Fourth Edition

CONSTRUCTION MATERIALS, METHODS, AND TECHNIQUES
Building for a Sustainable Future

EVA KULTERMANN | WILLIAM P. SPENCE

CENGAGE Learning

Australia • Brazil • Mexico • Singapore • United Kingdom • United States

Construction Materials, Methods & Techniques, Fourth Edition
Eva Kultermann & William P. Spence

SVP, GM Skills & Global Product Management: Dawn Gerrain

Product Director: Matthew Seeley

Senior Product Manager: Vanessa Myers

Senior Director, Development: Marah Bellegarde

Senior Product Development Manager: Larry Main

Associate Content Developer: Jennifer Alverson

Product Assistant: Jason Koumourdas

Vice President, Marketing Services: Jennifer Ann Baker

Marketing Manager: Scott Chrysler

Senior Production Director: Wendy Troeger

Production Director: Andrew Crouth

Senior Content Project Manager: Glenn Castle

Managing Art Director: Jack Pendleton

Cover image(s): © GettyImages, Inc./Allan Baxter

Notice to the Reader

Publisher does not warrant or guarantee any of the products described herein or perform any independent analysis in connection with any of the product information contained herein. Publisher does not assume, and expressly disclaims, any obligation to obtain and include information other than that provided to it by the manufacturer. The reader is expressly warned to consider and adopt all safety precautions that might be indicated by the activities described herein and to avoid all potential hazards. By following the instructions contained herein, the reader willingly assumes all risks in connection with such instructions. The publisher makes no representations or warranties of any kind, including but not limited to, the warranties of fitness for particular purpose or merchantability, nor are any such representations implied with respect to the material set forth herein, and the publisher takes no responsibility with respect to such material. The publisher shall not be liable for any special, consequential, or exemplary damages resulting, in whole or part, from the readers' use of, or reliance upon, this material.

The MasterFormat® Groups, Subgroups, and Divisions and cover used in this book are from *MasterFormat®*, published by CSI and Construction Specifications Canada (CSC), and are used with permission from CSI. For those interested in a more in-depth explanation of *MasterFormat®* and its use in the construction industry visit www.csinet.org/masterformat or contact:

CSI
110 South Union Street, Suite 100
Alexandria, VA 22314
800-689-2900: 703-684-0300
www.csinet.org

Portions of this publication reproduce excerpts from the *2015 International Building Code*, International Code Council, Inc., Washington, D.C. Reproduced with permission. All rights reserved. www.iccsafe.org

For product information and technology assistance, contact us at
Cengage Learning Customer & Sales Support, 1-800-354-9706

For permission to use material from this text or product,
submit all requests online at **www.cengage.com/permissions**
Further permissions questions can be e-mailed to
permissionrequest@cengage.com

Library of Congress Control Number: 2015959512

ISBN: 978-1-305-08627-2

Cengage Learning
20 Channel Center Street
Boston, MA 02210
USA

Cengage Learning is a leading provider of customized learning solutions with employees residing in nearly 40 different countries and sales in more than 125 countries around the world. Find your local representative at **www.cengage.com**

Cengage Learning products are represented in Canada by Nelson Education, Ltd.

To learn more about Cengage Learning, visit **www.cengage.com**

Purchase any of our products at your local college store or at our preferred online store **www.cengagebrain.com**

Printed in Mexico
Print Number: 05   Print Year: 2019

Case Studies

INTENT OF THIS BOOK

This book is designed to support introductory courses on construction materials and methods, construction management, and professional programs in architecture at two-year and four-year universities. The text follows the logical progression of the construction sequence, with reference to the 2004 edition of the MasterFormat™, developed by the Construction Specifications Institute and Construction Specifications Canada. Students are able to develop a foundation of knowledge pertaining to individual building components and details, their performance requirements, process of manufacture and assembly, and systematic organization into various building assemblies. The text also gives insight into current best practices in the construction industry.

PREFACE TO THE FOURTH EDITION

This fourth edition of *Construction Materials, Methods and Techniques* has been updated to incorporate the latest advances in materials science, and construction methods and techniques. This revision continues to integrate sustainable construction innovations holistically into the larger construction context. Green, or High-Performance Construction seeks to mitigate the negative impacts of inefficient construction practices by promoting strategies that will conserve natural resources, advance energy and resource efficiency, deal responsibly with waste, and create healthy environments.

Sustainable construction is no longer a unique means of building, but rather an integral part of how the industry is now operating. Cutting-edge building materials, systems, and construction practices have been added to the existing knowledge in each chapter. In addition, new case studies of individual buildings that exemplify sustainable construction materials and technologies have been added. This book has been thoroughly updated and revised to remain current with industry technologies and standards:

- The U.S. Green Building Council and LEED V.4 rating system

- Effective construction planning for efficient material use
- Environmentally friendly building materials with emphasis on recycled content and materials that promote manufacturer, contractor, and occupant health
- Energy-efficient environmental systems
- Sustainable construction operations and practices
- New organizations and resources actively providing metrics and research

HOW TO USE THIS BOOK

This text provides a detailed view of modern construction methods and building systems and the vast array of materials and products provided by manufacturers supplying the construction industry.

The first part of the text gives a brief overview of the basic characteristics and procedures of the construction industry. Pre-construction activities, the role of design professionals, various project delivery methods, and the MasterFormat™ specification system are covered. A discussion of zoning and building codes is followed by information about some of the industry's major professional and technical organizations. A section on sustainable design and construction discusses the environmental impact of the construction industry and provides an introduction of LEED and other sustainable building certification systems.

Early in the book, the physical properties of construction materials are outlined. This overview relates to all of the materials in the remainder of the book. New materials are developed each year, and the architect, contractor, and engineer must be informed of their properties in order to use these materials in the most effective and safe manner.

The remainder of the text is organized following selected divisions of the 2004 Edition of the MasterFormat™.

Division 1 addresses general requirements, including activities such as contractor selection, price and payment procedures, administrative requirements, and various legal controls and requirements. Additional details are given in Appendix B.

Division 2 outlines the existing conditions present on a project site, their assessment, and technologies available for remediation.

Division 3 presents a detailed study of the manufacture, types, characteristics, and properties of concrete. Consideration of the impact of admixtures, proportions, water, mixing, placing, and curing is included. Emerging technologies, such as smog-eating and bio-concrete, have been added. Drawings and photos are used extensively to illustrate cast-in-place and precast concrete structural systems.

Division 4 includes detailed information on mortar, the key to durable masonry construction. The materials, manufacturing processes, and construction techniques utilized in clay brick, concrete masonry, and stone construction are explained in detail and generously illustrated.

Ferrous and nonferrous metals are presented in Division 5. Their characteristics, mechanical properties, and practical applications are discussed. Steel frame and pre-fabricated construction systems and details are illustrated.

One of the largest divisions in the book, Division 6, covers the vast array of wood and plastic materials that are used in construction. Their properties, characteristics, and recommended applications are explained. Several chapters detail wood structural framing systems, including light wood frame, heavy timber, and innovative new panelized assemblies. The remainder of this division is used to present information on composites and plastic materials that find frequent use in modern construction.

Insulating, waterproofing, and sealing buildings against the weather are covered in Division 7. Structures are exposed to a variety of weather extremes, including temperature swings, driving rain and wind, and subsurface water. Walls, ceilings, and floors must be properly insulated and sealed against moisture penetration. Bituminous materials, bonding agents, sealers, and sealants are covered. Details, components, and materials commonly used for both residential and commercial roofing systems conclude this division.

The types, styles, methods of operation, and materials used for doors and windows are expanding continuously. Many of the products currently employed are illustrated in Division 8, in addition to stock and custom-made storefronts. Glass is used extensively in façade and window assemblies, so the types, properties, and uses of the numerous glass products available are discussed. New and innovative glass products are increasing the range of uses and applications of the material in modern commercial construction. Finally, an entire chapter is devoted to discussing and illustrating various types of cladding systems that find common use in commercial buildings.

Finishing the interior of a building involves an incredibly diverse range of products in any area of construction. Division 9 includes interior finishes; decorative and protective coatings; gypsum, lime, and plaster products; acoustical finishes and materials; and all types of finish flooring. The construction and finish of interior walls, partitions, and ceilings, and their properties and fire resistance are integrated.

Division 10, 11, and 12 cover the variety of specialty products, equipment, and furnishings that find use both within buildings and on project sites. Examples include visual displays, telephone enclosures, vending equipment, casework and furniture.

A most interesting compilation of special construction assemblies are found in Division 13. Special structural systems refer to innovative long-span structures that are used to provide enclose for a variety of activities. This division discusses and illustrates a diverse offering, including air-supported and fabric structures, geodesic domes, and other pre-engineered assemblies.

Division 14 discusses and illustrates the range of conveying systems in current use, including conveyors, elevators, escalators, moving walks, and material-handling systems.

Division 21 organizes the many complex factors related to the regulation, operation and maintenance of fire-suppression systems, including installation, instrumentation, and control. Water, carbon dioxide, clean-agent, wet chemical, and dry chemical fire-extinguishing systems are explained.

A basic need for a successful building is a well-functioning plumbing system. Division 22 summarizes the system of pipes and other devices installed in a building for the distribution of potable water and the removal of waterborne wastes. Innovative new technologies and fixtures for a variety of applications are outlined.

Division 23 is a very large division including heating, ventilating, and air-conditioning systems. Basic methods of heat transfer, the fuels employed, and the types of heating and cooling systems and equipment are explained and illustrated.

Electrical distribution systems inside a building supply electrical power as needed and transmit information through an internal communications system. Division 26, Electrical Systems, carries the discussion from the generation and transmission of electrical power to the service components in both residential and commercial buildings. Renewable power systems are illustrated, and extensive information on lighting technologies is accessible. Equipment for controlling and operating the electrical system, as well as equipment used for communication, such as alarm, television, public address, and other communication systems, is presented.

Division 28 covers the various electronic safety and security systems in buildings and related exterior areas where security is important. Video surveillance and personal protection systems are included, as are electronic detection and alarm systems for fire detection.

Division 31 encompasses the wide range of site construction activities, including soils and subsurface investigations, site preparation, excavations, earthwork, foundations, and the installation of utilities and drainage systems.

At the end of each chapter are materials providing the reader with a means of reviewing what has been read and reinforcing the learning experience. Key terms are listed. Sources of additional information are provided to enable the reader to explore areas in greater depth.

A master glossary at the end of the book provides an additional means of locating definitions. Appendix A provides the MasterFormat numbering system in its entirety. Appendix B provides details concerning the information in Division 1, General Requirements. Appendix C provides an extensive listing of U.S. and Canadian professional and trade organizations. These organizations provide information that is vital to the continual improvement of materials and construction techniques. They develop materials standards and installation specifications, and publish building codes, manuals, and technical reports on an ongoing basis. Readers can reference this information to provide additional insight as the chapters in this book are read. Detailed metric information is available in Appendix D. This information is essential because construction is moving toward using the metric system in the future. Appendix E lists the weights of commonly used construction materials. Appendix F gives the names and atomic symbols of selected chemical elements, and Appendix G provides data on the coefficients of thermal expansion of selected construction materials.

FEATURES OF THIS BOOK

This book includes many features to assist students as they progress through the chapters:

- Correlations to the MasterFormat™ are included at the beginning of each chapter, providing students with a quick reference to this essential manual.
- Individual case studies provide examples of the application of new construction materials and strategies.
- Inserts on Construction Materials, Methods, and Techniques provide detailed information on selected areas of interest.
- Learning Objectives and Review Questions open and close each chapter to provide students with a framework for study to ensure full comprehension of the material.
- Key Terms are in color throughout the text and are listed at the end of each of the chapters, highlighting essential terminology. Complete definitions are provided in the Glossary.
- Activities encourage students to apply what they have learned in the chapter and to build experience for actual on-the-job tasks.
- Additional Resources are listed at the end of each chapter that point students in the direction of organizations, periodicals, Web links, and other references to further learning on selected topics.

NEW TO THIS EDITION

This new edition emphasizes innovative materials and technologies that encourage the student of construction to aggressively deal with the challenges of a changing industry. In light of advances in the field, many topics have been updated whereas others have been reorganized. New topics include:

- An expanded discussion of moisture transfer within assemblies and the technologies available to effectively control it.
- The addition of door and window nomenclature and the hardware associated with their installation.
- Additional information on new cladding technologies, including structural glazing, structural standoff systems, and new metal panel assemblies.
- A new section on fire and smoke management in buildings.

- New information on recent advances in telecommunications and data systems.
- Additional Case Studies, including the Omega Center, The Passive House, and the Lewis Center for Environmental Studies, provide detailed information on the application of emerging sustainable construction technologies.
- The entire text has been updated with extensive new photos and revised detail drawings of the construction process for various assemblies.

SUPPLEMENTS

Instructor Companion Site

Spend less time planning and more time teaching with the **Instructor Companion Site**: Everything you need for your course in one place! This collection of book-specific lecture and class tools is available online via www.cengage.com/login. Access and download PowerPoint presentations, images, instructor's manual, and more.

Cengage Learning Testing Powered by Cognero

The Instructor Companion Site features **Cengage Learning Testing Powered by Cognero.** This flexible, online system allows you to:

- author, edit, and manage test bank content from multiple Cengage Learning solutions
- create multiple test versions in an instant
- deliver tests from your LMS, your classroom or wherever you want

Start right away!
Cengage Learning Testing Powered by Cognero works on any operating system or browser.

- No special installs or downloads needed
- Create tests from school, home, the coffee shop—anywhere with Internet access

What will you find?
- Simplicity at every step. A desktop-inspired interface features drop-down menus and familiar intuitive tools that take you through content creation and management with ease.
- Full-featured test generator. Create ideal assessments with your choice of fifteen question types (including true/false, multiple choice, opinion scale/likert, and essay). Multi-language support, an equation editor

and unlimited metadata help ensure your tests are complete and compliant.
- Cross-compatible capability. Import and export content into other systems.

MindTap

MindTap is a personalized teaching experience with relevant assignments that guide students to analyze, apply, and improve thinking, allowing you to measure skills and outcomes with ease.

- Personalized Teaching: Becomes yours with a Learning Path that is built with key student objectives. Control what students see and when they see it. Use it as-is or match to your syllabus exactly—hide, rearrange, add, and create your own content.
- Guide Students: A unique learning path of relevant readings, multimedia, and activities that move students up the learning taxonomy from basic knowledge and comprehension to analysis and application.
- Promote Better Outcomes: Empower instructors and motivate students with analytics and reports that provide a snapshot of class progress, time in course, engagement and completion rates.

ABOUT THE AUTHORS

Eva Kultermann is a licensed architect, and associate professor and associate dean of curriculum at the Illinois Institute of Technology, College of Architecture. She has a background in construction and is currently involved in developing, reviewing and revising the institution's architectural curriculums.

William P. Spence was Dean of the College of Technology and remains Professor of Construction Engineering Technology emeritus at Pittsburg State University, KS. He currently is active as a full-time author of more than thirty construction technical books.

ACKNOWLEDGMENTS

A major factor in the organization, writing, and illustrating of this book was the help given by hundreds of manufacturer representatives. Representatives of many of the professional and technical organizations supporting the construction industry also made important contributions. A special note of appreciation is due the consultants located at universities across the country for their assistance in reviewing the manuscript and the illustrations.

Disclaimer

The information presented in this book was secured from a wide range of manufacturers, professional and trade associations, government agencies, and architectural and engineering consultants. In some cases, generalized or generic examples are used. Every effort was made to provide accurate presentations. However, the author and publisher assume no liability for the accuracy of applications shown. It is essential that appropriate architectural and engineering staff be consulted and specific information about products be obtained directly from manufacturers.

Procurement and Contracting Requirements CSI MasterFormat™

General Requirements
CSI MasterFormat™

The Construction Industry: An Overview

Upon completion of this chapter, the student should be able to:

- Gain an understanding of the scope of the construction industry.
- Identify the different phases and activities that make up the building design and construction planning process.
- Recognize the types of drawings and specifications used in a set of construction documents.

- Be familiar with the roles and responsibilities of owners, architects, and contractors.
- Recognize the make-up and organization of construction documents.
- Categorize the types of project delivery methods used in commercial construction and how they are administered.
- Describe the phases and administrative procedures that serve to organize the on-site construction phase.

The construction industry is one of the largest business activities in the United States, encompassing establishments engaged in the construction of buildings and larger engineering projects, such as roadways and utility infrastructure. Construction with all of its related and supporting industries is vital to the economic health of the country, accounting for an annual average of 5 to 10 percent of our gross domestic product.

Compared with other large industries, some characteristics are unique to the planning of buildings and infrastructure. Most new buildings are custom designed and constructed, involving long development schedules. Because each project is site specific, its execution is influenced by physical, social, and regulatory conditions, such as weather, availability of skilled workers, and local building codes. Since the service life of a commercial building can be more than 50 years, future technical requirements must be considered throughout the planning process. Because of the technological complexity of construction, building plans must be flexible enough to allow for changes and adjustments both during construction and after.

The process by which a building or other project is designed and constructed can be divided into four discreet project phases (Figure 1.1). The *pre-design phase* sets the initial objectives and criteria under which a building will be planned. The *design phase* determines the actual geometry, materials, and performance characteristics of the finished structure. The preconstruction phase selects the

Figure 1.1 The process by which a building or other project is designed and built can be divided into four discreet project phases.

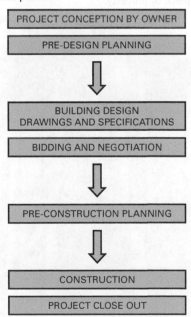

various project participants and outlines their contractual agreements. Finally, the *construction phase* is involved with the physical realization of the finished building.

CONSTRUCTION: A DYNAMIC INDUSTRY

The construction industry is divided into three major areas: building construction, heavy or infrastructure construction, and industrial construction (Figures 1.2 through 1.4). *Building construction* involves the erection of a building on a piece of property. This includes residential, commercial, civic, educational, religious, and agricultural buildings, among others. While many building projects consist of new construction, a considerable amount of construction activity is concerned with the renovation, repair, or the provision of additional space for existing buildings. *Heavy construction* is the term used for larger infrastructure projects, such as highways, bridges, canals, dams, subways, tunnels, utility piping systems, water control construction, and communications networks. Heavy construction projects are usually financed by governmental agencies and other institutions, or are incorporated into master plans (such as universities) to serve the public good. *Industrial construction* refers to the building of large-scale manufacturing, and processing plants or utility generation installations. Regardless of the type

Figure 1.3 "Heavy construction" is the term used for larger infrastructure projects, such as highways, bridges, and canals.

Figure 1.2 Building construction involves the erection of a building on a piece of property.

Figure 1.4 "Industrial construction" refers to the building of large-scale manufacturing, processing, and other mechanized plants, or utility generation installations.

of construction, each requires planning, financing, and compliance with regulatory constraints.

All construction involves the confluence of skilled workers, architects, engineers, and a vast array of materials and equipment to execute a carefully conceived plan. Supporting them is a broad range of industries that manufacture the materials and components designated for a project. These can include suppliers of aggregate and cement products, manufacturers of lumber, structural beams, doors and windows, siding, roofing, and finish materials, and dealers of appliances and mechanical equipment (Figure 1.5). Many manufacturing endeavors involve the cutting, shaping, and assembling of various materials into a useful product. Each product area has a tremendous variety of materials and installation methods, requiring extensive technical expertise. Consider just the electrical and mechanical components involved in servicing a building and the procedural knowledge required to design, specify, manufacture, and install them.

Figure 1.5 The construction industry is supporting by a broad range of industries that manufacture the materials and components designated for a project.

Construction materials and methods are constantly changing fields with new products and equipment being developed continuously. The people involved in construction must be educated and constantly kept up to date as new materials and methods of construction are introduced. They need to maintain membership in professional organizations, attend conferences, and read professional journals to recognize the value of new technical developments and their proper application.

The industry is currently in the midst of profound changes caused by unprecedented environmental forces that are reshaping all areas of production. The most important challenge of the twenty-first century is to develop strategies of using natural resources in ways that will sustain the natural environment for generations to come. The importance of the construction industry instituting changes that promote both energy and resource efficiency cannot be overstated and will be discussed in detail in Chapter 2.

PRE-CONSTRUCTION ACTIVITIES

Owners, both public and private, plan buildings and other construction projects to accommodate desired functional and spatial needs. The design and construction of a building project is the result of the collective ideas, talents, and services of a large and diverse group of professionals. A vast array of planning decisions must be made at the onset of any construction project. These include an evaluation of the vision for a project, its scope, functional requirements, the determination of a site, budgets and financing, project scheduling, and assurance that regulatory constraints such as zoning and code regulations will be satisfied.

One of the first tasks in commencing larger construction project is in the selection of the project team participants. Members of this multi-disciplinary group are selected based on their professional standing, experience with projects of similar size and scope, and sometimes, previous working collaboration among the various team members (Figure 1.6). The established project team works together in developing a project plan that addresses the roles and responsibilities of each member, and identifies at what point in the process each member will join in the work.

Figure 1.6 The design and construction of a building project is the result of the collective ideas, talents, and services of a large and diverse group of professionals.

Regardless of size, the three main players in any construction team are:

- The owner that initiates the project
- The architect and engineers who design the building
- The general contractor who completes the actual construction

Supporting these main players is a team of other professionals, each involved with particular aspects of the larger project. These include:

- Consultants that provide professional services to the architect, contractor, or owner

- Subcontractors who perform portions of the work under the direction of the general contractor. Product representatives who assist with submittals and furnish field services such as providing consulting and inspecting the installed work

- Manufacturers who produce materials or products, and material suppliers who furnish materials and products of construction

- Testing laboratories and inspection agencies provide quality control services to the owner and contractor

- Financial advisers and institutions that arrange the project financing, and attorneys who coordinate the legal and contractual issues of the project

- Insurance companies that provide risk coverage to the owner, contractor, and design consultants

- Authorities and regulatory agencies that have jurisdiction over the construction and establish standards in the form of codes, ordinances, and permits

Once the project team is established, it begins the investigation of the multitude of project parameters required for the development of project definition and feasibility, referred to as the pre-design phase.

Pre-Design and Design Development

Owners most often procure the services of an architect or an architect/engineering/contracting firm (AEC) to help in the development of project definition and feasibility. Once an architect or AEC firm has been identified, the pre-design portion of the work can begin. In this phase, the architect, other design consultants, and the owner mutually determine the goals and objectives of the project. Most projects begin with a thorough examination of the existing needs, goals, and constraints that will influence a design. *Programming* is the research and decision-making process that determines the specific set of needs that a building is expected to fulfill, and identifies the scope of work to be designed.

Design sketches and feasibility studies are developed as a series of alternatives for approval by the owner. Early design efforts focus on establishing what is known as the design intent. *Design intent* is a statement that defines the anticipated aesthetic, functional, and performance characteristics of the finished building or project.

Functional and spatial requirements, including sizes and adjacencies of spaces, are defined and tabulated. The architect will conduct initial checks that zoning and building code requirements can be met within the design concept. Surveys and drawings of existing conditions are obtained or prepared. The result of pre-design work will be a written program. A program is a written document that explains design intentions, controls, and standards for a project, including detailed space requirements and the types of equipment and systems to be used. This phase is complete when the owner and architect agree that the scope of work, anticipated construction cost, and time schedule are well defined.

During design development, initial design ideas are further developed into detailed drawings of the building, indicating exact sizes and relationships between building elements. Architects use a system of orthographic drawings of a building or structure to simplify the graphic understanding of complex forms. The *floor plan* is a representation of a building looking down after a horizontal plane has been cut through it and the top portion removed (Figure 1.7). A *building section* gives a view of a building after a vertical plane has been cut through it and the front portion removed (Figure 1.8). An *elevation* drawing shows the exterior façade of a building, delineating geometries and the materials of construction (Figure 1.9).

Since plans, sections, and elevations cannot show all the specific aspects of the construction, larger-scale detail drawings are used to explain the joining of materials, elements, and components of the various building assemblies. Detail drawings are keyed into plans and sections drawings to give more information on the specific means of assembly. Other information included in the drawings are schedules that reduce the amount of information that must be placed directly on the drawings themselves. Door and window schedules, for instance, use graphic symbols on the plan drawings that are keyed into a schedule, a table format that lists the type, size, and operating characteristics of each element. Schedules are also used for interior finishes, partition types, and structural elements of a building.

Most architectural drawings generated today use a system of *computer-aided drafting*, known as CAD. Before the advent of CAD, drawings were painstakingly generated by hand, and making significant changes to the drawings was difficult and time consuming. CAD software enables the design team to efficiently create and manage modifications to the architectural drawing set. The use of the software also facilitates effective communication between the various consultants on a project through digital file sharing.

Figure 1.7 The floor plan is the most-often-referred-to drawing in a set of construction documents showing the relationship of spaces and overall building dimensions.

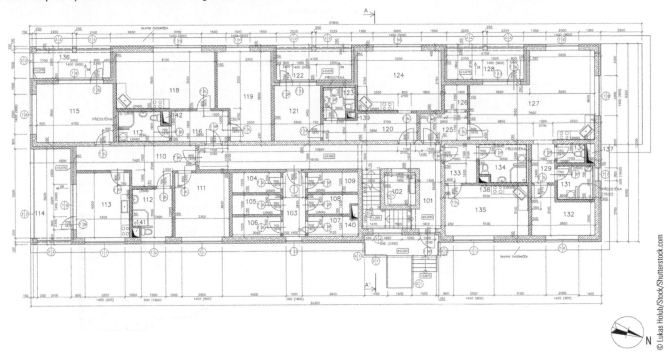

Figure 1.8 A building section graphically illustrates a cut through the building from foundation to roof, showing interior spaces and the materials of construction.

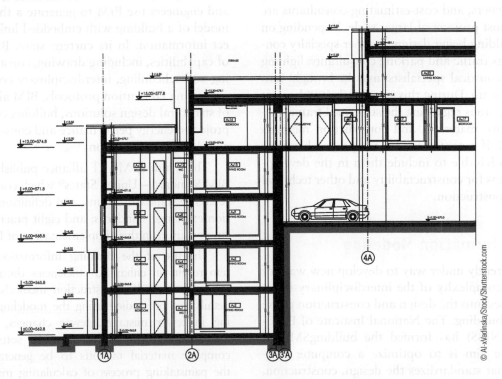

Figure 1.9 An elevation drawing shows the exterior façade of a building, delineating geometries, building heights, and the materials of construction.

Because of the multitude of systems under consideration at this stage, the architect will normally hire the services of other design consultants in the development of systems. Structural, mechanical, electrical, plumbing and civil engineers, fire protection consultants, interior and landscape designers, and cost-estimating consultants are common on most projects of larger scale. Depending on the type of building being designed, other specialty consultants, such as traffic and parking consultants, lighting designers, or acoustical specialists, may be brought into work with the team. During this stage of the work, most substantive decisions regarding the structural frame, mechanical systems, materials, and construction methods are determined. If a contractor for the project has been selected, it is advisable to include them in the development of solutions for constructability and other technical aspect of the construction.

Building Information Modeling

Efforts are currently under way to develop new ways of managing the complexity of the interdisciplinary information that goes into the design and construction phase activities of a building. The National Institute of Building Sciences (NIBS) has formed the buildingSMART alliance, whose aim is to optimize a computer software system that standardizes the design, construction, operation, and maintenance process for buildings by compiling all of the related information in a single electronic format. The software, known as *Building Information Modeling (BIM)*, has revolutionized the way buildings are designed, constructed, and maintained. Architects and engineers use BIM to generate a three-dimensional model of a building with embedded links to other project information. In its current state, BIM has a range of capabilities, including drawing, construction estimating and scheduling, interdisciplinary coordination, and generating fabrication protocols. BIM also runs analysis of structural design solutions, building code compliance, projected energy performance and consumption models, and daylight penetration (Figure 1.10).

The buildingSMART alliance publishes the National BIM Standard – United States® which consists of nineteen reference standards, terms, and definitions; nine information exchange standards; and eight practice guidelines to support users in their implementation of BIM .

The use of the Building Information Modeling system results in enhanced efficiencies during the construction process itself. Construction methods, sequences, and details can be studied using the modeling process, resulting in fewer mistakes, fewer changes, and, ultimately, shorter construction schedules. The software allows for complete material takeoffs to be generated, eliminating the painstaking process of calculating material quantities

Figure 1.10 A BIM model illustrating the capability of visualization and analysis of various building systems.

Image courtesy of Mortenson Construction and McKinstry Company

by hand. By designing in a three-dimensional model systems, integration can be visualized and better coordination realized. Information from models can be fed directly to material fabricators, eliminating the need for shop drawings and ensuring tight dimensional tolerances. Students of construction should become familiar with the BIM software, because its use in the construction industry is becoming widespread.

Construction Documents

Once all basic decisions have been approved by the owner, the architect and consultants will commence with the final phase of the design process, the preparation of construction documents. *Construction documents (CD)* consist of two interdependent components: the drawings and the specifications.

Construction drawings, sometimes referred to as working drawings, visually outline the dimensional relationships between all aspects of the building including their location, form, size, and quantities, as well as the materials used, and their colors and textures. Working

drawing sets are organized from the general overall building plans, sections, and elevations, down to the level of the detail, indicating fasteners and connections. Full drawing sets are collated according to discipline and include civil, architectural, structural, mechanical, electrical, and plumbing (MEP), and fire protection sheets, among others (Figure 1.11). The drawings are the basis on which the contractor generates cost estimates, and are used to guide the actual construction of the building on the site.

Since working drawings cannot give all of the details involved in describing the quality and performance characteristics of specific materials, construction methods, and project procedures, the construction documents include a written manual called the specifications. The *specifications* describe in writing more detailed information on the exact types of materials to be used and the ways in which construction processes are conducted. The drawings graphically indicate the relationship between elements, while the specifications spell out the specific types, qualities, and properties of those materials; their installation; and other information not shown in the drawings.

Figure 1.11 Construction drawings are organized sequentially from the general to the detail and in order of design discipline.

INDEX OF DRAWINGS

COVER	CODE MATRIX/SITE LOCATION
C1.1	SITE PLAN
ARCHITECTURAL	
A1.1	BASEMENT AND 1ST FLOOR DEMOLITION PLANS
A1.2	2ND FLOOR AND ROOF DEMOLITION PLANS
A2.1	BASEMENT AND 1ST FLOOR PLANS
A2.1	2ND FLOOR PLAN/DOOR AND FINISH SCHEDULE
A3.1	ENLARGED INTERIOR PLANS
A4.1	1ST AND 2ND FLOOR REFLECTED CEILING PLANS
A5.1	EXTERIOR ELEVATIONS
A6.1	BUILDING SECTIONS
A7.1	STAIR SECTIONS AND DETAILS
A8.1	ENLARGED WALL SECTIONS
A9.1	DETAILS
A9.2	SOLAR CHIMNEY DETAILS
A9.3	MILWORK AND TRIM DETAILS

STRUCTURAL	
S1.1	FOUNDATION PLANS
S1.2	1ST AND 2ND STRUCTURAL FLOOR PLANS
S1.3	STRUCTURAL DETAILS
MECHANICAL	
M1.1	1ST AND 2ND FLOOR MECHANICAL PLANS
M1.2	MECHANICAL DETAILS AND SCHEDULES
PLUMBING	
P1.1	PLUMBING RISER DIAGRAMS
P1.2	PLUMBING SCHEDULES AND DETAILS
ELECTRICAL	
E1.1	BASEMENT ELECTRICAL PLANS AND NOTES
E1.2	1ST AND 2ND FLOOR POWER PLANS
E1.3	1ST AND 2ND FLOOR LIGHTING PLANS

Specifications and the MasterFormat

Construction specifications for buildings and other projects are written using the MasterFormat. The MasterFormat system was developed by the Construction Specifications Institute (CSI) and Construction Specifications Canada. It is used by U.S. and Canadian construction companies and material suppliers, the McGraw-Hill Information Systems as a basis for their Sweet's Catalog Files, and the R. S. Means Company construction cost data publications. *MasterFormat* provides a standard for writing specifications using a system of descriptive titles and numbers to organize construction activities, products, and requirements into a standard order that facilitates the retrieval of information and serves as a means for all participants within the construction industry to communicate (Figure 1.12). Since the design and completion of construction projects involves individuals in many technical fields, the ability to communicate effectively by having a standard sequence for identifying and referring to construction information is essential.

The numbers and titles in MasterFormat are divided into 50 basic groupings called divisions, 34 of which are active, with 16 reserved for future expansion. Each division has a title and identifying number (Table 1.1). Division 1 describes the general requirements of the contract, outlining administrative requirements for a construction project, such as project management procedures,

Figure 1.12 The 2014 MasterFormat is used by U.S. and Canadian construction companies and material suppliers to organize construction activities, products, and requirements into a standard order.

MasterFormat®

Master List of Numbers and Titles for the Construction Industry

2014 Update

Table 1.1 Divisions 2–19 MasterFormat Facilities Subgroup

DIVISION 02	EXISTING CONDITIONS
DIVISION 03	CONCRETE
DIVISION 04	MASONRY
DIVISION 05	METALS
DIVISION 06	WOOD, PLASTICS AND COMPOSITES
DIVISION 07	THERMAL AND MOISTURE PROTECTION
DIVISION 08	OPENINGS
DIVISION 09	FINISHES
DIVISION 10	SPECIALTIES
DIVISION 11	EQUIPMENT
DIVISION 12	FURNISHINGS
DIVISION 13	SPECIAL CONSTRUCTION
DIVISION 14	CONVEYING SYSTEMS
DIVISION 15–19	RESERVED FOR FUTURE EXPANSION

© Construction Specifications Institute

and construction facilities and controls. See Appendix B for more information on Division 1. Divisions 2 through 19, the *Facility Construction Subgroup*, deal mainly with the materials of construction for buildings. The *Facilities Services Subgroup*, divisions 20 through 29, deal with mechanical, electrical, plumbing, fire protection, and communications equipment. Divisions 30 through 39, the *Site Facilities Subgroup*, are concerned with earthwork, transportation, and marine construction topics, while the final subgroup, *Process Equipment*, deals with larger industrial processes. A complete listing of the 2014 MasterFormat™ level-two numbers and titles is given in Appendix A.

Divisions 1 through 16, which previously encompassed the entire MasterFormat, are primarily concerned with the construction of buildings. The division ordering is loosely arranged to follow the sequence of the construction process itself, helping to facilitate easy recall of where information is located. Division 3, concrete, is the material from which most foundations are built. The divisions move on through major structural and cladding materials, to interior furnishings and components, and finally mechanical and electrical systems.

Under each division, section numbers are generally composed of six digits, although they may be extended to eight digits. Each section should be read as pairs of numbers, for example: Section 07 32 13 describes Clay Roof Tiles. The first two numbers indicate the division, the second pair, the broad topic scope, and the third pair, the narrow scope.

Level 1: Division 07: Thermal and Moisture Protection

Level 2: Broad scope number 32: Roof Tiles

Level 3: Narrow scope number 13: Clay Roof Tiles

Each Level 3 heading is further subdivided into three parts. Part one describes the general requirements for the category dealing with definitions, referencing standards, quality control, and warranties. Part two describes the actual materials and products, their physical properties, manufacturers, and performance requirements. Part three describes how and under what circumstances the product must be installed in the field. Within this system, every building component can be easily found under its division and number.

In addition to purely technical information, Division 00 of the specifications outlines the guidelines for the pre-construction phase of the project. These include the General Conditions of the contract, a blueprint for the construction delivery process. Division 00 addresses such issues as how the bidding process will be conducted, what kinds of bonds and insurance the contractor is required to hold, and when the actual on-site construction may commence. It includes the bid forms that the contractor will use in submitting the proposal for construction. The completed specifications are known as the *project manual*. The completed *contract documents* then consist of the construction drawings, and the project manual consisting of the final specifications.

GreenFormat, a newly released version of the MasterFormat, was introduced to incorporate new environmentally friendly products and construction procedures that were previously not included. GreenFormat is a new CSI framework that provides designers, contractors, and building operators with basic information to help meet green building requirements. The new format is entirely web based and gives building product manufacturers a process by which to describe the sustainable aspects of the products for incorporation in the MasterFormat sequence.

THE PROJECT DELIVERY PROCESS

Crucial for the successful delivery of a complex building project is the selection of a qualified contractor and an effective project delivery process. A project delivery method governs the conditions under which a construction project will be completed, and defines the relationships between the owner, architect or AEC firm, and contractor.

Construction Contractors

Construction contractors perform the on-site construction of a project. They are generally divided into general contractors and specialty contractors. *General contractors (GC)* assume the responsibility for the construction of an entire project at a specific cost and by a specified date. They are responsible for developing project schedules and sequencing, and coordinating the work of all subcontractors. The general contractor also determines the actual methods and techniques of construction, and implements safety precautions on the building site. A general contractor signs contracts with *subcontractors* who perform the focused labor within their technical areas, such as the electrical, plumbing, roofing, bricklaying, carpentry, or concrete work.

Specialty contractors do the work required in a limited area, like elevator or communication equipment installation, for example (Figure 1.13). Both specialty and subcontractors are independent contractors employed by the general contractor to perform specific work at a specified cost. While subcontractors work independently on a site, bringing their own employees, supervisors, and tools to the job, their work is overseen by a project superintendent. The *superintendent* is the general contractor's on-site representative responsible for continuous field supervision, coordination, and completion of the work. The superintendent makes sure that work proceeds according to the project schedule, and that the activities of the various subcontractors working on the site will not interfere with one another.

Figure 1.13 General contractors assume the responsibility for the construction of an entire project at a specific cost and by a specified date. The general contractor signs contracts with subcontractors, who perform the required work within their technical areas, such as this electrician.

© Bunwit Unseree/Stock/Shutterstock.com

Selecting the Contractor

A number of different options are available in the selection of a contractor for construction. For publicly financed projects, construction contractors are generally sought through a process of *competitive bidding*. In this method, qualified construction contractors are invited to bid on the project on a competitive basis, with the contract often being awarded to the contractor that submits the lowest bid. A variation of this process used more on privately funded projects is called *invitational bidding*, wherein only preselected contractors are asked to provide bids on a project.

An owner may decide on a contractor with whom they have worked on previous projects without seeking other bids and agree upon a negotiated contract. A *negotiated contract* can have the advantage of bringing a contractor into the project in the pre-construction phase of the work, encouraging a team approach that incorporates design, construction, and budget planning with owner involvement in a cooperative and coordinated effort. This integration, which incorporates estimating and cost control throughout the design and development effort rather than at its conclusion, often facilitates the best overall solutions.

Most projects select a single contractor to oversee and complete the work on an entire project. The single prime contractor is responsible for all of the work, including that which has been subcontracted. An alternative method is to divide portions of the work among more than one entity by employing the services of multiple prime contractors. *Multiple prime contracts* are often used on large and complex projects where a number of specialty contractors are engaged to complete different parts of the work. A high-rise building for instance may use one contractor for the foundations and site work, another for the building structure, and a third for the interior build out and finishes. In this case, each will enter into a separate contractual agreement with the owner. The use of multiple prime contracts also has the advantage of reducing construction costs by eliminating the general contractor's fee.

Multiple prime contracts are often used for projects that are fast tracked. *Fast tracking* a project shortens the overall construction time by beginning work on site prior to the completion of the construction documents. Construction documents for fast-track projects are often completed in sequenced packages. A first package may be for site preparation and foundations, a second for the structural frame, and a third for the installation of the exterior envelope. Careful coordination between the design team and multiple contractors is essential for a fast-track project.

Project Delivery Methods and Types of Construction Contracts

A project delivery method determines how a construction project will be completed; delegates the responsibilities, rewards, and risks between participants; and regulates the relationship between the owner and the contractor. Division 00 of the written specifications, General Requirements, outlines the general conditions of the construction contract. This information provides the basis upon which the contractual relationships between all parties involved in a project will be administered. Standard forms for construction contracts, which can be modified to suit conditions, are available from the American Institute of Architects (AIA). A number of options exist to define the contractual relationship between the owner and contractor on a particular project. The three most common project delivery methods currently in use are:

- Design-Bid-Build (DBB)
- Design-Build
- Construction Management (CM)

Design-Bid-Build

Design-Bid-Build (DBB) is the traditional method of project delivery that moves sequentially from conception of a project through its construction. The owner contracts initially with an architect to define the project scope, perform pre-design services, and produce the construction documents. Once the construction documents are complete, the project is sent out for proposals to qualified contractors. Division 00 of the written specifications, Procurement and Contracting, gives instructions to bidders for how to prepare the proposal and bidding forms. The contract is usually awarded to the most-qualified bidder who submits the lowest cost estimate. The general contractor solicits bids from subcontractors using the same bidding procedure. The project responsibilities are divided among the various team members with whom the owner establishes separate contractual relationships (Figure 1.14). Because the design, bidding, and construction phases occur sequentially, the design-bid-build delivery process tends to produce longer overall project schedules. The delivery model generally results in the lowest construction cost by offering the owner the financial advantage of an open competitive bidding process. Many state and local government institutions dictate that competitive bidding be used in public construction projects.

Figure 1.14 The traditional design-bid-build project delivery method. Solid lines in this image indicate the contractual relationships that are held between project participants, whereas dashed ones indicate lines of communication.

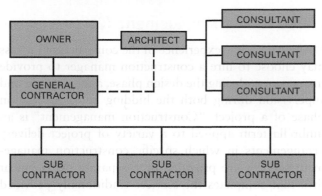

Design-Build

In the *design-build* contracting method, the owner contracts with a single party that completes all portions of the work from design through construction. Rather than having multiple contracts with the architect and contractor, the design-build method assigns a single entity with full responsibility (Figure 1.15). The entity may be a construction company that has contracted with an architect or engineering firm for design services, or one that maintains design professionals as permanent members of their staff. The design-build organization usually provides a bid price for both design and construction at the onset of a project, resulting in earlier and more tightly constrained cost controls. Because the bidding portion of the traditional design-bid-build delivery method is eliminated under this procedure, design-build contracts tend to produce shorter overall project schedules.

Figure 1.15 In the design-build delivery process, the owner contracts with a single entity that provides both design and construction services for the project.

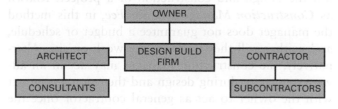

Many owners choose Design-Build in order to reduce the risks associated with a project. By combining the responsibility for design and construction under a single provider, coordination, construction process, and project costs are improved. Most of the risk is borne by the design-builder in exchange for retaining some or all of any savings identified.

Construction Management (CM)

Owners with less experience in the construction process may choose to hire a construction manager to provide management during the design phase, and oversight and supervision during both the bidding and construction phase of a project. "Construction management" is an umbrella term applied to a variety of project delivery arrangements in which specific construction management services are provided by a separate management firm that completes the services traditionally provided by a general contractor (Figure 1.16).

Figure 1.16 In the construction manager delivery method, the owner contracts with a construction manager to oversee both the design and construction process. Solid lines in this image indicate the contractual relationships that are held between project participants, whereas dashed ones indicate lines of communication.

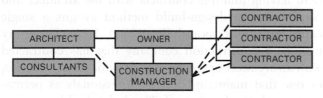

A construction manager is hired early in the design process to consult on issues of design, cost, and scheduling, and continues to supervise the building planning and construction. This method is particularly advantageous on large, complicated projects where ongoing oversight can provide innovative and cost-saving solutions to complex problems. Construction management is contracted in one of two ways.

The construction manager may act purely as an advisor, providing guidance and expertise throughout the design and construction of a project. Known as *Construction Management for Fee*, in this method the manager does not guarantee a budget or schedule, and neither will they perform any work on site. Alternatively, the construction manager may act in an advisory capacity during design and then sign a contract with the owner to act as general contractor once the construction documents are complete. When the management firm acts as general contractor, it assumes all the liability and responsibility of a general contractor, which is why this delivery method is known as *Construction Management at Risk*.

In both cases, the construction manager becomes a representative of the owner in overseeing the work of both the architect and the contractor. In the construction management delivery method, the contractor is involved in the planning of a project prior to the start of construction to help oversee scheduling, cost control, constructability, and additional expertise.

Integrated Design Process

Integrated Design Process (IDP) is a relatively new project delivery approach that attempts to bring together all the diverse participants of a project at the beginning of the design phase. A project team including the owner, architect or AEC firm, engineers, energy and green design consultants, contractor, construction manager, material suppliers, facilities manager, and regulatory officials is established at the onset of building planning and design. The team meets for a brainstorming session at the beginning of a project and continues to work collaboratively in establishing design intent and functional requirements. By including all stakeholders early in the design process, many coordination issues can be solved efficiently throughout the design and construction sequence. This approach considers the multifaceted systems involved in a project as acting interdependently, and seeks to identify synergies throughout the process. In general, integrated design is an approach that seeks to achieve high performance on a wide variety of well-defined environmental and social issues while staying within budgetary and scheduling constraints. It follows the design through the entire project life, from pre-design through occupancy and into operation.

THE CONSTRUCTION PROCESS

After the execution of an agreement, the construction stage of a project involves the contractors planning, administration, and completion of the actual construction of the final building. Construction is without question the most demanding stage of any project.

The *notice to proceed* is a written directive from the owner to the contractor that sets the date that the contractor can begin the work under the conditions of the contract. The directive identifies the contract delivery method and establishes the start of the construction schedule. The architect and other members of the design team are typically responsible for monitoring the contractor's compliance with the contract documents by conducting site visits, evaluating submittals, reviewing

Construction Methods

Subcontractor Agreements

Seldom does a general contractor have the personnel to complete all of the tasks required to construct a building. Many activities, such as mechanical, electrical, and plumbing (MEP) work, are handled by subcontractors employed by the general contractor. It is essential that the general contractor and subcontractor have a detailed written contract, stipulating what portion of the work is to done by which entity. Following are some of the basic topics covered in subcontractor agreements.

Contracts should include a schedule that includes the starting, interim, and completion dates. Clauses outlining who is to pay for expenses caused by any delays are included. These may define what type of delay would constitute a breach of contract that would permit the general contractor to employ a new subcontractor. Who assumes responsibility for obtaining required permits and notifying inspectors when the job has progressed to benchmark points should be noted. Some statement regarding the failure of a subcontractor to complete the work to the required quality expected is necessary.

Of great importance is a clearly specified payment schedule. This includes the costs of materials and labor provided, and the terms of payment. Materials suppliers for major portions of the work can be specified. Typically, a plumber supplies the pipe material, but the general contractor may prefer to purchase the plumbing fixtures. Drywall installers typically prefer to do the work but expect the general contractor to have the materials available when they appear on the job. If changes are to be made, the agreement should specify that the general contractor review these changes with the owner. The subcontractor should not proceed with changes unless they are approved by the general contractor.

All trades are subject to meeting safety standards, and even though the general contractor may have the overall responsibility, the contract should indicate these safety standards must be met by all subcontractors. Another concern on all construction jobs is site cleanup. The contract must specify daily and final cleanup responsibilities in detail.

All contracts should require that the subcontractor is responsible for having all the required licenses for his or her trade and for having a comprehensive, up-to-date insurance policy. Although legal differences exist across the country, it is advantageous to have the subcontractor agree to indemnify the general contractor if the subcontractor fails to pay workers compensation or injury claims, and the general contractor is held responsible for paying them. The subcontractor should provide a written warranty for the work done so the owner can contact the subcontractor if something fails or needs repair.

---

requests for contract changes or substitutions, and certifying applications for payments.

The construction phase typically ends when the owner submits the final payment to the contractor. During the construction phase of a project, the contractor is the primary participant of the ongoing work.

Construction Scheduling

The construction schedule assigns progressive dates to all project activities and phases. Project scheduling is used to match the resources of equipment, materials, and labor with the project construction phases over time. It indicates the estimated start and completion dates of various components, establishing milestones and sequences that help in coordinating the numerous subcontractors on the site. Poor scheduling can result in considerable waste as workers and equipment wait for needed resources or the completion of preceding tasks by other trades.

The most widely used scheduling technique is the *critical path method (CPM)* for scheduling, sometimes referred to as "critical path scheduling." This method calculates the minimum time required for a certain task, along with the possible start and finish times for the project activities. The duration of the critical path represents the minimum time required to complete a project. Computer programs for critical path scheduling are widely available and can efficiently schedule large and complex building projects.

Permitting

Prior to any work commencing on site, the constructor must secure a variety regulatory permits. The *building permit* is a certificate issued by the local governing *authority having jurisdiction (AHJ)* authorizing the construction of a project after a thorough review of the construction documents to ensure compliance with local building, safety, and fire codes. The building permit must be posted in a clearly visible location until the project is completed (Figure 1.17). Specialty subcontractors are responsible for securing permits to allow their own portion of the work to proceed.

Figure 1.17 A building permit is issued by the governing authority having jurisdiction that approves the construction of a project after a thorough review of the construction documents. The permit must be displayed throughout the duration of construction.

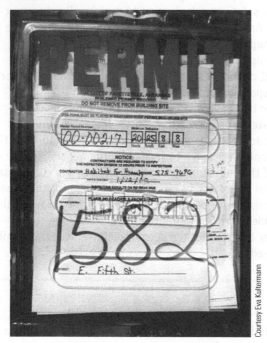

Courtesy Eva Kultermann

Most commercial projects will require a general building permit, as well as electrical, plumbing, HVAC, and other permits. Additional certifications for the use of cranes or the use of other heavy equipment may also need to be secured.

Inspectors employed with the jurisdiction will conduct periodic inspections of the work throughout the progress of construction to ensure that the installed work adheres to all applicable codes and standards. Permit fees are usually based on a percentage of the total construction costs. Any work involving obtaining utilities from the public right of way for the site and building typically requires a Public Way construction permit obtained from the local government. A more thorough discussion of building codes follows in Chapter 2.

Pre-Construction Planning and Temporary Facilities

Once a building permit has been secured and the contract between the owner and contractor is signed, the actual on-site construction can begin. The formal notice to proceed indicates that all pre-construction requirements have been met. Requirements for temporary facilities are outlined in Division 01 of the contract documents. On larger projects, the general contractor may be required to establish a field office on the construction site. Field offices are often housed in mobile trailers that are fully furnished and provided with modern communication equipment (Figure 1.18). The office provides a central command post for the day-to-day site activities, housing the construction documents, project files, and providing an office for the project superintendant. Temporary utilities must be brought to the site to provide the power and sanitation required by workers and equipment. Access for deliveries and equipment must be arranged, along with designated space for the storage and staging of materials.

Figure 1.18 A field office provides a central command post for the day-to-day site activities, housing the construction documents, project files, and provides an office for the project superintendent.

Courtesy Eva Kultermann

Many municipalities have instituted construction and demolition recycling ordinances requiring construction sites to recycle a percentage of the waste that is generated. This requires the contractor to identify a portion of the site to be used for the sorting and storage of construction waste materials to be picked up by a certified recycling agency. Materials such as wood, aluminum, steel, bricks and tile, and packaging materials can be easily recycled in today's expanding recycled materials market.

A preconstruction conference, attended by the owner, architect, general contractor or construction manager, and invited subcontractors is held to communicate the ongoing management procedures of the project. These conferences establish efficient communication channels, discuss payment procedures, the sequencing of work, and set a schedule for ongoing progress meetings. The completed construction documents, both drawings and specifications, provide the guidelines under which the project will be completed, but a number of additional submittals must be generated to fully define all portions of the work.

Shop Drawings, Submittals, and Mock-Ups

While the construction documents give detailed information as to how components and assemblies will be constructed, other more-specific drawings are required on most projects. A *shop drawing* gives precise directives for the fabrication of certain components, such as structural steel work, concrete reinforcing, or pre-cast concrete components. It is generated to explain the fabrication and sometimes installation procedures of the items to the manufacturer's production crew or contractor's installation crews (Figure 1.19). Shop drawings are substantially different from architectural drawings in

style and content, including manufacturing conventions, and special fabrication instructions. They are generated by the manufacturer of the material, reviewed by the contractor, and submitted to the architect for final approval. The fast processing of shop drawings is crucial to the start of any project, as the ordering of materials is often dependent on their completion.

The written specifications of the construction documents typically require the contractor to submit product data and material samples to the architect for final approval. The submittal process is an important and time-consuming step at the beginning of managing any construction project. Product data submittals are drawings, schedules, performance data, and brochures that give manufacturer's information on the characteristics of a material and allow the architect to verify that the product under consideration will satisfy the requirements listed in the specifications. A product sample is an actual physical example of a material that can be examined to assure that colors, textures, and other characteristics adhere to the original design intent.

Some contracts will ask for a sample of an entire building component, such as a wall panel, to be built for on-site evaluation. A *construction mock-up* is a full-size model of a proposed construction system built to judge the appearance of an assembly, examine its construction

Figure 1.19 A shop drawing outlining the fabrication of a steel beam and column system.

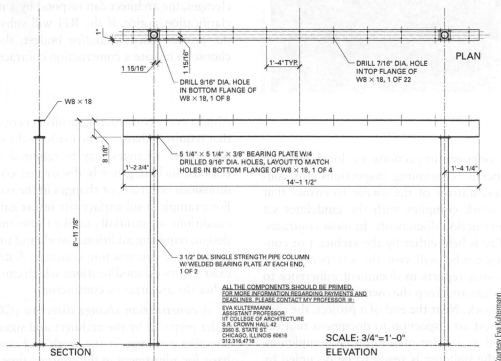

PLAN

1"

1 15/16"

1 15/16"

DRILL 9/16" DIA. HOLE
IN BOTTOM FLANGE OF
W8 × 18, 1 OF 8

1'–4" TYP.

DRILL 7/16" DIA. HOLE
IN TOP FLANGE OF
W8 × 18, 1 OF 22

W8 × 18

8 1/8"

1'–2 3/4"

5 1/4" × 5 1/4" × 3/8" BEARING PLATE W/4
DRILLED 9/16" DIA. HOLES, LAYOUT TO MATCH
HOLES IN BOTTOM FLANGE OF W8 × 18, 1 OF 4

14'–1 1/2"

1'–4 1/4"

8'–11 7/8"

3 1/2" DIA. SINGLE STRENGTH PIPE COLUMN
W/ WELDED BEARING PLATE AT EACH END,
1 OF 2

ALL THE COMPONENTS SHOULD BE PRIMED.
FOR MORE INFORMATION REGARDING PAYMENTS AND
DEADLINES, PLEASE CONTACT MY PROFESSOR @:

EVA KULTERMANN
ASSISTANT PROFESSOR
IIT COLLEGE OF ARCHITECTURE
S.R. CROWN HALL 42
3360 S. STATE ST.
CHICAGO, ILLINOIS 60616
312.316.4718

SCALE: 3/4"=1'–0"

SECTION   ELEVATION

Courtesy Eva Kultermann

details, and test for performance under actual site conditions (Figure 1.20). These models allow the contractor to investigate different means of achieving the level of workmanship required by the construction documents. Mock-ups often remain on the site throughout construction, to provide a basis against which to judge the finished construction as it progresses.

Figure 1.20 An example of a full-scale mock-up used to study façade details on the Sacramento Civic Center.

Courtesy Fentress Architects

Construction Observation

In addition to ongoing inspections by local building department officials, continuing inspections are conducted by representatives of the owner to ensure that the completed work complies with the guidelines set forth in the construction documents. In most contracts, this responsibility is held either by the architect or construction manager, who will visit the site periodically and prepare written reports to document adherence to the project schedule and keep the owner updated on the progress of the work. Near the end of a project, the contractor will request an inspection to document that the work has arrived at a point of substantial completion, indicating that the building is ready to be occupied by

the owner. Some owners may also secure the services of an independent inspection firm to conduct testing and inspections in an effort to ensure quality control. If inspections find either materials or workmanship to be in noncompliance, the contractor is required to repair or replace the work at their own expense.

Contractor Requests for Information (RFI)

During the construction process, it is not uncommon for the contractor to encounter portions of the work for which the contract documents do not provide clear and comprehensive directives. The *request for information (RFI)* is used by the contractor to obtain clarification on specific components and assemblies that are not fully detailed or understood through a comprehensive review of the drawings and specifications. These requests often occur during the bidding phase, as the contractor examines the documents in detail for the first time.

A contractor could ask for instance for more information on the details of an interior finish application. In order to obtain the answer, the contractor can submit a request for information to the architect. The RFI should not be used to request that an alternate product model or manufacturer be substituted for the one originally specified.

The RFI must be processed quickly by the architect, since both the project budget and schedule may be affected. If the response to the RFI involves no time or cost changes, the architect can respond by a memorandum or clarification notice. If the RFI will substantially change the project schedule and/or budget, the architect may choose to initiate a construction contract modification.

Modifications to the Construction Contract

Almost every construction project encounters conditions that require a change to be made to the original contract documents. Changes may be required for a number of reasons, including newly discovered conditions, design omissions or errors, or changes in the scope of the work. For example, a subsurface site investigation may expose conditions substantially unlike those anticipated during design, requiring additional work and time from that set forth in the construction contract. A number of means exist to process modifications which can be requested by either the architect or contractor.

A *construction change directive (CCD)* is a written order prepared by the architect and signed by the owner directing a change in the work and stating a proposed basis for adjustment in the contract time, cost, or both.

A *change order* is a written directive to the contractor issued after the execution of an agreement that authorizes an addition, deletion, or revision to the project, along with the related adjustments in contract budget and time.

Minor changes in the work can be addressed with a written document issued by the architect to document minor changes in the work that will not involve adjustments to the project cost or schedule.

Standard forms available from the American Institute of Architects are used to describe modifications, along with additional drawings and specifications when necessary. After completion, the order is approved by the owner and architect and becomes a modification of the construction contract. It authorizes the contractor to do the work and obligates the owner to cover the expense.

Project Close Out

Prior to all construction being concluded, a number of administrative benchmarks must be met to complete the contractual obligations between the owner and the contractor. When the contractor determines that the work is nearing completion, the architect or construction manager is asked to return to the site and conduct an inspection of substantial completion. The inspection occurs at a point when all major portions of the work have been completed but a number of final details have yet to be finished. A listing of the remaining items to be installed or repaired is called a *punch list*. The punch list is initially prepared by the contractor and then added to by the architect and engineering consultants during their inspections.

During project close out, the contractor is obligated to collate and submit all product and equipment warranties for the installed work to the owner. Warranties are common for such items as windows; roofing; appliances; and electrical, heating, and other mechanical equipment. The contractor also typically provides a warranty that the finished product is free of all defects and agrees that any work that requires it will be corrected within a one-year period. Operations and maintenance manuals are generated to educate building staff in the proper use and care of all building systems and equipment. Finally, record drawings are completed, indicating where modifications were made to the original contract documents.

Before a final certificate of completion can be issued, the inspecting authority must substantiate that all work is installed and complete, and that the quality of workmanship meets the specified standards. All equipment and appliances must have undergone functional testing and be working properly. The site must be thoroughly cleaned, surplus materials and temporary equipment removed, and surrounding streets and sidewalks returned to their pre-construction condition. The contractor will apply to the local jurisdiction for a *certificate of occupancy*. The certificate is issued by the local building department indicating that the building is compliant with locally adopted building codes and is in proper condition to be occupied.

Construction Safety

Throughout any construction project, a thorough accident prevention and safety training program must be implemented and maintained. The construction industry has one of the most hazardous work environments in the country, accounting for hundreds of fatalities and countless injuries each year. In 1970, the U.S. Congress passed the Williams-Steiger Occupational Health and Safety Act (OSHA) that initiated the regulation of safety standards for the construction industry. OSHA develops safety guidelines and administers their enforcement through a system of recurring site inspections. OSHA Standards-29 CFR–1926 Safety and Health Regulations for Construction, give extensive guidelines for construction means, methods, and materials handling. The standards cover every conceivable aspect of construction safety, from personal protective and life-saving equipment (Figure 1.21), tool and equipment safety, fire protection and prevention, materials storage, use, and disposal, to the signs, signals, and barricades that are required on a site.

Figure 1.21 Examples of common personal protective equipment used on the construction site.

The general contractor is responsible for administering safety procedures on the construction site. Detailed safety procedures are often specified in the general requirements division of the construction documents. An accident prevention program, including training and education, must be implemented at the start of any construction project. Workers must be trained in recognized safe working practices and the proper use of mandated personal protective gear. Copies of the current OHSA standards are required by law be kept in the field office along with material data sheets outlining the hazards associated with certain construction materials. All safety equipment and barriers must undergo regular inspections and be documented in a written log. With knowledge and expertise in construction safety, general contractors must provide the quality guidance that workers must have in order to prevent injury and the loss of life.

Review Questions

1. What are the three major divisions of the construction industry?

2. What professionals are responsible for the preparation of construction documents?

3. What is the difference between construction drawings and specifications?

4. How does the MasterFormat help all participants in the industry communicate?

5. What is the difference between a general contractor and subcontractors?

6. What are the three most commonly used types of project delivery methods?

7. How are contractors selected to work on a construction project?

8. What is the purpose of a shop drawing?

9. When is a punch list generated and what does it contain?

10. What purposes are served by OSHA?

11. What are material data sheets?

Key Terms

Authority Having Jurisdiction (AHJ)	Construction Phase	Invitational Bidding
Building Construction	Contract Document	MasterFormat
Building Information Modeling (BIM)	Critical Path Method (CPM)	Multiple Prime Contract
Building Permit	Design-Bid-Build (DBB)	Negotiated Contract
Building Section	Design-Build	Notice to Proceed
Certificate of Occupancy	Design Intent	Pre-design Phase
Change Order	Design Phase	Programming
Competitive Bidding	Elevation	Project Manual
Computer-Aided Drafting (CAD)	Fast-Track Project	Punch List
Construction Change Directive (CCD)	Floor Plan	Request for Information (RFI)
Construction Document (CD)	General Contractor (GC)	Shop Drawing
Construction Drawing	GreenFormat	Specifications
Construction Management at Risk	Heavy Construction	Subcontractor
Construction Management for Fee	Industrial Construction	Superintendent
Construction Mock-Up	Integrated Design Process (IDP)	

Activities

1. Contact the office of a local architect and ask them to conduct a tour of their facilities for the class. Request to see a completed set of construction documents while in the office.

2. Visit the Construction Specifications Institute website and investigate the online forums for current discussions and developments.

3. Ask a local contractor and one of their subcontractors to address the class and explain the process of competitive bidding and the preparation of bid documents.

Additional Resources

The Architects Handbook of Professional Practice, AIA Press, Washington, DC.

The Association of General Contractors, www.agc.org/

The Construction Specifications Institute, www.csinet.org/s_csi/index.asp

The Project Resource Manual, The CSI Manual of Practice, McGraw-Hill Publishing.

Fisk, Edward, 1997, *Construction Project Administration*, Upper Saddle River, NJ: Simon and Shuster.

Regulatory Constraints, Standards, and Sustainability

LEARNING OBJECTIVES

Upon completion of this chapter, the student should be able to:

- Discus the purposes served by zoning ordinances.
- Describe the content and application of the International Building Code.
- Understand the function of the Americans with Disabilities Act.
- Become familiar with professional and trade organizations and their contribution to the construction industry.

- List the impacts of the construction industry on global warming and resource depletion.
- Understand the concept of sustainable development and construction.
- Discuss the purposes served by green building rating and certification systems.

The primary objective of any construction project is that the completed building will be safe, provide a healthy environment for its occupants, use resources and energy as efficiently as possible, and preserve productive natural systems whenever possible. The design and construction of buildings is controlled by a variety of federal, state, and local regulations that have been enacted by law to ensure the safety of the public and the environment. Building and zoning codes regulate the design and construction of buildings by providing minimum standards established to safeguard both life and property. The codes are supported by standard writing organizations that develop additional best-practice guidelines. Further regulatory controls include the Americans with Disabilities Act (ADA) and a variety of environmental regulations. Chapter 2 examines the range of regulatory and additional standards and guidelines that influences the construction industry.

The chapter then examines the challenge shared by the construction industry of reducing resource consumption and protecting our environment. Construction activities consume large amounts of fossil fuels and other natural resources. Today, the industry is keenly aware of the potential to make a significant contribution in meeting this challenge.

THE REGULATORY ENVIRONMENT— ZONING AND CODES

Construction is regulated and supported by diverse local, state, and federal authorities to ensure that buildings and infrastructure will be built to industry-determined safety standards. Architects manage code issues during the design phase and contractors ensure that code-mandated requirements are met during the construction process.

Zoning Ordinances

Zoning ordinances are local land use regulations written into municipal law that regulate the degree of building development according to land use in order to promote the health, safety, and general welfare of a municipality and its citizens. The specific goals of zoning ordinances are varied but usually include the following components:

- Separating incompatible uses
- Providing for safe and efficient transportation

- Satisfying recreational needs by designating parks and open space
- Promoting aesthetic values
- Controlling density for safety and welfare reasons

Most zoning ordinances group buildings together into zoning districts that house similar activities in order to separate incompatible uses and promote cohesive city planning. Zoning districts are typically designated by a letter indicating the general use classification, such as R for residential and C for commercial zones. Typical zoning includes the following types of uses: single and multi-family residential zones, business and commercial zones, parks and open lands, and industrial zones (Figure 2.1). Within districts, the ordinances set guidelines for the density of population, height and area of buildings, building coverage on a property in relation to open space, setbacks from property lines, and ancillary facilities, such as parking. In certain jurisdictions, zoning ordinances may also stipulate building style and the types of materials or colors that can be used on building exteriors. Zoning ordinances are usually based on a comprehensive master plan. A master plan is a long-range plan for the growth of an entire municipality over time. The goal of a master plan is to direct future growth of a municipality in terms of the provision of public services, transportation, and utility infrastructure. Zoning ordinances and master plans can be obtained from the local department of buildings or on their municipal website.

Building Codes

Building codes regulate the design and construction of buildings and structures by providing minimum standards that are established to safeguard the life, health, property, and general welfare of the public. Within their jurisdiction, the codes prescribe the design of buildings, their permitted means of construction, and maintenance requirements. Their intent is to establish provisions for structural integrity, adequate light, ventilation and sanitation, energy conservation, and safety from fire and other hazards. Building codes control not only the initial construction of a building, but also any alteration, enlargement, repair, or demolition of buildings and other structures. The codes provide a basis upon which architects and engineers design buildings, and material and equipment suppliers fabricate products. They ensure the stability and safety of buildings by providing minimum construction standards, and ensuring that they are adhered to through inspection and verification of compliance.

Building codes are legally binding documents and, as such, are voted upon and approved by a municipalities governing body. They are adopted and administered by either local, county, or state authorities that review construction documents, issue building permits, and conduct periodic inspections of the work (see Chapter 1). Most jurisdictions adopt model codes, although some cities choose to write and maintain municipal codes that are tailored to address the specifics of local conditions.

The Development of Building Codes

Regulations regarding construction safety have been in use since the dawn of humanity. The ancient Babylonians decreed in writing that builders whose constructions led to occupant injury could be punished. The code of Hammurabi from the second century BC stated: *"If a builder build a house for someone, and does not construct it properly, and the house which he built fall in and kill its owner, then that builder shall be put to death."*

In early America, Thomas Jefferson promoted the development of building regulations to provide minimum standards that would ensure health and safety. The growth of modern building codes originates with early efforts to prevent the outbreak of fires that were commonplace in the predominantly wood-constructed cities of the nineteenth century. Around the same time, the first laws intended to assure adequate light and ventilation in residential buildings were enacted. While still concerned primarily with fire safety, the codes have expanded over time to include other safety issues, such as structural design and the protected exiting of a building.

Until 1995, most codes were written, maintained, and revised by four model code organizations, each adopted in different regions of the country. These organizations relied heavily on standards developed by groups such as the American Society for Testing and Materials (ASTM), the American Society of Heating, Refrigerating, and Air-Conditioning Engineers (ASHRAE), and the National Fire Protection Association (NFPA). Local and state building ordinances typically adopted one of the four model codes: The Uniform Building Code (UBC) published by the International Conference of Building Officials (ICBO), the Standard Building Code (SBC) published by the Southern Building Code Congress International (SBCCI), the Boca National Building Code (BOCA/NBC) published by the Building Officials and Code Administrators International (BOCA), or the International Code (IC) published by the International Codes Council (ICC).

Figure 2.1 This municipal zoning map defines residential, commercial, parkland, and industrial districts.

Because of the difficulties associated with inconsistent codes regulating construction across different geographical regions of the country, the ICC moved to develop one comprehensive model code by working cooperatively with the three model code organizations. After years of intensive research, gathering input from design professionals, building contractors, material manufacturers, government organizations, and trade associations, the new *International Building Code (IBC)* was unveiled in 1997. The new IBC provides a model code for adoption by all government agencies across the United States and other countries. The use of this code places all construction in the country under the same regulations related to design, materials, and construction techniques. While the original four model codes are still available for adoption, their use has diminished over the years. In addition to the Building and Residential codes, the ICC publishes and maintains a suite of 13 model codes, each addressing different construction aspects (Table 2.1).

Table 2.1 International Code Council Suite of Codes

The ICC Model Codes
International Building Code: Regulates all structures except one- and two-family dwellings.
International Energy Conservation Code: Outlines requirements for energy performance.
International Existing Building Code: Concerned with alterations to existing buildings.
International Fire Code: Deals specifically with fire prevention and suppression.
International Fuel Gas Code: Concerned with fuel gas used for heating.
International Mechanical Code: Relates to heating, ventilating, and air-conditioning systems.
ICC Performance Code: A rewriting of the ICC code as performance provisions.
International Plumbing Code: Outlines requirements for water supply and waste systems.
International Private Sewage Disposal Code: Regulates individual sewage systems.
International Property Maintenance Code: Relates to the upkeep of existing structures.
International Residential Code: Regulates construction for residential dwellings.
International Wild Land Urban Interface Code: Addresses the mitigation of fire in the urban-wild land interface.
International Zoning Code: Promotes uniformity in zoning for city planners and code officials.

Each code references national consensus standards and is comprehensive in itself, and all are compatible with each other. For example, the International Energy Conservation Code (IECC) addresses energy efficiency by regulating and promoting reduced energy usage and the conservation of natural resources. The codes are not static documents but are continuously updated, amended, and expanded for republication on a three-year cycle.

The International Residential Code

The International Building Code regulates all structures within the jurisdictions where it has been adopted, except one- and two-family residential dwellings. To deal with residential buildings, the ICC Board of Directors teamed with the National Association of Home Builders (NAHB) to form the ICC/NAHB Task Force. Input from this partnership led to the recommendation that the ICC develop and maintain a standalone residential code for one- and two-family dwellings and multiple single-family dwellings, such as townhouses. The task force developed a comprehensive residential code that is consistent with and inclusive of the scope and content of the existing codes promulgated by the BOCA, ICBO, and SBCCI. The resulting cooperative work led to the *International Residential Code for One- and Two-Family Dwellings*, first published in 2000.

The International Building Code (IBC)

The International Building Code establishes minimum regulations for commercial building systems using both prescriptive- and performance-related provisions (Figure 2.2). A prescriptive code specifies exactly what type of assembly or material may be used and gives explicit guidelines for its construction. A performance code lists the exact operating characteristics of a component and references industry standards to describe a minimum level of performance that an assembly must provide, without stipulating the exact materials or implementation strategies. Performance codes allow for a greater amount of flexibility in how requirements are met.

The written text of building codes is sometimes technically complex and difficult to interpret. The authority having jurisdiction determines the intent and interpretation of a particular code passage. The designer, contractor, and material supplier must work closely with the local building official who has ultimate authority in how the wording of the code will be interpreted.

Many cities have municipal building codes that may be significantly different from the IBC, or may exceed

Figure 2.2 The International Building Code (IBC) establishes minimum regulations for commercial building systems using prescriptive and performance-based provisions.

© International Code Council

the IBC's requirements in certain instances. A check with the local government buildings department is required to understand what codes and standards must be followed.

Organization and Content of the International Building Code

The IBC is composed of 34 chapters, each pertaining to a particular construction component or material. The code begins with an administrative introduction that provides instructions as to how the code is to be adopted, administered, and enforced by a jurisdiction. The first chapter outlines the applicability of the code, clarifies the duties of building officials, dictates permitting and inspection procedures, and assigns fee schedules. A second chapter gives a listing of definitions for the language used throughout the text of the code.

Use and Occupancy Classification Buildings and structures in the IBC are controlled by a classification system according to their intended function. The use that a facility is intended to accommodate determines

the occupancy classification it is assigned under the code. Structures are classified with special requirements based on use and occupancy in one or more of the following groups (Table 2.2):

Table 2.2 Occupancy Classifications

Group A	Assembly
Group B	Business
Group E	Educational
Group F	Factory and Industrial
Group H	High hazard
Group I	Institutional
Group M	Mercantile
Group R	Residential
Group S	Storage
Group U	Utility and Miscellaneous

Occupancy classifications are concerned with how many people are expected to occupy an area according to the potential hazards to their safety from within and outside of the building. A building that accommodates more than one type of use is classified as mixed-occupancy and must conform to the code requirements of both, or whichever is more restrictive. In addition to the main occupancy categories listed, the code specifies individual requirements for what are termed special uses—such as high-rise buildings or covered malls—that may require additional clarification. For instance, a building that houses hazardous materials will require supplementary fire protection equipment than required of normal occupancies. The occupancy group classification of a building governs much of the requirements covered in the remaining topics of the code, including allowable height and area, and the type of construction that may be utilized.

Types of Construction The IBC classifies building construction according to five categories based on the types of materials used and their degree of combustibility. The construction type of a building gives an indication the amount of time required for emergency response in case of a fire.

Chapter six of the IBC designates five types of construction by Roman numerals, with Type I being the most fire resistant and Type V being the least. Type I and II construction are inherently noncombustible and utilize materials such as concrete and steel (Figure 2.3). Type III construction utilizes exterior building materials that are noncombustible, with structural (load-bearing)

Figure 2.3 The IBC classifies building construction according to five categories based on the materials used and their combustibility. Type I and II construction are inherently noncombustible and utilize materials such as concrete and steel.

© Racheal Grazias/iStock/Shutterstock.com

materials that may be of varying degrees of combustibility. Type IV construction defines exterior walls that are of noncombustible materials, with interior building elements of solid or laminated wood without concealed spaces. This type is often used to designate heavy timber construction. Type V is that type of construction in which the structural elements, exterior walls, and interior walls are of any material. Most wood frame buildings are designated as Type V construction.

Each of these broad categories is further subdivided into A–protected, and B–unprotected. "Protected" describes a building in which all structural members have an additional fire-rated coating or cover that extends the fire resistance of the assembly. "Unprotected" means that all structural members of a building or structure have no additional fire-rated protective cover. Protected buildings are generally allowed greater building heights and areas. Further subcategories account for the effects of a building being provided with an automatic sprinkler system.

Building Area and Height   The height and area limitation for buildings of different construction types is governed by their intended use and construction type. Table 504.3 in chapter five of the IBC specifies the maximum floor areas allowed according to occupancy and construction type for various building types (Figure 2.4). Buildings of Type I and II construction are generally allowed unlimited areas due to their inherent

incombustibility. Construction types of less fire resistance must conform to the maximum area allowed.

The permissible height and number of floors of a building are also tabulated in Table 504.3 of the IBC. The building height is considered the vertical dimension from the ground plane to the average height of the highest roof surface. For example, based on Table 504.3, an A-1 assembly occupancy using Type I–A (protected) construction is allowed both unlimited height and area. If the same building uses Type 5–A construction, its area is limited to 11,500 square feet and its height to two stories. Certain building types, such as low-hazard industrial processes that require large areas and unusual heights, are exempt from area and height limitations. Code requirements must be balanced with local zoning ordinances, which may be more or less restrictive.

Means of Egress   Chapter 10 of the International Building Code is concerned with providing both safe ways for occupants to exit a building during a fire, and safety to firefighters and first responders during emergency operations. The IBC defines a *means of egress* as, "*A continuous and unobstructed path of vertical and horizontal egress travel from any occupied portion of a building or structure to a public way. A means of egress consists of three separate and distinct paths: the exit access, the exit, and the exit discharge*" (Figure 2.5). The required number and types of exits from a building, as well as the width of the egress path, vary according to building type, occupant load (the number of occupants), construction type, and whether or not a building has sprinklers. The code goes on to define the physical requirements for egress paths, including minimum width and headroom, allowable projections, and the surface characteristics of the path. Travel distance, the distance a person must travel from any point in a building to a public way, is also specified. For most non-sprinkler occupancies, travel distances are set at 200 feet. Specific guidelines are given for exit signage, illumination, and the configuration of stairs, doors, gates, and turnstiles.

Engineering Requirements   Subsequent chapters of the IBC define specific engineering guidelines for the structural adequacy of a building structure, enclosure, and finishes. Chapters 14 and 15 of the IBC outline requirements for the exterior envelope of a building, covering walls and roofs including building openings, architectural trim, and projections such as balconies and canopies. Chapters 16 through 18 of the IBC set forth design criteria for different types of structural

Figure 2.4 Table 503 of the ICC specifies maximum building height and area allowed according to occupancy and construction type for different occupancy types.

		TYPE OF CONSTRUCTION								
		TYPE I		TYPE II		TYPE III		TYPE IV	TYPE V	
		A	B	A	B	A	B	HT	A	B
HEIGHT (feet)		UL	160	65	55	65	55	65	50	40
GROUP		STORIES(S) AREA(A)								
A-1	S	UL	5	3	2	3	2	3	2	1
	A	UL	UL	15,500	8,500	14,000	8,500	15,000	11,500	5,500
A-2	S	UL	11	3	2	3	2	3	2	1
	A	UL	UL	15,500	9,500	14,000	9,500	15,000	11,500	6,000
A-3	S	UL	11	3	2	3	2	3	2	1
	A	UL	UL	15,500	9,500	14,000	9,500	15,000	11,500	6,000
A-4	S	UL	11	3	2	3	2	3	2	1
	A	UL	UL	15,500	9,500	14,000	9,500	15,000	11,500	6,000
A-5	S	UL	UL	UL	UL	UL	UL	UL	UL	UL
	A	UL	UL	UL	UL	UL	UL	UL	UL	UL
B	S	UL	11	5	3	5	3	5	3	2
	A	UL	UL	37,500	23,000	28,500	19,000	36,000	18,000	9,000
E	S	UL	5	3	2	3	2	3	1	1
	A	UL	UL	26,500	14,500	23,500	14,500	25,500	18,500	9,500
F-1	S	UL	11	4	2	3	2	4	2	1
	A	UL	UL	25,000	15,500	19,000	12,000	33,500	14,000	8,500
F-2	S	UL	11	5	3	4	3	5	3	2
	A	UL	UL	37,500	23,000	28,500	18,000	50,500	21,000	13,000
H-l	S	1	1	1	1	1	1	1	1	NP
	A	21,000	16,500	11,000	7,000	9,500	7,000	10,500	7,500	NP
H-2[d]	S	UL	3	2	1	2	1	2	1	1
	A	21,000	16,500	11,000	7,000	9,500	7,000	10,500	7,500	3,000
H-3[d]	S	UL	6	4	2	4	2	4	2	1
	A	UL	60,000	26,500	14,000	17,500	13,000	25,500	10,000	5,000
H-4	S	UL	7	5	3	5	3	5	3	2
	A	UL	UL	37,500	17,500	28,500	17,500	36,000	18,000	6,500
H-5	S	4	4	3	3	3	3	3	3	2
	A	UL	UL	37,500	23,000	28,500	19,000	36,000	18,000	9,000
I-1	S	UL	9	4	3	4	3	4	3	2
	A	UL	55,000	19,000	10,000	16,500	10,000	18,000	10,500	4,500
I-2	S	UL	4	2	1	1	NP	1	1	NP
	A	UL	UL	15,000	11,000	12,000	NP	12,000	9,500	NP
I-3	S	UL	4	2	1	2	1	2	2	1
	A	UL	UL	15,000	10,000	10,500	7,500	12,000	7,500	5,000
I-4	S	UL	5	3	2	3	2	3	1	1
	A	UL	60,500	26,500	13,000	23,500	13,000	25,500	18,500	9,000
M	S	UL	11	4	2	4	2	4	3	1
	A	UL	UL	21,500	12,500	18,500	12,500	20,500	14,000	9,000
R-1	S	UL	11	4	4	4	4	4	3	2
	A	UL	UL	24,000	16,000	24,000	16,000	20,500	12,000	7,000
R-2	S	UL	11	4	4	4	4	4	3	2
	A	UL	UL	24,000	16,000	24,000	16,000	20,500	12,000	7,000
R-3	S	UL	11	4	4	4	4	4	3	3
	A	UL	UL	UL	UL	UL	UL	UL	UL	UL
R-4	S	UL	11	4	4	4	4	4	3	2
	A	UL	UL	24,000	16,000	24,000	16,000	20,500	12,000	7,000
S-1	S	UL	11	4	2	3	2	4	3	1
	A	UL	48,000	26,000	17,500	26,000	17,500	25,500	14,000	9,000
S-2[b,c]	S	UL	11	5	3	4	3	5	4	2
	A	UL	79,000	39,000	20,000	39,000	26,000	38,500	21,000	13,500
U[c]	S	UL	5	4	2	3	2	4	2	1
	A	UL	35,500	19,000	8,500	14,000	8,500	18,000	9,000	5,500

Figure 2.5 The IBC defines a means of egress as a continuous path of travel from any occupied portion of a building to a public way.

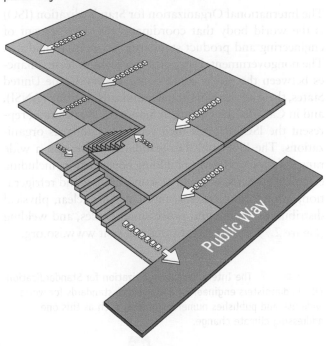

loading, structural tests and inspections, and the construction of foundation systems. The code defines two types of loading conditions that act on a structural system. The weight of the actual materials of construction, walls, floors and roofs, and finishes is known as the *dead load*. Conversely, a *live load* is a result of occupancy and use, consisting of people, furnishings, and environmental forces such as wind, snow, and seismic (earthquake) activity. Chapters 19 through 26 of the IBC give criteria for the use of specific materials, including concrete, aluminum, masonry, steel, wood, glass and glazing, gypsum board and plaster, and plastics. References to the ICC Electrical and Mechanical codes are given in lieu of detailed regulations for these systems. Final chapters of the code specify requirements for elevators, conveying systems, and other special constructions.

AMERICANS WITH DISABILITIES ACT (ADA)/UNIVERSAL DESIGN

The Americans with Disabilities Act (ADA) is federal legislation enacted in 1990 that prohibits discrimination on the basis of disability in employment, state and local government, public accommodations, commercial facilities, transportation, and communications. The ADA is not a code, but a federal law that is enforced by legal action through the courts. Title III: of the law—Public Accommodations—has a significant impact on the design of buildings by requiring that all public accommodations and commercial facilities be accessible and usable by individuals with disabilities. The majority of the title III regulations are concerned with ensuring access to people in wheelchairs, but the legislation also addresses the needs of the blind, hearing and cognitive impaired, and the elderly.

The ADA regulates both new construction and alterations to existing buildings. New commercial building designs must comply with the ADA immediately, while existing structures are required to provide accessibility when they are remodeled. For existing buildings, the language of the legislation stresses that building modifications must be "readily achievable," meaning that compliance should not create an undue burden to the building owner.

Specific building components that are impacted by the ADA include building access, both inside and outside; parking requirements; and the design of facilities such as bathrooms and drinking fountains. A detailed explanation of the ADA guidelines for building structures can be found in the *ADA Accessibility Guidelines for Buildings and Facilities* (ADAAG) (Figure 2.6). In recent years, ADA guidelines have included issues in relation to the public right of way. Developers may

Figure 2.6 A graphic from the ADA Accessibility Guidelines illustrating the minimum clear width passage for a wheelchair.

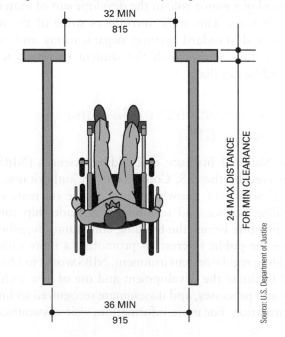

Source: U.S. Department of Justice

be required to provide ADA ramps with detectable warning strips, and sidewalks and drives with ADA compliant slopes. Local building departments should be consulted for the latest ADA regulations and standard details.

In moving beyond the barrier-free guidelines of the ADA, *Universal Design* advocates the design of products and environments that are usable by all people, to the greatest extent possible, without the need for adaptation or specialized design. The concept encourages architects and product designers to make all spaces and fixtures accessible and comfortable to users with a variety of needs and abilities, including the disabled, elderly, expectant women, and individuals with temporary injuries. Universal design accommodates the widest range of individual abilities by creating spaces that are easy to understand regardless of the user's experience, knowledge, or language skills.

STANDARD DEVELOPMENT ORGANIZATIONS

In addition to the building codes, many professional, private, and governmental organizations make major contributions to the safe and efficient use of construction materials. These organizations support the construction industry by conducting research and testing of materials and components and widely disseminating the resulting information. Many of the provisions cited in the building codes reference standards produced by some of the following organizations. Trade associations play a major role in the development of materials and systems. This topic introduces some of the most influential standard writing organizations and trade associations with which the student of construction should be familiar.

National Institute of Building Sciences (NIBS)

The National Institute of Building Sciences (NIBS) is chartered by the U.S. Congress as an authoritative national source of knowledge and advice on matters of building science and technology. A leadership forum, the institute brings the building community together to serve the public interest by promoting a more rational building regulatory environment. NIBS works to identify and facilitate the development and use of new technology and processes, and disseminate recognized technical information. For more information, visit www.nibs.org.

International Organization for Standardization (ISO)

The International Organization for Standardization (ISO) is the world body that coordinates the development of engineering and product standards for international use. The nongovernmental organization seeks to form alliances between the public and private sectors. In the United States, the American National Standards Institute (ANSI), and in Canada, the Canadian Standards Association, represent the ISO through their national standards organizations. The ISO publishes standards addressing a wide range of topics related to building construction, including electrical systems, heating, air-conditioning and refrigeration, information systems, mechanical, nuclear, physical distribution, piping and processing, textiles, and welding (Figure 2.7). For more information, visit www.iso.org.

Figure 2.7 The International Organization for Standardization (ISO) administers engineering and product standards for worldwide use and publishes numerous guides, such as this one addressing climate change.

American National Standards Institute (ANSI)

The American National Standards Institute (ANSI) is the coordinating organization for the national standards system in the United States. ANSI serves the nation in a variety of ways. Through cooperative efforts of its member organizations, ANSI's councils, boards, and committees coordinate the efforts of the hundreds of organizations in the United States that develop standards. ANSI helps identify what standards are needed and establishes timetables for their completion. It arranges for competent organizations to develop them or, if a standard-developing organization does not exist, it forms teams who have the necessary competencies to develop the standard. ANSI provides and administers the only recognized system in the United States for establishing standards, regardless of the originating source, as American National (Consensus) Standards.

There are now more than 10,000 American National Standards in virtually every field and discipline. Standards deal with dimensions, ratings, terminology and symbols, test methods, and performance and safety specifications for materials, equipment, components, and products in the following fields: construction, electrical and electronics, heating, air-conditioning and refrigeration, information systems, medical devices, mechanical, nuclear, physical distribution, piping and processing, photography and motion pictures, textiles, and welding. An American National Standard is designated by code numbers beginning with the letters ANSI followed by letters indicating the organization that formulated the standard, such as ACI (American Concrete Institute). These letters are followed by an identification number and the date the standard was issued or released. A typical example is ANSI/ACI 308-92, *Standard Practice for Curing Concrete*. For more information, visit www.ansi.org.

ASTM International

ASTM International, previously known as the American Society for Testing and Materials (ASTM), is a nonprofit corporation formed for the development of standards on the characteristics and performance of materials, products, systems, and services, and for the promotion of related knowledge. The standards include test methods, definitions, recommended practices, classifications, and specifications. Standards developed by the society's committees are published annually in an 80+ volume publication, *Annual Book of ASTM Standards* (Figure 2.8). ASTM International specifications are designated by the initials ASTM followed by an identification number and the year of last revision. One example is *ASTM A325-79, High-Strength Bolts for Structural Steel Joints*.

ASTM also publishes *Standards Adjuncts*, which include supporting material such as reference charts, tables, and drawings used in conjunction with ASTM test methods. They are identified by a single letter of the

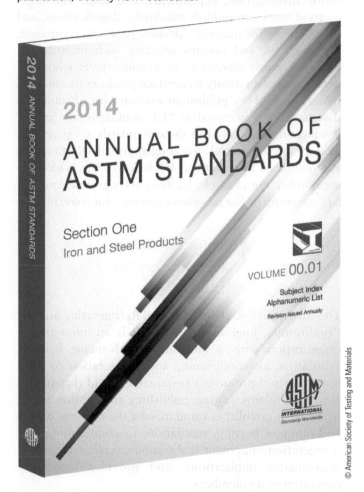

Figure 2.8 Standards developed by the American Society of testing and materials are published annually in a 48-volume publication, *Book of ASTM Standards*.

© American Society of Testing and Materials

alphabet followed by a code number and the word "Adjunct." An example is *Adjunct C856 Practice for Petrographic Examination of Hardened Concrete*. ASTM also publishes various technical publications such as: Manuals, Data Series, and *Special Technical Publications*, which are papers presented on specific technical, scientific, and standardization topics. They are identified by the letters STP followed by a code number. An example is STP 691 Durability of Building Materials and Components. For more information, visit www.astm.org.

Underwriters Laboratories Inc. (UL)

Underwriters Laboratories Inc. (UL) is a nonprofit organization that establishes, maintains, and operates laboratories for the examination and testing of devices and materials to determine their potential for hazards to life and property. UL conducts research to determine the probability of various materials, products, equipment, assemblies, and systems to hazards related to their use affecting life and property. This is accomplished by scientific investigation, experiments, and tests. The organization works to publish standards, classifications, and specifications for materials, devices, products, equipment, constructions, and systems affecting such hazards. UL publishes yearly directories of manufacturers who have demonstrated an ability to produce products meeting UL requirements. They publish an extensive list of standards that have been accepted as "UL Standards for Safety." Products manufactured to these standards are permitted to carry the UL symbol indicating approval. UL standards are designated by the initials UL followed by an identifying number. For example, UL943 covers ground-fault circuit interrupters. For more information, visit www.ul.com.

American Society of Heating, Refrigerating and Air Conditioning Engineers (ASHREA)

The American Society of Heating, Refrigerating and Air Conditioning Engineers (ASHRAE) is an international organization with a mission of "advancing heating, ventilation, air conditioning and refrigeration to serve humanity and promote a sustainable world through research, standards writing, publishing and continuing education." It establishes standards for the uniform testing and rating of heating, ventilation, air conditioning, and refrigeration equipment. It also conducts related research, disseminates publications, and provides continuing education to its members.

The *ASHRAE Handbook* is an influential four-volume publication that is considered the practical store of knowledge on the various topics that form the field of heating, ventilation, air-conditioning, and refrigeration (HVAC&R). The four volumes are *Fundamentals, Refrigeration, HVAC Applications,* and *HVAC Systems and Equipment.* For more information on ASREA, visit www.ashrae.org.

National Association of Home Builders of the United States (NAHB)

The National Association of Home Builders (NAHB) is a federation of approximately 800 state and local builders associations in the United States. NAHB established and staffs the National Housing Center in Washington, DC; manages a National Housing Center Library; and publishes a number of newsletters and magazines. It issues reports pertaining to economic research and analysis and offers nationwide seminars, workshops, and conferences.

The NAHB Research Foundation is a subsidiary of the National Association of Home Builders. Its four divisions serve as a center of technical research for the residential and light frame construction industries. The Laboratory Services Division performs tests on materials and components under contract from manufacturers. The Industrial Engineering Division designs and performs cost and economic studies of alternative construction techniques and materials. A Building Systems Division provides technical studies of functional systems in small buildings. The Special Studies Division initiates projects in new areas of interest to the residential construction industry. For more information, visit www.nahb.org.

TRADE ASSOCIATIONS

Other organizations are devoted to the advancement of knowledge about individual materials and methods used in construction. Some of these have developed guidelines that influence the content of the building codes and establish industry-wide standards. Trade associations are a major factor in materials development. A *trade association* is an organization whose membership comprises manufacturers or businesses involved in the production or supply of materials and services in a particular area. An example is the American Institute of Steel Construction (AISC). The goal of a trade association is to promote the interests of its membership. This includes dissemination of information on the proper use of the materials, products, and services of the members. Some trade associations support research into the proper use and improvement of construction materials and better construction methods. As they do this, they develop material specifications and performance procedures for a particular trade area. Often, these become accepted national standards. Some trade associations are involved in programs of certification. Products made according to their specifications and standards are identified with a seal, label, or stamp. An example is the certification stamp on plywood issued by APA–The Engineered Wood Association. Following are brief introductions to some of the most commonly referenced trade associations in the construction industry.

American Concrete Institute (ACI)

The American Concrete Institute (ACI) is a technical and educational organization dedicated to improving the design, construction, manufacture, and maintenance of concrete products and reinforced concrete structures (Figure 2.9). ACI disseminates its information through

Figure 2.9 The American Concrete Institute helps to develop standards dedicated to improving construction of reinforced concrete structures.

© sima/iStock/Shutterstock.com

conventions, seminars, publications, and volunteer committee work. The ACI series of technical publications is extensive, and one, *The ACI Manual of Concrete Practice,* is a primary reference for the industry. For more information, visit www.concrete.org.

Portland Cement Association (PCA)

The Portland Cement Association (PCA) is a nonprofit organization dedicated to improving and promoting the use of Portland cement and concrete. Supported by the voluntary financial contributions of its member companies, the PCA serves as a clearinghouse for concrete design practices and construction methods. PCA provides research and development/engineering services that include troubleshooting field problems, engineering investigations, development services, and assistance with the manufacture of Portland cement. It has an extensive service program that

offers publications, motion pictures, and slides depicting up-to-the-minute developments. For more information, visit www.cement.org.

The Masonry Institute of America

The Masonry Institute of America was founded in 1957 as a technical and research organization established to improve and extend the use of masonry. The Masonry Institute of America is active in the development of national building codes, promoting new ideas and masonry work, improving existing building codes, conducting research projects, presenting design, construction and inspection seminars, and writing technical and non-technical papers, in order to improve the masonry industry. For more information, visit www.masonryinstitute.com.

American Institute of Steel Construction (AISC)

The American Institute of Steel Construction (AISC) is a nonprofit association serving the fabricated structural steel industry. Its objectives are to improve and advance the use of fabricated structural steel through research and engineering studies to develop the most efficient and economical design of structures. The institute publishes manuals, textbooks, specifications, and technical books. The most widely used publication is the *Manual of Steel Construction.* For more information, visit www.aisc.org.

American Wood Council (AWC)

The American Wood Council (AWC) provides an organizational structure for wood products companies and associations to work together on building codes and standards, green building policy issues, and a focused set of environmental regulations. AWC's engineers, scientists, and building code experts develop engineering data, technology, and standards on structural wood products for use by design professionals, building officials, and wood products manufacturers to assure the safe and efficient design and use of wood structural components. For more information, visit www.awc.org.

SUSTAINABLE DESIGN AND CONSTRUCTION

Humanity faces perhaps its greatest challenge ever, to support a continued growth in living standards worldwide within the limits of diminishing natural resources and escalating levels of environmental pollution. This challenge is driving transformation at all levels of government, industry, and private society. The primary goal of sustainable construction is to reduce the damage caused by inefficiently designed structures through the application of traditional building design strategies used in clever combination with new technological advances. This section introduces some of the environmental issues that are relevant today and outlines specific technologies to lessen their impact in the design and construction of buildings.

The Environmental Context

At the dawn of the twenty-first century, the global population reached the 6 billion mark and is expected to continue growing at an annual rate of 1.2 percent. In an attempt to satisfy the needs of this ever-expanding population, natural resources are being depleted at unprecedented rates. Land that previously supported natural systems and agricultural production is coming under increasing development pressure. In the United States alone, some 34 million acres, an area roughly equal to the state of Illinois, was developed between 1982 and 2001. Sometimes referred to as *urban sprawl*, this unchecked land development places tremendous burdens on natural systems (Figure 2.10).

Figure 2.10 This housing development is an example of the rapid unchecked municipal growth known as urban sprawl.

For the past half-century, global energy consumption patterns have followed hand in hand with economic development. During this period, energy consumption doubled to keep up with the demands of the growing population and an ever-increasing array of technological developments. The fossil fuels we use today are the result of millions of years of forces and pressures on biological materials being broken down into pure carbons (coal)

and petroleum oils. These fossil fuels are a nonrenewable resource; once they have been consumed, they cannot be immediately replaced. Today, nearly all of the known petroleum reserves have been discovered and most are fully explored. Geologists predict that in the near future, global oil producers will reach "peak production," a point after which securing more oil will become increasingly difficult and expensive (Figure 2.11).

Figure 2.11 Geologists predict that in the near future securing more oil will become increasingly difficult and expensive.

The larger problem of the increased use of fossil fuels is the resulting impact on the environment. Scientists now believe that climate change, which for decades had been a hotly contested theory, has become a documented fact. Recent findings by the Intergovernmental Panel on Climate Change (IPCC) suggest, *"there is new and stronger evidence that most of the warming observed over the last fifty years is attributable to human activities"* (Figure 2.12). As human societies have adopted increasingly automated lifestyles, the amount of heat trapping gases present in the atmosphere has multiplied. The greenhouse effect is a process by which the earth's atmosphere traps solar radiation due to the existence of gases, such as carbon dioxide (CO_2), water vapor, and methane, that allow incoming sunlight to pass through but absorb the heat radiated back from the earth's surface. The burning of fossil coal, oil, and natural gas, coupled with widespread deforestation, has raised the total amount of atmospheric CO_2 by 30 percent since the beginning of the industrial revolution (Figure 2.13). Even within conservative estimates, global warming is predicted to have detrimental consequences for forest productivity, agricultural production, and air and water quality. Additional long-term consequences include a continued melting of polar ice, resulting in rising sea

Figure 2.12 This figure shows how annual average temperatures worldwide have changed since 1901.

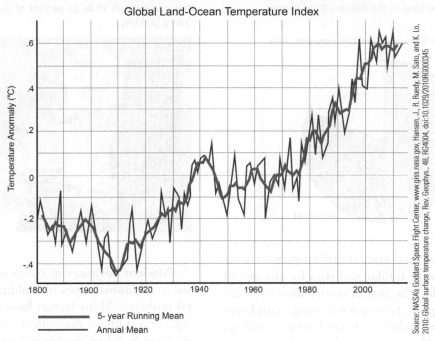

Global Land-Ocean Temperature Index

Source: NASA's Goddard Space Flight Center, www.giss.nasa.gov, Hansen, J., R. Ruedy, M. Sato, and K. Lo, 2010: Global surface temperature change. Rev. Geophys., 48, RG4004, doi:10.1029/2010RG000345

Figure 2.13 The burning of fossil coal, oil, and natural gas coupled with widespread deforestation has raised the total amount of atmospheric CO_2 by 30 percent since the beginning of the Industrial Revolution.

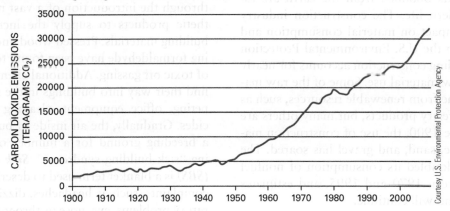

Courtesy U.S. Environmental Protection Agency

levels, and more frequent extreme weather events. The IPCC, which for decades had been concerned with developing strategies for climate change mitigation, is today advancing policies for climate change adaptation.

The Environmental Impact of the Construction Industry

Buildings, both commercial and residential, are a major contributor to energy and resource depletion. Each year, vast amounts of the raw materials and energy produced in the world are used in the construction industry,

generating millions of tons of greenhouse gasses and additional concentrations of solid wastes. According to the U.S. Energy Information Administration (EIA), the amount of energy consumed by buildings of all types accounts for around 40 percent of our total annual primary energy consumption, more than is used by industry or transportation (Figure 2.14). More than half of that total was attributed to residential building use, with the balance ascribed to commercial buildings. Around 65 percent of total electricity use is used in the heating, cooling, and lighting needs of buildings. Combined, this translates into buildings contributing

Figure 2.14 Buildings, both commercial and residential, account for nearly 40 percent of the nation's total energy consumption.

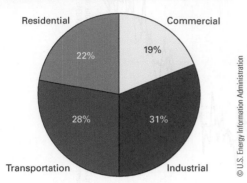

End-Use Sector Shares of Total Consumption, 2011

© U.S. Energy Information Administration

Figure 2.15 The amount of waste generated during construction can be from 10 to 15 percent of the total materials that go into a building.

© Robert Asento/iStock/Shutterstock.com

a solid 30 percent of total annual carbon emissions for the United States, with similar statistics for the rest of the developed world. The largest contributor to building energy consumption is in space heating, with lesser amounts utilized by lighting, water heating, cooling, and refrigeration. The mining and manufacturing of construction materials requires additional amounts of energy input.

Natural materials obtained from the earth are essential to our modern life. The construction industry has an extensive impact on material consumption and waste. According to the U.S. Environmental Protection Agency (EPA), building construction accounts for nearly 30 percent of all raw material use. Some of the raw materials required come from renewable resources, such as agricultural and forestry products, but many others are nonrenewable. Since 1900, the use of construction materials such as stone, sand, and gravel has soared. The world population doubled its consumption of nonfuel raw materials between 1970 and 1995, and estimates predict that this trend will continue.

Construction and demolition waste accounts for more than 100 million tons disposed in U.S. landfills annually, representing almost 40 percent of the total amount of municipal solid waste produced (Figure 2.15). *Construction and Demolition Debris (C&D)* are waste materials generated by building and demolition activities, such as scrap, damaged or temporary materials, packaging, and others wastes. Studies haves shown that the amount of waste can be from 10 to 15 percent of the total materials that go into a building, much higher than the 3–5 percent generally assumed by material estimators. Especially problematic are wastes from sources such as solvents, paints, or chemically treated wood that can result in soil and water pollution.

Modern advances in component technologies have led to additional types of building-related environmental problems. Many factors have contributed to increasing complaints of discomfort associated with building occupancy. When mechanical systems are not properly maintained, the exchange of contaminated air with fresh air is reduced and bacterial agents can enter the air stream. Indoor air quality is further compromised through the introduction of a vast number of new synthetic products to supply the increased demand for building materials. Pressed wood and fiber panels utilizing formaldehyde have been found to be a major culprit of toxic off gassing. Additional chemical concentrations find their way into buildings in the form of paints, carpeting, office equipment, cleaning agents, and pesticides. Gradually, the air inside of buildings has become a breeding ground for a number of maladies, including "sick building syndrome." *Sick building syndrome (SBS)* is a blanket term used to describe building-related complaints, such as headaches, dizziness, nausea, respiratory problems, eye, nose or throat irritations, and skin problems, to name a few.

Energy efficiency, renewable energy technologies, and the use of healthy, renewable materials in buildings have an enormous potential to mitigate climate change, improve our quality of life, increase productivity through better occupancy conditions, and reduce operating costs.

SUSTAINABLE CONSTRUCTION

The challenge of reducing resource consumption and protecting our environment is not an optional one, but an imperative for governments, corporations, and all levels of industry. Today, the construction industry is keenly aware of the potential to make a significant contribution in meeting this challenge. The primary goal of sustainable building is to lessen the negative impact related to construction by combining the best traditional building design strategies in clever combination with new technological advances (Figure 2.16). The ultimate purpose is to create buildings, communities, and urban areas that support healthy living through clean water and air, while protecting natural resources and habitats. The commonly accepted definition for sustainable development was formulated in 1987 by the United Nations World Commission on Environment and Development as: *"Sustainable development is development that meets the needs of the present without compromising the ability of future generations to meet their own needs."*

Figure 2.16 Sustainable building seeks to combine the best traditional building strategies in clever combination with new technological advances.

Sustainable development is a far-reaching concept, encompassing social and economic in addition to environmental concerns.

Sustainable Design

A number of terms are used to describe sustainable building including *green building*, high-performance building, and climate-responsive building. The American Society for Testing and Materials (ASTM) defines green building as, *"a building that provides the specified building performance requirements while minimizing disturbance to and improving the functioning of local, regional, and global ecosystems both during and after its construction and specified service life."* The organization continues, advocating a design and construction process that emphasizes a reduction in the use of materials, resources, and energy. Green building seeks to achieve resource and energy efficiency, integrate healthy materials, and promote ecologically and socially sensitive land use (Figure 2.17).

Figure 2.17 Green building seeks to achieve resource and energy efficiency, integrate healthy materials, and promote ecologically and socially sensitive land use.

The basic strategies that define sustainable building design and construction are outlined in Table 2.3.

While much of the work to be done in the creation of sustainable buildings rests in the domain of architects

Table 2.3 Sustainable Building Design and Construction Guidelines

Site	Water
Develop building sites to restore and enhance natural ecosystems.	Use water-efficient fixtures, appliances, and equipment.
Develop and restore previously disturbed sites instead of unspoiled rural or natural areas.	Minimize the use of potable water for landscape irrigation.
Develop native and drought-tolerant landscape systems.	Utilize rainwater harvesting for both building and landscape uses.
Minimize site paving and use pervious paving systems.	**Material Resources**
Minimize construction activity-related pollution.	Reuse existing buildings and materials whenever possible.
Energy	Design buildings for long service life and eventual recycling of components.
Promote energy conservation by avoiding energy-intensive operations.	Use renewable and recycled-content materials.
Use daylight and energy-efficient electric lighting systems.	Practice waste management during construction to minimize waste.
Use renewable energy sources, such as solar, wind, and geothermal energy.	Use locally or regionally sourced materials to minimize transportation energy.
Provide high levels of thermal insulation and an air-tight building envelope.	**Indoor Environment**
Use natural ventilation in place of mechanical ventilation.	Use non-toxic, natural materials and finishes.
Assure that all installed systems are calibrated for optimal functioning.	Curtail sources of volatile organic compounds and other pollutants.
Utilize energy and heat recovery.	Increase the use of natural daylight.
	Provide high levels of natural ventilation.

and engineers, the importance of considering sustainability during the construction process is crucial. Green building design is implemented on the construction site. Decisions made in the field regarding material procurement, site preservation during construction, equipment use, and the sequence and organization of the work are the responsibility of the contractor. By implementing a combination of traditional techniques and new technologies, the construction industry can reduce and sometimes eliminate a great deal of the environmental damage associated with construction processes (Figure 2.18).

Sustainable Building Certification Systems

In order to both implement and market sustainable design, it is important to be able to make definitive assessments so that building designs can be certified for performance. A number of organizations provide a means to both guide and quantify sustainable building through a recognized system of certification. Most are voluntary rating systems that have been developed by both public and private organizations. In addition to certifying the environmental performance of a building, the systems provide a means to compare and rate the success of different green building strategies.

The U.S. Green Building Council (USGBC) and LEED Rating System

The U.S. Green Building Council (USGBC) was founded in 1993 to identify and promote strategies with which the design and construction industry could advance the creation of energy- and resource-efficient buildings. The nonprofit organization is made up of a diverse membership including architects, contractors, material manufacturers, financial and insurance firms, research institutions, and government organizations. Council members cultivate new standards, practices, and guidelines for the design, construction, and operating procedures of high-performance sustainable buildings. Working in a committee-based, consensus-driven process, the USGBC developed the *Leadership in Energy and Environmental Design (LEED)* green building rating system. The aim of LEED is to define principles for sustainable building by providing a common standard of measurement. LEED is a third-party certification program that provides a nationally accepted standard for the design, construction, and maintenance of high-performance buildings. The rating system promotes an integrated, whole-building design process that stimulates green competition while also helping to raise public awareness of green building. After years of research and development, LEED version 1 was launched in 1998, with only a handful of projects

Figure 2.18 A diagrammatic view of the Culver House project, illustrating the strategies that define sustainable building design and construction.

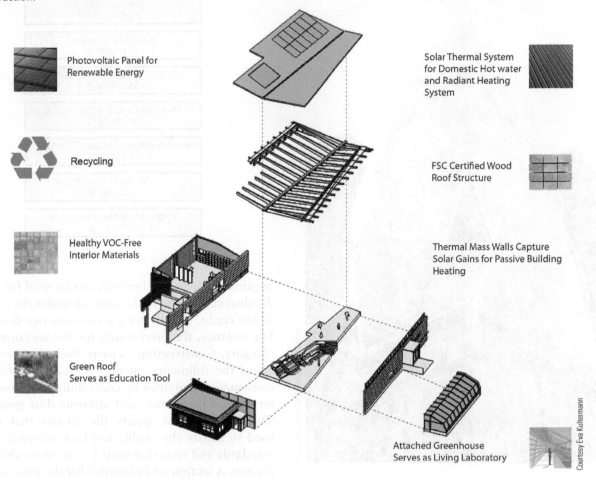

Photovoltaic Panel for Renewable Energy

Solar Thermal System for Domestic Hot water and Radiant Heating System

Recycling

FSC Certified Wood Roof Structure

Healthy VOC-Free Interior Materials

Thermal Mass Walls Capture Solar Gains for Passive Building Heating

Green Roof Serves as Education Tool

Attached Greenhouse Serves as Living Laboratory

Courtesy Eva Kultermann

achieving certification. Continued development and the incorporation of regional concerns has led to the current standard, LEED version 4, launched in 2013.

In order to address a variety of building and construction types, the USGBC has developed a suite of rating systems that are tailored to specific project types (Figure 2.19). The currently available LEED standards are outlined in Table 2.4.

LEED NC Version 3

The LEED rating system provides a tool that simultaneously guides and evaluates a building over its entire life cycle, from project inception through building operation. Certification is based on achieving a certain number of points, or credits for each building standard that a project adopts. The standards are organized into five major topics: Sustainable Sites, Water Efficiency, Energy and

Atmosphere, Materials and Resources, and Indoor Environmental Quality. Each category contains a number of base criteria, or prerequisites, which must be achieved in order for other credits in the category to be counted. The collected LEED building strategies amount to 100 credits, with two additional categories, Innovation in Design and Regional Bonus Credits, bringing the total number of points to 110 (Figure 2.20). The Innovation in Design category awards credits to projects that exhibit exceptional performance in achieving energy and resource efficiency. Regional and bonus credits can be awarded by local chapters to recognize specific regional issues. The USGBC certifies individuals who have completed a comprehensive training program as *LEED accredited professionals (LEED AP)*. The inclusion of a LEED AP on a project team to support the design and application process adds one credit toward certification. Depending on the number of points achieved, a project

Figure 2.19 The Hearst Tower in New York achieved a Gold rating under LEED Core and Shell rating system.

© Michel Stevelmans/Stock/Shutterstock.com

Figure 2.20 LEED categories and possible credits.

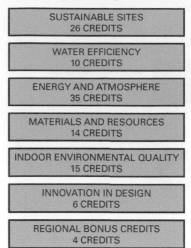

SUSTAINABLE SITES
26 CREDITS

WATER EFFICIENCY
10 CREDITS

ENERGY AND ATMOSPHERE
35 CREDITS

MATERIALS AND RESOURCES
14 CREDITS

INDOOR ENVIRONMENTAL QUALITY
15 CREDITS

INNOVATION IN DESIGN
6 CREDITS

REGIONAL BONUS CREDITS
4 CREDITS

Table 2.4 LEED Rating Systems

LEED for New Construction: for buildings that are undergoing new construction or major renovations that include a complete interior fit-out

LEED for Existing Buildings: Operations & Maintenance

EED Commercial Interiors

LEED Core & Shell

LEED for Schools

LEED Retail

LEED Healthcare

LEED Homes

LEED Neighborhood Development

requirements, and technologies to be used for each individual credit. The basic intent identifies the main goal of the credit, usually with a one-sentence description. For instance, the prerequisite for the sustainable sites category, Construction Activity Pollution Prevention, states the following intent: "Reduce pollution from construction activities by controlling soil erosion, waterway sedimentation, and airborne dust generation." The *Requirements* specify the criteria that must be used to satisfy the credit, and lists accepted industry standards and tests that must be met to establish compliance. A section on *Submittals* list the basic submittal requirements, templates, calculations, written summaries, drawings, and diagrams that must be sent to the USGBC for verification. Finally, a summary of references and standards cites the referenced standards on which the credit is based. The reference guide accompanies each credit explanation with a section entitled *Green Building Concerns* that provides a short narrative overview into the subject listing environmental, economic, and community concerns.

Certifying a project is a three-step process consisting of project registration, submittal of technical documentation, verification, and final certification by the USGBC LEED certification provides a number of tangible benefits including recognition of environmental stewardship by third-party certification, a plaque and official certificate for display, and a variety of marketing opportunities (Figure 2.21). Certification under LEED adds to project costs, requiring additional consultants to manage the documentation of credits and perform energy modeling. Costs are currently estimated at anywhere between 3 to 5 percent of total project expenditures. The LEED

can be designated as either: certified (40–49 credits), silver (50–59 credits), gold (60–79 credits), or platinum (80 credits and above), the highest rating possible.

The rating system is explained by a reference guide that lists in detail the intent, implementation, and submittal

Figure 2.21 LEED certification provides benefits to building owners including recognition of environmental stewardship and a plaque for display.

system is growing in popularity, with more federal and local municipalities requiring certification for all public buildings within their jurisdiction.

Green Globes

The Green Building Initiative (GBI) is a nonprofit organization that defines its mission as: *"to accelerate the adoption of building practices that result in energy-efficient, healthier and environmentally sustainable buildings by promoting credible and practical green building approaches for residential and commercial construction."*

The Green Globes Rating System provides a third-party green building verification procedure that is administered entirely online. The system delivers an assessment procedure that gives guidance for the design, construction, and management of a building. Like LEED, the system addresses specific building and project types. Point categories in the Green Globes system include project management, site, energy, water, resources, emissions and other impacts, and the indoor environment. The system is interactive and flexible, providing project teams with immediate feedback and a less-costly certification option.

NAHB National Green Building Program™

The National Association of Home Builders (NAHB) and the International Code Council (ICC) have collaborated to establish a nationally administered green certification program for single- and multifamily homes, residential remodeling, and site development projects. The ICC 700-2012 National Green Building Standard provides a flexible third-party rating system for residential green building certification. The guidelines were developed through a public process that included an extensive review of local green home builder programs, existing energy-efficiency programs endorsed by NAHB, and of the leading life cycle analysis (LCA) tools available for use in residential design and construction. Similar to the NAHB Model Green Homebuilding Guidelines, the main purpose is to provide a framework for builders to reduce the energy- and resource-related impacts of a home. Each topic addressed in the program has a point value attributed to it that is evaluated according to environmental impact, building science and best building practices, and ease of implementation.

The NAHB Research Center provides third-party certification for the program. There are four different levels of certification available to builders wishing to rate their projects: bronze, silver, gold, and emerald. A minimum number of points are required for each of the six guiding principles to ensure that green building strategies are addressed in a balanced and integrated way. After reaching the minimum requirements, an additional 100 points must be achieved by implementing any of the remaining line items.

The program is administered through a Green Scoring Tool available on the NAHB website. Contractors can compile information about the products and strategies used in a residential project. The tool specifies points that are awarded per the criteria of either the Guidelines or the Standard. The NAHB Research Center Green Certified mark identifies a project has been inspected at least twice by an independent verifier to confirm that every green point earned has been correctly implemented.

Energy Star

ENERGY STAR® is a U.S. Environmental Protection Agency (EPA) voluntary program for products, buildings, and homes. Established in 1992, the ENERGY STAR label is widely used to certify appliances, office equipment, lighting fixtures, home electronics, and, more recently, residential, commercial, and industrial buildings. The program administers a third-party verification system to certify the energy performance of new homes and remodeling projects, and publishes a series of builder's guides to help explain sustainable technologies. Unlike the LEED and Green Globes rating systems,

Heifer International Headquarters

Architect: Polk Stanley Rowland Curzon Porter Architects, Ltd.	Building type: Commercial office	Rating: U.S. Green Building Council
Contractor: CDI Contractors, LLC	Size: 94,000 sq. feet	LEED Platinum (52 points)
Location: Little Rock, Arkansas	Completed February 2006	

Overview

Heifer International is a nonprofit organization dedicated to fighting global hunger and poverty. It provides education in sustainable agriculture and gives donations of farm animals and plants to disadvantaged families around the world. The design intent of the new building was to provide a high performance building that would visibly communicate the sustainable building features utilized. The architects and construction team held frequent meetings early in the design process to communicate the implementation and documentation requirements of LEED criteria. The contractors played an active role in the process, suggesting alternate materials and construction methods to ensure compliance with sustainability goals.

The Heifer International Headquarters building was designed to clearly express its sustainable features.

Courtesy of Tarek Zohdi, University of California

Site Design

The building occupies a former railroad switching yard that required environmental remediation to remove leftover petroleum and other chemicals. New parking, paved with permeable materials and marked by bioswales between rows, leave the majority of the site free for walking paths, native vegetation, and a constructed wetland. A *bioswale* is a linear drainage ditch with gently sloped sides covered in vegetation designed to remove silt and pollution from surface runoff water. Storm water runoff is filtered through the bioswales before being reused for landscape irrigation.

Energy

The building minimizes energy use in a variety of ways, each designed to be easily understood. Large expanses of clear glass ensure that sunlight will provide the majority of lighting required during working hours. The building's 62-foot width allows every employee access to light and views of the adjacent river and parkland. Electric lights are set on automatic dimmers in response to daylight levels, and occupancy sensors ensure that lighting activates only when needed. Extensive shading systems prevent unnecessary heat gain from reaching the interior. The building envelope, both opaque walls and glazing, utilizes high levels of thermal insulation. The completed building uses 55 percent less energy than a conventional office building of the same size.

Materials

Construction materials for the project were sourced locally to minimize the energy used in transporting them to the site. The materials for the structural steel (97 percent recycled) and the aluminum and glass façade were fabricated by adjacent plants. The architects selected high-recycled content, renewable, low-toxicity, and durable materials for the interior finishes including flooring of bamboo and cork, insulation of recycled cotton and spray-on foam made from soybean by-products, recycled brick, and countertops made of recycled glass. To ensure good indoor air quality, only materials with low levels of volatile organic compounds (VOCs) were utilized. During construction, the contractor recycled 75 percent of the building's construction waste, by weight.

The steel structure and aluminum and glass facades were both fabricated by adjacent manufacturing plants reducing energy use in the long distance transportation of materials.

Water

The roof shape allows rainwater to be captured in a 42,000-gallon water tower wrapped with a glass-enclosed fire stair. The water is reused in flushing toilets, in floor-radiant heating installation, and in the building cooling system. Low-water-use plumbing fixtures, together with waterless urinals, greatly reduce the amount of potable water required to operate the building.

In addition to achieving the LEED platinum rating, the building has won numerous awards, including a U.S. Environmental Protection Agency award for Brownfield remediation (see Chapter 4). Heifer's headquarters provides a showcase for high-performance, energy-efficient sustainable construction, and is an inspiration for other project teams.

certification under ENERGY STAR entails a series of site inspections during the construction process. These site visits are useful in identifying construction errors, allowing corrections to be made to ensure full compliance with the ENERGY STAR guidelines. The latest version of LEED now incorporates minimum benchmarks that align ENERGY STAR with the categories in LEED as part of the certification process. New homes certified under ENERGY STAR receive an official certificate that identifies the contractor and third-party verification entity (Figure 2.22).

Other Assessment and Certification Programs

In the United Kingdom, the British Research Establishment's Environmental Assessment Method (BREEAM) offers a verification standard that is similar to the LEED system in content and organization. The BREEAM rating system has been approved for use in Canada and several European countries. The Living Building Challenge (LBC) is perhaps the most stringent certification of sustainability in the built environment possible today. The LBC requires that all of a building's energy needs must be supplied by on-site renewable energy, and that all storm water and building water discharge

Figure 2.22 New homes certified under ENERGY STAR receive an official certificate that is posted on a buildings breaker box.

must be managed on site and used to supply the building's water needs. The LBC also mandates a strict approach to the use of materials "Endorsing products and processes that are safe for all species through time." Under the LBC, a "Red list" contains materials and chemicals that cannot be used on a project, including asbestos, mercury, lead, and polyvinyl chloride (PVC) among others. Their strict criterion makes the LBC the most difficult certification to obtain of the certification systems reviewed.

Review Questions

1. Describe the purpose of local zoning ordinances.
2. What is the main intent of a building code?
3. What model building code is currently available for adoption in the United States and Canada?
4. What is the purpose of the American with Disabilities Act?
5. What is universal design?
6. What is the mission of the American Society of Testing and Materials?
7. What purposes are served by trade associations?
8. What are the environmental impacts of the construction industry?
9. What is urban sprawl?
10. What is meant by sustainable development?
11. What is the purpose of the LEED rating system?
12. What are the five broad topics addressed by the LEED rating system?
13. What are the purposes of the ENERGY STAR program?

Key Terms

Bioswale

Building Code

Construction and Demolition Debris (C&D)

Dead Load

Green Building

LEED (Leadership in Energy and Environmental Design)

Live Load

Means of Egress

Sick Building Syndrome

Trade Association

Universal Design

Urban Sprawl

Zoning Ordinance

Activities

1. Visit a municipal zoning department and obtain a copy of local zoning districts for study.
2. Ask a local building inspector to address the class and review local building code requirements and the inspection process.
3. Contact one of the trade organizations and request information concerning its activities and publications.
4. Contact your local U.S. Green Building Council chapter to identify a LEED certified building in your area. Set up a tour for the class to visit the building.

Additional Resources

ENERGY STAR, www.energystar.gov/index.cfm?c=home.index

International Code Council, 2006 International Building Code

U.S. Green Building Council, www.usgbc.org

United Nations World Commission on Environment and Development (1987), *Our Common Future*, Oxford: Oxford University Press.

Center for Universal Design (2008) *Universal Design Principles*, www.design.ncsu.edu/cud/

See Appendix C for addresses of professional and trade associations and other technical information.

Properties of Materials

Upon completion of this chapter, the student should be able to:

- Identify the properties of materials that must be considered when specifying construction products.
- Discuss the technical aspects of mechanical, thermal, acoustical, and chemical properties.

- List the criteria for the selection of environmentally preferable materials.
- Explain the concepts of embodied energy and life cycle analysis.

For a material to be suitable for a particular application, it must have predictable behavior. When the properties of the material are defined, the designer can calculate its behavior and verify performance against guidelines specified by codes and standards. As new materials are developed each year, the architect, engineer, and contractor must be informed of their properties in order to use these materials in the most effective and safest manner. The process of selecting materials for construction is far more complex than you might think. In addition to a material's availability, strength, durability, and cost implications, fundamental new questions arise as the construction industry attempts to define what materials are sustainable and how to evaluate them.

Material selection involves researching potential materials and evaluating their characteristics in order to employ the materials according to their qualities and properties.

MATERIAL GROUPS

For the purposes of discussing material properties, construction materials can be grouped into four broad categories: metals, non-metallic inorganic materials, polymers, and organic materials. *Metals* are refined from ores that have been extracted from the earth (Figure 3.1). Due to the addition of a mixture of materials, metals have a wide variation in properties.

Figure 3.1 Construction materials can be grouped into four broad categories: metals, non-metallic inorganic materials, polymers, and organic materials. Metals are refined from ores that have been extracted from the earth.

© hxdbxyz/Shutterstock.com

Most metals are ductile, strong, and are especially useful when tensile forces are expected to occur.

Non-metallic inorganic materials are also extracted from the soil. Often referred to as ceramics, they are refined into a variety of products (Figure 3.2). Typical inorganic materials include sand, limestone, glass, brick, cement, gypsum, mortar, and mineral wool insulation. Most are hard, rigid, brittle, and heavy, and are especially useful when compression forces are expected to occur.

Figure 3.2 Non-metallic inorganic materials include sand, limestone, glass, brick, cement, and gypsum.

Polymers consist of large molecules composed of repeating structural units connected by chemical bonds. While "polymer" in common usage suggests plastic, the term actually refers to a large class of natural and synthetic materials that can be shaped when heated and harden when cooled. Polymers are used in a variety of construction applications, including pipes and conduit, wire and cable, textiles, adhesives, roofing, and insulation, among others (Figure 3.3).

Figure 3.3 Polymers are used in a variety of construction applications including pipes and conduit, wire and cable, textiles, adhesives, and roofing.

Organic materials include wood, grasses, bitumen, and many synthetic materials based on a chemical compound containing carbon. Examples include wood and paper products, asphalts and rubber (Figure 3.4).

Because all of these materials vary a great deal in their properties, it is important to know how they will respond, according to established design requirements

Figure 3.4 Organic materials include wood, grasses, bitumen, and many synthetic materials based on a chemical compound containing carbon.

and tabulated test results. The material properties that most influence decisions in construction can be divided into five categories: mechanical, thermal, acoustical, chemical, and environmental.

MATERIAL PROPERTIES

Materials chosen for construction applications must be able to perform under a variety of loadings, stresses, and environmental conditions. All materials respond to changes in applied load, temperature, and moisture content. A building structure consists of numerous construction components, such as beams, columns, walls, and floors. The dead load of a building is defined as the structural load that results from the actual materials of construction, including walls, floors, and ceilings that are permanently fixed. In addition to supporting its own weight, a building structure must also resist a variety of additional short-lived loads. Live loads are the structural forces acting on a building resulting from occupancy and use, such as those caused by people, furnishings, storage items, and environmental loads, including wind, snow, and seismic (earthquake) activity. Dead loads are invariable, and are considered static loads because they are unchanging and stay constant. Live loads, on the other hand, are defined as dynamic loads, those that can change rapidly and abruptly in force and duration. Figure 3.5 illustrates the basic loads and forces that act upon a building structure.

Mechanical Properties

Mechanical properties are a measure of a material's ability to resist the variety of mechanical forces just described. They are related to the response of a material to

Figure 3.5 Building structures and materials must resist a variety of loads and forces.

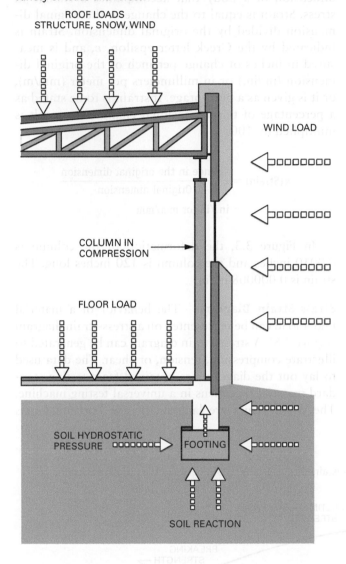

Figure 3.6 Stresses are created by tension, compression, and shear forces.

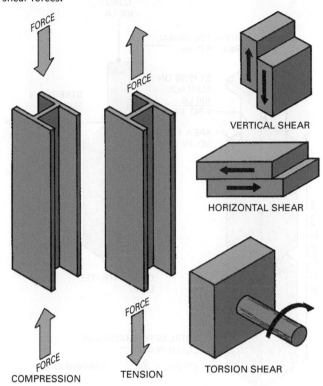

both static (dead) and dynamic (live) loads. Mechanical properties include tensile strength, compressive strength, shear strength, elasticity, ductility, hardness, impact resistance, fatigue strength, and permeability.

Stresses: Tension, Compression, Shear When a building load, or force, acts on a building component, such as a column, the component takes on an internal resistance to the force, known as "stress." As shown in Figure 3.6, there are three possible types of stresses: tension, compression, and shear. *Tensile stresses* are created when forces pull on a member and tend to increase its length. *Compressive stresses* push on a member and tend to shorten it. *Shear stresses* produce forces that work in opposite directions parallel with the plane of the force, causing adjacent parts of a material to slide past one another.

How a material behaves under load and the amount of stress a member can withstand is determined through standardized tests, as specified by the American Society for Testing and Materials (ASTM).

Stress and Strain When a load is applied to a member, it causes a stress-strain condition. *Stress* is the intensity of internally distributed forces that resist a change in the form of a body. It is measured in pounds per square inch (psi), mega-Newtons per square meter (MN/m²), or mega-Pascals (MPa). Stress is equal to the load divided by the area upon which it is acting. Normal stress is indicated by the Greek letter sigma, σ.

$$\sigma \text{ (Stress)} = \frac{\text{Load}}{\text{Area}} = \text{psi, MN/m}^2, \text{MPa}$$

The formula applies to materials that have a force applied parallel with the axis of the member. It is important to know if a load is to be distributed over an area or is concentrated on a small point. For example, if a 100-pound load is distributed over a 2-inch square column, the stress is 25 psi (100 divided by 2 × 2). If the same load is concentrated on the end of a rod with a cross-sectional area of 1 inch squared, the stress would be 100 psi (Figure 3.7).

Figure 3.7 Columns are subjected to stress and strain forces.

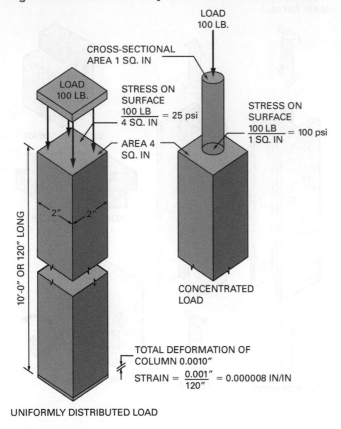

UNIFORMLY DISTRIBUTED LOAD

Strain is the change per unit length in the linear dimension of a body that accompanies a change in stress. Strain is equal to the change in the original dimension divided by the original dimension. Strain is indicated by the Greek letter epsilon, ε, and is measured in inches of change per inch of the original dimension (in./in.) or in millimeters per meter (mm/m), or it is given as a percentage. If strain is to be stated as a percentage of the dimensional change, the result is multiplied by 100.

$$\varepsilon\,(\text{Strain}) = \frac{\text{Change in the original dimension}}{\text{Original dimension}}$$

$$= \text{in./in. or mm/mm}$$

In Figure 3.3, the deformation of the column is 0.0010 inches and the column is 120 inches long. The strain is 0.000008 in./in.

Stress-Strain Diagrams The behavior of a material under load can be represented on a stress-strain diagram (Figure 3.8). A stress-strain diagram can be generated to illustrate compression, tension, or shear. The data used to lay out the diagram are obtained from testing standard material specimens in a universal testing machine. The strain values are laid out on the horizontal axis

Figure 3.8 A typical stress-strain diagram. Note that this material sustained considerable strain between the ultimate strength and before it fractured.

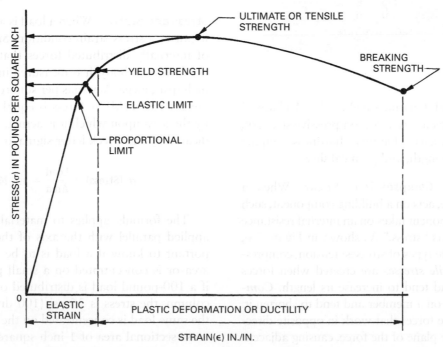

(abscissa) and the stress values on the vertical axis (ordinate). An increase in stress will produce a proportionate increase in strain, up to the elastic limit. The *elastic limit* is the highest stress that can be imposed on a material without permanent deformation. This produces a straight line on the diagram from zero to the elastic limit stress value.

Modulus of Elasticity The *modulus of elasticity (E)* is a proportional constant between stress and strain. For stresses within the elastic range, E, sometimes referred to as E-value, equals stress divided by strain. The modulus of elasticity is basically a measure of stiffness and rigidity. For beams, E-values are an indication of their resistance to deflection (bending). The modulus of elasticity then defines the stiffness of a material, governs deflections, and influences buckling behavior. Table 3.1 lists E-values for selected construction materials. Remember that there is no single value of E for a particular material, because of the many variations in the makeup of materials. For example, there are many types of plastics, each having their own stress-strain behavior. Within materials, such as steel, there are variations in the alloying ingredients used. Steel demonstrates high E-values, and provides a strong material where loads must be carried with

minimal deflection. The E-value of wood is higher than that of plastics but lower than that of concrete. Wood is useful in carrying light structural loads over short distances, as is evidenced in the use of residential wood joists and rafters. A comparison of the relative stiffness of selected construction materials is shown in Figure 3.9.

Table 3.1 Modulus of Elasticity for Selected Materials

Material Group	Material	10⁶ psi
Metals	Aluminum alloys	25
	Cast iron, malleable	25
	Copper alloy	13–17
	Steel, carbon	29
Ceramic Materials	Brick	1.8–2.2
	Concrete	1.0–5.0
	Glass	9–11
	Stone	10–15
Organic Materials	Particleboard	0.9–2.0
	Plastics, molded	0.1–1.9
	Plastics, reinforced	0.1–2.2
	Wood, 2", visually graded	0.9–1.6

(These are approximate. Values will vary with specific samples.)

Figure 3.9 A comparison of modules of elasticity for selected construction materials.

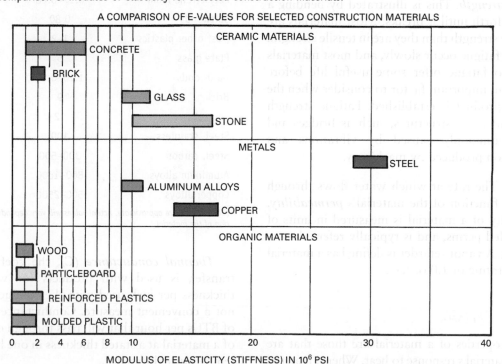

Ductility *Ductility* refers to the ability of a material to be deformed plastically without actually breaking or fracturing. Copper, for example, is a very ductile material. When tested under tension, it stretches for some distance without breaking.

Hardness *Hardness* is a measure of the ability of a material to resist indentation or surface scratching. It is the result of several properties of a material, such as elasticity, ductility, brittleness, and toughness. Hardness is related to the tensile strength of the material. Generally, a material that has a high hardness rating also will have a high tensile strength. This is not true for all materials; concrete, for example, has high hardness and low tensile strength.

Impact Strength *Impact strength* is the ability of a material to resist a rapidly applied load, such as the strike of a hammer. It is an indication of the toughness of the material. A material with high impact strength will absorb the energy of impact without fracturing. Impact strength is affected by strength and ductility. Metals that are strong and ductile have high impact strength. Ceramics are strong in compression but are brittle (lack ductility) and break under impact. In general, plastics are very ductile but low in strength and do not absorb impact well.

Fatigue Strength Resistance of a material to a cyclic load, one that varies in direction and/or magnitude, is called *fatigue strength*. This is illustrated by bending a wire back and forth until it breaks. Most materials are lower in fatigue strength than they are in tensile strength. Failures due to fatigue occur slowly, and most materials that fail due to fatigue offer some useful life before failure. This is an important factor to consider when the life cycle of a product is established. Fatigue strength in buildings and other structures, such as bridges and elevators, is commonly caused by vibrations and rotational motion produced by machinery.

Permeability The rate at which water flows through a material is a function of the material's *permeability*. The permeability of a material is measured in units of permeance, called perms, and is typically referred to as the perm rating. A vapor retarder is defined as a material having a perm rating of 1.0 or less.

Thermal Properties

The thermal properties of a material are those that are related to the material's response to heat. When a material is subjected to a change in temperature, it may expand, contract, conduct, or reflect heat. Construction materials can be classified either as insulators, materials that resist the transfer of heat, or conductors, materials that encourage the transfer of heat. Fibrous materials, such as rock wool, cotton, and cork, are good insulators, whereas most metals are good conductors.

Thermal Conductivity and Thermal Conductance *Thermal conductivity (k)* denotes the ability of a material to transfer heat from an area of high temperature to an area of lower temperature. Thermal conductivity is measured in the number of BTUs (British thermal units) that can pass through one square foot of a material one inch thick when the temperature difference between the two surfaces is one degree Fahrenheit. It is expressed as (Btu. in)/(hr. ft.^{2} °F) or W/(m^2 °C). Metals tend to have the highest values of thermal conductivity. Non-metallic inorganics have much lower conductivities, and organic materials have the lowest (Table 3.2). Materials with low thermal conductivity resist the transfer of heat and tend to have excellent insulation qualities.

Table 3.2 Thermal Conductivity (k) of Selected Materials

Material	(Btu/in.)/ (hr./ft.^{2}/°F) (English units)	W/(m^2°C) (metric units)
Polyurethane foam	0.11–0.14	0.016–0.020
Polystyrene foam	0.17–0.28	0.024–0.040
Softwoods	0.80	0.12
Most other plastics	1–2	0.14–0.28
Plate glass	5	0.72
Hardwoods	1.10–0.16	0.16–0.02
Brick	9	1.3
Concrete	12	1.7
Steel, stainless	100	14
Steel, carbon	300–500	43–70
Aluminum alloys	800–1600	115–230
Copper alloys	500–2500	70–360

(These values are approximate. Actual values will vary depending on the composition of the material.)

Thermal conductance (C), the coefficient of heat transfer, is used with products for which the unit thickness per inch used with the conductivity values is not a convenient measure. Conductance is the number of BTUs per hour that will pass through one square foot of a material at a stated thickness in one hour per degree temperature difference.

Composite Thermal Performance A single expression is used to describe the overall rate of heat transfer through a layered assembly of materials. The U-value is the overall *coefficient of thermal expansion* in Btu/(hr.ft² °F) or W/m² K. Thermal resistance, or R-value, is a measure of an individual material's resistance to heat flow, while U-value is used as a summary factor for the conductive energy potential of building assemblies. U-values are determined by calculating the thermal resistance of each layer of an assembly, summing them to arrive at a total resistance, and taking their reciprocal.

$$U = \frac{1}{\Sigma R}$$

Both R-values and U-values are frequently used in the building codes to specify the minimum thermal criteria for an exterior building component. The insulating potential of building materials, such as insulation and windows, is often identified by either their U-value or R-value.

Changes of State Freezing and melting indicate the temperature at which a material is said to undergo a change of state. Since most construction materials are solid ("frozen") at normal service temperatures, a material's melting point is of greater importance. The *melting temperature* is the point at which a material turns from a solid to a liquid form. In general, materials with high melting temperatures, such as non-metallic inorganics, perform better at high temperatures. They also tend to retain their mechanical properties over a wider range of temperatures. The melting point of metals is below that of non-metallic inorganics, so metals do not perform as well under service conditions having high temperatures. This is why steel structural members must be protected with a fire-resistant material. Although steel will not burn, it does lose strength as its temperature rises. Organic materials perform the poorest in conditions of high temperature.

Heat Capacity The ability of a material to store and release heat is an important thermal property for building construction. Through natural heat transfer, heat always flows from hot to cold. Solid and liquid materials with a high specific *heat capacity* can store heat from the sun during the day and then discharge that heat by natural heat transfer during the night. Sometimes referred to as *thermal mass,* the process is an excellent way to provide low-cost heating and energy management. The subsequent transfer of heat occurs naturally, as the material with an elevated temperature cools to the temperature of the surrounding environment, thus giving off heat. Heavy mass materials, such as masonry and concrete, have excellent thermal mass properties.

Heat can also be stored in a gaseous material, such as the air flowing through a central heating system, to provide thermal comfort in winter. The specific heat capacity of a solid, liquid, or gas is defined as the heat required to raise a unit mass of substance by one degree of temperature and is expressed in terms of Btu/(lb.°F) in the Imperial system, or Joules per kilogram per degree centigrade (J/(kg.°C)) in the metric system.

Acoustical Properties

Acoustics is the branch of physics that deals with the generation, transmission, and control of sound waves. It considers the ability of a material to either absorb or reflect sound waves within a room. The acoustical properties of interior finish materials directly affect occupants by influencing the quality of speech, music, and other audible sounds projected in a space. Acoustical materials that perform well as sound absorbers include soft materials (such as fabrics), rigid but soft materials, and rigid but hard materials that have the exposed surface perforated with holes or slots of varying sizes and placement. Various construction materials can be tested to ascertain their sound control properties. A more detailed discussion is presented in Chapter 31.

Chemical Properties

The chemical properties of a material describe its tendency to undergo a chemical change or reaction due its composition and interaction with the environment. A chemical change can alter the original composition of a material and thereby affect its properties. Iron, for instance, has a tendency to oxidize or corrode under certain conditions. In addition to corrosion, other chemical properties include ultraviolet degradation and fire resistance.

Construction materials are chemically degraded by the environment in which they are placed. This degradation is called atmospheric corrosion. The most common form of atmospheric corrosion is oxidation. *Oxidation* is the reaction of a material with the oxygen in the atmosphere, such as iron rusting. Other corrosive chemicals are present in the air, water, and soil to which materials are exposed, including sulfate ions in groundwater, soils, and seawater; sulfur dioxide gas in the air; and alkali in soils.

Metals usually corrode due to electrochemical action. Corrosion occurs when minute amounts of electricity that occur naturally in the atmosphere or soil flow from one metal called an *anode* to a dissimilar metal called a *cathode* through a current-carrying medium (moisture) called the electrolyte. The *anode* deteriorates and the *cathode* remains unaffected by the electrochemical action. The electrolyte is usually a water or gas, such as carbon dioxide or sulfur dioxide. Metals are ranked by their tendency to be anodic or cathodic. In Table 3.3, selected metals are ranked from those most anodic to those most cathodic. When a material near the top of the list, such as steel, is placed in contact with a material near the bottom of the list, such as copper, and an electrolyte is present, corrosion will occur. The metal that is destroyed (corrodes) is one that is high on the galvanic table. Metals that are cathodic are said to be nobler, while anodic materials are said to be less noble.

Table 3.3 Galvanic Series of Selected Metals and Alloys

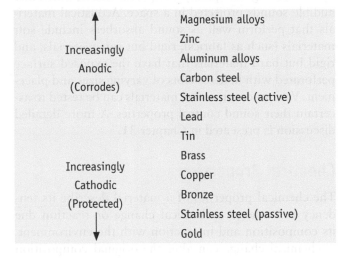

Increasingly Anodic (Corrodes)	Magnesium alloys
	Zinc
	Aluminum alloys
	Carbon steel
	Stainless steel (active)
	Lead
	Tin
Increasingly Cathodic (Protected)	Brass
	Copper
	Bronze
	Stainless steel (passive)
	Gold

Metals that are widely separated on the galvanic table should not be in contact with each other. Those close together, such as cadmium-plated steel and aluminum, have practically no difference in electrolytic potential and, therefore, almost no galvanic action. Corrosion can be prevented by breaking up the electrical circuit through separation of dissimilar metals. This can be done by painting or covering the metal surfaces with a material that will not conduct electricity and that protects the metal from other corrosive materials in the air or soil. Another means of preventing galvanic action is by specifying only one metal in all parts of an assembly. For example, aluminum gutters can be hung with aluminum nails rather than copper nails.

Sometimes a metal, such as steel, is coated with another metal, such as zinc. The zinc, which is high on the galvanic series, is purposely sacrificed to save the steel. Aluminum and stainless steel are anodic and tend to corrode. However, they do develop some natural corrosion resistance when exposed to oxygen, which forms a transparent oxide film that protects the metal.

Ultraviolet Degradation   The reduction in performance of some building materials is related to their exposure to ultraviolet (UV) light, or that portion of sunlight in the ultraviolet spectrum. UV radiation tends to break down solvents in many plastics, causing the material to become brittle, fade, and ultimately fail. Paint designed to be UV resistant is often used to prevent the adverse effects of continued exposure to direct sunlight. Data regarding a product or material's resistance to ultraviolet light should be consulted during the material selection process.

Fire Resistance   Combustibility is of prime importance in the selection of building materials. A material of low combustibility is said to be fire resistant. For example, a firebrick in a fireplace must withstand temperatures of up to 2,000 degrees Fahrenheit. An emphasis on the resistance to fire is a large component of building code regulations (see Chapter 2). Fires create indoor air-quality problems and can result in structural failure. The fire-related properties of specific materials are discussed in more detail in the various chapters related to each material.

ENVIRONMENTALLY PREFERABLE PRODUCTS (EEPS)

In any construction project, building materials are evaluated and chosen based on performance, aesthetics, and cost. For sustainable building materials, these established selection criteria are expanded to include both human health and the larger environmental impacts associated with the use and manufacture of a matrial. In recent years, a wide range of terminology has developed to aid in defining the attributes of green and sustainable construction materials.

The federal government, in an effort to help federal purchasers and simultaneously stimulate market demand for green products and materials, mandates the use of environmentally preferable products in all new federally financed construction projects. *Environmentally preferable products (EPPs)* are defined by the federal government in Executive Order 13101 as those that have "a lesser or reduced effect on human health and

Construction Materials

Composite Materials

Composite materials, or just composites, are engineered materials derived from two or more constituent materials with significantly different physical or chemical properties. They utilize the combination of two compounds, a matrix, and reinforcement. The matrix surrounds the reinforcement material, and the reinforcing imparts its special mechanical and physical properties to support the matrix properties. The combination results in material properties that can be engineered to produce a product or structure with optimum characteristics for a required application. Engineered composite materials are typically formed to their final shape by molding or tooling. Composites are used widely in the automobile, space, aeronautics, and construction industries.

An example of a historical composite is the common brick, made from mud and straw. The most visible composite today is the steel- and aggregate-reinforced Portland cement or asphalt concrete that paves our roadways. Another type is glass-fiber reinforced concrete. Glass-fiber concretes are mainly used in exterior building façade panels and as architectural precast concrete. The material is well suited for creating complex shapes for building façades, canopies, or roofs and is less dense than steel.

A variety of composite materials are used for architectural cladding.

the environment when compared to competing products that serve the same purpose."

A variety of different factors are considered in the determination of the environmental impacts of construction materials.

A *Life Cycle Analysis (LCA)* is a detailed procedure for compiling and analyzing the inputs and outputs of resources and energy and their associated environmental impacts due to the use of a material throughout its life cycle. Life cycle assessment takes into account all of the resource and energy inputs that go into a material, while simultaneously calculating the resulting airborne emissions, solid and water-borne wastes, and any other by-products and releases.

During a life cycle analysis, data compiled for every phase of a product's life is evaluated to determine the environmental impact, in terms of global warming potential, primary energy consumption, air and water pollution, and the use of natural resources.

Embodied Energy

Embodied energy is defined as the energy consumed by all processes associated with the production and use of a material or assembly, from the acquisition of natural resources to its final reuse or demolition (Figure 3.10).

Figure 3.10 The stages considered when determining the embodied energy content of a building material or component.

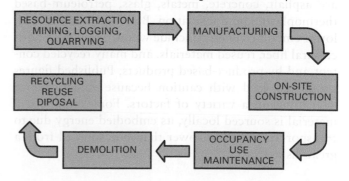

Whereas the energy used in operating a building can be readily measured, the embodied energy contained within building materials is more difficult to calculate. This energy use is often unknown and can only be fully quantified through a complete life cycle analysis (LCA).

The life cycle energy cost of most building products starts with the extraction of raw materials. Here the energy consumed in the actual harvesting, mining, or quarrying of a resource, the building of access roads, and the transportation of raw resources to the mill or plant is tabulated. A second stage is initiated with the delivery of raw materials to the processing plant and ends with the finished product ready for shipment. The manufacturing stage typically accounts for the largest portion of embodied energy and emissions associated with the life cycle of a building product.

The on-site construction is like an additional manufacturing step where individual products and components come together in the assembly of an entire building. Energy is consumed in the transportation of individual products and assemblies from manufacturing plants to regional distribution centers, as well as in the use and transport of equipment, temporary heating, and electricity used during construction. Once the building is completed and occupied, operating energy is calculated, taking into account functions such as heating, cooling, lighting, and water use. Demolition marks the end of a building's life cycle, although it is not necessarily the end for all individual materials, some of which may face subsequent recycling or reuse.

Typically, embodied energy is measured as a quantity of nonrenewable energy per unit of building material or component in either megajoules (MJs) or gigajoules (GJs) per unit of weight (ton or kg) or area (square foot or meter) of a material. Table 3.4 gives embodied energy values for some common building materials. Examples of materials with high-embodied energy are asphalt, concrete, metals, glass, petroleum-based thermoplastics, and insulation. Building products with lower embodied energy include wood, wood fiber, agricultural fiber, reused materials, and many recycled content and by-product-based products. Published figures should be used with caution because values change depending on a variety of factors. For example, if a material is sourced locally, its embodied energy due to transportation will be lower than one sourced from a great distance.

Table 3.4 Embodied Energy of Selected Materials

Material	Embodied Energy in MJ/Kg
Fly ash (In concrete)	< 0.1
Gravel	0.2
Sand	0.6
Brick	2.7
Ceramics and quarry tiles	5.5
Glass	6.8
Wood	10.8
Steel	28.0
Cast iron	32.8
Asphalt	50.2
Polyethylene	79.5
EPDM Roofing	183
Aluminum (Not recycled)	207
Brass	239

Natural Resources/Habitat Degradation

The extraction, manufacture, and use of a construction material have profound effects on ecological systems and the supply of natural resources. To avoid habitat degradation, materials whose use results in reduced effects on erosion, salinity, vegetation loss, changes in the nutrient characteristics of soil, and aesthetic damage to landscapes are preferred. Impacts on water systems, such as the release of nutrients, salts, toxins, or suspended solids into aquatic systems, are also considered.

Renewable Materials

Renewable materials are derived from a living plant, animal, or ecosystem that has the ability to regenerate itself. The USGBC defines *rapidly renewable materials* as those that are able to regenerate themselves in ten years or less. Renewable materials are preferred because, when managed properly, their supply will not be depleted. Most result in biodegradable waste and tend to have reduced net emissions of CO_2 across their life cycle compared to materials derived from fossil fuels.

Wood harvested from sustainably managed forests is a perfect example of the efficient use of a renewable material. It cleans the air and water as it matures (trees bind carbon during their growth), we are able to utilize nearly all of the raw material, its production uses low energy processes, and it is readily renewable and recyclable. Other examples of renewable materials include bamboo, cork, and various composite materials derived from agricultural products.

Recycled and Post-Consumer/Post-Industrial/Agricultural Materials

The use of materials that utilize the waste products from a variety of processes results in a reduced use of natural resources and energy consumption. Recycled content is generally classified in two categories: post-consumer or post-industrial. *Post-consumer recycled materials* have served their intended use, been collected from the consumer, and reprocessed (Figure 3.11). Consumer wastes such as clear plastic water and soda bottles made from PET (polyethylene terephthalate) can be repurposed for carpeting, fabrics, and other building materials.

Figure 3.11 An example of a post-consumer recycled material is cellulose insulation made from recycled post-consumer newsprint and paper stock.

© Stocksnapper/Shutterstock.com

Post-industrial recycled content materials are created using an industrial waste or by-product, diverting it from landfill, and integrating it into new products. An example is the use of fly ash, a glass-like powder recovered from the gases of coal-fired electricity production, as an inexpensive replacement of Portland cement in concrete.

A good resource for selecting recycled content materials is the U.S. EPA's Comprehensive Procurement Guideline program, which exists to promote the use of materials recovered from solid waste. More and more building products with recycled content are becoming available, and most are competitive with non-recycled products in both cost and performance.

Toxicity to the Environment

Studies have found that humans now spend more than 80 percent of their time indoors. The EPA ranks *indoor air quality (IAQ)* as one of the most prominent environmental problems today. ASHRAE defines acceptable indoor air quality as air in which there are no known contaminants at harmful concentrations. Poor indoor air quality can result from inadequately maintained HVAC systems, as well as the material makeup of interior finish materials. Many synthetic products containing *volatile organic compounds (VOCs)* tend to off-gas from interior materials, furnishings, and equipment. Off-gassing occurs when solid but chemically unstable materials evaporate at room temperature and slowly release contaminants into the air. More than 500 pollutants identified in common building materials are known to cause asthma, allergies, cancer, and reproductive and nervous system disorders. Builders receive both short-term and long-term exposures to these chemical hazards from the off-gassing of solvent-based products and cleaners, as well as from sawdust and dust particulates from construction materials.

Certain materials used frequently in the last few decades that carry risk of contamination include polyvinyl chloride (PVC), used in flooring, wall coverings, plumbing, carpet backing, and electronics; formaldehyde products, such as plywood and particleboard; synthetic carpets; and sheet vinyl. A number of organizations now provide official recognition and labeling that certifies the non-toxicity of building materials. More detailed information pertaining to the toxicity of specific materials is given in subsequent chapters.

Sustainable Construction Materials Assessment Criteria

There is a rapid increase in the types and numbers of materials that respond to environmental issues in the construction industry. Designers and builders should strive to specify materials that protect the natural environment, are renewable and recyclable, minimize resource and energy consumption, and are healthy and nontoxic. A complete life cycle assessment of proposed choices is recommended for a thorough understanding of the larger impacts of material selection.

Review Questions

1. What are the major material groups?

2. List the parameters to be considered when selecting a construction material.

3. What are the mechanical properties that define a material?

4. How does a tensile stress differ from a compressive stress?

5. What are the major thermal factors to consider when selecting a material?

6. Which of the four major material groups has the lowest thermal conductivity?

7. What factor describes the overall thermal performance of a building assembly?

8. What is the most common form of chemical degradation?

9. What stages in a material's life cycle are considered when calculating embodied energy?

10. What inputs and outputs does a life cycle analysis (LCA) of a material consider?

11. Give examples of renewable building materials.

Key Terms

Coefficient of Thermal Expansion

Compressive Stress

Ductility

Elastic Limit

Embodied Energy

Environmentally Preferable
    Products (EPPs)

Fatigue Strength

Hardness

Heat Capacity

Impact Strength

Indoor Air Quality (IAQ)

Life Cycle Analysis (LCA)

Melting Temperature

Metals

Modulus of Elasticity (E)

Non-Metallic Inorganic

Oxidation

Permeability

Polymer

Post-Consumer Recycled Material

Post-Industrial Recycled Content
    Material

Rapidly Renewable Materials

Renewable Material

Shear Stress

Strain

Stress

Tensile Stress

Thermal Conductance (C)

Thermal Conductivity (k)

Thermal Mass

Volatile Organic Compound (VOC)

Activities

1. Prepare samples of various materials and test for compression and tensile strength. Keep an accurate record of what each sample contains and the conditions of its preparation. For example, prepare concrete samples, keeping a record of the ingredients, curing time, and how it was contained during curing. Vary certain ingredients in various samples. Test each for compression strength, and compare the results to try to verify how the differences in mixture, curing time, and methods of curing influenced the strength. Similar tests can be made on other materials, such as various species

of wood and different types of metals. Run some experiments on thermal conduction by applying heat to identical size samples of various materials and recording the time it takes to feel heat at the end opposite the end exposed to heat. Although this is rather crude, it will illustrate the property of thermal conduction.

2. Most schools have material-testing equipment or even an entire testing laboratory. Have the instructor in this area help the class devise some more accurate, scientifically based experiments and help the class run these in the laboratory. For

example, with the proper equipment, the class could possibly test a sample, such as a metal, and record the elastic limit, yield point, and breaking strength.

3. Try to set up a device that will subject a material, such as a metal, to constant rapid flexing and see which material has the best fatigue strength.

4. Compare the environmental impacts between using a wood-framed wall and a metal-framed wall using the Building for Environmental and Economic Sustainability (BEES) tool. The BEES tool provides a simple and clear tool for selecting cost-effective green building products. Visit the BEES website at: www.nist.gov/el/economics/BEESSoftware.cfm, and select the BEES on line version.

Select: analyze building products

Under: Environmental Impact Category Weights;

Select: EPA Science Advisory Board Based

Under: Building Element for Comparison

Select: Shell as the major group element

Select: Exterior enclosure under group element

Select: Framing as individual element

Under: Select alternatives:

Select: Generic steel framing and wood frame (untreated)

Select: Compute and View Results to determine which wall assembly performs better in regard to environmental considerations.

Additional Resources

American Society for Testing and Materials, www.astm.org

The Athena Sustainable Materials Institute, www.athenasmi.org

U.S. EPA's Comprehensive Procurement Guideline program, www.epa.gov/epawaste/conserve/tools/cpg/index.htm

Shiers, H., Howard, S., Sinclair, M. (1996), *The Green Guide to Specification*. Oxon: GTI Specialist Publishers.

Existing Conditions
CSI MasterFormat™

© Dmitry Kalinovsky/Shutterstock.com

DIVISION

31

Earthwork
CSI MasterFormat™

CHAPTER 4

The Building Site

Upon completion of this chapter, the student should be able to:

- Define the types of site surveys that are required to document site conditions.
- Be aware of the problems that may occur during excavation and techniques to control them.

- Discuss the types of structures used to control surface and subsurface storm run-off.
- Be familiar with paving materials and applications.
- Understand the planning and construction of different earthwork components and techniques.

Site planning involves the development of land in terms of zoning, access, built elements, land drainage, and additional factors. The coordinated effort of architects, landscape architects, civil engineers, and construction managers is required to ensure that a development will optimize the use of the site, and work efficiently within existing natural systems.

Site work includes a wide range of preconstruction activities encompassing subsurface investigations, preparing the site for construction, dewatering excavations, shoring and underpinning, earthwork, installing various types of utility infrastructure, and constructing site improvements, such as roads, walks, and landscaping (Figure 4.1). There is a complex set

Figure 4.1 Site preparation involves the work of a number of different construction trades.

© Leah-Anne Thompson/Shutterstock.com

of legislation, polices, and guidance applying to the design of site construction, both within and outside of a project location. Federal and state regulations, as well as local planning and zoning codes, must be reviewed in order to identify all relevant issues early in the planning process.

SITE ASSESSMENT

Before any construction project can begin, the building site must be thoroughly analyzed and understood to ensure an efficient design and construction process. A series of preconstruction investigations are conducted, including land and topographical surveys, subsurface soil testing, and environmental studies.

Land Surveying

The site documentation process begins with the acquisition of the deed and plat of a property from the local authority. The deed documents ownership and provides a legal description of the property. A plat is a map of a property that locates property lines and adjacent streets and buildings, as well as easements and right of ways.

Land surveys provide the detailed information required to assure that site preparation and construction effectively coordinate the new work with existing conditions. Surveying is an essential element in site development and is required in the planning and execution of nearly every form of construction. Land surveying is the science of using measurements to determine the relative position of points or physical details, relative to the earth, and to depict them in a usable format. It involves the establishment of land boundaries based on documents of record and historical evidence, as well as site mapping and related data accumulation.

The basic tools used in land surveying are measuring devices for determining horizontal distances, a level to determine height or elevation differences, and a theodolite, or transit, to measure angles. Starting from a position with known location and elevation, the distance and angles to the unknown points are measured. Modern surveying equipment called total stations are fully robotic, and are able to electronically send data to office computers and connect to Global Positioning Systems (Figure 4.2).

Figure 4.2 Surveying is required in the planning and execution of nearly every form of construction.

Survey types typically encountered during commercial construction projects are outlined below.

Property Survey   Sometimes called a boundary survey or cadastral survey, the property survey establishes the extent and geometry of property boundaries.

Existing Conditions Survey   The *Existing Conditions Survey* is a survey of all utilities, infrastructure, built structures, and significant natural conditions existing above or below grade.

Topography Survey   A survey of an areas surface features in relief, showing land form or the shape of the ground and giving elevations.

Construction Survey   During the on-site construction, surveys are used to establish the exact locations of building points and elevations for civil engineering and architectural built elements. Construction surveys are used for everything from locating foundation outlines to laying out roadways. All surveys must be conducted by state-licensed surveyors.

Environmental Assessment

An *Environmental Site Assessment (ESA)* is a report prepared for a building site that identifies potential or existing environmental contamination. A Phase I – ESA, defined by ASTM Standard E-1527-05, reviews the history of a site and addresses the land itself as well as any physical improvements existing on a property, such as buildings, wells, or submerged tanks. The assessment looks for contamination to soil, subsurface materials, and the existence of hazardous substances, such as asbestos, within buildings.

If a site is considered contaminated, a more detailed investigation involving chemical analysis for hazardous substances is conducted. A Phase II – ESA may include the Phase I portion of the assessment, and also soil and groundwater and surface sampling in suspect areas (Figure 4.3). Soil samples are collected using hand augers

Figure 4.3 An Environmental Site Assessment (ESA) may include soil and groundwater and surface sampling in suspect areas.

or a drilling rig and shipped to a laboratory according to standard industry methods. Groundwater samples are collected from the borings or permanent monitoring wells located on a property. The Phase II – ESA report describes the soil borings completed, soil texture, soil and groundwater analytical results, and presents the data in tabular format with a map illustrating the sampling locations and plan of site.

Sites found to contain environmental contamination are referred to as brownfield sites. The EPA defines a *brownfield site* as "a property, the expansion, redevelopment, or reuse of which may be complicated by the presence or potential presence of a hazardous substance, pollutant, or contaminant." Sites containing high levels of hazardous contamination are known as EPA superfund sites. Properties that contain hazardous contamination must be remediated prior to the commencement of new construction activities. Environmental assessments require the services of professionals with knowledge in chemistry, physics, geology, microbiology, and botany.

Subsurface Investigation

A key factor in the design of any structure is an investigation of the characteristics of the soil upon which it is built. This subsurface exploration, also called a geotechnical investigation, requires an examination of site-soil samples by field-test procedures or in a testing laboratory. Certified geotechnical engineers perform the tests to determine the engineering properties of the soils and determine how those properties will interact with the proposed construction.

The resulting soil report is used by the architect and design engineers to determine how the building must be designed and constructed. Subsurface site conditions influence the design not only of foundations, but also of underground utilities, site paving, retaining walls, and a host of other factors. Characteristics of the soil will also have an influence on the planning of excavation methods. More information on soil testing activities is outlined in Chapter 5.

There are times when unexpected site conditions are identified after contracts have been signed and site work begins. For example, say a site is being excavated for a four-story masonry building and massive foundations from a previous building were found. Who should pay to excavate and remove the material? To prevent these kinds of potential problems, unexpected findings clauses can be written into contracts and thorough subsurface investigation should be conducted.

Site Plans

A *site plan* is a detailed drawing that shows the proposed improvements for a site development project. Site plans illustrate the location arrangement, dimensions, and finish floor elevations of proposed buildings and structures. Site entrances and exits, vehicular drives including fire and service lanes, parking and loading areas, and pedestrian walkways are located. The plan notes existing and proposed underground and overhead utilities, noting line sizes, grades, and the location of public or private site lighting and fire hydrants. Existing contours and proposed elevation changes are located as needed to verify drainage, along with existing and new landscaping materials.

Site planning components may be shown on a single drawing or multiple drawings, depending on the complexity of the project. It is common to have a drainage plan, utility plan, or landscape plan separate from the site plan. Site development plans must be prepared by a licensed architect, engineer, or surveyor. A typical site plan is shown in Figure 4.4.

SITE WORK ACTIVITIES

Site work includes many other activities, such as erosion control, demolition, earthwork and site remediation. Foundations are a big part of site work and include spread footings, piles, and caissons (see Chapter 6). Building in wet environments may necessitate dredging, underwater work, and constructing seawalls, jetties, and docks. Utilities require site trenching for water, sewer, gas, oil, and steam distribution systems. The construction of ponds, drainage reservoirs, and sewage lagoons may also be required. Finally, extensive site improvements are often needed in the form of walks, fences, irrigation systems, and landscaping.

Activities on a construction site should make every effort to conserve natural features, prevent the escape of pollutants from the site, and restore existing areas that may be damaged. Earthwork and other site disturbance should be kept as close as possible to the perimeter of built structures, roads and walkways, and utility trenches. Existing vegetation that is to remain must be protected during construction and restored at the end of a project. Topsoil must be carefully stripped, protected, and stockpiled for reuse. Discharge of toxic waste materials must be prohibited. The existence of flood zones must be acknowledged and decisions made concerning their impact on the site.

Figure 4.4 A typical site plan showing property lines, utility locations, landscape elements, and proposed built improvements.

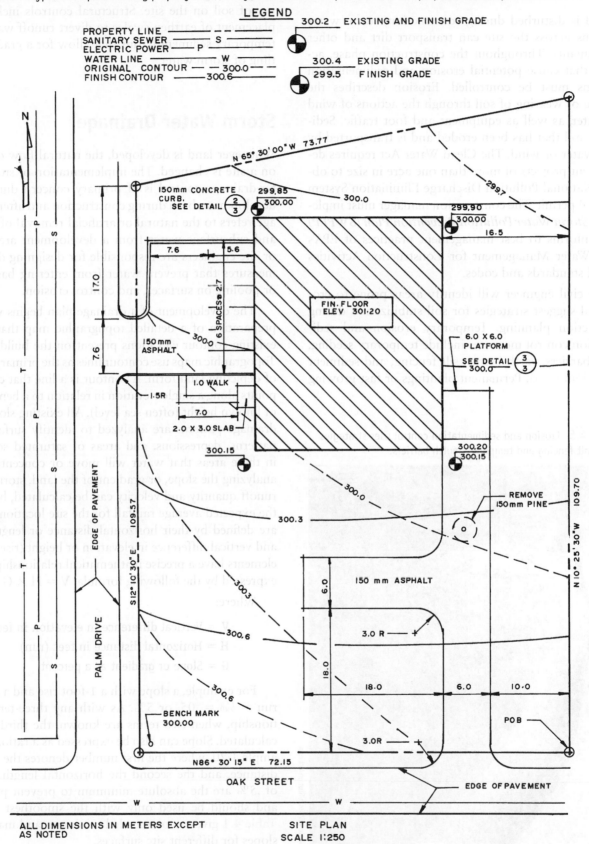

LEGEND

PROPERTY LINE	
SANITARY SEWER	S
ELECTRIC POWER	P
WATER LINE	W
ORIGINAL CONTOUR	300.0
FINISH CONTOUR	300.6

300.2 EXISTING AND FINISH GRADE

300.4 EXISTING GRADE
299.5 FINISH GRADE

N 65° 30' 00" W 73.77

150mm CONCRETE CURB. SEE DETAIL (2/3)

299.85 / 300.00

300.0

299.90 / 300.00

16.5

7.6   5.6

17.0

6 SPACES @ 2.7

FIN. FLOOR ELEV 301·20

7.6

150 mm ASPHALT

300.0

6.0 X 6.0 PLATFORM
SEE DETAIL (3/3)
300.0

1.0 WALK
1.5R
7.0
2.0 X 3.0 SLAB

300.15

300.20 / 300.15

300.0

300.3

REMOVE 150mm PINE

EDGE OF PAVEMENT

109.57

S 12° 10' 30" E

N 10° 25' 30" W   109.70

300.3

150 mm ASPHALT

6.0

3.0 R +

18.0

PALM DRIVE

300.6

300.6

18.0   6.0   10·0

BENCH MARK 300.00

3.0 R +

POB

N86° 30' 15" E   72.15

OAK STREET

EDGE OF PAVEMENT

W   W

N

T

P

S

ALL DIMENSIONS IN METERS EXCEPT
AS NOTED

SITE PLAN
SCALE 1:250

Sedimentation and Erosion Control

As land is disturbed during construction, storm water that runs across the site can transport dirt and other contaminants. Throughout the construction phase, activities that cause potential erosion and sedimentation problems must be controlled. Erosion describes the washing or wearing of soil through the actions of wind and water, as well as equipment and foot traffic. Sediment is soil that has been eroded and is transported by either water or wind. The Clean Water Act requires development projects of more than one acre in size to obtain a National Pollution Discharge Elimination System (NPDES) permit. The construction manager must implement a *Storm Water Pollution Prevention Plan (SWPPP)* that conforms to best management practices of EPA's Storm Water Management for Construction Activities or local standards and codes.

The civil engineer will identify areas prone to erosion and suggest strategies for soil stabilization during construction planning. Temporary erosion and sedimentation control measures include temporary seeding, mulch barriers, earth berms, silt fencing, and sediment traps (Figure 4.5). Permanent plantings or fast-growing

Figure 4.5 Erosion and sedimentation control uses strategies such as silt fencing and temporary mulch barriers.

© Gary Whitton/Shutterstock.com

temporary grasses can be utilized to prevent the eroding of soil on the site. Structural controls include the placement of earthen swales to divert runoff water into temporary sediment basins that allow for a gradual settling of storm water.

Storm Water Drainage

Whenever land is developed, the natural flow of water on a site is changed. The implementation of storm water-drainage control is of primary concern during site development, both during construction and after. Drainage refers to the natural or artificial removal of surface and subsurface water from a development area. Civil or site engineers are responsible for designing drainage measures that prevent water from entering basements or pooling on surfaces, and control erosion.

The development of a drainage plan begins with the preparation of a detailed topographic map that shows existing contour elevations present on the building site. Topographic maps use contour lines as the primary means of depicting land form. A contour is a line that connects points along a single elevation in relation to a benchmark of known height (often sea level). All existing slopes and drainage patterns are analyzed to identify surface flow patterns, depressions, and areas of saturated soil. It is in these areas that water will move or concentrate. By analyzing the slope, or gradient of the land, storm water runoff quantity and velocity can be calculated, based on the expected average rainfall for the site location. Slopes are defined by their horizontal distance or length (run) and vertical difference in elevation or height (rise). These elements have a precise mathematical relationship that is expressed by the following formula: $V = H \times G$

Where:

> V = Vertical difference in elevation in feet (rise)
> H = Horizontal distance in feet (run)
> G = Slope or gradient as a percent

For example, a slope with a 1-foot rise and a 20-foot run = $\frac{1}{20}$ = .05, or 5%. As with any three-term relationship, when two terms are known, the third can be calculated. Slope can also be expressed as a ratio, for example 1:20, where the first number denotes the vertical distance, and the second the horizontal length. Slopes of .5% are the absolute minimum to prevent ponding, and should be used only with the smoothest paving. Table 4.1 gives recommended minimum and maximum slopes for different site surfaces.

Table 4.1 Recommended Slopes for Site Development

Site Surface	Minimum Slope	Maximum Slope
Walkways	0.5%	5% (walks >5% are ramps)
ADA Accessible Ramps	5%	8.33% (1ft. rise :12ft. run)
Parking Areas	1%	5%
Lawn Areas	1–2%	25%
Stabilized Slopes (Ground cover, rock)	1%	10%

Figure 4.6 The installation of a catch basin that conducts runoff into subsurface drainpipes.

© TFoxFoto/Shutterstock.com

Drainage systems can achieve storm-water management goals by using one of two basic elements, either alone or in combination, depending on the particulars of a development site. Surface drainage systems collect and dispose of storm water runoff with ditches, swales and detention basins and are generally used on sites with sufficient open areas for the flow of water above grade to be accommodated. Subsurface storm drainpipes made from concrete, aluminum, or polyvinyl chloride (PVC) can be used to convey rainwater to the storm-water system or other catchment area. French drains, similar to foundation drain systems, can be installed anywhere on a site that collects water. Designed in either a random, parallel line or grid system, the piping is located at low points in the site topography to collect subsurface water. The depth and spacing of the pipes depends on the existing soil conditions and requires the expertise of a geotechnical engineer. Different inlet types include trench drains, yard and curb inlets, and catch basins that conduct the runoff into subsurface drainpipes (Figure 4.6). Piped systems are designed with cleanouts at easy access points for maintenance of the storm-water drainage system.

Best Management Practices (BMPs) are alternatives to conventional drainage techniques and that can substantially reduce surface runoff quantities and resultant pollutant loadings. BMPs reduce the amount of impervious surface areas on a site to reduce storm water runoff. They utilize the landscape and soils to naturally move, store, and filter storm water runoff before it leaves the development site.

Demolition

On some development sites, preparing for construction can require the demolition of existing structures. This may include the breakdown and removal of existing buildings, roadways or parking areas, and existing elements below grade, such as submerged tanks or abandoned utility lines. Site clearing includes the removal of trees and other vegetation in areas to be occupied by new development. Grubbing involves the removal of nonvisible subsurface obstructions, such as root systems, stumps, and other vegetation material, to prevent potential voids in the ground due to the decay of organic material. For building demolition, asbestos abatement and the removal of other hazardous or regulated materials may be required.

Demolition work is subject to control by the local building codes and safety practices must be followed. Many municipalities maintain and enforce construction waste regulations stipulating what portion of material must be recycled and how the remaining material is disposed of in an approved manner. No demolition can occur until the local building authority approves the construction documents.

Site Remediation

Remediation activities result in long-term benefits for building sites by preventing erosion, generating productive soils, restoring natural habitats, and removing toxic substances. On previously polluted sites, restoration of structures, soils, and vegetation may be required. Both urban and rural sites can be problematic in a variety of ways, from minor damage and neglect to more complicated contamination. Depending on the type, extent, and toxicity of contamination found on a site, cleanup activities might include soil, surface water, and groundwater remediation. Historically, the main remediation strategy for damaged sites was the removal of contaminated material to special contained landfills, a procedure known as dig and dump.

Means for remediating contaminated soils on-site, rather than dumping contaminated soil, are now in common use. New technologies include phyto-remediation, which utilizes plants to metabolize organic contaminants. Another method, known as *bio-remediation*, uses bacteria that are able to digest and remove toxic chemicals over time. The U.S. Environmental Protection Agency has a number of programs and resources to aid in the remediation of polluted sites and provides grants for assessment and cleanup, in addition to technical advice and training.

EARTHWORK

Earthwork refers to construction work that includes soil excavation, transportation, and relocation in a predetermined location. Earthwork involves a variety of activities, including grading, excavating, backfilling, compacting, laying base courses, stabilizing the soil, setting up slope protection, and establishing erosion control.

Grading

Grading is the alteration of a site's land form, usually by means of heavy equipment, to provide a level area for a structure, create movement paths, and create drainage and landscape features. Rough grading involves adjusting the level of the ground to facilitate the excavation and construction of the building. Equipment for grading operations includes graders, front-end loaders, and backhoes. Prior to any grading activities, valuable topsoil is carefully removed, covered, and stockpiled during construction. When construction is complete, the topsoil is redistributed with compost amendments to support plantings.

Grading to create a level plane for a building can be accomplished in three ways: by cutting, by filling, or by using a combination of cut and fill (Figure 4.7). Cutting involves removing soil from a slope to create a large enough area for construction activities. The slope behind

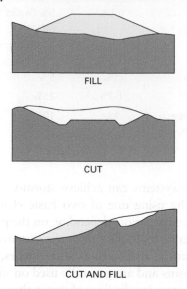

Figure 4.7 Grading to create a level plane for a building can be accomplished in three ways: cutting, filling, or a combination of cut and fill.

the cut must be formed to allow the earth to retain itself, or be held with elements such as retaining walls. Filling a slope involves importing soil, forming the level area, and compacting it to support construction loads. Cutting and filling is the preferred method of grading, as it balances the soil quantities on a site and eliminates the need to remove or import soil, which is costly. Cut and fill volumes are calculated using a grid analysis that divides a site into smaller square areas. By determining the average depth of cut and fill in each square and multiplying the average depth by the surface area, the volume of cut and fill can be calculated by averaging the depth at the corner of each area. The volume of soil changes as it is excavated, hauled, placed, and compacted. Additional calculations must be made to account for this change in volume due to excavation activities. A variety of cut and fill calculation software available today simplify the calculation process. Finish grading occurs after the building and other improvements are complete, bringing the site elevations up to the levels shown on the site or grading plan.

Grading operations are designated on the site plan or grading plan. Natural grade, designated by existing contours, are shown in a lightly dashed line. Finish grade, designated by proposed contours, are shown in a solid line. Construction costs can be lessened by keeping earthwork activities to a minimum, working with existing site grades, and avoiding large-scale grading, which creates dust and the potential for erosion.

Excavation

Excavations are required for footings, foundations, underground utilities, and other aspects of site development. A wide variety of equipment can be used, including power shovels, backhoes, draglines, clamshells and cranes, trenching machines, wheel-mounted belt loaders, and bulldozers.

If foundations are shallow, they are usually dug with a backhoe. The depth of a foundation footing is determined by geographic location, and its base must always be placed below the frost line. Excavations must reach to the depth at which tests have shown the soil to have the required load-bearing capacity. A building with a basement will have a large excavation, and a considerable amount of soil must be removed and stockpiled.

Multistory buildings require extensive excavations that can reach several stories underground to access bedrock or other adequate supporting soil. If solid rock must be removed, the cost of excavation increases rapidly. Weak and thinly layered rock can be loosened with power shovels, tractor-mounted rippers, backhoes, or pneumatic hammers. Larger rock formations can be fractured by explosives and removed from the excavation by a front-end loader. Like soil, rock must be stockpiled or removed from the site.

Deep excavations present an ever-present danger of collapsing sides. The depth and type of soil and the location of the site influence what strategies should be used. If the excavation is not too deep and the site is large enough, excavation sides can be sloped. The angle of the sloped surface will vary with the type of soil. The *angle of repose* defines the steepest angle of a surface at which loose material will remain in place before sliding. This angle will range from 34° for cohesionless soil to 90° for stable rock. Deeper excavations can use a bench system, as shown in **Figure 4.8**. The particles in cohesionless soils tend to dispersion, which can result in slope failure unless the angle of repose is small. Cohesive soil grains tend to bond and provide good shear strength. Maximum allowable excavation slopes for different materials are specified by OSHA regulations.

Sheeting

If the sides of the excavation cannot be sloped to provide protection from slides, some form of sheeting must be used. *Sheeting*, in the form of sheet piling, lagging, and slurry walls, is used to support the face of an excavation.

Sheet piling may be wood, aluminum, steel, or precast concrete, placed vertically into the ground

Figure 4.8 Ways of containing the sides of an excavation to prevent collapse.

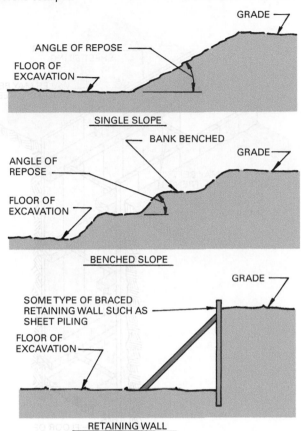

(Figure 4.9). While wood pilings are usually removed, steel and concrete sheet pilings are sometimes left in place after construction is finished. Another sheeting technique uses vertical steel columns, called soldier piles, that are driven into the soil. Horizontal wood lagging is then placed between them to contain the excavation (Figure 4.10). Steel sheet piling uses interlocking steel sheets that are typically driven or vibrated into the earth before excavation begins (Figure 4.11).

Since sheeting is subject to soil and subsurface water pressures, some form of bracing is often required as the excavation gets deeper. In shallow excavations, the sheeting may be driven deep enough below the bottom of the excavation so that bracing is not required. Vertical sheeting and narrow excavations can use horizontal beams, called wales, placed at regular intervals to resist soil pressures.

Wider excavations require the use of more substantial cross-lot bracing supported by vertical steel posts driven into the ground. Rakers are used on wide excavations where cross-bracing would be impractical. The rakers are set on an angle and transfer the forces to a

Figure 4.9 Vertical wood sheet piling can be used to support the sides of an excavation.

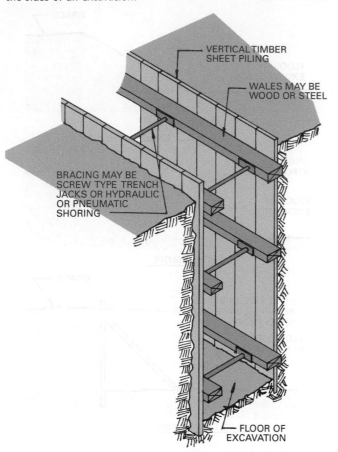

VERTICAL TIMBER
SHEET PILING

WALES MAY BE
WOOD OR STEEL

BRACING MAY BE
SCREW TYPE TRENCH
JACKS OR HYDRAULIC
OR PNEUMATIC
SHORING

FLOOR OF
EXCAVATION

Figure 4.10 This form of sheeting uses steel soldier piles and horizontal timber lagging.

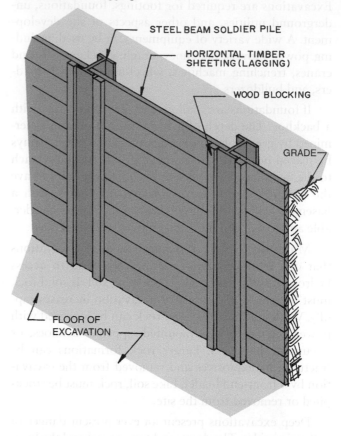

STEEL BEAM SOLDIER PILE

HORIZONTAL TIMBER
SHEETING (LAGGING)

WOOD BLOCKING

GRADE

FLOOR OF
EXCAVATION

Figure 4.11 Steel sheet piling is driven into the ground before the excavation begins.

footing set in the bottom of the excavation (Figure 4.12). Both types block the open space in the excavation and interfere with any work that occurs within it.

Some soils permit the use of tiebacks. Tiebacks are steel cables, or tenons, that are inserted into holes drilled through the sheeting and into the rock or subsoil. The drilled hole is filled with concrete grout. When the grout has set, the cables are tightened with hydraulic jacks and fastened to the wales to provide an excavation free of barriers (Figure 4.13). A second type of tieback uses a screw anchor that is installed with the same rotary drilling equipment used in the grouted anchor construction. The screw anchor can be installed quickly and does not require the drilling of an anchor hole or the injection of grout. They are especially useful in loose sandy and clay soils where a drilled hole may collapse. Screw anchors can be withdrawn and reused.

A *slurry wall* serves as sheeting that protects the excavated area and becomes a part of the permanent foundation. It may be cast in place or built from precast concrete panels. The excavation is dug with a narrow

Figure 4.12 Steel pilings on wide excavations can be braced with rakers or cross-lot bracing.

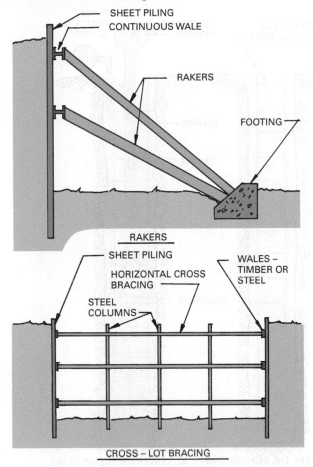

Figure 4.13 Sheeting can be braced against hydrostatic pressures by tiebacks anchored in stable earth or rock.

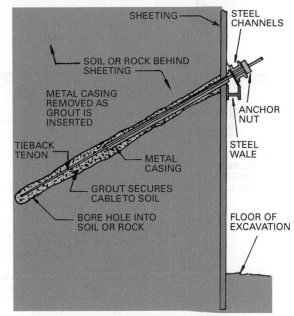

1. BORE A HOLE IN SOIL OR ROCK WITH A ROTARY DRILL. INSERT A METAL CASING TO KEEP THE SOIL FROM FILLING THE HOLE.
2. INSERT STEEL TIEBACK TENONS INTO THE HOLE. FILL THE HOLE WITH GROUT AS THE STEEL CASING IS REMOVED. LET THE GROUT SET.
3. WHEN THE GROUT HAS SET POST-TENSION THE TENON WITH A HYDRAULIC JACK AND ANCHOR TO THE WALE.

clamshell bucket that establishes the width of the wall. To help prevent collapse, the excavation is filled with *slurry* composed of bentonite clay and water, which stabilizes the walls. The clamshell bucket moves through the slurry to continue digging. After the required depth is reached, a welded cage of steel reinforcing is lowered into the cavity. Concrete is poured from the bottom of the excavation, filling it to the top (Figure 4.14). The slurry rises above the concrete before being pumped off. Once the entire wall is poured and has cured to sufficient strength, excavations can begin from the inside. As the excavation deepens, steel tiebacks are set in holes drilled through the wall and into the subsoil .

A second technique is to insert precast concrete wall sections into the slurry instead of site casting the concrete wall. The surfaces of the precast sections on the excavation side are coated with a release agent to prevent the slurry from bonding to them. A tongue-and-groove joint or rubber gasket is used to seal adjacent sections. After the slurry remaining in the excavation

has hardened, the soil can be dug away. The slurry on the backside of the wall remains and aids in waterproofing. The slurry on the excavated side falls free, leaving the surface of the precast units exposed.

Soil anchors are metal shafts grouted into holes to stabilize sides of an excavation. This involves drilling holes into the soil embankment, removing the drill, and filling the hole with grout (Figure 4.15). Then a soil nail, typically a section of steel reinforcing bar, is pushed into the grouted hole. A centering device is attached to the nail so it remains in the center of the grouted hole. (Additional information can be found in Chapter 6.)

Cofferdams and Caissons

Cofferdams are temporary watertight enclosures used either in water-bearing soil or directly in water. They prevent water from entering an interior area, allowing construction to proceed in a dry and stable environment. Water is pumped from within the cofferdam, and pumps are maintained during construction to remove any leakage that may occur. Cofferdams are typically built using sheet piling, soldier beams with lagging, or as

Figure 4.14 The typical procedure for constructing a slurry wall.

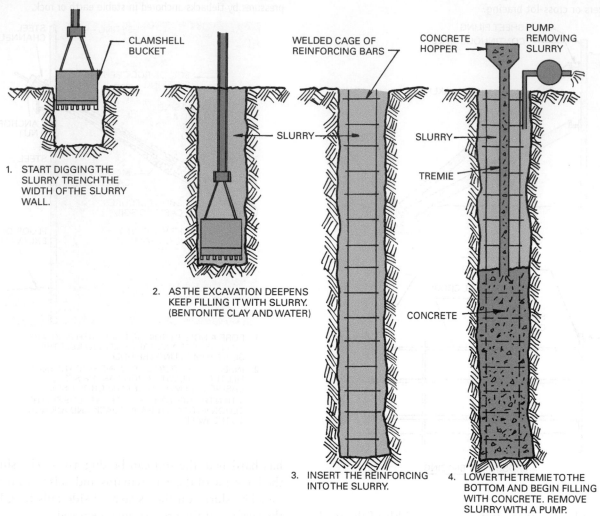

CLAMSHELL BUCKET

WELDED CAGE OF REINFORCING BARS

CONCRETE HOPPER

PUMP REMOVING SLURRY

SLURRY

SLURRY

TREMIE

CONCRETE

1. START DIGGING THE SLURRY TRENCH THE WIDTH OF THE SLURRY WALL.

2. AS THE EXCAVATION DEEPENS KEEP FILLING IT WITH SLURRY. (BENTONITE CLAY AND WATER)

3. INSERT THE REINFORCING INTO THE SLURRY.

4. LOWER THE TREMIE TO THE BOTTOM AND BEGIN FILLING WITH CONCRETE. REMOVE SLURRY WITH A PUMP.

Figure 4.15 Construction activities proceed inside a slurry wall sheeting anchored by tiebacks.

© Alison Hancock/Shutterstock.com

a double-wall structure. A typical double-wall cofferdam is shown in **Figure 4.16**. The cofferdam extends through the areas of permeable water-bearing soil formations, through any impervious rock formations that have low bearing capacity, and on to solid bedrock that is used to support the foundation.

A *caisson* is a watertight shell in which construction work is carried out below water level. Caissons may be open or pneumatic. The top of the open caisson is exposed to the weather, and work is performed under normal atmospheric pressure. Pneumatic caissons are air and watertight. They are open on the bottom so soil excavation can be accomplished. As an excavation proceeds, the caisson is consistently filled with pressurized air to keep water from entering (Figure 4.17). Workers must go through decompression after working in pressurized caissons to avoid suffering from the bends,

Construction Techniques

Blasting

Many parts of the country have considerable rock formations extending below the surface of the earth. Before a foundation, utility trenches, or other subsurface work can begin, it is sometimes necessary to blast out rock located in an excavation's path. Blasting requires the services of a qualified construction blaster who has a record of successful projects, is licensed as required by the state, and has adequate insurance coverage.

Drilling and blasting is expensive, and a careful analysis of the site must be made during the bidding phase of a project. When an area is known to contain rock, soil testing must be conducted to check for subsurface rock and determine whether blasting is needed prior to the start of construction. If rock is encountered once construction has begun, the material will have to be removed with air hammers, a time-consuming process. Subsurface investigations can reveal the presence of solid rock beneath a site. The project team may decide that structures and utility lines should be relocated to avoid the expense of blasting.

Explosives are employed in blasting for mining, roadwork, excavations demolition, and tunneling. They are made from materials that are extremely sensitive to stimuli from impact, friction, heat, or electrostatic sources of initiation. Explosives such as lead azide, lead styphnate, and TNT have a high rate of energy release and produce a forceful blast pressure.

Once subsurface conditions are known, a number of holes are drilled into the rock and then filled with explosive. Detonating the explosive will cause the rock to collapse and loosen, allowing for removal. Special care must be exercised if there are existing structures, such as buildings, wells, or submerged tanks, nearby. This may require more drilled holes and smaller charges or detonating fewer holes at one time. Heavy blasting mats can be utilized to cover the charged area and reduce the distribution of material.

A rock drill working a site in preparation for a blasting operation in shale rock.

© TFoxfoto/Shutterstock.com

---

a painful and sometimes fatal disorder. The structure forming the caisson is frequently left in place and filled with concrete, forming a caisson pile.

DEWATERING TECHNIQUES

Dewatering techniques involve lowering the level of subsurface water on a site to allow excavation to occur in a dry and stable environment. Dewatering usually begins before excavation and continues as the excavation proceeds. Subsurface water is removed using a system of pumps, pipes, and well points. A well point is a perforated unit placed at the bottom of a pipe that is driven into the soil. A series of well points connected to horizontal pipes above the ground, called headers,

are driven in and around the area to be excavated. The headers are connected to centrifugal pumps that draw water up into the header pipes and eject it for drainage away from the excavation (**Figure 4.18**). More than one series of well point, header, and pump assembly is usually required, especially as the excavation gets deeper.

Water can also be kept out of an excavation by constructing a watertight wall. This could be a slurry wall or one constructed with sheet piling, as discussed under the Cofferdams heading. The wall must resist the hydrostatic pressure of the subsurface water and the pressure generated by the soil itself. It has to reach a depth where it penetrates a stratum of impermeable material so a watertight seal can be achieved. Pumps should be kept available to remove any water that may seep into the excavation.

Figure 4.16 A cofferdam is used to provide a watertight area for the construction of footings and foundations.

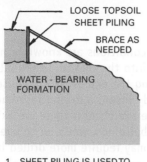

1. SHEET PILING IS USED TO HOLD THE LOOSE TOPSOIL FORMATION

LOOSE TOPSOIL
SHEET PILING
BRACE AS NEEDED
WATER – BEARING FORMATION

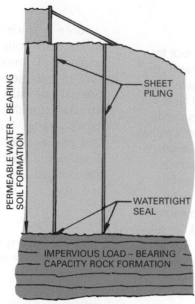

2. SHEET PILINGS ARE DRIVEN THROUGH THE PERMEABLE WATER – BEARING SOIL FORMATION TO IMPERVIOUS ROCK. THE PILING FORMS A WATER-TIGHT SEAL WITH THE ROCK.

PERMEABLE WATER – BEARING SOIL FORMATION
SHEET PILING
WATERTIGHT SEAL
IMPERVIOUS LOAD – BEARING CAPACITY ROCK FORMATION

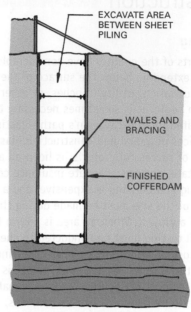

3. EXCAVATE THE AREA BETWEEN THE SHEET PILING AND BRACE WITH WALES AND CROSS BRACING AS THE EXCAVATION PROCEEDS

EXCAVATE AREA BETWEEN SHEET PILING
WALES AND BRACING
FINISHED COFFERDAM

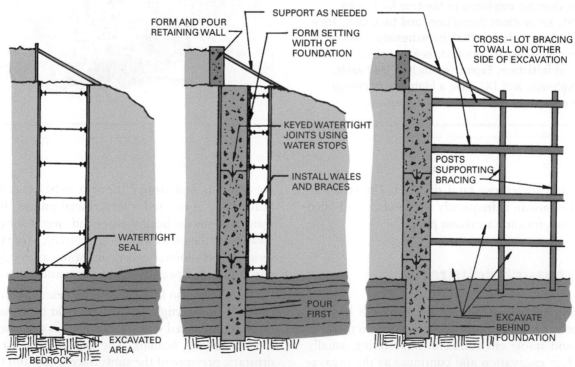

4. EXCAVATE FOR THE FOUNDATION THROUGH THE IMPERVIOUS ROCK DOWN TO BEDROCK. NO SHEET PILING NEEDED IN THIS AREA.

WATERTIGHT SEAL
EXCAVATED AREA
BEDROCK

5. INSTALL THE FOUNDATION WALL FORM IN THE EXCAVATED AREA. INSTALL REINFORCING AND FILL WALL CAVITY WITH CONCRETE. POUR IN SECTIONS WITH WATER STOPS BETWEEN THE SECTIONS.

FORM AND POUR RETAINING WALL
SUPPORT AS NEEDED
FORM SETTING WIDTH OF FOUNDATION
KEYED WATERTIGHT JOINTS USING WATER STOPS
INSTALL WALES AND BRACES
POUR FIRST

6. BEGIN EXCAVATION OF THE AREA BEHIND THE FOUNDATION. BRACE THE FOUNDATION AS THE EXCAVATION PROCEEDS. WHEN COMPLETELY EXCAVATED THE WALL CAN BE SUPPORTED WITH CROSS – LOT BRACING.

CROSS – LOT BRACING TO WALL ON OTHER SIDE OF EXCAVATION
POSTS SUPPORTING BRACING
EXCAVATE BEHIND FOUNDATION

Figure 4.17 Pneumatic caissons operate under elevated air pressure.

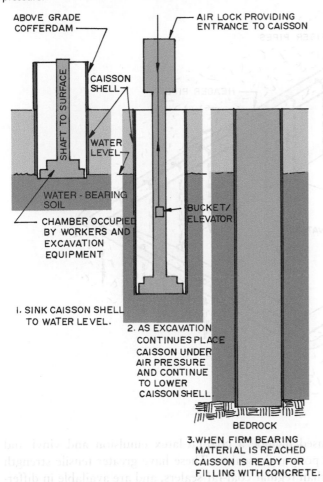

ABOVE GRADE
COFFERDAM

AIR LOCK PROVIDING
ENTRANCE TO CAISSON

CAISSON
SHELL

SHAFT TO SURFACE

WATER
LEVEL

WATER - BEARING
SOIL

CHAMBER OCCUPIED
BY WORKERS AND
EXCAVATION
EQUIPMENT

BUCKET/
ELEVATOR

1. SINK CAISSON SHELL
TO WATER LEVEL.

2. AS EXCAVATION
CONTINUES PLACE
CAISSON UNDER
AIR PRESSURE
AND CONTINUE
TO LOWER
CAISSON SHELL.

BEDROCK

3. WHEN FIRM BEARING
MATERIAL IS REACHED
CAISSON IS READY FOR
FILLING WITH CONCRETE.

UNDERPINNING TECHNIQUES

Occasionally, the excavation for a new building is so close to another building that it becomes necessary to support the existing building during construction by *underpinning*. This underpinning requires considerable engineering expertise, and special design considerations are needed for each situation.

The existing soil conditions must be evaluated to ensure that the new footings will be bearing on a suitable stratum. There are various ways to underpin a foundation. Two frequently utilized techniques are the use of trenches dug below the foundation of the existing building or the use of needles.

The first technique uses trenches dug at intervals beneath the existing foundation. The building is supported by the remaining undisturbed soil. The required underpinning is

installed in the trenched area and additional trenches are dug through the unexcavated area. This is repeated until the extant building has adequate underpinning.

Needles are heavy wooden timbers or steel beams that are run horizontally through the wall of a building and supported on each end. An engineer determines the spacing between needles in a wall. In Figure 4.19, the building is supported by needles run through the foundation, then the area under the foundation is excavated and a new footing and foundation are built. Notice that this system uses sheeting under the building to prevent cave-in of the excavation wall.

PAVING

A variety of surfacing types are in common use for vehicular and pedestrian traffic systems. For roadways and parking, paving may be stone aggregate bound with asphalt or Portland cement, gravel, or stabilized soil. Unit pavers are generally used for walkways, patios, building entrances, and plazas.

The most commonly used paving materials for roads and parking areas are asphaltic concrete or concrete. Finish paving occurs after a sub-base and base have been laid. A typical situation with a flexible pavement, such as asphaltic concrete, is shown in Figure 4.20. The lowest layer is the natural subgrade, which is consolidated by compaction. The sub-base course, consisting of a natural soil of higher quality, is laid on top of the natural subgrade to provide additional support for the distribution of loads imposed on the finished surface. The base course is laid over the sub-base to provide the surface on which the finished wearing surface is laid. It is a granular base of high quality, such as crushed stone, gravel, slag, or some combination of these, sometimes mixed with sand. The material is often treated with bitumen to bind the materials. The sub-base and base courses are extended beyond the edge of the next layer to provide a cone disbursement for the loads.

Figure 4.21 shows a section through the base material for a typical rigid pavement, made of concrete. In this design, the concrete slab has to be thick, reinforced, and strong enough to resist the imposed traffic loads. The base material helps support the loads, but the concrete slab actually carries most of the weight. The loads are transmitted over the entire width of the slab, which reduces the forces on the base materials.

Figure 4.18 This excavation is kept dry with a well point system as the base is covered with aggregate delivered by a boom conveyer.

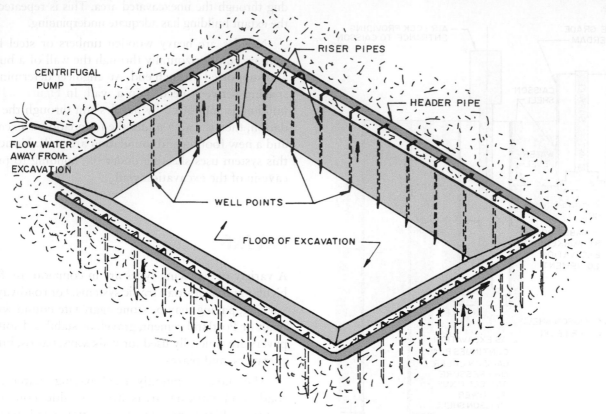

Asphalt Paving

Asphalt concrete consists of asphalt cement and graded aggregates carefully proportioned and mixed in an asphalt plant at controlled temperatures. Four types of asphalt paving are in common use, made with either cutback or emulsified asphalt. The mix is transported to the site, spread by an asphalt-paving machine, and rolled while it is still hot (Figure 4.22). Cold-laid asphalt is similar to asphalt concrete but is laid when the asphalt liquid is cold.

Asphalt macadam is laid using a penetration method. A coarse aggregate is laid over the base, compacted to a smooth surface, and sprayed with an asphalt emulsion or hot asphalt cement. This is covered with fine aggregates and rolled, forcing the fine aggregate into the voids between the larger aggregate.

Old asphalt surfaces can be restored by applying a surface treatment consisting of a 1-inch (25.4 mm) asphalt sealer. The treatment may consist of several layers of liquid asphalt covered with a mineral aggregate or a single layer of cold or hot asphalt mix that is plant-mixed and rolled after being machine laid. Other sealers

used include a coal-tar-latex emulsion and vinyl and epoxy resin sealers. These have greater tensile strength than regular coal-tar sealers, and are available in different colors. Additional information on bitumens can be found in Chapter 25.

Concrete Paving

Roads, driveways, and parking lots are often paved with concrete. The subgrade, sub-base, and base course preparation is much like that described for asphalt paving. Side forms are placed along the edges of the paving to contain the plastic concrete and act as rails on which the concrete placing spreader rides. Some placing spreaders are designed to ride inside the side forms, and others ride outside the forms.

The paving machinery may be either form or slip-form types. The form-type paver rides on metal side forms, places the concrete, strikes it off, and consolidates it. Slip-form pavers place a slab without the use of side forms. The concrete is of a consistency that it develops sufficient strength to be self-supporting as it leaves the paver. The slip-form paver spreads, consolidates,

Figure 4.19 A typical underpinning installation using needles to support the existing building.

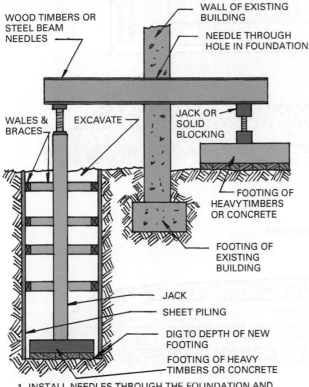

WOOD TIMBERS OR STEEL BEAM NEEDLES

WALL OF EXISTING BUILDING

NEEDLE THROUGH HOLE IN FOUNDATION

WALES & BRACES

EXCAVATE

JACK OR SOLID BLOCKING

FOOTING OF HEAVY TIMBERS OR CONCRETE

FOOTING OF EXISTING BUILDING

JACK

SHEET PILING

DIG TO DEPTH OF NEW FOOTING

FOOTING OF HEAVY TIMBERS OR CONCRETE

1. INSTALL NEEDLES THROUGH THE FOUNDATION AND RAISE TO CARRY THE WEIGHT OF THE BUILDING.

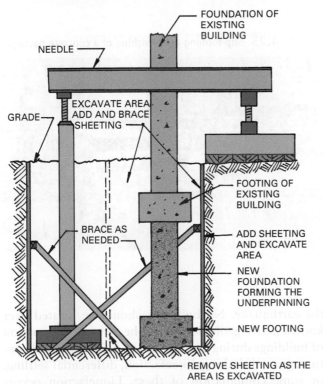

FOUNDATION OF EXISTING BUILDING

NEEDLE

GRADE

EXCAVATE AREA ADD AND BRACE SHEETING

FOOTING OF EXISTING BUILDING

BRACE AS NEEDED

ADD SHEETING AND EXCAVATE AREA

NEW FOUNDATION FORMING THE UNDERPINNING

NEW FOOTING

REMOVE SHEETING AS THE AREA IS EXCAVATED

2. EXCAVATE THE AREA BELOW THE BUILDING FOOTING AND INSTALL THE UNDERPINNING

and finishes the concrete slab (Figure 4.23). Special machinery is used to create curb and gutter profiles (Figure 4.24). Information on concrete is presented in Chapter 7.

Unit Paving

Unit paving systems are used for sidewalks, building entrances, patios, and, sometimes, parking areas or streets. Paving units of brick, stone, or concrete are selected according to their shape, color, strength, and resistance to abrasion and weathering. Unit paving systems are classified by the type of base layer employed. Base layers may be rigid, bonded by mortar, or mortarless and flexible. Heavy vehicular loads necessitate the use of a continuous rigid base. Pedestrian paving can use either a rigid, semi-rigid, or a flexible base (Figure 4.25). Rigid base materials for unit paving include reinforced or unreinforced concrete. Asphalt or bituminous concrete are used for semi-rigid applications, and gravel or sand are used for flexible paving. A setting bed, laid between the base material and the paving units, serves to level the pavers and compensate for any irregularities in the base and the units. Setting beds can be of a 1–2 inch layer of sand, pea gravel, or finely crushed stone.

Permeable paving refers to paving materials, typically concrete, stone, or plastic, that promote the absorption of rain and snowmelt. Materials that are pervious to water—such as gravel, crushed stone, open paving blocks, or pervious concrete—minimize runoff from a site and increase the natural infiltration of storm water. Pervious materials can be used for drives, parking areas, walkways, and patios, and are able to withstand both foot and vehicular traffic. In order to ensure proper drainage, pervious systems will function best with a similarly porous subgrade or permeable backfill.

SEISMIC CONSIDERATIONS

Seismic forces are destructive forces caused by earthquakes. Although accurate prediction of earthquakes is difficult, certain *seismic areas* of the country are known to have subsurface conditions that make them possible earthquake zones, as shown in Figure 4.26. These areas have special building code requirements regulating the design and construction of buildings and other structures.

The magnitude of an earthquake is measured by the Richter scale, which is based on the maximum single movement recorded on a seismograph. The

Figure 4.20 The base course and wearing surface for a typical flexible pavement.

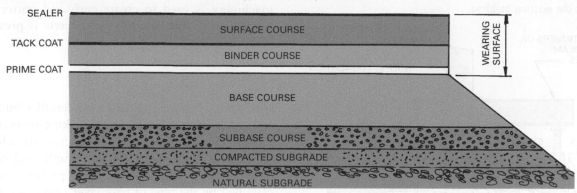

Figure 4.21 The base courses and concrete paving for a typical rigid pavement.

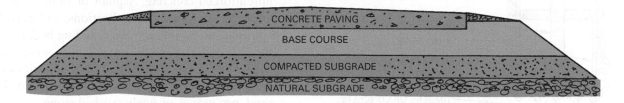

Figure 4.22 Asphalt paving is transported to the site, spread by an asphalt paving machine, and rolled while it is still hot.

Figure 4.23 Slip forming and finishing of a concrete paving.

measurement most widely used by seismologists is the moment magnitude. It is based on the size of the fault on which the earthquake occurs and the amount of earth slippage. The larger the fault and the larger the actual slip, the higher the moment magnitude of

the earthquake. No buildings should be located over known active geologic faults. The types of destruction of buildings during an earthquake include ground rupture, ground shaking, liquefaction, differential settling, or some combination of these. Liquefaction occurs

Figure 4.24 Special machinery creates curb and gutter profiles.

Copyright © GOMACO Corporation

Figure 4.25 Brick pavers on a flexible base are suitable for light traffic applications.

© Sylvie Bouchard/Shutterstock.com

Figure 4.26 Seismic zone map of the United States.

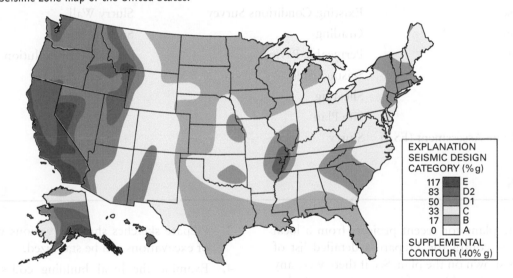

EXPLANATION
SEISMIC DESIGN
CATEGORY (% g)

117	E
83	D2
50	D1
33	C
17	B
0	A

SUPPLEMENTAL
CONTOUR (40% g)

when the motions of the earthquake transform the soil into a semi-liquefied state much like quicksand. It occurs in areas with loose sands and silts that have high water tables.

The design of footings for buildings in seismic areas is regulated by local codes and varies with the soil classification. Masonry construction must meet the requirements for seismic design set forth by the American Concrete Institute, American Society of Civil Engineers, and the Masonry Council. Seismic requirements for reinforced concrete are set forth by the American Concrete Institute and include any additional requirements specified by local codes. Structural steel seismic requirements are usually those set forth by the American Institute of Steel Construction, Inc., in their publication *AISC Seismic Provisions for Structural Steel Buildings* (plus other requirements specified by the local code). Seismic requirements for wood and timber construction follow the recommendations of the National Forest and Paper Association and additional local code specifications.

Review Questions

1. What site development components are shown on a site plan?

2. What types of activities are required to prepare the site for construction?

3. What different methods are used in controlling site drainage?

4. What information is gained through subsurface investigations?

5. What types of technologies can be used for site remediation?

6. What type of work is conducted under the earthwork section of a construction contract?

7. What are the various ways the sides of an excavation can be supported to protect workers from caving earth?

8. What purpose does dewatering serve?

9. When would underpinning be necessary?

10. What are the layers of material used to pave a parking area with asphaltic concrete?

11. What is the difference in laying a concrete paving between form type pavers and slip-form pavers?

12. Describe what happens in the earth when an earthquake occurs.

Key Terms

Angle of Repose

Bio-remediation

Brownfield Site

Caisson

Cofferdam

Dewatering

Environmental Site Assessment (ESA)

Existing Conditions Survey

Grading

Permeable Paving

Seismic Area

Sheeting

Site Plan

Slurry

Slurry Wall

Soil Anchors

Storm Water Pollution Prevention Plan (SWPPP)

Underpinning

Activities

1. Secure site plans for recent projects from a local architect or contractor. Prepare a detailed list of the things shown on the plan. See if there were any environmental problems; if they exist, record what these were and the actions taken.

2. Visit local building sites where excavation is underway or just completed. Prepare a written and photographic report of the things you saw. Did you notice any especially good practices or any unsafe conditions?

3. Prepare sketches showing various ways the sides of excavations can be stabilized.

4. Examine the local building codes and list the requirements for site control of erosion, tree removal, and spoil banks.

5. If your locale is in an identified seismic area, cite the building code requirements specified because of possible seismic activity.

Additional Resources

Ambrose, J., *Building Construction: Site and Below-Grade Systems*, Van Nostrand Reinhold, New York.

Hosack, W., *Land Development Calculations: Interactive Tools and Techniques for Site Planning, Analysis, and Design*, McGraw-Hill, New York.

DeChiara, J., and Koppelman, L. E., *Time-Saver Standards for Site Planning*, McGraw-Hill, New York.

Publications from the Geo-Institute of the American Society of Civil Engineers, Reston, VA.

Soils

Upon completion of this chapter, the student should be able to:

- Define the major types of soils and their classification systems.
- Be familiar with commonly performed soil tests.

- Understand how factors related to soils can affect construction activities.
- Describe various techniques available to stabilize soils.

Soils are created through mechanical, chemical, and biological forces acting on the earth's crust of rock. Naturally occurring forces involving heat and gravity wear down large masses of rock into smaller particles, producing gravels, sands, and fine silts. Over time, weathering and erosion of these particles results in a natural material deposited in layers of variable size and composition known as soil horizons. Soil formation depends significantly on local conditions, and soils from different climate zones show distinctive characteristics. The activities of plants, animals, bacteria, and humans also affect soil formation.

Soil structure affects bearing capacity, water movement, resistance to erosion, and the soil's ability to support vegetation. An understanding of soils, their characteristics, and properties is essential for those who design foundations. Building codes specify maximum design loads for various types of soils, and soil investigations on a site are required to produce the information needed for design.

TYPES OF SOILS

Soils are classified by the sizes of their particles and their physical properties. A sieve analysis is performed to separate soils according to the size of soil particles. Most soils are a mixture of the following five types of soils:

- *Gravel* is a hard rock material with particles larger than ¼ in. (6.4 mm) in diameter but smaller than 3 in. (76 mm).

- *Sand* consists of separate particles ranging from .002 in. to ¼ in. (.050 mm to 6.3 mm) in size that are too small to handle individually. Sand is neither plastic nor cohesive. It is permeable to water and, therefore, has excellent drainage characteristics.
- *Silt* is fine sand with particles smaller than 0.002 in. (0.05 mm) and larger than 0.00008 in. (0.002 mm).
- *Clay* is a very cohesive material with microscopic particles (less than 0.00008 in. [0.002 mm].). Clay exhibits deformation when mixed with water. Relatively impervious, it expands when it absorbs water and shrinks when dry. Clay is one of the least stable and most unpredictable soils for the support of building foundations.
- *Organic matter* is partly decomposed animal and vegetable matter.

Rock particles larger than 3 in. (76 mm) are called cobbles or boulders, and are not classified as soil. Non-problem soils are defined as those that are dense, coarse-grained, and properly drained. They possess excellent bearing capacity because they are not subject to significant volume change. Rock and gravel are generally considered non-problem soils.

Problem soils are defined as those that do not provide adequate bearing characteristics for new construction, or those that expand and contract with the addition of water. Problem soils include highly compressible or expansive clays, highly plastic (liquid like) loose soils, and unconfined sands and silts.

SOIL CLASSIFICATION

Soils are classified based on the size of the soil particles, their distribution and characteristics, and their moisture content. A sieve analysis is used to separate soils into different grain sizes. Dried soil samples are passed through a series of woven screens (sieves) with increasingly smaller openings (Figure 5.1). The results of a sieve analysis allow soils to be classified as either coarse-grained or fine-grained.

Figure 5.1 Standard sieves are used to separate soil samples into soil types based on the size of individual particles.

Moisture content, or the amount of water present in the soil, is a feature in understanding soil classification. Important are the liquid limit, plastic limit, plasticity index, and the shrinkage limit of soils. These states of soil consistency are indicated in terms of water content (Figure 5.2). The *liquid limit (LL)* of a soil is the water content expressed as a percentage of the dry weight at

which the soil will start to flow when subjected to a shaking test. The *plastic limit (PL)* of a soil is the moisture content percentage at which the soil begins to crumble when it is rolled into a thread ⅛ in. (3 mm) in diameter. The *plasticity index (PI)* is the difference between the liquid limit and the plastic limit, and indicates the range in moisture content over which the soil will remain in a plastic condition. The *shrinkage limit (SL)* is the water content at which soil volume is at its minimum.

Also referred to as the Atterberg Limits, the shrinkage limit, plastic limit, and liquid limit are used mainly with clayey or silty soils since these are most prone to expand and shrink due to moisture content. They are used to not only identify a soil's classification, but also define other soil properties, such as compressibility, permeability, and strength.

Test methods, practices, and guides for classifying soils and preparing test specimens are detailed in the American Society of Testing and Materials publication *ASTM Standards on Soil Stabilization with Admixtures*.

The two commonly used soil classification systems are the Unified Soil Classification System (USCS) and the system of the American Association of State Highway and Transportation Officials (AASHTO).

Unified Soil Classification System

The Unified Soil Classification System was developed by the U.S. Army Corps of Engineers to classify soils for use in roads, embankments, and foundations. Soils are classified according to the percentage of grain-size particles in each of the established soil grain sizes in the soil distribution index, the plasticity index, the liquid limit, and the organic-matter content. Soils are grouped in 15 classes. Eight of these are coarse-grained soils that are identified by the letter symbols GW, GP, GM, GC, SW, SP, SM, and SC. Six classes are fine-grained and identified as ML, CL, OL, MH, CH, and OH. The one class of highly organic soils is identified as PT. Soils on the borderline between two classes are given a dual classification, such as GP or GW. Descriptive details are given in Table 5.1.

American Association of State Highway and Transportation Officials System

The AASHTO soil classification system classifies soils according to those properties that affect their use in highway construction and maintenance. In this system, the mineral soil is classified in one of seven basic groups, ranging from A-1 through A-7, on the basis of grain-size distribution, liquid limit, and plasticity index. Soils in

Figure 5.2 As moisture content increases, soils move from a solid to a plastic or liquid state.

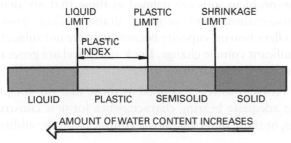

Table 5.1 Classification of Soils Using the Unified Classification System

Type	Letter Symbol	Description	Rating as Subgrade Material	Rating as Surfacing Material
Gravel and gravelly soils	GW	Well-graded gravel; gravel-sand mixture; little or no fines	Excellent	Good
	GP	Poorly graded gravel; gravel-sand mixture; little or no fines	Good	Poor
	GM	Gravel with silt; gravel-sand-silt mixtures	Good	Fair
	GC	Clayey gravels; gravelly sands; little or no fines	Good	Excellent
Sand and sandy soils	SW	Well-graded sands; gravelly sands; little or no fines	Good	Good
	SP	Poorly graded sands; gravelly sands; little or no fines	Fair	Poor
	SM	Silty sands; sand-silt mixtures	Fair	Fair
	SC	Clayey sands; sand-clay mixtures	Fair	Excellent
	ML	Inorganic salts; fine sands; rock flour; silty and clayey fines	Fair	Poor sands with slight plasticity
Silts and clays with liquid limit greater than 50[a]	CL	Inorganic clays of low to medium plasticity; gravelly clays; silty clays; lean clays	Fair	Fair
	OL	Organic silts of low plasticity	Poor	Poor
Silts and clays with liquid limit less than 50[a]	MH	Inorganic silts; micaceous or diatomaceous fine sandy or silty soils; elastic silts	Poor	Poor
	CH	Inorganic clays of high plasticity	Very poor	Poor
	OH	Organic clays of medium to high plasticity; organic silts	Very poor	Poor
Highly organic soils	Pt	Peat and other highly organic soils	Unsuited for subgrade surfacing	Unsuited for material

Source: U.S. Department of Interior, Bureau of Reclamation

[a]The liquid limit is the water content, expressed as a percentage of the weight of the oven-dried soil, at the boundary between the liquid and plastic states of the soil.

the A-1 through A-3 groups are sands and gravels. They have a low content of fines (very small particles). Soils in the A-4 through A-7 groups are silts and clays and are, therefore, fine-grained soils. A special grade, A-8, is reserved for highly organic soils.

When laboratory tests are made on soil samples, the A-1, A-2, and A-7 groups can be further classified into groupings such as A-1-a, A-1-b, A-2-4, A-2-5. Details are shown in Table 5.2.

SUBSURFACE SOIL TESTING

For most commercial construction projects, soil samples must be obtained on the building site for testing and analysis. The design team designates the exact locations where soil borings are to be taken, normally along the building perimeter, at building corners, and at critical interior column loads.

A number of soil testing methods are in common use:

Soil Load Tests use incremental loads that are applied to a platform erected on an area of undisturbed soil and measured to determine the settlement characteristics of the soil.

A *Standard Penetration Test (SPT)* uses a pipe-like weight of around 140 lbs placed on top of a drive rod. The weight is repeatedly raised and dropped a distance of 30 inches, driving it into the soil. The number of hits or blows required to drive the weight every 6 inches allows for a determination of bearing capacity.

Auger borings bring up soil samples using a 2-in. or 2½-in. auger fastened to a long rotating pipe (Figure 5.3). Limited to depths of 50 feet, auger borings are most effective in sand or clay, because the auger will usually falter at obstructions such as boulders, rock, or even tree stumps.

Samples taken by *core borings* are more costly than any of the other methods, but result in the most reliable testing by allowing penetration through all materials and

Table 5.2 The AASHTO System of Soil Classification

	A-1		A-2				A-3	A-4	A-5	A-6	A-7
	Sand and Gravel		**Gravel or Silty or Clayey Sand**								
Typical Material	A-1-a	A-1-b	A-2-4	A-2-5	A-2-6	A-2-7	**Fine Sand**	**Silt**	**Silt**	**Clay**	**Clay**
No. 10 sieve	50% max.										
No. 40 sieve	30% max.	50% max.					51% min.				
No. 200 sieve	15% max.	25% max.	35% max.	35% max.	35% max.	35% min.	10% max.	36% min.	36% min.	36% min.	36% min.
Fraction passing No. 40 sieve											
Liquid limit	6% max.	6% max.	40% max.	41% min.	40% max.	41% min.	—	40% max.	41% min.	40% max.	41% min.
Plasticity index	—	—	10% max.	10% max.	11% min.	11% min.	—	10% max.	10% max.	11% min.	11% min.

Source: Standard Specifications for Transportation Materials and Methods of Sampling and Testing, 2004, American Association of State Highway and Transportation Officials, Washington, D.C. Used by permission.

Figure 5.3 Soil samples taken by core borings are obtained on the building site for testing and analysis.

Photo courtesy of constructionphotographs.com

to great depths. Soil samples are brought up using an inner and outer sampling pipe that can be opened along the side. Samples are collected in the inner pipe at 5-in. intervals and give a detailed picture of a site's soil horizons.

The subsurface boring samples are taken from the building site to a soils testing facility to be analyzed. The soil analysis, or soil test report, is prepared by a licensed geotechnical or civil engineer experienced in soils engineering. The report includes a detailed geologic description

of the site subsurface conditions, along with a graphic representation of the location and depth of soil strata (Figure 5.4). Issues that could affect the integrity of foundations, such as active faulting, slope instability, the presence of expansive or corrosive soils, and groundwater seepage, are addressed. The soils report provides the necessary data for the project civil engineer, architect, or structural or foundation engineer to design the project appropriately in consideration of the site geotechnical conditions.

FIELD CLASSIFICATION OF SOIL

When it is not possible to have soil samples tested in a laboratory, a number of field tests can be made. Field testing involves separating soil particles according to size by screening them through sieves of various sizes. The soil is separated as smaller grains fall through the sieve, while larger particles are retained. The openings in the sieves are designed to separate samples into the basic soil types of gravels, sands, silts, and clays. Sieve sizes are denoted either by the size of the screen openings, or by indicating the number of screen openings per square inch. Gravels, for instance, are retained by a number 4 sieve (four openings per inch) but flow through a 3-in. sieve. Frequently used sieve sizes are shown in Table 5.3. Following are generalized descriptions of field tests made by a soils engineer. More detailed information can be obtained from the U.S. Department of Interior, Bureau of Reclamation.

Table 5.4 shows the division of soil into various fractions based on the size of the soil particles. Although this is important when considering the properties of a soil,

Figure 5.4 A typical soils report on the findings of a subsurface soil test.

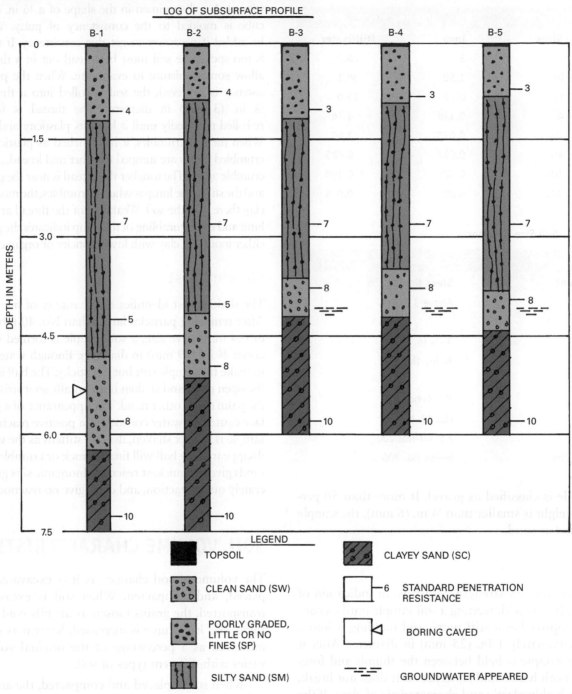

LOG OF SUBSURFACE PROFILE

LEGEND

TOPSOIL

CLEAN SAND (SW)

POORLY GRADED, LITTLE OR NO FINES (SP)

SILTY SAND (SM)

CLAYEY SAND (SC)

STANDARD PENETRATION RESISTANCE

BORING CAVED

GROUNDWATER APPEARED

other properties, such as the shape of the particles (angular or sharp distinct edges versus sub-angular or rounded edges), also influence the ability of a soil to interlock particles.

Testing for Soil Coarseness

To test for coarseness, all soil particles larger than 3 in. (7.6 cm) in diameter are removed from the sample.

The remaining particles are separated through a No. 200 sieve, which permits the smallest particles that can be seen by the naked eye to pass through. If more than 50 percent of the sample by weight does not pass through the sieve, the sample is a coarse-grained soil. The coarse sample is then divided into particles larger and smaller than ¼ in. (6 mm). If more than 50 percent of this sample by weight is larger than ¼ in. (6 mm),

Table 5.3 Some Frequently Used Standard Sieves

U.S. Standard	Opening Size	
Sieve Sizes	Inch	Millimeter
3 in.	3	76.2
1½ in.	1.50	38.1
¾ in.	0.75	19.0
No. 4	0.186	4.76
No. 10	0.078	2.00
No. 40	0.017	0.425
No. 100	0.006	0.150
No. 200	0.003	0.075

Table 5.4 Soil Size Fractions

Constituent	U.S. Standard Sieve No.
Cobbles	Above 3 in.
Gravel	
Coarse	3–¾ in.
Fine	¾ in.–No. 4
Sand	
Coarse	No. 4–No. 10
Medium	No. 10–No. 40
Fine	No. 40–No. 200
Silts and clays	Below No. 200

the sample is classified as gravel. If more than 50 percent by weight is smaller than ¼ in. (6 mm), the sample is classified as sand.

Dry Strength Test

A field test for dry strength, which is an indication of plasticity, begins with wetting a soil sample until its consistency approaches a stiff putty, and molding it into a ball approximately 1 in. (25 mm) in diameter. After it dries, the sample is held between the thumb and forefinger of both hands and squeezed. If it does not break, the soil is highly plastic and characteristic of clays. If the sample breaks but it is difficult to cause the sections to powder when rubbed between the fingers, it has a medium plasticity. If the sample easily breaks down into a powder, it has low plasticity.

Toughness Test

A toughness test ascertains the consistency of a soil sample near the plastic limit. All soil particles larger than the

opening in a No. 40 (about ¹⁄₆₄ in. or 0.4 mm) sieve are removed. A soil specimen in the shape of a ½ in. (12.5 mm) cube is molded to the consistency of putty. Water may be added if necessary to get this consistency. If the sample is too sticky, the soil must be spread out in a thin layer to allow some moisture to evaporate. When the proper consistency is achieved, the soil is rolled into a thread about ⅛ in. (3 mm) in diameter. The thread is folded and re-rolled repeatedly until it loses its plasticity and crumbles. When the soil crumbles, it has reached its plastic limit. The crumbled pieces are lumped together and kneaded until they crumble again. The tougher the thread is near the plastic limit and the stiffer the lump is when it crumbles, the more colloidal clay there is in the soil. Weakness of the thread at the plastic limit and easy crumbling of the lump indicates the presence of either inorganic clay with low plasticity or organic clay.

Shaking Test

The shaking test identifies the character of fines in a soil. After removing particles larger than No. 40 (about ¹⁄₆₄ in. or 0.4 mm) sieve size, a soil sample is formed into a ball about ¾ in. (19 mm) in diameter. Enough water is added to make the sample soft but not sticky. The ball is placed in the open palm and shaken horizontally so it strikes against the palm of the other hand. The appearance of a glossy surface caused by water constitutes a positive reaction. As the sample is further shaken, the ball stiffens as the water gloss disappears. The ball will finally crack or crumble. Very fine sands give the quickest reaction. Inorganic silts give a moderately quick reaction, and clays give no reaction.

SOIL VOLUME CHARACTERISTICS

The volume of soil changes as it is excavated, hauled, placed, and compacted. When soil is excavated and transported, the grains loosen as air fills voids between them, and its volume is increased. Known as swell, it is measured as a percentage of the original volume and varies with different types of soil.

When soil is placed and compacted, the air between the grains is forced out and the soil's volume decreases. This is called shrinkage. The amount of shrinkage varies with the type of soil and is measured as a percentage of the original volume.

The volume of one cubic yard of soil *in situ* (undisturbed soil) is referred to as **bank measure**. When disturbed, it will produce more than 1 cubic yard and when compacted, less than 1 cubic yard of soil. The weight and volume changes for typical types of soil are shown in Table 5.5.

Table 5.5 Typical Weights, Swell, and Shrinkage for Selected Types of Soils

| | Weight[a] | | | | | | | |
| | Loose | | Bank Measure | | Compacted | | | |
	lb./yd.³	kg/m³	lb./yd.³	kg/m³	lb./yd.³	kg/m³	Swell	Shrinkage
Clay, natural	2,300	1,300	2,900	1,720	3,750	2,220	38%	20%
Common earth, dry	2,100	1,245	2,600	1,540	3,250	1,925	24%	10%
Sand with gravel, dry	2,900	1,720	3,200	1,900	3,650	2,160	12%	12%
Sand, dry	2,400	1,400	2,700	1,600	2,665	1,580	15%	12%

[a]Weights, swell, and shrinkage of actual samples may be larger or smaller.

Soil volume also changes due to fluctuations in moisture content. Clays are especially susceptible to changes in moisture. As moisture content increases, the soil expands, sometimes to the point of lifting a footing or slab. If the moisture decreases, the soil will shrink, sometimes causing a footing or slab to settle.

SOIL WATER

The two major types of water present in soils are capillary water and gravitational water. Capillary water is retained within the microscopic pores of individual soil particles, enabling them to bind together. Gravitational water flows within soil in the voids between particles. The water table, or *groundwater level*, is the level of free-flowing water below the ground surface. Since it is free flowing, water levels below the soil surface can be varied by pumping water from wells, by dry or rainy seasons, or by installing drainpipes to remove water from the construction area.

The moisture content of soil is an important factor during soil compaction. The density (p/cf or kg/m³) varies with moisture content. A moisture content in optimal amounts enables soil to be compacted to its maximum density. This is called the optimum moisture content and is determined by laboratory testing of soil samples.

Another factor to consider when examining soil on a site is its permeability. Permeability measures the ability of water to penetrate through the soil. For example, clay has low permeability, whereas sand has a high permeability rating. Soil permeability is measured by testing samples with a permeameter.

SOIL-BEARING CAPACITIES

The required load-bearing capacities of the various types of soils are specified by local building codes. Table 5.6 shows the maximum allowable load-bearing values for

Table 5.6 Presumptive Load-Bearing Values of Foundation Materials

| | Load-Bearing Pressure | |
Class of Material	lb./ft.²	kg/m²
Crystalline bedrock	12,000	58,560
Sedimentary rock	6,000	29,280
Sandy gravel or gravel	5,000	24,400
Sand, silty sand, clayey sand, silty gravel, and clayey gravel	3,000	14,640
Clay, sandy clay, silty clay, and clayey silt	2,000	9,760

Source: The BOCA National Building Code/1993.
Source: Building Officials and Code Administrators International, Inc. Copyright 1993. All rights reserved.

supporting soils under spread footings placed at or near the surface of a site. In many cases, it is necessary to test soil samples to determine the maximum allowable pressure that a soil can support. These surface values must be adjusted for deep footings and load-bearing strata under piles, as specified by the building code.

If the soil in the area of a structure is uniform in composition and the footing design is adequate, the building will remain stable. While some settlement may occur over time, it will be uniform. If the soil composition varies, and is not compensated for in foundation design, the building could experience differential settlement. Other actions can also affect the stability of supporting soils. Extensive pumping of subsurface water, for example, can cause soil to settle because of a lowering of the water table.

SOIL ALTERATIONS

After soil tests have been made and the soil's properties determined, the soil can be modified to improve its load-bearing characteristics. Commonly used processes

Chicago Center for Green Technology

Architect: Farr Associates Architecture and Urban Design	Building type: Educational Center, Commercial Office	Completed 2003
Location: Chicago, Illinois	Size: 40,000 sq. feet (3,720 sq. meters)	Rating: U.S. Green Building Council LEED Platinum (37 points)

Overview

The Chicago Center for Green Technologies (CCGT) building is open for visitors to gain firsthand knowledge of how green buildings contribute to the good of the environment. The center provides green technology educational resources to the public, including green building standards and construction guides, reference books, samples of green building materials, and staff and volunteers to guide the research process. The building also houses a recycling center, educating occupants and residents in everyday recycling strategies.

The structure is a renovation of an existing 1952 office building that was previously owned by a construction materials recycling company. The business was closed after it was discovered that the 17-acre site contained 600,000 cubic yards of illegally dumped debris stacked up to seven stories high. The Chicago Department of the Environment purchased and cleaned the brownfield site, recycling the discarded stone and concrete into public infrastructure projects throughout the city.

Site Assessment and Remediation

The remediation of the brownfield site required detailed assessments and multiple reports to ascertain the level of contamination. Prior to the environmental assessment of the site, the removal of over 500,000 cubic yards of illegally stockpiled C&D debris was required. A Phase I Environmental Site Assessment was conducted that identified that the former site was used as a foundry, with a former gas plant located directly next door. Both former uses suggested the presence of products of environmental concern such as cyanide, sulfur, iron, and heavy petroleum products such as oils and tars.

A Phase II Environmental Site Assessment was performed to evaluate the recognized environmental conditions (RECs) identified in the Phase I ESA. The Phase II ESA revealed that in the uppermost levels of soil, numerous toxic substances at levels exceeding the allowable levels for commercial land use were present. Once the full extent of soil contamination was defined, a Remedial Action Plan (RAP) was developed to identify the required actions needed to remediate the site.

The remediation consisted of partial excavation and removal of soil that was subsequently replaced

View from the green roof of the Chicago Center for Green Technology.

with clean fill in the most polluted areas of the site. Because the soil impacts were limited, the majority of the remaining polluted soils were managed *in situ* by installing an engineered barrier consisting of impermeable surfaces, such as paved parking lots that prevented human exposure to the underlying pollutants.

Site Design

The landscaping components of the center were designed to showcase sustainable site-design technologies. A thorough sedimentation and storm water control system was implemented. A constructed wetland and system of bioswales slows the volume of rainwater so that pollutants can settle out of the water before it reaches the ground or sewers. Native drought-resistant landscaping serves to filter and clean a majority of runoff directly on the site. All walks and drives were constructed using a system of permeable pavings. The paving matrix prevents contamination of nearby water systems by limiting urban storm-water runoff. Rather than allowing storm water to flow over impervious surfaces, picking up contaminants along the way, all storm water is retained on site and slowly percolated back into the ground.

A large part of the roof at the Center for Green Technology consists of a green roof system utilizing sedum-based plantings that reduces the cooling load of the building, while simultaneously protecting the roof's waterproof membrane. The planting species of the green roof absorb large portions of rainwater. The remaining storm water is collected in a series of cisterns and reused in landscape irrigation. This recycling of storm water for landscaping also helps reduce the center's water usage.

to alter soil properties include compaction, stabilization, and dewatering.

Compaction

Compaction of soil refers to increasing its density by mechanically forcing the soil particles closer together. This expels the air present in the voids between particles. Soil density can also be increased by consolidation, which involves forcing the particles closer together through the removal of water from the voids between soil particles. Compaction tends to produce immediate results, while consolidation requires longer time periods to increase the soil density.

The amount of compaction that can be obtained depends on the physical and chemical properties of the soil, its moisture content, the compaction method used, and the thickness of the soil layer being compacted. As compaction occurs, the air and water between particles is forced out of the soil and must be drained away. The soil engineer will use the results of soil tests to decide the best way to achieve the required compaction density.

Soil compaction can be achieved in a variety of ways, including a kneading action, vibration, static weight, explosives, and impact. Pneumatic-tire rollers are machines with multiple tires that provide compaction of the soil through a kneading action. The rows of tires are staggered to give wide coverage, and some units have wheels mounted to give a wobbly effect that increases the kneading action. The weight of pneumatic rollers can be varied by adding ballast. They are effective on most soils but least effective on sands and gravels.

Tamping foot rollers use static weight to compact soil (Figure 5.5). Tamping foot rollers are available in a range of foot sizes and shapes. As they pass over the soil, the feet sink into it, compacting material below the surface. With repeated passes, the feet do not penetrate as deep and eventually walk on the top surface. Tamping rollers are most effective on cohesive soils.

Figure 5.5 This compactor uses sheepsfoot rollers to produce a tamping action.

Another type of static-weight compactor uses smooth steel wheels or drums (Figure 5.6). It is best used to compact granular bases, asphalt bases, and asphalt pavements. It is not effective on cohesive soils, because it tends to compact the surface, forming a crust over a loose, non-compacted subsurface.

Figure 5.6 Smooth-wheel rollers are used to compact granular materials and asphalt pavements.

© Dmitry Kalinovsky/Shutterstock.com

A variety of compaction devices use vibration. Small hand-operated vibratory compactors are useful in compacting areas where it may be difficult to fit larger equipment. Larger vibratory compactors include tamping foot rollers and smooth drum rollers. The vibratory action provides compaction in addition to static weight.

Impact rammers are used to provide compaction in very tight areas. The vertical movement of the rammer provides considerable compaction. Units are also available for mounting on the end of a backhoe boom.

Loose, saturated, granular soil can be compacted by subjecting it to a sudden shock and vibration. This causes the soil particles to fall into a denser pattern, displacing water from between them. The weight of the particles forces the water to flow from the soil. Explosives are typically used for this purpose. The spacing, depths, and sizes of the explosive charges are determined by experienced soil engineers.

Another way to increase the density of cohesionless soils is to use vibro-compaction. This employs a vibratory probe, a large-diameter tube with a vibrating device mounted on one end. The probe is lifted by a crane and driven into the soil by the weight of the tube and the vibrations. The frequency and duration of vibration can be monitored to achieve maximum compaction for the prevailing soil conditions. The degree of compaction depends on the soil type, spacing of probes, frequency, and the duration of vibration.

Compaction of soil can also be achieved through a process called vibro-flotation. As shown in Figure 5.7, a probe with a powerful cylindrical vibrator penetrates the soil via vibration and is assisted by compressed air or water jets. When the probe reaches the desired depth, the space created by the vibroflot is filled with gravel by the simultaneous introduction of the material and withdrawal of the probe. Vibration during withdrawal ensures the efficient compaction of the added material and the neighboring soil.

Soil Stabilization with Admixtures

Test methods and specifications on soil stabilization with admixtures are detailed in the publication *ASTM Standards on Soil Stabilization with Admixtures*. Following are explanations of some of these. Soil remediation is discussed in Chapter 4.

Blending Soils   Soils can be blended by mixing imported material with the original using a motor grader or disc. Power shovels or deep-cutting belt loaders can be used to cut deeply through the soils and provide consistent mixing.

Lime–Soil Stabilization   Clays and silty clay soils can be stabilized through the addition of lime, which produces a chemical reaction. Clays expand and contract according to moisture content and can cause damage to the structures resting on them. A concrete slab, for instance, can heave and crack due to the expansion of clay. The addition of lime to clay reduces the amount of expansion and forms a moisture barrier that protects the expansive clay from subsurface water. In some cases, lime can eliminate the need to excavate and replace unsatisfactory soil. Typically, slaked, hydrated lime, or quicklime, is spread over the base material and blended into it with a pulverizing machine. Water may be added if conditions warrant. The blending layer is generally 12 to 18 in. (300 to 450 mm) deep and compacted to the required thickness. The compacted layer will continue to gain strength for many months.

Asphalt–Soil Stabilization   Asphalts blended with granular soils produce a durable, stable soil that can be used as a finished surface for low-traffic roads or

Figure 5.7 The vibro-flotation method inserts a vibrating probe into the soil. Penetration is assisted by jets of compressed air or water.

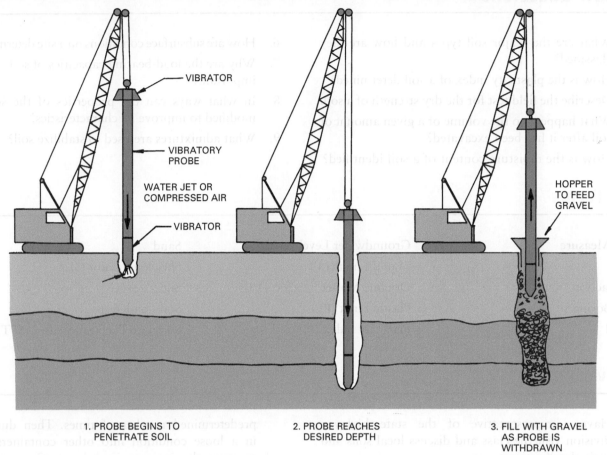

VIBRATOR

VIBRATORY PROBE

WATER JET OR COMPRESSED AIR

VIBRATOR

HOPPER TO FEED GRAVEL

1. PROBE BEGINS TO PENETRATE SOIL

2. PROBE REACHES DESIRED DEPTH

3. FILL WITH GRAVEL AS PROBE IS WITHDRAWN

a stabilized base course for higher-quality pavements. Some soils require the addition of fines with the asphalt to fill the spaces between the soil particles.

Cement–Soil Stabilization Predominantly granular soils that have minute amounts of clay particles can be stabilized by blending them with Portland cement. In some cases, fly ash is used to replace some of the Portland cement. The cement is uniformly spread over the surface and blended with the soil using a pulverizing machine. After this has been accomplished to the specified depth, the surface is fine-graded and compacted. Water can be sprinkled on the surface during blending if the soil moisture content is low. Since Portland cement sets up quickly, compaction immediately after blending

is required. Initial compaction can be accomplished with tamping or pneumatic rollers and followed by a steel smooth-wheel roller.

Salt–Soil Stabilization Coarse crushed rock salt can be used to stabilize well-graded clay or loamy soil containing some limestone fines. The salt can be blended in the soil in dry form or mixed with water to form brine. The base soil is pulverized to the specified depth, the salt added, and the mixture blended. This layer is then compacted, fine-graded, watered, and finished with a steel smooth-wheel roller. The salt and water form a bond with soil particles, creating a stable soil. It can take up to several weeks before the soil is cured.

Review Questions

1. What are the major soil types and how are they classified?

2. How is the plasticity index of a soil determined?

3. Describe the field test for the dry strength of a soil.

4. What happens to the volume of a given amount of soil after it has been excavated?

5. How is the moisture content of a soil identified?

6. How are subsurface conditions on a site determined?

7. Why are the load-bearing capacities of soil important?

8. In what ways can the properties of the soil be modified to improve its characteristics?

9. What admixtures are used to stabilize soil?

Key Terms

Bank Measure

Clay

Compaction

Core Boring

Gravel

Groundwater Level

Liquid Limit (LL)

Organic Matter

Plastic Limit (PL)

Plasticity Index (PI)

Sand

Shrinkage Limit (SL)

Silt

Soil Load Test

Standard Penetration Test (SPT)

Activities

1. Have a representative of the state highway division visit the class and discuss local soils and their classifications.

2. Arrange for a visit to a local soils testing laboratory.

3. Secure soil-testing sieves of various sizes and run some tests on local samples for classification.

4. Run dry strength and toughness field tests on soil samples from various sites.

5. Pack soil samples in a container of known volume, such as a quart or gallon. Ram each sample a predetermined number of times. Then dump it in a loose condition into other containers and measure the increase in volume. Compute the percentage of swell for each sample. Although this is a crude test, it will possibly give some interesting results if the soil samples are chosen from a wide geographic area.

6. Visit a road construction site, observe the grading and compaction process, and report on the compaction and soil stabilization techniques being used.

Additional Resources

PCA Soil Primer, Portland Cement Association, Skokie, IL.

Standards on Soil Stabilization with Admixtures, American Society for Testing and Materials, Philadelphia, PA.

Standard Specifications for Transportation Materials and Methods of Sampling and Testing, American Association of State Highway and Transportation Officials, Washington, DC.

See Appendix C for addresses of professional and trade organizations and other sources of technical information.

Foundations

Upon completion of this chapter, the student should be able to:

• List the factors to be considered when designing foundations.

• Identify various types of shallow and deep foundation systems.

• Discuss the features and construction of common types of foundations.

The function of a foundation is to safely support the loads of the structure above, and transmit them to the ground below. While foundation systems are found in all buildings, their construction can vary dramatically in scale and cost. A small commercial building may utilize a simple slab on grade, while large high-rise buildings can require foundations that are hundreds of feet deep. Foundations are classified as either shallow or deep (Figure 6.1). Shallow foundations, such as spread footings and slabs, transmit structural loads directly beneath the substructure and close to existing grade. Deep foundations, such piles, reach far below grade to access materials suitable for the required bearing. The choice of foundation systems is determined by the size and type of building to be constructed, the soil and sub-surface water conditions, and building code requirements.

FOUNDATION DESIGN

Buildings can be conceived of as having three major elements: the *superstructure*, the portion of the building above the ground; the *substructure*, the enclosed and conditioned portion below grade; and the *foundation*, the element that supports the loads superimposed on it through transmitting elements into solid bearing below. Foundations also serve to resist uplift or overturning forces caused by wind and ground swelling (Figure 6.2). The design of a building's foundation begins as soon as the initial architectural plans are developed. The architect is responsible for resolving the shape and size of a building, its distribution of functional spaces, and the selection of structural and other materials.

The initial building layout is sent to the structural engineer who reviews the existing site conditions, completes the necessary calculations, and develops designs for the structural frame of the building. A civil or foundation engineer, specializing in subsurface construction may be consulted to work with both the architect and structural engineer as the foundation design is developed. The final foundation plan is the result of the cooperative effort of these three design disciplines. The foundation engineer considers the size of the building, loads to be

Figure 6.1 Foundations are classified as either shallow or deep.

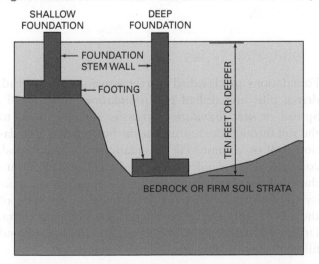

Figure 6.2 Differential settlement can occur if building loads are unevenly distributed or variations exist in subsoil properties.

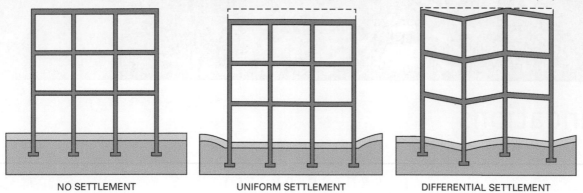

NO SETTLEMENT   UNIFORM SETTLEMENT   DIFFERENTIAL SETTLEMENT

carried, conditions on the site, the climate, type of structural system to be used, presence of water, winds, potential earthquakes, and other factors. Building codes set a number of guidelines that heavily influence the design process and may dictate the type of foundation used.

An important consideration in foundation design is the prevention of settlement. Settlement refers to the amount a structure will sink during and after construction. All buildings will experience a certain amount of settling as the ground adjusts to the newly applied weight. Most building settling is negligible but in some cases, it can be excessive and cause building failure. Total settlement refers to the uniform settlement of the entire structure resulting from the weight of the structure and imposed loads. Differential or uneven settlement can occur if the loads on the structure are unevenly distributed or variations are present in the soil properties (see Figure 6.2).

The depth of a foundation and its footings is also impacted by the climate conditions of the site. The base of a foundation must always be placed below the frost line to prevent possible heaving of the building structure due to freeze (contracting) and thaw (swelling) movement of the ground. The building site, its access and room for the equipment associated with constructing the various types of foundations, as well as proximity to neighboring buildings are considered.

Hydrostatic forces, or the amount of water pressure the foundation will be subjected to, are also planned for during foundation design. The amount of expected precipitation as well as the ability of soils to absorb water can potentially cause a foundation system to heave upward. All ground and surface water must be diverted from the building foundation in order to maintain soil conditions that will support the building without failure.

Building codes mandate that the slope of the finish grade slant away from the building perimeter at a minimum gradient of ½ in. per foot for the first 6 feet.

Foundation and basement walls are waterproofed with asphaltic or membrane materials to ensure a watertight wall system. Gravel or other coarse-grained material is then backfilled against the sides of foundations in order to provide a permeable drainage channel for both building and surface water runoff. A perforated drain is usually incorporated at the base of the gravel fill. The drain surrounds the perimeter of the building and conducts water toward appropriate site drainage channels.

The final design of foundations is described in the building foundation plan and details. As drawing sets are organized according to the construction sequence, foundation drawings are generally at the beginning of the construction drawing set. The foundation plan is a scale drawing showing the location, depth, and size of footings, size and dimensions of piers, and the construction measurements and details of the subfloor area (Figure 6.3).

Foundation Types

Foundations are classified into three major types: spread, driven pile, and drilled *pile foundations* (Figure 6.4). Spread or *mat foundations* transfer building loads to the soil through the footings at the bottom of a foundation wall or column. Pile foundations use long wood, concrete, or steel piles that are driven or drilled into the earth. Driven piles get their load-carrying capacity through frictional forces developed on the sides of the pile or from the end resting on load-bearing strata. Drilled piles are formed by boring holes in the earth and filling them with concrete and reinforcing steel.

Figure 6.3 Foundation plans note only the elements directly associated with subsurface construction.

Foundation Plan
Scale: 1/4" = 1'-0"

NORTH

Figure 6.4 Types of shallow and deep foundations.

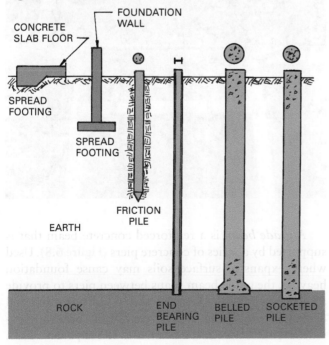

SHALLOW FOUNDATIONS

Shallow foundation usually refers to those supported on stratum with adequate bearing capacity close to the surface of the ground. Common examples include pad, strip (continuous footing), or raft foundations. A combination of two or three types of shallow foundations in a single structure is not uncommon. Shallow foundations are generally more affordable than deep foundations and utilize simple construction procedures that do not require extensive expertise.

Shallow foundations consist of a *footing*, the base of the foundation that transmits building loads directly to the soil, and a stem wall, the vertical portion of the foundation that reaches to grade and provides the base for the superstructure. Stem walls can be constructed of poured in place concrete or reinforced masonry.

Spread Footings

The most common shallow foundation type is the continuous *spread footing*, or strip footing. A strip footing provides a stable and continuous foundation around the full perimeter of a structure. The soil below the footing must be undisturbed or compacted to provide a solid substrate. Compaction is achieved using heavy rollers or vibratory tamping that reduces the volume of air voids between individual soil particles. If soil conditions allow for trenching without the need for stabilization, the excavated trench can act as the formwork for the footing concrete. In loose soils and unconfined sands, wood or steel formwork is constructed to contain the wet concrete during the concrete placement. When used on sloping sites, the footings are stair-stepped with the slope to accommodate the change in grade (Figure 6.5). Buildings with spread footings often utilize interior independent footings to support point or concentrated loads, such as columns or piers within the building footprint. *Independent footings*, also known as footers, are an enlargement at the bottom of a column or interior bearing wall that spreads the applied structural loads over a sufficiently large soil area (Figure 6.6).

Figure 6.5 On sloping sites, footings are stair-stepped with the slope to accommodate changes in grade.

© Christian Delbert/Shutterstock.com

Concrete has no tensile strength of its own, so steel reinforcing with high-tensile strength must be added. The type, spacing, and placement of the reinforcing are specified on the foundation drawings. The steel bars, located in the bottom of the footings where flexural forces are the greatest, are either hung from formwork bracing or supported by special wire bar supports (Figure 6.7). The bars are bent at corners and wall intersections and overlapped to establish structural continuity throughout the geometry of the footing construction. Prior to pouring the wet concrete, elevation stakes are driven into the formwork to act as leveling guides during the concrete placement. Vertical steel reinforcing bars that project upward out of the formwork is used to connect the footing with the stem wall that will be built on it.

Figure 6.6 Strip footings provide a continuous foundation around the full perimeter of a structure. Independent footings carry point loadings of interior column loads.

Figure 6.7 Concrete has no tensile strength of its own, so steel reinforcing with high-tensile strength is added.

A *grade beam* is a reinforced concrete beam that is supported by a series of concrete piers (Figure 6.8). Used where expansive surface soils may cause foundation heaving, the grade beam spans between piers to provide a solid base over the unstable soil. Cardboard that will decompose with time is laid beneath the beam to provide a space between the beam and the expansive soils.

Figure 6.8 Grade beams distribute loads between deep foundation elements, such as piles.

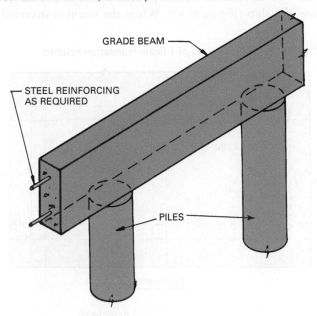

Grade beam footings differ from continuous spread footings in how they distribute loads. The depth of a grade beam footing is designed to distribute loads to bearing points, while the width of a continuous spread footing is designed to transfer loads to the ground.

Slab and Raft Foundations

A *slab on grade* is a reinforced concrete foundation that is placed directly on the ground to provide the structural support for a building (Figure 6.9). Slabs can be constructed as an extension of a stem wall to provide a

Figure 6.9 A reinforced concrete slab on grade is a foundation that is placed directly on the ground to provide the structural support for a building.

Photo courtesy of constructionphotographs.com

monolithic construction. In very warm climates, deep excavations can be eliminated and the slab constructed with a thickened edge at its perimeter and at bearing points within the building required to support interior load-bearing partitions and columns. Most codes require that slabs on grade be a minimum of 4 inches (10.1 cm) thick, although heavier commercial construction often utilizes slabs up to 6 inches (16.2 cm) in thickness. Regardless of its depth, the slab acts as a floor system only and is not designed to carry additional vertical building loads.

Formwork, similar to that used for spread footings, is built at the slab edge. The ground below the slab must first be compacted to assure a firm bearing surface.

With the concrete formwork in place, a 4- to 6-inch bed of compacted fill gravel is laid into the forms to provide a solid and well-drained base. To protect the slab from ground moisture, a layer of polyethylene sheeting is installed between the gravel and the slab. Where ground temperatures fall significantly below interior air temperature, the potential for large heat losses necessitates the use of insulation below the slab. Rigid insulation can be laid directly under the slab and along the vertical perimeter stem walls (Figure 6.10). Slabs on grade are reinforced using steel bars (often called rebar) and steel mesh materials. Before the slab is poured, plumbing, heating, and other utility lines are placed in their appropriate locations to stub out of the finished slab. This procedure, referred to as the utility rough in, requires close coordination between all subcontractors.

Figure 6.10 The component materials of a typical slab on grade.

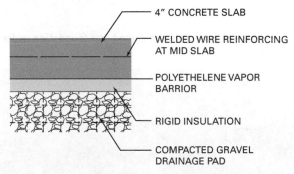

Post-tensioned slabs on grade are used to overcome problems associated with unstable soils for ground-supported residential foundations. Construction of post-tensioned slabs on grade uses high strength reinforcing steel placed inside plastic ducts with tension anchors located in pockets embedded into the slab edge. Once the concrete pour has gained sufficient strength, usually after two to three days, the tendons are stressed. After stressing, the tendon ends are cut off and the anchor pockets filled with grout to protect them from corrosion. A typical residential post-tensioned concrete slab ranges from 8 to 12 in. (20.3 to 30.4 cm) in thickness and uses high-strength concrete. Post-tensioned slabs are also used in areas with stable soils to reduce slab cracking, decrease or eliminate control joints, and increase flexural capacity.

Another method of dealing with expansive soils on project sites is through the use of under-slab voids that are created by pouring slabs on top of void forms, also known as carton forms. Laid directly on the earth, the cardboard void forms absorb moisture from the soil and degrade over time, while the concrete slab floats above. The void created between the bottom of the slab and the top of the soil allows expansive soils to swell without influencing the slab. In addition to cardboard, materials used to create the void include polystyrene and plastic form components.

Raft or mat foundations are reinforced concrete slabs several feet in thickness that cover the total footprint of a building (Figure 6.11). The concrete mat spreads the weight of the building evenly over the entire area below the building, thereby reducing the load per square foot on the soil. This is especially useful when the soil characteristics have low bearing capacity or vary across the excavated area. As a general rule of thumb, if spread footings would cover more than 50 percent of the building footprint area, a mat or some type of deep foundation will usually be more economical. To avoid construction joints, the entire slab is poured monolithically and continuously.

Figure 6.11 A monolithically poured mat foundation under construction.

A T-beam raft foundation uses a solid poured slab and two-directional beams poured either above or below the slab (Figure 6.12). When the beam is inverted,

Figure 6.12 Two types of T-beam foundation systems.

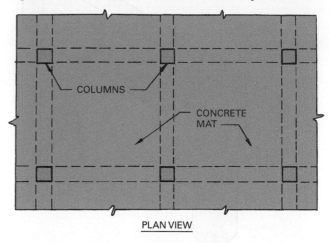

PLAN VIEW

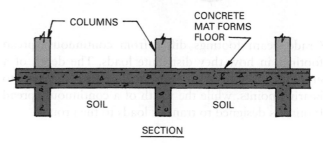

SECTION

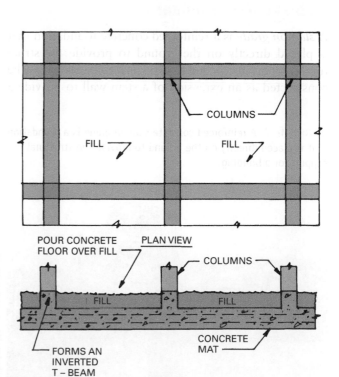

PLAN VIEW

SECTION

the slab becomes the basement floor. The beams' cavities are dug in the soil, which must be able to stand without caving in as the beams and slab are poured. When the beams are on top of the slab, the spaces between them must be filled before pouring a concrete floor.

For heavy loading conditions, steel grillage footings incorporate steel beams to reinforce the foundation support (Figure 6.13). A base layer of steel is sized to the area needed to support the load. The upper layer of beams is joined at right angles to the base level, and the entire construction is encased in concrete.

Figure 6.13 A steel grillage footing.

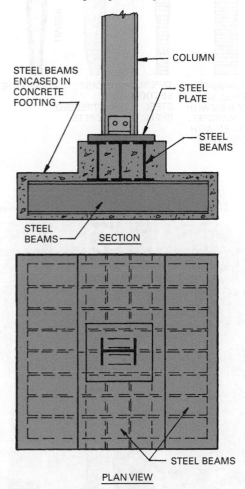

PLAN VIEW

Combined footings are used to support walls and columns near the property line, where it is not possible for the footing to cross the line onto a neighboring property. The combined footing ties the outside row of columns to the next row within the building. If the outer row of columns were to be placed on the edge of individual footings, the loads would not be symmetrical, and the footings would tend to settle unevenly or rotate (Figure 6.14).

Figure 6.14 Footings close to a property line can be reinforced with a combined footing.

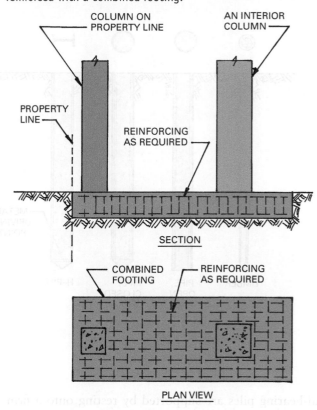

Cantilever footings are a form of combined footing that permits erection of wall columns near the edge of the footing. A beam rests on independent footings and cantilevers over the outer footing to support the column. The footings may be reinforced concrete or steel grillage.

DEEP FOUNDATIONS

When it becomes impossible to provide a suitable bearing surface for a structure close to the surface of a site, the use of deep foundations becomes necessary. Deep foundation systems reach far below the surface of the ground to find suitable bearing for larger structures. They are commonly used for high loading conditions, when soft or compressible soils are found near grade, or when firm soil is underlain by soft or compressible soil below. Deep foundations generally make use of piles. A pile foundation employs long and slender members that bypass shallow soil of low bearing capacity to transfer building loads to deeper soil or rock of high bearing capacity and (Figure 6.15).

The load from a superstructure to a pile foundation is transmitted to the subsoil in one of two ways.

Figure 6.15 Commonly used pile types.

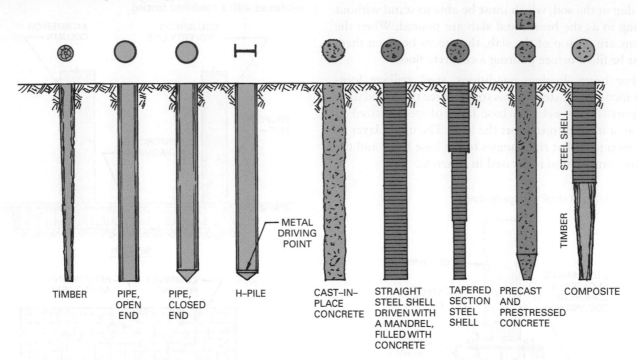

TIMBER | PIPE, OPEN END | PIPE, CLOSED END | H–PILE | CAST–IN–PLACE CONCRETE | STRAIGHT STEEL SHELL DRIVEN WITH A MANDREL, FILLED WITH CONCRETE | TAPERED SECTION STEEL SHELL | PRECAST AND PRESTRESSED CONCRETE | COMPOSITE

METAL DRIVING POINT

End-bearing piles are supported by resting onto a firm stratum, such as bedrock or suitable subsoil with the required bearing capacity. Skin-friction piles are supported by the frictional resistance created between the contact surface of the pile and the embracing soil. Piles for deep foundations are constructed in two very different ways.

Displacement Piles

Displacement piles (driven) dislodge soils radially and vertically as a slender wood, steel, or pre-cast concrete pile shaft is driven or mechanically vibrated into the ground. Sometimes referred to as driven piles, these typically develop bearing strength through friction between the pile and the soil, but may utilize end bearing as well.

Timber piles are tree trunks trimmed, stripped of bark, and treated with preservatives. They are driven with the small end down and may have a pointed metal shoe on the tip. The minimum tip is 6 in. (152 mm) in diameter. The top or butt end, typically with a minimum diameter of 12 in. (305 mm), is fitted with a metal pile ring to protect it during driving. Timber piles work best in relatively rock-free soils.

H-piles are structural steel HP shapes that are especially well suited for driving into soil containing rock or thin rock strata (Figure 6.16). Since they have

Figure 6.16 A steel HP pile being driven by a pile driver.

a small surface area, they cannot rely solely on friction but must be driven to end bearing on rock. They are available in widths from 8 to 14 in. (200 to 355 mm) and can be driven to depths of 150 to 200 ft. (45 to 61 m). These longer lengths are developed by driving a series of shorter sections and welding them to one another.

Pipe piles utilize heavy-gauge steel pipes that are driven with an open end. Small diameter piles generally have the end closed. As open-end piles are driven, the earth inside is gradually removed to decrease resistance. The soil can be removed by inserting a pipe and blowing it out or adding water to help loosen it. Once the required depth has been reached, concrete is placed inside the pipe, starting the fill from the bottom. Pipe diameters may be from 8 to 18 in. (203 to 457 mm) in width, with wall thicknesses from ¼ to ½ in. (6 to 12 mm).

Pre-cast and pre-tensioned concrete piles are made in a variety of ways. They may be cylindrical, square, or octagonal and are available either tapered or uniform in size for their entire length (Figure 6.17). Although most are solid in cross-section, some cylindrical types are hollow. The piles are reinforced with steel to withstand the installation stresses, as well as the building loads. Piles driven in clay soils have a blunt point, while a tapered point is used in sand and gravel. Pre-cast concrete piles can be lengthened by welding together the reinforcing bars of two adjoining sections. Some are installed by jetting the excavation with water. A pipe is cast in the center of the pile and high-pressure water is forced through it, blasting the soil as the pile penetrates it.

Varieties of composite piles are used to reduce costs or to achieve other advantages. For example, a wood pile (low in cost) could be driven into the area below ground water and a concrete pile could be placed on top.

Structural foundation drawings will specify the site locations where the piles are to be placed. Piles are usually grouped in clusters with a cast-in-place reinforced concrete *pile cap* that distributes loads equally among the piles below the cap (Figure 6.18). The footings can be supported by piles, or grade beams can be run over

Figure 6.18 A pile cap distributes structural loads equally among the piles below the cap.

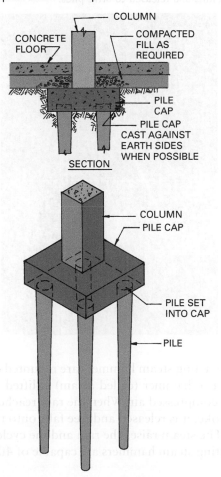

Figure 6.17 Pre-cast and pre-tensioned concrete piles may be cylindrical, square, or octagonal.

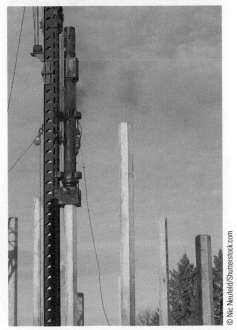

the pile caps to provide a foundation for construction of a wall. Mat foundations can also be supported on piles.

Individual pile caps are often interconnected by steel ties that carry tension and compression loads as specified by the building code.

Methods of Driving Displacement Piles

Piles are driven into the earth with *pile hammers*. The most commonly used types include drop, single-acting steam, double-acting steam, diesel, vibratory, and hydraulic.

Drop hammers use a heavy weight lifted up vertical rails called leads (Figure 6.19). When the weight reaches the top of the rail, it is released and falls onto the pile, driving it into the earth. Repeated cycles drive the pile to the required depth. A drop hammer can weigh from 500 to 3,000 lb. (226 to 1,360 kg), and be released from a height of 5 to 20 ft. (1.5 to 6.0 m), depending on the type of pile and soil conditions.

Figure 6.19 Drop hammers use a heavy weight that is lifted up vertical rails and released to drive piles.

Single-acting steam hammers are mounted on top of the pile. The hammer (called a ram) is lifted by either steam or compressed air. When the ram reaches the top of the stroke, it is released and free falls onto the top of the pile. The steam raises the ram and the cycle repeats. Single-acting steam hammers are capable of 40 or more

blows per minute. Double-acting steam hammers operate in the same manner as single-acting hammers except that, on the down stroke, the steam pressure is applied to the backside of a piston, which increases the energy per blow. Hydraulic pile hammers operate much the same, but hydraulic fluid is used to move the ram.

Diesel pile hammers are self-contained driving units that include a vertical cylinder, a ram, an anvil, and a diesel power unit. The ram is placed on top of the pile and released. As the ram falls, the fuel pump injects diesel fuel into the combustion chamber. As the ram continues to fall, the fuel is compressed and the heat generated causes an explosion. This drives the anvil hard against the pile and causes the ram to rise, thus starting a new cycle.

Vibratory pile hammers are held by a crane and mounted on the end of the pile (Figure 6.20). A motor drives eccentric weights in opposing directions, producing vibrations that are transmitted to the pile. The vibrations agitate the soil, reducing the friction between the pile and the soil and allowing the pile to penetrate the soil. Vibratory pile hammers are especially effective for driving into water-saturated non-cohesive soils. They have difficulty in dry sand and will not penetrate tight and cohesive soils. Vibratory pile hammers are also used in driving sheet piling for excavations.

Figure 6.20 Vibratory pile hammers reduce friction between the pile and the soil, allowing the pile to penetrate the soil.

Non-Displacement Piles

Non-displacement piles (drilled) are end-bearing units that are drilled using large-diameter drilling rigs. Bored piles are of a replacement nature, forming an *in situ* pile by drilling and filling the excavation with concrete. Soils such as stiff clays are particularly amenable to the formation of bored piles, since the borehole walls do not require temporary support except close to the ground surface. In unstable ground, such as gravel the ground requires temporary support from casing or the use of a bentonite slurry. Unlike displacement piles, non-displacement piles carry building loads solely through end bearing (Figure 6.21).

Figure 6.21 Non-displacement piles carry building loads solely through end bearing.

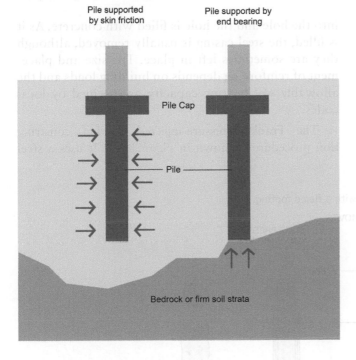

Three main factors are considered in constructing bored piles:

1. How to drill the hole, using what kind of machine and method to form the excavation in the ground.

2. How to protect the soil from collapsing into the borehole during drilling, usually done by inserting a steel casing or using a drilling fluid.

3. How to take the soil out of the borehole during drilling, either by a bucket, drilling fluid, or compressed air.

Non-displacement piles are bored through weaker surface materials until rock or other acceptable bearing soil is found. Various forms of drilling rigs are used to drill non-displacement piles.

A *continuous flight auger* is like a large drill bit that can form a bored pile ranging from 20 to 48 inches (20 to 121 cm) in diameter (Figure 6.22). The pile is drilled to the specified depth in one continuous process. While the auger is drilled into the ground, the flights of the auger are filled with soil, which provides lateral support and maintains the stability of the hole. At the same time the augers is withdrawn from the hole, concrete or grout is placed by pumping the concrete/grout mix through the hollow center of the auger pipe to the base. Once the excavation is complete, reinforcement is placed into the fluid and the hole filled with concrete to complete the pile. For successful operation of a rotary auger, the soil must be reasonably free of tree roots, cobbles, and boulders, and it must be self-supporting. When boulders or decomposed rock is encountered, the auger head can be fitted with a cutting bit to aid in drilling.

Figure 6.22 A continuous flight auger is like a large drill bit that can form a bored pile in one continuous process.

Another means of drilling for bored piles involves the use of a steel casing that is lowered into the hole as drilling continues (Figure 6.23). Shaft diameters can vary from 18 in. (460 mm) up to 5 or 6 ft. (1.5 to 1.8 m)

Figure 6.23 Drilling for bored piles can utilize a steel casing that is lowered into the hole as drilling continues.

or more. Bored piles are drilled through the weaker surface soils until they strike rock or other acceptable bearing soil. The soil must be able to stand without collapsing when the casing is removed. In cohesive soils, the bottom of the pile is flared outward using a belling bucket on the end of the drill to provide a larger end-bearing surface. Once the excavation is complete, the required reinforcing steel is lowered

into the hole and the hole is filled with concrete. As it is filled, the steel casing is usually removed, although they are sometimes left in place. The size and placement of reinforcing depends on building loads and the allowable soil bearing capacity as specified by local codes.

The Franki pressure-injected footing construction procedure is shown in Figure 6.24. It uses a steel

Figure 6.24 The Franki pressure-injected process produces a pile with a flared footing.

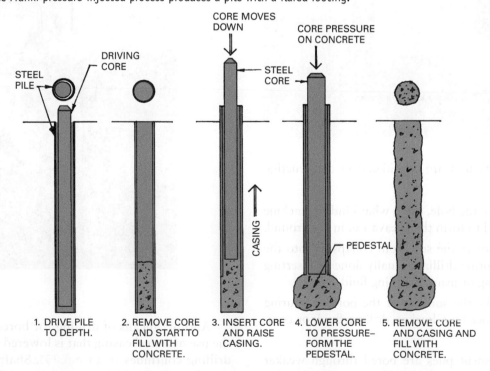

1. DRIVE PILE TO DEPTH.
2. REMOVE CORE AND START TO FILL WITH CONCRETE.
3. INSERT CORE AND RAISE CASING.
4. LOWER CORE TO PRESSURE–FORM THE PEDESTAL.
5. REMOVE CORE AND CASING AND FILL WITH CONCRETE.

casing into which zero-slump concrete is rammed to produce a pedestal footing. Another means of drilling for bored piles involves the use of a bucket barrel. The bucket barrel uses a cylindrical rotary drilling tool with a hinged bottom containing a soil-cutting blade. Spoil enters the bucket, is lifted out of the hole, swung aside, and dumped by releasing a latch on the hinged bottom. A steel casing is sometimes lowered into the hole as drilling continues to prevent collapse.

Helical Pier® Foundation Systems

Helical Pier foundations use helical (screw) steel piers installed at intervals between the footing forms. They are installed by screwing the steel into the soil (Figure 6.25). The torque required to drive each pier correlates to the bearing capacity of the soil below and is used to ascertain the anchor depth needed to support the required structural load. Helical Piers can also be used to underpin existing foundations that are experiencing settling by using a special support bracket.

Figure 6.25 The Helical Pier foundation uses steel helical piers drilled into the soil before the footing is poured.

Review Questions

1. What are the factors considered during foundation design?

2. How are foundation and basement walls waterproofed?

3. What are the major types of foundations?

4. What types of materials are used for pile foundation construction?

5. What distinguishes displacement piles from non-displacement piles?

6. What types of equipment are used to drive piles?

7. How are non-displacement piles constructed?

Key Terms

Continuous Flight Auger	Independent Footing	Pile Hammer
Displacement Piles	Mat Foundation	Post-tensioned Slabs on Grade
Footing	Non-displacement Piles	Raft Foundation
Foundation	Pile Cap	Slab on Grade
Grade Beam	Pile Foundation	Spread Footing

Activities

1. Secure foundation drawings from local architects and general contractors and identify the types of foundation specified, reinforcing, and other details. See if a log of test borings was made and identify the characteristics of the soils as the borehole goes deeper.

2. Visit construction sites and observe the excavation, forms, piers, reinforcing, and other materials used to construct foundations.

3. If any buildings in your area are being built on piles, try to visit the site and take photos to illustrate the process, from locating the piles to establishing the finished elevation of the tops of the piles. What types of piles were being used?

4. When visiting foundation construction, report on any actions taken to keep the excavation dry.

5. As you visit construction sites, observe situations you think are not safe and prepare remarks so you can lead a class discussion on safety. (While on the site, do not charge the supervisor or workers with allowing unsafe practices. Simply observe and remember.)

Additional Resources

Builders Foundation Handbook, Oak Ridge, TN: U.S. Department of Energy, http://web.ornl.gov/sci/roofs+walls/foundation/ORNL_CON-295.pdf

Concrete Masonry Basement Wall Construction, Herndon, VA: National Concrete Masonry Association, www.ncma.org/etek/Pages/ManualViewer.aspx?filename=TEK%2003-11.pdf&apf=1

Other resources include: Many publications available from the U.S. Department of Housing and Urban Development, HUD USER, Rockville, MD.

See Appendix C for addresses of professional and trade organizations and other sources of technical information.

Concrete
CSI MasterFormat™

Courtesy Portland Cement Association

Concrete

Upon completion of this chapter, the student should be able to:

- Identify the various types of Portland cement and cite their purposes and uses.

- Be familiar with the properties of Portland cement and their effect on concrete.

- Discuss the role of water in concrete and its desired properties.

- List the types of aggregate used in concrete and their desirable characteristics.

- Understand the purposes of various concrete admixtures and their effect on concrete.

- Be aware of how commonly used concrete tests are conducted and what they reveal about a mix.

Build Your Knowledge

For further information on these materials and methods, please refer to:

Chapter 10 Mortars for Masonry Walls

Chapter 12 Concrete Masonry

Concrete is one of the most widely used construction materials, with a long history of use. In the United States, almost twice as much concrete is used as any other material. Concrete is strong and durable, long lasting, and possesses plastic qualities for infinite formal possibilities. Its constituent ingredients are derived from a wide variety of naturally occurring materials that are readily available in most parts of the world. Concrete can be made by simple hand-mixing methods or in large quantities at a computer-controlled batching plant.

Concrete is a solid, hard material produced by combining Portland cement, coarse (gravel) and fine (sand) aggregates, and water in proper proportions (Figure 7.1). The mix is composed of paste and aggregates. The paste, composed of Portland cement and water, coats the surface of the fine and

Figure 7.1 Concrete components: Cement, water, sand, and coarse aggregate next to a cut section of hardened concrete.

Courtesy Portland Cement Association

coarse aggregates. Through a chemical reaction called **hydration**, the paste hardens and gains strength to form the rock-like mass known as concrete. The material is inexpensive, but has no form of its own and possesses no tensile strength. Forms must be built to shape it into useful structures, and reinforcing steel must be added to provide tensile strength.

Modern concrete engineering is the result of more than 2,000 years of development. The ancient Romans used concrete made from a mix of lime mortar and pozzolana, known as *opus caementitium* (Figure 7.2). This early concrete was waterproof and used stone chips as the aggregate. With the fall of the Roman Empire, the knowledge of making concrete was lost until the nineteenth century. It was rediscovered by the Englishman Joseph Aspdin, who named his 1824 patent Portland cement after the color of the limestone on his native Isle of Portland. Over time, fundamental methods of engineering analysis have led to a series of applicable design standards for modern reinforced concrete. Concrete is produced in three basic types: redi-mix concrete, pre-cast concrete, and concrete masonry units, each with unique applications and properties. This chapter introduces the constituent materials of concrete, their mixing, properties, and testing procedures.

Figure 7.2 The unreinforced concrete dome of the Pantheon has survived for two millennia.

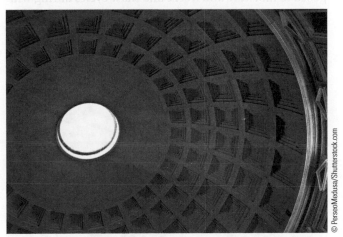

© PerseoMedusa/Shutterstock.com

PORTLAND CEMENT

Portland cement, the binding compound of concrete, is a fine, pulverized material consisting of compounds of lime, iron, silica, and alumina. The manufacture of Portland cement produces a combination of these elements in proper proportion under carefully controlled conditions. The exact composition of different types of Portland cement varies, but the composition of Type 1, Normal, is representative (Table 7.1). Current manufacturing techniques emit large amounts of CO_2 into the atmosphere, and research is underway to reduce the amount of Portland cement in concrete through the addition of other ingredients. (See "Construction Materials: Supplementary Cementing Materials" on page 111.)

Table 7.1 Oxide Composition of Type 10 (ASTM Type I) or Normal Portland Cement

Oxide Ingredient	Range, %
Lime, CaO	60–66
Silica, SiO_2	19–25
Alumina, AL_2O_3	3–8
Iron, Fe_2O_3	1–5
Magnesia, MgO	0–5
Sulfur trioxide, SO_3	1–3

Source: Portland Cement Association

The lime in Portland cement is commonly derived from limestone, marble, marl, or seashells. Iron, silica, and alumina are obtained from mining clays containing these elements. Increasingly, iron furnace slag and flue dust are also used, as are sand, chalk, and bauxite. These ingredients are crushed in a primary crusher and sent through a vibrating screen. After crushing, either a wet or a dry process is used to produce the final cement.

The dry process grinds the raw materials to a powder, blends them in mixing silos, and moves them to a kiln. In the wet process, the ground materials are mixed with water to form a slurry that is blended, and moved to a kiln. The kiln is a rotating cylinder operating at 2600–3000°F (1600–1780°C) that burns the materials into a clinker. The clinker is cooled, a small amount of gypsum is added, and ground into a powder so fine it can pass through a 40,000-openings-per-square-inch sieve. The gypsum serves to retard the curing process.

Most Portland cement is shipped in bulk, in railroad cars or large trailers designed especially for the material. Bulk cement is sold by the ton (2,000 lb. or 907.2 kg), and smaller quantities are bagged at a volume of one cubic foot (0.028 m³) weighing 94 pounds (42 kg).

ASTM-Designated Types of Portland Cement

Cement is an easily molded temporary paste when mixed with water. After a short time, it hardens, or sets, to form a rigid mass. Hardening is not simply a process of evaporation or drying; hydration is the result of a chemical reaction between the Portland cement and water. The term "heat of hydration" is applied to the heat generated by the chemical reactions, which occur in setting concrete between the water and cement. The heat causes the concrete first to expand and then to shrink as it cools. Portland cement is made following ASTM Designation C150, which defines eight types of cement (Table 7.2).

Table 7.2 Types of Portland Cement

Canadian Designation	United States Designation	Type
10	I	Normal
20	II	Moderate
30	III	High early strength
40	IV	Low heat of hydration
50	V	Sulfate resisting
Air-Entrained Types		
10A	IA	Normal air-entrained
20A	IIA	Moderate air-entrained
30A	IIIA	High early strength air-entrained

Source: Portland Cement Association

Type I: Normal Type I is a general purpose Portland cement used whenever other special properties are not necessary. It is used for pavements, sidewalks, reinforced concrete structural members, bridges, tanks, water pipes, and masonry building units, among other components.

Type II: Moderate Type II cement is used when protection against moderate sulfate attack, found in some soils and groundwater, is required. It generates less heat by hydrating at a slower rate than Type I. This makes Type II useful in structures that have larger masses of concrete, such as abutments, piers, and large retaining walls.

Type III: High Early Strength Type III Portland cement provides higher strength in a shorter time than Types I or II. It reduces the curing period, allowing for quick removal of forms, especially in cold-weather situations when freezing may pose problems. High early strength normally reduces the required curing time to one week or less.

Type IV: Low Heat of Hydration Type IV reduces the rate and amount of heat generated by hydration. It develops strength slower than Type I, and is mainly used for structures having very large concrete masses, such as dams and nuclear plants. The use of type IV has diminished over the years.

Type V: Sulfate Resisting Type V Portland cement is used only in concrete that is exposed to severe sulfate

conditions. This most frequently occurs in soils and groundwater in areas where sulfate concentrations are high.

White Portland Cement White Portland cement is a true Portland cement except that during the manufacturing process the components are controlled so the finished products will be white. It is used mainly for architectural purposes, such as in precast curtain-wall and facing panels, terrazzo floors, stucco, finish-coat plaster, and tile grout.

Types IA, IIA, IIIA: Air-Entrained There are three types of *air-entrained Portland cement*: IA, IIA, and IIIA. These are similar to Types I, II, and III, except that small quantities of air-entraining materials are mixed with the clinker and gypsum.

Air-entraining encapsulates millions of microscopic air bubbles into the concrete mix (Figure 7.3). This greatly improves the durability of concrete that is exposed to moisture and freeze and thaw cycles during winter. It also increases the concrete's resistance to surface

Figure 7.3 A polished section of air-entrained concrete as seen through a microscope.

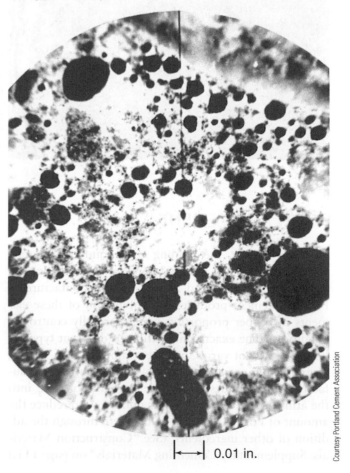

|← →| 0.01 in.

Courtesy Portland Cement Association

scaling, which is caused by salts used to remove winter ice. Air-entraining improves the workability of concrete and reduces segregation of the aggregate, bleeding, and the amount of water needed.

Blended Hydraulic Cements Hydraulic cements are capable of hardening under water. There are two blended hydraulic cements: Portland blast-furnace-slag cements and Portland pozzolan cements. ASTM C595 sets requirements for two types of Portland blast-furnace-slag cements: IS and IS-A (air-entrained). Blast-furnace slag is ground with the Portland cement clinker or separately ground and blended with Portland cement (Figure 7.4). The blast-furnace-slag content of these cements is 25 to 65 percent of the cement's weight.

Figure 7.4 Blended cements use a combination of portland cement or clinker and gypsum blended or interground with pozzolans, slag, or fly ash.

Courtesy Portland Cement Association

The four types of Portland pozzolan cements are IP, IP-A, P, and P-A. The "A" indicates air-entraining. They are manufactured by mixing Portland cement clinker with a suitable pozzolan. *Pozzolan* is a siliceous and aluminous material that chemically reacts with calcium hydroxide to form compounds possessing cementitious properties. Natural pozzolans include pumicite, volcanic ash, and volcanic tuff. Tuff is a porous rock formed by the consolidation of volcanic ashes and dust. Processed natural pozzolans include clay and shale that are burned or calcined in a kiln, then crushed and ground. All four types can be used for general concrete construction. Types P and P-A are used in massive concrete structures that do not require high early strengths.

Masonry Cements Masonry cements, covered under ASTM C91, are used for the manufacture of masonry mortars. They typically contain some of the following: Portland cement, slag cement, and hydraulic lime. They may also contain hydrated lime, limestone, chalk, calcareous shell, talc, slag, or clay. Mortar cements must have excellent workability, plasticity, and water-retention properties.

Waterproof Portland Cement Waterproof Portland cement has a small amount of calcium stearate or aluminum stearate added to the Portland cement clinker during grinding.

Plastic Cements Plastic cements have plasticizing agents added to Type I or Type II Portland cement during the milling operation. They are used in stucco and plaster.

Expansive Cement Expansive cement is hydraulic cement that expands during the early hardening period. The three types available are K, M, and S. They can be used to compensate for the effects of drying shrinkage.

Regulated Set Cement Regulated set cement is hydraulic cement that can be compounded to have a set time from one to two minutes to as long as fifty or sixty minutes.

Properties of Portland Cement

The specifications for Portland cement place limits on its chemical composition and physical properties. Knowledge of these properties is necessary to evaluate the results of tests on the cement itself and on concrete made with the cement.

Fineness As cement increases in fineness, the rate of hydration increases, which accelerates the strength development of the concrete. The effects of greater fineness on strength are most apparent during the first five to seven days. The compound composition and fineness of Portland cements are listed in Table 7.3.

Table 7.3 Potential Compound Composition and Fineness of Cements

Type of Portland Cement	Compound Composition, Percentage				Wagner Fineness, m^2/g
	C_3S	C_2S	C_3A	C_4AF	
Type I	55	19	10	7	0.18
Type II	51	24	6	11	0.18
Type III	56	19	10	7	0.26
Type IV	28	49	4	12	0.19
Type V	38	43	4	9	0.19
White	33	46	15	2	

Source: Portland Cement Association

Soundness   Soundness describes the ability of a hardened paste to retain its volume after setting. A major problem is delayed destructive expansion after the paste has hardened caused by excessive amounts of hard-burned free lime or magnesia. Cement can be tested following ASTM standards to determine soundness.

Consistency   Consistency is the relative mobility of a fresh mixture, in other words, its ability to flow and its workability. Cement-paste tests for consistency are made using the tests specified by ASTM C191.

Setting Time   Setting-time tests are made to determine how the cement paste undergoes setting and hardening during the first few hours. The initial set must not occur before the concrete surface finishing operations can be completed. After finishing, the final set should occur without unnecessary delay. Setting times of cement pastes and concrete do not correlate directly because of water loss, the surface on which the concrete has been poured, and temperature differences.

False Set   False set results in the loss of plasticity without the development of much heat shortly after the concrete has been mixed. False set can be countered through the addition of water to the concrete mix. Chemical admixtures can also be added to delay its occurrence.

Compressive Strength   Compressive strength, a major physical property of cement, is found by testing 2 in. (50 mm) cubes of mortar, as specified by ASTM C150. Compressive strength is influenced by the type of cement used, its compound composition, and fineness. Due to variances in concrete mixtures, cement compressive strengths cannot be used to determine concrete compressive strengths. Actual concrete samples must be tested to determine the compressive strength of concrete. Compressive strengths of cements are shown in Table 7.4.

Heat of Hydration   When cement and water chemically react, the heat generated is known as the heat of hydration. The amount of heat generated depends on the chemical composition of the cement. The rate of heat generated is affected by the fineness of the cement, its chemical composition, and the temperature during hydration. Structures with a large concrete mass may experience a significant rise in temperature unless the heat can be rapidly dissipated. If not controlled, it may create undesirable stresses. In cold weather, the excessive heat can be beneficial because it helps to maintain favorable curing temperatures. The approximate amounts of heat of hydration that is generated are shown in Table 7.5.

Table 7.4 Relative Compressive Strength Requirements as Affected by Type of Cement[a]

Type of Portland Cement	Compressive Strength, Minimum, Percentage of Strength of Type 1 at 7 days			
	1 day	3 days	7 days	28 days
I	—	64	100	143[c]
IA	—	52	80	114[c]
II[b]	—	54	89	143[c]
IIA[b]	—	43	71	114[c]
III	64	125	—	—
IIIA	52	100	—	—
IV	—	—	36	89
V	—	43	79	107
IS	—	64	100	125
IS-A	—	52	80	100
IS(MS)	—	36	64	125
IS-A(MS)	—	27	50	100
IP	—	64	100	125
IP-A	—	52	80	100
P	—	—	54	107
P-A	—	—	45	89

[a]ASTM Designations: C150-77 and C595-77.
[b]A strength reduction of one-fifth to one-third is allowed if the optional heat of hydration or the chemical limit on the sum of the C3S and C3A is specified. See ASTM specification.
[c]Optional specification requirement.
Note: When suffixes MH or LH are added to any of last eight cement types listed, the strength requirement is 80% of the values shown.

Source: Portland Cement Association

Table 7.5 Approximate Relative Amounts of Heat Generated During Hydration[a]

Portland Cement Type	Percent of Hydration, Relative to Type 1, Normal
Type 1, Normal	100
Type 2, Moderate	80 to 85
Type 3, High early strength	up to 150
Type 4, Low heat of hydration	40 to 50
Type 5, Sulfate resisting	60 to 75

[a]First seven days.

Source: Portland Cement Association

Specific Gravity   The specific gravity of Portland cement is about 3.15. Portland blast-furnace-slag and Portland pozzolan cements may have values as low as

2.90. The specific gravity is not an indication of the quality of cement, but is used when calculating mix designs.

Storing Portland Cement

Portland cement is moisture sensitive, so it must be protected from dampness. Sacked material must be stored on pallets, whether it is in a warehouse with a concrete floor or on a site in the open. Warehouse storage must be watertight, and bags must not touch the exterior walls. Bags should be packed closely to reduce airflow and covered with plastic or tarpaulins if stored for long periods. Bagged cement tends to pack if it is stored a long time. This can be corrected by rolling the bags on the floor or ground before opening them. Bulk cement is stored in watertight bins or silos. Dry low-pressure aeration or vibration should be used to make the cement flow better. When cement is loaded into bins or silos, it swells, so a unit will store only about 80 percent of its rated capacity.

WATER

Water used in making concrete should be clear and free of sulfates, acids, alkalis, and humus. Potable water from municipal water systems or wells provides water suitable for use. Water from lakes, ponds, or rivers should be carefully checked for suitability before use.

Water of questionable quality can be used for concrete if test mortar cubes have seven-day and twenty-eight-day strengths equal to 90 percent of samples made with drinkable water. ASTM C191 specifies how to test the samples to see if impurities in the water adversely shorten or lengthen the setting time.

Alkali carbonate and bicarbonate, chloride, sulfate, carbonates of calcium and magnesium, iron salts, inorganic salts, acid waters, and alkaline waters all have an effect on concrete. Additionally, sugar, silt, suspended particles of clay or fine rock, oils in suspension, and algae in water can make it unsuitable for concrete mixing.

Construction Materials

Supplementary Cementing Materials (SCMs)

The manufacturing of Portland cement is a major source of environmental pollution. Per ton of clinker produced, 3 to 6 million BTU of energy and 1.7 tons of raw materials, chiefly limestone, are required. Significant emissions, including carbon dioxide (CO_2), nitrogen oxides, sulfur oxides, and particulates, are produced. The industry is now looking at the use of both alternative fuels and supplementary cementing materials. *Supplementary cementing materials (SCMs)* are mineral admixtures consisting natural or by-products of other manufacturing processes that are used to reduce the amount of Portland cement in a concrete mix.

Fly ash is a fine, glass-like powder recovered from the gases created by coal-fired electric power generation. U.S. power plants produce millions of tons of fly ash annually, much of it discarded in landfills. Fly ash can provide an economical and environmentally preferable substitute for the Portland cement used in concrete, brick, and block production.

Consisting mostly of silica, alumina, and iron, fly ash is a material that forms cement in the presence of water. When mixed with lime and water, it forms a compound with properties similar to Portland cement.

The circular shape of fly ash particles reduces friction and increases the concrete's consistency and workability. Improved workability means less water is needed, resulting in less segregation of the mixture and higher ultimate strength. The resulting concrete also tends to be denser, with a smoother surface and sharper detail. Fly ash can substitute for up to 20 to 35 percent of the Portland cement used to make concrete.

Class F fly ash reduces the risk of expansion due to sulfate attack, which can occur in certain soils or near coastal areas. It is resistant to expansion from chemical attack, has a higher percentage of calcium oxide, and is commonly used for structural concrete. Fly ash used in Portland-cement concrete is produced according to the requirements of ASTM C618, *Standard Specification for Fly Ash and Raw or Calcined Natural Pozzolan: Class C Fly Ash for Use As a Mineral Admixture in Portland Cement*.

As a recycled, post-industrial material, fly ash use in concrete offers environmental advantages by diverting material from the waste stream, conserving virgin materials, and reducing the energy investment in processing. The use of fly ash is eligible for one recycled content credit in the LEED rating system, under the materials and resources category.

Seawater containing up to 35,000 ppm (parts per million) of dissolved salts is generally suitable as mixing water for unreinforced concrete. Concrete made with seawater has a higher early strength than normal concrete and usually has lower strength after twenty-eight days. Seawater may be used for reinforced concrete, but the reinforcement must be protected from corrosion, and the concrete must contain entrained air. It is also necessary to check for sea salt on aggregates that were exposed to salt water. Not more than 1 percent of sea salt on aggregates is acceptable. The Environmental Protection Agency (and some state agencies) prohibit discharging untreated wash water from returned concrete or mixer washout operations into rivers, streams, and lakes.

AGGREGATES

Aggregates are inert granules such as crushed stone, gravels, and minerals that are mixed with cement and sand to form concrete. Approximately 60 to 80 percent of the concrete mix is made up of aggregates. The cost of concrete and its properties are directly related to the aggregates used. Natural aggregates are composed of rocks and minerals. Minerals are naturally occurring inorganic substances that have distinctive physical properties and a composition that can be expressed by a chemical formula. Rocks are usually composed of several minerals. For example, limestone is basically calcite (a mineral) with small amounts of quartz, feldspar, and clay (all minerals). The crushing and weathering of rock produces stone, gravel, sand, silt, and clay. Aggregates derived from the recycling and crushing of demolished concrete are also finding expanded use. Rocks and minerals commonly found in aggregates are shown in Table 7.6.

Aggregates must conform to ASTM specifications. They must be clean, strong, free of absorbed chemicals, and devoid of coatings of clay, humus, and other fine materials. Aggregates containing some shale, shaly rocks, soft or porous rocks, and some types of chert are not suitable because they do not weather well and can cause pop-outs on the exposed concrete surface.

Characteristics of Aggregates

The major characteristics for aggregates, listed in Table 7.7, are described below.

Resistance to Abrasion and Skidding Abrasion resistance is important when the aggregate is to be used in an

Table 7.6 Rock and Mineral Constituents in Aggregates

Minerals		Sedimentary Rocks
Silical	Pyrite	Conglomerate
Quartz	Marcasite	Sandstone
Opal	Pyrrhotite	Quartzite
Chalcedony	Iron Oxide	Graywacke
Tridymite	Magnetite	Subgraywacke
Cristobalite	Hematite	Arkose
Silicates	Goethite	Claystone, Siltstone,
Feldspars	Ilmenite	Argillite and Shale
Ferromagnesian	Limonite	Carbonates
Hornblende	**Igneous Rocks**	Limestone
Augite	Granite	Dolomite
Clay	Syenite	Marl
Illites	Diorite	Chalk
Kaolins	Gabbro	Chert
Chlorites	Peridotite	
Montmorillonites	Pegmatite	**Metamorphic Rocks**
Mica	Volcanic Glass	Marble
Zeolite	Obsidian	Metaquartzite
Carbonate	Pumice	Slate
Calcite	Tuff	Phyllite
Dolomite	Scoria	Schist
Sulfate	Perlite	Amphibolite
Gypsum	Pitchstone	Hornfels
Anhydrite	Felsite	Gneiss
Iron Sulfide	Basalt	Serpentinite

Note: For brief descriptions, see "Standard Descriptive Nomenclature of Constituents of Natural Mineral Aggregates" (ASTM C294).

Source: Portland Cement Association

area that is subject to heavy abrasive use, such as a factory floor. Aggregate for abrasion resistance is tested following ASTM C131 standards. To give skid resistance, the siliceous particle content of the fine aggregate should be 25 percent or more.

Resistance to Freezing and Thawing Freeze–thaw properties of an aggregate are important in concrete that will be exposed to a wide range of temperatures. Significant considerations are the porosity, absorption, permeability, and pore structure of the aggregate. Suitable aggregates may be chosen from those used in the past that have given good results. Unknown materials should be evaluated using ASTM tests.

Compressive Strength The strength of aggregates under compression is an important factor to consider when

Table 7.7 Characteristics and Tests of Aggregates

Characteristic	Significance	Test Designation	Requirement or Item Reported
Resistance to abrasion	Index of aggregate quality; wear resistance of floors, pavements	ASTM C131 ASTM C295 ASTM C535	Maximum percentage of weight loss
Resistance to freezing and thawing	Surface scaling, roughness, loss of section, and unsightliness	ASTM C295 ASTM C666 ASTM C682	Maximum number of cycles or period of frost immunity; durability factor
Resistance to disintegration by sulfates	Soundness against weathering action	ASTM C88	Weight loss, particles exhibiting distress
Particle shape and surface texture	Workability of fresh concrete	ASTM C295 ASTM D3398	Maximum percentage of flat and elongated pieces
Grading	Workability of fresh concrete; economy	ASTM C117 ASTM C136	Minimum and maximum percentage passing standard sieves
Bulk unit weight or density	Mix design calculations; classification	ASTM C29	Compact weight and loose weight
Specific gravity	Mix design calculations	ASTM C127, fine aggregate ASTM C128, coarse aggregate ASTM C29, slag	—
Absorption and surface moisture	Control of concrete quality	ASTM C70 ASTM C127 ASTM C128 ASTM C566	—
Compressive and flexural strength	Acceptability of fine aggregate failing other tests	ASTM C39 ASTM C78	Strength to exceed 95% of strength achieved with purified sand
Definitions of constituents	Clear understanding and communication	ASTM C125 ASTM C294	—

Source: Portland Cement Association

choosing materials. This is tested by standard compression tests on hardened concrete samples.

Shape and Texture of Particles   The shape and texture of aggregate particles influence the properties of fresh concrete more than cured concrete. Rough-textured, angular, elongated particles require more water than do smooth, rounded aggregates. Therefore, angular particles require more cement to maintain the required water–cement ratio. Cement tends to bond better to angular particles than to smooth particles. This must be considered when flexural strength or high compressive strength is specified.

Specific Gravity   Specific gravity is a measure of the relative density of an aggregate. It is a ratio of an aggregate's weight to the weight of an equal volume of water and is used to determine the absolute volume occupied

by the aggregate. Most natural aggregates have specific gravities between 2.4 and 2.9.

Absorption and Surface Moisture   The absorption and surface moisture conditions of an aggregate are tested in order to control the net water content of the concrete and determine suitable batch weights.

Moisture conditions are designated in four categories:

1.   Oven dry. Completely dry and fully absorbent.

2.   Air dry. Dry on the surface but with some interior moisture.

3.   Saturated surface dry. Neither absorbing water nor contributing water to the mix. The surface is dry but the voids and interior of the aggregate are fully saturated.

4.   Damp. Contains an excess of surface moisture.

Surface moisture can increase the bulk (volume) of average and fine sands more than coarse sands. Since moisture in sands delivered on a job can vary, it is necessary to weigh the aggregate rather than using volume measures when portioning concrete ingredients.

The test to determine the amount of moisture on fine aggregates is made according to ASTM C70, *Surface Moisture in Fine Aggregate*, which depends on the displacement of water by a known weight of moist aggregate. The density of the aggregate must be known to use this method. Another method, ASTM C566, *Total Moisture Content of Aggregate by Drying*, involves weighing a sample, heating it until dry, and reweighing it. The difference in weights is used to calculate the moisture content.

Chemical Stability   Chemically stable aggregates do not react chemically with cement, which could cause harmful reactions. Some aggregates contain minerals that do react with the alkalies in cement, causing abnormal expansion and cracking in concrete. Field records of aggregates provide evidence of their chemical stability. If this is unknown, a laboratory test based on ASTM specifications should be made.

Grading Aggregates

Aggregates are graded into standardized sizes by passing them through a sieve, as specified by ASTM C136 and CSA A23.2.2. Sieves used for fine aggregates consist of seven sizes; there are ten sieve sizes for coarse aggregates (Table 7.8).

The limits used are specified for the percentage of material passing each sieve. Figure 7.5 shows the grading limits for fine aggregates. Notice that almost all of the particles of fine sand pass through the No. 4 sieve, but only 10 to 20 percent pass through the No. 100 sieve. The coarse sand graph shows that about 98 percent of particles pass through the No. 4 sieve, and only about 3 percent pass through the No. 100 sieve. Each product is, therefore, an accumulation of the particles of various percentages passing through the various sieves.

When concrete mixes are designed, limits are specified for the percentage of material passing each sieve. This, plus specifying the maximum aggregate size, affects the relative proportions of aggregate, as well as the cement and water requirements, workability, economy, porosity, shrinkage, and durability of concrete. The best mixes are those that do not have a large deficiency or an excess of any size. Very fine sands are uneconomical,

Table 7.8 Fine and Coarse Sieve Sizes

Fine Sieve Sizes
³/₈ in. (9.5 mm)
No. 4 (4.75 mm)
No. 8 (2.36 mm)
No. 16 (1.18 mm)
No. 30 (600 µm)
No. 50 (300 µm)
No. 100 (150 µm)

Coarse Sieve Sizes	
Size Number	**Nominal Size (sieves with square openings)**
1	3½ : to 1½ in. (90 to 37.5 mm)
2	2½ to 1½ in. (63 to 37.5 mm)
357	2 in. to No. 4 (50 to 4.75 mm)
467	1½ in. to No. 4 (37.5 to 4.75 mm)
57	1 in. to No. 4 (25.0 to 4.75 mm)
67	¾ in. to No. 4 (19.0 to 4.75 mm)
7	½ in. to No. 4 (12.5 to 4.75 mm)
8	³/₈ in. to No. 8 (9.5 to 2.36 mm)
3	2 to 1 in. (50.0 to 25.0 mm)
4	1½ to ¾ in. (37.5 to 19.0 mm)

Source: Portland Cement Association

Figure 7.5 The graph shows the grading limits for fine aggregates.

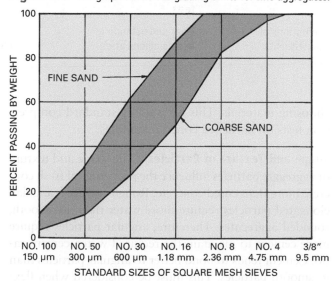

Courtesy Portland Cement Association

while very coarse sands produce a harsh, unworkable mix. A wide range of particle sizes is necessary to fill the voids between aggregates (Figure 7.6). As the range of sizes fill the voids, voids constitute less volume, which decreases the cement needed.

Figure 7.6 Separation of concrete aggregates into various sieve sizes gives a visual illustration of why control of aggregate gradation is important in maintaining uniformity in the concrete.

Courtesy Portland Cement Association

Sand is graded using a layered set of sieves inside a shaker. Each sieve layer has a fine mesh. The sand remaining on each sieve can be weighed to get the percent of the sample for each sieve. Following is an example of a typical test using 1,000 g of sand.

Sieve	Weight of Sand Retained in Grams	Percent Retained	Percent Cumulative
4 (5 mm)	4	4	115.0
8 (2.5 mm)	115	11.5	115.5
16 (1.25 mm)	210	21.0	136.5
30 (630 mm)*	262	26.2	162.7
50 (315 mm)	201	20.1	182.8
100 (160 mm)	172	17.2	100.0
		Total	302.5

*micro millimeter

The fineness modulus of the sand sample can be found by adding the cumulative-percent-retained figures and dividing by 100. In the previous illustration, this total was 302.5, which gives a fineness modulus of 3.02. A range of 2.30 to 2.60 denotes fine sands, 2.61 to 2.90 medium, and 2.91 to 3.10 coarse. The sand in the example above was coarse.

Structural Lightweight Aggregates

Lightweight aggregates are used to produce lightweight structural concrete. Lightweight concrete has a density ranging from 90 to 115 lb./ft.³ (1,440 to 1,850 kg/m³), depending on the aggregate. Pelletized clay or crushed shale or slate is placed in a rotary kiln and heated to about 1,800°F (1,000°C). Gases in the material cause it to expand, forming small air cells in the particles. Another process, known as sintering, involves crushing the aggregate, mixing it with a small amount of finely ground coal or coke, spreading it over a grate, and igniting the coke. Blowers increase the temperature of the burning, causing the material to expand and bond into large pieces that are then crushed to form the smaller aggregate.

Expanded blast-furnace-slag aggregate is produced by treating the slag with steam or water while it is in a molten state, causing it to expand and bond into large particles that are then crushed to the desired size.

Insulating Lightweight Aggregates

Perlite, diatomite, vermiculite, pumice, and scoria are commonly used as insulating lightweight aggregate. Perlite is siliceous volcanic rock that contains moisture. It is crushed and heated, causing it to expand and form a honeycomb structure. In addition to aggregate, it is used for loose-fill insulation and plaster. Diatomite is a diatomaceous earth, composed of silicified skeletons of microscopic one-celled animals. Vermiculite is a hydrated magnesium-aluminum-iron silicate that occurs in thin layers with moisture between them. When vermiculite is crushed and heated, the layers expand, forming dead air cells. Pumice is a type of volcanic rock that is porous and lightweight. Scoria is a type of volcanic slag; scoria also is used to describe a form of blast-furnace slag that is crushed to form aggregate of various sizes.

Heavyweight Aggregates

Heavyweight aggregates are used to produce heavyweight concretes with densities of up to 400 lb./ft.³ (6400 kg/m³) that are used for radiation shielding and other applications that necessitate great weight. Frequently used heavyweight aggregates include ferrophosphorus, barite, goethite, hematite, ilmenite, limonite, magnetite, and steel shot and punchings.

Testing Aggregates for Impurities

Aggregates must be free of clay, silt, and organic impurities. If present, these could influence the composition of the cement paste and, consequently, the strength and setting of the concrete. Organic impurities in fine aggregate are determined by ASTM C40, *Organic Impurities in Sands for Concrete.* Excess clay and silt in an aggregate can cause an increase in shrinkage, affect durability, and cause separation from the other aggregate particles. Aggregate specifications usually limit clay and silt passing through a No. 200 (75 mm) sieve to 2 or 3 percent for sand and less than 1 percent in coarse aggregate. Tests are made based on ASTM C117, *Sieve in Mineral Aggregates by Washing*. Testing for clay lumps is made following ASTM C142, *Clay Lumps and Friable Particles in Aggregates*.

Handling and Storing Aggregates

Aggregates should be stored in layers of uniform thickness. In stockpiles, each truckload should be discharged tightly against the previous load. Aggregate is removed from stockpiles with a front-end loader. If removing it from a tall, conical pile, the loader should move back and forth across the face of the pile to re-blend sizes. Washed aggregates must have sufficient time in storage to drain to uniform moisture content. Fine aggregates should not be dropped from a bucket or conveyor because the wind tends to blow away the very fine particles. Grades should be kept from intermingling by building dividers or storing in completely separate locations. Above all, aggregates must be stored and handled so they are not contaminated by unwanted substances, such as earth and leaves.

ADMIXTURES

Admixtures are ingredients added to concrete, other than Portland cement, aggregates, and water. They may be added before or during mixing. Admixtures change the properties of concrete, so should be used sparingly and only on the advice of a concrete specialist.

Concrete should be workable, finishable, strong, durable, watertight, and wear resistant. Whenever possible, these properties are obtained by careful selection of suitable types of aggregate, Portland cement, and the water–cement ratio. If this is not possible, or special circumstances such as freezing weather exist, admixtures can be of benefit. The following discussion outlines the main admixtures typically used in concrete construction. A summary of admixtures and their uses is given in Table 7.9.

Air-Entraining Agents

Air-entraining admixtures are used to entrain microscopic air bubbles in concrete. Entrainment can be produced by using air-entrained Portland cement or by adding an

Table 7.9 Admixtures by Classification

Desired Effect	Type of Admixture	Material
Improve durability	Air entraining (ASTM C260)	Salts of wood resins
		Some synthetic detergents
		Salts of sulfonated lignin
		Salts of petroleum acids
		Salts of proteinaceous material
		Fatty and resinous acids and their salts
		Alkylbenzene sulfonates
Reduce water required for given consistency	Water reducer (ASTM C494, Type A)	Lignosulfonates
		Hydroxylated carboxylic acids
		(Also tend to retard set so accelerator is added)
Retard setting time	Retarder (ASTM C494, Type B)	Lignin
		Borax
		Sugars
		Tartaric acid and salts
Accelerate setting and early strength development	Accelerator (ASTM C494, Type C)	Calcium chloride (ASTM D98) Triethanolamine
Reduce water and retard set	Water reducer and retarder (ASTM C494, Type D)	(See water reducer, Type A, above)
Reduce water and accelerate set	Water reducer and accelerator (ASTM C494, Type E)	(See water reducer, Type A, above. More accelerator is added)

Source: Portland Cement Association

(Continued)

Table 7.9 Admixtures by Classification (*Continued*)

Improve workability and plasticity	Pozzolan (ASTM C618)	Natural pozzolans (Class N)
		Fly ash (Class F and G)
		Other materials (Class S)
Cause expansion on setting	Gas former	Aluminum powder
		Resin soap and vegetable or animal glue
		Saponin
		Hydrolyzed protein
Decrease permeability	Dampproofing and waterproofing agents	Stearate of calcium, aluminum, ammonium, or butyl
		Petroleum greases or oils
		Soluble chlorides
Improve pumpability	Pumping aids	Pozzolans
		Organic polymers
Decrease air content	Air detrainer	Tributyl phosphate
High flow	Superplasticizers	Sulfonated melamine formaldehyde condensates
		Sulfonated naphthalene formaldehyde condensates

air-entraining admixture to concrete as it is being mixed. The entrained air bubbles are distributed uniformly throughout the cement paste, constituting 2 to 8 percent of the volume of the concrete. Air-entrained Portland cement has the air-entraining material ground in during manufacture and can be added to the concrete mix before or during the mixing process. Numerous commercial air-entraining admixtures are manufactured from a variety of materials. Ingredients include polyethylene oxide polymers, fats and oils, sulfonated compounds, and detergents. Air-entraining admixtures are specified by ASTM C226.

Entrained air bubbles improve the durability of concrete, which increases resistance to damage due to freeze–thaw cycles and deicers, which can cause scaling. Air-entraining gives improved workability during placement and superior water-tightness. It also improves resistance to sulfate attack from soil water and seawater. Another important feature is that properly proportioned air-entrained concrete requires less water per cubic yard than non-air-entrained concrete of the same slump, resulting in an improved water–cement ratio. Air-entrained concrete is used in cold climates where concrete, such as paving and architectural concrete, is exposed to the freeze–thaw cycle. It is also effective for concrete exposed to soil and water, where sulfate attack is possible.

Retarders

Retarding admixtures, or *retarders*, are used to slow the setting time of cement paste in concrete. They are often employed in hot weather, where hydration is accelerated by excessive heat. Without a retarder in hot weather, more water is required to achieve the desired slump, which produces lower-strength concrete. Retarding admixtures tend to reduce the water required, resulting in a better water–cement ratio and ultimately increased concrete strength. Retarders also help when it is necessary to pour large amounts of concrete or where placement is difficult. For example, if concrete must be pumped a considerable distance, retarders will enable it to be moved, placed, and finished before premature setting occurs. They are sometimes used in concrete mix trucks that have to travel an unusually long distance to a job. They also reduce increased temperatures caused by the heat of hydration in large concrete masses.

A variety of chemicals are used as retarders, and their use results in a reduction in strength during the first one to three days. Retarders can also cause shrinkage, which may cause cracking. Before retarders are used, tests should be made with job materials and conditions. Set times can often be slowed by other means. For example, higher air temperatures during hydration often cause increased hardening rates. A simple means for retarding set time in hot weather without admixtures is cooling the mixing water, aggregates, or both. Set time can also be reduced by shading the concrete so it is not directly exposed to sunlight.

Water Reducers

Water-reducing admixtures lower the amount of water needed to produce concrete of a given consistency. They can also be used to increase the amount of slump without requiring additional water. This makes for a lower water–cement ratio, resulting in greater concrete strength. Some water-reducing admixtures shorten set time and can cause increased drying shrinkage. Lignin solfonic acids and metallic salts are common water-reducing agents.

Accelerators

An *accelerator* admixture speeds up the strength development of concrete. Strength development can also be accelerated by using Type III high-early-strength Portland cement, by increasing the amount of cement to lower the water–cement ratio, or by curing at higher temperatures. Accelerators are used in cold weather to develop strength faster in order to offset freeze damage.

Calcium chloride is a frequently used accelerator. It should be added to the concrete mix in solution rather than dry form. The amount added varies, depending on conditions and the desired set time, but never exceeds 2 percent of the cement's weight. Commercially available calcium chloride ($CaCl_2$) is produced in regular and concentrated flake types. Regular flake contains at least 77 percent $CaCl_2$, while concentrated flake has 94 percent $CaCl_2$.

Experience has shown that using 2 percent or less of calcium chloride has no significant corrosive effect on steel reinforcing materials if the concrete is of high quality. Calcium chloride is not recommended for use in (1) prestressed concrete, because of the possibility of corrosion; (2) in concrete containing embedded aluminum, such as electrical conduit; (3) in concrete subject to alkali–aggregate reaction or soils or water containing sulfates; (4) in nuclear shielding concrete; (5) in floor slabs to receive dry-shake metallic finishes; and (6) in hot weather.

Additional commercial accelerators are made from chemicals other than calcium chloride. Some take a non-hygroscopic powder form and others are concentrated liquids. Since they are free of chlorides, they are useful in concrete in which steel is embedded. They produce a high early set, reducing normal set time from three hours to one hour, and the concrete develops higher-than-normal strength during the initial three-day curing period.

Pozzolans

A pozzolan is a siliceous and aluminous material that, when finely ground in the presence of moisture, chemically reacts with calcium hydroxide at ordinary temperatures to form compounds possessing cementitious properties. (See the earlier discussion under "Blended Hydraulic Cements" for more information.) Pozzolan materials are sometimes added to concrete to help reduce internal temperatures, helpful when pouring large masses of concrete. Some pozzolans are used to reduce or eliminate potential concrete expansion from alkali-reactive aggregates, while others improve resistance to sulfate attack.

Pozzolans can replace 10 to 35 percent of the cement. They substantially reduce the twenty-eight-day strength of concrete and require continuous wet curing and the maintaining of favorable temperatures for a longer period than that needed for normal concrete.

Workability Agents

If fresh concrete is harsh due to improper aggregate grading or incorrect mix proportions, agents can be added to improve workability (Figure 7.7). *Workability* is a term used to describe the ease with which concrete can be placed and consolidated. Improved workability

Figure 7.7 Workable concrete should flow sluggishly into place without segregation.

© Don Cline/Shutterstock.com

may be needed if the concrete requires pumping or placing in forms containing considerable reinforcing. If concrete needs a troweled finish, workability is important.

The addition of entrained air is the best workability agent. Some organic materials, such as alginates and cellulose derivatives, will increase slump. Finely divided materials can be used as admixtures to improve the workability of mixes deficient in aggregates passing through the No. 50 and No. 100 sieves. When added to mixes not deficient in fines, additional water is usually required. This may reduce strength, increase shrinkage, and adversely affect other properties of concrete. Fly ash and pozzolans used as workability agents must meet ASTM specifications. These also tend to reduce early strength and entrained air, so additional air-entraining admixture must be used.

Superplasticizers

When cement and water mix, the wet cement particles can form small clumps that inhibit proper mixing of cement and water. This reduces workability and inhibits hydration. Superplasticizers are admixtures that coat the cement particles, causing them to break away from the lumps and disperse in water. Superplasticizers give each cement particle a negative charge, causing them to repel each other, thus providing more thorough dispersement. Superplasticizers can be used to:

- Reduce water and cement at a constant water–cement ratio, giving a concrete the same strength as a normal mix but reducing the amount of cement used.

- Produce normal concrete at normal water–cement ratios that is so workable it can be placed with little or no vibration or compaction and not have excessive bleed or segregation.

- Produce a concrete of higher strength by reducing the water content required while maintaining the normal cement content.

Some superplasticizers are formulated to give slightly accelerated or high early strength. Others are used with water-reducing admixtures. These tend to retard and lower early strength but increase ultimate strength. It is recommended that superplasticizers be added directly to the mix in the concrete mixer truck to enable the slump to be maintained during long hauls or delivery delays. Commonly used superplasticizers include a higher-molecular-weight condensed sulfonate naphthalene formaldehyde, sulfonated melamine formaldehyde, and modified ligninsulfonates, all of which come in liquid form. The effectiveness of superplasticizers has a short duration, from thirty to sixty minutes before the concrete has a rapid loss in workability.

Permeability-Reducing and Damp Proofing Agents

Sound, dense concrete that has a water–cement ratio of 0.50 by weight and is properly placed and cured will be watertight. Concretes that have low cement contents, a deficiency in fines, or a high water–cement ratio can have permeability reduced by adding permeability-reducing agents, such as certain soaps, stearates, and petroleum products.

Permeability is a measure of the amount of water that passes through channels running between the outer faces of the concrete. A permeability-reducing agent reduces the flow of water through these channels. They should generally not be used in well-proportioned mixes because they increase the amount of mixing water required, thus increasing rather than decreasing permeability.

Other permeability-reducing strategies, which may avoid the need for a permeability reducing agent, can be employed. Concrete that is properly placed, compacted, and cured will have reduced permeability. Air-entraining admixtures also reduce permeability by increasing the plasticity and reducing the amount of water needed. In effect, air-entraining is an excellent permeability reducing admixture. Superplasticizers also disperse cement particles throughout the mix and help reduce permeability.

Damp-proofing admixtures are used to reduce moisture that is transferred by capillary action. This occurs when one side of the concrete is exposed to moisture and the other to air, as in a slab on-grade. The surface exposed to air tends to dry. Capillary action occurs as moisture flows to a dry surface. The effectiveness of damp-proofing admixtures varies, and manufacturers' test data must be studied.

There are other ways to damp-proof concrete. Various types of film can be applied to the damp side. A typical example involves putting a plastic membrane on the ground below a concrete slab floor. Various coatings, such as asphalt, sodium silicate, and metallic aggregate mixed with Portland cement, can also be applied. Above-grade concrete and masonry can be damp proofed by applying a coat of silicone sealer. This also protects against freeze–thaw cycles, efflorescence, weathering, and staining.

Bonding Agents

Fresh concrete must often be placed over a previously poured surface that has already set. When fresh concrete is poured over hardened concrete, the fresh concrete shrinks, breaking its bond to the hardened concrete. The hardened surface must be prepared so the fresh concrete will firmly bond to the aggregate in the hardened surface. The condition of the surface is of great importance. It must be dry and clean (free of dirt, dust, grease, paint, etc.), and the proper temperature should be maintained.

Bonding admixtures can be added to Portland cement mixtures or applied to the surface of old concrete to increase bond strength. These admixtures are usually water emulsions of certain organic materials, such as a liquid acrylic polymer, which may be added to the Portland cement with or without mixing water. The pores of the concrete absorb the water and the resin unites into a mass that bonds the two layers of concrete.

Bonding is also accomplished by exposing the surface of the aggregate in the hardened concrete, applying cement paste slurry to the hardened surface, and immediately pouring the new layer of concrete over it. Another type of bonding agent that also forms a waterproof membrane is a two-compound moisture-insensitive epoxy adhesive applied to hardened concrete and immediately covered by fresh concrete.

Coloring Agents

Concrete can be colored by mixing pure, finely ground mineral oxides with dry Portland cement. Thorough mixing is necessary to produce a uniform color. Oxides added to normal Portland cement are usually limited to earthy colors and pastels because of the cost and graying effect of the cement (Figure 7.8). White Portland cement produces clearer, brighter colors and is preferred. Some color agents are compounded to provide water-reducing and set-controlling properties.

Concrete can also be colored by exposing the aggregate. Colored aggregates are spread on the surface of the freshly cast concrete and floated in place. At the proper degree of set, the un-hydrated paste is washed away. Color can also be placed on the surface of concrete before it sets. One way involves using natural mineral oxides of cobalt, chromium, iron, or ochres and umbers ground to fine powder. These are combined with Portland cement into a topping mix that is troweled over the surface of the concrete.

Figure 7.8 This concrete slab was colored by combining mineral oxides with dry Portland cement in the concrete mix.

Another method uses synthetic oxides mixed with fine silica sand. This is spread over the unset but floated concrete surface. The mixture is then floated and troweled into the uncured surface and cured in the normal way. Colors can also be mixed with a metallic aggregate and dry Portland cement. This mixture is applied as a dry shake to the surface, then is floated, troweled, and cured. Some dry shakes also give additional hardness to the surface.

Color can also be achieved by simply painting or staining the concrete surface after it has been completely cured and neutralized. Neutralization can occur naturally via aging and weathering or it can be accomplished with a neutralizing agent, such as zinc sulfate.

Hardeners

When a concrete surface is subject to heavy wear, such as that of a factory or warehouse floor, its life can be extended by using a liquid-chemical or dry-powder hardener. One form of chemical hardener is a colorless, nontoxic, nonflammable liquid containing magnesium and zincfluosilicates with a wetting agent. The wetting agent reduces the surface tension of the liquid hardener, which makes it easier for it to enter the pores of the concrete. The hardening agent produces a chemical reaction with the free lime and calcium carbonates in the Portland cement. This reaction densifies the surface, making it more wear resistant and impenetrable to many liquids and chemicals. The liquid hardener is applied to the surface of the cured concrete in several coats. Similar hardeners are made using Baume sodium silicate that reacts with the lime to form an insoluble crystal within the pores of the concrete.

Concrete surfaces can also be hardened using dry powder hardeners. One product uses quartz silica aggregates and alkali-fast inorganic oxides, which color the concrete. These two are mixed with Portland cement and plasticizing agents, giving a dry shake that is applied to freshly poured concrete. This provides a high-strength surface with color provided by the oxides. Another type uses a finely ground iron aggregate and, if desired, inorganic oxides for color. These are mixed with Portland cement and plasticizing agents and applied as a dry shake on the surface of freshly laid concrete.

Grouting Agents

Portland cement grouts are widely used for stabilizing foundations, filling cracks in concrete walls, filling joints, grouting tenons or anchor bolts, and in other applications. Grout properties can be altered using the various admixtures discussed in this chapter.

Gas-Forming Agents

Gas-forming agents are added to concrete or grout to cause a slight expansion in it before it hardens. Aluminum powder, one of several gas-forming agents in use, reacts with the hydroxides in hydrating cement and produces small hydrogen-gas bubbles. This also helps eliminate voids caused by settlement of the concrete or grout. A typical application consists of expanding the grout under a difficult-to-reach machine base or column to be certain the area is fully grouted. After concrete or grout has hardened, gas-forming agents will not overcome shrinkage. When they are used in large amounts, gas-forming agents produce lightweight cellular concrete.

BASICS OF CONCRETE

Concrete is made up of two parts: aggregates and a paste. The paste is a Portland cement and water mixture with some entrapped air that coats the particles of aggregate and binds them. The paste hardens due to hydration, a chemical reaction between the Portland cement and water. The aggregates used range from fine to course, with various degrees of size within each.

Figure 7.9 shows the range in proportions of materials used in typical designs for rich and lean mixes. Notice that the volume of both cement and water varies from 7 to 15 percent and 14 to 21 percent, respectively. A small percentage consists of air and the rest, the major volume, is aggregate. This means that the selection of aggregates

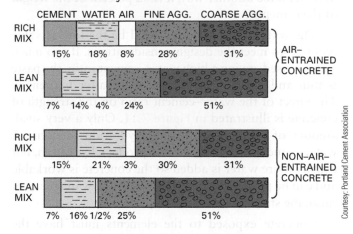

Figure 7.9 The graph shows the range in proportions of materials used in concrete by volume.

Courtesy: Portland Cement Association

is vital for the production of quality concrete. It is essential that a continuous graduation of particle size be maintained. Figure 7.10 shows a cross-section cut through a sample of hardened concrete. The paste must completely fill the spaces between the particles and coat each particle.

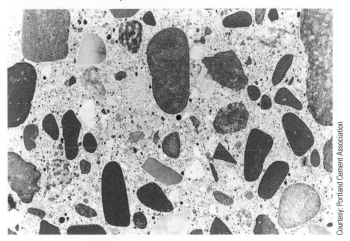

Figure 7.10 A cross-section of hardened concrete showing how the cement and water ratio coats each particle of aggregate and fills the voids between particles.

Courtesy: Portland Cement Association

Water–Cement Ratio

If all conditions are held constant, the quality of the hardened concrete is determined by the *water–cement ratio*. The water–cement ratio is the ratio by weight between the water and cement used to make paste. For example, a mix that uses 0.62 lb. of water per one pound of cement has a water–cement ratio of 0.062/100, or 0.62. In metric measure, this would be 620 g of water per kilogram of cement, for a metric water–cement ratio of 620. Stated another way, it is 62 percent of the weight of the cement.

The water–cement ratio should be the lowest value required to meet the design considerations. If too much water is used, giving a high water–cement ratio, the paste is thin and will be porous and weak after hardening. The effect of the water–cement ratio on the strength of concrete is illustrated in Figure 7.11. Only a very small amount of water is needed for hydration to occur. A water–cement ratio of 0.31 will produce hydration, but usually more water is added so the concrete is workable and can be properly placed. The lower the water–cement ratio, the stronger the concrete (Table 7.10).

Concrete exposed to the elements must have the following for durability:

- Air entrainment
- Low water–cement ratio
- Quality cement and aggregate
- Proper curing
- Proper construction practices

Table 7.10 Maximum Permissible Water–Cement Ratios for Concrete When Strength Data from Trial Batches or Field Experience Are Not Available

Specified Compressive Strength F'_c, psi[a]	Maximum Absolute Permissible Water–Cement Ratio, by Weight	
	Non-air-entrained Concrete	Air-entrained Concrete
2,500	0.67	0.54
3,000	0.58	0.46
3,500	0.51	0.40
4,000	0.44	0.35
4,500	0.38	[b]
5,000	[b]	[b]

[a]28-day strength. With most materials, the water cement ratios shown will provide average strengths greater than required.
[b]For strengths above 4500 psi (non-air-entrained concrete) and 4000 psi (air-entrained concrete) proportions should be established by the trial batch method. 1000 psi ≈ 7 MPa.

Source: Portland Cement Association

Figure 7.11 The graph demonstrates the effect of the water–cement ration on the strength of concrete.

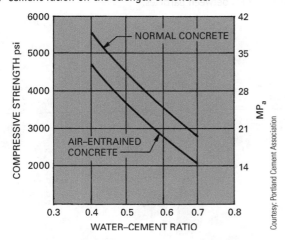

Courtesy: Portland Cement Association

Table 7.11 Maximum Water–Cement Ratios for Various Exposure Conditions

Exposure Condition	Normal-Weight Concrete, Absolute Water–Cement Ratio by Weight
Concrete protected from exposure to freezing and thawing or application of deicer chemicals	Select water–cement ratio on basis of strength, work ability, and finishing needs
Watertight concrete	
In fresh water	0.50
In seawater	0.45
Frost-resistant concrete	
Thin sections; any section with less than 2-in. cover over reinforcement and any concrete exposed to deicing salts	0.50
All other structures	0.45
Exposure to sulfates	
Moderate	0.50
Severe	0.45
Placing concrete under water	Not less than 650 lb. of cement per cubic yard (386 kg/m³)
Floors on-grade	Select water–cement ratio for strength, plus minimum cement requirements

Source: Portland Cement Association

The required water–cement ratios for various conditions are shown in Table 7.11. These data do not give consideration to relative strength. When durability is not the major consideration, the water–cement ratio is selected based on compressive strength. This often requires that tests be made using the actual job-site materials. If flexural strength is the basis for concrete design, tests are conducted to find the relationship between the water–cement ratio and flexural strength.

Minimum Cement Content

In addition to the specification of the water–cement ratio, the minimum cement content is also specified. This ensures that the concrete will have good finishability, good wear-resistance, and good appearance (Table 7.12).

Table 7.12 Minimum Cement Requirements for Concrete Used in Flatwork

Maximum Size of Aggregate, In.	Cement, lb. per Cubic Yard
1½	470
1	520
¾	540
½	590
⅜	610

1 in. ≈ 25 mm
100 lb./yd.³ ≈ 60 kg/m³

Source: Portland Cement Association

Aggregates

Aggregates must be properly graded and of proper quality. Grading pertains to the particle size and the distribution of particles. Properly graded aggregate produces the most economical concrete because it allows the use of the larger size coarse aggregate. This reduces the amount of water and cement required, and a reduction in cement reduces the cost.

The maximum size of aggregate used depends on the size and shape of the members being formed and the amount of reinforcing steel required. The maximum-size aggregate acceptable can be no more than one-fifth the narrowest dimension between the sides of the forms or three-fourths the clear space between steel, such as reinforcing bars, ducts, conduit, or bundles of bars. Aggregate in unreinforced slabs on the ground should not exceed one-third the slab thickness. For leaner concrete mixes, a finer grade of sand is used to improve workability. Richer mixes use a coarse grade of sand for greater economy.

Entrained Air

Entrained air should be used in all concrete paving regardless of the temperatures but especially for those exposed to freeze and thaw cycles. The required percentage of entrained air is shown in Table 7.13. The amount of entrained air required decreases as the maximum size of the aggregate increases. Entrained air reduces the amount of water required, producing a lower water–cement ratio.

Slump

Slump is a measure of the consistency of concrete, defining the ability of fresh concrete to flow. This is measured by a slump test explained later in this chapter. Slump is the decrease in height of a molded mass of fresh concrete that occurs immediately after it is removed from a standard metal slump cone. The higher the slump, or the more the sample lowers, the wetter the concrete mixture. A measure of slump can only be used to compare mixes of identical design. Slump is usually specified in the concrete specifications. Recommended slumps for concrete consolidated by mechanical vibration are shown in Table 7.14.

CONCRETE DESIGN

Concrete mixes are specially designed to provide the needed characteristics for a particular application. A mix designed for structural purposes will be higher in strength, while one intended for architectural finishes may have superior surface qualities. The unit of measure for concrete is the cubic yard, which contains 27 cubic feet.

Table 7.13 Recommended Average Total Air-Content Percentage for Level of Exposure

	Nominal Maximum Sizes of Aggregates					
Exposure	⅜ in. (10 mm)	½ in. (13 mm)	¾ in. (19 mm)	1 in. (25 mm)	1½ in. (40 mm)	2 in. (50 mm)
Mild	4.5	4.0	3.5	3.0	2.5	2.0
Moderate	6.0	5.5	5.0	4.5	4.5	4.0
Extreme	7.5	6.0	6.0	6.0	5.5	5.0

1 in. ≈ 25 mm

Source: Portland Cement Association

Table 7.14 Recommended Slumps for Various Types of Construction

Concrete Construction	Slump, In.	
	Maximum[a]	Minimum
Reinforced foundation walls and footings	3	1
Plain footings, caissons, and substructure walls	3	1
Beams and reinforced walls	4	1
Building columns	4	1
Pavements and slabs	3	1
Mass concrete	2	1

Source: Portland Cement Association

[a]Slumps shown are for consolidation by mechanical vibration. May be increased 1 in. for consolidation by hand methods, such as rodding and spading.

To determine the amount of concrete needed, the volume of concrete in cubic feet of the area is calculated and divided by 27.

As discussed earlier, the water–cement ratio is a basic premise used when designing normal-weight concrete. Based on research and experience, water–cement ratios for various applications can be recommended, as shown in Table 7.11. Another way mix design is accomplished involves using the absolute volume of material amounts. The absolute volume of a loose material, such as aggregate, is the total volume, including the particles and air spaces between them. This includes the absolute volume of the cement, aggregate, water, and trapped air.

$$\text{Absolute volume} = \frac{\text{weight of dry material}}{\text{specific gravity} \times \text{unit weight of water}}$$

For example, the absolute volume of 100 pounds of aggregate having a specific gravity of 2.5 would be:

$$\text{Absolute volume} = \frac{100 \text{ lb.}}{2.5 \times 62.5 \text{ (one cubic foot of water)}}$$
$$= 0.64 \text{ ft}^3$$

Sometimes, concrete mixes are still called out based on the number of sacks of 94 lb. cement used in a cubic yard of concrete, for example a 4-bag mix, or a 6-bag mix. Sack counts do not accurately describe the desired properties of the finished concrete and detailed proportioning specifications should be followed. Publications of the Portland Cement Association give complete instructions on concrete design by water–cement ratio and absolute volume.

Normal-Weight Concrete

Normal-weight concrete weighs about 135 to 160 lb. per ft.^{3} (2,150 to 2,550 kg/m^3). It uses sand, gravel, crushed stone, and air-cooled blast-furnace-slag aggregates, as specified by ASTM C33. The design of normal-weight concrete depends primarily on the required strength and durability. Consideration of workability, plasticity, and cost are factored secondarily.

Lightweight Insulating Concrete

Insulating concrete, specified in ASTM C332, weighs about 15 to 90 lb. per ft.^{3} (250 to 1450 kg/m^3) and uses pumice, scoria, perlite, vermiculite, and diatomite aggregates. The design of lightweight insulating concrete depends on the aggregate used and the desired compressive strength. The amount of water required varies greatly, depending on the circumstances. An air-entraining agent is recommended for some mixes. Detailed information is available from the Portland Cement Association.

Lightweight Structural Concrete

Structural lightweight concrete weighs about 85 to 115 lb. per ft.^{3} (1350 to 1850 kg/m^3) and uses expanded shale, clay, slate, and slag aggregates. It is specified in ASTM C330. Lightweight structural concrete can be designed to produce structural members that are 25 to 35 percent lighter than members made with normal-weight concrete and have no loss in strength. Since the aggregate is cellular, and weights are different from normal aggregate, its design is usually derived from testing trial batches, experience, and other reliable test data.

Heavyweight Structural Concrete

Heavyweight concrete weighs 175 to 400 lb. per ft.^{3} (2800 to 6400 kg/m^3) and uses barite, limonite, magnetite, ilmenite, hematite, iron, and steel slugs as aggregates. It is specified in ASTM C637. The main application for heavyweight concrete is for radiation shielding in medical or nuclear applications.

CONCRETE TESTS

Both freshly mixed and cured concrete are tested to make certain they meet the written specifications for the concrete. This is especially important when working in different parts of the country where aggregates and water differ and the mix must be adjusted accordingly. Concrete testing is conducted both on the building site and in certified testing laboratories. Certified testing laboratories are organizations that have demonstrated the knowledge and ability to properly perform, record, and report the results of laboratory procedures related to the determination of concrete strength and other characteristics.

Tests with Fresh Concrete

When testing fresh concrete, it is essential to take samples that are representative of the batch. Samples must be taken and handled following the specifications in ASTM C172. Except for slump and air-content tests, the sample must be at least 1 ft.^{3} (0.030 m^3) in volume. After it has been taken from the batch, the sample must be used within fifteen minutes and protected from sources of rapid evaporation during the test. Samples taken from the beginning and the end of a batch are not representative.

Slump Test   A *slump test* is made on site to ensure that the required consistency of a concrete mix has been achieved. Each load of transit-mixed concrete has a certificate listing its ingredients and their proportions. A mix with a high slump may be too wet, and one with a low slump may be too stiff. The slump test is made following ASTM C143 specifications. A standard slump cone is 8 in. (200 mm) in diameter at the bottom and 12 in. (305 mm) high. The cone is placed on a flat surface and held still by standing on the foot support (Figure 7.12). It is filled full and rodded twenty-five times with a ⅝ in. (16 mm) diameter, 24 in. (600 mm) long rod with a rounded tip. A second layer is poured and rodded as above, making certain the rod penetrates the surface of the layer below. After the top layer has been rodded, the excess concrete is struck off, leveling the top surface, and the mold is carefully lifted. The amount the concrete will slump is measured from the top of the cone (Figure 7.13). For example, if the top of the concrete is 4 in. below the top of the cone, the slump is 4 in.

Another method for testing slump is the ball penetration test, as specified by ASTM C360. The depth to which a 30 lb. (13.6 kg), 6 in. (150 mm) diameter hemisphere will sink into fresh concrete is measured. When calibrated for a particular set of materials, the results can be related to the slump. The concrete is placed in a container at least 18 in. (450 mm) square and at least 8 in. (200 mm) deep.

Figure 7.12 A concrete sample is rodded during a slump test.

Figure 7.13 The cone is lifted and the amount the concrete slumps below the top of the cone is measured.

Unit Weight Test   The unit weight test involves weighing a properly consolidated specimen in a calibrated container following ASTM C138 standards. It can determine the quantity of concrete produced per batch and give indications of air content.

Air-Content Test Methods for measuring air content include the pressure method (ASTM C231), the volumetric method (ASTM C173), and the gravimetric method (ASTM C138). The pressure method requires the sample be placed in a pressure air meter and subjected to an applied pressure (Figure 7.14). The air content can be read directly. When lightweight aggregates are used, this method also compresses the air in them; therefore, tests by the pressure method are not recommended for concrete with lightweight aggregates.

Figure 7.14 An engineer using pressure testing to measure air content.

© TFoxFoto/Shutterstock.com

The volumetric method measures air content by agitating a known volume of concrete in an excess of water. This method is suitable for concrete containing all types of aggregate. The gravimetric method uses the same test used for the unit weight test of concrete. The actual unit weight of the sample is subtracted from the theoretical unit weight, as determined from the absolute volumes of the ingredients, assuming no air is present. The mix proportions and specific gravities of the ingredients must be known. The difference in weight is given as a percentage, as is the air content. This method requires laboratory control and, thus, is not suitable for on-site use. A quick check for air content can be made using a pocket-size air indicator. This suffices only for a quick on-site test and is not suitable for a standard ASTM test.

Cement Content Test The cement content test is used to determine the water and cement content of fresh concrete. The water–cement ratio has a major influence on strength. Therefore, this test gives an estimate of the strength potential without waiting for samples to harden and cure, which usually takes seven to twenty-eight days.

Strength Specimens The compressive strength of concrete is one of the most important performance measure used by engineers in the design of a structure. Strength is determined by compression testing, and is expressed in pounds per square inch (PSI). Normal concrete mixes range from 2,500 PSI to 5,000 PSI, with engineered design mixes of over 10,000 PSI available for specialized applications. 2,500 PSI concrete can be used in driveways, walkways, and even floor slabs on grade. This concrete is usually the most economical available from batch plants. 3,000 PSI concrete is a standard multipurpose mixture for general use in construction. It is durable, has sufficient cement to give it good finishing characteristics, and can be placed fairly wet without sacrificing quality. 4,000 PSI concrete is used for heavy-traffic pavement, severe-use floor slabs like shops and warehouses, and concrete footings designed to support heavy loads. 5,000 PSI and higher concrete mixes are usually used for specialized construction projects where high impact resistances, very low wear rates, or extreme temperature conditions are expected.

Compressive strength is measured by breaking cylindrical samples of wet concrete, called *test cylinders*, taken on the building site. Specimens of freshly poured concrete may be field molded or laboratory molded. The molding of test cylinders should be started within fifteen minutes after the specimens are obtained. Field-molded specimens (Figure 7.15) should be made and cured as specified by ASTM C31 or AASHTO T23. Laboratory-molded specimens should be made and cured as specified by ASTM C192 or AASHTO T126. The size of the test cylinder depends upon the aggregate size. For

Figure 7.15 Concrete cylinders being cast on the job site for compressive strength testing.

Courtesy: Portland Cement Association

Tests with Hardened Concrete

Compressive Strength Test   After the test cylinder has cured as required, it is ready for the compression strength test. The compression test is made according to ASTM specifications and is one of the most frequently required tests. The cylinders are placed in a compression-testing machine (Figure 7.16). As pressure is applied, the compressive strength is recorded up to the point when the cylinder fractures.

Figure 7.16 Compression strength testing of a capped test cylinder.

Courtesy: Portland Cement Association

example, a specimen with a maximum aggregate size of 2 in. (50 mm) or smaller can be made in a cylinder 6 in. (150 mm) in diameter.

The size of the mold for specimens used to test beams for flexural strength also varies depending on the size of the aggregate. For example, a mold should be 6 × 6 in. (150 × 150 mm) in cross-section for specimens with aggregates 2 in. (30 mm) or smaller. The cylinders are rodded and filled as specified.

After casting, the tops of the specimens are covered with an oiled glass or metal plate, a special cylinder cap, or a plastic bag. The strength of the test specimen can be greatly affected by changes in temperature, jostling of the mold, and exposure to drying. Test specimens should be cast where they can be properly protected and movement is not necessary. Specimens taken and cured on the site in the same manner as the cast structure more closely represent the actual strength of the concrete in the structure at the time of testing. Detailed information is available from the Portland Cement Association.

Flexural Strength Test   The flexural strength test is used to determine the flexural, or bending strength, of concrete. The concrete sample is formed in a mold in the shape of a beam. Samples with aggregates up to 2 in. (50 mm) should have a minimum cross-section of 6 × 6 in. (150 × 150 mm). Large aggregate samples should have a minimum cross-section dimension of three times the maximum size of the aggregate. The span of the test beam should be three times the depth of the beam plus two additional inches. For example, a 6 × 6 in. (150 × 150 mm) beam would be 20 in. (508 mm) in length. The mold is filled in two layers with one rodded stroke for every 2 in.^{2} (13 cm^2) of area. The top is struck flush with the mold and the sample is cured with controlled temperature and moisture. The cured specimen is tested as shown in Figure 7.17. It is supported on each end, and

Figure 7.17 A concrete beam undergoing flexural strength testing.

Courtesy: Portland Cement Association

Figure 7.18 Test apparatus for measuring abrasion resistance of concrete. The machine can be adjusted to use either revolving disks or dressing wheels.

Courtesy: Portland Cement Association

curing period. ASTM C684 outlines three methods for making accelerated strength tests.

Nondestructive Tests

Nondestructive tests are used to evaluate the strength and durability of hardened concrete. Commonly used tests are rebound, penetration, pull-out, and dynamic or vibration tests.

Rebound Tests   Rebound tests are made with a Schmidt rebound hammer. It measures the distance a spring-loaded plunger rebounds after striking the concrete surface. The reading is related to the compressive strength of concrete.

Penetration Tests   The penetration method uses a Windsor probe, which is a power-activated gun that drives a hardened alloy probe into the concrete. The exposed length of the probe is measured and related by a calibration table to the compressive strength of the concrete. This leaves a small indentation in the concrete surface.

Pull-Out Tests   A pull-out test requires that a steel rod with an enlarged end be cast in the concrete. A device used to pull the rod from the concrete measures the force required. This gives the shear strength of the concrete. It has the disadvantage of damaging the surface of the concrete.

pressure is applied to the midpoint until the specimen breaks. The ultimate flexural strength is read on a dial in pounds per square inch (kilopascals).

Abrasion Test   The abrasion test is used to ascertain the resistance to wear of hardened concrete samples. A hardening admixture or surface coating is used with the sample concrete mix. The test is made on a machine that rolls steel balls under pressure in a circular motion on the surface of the specimen. The specimen is weighed before and after the test. The loss in weight determines the ability to resist abrasion (Figure 7.18).

Freeze–Thaw Test   Cured concrete specimens are placed in a freeze–thaw tester, a cabinet much like a freezer, and run through a series of freeze–thaw cycles. The loss between the original weight and final weight of the specimen is used to determine which samples withstand the freeze–thaw cycle best.

Accelerated Curing Tests   Accelerated curing tests are used when it is desirable to determine acceptance of structural concrete without the usual twenty-eight-day

Dynamic or Vibration Tests A dynamic or vibration test uses the principle that the velocity of sound in a solid can be measured either by recording the time it takes short impulses of vibrations to pass through a sample, or by determining the resonant frequency of a specimen. High velocities indicate a very good concrete, whereas very low velocities indicate a poor concrete.

EMERGING TRENDS IN CONCRETE

Reactive Powder Concrete (RPC) is a high-performance concrete able to achieve compressive strengths of up to 30,000 psi. Made by combining fine silica sand, cement, silica fume, and crushed quartz with steel and synthetic fibers, RCP is highly ductile, extremely workable, and relatively self-placing, requiring only minimal vibration.

Self-Consolidating Concrete (SCC), also known as self-compacting concrete, is a highly-flowable, non-segregating concrete that fills formwork and the most congested reinforcement without any mechanical vibration. SSC uses new super-plasticizers that eliminate the need for vibration and provides a smooth surface finish with no voids.

Smog-eating concrete uses cement that has been treated with titanium dioxide. The titanium dioxide sets off a photo-catalytic reaction with ultra violet light that accelerates natural oxidation and prevents bacteria and dirt from accumulating on a surface. The material can also break down nitrogen oxides emitted in the burning of fossil fuels.

Permeable concrete is made with narrowly graded coarse aggregate, little or no fine aggregate, and a very low water–cement ratio. The result is a stiff, pebbly mixture with 15 to 25 percent of its volume composed of interconnecting pores through which water can flow. Pervious concrete is used for drives and walkways that allow water to percolate through a surface and thereby educe stormwater runoff.

Bioconcrete is a revolutionary new concrete technology that heals itself by using bacteria. No matter how carefully it is mixed or reinforced, all concrete eventually develops cracks that lead to water penetration and under some conditions, structural failure. The bacteria are blended into the fresh concrete in small biodegradable capsules that are activated only in the presence of water. When cracks and water penetration occur in the concrete, the bacteria is activated by the water and feed on calcium lactate present in the capsules to form a limestone like material that fills the cracks.

Rammed earth, or *pisé de terre*, is a historic construction material that can be classified as a green building material because it utilizes locally available materials with little embodied energy (Figure 7.19). Rammed earth walls are made by compressing a damp mixture of earth combined with sand, gravel, clay, and a small amount of Portland cement into conventional formwork. The damp material is poured into the forms 4 to 10 in. (100 to 250 mm) at a time and is "rammed" with a pneumatically powered backfill tamper that compacts the mix to around 50 percent of its original height. Subsequent layers of the material are added and the process is repeated until the wall has reached the desired height. Like all solid masonry construction, a significant benefit of rammed earth construction is its excellent thermal mass qualities; it heats up slowly during the day and releases its heat during the evening. This can even out daily temperature variations and reduce the need for mechanical air conditioning and heating.

Figure 7.19 The striations, natural earth tone colors, and textures of a rammed earth wall.

Review Questions

1. What are the basic ingredients in concrete?

2. How does concrete cure and harden?

3. What ingredients normally make up Portland cement?

4. What are the designations for the different types of Portland cement?

5. What is meant by air-entraining?

6. How does the addition of air-entraining materials to Portland cement improve the concrete?

7. What ingredients may be used in masonry cements?

8. How does the fineness of the cement affect the concrete?

9. Explain how Portland cement should be stored.

10. What natural aggregates are used in making concrete?

11. What aggregates should be avoided when making concrete?

12. What aggregates are used in insulating concrete?

13. What are the commonly used admixtures?

14. What is the recommended water–cement ratio?

15. What five factors influence the quality of finished concrete exposed to the elements?

16. What can be done in hot weather to help retard rapid setting times?

Key Terms

Accelerator

Admixture

Aggregate

Air-Entrained Portland Cement

Concrete

Hydration

Portland Cement

Pozzolan

Rammed Earth

Retarder

Slump

Slump Test

Supplementary Cementing Materials (SCMs)

Test Cylinders

Water–Cement Ratio

Workability

Activities

1. Visit a local concrete batch plant. Ask the supervisor to show the process for producing concrete and to explain instructions given to drivers of the delivery trucks.

2. If the concrete plant has a test lab, see if you can observe some of the tests made for the contractor.

3. Mix samples of concrete using various types of cement, varying the proportions of the ingredients and using different aggregates. Cure the samples and conduct compression tests. Compare the results and explain the differences that occur.

4. Prepare identical concrete samples and cure each the same number of days but under different temperatures (for example, at room temperature, above 100°F, and below freezing (use a freezer)). Test for compression strength. Examine the broken samples and report any differences in the appearance of the surfaces.

5. Prepare identical concrete samples, cure them under ideal conditions, and then expose them to various conditions to see how they perform. For example, immerse in salt water, fresh water, gasoline, oil, and other materials to which concrete is often exposed. Report your findings in a written paper.

6. On construction sites visited, observe how the aggregates are stored. Report the good and poor practices observed.

7. Prepare a list of admixtures and tell what purposes they serve.

8. Mix standard proportion concrete ingredients but vary the amount of water used. Measure the slump of each sample.

Additional Resources

Canadian Design and Control of Concrete, and other publications, Cement Association of Canada Headquarters, Ottawa, Ontario, Canada K1P 5Y7.

Concrete in Practice, National Ready Mixed Concrete Association, Silver Spring, MD.

The Contractors Guide to Quality Concrete Construction, Concrete Fundamentals, Cast-In-Place Walls, and numerous other technical concrete publications, American Concrete Institute, Farmington Hills, MI.

Design and Control of Concrete Mixtures, U.S. and Canadian metric editions, Portland Cement Association, Skokie, IL. Many other publications available.

Manual of Concrete Practice, and other publications, American Concrete Institute, Farmington Hills, MI.

Seismic and Wind Design of Concrete Buildings, International Code Council, Falls Church, VA.

Cast-in-Place Concrete

Upon completion of this chapter, the student should be able to:

- Discuss the processes for preparing, transporting, handling, and placing cast-in-place concrete.
- Explain how cast-in-place concrete walls, beams, and columns are formed, reinforced, and poured.
- Discuss the finishes used on concrete surfaces.
- Know how concrete is cured.

- Cite various types of formwork used for cast-in-place concrete.
- Identify and describe types of concrete-reinforcing materials.
- Prepare sketches illustrating the various types of monolithically cast slab and beam floors and roofs.
- Describe the procedure for casting and erecting tilt-up concrete walls.
- Explain briefly what is meant by lift-slab construction.

Build Your Knowledge

For further information on these materials and methods, please refer to:

Chapter 7   Concrete

Topic: Basics of Concrete

Cast-in-place concrete is a widely used and robust construction material. Also known as site-cast concrete, the material is produced on the building site by setting wood, metal, or molded plastic forms in place; placing reinforcing material in the forms; pouring the concrete over the reinforcing to fill the form; and letting it cure into a hardened material.

An engineer can design cast-in-place concrete members in a wide range of sizes and shapes, with a variety of surface textures and colors. Although some concrete structural members can be pre-cast and shipped to the site (see Chapter 9), cast-in-place parts of a structure are cast on the building site. These include spread footings, foundation caissons, pilings, piers, slabs on-grade, and any members too large to pre-cast and move to the site. Some designs have irregular shaped features that are difficult to pre-cast and transport, so the pieces are cast-in-place. For example, construction of a dam requires

huge intricate forms and massive amounts of concrete, all site built and cast in place.

Cast-in-place concrete structural members usually are heavier than steel, wood, or pre-cast concrete members, resulting in increased loads on the foundation. Because the concrete is poured and cured under the environmental conditions present at the time, site-cast concrete may be difficult to complete during extreme weather conditions. There are continuing developments to make cast-in-place concrete faster and easier. A wide range of forms are available, in addition to equipment, such as concrete pumps and power finishing machines needed to speed up the process. Cast-in-place concrete is a versatile and effective construction material.

BUILDING CODES

Reinforced concrete structural members and pre-stressed concrete must be designed and constructed according to the provisions in the building code. This includes provisions to resist seismic forces if they are a factor in the area. Codes specify how to bend reinforcement, what the surface conditions must be, how to place the concrete in the forms, and what the coverage of the reinforcing within the concrete must be. For example, concrete that is cast against and will remain on the earth requires a minimum 3 in. (76.2 mm) of concrete cover over the reinforcement, while concrete walls, joists, and slabs not exposed to ground or earth require only ¾ to 1.5 in. (19–38 mm) of concrete cover. Codes also include specifications for placing concrete in corrosive environments, for thicknesses over reinforcing, fire protection, resistance to frost action, and for vertical and lateral loads.

PREPARING CONCRETE

Batching

Concrete is usually prepared in batches. A *batch* is the amount of concrete mixed at one time. The quantities of dry ingredient are usually weighed, while water and admixtures can be specified by either weight or volume. The use of volume measurements is discouraged, as they tend to be inaccurate because moisture in the aggregate moisture changes the weight, which is not accounted for in volume measures. Aggregates, especially sand, tend to fluff when handled, so the actual volume of sand can vary from batch to batch. When concrete is produced by a continuous mixer, volumetric measure is used. The job specifications usually establish the percentage of accuracy allowed when measuring ingredients. These typically range from 1 to 2 percent, so accurate weighing facilities are essential.

Mixing

Concrete is mixed until it is uniform in appearance and all ingredients are evenly distributed. If an increased amount of concrete is needed, an additional mixer should be used, rather than overloading or speeding up those in operation. It is important to follow the manufacturer's recommendations and to keep the mixing blades clean. Bent or worn blades should be replaced.

Stationary Mixing

On a large job, the concrete is often mixed on site using a stationary mixer. This can be a tilting or non-tilting type and may be manually, semi-automatically, or automatically controlled. Automatically controlled mixers have the ability to produce various mix designs that are stored in the machine's computer memory. Generally, the batch is mixed one minute for the first cubic yard and an additional fifteen seconds for each additional cubic yard (0.35 m³) or fraction thereof. Mixing time is measured

from the moment all ingredients are placed in the mixer. All water must be added before one-fourth of the mixing time has elapsed. About 10 percent of the mixing water is placed in the drum before dry ingredients are added. The remaining water is combined uniformly with the dry components saving 10 percent for addition after all dry ingredients are in the drum.

Ready-Mix Concrete

Ready-mix concrete may be fully mixed in a central mixing plant and delivered to the site in a truck mixer that operates at agitating speed (Figure 8.1). The concrete may be partially blended in a central mixer and completed in a truck as it is moved to the site, or the dry ingredients may be placed in a transit truck mixer and the entire process done by the truck after the addition of water. When the entire batch is made in the truck mixer, seventy to one hundred rotations of the drum at the rotating speed specified by the manufacturer are enough to produce a uniform mix. All revolutions after one

Figure 8.1 A truck mixer being loaded with dry ingredients and water at a central mixing plant.

hundred should be at a slower agitating speed so as not to over-mix the batch. Concrete must be delivered and discharged within hours or before the drum has revolved three hundred times after the introduction of water.

Remixing

Fresh concrete in the drum tends to stiffen even before the concrete has hydrated to initial set. It can be used if remixing will restore sufficient plasticity for compaction in the forms. Under special conditions, a small amount of water can be added, but it must not exceed the allowable water–cement ratio, designated slump, and allowable drum revolutions, and must be re-mixed at least half the minimum required mixing time or number of revolutions.

TRANSPORTING, HANDLING, AND PLACING CONCRETE

Before the fresh concrete arrives on the job, preparations for moving it to its point of placement must be complete. Delays in placing the concrete can cause a loss of plasticity. In addition, the method of moving the concrete must not result in the segregation of concrete materials. *Segregation* is the tendency of the coarse aggregate to separate from the sand-cement mortar. In some cases, the heavy aggregate settles to the bottom and the sand-cement mortar rises to the top, producing unsatisfactory results.

Concrete can be moved and placed in a variety of ways. Truck agitators, or non-agitating trucks are commonly used to deliver the fresh concrete. Once on site, the material is moved to its point of placement with cranes using concrete buckets, barrows and buggies, chutes, belt conveyors, and concrete pumps (Figure 8.2). A special placement technique known as "shotcrete" is placed by pneumatic guns.

Moving Concrete

If transporting units are to be filled from a hopper, the concrete should pour straight into them. Concrete coming from a conveyor should be directed straight down by using deflector plates and a down pipe. If the concrete is moved with a chute, the chute should also be placed perpendicular to the receiving unit through a down pipe to prevent segregation of the mix. For many small jobs, such as pouring a garage floor or residential basement walls, wheelbarrows or manually pushed buggies can provide an adequate delivery of material. To speed up the flow of concrete, powered buggies can also be used.

Figure 8.2 Concrete pump trucks use heavy-duty piston pumps to force concrete through a pipe mounted on a mast.

Often a slab or foundation can be poured directly from the ready-mix concrete truck. If the truck can back up close enough, and the pour is not too high above grade, much of the pour can come directly from the truck (Figure 8.3). In some cases, wheelbarrows or buggies may be used for part of a pour, with the rest pouring directly from the truck.

Figure 8.3 Ready-mixed concrete can be discharged directly from a truck mixer to its final location.

Multistory slabs, beams, and columns are poured from buckets lifted by cranes or from concrete pumps. The buckets are filled on the ground, lifted over the point of pour, and opened to deposit the concrete as shown in Figure 8.4. Buckets are available in 2 yd.³ (1.5 m³) capacities and may be round or square. They are opened by hand and poured from the bottom.

Figure 8.4 Ready-mixed concrete being lifted by bucket and crane to the top of a high-rise building.

Concrete pumps are heavy-duty piston pumps that force concrete through a pipe ranging from 6 to 8 in. (152 to 203 mm) in diameter (see Figure 8.2). They can place concrete over long distances, ranging up to about 100 ft. (30.5 m) vertically and 800 ft. (244 m) horizontally. Pumps can also place concrete below grade, as required for foundations of multistory buildings, whose excavations must be several stories below grade. The maximum aggregate size for an 8 in. (203 mm) pipeline is 3 in. (75 mm); for a 7 in. (178 mm) pipeline it is 2 in. (64 mm); and for a 6 in. (152 mm) pipeline it is 2 in. (50 mm). The pump requires an uninterrupted

flow of plastic concrete to mitigate premature set and the production of unwanted joints in the slab, beams, or column. On the end of the discharge line is a choke that controls the flow of concrete. The pump and pipeline are thoroughly flushed after each use. Concrete is also moved with a conveyor.

Placing Concrete

Concrete placement occurs after the base for on-grade pours is ready or the forms for walls, columns, and beams are erected and reinforcing is in place (Figure 8.5). Concrete should be placed continuously and as near as possible to its final location. In slab construction, work starts along one end, and each batch is discharged against the one previously placed. If the concrete is to be thick, like in a foundation wall, it should be placed in layers 6 to 20 in. (150 to 500 mm) deep for reinforced members and 15 to 20 in. (400 to 500 mm) deep for mass work. Each layer should be consolidated before a second layer is placed on it. It is necessary to work fast so the first layer is still plastic when the next layer is poured.

Figure 8.5 Before the fresh concrete arrives on the site, all formwork, reinforcing, and bracing must be in place and inspected for compliance with the design specifications.

Consolidation is the process of compacting freshly placed concrete in forms and around reinforcing steel to remove air and aggregate pockets. Consolidation can be done by manually pushing a rod into the concrete or with a mechanical vibrator (Figure 8.6). The vibrator should be lowered vertically into the concrete. It should never be used to move concrete along in a form. Rodding or vibrating should extend through the layer being consolidated and about 6 in. (150 mm) into the layer below. It is important not to over consolidate. Too much consolidation tends to force heavy aggregate to the bottom and lighter cement paste to the top.

Figure 8.6 A mechanical vibrator assures proper placement and consolidation of concrete even in areas of dense reinforcing.

Placing Concrete in Cold Weather

Concrete placed in cold weather gains strength slowly. The critical period after which concrete is not seriously damaged by several freezing cycles depends on ingredients, conditions of mixing, placing, curing, and long-term drying. The concrete designer must consider the heat of hydration, the use of special cements and admixtures, and the temperature of the concrete, which is influenced by heating the aggregates and water. A formula is used to determine the temperature of fresh concrete. The final temperature of the combined ingredients should be well below 100°F (38°C) and most batches are kept within a 60°F to 80°F (15°C to 27°C) range. The temperatures of all batches should be about the same. If the overall temperature of the concrete exceeds 100°F (38°C), the concrete may flash set. When water is heated, most of the cement is not added until the bulk of the water and aggregates have been mixed in the drum. After placement, concrete must be protected from freezing. The length of time depends on the concrete mix and local conditions. Common methods of protection include insulated blankets and air heaters.

Placing Concrete in Hot Weather

In addition to maintaining low concrete temperature by cooling the aggregates and water, precautions must be taken to maintain a low temperature while placing concrete. This can involve shading and painting the mixers, chutes, hoppers, pump lines, and other concrete handling equipment white. Forms can be cooled with water, and the sub-grade can be moistened before the concrete is placed. Concrete must be transported from the mixing station to the point of placement as quickly as possible. In hot weather, the mix should be in place 45 to 60 minutes after mixing.

Pneumatic Placement

Concrete can be placed by pneumatically forcing a dry mixture of sand, aggregate, and cement through a hose and mixing it with water at a nozzle. This is referred to as pneumatically placed concrete, or shotcrete. It is used to form thin sections in difficult locations and to cover large areas. Shotcrete is ideal for placing concrete in free-form shapes, such as domes and shells; for applying protective coatings; and for repairing concrete surfaces. Typical applications include forming swimming pools, covering rock outcroppings along highways to prevent rock falls, and providing underground support, as in tunnel and coal mine shaft linings. A surface can be covered with metal lath and concrete sprayed over it, thereby eliminating the construction of expensive forms.

FINISHING CONCRETE

After concrete has been placed and consolidated, it is screeded (Figure 8.7). An initial *screeding*, also called strike-off, involves removing excess concrete with a screed to scrape it flush with the top of the form work. Immediately after strike-off, the surface is bull floated to lower high spots, fill low spots, and embed large aggregate that may be on the surface. A *bull float* has a long handle connected to a float (Figure 8.8). A darby has a shorter handle and is used for shorter distances. This work must be done before any bleed water appears on the surface. *Bleed* refers to water that rises to the surface very soon after concrete is placed in forms.

When the bleed water sheen has evaporated, the surface is ready for final finishing. Any finishing operation performed on the surface of the concrete while bleed water is present will cause it to scale and dust. Final finishing includes one or more of the following: edging, jointing, floating, troweling, and brooming.

Figure 8.7 Vibratory screeds units reduce the work of strike-off while consolidating the concrete.

Copyright © ConstructionPhotographs.com

Figure 8.8 A bull-float is used to finish the concrete flatwork by lowering high spots and filling in low spots.

© Richard Thornton/Shutterstock.com

Edging rounds off the edges of a slab to prevent chipping. *Jointing* forms control joints in a slab. A groove with the thickness of the slab is formed across it at intervals specified by the architect. Its purpose is to provide a weak spot where the slab can crack when stresses exceed the strength of the concrete. Control joints can be formed in wet concrete, sawed after concrete has hardened, or formed by inserting plastic or hardwood strips in the concrete. Isolation and construction are other joints used. Isolation joints provide a space between a slab and a wall, allowing each to move without disturbing the other. Construction joints are formed where one pour ends and a joining one meets it (Figure 8.9).

Figure 8.9 Control joints provide a place for the slab to crack without being visible. Construction joints occur where two pours meet. Isolation joints separate a pour from a wall, column, or other abutting form.

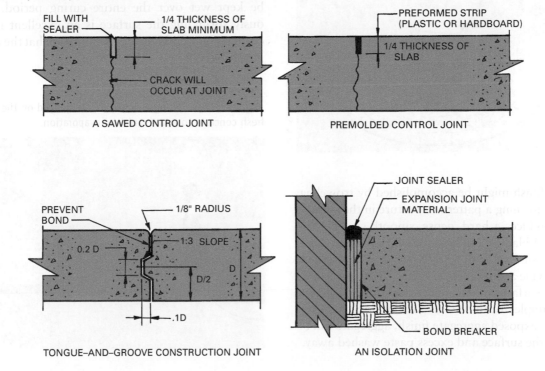

FILL WITH SEALER

1/4 THICKNESS OF SLAB MINIMUM

CRACK WILL OCCUR AT JOINT

A SAWED CONTROL JOINT

PREFORMED STRIP (PLASTIC OR HARDBOARD)

1/4 THICKNESS OF SLAB

PREMOLDED CONTROL JOINT

PREVENT BOND

1/8" RADIUS

1:3 SLOPE

0.2 D

D/2

D

.1D

TONGUE–AND–GROOVE CONSTRUCTION JOINT

JOINT SEALER

EXPANSION JOINT MATERIAL

BOND BREAKER

AN ISOLATION JOINT

After concrete is edged and jointed, it can be floated. This is done with a wood or metal handheld float or a finishing machine with float blades (Figure 8.10). Floating embeds aggregate slightly below the surface, removes imperfections, compacts mortar at the surface for final finishing, and keeps the surface open, allowing excess moisture to escape. Marks left by edging and jointing are removed by floating. If they are wanted for decorative purposes, they need to be reformed after floating.

Figure 8.10 A power float is used after the bleed water sheen has evaporated and the concrete will sustain foot pressure with only slight indentation.

The final finish might be accomplished by troweling, brooming, or forming a pattern or texture in the surface. *Troweling* produces a hard, dense, smooth surface. The trowel is a steel-bladed handheld tool. Brooming involves roughing the surface with a steel-wire or coarse-fiber broom to provide a slip-resistant surface. Patterns can be formed in the surface by placing divider strips in the concrete. For example, it can be segregated to look like flagstones. For an exposed aggregate finish, aggregate can be embedded in the surface and excess paste washed away.

CURING CONCRETE

As explained in Chapter 7, adding water to Portland cement produces a chemical reaction called *hydration*. This reaction produces a hard cement paste that bonds the aggregate into a solid mass. Hydration continues for an indefinite period at a decreasing rate as long as water is in the mix and the temperature is favorable (73°F [23°C] is recommended). The material reaches its specified design strength after 28 days of *curing*. Concrete should be protected so moisture remains in the mix during the early hardening period and the temperature is maintained. Protected concrete that is kept moist for seven days has about twice the compressive strength of unprotected concrete (i.e., concrete exposed to air with no attempt to keep moisture in the mix). The curing process is essential to producing concrete members with the expected compressive strength. The length of the curing period depends on the type of cement, the design of the mix, the required strength, the size of the member poured, and weather conditions.

Forms may be left in place and the exposed concrete surfaces kept moist. Curing compounds are sprayed on the surface to retard moisture evaporation (Figure 8.11). After the forms have been removed, the exposed concrete can be sprayed with curing compound. To hold moisture, exposed concrete can be covered with waterproof curing paper or plastic film. Wet covering materials, such as burlap or moisture-retaining fabrics, can be placed over the concrete. They should be kept wet over the entire curing period. Continuously sprinkling the surface is an excellent method of curing. The sprinkling must be done so that the concrete is always wet.

Figure 8.11 Curing compounds are sprayed on the surface of fresh concrete to retard moisture evaporation.

FORMWORK

Formwork is the term given to either temporary or permanent molds into which concrete is poured. *Falsework* is any additional temporary bracing that serves to support the formwork and permanent structure until it is able to support its own weight. A variety of manufactured metal forms (steel and aluminum) are available in a range of sizes, along with assembly and bracing systems (Figure 8.12). Except for very large projects, the responsibility for formwork design normally lies with the general contractor. The contractor may hire an engineer, if the specifications or building codes require one, or design the temporary structures based on past experience. The design and construction of formwork should maximize economy without sacrificing quality or safety. Size, shape, and alignment of slabs, beams, and other concrete structural elements depend on the accurate construction of the forms. Formwork must be sufficiently rigid under construction loads to maintain the designed shape of the concrete, stable and strong enough to maintain large members in correct alignment, and designed and constructed to withstand handling and reuse without losing their dimensional integrity. Many technical manuals and publications are available to assist in the design of formwork. Among the most commonly used are: *Formwork for Concrete*, and *Recommended Practice for Concrete Formwork*, both published by the American Concrete Institute (ACI).

The inside of formwork is coated with a release agent that prohibits the concrete from bonding to the formwork surface and facilitates the easy removal of the forms. Once the concrete has been poured into formwork and has set or cured, the formwork is removed to expose the finished concrete. The time between pouring and formwork stripping depends on the job specifications, the cure required, and whether the form is supporting any weight, usually at least 24 hours after the pour is completed.

Wooden site-constructed forms are built in place by assembling individual components for one-time use (Figure 8.13). Lumber suitable for concrete formwork is

Figure 8.12 Panelized or modular forms prefabricated of aluminum or steel are easily bolted together at the jobsite.

Figure 8.13 Site-built formwork for a concrete wall made of plywood panels with lumber frames and bracing.

available in a range of sizes, grades, and species. Doug-las Fir, southern pine, and spruce/pine/fir are among the strongest species that provide economy and are, there-fore, often used for constructing forms. Plywood is used extensively for concrete forms because it is economical, available in large panels and a variety of thicknesses, and has a predictable strength. Plywood is manufactured with numerous types of surface textures that can pro-vide various architectural finishes on the cured concrete. While many types of exterior grade plywood find use in formwork, the plywood industry produces a special product for concrete forms called Plyform. The product is available in three strength classes and can be specified with a resin-impregnated surface finish that results in an extremely smooth finish. Wafer board can be used in concrete form construction, and a number of molded plastic and waxed cardboard forms are also available.

While site-constructed forms for one-time use are still used on many projects, their associated labor costs and the need for precision and economy have increased the use of prefabricated reusable form panels and com-ponents. Panelized, or modular, forms prefabricated of plywood, aluminum, or steel are easily bolted together at the jobsite to rapidly form large areas of concrete walls, floors, or columns. They offer a number of dis-tinct advantages. Components can be assembled for al-most any size or shape with less need for on-site skilled labor. The forms can be used and reused in both large sections or as individual units. Panelized prefabricated forms are generally manufactured in modular sizes. The 2- and 4-ft (.60–1.2 M) widths are the most common, with heights ranging from 2 to 12 ft (.60–3.6 M).

Both wood and metal forms are generally assembled in inside/outside pairs. The face of the form is oiled or treated with a chemical release agent to facilitate easy stripping. One form side is set, and the reinforcing bars are placed inside before the second side is assembled (Figure 8.14).

Snap ties hold forms together at the desired width. They support the forms' sides against lateral pressure from the wet concrete. Walers are used to provide addi-tional reinforcing to form work. Once concrete has cured and the forms are removed, the snap tie ends are broken off just inside the surface of the concrete. The remaining holes are either grouted or capped (Figure 8.15).

Other forms are used to construct cast-in-place structural members. Columns are usually round, square, or rectangular.

A variety of specially made forms are offered by sup-pliers of concrete forms. These forms are custom-made based on architectural designs. An example of one-sided

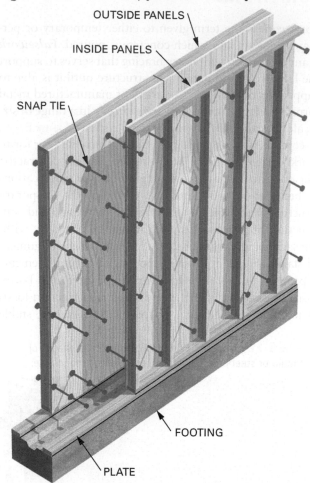

Figure 8.14 A site-built plywood form assembly.

OUTSIDE PANELS

INSIDE PANELS

SNAP TIE

FOOTING

PLATE

formwork is shown in Figure 8.16. This form uses verti-cal truss-like structural members against which the form sheets are placed. The bracing is set on footings and is positioned with a number of screw-adjusted feet. Cir-cular formwork uses horizontal walers made of curved metal units supported with round steel braces bolted to the concrete floor.

Climbing formwork is a special type of formwork for vertical concrete structures that rises with the building process. While relatively complicated and costly, it is ideal for buildings that are repetitive in form or that re-quire a seamless wall structure. An example of climb-ing formwork is shown in Figure 8.17. After a section is poured and reaches sufficient strength, the form is raised to the next level and positioned, reinforcing is set in place, and the pour is repeated.

Large projects use a variety of forms. Figure 8.18 shows forms being installed for part of a dam construc-tion, including climbing, curved concrete wall forms and forms for slanting concrete surfaces. Form construction

Figure 8.15 Snap tie caps on a completed concrete wall.

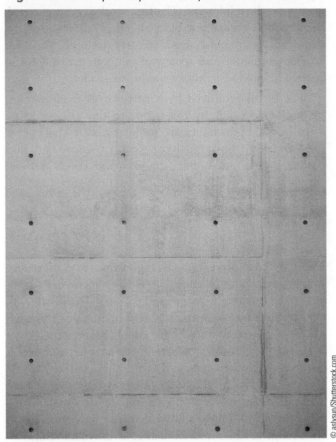

Figure 8.16 A one-sided form is supported by heavy tubular steel bracing.

Figure 8.17 Climbing formwork used to pour exterior wall panels is raised from one level to the next.

Figure 8.18 A concrete dam under construction.

Figure 8.19 These polystyrene pans with steel reinforcing are used to form cast-in-place slabs with coffered ceilings.

and installation is a major consideration in large projects such as this.

Concrete floors and roofs are constructed in a number of ways. Figure 8.19 shows pans set for pouring a waffle floor. Reinforcing bars needed in the ribs run between the pans. The reinforcing for the floor is placed over the pans. Larger areas without pans form column heads cast over each column below the floor. These are discussed in greater detail later in this chapter.

Flying formwork is built in large table-like sections that are lifted by crane from floor to floor and reused (Figure 8.20). The formwork is supported by metal posts, providing a strong and rigid unit. The reusability of the forms and bracing reduces costs by eliminating some of the labor required to strip and rebuild form-work on the next floor. The tables are supported by shoes attached to previously poured columns and walls. Climbing formwork works in a similar fashion by using pre-fabricated forms that are lifted from one level to the next for wall construction.

Figure 8.20 Flying formwork is built in large table-like sections that are lifted by crane from floor to floor.

Another type of form used for residential and light commercial foundation construction is an ***insulated con-crete form (ICF)*** with integral stay-in-place insulation (Figure 8.21). The forms are made from expanded polysty-rene in units that are notched to facilitate stacking to form the cavity for cast-in-place concrete walls. Plastic form ties at

Figure 8.21 Insulated concrete forms use rigid thermal insulation that stays in place as a permanent substrate for walls, floors, and roofs.

regular intervals hold the foam forms in position and serve to support the reinforcing. The formwork is nonstructural, and although it remains in place after the wall is poured, the reinforced concrete is designed to carry the loads. The form provides insulation and acoustical values to the wall. It can be used to build walls above and below grade.

The finish produced by forms ranges from the un-treated surface left by the forms to one produced by form liners. Form liners are molded plastic sheets that are modelled from actual concrete, masonry, or wood patterns. They are bonded to plywood sheets and se-cured inside a form, producing a textured surface.

CONCRETE REINFORCING MATERIALS

The material concrete exhibits high values of compressive strength but no useful tensile strength on its own. By com-bining the concrete with a material of high tensile strength like steel, the best properties of both materials can be ex-ploited. Concrete and steel have about the same coefficient of thermal expansion, concrete bonds to steel, and steel is not corroded by concrete, so they work together to provide an efficient structural system. The two commonly used steel reinforcing materials are reinforcing bars with associ-ated hooks and stands, and welded wire reinforcement. A wide range of fibers also serve as concrete reinforcing.

Steel Reinforcing Bars

Reinforcing bars (also called rebar) are hot rolled steel rods that may be plain (smooth) or deformed. The deformed type has surface ridges, which provide better bond-ing to the concrete (Figure 8.22). The smooth type is used for special applications. The bars are available in

Figure 8.22 Deformed steel reinforcing bars use surface ridges to establish better bonding to concrete.

60 ft. (18.3 m) lengths and in eleven standard diameters. The metric bars are made in eight diameters (Table 8.1). The diameters are identified by the bar size designation. Inch-size bar designations represent inches of bar diameter. For example, a No. 4 bar is ⁴⁄₈ in. or ½ in. in diameter. Metric designations give the diameter in millimeters.

Table 8.1 Steel Reinforcing Bar Sizes

ASTM Inch-Size Steel Reinforcing Bars			
Bar Size Designation	Weight in Pounds Per Foot	Nominal Dimensions Diameter in Inches	Cross-Sectional Area in Square Inches
#3	0.376	0.375	0.11
#4	0.668	0.500	0.20
#5	1.043	0.625	0.31
#6	1.502	0.750	0.44
#7	2.044	0.875	0.60
#8	2.670	1.000	0.79
#9	3.400	1.128	1.00
#10	4.303	1.270	1.27
#11	5.313	1.410	1.56
#14	7.650	1.693	2.25
#18	13.60	2.257	4.00

ASTM Metric-Size Steel Reinforcing Bars			
Bar Size Designation	Mass (kg/m)	Nominal Dimensions Diameter (mm)	Area (mm²)
#10M	0.785	11.3	100
#15M	1.570	16.0	200
#20M	2.355	19.5	300
#25M	3.925	25.2	500
#30M	5.495	29.9	700
#35M	7.850	35.7	1,000
#45M	11.775	43.7	1,500
#55M	19.625	56.4	2,500

Source: American Society for Testing and Materials.

Reinforcing bars are manufactured to ASTM standards A615, A616, A617, and A706. They are made in grades 40, 50, 60, and 75. These refer to the minimum yield strength of the steel, 40,000, 50,000, 60,000, and 75,000 psi (276, 345, 414, 517 MPa) (Table 8.2).

Reinforcing bars are made from three types of steel: rail steel, axle steel, and billet steel. The bars are available

galvanized or coated with epoxy to prevent corrosion. Examples of the markings stamped into the rebar are shown in Figure 8.23. The type of steel, bar size, grade mark, and identification of the production mill are given. Higher strength bars are used where space is tight and smaller diameter higher-strength bars provide the strength needed. Concrete columns and beams are members for which high-strength rebar is often used.

The amount and placement of reinforcing steel for concrete structural members is specified on the structural drawings. The engineering drawings give the size, location, and bending information (Figure 8.24). The steel fabricator cuts the bars to length and makes the required bends (Figure 8.25). The bends are specified by the engineer on the engineering drawings. Information pertaining to the bend designs can be obtained from the Concrete Reinforcing Steel Institute.

Bend types are standardized and identified by number. A few of these are shown in Figure 8.26. Some of the standard hook bends are shown in Figure 8.27. Reinforcing bars must be bent cold unless specific approval is given by the design engineer. The formed bars are wired together into bundles and tagged. The tag has the name of the fabricator, the address of the job, fabrication data, and the mark that locates their place on the structural drawing.

Some reinforcing is preassembled before being placed in the forms. Examples include column spirals, column ties, and footing bars (Figure 8.28). The steel in beams usually involves a set of bottom bars and stirrups. The bottom bars resist tension forces that exist in the bottom of the beam. The stress is dissipated by the bars and into the concrete through the bond between the concrete and the bars. The concrete in the top of the beam is in compression, which the concrete is able to resist. The ends of the bottom bars are bent into hooks that help dissipate tension forces at the support points of the beam. Tension forces remaining in the bearing ends of the beam are resisted by stirrups. Most applications use U-stirrups, but in some cases closed stirrup ties are required (Figure 8.29).

The bottom bars in the beam are raised above the bottom of the form with one of several types of bar supports. These can be bolsters or chairs. Bars in the top of the slab are supported with high chairs. Bar supports are available in wire, pre-cast concrete, reinforced cementitious fiber, and all-plastic types (Figure 8.30). Reinforcement that rests directly on the ground can be supported by bar supports made from concrete, which become completely sealed from the earth by the concrete. Steel wire chairs tend to provide a passage for moisture and rust to reach the bottom reinforcing bars.

Table 8.2 Grades, Strengths, and Types of Reinforcing Steel

Type of Steel	Steel Type Symbol	Yield Strength (psi)	Tensile Strength (psi)	Yield Strength (MN/m²)	Tensile Strength (MN/m²)
ASTM A615					
Billet steel					
Grade 40	S	40,000	70,000	276	483
Grade 60		60,000	90,000	414	621
Grade 75		75,000	100,000	517	690
ASTM A616					
Rail steel					
Grade 50	R	50,000	80,000	345	552
Grade 60		60,000	90,000	414	621
ASTM A617					
Axle steel					
Grade 40	A	40,000	70,000	276	483
Grade 60		60,000	90,000	414	621

Source: Concrete Reinforcing Steel Institute

Figure 8.23 Markings stamped into reinforcing bars give the bar size, type of steel, and a grade mark identifying the mill that produced the bar.

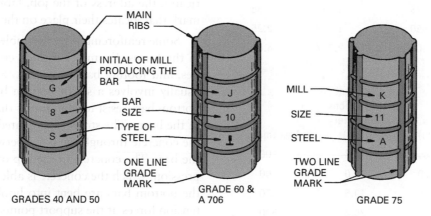

MAIN RIBS

INITIAL OF MILL PRODUCING THE BAR

BAR SIZE

TYPE OF STEEL

ONE LINE GRADE MARK

GRADES 40 AND 50

GRADE 60 & A 706

MILL

SIZE

STEEL

TWO LINE GRADE MARK

GRADE 75

LINE SYSTEM TO INDICATE GRADE MARKS

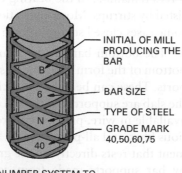

INITIAL OF MILL PRODUCING THE BAR

BAR SIZE

TYPE OF STEEL

GRADE MARK 40,50,60,75

NUMBER SYSTEM TO INDICATE GRADE MARKS

FIELD IDENTIFICATION SYMBOLS FOR STEEL TYPE

S – BILLET STEEL (ASTM A 615)
⊥ – RAIL STEEL (ASTM A 616)
A – AXLE STEEL (ASTM A 617)
W – LOW ALLOY STEEL (ASTM A 706)

Figure 8.24 Examples of engineering plan and section details for the cast-in-place structural members of a building.

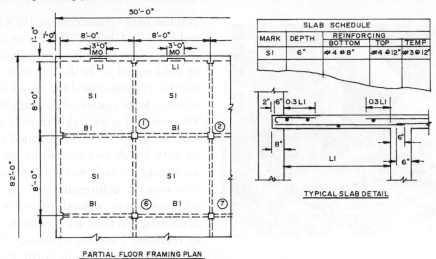

SLAB SCHEDULE				
MARK	DEPTH	REINFORCING		
		BOTTOM	TOP	TEMP
S1	6"	#4 @ 8"	#4 @ 12"	#3 @ 12"

PARTIAL FLOOR FRAMING PLAN

TYPICAL SLAB DETAIL

BEAM AND GIRDER SCHEDULE							
MARK	SIZE		BOTTOM		TOP	STIRRUPS	
	W	D	"A" BARS	"B" BARS		NO.– SIZE	SPACING FROM FACE OF SUPPORT
G1	12"	30"	3 # 8	3 # 8	2 # 9 NON-CONTINUOUS ENDS 3 # 8 AT COLUMNS # 8 IN TOP LAYER	20 # 3	1 @ 2", 3 @ 6", 2 @ 9", 2 @ 10", 2 @ 12"
B1	10"	20"	3 # 8	3 # 6	3 # 10 AT COLUMNS	10 # 3	1 @ 2", 2 @ 12", 2 @ 18"
L1	8"	12"	2 # 6	—		NONE	

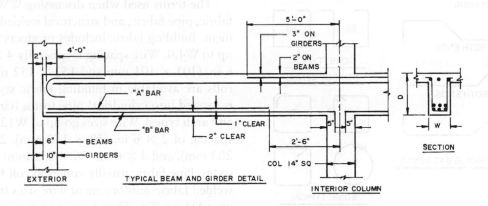

TYPICAL BEAM AND GIRDER DETAIL

SECTION

INTERIOR COLUMN

BENDING DETAILS							
MARK	SIZE	LENGTH	TYPE	A	B	C	D
B1	8	8'-6"	2	8	8'-0"		
S1	4	4'-6"	1	6	4'-0"		

TYPE 1

TYPE 2

TYPE 3

Figure 8.25 Bars are cut to length and bent by the steel fabricator.

Figure 8.26 Some of the standard bends used with steel reinforcing bars.

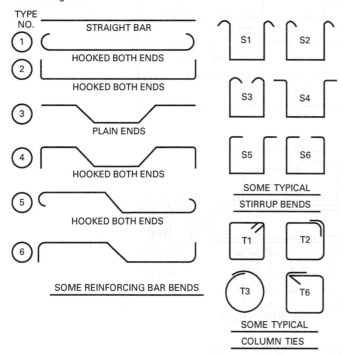

If a member is formed from several pours, the reinforcing bars are extended beyond the end of the first pour into the next one. The bars in the second pour overlap those from the first at a distance specified by codes. Figure 8.31 shows concrete columns with the steel extended. Connections between reinforcing bars in columns require that they be spliced end-to-end and secured with a mechanical splicing device or by welding. Connections should be made as specified by the structural engineer.

Welded Wire Reinforcement

Welded wire reinforcement (WWR), sometimes called welded wire fabric (WWF), is an assembly of steel reinforcing wires made from rod that is either cold drawn or cold-rolled or both. The transverse wires are electrically resistance-welded to each of the longitudinal wires to form square or rectangular grids (Figure 8.32). The material is available in rolls and sheets. The wires may be plain (W) or deformed (D). The plain welded wire bonds to concrete by the positive mechanical anchorage at each welded intersection. The deformed wire uses the deformations in the surface of the wire in addition to the welded intersections for bonding and anchoring. Welded wire reinforcement is widely used to reinforce concrete structures. The smaller diameter wires provide more uniform stress distribution and crack control. The selection of wire sizes by the engineer provides the needed cross-sectional area of the reinforcing steel.

In addition to uncoated wire, two coatings are available. One is a hot-dipped galvanized coating specified by ASTM A641 or A153. It is usually applied to the wire before welding. The other is an epoxy coating specified by ASTM A884. This is applied after the sheets have been welded.

The terms used when discussing WWR are building fabric, pipe fabric, and structural welded wire reinforcement. Building fabric includes products with wire sized up to W4.0. Wire spacing is generally 4 × 4 in. and 6 × 6 in. (101 × 101 mm and 152 × 152 mm). Sheets and rolls are available in building fabric styles. Pipe fabric is formed into cylindrical pipe forms (circular, elliptical, and arch types). Wire sizes go up to W12, generally with spacing of 2 × 6 in. (50 × 152 mm), 2 × 8 in. (50 × 203 mm), and 3 × 6 in. (76 × 152 mm) in the standard styles. Pipe fabric usually comes in roll form. Structural welded fabric-reinforcement wire sizes include anything over D4 or W4. They have a variety of wire spacing, from 3 in. to 18 in. (76 to 457 mm) in both directions. Generally, structural welded wire is furnished in sheet or mat form.

Welded wire reinforcement is specified by listing the longitudinal wire spacing, transverse wire spacing, longitudinal wire size, and the transverse wire size. An example for WWR style is 12 × 12 – W12 × W5. This means that the longitudinal and transverse wire spacing is 12 in. apart. The longitudinal wire type and area: W12 denotes a plain wire with an area of 12 in.²/ft. The transverse wire type and area: W5 means plain wire, with an area of .05 in.²/ft.

Figure 8.27 Standard hook forms used with steel reinforcing bars.

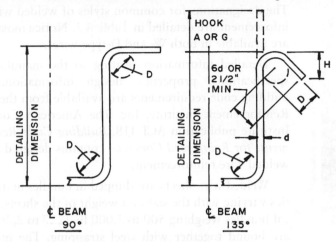

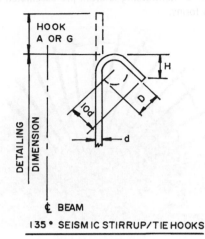

135° SEISMIC STIRRUP/TIE HOOKS

STIRRUP AND TIE HOOKS

Bar Size	D, In.	Stirrup and tie hook dimensions, in.*			
		90-deg hook		135-deg hook	
		A or g		A or G	H, approx
#3	1¹/₂	4		4	2¹/₂
#4	2	4¹/₂		4¹/₂	3
#5	2¹/₂	6		5¹/₂	3³/₄
#6	4¹/₂	1-0		7³/₄	4¹/₂
#7	5¹/₄	1-2		9	5¹/₄
#8	6	1-4		10¹/₄	6

Bar Size	D, In.	135 deg seismic stirrup/tie hook dimensions, in.*	
		135-deg hook	
		A or G	H, approx
#3	1¹/₂	5	3
#4	2	6¹/₂	4¹/₂
#5	2¹/₂	8	5¹/₂
#6	4¹/₂	10³/₄	6¹/₂
#7	5¹/₄	1-0¹/₂	7³/₄
#8	6	1-2¹/₄	9

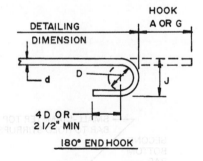

180° END HOOK

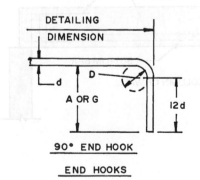

90° END HOOK

END HOOKS

Bar size	RECOMMENDED END HOOKS ALL GRADES			
	Finished bend diameter D, in.	180-deg hooks		90-deg hooks
		A or G, in.	J, in.	A or G, in.
#3	2¹/₄	5	3	6
#4	3	6	4	8
#5	3³/₄	7	5	10
#6	4¹/₂	8	6	1-0
#7	5¹/₄	10	7	1-2
#8	6	11	8	1-4
#9	9¹/₂	1-3	11³/₄	1-7
#10	10³/₄	1-5	1-1¹/₄	1-10
#11	12	1-7	1-2³/₄	2-0
#14	18¹/₄	2-3	1-9³/₄	2-7
#18	24	3-0	2-4¹/₂	3-5

Figure 8.28 Reinforcing is often pre-assembled before being placed into forms.

Longitudinal wire spacings are available in 2, 3, 4, 6, 8, 10, 12, and 16 in., and in some cases, spacing to 24 in. is available. Soft metric conversions are 51, 76, 102, 152, 203, 254, 305, and 406 mm. Transverse wire spacings are 3, 4, 6, 8, 12, and 16 in., with soft metric

conversions of 76, 102, 152, 203, 305, and 406 mm. The designations for common styles of welded wire reinforcement are detailed in Table 8.3. Notice most sizes are available in both W- and D-type wires.

Detailed information relating to the manufacture, specifications, properties, design information, and building code requirements are available from the Wire Reinforcement Institute, Inc. The American Concrete Institute publication ACI 318, *Building Code Requirements for Reinforced Concrete*, contains design data on welded wire reinforcement.

Welded wire sheets are shipped in bundles in quantities varying with the size and weight of the sheets. Typical bundles weighing 500 to 5,000 lb. (227 to 2,268 kg) are bound together with steel strapping. The bundles should never be lifted off a truck by the steel strapping. Lifting eyes can be specified when bundles are lifted by crane.

The engineer specifies the amount of reinforcement required and the correct placement of it within a

Figure 8.29 Typical reinforcing for a concrete beam.

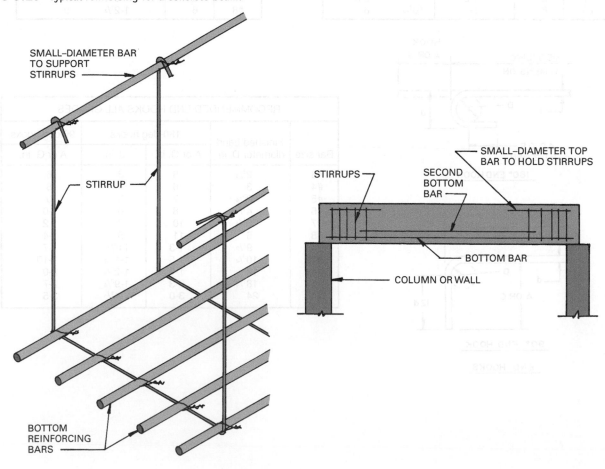

Figure 8.30 Types and sizes of reinforcing bar supports.

TYPICAL TYPES AND SIZES OF WIRE BAR SUPPORTS

SYMBOL	BAR SUPPORT ILLUSTRATION	BAR SUPPORT ILLUSTRATION PLASTIC CAPPED OR DIPPED	TYPE OF SUPPORT	TYPICAL SIZES
SB	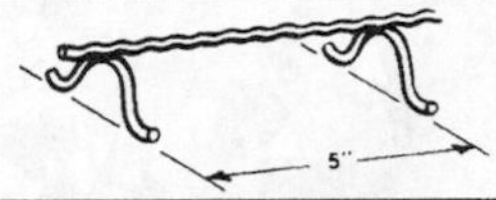	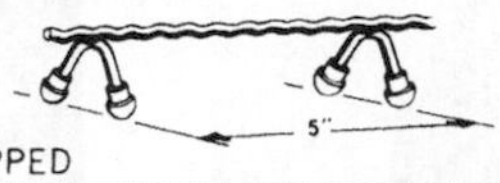 CAPPED	Slab Bolster	¾, 1, 1½, and 2 inch heights in 5 ft. and 10 ft. lengths
SBU*	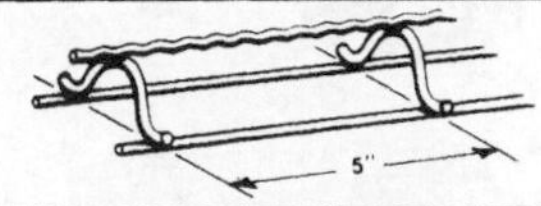		Slab Bolster Upper	Same as SB
BB	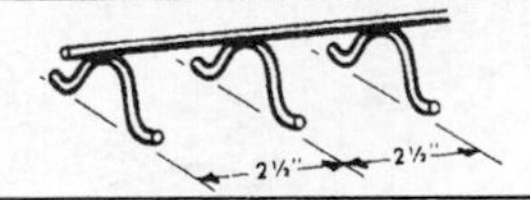	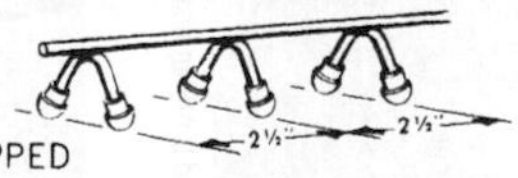 CAPPED	Beam Bolster	1, 1½, 2, over 2" to 5" heights in increments of ¼" in lengths of 5 ft.
BBU*	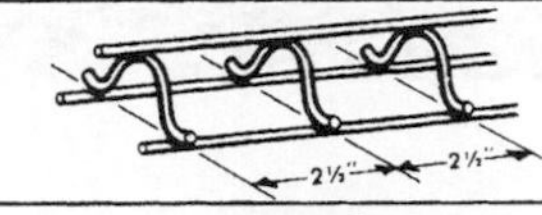		Beam Bolster Upper	Same as BB
BC		 DIPPED	Individual Bar Chair	¾, 1, 1½, and 1¾" heights
JC	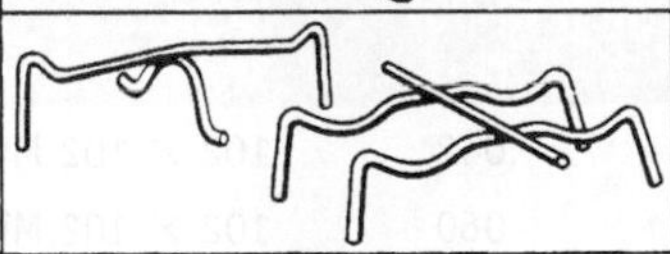	 DIPPED DIPPED	Joist Chair	4, 5, and 6 inch widths and ¾, 1 and 1½ inch heights
HC	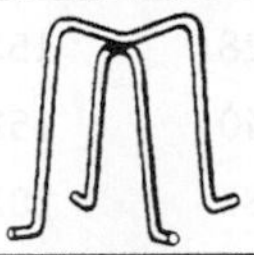	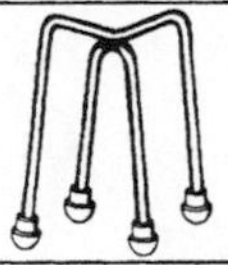 CAPPED	Individual High Chair	2 to 15 inch heights in increments of ¼ inch
HCM*	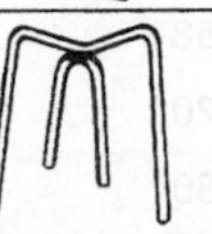		High Chair for Metal Deck	2 to 15 inch heights in increments of ¼ in.
CHC	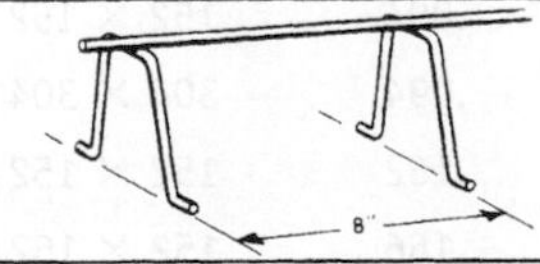	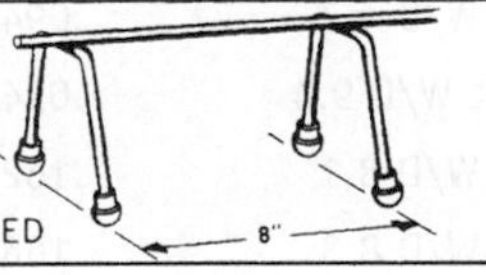 CAPPED	Continuous High Chair	Same as HC in 5 foot and 10 foot lengths
CHCU*	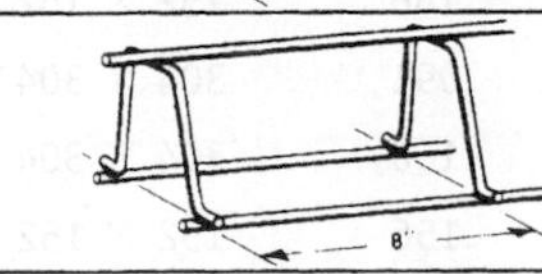		Continuous High Chair Upper	Same as CHC
CHCM*	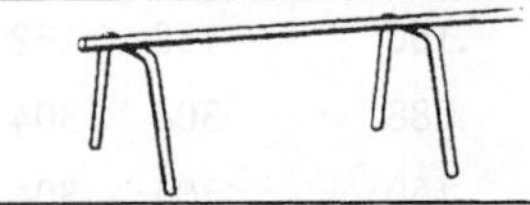		Continuous High Chair for Metal Deck	Up to 5 inch heights in increments of ¼ in.
JCU**	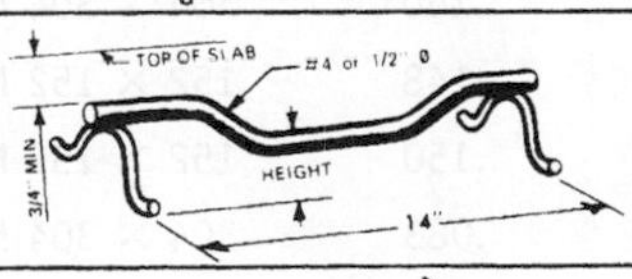	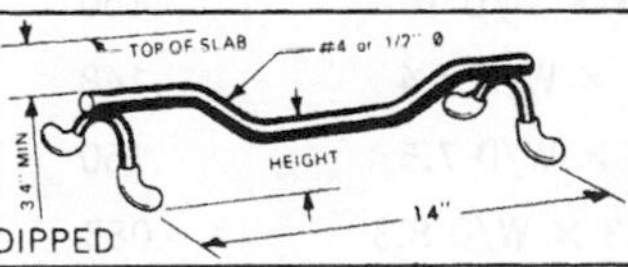 DIPPED	Joist Chair Upper	14" Span Heights −1" thru +3½" vary in ¼" increments
CS	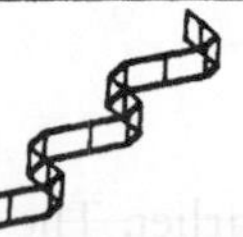		Continuous Support	1½" to 12" in increments of ¼" in lengths of 6'-8"

*Usually available in Class 3 only, except on special order.
**Usually available in Class 3 only, with upturned or end bearing legs.

Figure 8.31 Reinforcement in columns extends upward for attachment to subsequent pours.

© Eldred Lim/Shutterstock.com

Figure 8.32 Welded wire reinforcement uses a grid assembly of steel reinforcing wires to reinforce concrete elements.

© Lakeview Images/Shutterstock.com

Table 8.3 Common Styles of Welded Wire Fabric

Yield Strength (min.) (fy in psi)	Style Designation (W = Plain, D = Deformed)	Steel Area (in.²/19-00) Longit.	Trans.	Metric[a] Style Designation
65,000 (W only)	4 × 4-W1.4 × W1.4	.042	.042	102 × 102 MW 9.1 × MW 9.1
	4 × 4-W2.0 × W2.0	.060	.060	102 × 102 MW 13. × 3 MW 13.3
	6 × 6-W1.4 × W1.4	.028	.028	152 × 152 MW 9.1 × MW 9.1
	6 × 6-W2.0 × W2.0	.040	.040	152 × 152 MW 13.3 × MW 13.3
	4 × 4-W2.9 × W2.9	.087	.087	102 × 102 MW 18.7 × MW 18.7
	6 × 6-W2.9 × W2.9	.058	.058	152 × 152 MW 18.7 × MW 18.7
70,000 (W & D)	4 × 4-W/D 4 × W/D 4	.120	.120	102 × 102 MW 25.8 × MW 25.8
	6 × 6-W/D 4 × W/D 4	.080	.080	152 × 152 MW 25.8 × MW 25.8
	6 × 6-W/D 4.7 × W/D 4.7	.094	.094	152 × 152 MW 30.3 × MW 30.3
	12 × 12-W/D 9.4 × W/D 9.4	.094	.094	304 × 304 MW 60.6 × MW 60.6
72,500 (W & D)	6 × 6-W/D 8.1 × W/D 8.1	.162	.162	152 × 152 MW 52.3 × MW 52.3
	6 × 6-W/D 8.3 × W/D 8.3	.166	.166	152 × 152 MW 53.5 × MW 53.5
	12 × 12-W/D 9.1 × W/D 9.1	.091	.091	304 × 304 MW 58.7 × MW 58.7
	12 × 12-W/D 16.6 × W/D 16.6	.166	.166	304 × 304 MW 107.1 × MW 107.1
75,000 (W & D)	6 × 6-W/D 7.8 × W/D 7.8	.156	.156	152 × 152 MW 100.6 × MW 100.6
	6 × 6-W/D 8 × W/D 8	.160	.160	152 × 152 MW 51.6 × MW 51.6
	12 × 12-W/D 8.8 × W/D 8.8	.088	.088	304 × 304 MW 56.8 × MW 56.8
	12 × 12-W/D 16 × W/D 16	.160	.160	304 × 304 MW 103.2 × MW 103.2
80,000 (W & D)	6 × 6-W/D 7.4 × W/D 7.4	.148	.148	152 × 152 MW 95.5 × MW 95.5
	6 × 6-W/D 7.5 × W/D 7.5	.150	.150	152 × 152 MW 48.4 × MW 48.4
	12 × 12-W/D 8.3 × W/D 8.3	.083	.083	304 × 304 MW 53.5 × MW 53.5

[a]These are soft conversions from the inch sizes.
Source: Wire Reinforcement Institutes

concrete component. The sheets must be placed on supports, or, in the case of walls, firm support spacers are used to maintain their position as the concrete is placed. The supports are usually concrete, steel, or plastic chairs, as discussed earlier. The amount of splice for sheets is also specified by the engineer. Slab-on-grade splices can generally be less than a structural splice, since the steel reinforcement is used primarily for crack control.

Fiber Reinforcement

In addition to steel reinforcing bars and welded wire fabric, a number of fibers are used to reinforce concrete. They are added to concrete as it is prepared in the mixer. A number of manufacturers produce these products, and research will probably increase their effectiveness and use in the future. Under some conditions, they may reduce the amount of reinforcing bars required and replace welded wire fabric in some installations. Fibers also are used, along with rebar and welded wire, to produce more desirable properties in concrete.

The types of fibers available include glass fibers, polymeric (polypropylene, polyethylene, polyester, acrylic, and aramid), steel, asbestos, carbon, and natural fibers (wood, sisal, coconut, bamboo, jute, okwara, and elephant grass). Each of these fibers has different characteristics, and consultation with a concrete specialist should occur before using them. The design of the mix, strength, fatigue resistance, durability, shrinkage control, and other factors must be considered.

In addition to employment in batch-mixed concrete, fiber reinforcement is also used in pneumatically placed concrete (shotcrete). The fiber is added with the dry concrete mixture and fed through a hose to a nozzle where water is injected. The mix is then sprayed onto the desired surface. This is referred to as SFRC (sprayed fiber reinforced concrete).

CAST-IN-PLACE CONCRETE ELEMENTS

Casting On-Grade Slabs

On-grade concrete slabs require preparation of the slab base. This varies with the type of soil, but often a layer of compacted gravel is required over the soil. Some soils only require compacting before pouring the slab. To control moisture penetration, a plastic sheet is laid over the base. Sometimes 2 or 3 in. (50 to 75 mm) of compacted sand is placed over the plastic sheet. Rigid insulation should also be placed below the slab for conditioned spaces. Finally, the reinforcing is placed over this base and held the required distance above the surface with bar supports (Figure 8.33).

Lightly reinforced slabs are constructed with welded wire fabric, fiber reinforcement, or both, and depend on the earth for uniform support. This helps hold together surface cracks that occur during curing. Structurally reinforced slabs contain steel reinforcing bars and often welded wire fabric and/or fiber reinforcing. The design of

Figure 8.33 A concrete slab with insulation and reinforcing in place.

© swelsh1/Shutterstock.com

the reinforcing and thickness of the slab varies with the loads to be carried. Some designs depend on the base for support, while the reinforcing helps control tensile stress. Other designs have sufficient reinforcing so that the slab can extend from one support, such as a foundation, to another without depending on the earth for support.

Joints are necessary when building concrete slabs on-grade to help control cracking, reduce the size of the pour, and separate the slab from surfaces to which it should not bond. Control joints are used to provide a weakened place in the slab where it can crack without causing problems, preventing cracks from running across slabs at all angles. Control joints are spaced 15 to 20 ft. (4.6 to 6.1 m) apart. They are formed by sawing into the slab after it has begun to harden or by placing molded strips in the concrete before it hardens.

Construction joints are used to separate a large area to be poured into smaller, more manageable areas. They also help prevent cracking because each joint serves as a control joint, allowing for expansion and contraction within a large slab.

Isolation joints are used to keep a slab from bonding to some abutting part of the building. This permits each part to move independently. Typical isolation joints are ⅛ to ¼ in. thick (3 to 6 mm) asphalt-impregnated fiber or molded plastic strips (Figure 8.34).

Cast-in-Place Reinforced Concrete Walls

Reinforced concrete cast-in-place walls may rest on a continuous concrete footing below grade, such as a basement wall, and extend above grade to form the exterior or interior walls of a building. When the footing is

Figure 8.34 Commonly used construction joints.

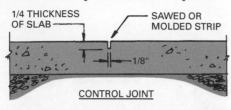

CONTROL JOINT

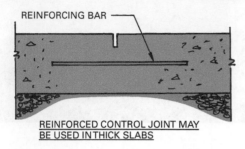

REINFORCED CONTROL JOINT MAY
BE USED IN THICK SLABS

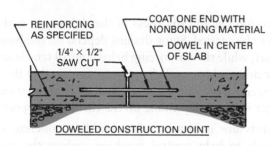

DOWELED CONSTRUCTION JOINT

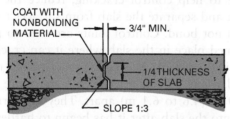

TONGUE–AND–GROVE CONSTRUCTION JOINT

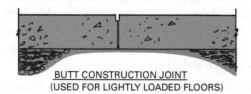

BUTT CONSTRUCTION JOINT
(USED FOR LIGHTLY LOADED FLOORS)

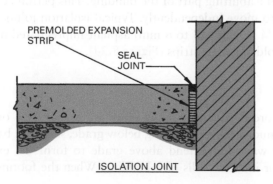

ISOLATION JOINT

poured, vertical reinforcing dowels are inserted that project above the top of the footing. Some footings have a key cast into them that forms a tie at the bottom of the wall (Figure 8.35). One side of the wall form is set on the footing and braced. The vertical bars are wired to the foundation dowels. The top of each vertical bar is wired to a horizontal bar, which maintains the required spacing. Additional horizontal bars are located and wire tied to the vertical bars, to form a grid known as a reinforcing wall mat. The mat is secured to the top of the form so that it remains in a vertical position during pouring. Sometimes the mat is built on the ground and then lifted and placed in the form with a crane. If the wall is to be topped with a concrete slab floor or roof, the working drawings will show the locations of required reinforcing for the connection. Typically, vertical reinforcing extends beyond the top of a wall and is bent to be cast into the slab. Corners between walls are also tied together using some form of a hook or elbow bar (Figure 8.36). After all steel is in place, it should be rechecked before the release agent is applied and the enclosing side of the form is set and braced.

Figure 8.35 Vertical reinforcing bars are tied to dowels cast in the footing.

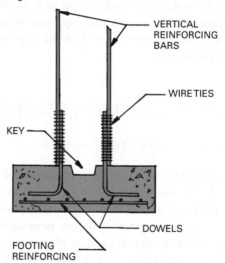

Wall ties, or snap ties, are used to hold the sides of the formwork in the correct position under the fluid pressure of the wet concrete. Some ties are designed to spread the forms and hold them at the specified width before the concrete is poured. They are loaded in tension to support the forms' sides against lateral pressure from the wet concrete. Once the concrete has cured and the forms are removed, the snap tie ends are "snapped" off just inside the surface of the concrete. The remaining holes are either grouted or capped.

Figure 8.36 Typical corner reinforcing for cast-in-place walls.

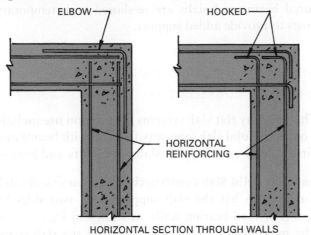

HORIZONTAL SECTION THROUGH WALLS

Reinforcing Cast-in-Place Beams

A simple single-span cast-in-place beam rests on end supports. The typical arrangement for placing the reinforcing was shown in Figure 8.29. The stirrups are held in place with a small-diameter top bar or the top member of a truss bar.

Continuous-beam casting is common in structures with a cast-in-place concrete structural system. The beam ties together the structure from column to column. The bottom of the beam at mid-span is in maximum tension, and this force dissipates toward the ends of the beam. The stirrups at the end transfer these tension forces to the concrete. Over the column, the top of the beam is subject to tension forces due to bending, so appropriate top bars are placed over each column (Figure 8.37).

Reinforcing Cast-in-Place Columns

Cast-in-place concrete columns are typically round, square, or rectangular. They are reinforced with vertical bars, which help carry compressive loads and resist tension forces from lateral loads, such as those caused by a high wind. The vertical bars may be arranged in square, rectangular, or circular patterns. Either pattern can be used for square and circular columns.

Tied columns have vertical bars tied together with small-diameter smooth steel bars placed horizontally and wired to the vertical bars (Figure 8.28). These ties hold the vertical bars in position during the pour and help them resist outward buckling when under load. Spiral columns act in the same manner as tied columns except they also restrain the concrete inside them. Rather than cracking, this enables the column to bend or bow under load.

CAST-IN-PLACE CONCRETE FRAMING SYSTEMS

A typical cast-in-place concrete framing system utilizes cast-in-place columns; one-way or two-way concrete slabs; and cast-in-place joists, beams, and girders. The exterior can be finished in a number of ways, including pre-cast concrete panels and curtain wall systems. The flat slab concrete floor and roof slabs are among the simplest to form, reinforce, and pour. A generic example is shown in Figure 8.38.

Several types of reinforced concrete framing systems are used with cast-in-place concrete construction. In some cases, the beams and girders are cast with the floor or roof slab. The two commonly used reinforced concrete floor and roof systems are the flat slab and the flat plate. The flat slab is a concrete slab reinforced in two or more directions and supported by columns with dropped panels and capitals, which enlarge the columns at the top, or by beams or joists. When two-way reinforcement is used in the slab, it is called a two-way flat slab. A flat

Figure 8.37 Typical reinforcing for cast-in-place beams.

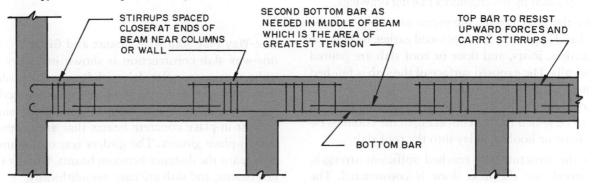

STIRRUPS SPACED CLOSER AT ENDS OF BEAM NEAR COLUMNS OR WALL

SECOND BOTTOM BAR AS NEEDED IN MIDDLE OF BEAM WHICH IS THE AREA OF GREATEST TENSION

TOP BAR TO RESIST UPWARD FORCES AND CARRY STIRRUPS

BOTTOM BAR

Figure 8.38 A cast-in-place concrete framing system utilizing cast-in-place columns, slabs, joists, beams, and girders.

© Pavel L Photo and Video/Shutterstock.com

plate floor or roof is much like the flat slab. Its reinforcing runs in two directions, but it is supported by columns that are not enlarged where they meet the slab.

General Construction Procedures

After the foundation is in place, the load-bearing walls and columns are poured. When they have cured enough to carry the floor or roof load, form work for the slab is assembled. Generally, the beams are formed and poured with the floor, resulting in a monolithic unit. The slab forms are supported with temporary joists and beams of wood or metal, which are supported by adjustable length temporary shores.

Sharp corners on concrete members are hard to cast and tend to break away if struck, leaving a ragged edge. Therefore, wood or plastic inserts can be placed in the form to produce a beveled or rounded corner. The interior surfaces of the forms are coated with a form-release compound, which prevents the concrete from bonding to the form and aids in removing it with minimum damage to the concrete member and the form. Forms are expensive and must be preserved for reuse. Finally, the reinforcing steel is placed as specified by the engineer's placing drawings.

Before the pour begins, the entire assembly must be inspected and approved. As mentioned earlier, usually the beams, girders, joists, and floor or roof slab are poured monolithically. The exposed surface of the slab is finished as specified. If additional floors are to be built above the one poured, there will be reinforcing protruding through the slab to be joined to the reinforcing in the columns on the next floor or hooked to lay into the roof slab.

After the structure has reached sufficient strength, the formwork for the next floor is constructed. The formwork for the lower floor is stripped, and the freshly cured beams and slabs are re-shored with temporary posts to provide added support.

One-Way Flat Slab Floor and Roof Construction

The one-way flat slab systems in common use include a one-way solid slab, one-way flat slab with beams and girders, and one-way flat slab with joists and beams.

One-Way Solid Slab Construction One-way solid slab construction has the slab supported on two sides by beams or load-bearing walls, as shown in Figure 8.39. The main beams are on one axis, and the slab spans the distance between them. The reinforcing bars or pre-stressed tendons are placed perpendicular to the supporting walls or beams.

Figure 8.39 One-way solid slab with beam construction has the main beams on one axis.

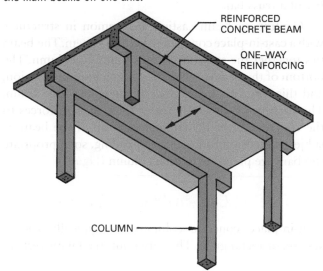

REINFORCED
CONCRETE BEAM

ONE–WAY
REINFORCING

COLUMN

ONE–WAY SOLID SLAB WITH BEAMS

One-Way Flat Slab with Beams and Girders Another one-way slab construction is shown in Figure 8.40. It utilizes a one-way flat slab with beams and girders. The construction follows the same procedures described for one-way solid slab construction, but this slab is supported by cast-in-place concrete beams that are supported by cast-in-place girders. The girders rest on columns. The slab spans the distance between beams. Usually the girders, beams, and slab are cast monolithically.

Figure 8.40 A one-way flat slab with beams and girders has a slab that spans the beams that are supported by girders.

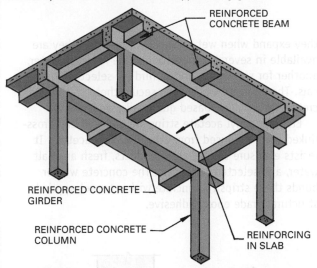

ONE–WAY FLAT SLAB WITH BEAMS AND GIRDERS

One-Way Flat Slab with Joists and Beams This assembly follows the same procedures described for the one-way solid slab but adds cast-in-place concrete beams and girders to support the slab. The girders are supported directly on columns that in turn support secondary beams that hold up the slab (Figure 8.41). As the span increases, the thickness of the slab increases, which adds to the total weight.

Figure 8.41 A one-way slab with joists and beams has joists that span the distance between beams.

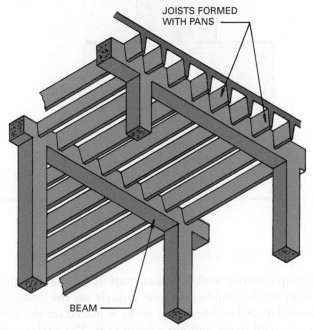

ONE–WAY FLAT SLAB WITH JOISTS AND BEAMS

The beams, joists, and slab are cast monolithically. The joists are formed with molded fiber-glass or metal domes that rest on the temporary framing and shoring. When the forms are removed, the exposed ceiling appears, as shown in **Figure 8.42**. This can be painted, left natural, sprayed with acoustical plaster, or covered with a suspended ceiling.

Two-Way Flat Slab Floor and Roof Construction

Two-way flat slabs have reinforcing running in two directions perpendicular to one another. Two-way concrete systems provide economy by utilizing a square bay system. The systems in general use are the two-way flat slab, two-way solid slab, two-way flat plate, and the two-way joist (or waffle) slab.

Two-Way Flat Slab Construction Two-way flat slab construction utilizes a flat slab supported by thickening the slab at each column with a drop panel or a drop panel and a capital (Figure 8.43). The slab reinforcing runs in two directions. The increased thickness at each column helps resist the shear forces at the top of the column. This system is useful for buildings with heavy loadings. In buildings designed for lighter loads, the thickness of the slab above the column can be reduced. A simplified partial-plan view and sections are shown in Figure 8.44.

The slab reinforcing is divided into column strips and middle strips. Each is about one-half span. The column strips carry the bend forces developed in the area over the column, while the middle strips reinforce the slab in the area between columns. The slab reinforcement is shown on the engineering drawings. The drop panel usually has no additional reinforcing.

Two-Way Solid Slab The two-way solid slab has the flat slab supported by beams that run between columns (Figure 8.45). The slab and beams are cast monolithically. This construction can be used for buildings designed to carry heavy loads. The slab is reinforced in two directions.

Two-Way Flat Plate The two-way flat plate uses the same general construction as the flat slab, but it has no drop panel or capital on the top of each column. The design is very similar to the flat slab.

Two-Way Joist (Waffle) Flat Slab The two-way joist framing system consists of a series of concrete joists cast at right angles to each other to form a grid. The

Construction Materials

Strip-Applied Waterstops

Waterstops in concrete construction are used to install construction, contraction, and expansion joints as well as for creating the perimeter of any penetrations in a concrete wall. Strip-applied waterstops are typically ½ × ¾ in. (25 × 19 mm), although some are larger. They are made from a variety of materials, which have varying characteristics and, therefore, different applications and installation procedures. Types commonly available include bituminous strips, bentonite-based strips, hydrophilic rubber strips and vinylester gaskets, and ethylene-vinyl acetate strips. The concrete worker must know the proper way to install each type, including the method of bonding each to concrete. The engineer designing the concrete structure must choose the product best suited for the conditions that will exist.

Bituminous strips are typically some blend of refined bituminous hydrocarbon resins and plasticizing compounds reinforced with inert mineral fillers. They resist fresh and salt water and acids but not oil or oil byproducts. They are bonded with an asphalt primer.

Bentonite-based strips are a blend of sodium-bentonite clay and butyl rubber, or sodium-bentonite clay with various binders and fabrics. Various grades are available that resist exposure to salt water, fresh water, and certain chemicals. They are bonded to concrete with a water-based latex adhesive. One type has an adhesive back that is simply pressed onto the concrete surface.

Hydrophilic rubber strips are made from various types of rubber that have been modified to make them hydrophilic. "Hydrophilic" refers to the product having a strong affinity for water. These strips swell when exposed to water and shrink to normal size when dry. They can resist fresh and salt water and various chemicals. Epoxy or polyurethane sealants are used to bond them to concrete.

Hydrophilic vinylester gaskets are made from a vinylester that is hydrophilic. Like hydrophilic rubber strips,

they expand when wet and shrink when dry. They are available in several compositions, one for fresh water, another for salt water, and a third for selected chemicals. The concrete worker can secure these to the concrete with a xylene-based glue, nails, or screws.

Ethylene vinyl acetate strips are a closed-cell cross-linked foam produced from ethylene vinyl acetate. It resists exposure to oil-based products, fresh and salt water, and selected chemicals. The concrete worker bonds these strips to concrete with a trowel-applied structural grade epoxy adhesive.

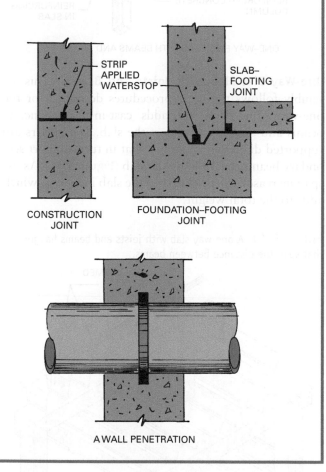

CONSTRUCTION JOINT   FOUNDATION–FOOTING JOINT

A WALL PENETRATION

slab spans the distances between the joists and both are poured monolithically (Figure 8.46). The system is designed around a series of columns, usually equally spaced. The joists are formed by domes placed on a wooden deck that is supported by shoring. Areas over the columns are left open so a solid concrete slab area called a column head can be poured.

Tilt-Up Wall Construction

Tilt-up concrete wall panels are cast in a horizontal position on the building site, or directly on the concrete floor slab. A bond-breaking agent is placed on the slab so the wall panel will not adhere to it. The surfaces of the panel can be finished in many ways: textured (from

Figure 8.42 The beams, joists, and flat slab are cast monolithically using domes to form the joists and the base for the floor.

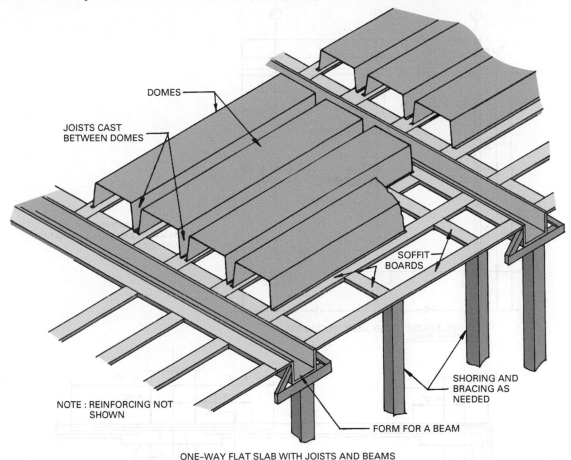

DOMES

JOISTS CAST
BETWEEN DOMES

SOFFIT
BOARDS

SHORING AND
BRACING AS
NEEDED

NOTE : REINFORCING NOT
SHOWN

FORM FOR A BEAM

ONE–WAY FLAT SLAB WITH JOISTS AND BEAMS

Figure 8.43 Two-way flat slab construction with columns using drop panels and a column capital.

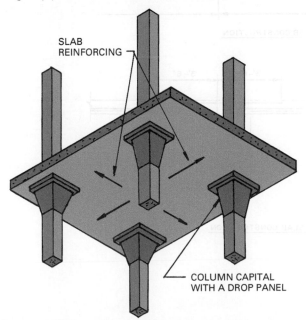

SLAB
REINFORCING

COLUMN CAPITAL
WITH A DROP PANEL

TWO–WAY FLAT SLAB WITH DROP PANELS AND COLUMN CAPITALS

a form liner), washed to expose embedded aggregate, troweled smooth, or painted.

After the wall panel has developed sufficient strength, it is lifted into position with a crane and secured to the footing (Figure 8.47). The panels are temporarily braced until all wall and roof structural members are in place. The panels are secured to columns, which may be pre-cast concrete, steel, or concrete that is cast-in-place after the walls are up.

Panels generally range in thickness from about 5 to 7 in. (140 to 190 mm) and can be formed with openings for windows and doors; they can be several stories high. A key to success is the selection and placement of steel reinforcing, which not only must serve when the wall is erect and carrying a roof or floor load but also must withstand the bending loads generated when the panel is tilted and lifted into place. Pick-up hangers are carefully located in the panel for the crane can attach to and lift the wall.

Generally, the wall panels are load bearing and will support floors and roofs. Typical roof deck materials include hollow-core units, single or double tees, and metal open-web joists.

Figure 8.44 A partial drawing showing plan and section details for a two-way flat slab with drop panels.

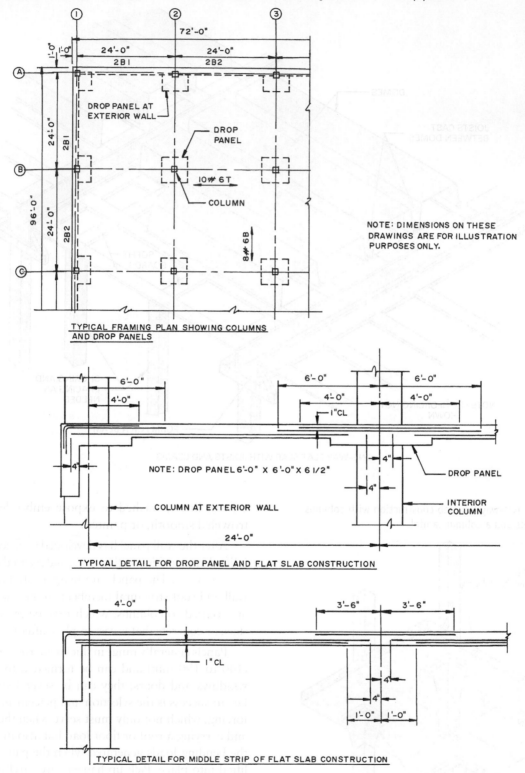

NOTE: DIMENSIONS ON THESE DRAWINGS ARE FOR ILLUSTRATION PURPOSES ONLY.

TYPICAL FRAMING PLAN SHOWING COLUMNS AND DROP PANELS

NOTE: DROP PANEL 6'-0" X 6'-0" X 6 1/2"

TYPICAL DETAIL FOR DROP PANEL AND FLAT SLAB CONSTRUCTION

TYPICAL DETAIL FOR MIDDLE STRIP OF FLAT SLAB CONSTRUCTION

Figure 8.45 A two-way solid slab and beam is supported by beams running between the columns.

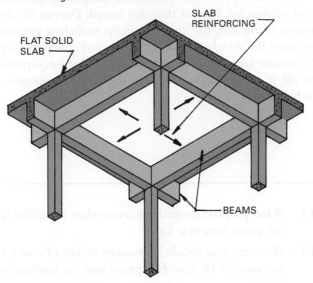

SLAB REINFORCING

FLAT SOLID SLAB

BEAMS

TWO–WAY SOLID SLAB AND BEAM

Figure 8.46 A waffle slab is a two-way joist framing that casts joists run at right angles to each other creating a coffered ceiling.

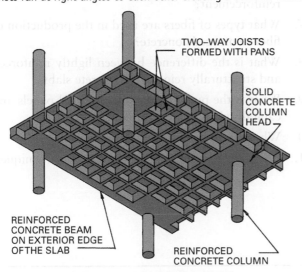

TWO–WAY JOISTS FORMED WITH PANS

SOLID CONCRETE COLUMN HEAD

REINFORCED CONCRETE BEAM ON EXTERIOR EDGE OF THE SLAB

REINFORCED CONCRETE COLUMN

TWO–WAY JOISTS WITHOUT BEAMS (WAFFLE FLAT PLATE CONSTRUCTION)

Panels have ¾ in. (18 mm) chamfers around the outside edges. Sharp corners are hard to cast and break off easily when moved. The wall panels usually are not connected to each other but have a ¾ in. (18 mm) space sealed with backer rods and a weatherproof sealant. A backer rod is a flexible, compressible rod made of foam plastic. It is placed in a joint to limit the depth to which a sealant must fill the joint. The wall panels are tied together by the roof framing and floor slab. Each

Figure 8.47 A tilt-up wall being lifted into place.

panel acts as an independent member. A typical wall-floor connection is made by leaving out a section of the floor along the wall when the on-grade slab is cast. The wall panels may be supported by continuous footings, isolated footings, or piers. They are aligned with dowels or metal angles. The void between the bottom of the panel and the top of the footing is filled with grout.

Lift-Slab Construction

Lift-slab construction involves casting a reinforced concrete slab for each floor of a building and the roof, one on top of the other. The ground floor on-grade slab is cast first. It is coated with a bond-release compound that prevents the next slab from bonding to it. Slabs are cast for each floor and the roof slab. Each is coated with a bond-release compound. The slabs, usually two-way flat slabs, can be reinforced concrete flat slabs or pre-stressed. The slabs are lifted by hydraulic jacks mounted on the tops of the columns. They slide up to the desired elevation and are secured to each column by welding, bolting, or a series of pins. This is repeated for each floor.

Slip Forming

Slip-forming is used to cast-in-place tall structures, such as grain elevators and stairwells. It involves constructing formwork, which is then lifted by jacks supported by vertical steel rods. The forms are filled with concrete and reinforcing bars as they are raised. During the duration of the lift, the concrete gains sufficient strength to carry the weight of the newly poured concrete. Keys to success are a proper concrete mix and the equipment to lift it to successive levels as the structure increases in height.

Review Questions

1. How long can ready-mix concrete remain in the mixer and still be used?
2. What devices are used on site to move concrete to point of placement?
3. How is concrete consolidated?
4. What may happen if the temperature of a concrete mix exceeds 100°F (38°C)?
5. Why are freshly poured concrete slabs bullfloated?
6. What is the difference between an isolation joint and a construction joint?
7. What is meant by "hydration"?
8. What are the curing methods in common use?
9. What is the purpose of form liners?
10. How is the tensile strength of concrete beams increased?
11. What two styles of reinforcing bars are used?
12. What is the diameter in inches of a No. 8 bar?
13. What purpose do stirrups serve when installing reinforcing bars in a form?
14. How can you decide how many inches of concrete are needed to give fire protection to reinforcing bars?
15. What styles of wire are used to make welded wire reinforcement?
16. What types of coatings are used on welded wire reinforcement?
17. What types of fibers are used in the production of fiber-reinforced concrete?
18. What is the difference between lightly reinforced and structurally reinforced concrete slabs?
19. What is the purpose of using dropped panels and capitals when casting flat concrete slabs?
20. What is meant by tilt-up construction?
21. What is unique about the slip-forming technique?

Key Terms

Batch

Bleed

Bull float

Cast-in-Place Concrete

Concrete Pump

Consolidation

Curing

Falsework

Formwork

Hydration

Insulated Concrete Form (ICF)

Jointing

Screeding

Segregation

Troweling

Activities

1. Visit a construction site where a cast-in-place concrete structural frame is being erected. Prepare a photographic report giving full details on the types of forms and reinforcing used, and how the concrete was delivered to its point of placement and poured. How often are concrete samples taken from the delivery truck and on-site mixed batches? Who does the testing? What type of structural system was being utilized?

2. Review the local building codes and list briefly the regulations relating to cast-in-place concrete structural systems.

3. Observe the pouring of floor and roof slabs and report on the use of various joints observed, the surface finish, and the protection used during the curing phase.

4. Collect small samples of concrete-reinforcing materials. Prepare a display and lead a discussion on reinforcing techniques.

5. Build forms and pour small samples of a reinforced concrete slab.

Additional Resources

ACI Building Code Requirements for Structural Concrete and Commentary, American Concrete Institute, Farmington Hills, MI.

Builders Foundation Handbook, U.S. Department of Energy, Oak Ridge, TN.

Building Code Requirements for Structural Concrete (2005), Construction Practices (2004), *Essential Requirements for Reinforced Concrete Building (2002),* ACI Manual of Concrete Practice (2005), Farmington Hills, MI.

Cast-In-Place Home Building, Portland Cement Association, Skokie, IL. Many other publications available.

Chemical Admixtures for Concrete, Cold Weather-Concreting, Guide for Consolidation of Concrete, and many other concrete-technical publications, American Concrete Institute, Farmington Hills, MI.

American Concrete Institute Student Page: www.concrete.org/Students.aspx

Other resources include: Publications of the Wire Reinforcement Institute, Findley, OH.

Technical publications from the Concrete Reinforcing Steel Institute, Schaumburg, IL.

See Appendix C for addresses of professional and trade organizations and other sources of technical information.

Pre-cast Concrete

Upon completion of this chapter, the student should be able to:

- Describe and identify the major types of pre-cast concrete units.
- Explain how pre-cast units are manufactured.
- Cite the advantages and limitations of using pre-cast concrete structural units.
- Explain the differences between pre-stressed and non-pre-stressed pre-cast concrete units.

- Discuss the differences between pre-tensioned and post-tensioned structural concrete units.
- Describe the various types of pre-cast concrete slab units and their applications.
- List standard types sizes of pre-cast concrete columns, beams, girders, and wall panels.
- Discuss the procedure for erecting pre-cast concrete units and the types of connections used.

Build Your Knowledge

For further study of these materials and methods, please refer to:

Chapter 12 Concrete Masonry

Pre-cast concrete units are cast and cured under factory-controlled conditions and then transported to the building site for assembly (Figure 9.1). Pre-cast concrete units can be classified into two major groups: pre-cast structural units and pre-cast architectural units. Typical structural units include floor and roof slabs, beams, girders, columns, and wall panels. Pre-cast architectural concrete units are used for façade wall panels, cast corners, screens, louvers, fins, and other visible façade elements. Pre-cast concrete is cast in permanent forms made of metal, wood, or fiberglass. The texture and shape of the form's surface lining produces the finished surface of the pre-cast unit.

ADVANTAGES OF PRE-CAST CONCRETE UNITS

Pre-casting provides some advantages over conventional site-cast concrete. Casting takes place in automated facilities where experienced crews produce units under controlled conditions (Figure 9.2). The control of materials, their mixing, and the placement of the concrete is automated and regularly tested per ASTM standards to produce a higher-quality concrete. A high-strength concrete mix (typically 5,000 psi) is combined with superior strength reinforcing steel. The concrete is mechanically vibrated in the forms and steam cured to speed up production (Figure 9.3). Units are finished and cured under carefully controlled environmental conditions. Once the concrete has achieved design strength, the units are removed from the casting bed, stored, and finally delivered to the job site. Very large pre-cast units can be cast by moving forms to the job and casting on site.

Figure 9.1 A structural system of pre-cast concrete members.

Figure 9.3 Pre-cast units are poured from a crane-operated hopper.

© sima/Shutterstock.com

Figure 9.2 A pre-cast concrete casting bed with reinforcing in place.

Additional advantages are realized in pre-casting concrete during the on-site construction stage. The pre-cast units are delivered in a kit of parts that is quickly erected without the need to wait for wet concrete to cure, or building and stripping formwork. Inclement weather does not slow pre-cast construction as easily as it does cast-in-place construction. The building site must have sufficient space to deliver and unload members and allow for lifting them into place. Careful hoisting and handling of pre-cast units is necessary so that surface finishes are not damaged.

BUILDING CODES

Building codes have extensive specifications pertaining to the design and installation of pre-cast concrete members. In addition to engineering design data, the codes provide guidelines for reinforcing, connections, lifting devices, fabrication, shop drawings, tendon anchorage, and grouting procedures. The codes also specify the requirements for meeting fire codes.

PRE-STRESSING PRE-CAST CONCRETE

Non-pre-stressed units are cast in molds in a plant, then cured and shipped to the job site. They are reinforced in the same manner as cast-in-place concrete (Figure 9.4). Some large or unusual beams, girders, and many columns and lintels are cast this way. The architect works directly with the casting company to establish the design, secure samples of the proposed finish for approval, and receive testing evidence that the units will meet quality standards.

Figure 9.4 A typical reinforcing cage for a pre-cast, non-pre-stressed concrete beam.

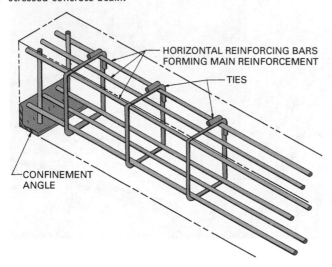

Pre-stressed concrete units have stresses introduced before they are placed under a load. A beam transfers its load laterally along its length to supporting elements. A simple beam with a load applied experiences bending. Under loading, the beam deflects or displaces in a curved downward direction. As bending occurs, the concrete fibers in the lower part of the beam become stretched in tension, while those in the upper part of the beam are in compression. As the steel in the lower part is elongated, the concrete around it begins to crack under the tension forces. This cracking, sometimes noticeable to the naked eye, demonstrates that the concrete is not carrying any loading but only serving to hold the steel in place. By tightening high-strength reinforcing bars into the concrete at high tension before it is loaded, and then releasing them against the concrete, the concrete at the bottom of the beam is placed in compression, a force it is able to counteract.

A comparison of non-pre-stressed (Figure 9.5) and pre-stressed (Figure 9.6) concrete shows that the non-pre-stressed unit has no *camber* (arch) and deflects slightly

Figure 9.5 A cast-in-place concrete beam will deflect under load and possibly crack on the bottom because concrete cannot resist the tension forces.

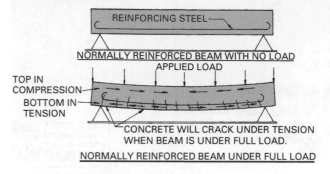

Figure 9.6 A pre-stressed concrete beam has a slight camber, which places the concrete in compression when subjected to a load.

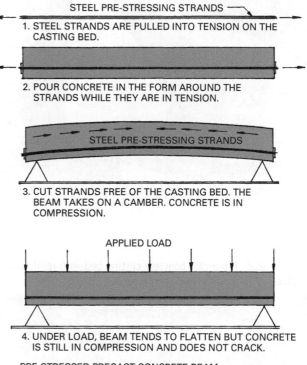

under load. The pre-tensioned member has a camber due to the tension in the strands. When under load, a pre-tensioned member will settle in a level position. The stressed strands produce high compressive stresses in the lower part of the unit and a tensile stress in the upper part. As the unit is loaded, these stresses are reduced.

Pre-stressing concrete provides a means to overcome the material's inherent weakness in tension. Both pre-stressing and post-tensioning are used to produce beams, slabs, or bridge girders with a longer span than is practical with ordinary reinforced concrete.

Pre-Tensioned Units

Pre-tensioning places the steel reinforcing in tension before the wet concrete is poured. They are cast in forms, some as long as 800 ft. (250 m) or more. High-tensile-strength steel reinforcing strands are stretched from one end of the form to the other. One end is anchored to a large fixed abutment. The other end passes through the abutment on the opposite side and has hydraulic stressing equipment fastened to it. The desired design stress is applied to the cable by pulling it. When the design tension is reached, the cable is clamped to the abutment.

After the strand is stressed, the concrete is placed in the form. As it hardens, it bonds to the strands. After twenty-four hours, samples of the concrete are taken and tested. If they have attained the required strength, the strands are cut from the abutments. The resulting forces produced by the strands pre-stress the concrete. The units are lifted from the bed and moved to storage, from where they are shipped to job sites. The forms are now ready for laying up another pre-tensioned member.

Post-Tensioned Units

Post-tensioning applies stresses to the concrete unit after it has been cast and hardened. It is applied frequently with cast-in-place concrete but is sometimes used to tension pre-cast concrete units (Figure 9.7).

Figure 9.7 Post-tensioning is used frequently in cast-in-place concrete slabs.

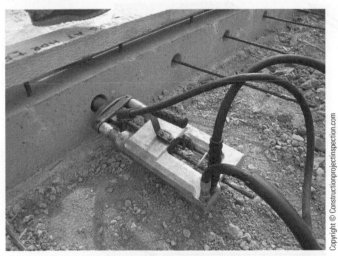

Copyright © Constructionprojectinspection.com

Post-tensioning involves casting the unit using normal reinforcement and, in addition, running several post-tensioning strands through the unit in such a way that the concrete does not bond to them as it sets. The strands may be greased or placed in thin-walled metal tubes to prevent bonding (Figure 9.8). Once the concrete has hardened, the post-tensioning strands are anchored firmly on one end and a hydraulic jack is used on the other to apply tension to the strand. Once the desired tension is reached, this end is also anchored. The steel strands may be left un-bonded. If they are in steel tubes, they can be bonded by filling the space between the strands and the tube with high-pressure grout. Figure 9.9 shows the abutment and tenons in a post-tensioned concrete casting bed.

Figure 9.8 Post-tensioned pre-cast concrete members develop camber by applying tension after the member is cast and cured.

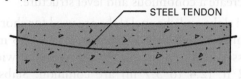

STEEL TENDON

1. CONCRETE IS POURED OVER A STEEL TENDON THAT IS GREASED OR IN A TUBE PREVENTING IT FROM BONDING TO THE CONCRETE. THE TENDON IS DRAPED.

2. THE CONCRETE IS ALLOWED TO CURE. THEN THE TENDONS ARE TENSIONED AND ANCHORED TO THE ENDS OF THE BEAM PRODUCING SOME CAMBER.

Figure 9.9 A post-tensioned bed with a hydraulic jack tensioning the cables in each tendon.

Courtesy Portland Cement Association

PRE-CAST CONCRETE ELEMENTS

Pre-Cast Concrete Slabs

A number of standardized pre-cast slabs are used for floor and roof construction (Figure 9.10). They may be supported by frames of steel, site cast, or pre-cast concrete, or bearing walls of concrete or masonry. All are one-way spanning elements and are typically pre-stressed of either normal or lightweight concrete. A topping of 1½ – 2½ in. (38–63 mm) thick concrete is applied after installation to increase structural performance and fire rating. The topping joins with the rough surface of the individual members to create a continuous and level structure.

Solid slabs and channel slabs are used for short spans and minimum slab depth. Spans of 25 ft. (7.6 m) with thicknesses of 2 to 8 in. (50 to 200 mm) and widths of 4 to 12 ft. (2.4 to 3.7 m) are common. Slabs made thicker than this to span longer distances become very heavy and are not economical.

Hollow-core slabs are used for spans ranging up to 40 ft. (12 m) in length and are 6 to 12 in. (150 to 300 mm) thick. The hollow core, created by removing concrete from an area of the slab that has little influence on strength, lightens the member. *Double Ts* and single Ts are used to span the longest distances. Single Ts come in widths of 6, 8, 10, and 12 ft. (1.8, 2.4, 3.0, and 3.7 m) and depths of 16 to 48 in. (406 to 1219 mm). Double Ts come in widths

of 4, 8, 10, and 12 ft. (1.2, 2.4, 3.0, and 3.7 m) and depths from 10 to 40 in. (254 to 1016 mm). Standard lengths range to over 100 ft. (30.5 m), although lengths over 60 ft. (18.3 m) present special transportation problems.

Tees are cast with a rough or smooth surface on top. The rough surface is used when a concrete topping is to be applied over them. This topping may be from 2 to 4 in. (50–100 mm) thick. Structural continuity across the units can be provided by placing steel reinforcing bars in the topping. Normal-weight and lightweight concrete are used to cast these units. Lightweight concrete is more expensive than normal weight, but it reduces the weight on the structure. Since double Ts do not require temporary support to prevent tipping, they are easier and more economical to erect than single Ts.

Pre-cast Concrete Columns

Pre-cast columns can be connected with pre-cast concrete beams, to form a post-and-beam structural framework. Most pre-cast concrete columns are reinforced conventionally. If they are pre-stressed, this is done mainly to reduce stresses on the column during transportation and handling. Square sizes range from 10 to 24 in. (254–619 mm). Rectangular columns vary, depending on the design. Columns up to 60 ft. (18 m) in length can be transported. Those above that size need special arrangements. Columns are often designed to incorporate

Figure 9.10 Pre-cast concrete spanning units used for floors and roof decks.

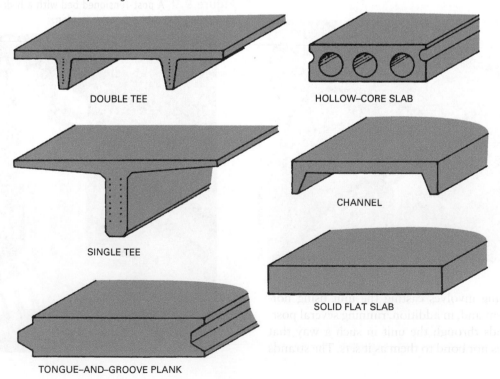

DOUBLE TEE

HOLLOW–CORE SLAB

SINGLE TEE

CHANNEL

SOLID FLAT SLAB

TONGUE–AND–GROOVE PLANK

haunches, projections used to support beams, on two or three sides (Figure 9.11). Concrete quality may range from 5,000 to 11,000 psi (48 to 76 MPa).

Figure 9.11 Columns are cast with projecting haunches on opposite sides to support beams.

Beams and Girders

Pre-cast pre-stressed concrete beams and girders are efficient load-bearing members able to achieve long spans in small profiles. They are made in several standard shapes, and custom designs can be produced (Figure 9.12). The L-shaped and inverted-T beams provide a bearing surface for pre-cast floor units, such as single- and double-T units. This results in a lower overall floor-to-ceiling height than does resting floor and roof units on top of rectangular beams.

Rectangular beams range in depth from 18 to 48 in. (457 to 1,219 mm) and are 12 to 36 in. (305 to 914 mm) wide. Inverted-T and L-beams range in depth from 18 to 60 in. (457 to 1,524 mm) and in width from 12 to 30 in. (305 to 762 mm). Beam ledges are usually 6 in. (152 mm) wide and 12 in. (305 mm) deep. Beams are available in increments of 2 or 4 in. (50 or 100 mm).

Pre-cast Concrete Wall Panels

Pre-cast concrete wall panels can be manufactured in an endless variety of designs. They may be pre-stressed or conventionally reinforced and are used for both

Figure 9.12 Typical precast pre-stressed concrete beams.

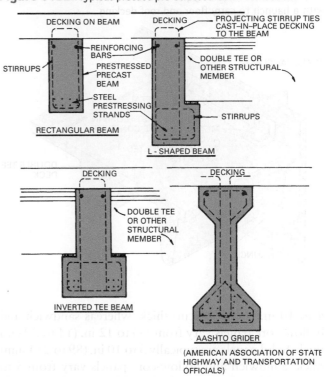

(AMERICAN ASSOCIATION OF STATE HIGHWAY AND TRANSPORTATION OFFICIALS)

load-bearing and non-load-bearing applications. They may be hollow, solid, or of an insulated sandwich construction with a flat, ribbed, or other surface configuration (Figure 9.13). Their design varies considerably and customized architectural panels are common. Form liner molds can provide a variety of flat, ribbed, or other surface shapes and textures. Pre-cast wall panels are used in connection with a pre-cast or cast-in-place concrete, or steel framing system. Panels may be single or multistory, horizontal spandrel, or vertical column covers. Multistory load-bearing panels can be cast with a haunch or other type of support to carry floor and roof slabs (Figure 9.14). Solid panels are typically from

Figure 9.13 A wide variety of pre-cast panels with openings or without can be manufactured.

Figure 9.14 Multistory pre-cast concrete panels can be cast with a haunch to carry floor and roof decking.

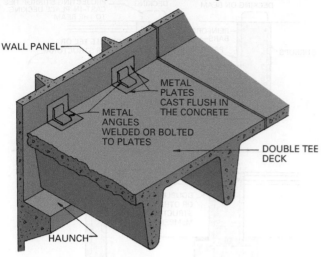

4 to 10 in. (10 to 24 cm) thick, whereas sandwich and hollow-core panels vary from 5½ to 12 in. (14 to 30 cm) thick. Solid panels are typically 3 to 10 in. (89 to 254 mm) thick. Sandwich and hollow-core panels vary from 5 to 12 in. (140 to 305 mm) thick. Ribbed panels are made in thicknesses from 12 to 24 in. (305 to 610 mm). Most panels are designed with an 8 ft. (2.4 in.) width.

The panels are installed with metal anchors cast into them and bolted to the structural frame. Panel connections require special sealants to provide a watertight surface.

Other wall systems use a double-wall insulated pre-cast concrete sandwich panel for both walls and floors. Common wall thickness is 8 in. (20 cm), consisting of two wythes of reinforced concrete 2 in. (5.08 cm) thick, sandwiched around 3 in. (7.62 cm) of high R-value insulating foam. The two concrete layers are poured in sequence and held together with steel truss reinforcing

CONNECTING PRE-CAST UNITS

The types of connections for pre-cast units include bolted, welded, post-tensioned, and doweled connections. The main connecting elements typically involve metal plates or anchor rods that are cast into the pre-cast elements.

Types of Connections

Bolted connections speed up the construction process by providing immediate joining of elements and allowing final alignments to be made after the crane has placed the member. Generally, ½, ¾, and 1 in. (12, 18,

and 25 mm) bolts with national coarse threads and steel washers are used, tightened to the recommended torque. Steel washers may be required under certain conditions.

Welded connections are strong and easy to make on site. They should be made following details on erection drawings, which include the size, type, length of the weld, type of electrode, and required temperatures. Welded connections are made before a unit is released by the hoisting device. This tends to tie up the hoisting device for longer periods than for bolted connections. The amount of heat generated by welding activities may cause the concrete to crack and metal anchors to distort. Large, long, and continuous welds should be used sparingly, and welding temperature must be carefully controlled.

Post-tensioned connections are made using either bonded or un-bonded tendons. Bonded tendons are installed in holes cast in the member and bonded to the member after tensioning by filling the hole with grout. Un-bonded tendons are not grouted in place but are provided with an organic coating to prevent corrosion. Doweled connections use reinforcing bars grouted into dowel holes in a pre-cast member (Figure 9.15). The

Figure 9.15 An example of a doweled connection made using a metal sleeve.

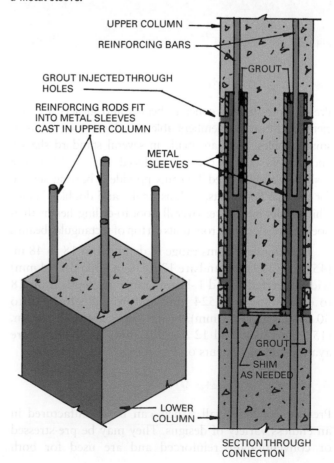

strength of the connection depends on the dowel diameter, its depth in the member, and the bond developed by the grout. A number of manufactured dowel systems use a sleeve with the dowel.

Grout, Mortar, and Drypack

Grouts and mortars are used to transfer loads between members and to fill noncritical voids. Load-bearing grouts should be non-shrinking, and only grouts manufactured for this specific purpose should be used (Figure 9.16). *Drypack* describes a method where the dry grout material has only enough water added to produce a stiff granular mix, which is packed into an opening.

Figure 9.16 Non-shrink load-bearing grout is placed below pre-cast columns and pre-cast walls.

Courtesy Precast/Prestressed Concrete Institute

Connection Details

The following detail drawings illustrate typical connections used in pre-cast concrete construction. Member sizes, shapes, their reinforcing, and connections are designed by a professional engineer to meet the requirements of each specific situation. Typical column-to-column connections are shown in Figure 9.17 and Figure 9.18. The columns are shimmed as needed to achieve vertical plumb. Steel base plates connected to the column are bolted and the voids filled with drypack.

Figure 9.17 A typical column-to-column connection.

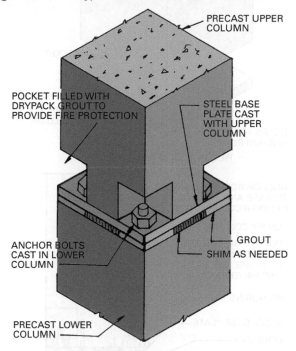

PRECAST UPPER COLUMN

POCKET FILLED WITH DRYPACK GROUT TO PROVIDE FIRE PROTECTION

STEEL BASE PLATE CAST WITH UPPER COLUMN

ANCHOR BOLTS CAST IN LOWER COLUMN

GROUT

SHIM AS NEEDED

PRECAST LOWER COLUMN

A typical column-to-foundation connection is shown in Figure 9.19. Steel anchor bolts are cast in the foundation. The column base plate rests on leveling nuts, which are used to plumb the column. The anchor bolts area fastened once correctly shimmed and tightened, and the area below the base plate is filled with grout.

Typical beam-to-column connections include the use of dowels or haunches. Figure 9.20 shows metal dowels that are cast with the column and fit into holes cast in the beam. Figure 9.21 shows pre-cast columns with haunches on two sides designed to support the spanning beams. Typically, beams are connected to columns by welding an angle to metal anchor plates cast into the beam and column. Bearing pads on the column are shimmed as needed before the dowel pockets are grouted.

Connections of pre-cast walls to a foundation are shown in Figure 9.22. This typically involves metal plates that are welded or an anchor rod that is post-tensioned. One way to connect a pre-cast concrete interior wall to a floor in multistory construction is shown in Figure 9.23. The wall has metal anchor plates cast in

Figure 9.18 This column is anchored to the foundation with bolts located within the base of the column.

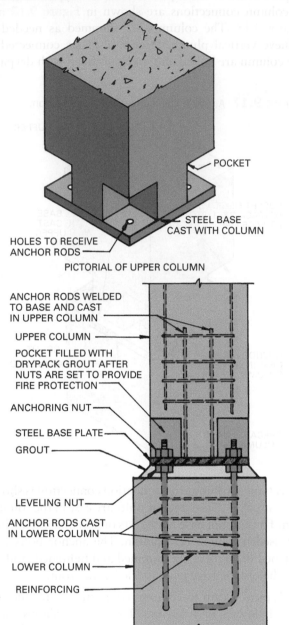

PICTORIAL OF UPPER COLUMN

POCKET

STEEL BASE CAST WITH COLUMN

HOLES TO RECEIVE ANCHOR RODS

ANCHOR RODS WELDED TO BASE AND CAST IN UPPER COLUMN

UPPER COLUMN

POCKET FILLED WITH DRYPACK GROUT AFTER NUTS ARE SET TO PROVIDE FIRE PROTECTION

ANCHORING NUT

STEEL BASE PLATE

GROUT

LEVELING NUT

ANCHOR RODS CAST IN LOWER COLUMN

LOWER COLUMN

REINFORCING

it with a pocket above. Anchor bolts in the wall below are secured to the wall above by bolting.

There are a number of ways to make floor and roof panel connections. The installation of hollow core deck units to a beam is shown in Figure 9.24. Reinforcing rod is cast in concrete between butting units. Additional rods may be placed in the cores on either side of the joint. Figure 9.25 shows a detail for connecting hollow-core units to an interior pre-cast concrete wall.

Figure 9.19 This column is anchored to the foundation with bolts located outside the base of the column.

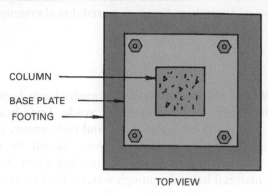

COLUMN

BASE PLATE

FOOTING

TOP VIEW

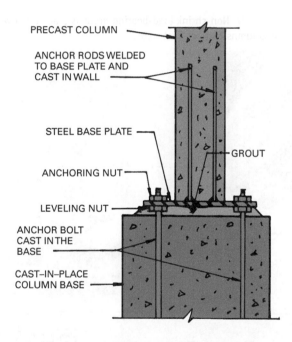

PRECAST COLUMN

ANCHOR RODS WELDED TO BASE PLATE AND CAST IN WALL

STEEL BASE PLATE

GROUT

ANCHORING NUT

LEVELING NUT

ANCHOR BOLT CAST IN THE BASE

CAST–IN–PLACE COLUMN BASE

Single and double Ts can be installed on a pre-cast L-shaped beam that is supported on column haunches (Figure 9.26). The beam is connected to the column by welding metal plates cast into each unit. The T member is welded to the L-beam. Another technique involves supporting inverted-T beams on a pre-cast column haunch and placing the pre-cast T decking units as shown in Figure 9.27. A view of this construction is shown in Figure 9.28.

ERECTING PRE-CAST CONCRETE

Construction Planning

An example of a pre-cast concrete floor *framing plan* for a small building using concrete bearing walls and

Figure 9.20 A doweled connection for a pre-cast rectangular beam to a pre-cast column.

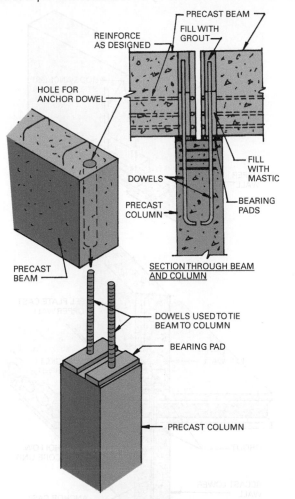

hollow-core slabs is shown in **Figure 9.29**. The plan is accompanied by construction details and a schedule giving sizes of the types of units used.

After the structural design work has been completed and reviewed, the plans are sent to the pre-cast manufacturer that prepares shop drawings and erection specifications. The work is planned using standard molds when possible and custom-built molds as needed. The units are produced through cycles of casting and marked to identify their location within the building.

A major factor in planning for pre-cast construction is the sequencing of the manufacture and delivery of members to the job site. A production and delivery schedule must be planned so units can be delivered to the site as they are needed. Adequate time for the units to cure and gain sufficient strength to permit movement to the job site must be included.

Consideration of the erection procedures also begins before the units are cast. The design engineer must consider the use of lifting hardware and connections. Pre-cast units normally have lifting hooks cast in them so the unit can be hoisted without damage and leveraged into the position needed for connection. Special erection plans show how the pre-cast units are to be raised and connected. They specify dimensional tolerances, hardware sizes, weld lengths, and the required torque on bolts. Connections should always be planned to allow workers to install them from a stable work platform. Planning should include provision for activities such as welding, post-tensioning, and grouting. Temporary bracing may

Figure 9.21 Pre-cast rectangular beams can be supported on column haunches.

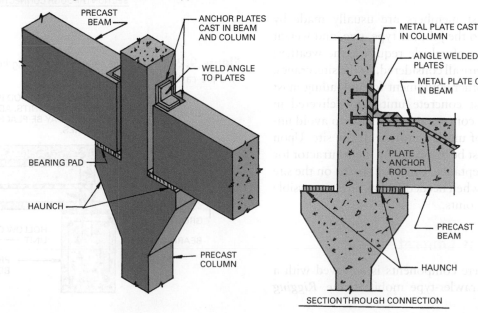

Figure 9.22 Two ways to connect a pre-cast wall to the foundation.

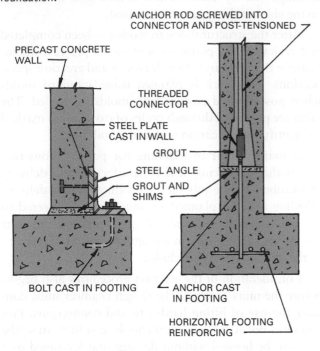

ANCHOR ROD SCREWED INTO
CONNECTOR AND POST-TENSIONED

PRECAST CONCRETE
WALL

THREADED
CONNECTOR

STEEL PLATE
CAST IN WALL

GROUT

STEEL ANGLE

GROUT AND
SHIMS

BOLT CAST IN FOOTING

ANCHOR CAST
IN FOOTING

HORIZONTAL FOOTING
REINFORCING

be required during construction. The erection procedure should include plans to maintain the stability and equilibrium of the structure during construction as well as after the structure is completed. For example, when an L-shaped or inverted-T ledger beam is subjected to eccentric loading (unsymmetrical load), the beam experiences torsion and may tilt slightly on the column below. The connections should be designed to resist this motion.

Transporting Units to the Site

Deliveries of pre-cast members are usually made by truck. The distance to the job; the type, size, and weight of members; the type of vehicle required; the weather; and road conditions are all considered. At the site, cranes, lifting devices, and other equipment for unloading must be in place. Pre-cast concrete units are delivered in accordance with the construction schedule to avoid unnecessary build-up of units in storage at the site. Upon delivery, all units must be inspected by the contractor for quality and final acceptance. Units are stored on the site to be easily reached when needed and as near as possible to their destination points.

Erecting Pre-cast Concrete

Lifting of the concrete components is achieved with a truck mounted or crawler-type mobile crane. *Rigging*

Figure 9.23 Anchor bolts used to connect pre-cast interior walls when hollow core decking is used.

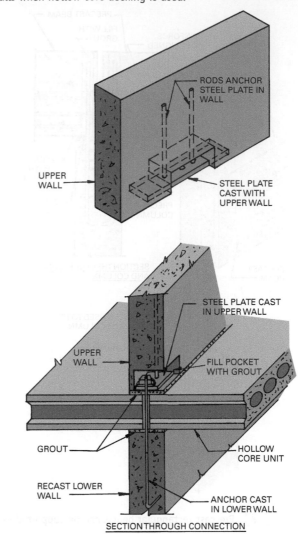

RODS ANCHOR
STEEL PLATE IN
WALL

UPPER
WALL

STEEL PLATE
CAST WITH
UPPER WALL

STEEL PLATE CAST
IN UPPER WALL

UPPER
WALL

FILL POCKET
WITH GROUT

GROUT

HOLLOW
CORE UNIT

RECAST LOWER
WALL

ANCHOR CAST
IN LOWER WALL

SECTION THROUGH CONNECTION

Figure 9.24 A standard detail for installing hollow core units on a pre-cast rectangular beam.

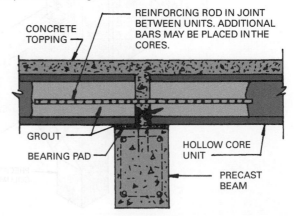

REINFORCING ROD IN JOINT
BETWEEN UNITS. ADDITIONAL
BARS MAY BE PLACED IN THE
CORES.

CONCRETE
TOPPING

GROUT

BEARING PAD

HOLLOW CORE
UNIT

PRECAST
BEAM

Figure 9.25 A typical detail for supporting hollow core units on a pre-cast interior wall.

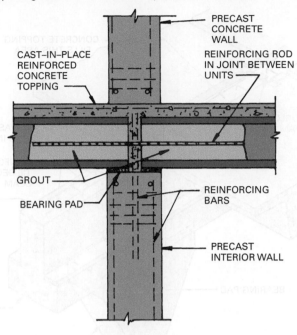

Figure 9.26 Double tees can be supported on a ledger beam that is seated on a haunch of a pre-cast concrete column.

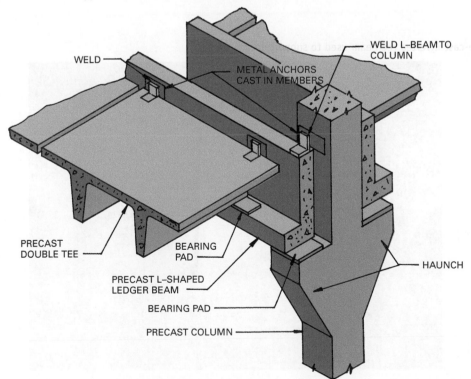

Figure 9.27 Double tees can be supported by pre-cast inverted tee beams that rest on the haunches of pre-cast concrete columns.

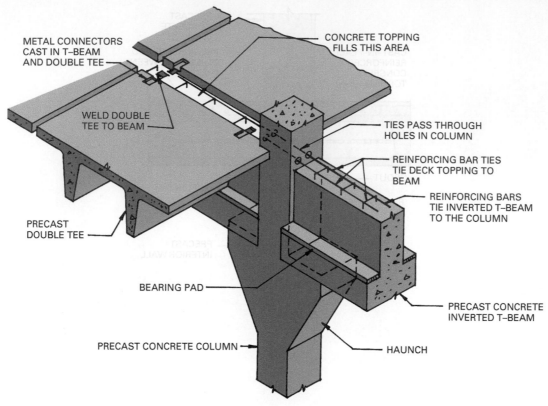

METAL CONNECTORS
CAST IN T–BEAM
AND DOUBLE TEE

CONCRETE TOPPING
FILLS THIS AREA

WELD DOUBLE
TEE TO BEAM

TIES PASS THROUGH
HOLES IN COLUMN

REINFORCING BAR TIES
TIE DECK TOPPING TO
BEAM

REINFORCING BARS
TIE INVERTED T–BEAM
TO THE COLUMN

PRECAST
DOUBLE TEE

BEARING PAD

PRECAST CONCRETE
INVERTED T–BEAM

PRECAST CONCRETE COLUMN

HAUNCH

Figure 9.28 A pre-cast inverted tee beam used to support double tees.

Figure 9.29 A framing plan for a building with hollow core floor units. Included are drawings showing construction details and a schedule giving information about the units required.

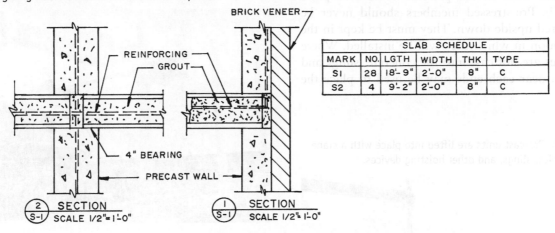

SLAB SCHEDULE					
MARK	NO.	LGTH	WIDTH	THK	TYPE
S1	28	18'-9"	2'-0"	8"	C
S2	4	9'-2"	2'-0"	8"	C

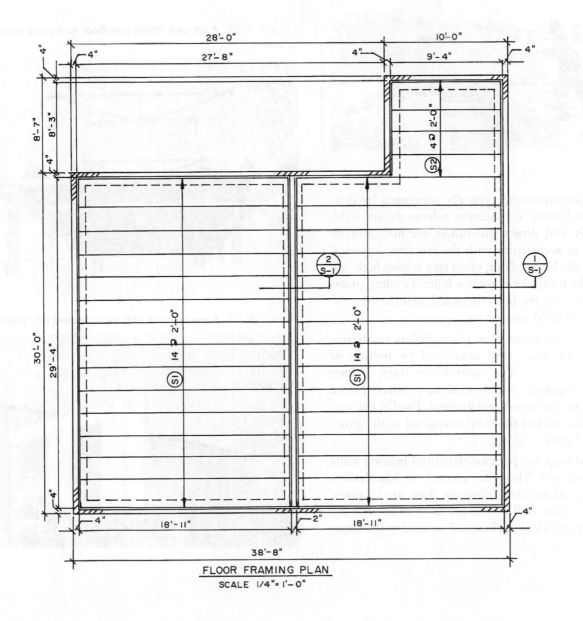

FLOOR FRAMING PLAN
SCALE 1/4"= 1'-0"

refers to the equipment, cables, slings, and other hoist-ing equipment used to lift pre-cast units into place (Figure 9.30). Pre-stressed members should never be lifted or stored upside down. They must be kept in the upright position in which they will be installed. Where lifting devices are not cast into the concrete, slings and ropes placed near each ends may be used to place the member.

Figure 9.31 Pre-cast pre-stressed beams are connected to columns.

Figure 9.30 Pre-cast units are lifted into place with a crane and proper cables, slings, and other hoisting devices.

Figure 9.32 A pre-cast hollow core floor deck being secured to beams.

Pre-cast structures are typically supported by cast-in-place foundations. Connection tolerances are tight, and assembly will slow if tolerances are not carefully maintained. In accordance with the erection drawings, columns for the lowest floor, often two stories high, are lifted over the foundation anchor bolts. Leveling plates are set and aligned, the bolts tightened, and the area un-der the column filled with grout.

After the columns are set in place, girders and beams are hoisted into place and connected by bolting or welding (Figure 9.31). The entire lower frame is then straightened, squared, and braced as required before connections are tightened and grouted. Finally, the pre-cast floor slabs are installed and connected to the struc-tural frame (Figure 9.32).

Some buildings use pre-cast structural bearing walls instead of columns. These are erected on foundations and braced until members abutting them are in place. If pre-cast exterior wall panels are used to enclose the building, they are lifted in place and secured to the struc-tural frame (Figure 9.33).

Figure 9.33 A pre-cast wall unit being hoisted into position.

Review Questions

1. Give some examples of pre-cast concrete structural units.

2. How do non-pre-stressed structural units differ from pre-tensioned units?

3. How does camber in a beam serve to resist loads?

4. What are the commonly used types of pre-cast pre-stressed floor and roof slabs?

5. Which type of pre-cast pre-tensioned slabs span the greatest distances?

6. What is the main value in pre-stressing pre-cast concrete columns?

7. What is the difference between post-tensioning and pre-tensioning?

Key Terms

Camber

Double Ts

Drypack

Framing Plan

Grout

Haunch

Non-Pre-Stressed Unit

Post-Tensioning

Pre-Stressed Unit

Pre-Tensioning

Rigging

Activities

1. If there is a pre-casting plant within reasonable driving range, it is well worth the effort to make a visit. Ask the supervisor to explain and show the process from beginning to end. Take photos and prepare an exhibit detailing the process.

2. Explain how a pre-cast plant prepares the concrete and how the quality control tests are taken.

3. See if you can get the structural plans for a pre-cast concrete structure from a local architect or contractor. Prepare a report citing the size and spacing of columns, beams, and girders. What type of floor and roof decking were specified, and what are the sizes of each? Is there anything else on the plan that you would like to report?

4. Make a model with wood pieces representing pre-cast concrete beams, girders, columns, wall panels, and decking.

5. Examine the local building code and briefly cite the specific areas related to pre-cast concrete structures.

6. If you are fortunate enough to be able to visit the site of a pre-cast concrete structure being erected, take photos and notes describing what you see. For example, what types of connectors are used? How are members lifted into place? What safety devices are the workers wearing?

Additional Resources

Architectural Pre-cast Concrete, and numerous other publications, Pre-cast/Pre-stressed Concrete Institute, Chicago, IL. www.pci.org/

PCI Design Handbook, and numerous other publications, Pre-cast/Pre-stressed Concrete Institute, Chicago, IL.

Other resources include: Many related publications available from the American Concrete Institute, Farmington Hills, MI.

APA The Architectural Pre-cast Association. www.archprecast.org/

See Appendix C for addresses of professional and trade organizations and other sources of technical information.

Masonry
CSI MasterFormat™

Mortars for Masonry Walls

CHAPTER 10

LEARNING OBJECTIVES

Upon completion of this chapter, the student should be able to:

- Describe the various materials used to produce mortar.
- Classify the ingredients used to manufacture masonry cements.
- Identify and cite the uses of the types of standard mortars.

- Explain the desirable properties of plastic mortar.
- Discuss the properties of quality-hardened mortar.
- Name the types of frequently used mortar tests.
- Specify how to cure masonry mortar laid in cold weather.

Build Your Knowledge

For further information on these materials and methods, please refer to:

Chapter 7   Concrete

Masonry construction boasts a long tradition of providing strong, durable, and aesthetically flexible assemblies. The term masonry is used to describe either solid stone or man made modular units that are formed into building blocks. The modular units are manufactured from ceramic materials and hardened either by heat as in brick, clay tiles, or terra cotta, or by chemical reaction in the case of concrete block. By themselves, masonry units are relatively weak and rely on mortar to join them together into strong structural assemblies.

Mortar is the bonding agent used to bind masonry units into an integral structure. In addition to bonding the units together, mortar also seals the spaces between the units so they are not penetrated by air or moisture; provides a place to secure steel reinforcement, ties, and anchor bolts into walls; and allows for adjustment due to the slight variations in size and surface that occur in masonry units. Historically, mortars were made of mud and clay, the materials that were readily available on a site. Later developments in mortar

composition utilized a combination of lime and sand that developed their strength slowly. Modern mortar ingredients include Portland cement as the binding agent, hydrated lime to impart workability, and sand and water. Mortar accounts for around 17–20% of a finished wall surface (Figure 10.1).

Figure 10.1 The primary function of mortar is to bond the masonry units into an integral element.

© 1000 Words/Shutterstock.com

MORTAR COMPOSITION

Mortar is composed of *cementitious materials*, clean carefully graded mortar sand, clean water, and sometimes a coloring agent or admixture. The actual composition of mortars varies according to intended use. For general masonry applications, mortars may contain Portland cement or masonry cement, sand, hydrated lime or lime putty, and water. Mortars for special applications, such as prefabricated masonry units that must be moved into place, require additives that increase compressive and tensile strength and have greater bonding capabilities.

Cementitious Materials

Cementitious materials provide the bonding ingredient in mortars. They are made according to ASTM specifications, as shown in the following list:

> Masonry cement—ASTM C91
>
> > (Types M, S, or N), CSA A8 (Types S or N)
>
> Portland cement—ASTM C150
>
> > (Types I, IA, II, IIA, III, or IIIA), CSA A5 (10, 20, 30, or 50)
>
> Blended hydraulic cement—ASTM C595
>
> > (Types IS, IS-A, IP, IP-A, I(PM), or I(PM)-A) (Pozzolan modified Portland cement), CSA A362 (Type 10S)
>
> Hydrated lime for masonry purposes—ASTM C207
>
> > (Types S, SA, N, or NA)
>
> Quicklime for structural uses (for lime putty)—ASTM C5

Detailed information on various cements is given in Chapter 7. Most mortars are now made with mortar cement. Masonry cement mortars are made by combining masonry cement, clean carefully graded sand, and enough clean water to produce a plastic, workable mix.

Masonry Cements

Masonry cements are hydraulic cements used in mortars for masonry construction. They produce a mortar that has greater plasticity and water retention than Portland cement. Typically, they include Portland cement, slag cement, blast-furnace slag cement, Portland–pozzolan cement, natural cement, and hydraulic lime. Manufacturers may also include chalk, clay, talc, calcareous shell, or limestone to add the properties desired.

Masonry cements meet the specifications found in ASTM C91. This standard specifies three types of masonry cement: Type N, Type S, and Type M. They are used to produce mortars: Types N, O, S, and M, as specified in ASTM C270. Type N masonry cement is used in Type N and O mortars. It can be blended with Portland cement for use in Type S and M mortars.

Other Mortar Ingredients

Lime is used to help stabilize the volume of mortar by controlling shrinking and expansion. Masonry sand may be natural or a manufactured product that meets the requirements of ASTM C144, *Standard Specification for Aggregate for Masonry Mortar*. In Canada, the standard is CSA A82.56, *Aggregate for Masonry Mortar*. Since sand makes up the major portion of a mortar, the quality of the finished mortar depends heavily on clean, quality sand. The gradation requirements for sand specified in ASTM C144 are shown in Table 10.1.

Table 10.1 Aggregate Gradation for Masonry Mortar

Sieve Size No.	Gradation Specified, Percent Passing ASTM C144[a]	
	Natural Sand	Manufactured Sand
4	100	100
8	95 to 100	95 to 100
16	70 to 100	70 to 100
30	40 to 75	40 to 75
50	10 to 35	20 to 40
100	2 to 15	10 to 25
200	—	0 to 10

[a]*Additional requirements: Not more than 50% shall be retained between any two sieve sizes, nor more than 25% between No. 50 and No. 100 sieve sizes. Where an aggregate fails to meet the gradation limit specified, it may be used if the masonry mortar will comply with the property specification of ASTM C270 (Table 2)*

Source: American Society for Testing and Materials

Water used to produce mortar must be clean and free of acids, alkalies, and other organic materials. Water containing soluble salts will cause the mortar to effloresce. *Efflorescence* appears on the finished masonry wall as a white powdery substance that discolors the mortar and masonry units.

Mortar can be colored by adding various organic pigments. White mortar is made with white masonry cement or white Portland cement, lime, and white sand. Colored mortars are made with white masonry cement and pigments. The pigments typically include some type of mineral oxide compound, such as iron, chromium, cobalt, or manganese oxides. Carbon black is used to produce a dark gray or black mortar.

Admixtures are sometimes used to alter the properties of a mortar. However, the possibility of their creating unforeseen problems is high. Admixtures are typically not used unless laboratory tests have been made to verify the effects they will have on the mortar. For example, certain admixtures may affect the hydration process and hardening properties.

Table 10.2 lists some of the commonly available admixtures (modifiers), their benefits, and possible

Table 10.2 Admixtures—Benefits and Concerns

Admixture	Primary Benefits	Possible Concerns
Air-entraining	Freeze-thaw durability, workability	Effect on compressive and bond strengths
Bonding	Wall tensile (and flexural) bond strength	Reduced workability, bond strength regression upon wetting, corrosive properties
Plasticizer	Workability, economy	Effect on hardened physical properties under field conditions Effectiveness at cold temperatures, corrosive properties, effect on efflorescence potential of masonry
Set accelerator	Early strength development	
Set retarder	Workability retention	Effect on strength development, effect on efflorescence potential of masonry
Water reducer	Strength, workability	Effect on strength development under field conditions with absorptive units
Water repellent	Weather resistance	Effectiveness over time
Pozzolanic	Increase density and strength	Effect on plastic and hardened physical properties under field conditions
Color	Esthetic versatility	Effect on physical properties, color stability over time

Source: Portland Cement Association

problems they may cause. More information on admixtures is detailed in Chapter 7.

TYPES OF MORTAR

Four types of mortar are specified for use in the United States. These are M, S, N, and O, as specified by ASTM C270 *Standard Specification for Mortar for Unit Masonry*. The Canadian Standards Association Standard A179 specifies two types, S and N.

Mortar types are identified by either proportion or property specifications. Table 10.3 lists the specifications for mortar proportion indicated by the combinations of Portland cement or blended cement, masonry cement, hydrated lime or lime putty, and aggregate. Blended cement is a combination of Portland cement and masonry cement. Mortar types, classified under property specifications, are based on compressive strength, water retention, and air content (Table 10.4). This data is developed by testing 2 in. (50.8 mm) cubes of hardened mortar to find the compressive strength. Type M mortar is a high-strength mortar with a compressive strength of 2,500 psi (17 MPa). It has better durability than the other types. Type M is recommended for masonry below grade in contact with earth and for conditions of severe frost action. It is also used for reinforced masonry.

Type S mortar is a medium-high-strength mortar with a compressive strength of 1,800 psi (12.5 MPa). It is often permitted in wall construction instead of Type M because it has almost the same allowable strength, and

can be used above or below grade. Type S has better workability and more water retention than Type M. It is used for reinforced and unreinforced masonry and has high tensile bond strength.

Type N mortar is a medium-strength mortar with a compressive strength of 750 psi (5 MPa). It is used for above-grade general construction involving exposed masonry load-bearing walls where compressive strength and lateral strength requirements are not high.

Type O mortar is a medium-strength mortar with a compressive strength of 350 psi (2.5 MPa). It is used for general interior purposes, such as for non-load-bearing walls where the compressive strength required does not exceed 100 lb. per sq. in. It must not be used in contact with soil or exposed to freezing conditions. A guide for selecting masonry mortars for various uses above and below grade is given in Table 10.5.

Sources of Mortar

Mortars are available ready-mixed, dry-batch, and on-site mixed. Ready-mixed mortar is prepared at a central batch plant and transported to the job site, where a slump test is made. If required, additional water is added to get the desired slump. This mortar contains a retarding, set-controlling admixture that keeps the mortar workable and in a plastic condition for more than twenty-four hours. The mix is delivered by a ready-mix truck and stored on site in large tubs. Ready-mixed mortar must meet the requirements of ASTM

Table 10.3 Proportion Specifications for Mortar

Mortar type	Portland cement or blended cement	Masonry cement type			Hydrated lime or lime putty	Aggregate[a]
		M	S	N		
	1	—	—	1	—	4½ to 6
M	—	1	—	—	—	2¼ to 3
	1	—	—	—	¼	2¹³/₁₆ to 3¾
	½	—	—	1	—	3⅜ to 4½
S	—	—	1	—	—	2¼ to 3
	1	—	—	—	Over ¼ to ½	.
	—	—	—	1	—	2 to 3
N	1	—	—	—	Over ½ to 1 ¼	.
	—	—	—	1	—	2 to 3
O	1	—	—	—	Over 1¼ to 2 ½	.

United States—ASTM C270 / Parts by Volume

Mortar type	Portland cement	Masonry cement type		Hydrated lime or lime putty	Aggregate[a]
		S	N		
S	—	—	1	—	2¼ to 3
	½	—	1	—	3½ to 4½
	1	—	—	½	3½ to 4½
N	—	—	1	—	2¼ to 3
	—	—	—	1	4½ to 6

Canada—CSA A179M / Parts by Volume

[a]The total aggregate shall be equal to not less than 2¼ and not more than 3 times the sum of the volumes of the cement and lime used.

Notes: 1. Under both ASTM C270, Standard Specification for Mortar for Unit Masonry, and CSA A1 79, Mortar and Grout for Unit Masonry, aggregate is measured in a damp, loose condition and 1 cu ft. of masonry sand by damp, loose volume is considered equal to 80 lb. of dry sand (in SI units 1 cu m of damp, loose sand is considered equal to 1,280 kg of dry sand).

2. Mortar should not contain more than one air-entraining material.

Source: Portland Cement Association and the Canadian Standards Association.

Table 10.4 Property Specifications for Laboratory-Prepared Mortar[a]

Mortar specification	Mortar type	Minimum 28-day compressive strength, psi	Minimum water retention, %	Maximum air content, %[b]
	M	2,500	75	12[c]
	S	1,800	75	12[c]
ASTM C270	N	750	75	14[c]
	O	350	75	14[c]

United States

Mortar specification	Mortar type	Minimum compressive strength, MPa		Minimum water retention, %
		7-day[d]	28-day	
	S	7.5	12.5	70
CSAA179	N	3	5	70

Canada

[a]The total aggregate shall be equal to not less than 2¼ and not more than 3½ times the sum of the volumes of the cement and lime used.

[b]Cement-lime mortar only (except where noted).

[c]When structural reinforcement is incorporated in cement-lime or masonry cement mortar, the maximum air content shall be 12% or 18%, respectively.

[d]If the mortar fails to meet the 7-day requirement but meets the 28-day requirement, it shall be acceptable.

Source: Portland Cement Association and the Canadian Standards Association.

Table 10.5 Guide for the Selection of Masonry Mortars (United States)[a]

Location	Building segment	Mortar type Recommended	Alternative
Exterior, above grade	Load-bearing walls	N	S or M
	Non-load-bearing walls	O[b]	N or S
	Parapet walls	N	S
Exterior, at or below grade	Foundation walls, retaining walls, manholes, sewers, pavements, walks, and patios	S[c]	M or N[c]
Interior	Load-bearing walls	N	S or M
	Non-load-bearing partitions	O	N

Source: American Society for Testing and Materials

[a]Adapted from ASTM C270. This table does not provide for specialized mortar uses, such as chimney, reinforced masonry, and acid-resistant mortars.
[b]Type O mortar is recommended for use where the masonry is unlikely to be frozen when saturated or unlikely to be subjected to high winds or other significant lateral loads. Type N or S mortar should be used in other cases.
[c]Masonry exposed to weather in a nominally horizontal surface is extremely vulnerable to weathering. Mortar for such masonry should be selected with due caution.

C1142, *Specifications for Ready-Mixed Mortar for Unit Masonry*.

Dry-batch mortar has the required ingredients blended and packaged in bags that are delivered to the site and stored in a sealed hopper. When mortar is needed, the dry ingredients are moved to an on-site mixer and the required water is added. Dry-batching provides control over the proportions of the various ingredients in the mortar. Premixed dry mortar ingredients are also available packaged in waterproof bags.

On-site mixed mortar can be made by storing dry ingredients in a silo-type mixer in which an auger mixing device receives the correct proportions of dry ingredients and blends them together (Figure 10.2).

The dry ingredients are stored in separate chambers, from which they are fed to the mixer. The water is then injected from a pressurized water source, such as a city water main, and the mixer produces the finished mortar. The process is usually computer controlled. The mortar is discharged into tubs or wheelbarrows for distribution on site. This produces a very accurately proportioned mortar.

For small jobs, the ingredients typically are delivered to the site separately. The mortar cement is in bags, and the sand in bulk. The sand must be protected from moisture and dirt. The mortar is often proportioned by placing shovelfuls of sand into a small power-operated mixer and adding the required number of bags of mortar cement with water (Figure 10.3). The gasoline-powered

Figure 10.2 Sand and cement stored in a computer-controlled silo are proportioned, discharging the desired mix and amount of mortar.

© Kenneth Johansson/Corbis News/Corbis

Figure 10.3 Small quantities of on-site mortar can be mixed in a portable mixer.

© LUCARELLI TEMISTOCLE/Shutterstock.com

mixer produces the mortar, which usually is moved to the masons by wheelbarrow. The proportions of the ingredients can vary considerably, as can be evidenced after mortar has cured and joints over a wall do not have a uniform shade. To be most effective, proportions should be carefully measured and the power mixer run from three to five minutes. Under-mixing produces a non-uniform mortar, and over-mixing may reduce a mortar's strength.

PROPERTIES OF PLASTIC MORTAR

The properties that affect mortar in a plastic condition are workability and water retention.

Workability

"Workability" is a term used to describe the condition of a mortar that will spread easily, cling to vertical surfaces, extrude easily from joints but not drop off, and permit easy positioning of masonry units (Figure 10.4). It is a condition of the mortar that is influenced by other properties, such as consistency, water retention, setting time, weight, adhesion, and penetrability. Masons judge workability by observing how the mortar slides or adheres to their trowel. Mortar is made in amounts that can be used in a short time. After a couple of hours, workability is reduced and new batches must be mixed.

Figure 10.4 Mortar must be workable, adhere to vertical surfaces, and extrude from joints without falling off.

© John Leung/Shutterstock.com

Water Retention

Mortar must have good *water retention* to resist rapid loss of mixing water to the air or an absorptive masonry unit. Precipitous mix water loss causes mortar to stiffen, making it difficult to achieve a good bond or watertight joint. Mortar that has good water retention remains workable, enabling masons to properly place masonry units. Low-absorption masonry units may float when placed on a mortar with too much water retention. This causes the mortar joint to bleed. Water retention is increased by entrained air, very fine aggregate, or cementitious materials.

Mortar Flow Test   The water retention limit of mortar is measured by both initial flow and flow following laboratory suction tests, as described in ASTM C91 *Standard Specification for Masonry Cement*. The initial flow test is made by placing a truncated cone of mortar with a 4 in. (100 mm) diameter on a metal flow table. The table is mechanically dropped ½ in. (12 mm) twenty-five times in fifteen seconds. The mortar will flow into an enlarged circular shape. The diameter of this shape is measured and compared with the original diameter. Allowable initial flow should be in a range of 100 to 115 percent.

The flow-after-suction test is used to determine flow after loss of water to an absorbent masonry unit. The mortar sample is placed for one minute in a vacuum device that removes some water. The mortar is then tested as described above and the flow measured. The flow after suction should be in a range of 70 to 75 percent.

PROPERTIES OF HARDENED MORTAR

The properties essential to quality-hardened mortar are bond strength, durability, compressive strength, low volume change, appearance, and rate of hardening.

Bond Strength

Bond strength refers to the degree of contact between mortar and masonry units as well as the *tensile bond strength* available for resisting forces that tend to pull masonry units apart.

The degree of contact between masonry units is essential to watertight joints and tensile bond strength. Good bond strength requires workable, water-retentive mortar, good workmanship, full joints, and masonry units with a medium rate of absorption (*suction rate*).

Tensile bond strength is necessary to withstand forces such as wind, structural movement, expansion of clay masonry units, shrinkage of mortar or concrete masonry units, and temperature changes. Tensile bond strength is tested by bonding together samples of concrete masonry units with mortar, curing the mortar, and pulling the units apart on a tensile testing machine.

The major factors affecting the bond strength are the characteristics (strength) of the masonry units, the quality of the mortar, the workmanship, and curing conditions.

Bond strength tends to be high on textured surfaces and low on smooth surfaces. Suction rates of masonry units influence bond strength. Concrete masonry units tend to retain moisture after curing and have relatively low suction rates. Some clay bricks have very high suction rates and, unless wetted before use, will pull water from the mortar resulting in a poor bond. After bricks are wetted, surfaces should be permitted to dry before use.

Mortar flow influences tensile bond strength. As the water content increases, the bond strength increases. This indicates that it is wise to use the highest water content possible while retaining a workable mortar. As water content increases, mortar compressive strength decreases. Bond strength takes precedence over compressive strength.

Good workmanship requires a minimum of elapsed time between spreading the mortar and placing the masonry unit. Some water in the mortar will evaporate and some will be sucked away by the masonry unit on which it is placed, leaving insufficient water to form a good bond on the next masonry unit. After placing a unit on mortar and getting its initial alignment, it should not be moved, tapped, or slid in any way. This can break the initial bond, which cannot be reestablished. The mortar must be replaced if this happens.

Good curing conditions require the maximum amount of water possible be in the mortar, because it is needed for hydration. The laid units should be covered with plastic to retain moisture while curing. Under severe dry conditions, it may be necessary to keep the wall wet with a fine mist spray for several days. Walls must also be protected from freezing with insulating blankets.

Durability

Mortar needs the specified durability to withstand weathering forces. Frost or freezing will not damage mortar joints unless they are water soaked and leak. High-compressive-strength mortar usually has good durability. Air-entrained mortar also provides protection against freeze–thaw cycles.

Compressive Strength

The compressive strength of mortar depends mainly on the type and quantity of cementitious material used. It increases as cement content increases and decreases as air-entrainment, lime, or water content increases. The compressive strength of mortar is found by testing standard cured 2 in. (50.8 mm) square cubes in a laboratory compression-testing machine following ASTM C270 standards. Mortar can be tested in the field using ASTM C780 *Standard Test Method for Preconstruction and Construction Evaluation of Mortars for Plain and Reinforced Unit Masonry* standards.

The compressive strength of a wall depends not only on the mortar but also on the masonry unit, design of the structure, workmanship, and curing.

Low Volume Change

Mortars have low volume change. The actual shrinkage during curing of a mortar joint is negligible.

Appearance

The mortar joints in a wall should have a uniform color or shade. Each batch of mortar should have exactly the same proportions of ingredients. Time of tooling also causes variances in the shade of a joint. If mortar is tooled when fairly hard, a darker shade will occur; if fairly soft, then tooling produces a lighter shade. White mortar cement should be tooled with a glass or plastic joint tool, as metal tools will darken the joint.

Any pigments added to color the mortar must be carefully measured. Typically, the color is premixed with enough mortar cement to do the entire job, rather than one batch at a time. A joint's color and shade is also affected by atmospheric conditions, admixtures, and moisture content of masonry units.

Rate of Hardening

The rate of hardening of a mortar due to hydration is the speed at which it develops the strength to resist an applied load. Mortar that hardens very slowly can delay construction because it will not support brick courses laid above it. Mortar with a very high rate of hardening can be difficult for the mason to work with. The rate of hardening must be consistent from batch to batch, allowing sufficient time to place masonry units and tool joints.

Loss of water can also influence the stiffening of mortar. This is of particular concern in hot weather. In high temperatures, masons frequently lay fewer masonry units on shorter beds before stopping to tool joints.

COLORED MORTAR

Architects frequently specify the color of mortar to be used. White mortar is made using white masonry cement or white Portland cement, lime, and sand. Colored mortars are made using white masonry cement or white Portland cement and adding color pigments, colored sand, or colored masonry cements (Figure 10.5). The final color achieved is the result of blending these ingredients. Trial batches are made and cured until the desired color is developed.

Figure 10.5 A colored mortar being made from white masonry and colored sand.

Color pigments are a form of mineral oxide, such as iron, manganese, chromium, and cobalt oxides. Carbon black is used to produce a dark gray to black mortar.

Following are recommended pigments.

Color Used*	Pigment	Maximum Amount Used*
Gray to black	Carbon	3
Green	Chromium oxide	10
Blue	Cobalt oxide	10
Reds, yellows, browns, blacks	Iron oxide	10

*Maximum amount of pigment as a percent of the weight of the Portland cement.

MORTAR IN COLD WEATHER

When mortar is placed and cured, its temperature should be kept in the range of 60°F to 80°F (15.7°C to 26.9°C). The water in the mortar is needed for hydration (the chemical reaction between masonry cement and water). This leads to the mortar's hardening, which slows or stops if the cement paste in the mortar drops below 40°F (4.5°C).

Mortar can be laid when air temperature is above 40°F (4.5°C) using normal procedures. When air temperature is below 40°F (4.5°C), the mortar water requires heating. However, the temperature of the mortar should never exceed 120°F (50°C) because higher temperatures will cause the mortar to set up too fast, resulting in a loss of compressive and bond strength. If air temperature falls as the mortar is laid, the minimum mortar temperature should be 70°F (21°C). The recommended mortar temperatures for various air temperatures are given in Table 10.6 and Table 10.7.

The finished laid masonry should be covered when work stops. When air temperatures top 40°F (4.5°C), walls must be covered with plastic sheets to protect them from rain or light snow. Below 40°F (4.5°C), the walls should be covered with plastic or canvas to prevent their freezing or becoming wet. Some masons use insulated plastic or canvas-covered blankets. When air temperatures fall below 32°F (0°C), walls must be covered with insulated blankets and a source of heat provided to keep the mortar from freezing for at least twenty-four hours.

SURFACE-BONDING MORTARS

A variety of surface-bonding mortars are available from various manufacturers. These are applied by trowel, brush, or spray to any masonry surface and provide

Table 10.6 Recommendations for All-Weather Masonry Construction[a]

Air temperature, °F	Construction Requirements	
	Heating of materials	Protection
Above 100 or above 90 with wind velocity greater than 8 mph	Limit open mortar beds to no longer than 4 ft. and set units within one minute of spreading mortar. Store materials in cool or shaded area.	Protect Wall from rapid evaporation by covering, fogging, damp curing, or other means.
Above 40	Normal masonry procedures. No heating required.	Cover walls with plastic or canvas at end of work day to prevent water entering masonry.
Below 40	Heat mixing water. Maintain, mortar temperatures between 40°F and 120°F until placed.	Cover walls and materials to prevent wetting and freezing. Covers should be plastic or canvas.
Below 32	In addition to the above, heat the sand. Frozen sand and frozen wet masonry units must be thawed.	With wind velocities over 15 mph, provide windbreaks during the work day and cover walls and materials at the end of the work day to prevent wetting and freezing. Maintain masonry above 32°F using auxiliary heat or insulated blankets for 24 hours after laying units.
Below 20	In addition to the above, dry masonry units must be heated to 20°F.	Provide enclosure and supply sufficient heat to maintain masonry enclosure above 32°F for 24 hours after laying units.

[a]Adapted from recommendations of the International Masonry Industry All Weather Council and requirements of ACT530 1/ASCE 6/TMS 602. (References 4 & 13).

Table 10.7 Canadian Protection Requirements[a]

Mean daily air temperature, °C	Protection
0 to 4	Masonry shall be protected from rain or snow for 48 h.
24 to 0	Masonry shall be completely covered for 48 h.
27 to 24	Masonry shall be completely covered with insulating blankets for 48 h.
27 to below	The masonry temperature shall be maintained above 0°C for 48 h by enclosure and supplementary heat.

Note: The amount of insulation required to properly cure masonry in cold weather shall be determined on the basis of the expected air temperature and wind velocity and the size and shape of the structure.

Source: Canadian Standards Association

a base to which plaster, stucco, concrete, and cement-based paints will adhere. Surface bonding is the application of a cement mortar, reinforced with glass fibers, to both surfaces of concrete block walls laid up without mortar. This bonds them into a solid wall and provides a waterproof coating. Tests indicate that surface-bonded walls are as strong in bending flexure as walls laid with conventional mortar joints. If the surface between the blocks is not flat and smooth, vertical compressive strength is reduced.

Review Questions

1. What are the basic materials used to produce mortar?

2. What is meant by the workability of a mortar?

3. Why are mortar water-retention properties important?

4. What steps are performed to make a mortar flow test?

5. What does a flow-after-suction test reveal?

6. What is meant by the tensile bond strength of mortar?

7. What type of masonry surfaces produce the best mortar bond?

8. How does the water content in mortar influence bond and compressive strengths?

9. How does air-entrainment influence the compressive strength of mortar?

10. How does the volume change in mortar as it cures affect a mortar joint?

11. What things can cause a change in the shade of a mortar joint?

12. What are the types of mortar used in the United States and Canada?

13. Which type of mortar has the highest strength and where is it used?

14. Where would you use the lowest-strength mortar?

15. Which type of mortar is used for general construction of exposed masonry load-bearing walls?

16. What materials can be used to produce colored mortar?

17. How can you keep prepared mortar from having excessive moisture loss?

18. What should you do with mortar that becomes stiff from hydration while waiting for use?

19. What is the lowest temperature at which mortar can be laid before requiring protection from freezing?

Key Terms

Cementitious Material	Mortar Flow	Water Retention
Efflorescence	Suction Rate	
Mortar	Tensile Bond Strength	

Activities

1. Prepare a number of mortar samples using different variations of ingredients, including Portland cement mortar and masonry cement mortar. Cure each under identical conditions and test their compressive strength.

2. Get a supply of clay masonry and concrete masonry units. Mix mortars of different consistencies and use them to lay up a small wall. Observe and record the ease or difficulty in using the mortar and maintaining the desired-size mortar joint. After the walls have cured, try to separate the masonry units. Do some seem to have greater bonding strength?

3. Prepare several samples of mortar, varying the water content slightly. Then conduct an initial mortar flow test and report your results.

4. Prepare several samples of mortar using various ingredients. After they have cured, make compression tests and record your results.

Additional Resources

Cement, Lime, Gypsum, Volume 04.01, American Society for Testing and Materials, West Conshohocken, PA.

Other resources include: Numerous publications from the Portland Cement Association, Skokie, IL.

Publications from the Canadian Standards Association, Etobicoke, Ontario, Canada.

Clay Masonry

LEARNING OBJECTIVES

Upon completion of this chapter, the student should be able to:

- Describe how clay bricks are made.
- Identify various clay masonry products and explain how they are typically used.

- Be aware of the grades, types, and classes of clay masonry products.
- Use the properties of clay masonry products to select materials for various applications.

Build Your Knowledge

For further study on these materials and methods, please refer to:

Chapter 33 Interior Walls, Partitions, and Ceilings

Chapter 34 Flooring

Brick is an ancient building material. More than 6,000 years ago, sun-baked clay bricks were used in the construction of palaces and fortifications. Masonry structures work in compression and provide a load-bearing material with a durable surface finish that requires little maintenance. Clay brick is made from surface- or deep-mined clays that possess the necessary plasticity when mixed with water to permit molding to desired shapes. The clay must have the tensile strength to hold a shape while in a plastic condition and contain clay particles that will fuse together when subjected to high temperatures. Properly manufactured brick is fire resistant.

Brick manufacturing is one of the most efficient uses of materials to produce a product. Clay has been termed an abundant resource, and brick plants are typically located close to raw material sources.

CLAYS

Clay is found in three forms: *surface clay*, shale, and fireclay. Most bricks are made from surface-mined clays that reside near the surface of the earth and are strip-mined. *Shales* are clays that have been subjected to high pressures, causing them to be relatively hard. *Fireclays* are found at deeper levels and have more uniform physical and chemical properties. They can withstand higher temperatures and are used to produce fire-bricks for high-temperature applications.

Clays contain a variety of materials, but they are predominantly silica and alumina, with smaller amounts of metallic oxides and other ingredients. Clays are divided into two classes, calcareous and non-calcareous. Calcareous clays contain about 15 percent calcium carbonate and have a yellow color when burned. Non-calcareous clays contain silicate of alumina, feldspar, and iron oxide. The color varies depending on the amount of iron oxide from a buff, red, or salmon color when burned.

MANUFACTURING CLAY BRICKS

The manufacture of clay bricks requires a seven-step manufacturing process (Figure 11.1).

Figure 11.1 The flow of finished bricks in a modern brick manufacturing plant.

Figure 11.2 The stiff mud process forces clay through a die and produces a column of brick that is cut to size with a wire cutter.

©iStock.com/bucky_za

Winning and Storage

Clay masonry is manufactured in a multistep process. *Winning* is a term used to describe the mining of clay. Most bricks are made from surface-mined clays dug from open pits, although fireclays are obtained from underground mines. The clays are moved by truck or rail to a plant where they are crushed and blended to produce the desired chemical composition and physical properties. Once blended, the clays are moved to crushers where stones are removed and the clay lumps are reduced to a maximum of about 2 in. (50mm) diameter. This material is then moved by conveyor to grinders, where it is ground to a fine powder and passed over vibrating screens.

Forming the Bricks

The three major methods for forming bricks are the soft-mud process, the stiff-mud process, and the dry-press process.

The *stiff-mud process* is most widely used. This highly automated production procedure passes clay of 12 to 15 percent moisture through a vacuum that removes air pockets. The clay is then forced by an auger through a die, producing a continuous column of the desired size and shape (Figure 11.2). As the clay leaves the die, a surface texture is applied by attachments that scratch, brush, roll, or in some way leave markings on the brick's face. Common textures for stiff-mud extruded bricks include smooth, matt, rugs, barks, and stippled.

The column then passes through a wire cutter that trims the bricks to size, then move via conveyor to an inspection area, where imperfect bricks are removed and returned for reprocessing. The good bricks are placed on drier cars for transfer to a drier kiln.

The *soft-mud process* is used for brick making from clays having too much natural water to permit the use of the stiff-mud process. The bricks are shaped in molds lubricated with water or sand (Figure 11.3). Bricks formed in water-lubricated molds are called water-struck; those made in sand-lubricated molds are called sand-struck. Water-struck bricks have a relatively smooth finish, whereas sand-struck bricks have a matte textured surface.

Figure 11.3 In the soft-mud process, the mixture is placed in molds, formed, removed, and placed in the kiln.

The *dry-press process* is used with clays having 10 percent or less moisture. The mix is formed into bricks in steel molds under high pressure.

Drying

The moisture content of green bricks (unfired, newly formed bricks) varies depending on the clay and the process used. Once formed, the bricks are placed in a low-temperature drier kiln for one to two days. The temperature and humidity are carefully controlled to prevent rapid shrinkage and possible cracking. They are then glazed, if required, or moved directly to high-temperature kilns.

Glazing

Some bricks have a ceramic glaze applied to one or more surfaces after the brick has been dried. *Glaze* is a sprayed coating of mineral ingredients that melts and fuses to the brick when subjected to the required temperature. The glaze forms a smooth, glasslike coating and is available in a wide range of colors.

Burning and Cooling

Burning involves raising the temperature of dried bricks to a predetermined level. The two common types of kilns in use are a periodic kiln and a tunnel kiln. The periodic kiln is filled with bricks stacked so air can circulate between them. The temperature in the kiln is raised, held, and lowered. The tunnel kiln is a long, narrow structure through which the bricks move on cars. The bricks enter the kiln on one end and are fired and cooled as they move through to the other. The burning process for both methods takes from 40 to 150 hours, depending on desired results (Figure 11.4).

Figure 11.4 Masonry units are fired at temperatures from 400 to 2,400 degree Fahrenheit.

The burning process itself involves several stages: water-smoking, dehydration, oxidation, vitrification, flashing, and cooling.

Water-smoking removes free water by evaporation and requires temperatures up to 400°F (204°C). Dehydration removes additional moisture and requires temperatures ranging from 300 to 1,800°F (150 to 980°C). Oxidation temperatures range from 1,000 to 1,800°F (540 to 980°C) and vitrification from 1,600 to 2,400°F (870 to 1,315°C). These last processes transform clay into a solid, ceramic material. Flashing, if required, follows at this point. It is accomplished by adjusting the fire to reduce the atmosphere in the kiln (resulting in insufficient oxygen to support combustion). This produces a variation in the colors and color shading of the bricks.

As bricks are burned, they undergo considerable shrinkage. This is taken into account when they are cut to size or molded. The higher the burning temperature, the more the shrinkage and the darker the color of the brick. Therefore, dark bricks are usually slightly smaller than are light-colored bricks. Some size variation is always possible, and slight distortion caused by the burning process is normal.

After bricks have been burned and flashed as required, a cooling period of forty-eight to seventy-two hours begins. The rate of cooling affects a brick's color and controls cracking and checking. The color of brick is related to the chemical composition of the clay or shale used and the temperature during the burn. The iron in clay turns red in an oxidizing fire and purple in a reducing fire.

Drawing and Storage

After the cooling stage is complete, bricks are removed from the kiln, sorted, graded, stacked on wood pallets, and wrapped in plastic to keep the bricks dry. Brick not meeting standards after firing are culled from the process and ground to be used as grog in manufacturing brick, or crushed and used as landscaping material.

CLAY MASONRY UNITS

Structural clay masonry units are classified as either solid masonry or hollow masonry.

Solid Masonry

Bricks are classified as *solid masonry* if they have cores whose area does not exceed 25 percent of the gross cross-sectional area of the brick. The cores help during the drying and burning of the unit and reduce its weight. Bricks are made in a variety of types and sizes varying in length, width, and height. Commonly manufactured modular bricks are shown in Figure 11.5.

Figure 11.5 Some of the commonly used types of modular and non-modular bricks.

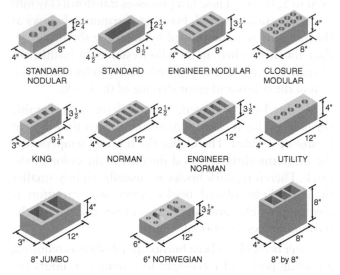

STANDARD NODULAR STANDARD ENGINEER NODULAR CLOSURE MODULAR

KING NORMAN ENGINEER NORMAN UTILITY

8" JUMBO 6" NORWEGIAN 8" by 8"

Modular bricks are those whose actual size plus a mortar joint can be assembled on a 4-inch standard module. The actual size of a standard modular brick is 3⅝ × 2¼ × 7⅝ in. The nominal size of the unit is the actual size plus the thickness of a mortar joint (4 × 2⅔ × 8 in.). Actual and nominal sizes for modular inch bricks are shown in Table 11.1. Brick sizes are specified by three dimensions: width, thickness, and length, given in that order. Figure 11.6 shows the application of the modular size of a standard inch modular brick. Using the 4 in. module, three courses of brick produces an 8 in. module. Engineered or oversize brick have the same length and width as the standard modular but the face is 2¾ in. high. Economy, or jumbo utility brick have a length of 11⅝, a face of 3⅝, with a width of 3½ in. The use of utility brick reduces construction costs as less units must be laid. The actual sizes of non-modular inch bricks are shown in Table 11.2.

The actual size of modular metric brick is that of the manufactured product. The modular size is the size stated when specifying the brick. For example, when ordering an 89 × 57 × 190 mm actual-size modular

Table 11.1 Modular Inch-Size Common Brick Sizes

Unit Name	Actual Dimensions[a]		Nominal Dimensions[b]	Joint Thickness	Modular Courses
Modular	w	3½	4	½	3C = 8″
	h	2¼	2⅔		
	l	7½	8		
Engineer modular	w	3½	4	½	5C = 16″
	h	2¾	3⅕		
	l	7½	8		
Closure modular	w	3½	4	½	1C = 4″
	h	3½	4		
	l	7½	8		
Roman	w	3½	4	½	2C = 4″
	h	1⅝	2⅔		
	l	11½	12		
Norman	w	3½	4	½	3C = 8″
	h	2¼	2		
	l	11½	12		
Engineer norman	w	3½	4	½	5C = 16″
	h	2¾	3⅕		
	l	11½	12		
Utility	w	3½	4	½	1C = 4″
	h	3½	4		
	l	11½	12		

Source: Brick Institute of America

[a]Actual size unit as manufactured.
[b]Specified unit size plus intended joint size.

Table 11.2 Specified Inch-Size Non-modular Common Brick Sizes

Unit Name		Actual Dimension[a]
Standard	w	3⅝
	h	2¼
	l	8
Engineer standard	w	3⅝
	h	2⅔
	l	8
King	w	3
	h	2¾
	l	9⅝
Queen	w	3
	h	2¾
	l	8

Source: Brick Institute of America

[a]Anticipated manufactured dimension.

Figure 11.6 Modular bricks are sized for an 8-in. module with ½ or ⅜ in. mortar joints.

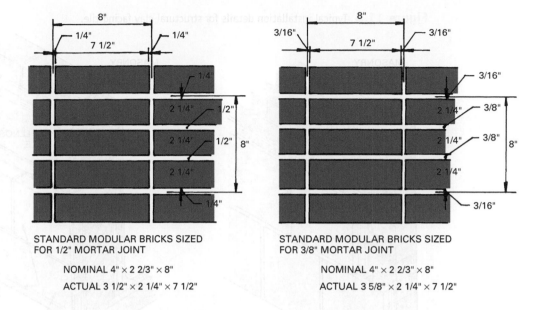

STANDARD MODULAR BRICKS SIZED
FOR 1/2" MORTAR JOINT

NOMINAL 4" × 2 2/3" × 8"

ACTUAL 3 1/2" × 2 1/4" × 7 1/2"

STANDARD MODULAR BRICKS SIZED
FOR 3/8" MORTAR JOINT

NOMINAL 4" × 2 2/3" × 8"

ACTUAL 3 5/8" × 2 1/4" × 7 1/2"

metric brick, 90 × 57 × 190 mm is specified. The nominal size is the modular size plus the 10 mm mortar joint.

Brick Pavers

Brick pavers are solid clay units whose selection is based on weather resistance, abrasion resistance, and appearance relating to bond patterns, size and tolerances. Pavers come in two thicknesses: 1½ in., commonly used for pedestrian situations, and 2¼ in. for heavier traffic/load conditions.

Fire Brick

Firebrick are made from special clays containing alumina-silica with traces of iron oxide, and alkalis that are very uniform in structure and therefore able to withstand high temperatures. Firebricks are generally larger than other structural bricks and are often hand molded. They find use in fireplaces, chimneys, and kiln construction.

Hollow Masonry

Hollow clay masonry comprise units whose net cross-sectional area in the plane of the bearing surface is not less than 60 percent of the gross cross-sectional area of that face. A hollow brick may have a cored area from

25 to 40 percent of the gross cross-sectional area of the bearing surface.

Structural clay tiles are hollow clay units whose cores exceed 40 percent of the gross cross-sectional area (Figure 11.7). They are used in load-bearing and non-load-bearing walls and are available in a variety of glazed and unglazed surface finishes. A smooth, colored glaze is most frequently used, but they also come in matte, speckled, and mottled finishes. Tiles with rough textures or those covered with small-diameter holes of varying sizes are used for acoustical treatment. Clay facing tiles must meet or exceed ASTM requirements for imperviousness and resistance to fading and scratching and can be installed over wood and metal stud walls and any type of masonry or concrete wall (Figure 11.8).

Figure 11.7 Structural clay tile is used for both load-bearing and non-load-bearing applications.

Figure 11.8 Typical installation details for structural clay facing tile.

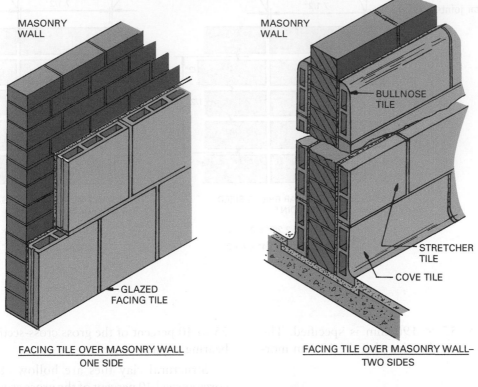

MASONRY
WALL

GLAZED
FACING TILE

FACING TILE OVER MASONRY WALL
ONE SIDE

MASONRY
WALL

BULLNOSE
TILE

STRETCHER
TILE

COVE TILE

FACING TILE OVER MASONRY WALL–
TWO SIDES

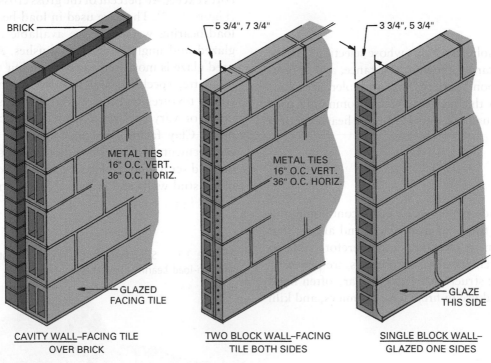

BRICK

METAL TIES
16" O.C. VERT.
36" O.C. HORIZ.

GLAZED
FACING TILE

CAVITY WALL–FACING TILE
OVER BRICK

5 3/4", 7 3/4"

METAL TIES
16" O.C. VERT.
36" O.C. HORIZ.

TWO BLOCK WALL–FACING
TILE BOTH SIDES

3 3/4", 5 3/4"

GLAZE
THIS SIDE

SINGLE BLOCK WALL–
GLAZED ONE SIDES

MATERIAL SPECIFICATIONS FOR CLAY MASONRY

Specifications for clay masonry units are developed by the American Society for Testing and Materials (ASTM) and are based on the weathering index.

Weathering Index

The *weathering index* reflects the ability of clay masonry to resist the effects of weathering according to the number of freeze cycle days coupled with the annual winter rainfall. A *freeze cycle day* is a day when the temperature of the air rises above or falls below 32°F (0°C). The weathering index ranges from 0 to more than 500. Regions rated higher than 500 are considered severe weathering regions, those between 50 and 500 are moderate regions, and areas below 50 are negligible regions (Figure 11.9).

Figure 11.9 The weathering index is an indication of the ability of clay masonry units to resist the effects of weather.

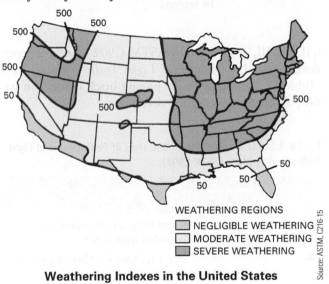

WEATHERING REGIONS
NEGLIGIBLE WEATHERING
MODERATE WEATHERING
SEVERE WEATHERING

Source: ASTM, C216-15

Weathering Indexes in the United States

Grades, Types, and Classes

Building (common), facing, and hollow brick are available in a variety of grades based on the degree of weathering and appearance (Tables 11.3 to 11.5).

Load-bearing structural clay tile is available in two grades, LBX and LB, as specified by ASTM C34 *Standard Specification for Structural Clay Load-Bearing Wall Tile* (Table 11.6). Load-bearing tile will carry building loads in addition to its own weight.

Table 11.3 Grades and Uses of Solid Building Brick (ASTM C62)

Grade (based on weathering index)	Use
SW Severe weathering	For wet locations below grade where bricks may be frozen, such as in foundations
MW Moderate weathering	For vertical masonry surfaces exposed to the weather in relatively dry conditions where freezing can occur
NW Negligible weathering	For use as backup or interior masonry where no freezing occurs

Source: American Society for Testing and Materials

Table 11.4 Grades, Types, and Uses of Solid Facing Brick (ASTM C216)

Grade (based on weathering index)	Use
SW Severe weathering	Masonry in wet locations, in contact with the ground, and subject to freezing
MW Moderate weathering	Exterior walls and other exposed masonry above grade where freezing can occur
Type (based on appearance of the finished wall)	Use
FBX	High degree of physical perfection, minimum variation in color, minimum variation in size
FBS	Wider color range and size variations than permitted in type FBX
FBA	Nonuniform in size, color, and texture

Source: American Society for Testing and Materials

Non-load-bearing structural clay tile is available in one grade, NB, as specified by ASTM C56 *Standard Specification for Structural Clay Nonloadbearing Tile* (Table 11.7). Non-load-bearing tile carries only its own weight.

Unglazed structural facing tile is available in two grades, FTX and FTS, as specified by ASTM C212 *Standard Specification for Structural Clay Facing Tile*. These grades are based on face-shell thickness and factors affecting the appearance of the finished wall (Table 11.8).

Ceramic glazed facing brick and structural clay facing tile are available in two grades, S and SS, and two types, I and II, as specified by ASTM C126

Table 11.5 Grades, Types, and Uses of Hollow Brick (ASTM C652)

Grade (based on weathering index)	Use
SW Severe weathering	High degree of resistance to disintegration by weathering when brick may be permeated with water and frozen
MW Moderate weatheringv	Moderate degree of resistance to frost action where brick is not likely to be permeated with water when it is exposed to freezing temperatures

Type (based on appearance of the finished wall)	Use
HBS	Visible interior and exterior walls where graded variations in color and size than specified for HBX are acceptable
HBX	Visible interior and exterior walls where a small variation in color and size are acceptable
HBA	Nonuniform in size, color, and texture
HBB	Color and texture are not a consideration
	Size variation greater than specified for HBX is acceptable

Source: American Society for Testing and Materials

Table 11.6 Grades and Uses of Structural Clay Load-Bearing Tile (ASTM C34)

Grade (based on weathering index)	Use
LBX	Tile exposed to the weather and as a base for the applications of stucco
LB	Tile not exposed to frost or earth May be used in exposed masonry if covered with 3 in. or more of other masonry

Source: American Society for Testing and Materials

Table 11.7 Grade and Use of Structural Clay Non-Load-Bearing Tile (ASTM C56)

Grade	Use
NB	Non-load-bearing walls, partitions, fireproofing, and furring

Source: American Society for Testing and Materials

Standard Specification for Ceramic Glazed Structural Clay Facing Tile, Facing Brick, and Solid Masonry Units (Table 11.9).

Pedestrian and light traffic paving brick is available in three classes, SX, MX, and NX, and three types,

Table 11.8 Grades and Uses of Unglazed Structural Clay Facing Tile (ASTM C212)

Grade	Use
FTX	Exposed masonry with minimum variation in color and dimensions, smooth face, mechanically perfect
FTS	Smooth or rough textured, with moderate absorption and variation in dimensions, medium color range, and minor surface finish defects

Source: American Society for Testing and Materials

Table 11.9 Grades, Types, and Uses of Ceramic Glazed Structural Clay Facing Tile (ASTM C126)

Grade	Use
S Select	For masonry with narrow mortar joints (")
SS Select sized or ground edge	For masonry where the face dimension variation is small
Type	**Use**
I Single-face units	Where only one finished face is to be exposed
II Two-faced units	Where two opposite finished faces are to be exposed

Source: American Society for Testing and Materials

I, II, and III, as specified by ASTM C902 *Standard Specification for Pedestrian and Light Traffic Paving Brick* (Table 11.10). Light traffic paving brick is used on patios, walkways, floors, and driveways.

Table 11.10 Classes, Types, and Uses of Pedestrian and Light Traffic Paving Block (ASTM C902)

Grade (based on weathering index)	Use
SX	Where brick may be frozen when saturated with water
MX	Where resistance to freezing is not necessary
NX	Interior use when an effective sealer or water-resistant surface coating will be applied
Type (based on traffic)	**Uses**
I	Where brick will be exposed to extensive abrasion, as in driveways
II	Where brick will be exposed to intermediate traffic, as on floors in stores
III	Where bricks will be exposed to low traffic, as in residences

Source: American Society for Testing and Materials

PROPERTIES OF CLAY BRICK AND TILE

Finished clay units vary considerably in their physical properties. The types of raw materials used and the effects of the production process greatly influence these characteristics. The most important properties are compressive strength, durability, absorption, color, and texture.

Compressive Strength

Compressive strength depends on the type of clay used, how the units are made, and the temperature and duration of firing. In general, plastic clays used in the stiff-mud process produce units with higher compressive strengths. Higher burn temperatures produce higher compressive strengths in almost any clay or process used. Bricks vary in compressive strength from 1,500 psi (10.35 MN/m²) to more than 5,000 psi (34.50 MN/m²).

The compressive strength in a wall of brick depends not only on the compressive strength of the brick, but also on that of the mortar. For example, a brick may have a compressive strength of 4,000 psi, but when it is combined with a mortar, the allowable compressive strength will be limited by the mortar's compressive strength.

Durability

Durability refers to the ability of a clay masonry unit to resist damage from freeze and thaw cycles and moist conditions. Durability is a result of the degree of fusion of clays during burning. High burn temperatures tend to produce a harder, more durable unit.

Absorption

Clay units absorb a certain amount of water that can affect the bond strength of mortar to brick. The properties of the clays, the process used to make the brick, and the burn temperature all influence absorption. In general, plastic clays and high-temperature burns produce units that have lower absorption. Units made with the stiff-mud process usually have lower absorption than those made with other processes. The rate at which a clay unit absorbs moisture is called the *suction rate*. Suction refers to the tendency of a brick to take up moisture in pores and small openings in its surface by capillary action. It refers to surface water that penetrates the brick itself. Suction affects the bond strength of mortar to brick. The strongest bond results when units have a suction rate that does not exceed 0.7 ounces (20 grams) of water per minute. If a brick has a greater suction rate than this, it should be wetted and the surface allowed to dry before the brick is laid.

The initial rate of absorption (suction) can be determined in a laboratory using ASTM 67 testing procedures. The masonry unit is immersed in ⅛ in. (3 mm) of water for one minute, removed and weighed. The gain in weight from its dry condition is an indication of the initial rate of absorption (IRA) and is measured in grams per minute (gpm) per 30 in.² Rate of absorption values range from 1 to 50 gpm or more.

Color

The color of clay masonry units depends on the clays used, burning temperature, and the method of controlling color during the burn. Burned clays vary widely in color, from creams and grays through reds and purples. Iron oxide, commonly found in clay, is the dominant oxide that influences color. It produces a red unit in an oxidizing fire and becomes purple if burned in a reduced atmosphere (called flashing). Under-burning produces a salmon-colored unit that is softer, has lower compressive strength, and a higher absorption rate than red bricks. Bricks that are over-burned are called clinkers and tend toward dark red to black when made from clays high in iron oxide. Buff clays are used to produce bricks ranging from yellow-brown to dark tan.

Since clay composition varies, the color of produced units has some variation within each unit. The application of surface coatings, such as glazes, enables units to be produced with almost any color desired. Chemicals that produce a range of colors when they vaporize can also be introduced into the clay.

Texture

The surface texture of a finished clay masonry unit is produced by the surface of the die or mold used to form the unit or by attachments that cut, scratch, roll, or in some other way alter the surface as the clay unit leaves the die. Some of the standard textures are smooth, matte (with horizontal or vertical markings), barks, rugs, sand-molded, stippled, water-struck, and sand-struck (Figure 11.10).

Figure 11.10 Clay masonry units are available in a wide variety of colors and textures.

©iStock.com/sndr

Heat Transmission

The ability of clay brick walls to facilitate or resist the transmission of heat directly influences the surface temperature of interior walls and rooms. In most applications, resistance to heat transmission is very important. In others, such as the storage and conduction of solar heat, the ability to transmit heat is important. Selected examples of heat transmission (U-value) and heat resistance (R-value) are given in Table 11.11. These figures vary for different types of brick because of the differences in both clay used and unit density.

Table 11.11 Coefficients of Heat Transmission and Resistance of Solid and Cavity Clay Brick[a]

Brick	Transmission (U)	Resistance (R)
4″ (100 mm) solid brick	2.27	0.44
6″ (150 mm) solid brick	1.52	0.66
8″ (200 mm) solid brick	1.14	0.88
12″ (300 mm) solid brick	0.76	1.32
10″ (250 mm) cavity brick	0.34	2.96

[a]Values reflect an average rating. Get actual values from the manufacturer of the masonry units to be used.

Fire Resistance

The fire resistance of a material is an indication of its ability to prevent materials behind it from igniting. Fire resistance is stated in hours and is usually called the "fire rating." These ratings are developed by testing units under actual fire conditions set up in a laboratory. Most floors, walls, and ceilings are made up of several materials, so fire ratings of these assemblies are also known. Table 11.12 lists the fire ratings for selected clay masonry units.

Table 11.12 Fire Resistance of Selected Brick and Tile Walls

Solid Brick	Fire Rating in Hours
4″ (100 mm) brick	1
6″ (150 mm) brick	3
8″ (200 mm) brick	4
Hollow Core Brick	
8″ (200 mm) brick wall	2
8″ (200 mm) brick wall plastered both sides	4
Structural Facing Tile	
4″ (100 mm) tile wall plastered one side	1
6″ (150 mm) tile wall	2
8″ (200 mm) tile wall plastered both sides	4

University of Florida Rinker Hall

Architect: Croxton Collaborative
Architects, P.C.

Contractor: Centex Rooney Construction
Co., Inc.

Location: Gainesville, Florida

Building type: Higher Education

Size: 47,300 sq. feet (4,390 sq. meters)

Completed March 2003

Rating: LEED Gold (39 points)

Rinker Hall is home to the University of Florida's College of Design and School of Building Construction, providing classrooms, construction teaching labs, and administrative facilities for 1,600 students and faculty. The new building incorporates a range of green building features and in 2004 became the first building in Florida to achieve a LEED Gold rating from the U.S. Green Building Council. Located on the site of a former parking lot, Rinker Hall was designed to maximize daylighting, collect rain water (used for flushing toilets), and features a white roof to reflect heat instead of absorbing it.

For the school of building construction, great care was taken in the selection and procurement of materials for the new building. Material selections were based on stringent prerequisites, including: proximity of manufacturing, recycled content, renewable-resource content, durability, low-maintenance attributes, low toxicity, and potential for recycling and reuse at the end of the building's life. A material proximity and "optimum pathways" exercise involving construction students supported the design team in the procurement of building materials.

When an existing building on the campus was demolished, its wall brick was carefully recycled for reuse on the new building. The brick salvaged from the demolished Hume Hall provided historical significance and a rustic appearance in addition to low cost. Salvaged brick is about $30 to $70 less expensive per 500 bricks than the cost of new bricks. The other main advantage to salvaging brick is the reduced volume of waste that otherwise goes to a landfill.

For bricks to be effectively recycled, they must be properly cleaned. If old mortar is not completely removed from a brick's surface, the new bond can be negatively impacted. Removing bricks in mortar can be easy or difficult, depending on how hard the mortar is. Typically, bricks at least fifty years old are used; otherwise the mortar is too difficult to remove. For the Rinker Hall project, students volunteered their time to clean the bricks and store them on pallets for later use.

Walls made from salvaged brick are generally less durable than those composed of new brick masonry. A lime mortar rather than one of Portland cement is recommended for laying reclaimed brick. Lime mortar is composed of lime and sand and is generally low in salt

content that can cause efflorescence on the brickwork. Lime mortar is also highly plastic and more likely to achieve a good bond with porous brick.

Reused brick on the project was utilized in retaining walls and other non-critical load-bearing elements of the structure. The project allowed students to gain first-hand experience in sustainable construction principles as applied to planning, design, and reconstruction of the built environment.

OTHER FIRED CLAY PRODUCTS

Clay tile is a vitrified clay product, such as clay pipe, used for drainage and sewer systems. Roof tile is burnt clay used for finished roofing (Figure 11.11). Tiles are made in flat, plain shingles, single-lap tiles, and interlocking tiles. They are fashioned in various styles, reflecting the countries where they have been used for centuries (Spanish, French, Roman, and English).

Figure 11.11 Clay roof tile has been used for centuries to provide a fire-proof and durable roofing material.

©iStock.com/phototropic

Architectural Terra-Cotta and Ceramic Veneer

Architectural terra-cotta is a hard-fired clay that has been used for hundreds of years for decorative purposes. Modern terra-cotta products are extruded and molded or pressed to shape. Most of these products are custom made to an architect's designs and specifications. Figure 11.12 shows a sun-shading façade panel made from architectural terra-cotta.

Figure 11.12 Architectural terra-cotta used as a screen over a glazed façade.

Architectural terra-cotta is a red earth color when left unglazed. It is available with a smooth, plain surface that may be unglazed, or it can have a transparent glaze, a non-lustrous glaze that gives a satin or matte finish, a ceramic color glaze, or a polychrome finish. Sculptural and decorative reproductions of classic architectural ornamentation can be made from terra-cotta and are generally handmade in special molds.

A machine-made unit having a flat or ribbed back and a flat face is called "ceramic veneer."

Ceramic veneer may be an earth red if unglazed but may have a transparent ceramic glaze, a non-lustrous glaze with a satin or matte finish, or a ceramic color glaze, either solid or a mottled blend of colors. A polychrome finish is used, which means two or more colors are applied to separate areas and each color is burned separately. The surface may be smooth, scored, combed, or roughened. Ceramic veneer is installed by mortaradhesion or through a system of mechanical anchoring.

Review Questions

1. What are the forms in which clay is found?
2. How do fireclays differ from surface clays?
3. How do the colors of calcareous and non-calcareous clays differ?
4. What methods are used to form bricks?
5. What are some of the common surface textures used on face bricks?
6. Why are the temperature and humidity carefully controlled when drying bricks?
7. How does the burning temperature influence the size and color of bricks?
8. What are the classifications of structural clay masonry units?
9. How can you identify solid masonry clay units?
10. What is the difference between modular and actual brick dimensions?
11. What is the allowable hollow cored area for a hollow brick?
12. What two factors are considered when establishing the weathering index?
13. What is the range of compressive strength for common bricks?
14. What unit of suction provides conditions for the best mortar bond?

Key Terms

Architectural Terra-Cotta	Glaze	Stiff-Mud Process
Brick Pavers	Hollow Clay Masonry	Suction Rate
Burning	Modular Bricks	Surface Clay
Dry-Press Process	Shale	Weathering Index
Fireclay	Soft-Mud Process	Winning
Freezing Cycle Day	Solid Masonry	

Activities

1. Collect samples of clay masonry products for use in the classroom. Label each, giving as much information about their properties as you can collect.

2. Design and run some tests on samples of clay masonry units, such as compression tests, hardness tests, and moisture absorption tests.

Additional Resources

Architectural Graphic Standards, the American Institute of Architects, John Wiley & Sons, New York, NY.

Jaffe, R., *Masonry Basics*, the Masonry Society, Boulder, CO.

Technical Notes on Brick Construction on CD-ROM, the Brick Industry Association, Reston, VA.

Numerous publications from the Portland Cement Association, Skokie, IL.

Other resources include:

Publications from the Brick Industry Association, Reston, VA.

See Appendix C for addresses of professional and trade organizations and other sources of technical information.

Concrete Masonry

Upon completion of this chapter, the student should be able to:

- Explain how concrete masonry units are manufactured.
- Use information about the physical properties of concrete masonry units when making material selections.

- Recognize the many types of concrete masonry units and be able to choose those suitable for various applications.

Build Your Knowledge

For further study on these materials and methods, please refer to:

Chapter 7   Concrete

Concrete masonry blocks, first introduced in the 1880s, find use in virtually every type of construction including commercial and residential buildings, fireproof enclosures and landscape retaining walls. They are manufactured in a wide range of standard sizes and custom-designed architectural units. As one of the most widely used modern construction materials, concrete blocks serve in both structural and nonstructural applications (Figure 12.1).

Figure 12.1 Concrete masonry blocks find use in virtually every type of construction.

© bodhichita/Shutterstock.com

MANUFACTURE OF CONCRETE MASONRY UNITS

Concrete masonry units (CMUs) are made of a relatively dry mix of Portland cement, aggregates, water, and, in some cases, admixtures. The dry materials are weighed and moved to a mixer. The mixer adds the required water and agitates the batch for a predetermined time. The mixed batch is discharged into the hopper of a block machine, fed into molds, and consolidated by pressure and vibration. The freshly molded "green blocks" are moved from the block machine on steel pallets to a curing rack.

The curing rack full of green units travels to either a low-pressure steam kiln or an autoclave for hardening. In a low-pressure steam *kiln*, the green units take on an initial set before steam is introduced. This takes from one to three hours with the temperature kept at 70°F to 100°F (22°C to 38°C). Then steam is introduced, providing heat and moisture. The temperature is gradually raised from 150°F to 180°F (66°C to 82°C), depending on the composition of the concrete. This condition is maintained for ten to twenty hours until the units reach the required strength. After removal, the blocks are stored in a protected condition and achieve almost full strength in two to four more days.

An *autoclave* uses high-pressure steam. The molded units are placed in the autoclave, allowed to set for two to five hours, and then gradually heated with saturated steam under a pressure of 150 psi (1,035 kPa) for two to three hours. Once the maximum temperature (350°F (178°C)) and pressure are reached, the units soak for five to ten hours. The pressure is gradually released over a 30-minute period before they are moved to storage. They can be used twenty-four hours after leaving the autoclave. High-pressure steamed units have greater stabilization against volume changes caused by moisture than the low-pressure units do.

PHYSICAL PROPERTIES OF CONCRETE MASONRY UNITS

The physical properties of concrete, discussed in Chapter 7, apply to concrete masonry units as well. The properties of the units are determined by the cement paste and aggregates. Differences that do exist between concrete and concrete masonry units are caused by different mix compositions, methods of consolidation, and curing processes.

Concrete masonry units are generally made with less cement per cubic yard and a lower water-cement ratio than redi-mix concrete. The aggregate is finer, with ⅜ in. (10 mm) being the largest size. Concrete masonry units have a large volume of void spaces between aggregate particles, while normal concrete should have no voids.

The important properties to consider when specifying concrete masonry units are weight, compressive strength, water absorption, and the coefficient of thermal expansion (Table 12.1). The required properties are established by national building codes and ASTM standards. The weight of a concrete masonry unit varies with the design of the block and the mix used to make it. It is necessary to know weights so the dead loads of a structure can be calculated.

Compressive strength data provide a means of determining a unit's ability to carry loads and withstand structural stresses. The strength varies depending on the wetness of the mix. Wetter mixes give the highest strength, but cause difficulties in manufacturing the units. The manufacturer develops units and tests them to determine

Table 12.1 Properties of Concrete Blocks with Various Aggregate

Customary Units			
Aggregate	Compressive Strength in psi (gross area)	Coefficient of Thermal Expansion per °F × 10⁻⁶	Water Absorption (lb./ft.³ of concrete)
Sand and gravel	1,200–1,800	5.0	7–12
Limestone	1,100–1,800	5.0	8–12
Air-cooled slag	1,100–1,500	4.6	9–13
Expanded shale	1,000–1,500	4.5	12–15
Cinders	700–1,000	4.5	12–18
Expanded slag	700–1,200	4.0	12–18
Pumice	700–900	4.0	13–20
Scoria	700–1,200	4.0	12–18
Metric Units			
Aggregate	Compressive Strength in kg/cm² (gross area)	Coefficient of Thermal Expansion mm/mm/°C × 10⁻⁶	Water Absorption (kg/m³)
Sand and gravel	84.4–127	9.0	128–190
Limestone	77.3–127	9.0	128–190
Slag	77.3–105.5	8.3	144–208
Expanded shale	70.3–105.5	8.1	192–240
Cinder	49.2–70.3	4.5	192–288
Expanded slag	49.2–84.4	7.2	192–288
Pumice	49.2–63.3	7.2	208–320
Scoria	49.2–84.4	7.2	192–288

the compressive strength of the products available. Compressive strengths, given in Table 12.1, are based on the gross bearing area of the block, including the core spaces. Compressive strength of the net area (actual surface, excluding the cores) is 1.8 times the values shown.

Using the graph in Figure 12.2, design compressive data can be obtained if the compressive strength of the concrete masonry unit (based on actual area) and the type of mortar are known.

Figure 12.2 The compressive strength of a wall made from concrete masonry units depends on the type of mortar and the strength of the concrete masonry units.

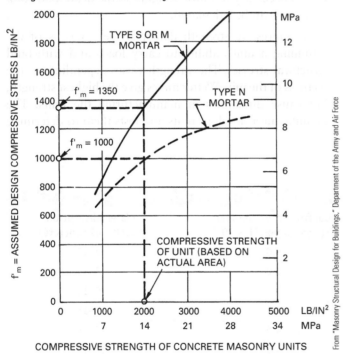

COMPRESSIVE STRENGTH OF CONCRETE MASONRY UNITS

Tensile strength is about 5 to 10 percent of compressive strength. Flexural strength is about 15 to 20 percent of compressive strength, and the modulus of elasticity ranges from 300 to 1,200 times the compressive strength.

Water absorption varies with the density of the concrete masonry unit. Absorption is a measure of the pounds of water per cubic foot of concrete. Units made with a dense aggregate have much lower absorption rates than those made with lightweight aggregate. High water absorption is not acceptable for many applications. Units having a large number of interconnected pores and voids have a high absorption rate (also called suction rate). Units with a high suction rate and high absorption by aggregate have a high permeability to water, air, and sound, and are more likely to be damaged by

freezing. Units with unconnected pores, such as those found in lightweight aggregate and in the air-entrained cement paste, provide for sound absorption, reduce thermal conductivity, and minimize permeability to water.

Units used for exterior walls that will not be painted should be low-absorption units. Painting and other waterproof coatings greatly reduce water absorption. It should be remembered that concrete masonry units ought to be in equilibrium with the surrounding air before installation. If they are used before this occurs, they will shrink slightly as moisture is lost, causing tensile and shearing stresses that may produce cracking.

The coefficient of thermal expansion is used to calculate the amount of expansion that can be expected as temperatures change. Concrete masonry units expand when heated and contract when cooled. These changes are reversible, and the unit returns to its original size after the temperature returns to the point at which the change started. Since the aggregate makes up about 80 percent of a unit, the coefficient of expansion of the aggregate is a major factor influencing expansion and contraction of the concrete masonry unit.

Other properties that are a result of those just discussed include insulation value, coefficient of heat transmission, sound-absorbing properties, and fire resistance.

The insulation value of units made with porous, lightweight aggregate is better than those using denser material. The units with lightweight aggregate also have a lower coefficient of heat transmission. Recall that heat transmission, U, is stated in British thermal units (BTU) per hour per square foot per degree Fahrenheit for each degree difference in temperature between the air on the cool and warm sides of a wall. Concrete masonry walls are poor insulators and good heat transmitters, and often require the addition of insulation materials to be energy efficient (Table 12.2).

Concrete masonry units resist the transmission of sound. Hollow block made with lightweight aggregate provides the best sound isolation. The addition of a

Table 12.2 Coefficients of Heat Transmission for Selected Concrete Masonry Units

Hollow Masonry			
Size	Aggregate	Btu/hr/ft²/8F	Watts/m²/8C
8″	Limestone	0.53	3.0
8″	Sand and gravel	0.53	3.0
8″	Cinders	0.37	2.1
8″	Expanded clay, slag, or shale	0.31	1.8

plaster interior or exterior finish increases this property. Concrete masonry units that have an open, porous surface absorb sound better than do denser units with smooth surfaces. Painting the surface fills these pores and reduces sound-absorbing properties. Acoustical concrete block units are manufactured to combine a sound-deadening liner panel with a concrete unit with open slots on the face (Figure 12.3).

Figure 12.3 The sound-absorbing qualities of acoustical blocks improve the acoustical properties of interior spaces.

Courtesy of Trenwyth Industries, Inc.

Building codes stipulate the required fire-resistance ratings for various assemblies in a building. Products such as concrete masonry units must be carefully tested before they are given a fire-resistance rating. The rating specifies the number of hours the material can be exposed to a flame before it fails. The fire-resistance of concrete masonry units varies depending on the aggregate used. Plaster on a concrete unit is an effective way of increasing its fire resistance. A method for estimating the fire-resistance rating of concrete masonry is shown in Table 12.3.

TYPES OF CONCRETE MASONRY UNITS

Concrete masonry units, often called concrete blocks, are made in a wide range of sizes and types. Some are made based on custom designs prepared by an architect. They can be divided into three classifications: concrete brick (solid units), concrete block (hollow and solid units), and special units.

Modular Size Blocks

Concrete masonry units are made in modular sizes and are specified by their nominal or modular size. They are produced on a 4 in. module. The metric modular unit is 100 mm.

The nominal or modular size is the theoretical size without allowance for a mortar joint. The actual size is ⅜ in. less than the nominal size, so the actual unit plus a ⅜ in. mortar joint gives the modular size (Figure 12.4). Concrete units are also related in size to standard bricks, allowing composite assemblies to be realized. For example, as shown in Figure 12.5, a standard concrete block is two bricks wide, three brick courses high, and two bricks long.

True metric modular concrete block modular size is 200 × 200 × 400 mm. Actual modular size is 190 × 190 × 390 mm (7½ × 7½ × 15⅜ in.). American metric modular blocks soft-converted from inch sizes are (actual size) 194 × 194 × 397 mm (7⅝ × 7⅝ × 15⅝ in.), which is quite similar to a true metric block. The mortar bed is 10 mm. (See Figures 12.4 and 12.5).

Concrete Brick

Concrete bricks are made either solid or with a depressed area called a "frog." The frog reduces the weight. They are laid with a ⅜ in. mortar joint. A standard modular brick is 4 × 2⅜ × 8 in. nominal. Some block manufacturers make other sizes.

Slump Block

Slump block is a concrete masonry block that is removed from the mold before it has had a chance to set completely. This causes the concrete block to retain a slightly slumped appearance. The units have irregular faces and some differences in height and surface texture, giving a rustic appearance when laid into a wall.

Concrete Block

A wide variety of types of standard concrete blocks are available, many designed for special uses. Some are shown in Figure 12.6. The dominant unit manufactured today is the 8 × 8 × 16 in. nominal module. (Actual size 7⅝ × 7⅝ × 15⅝, allowing for a ⅜ in. mortar joint.) They are produced in three major groups: solid load-bearing, hollow load-bearing, and non-load-bearing units. A solid load-bearing unit is one whose cross-sectional area in every plane parallel to the bearing surface is not less than

Table 12.3 Estimated Fire Rating in Hours for Concrete Masonry

	Minimum Equivalent Thickness in Inches for Fire Ratings in Hours			
Type of Aggregate	1 hour	2 hours	3 hours	4 hours
Pumice or expanded slag	2.2*	3.2	4.0	4.7
Expanded shale, clay, or slate	2.6	3.6	4.2	5.1
Limestone, cinders, or unexpanded slag	2.7	4.0	5.0	5.9
Calcareous gravel	2.8	4.2	5.3	6.2
Siliceous gravel	3.0	4.5	5.7	6.7

How to Figure Equivalent Thickness for Cored Blocks

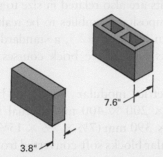

Equivalent thickness is the solid thickness that would be obtained if the same amount of concrete contained in a hollow unit were re-cast without core holes.

Calculating Estimated Fire Resistance. Example: A 7.6-in. (actual size) hollow masonry wall is constructed of expanded slag units reported to be 55%* solid. What is the estimated fire resistance of the wall? (modular units)

Eq Th = 0.55×7.6 in. = 4.2 in. From table: 3-hr fire resistance requires 4.00 in. Use 3-hr est. resistance.

	Minimum Equivalent Thickness in Millimeters for Fire Ratings in Hours			
Type of Aggregate	1 hour	2 hours	3 hours	4 hours
Pumice or expanded slag	55.9	76.2	101.6	119.4
Expanded shale, clay, or slate	66.0	91.4	106.7	129.5
Limestone, cinders, or unexpanded slag	68.6	101.6	127.0	149.9
Calcareous gravel	71.1	106.7	134.6	157.5
Siliceous gravel	76.2	144.8	114.8	170.2

How to Figure Equivalent Thickness for Cored Blocks

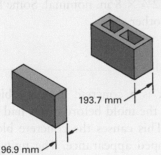

Equivalent thickness is the solid thickness that would be obtained if the same amount of concrete contained in a hollow unit were re-cast without core holes.

Calculating Estimated Fire Resistance. Example: A 193.7-mm (actual size) hollow masonry wall is constructed of expanded slag units reported to be 55%* solid. What is the estimated fire resistance of the wall? (modular units)

Eq Th = 0.55×193.68 mm = 106.43 mm. From table: 3-hr fire resistance requires 101.6 mm. Use 3-hr est. resistance.

Source: "Fire Safety with Concrete Masonry," 35C, with permission, National Concrete Masonry Association

75 percent of the gross cross-sectional area measured in the same plane. A hollow concrete block is one whose cross-sectional area in every plane parallel to the bearing surface is less than 75 percent of the gross cross-sectional area measured in the same plane.

Block cores are typically tapered so that the top surface of the block has a greater surface on which to spread a mortar bed. There may be two, three, or four cores, although two cores are the most common configuration. The core configuration allows steel reinforcing

to be inserted into the assembly, greatly increasing its strength. Reinforced cores are filled with grout to secure the reinforcing in proper relationship to the structure, and to bond the block and reinforcing. The reinforcing is primarily used to impart greater tensile strength to the assembly, improving its ability to resist lateral forces such as wind load and seismic forces.

Many units are made with two rather than three cores. The two-core unit has the advantage of having an increase of the face shell thickness and the center web.

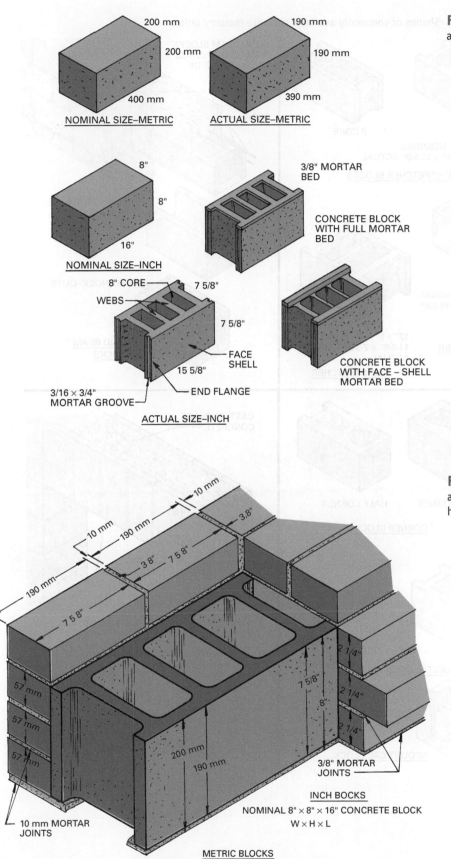

Figure 12.4 Standard inch and metric concrete blocks.

200 mm

200 mm

400 mm

NOMINAL SIZE–METRIC

190 mm

190 mm

390 mm

ACTUAL SIZE–METRIC

8"

8"

16"

NOMINAL SIZE–INCH

3/8" MORTAR BED

CONCRETE BLOCK WITH FULL MORTAR BED

8" CORE

WEBS

7 5/8"

7 5/8"

FACE SHELL

15 5/8"

3/16 × 3/4" MORTAR GROOVE

END FLANGE

CONCRETE BLOCK WITH FACE – SHELL MORTAR BED

ACTUAL SIZE–INCH

Figure 12.5 Standard brick and blocks are compatible in height and width.

10 mm

10 mm

190 mm

3 8"

7 5 8"

3 8"

190 mm

7 5 8"

57 mm

57 mm

57 mm

200 mm

190 mm

10 mm MORTAR JOINTS

7 5/8"

8"

2 1/4"

2 1/4"

2 1/4"

3/8" MORTAR JOINTS

INCH BOCKS

NOMINAL 8" × 8" × 16" CONCRETE BLOCK
W × H × L

METRIC BLOCKS
NOMINAL SIZE 200 × 200 × 400 mm

Figure 12.6 Shapes of commonly available concrete masonry units.

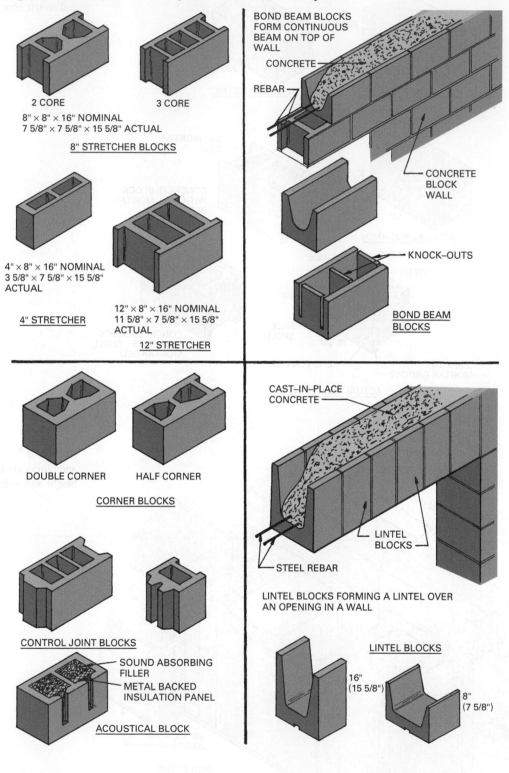

2 CORE   3 CORE

8" × 8" × 16" NOMINAL
7 5/8" × 7 5/8" × 15 5/8" ACTUAL

<u>8" STRETCHER BLOCKS</u>

4" × 8" × 16" NOMINAL
3 5/8" × 7 5/8" × 15 5/8"
ACTUAL

<u>4" STRETCHER</u>

12" × 8" × 16" NOMINAL
11 5/8" × 7 5/8" × 15 5/8"
ACTUAL

<u>12" STRETCHER</u>

DOUBLE CORNER   HALF CORNER

<u>CORNER BLOCKS</u>

<u>CONTROL JOINT BLOCKS</u>

SOUND ABSORBING
FILLER
METAL BACKED
INSULATION PANEL

<u>ACOUSTICAL BLOCK</u>

BOND BEAM BLOCKS
FORM CONTINUOUS
BEAM ON TOP OF
WALL

CONCRETE

REBAR

CONCRETE
BLOCK
WALL

KNOCK–OUTS

<u>BOND BEAM
BLOCKS</u>

CAST–IN–PLACE
CONCRETE

LINTEL
BLOCKS

STEEL REBAR

LINTEL BLOCKS FORMING A LINTEL OVER
AN OPENING IN A WALL

<u>LINTEL BLOCKS</u>

16"
(15 5/8")

8"
(7 5/8")

Figure 12.6 *Continued*

COLUMN BLOCK
WITH WALL INSET

OPEN END BLOCK FOR
PLUMBING STACKS
AND STEEL COLUMNS

HEADER BLOCK

WINDOW

ANOTHER TYPE
OF COLUMN BLOCK

SILL
BLOCK

JAMB BLOCK

SASH BLOCK

COPING BLOCK
ON TOP OF CONCRETE
BLOCK WALL

CORNER RETURN
BLOCK

PILASTER INSERT
BLOCK

ONE TYPE OF CHIMNEY
BLOCK

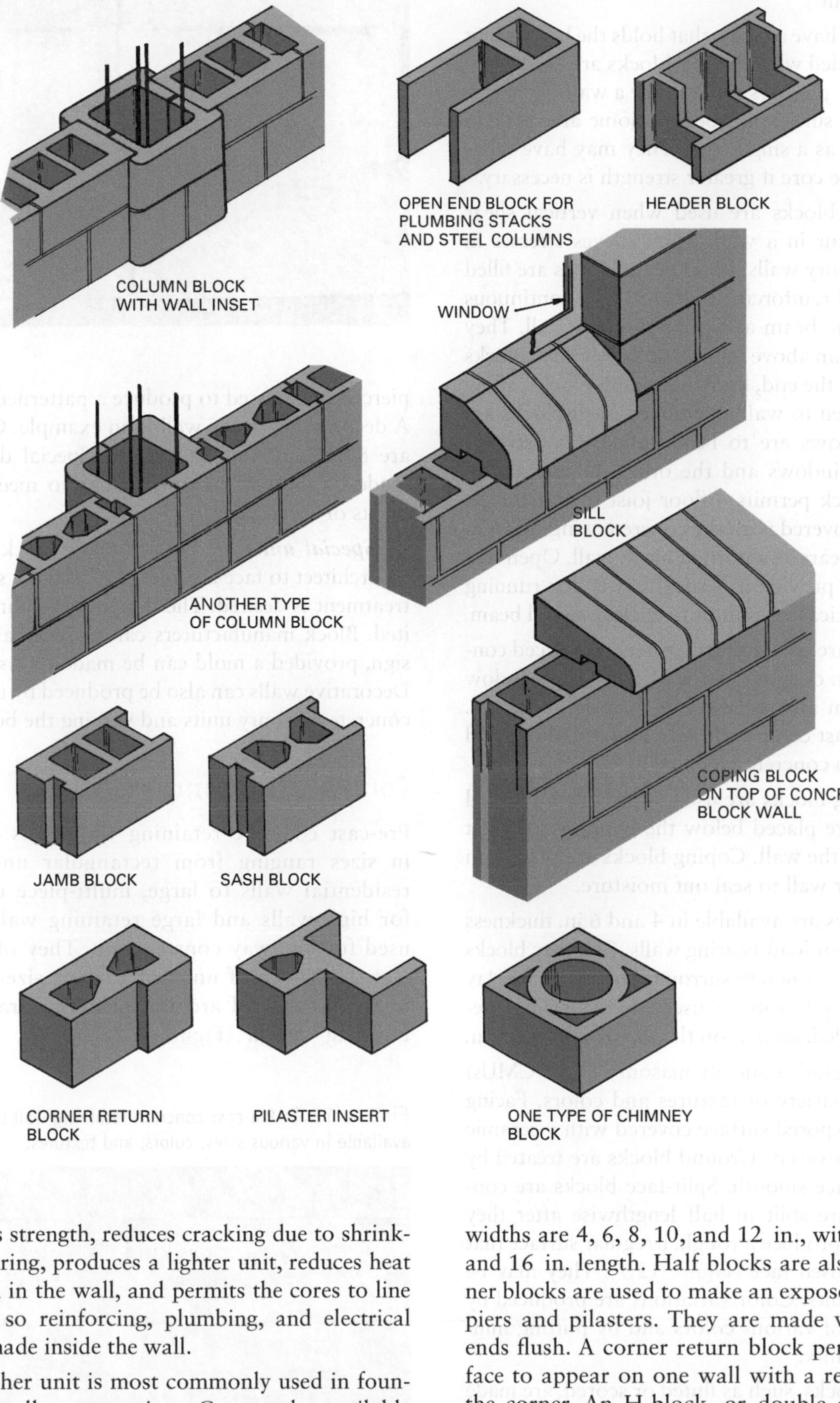

This increases strength, reduces cracking due to shrinkage during curing, produces a lighter unit, reduces heat concentration in the wall, and permits the cores to line up vertically so reinforcing, plumbing, and electrical runs can be made inside the wall.

The stretcher unit is most commonly used in foundation and wall construction. Commonly available widths are 4, 6, 8, 10, and 12 in., with an 8 in. height and 16 in. length. Half blocks are also available. Corner blocks are used to make an exposed corner, and for piers and pilasters. They are made with one or both ends flush. A corner return block permits a full block face to appear on one wall with a recess to help turn the corner. An H-block, or double open-end unit, is open on both ends which increases the space available for rebar and grout.

Header blocks have a recess that holds the header unit in a masonry-bonded wall. Pilaster blocks are used when the design calls for pilasters to reinforce a wall. They also provide a bearing surface for beams. Some are made in halves and others as a single unit. They may have rebar and concrete in the core if greater strength is necessary.

Control joint blocks are used when vertical shear control joints occur in a wall. This relieves stresses in long runs of masonry walls. Bond beam blocks are filled with concrete and reinforcing bar to form a continuous reinforced concrete beam along the top of a wall. They also are used to span above lintels and below sills. Blocks with a channel on the end, known as jamb blocks, allow doors to be secured to wall assemblies. Sash blocks are used where windows are to be installed. One style is used for wood windows and the other for metal windows. A joist block permits a floor joist to rest on the wall. The end is covered with the concrete wing, so from the exterior it appears as a normal block wall. Open-end units are used to provide a vertical cavity for running plumbing or electrical conduit or to enclose a steel beam.

Lintel blocks are used to form a steel reinforced concrete lintel over an opening in a wall, such as a window opening. They can also be used to form bond beams. Alternately, pre-cast concrete lintels are used above and below openings in concrete block walls.

Sill and coping blocks are used to provide a finished cap. Sill blocks are placed below the window to direct water away from the wall. Coping blocks are placed on top of the exterior wall to seal out moisture.

Partition blocks are available in 4 and 6 in. thickness and are used for non-load-bearing walls. Chimney blocks are used to form a concrete surround around a fireclay flue lining. Solid top blocks are used where a solid, coreless surface is needed, such as on the top of a foundation.

Architectural-grade concrete masonry units (CMUs) are produced in variety of textures and colors. Facing blocks have the exposed surface covered with a ceramic glaze or a plastic overlay. Ground blocks are treated by grinding the surface smooth. Split-face blocks are concrete units that are split in half lengthwise after they have hardened to produce a rough, irregular surface that serves as the exposed face (Figure 12.7). They may be hollow or solid units. Color variations are produced by using aggregates of various colors and by putting mineral colors in the mix.

Decorative blocks, such as fluted or scored, are made with many different face designs. Typically, they are

Figure 12.7 Split face block adds shadow and texture to a wall.

pierced or recessed to produce a patterned wall surface. A decorative privacy wall is an example. Common sizes are 8×16 in. and 12×12 in. Special designs can be made by the block manufacturer to meet the requirements of an architect.

Special units are custom-made blocks designed by an architect to face a project for which a special surface treatment is desired. The design possibilities are unlimited. Block manufacturers can produce almost any design, provided a mold can be made to fashion the unit. Decorative walls can also be produced by using standard concrete masonry units and varying the bond pattern.

Concrete Retaining Walls

Pre-cast concrete retaining wall units are available in sizes ranging from rectangular units for small residential walls to large, multi-piece units suitable for high walls and large retaining walls, like those used for highway construction. They offer an architect the choice of units of various sizes, colors, and textures to blend architectural appearance with the building design (Figure 12.8). Most are made of

Figure 12.8 Pre-cast concrete retaining wall units are available in various sizes, colors, and textures.

interlocking units and some are designed for a battered wall configuration.

Autoclaved Aerated Concrete Block (AAC)

Autoclaved aerated concrete block uses Portland cement mixed with lime, silica sand, or recycled fly ash, water, and aluminum powder. The mix is poured into a mold where the reaction between aluminum and concrete causes microscopic hydrogen bubbles to form, expanding the mix to about five times its original volume. After evaporation of the hydrogen, aerated concrete is cut to size and formed by steam curing in a pressurized chamber (an autoclave). The result is a nontoxic, airtight material with excellent thermal, fire, and acoustical-resistance properties that can be used for wall, floor, and roof panels and blocks, and lintels to provide structural capacity (Figure 12.9).

Figure 12.9 Autoclaved aerated concrete block provide excellent thermal, fire, and acoustical-resistance properties.

© pryzmat/Shutterstock.com

MATERIAL SPECIFICATIONS

Concrete masonry units (CMUs) are manufactured to conform to ASTM C140, *Standard Test Methods for Sampling and Testing Concrete Masonry Units and Related Units*. ASTM has specifications for four types of CMUs (Table 12.4).

Table 12.4 ASTM[a] Standards for Concrete Masonry Units

Standard	Type of Unit
ASTM C55	Concrete building brick, solid veneer, and split block
ASTM C90	Hollow load-bearing concrete masonry units
ASTM CI29	Hollow non-load-bearing concrete masonry units
ASTM CI45	Solid[b] load-bearing concrete masonry units

[a]*American Society for Testing and Materials.*
[b]*Units with 75% or more net area.*
Source: *American Society for Testing and Materials*

Grades

Load-bearing concrete masonry units are available in two grades, S and N (Table 12.5). Grade S units are restricted to above-grade applications and, when used for exteriors, must have a protective coating. Grade N units can be used above and below grade, and may be exposed to the weather or moisture penetration. When used below grade, protective coatings are recommended and often are required by building codes.

Table 12.5 Grades, Types, Weights, and Uses of Concrete Masonry Units (ASTM C90)

Grade	Use
N	General use above and below grade
S	Above grade only in walls not exposed to weather or with weather-protective coating if exposed to the weather

Type	Use
I	Moisture controlled during manufacture
II	Non-moisture-controlled units

Weight	Use
Normal weight	Manufactured using concrete weighing more than 125 pcf (2,000 kg/m^3)
Medium weight	Manufactured using concrete weighing between 105 and 125 pcf (1,680 to 2,000 kg/m^3)
Lightweight	Manufactured using concrete weighing 105 pcf (1,680 kg/m^3) or less

Types

Two types of concrete masonry units are specified by ASTM C90 *Standard Specification for Loadbearing Concrete Masonry Units*, Type I and Type II. Type I units are manufactured with specific limits on moisture content. Type II has no moisture control limits. Moisture limits on block manufacturing have the advantage of minimizing shrinkage after units have been laid. Shrinkage can cause cracking of the wall. Uses of the various grades and types are shown in Table 12.6 and Table 12.7.

Table 12.6 Grades and Uses of Solid Load-Bearing Concrete Block, ASTM C145 and Hollow Load-Bearing Concrete Block, ASTM C90

Grade	Use
N-1, N-2	Exterior walls above or below grade that may be exposed to moisture and weather, and for interior walls and backup
S-1, S-2	Above-grade walls not exposed to weather or exterior walls if covered with weather- protective coatings

Table 12.7 Grades and Uses of Concrete Building Brick, ASTM C55

Grade	Use
N-1, N-2	On interior and exterior walls where high strength and resistance to moisture and frost action are required
S-1, S-2	Where moderate strength and resistance to moisture and frost action are required

Weights

ASTM C90 establishes three weights of concrete masonry units: normal, medium, and lightweight. Normal weight units are made from concrete weighing more than 125 pcf (2,000 kg/m^3); medium weight uses concrete weighing 105 to 125 pcf (1,680 to 2,000 kg/m^3); and lightweight uses concrete 105 pcf (1,680 kg/m^3) or less. Due to the weight of concrete masonry units, erection is usually limited to ten courses or rows per day, because mortar needs time to set and gain strength before additional weight is placed on it.

Review Questions

1. What are the different ways concrete masonry units are cured?

2. What advantages are there to autoclaved concrete masonry units?

3. What two things influence the properties of concrete masonry units?

4. How does the tensile and flexural strength of concrete masonry units compare with compressive strength?

5. What affects the water absorption of concrete masonry units?

6. How is the insulation value of concrete masonry units improved?

7. What are the three major classifications of concrete masonry units?

8. What is the design module used when sizing concrete masonry units?

9. What are the grades established for load-bearing concrete masonry units?

Key Terms

Autoclave

Concrete masonry unit (CMU)

Kiln

Special Unit

Activities

1. If you are near a plant that produces concrete masonry units, arrange for a visit. Before the visit, prepare a list of specific things to look for and ask questions about, such as the actual proportions of the concrete used, curing procedures, and tests used to ensure quality control of products produced.

2. Secure a supply of concrete blocks and lay up a short wall and turn a corner. Vary the amount of water and mortar cement in several batches of mortar and prepare a written report citing what this does to the finished wall before and after curing.

Additional Resources

Beall, C., *Masonry*, Creative Homeowner Publishers, Upper Saddle River, NJ.

Building Code Requirements for Masonry Structures (ACI 530/ ASCE 5/ tms 402), American Concrete Institute, Farmington Hills, MI.

Concrete Masonry Handbook for Architects, Engineers, Builders, Portland Cement Association, Skokic, IL 50077. Many other publications available.

Ramsey, C. G., Sleeper, H. R., and Hoke, J. R., *Architectural Graphic Standards,* the American Institute of Architects, John Wiley & Sons, Hoboken, NJ.

Other resources include:

Numerous publications from the National Concrete Masonry Association, Herndon, VA.

See Appendix C for addresses of professional and trade organizations and other sources of technical information.

Stone

mortar and prepare a written report citing what this does to the finished wall before and during.

questions about, such as the actual proportions of the concrete used, curing procedures, and tests used to ensure quality control of products produced.

LEARNING OBJECTIVES

Upon completion of this chapter, the student should be able to:

- Discuss the characteristics and uses of various stones used in building construction.

- Identify the various types of commercially available stone.

- Understand the processes used to quarry and work stone to make it useful for building construction.

- Select stone for a project based on the requirements of the job.

- Define manufactured stone.

Build Your Knowledge

For further information on these materials and methods, please refer to:

Chapter 34 Flooring

For centuries, stone was used as a material to build structural load-bearing walls. Today it is used mainly as a veneer or facing, which greatly reduces the weight of a building while still enabling a designer to take advantage of stone's beauty as a finish material (Figure 13.1). **Rock** is solid mineral matter, occurring in individual pieces or large masses, such as an outcropping. **Stone** is rock that is quarried and shaped to size for construction purposes.

Figure 13.1 Stone is used as a finish material on the interior and exterior of buildings.

BASIC CLASSIFICATIONS OF ROCK

Rock is divided into three basic categories, depending on its origin: igneous, sedimentary, and metamorphic.

Igneous Rock

Igneous rock is formed, usually deep in the earth, when a molten (magma or lava) material changes from a liquid to a solid state. Commonly used forms include granite, serpentine, and basalt.

Granite   Granite is an igneous rock containing crystals or grains of visible size. It consists mainly of quartz, feldspar, mica, and other colored minerals. Colors include black, gray, red, pink, brown, buff, and green. It is hard, strong, nonporous, and durable. Granite is one of the most permanent building stones. It can be used under severe weather conditions and in contact with the ground. Granite can be finished with a range of surface textures, from rough to highly polished. Granite is used for windowsills, cornices, columns, floors, countertops, and wall veneers (Figure 13.2).

Figure 13.2 Granite is a hard, strong, and durable stone used for both interior and exterior applications.

© Vladislav Gurfinkel/Shutterstock.com

Serpentine   Serpentine is an igneous rock named after its major ingredient. It ranges from olive-green to greenish black, has a fine grain, and is dense. Since some types deteriorate due to weathering, its major uses are for interior applications. It can be cut into thin sections, to ⅞ to 1¼ in. (22 to 32 mm), and is used for paneling, windowsills, stools, stair treads and risers, and landings.

Basalt   Basalt is an igneous rock that ranges in color from gray to black. It has a fine grain and is used mainly for paving stones and retaining walls.

Sedimentary Rock

Sedimentary rock is formed of materials (sediments) deposited on the bottom of bodies of water or on the surface of the earth. Major types include sandstone, shale, and limestone.

Sandstone   Sandstone is a sedimentary rock composed of sand-sized grains cemented together by naturally occurring mineral materials, such as silica, iron-oxide, and clay. Quartz grains predominate in sandstone used for building construction. The two most familiar forms are brownstone, used mainly in wall construction, and bluestone, used for paving and wall copings. Available colors include gray, brown, light brown, buff, russet, red, copper, and purple.

Since sandstone's hardness and durability depend on the cementing material, there is a wide range of weight and porosity. Sandstone is used for wall facing panels and can have a variety of surface finishes, including chipped, hammered, and rubbed.

Shale   Shale is a sedimentary rock derived from clays and silts. It forms in thin laminations and tends to be weak along planes. It is not suitable as a concrete aggregate because it may expand in the presence of water, causing concrete failure. Shale that is high in limestone is ground into small particles and used in making cement, bricks, and tiles. It is predominantly gray in color but can be found in hues ranging from black to red, yellow, and blue.

Limestone   Limestone is a sedimentary rock composed mainly of calcite and dolomite. There are three types of limestone. Oolitic is a calcite-cemented calcareous stone formed from shells that are uniform in composition and structure. Dolomitic limestone consists mainly of magnesium carbonate and has a greater compressive strength than oolitic. Crystalline limestone consists mainly of calcium carbonate crystals. It has high tensile and compressive strengths.

Limestone is used for building stones and is available as dimension (cut), ashlar, and rubble stones. It is used for paneling, veneer, window stools and windowsills, flagstone, mantels, copings, and facings (Figure 13.3). It is also pulverized to form crushed stone aggregate and burnt to produce lime.

Figure 13.3 This limestone facing exposes the sedimentary formation of the natural rock.

© alb_photo/Shutterstock.com

Metamorphic Rock

Metamorphic rock is either igneous or sedimentary rock that has been altered in appearance, density, and crystalline structure by high temperature and/or high pressure. Major types used in construction include marble, quartzite, schist, and slate.

Marble   Marble is a metamorphic rock made up largely of calcite or dolomite that has been recrystallized. There are a number of types of marble, with colors varying from white through gray and black. The presence of oxides of iron, silica, graphite, carbonaceous matter, and mica produce other color variations, including red, violet, pink, yellow, and green. Marble is used for wall panels and column facings, as well as window stools, windowsills, and floors. Some types are used on building exteriors, and others are limited to interior applications. Its surface can be ground to a fine, polished condition.

Quartzite   Quartzite is a metamorphic rock that is often confused with granite. It is a type of sandstone composed mainly of granular quartz that is cemented by silica, producing a coarse, crystalline appearance. It has high tensile and crushing strengths and is available in brown, buff, tan, ivory, red, and gray. Quartzite is used for building stone, gravel, and aggregate in concrete.

Schist   Schist is a metamorphic rock generally made up of silica with smaller amounts of iron oxide and magnesium oxide. Its color depends on mineral makeup, but blue, green, brown, gold, white, gray, and red are common. It is commonly available in rubble veneer and flagstone and is used for interior and exterior wall facing, patios, and walks.

Slate   Slate is a hard, brittle metamorphic rock consisting mainly of clays and shales. Its major ingredients are silicon dioxide, aluminum oxide, iron oxide, potassium oxide, magnesium oxide, and, sometimes, titanium, calcium, and sulfur. Slate is found in parallel layers, which enables it to be cut into thin sheets.

Slate is produced in three textures: sand-rubbed, honed, and natural cleft. It is cut into three types: roof tiles, random flagging, and dimension slate (cut to size) (Figure 13.4). It is commonly used for interior and

Figure 13.4 Natural slate is used for wall facings, flooring, and countertops.

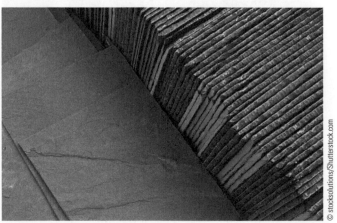

exterior wall facing, flooring, flagstones, countertops, coping, and windowsills and window stools.

TYPES AND USES OF STONE

Commercially, stone is used in several types: rubble stone, rough stone, monumental stone, dimension stone, flagstone, broken and crushed stone, and stone powder and dust.

Rubble stone consists of irregular fragments from a quarry that have one good face. The pieces are irregular in shape and sized usually in pieces 12 in. (300 mm) by 24 in. (600 mm) that are cut and fitted by a mason.

Rough building stone, sometimes called fieldstone, occurs in naturally found rock masses. The stone is generally used in its natural shape.

Dimension stone, also referred to as cut stone, is cut to size at a stone mill and shipped to site. Its surface may be rough, as occurs when it is split, or polished. It is used as veneer on interior and exterior walls, floors, copings, stair treads, and for other similar applications. Ashlar, a form of dimension stone, is a cut rectangular stone with square corners and faces that is smaller than other dimension stone (Figure 13.5).

Figure 13.5 Dimensional sandstone used in a random ashlar pattern.

Flagstone is thin, flat stone from ½ to 4 in. (12 to 100 mm) in thickness. Flagstones laid over a concrete base are usually ¾ to 1 in. (18 to 25 mm) thick. If laid over a sand or loam base, pieces 1¼ to 1½ in. (31 to 37 mm) thick are required. Its surface may be left rough or polished. Random flagstones, with the exception of minor shaping, are left in their natural shape. Trimmed flagstones are random, shaped pieces with several edges sawed straight. Trimmed rectangular flagstones are pieces with four sides sawed, forming square and rectangular pieces (Figure 13.6).

Figure 13.6 Flagstone is thin, flat stone that can be laid over a concrete, sand, or loam base.

© Christina Richards/Shutterstock.com

Broken and crushed stone includes irregular shapes and crushed pieces of one type of stone that are graded for hardness and size and used as aggregate in concrete for surfacing roads and driveways and as aggregate for surfacing fiberglass asphalt shingles and built-up roofing. Stone powder and stone dust are used as fill in paints and asphalt paving surfaces.

QUARRYING AND PRODUCING NATURAL STONE

A quarry is an excavation from which stone used for building is taken via blasting or cutting (Figure 13.7). Broken stone is produced by blasting the rock. Larger pieces can be broken or cut into smaller units for use as an exterior finish material. The rest is crushed and sorted into various sizes for use as aggregate.

Most stone used in building construction is dimensional and produced by cutting large blocks from a quarry, often with a channeling machine, which makes an incision from 1 to 3 in. wide. Some machines use a rotating chisel cutter, and others use a wire that runs over pulleys and moves a quartz-sand cutting agent over the stone, producing a saw-type cut.

Large blocks are removed from the quarry to a mill, where they are cut to the sizes and thicknesses needed (Figure 13.8). The architectural drawings specify the shape and size of each stone. Water-jet cutting technology

Figure 13.7 Stone used in building construction is produced by cutting large blocks from a quarry with channeling machines, chisel cutters, or wire saws.

© vallefrias/Shutterstock.com

Figure 13.8 Rough-cut blocks are moved to a mill, where they are cut to size and fitted with holes or anchors, as required.

© vallefrias/Shutterstock.com

uses high-pressure water jets to execute complex shapes and designs on stone materials, such as granite, marble, quartz. Holes are drilled in each block, as required for lifting and anchoring it in place.

CHOOSING STONE

When selecting the stone to be used in a building, an architect must consider a variety of parameters, including cost, strength, durability, hardness, grain and color, and texture and porosity (Figure 13.9).

Figure 13.9 Stone selected for carving is carefully chosen for proper grain, color, durability, and hardness.

© pzAxe/Shutterstock.com

The ease with which a stone can be quarried and shaped influences economic considerations. Soft stones have lower production costs. Accessibility is a second cost factor, because costs rise as stone transportation distances increase.

If stone is exposed to detrimental weather, it must have sufficient durability to withstand freeze–thaw cycles and erosive conditions. Hardness is very important in stones used for floors, steps, patios, and other areas exposed to traffic. Grain and color are considered when the appearance of stone in its finished position is decided. As colors vary widely, they are usually selected by inspecting actual material samples. All types of stone is ordered by the percentage of colors desired to make the blend, and by quantity of flat stones, corners, and other specialty units needed. The texture of the stone has a great influence on the finished appearance. Fine-grained stones can have a smoother, polished surface, while coarse-grained stones present a more open face. Porosity pertains to the ability of stone to resist moisture penetration. Porous rock tends to permit some minerals to dissolve that can cause staining of the exposed face. It also is not durable and will be damaged by freeze–thaw cycles.

MANUFACTURED STONE

Manufactured stone is an architectural building unit manufactured to simulate natural cut stone used in masonry applications (Figure 13.10). While technically a concrete masonry unit, cast stone is included here because its use more closely resembles that of natural stone. Also referred to as cultured stone, the material is cast of Portland cement mixed with high-quality fine aggregates and mineral coloring pigments to achieve the desired color and appearance.

Figure 13.10 Manufactured stone can provide an economical substitute to natural cut stone.

© Greg Henry/Shutterstock.com

Cast stone frequently is produced with a low water-to-cement ratio mixture. The mix is consolidated into molds using an air- or electric-driven tamping device, or vibration under pressure, to replicate the formation of natural sedimentary rock. Products manufactured in this manner are referred to as vibrant-dry-tamped (VDT) cast stone. Physical properties and raw materials ingredients are specified by ASTM C 1364, *Standard Specification for Architectural Cast Stone*.

The product can be reinforced as needed to increase its structural integrity and is made from molds with precise dimensional properties. Lifting inserts, anchors, and drips can be cast into the stone enabling installation that is more efficient.

Review Questions

1. What is the main use for stone in buildings built today?
2. What are the three basic categories of rock?
3. How is igneous rock formed?
4. What colors of granite rock are commonly available?
5. For what construction applications is serpentine rock used?
6. How is sedimentary rock formed?
7. Shale that has high limestone content is used to make what masonry materials?
8. What are the three types of limestone?
9. What are the main differences in the three types of limestone?
10. What are the major types of metamorphic rock used in building construction?
11. In what colors is marble found?
12. Where is marble commonly used in building construction?
13. In what textures is slate produced?
14. What type of machine is used in a quarry to cut rock into manageable sizes?
15. What factors must a designer consider when choosing stone for a building?

Key Terms

Igneous Rock
Manufactured Stone

Metamorphic Rock
Rock

Sedimentary Rock
Stone

Activities

1. Collect samples of various types of stone typically used in building construction. Write a report citing the characteristics of each type. Your building materials supplier can help with samples.

2. Test selected stone samples for compressive strength. Note variances between samples of the same stone and between the different types of stone. Report why you think these variances may have occurred.

Additional Resources

Dimension Stone Cladding: Design, Construction, Evaluation, and Repair, American Society for Testing and Materials, West Conshohocken, PA.

Other resources include:

Publications of the Building Stone Institute, Purdys, NY. www.buildingstoneinstitute.org/

Publications of the Cast Stone Institute, Winter Park, FL. www.caststone.org/default.htm

Publications of the Indiana Limestone Institute of America, Bedford, IN.

See Appendix C for addresses of professional and trade organizations and other sources of technical information.

Masonry Construction

Upon completion of this chapter, the student should be able to:

- Name the various brick positions and bond patterns used in masonry wall construction.
- Identify how openings are spanned in masonry walls.
- Understand the purpose of expansion and control joints in masonry construction.
- Recognize the types of brick and concrete masonry wall constructions and the properties associated with them.

- Explain the advantages and disadvantages of the various mortar joints.
- Describe the process used to lay brick and concrete masonry units.
- Evaluate the quality of stone wall construction.
- Recognize proper construction of structural clay-tile walls.

Build Your Knowledge

For further study on these materials and methods, please refer to:

Chapter 10   Mortars for Masonry Walls
            Topic: Types of Mortar
Chapter 11   Clay Masonry
Chapter 12   Concrete Masonry
Chapter 13   Stone

Masonry construction encompasses the use of clay brick, concrete masonry units, stone, or structural tile. While each masonry material has specific requirements associated with its use, construction techniques are similar for all masonry assemblies. Load-bearing masonry walls carry floor and roof loads on the exterior and interior of a building (Figure 14.1). Non-load-bearing masonry walls are used for interior partitions and on exterior building skins as protection from the elements. Masonry walls are easy to design and construct, and often produce a more economical building than do other construction methods. The individual masonry units fit easily into the mason's hand, requiring no automated or expensive equipment.

Figure 14.1 Masonry can provide structural support or be used as a non-load-bearing veneer for exterior finishes.

© Vladimirs Koskins/Shutterstock.com

Stonemasonry is one of the oldest building professions in history. As such, it is regarded as a traditional skill, and is one that is in heavy demand. Masons are trained either by an apprenticeship program, union program, or by another masonry contractor.

Because of masonry's high thermal mass, the material is frequently used in passive solar designs that capture and use solar radiation to help heat a building. While most building materials absorb heat, masonry's high density and mass allow it to absorb heat slowly and retain it longer. Masonry floors and walls store solar heat during the day, and radiate it back into the interior when temperatures drop at night.

MASONRY WALLS

Masonry units are generally rectangular in shape, allowing them to be placed into a wall in a variety of positions. Varying the orientation and positioning of units enables designers to not only satisfy the wall's structural requirements, but also to create distinctive effects and patterns. Masonry walls are laid in courses, horizontal rows of masonry units. How a masonry unit is oriented within a wall construction is defined by six distinct positions (Figure 14.2).

Masonry Unit Positions

The most common position is the *stretcher*, a unit that is positioned with its bottom side in mortar and its face in the plane of the wall. Bricks are often manufactured with only one side that is textured or colored to act as face brick, and the masons must be careful to place each unit correctly. A header is a unit that has its bottom bedded in mortar and its end placed in the plane of the wall. Brick headers are used to join two single-wythe walls by bridging across them.

Wythe is a term used to define a continuous vertical wall of masonry one unit in thickness. The rowlock position has its end in the face of the wall and its back side bedded in mortar. A rowlock is like a header that has been turned vertically. The position is often used to cap a double-wythe wall with the face of the unit providing the finished top. A soldier is a unit with its end bedded in mortar and its face positioned vertically in the plane of the wall, while a sailor is a has its end bedded in mortar, and either its top or bottom in the plane of the wall. The sailor position finds frequent use in the construction of masonry arches. Finally, the shiner position denotes a unit that has its face or back bedded in mortar, and

Figure 14.2 Terms used to describe brick positions in a wall.

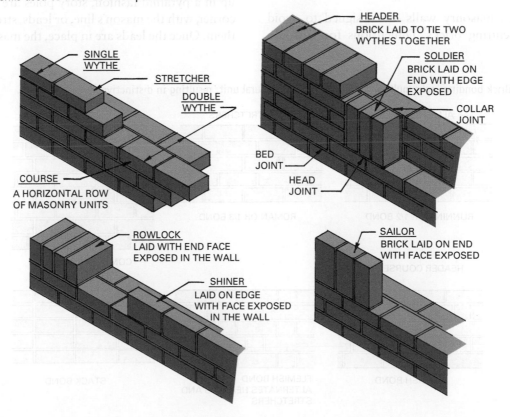

top or bottom oriented horizontally with the face of the wall. Both the sailor and shiner positions assume the use of a solid rather than cored unit.

Masonry Bond Patterns

Solid masonry walls without steel reinforcing have been used for thousands of years. While they tend to have limited applications in modern wall construction, solid masonry can be quite economical and suitable for low walls. The question of how to bond multiple wythes of brick into an efficient structural system has been solved in many ways in different regions of the world, resulting in a variety of standard patterns.

A *bond pattern* describes the arrangement of masonry units in a masonry wall. Bonding in solid masonry walls is achieved by overlapping individual units with previously laid courses. A running bond is a pattern made using only stretchers. A full course of masonry is laid, with the next course overlapping the brick below at either a half, quarter, or third of a brick. The running bond is not structurally strong and is used for veneer walls one unit in thickness. A number of other bonds use alternating headers to structurally bond to adjacent wythes of brick by stretching across them (Figure 14.3).

Laying Brick Masonry Walls

When possible, masonry walls are designed to avoid the need for cutting bricks. Dimensions for masonry structures are based on the actual dimensions of the units and mortar joints to be used (Figure 14.4). The actual size of the unit plus the thickness of a mortar joint (⅜ to ½ in.) are calculated to arrive at building dimensions that require no cutting of units.

Masonry walls must be placed on properly engineered foundation systems. Depending on structural design requirements, vertical reinforcing is often embedded in the foundation to extend into the open cells of the masonry. By grouting the reinforcing inside of the masonry, a solid anchorage is established between the wall and its foundation.

The basic process for laying a brick masonry wall is shown in Figure 14.5. The outer edge of the wall is located on its footing by snapping a chalk line. Starting with the corners, a first level, or course of masonry is laid without mortar. This allows for a check between the dimensions on the floor plan and how the first course actually fits the lay out. This dry run allows the masons to plan for units that will need to be cut to fit openings. Once all dimensions are verified, a brick corner is laid in mortar and leveled at each end of the wall.

Masons utilize a *story pole*, a measuring stick that can be braced into a true vertical position to hold a taut mason's line without bending. As the corners are laid up in a pyramid fashion, story poles are set up on each corner, with the mason's line, or leads, stretched between them. Once the leads are in place, the mason can quickly

Figure 14.3 Brick bonding locks multi-wythe walls into a structural unit, resulting in distinctive surface patterns.

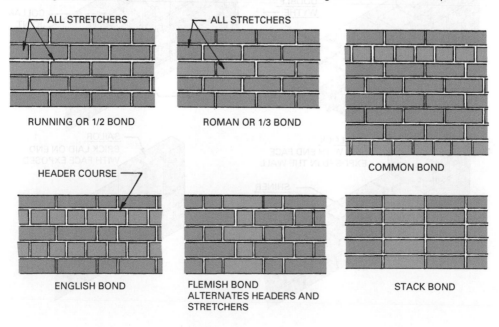

RUNNING OR 1/2 BOND   ROMAN OR 1/3 BOND   COMMON BOND

ENGLISH BOND   FLEMISH BOND ALTERNATES HEADERS AND STRETCHERS   STACK BOND

Figure 14.4 Masonry wall lengths are dimensioned to fit the module of the masonry units and avoid the need for cutting.

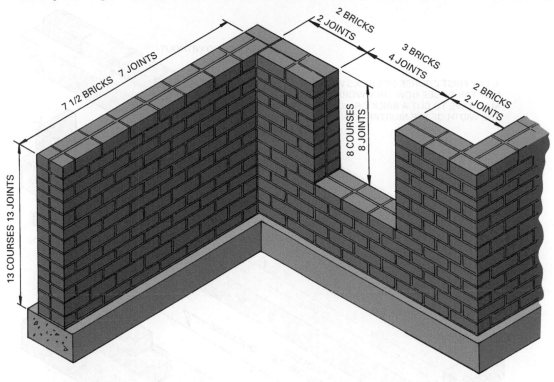

lay the infill bricks between corners. Bricks that need to be fitted can be cut with a water-cooled diamond blade cutter (Figure 14.6), or rough-cut with a sharp blow from a mason's hammer. The mason must work rapidly yet carefully in placing the bricks and maintaining a uniform mortar line. Because bricks are laid from each corner to the center of a wall, closure is made by placing the final brick in the last opening. Each course must be carefully executed to assure that the closure brick can be laid without requiring its being cut.

Mortar Joints

After the mortar has set sufficiently, various-shaped tools are used to profile the joints (Figure 14.7). The appearance and weather tightness of a finished masonry wall depends on the type of mortar joint employed (Figure 14.8). Mortar joints are finished by either troweling or tooling. *Troweling* involves striking off excess mortar with a mason's trowel, and does not produce a watertight joint. Tooled joints, including weathered, concave, and V-joints are the best at resisting leakage. Tooling joints result from compressing mortar into a joint and against the faces of adjacent units. Flush, ruled, and rodded joints are not as watertight. Concave joints are very effective in resisting

rain penetration and are recommended for use in walls exposed to wind-driven rain.

Once tooled, the wall is lightly scrubbed with a fiber brush to remove loose mortar particles. Some weeks after the wall has cured, it is usually washed with muriatic acid (HCl) and thoroughly rinsed with clean water. This removes any remaining mortar stains on the brick face.

Laying Concrete Masonry Walls

Concrete masonry wall lengths are established so that standard-size concrete masonry units can be used. These include full 16 in. blocks and half blocks (8 in.). Mortar joints of ⅜ or ½ in. in thickness are common for concrete block.

The procedure for laying concrete masonry walls is very similar to that described for clay masonry units. First, blocks without mortar are spaced along a footing to establish spacing and determine if blocks need cutting. The mason begins by laying a full mortar bed on the footing and building up each corner to several courses high. The height of the courses is checked with a story pole. A chalk line is run the length of the wall to establish the level height of the first course, which is laid on a

Figure 14.5 The procedure for laying a brick wall.

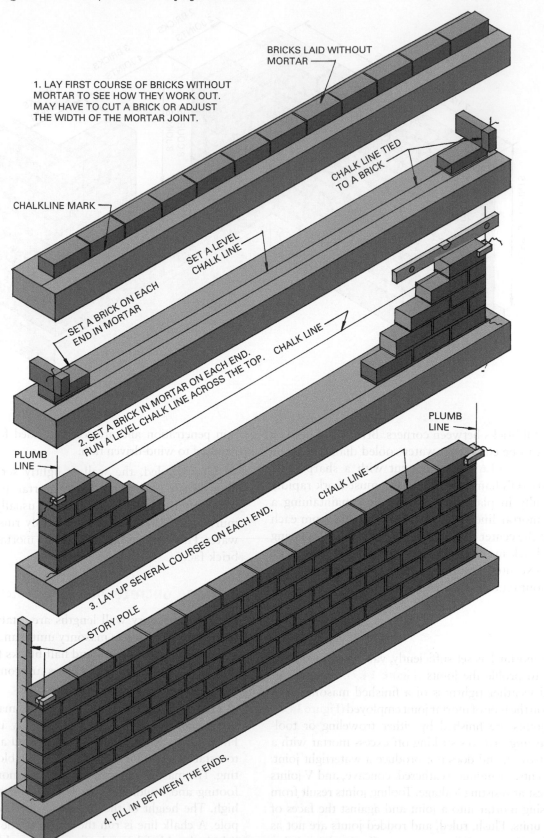

BRICKS LAID WITHOUT MORTAR

1. LAY FIRST COURSE OF BRICKS WITHOUT MORTAR TO SEE HOW THEY WORK OUT. MAY HAVE TO CUT A BRICK OR ADJUST THE WIDTH OF THE MORTAR JOINT.

CHALK LINE TIED TO A BRICK

CHALKLINE MARK

SET A LEVEL CHALK LINE

SET A BRICK ON EACH END IN MORTAR

CHALK LINE

2. SET A BRICK IN MORTAR ON EACH END. RUN A LEVEL CHALK LINE ACROSS THE TOP.

PLUMB LINE

PLUMB LINE

CHALK LINE

3. LAY UP SEVERAL COURSES ON EACH END.

PLUMB LINE

STORY POLE

4. FILL IN BETWEEN THE ENDS.

Figure 14.6 Masonry units that need to be fitted can be cut with a diamond blade saw or rough broken with a masonry hammer.

Figure 14.7 A mason tooling a concave mortar joint.

© Susan Law Cain/Shutterstock.com

Figure 14.8 Commonly used mortar joint profiles.

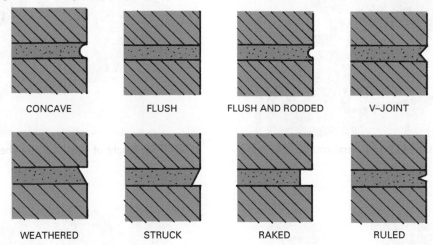

CONCAVE   FLUSH   FLUSH AND RODDED   V–JOINT

WEATHERED   STRUCK   RAKED   RULED

full mortar bed. Courses above it are usually laid without placing mortar on the webs. Mortar is placed on the face shells, and each course is laid from the corners to the center of the wall (Figure 14.9). After the mortar has set properly, joints are tooled to the desired contour, and mortar crumbs are scrubbed off with a soft brush. Lintels used in concrete masonry walls include the use of special *lintel blocks* that allow the steel reinforcing to be placed and grouted inside of the horizontal void.

Openings in Masonry Walls

Openings in masonry walls must be bridged over by some method to support the masonry above. Beams, lintels, and arches are used to span openings in masonry walls. These elements work to transfer the loads above the opening to the wall and foundations below.

Large openings are typically spanned with steel or cast-in-place, reinforced concrete lintels (Figure 14.10). To span small openings, steel angle, reinforced concrete masonry units (CMU), and pre-cast concrete are most commonly used. Lintels should generally overlap the sides of the opening by a minimum of 4 in. for openings up to 6 ft.; 6 in. for openings between 6 and 9 ft; and 12 in. for openings over 9ft. in width. The height of the lintel is normally sized to correspond to the vertical masonry module. Reinforced brick lintels use steel reinforcing that is bonded to the masonry to form a beam.

Figure 14.9 The procedure for laying a concrete block wall.

1. Lay the first course on a full mortar bed.

2. The blocks for the first corner are laid.

3. The corner blocks are laid up several courses.

4. The height of each course is checked with a story pole.

5. Blocks are laid up on each corner and worked toward the center of the wall.

6. The first course has been laid across the footing and the corners have been raised several courses. The mason is laying the second course.

Figure 14.9 (*Continued*)

7. A level is used to check the wall for plumb.

8. A level is used to check the wall for level.

9. The mortar is tooled to the desired contour when it has properly set.

10. Mortar crumbs are removed with a soft brush after tooling is finished.

11. The closure opening has mortar buttered on all sides.

12. The closure block has mortar on the ends and is slid into the opening. Excess mortar is cut away.

Figure 14.10 Beams, lintels, and reinforced brick are used to span openings in masonry walls.

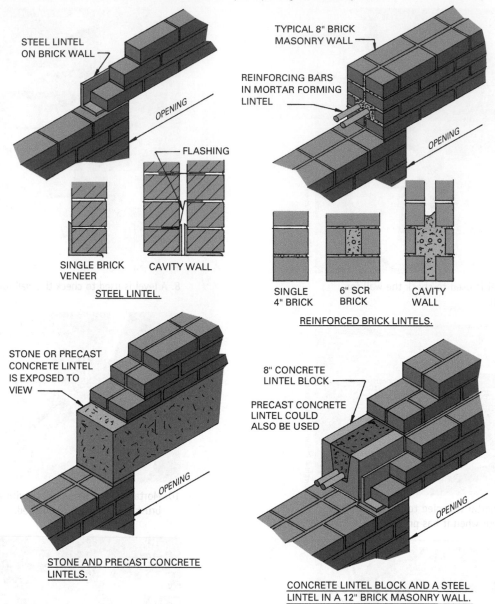

STEEL LINTEL
ON BRICK WALL

OPENING

FLASHING

SINGLE BRICK VENEER

CAVITY WALL

STEEL LINTEL.

TYPICAL 8" BRICK MASONRY WALL

REINFORCING BARS IN MORTAR FORMING LINTEL

OPENING

SINGLE 4" BRICK   6" SCR BRICK   CAVITY WALL

REINFORCED BRICK LINTELS.

STONE OR PRECAST CONCRETE LINTEL IS EXPOSED TO VIEW

OPENING

STONE AND PRECAST CONCRETE LINTELS.

8" CONCRETE LINTEL BLOCK

PRECAST CONCRETE LINTEL COULD ALSO BE USED

OPENING

CONCRETE LINTEL BLOCK AND A STEEL LINTEL IN A 12" BRICK MASONRY WALL.

Pre-cast concrete and concrete masonry lintels generally carry the loads imposed on the wall, and steel lintels carry the face brick. Some classic architectural styles use exposed concrete or stone lintels. In general, building codes do not offer detailed prescriptions for lintels, so most lintel manufacturers offer supporting calculations.

Masonry Arches Masonry arches are a curved structural construction that has been used to span openings in masonry for centuries. The *arch* uses compressive strength of masonry or stone to span an opening and carry imposed loads. The horizontal thrust generated is resisted by the masonry mass of the wall against which

it butts. Two equal arches close together provide opposite but equal thrust, thus resisting the horizontal forces. A masonry arch may be constructed using soldier courses, two or three rowlock courses, or alternating row-lock and soldier courses. Arches are constructed with temporary bracing called centering that supports the wet mortar until it has hardened. Special wedge-shaped units called voussoirs are manufactured to specifically to create the curvature in arches. When rectangular units are used, the mortar between them assumes a wedge shape. Arches vary in shape, from flat horizontal openings (Jack) through semicircular (Roman) and semi-elliptical (segmented) forms, to pointed (Gothic) **(Figure 14.11)**.

Figure 14.11 Arches vary in shape, from flat horizontal openings (A: Jack) through semicircular (B: Roman) and semi-elliptical (C: segmented) forms, to pointed (D: Gothic).

A

B

C

D

©iStock.com/eugenesergeev

CONTROL AND EXPANSION JOINTS

Within masonry walls, movements can occur that produce stresses that may cause cracking. These stresses are created by expansion and contraction due to changes in temperature, changes in moisture content, and structural movements due to settling of the foundation, and concentration of stresses at openings in the wall. Masonry walls often incorporate a number of different materials that expand and contract at different rates and can create additional stresses. A typical example is a wall using brick veneer over concrete block with metal lintels over openings. Control and expansion joints are used to relieve these stresses and must be carefully determined as part of the wall design.

Expansion joints are intentionally created slots that close to accommodate expansion of masonry materials. *Control joints* are intentionally created cracks that open to accommodate shrinkage of masonry by providing tension relief between parts of a masonry wall that may change from their original dimensions.

Masonry walls tend to contract directly after laying and drying. The tensile strength of the wall tries to resist these stresses, and the wall cracks if its strength is exceeded. This cracking is controlled by carefully locating control and expansion joints in the wall. Vertical joints are built into the wall as units are laid (Figure 14.12). Joints are made using bond breakers of asphalt felt or pre-formed rubber gaskets between units. Generally, they are located where there are

Figure 14.12 Expansion joints details and locations.

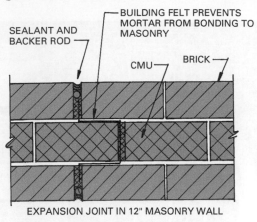

EXPANSION JOINT IN 12" MASONRY WALL

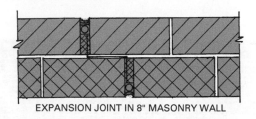

EXPANSION JOINT IN 8" MASONRY WALL

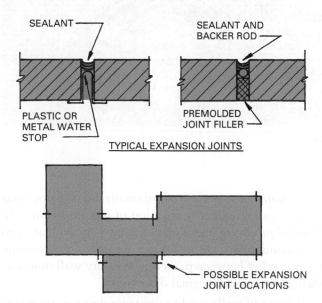

changes in wall height or wall thickness and at openings, wall intersections, building corners, and at junctions between walls and columns (Figure 14.13). Typical control joints for concrete masonry walls are shown in Figure 14.14.

Horizontal joints are placed above masonry walls that butt structural frames or on the bottom of floor or roof

structures. Required above 30 feet of wall height, they use a minimum ¼ in. space or compressible material directly below a shelf angle. The angle supports the masonry above, while the masonry below has space to expand.

The joints are sealed with elastic joint sealers of rubber or polymer-based materials. Joints should try to match the color of brick for vertical joints, and the color of mortar for horizontal joints. Regular inspection and maintenance of joints is required to ensure the seal is kept watertight.

TYPES OF MASONRY WALLS

A number of different masonry wall assemblies are commonly used for building construction (Figure 14.15). Masonry bearing walls are used to carry floor and roof loads of wood, steel, or concrete construction while simultaneously serving as the exterior enclosure for a building. A masonry load-bearing wall may be solid, unreinforced, or reinforced. *Reinforced masonry* is used when the flexural, compressive, and shear stresses exceed those permitted for partially reinforced or unreinforced masonry.

Reinforced Masonry

Reinforcing masonry enables the designer to optimize the natural property of masonry being strong in compression, and steel, that is strong in tension, in combination. Reinforcing can be applied in both single and multiple wythe walls. Reinforced masonry uses steel reinforcing that is placed either in the cores of the masonry units or between two wythes of masonry and filled with grout (a fluid mix of Portland cement, aggregate, and water) to produce an integral structural unit (Figure 14.16). The reinforcing increases the strength of the wall to resist lateral loads and reduce buckling. Building codes specify the amount of steel reinforcing required according to use and loading.

Grouted reinforced brick masonry walls use both vertical and horizontal reinforcing bars in the wall cavity. Vertical reinforcing bars of ¾ in. diameter minimum are placed at regular intervals, with a maximum spacing of 48 in. in longer walls. Reinforcing is also used around openings, at the top of foundations and parapet walls and at connections to floors and roofs. The vertical reinforcing bars can be placed before the masonry, or be positioned afterward. A typical reinforcing layout is shown in Figure 14.17. Once reinforcement and units are in place and the mortar has cured for at least four

Figure 14.13 A typical control joint in a masonry wall and recommended layout locations.

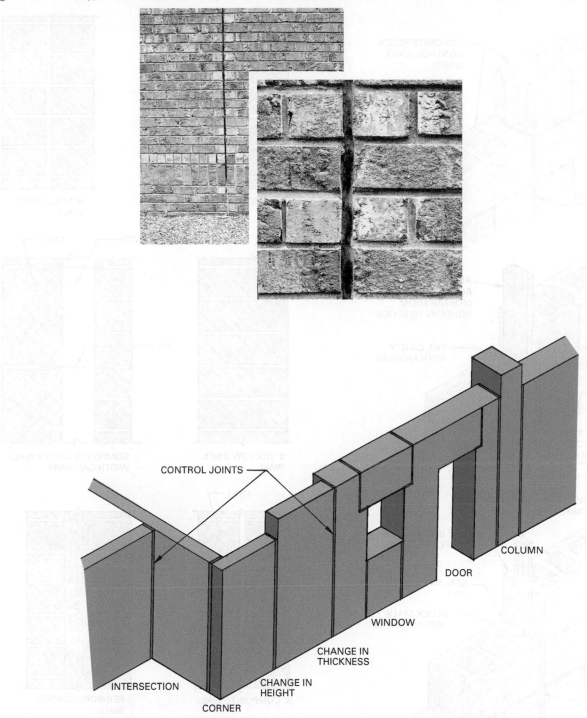

CONTROL JOINTS

COLUMN

DOOR

WINDOW

CHANGE IN THICKNESS

INTERSECTION

CHANGE IN HEIGHT

CORNER

hours, grouting can begin. Grout is placed in lifts, the amount of grout placed in a single continuous operation by either hand buckets or pumps. Connections and anchors for floor and roof assemblies can be bolted directly to the wall, or be made with mechanical anchors, grouted into the wall cavity.

Grouted reinforced masonry offers expanded structural possibilities in masonry construction. While its most common application is in single-story and low-rise construction, it can also be employed in high-rise design, allowing for the construction of taller, thinner walls.

Figure 14.14 Typical control joint details for concrete block construction.

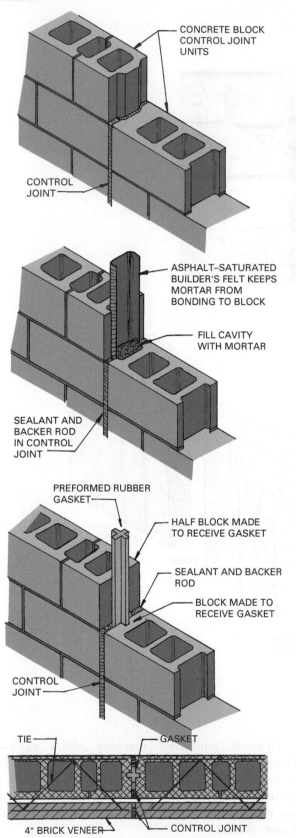

CONCRETE BLOCK
CONTROL JOINT
UNITS

CONTROL
JOINT

ASPHALT–SATURATED
BUILDER'S FELT KEEPS
MORTAR FROM
BONDING TO BLOCK

FILL CAVITY
WITH MORTAR

SEALANT AND
BACKER ROD
IN CONTROL
JOINT

PREFORMED RUBBER
GASKET

HALF BLOCK MADE
TO RECEIVE GASKET

SEALANT AND BACKER
ROD

BLOCK MADE TO
RECEIVE GASKET

CONTROL
JOINT

TIE   GASKET

4" BRICK VENEER   CONTROL JOINT

Figure 14.15 Types of masonry wall construction.

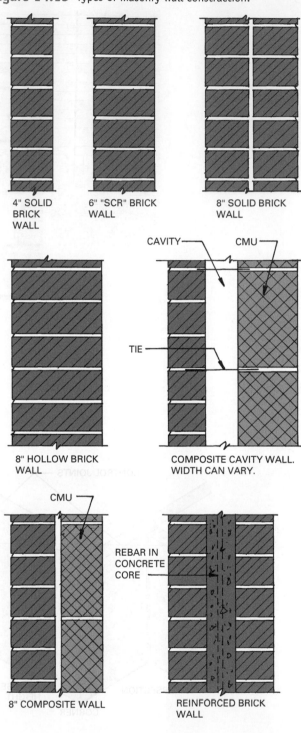

4" SOLID
BRICK
WALL

6" "SCR" BRICK
WALL

8" SOLID BRICK
WALL

8" HOLLOW BRICK
WALL

CAVITY   CMU

TIE

COMPOSITE CAVITY WALL.
WIDTH CAN VARY.

CMU

8" COMPOSITE WALL

REBAR IN
CONCRETE
CORE

REINFORCED BRICK
WALL

Masonry Composite Walls

Composite masonry walls consist of two wythes of different types of masonry units with wire reinforcing that binds them together to form a single structural unit able to resist forces and loads. A commonly used example is a brick outer wall with a CMU backup (**Figure 14.18**).

Figure 14.16 A grout-filled reinforced brick masonry wall with horizontal and vertical steel reinforcing.

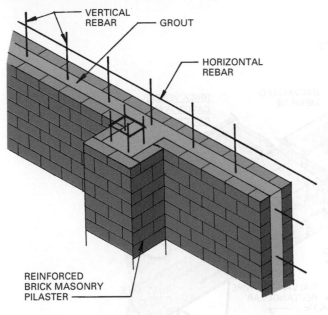

VERTICAL REBAR — GROUT

HORIZONTAL REBAR

REINFORCED BRICK MASONRY PILASTER

Figure 14.18 A composite masonry wall of concrete block with a brick facing.

© hans engbers/Shutterstock.com

Composite masonry walls often utilize a higher-quality masonry unit, such as brick, tile, or stone at the exterior, and a more economical one, such as concrete block or *structural clay tile*, at the hidden interior portion of the assembly. The design of composite walls must accommodate differences in thermal expansion, moisture absorption, and load-bearing capabilities of the individual masonry units.

The inner and outer wythes of composite walls are connected with either continuous horizontal or unit ties (Figure 14.19). Horizontal wire reinforcement comes in a variety of types with ladder, truss, and adjustable forms being the most common. Ladder-type and truss-type reinforcing use longitudinal rods welded to perpendicular or diagonal cross-rods. Adjustable ties use ladder- or truss-type design in the inner wythe, with adjustable tabs or eye-hooks connected to the outer reinforcing. The use of adjustable ties allows each wythe to be built separately, and accommodates movement between the wythes. All horizontal reinforcing types should be slightly narrower than the final wall thickness and must have a minimum of ⅝ in. mortar cover to prevent contact of moisture. For exterior walls and corrosive environments, the reinforcing used must be corrosion resistant hot-dipped galvanized or stainless steel. The reinforcing is placed continuously in horizontal mortar joints as the wall is being laid, usually at a maximum spacing of 16 in. vertically.

The inner wythe is normally laid prior to the outer one when adjustable ties are used. By completing the inner wall first, other trades are able to begin their work at the interior of the building. The face of the interior wythe exposed

Figure 14.17 Typical reinforcing in brick masonry load-bearing walls with openings.

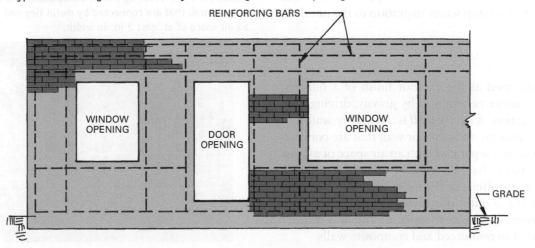

REINFORCING BARS

WINDOW OPENING

DOOR OPENING

WINDOW OPENING

GRADE

Figure 14.19 Masonry composite walls use various types of metal reinforcing.

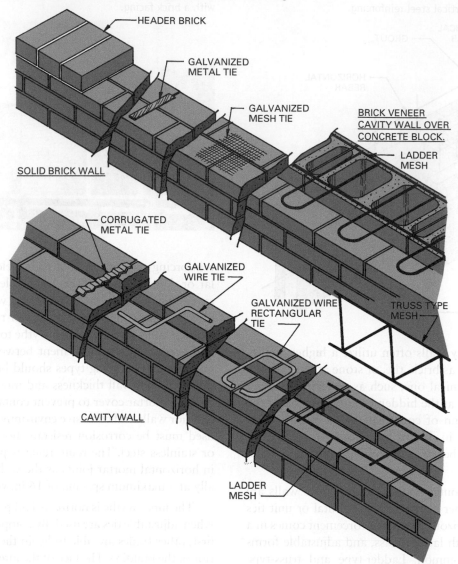

HEADER BRICK
GALVANIZED METAL TIE
GALVANIZED MESH TIE
BRICK VENEER CAVITY WALL OVER CONCRETE BLOCK.
LADDER MESH
SOLID BRICK WALL
CORRUGATED METAL TIE
GALVANIZED WIRE TIE
GALVANIZED WIRE RECTANGULAR TIE
TRUSS TYPE MESH
CAVITY WALL
LADDER MESH

to the cavity is usually sealed with an applied bituminous or acrylic coating to stop water migration to the interior.

Masonry Cavity Walls

Masonry walls used as the exterior finish of a building must resist moisture penetration by gravity, driving rain, and capillary action. A *cavity wall* is a masonry wall constructed of an interior and exterior wall that are connected by metal ties and separated with an air space of at least 2 in., and no more than 4½ in. in width (Figure 14.20). Cavity walls most often use the inner wall to carry structural loads and the outer one to resist weathering. They provide improved resistance to moisture and heat transfer when compared to reinforced and composite walls.

Figure 14.20 A cavity wall is constructed of an interior and exterior wall that are connected by metal ties and separated with an air space of at least 2 in. in width.

The air space acts as a drainage plane to control moisture and can be used to accommodate insulation for improved thermal performance of the assembly. Cavity walls are considered drainage wall systems because moisture that enters the outer wythe is drained down its inner surface and conducted back to the exterior by base flashing and weep holes. Flashing is required at the base of the wall, around 8 in. above its base, below and above all wall openings and shelf angles, as well as the cap, or top course, of units in the wall. *Weep holes* are located just above the bottom of the flashing. Weep holes are openings left in the head joints of masonry to permit water to escape and allow air circulation in the air space (Figure 14.21). Moisture that enters the cavity moves to the flashing and is directed out of the building through the weep holes (Figure 14.22).

Figure 14.21 Weep holes connected to flashing at the bottom of cavity walls allow moisture to escape from the cavity.

Figure 14.22 Metal flashing is located at the bottom of walls, and below and above windows and other openings.

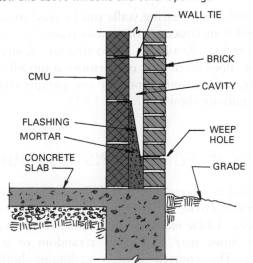

The flashing used to transport the water within the cavity is usually made of copper, galvanized metal or plastic sheet material. Aluminum flashing is not used because it has an unfavorable chemical reaction with mortar. Through wall flashing extends a minimum of 8 in. vertically from the outer wythe upward to the inner wythe. Flashing used at openings is continued past the opening at least 4 in. and turned up 1 in.

Weeps are formed by omitting mortar in head joints, with pieces of plastic or metal pipe located in the bottom of a joint, or by placing oiled rope in the joint and pulling it out after the mortar begins to set. Weep holes should be spaced at 24 in. horizontally in the outer wythe of the wall, although a maximum spacing of 16 in. is recommended for weeps using wick ropes. Mesh-type mortar dropping collection materials are used behind the weep to prevent mortar droppings from blocking the opening as the wall is laid. Masonry cavity walls used commonly used in commercial construction are outlined below.

Brick/Concrete Masonry Cavity Wall   The ubiquitous brick/CMU cavity wall is popular because it provides an extremely durable, long-lasting wall system with minimal long-term maintenance. It offers great design flexibility through different colors, patterns, sizes, and surface textures. Brick CMU walls can provide fire ratings of up to four hours and are ideal for noisy environments, as they naturally absorb sound.

Brick Veneer/Reinforced Concrete   Brick veneer on a cast-in-place concrete wall generally entail both higher weights and costs but result in a durable wall system. This substantial wall system may require larger foundations (Figure 14.23).

Figure 14.23 Brick veneer on a cast-in-place concrete wall provides a durable wall system.

Brick Veneer/Steel Stud This assembly looks like a solid masonry building but a steel stud backup system usually costs less and is lighter in weight than the brick on a CMU cavity wall. Steel studs flex more than masonry veneer and can lead to cracks developing in the masonry.

Insulated Masonry Cavity Walls

Solid masonry walls are poor insulators, readily conducting heat and cold through their massive construction. Double masonry cavity walls are insulated using rigid insulation, normally adhered to the outside face of the interior wythe of masonry, facing the air space. Rigid insulation consists of rigid sheets of expanded or extruded polystyrene, polyisocyanurate, as well as sprayed and batt materials. Rigid insulation for cavity walls is manufactured to correspond to masonry module sizes, often in 16 and 24 in. widths. The material is bonded to the inner masonry between the wall ties with a construction adhesive, insulation grips that snap onto the wall ties, or a combination of both (Figure 14.24).

Figure 14.24 A brick outer wythe with CMU backup cavity wall showing the adjustable eye-hooks and insulation.

For single-wythe walls or for additional insulation, the cores of hollow masonry units can be filled with a granular insulation. Granular fill insulations uses lightweight perlite or vermiculite that are poured into the cells.

Masonry Veneer

A veneer wall is a type of cavity wall that is non-load-bearing, as it supports no weight other than its own. *Masonry veneer* consists of a single wythe of masonry that is anchored to a load-bearing backup construction

with an air space between them. The veneer may be anchored to a steel frame, CMU, or cast-in-place concrete wall. While wood backup walls are still common in residential construction, most commercial projects are replacing them with steel stud walls backed with exterior gypsum board for a more-fire-resistant construction.

Masonry veneer walls are usually supported by concrete or CMU foundations. Typically, the foundation wall will utilize a wider dimension below, with an inset at grade that supports the masonry veneer called the brick shelf. The face of the veneer supported by the shelf is set beyond the edge of the foundation to create the cavity. The overhang may be more than a third of a brick width. This creates a cavity that helps to reduce water condensation by accommodating air circulation. The cavity width is determined by the designer according to the unit width, insulation requirements, and door and window widths. Cavity walls incorporate a moisture/air/ vapor barrier at the sheathing plane facing the unobstructed, cavity anchors and ties, and flashing and weeps.

BRICK MASONRY COLUMNS

Brick masonry columns are typically designed for applications that do not require brick cutting. The interior core of the column has reinforcing bars tied to the foundation, and is filled with concrete (Figure 14.25). *Pilasters* are built in the same manner as columns but become an integral part of a load-bearing wall.

MASONRY CONSTRUCTION DETAILS

Brick and block bearing walls can be used to support roof and floor structures of various types. Figure 14.26 shows various details for supporting wood, steel, and concrete floor assemblies on masonry composite walls. Examples of common roof edge and parapet construction details are shown in Figure 14.27.

STONE MASONRY CONSTRUCTION

Stone laid in mortar may be rubble or ashlar masonry. *Rubble* refers to stone found naturally in irregular shapes and sizes. *Ashlar masonry* uses units cut into squared shapes. Stone may be laid in a random or coursed pattern. The coursed pattern maintains horizontal

Figure 14.25 Reinforced brick masonry column and pilaster construction.

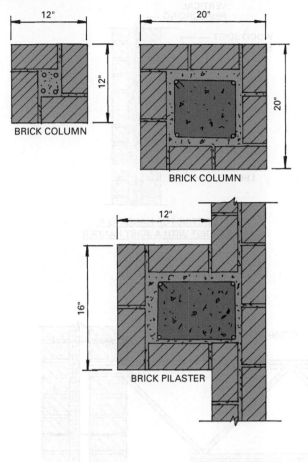

12"

BRICK COLUMN

20"

12"

20"

BRICK COLUMN

12"

16"

BRICK PILASTER

lines, while the random pattern uses irregular courses (Figure 14.28). One type of rubble stone masonry, ledge-rock, is a rock formed naturally into thin layers of varying thicknesses.

When stone is laid, the grain in each unit should run in a horizontal direction. The stone is stronger in this position and tends to better resist weathering. Rubble stone is more difficult to assemble because a mason must select pieces that fit open spaces and work in with the shapes of adjoining stones. Sometimes it is necessary to trim stone with a hammer. The resulting mortar joints are irregular and often quite large. Squared ashlar stone is easier to lay than rubble but requires the mason to select stones that will fit together as needed. Since the stones are squared, the mortar joint can be of relatively uniform width over the entire wall. Joints in ashlar masonry are usually to ⅜ to ¾ in. in width.

Large squared blocks are called *dimension stone*. They usually require a hoist to set them in place. Mortar joints in ashlar masonry are usually raked after setting and allowed to set thoroughly. Then the raked space is

filled with mortar (called *pointing*) and tooled to the specified shape. Great care is taken to avoid getting mortar on the face of the stone. The face is cleaned with a mild soap and soft brush and flushed clean with water.

Stone masonry is usually applied as a veneer over a backup wall, composed of concrete masonry units, for example. Metal ties are used to bond the stone veneer to the backup wall (Figure 14.29), much the same as with brick masonry construction. Anchors should be chromium-nickel, stainless steel, or a zinc alloy. Building codes often ban the use of galvanized steel. Copper, brass, and bronze ties may cause some staining.

Some stone is cut into large, thin panels that are scored to accept some form of metal tie that secures them in place to a backup wall. The manufacturers of various types of stone panels have engineered systems for securing panels in place. A number of connection methods are used. The first veneer panel on a base can be set on the foundation or other supporting masonry, or the panel can hang on the supporting structure and be secured with metal connectors, as shown in Figure 14.30. As panels are placed up the wall, a metal connector is used to secure them to the backup wall. Typically, metal connectors tie the stone veneer to a column. Preassembled column covers are available bonded with high-strength epoxy adhesives.

Stone veneered walls usually use stone windowsills (Figure 14.31) and various types of decorative moldings, balusters, and modillions, and consoles. A *modillion* is a scroll supporting the *corona* (overhanging member of a cornice) under a cornice. A *console* is a decorative vertical scrolled bracket that projects from a wall to support a cornice, door, or window. Balusters are vertical members in a stair rail used to support the handrail.

STRUCTURAL CLAY TILE CONSTRUCTION

Structural clay tile may be load bearing or non-load bearing. Load-bearing units can form an unfaced wall, or they may be used as a backup wall and be faced with another material. Non-load-bearing tile walls are used as interior partitions, fireproofing barriers, and masonry screens.

Examples of commonly used structural tile construction are shown in Figure 14.32. It should be noted that there are many possible variations, depending on expected loads, and desired surface characteristics. Some tile units are utility block used to construct backup walls only, while others have glazed exposed faces available in a variety of colors and textures.

Figure 14.26 Common roof details for masonry bearing wall construction.

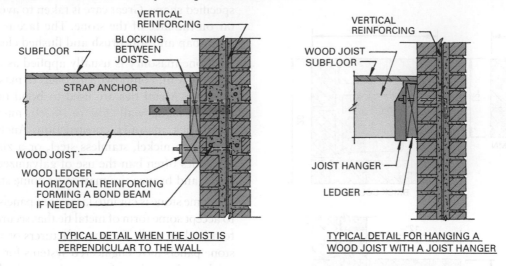

TYPICAL DETAIL WHEN THE JOIST IS
PERPENDICULAR TO THE WALL

TYPICAL DETAIL FOR HANGING A
WOOD JOIST WITH A JOIST HANGER

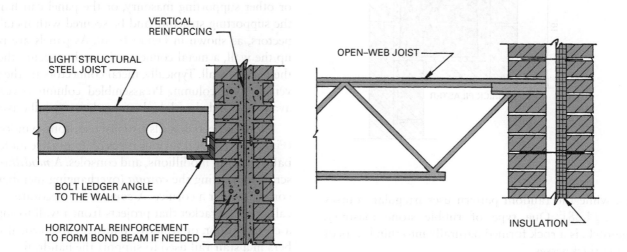

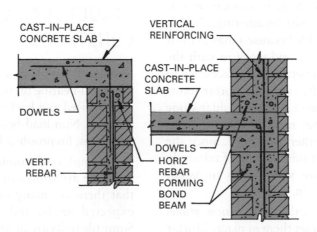

Figure 14.27 Examples of masonry roof and parapet details.

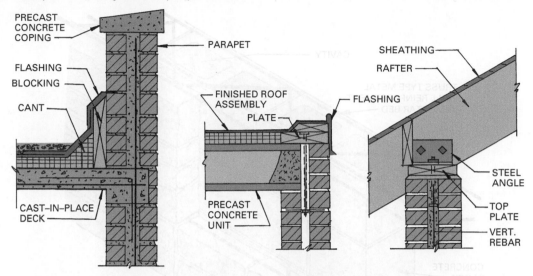

Figure 14.28 Patterns for laying ashlar and rubble stone.

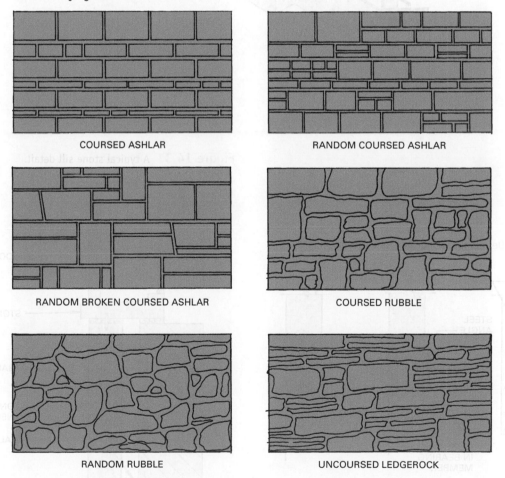

COURSED ASHLAR

RANDOM COURSED ASHLAR

RANDOM BROKEN COURSED ASHLAR

COURSED RUBBLE

RANDOM RUBBLE

UNCOURSED LEDGEROCK

Figure 14.29 Stone masonry veneer can be tied to a CMU backup wall with truss-type metal reinforcing.

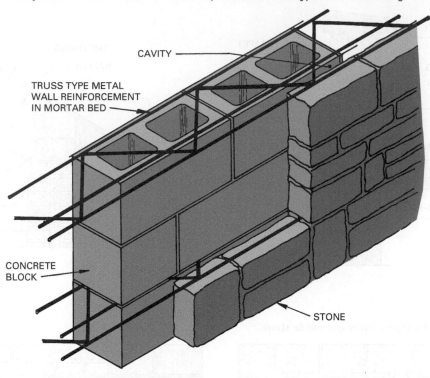

Figure 14.31 A typical stone sill detail.

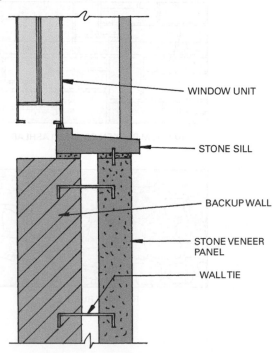

Figure 14.30 Details for anchoring stone facing panels.

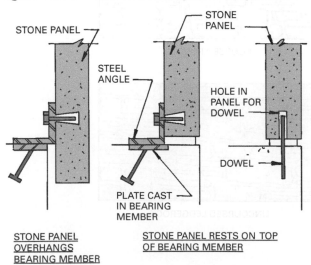

Figure 14.32 Typical structural clay tile wall and partition types.

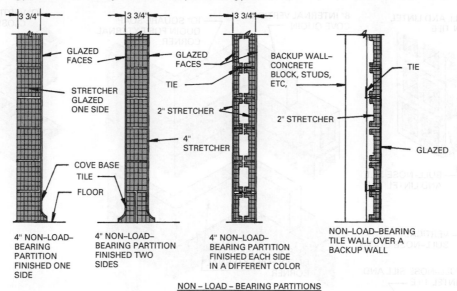

GLAZED FACES

STRETCHER GLAZED ONE SIDE

COVE BASE TILE

FLOOR

4" NON–LOAD– BEARING PARTITION FINISHED ONE SIDE

GLAZED FACES

TIE

2" STRETCHER

4" STRETCHER

4" NON–LOAD– BEARING PARTITION FINISHED TWO SIDES

BACKUP WALL– CONCRETE BLOCK, STUDS, ETC,

2" STRETCHER

4" NON–LOAD– BEARING PARTITION FINISHED EACH SIDE IN A DIFFERENT COLOR

TIE

GLAZED

NON–LOAD–BEARING TILE WALL OVER A BACKUP WALL

NON – LOAD – BEARING PARTITIONS

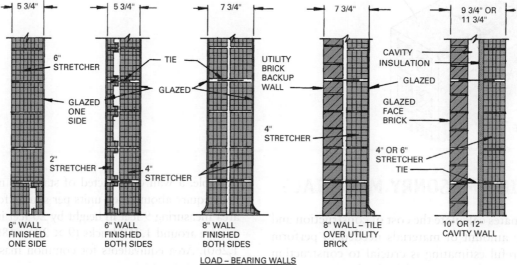

6" STRETCHER

GLAZED ONE SIDE

2" STRETCHER

6" WALL FINISHED ONE SIDE

TIE

GLAZED

4" STRETCHER

6" WALL FINISHED BOTH SIDES

UTILITY BRICK BACKUP WALL

4" STRETCHER

8" WALL FINISHED BOTH SIDES

CAVITY INSULATION

GLAZED

GLAZED FACE BRICK

4" OR 6" STRETCHER

TIE

8" WALL – TILE OVER UTILITY BRICK

10" OR 12" CAVITY WALL

LOAD – BEARING WALLS

Cavity walls of structural tile use flashing and weep holes as described for brick construction. They may be insulated with rigid plastic foam insulation. Floor joists and rafters are also installed much like those on brick and concrete masonry. Structural clay tile can be used as a veneer over a backup wall of a different material, such as concrete masonry units or stud walls. Openings in structural clay tile are usually spanned using reinforced tile lintel units.

Laying Structural Clay Tile Walls

As with brick and concrete masonry, structural clay tile walls are designed to use standard size units that avoid custom cutting. The face sizes of a standard stretcher block are 5 in. high by 11 in. long (5 in. nominal) and 7 in. high by 15 in. long (8 × 16 in. nominal). Buildings are often designed to meet these sizes, although other special units are also available.

High-strength mortars are used with structural tile. Local codes and information from tile manufacturers are used when specifying mortar for structural clay tile. Cavity walls utilize metal ties to integrate tile with its backup material, similar to how brick and concrete masonry are handled.

Corners can be built in several ways. Typical corner constructions and sills at openings can be fashioned using special bullnose units, as shown in **Figure 14.33**.

Figure 14.33 Several ways to construct corners and openings with structural clay tile.

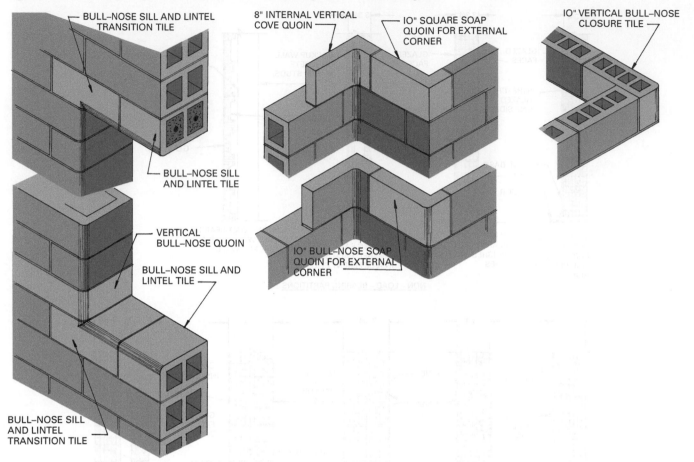

BULL–NOSE SILL AND LINTEL TRANSITION TILE

8" INTERNAL VERTICAL COVE QUOIN

IO" SQUARE SOAP QUOIN FOR EXTERNAL CORNER

IO" VERTICAL BULL–NOSE CLOSURE TILE

BULL–NOSE SILL AND LINTEL TILE

VERTICAL BULL–NOSE QUOIN

BULL–NOSE SILL AND LINTEL TILE

IO" BULL–NOSE SOAP QUOIN FOR EXTERNAL CORNER

BULL–NOSE SILL AND LINTEL TRANSITION TILE

ESTIMATING MASONRY MATERIALS

Material estimates calculate the cost of construction and determine the amount of materials needed to perform a project. Careful estimating is crucial to construction productivity by reducing time lost by having to reorder materials or remove excess material at the end of a project. There are various methods to estimate material quantities. Hand calculations and computer programs can be used, depending on the complexity of the building. Masonry units are delivered to the site in bands of units strapped into blocks containing a set number of units. The amount of materials required depends on the size of the wall and the size of the masonry unit.

The wall area method is the most widely used estimating procedure in masonry construction. Estimators begin by calculating the area of masonry to be assembled. The height and width of a wall is multiplied, with window and door areas subtracted out of the total. To estimate the required number of units, the calculated wall area is multiplied by known quantities of material required. For example, a wall constructed of standard modular bricks will require about seven units per square foot of area. An area measuring 9 feet in height by 20 feet in width would require around 1,260 bricks (9 × 20 = 180 sq. ft. × 7 = 1,260). Area equivalents for common masonry units are given in Table 14.1.

Table 14.1 Number of Masonry Units Required Per Square Foot

Unit Name	Actual Size	Units per Square Foot Area
Standard Modular Brick	(3-5/8 × 2-1/4 × 7-5/8)	7
Utility Brick	(3-5/8 × 3-5/8 × 11-1/2)	3
King	(3 × 2-5/8 × 9-5/8)	4.8
Jumbo	(3-5/8 × 2-3/4 × 8)	5.5
Oversize Brick	(3-5/8 × 2-1/4 × 7-5/8)	6
2-1/2 in. High Block	(2-1/2 × 8 × 16")	3.1
Half High Block	(3-1/2 × 7-5/8 × 15-5/8)	2.25
Standard Block	(7-5/8 × 7-5/8 × 15-5/8)	1.125

To determine mortar quantities, masons frequently use a rough estimate that eight bags of masonry cement will lay around 1,000 modular brick. In the example above, 1,260 bricks would require approximately ten bags of masonry cement (1,260 × 8/1,000 = 10.08). For block, three bags of mortar are estimated for every 100 blocks. Sand is estimated by volume or weight. The ratio of sand to masonry cement is about one part cement to 2.25 to 3 of masonry cement. Rough estimating calls for roughly 1 cubic yard of sand (1.25 tons) per 1,000 brick and per 200 block. Approximately 5 percent to 10 percent should be added to all material quantities to compensate for breakage, spillage, and errors.

Review Questions

1. What are the commonly used types of masonry load-bearing walls?
2. Why are control and expansion joints required in masonry walls?
3. Why are weep holes used?
4. What can be done to reduce moisture penetration through masonry walls?
5. What types of metal flashing can be used with masonry wall construction?
6. How are masonry units supported over openings in a wall?
7. How can the heat and cold conduction of masonry walls be reduced?
8. What are the commonly used brick patterns used in wall construction?
9. What are the advantages and disadvantages of the various types of mortar joints?
10. What size mortar joint is commonly used?
11. What is the difference between rubble and ashlar stone used for wall construction?
12. What types of structural clay tile are used for wall construction?

Key Terms

Arch

Ashlar Masonry

Bond Pattern

Cavity Wall

Console

Control Joint

Corona

Dimension Stone

Expansion Joint

Lintel Block

Masonry Veneer

Modillion

Pilaster

Pointing

Reinforced Masonry

Rubble

Story Pole

Stretcher

Structural Clay Tile

Troweling

Weep Hole

Wythe

Activities

1. Check your local building code and make a record of requirements pertaining to the construction of masonry load-bearing walls.
2. Using standard 8 × 8 × 16 in. concrete blocks, design a building so it will be approximately 10 ft. wide and 25 ft. long. Size the walls so it is not necessary to cut any concrete blocks. If the wall should be about 10 ft. high, how many blocks will it take to lay the back and one side?
3. Secure a quantity of clay bricks and mortar mix. Prepare a story pole using 1 in. mortar joints. Lay up a length of wall about 6 ft. long and 4 ft. high. Try making several types of finished mortar joints.

4. Collect samples of all the masonry units you can find. Most clay brick manufacturers have sample boards with thin sections of the bricks they make glued onto a strong backing.

5. Pick an exterior wall of a brick building and measure its width and height. Calculate the number

of standard and oversize bricks in the wall. Get the local cost of bricks and figure the cost of the masonry units in this wall. Refer to Chapter 11 for sizes of masonry units.

Additional Resources

ASTM Standards on Masonry, American Society for Testing and Materials, West Conshohocken, PA.

Beall, C., *Masonry Design and Detailing*, the Masonry Society, Boulder, CO.

Chrysler, J., and Amshein, J. E., *Reinforcing Steel in Masonry*, the Masonry Society, Boulder, CO.

Concrete Masonry Handbook, Portland Cement Association, Skokie, IL.

Jaffe, R., *Masonry Basics*, and many other technical publications, the Masonry Society, Boulder, CO.

Masonry Code and Specifications, the Masonry Society, Boulder, CO.

Masonry Designers Guide, the Masonry Society, Boulder, CO.

NCMA Masonry Design Software, National Concrete Masonry Association, Herndon, VA.

See Appendix C for addresses of professional and trade organizations and other sources of technical information.

05

Metals
CSI MasterFormat™

052000 Metal Joists
053000 Metal Decking
054000 Cold-Formed Metal Framing
055000 Metal Fabrications
057000 Decorative Metal

© Dwight Smith/Shutterstock.com

Ferrous Metals

Upon completion of this chapter, the student should be able to:

- Explain the processes for mining and processing iron ore and for producing pig iron and steel.

- Develop knowledge of the properties of ferrous metals to consider when making material selection decisions.

- Be familiar with the various steel identification systems and the Unified Numbering System for Metals and Alloys.

Build Your Knowledge

Metals had little use in construction until the late eighteenth century, when cast iron found limited use as a structural material. By the early nineteenth century, wrought iron and cast iron were used for structural purposes but not extensively, because cast iron was brittle, and the production of wrought iron was limited and expensive. A breakthrough occurred in the 1850s when the Bessemer process for removing impurities from molten iron was developed. The process enabled steel, a greatly improved material, to be produced quickly and in large quantities, thus reducing its cost.

Since it became widely available in the late nineteenth century, steel has had tremendous influence on the evolution of construction. Previous materials worked mostly in compression, restricting the spans that could be achieved. The stiffness of steel enables designers to achieve much greater spans and heights than is possible in either wood or masonry construction. In engineering terms, there are is almost no limit to what steel can achieve.

Ferrous metals are those in which the chief ingredient is the chemical element iron (ferrum). Iron (chemical symbol Fe), mixed with other minerals, is found in large quantities in the earth's crust. To be useful, iron must be extracted from mined ore, have impurities removed and ingredients added to alter its properties, and then be formed into usable products.

Ferrous metal products are widely used in the construction industry. They are a major construction material, and architects, engineers, and contractors should be familiar with the various types, their properties, and the proper applications for each. Although a ferrous metal product may fail, this usually does not occur because it is a poor material but rather because the type of ferrous metal chosen for a particular application was incorrect. When the properties of a ferrous metal are known, its performance can be accurately determined during the engineering design process.

IRON

Iron is found in large quantities in the earth's crust. Pure iron, free from impurities and other elements, is ductile and soft but generally not strong enough for structural purposes. It has good magnetic properties but oxidizes (rusts) easily and does not resist attack by acids and some chemicals. For commercial purposes, iron must have *alloying elements* added to improve its characteristics. Iron can be hardened by heating and rapid cooling, and can be made more workable by *annealing*, that is, heating it and then allowing it to cool slowly.

The starting point for commercial iron products is called *pig iron*. It contains 3 to 5 percent carbon and traces of other elements, such as manganese, sulfur, silicon, and phosphorus. Pig iron is the base material used to produce various types of iron and steel. Iron is used for a wide range of purposes in construction. It is the main ingredient in cast iron, steel, stainless steel, and iron alloys. Iron particles may be used as an abrasive for sandblasting and sometimes as aggregate for specialized concretes. They also form the basis for some color pigments.

Mining and Processing Iron Ore

Iron is found in rock, gravel, sand, clay, and mud that is mined in open pit and underground mines. Common iron-bearing minerals are pyrite (FeS_2), siderite ($FeCO_3$), and hematite (Fe_2O_3), which contain up to 70 percent iron. Jasper and taconite rock contain 20 to 30 percent iron.

Surface ores are mined, loaded onto trucks, trains, and ships, and moved to blast furnaces. If the iron content of the ore is less than 50 percent, it is too costly to ship any distance. These lower-content ores are processed on site by a *beneficiation* process that removes some of the unwanted elements, leaving an ore with high iron content. Beneficiation involves grinding ore to remove unwanted elements and then increasing the size of the ground particles by a process called *agglomeration*.

Agglomeration involves pelletizing the ore particles, which are then easier to ship than finely ground ore. Iron ore dust produced during this process is recycled through sintering. *Sintering* consists of fusing iron ore dust with coke and fluxes into a clinker that is high in iron content.

Ores that have very high iron content are made into pellets and briquettes containing more than 90 percent iron. These pellets are so pure they are not used in pig iron production but go directly into steelmaking. The method of producing these pure pellets is called *direct reduction*.

Producing Iron from Iron Ore

Iron ore is converted into pig iron in a blast furnace by smelting and reduction. *Smelting* is a process in which the ore is heated, permitting the iron to be separated from impurities that may be chemically or physically mixed in. *Reduction* is a process that separates the iron from oxygen with which it is chemically mixed.

The Blast Furnace The blast furnace separates the iron from the waste materials and sinters the ore and flue dust. A very large blast furnace is shown in Figure 15.1. It is computer controlled, fed by a conveyor, and operates continuously.

Figure 15.1 A large blast furnace is capable of producing 8,000 tons of molten iron per day.

© Enrique Izquierdo/Shutterstock.com

Coal, oil, and natural gas are the commonly used fuels in a blast furnace. Most operate on coke, which is produced from coal. The coke not only provides heat to melt ore, but also helps separate iron from its oxides.

A *flux* is a mineral added to molten ore in a blast furnace. The flux combines with impurities and forms a *slag* that floats on top of the molten material. Limestone and dolomite are typical basic fluxes, whereas sand, gravel, and quartz are acid fluxes.

The heart of the blast furnace is a tall, cylindrical shaft, 150 to 200 ft. (45.8 to 61 m) tall, and about 30 ft. (9.2 m) in diameter at the base. It is usually smaller in diameter at the top. It is lined with a refractory brick made of a material, such as magnesia, that can withstand temperatures approaching 3,000°F (1,662°C) (Figure 15.2).

Figure 15.2 A blast furnace receives the charge at the top and hot air at the bottom, causing combustion that frees the iron from the oxide-forming pig iron.

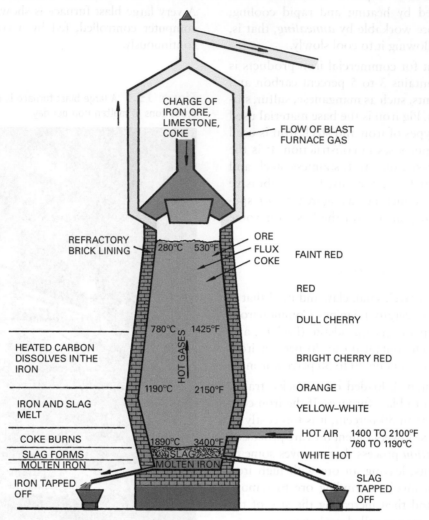

CHARGE OF IRON ORE, LIMESTONE, COKE

FLOW OF BLAST FURNACE GAS

REFRACTORY BRICK LINING — 280°C   530°F

ORE
FLUX
COKE

FAINT RED

RED

DULL CHERRY

780°C   1425°F

HEATED CARBON DISSOLVES IN THE IRON

BRIGHT CHERRY RED

1190°C   2150°F

ORANGE

IRON AND SLAG MELT

YELLOW–WHITE

COKE BURNS

1890°C   3400°F

HOT AIR   1400 TO 2100°F
760 TO 1190°C

SLAG FORMS
MOLTEN IRON

SLAG
MOLTEN IRON

WHITE HOT

IRON TAPPED OFF

SLAG TAPPED OFF

HOT GASES

A charge of ore, hot coke (fuel), and limestone (flux) are loaded at the top. A blast of hot air is injected at the bottom. As the air works its way up through the charge, oxygen in the hot air combines with the hot coke. This causes combustion, which produces the heat necessary to melt the ore. As the gases from combustion pass through the heated ore, a chemical reaction occurs that frees iron from its oxide. Gases removed from the top of the furnace are passed through filters that retain the dust. The cleaned hot air passes through heat exchangers that in turn heat new, incoming air. A blast furnace must operate continuously.

Molten pig iron settles to the bottom of the furnace and is tapped off every few hours. At this point, it is impure and brittle, containing about 4 to 5 percent carbon. This high carbon content is what makes pig iron brittle and not as useful as steel. It has no ductility and cannot be rolled into useable products like beams or studs. Pig iron must undergo additional processing in a steelmaking furnace to reduce carbon content. The impurities, flux, and oxygen combine to form a slag on top of the molten iron, which is periodically tapped off. Slag is used as an aggregate for concrete, loose fill, and road surfacing.

The molten pig iron is moved to furnaces called *mixers* where it is blended to equalize the chemical composition and kept in a liquid state until used. The molten iron sometimes goes into a casting machine where it is molded into ingots (*pigs*) and allowed to cool. In other operations, it is moved directly to basic oxygen steelmaking furnaces. In this case, energy is saved because the iron is already in molten form. A diagram of the total process is shown in Figure 15.3.

Pig iron ingots are usually made to specifications based on their end use. The elements that are carefully controlled include carbon, sulfur, silicon, phosphorus, and manganese.

Figure 15.3 The steelmaking process, from preparing the ore, through processing into steel, to producing steel beams.

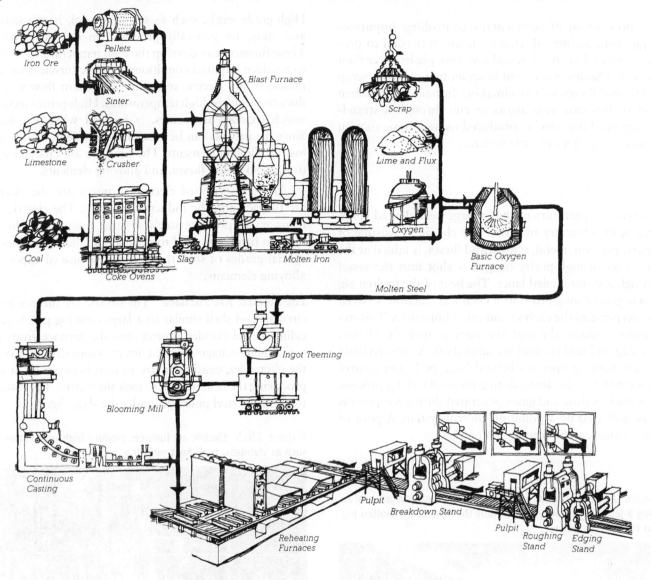

Iron Ore

Pellets

Sinter

Limestone

Crusher

Coal

Coke Ovens

Blast Furnace

Slag

Molten Iron

Scrap

Lime and Flux

Oxygen

Basic Oxygen Furnace

Molten Steel

Ingot Teeming

Blooming Mill

Continuous Casting

Reheating Furnaces

Pulpit

Breakdown Stand

Pulpit

Roughing Stand

Edging Stand

Cast Irons

Iron containing almost no carbon is identified as a wrought iron. A mixture of low-carbon iron and a large amount of slag, wrought iron is soft, tough, and ductile (easily worked). Ingot iron is a very low-carbon iron that has no slag and is also tough, ductile, and soft.

Cast irons have carbon contents above 1.7 percent and include white, gray, and malleable types. White cast irons have low silicon content and are cooled rapidly. They are hard and brittle and have few applications for construction uses. Gray cast irons are produced by increasing silicon content and cooling the molten metal slowly. They are tougher and softer than white cast iron,

and gray in color. They may have additional elements, such as nickel, copper, and chromium. Gray cast irons are widely used for all types of castings, such as sewer pipe, ornamental railings, and decorative lamp posts.

Malleability is a property of metal that allows it to be formed mechanically, by rolling or forging for example, without fracturing. Malleable cast iron is produced by reheating white cast iron, maintaining the required temperature for a long time, and then cooling it slowly. It is not as brittle as white and gray cast iron, because it has a lower carbon content and greater ductility. Malleable cast iron is used for hardware and other cast items requiring toughness and breakage resistance.

STEELMAKING

The production of steel entails controlling impurities in pig iron, adding alloying elements as needed to produce required properties, and lowering pig iron's carbon content. This involves combining molten pig iron, scrap metal, and fluxes in a steelmaking furnace. The molten steel is then cast into ingots or run through a strand-casting machine. Steel is produced using a basic oxygen process or with an electric furnace.

Basic Oxygen Process

The basic oxygen furnace has a large pear-shaped vessel lined with refractory material. The charge, consisting of molten pig iron, metal, scraps, and fluxes, is added at the top. A jet of high-purity oxygen is shot into the vessel through a water-cooled lance. The heat of the molten pig iron is great enough to start a chemical reaction between the oxygen and the carbon and other impurities. This *oxidation* produces the heat necessary to melt the charge. The slag and molten steel are tapped off. A modern basic oxygen furnace operates behind huge pollution control doors that remain closed during the steelmaking process. This enables dust and gases generated during the process to be collected by a pollution-control system. A pour of molten steel can be seen in Figure 15.4.

Figure 15.4 Molten iron is poured in the basic oxygen process using a jet of high-purity oxygen plus the heat of the molten pig iron to start the process.

Electric Steelmaking Processes

High-grade steels, such as stainless, tool, heat-resisting, and alloy, are generally produced in electric furnaces. These furnaces can develop the high temperatures needed to produce required conditions. Electric furnaces use arc radiation and electric resistance to current flow to produce the required high temperatures. High-purity oxygen, which oxidizes impurities, is injected when necessary. Since the process can be carefully controlled, there is less loss of alloying elements. The furnaces are charged with iron, scrap metal, fluxes, and alloying elements.

The two types of electric furnaces are the electric arc furnace and the induction furnace. The electric arc furnace can produce steels in large quantities. The induction furnace is used to produce smaller quantities of special grades of steel that require the use of expensive alloying elements.

The Electric Arc Furnace   The electric arc furnace has a circular steel shell similar to a large cooking pot. Several cylindrical electrodes project into the furnace from the top. A high-voltage electric current is introduced through the electrodes, causing an arc to pass between them and producing the heat needed to melt the charge. The furnace tilts and the steel pours off under the slag (Figure 15.5).

Figure 15.5 Electric arc furnaces produce high-grade steels, such as stainless, tool, heat-resisting, and alloy.

The Induction Furnace   This furnace has a cylindrical vessel made of magnesia and insulated with refractory materials. Outside the vessel are windings of copper tubing through which high-voltage alternating current passes, creating an induction current resisted by the charge. This resistance generates the heat needed for melting the charge. The vessel tilts and the steel pours off below the slag.

Casting the Steel   When steel is ready to be poured from a furnace, it is either cast into ingot molds or run through a *strand-casting machine* (Figure 15.6).

Figure 15.6 Molten steel can be cast into a continuous slab by a strand-casting machine.

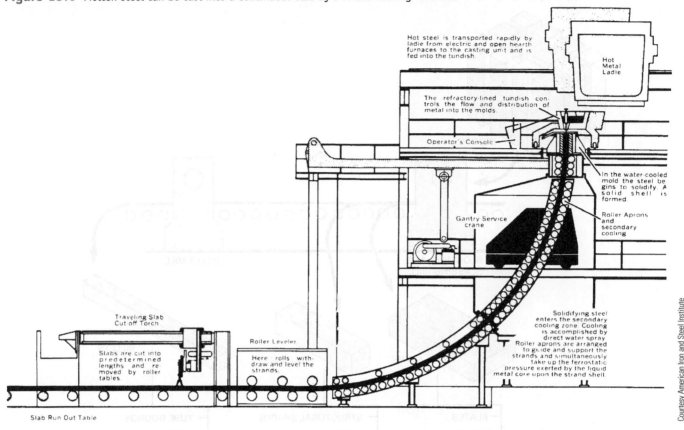

Ingot molds can be several feet in diameter and six to eight feet high. The molds are coated inside to prevent surface damage from steel. After an ingot begins to harden, it is removed from the mold and taken to a soaking pit where the molten steel inside solidifies. Strand-casting machines produce a continuous ribbon of steel that begins to harden as it passes through a series of rollers (**Figure 15.7**). Once hardened, it reaches a horizontal conveyor and is cut to required lengths. This process takes about thirty minutes.

MANUFACTURING STEEL PRODUCTS

Steel products are manufactured by rolling, extruding, cold-drawing, forging, and casting. Items produced by casting are referred to as *cast steel products* and all of the others are called *wrought products*. The most frequently used products are produced by hot-rolling and cold finishing. Cast steel products are made by pouring molten steel into sand molds. The molten steel goes through a process to alter its properties to suit the purpose served by the finished product.

Figure 15.7 These steel strands are produced by a continuous-casting machine. The slabs are cut into predetermined lengths by an automatic torch machine.

Most wrought products are produced by hot-rolling and cold-rolling. The ingots fashioned in steel mills are used to produce slabs, blooms, and billets, which can be seen in **Figure 15.8**. Slabs, blooms, and billets are semi-finished forms of steel. When semi-finished steel is fabricated directly into a finished product or

Figure 15.8 The steps in producing ingots into semi-finished steel used to produce finished steel products.

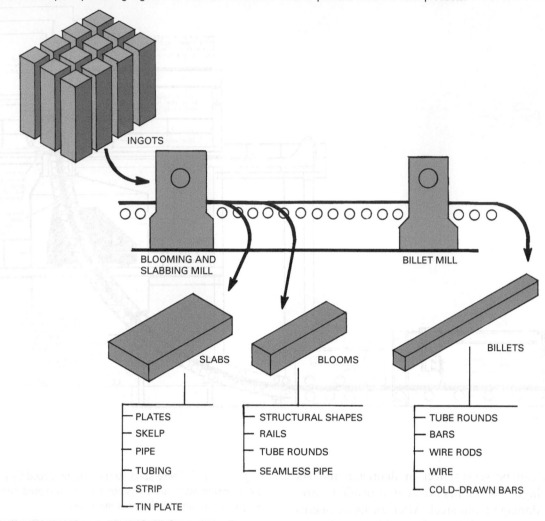

INGOTS

BLOOMING AND SLABBING MILL

BILLET MILL

SLABS

BLOOMS

BILLETS

PLATES
SKELP
PIPE
TUBING
STRIP
TIN PLATE

STRUCTURAL SHAPES
RAILS
TUBE ROUNDS
SEAMLESS PIPE

TUBE ROUNDS
BARS
WIRE RODS
WIRE
COLD–DRAWN BARS

into a product used to produce a finished product, it is considered finished steel.

Slabs are large rectangular semi-finished steel pieces made in a number of sizes. Their width is typically more than twice their thickness (Figure 15.9). The thickness is usually 10 in. (254 mm) or less. A bloom is an oblong of semi-finished steel with a square cross-section that is usually larger than 6 in. (152 mm) square. Billets are also oblongs of semi-finished steel, but they are smaller in cross-section and longer than blooms.

Figure 15.8 lists typical finished products that use these semi-finished steel products. Notice that plates and skelp (skelp is used to produce pipe and tubing) are rolled from slabs. Finished steel plates resting on cooling beds can be seen in Figure 15.10. The plate stock enters a four-tiered finishing stand, where it is rolled to

Figure 15.9 A continuous caster moving slabs after they have been cut to length.

Figure 15.10 These perfectly formed steel plates are resting on a cooling bed.

Figure 15.11 Structural steel beams that have been roll-formed hot being transferred to a "walking" cooling bed before they are cut to the desired lengths.

the final desired length and thickness by passing back and forth through finishing stand rollers.

Hot-rolling is used for most structural shapes such as I-beams, wide-flange beams, channel iron and angle iron. The bloom is taken to a structural mill still at a temperature of about 2200 F. It is passed through a succession of rollers that form the metal into the desired shape and size. A typical wide flange beam for instance, will pass through the rollers around 40 times before completion. Because the rollers are massive and expensive, hot-rolled shapes are available only in standard sizes.

Cold-rolling is a continuous bending operation done at room temperature in which sheet or strip metal is plastically deformed. Tandem sets of rolls shape the metal in a series of progressive stages until the desired cross-sectional configuration is obtained. Roll forming is ideal for producing parts with long lengths or in large quantities, and produces tight dimensional consistency.

Cold-drawing is to form steel profiles of complex cross-sections with high precision. Special profile bars are drawn through a forming die. Cold-drawing of hot-rolled and hot-extruded special pre-shape steel profiles tightens the cross sectional tolerances, leading to significant improvement in dimensional accuracy and surface quality. Cold-drawn steel special profiles offer the same precision achieved by machining but without the waste.

All types of structural shapes, including rods, rails, strip, and seamless pipe, are manufactured using blooms. Finished structural beams are shown in Figure 15.11. Billets are used to produce solid steel bars and rounds, wire rods and wire, seamless pipe, and cold-drawn bars.

Steel Recycling

Steel is one of the most widely recycled materials in the world (Figure 15.12). The steel industry has been engaged in recycling for more than a century because it is economically advantageous. It is less costly to recycle steel than to mine iron ore and process it to form new steel. Steel loses none of its inherent physical properties during the recycling process, and the energy and material requirements for recycling are far less than that for iron ore refining. Today, 60 to 90 percent of all steel products contain recycled content, and the energy saved by recycling reduces the industry's annual energy consumption by about 75 percent.

Architects and engineers are increasingly not only utilizing recycled steel for their projects, but also considering the potential for material reuse at the end of building's useful life. *Design for Disassembly (DFD)* is

Figure 15.12 Steel scrap stockpiled for reprocessing in a steel mill.

an emerging subfield of sustainable construction that promotes disassembly over demolition in an effort to maximize material recycling. Steel frame buildings are excellent candidates for disassembly due to the durability, versatility, and the standardized sizing of steel framing members.

STEEL IDENTIFICATION SYSTEMS

Nationally used metal and alloy numbering systems are administered by societies, trade associations, and individual users and producers of metals and alloys. Among these are numbering systems by the American Society for Testing and Materials (ASTM), SAE/Aerospace Materials Specifications (AMS), American Welding Society (AWS), American Iron and Steel Institute (AISI), Society of Automotive Engineers (SAE), Federal Specifications, Military Specifications (MIL-SPEC), and the American Society of Mechanical Engineers (ASME). Following are brief descriptions of several of these systems.

The American Iron and Steel Institute (AISI) and the Society of Automotive Engineers (SAE) series of identifying numbers for carbon and alloy steels use four- or five-digit numbers. The first two digits indicate type of steel (carbon or alloy) and the last two digits indicate carbon content. For example, in AISI 1030, the 10 indicates a carbon content of 0.30 (actual range is 0.28 to 0.34). An AISI 4012 indicates a molybdenum steel alloy composed of 0.15 to 0.25 molybdenum. It has 0.09–0.14 carbon, 0.75–1.00 manganese, 0.035 phosphorus maximum, 0.040 sulfur maximum, and 0.15–0.35 silicon.

The American Society for Testing and Materials (ASTM) sets standards for steel designations using an arbitrary number to indicate chemical compositions and specify minimums for strength and ductility. These specifications regulate specific chemical elements that directly affect the fabrication and erection of the steel. This makes it possible for various proprietary steels of different chemical compositions to conform to ASTM performance standards. Structural steels used in construction are designated by ASTM standards. Typical steels used for construction purposes are identified in Table 15.1.

One system, the Unified Numbering System for Metals and Alloys (UNS), was developed jointly by the SAE and the ASTM. It provides a means for describing the composition of several thousand metal designations and cross-references the systems of the organizations just mentioned.

Table 15.1 Identification Number, Type, and Yield Point of Selected Structural Steels

Unified Number	ASTM Number	Type	Yield Point, ksi[a]
K02600	A36	Carbon steel	36
K11510	A242	High-strength low-alloy, corrosion resistance 5 to 8 times carbon steel	42–50
K11630	A514	High-yield-strength, quenched and tempered alloy steel	90–100
K02703	A529	Carbon steel	42
K02303	A572	High-strength low-alloy, niobium-vanadium, structural quality, corrosion resistance four times carbon steel	42–65
K11430	A588	High-strength low-alloy, corrosion resistance four times carbon steel	42–50
K01803	A633	Normalized high-strength low-alloy steel	42–60
K01600	A678	Quenched and tempered steel	50–75
K12043	A852	High-strength quenched and tempered alloy steel	70

[a]Kips per square inch (ksi). One ksi equals 1,000 lb. per square inch.

The Unified Numbering System for Metals and Alloys

The Unified Numbering System (UNS) provides the uniformity required for indexing, record keeping, data storage and retrieval, and cross-referencing of metals. The UNS is not a specification but is used as an identifier of a metal or alloy for which controlling limits have been established in specifications published elsewhere.

The UNS has eighteen series of designations for metals and alloys, seventeen of which are active. Each UNS designation has a single-letter prefix followed by five digits. The letter is used to identify the family of metals, such as S for stainless steels and A for aluminum. The prefixes are listed in Table 15.2.

The significance of the digits can vary with the UNS series. Therefore, their meaning for each series can serve a different purpose. Effort has been made to relate UNS digits to those established by related trade organizations. In the published UNS manual *Metals and Alloys in the Unified Numbering System*, UNS numbers are cross-indexed with systems of other organizations whose documents describe materials that are the same as or similar to those covered by UNS numbers. These systems are identified by the letters for each organization, shown in Table 15.3.

Table 15.2 The Unified Numbering System (UNS) for Identifying Metals and Alloys

Series Designation	Family of Metals
Axxxxx	Aluminum and aluminum alloys
Cxxxxx	Copper and copper alloys
Exxxxx	Rare earth and similar metals and alloys
Fxxxxx	Cast irons
Gxxxxx	AISI and SAE carbon and alloy steels
Hxxxxx	AISI and SAE H-steels
Jxxxxx	Cast steels (except tool steels)
Kxxxxx	Miscellaneous steels and ferrous alloys
Lxxxxx	Low melting metals and alloys
Mxxxxx	Miscellaneous nonferrous metals and alloys
Nxxxxx	Nickel and nickel alloys
Pxxxxx	Precious metals and alloys
Rxxxxx	Reactive and refractory metals and alloys
Sxxxxx	Heat and corrosion-resistant steels (including stainless), valve steels, and iron-base "superalloys"
Txxxxx	Tool steels, wrought and cast
Wxxxxx	Welding filler metals
Zxxxxx	Zinc and zinc alloys

Source: Society of Automotive Engineers

Table 15.3 Numbering Systems Cross-Referenced to UNS

Cross-Reference Prefix	Specifying Organization
AA	(Aluminum Association) numbers
ACI	(Steel Founders Society of America) numbers
AISI	(American Iron and Steel Institute) including SAE (Society of Automotive Engineers) numbers (carbon and low-alloy steels)
AMS	(SAE/Aerospace Materials Specification) numbers
ASME	(American Society of Mechanical Engineers) numbers
ASTM	(American Society for Testing and Materials) numbers
AWS	(American Welding Society) numbers Federal Specification Numbers
MIL	(Military Specification) numbers
SAE	(Society of Automotive Engineers) "J" numbers

Source: Society of Automotive Engineers

The base elements in the specification of composition of a metal are identified by standard chemical symbols. Some of these are shown in Appendix (F).

STEEL AND STEEL ALLOYS

"Steel" is a term generally applied to plain carbon steels that are alloys of iron and carbon (with a carbon content of less than 2 percent). Other steel products include stainless and alloy steels.

Plain Carbon Steels

Plain carbon steels have iron as the major element (more than 95 percent), but they also contain impurities, such as sulfur, nitrogen, and oxygen. Other elements may be present as residual impurities or as alloys added to change the properties of the steel. These can include phosphorus, nickel, aluminum, copper, silicon, and manganese.

The properties of carbon steel are varied not only by the elements that are added to alter their chemical composition but also by the type of mechanical and heat treatment used in producing the steel. For example, during production, carbon steel could be hot or cold-rolled, cast, cooled slowly, or cooled rapidly, all of which influence final properties.

Control of carbon content is the major factor in establishing the properties of carbon steel. As the carbon content increases, so does the strength and hardness, while ductility decreases. Differences in the amount of carbon among steel types is very small, usually hundredths of a percent. The percentage of carbon in several types of carbon steel is shown in Table 15.4.

Table 15.4 Carbon Content of Carbon Steels

Type of Steel	Percent Carbon Content	Characteristics
Extra soft grade	0.05–0.15	Ductile, soft, tough Used for wire, rivets, pipe, sheets
Mild structural grade	0.15–0.25	Ductile, strong Used for boilers, bridges, buildings
Medium grade	0.25–0.35	Harder than mild structural Used for machinery and general structural purposes
Medium hard grade	0.35–0.65	Harder than medium grade Resists abrasion and wear
Spring grade	0.85–1.05	Strong and hard Used to make springs
Tool steel	1.05–1.20	Hardest and strongest of carbon steel

Carbon steels contain other elements, such as manganese, phosphorus, and sulfur, and amounts vary with the chemical design of the metal. For example, AISI/SAE 1008 carbon steel contains 0.10 percent maximum carbon, 0.30 to 0.50 percent manganese, 0.040 percent maximum phosphorus, and 0.050 percent maximum sulfur. Carbon steels are used for numerous construction products, including structural shapes, bars, sheet and strip products, plate, pipe, tubing, wire, nails, rivets, and screws. Carbon steel is also used to produce cast products, typically from medium-grade carbon steel. The castings are often heat treated to relieve internal strain developed during the casting process.

Alloy Steels

An alloy steel contains one or more alloying elements other than carbon (such as chromium, nickel, or molybdenum) that have been added in amounts exceeding a specified minimum to produce properties not available in carbon steels. These elements give particular physical, mechanical, and chemical properties to the steel. Stainless steels, specialty steels, and tool steels are not considered alloy steels, although they do contain alloying elements.

Standard alloy steels generally have elements of carbon, manganese, and silicon occurring naturally. Alloy steel that has one additional alloying element is identified as a single alloy steel. Adding two alloying elements produces a double, or binary, alloy steel, and three elements produce a triple, or ternary, steel. The major alloying elements and the property changes they produce can be found in Table 15.5.

Table 15.5 also demonstrates that alloying elements are used to improve properties, such as hardness, performance of the material at high and low temperatures, strength, ductility, workability, wear resistance, electromagnetic properties, and electrical resistance or conductivity. Architects and structural engineers are especially concerned with those alloys having increased strength, wear resistance, and resistance to *corrosion*, expansion, contraction, and ductility.

There are many alloy steels specified by the various organizations mentioned earlier. Their uses range widely and include such types as alloy steel electrode, alloy steel welding wire, alloy steel, high-strength low-alloy steel, and others. Some of these have direct applications to products used in construction. Examples include high silicon content, which improves magnetic permeability, making it useful in transformers, motors, and generators. Nickel improves toughness, so nickel alloys are employed on cutting tools used in

Table 15.5 Properties Imparted to Steel Alloys by Alloying Elements

Element	Properties
Aluminum	A deoxidizer used to control the grain size within the structure of the steel, promotes surface hardening
Boron	Increases the depth of hardness
Carbon	Increases hardness, strength High percentages contribute to brittleness
Chromium	Increases hardness, corrosion, and wear resistance
Cobalt	Hardens or strengthens the ferrite, resists softening at high temperatures Used in high-speed tool steels
Copper	Increases resistance to atmospheric corrosion and increases yield strength
Manganese	Increases strength and resistance to wear and abrasion
Molybdenum	Increases corrosion resistance, raises tensile strength and elastic limit, reduces creep, improves impact resistance
Nickel	Increases elastic limit and internal strength Increases strength and toughness in heat treated steel In some steels increases hardness, fatigue, and corrosion resistance
Niobium	Retards softening during tempering operations, increases resistance to creep at high temperatures, increases ductility and impact strength
Phosphorus	Increases corrosion resistance and strength
Silicon	Increases hardenability, strength, and magnetic permeability in low-alloy steels
Sulfur	Improves machining properties Especially useful in mild steels
Titanium	Prevents intergranular corrosion of stainless steels, is a deoxidizer, increases strength in low-carbon steels
Tungsten	Used in tool steels to promote hardness In stainless steels helps maintain strength at high temperatures
Vanadium	Improves resistance to thermal fatigue and shock
Zirconium	Inhibits grain growth and is a deoxidizer

rock-drilling machines and air hammers. Since alloy steels are more expensive than carbon steels, they are typically not used for structural members unless special requirements exist, such as high temperature resistance or very high strengths. They are used in the manufacture of power tools and heavy construction equipment. Some common uses for carbon and alloy steels are shown in Table 15.6.

Table 15.6 Construction Uses for Selected Carbon and Alloy Steels

UNS Designation	ASTM Designation	Construction Use
K02600	A36	Structural steel members
K20504	A53	Welded and seamless pipe
K11510	A242	Structural members
K02303	A572	Structural members
K11430	A588	Structural members
K02703	A529	Structural members
K01803	A633	Structural members
K03000	A500	Cold-formed, welded, and seamless tubing
K10600	A678	Quenched and tempered steel plate
K03000	A501	Hot-formed, welded, and seamless tubing
K02706	A325	High-strength bolts
K03900	A490	High-strength alloy steel bolts
	A307	Machine bolts and nuts
	A328	Sheet piling
KJ0300	A27	Cast steel items
J31575	A148	High-strength cast steel items
	615	Billet steel bars for reinforcing concrete
	616	Rail-steel bars for reinforcing concrete
	606, 607	Sheet steel

Source: Society of Automotive Engineers

Types of Structural Steel

Steel specified for structural purposes is of major importance to architects and engineers. It has low-to-medium carbon content. The American Institute of Steel Construction (AISC) publication *Code of Standard Practice for Steel Buildings and Bridges* includes requirements that must be specified in construction documents, including columns, beams, trusses, bearing plates, and various fastening devices and connectors. The AISC publication *Specification for Structural Steel Buildings* details information on structural steels for use in building construction.

Structural steels fall into four major classifications:

1. Carbon steel (ASTM A36, A529, UNS K02600)

2. Heat-treated construction alloy steel (ASTM A514, UNS K11630)

3. Heat-treated high-strength carbon steel (ASTM A633, A678, A852, UNS K01803, K01600, K12043)

4. High-strength low-alloy steel (ASTM A242, A572, A588 UNS K11510, K02303, K11430)

Carbon steels must meet maximum content requirements for manganese and silicon. Copper requirements

have minimum and maximum specifications. There are no other minimums specified for other alloying elements. Heat-treated construction alloy steels have more stringent alloying element specifications than carbon steel. They produce the strongest general-use structural steel.

Heat-treated high-strength carbon steels are brought to desired strength and toughness levels by heat-treating. Heat-treating refers to the process of heating and cooling metals to produce changes in the physical and mechanical properties.

High-strength low-alloy steels are a group of steels to which alloying elements have been added to produce improved mechanical properties and greater resistance to atmospheric corrosion. Their carbon range is typically from 0.12 to 0.22 percent. The percent of alloying agents varies with the manufacturer of the steel and it is available from manufacturers under specific trade names. ASTM has specifications to cover all the trade name steels. One type, known by the general name *weathering steels*, best known under the trademark COR-TEN, is of special interest. They are designed for exterior architectural applications where exposure to the atmosphere causes the steel to form a natural, rust-colored, self-healing oxide coating. It is never painted but left to protect and color the exterior material (**Figure 15.13**).

Figure 15.13 Weathering steels form a natural, rust-colored, self-healing oxide when exposed to the atmosphere.

High-strength low-alloy steels can be worked by many metal processing operations, including hot- and cold-forming, punching, shearing, gas cutting, and welding. In construction, alloy steels find use in high-strength bolts, cables used in pre-stressed concrete, sheet and strip material, plates, various bars and structural shapes, wire, tubing, and pipe.

Stainless and Heat-Resisting Steels

Stainless steels have outstanding corrosion and oxidation resistance at a wide range of temperatures. Heat-resisting steels maintain their basic mechanical and physical properties when subjected to high temperatures. Typically, standard stainless steels have 10 to 25 percent chromium. The chromium alloying element gives stainless steel its corrosion-resistance qualities and produces a thin, hard, invisible film over the surface, which inhibits corrosion. Nickel and manganese increase strength and toughness and make the material easier to fabricate.

Other alloying elements that may be used to produce desired characteristics include zirconium, titanium, sulfur, silicon, selenium, phosphorus, molybdenum, and niobium. Refer to Table 15.5 for additional information on how each of these alters the characteristics of a product. The chemical compositions of several types of stainless steel frequently used in construction are shown in Table 15.7. Notice the small ranges allowed within each alloying element in each type of stainless steel.

Classifying Stainless and Heat-Resisting Steels

Stainless and heat-resisting steels are classified into three major groups based on their chemical composition and their reaction when heat-treated: ferritic steels, martensitic steels, and austenitic steels (Table 15.8).

Ferritic stainless steels have a chromium content of about 16 to 18 percent and a low carbon content of 0.12 maximum. These are non-hardenable steels, meaning they cannot be hardened by heat-treating. Ferritic steels have excellent corrosion resistance. Although they have especially good corrosion resistance at high temperatures, they do lose strength under these conditions. They are used when high temperatures exist, corrosion resistance is important, and when a low coefficient of thermal expansion is helpful.

Martensitic stainless steels have a chromium content of 11.50 to 13.50 percent and a carbon content of 0.15 percent maximum. They can be hardened by heat-treating and worked hot, forged, or formed cold. They are typically used when hardness, strength, and abrasion resistance are critical, such as in a steam turbine.

Austenitic stainless steels have a chromium content of about 17 to 19 percent, a nickel alloying element of 8 to 10 percent, and a maximum carbon content of 0.15 percent. They are nonmagnetic and harden when worked cold. At high temperatures, they have high strength and are very tough and corrosion resistant. They have a high coefficient of thermal expansion, which must be taken into consideration by an engineer when specifying the use of austenitic steels. Of all of the stainless steels, they have the least resistance to corrosion by attack of sulfur gases. Typical uses include curtain wall panels, railings, and door and window finished surfaces (Figure 15.14). Some types are used in the manufacture of food preparation equipment and various appliances in hospitals.

Designations for Stainless and Heat-Resisting Steels

The American Iron and Steel Institute (AISI) designations for heat-resisting and stainless steels are based on a three-digit numbering system ranging from 200 to 500,

Table 15.7 Chemical Compositions of Stainless Steels Used in Construction

UNS Designation	AISI/SAE Designation	Chemical Composition (percent)						
		Fe	Cr	Ni	Mn	C	Mo	Si
S20100	201[a]	67.51–73.51	16.0–18.0	3.5–5.5	5.5–7.5	0.15 max.		1.00 max.
S20200	202[a]	63.51–70.01	17.0–19.0	4.0–6.0	7.5–10.0	0.15 max.		1.00 max.
S30100	301[b]	70.84–74.84	16.0–18.0	6.0–8.0	2.0 max.	0.15 max.		1.00 max.
S30200	302[b]	67.84–70.84	17.0–19.0	8.0–10.0	2.0 max.	0.15 max.		1.00 max.
S30400	304[b]	66.28–70.77	18.0–20.0	8.0–10.5	2.0 max.	0.15 max.		1.00 max.
S31600	316[b]	61.83–68.83	16.0–18.0	10.0–14.0	2.0 max.	0.10 max.	2.0–3.0	1.00 max.
S43000	430[b]	79.81–81.05	14.0–18.0		1.0 max.	0.12 max.	0.75–1.25	1.00 max.

[a]Contains maximum 0.06% phosphorus, 0.03% sulfur, 0.25% nitrogen.
[b]Contains maximum 0.04% phosphorus, 0.03% sulfur.
Source: American Iron and Steel Institute

Table 15.8 AISI/SAE and UNS Designations and Characteristics of Heat-Resisting and Stainless Steels

Grain Structure	Chief Alloying Elements	Series	AISI/SAE Designation	UNS Designation	Characteristics	Steel Type
Martensitic (ferromagnetic and hardenable by heat treatment)	Chromium 4%–6%	500	502	S50200	Retain their mechanical properties at high temperatures Moderate corrosion resistance, high strength and hardness	Heat-resisting steels
	Chromium 11.50–13.50%	400	410	S41000		
Ferritic (ferromagnetic and nonhardenable)	Chromium 16%–18%	400	430	S43000	Very good corrosion resistance, particularly at high temperatures	
Austenitic (nonmagnetic and hardenable by cold working)	Chromium 17%–19% Nickel 8%–10% Manganese 2.0% Max.	300	302	S30200	Excellent corrosion resistance, high strength and ductility Suitable for many fabrication techniques	Stainless steels
	Chromium 17%–19% Nickel 4%–6% Manganese 7.5%–10%	200	202	S20200	Excellent corrosion resistance at high temperatures, high strength and toughness	

Source: American Iron and Steel Institute

as shown in Table 15.7 and Table 15.8. The first digit indicates the group, and the last two digits indicate the type within the group. Some types have the prefix TP, which refers to tubular grades.

Stainless Steel Uses in Construction Typical applications for stainless steels frequently specified by architects and engineers are shown in Table 15.9. Of the ferritic stainless steels, Type 430 is most commonly used in

Figure 15.14 The Lloyds of London building utilizes a stainless steel exterior cladding.

© Elena Elisseeva/Shutterstock.com

Table 15.9 Stainless Steels Used in Construction

UNS Designation	AISI Designation	Typical Applications
S30100	301	Gutters, trim, flashing, household and industrial appliances
S30200, S30400	302, 304	Storefronts, curtain walls, doors, windows, railings, household and industrial appliances
S31600	316	All exterior marine uses, on seacoast building exteriors, in chemical, petroleum, and paper manufacturing facilities
S43000	430	Exterior applications in areas exposed to salt water atmosphere, column covers, trim, grills, gutters
S20100	201	Gutters, trim, flashing, household and industrial appliances
S20200	202	Storefronts, curtain walls, doors, windows, railings, household and industrial appliances

construction for such applications as column covers, trim, grills, and gutters. Of the austenitic stainless steels, Types 301, 302, and 304 are used for applications such as storefronts, doors, windows, and railings. Types 201 and 202 are used for the same applications as 301 and 302. Other uses include drinking fountains, kitchen sinks, flatware, and cooking utensils.

STEEL PRODUCTS

Steel of various types is used extensively in construction, ranging from something as small as a nail to large beams and columns. Following are a few of the products in common use.

Structural Steel Products

Rolled structural steel members are made up of standard structural shapes manufactured in a wide variety of sizes and cross-sections (Figure 15.15). The three most

common types of structural members are the W-shape (wide flange), the S-shape (American Standard I-beam), and the C-shape (American Standard channel). These three types are identified by their nominal depth in inches, and the weight per foot of length in pounds. For example, a W 16 × 31 indicates a W-shape with a web 16 inches deep and a weight of 31 pounds per linear foot. The difference between the W-shape and the S-shape is in the form of the inner surfaces of the flange. The W-shape has parallel inner and outer flange surfaces with a consistent thickness, while the S-shape has a slope of approximately 17 degrees on the inner flange surfaces. The C-shape is similar to the S-shape in that its inner flange surface is also sloped.

The most widely used structural steel member is the W-shape, whose cross-section forms the letter H. It is designed so that its flanges provide strength in a horizontal plane, while the web gives strength in a vertical plane. W-shapes are used as beams, columns, truss members, and in other load-bearing applications (Figure 15.16). The bearing pile, or HP-shape, is almost identical to the

Figure 15.15 Commonly available structural steel shapes.

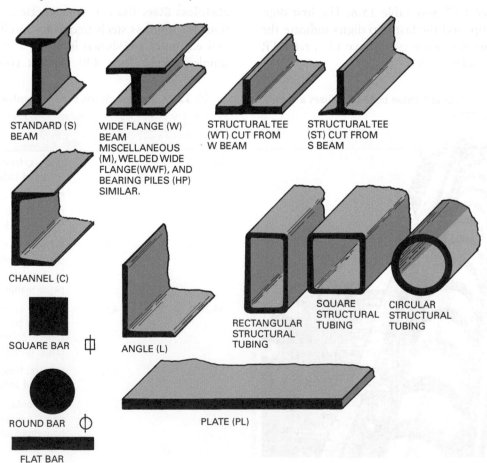

STANDARD (S) BEAM

WIDE FLANGE (W) BEAM MISCELLANEOUS (M), WELDED WIDE FLANGE(WWF), AND BEARING PILES (HP) SIMILAR.

STRUCTURAL TEE (WT) CUT FROM W BEAM

STRUCTURAL TEE (ST) CUT FROM S BEAM

CHANNEL (C)

SQUARE BAR

ANGLE (L)

RECTANGULAR STRUCTURAL TUBING

SQUARE STRUCTURAL TUBING

CIRCULAR STRUCTURAL TUBING

ROUND BAR

PLATE (PL)

FLAT BAR

Figure 15.16 Structural steel H beams are available in a wide range of sizes.

© t.peter photodesign-tp de/Shutterstock.com

W-shape except that the flange thickness and web thickness of the bearing pile are equal, whereas the W-shape has different web and flange thicknesses.

The S-shape (American Standard I-beam) is distinguished by its cross-section being shaped like the letter I. S-shapes are used less frequently than are W-shapes, since the S-shapes provide less strength. The C-shape is called the American Standard channel and has a cross-section similar to the letter C. It is especially useful in locations where a single flat face without a flange on one side is required. The C-shape is not very efficient for a beam or column when used alone, but can be efficiently built up of channels welded together with other structural shapes.

A steel angle is a structural shape whose cross-section resembles the letter L. Angles are available in an equal leg angle and an unequal leg angle. The angle is identified by the length and thickness of its legs; for example, 8 in. × 6 in. × ½ in. When an angle has unequal legs, the dimension of the wider leg is given first. The third dimension applies to the thickness of the legs, which always have equal thickness.

Structural tees have a cross-section that resembles the letter T. They are produced by splitting the webs of beams with rotary shears. Structural tees are designated by their depth and weight per liner foot. Steel pipe and structural tubing are available in square, rectangular, and circular shapes and are commonly used for columns and other load-bearing applications. They are designated by their cross-sectional dimensions. Steel plate is a structural shape whose cross-section is in the form of a flat rectangle that has a width of greater than 8 in. and a thickness of ¼ in. or greater. Plates frequently are referred to by their thickness and width in inches, as plate ½ in. × 24 in. Plates are frequently used to make connections between other structural members or as component parts of built-up structural members. Shapes are specified on architectural drawings as shown in Table 15.10.

Open-web steel joists are a widely used floor and roof framing material. They are lightweight and produced by welding structural steel shapes, such as angles and bars, into a Warren truss unit. They are manufactured in short-span, long-span, and deep long-span series. The short-span joists are manufactured to span clear openings from 8 ft. (2.4 m) to 60 ft. (18.3 m) and in depths from 8 in. (203 mm) to 30 in. (762 mm). The long-span joist series is heavier, made to span clear openings from

Table 15.10 Symbols and Abbreviations for Structural Steel Members

Structural Shape	Symbol	Order of Presenting Data	Sample of Abbreviated Note
Square bar	⊡	Bar, size, symbol, length	Bar 1 ⊡ 5' − 9"
Round bar	Φ	Bar, size, symbol, length	Bar 5/8 Φ 7' − 8"
Plate	PL	Symbol, thickness, weight	PL ½ × 24
Angle, equal legs	L	Symbol, leg 1, leg 2, thickness, length	L 2 × 2 × ¼ × 9' − 6"
Angle, unequal legs	L	Symbol, long leg, short leg, thickness, length	L 4 × 3 × ¼ × 6' − 2"
Channel	C	Symbol, depth, weight, length	C 6 × 10.5 × 12' − 7"
American Standard beam	S	Symbol, depth, weight, length	S 10 × 35.0 × 13' − 6"
Wide flange beam	W	Symbol, depth, weight, length	W 16 × 64 × 20' − 2"
Structural tee	ST, WT	Symbol, depth, weight	ST 10 × 50
Pipe	Name of pipe	Name, diameter, strength	Pipe 3 X-strong
Miscellaneous	M	Symbol, depth, weight, length	M 14 × 17.2 × 19' − 0"
Bearing pile	HP	Symbol, depth, weight, length	HP 14 × 117 × 36' − 0"
Welded wide flange	WWF	Symbol, depth, weight, length	WWF 14 × 84 × 15' − 6"

25 ft. (7.6 m) to 96 ft. (29.2 m) and depths from 18 in. (457 mm) to 48 in. (1219 mm). The deep long-span series spans a clear opening from 89 ft. (27 m) to 144 ft. (43.9 m) and in depths from 52 in. (13 mm) to 72 in. (1829 mm). They are used to support floor and roof loads. The openings in the web permits the passage of plumbing, heating ducts, and electrical runs (Figure 15.17).

Figure 15.17 Open-web joists used with a corrugated steel decking.

Sheet Steel Products

Many products are made by rolling them to shape from flat steel sheets. Common among these are roofing, siding, decking, and light-gauge steel framing systems. The thickness of steel sheets is given by gauge numbers, as shown in Table 15.11.

Many styles of steel roof and siding systems are available. A variety of coatings are used to protect the steel surfaces. Typical coatings include galvanizing, zinc-aluminum alloy with a covering paint, siliconized polyester in a variety of colors, fluoro-polymer paint finish, and weathering copper coating. A few of the available wall panel configurations are shown in Figure 15.18.

Steel floor decking is made of steel sheets that have been bent or corrugated to improve their strength. Corrugated steel floor decking is used with structural steel framing (Figure 15.19). Spanning capabilities of metal decking vary with the thickness of the metal and the geometry of the corrugations and range from 6 to 15 ft. (1.8 to 4.6 m). Cellular floor decking is made up of two sheets of metal, one flat and one corrugated, which are welded together. Cellular decking is advantageous in that it provides raceways in the floor assembly for running electrical conduit and data cables.

Table 15.11 Standard Thicknesses for Some Basic Steel Sheets

Minimum Thickness	
Equivalent Inches	**Millimeters**
0.3937	10.0
0.3543	9.0
0.3150	8.0
0.2756	7.0
0.2362	6.0
0.2165	5.5
0.1969	5.0
0.1890	4.8
0.1772	4.5
0.1654	4.2
0.1575	4.0
0.1496	3.8
0.1378	3.5
0.1260	3.2
0.1181	3.0
0.1102	2.8
0.0984	2.5
0.0866	2.2
0.0787	2.0
0.0709	1.8
0.0630	1.6
0.0551	1.4
0.0472	1.2
0.0433	1.1
0.0394	1.0
0.0354	0.90
0.0315	0.80
0.0276	0.70
0.0256	0.65
0.0236	0.60
0.0217	0.55
0.0197	0.50
0.0177	0.45
0.0157	0.40
0.0138	0.35

Steel roof decking may have a site-cast lightweight concrete or gypsum topping, or it may be covered with some form of insulation board and finished roofing system, such as tar and gravel. Examples of decking products are shown in Figure 15.20.

Figure 15.18 Metal siding is available in a variety of configurations and colors.

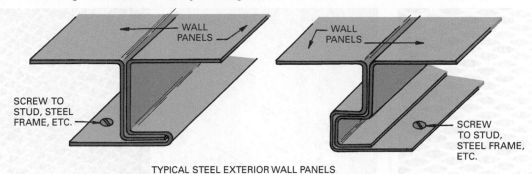

TYPICAL STEEL EXTERIOR WALL PANELS

Figure 15.19 Metal decking is made of steel sheets that have been corrugated to provide strength.

Light-gauge steel framing systems use studs, joists, channels, and runners to frame wall and floor systems. Metal studs are used in the same manner as wood studs. They have metal runners for top and bottom plates. Metal joists are welded to the top runner and have a perimeter channel for a header. The floor and roof deck can be metal or plywood (**Figure 15.21**). Additional information can be found in Chapter 20.

Figure 15.21 Construction details of lightweight steel framing.

Expanded steel mesh is made by slitting metal sheets and stretching them to form diamond-shaped openings. It is used on gratings, decks, partitions, as a base for troweled stucco and plaster application, and in many other applications. Expanded metal mesh is usually designated by the width of the mesh opening and the gauge of the steel sheet. For example, a ¾ in. (18 mm) expanded metal mesh has diamond ¾ in. (18 mm) wide on 16-gauge material. The length of the diamond is approximately twice the width, or 1½ in. (38 mm) in this example.

Metal lath is a form of expanded steel that has a flat or ribbed mesh ⅜ in. (9 mm) in height. It is used as a base for plaster and expanded metal corner bead. A wide variety of other expanded, slit, and woven products are available for both utilitarian and decorative purposes (**Figure 15.22**).

Figure 15.20 Several of the many types of steel roof decking available.

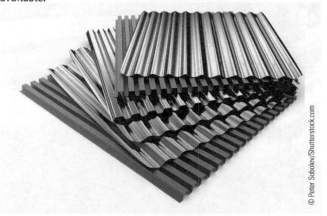

Figure 15.22 A variety of utilitarian and decorative metal mesh products are available.

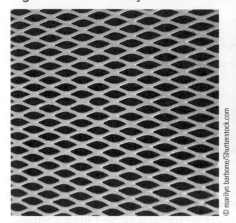

Other Products

Welded wire fabric is used to reinforce concrete slabs. Several commonly used sizes are listed in Table 15.12. More information is given in Chapter 8. Inch sizes of welded wire fabric are designated by two numbers and two letter-number combinations. The first two numbers, for example, 6 × 6, give the spacing of the wires in inches. The first number is the spacing of the longitudinal wires. The second number gives the spacing of the transverse wires in inches. The first letter-number combination gives the type and size of the longitudinal wire and the second for the transverse wire. "W" indicates a smooth wire, and the number following gives its cross-sectional area. A "D" instead of a "W" indicates the use of a deformed wire. The area is given in hundredths of an inch per foot. For example, a W8.0 wire is a smooth wire with a cross-sectional area of 0.08 in.²

Longitudinal wires are spaced at 2, 3, 4, 6, 8, and 12 in. Transverse wires are spaced 4, 6, 8, and 12 in. apart. Metric welded wire fabric sizes are given in millimeters. Typical sizes are also shown in Table 15.12. The first two numbers represent the wire spacing in millimeters. The last two numbers indicate the type of wire and give the cross-sectional area in square millimeters (mm²).

Reinforcing bars are placed in concrete members to improve the tensile strength of concrete (Figure 15.23). They are made in accordance with ASTM requirements for yield and tensile strength, cold bend, and elongation.

Table 15.12 Selected Sizes of Welded Wire Reinforcement (WWR) (Inch and Metric)

Inch Sizes		
	Steel Area (in.²/ft.)	
Wire Style Designation (W or D)	Longitudinal Wire	Transverse Wire
4 × 4–W1.4 × W1.4	0.042	0.042
4 × 4–W2.0 × W2.0	0.060	0.060
4 × 4–W2.9 × W2.9	0.087	0.087
6 × 6–W1.4 × W1.4	0.028	0.028
6 × 6–W2.0 × W2.0	0.040	0.040
6 × 6–W2.9 × W2.9	0.058	0.058
Metric Sizes		
	Steel Area (mm²/m)	
Wire Style Designation (MW or MD)	Longitudinal Wire	Transverse Wire
102 × 102–MW9 × MW9	88.9	88.9
102 × 102–MW14 × MW14	127.0	127.0
102 × 102–MW19 × MW19	184.2	184.2
152 × 152–MW9 × MW9	59.3	59.3
152 × 152–MW14 × MW14	84.7	84.7
152 × 152–MW19 × MW19	122.8	122.8

Figure 15.23 Reinforcing bars and welded wire fabric provide tensile strength for concrete members.

Three grades are available: structural, intermediate, and hard. Structural grade has the lowest yield point and tensile strength, and hard has the highest. The deformed bar has surface projections that provide a better bond to concrete. They are available in sizes 3 through 18. The number indicates the diameter in eighths of an inch. For example, a No. 5 bar has a diameter of ⁵/₈ in. Metric sizes are given in numbers shown in Table 15.13. A variety of accessories can hold reinforcing bars in place until concrete is poured. They are made from steel wire. Detailed information is available in Chapter 8.

Table 15.13 Metric and Inch Reinforcing Bar Sizes and Diameters

Metric Bar Number	Diameter (mm)	Imperial - Equivalent Bar Number	Diameter (inches)
10	9.5	3	.375
13	12.7	4	.500
16	15.9	5	.625
19	19.1	6	.750
22	22.2	7	.875
25	25.4	8	1.000
29	28.7	9	1.128
34	34.3	10	1.270
36	35.8	11	1.410
43	43.0	14	1.693
57	57.4	18	2.257

Many fasteners used in construction are made from steel. Commonly used fasteners include bolts, nails, rivets, and screws. Standard steel bolts are available in a wide range of sizes and head types (Figure 15.24). High-strength steel bolts are used where tensile strength is important, such as in structural steel framing. Additional

Figure 15.24 Standard steel bolts.

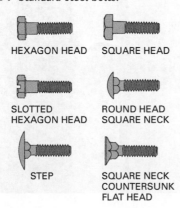

HEXAGON HEAD   SQUARE HEAD

SLOTTED HEXAGON HEAD   ROUND HEAD SQUARE NECK

STEP   SQUARE NECK COUNTERSUNK FLAT HEAD

details are given in Chapter 17. Commonly used nails are shown in Figure 15.25. Some nails are specified in length by inches and others by the term "penny" (d). Penny lengths are shown in Table 15.14.

Figure 15.25 Commonly used nails.

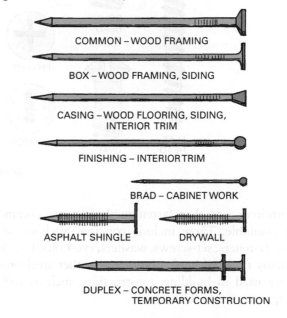

COMMON – WOOD FRAMING

BOX – WOOD FRAMING, SIDING

CASING – WOOD FLOORING, SIDING, INTERIOR TRIM

FINISHING – INTERIOR TRIM

BRAD – CABINET WORK

ASPHALT SHINGLE   DRYWALL

DUPLEX – CONCRETE FORMS, TEMPORARY CONSTRUCTION

Table 15.14 Lengths and Diameters of Common, Box, Casing, and Finishing Nails

Size	Length (in.)	American Steel Wire Gauge Number		
		Common	Box and Casing	Finishing
2d	1	15	15½	16½
3d	1¼	14	14½	15½
4d	1½	12½	14	15
5d	1¾	12½	14	15
6d	2	11½	12½	13
7d	2¼	11½	12½	12½
8d	2½	10¼	11½	12½
9d	2¾	10¼	11½	12½
10d	3	9	10½	11½
12d	3¼	9	10½	11½
16d	3½	8	10	11
20d	4	6	9	10
30d	4½	5	9	
40d	5	4	8	

Most wood screws used have flat, round, or oval heads. Their lengths are specified in inches and their diameters with wire gauge numbers. A variety of slots are used in the heads (Figure 15.26).

Figure 15.26 Common head types used on wood screws.

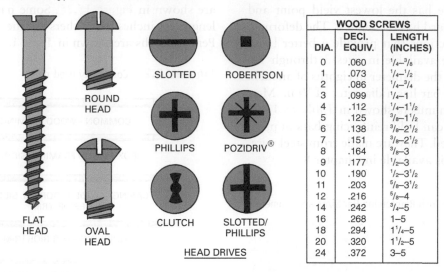

WOOD SCREWS		
DIA.	DECI. EQUIV.	LENGTH (INCHES)
0	.060	$1/4$–$3/8$
1	.073	$1/4$–$1/2$
2	.086	$1/4$–$3/4$
3	.099	$1/4$–1
4	.112	$1/4$–$1 1/2$
5	.125	$3/8$–$1 1/2$
6	.138	$3/8$–$2 1/2$
7	.151	$3/8$–$2 1/2$
8	.164	$3/8$–3
9	.177	$1/2$–3
10	.190	$1/2$–$3 1/2$
11	.203	$5/8$–$3 1/2$
12	.216	$5/8$–4
14	.242	$3/4$–5
16	.268	1–5
18	.294	$1 1/4$–5
20	.320	$1 1/2$–5
24	.372	3–5

Hundreds of other fastening devices are commercially available. These include concrete anchors, self-drilling fasteners, set-screws, washers, eye-bolts, U-bolts, and many types of hooks. Numerous other steel products are used in building construction, such as locks, hinges, gutters, and flashing.

TESTING METALS

Engineers and architects must know the capabilities of potentially usable metals. These are determined by standardized testing that follows the procedures of the American Society for Testing and Materials (ASTM). Some of the more frequently used tests include analyses of hardness, tensile strength, and fatigue.

Hardness testing determines the resistance of a metal to penetration, and is conducted using a Tinius Olsen Air-O-Brinell hardness tester. To conduct a test, the operator adjusts the air regulator until the desired Brinell load in kilograms is indicated. The specimen is placed on the anvil and the operator pulls the plunger valve. The plunger, which has a hardened ball on the bottom, impacts the metal and forms an impression in its surface. The plunger valve depresses and the plunger and ball retract from the metal. The impression diameter is measured, using optical instruments to provide an automatic and precise reading. A computer makes it possible for the operator to customize test parameters and test report formats. Test data can be recalled by individual test number, lot identifier, heat number, date, time, or other specific data.

Tensile testing determines the mechanical properties of a metal when it is subjected to a force that tends to pull it apart. The tensile test conducted on a Tinius-Olsen Universal Testing Machine (Figure 15.27) determines the ultimate strength, yield point, yield strength, and factors related to ductility. It is available

Figure 15.27 A tensile test on a Tinius-Olsen Universal Testing Machine determines the ultimate strength, yield point, and yield strength.

in capacities up to 600,000 lb. (2,000 kN). The test data is recorded by a four-channel digital indicating system. The testing sample is gripped with pinion-type wedge grips in the crossheads. Once the sample is secure in the grips, the lower crosshead moves down, exerting tension on the material. Data are recorded and the stress-strain curve values displayed.

Fatigue testing determines the stress level a metal can withstand without failure when subjected to an infinitely large number of repeated alternating stresses. Typical tests include subjecting a specimen, such as a beam, to rotating motion while under a bending movement. Another test involves bending the metal back and forth without rotating it.

Review Questions

1. What is the key element that influences the properties of iron?

2. What process is used to remove unwanted elements from freshly mined iron ore?

3. What fuels and fluxes are commonly used in a blast furnace?

4. What uses are made of slag removed from a blast furnace?

5. What are the commonly produced cast irons?

6. What are the main properties of white cast iron?

7. Why is malleable cast iron less brittle than white cast iron?

8. How does the alloying element nickel affect the properties of steel?

9. If steel has a carbon content less than 0.90 percent, how does this affect the properties of the steel?

10. How do the alloying elements copper and cobalt influence the properties of steel?

11. What are the three basic steelmaking processes?

12. How does carbon steel differ from alloy steel?

13. What are the two outstanding properties of stainless steel?

14. Explain the purpose of each part of the AISI steel designation code.

15. Why does stainless steel resist corrosion?

16. What are the three groups of stainless and heat-resisting steels?

17. Identify the following structural steel shapes: W, S, HP, ST, HSS, C, M.

18. How is welded wire fabric used in building construction?

19. What tests are used on metal to determine its hardness?

20. What does a tensile test show?

21. What does a fatigue test show?

Key Terms

Agglomeration

Alloying Element

Annealing

Beneficiation

Cast Iron

Cast Steel Products

Corrosion

Design for Disassembly (DFD)

Direct Reduction

Flux

Malleability

Mixers

Oxidation

Pig

Pig Iron

Reduction

Sintering

Slag

Smelting

Strand-Casting Machine

Weathering Steel

Wrought Products

Activities

1. Collect representative samples of ferrous metals and prepare a sheet for each one, identifying the type and its properties.

2. Make a tour of your campus and compose a list of all the applications of ferrous metals you can find. Take photos, if possible.

3. Test samples of ferrous metals for tensile strength and hardness and keep a record of your findings for each sample. Compare your findings with those of others who tested the same materials.

4. Find products made from ferrous metals produced by various manufacturing processes, such as casting and extruding.

Additional Resources

Metals and Alloys in the Unified Numbering System, and other related publications, American Society of Automotive Engineers, Inc., Warrendale, PA, and the American Society for Testing and Materials, West Conshohocken, PA.

Other resources include:

Numerous publications related to metals, American Society for Testing and Materials, West Conshohocken, PA.

Publications of the Wire Reinforcement Institute, Hartford, CT.

See Appendix C for addresses of professional and trade organizations and other sources of technical information.

Nonferrous Metals

Upon completion of this chapter, the student should be able to:

- Select nonferrous metals for a wide range of applications.

- Know the properties of nonferrous metals and how these will influence a material's performance.
- Understand the effects of galvanic corrosion and how to design to eliminate it.

Build Your Knowledge

For further study on these materials and methods, please refer to:

Chapter 3   Properties of Materials

       Topic: Mechanical and Chemical Properties

Chapter 27   Roofing Systems

       Topic: Roofing Materials

Chapter 39   Plumbing Systems

       Topics: Piping, Tubing, Fittings

Nonferrous metals are those containing little or no iron. In other words, all metals other than iron and steel are nonferrous. Nonferrous metals commonly found in construction include aluminum, copper, lead, tin, and zinc. Nickel is used mainly as an alloying element. Titanium, once used in aircraft, aerospace, and military applications, is now employed as a building cladding material.

GALVANIC CORROSION

Both ferrous and nonferrous metals are subject to *galvanic corrosion*. In the presence of an electrolyte (moisture in the atmosphere), dissimilar metals in contact will corrode more rapidly than similar metals in contact. For example, an aluminum gutter secured with copper nails produces galvanic action at the point of contact. The presence of moisture in the atmosphere sets up an electrolytic action that causes the aluminum to corrode. The metal that is higher on the table of electrolytic corrosion potential will corrode more (Table 16.1).

Table 16.1 The Galvanic Series

Electrolytic Potential	Metals and Alloys
High potential (anode +)	Magnesium
	Magnesium alloys
	Aluminum (pure and several cast and wrought alloys)
	Zinc
	Cadmium
	Aluminum (wrought 2024 and 356.0 cast)
	Iron or steel
Electric current flows from positive (+) to negative (−)	Cast iron
	Stainless steel
	Lead
	Tin
	Nickel (active)
	Brass
	Copper
	Bronze
Low potential (cathode −)	Chromium stainless steel
	Silver
	Titanium
	Platinum
	Gold

When aluminum is plated with zinc and electrolytic action corrosion occurs, the zinc corrodes and protects the aluminum. When steel is coated in zinc, it is called *galvanized*. When it is not possible to avoid contact between dissimilar metals, they can be given a coat of unleaded paint or separated with a plastic or other nonconducting material. A joint can also be caulked or otherwise sealed to keep out moisture, thus eliminating the electrolyte.

ALUMINUM

Aluminum (chemical symbol Al) is a versatile material used widely in building, and the construction industry is one of its largest consumers. Aluminum is lightweight, having a specific gravity of only 2.7 times that of water and approximately one-third that of steel. Pure aluminum melts at 1,220°F (665°C), considerably lower than the melting point of other structural metals. It also is relatively weak as far as mechanical properties are concerned. Aluminum elastically deforms about three times more than steel under comparable loading. It can be strengthened by alloying, cold-working, or *strain hardening*. Aluminum alloys do not lose ductility or become brittle at cryogenic (low) temperatures.

Aluminum is a good conductor of electricity. Compared to copper wire of the same diameter, aluminum's conductivity is roughly 65 percent that of copper.

Mining Aluminum

Aluminum is found in most rocks and clays, but concentrations of the aluminum oxide content must be about 45 percent to be viable for economic extraction. Aluminum ores are called *bauxites*. Most ores are secured via open-pit mining. After extraction, the ore is crushed, washed, screened, ground, and dried. The embodied energy impacts of aluminum production are high, though they are offset by its inherent recyclability. Aluminum is widely recycled, and reprocessing the material requires considerably less energy.

Refining Bauxite Ore

Aluminum is refined using a two-step process. The first, the Bayer process, produces a very pure alumina (Al_2O_3). Alumina is an oxide of aluminum in crystal form. The second step reduces the alumina to a metallic aluminum, which is about 99 percent pure. This process is referred to as the Hall-Héroult process. It is named after the two men who developed the electrolytic method of aluminum production.

The Bayer Process

Ground dried bauxite is mixed in a digester with soda ash, crushed lime, sodium hydroxide, and hot water.

Construction Materials

History of Aluminum

People were using alum, one of the aluminum compounds found in nature, as early as 500 BC. These naturally occurring compounds were used as astringents (a substance that contracts the tissues of the body), to fix dyes in cloth, and in the tanning of animal skins.

The Romans used a natural potassium sulfate they called *alumen*. This was purified into crystalline alum around AD 1200, and by the 1500s alum was produced from the clay in which it occurred.

The actual production of aluminum as we know it today happened much more recently; in fact, it is the most recently developed and used metal in construction. Various attempts were made by German, French, and Danish chemists and scientists to isolate aluminum. Around 1850, one chemist was successful in isolating a small quantity of the metal by decomposing anhydrous aluminum chloride with potassium.

In 1855, an aluminum ingot was exhibited at the Paris Exposition. It was a rarity worth more per pound than gold. Emperor Napoleon III commissioned Henri Sainte-Claire Deville to find a way to manufacture large amounts of aluminum to be used for military equipment. He successfully produced a few tons but not the huge commercial quantities required.

In 1886, an American, Charles Martin Hall, and a Frenchman, Paul Louis-Toussaint Héroult, independently developed the production method used today, in which the alumina is dissolved in molten cryolite and decomposed electrolytically. Since that time, industry has begun to produce large amounts of aluminum. It first appeared in construction products around 1925. Applications have expanded since then to include hundreds of interior and exterior uses.

Live steam and mechanical agitators stir the mixture, which is pumped to digester tanks and churned under high pressure. The chemical reaction forms sodium aluminate, and the insoluble impurities form a waste material that is removed. The sodium aluminate solution passes through a filter to a cooling tower and into a precipitator where aluminum hydrate is added. Compressed air agitates the concoction, and cooling continues, allowing the sodium aluminate to precipitate as aluminum hydrate. This is pumped into filter tanks that separate the aluminum hydrate from the solution. It is then calcined in a rotary kiln operating at about 2,000°F (1,100°C), which produces the alumina used to produce the metallic aluminum in the second step, the Hall-Héroult electrolytic process (Figure 16.1).

Figure 16.1 The first step in producing aluminum, the Bayer process, separates the alumina from the bauxite ore in large rotary kilns.

THE HALL-HÉROULT ELECTROLYTIC PROCESS

In the Hall-Héroult process, aluminum is produced from the oxide *alumina* by reduction. *Reduction* is an electrolytic procedure that uses a carbon-lined vessel containing molten cryolite and alumina. An electric current is passed through this liquid via large carbon anodes suspended in it. When the current hits the liquid, molten aluminum separates and settles on the bottom of the vessel, where it is siphoned off. Pure aluminum may be cast in molds for use in products requiring its properties or sent to a holding furnace where alloying elements are added to alter its physical properties (Figure 16.2).

Figure 16.2 Pure aluminum ready for processing into useful products rolls from a processing plant.

Aluminum Alloys

Produced aluminum is between 99.5 and 99.9 percent pure. In this form, it is relatively soft and ductile and has a tensile strength of around 7,000 psi (48,258 kPa). For many products, aluminum must have alloying elements added to alter its physical properties.

Aluminum alloys have two classifications: wrought and casting. *Wrought alloys* are those that are mechanically worked by processes, such as forging, drawing, extruding, or rolling, to form sheet material. *Cast alloys* are those used to produce a product for which the molten metal is cast in a finished shape, such as a grille, in a sand mold or permanent mold.

Wrought Aluminum Alloy Classifications The major alloying elements added to aluminum are manganese, copper, magnesium, silicon, and zinc. A wide variety of aluminum alloys are identified by a code system developed by the Aluminum Association, Inc. Each wrought aluminum alloy is specified by a four-digit code number, as shown in Table 16.2.

The first digit denotes the alloy series and indicates the major alloying element. The second digit in the 1xxx series represents a modification of impurity limits, and in the 2xxx through 9xxx series, it represents a modification of the alloy. The last two digits in the 1xxx series indicate aluminum purity above 99 percent. For example, a 1050 specifies an aluminum containing 0.50 percent more aluminum than the minimum 99 percent. In the 2xxx through 9xxx series, the last two digits are arbitrary numbers identifying the alloy in the series. Table 16.3 and Table 16.4 give a breakdown of these classification systems.

Table 16.2 Wrought Aluminum and Aluminum Alloy Designation System

Aluminum 99.00% Minimum 1xxx	
Aluminum Alloys Grouped by Major Alloying Elements	
Copper	2xxx
Manganese	3xxx
Silicon	4xxx
Magnesium	5xxx
Magnesium and silicon	6xxx
Zinc	7xxx
Other elements	8xxx
Unused series	9xxx

Source: The Aluminum Association

Table 16.3 Breakdown of 1xxx Aluminum Designations

1 X XX		
Indicates 99% pure commercial aluminum	0 indicates no special impurity control	Indicates pure aluminum content
	1–9 indicates special specific impurity controls	above 99% in hundredths of a percent

Table 16.4 Breakdown of 2xxx–9xxx Aluminum Designations

X X XX		
1 commercially pure	0 indicates original alloy developed	in 1xxx series indicates impurity limits
2–9 indicates major alloying element as shown in Table 16.2	2–9 indicate modifications of the original alloy	2xxx through 9xxx series— arbitrary numbers identifying the alloy in the series

Cast Aluminum Alloy Classifications Cast alloys are specified by a three-digit number followed by a decimal (Table 16.5). The first digit identifies the alloy series and the second and third digits the specific alloy or purity. The decimal indicates whether the alloy composition is for final casting (.0) or for ingot (.1 or .2).

Table 16.5 Cast Aluminum and Aluminum Alloy Designation System

Aluminum 99.00% minimum	1xx.x
Aluminum Alloys Grouped by Major Alloying Elements	
Copper	2xx.x
Silicon, with added copper and/or magnesium	3xx.x 4xx.x
Silicon	
Magnesium	5xx.x
Zinc	7xx.x
Tin	8xx.x
Other elements	9xx.x
Unused series	6xx.x

Source: The Aluminum Association

UNS Designations

The Unified Numbering System (UNS) is described in detail in Chapter 15. It uses a five-digit number with a letter prefix to classify and identify various types of aluminum and aluminum alloys specified by other organizations, such as the Aluminum Association (AA), American Society for Testing and Materials (ASTM), Society of Automotive Engineers/Aerospace Materials Specifications (AMS), Military Specifications (MIL-SPEC), and Federal Specifications (FS). The UNS pulls together those materials with like specifications from specifying organizations and gives them a unified number. Table 16.6 gives a few selected examples to illustrate how this system combines them.

Table 16.6 Selected Examples of Aluminum Alloy UNS Designations

UNS Designations	Material	Numbers Used by Various Organizations
A02400	Aluminum foundry alloy, casting	AA 240.0, AMS 4227
A91035	Wrought aluminum alloy, non-heat-treatable	AA 1035
A92011	Wrought aluminum alloy, heat-treatable	AA 2011, ASTM B210, FS QQ-A-22513, SAE J454

Source: Society of Automotive Engineers

Temper Designations

Temper is the degree of hardness and strength imparted to a metal by a process, such as *heat-treating* or cold-working. Tempering refers to the process that renders the proper degree of hardness and elasticity to a metal so it can be used for an intended purpose.

Wrought aluminum alloys fall into two classes: heat-treatable and non-heat-treatable. *Heat-treatable alloys* are those whose strength characteristics are improved by heat-treating. Their strength is also improved by adding alloying elements, such as copper, zinc, silicon, and magnesium.

Non-heat-treatable alloys have alloying elements added that do not cause an increase in strength when heat-treated. The strength of non-heat-treatable alloys depends on elements such as iron, magnesium, manganese, and silicon. These alloys are strengthened by cold-rolling or strain hardening.

Some temper specifications apply only to cast aluminum, and others apply only to wrought aluminum. Alloys specified as F, H, or O can be hardened by cold-working and may or may not be non-heat-treatable. Heat-treatable aluminum alloys use the T and W designations.

The specification of an aluminum alloy requires a designation of temper or metallurgical condition. The following temper designation system for aluminum alloys was developed by the Aluminum Association, Inc. The *temper designation* consists of letters and numbers placed after an alloy number and separated from it by a dash.

Temper designations include:

F (as fabricated): No special control over thermal or work-hardening conditions is used.

Q (annealed): Wrought products have been heated to effect recrystallization, which produces the lowest strength. Cast products are annealed to improve stability and ductility.

H (strain hardened): Wrought products are strain hardened through cold-working. The H is followed by one or two digits. See Table 16.7.

W (solution heat-treated): The alloy is heated to about 1,000°F (542°C) and then quenched.

T (thermally treated): The product has been heat treated and then strain hardened. The T is followed by one or two digits. See Table 16.8.

Some applications for aluminum alloys are shown in Table 16.9. Table 16.10 gives some detailed uses of aluminum in construction.

Table 16.7 Subdivision of H Temper: Strain Hardened

First Digit Indicates Specific Treatment:
H1—Strain hardened only
H2—Strain hardened and partially annealed
H3—Strain hardened and stabilized
H4—Strain hardened and lacquered or painted
Digit (0–8) Indicates the Degree of Strain Hardening as Identified by a Minimum Value of the Ultimate Tensile Strength[a]:
0—Annealed
2—Tempers whose ultimate tensile strength is midway between 0 and 4.
4—Tempers whose ultimate tensile strength is midway between 0 and 8.
6—Tempers whose ultimate tensile strength is midway between 4 and 8.
8—Tempers whose ultimate tensile strength exceeds that of 8 by 2 ksi or more.
1, 3, 5, 7—Tempers whose ultimate tensile strength falls between those defined above.

[a]*The tensile strength for each number is specified by Aluminum Association tables.*

Source: The Aluminum Association

Table 16.8 Subdivisions of T Temper: Thermally Treated

First Digit Indicates Specific Sequence of Treatments:
T1—Naturally aged after cooling from an elevated temperature shaping process
T2—Cold-worked after cooling from an elevated temperature shaping process and then naturally aged
T3—Solution heat treated, cold-worked, and naturally aged
T4—Solution heat treated and naturally aged
T5—Artificially aged after cooling from an elevated temperature shaping process
T6—Solution heat treated and artificially aged
T7—Solution heat treated and stabilized (overaged)
T8—Solution heat treated, cold-worked, and artificially aged
T9—Solution heat treated, artificially aged, and cold-worked
T10—Cold-worked after cooling from an elevated temperature shaping process and then artificially aged
Second Digit Indicates Variation in Basic Treatment:
Examples:
T42 or T62—Heat treated to temper by user
Additional Digits Indicate Stress Relief:
Examples:
TX51—Stress relieved by stretching after solution heat treating
TX52—Stress relieved by compressing after solution heat treating or cooling

Source: The Aluminum Association

Table 16.9 Typical Applications of Aluminum Alloys

Typical Applications of Non-Heat-Treatable Aluminum Alloys			
AA Alloy Series	**AA Typical Alloys**	**UNS Designation**	**Typical Applications**
1xxx	1350	A91350	Electrical conductors
	1060	A91060	Chemical equipment, tank cars
	1100	A91100	Sheet metal work, cooking utensils, decorative
3xxx	3003	A93003	Sheet metal work, chemical equipment, storage tanks
4xxx	4043	A4043	Welding electrodes
	4343	A94343	Brazing alloy
5xxx	5005	A95005	Decorative and automotive trim, architectural and anodized, sheet metal work, appliances
	5050	A95050	
	5454	A95454	
	5456	A95456	
5xxx (3% Mg)	5083	A95083	Marine, welded structures, storage tanks, pressure vessels, armor plate, cryogenics
	5086	A95086	
	5454	A95454	
	5456	A95456	
Typical Applications of Heat-Treatable Aluminum Alloys			
AA Alloy Series	**AA Typical Alloys**	**UNS Designation**	**Typical Applications**
2xxx (Al-Cu)	2011	A92011	Screw machine products
	2219	A92219	Structural, high temperature
2xxx (Al-Cu-Mg)	2014	A92014	Aircraft structures and engines, truck frames and wheels
	2024	A92024	
	2618	A92618	
4xxx	4032	A94032	Pistons
6xxx	6061	A96061	Marine, truck frames and bodies, structures, architectural furniture
	6063	A96063	
7xxx (Al-Zn-Mg)	7004	A97004	Structural, cryogenics, missile
	7005	A97005	
7xxx (Al-Zn-Mg-Cu)	7001	A97001	High-strength structural and aircraft
	7075	A97075	
	7178	A97178	

Source: The Aluminum Association

Table 16.10 Specific Construction Product Applications

UNS Designation	AA Alloy Number	Use
A93003	3003	Flashing, ducts, garage doors, curtain wall panels, grilles, louvers, siding, shingles, termite shields
A91235	1235	Vapor barriers, insulation
A04430	B443.0	Cast products, such as hardware, curtain wall castings, architectural letters
A96063	6063	Extruded products, such as curtain wall, door, and window frames; grilles, mullions; railings; thresholds

Aluminum Castings

The conditions involved with the production of aluminum castings influence their physical properties. The metallurgist has to consider not only the characteristics of an alloy and the use to which the product will be put but also how it will be cast. In general, aluminum cast in permanent (metal) molds is stronger than sand-cast products (Figure 16.3). The cost of producing metal molds is high but can be justified when many products need casting and the molds can be reused. A sand-casting mold produces only one casting.

Figure 16.3 Molten aluminum is cast in steel molds that can be reused repeatedly.

Heat-treated alloys typically are stronger and have greater ductility than non-heat-treatable alloys. After casting, heat-treatable alloy castings are subjected to a very high temperature (but below the alloy's melting point), then quenched in cold or hot oil or water and left to age at room temperature. This can help control the characteristics of the alloy.

Aluminum Finishes

As mentioned previously, when aluminum is exposed to oxygen it forms a natural protective *oxide layer*, so under ordinary circumstances no protective coating is necessary. However, various finishes are applied to aluminum products to improve their appearance and usefulness. The Aluminum Association specifies three categories of finishes: mechanical, chemical, and coatings. Aluminum can also have an *anodized* finish, or be left natural. The flow chart in Figure 16.4 illustrates these finishes and shows their relationships.

The Aluminum Association classifies finishes with a letter and a two-digit number. Mechanical finishes are designated with "M" and chemical finishes with "C." Coatings are designated by letters denoting a particular type, and anodic coatings use the letter A. The specifications and designations for aluminum finishes are given in Table 16.11, Table 16.12, and Table 16.13.

Figure 16.4 The finishing processes used on various aluminum products.

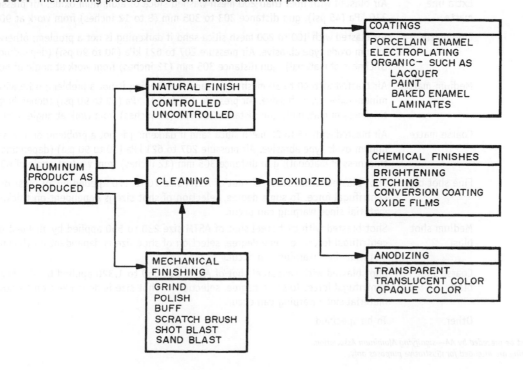

Table 16.11 Mechanical Finishes on Aluminum

Type of Finish	Designation[a]	Description	Examples of Methods of Finishing[b]
As fabricated	M10	Unspecified	
	M11	Specular as fabricated	
	M12	Nonspecular as fabricated	
	M1X	Other	To be specified.
Buffed	M20	Unspecified	
	M21	Smooth specular	Polished with grits coarser than 320. Final polishing with a 320 grit using peripheral wheel speed of 30 m/s (6,000 feet per min). Polishing followed by buffing, using tripoli based buffing compound and peripheral wheel speed of 36 to 41 m/s (7,000 to 8,000 feet per min).
	M22	Specular	Buffed with tripoli compound using peripheral wheel speed 36 to 41 m/s (7,000 to 8,000 feet per min).
	M2X	Other	To be specified.
Directional textured	M30	Unspecified	
	M31	Fine satin	Wheel or belt polished with aluminum oxide grit of 320 to 400 size; peripheral wheel speed 30 m/s (6,000 feet per min).
	M32	Medium satin	Wheel or belt polished with aluminum oxide grit of 180 to 220 size; peripheral wheel speed 30 m/s (6,000 feet per min).
	M33	Coarse satin	Wheel or belt polished with aluminum oxide grit of 80 to 100 size; peripheral wheel speed 30 m/s (6,000 feet per min).
	M34	Hand rubbed	Hand rubbed with stainless steel wool lubricated with neutral soap solution. Final rubbing with No. 00 steel wool.
	M35	Brushed	Brushed with rotary stainless steel wire brush, wire diameter 0.24 mm (0.0095 in.); peripheral wheel speed 30 m/s (6,000 feet per min); or various proprietary satin finishing wheels or satin finishing compounds with buffs.
	M3X	Other	To be specified.
Nondirectional textured	M40	Unspecified	
	M41	Extra fine matte	Air blasted with finer than 200 mesh washed silica or aluminum oxide. Air pressure 310 kPa (45 psi); gun distance 203 to 305 mm (8 to 12 inches) from work at 90° angle.
	M42	Fine matte	Air blasted with 100 to 200 mesh silica sand if darkening is not a problem; otherwise aluminum oxide type abrasive. Air pressure 207 to 621 kPa (30 to 90 psi) (depending upon thickness of material); gun distance 305 mm (12 inches) from work at angle of 60 to 90°.
	M43	Medium matte	Air blasted with 40 to 50 mesh silica sand if darkening is not a problem; otherwise aluminum oxide type abrasive. Air pressure 207 to 621 kPa (30 to 90 psi) (depending upon thickness of material); gun distance 305 mm (12 inches) from work at angle of 60 to 90°.
	M44	Coarse matte	Air blasted with 16 to 20 mesh silica sand if darkening is not a problem; otherwise aluminum oxide type abrasive. Air pressure 207 to 621 kPa (30 to 90 psi) (depending upon thickness of material); gun distance 305 mm (12 inches) from work at angle of 60 to 90°.
	M45	Fine shot blast	Shot blasted with cast steel shot of ASTM size 70 to 170 applied by air blast or centrifugal force. To some degree, selection of shot size is dependent on thickness of material since warping can occur.
	M46	Medium shot blast	Shot blasted with cast steel shot of ASTM size 230 to 550 applied by air blast or centrifugal force. To some degree, selection of shot size is dependent on thickness of material since warping can occur.
	M47	Coarse shot blast	Shot blasted with cast steel shot of ASTM size 660 to 1,320 applied by air blast or centrifugal force. To some degree, selection of shot size is dependent on thickness of material since warping can occur.
	M4X	Other	To be specified.

[a]The complete designation must be preceded by AA—signifying Aluminum Association.
[b]Examples of methods of finishing are intended for illustrative purposes only.

Table 16.12 Chemical Finishes on Aluminum

Type of Finish	Designation[a]	Description	Examples of Methods of Finishing[b]
Nonetched cleaned	C10	Unspecified	
	C11	Degreased	Organic solvent treated.
	C12	Inhibited chemical cleaned	Inhibited chemical type cleaner used.
	C1X	Other	To be specified.
Etched	C20	Unspecified	
	C21	Fine matte	Trisodium phosphate, 22–45 g/l (3–6 oz per gal) used at 60–71°C (140–160°F) for 3 to 5 min.
	C22	Medium matte	Sodium hydroxide, 30–45 g/l (4–6 oz per gal) used at 49–66°C (120–150°F) for 5 to 10 min.
	C23	Coarse matte	Sodium fluoride, 11 g/l (1.5 oz) plus sodium hydroxide 30–45 g/l (4–6 oz per gal) used at 54–66°C (130–150°F) for 5 to 10 min.
	C2X	Other	To be specified.
Brightened	C30	Unspecified	
	C31	Highly specular	Chemical bright dip solution of the proprietary phosphoric-nitric acid type used, or proprietary electrobrightening or electropolishing treatment.
	C32	Diffuse bright	Etched finish C22 followed by brightened finish C31.
	C3X	Other	To be specified.
Chemical coatings[c]	C40	Unspecified	
	C41	Acid chromate-fluoride	Proprietary chemical treatments used producing clear to typically yellow colored surfaces.
	C42	Acid chromate-fluoride-phosphate	Proprietary chemical treatments used producing clear to typically green colored surfaces.
	C43	Alkaline chromate	Proprietary chemical treatments used producing clear to typically gray colored surfaces.
	C44	Non-chromate	Proprietary chemical coating treatment employing no chromates.
	C45	Non-rinsed chromate	Proprietary chemical coating treatment in which coating liquid is dried on the work with no subsequent water rinsing.
	C4X	Other	To be specified.

[a]The complete designation must be preceded by AA—signifying Aluminum Association.
[b]Examples of methods of finishing are intended for illustrative purposes only.
[c]Includes chemical conversion coatings.

Source: The Aluminum Association

Natural Finishes

Natural finishes may be controlled or uncontrolled. Uncontrolled finishes are imparted to wrought and cast products by the surface conditions of the rollers, molds, or extruding dies used to form them. Hot-rolled products have a brighter surface. Controlled finishes result from varying the smoothness of rollers or mold surfaces. A sheet can have a controlled finish on one side, affected with smooth rollers, and an uncontrolled finish on its other side. Sheets may be embossed using rollers with surface designs, as in the case of imitation wood-grain aluminum siding.

Mechanical Finishes

Mechanical finishes are used to alter surface appearances of aluminum products. They are usually produced before surfaces are cleaned to receive other finishing processes.

Surfaces may be buffed with a coarse abrasive or receive a high luster polish via a fine abrasive. They may be ground with a dry grinding wheel. This is often necessary to clean up ridges on castings. It produces a rough, scratched finish. Rotating wire brushes produce scratched surfaces. The scratches can vary from very fine to coarse, depending on the size of the wire in the brushes. A matte surface is produced by blasting with sand or steel shot. Round pieces of abrasive cloth or steel wool can be rotated against a surface to produce a series of concentric circles. Code designations for mechanical finishes are given previously in Table 16.11.

Chemical Finishes

Chemical finishes are produced by the reaction of an aluminum surface to various chemicals. Conversion coatings are a major chemical finish. They prepare surfaces for

Table 16.13 Anodic Coatings on Aluminum

Type of Finish	Designation[a]	Description	Examples of Methods of Finishing[b]
General	A10	Unspecified	
	A11	Preparation for other applied coatings	3 μm (0.1 mil) anodic coating produced in 15% H_2SO_4 at 21° ± 1°C (70°F ± 2°F) at 129 A/m² (12 A/ft.²) for 7 min, or equivalent.
	A12	Chromic acid anodic coatings	To be specified.
	A13	Hard, wear and abrasion resistant coatings	To be specified.
	A1X	Other	To be specified.
Protective and Decorative Coatings less than 10 mm (0.4 mil) thick	A21	Clear coating	Coating thickness to be specified. 15% H_2SO_4 used at 21° ± 1°C (70°F ± 2°F) at 129 A/m² (12 A/ft.²).
	A211	Clear coating	Coating thickness—3 μm (0.1 mil) minimum. Coating weight—6.2 g/m² (4 mg/in.²) minimum.
	A212	Clear coating	Coating thickness—5 μm (0.2 mil) minimum. Coating weight—12.4 g/m² (8 mg/in.²) minimum.
	A213	Clear coating	Coating thickness—8 μm (0.3 mil) minimum. Coating weight—18.6 g/m² (12 mg/in.²) minimum.
	A22	Coating with integral color	Coating thickness to be specified. Color dependent on alloy and process methods.
	A221	Coating with integral color	Coating thickness—3 μm (0.1 mil) minimum. Coating weight—6.2 g/m² (4 mg/in.²) minimum.
	A222	Coating with integral color	Coating thickness—5 μm (0.2 mil) minimum. Coating weight—12.4 g/m² (8 mg/in.²) minimum.
	A223	Coating with integral color	Coating thickness—8 μm (0.3 mil) minimum. Coating weight—18.6 g/m² (12 mg/in.²) minimum.
	A23	Coating with impregnated color	Coating thickness to be specified. 15% H_2SO_4 used at 27°C ± 1°C (80°F ± 2°F) at 129 A/m² (12 A/ft.²) followed by dyeing with organic or inorganic colors.
	A231	Coating with impregnated color	Coating thickness—3 μm (0.1 mil) minimum. Coating weight—6.2 g/m² (4 mg/in.²) minimum.
	A232	Coating with impregnated color	Coating thickness—5 μm (0.2 mil) minimum. Coating weight—12.4 g/m² (8 mg/in.²) minimum.
	A233	Coating with impregnated color	Coating thickness—8 μm (0.3 mil) minimum. Coating weight—18.6 g/m² (12 mg/in.²) minimum.
	A24	Coating with electrolytically deposited color	Coating thickness to be specified. Application of the anodic coating, followed by electrolytic deposition of inorganic pigment in the coating.
	A2X	Other	To be specified.
Architectural Class II[c] 10 to 18 mm (0.4 to 0.7 mil) coating	A31	Clear coating	15% H_2SO_4 used at 21°C ± 1°C (70°F ± 2°F) at 129 A/m² (12 A/ft.²) for 30 min, or equivalent.
	A32	Coating with integral color	Color dependent on alloy and anodic process.
	A33	Coating with impregnated color	15% H_2SO_4 used at 21°C ± 1°C (70°F ± 2°F) at 129 A/ft.²) for 30 μin, followed by dyeing with organic or inorganic colors.
	A34	Coating with electrolytically deposited color	Application of the anodic coating followed by electrolytic deposition of inorganic pigment in the coating.
	A3X	Other	To be specified.

Table 16.13 Anodic Coatings on Aluminum (*Continued*)

Type of Finish	Designation[a]	Description	Examples of Methods of Finishing
Architectural Class I[c] 18 mm (0.7 mil) and thicker coatings	A41	Clear coating	15% H_2SO_4 used at 21°C ± 1°C (70°F ± 2°F) at 129 A/m² (12 A/ft.²) for 60 μin, or equivalent.
	A42	Coating with integral color	Color dependent on alloy and anodic process.
	A43	Coating with impregnated color	15% H_2SO_4 used at 21°C ± 1°C (70°F ± 2°F) at 129 A/m² (12 A/ft.²) for 60 min, followed by dyeing with organic or inorganic colors, or equivalent.
	A44	Coating with electrolytically deposited color	Application of the anodic coating followed by electrolytic deposition of inorganic pigment in the coating.
	A4X	Other	To be specified.

[a]*The complete designation must be preceded by AA—signifying Aluminum Association.*
[b]*Examples of methods of finishing are intended for illustrative purposes only.*
[c]*Aluminum Association Standards for Anodized Architectural Aluminum.*

the bonding of paints, organic coatings, and laminates. A natural oxide film does not always provide an adequate bonding surface. A frosty surface can be etched with chemicals. Designs can be etched on surfaces by masking all areas wexcept those to be etched. Chemical oxide films produce surfaces with greater corrosion resistance. Aluminum can be plated in a process called zincating, which produces a thin zinc coating. This coating protects aluminum from galvanic action and also prepares surfaces for electroplating. Aluminum can produce surfaces highly reflective of heat and light. These mirror-like finishes result from chemical brightening. Code designations for chemical finishes are given previously in Table 16.12.

Anodic Finishes

A widely used finish on aluminum is an anodized electrolytic oxide layer. Anodized films can be used as finishes on surfaces to be painted. Anodized coatings are most often employed as a finished protecting layer. A film is less than 0.1 mil thick, while a coating is 0.1 mil or thicker.

Following is a typical procedure for the anodizing process.

1. Alkaline and/or acid cleaners are used to remove grease and dirt from a surface.
2. Next, the surface receives an etching or brightening pretreatment. Etching involves producing a matte surface with hot solutions of sodium hydroxide. This process removes minor surface imperfections, and a thin layer of aluminum. Brightening produces a mirror-like surface with a concentrated solution of phosphoric and nitric acids. These smooth the surface through chemical reaction.
3. The third step is the actual anodizing process, in which the anodic film is built and combined with the aluminum by passing an electric current

through an acid electrolyte bath immersing the aluminum. The coating thickness and finished surface characteristics can be carefully controlled.

4. Coloring the anodized surface can occur in several ways. One method is to combine the coloring with the actual anodizing process (step 3), which simultaneously forms and colors the oxide cell wall in bronze and black shades. This produces a more abrasion-resistant coating than other methods but is more expensive because it requires more electricity.

A second coloring procedure involves a two-step electrolytic coloring process. After aluminum is anodized (step 3), it is immersed in a bath containing an inorganic metal salt and subjected to an electric current that deposits the metal salt at the base of the pores. Color depends on the metal salt used (for example, tin, copper, nickel, cobalt). This method provides the greatest variation of colors. A third process involves organic dyeing. This produces vibrant colors that are highly weather resistant.

One other coloring process is described as interference coloring. It involves modification of the pore structure produced in sulfuric acid. It results in lightfast colors ranging from blue to green and yellow to red.

Anodic coatings are classified into four groups: General, Protective and Decorative, Architectural I, and Architectural II. Coatings that are less than 0.1 mil thick are classified as General. Protective and Decorative coatings are less than 0.4 mil thick. These two classes are used for general industrial applications. The architectural classes are used on materials exposed to weather and wear. Details can be found in Table 16.13.

Architectural I coatings are recommended for exterior use where they will receive no regular maintenance. They are also used for interior purposes where extra protection is needed. They must be over 0.7 mil in thickness and weigh more than 27 mg per in.²

Architectural II coatings are recommended for interior applications not expected to receive heavy wear, and for exterior uses where a product receives regular maintenance. They must have a thickness from 0.4 to 0.7 mils and weigh 17 to 27 mg per in.[2] Code designations for architectural classes can be found in Table 16.13.

Coatings

Many aluminum products are painted to provide additional protection or for surface decoration. These products, such as aluminum gutters, are factory painted and delivered to site finished and ready to install. The factory-applied finish is electrostatically sprayed or roller applied with an organic paint and then passed through an oven where it is baked to a hard, uniform finish. The paint is flexible and will not crack if the product is bent or formed on the job.

Porcelain enamel coatings are produced by using a vitreous inorganic material that is bonded to the metal by fusing it at high temperatures. The coating is very resistant to corrosion, durable, and available in a wide range of colors. It is widely used on curtain wall panels, as seen on high-rise buildings, where regular maintenance is difficult.

Aluminum can be electroplated with chromium, copper, and other materials. If chromium is electroplated, the aluminum must first be zincated or plated with copper, brass, or nickel. Chromium plating gives a mirror-like surface and provides resistance to abrasion. Copper plating requires that the aluminum be zincated. It is mainly used where electrical connections must be soldered to an aluminum product. Tin and brass plating also provide good soldering surfaces. Zinc and cadmium plating improve corrosion resistance.

Aluminum surfaces can also have laminated finishes. This involves bonding another material, such as vinyl or polyvinyl chloride films, to the surface with an adhesive. Usually the aluminum surface must be chemically cleaned to provide the strongest bond.

Aluminum surfaces that serve as high-quality mirrors are treated by electro-polishing. This is an anodic smoothing of the surface and requires high-purity aluminum. Finished surfaces usually have a final anodic protective coating applied.

Protecting a Finished Aluminum Product

Aluminum products on a construction site need protection from damage during delivery, storage, and installation. Following are several coatings used for this purpose.

The best and most expensive protective coatings consist of paper or plastic sheet material bonded to the product with an adhesive that permits easy removal, leaving little or no sticky residue behind. Sometimes a coat of clear lacquer is sprayed on the aluminum's surface. Methacrylate lacquers chalk off after several years, but applying an automobile wax or a polish can afford added protection.

Routine Maintenance of Aluminum Surfaces

When cleaning aluminum surfaces, the type of protective film must be considered. A natural aluminum will develop its own oxide film that when scrubbed or rubbed with abrasive cleaners or steel wool will be damaged. Eventually, the oxide layer will reform. However, unless the surface is stained, abrasive cleaners should be avoided. Likewise, anodized surfaces, plated surfaces, and other coatings can suffer damage that is not easily repaired.

The simplest cleaner is clear water. Surfaces can be washed with water and liquid soaps, and then rinsed. If some discoloration needs removing, a mild polish, such as liquid auto polish, is acceptable. Cleaners containing abrasive materials, such as kitchen cleansers, will scratch a surface. Liquid or paste wax is a good final protective coating.

Joining Aluminum Members

Aluminum members can be joined by any of the standard fastening techniques. Mechanical fasteners include screws, bolts, rivets, and a variety of specially designed products. Sheet stock can be joined by stitching, which involves sewing sheets together with aluminum wire. Aluminum members can be attached by welding, brazing, and soldering. This includes gas, arc, resistance, and inert gas-shielded arc welding. Adhesive bonding is a technique gaining wide usage. A variety of construction adhesives are used, depending on materials and design considerations. Bonding produces a strong joint and increases the design possibilities for fusing aluminum and forming laminates.

Aluminum Products

Aluminum is used in a variety of common construction materials. One major application for aluminum is in the production of metal doors and windows. The members used in windows are extruded from an aluminum alloy, such as 6063. A wide range of interior and exterior aluminum doors and windows are made using framed glass units encased in an extruded aluminum frame (Figure 16.5). These can be panelized, flush, louvered, and have glass lights. A variety of other aluminum extrusions are available (Figure 16.6).

Figure 16.5 Aluminum frame windows are widely used in both residential and commercial construction.

© M.Khebra/Shutterstock.com

Figure 16.6 A detailed view of extruded aluminum shapes.

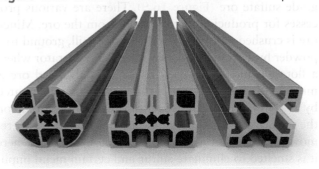

© Peter Sobolev/Shutterstock.com

Specifications developed by the Architectural Aluminum Manufacturers Association (AAMA) specify minimum frame strength, thickness, corrosion resistance, air infiltration, water resistance, wind load capacity, and condensation resistance. Units meeting these specifications have the AAMA seal attached.

Aluminum is used for residential and commercial siding and roofing. The specifications are developed by the AAMA. Residential siding is cold-rolled to shape and comes in several widths. Horizontal, vertical, and panelized siding sheets are available (Figure 16.7). The panels may or may not incorporate an insulative backer board. Backer boards can be made of fiberboard or foamed plastic, both of which provide higher insulation value.

Figure 16.7 Aluminum siding is durable and weather resistant.

© JOSEPH S.L.TAN MATT/Shutterstock.com

The same material is used for soffits, fascias, flashing, gutters, and frieze boards (Figure 16.8). Aluminum panels are secured to wood framing with aluminum nails.

Figure 16.8 Aluminum is widely used for soffits, fasciae, and gutters.

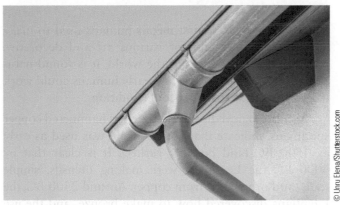

© Unru Elena/Shutterstock.com

Aluminum panels for siding and roofing commercial buildings are made in a variety of designs. Aluminum alloy 3004 is often used, resulting in panels that are one-third the weight of steel. Common finishes include mill, unpainted, painted, and stucco embossed.

Aluminum curtain wall systems include preformed insulated wall panels that are integrated with aluminum windows to form a weather-tight, maintenance-free exterior surface. They are available with anodized or factory-applied baked enamel finishes. More information is given in Chapter 29. A wide array of roof accessories, such as skylights, roof hatches, and smoke and fire vents and louvers, are also made with extruded and sheet aluminum.

Aluminum structural shapes are rolled in much the same configurations as those discussed for structural steel members. These shapes include S and W beams, channels (in a variety of special forms), tees, zees, bulb angles, round and square tubes, pipe, plate, and rods of varying aspects.

Aluminum is used for large-diameter electrical conductors, various sheet metal applications, and for situations involving corrosion, such as in food-processing and chemical plants.

COPPER

Copper (chemical symbol Cu) is a nonmagnetic reddish-brown metal with excellent electrical and thermal conductivity. It has the highest conductivity properties of all commonly used metals except silver. It is ductile, malleable, and easily worked. When alloyed, it offers a wide range of properties suitable for a variety of construction applications. Although copper products are initially more expensive than aluminum, they have properties that make them less costly over time, such as resistance to corrosion and other damaging conditions.

History of Copper

Copper is one of the first metals humans used to make tools and utensils and for various art and decorative purposes. In some parts of the world, it is found naturally in almost pure form, and early humans could work it into desired shapes in a cold condition.

Archaeological discoveries, such as hammered copper specimens, indicate that copper likely was used as early as 4500 BC (and possibly earlier). It is clear that by 3000 BC the Egyptians were making utensils, simple tools, and ornaments from copper. Around 2500 BC, the Egyptians discovered how to make bronze, and the use of copper and bronze spread over most of the Mediterranean area. Copper, named after the island of Cyprus, was first called cuprium, then cuprum, and, in English, became "copper." Through the years, with improvements in mining, refining, and alloying, copper has become one of the major metals used in construction products and tools.

Properties of Copper

The most important properties of copper and copper alloys as used in the construction industry are its electrical and thermal conductivity, resistance to corrosion, wear defiance,

ductility, and high temperature performance. Copper has a relatively low tensile strength, about 32,000 psi, which can be improved by heat-treating, cold-working, and alloying. High electrical conductivity makes it a good material for electrical wiring and parts in devices that conduct electricity. Copper's good thermal conductivity makes it useful in heat transfer situations, and its corrosion resistance properties make it ideal for plumbing pipe, gas lines, and components exposed to the atmosphere or corrosive chemical elements. Its ductility properties make it a material easily bent, stretched, stamped, machined, and otherwise formed into useful products. Copper has a melting point of 1,981°F (1,083°C) and a coefficient of thermal expansion of 0.0000168/°F (0.0000093/°C).

When unprotected copper is exposed to the atmosphere, it will organically develop a green coating over a period of years. This coating, called patina, provides a natural protection from additional corrosion. It takes many years of exposure to the atmosphere to build up a fully developed patina coating.

Production of Copper

Much of the copper mined in the United States is low-grade sulfate ore (Figure 16.9). There are various processes for producing metal copper from the ore. Mined ore is crushed and pulverized in a ball mill, ground to a powder by a grinder and moved to a concentrator where a flotation method is used. Here the pulverized ore is mixed with water, oil, and a foaming agent and agitated by air. The copper sulfate particles collect in the foam on the surface as waste materials settle to the bottom for removal. The froth flows to a reverberatory furnace where it is smelted to eliminate sulfur and certain metal impurities caused by oxidation. This produces a top layer of copper that is tapped off and moved to a converter.

Figure 16.9 Low-grade sulfate ore used to produce copper is mined in open pits.

© FCG/Shutterstock.com

The converting process is much like the Bessemer process used to produce steel. In the converter, the iron in the remaining material is oxidized, forming a layer of slag that is removed. The sulfur is oxidized, forming sulfur dioxide gas, and drawn off the converter. The remaining material, called blister copper, is about 99 percent pure. The blister copper is processed in a refining furnace, reducing its oxygen content and forming fire-refined copper. This fire-refined copper can be used to produce copper products by rolling, extruding, drawing, and casting.

The purest copper is produced by electrolytic processes. Fire-refined copper can be cast in molds that become anodes in an electrolytic process. The electrolytic process utilizes an acid-proof tank containing an electrolyte, warm diluted sulfuric acid, and copper sulfate. The fire-refined copper anodes are suspended in the tank. Thin sheets of pure copper are suspended between them and serve as cathodes. The system receives an electric current, causing the copper in the anodes to dissolve and bond to the cathodes. This produces copper approximately 99.9 percent pure. The cathodes are melted and used to produce the base metal for a wide variety of copper alloys.

After the copper has been fire-refined or purified by the electrolytic process, it is formed into sheets, bars, tubes, and wires used to produce various products. Examples of common plumbing fittings made from copper are shown in Figure 16.10.

Figure 16.10 A variety of copper fittings produced from copper alloys.

© B Calkins/Shutterstock.com

Classifying Copper and Copper Alloys

Copper and copper alloys are specified by the UNS. They are identified by a five-digit number code preceded by the letter C. Wrought materials are assigned UNS numbers from C10000 to C79999. Cast alloys are numbered from C80000 to C99999. A summary of these can be found in Table 16.14.

Table 16.14 UNS Designations for Coppers, Brasses, and Bronzes

Wrought Coppers and Copper Alloys	
C10000–C15760	Copper
C16200–C19900	High copper alloys
Cast Coppers and Copper Alloys	
C80100–C81200	Copper
C81400–C82800	High copper alloys
Wrought Brasses and Brass Alloys	
C21000–C28000	Brasses
C31200–C38500	Copper-zinc-lead alloys
C40400–C48600	Copper-zinc-tin alloys
Cast Brass and Brass Alloys	
C83300–C83810	Brasses
C84200–C84800	Copper-tin-zinc and copper-tin-zinc-lead alloys
C85200–C85800	Copper-zinc and copper-zinc-lead alloys
C86100–C86800	Manganese bronze and leaded manganese bronze alloys
C87300–C87800	Copper-silicon alloys
Wrought Bronzes and Bronze Alloys	
C50100–C54400	Copper-tin-phosphous alloys
C55180–C55284	Copper-phosphorus and copper-silver-phosphorus alloys
C60800–C64210	Copper-aluminum alloys
C64700–C66100	Copper-silicon alloys
C66400–C69710	Other copper-zinc alloys
C70100–C72950	Copper-nickel alloys
C73500–C79800	Copper-nickel-zinc alloys (nickel silvers)
Cast Bronzes and Bronze Alloys	
C90200–C91700	Copper-tin alloys
C92200–C92900	Copper-tin-lead alloys (leaded tin bronzes)
C93100–C94500	Copper-tin-lead alloys (high leaded-tin alloys)
C94700–C94900	Copper-tin-nickel alloys
C95200–C95900	Copper-aluminum-iron and copper-aluminum-iron-nickel alloys
C96200–C96900	Copper-nickel-iron alloys
C97300–C97800	Copper-nickel-zinc alloys (nickel silvers)
C98200–C98840	Copper-lead alloys
C99300–C99750	Special alloys

Source: Society of Automotive Engineers

Coppers are numbered from C10100 to C15999 and are pure or nearly pure, having a minimum copper content of 99.3 percent or higher. The coppers in this series are very similar in chemical composition. The numbering system identifies the production or refining processes used to produce the copper. For example, oxygen-free copper is C10200, electrolytic tough pitch copper is C11000, and phosphorus-deoxidized high-residual phosphorus copper is series C12200. Coppers containing less than 0.7 percent of specified alloying constituents include tellurium-bearing copper (C14500) and zirconium copper (C15000).

The numerical designations indicate the type of copper or copper alloy and the alloying elements and impurities. Comprehensive handbooks from the Copper Development Association give specific details.

Copper or copper alloys must be identified by their UNS number. Over the years, some frequently used types of copper were given trade names, as shown in the tables that follow. The trade name does not indicate the composition of the material and sometimes is used to describe similar materials that have slightly different amounts of various elements.

Deoxidized copper (UNS C122200) contains 99.9 percent copper and 0.025 percent phosphorus. It has better forming and bending qualities than electrolytic copper and resists brittleness at high temperatures. It is used for water and refrigeration piping, oil burner service, and in sheets and plates where welding constitutes the major joining method. Electrolytic tough pitch copper (UNS C11000) is 99.9 percent copper and has high electrical and thermal conductivity. It can be easily formed into useful shapes. Major uses include electrical conductors of all types, as well as flashing, gutters, roofing, and forgings (Figure 16.11).

Figure 16.11 Easily formed into useful shapes, copper uses include electrical conductors, flashing, gutters, and roofing.

© Inc/Shutterstock.com

Copper Alloys

The major copper alloying elements are tin, aluminum, zinc, nickel, silicon, manganese, lead phosphorus, and beryllium. The UNS designations shown are for wrought products. Designations for cast products are listed earlier in Table 16.14.

High-copper alloys (UNS C16200–C19199) are wrought alloys having specified copper contents from 96 to 99.3 percent. Cast high-copper alloys have a minimum of 94 percent copper. Wrought high-copper alloys have very high electrical and thermal conductivity, almost matching that of pure copper. However, they are much stronger than pure copper, which increases the number of possible uses. They also have good corrosion resistance.

Brasses (UNS C20000–C49999) are copper alloys with zinc as the major alloying element. Other elements, such as lead, phosphorus, nickel, silicon, iron, and aluminum, can be added in small amounts. They are extremely useful and find application in many products.

Bronzes (UNS C50000–C66399) are copper alloys in which neither nickel nor zinc is used as a major alloying element. The various types are used for electrical contacts and in corrosion-resistant applications.

Miscellaneous copper-zinc alloys (UNS C66400–C69999) are often referred to as manganese or nickel bronzes. Typically, the major alloying element is zinc, so they are much like some of the brasses.

Copper-nickel alloys (UNS C70000–C72999) contain from 3 to 33 percent nickel. Other elements may be added to improve corrosion resistance and strength. They are used in marine applications because of their outstanding capacity to resist corrosion. Typical applications include use in heat exchangers, condensers, piping, valve and pump parts, and relay and switch springs. Alloys with more than 50 percent nickel are called Monel® metals and form a separate class. Monels retain high strength at elevated temperatures.

Copper-nickel-zinc alloys (UNS C73000–C79999) are referred to as silver nickels because of their color. Zinc is the principle alloying element, and nickel is secondary, although other elements may be added to alter their properties. They have good electrical and mechanical characteristics, good corrosion resistance, and are used for fasteners and various electrical components.

Copper Alloy Finishes

Copper alloys are available with a variety of finishes. Some are supplied by the mill that manufactures the copper alloy, and others are supplied by the company fabricating the copper into a specific product or stock shapes. These finishes use the same three classifications used for aluminum products. They include mechanical, chemical, and coatings. A summary of coatings is given in Table 16.15. Notice the Copper Development Association finish designation. An "M" before the identifying digits denotes mechanical finishes, a "C" designates chemical finishes, and coatings use three-digit numbers.

Table 16.15 Copper Alloy Finishes

Finish	Copper Development Assn. Finish Designation
Mechanical	
As fabricated	M10 series
Buffed	M20 series
Directional textured	M30 series
Non-directional textured	M40 series
Patterned	M4X (specify)
Chemical	
Cleaned only	C10 series
Mane dipped	
Bright dipped	
Conversion coatings	C50 series
Coatings	
Organic:	
Air dry	060 series
Thermo-set	070 series
Chemical cure	080 series
Vitreous	
Laminated	L90 series
Metallic	

Source: Copper Development Association

Care of Copper and Copper Alloys

New copper products, especially sheet stock, have a bright, shiny, light-brownish color. A protective coating, such as a clear lacquer, is needed to maintain this color.

When left exposed to the atmosphere, it will first turn a darker brown and eventually take on a permanent light-green patina. This is considered highly desirable from an appearance standpoint and can be produced by artificial means if the natural aging is too slow.

Copper can be washed with liquid soap and water. If scrubbed with an abrasive cleaner, the brown or green color will be damaged and require time to return.

Uses of Copper

Copper and copper alloys are excellent for outdoor uses, such as siding, roofing, flashing, guttering, and screen wire. Alloys are used extensively for plumbing pipe in residential and commercial structures and in the manufacture of plumbing fittings, such as valves, drains, and faucets. Sewage treatment plants and industrial plants, such as chemical processing installations, utilize copper for many purposes, including lining vessels that contain corrosive materials. Various types of hardware and fasteners, such as nails, screws, and bolts, are made from copper alloys. A major use is in electrical wire, electrical conductors, and parts in electrical appliances that conduct electricity.

BRASS

Brasses (UNS C20000–C49999) are copper alloys having zinc as the principal alloying element, but variations are produced by adding small quantities of other elements. Zinc improves strength and ductility and produces changes in color. Adding lead improves machinability, and tin enhances strength, hardness, workability, and ductility.

Brasses are hardened by cold-working. However, hardness is also influenced by alloy composition. The compositions of several wrought brasses and brass alloys used in construction are shown in Table 16.16. Similar information is available for cast brasses and brass alloys. Notice that zinc is the major alloying element, with lead and iron present in much smaller amounts. The chemical symbols used to identify the various elements in all metals are shown in Appendix F.

Brasses fall into three general classes. White brasses contain less than 55 percent copper and are hard and brittle. They are used for cast products and cannot be hammered or worked without breaking. Alpha brasses contain 63 to 95 percent copper and are the easiest type to work. They are used to make radiator parts, springs, grilles, hardware, and decorative plating (Figure 16.12).

Table 16.16 Composition of Selected Wrought Brasses and Brass Alloys[a]

UNS Designation	Descriptive Name	Major Alloying Elements in Percent[b]					Other Named Elements
		Cu	Zn	Pb	Fe	Sn	
Copper-Zinc Alloys (Brasses)							
C22000	Commercial bronze	89.0–91.0	REM[c]	0.05 max.	0.05 max.		
C23000	Red brass	84.0–86.0	REM	0.05 max.	0.05 max.		
C26000	Cartridge brass	68.5–71.5	REM	0.07 max.	0.05 max.		
C28000	Muntz metal	59.0–63.0	REM	0.03 max.	0.07 max.		
Copper-Zinc-Lead Alloys (Leaded Brasses)							
C31400	Leaded commercial bronze	87.5–90.5	REM	1.3–2.5	0.10 max.		0.7 Ni
C35000	Medium-leaded brass	60.0–63.0	REM	0.8–2.0	0.15 max.		
C37700	Forging brass	58.0–61.0	REM	1.5–2.5	0.30 max.		
C38500	Architectural bronze	55.0–59.0	REM	2.5–3.5	0.35 max.		
Copper-Zinc-Tin Alloys (Tin Brasses)							
C44300	Admiralty, arsenical	70.0–73.0	REM	0.07 max.	0.06 max.	0.8–1.2	0.02–0.06 As
C46400	Naval brass, uninhibited	59.0–62.0	REM	0.20 max.	0.10 max.	0.5–1.0	
C48500	Naval brass, high lead	59.0–62.0	REM	1.3–2.2	0.10 max.	0.5–1.0	

[a]These are only a few of the many types of brass and brass alloys available.
[b]Cu copper, Zn zinc, Pb lead, Fe iron, Sn tin, Ni nickel, As arsenic.
[c]Remainder for the difference between elements specified and 100 percent.
Source: Standards Handbook, parts 5 and 6, Copper Development Association.

Figure 16.12 Brass is used for springs, grilles, hardware, and decorative plating.

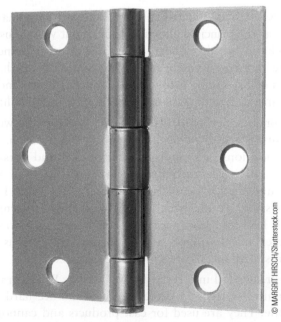

© MARGRIT HIRSCH/Shutterstock.com

Alpha-beta brasses contain 55 to 63 percent copper. They are stronger than alpha brasses and can be worked hot. They are used where strength is important, such as for rivets and screws. Table 16.14 gives the Unified Numbering System designation for wrought and cast brasses. Brasses are very important metals and are used in construction almost as much as copper.

History of Brass

The Romans may have been the first to deliberately add zinc to copper to produce brass. Used to manufacture coins and jewelry, by the Middle Ages, the production of brass in Europe was a major industry. It was widely used to cast religious objects, candlesticks, locks, and commemorative plates. By the nineteenth century, companies started producing brass utensils, lamps, and other household items.

Plain Brasses (Copper-Zinc Alloys)

Copper-zinc alloys are sometimes referred to as plain brasses. Several that find use in products related to construction are red brass, commercial bronze, cartridge brass, and Muntz Metal. The alloying elements for each can be found in Table 16.16.

Red brass contains 85 percent copper, 15 percent zinc, and small amounts of lead and iron. It has excellent resistance to corrosion and higher ductility and strength than copper. It is used for plumbing pipe, handrails, balusters, stair posts, tubing, and hardware (Figure 16.13).

Figure 16.13 Architectural bronze is used for decorative grilles, handrails, architectural trim, and hardware.

Commercial bronze contains 90 percent copper and 10 percent zinc. It has good ductility and cold-working properties. It is used for screws, forgings, and some types of hardware.

Cartridge brass contains 70 percent copper and 30 percent zinc. It has the best strength and ductility of all the brasses and is easily worked cold. It is widely used where copper products require extensive fabrication, such as stamping or deep drawing. It finds use in the manufacture of electric sockets, reflectors, rivets, and heating units.

Muntz Metal contains 60 percent copper and 40 percent zinc, which is the greatest percent of zinc alloyed with the copper. Muntz Metal has low ductility but high strength and is used for things such as sheet stock and exposed architectural features.

Leaded Brasses (Copper-Zinc-Lead Alloys)

Lead is added to brass to make it easier to machine. Leaded brasses are used in a variety of construction products, including architectural bronze, forging brass, and medium-leaded brass. The alloying elements for each can be found in Table 16.16.

Architectural bronze has the least copper and most lead of the three mentioned above. Widely used for forgings and products produced by machining, it contains 55 to 59 percent copper, 41 to 45 percent zinc, and 2.5 to 3.5 percent lead. Architectural bronze is used for decorative grilles, handrails, architectural trim, and hardware.

Forging brass contains about 60 percent copper, 38 percent zinc, and 2 percent lead. It has great plasticity when hot, and is used often for forgings. Because of its good corrosion resistance, it is used for plumbing and hardware items.

Medium-leaded brass contains about 62 percent copper, 34 percent zinc, and 2 percent lead. It is used when good machining properties are required, such as for keys, parts of locks, plaques, and various scientific instruments.

Tin Brasses (Copper-Zinc-Tin Alloys)

When tin is alloyed with copper, zinc, and other elements, the alloy has additional properties not present in plain brasses. The two types of tin brasses are admiralty and naval brasses.

Admiralty brass contains about 71 percent copper, 28 percent zinc, 1 percent tin, and traces of lead, iron, and arsenic. These alloying elements improve strength and ductility and, most important, increase resistance to corrosion. Admiralty brass is widely used in the manufacture of condenser and heat-exchanger plates, various tubes, and equipment in chemical and electrical power plants, as well as products needing resistance to seawater.

Naval brasses fall into several categories used for wrought products. These include uninhibited, arsenical, medium-leaded, and high-leaded. The copper content for all naval brass ranges from 59 to 62 percent. There is a wide range in the amounts of lead and tin alloying elements. Naval brasses are used in chemical, steam power plant, and marine equipment. They can withstand the corrosive effects of materials to which they are typically exposed.

BRONZE

Technically, the term "bronze" has been used to identify a product that is 90 percent copper and 10 percent tin. However, other alloying elements are now added to produce a wider range of materials in the bronze family of metals that have varied properties. Bronze now refers to copper alloys having elements of silicon, aluminum, manganese, and others. They may or may not have zinc. Some of the products in Table 16.16 are referred to as bronze but are actually brass alloys. These include commercial bronze, architectural bronze, and Muntz Metal.

The composition of a few selected wrought bronzes and bronze alloys are shown in Table 16.17. The

Table 16.17 Composition of Selected Bronzes and Bronze Wrought-Type Alloys[a]

UNS Designation	Pevious Trade Name	Major Alloying Elements in Percent[b]										Other Named Elements
		Cu	Pb	Fe	Sn	Zn	P	Al	Mn	Si	Ni	
Copper-Tin-Phosphorus Alloys (Phosphor Bronzes)												
C51800	Phosphor bronze	REM[c]	0.02 max.	—	4.0–6.0	—	0.10–0.35	0.01				
Copper-Tin-Lead-Phosphorus Alloys (Leaded Phosphor Bronzes)												
C53400	Phosphor bronze B-1	REM	0.8–1.2	0.10 max.	5.0–5.8	0.30 max.	0.03–0.35					
Copper-Aluminum Alloys (Aluminum Bronzes)												
C61300	Aluminum bronze	REM	0.01 max.	2.0–3.0	0.20–0.50	0.10 max.	—	6.0–7.5	2.0 max.	0.10 max.	0.15 max.	0.15 P
C64200	—	REM	0.05 max.	0.30 max.	0.20 max.	0.50 max.	—	6.3–7.6	0.10 max.	1.0–2.2	0.25 max.	0.15 As max.
Copper-Silicon Alloys (Silicon Bronzes)												
C65100	Low silicon bronze B	REM	0.05 max.	0.8 max.	—	1.5 max.	—	—	0.7 max.	0.8–2.0		

[a]These are only a few of the many alloys available.
[b]Cu copper, Pb lead, Fe iron, Sn tin, Zn zinc, P phosphorus, Al aluminum, Mn manganese, Si silicon, Ni nickel, As arsenic.
[c]Remainder for differences between elements specified and 100 percent.
Source: Standards Handbook, parts 5 and 6, Copper Development Association.

phosphor bronzes contain approximately 89 percent copper. C51800 has 4 to 6 percent tin and 0.1 to 0.35 percent phosphor. C53400 has higher lead content and is referred to as leaded phosphor bronze. Aluminum bronzes have 6 to 7.5 percent aluminum, while low silicon bronze has no tin or lead but 0.8 to 2.0 percent silicon. Cast bronzes have similar elements. Table 16.14 gives the UNS designations for bronzes. Bronze is used for various cast products and hardware, including screws, washers, nuts, bolts, and weather stripping.

LEAD

Lead (chemical symbol Pb) is a soft, heavy metal with good corrosion resistance that is easily worked. Its ability to resist penetration from radiation is a unique feature.

History of Lead

Archaeological findings indicate that lead may have been used as early as 6000 to 5000 BC. Through the centuries, it found use in pottery glazing and as a solder for joining copper. It was used in China for currency and as an alloying element in early bronzes. The Romans made wide use of lead pipes for water distribution, not realizing lead poisoning results from its ingestion. By 800 BC, lead was used in the manufacture of glass and, by AD 1100, as roofing and cames, the latticed framework that holds glass pieces in stained glass windows.

Properties of Lead

Other advantageous properties of lead are its high density and weight, softness and malleability, low melting point (620°F; 327°C), and good electrical conductivity. Potentially disadvantageous is lead's low strength and lack of elasticity.

Production of Lead

The major source of lead is the mineral galena (or lead sulfide). Other sources include cerussite (lead carbonate) and anglesite (lead sulfate). Lead ores frequently contain zinc, and some have gold, silver, and other metals.

Lead-bearing ore is first crushed and ground into fine particles. Metal-bearing material is separated from the rock particles via flotation. Flotation involves mixing water, oils, and chemicals with the ground ore. As air is blown into the mix from the bottom, the lead-bearing particles are wetted by the oil and float to the top in a froth of air bubbles. The lead particles are drawn off and the waste material (gangue) settles to the bottom and is removed.

The concentrated ore is roasted in the air, which changes the lead sulfide to lead oxide. Sulfur escapes as sulfur dioxide gas and is recovered and made into sulfuric acid. The lead oxide is then smelted in a blast furnace, which causes the lead to settle to the bottom. Any gold or silver settles with it. Waste materials float to the top, forming slag, and are removed. The lead is then processed to separate the gold and silver (Figure 16.14).

Figure 16.14 Processed lead bars ready to be formed into useful products.

Grades of Lead

Of the several grades of lead, chemical lead, desilverized lead, and corroding lead find use in some construction applications. Chemical lead and desilverized lead are used for pipes, sheets, and alloys. Corroding lead is used for white lead, red lead, and litharge (used in the manufacture of batteries, pottery, lead glass, and ink).

Lead Alloys

Lead is alloyed with antimony to improve hardness and strength. However, many other elements are also added. Among these are arsenic, nickel, zinc, copper, iron, and manganese.

UNS Designations

Lead and lead alloys are designated by the UNS numbers L50001 through L59999. For example, L50121 is described as a solder alloy containing 98.0 percent lead,

and L50770 is a battery grid alloy, lead-calcium, containing 99.6 percent lead.

Uses of Lead

Lead pipes and tank liners are used in installations that process highly corrosive materials, but they are never used for piping to carry drinking water. Since lead is a good self-lubricant, it is used when high-pressure lubricating is necessary. Lead solder works for electrical connections because it is a good conductor, but it is not used on water pipe connections. Hard solders have antimony added. High-temperature solders are alloyed with silver and are generally referred to as silver solder.

The use of lead in solder for joining copper water pipes has been banned because of the possibility of increasing lead content in water, which causes lead poisoning. Lead pipes are not used to move drinking water for the same reason. This danger is especially present when soft or distilled water is used. However, lead pipe and lead-lined tanks have high corrosion resistance and can find use in industrial production applications, such as in the chemical manufacturing industry (Figure 16.15). Sheet lead is used for roofing, flashing, and spandrels in areas where there is severe industrial air contamination or on seacoasts.

Figure 16.15 Lead pipe and lead-lined tanks have high corrosion resistance and find use in industrial applications.

Another interesting use for lead occurs in the form of lead azide. Lead azide is easily exploded by an electrically heated wire, so it is used in the manufacture of blasting caps, which set off other explosives.

Lead works in products such as adhesives, caulking, pigments, glazing, pipes and tanks, compounds, and protective coatings over steel and copper. It is an element added to produce brass, bronze, certain asphalt products, glazes, various fusible alloys, glass, porcelain enamel, iron, steel, certain plastics, solder, rust-resistant pre-painting primer coatings, tin. It also serves as an additive in certain wood preservative preparations. As mentioned in Chapter 31, the use of lead in paint has been banned to avoid the danger of children ingesting peeling pieces of paint.

Architects specify lead for waterproofing, soundproofing, reduction of vibration, and radiation shielding. Following are more detailed examples of two lead products in order to provide a closer look at its elements and properties.

Lead strip and sheet products are made from lead that is almost 100 percent pure and from an alloy with about 7 percent antimony, which improves strength and stiffness. Seams can be folded, soldered, or folded and soldered. Soldering is done with a 50 percent lead, 50 percent tin solder.

Red lead, a lead oxide, is a widely used primer applied generally to steel before it receives final coats of finish paint. It is available dry, in a paste, or as a liquid paint. Dry red lead is available in grades A, B, and C. Grade C has the highest lead content. Red lead paste in an oil is available in grades B and C. Again, grade C has a higher percentage of lead.

SOLDERS

Solders are nonferrous metals used to join metals in a waterproof joint and to make secure electrical connections.

Properties

Solders have low melting temperatures, typically around 375°F (192°C) to 595°F (315°C). However, some high-temperature solders range up to about 740°F (396°C). The low melting point enables a solder to join metals without melting them; therefore, the solder has no alloying action with the metals being joined.

Solders have little shear, tensile, or impact strength, so they are used when there is no load on a joint or on joints that have other fasteners, rivets, bolts, or interlocking seams to carry loads or other stresses.

Types of Solder

Solders are mainly an alloy of lead and tin, with small amounts of other elements included. The four major classifications of solder are tin-lead, tin-lead-antimony, silver-lead, and a variety of special alloys.

Tin-lead solders are general-purpose alloys used for joining metals. The tin-lead composition varies from an alloy with 70 percent lead and 30 percent tin through a series of ten or so combinations in which the amount of lead increases and the amount of tin decreases. The maximum lead solder has 90 percent lead and 10 percent tin. Tin-lead solders have less than 1 percent antimony. Typical 50–50 tin-lead solder is most commonly used.

Tin-lead-antimony solders have from 1 to 2 percent antimony, and of the five commonly available types, all have more lead than tin. Tin content varies from 20 to 40 percent.

Silver-lead solders contain more than 97 percent lead, from 0 to 1.25 percent tin, traces of antimony, and a silver content of 1.5 to 2.5 percent. They have melting points of about 580°F (307°C) and produce strong corrosion-resistant joints. They are used for soldering copper and brass using a torch-heating method. Special purpose solders are designed for a specific application, such as soldering copper roofing, or when a high-temperature solder is required. They typically contain various amounts of lead, zinc, silver, and cadmium.

Fluxes

A metal surface must be clean and free of oxides for solder to bond to it. In addition to mechanically cleaning the surface (washing, buffing, brushing, etc.), any oxides on the surface must be removed. Fluxes are materials used to remove these oxides. Three general types of fluxes are available: neutral, corrosive, and noncorrosive.

Neutral fluxes are mild and used on easily soldered metals, such as copper, lead, brass, and tin plate. They typically take the form of a mild acid that is wiped on a surface and need not be removed before soldering begins.

Corrosive fluxes include salt-type and acid-type. They are more effective for cleaning surfaces than neutral fluxes, but it is important that they, along with any oxide residue, be removed. Otherwise their corrosive action continues to attack the base metal being soldered. Corrosive fluxes (often called acid fluxes) cannot be used on electrical connections.

Noncorrosive fluxes tend to be used only on easily soldered metals and electrical connections. Typically, they contain rosin as the main flux. Noncorrosive fluxes are good for soldering electrical connections because they will not cause wires to corrode.

TIN

Tin (chemical symbol Sn) is produced from the ore containing the mineral cassiterite, a form of tin oxide. Since there is little cassiterite in North America, tin has to be imported from Malaysia, Brazil, Russia, Indonesia, Thailand, China, and Bolivia. Ore usually contains little tin, so an extensive refining process is required.

Properties and Working Characteristics of Tin

Tin is a soft metal that is malleable, ductile, and blue-white in color. It is corrosion resistant when exposed to air and moisture, has a low melting point of 450°F (232°C), and can be cast. It will take a high polish and has properties enabling it to coat other metals. Because tin is soft and malleable, it can be worked by rolling, spinning, extrusion, and casting.

Production of Tin

After ore is mined, impurities must be removed to get a concentrated ore. This involves a series of mechanical and chemical processes for crushing and cleaning the ore. Magnetic separators and screening operations are used to separate the ore and the tailings.

The concentrated cassiterite is refined, producing a high-purity metal of at least 99.8 percent tin. Several different processes are used, but all involve smelting ore in a furnace where crude tin is separated from slag. The crude tin is then heated to a predetermined temperature at which impurities with higher melting temperatures than tin remain. This process is called liquidation. This partly pure tin is drawn off and heated above its boiling point. As it is stirred, additional impurities rise to the surface and are drawn off. This continues until most impurities are removed, leaving an almost pure tin.

UNS Designations

Tin alloys are designated in the UNS by the numbers L13001 through L13999. For example, UNS L13630 is a lead-tin solder containing 37 percent lead and 63 percent tin.

Uses of Tin

Since tin and tin alloys have high corrosion resistance and excellent coating ability, they are used extensively to provide protective coatings on other metals, especially steel (Figure 16.16). One example is the coating on cans used to store food for retail. Tin is also used as an alloying element in other metals. Its low melting point makes it useful in some solders. It finds applications in mirrors, hardware, and fusible alloys. Tin compounds are used in the production of glazes, glass, and porcelain enamel.

Figure 16.16 Tin and tin alloys have high corrosion resistance and are used extensively to provide protective coatings on other metals, especially steel.

© Ouka/Shutterstock.com

TERNEPLATE

Terneplate is a mixture of lead and tin applied to copper-bearing steel sheet or stainless-steel sheet to produce a corrosion-resistant coating. Tin is added because lead alone will not alloy with the iron. Terneplate sheets are available in two types: short terne and long terne.

Short terne, used for roofing, is available with stainless-steel or copper-bearing steel bases. The stainless-steel type uses UNS S30400 stainless steel in 26 and 28 gauge thicknesses. Copper-bearing steel terneplate is made in 26, 28, and 30 gauge thicknesses. Both types use a coating on both sides consisting of 75 to 80 percent lead and 20 to 25 percent tin.

Long terne is used for various industrial purposes and is available in 14 to 30 gauge low-carbon steel with a coating on both sides consisting of 75 to 87 percent lead and 12 to 25 percent tin.

Uses of Terneplate

Short terne, often called roofing terne, is used for finished roofing, gutters, downspouts, and flashing. Terneplate roofing is installed with batten and standing seams. Flat-locked and horizontal seams can be used when it is installed over wood sheathing.

Long terne is used for fireproof doors and frames, other fireproofing items, and roofing. The various types are available in sheets and rolls. Terne on steel should be painted on both sides after installation to seal any pinholes. If pinholes remain in terne-coated steel, corrosion results. Frequently, a mill applies a red iron-oxide primer, but a high-quality finish coat of paint is required over this. Terne-coated stainless steel does not require a primer or painting.

Production of Terne Sheets

Sheets are chemically cleaned by passing them through a dilute solution of sulfuric or hydrochloric acid. Then they are fluxed by a heated solution of zinc chloride. Finally, the sheets pass through a molten solution of lead and tin. The terne-coated sheet is then moved between smooth metal rollers to produce the finished surface. After cooling, it is cleaned to remove any traces of oil. It is then ready for priming.

TITANIUM

Titanium (chemical symbol Ti) is found in large quantities in the earth's surface. It is a very light, strong, ductile, silvery metal, and is one of the most common elements.

History

The element titanium was discovered in England around 1790. It was in the form of black sand, which, upon analysis, yielded a white metallic oxide: titanium dioxide, a compound found in nature that produces a white powder when refined. This powder is used today in the paint industry. In 1910, a laboratory experiment proved that titanium in metallic form could be produced. It was not until 1946 that a process was patented that enabled titanium to be produced in commercial quantities. By 1950, an industrial titanium was available.

Properties

Titanium has low electrical conductivity and a low co-efficient of thermal expansion. It is also paramagnetic, meaning that when placed in a magnetic field, it possesses magnetization in direct proportion to the field's strength. It has a melting point of 3,300°F (1,820°C) and a coefficient of thermal expansion of 0.0000085/°C.

Two important characteristics of titanium are its high strength-to-weight ratio and its ability to resist corrosion by salt water and the atmosphere. These have an important impact on its use in various products.

Working Characteristics

Pure titanium is easier to use than are titanium alloys; however, in general, all types can be fabricated using standard manufacturing processes. These include machining, welding, riveting, drilling, punching, hot- and cold-rolling, extruding, forging, and drawing.

Production of Titanium

Titanium is produced by the Kroll process, named after William J. Kroll, the person who developed it. The process is chemically driven, using magnesium in an inert atmosphere of helium or argon to produce the reduction of the titanium tetrachloride. These elements react, releasing magnesium chloride, which is distilled off. Left is a sponge metal that is crushed and melted into ingots. The ingots (similar to pig iron ingots) are used to produce titanium and titanium alloys that are processed into various products.

Titanium Alloys

The strength of titanium varies depending on the purity of the metal. The higher the purity the weaker the metal; therefore, various elements are added to produce titanium alloys with greatly increased strength. Typically, vanadium, molybdenum, aluminum, iron, chromium, and manganese are used as alloying elements.

UNS Designations

Titanium alloys are designated in the UNS by the numbers UNS R50001 through R59999. For example, R56210 is a titanium alloy containing 90.2 percent titanium.

Uses of Titanium

The major demand for titanium comes from aircraft and aerospace industries and the military. Its strength and light weight make it a desirable material for these applications. It can also be formed in sheet, strip, pipe, and tube products, and it can be forged and cast. Titanium is beginning to find uses in construction as cladding, flashing, and guttering (Figure 16.17). Since titanium has a high melting point, it also works as a structural material.

Figure 16.17 Titanium, previously used mainly in the aircraft and aerospace industries, is finding increased use as a cladding material.

© David Woods/Shutterstock.com

NICKEL

Nickel (chemical symbol Ni) is a silver-colored metal mainly used as an alloying element. It provides increased resistance to atmospheric and chemical corrosion and increases an alloy's strength.

History

Nickel in various alloyed forms was used by prehistoric man because of its presence in the meteoric iron used to produce tools. Other metallic items from China and Asia Minor have been found to contain nickel, as well as copper and other alloying elements. In the mid-1700s, a crude nickel was first isolated from an ore in England, and by the early 1800s, refined nickel was being mass produced. Canada and New Caledonia are the largest producers of nickel-bearing ore.

Properties

Nickel is resistant to strong alkalis and many acids. It has good resistance to corrosion and oxidation and is strong and tough. It has a melting point of 2,651°F (1,455°C),

a coefficient of thermal expansion of 0.000013/°C, and is magnetic up to 680°F (360°C).

Working Characteristics

Nickel can be fabricated using most of the commonly used processes, such as hot- and cold-rolling, extruding, bending, forging, and spinning. It can be soldered, brazed, joined by some welding processes, or joined with mechanical fasteners.

Production of Nickel

Several processes are used to produce nickel from ore. The most recent is the Hybinette process developed in Canada. Ore is processed using the Bessemer process and the molten metal goes to a cooling chamber. It is then crushed, ground, and magnetically separated. The result is a nickel-copper platinum alloy that is treated by an electrolysis process to separate the nickel, copper, and platinum.

Nickel Alloys

A major use for nickel is as an alloying element. The alloying of nickel to other metals provides increased ductility, corrosion resistance, strength, hardness, and toughness. Nickel alloyed to nonferrous metals improves electrical resistance and magnetism and helps control expansion. An examination of steel alloys in Chapter 15 demonstrates that nickel is a commonly used alloying element. It is also widely used in Monel® metals and aluminum alloys. Monel alloy is about 66 percent nickel and 34 percent copper. Inconel 600® is a special nickel alloy containing about 75 percent nickel, 15 percent chromium, and 7 percent iron. Since it has excellent oxidation resistance, it is used in the food processing and chemical industries. Other nickel alloys include those used for electrical resistance coils, magnetic and nonmagnetic alloys containing iron, alloys used in the production of glass that are designed to have a high coefficient of thermal expansion, and copper-nickel alloys used for products exposed to marine conditions.

UNS Designations

Nickel and nickel alloys are designated in the UNS by the numbers N02001 through N99999. For example, UNS N02250 is a commercially pure nickel alloy having 99.0 percent minimum nickel.

Uses of Nickel

In addition to its use as an alloying element in ferrous and nonferrous metals, nickel is also an excellent material for electroplating and electroless plating. Electroplating is the process of depositing a coating of metal on another metal by electrolysis. Electrolysis is a method of plating a material by chemical means in which the piece to be plated is immersed in a reducing agent that, when catalyzed by certain materials, changes metal ions to metal, forming a deposit on the surface of the piece. Nickel is used in electric heating elements, lamp filaments, plumbing fittings, and hardware.

ZINC

Zinc (chemical symbol Zn) is a bluish-white metal that is brittle and has low strength. It is often referred to as a white metal and is widely used as a protective coating over steel to prevent corrosion.

Properties

Zinc has low strength and is brittle. Although it can be damaged by alkalis and acids, it resists corrosion by water and forms a protective oxide when exposed to air. Zinc has a melting point of 787°F (419°C). It is also subject to creep, a permanent dimensional deformation over time. Its tensile strength can be greatly increased by cold-working and alloying.

Working Characteristics

Since zinc is a soft material, it can be hot- and cold-rolled, drawn, extruded, cast, and machined. It can be joined by welding, soldering, and various mechanical fasteners.

Production of Zinc

Zinc is extracted from zincblende (sphalerite) ore. The mined ore is crushed and ground, and the ore particles are separated from the rock. Elements present in the ore, such as copper, lead, and iron sulfides, are separated by a flotation process. This involves dividing the elements in the finely grained ore by floating them on a liquid. The floating capacity of the elements varies. Lead and copper sulfates will float off the top of the liquid. Added chemicals cause the zinc sulfide to float, enabling it to be taken off. This concentrated zinc material is dried and ready to be refined into the metal zinc.

The electrolytic process involves roasting the zinc concentrate and removing soluble parts with a weak sulfuric acid solution. The solution is filtered to remove some of the other metals. Finally, this solution is moved to electrolytic tanks where cathodes of pure aluminum and anodes of lead or lead-silver are lowered into the tank and an electric current is passed between them through the solution. Pure zinc is attracted to and plates the cathodes. The layer is removed, melted, and poured in slabs for processing into various applications.

The vertical furnace method involves mixing coking coal briquettes and dried zinc concentrate in the top of a vertical furnace. They are then heated until the zinc concentrate vaporizes. The vapor is removed and condensed at the temperature at which the vapor turns into a solid. Solid zinc slabs are then ready for use in producing various products.

Zinc Alloys

Zinc alloys used for die casting consist of about 95 percent zinc and 4 percent aluminum and magnesium. Some copper may be present.

UNS Designations

Zinc and zinc alloys are designated in the UNS by the numbers Z00001 through Z99999. For example, Z13001 is identified by the name "zinc metal" and contains 99.90 percent zinc at minimum.

Uses of Zinc

Zinc's major use is in forming a protective coating over steel to prevent rust. This is referred to as galvanizing. *Galvanizing* involves placing the steel to be coated into a bath of molten zinc, which bonds to its surface (Figure 16.18). It is important that the coating be free of imperfections, such as pinholes, that permit moisture to reach the steel and cause it to rust. Both galvanized sheet and strip material are available.

Since zinc and zinc alloys have low melting temperatures, they are easy to cast and are used for some types of hardware and plumbing items. They are usually die cast and finished by polishing or plating with chromium, brass, or other materials.

Figure 16.18 Steel beams are immersed in a bath of molten zinc that provides a protective coating.

© the palms/Shutterstock.com

Zinc also finds use as an alloying element in brasses. Various zinc compounds serve in the production of paper, plastics, ceramics, rubber, abrasives, paint, and other products. Zinc is also used for specialized products in which corrosion resistance is important, such as anchors, flashing, screws, nails, expansion joints, and corner beads. Solid zinc strip material is used to produce a wide range of products, such as low-voltage buss bars, cavity wall ties, electric cable binders, electric motor covers, grading screens, and roofing and fascia material.

Zinc is high on the galvanic table of electrolytic potentials. This means it can be used to coat a material lower on the table to protect that material from galvanic action. The zinc is sacrificed, thus protecting the coated metal.

Zinc Galvanizing Processes

Zinc sheets, strips, coils, and wire are galvanized in a continuous process. The processes in use include electro-galvanizing, hot-dip galvanizing, metallic spraying, and sherardizing.

Electro-galvanizing involves placing cleaned steel or iron in an electrolyte solution of zinc sulfate. The electrolytic action deposits a layer of zinc on the material. The thickness of the coating can be controlled, but it is limited. Typical thicknesses range from 0.0001 to 0.0005 in. (0.0025 to 0.0127 mm). Hot-dip galvanizing involves immersing clean steel in a bath of molten zinc. It is a semiautomatic process. Metallic spraying involves coating sheet iron or steel by applying a fine spray of molten zinc. It can be applied after an installation is complete, thus also coating bolts, rivets, and welds. *Sherardizing* is a process in which cleaned iron or steel is placed in a container filled with zinc dust. The temperature in the container is raised and the objects to be coated are tumbled in the dust. The heated zinc bonds to the metal, forming a thin coating. This is usually used for small parts and provides minimum protection.

Review Questions

1. How does the specific gravity of aluminum compare with that of steel?

2. How does aluminum compare with copper as a conductor of electricity?

3. How is alumina produced from processed aluminum hydrate?

4. What is the purity of newly produced aluminum?

5. What do the aluminum temper designations F, O, and H mean?

6. What does solution heat-treating mean?

7. When galvanic corrosion occurs, what forms the electrolyte?

8. How can galvanic corrosion be reduced or eliminated?

9. What types of finishes are used on aluminum products?

10. What is a natural, controlled finish?

11. What is an anodized finish?

12. What are the types of anodic coatings used in industrial and architectural applications?

13. How are aluminum surfaces painted in a factory?

14. What is a porcelain enamel coating?

15. What is laminated finish?

16. How can aluminum members be joined?

17. What are the major properties of wrought copper alloys?

18. What are the two most widely used copper alloys?

19. What is copper patina?

20. What type of ore contains lead sulfide?

21. What are the major properties of lead?

22. What are the major properties of tin?

23. What are the major properties of zinc?

Key Terms

Alumina

Anodize

Bauxite

Cast Alloy

Galvanic Corrosion

Galvanize

Heat-Treatable Alloy

Heat-Treating

Non-Heat-Treatable Alloy

Nonferrous Metals

Oxide Layer

Reduction

Sherardizing

Strain Hardening

Temper Designation

Wrought Alloy

Activities

1. Collect samples of each of the commonly used nonferrous metals. Identify and list the properties of each.

2. How many uses of nonferrous metals can you find around your campus? Make a list of locations and types of material.

3. Test samples of a variety of nonferrous metals for tensile strength and hardness. Compare your findings with those of others who made the same tests.

4. Cite the location and name of a product made from nonferrous metals that were extruded, drawn, rolled sheet, and cast.

Additional Resources

Aluminum Design Manual, and many other technical publications related to aluminum, the Aluminum Association, Arlington, VA.

Metals and Alloys in the Unified Numbering System, and other related publications, American Society of Automotive Engineers, Inc., Warrendale, PA, and the American Society for Testing and Materials, West Conshohocken, PA.

Other resources include:

Numerous publications related to metals, American Society for Testing and Materials, West Conshohocken, PA.

Publications of the Copper Development Association, New York.

See Appendix C for addresses of professional and trade organizations and other sources of technical information.

Steel Frame Construction

LEARNING OBJECTIVES

Upon completion of this chapter, the student should be able to:

- Become acquainted with drawings prepared for structural steel buildings.

- Describe the procedure for fabricating and erecting the structural steel frame of a building.

- Be familiar with the fastening techniques used to join structural steel members.

- Describe fire-protection procedures for structural steel members required by building codes.

- Discuss steel framing systems using manufactured components.

Build Your Knowledge

For further study on these materials and methods, please refer to:

Chapter 15 Ferrous Metals

   Topics: Steel and Steel Alloys, Steel Products

Chapter 20 Wood and Metal Light Frame Construction

   Topic: Steel Framing

Chapter 33 Interior Walls, Partitions, and Ceilings

Steel-framed buildings utilize a skeleton frame construction in which the walls, floors, and roof are supported by a structural framework of steel beams, columns, girders, and related structural elements (Figure 17.1). Steel frames are strong, lightweight, and durable structures. Some designs rely partly on the skeleton frame for support while utilizing other systems, such as wall-bearing construction, where walls (masonry, panelized, etc.) form part of the overall structural system. The partition walls in a skeleton-framed building are non-load-bearing, permitting greater freedom in the layout of interior spaces. Frame members are manufactured to design sizes, transported to the construction site, and erected.

The design of a structural frame requires extensive engineering analysis. Design considerations include live and

Figure 17.1 Steel-framed buildings utilize a skeleton frame of walls, floors, and roof, supported by a structural framework of steel beams, columns, and girders.

© Stephen Finn/Shutterstock.com

dead loads, as well as lateral loads caused by earthquakes, wind, rain, and snow. Soil and hydrostatic pressures act horizontally below grade and must also be considered. The stresses to which steel is subjected while in use must be considered as the design process continues. Provisions also must be made for temporary stresses that occur during construction, such as those resulting from supporting a crane. Allowable working stresses are regulated by building codes. Steel frame construction is conducted according to the *Code of Standard Practice for Structural Steel Buildings and Bridges* (AISC 303-10). The code provides a framework for a common understanding of the acceptable standards when building in structural steel. The existing trade practices the code outlines are considered the standard custom and usage of the industry and are, therefore, often incorporated into the specifications and contract.

The economic use of structural steel members is another design consideration. The spacing of columns influences the span of beams and girders. Spans that are too short or too long result in the uneconomical use of framing members. The design of connections and the securing methods for rivets, bolts, or welding are other crucial design considerations.

The procedure for the design and construction of structural steel-framed buildings includes the engineering design of the structure, the preparation of shop drawings, the manufacture of structural members, and the erection of these structural members on site.

STRUCTURAL STEEL DRAWINGS

Several types of drawings are required for steel frames, including engineering design drawings, shop drawings, and erection plans. *Design drawings* are prepared by a structural engineer. They indicate the type of construction and give data on shears, loads, moments, and axial forces that must be resisted by each member and all connections. Figure 17.2 is a partial drawing that shows the elevation of a beam as (75' 0") above an established site datum, such as the finished floor. The sizes of the beams are given, and the forces they must resist are indicated in k (kips). Notes are used to provide additional information that allows a structural detailer to prepare shop drawings for the various members.

Shop drawings contain the necessary information for the fabrication of each member. They specify the size of the member and the exact location of holes for connections. Clearances must be allowed for so erection can proceed without interference between joining members. A typical example of a shop drawing is shown in Figure 17.3, which gives specifications and the mark (B1) identifying the beam. Shop drawings are submitted by the manufacturer or fabricator of the assembly, and must be approved by both the architect and contractor

Figure 17.2 A partial engineering design drawing contains the data needed by the steel fabricator.

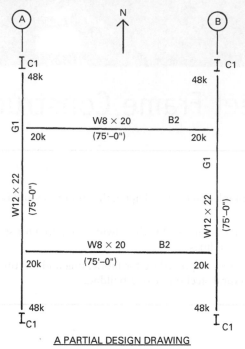

A PARTIAL DESIGN DRAWING

GENERAL NOTES
SPECIFICATIONS: LATEST AISC EDITION
MATERIAL: ASTM A36
FASTENERS: 3/40 A325 IN BEARING TYPE CONNECTIONS
THREADS IN SHEAR PLANES
CONNECTIONS: PER AISC MANUAL AND DEVELOP
INDICATED END REACTIONS

for conformance with the contract documents and field conditions.

Erection plans are assembly drawings used on site to direct the placement of each member in its desired location. An example is shown in Figure 17.4. Each piece, or any subassembly of pieces, and its assigned shipping and assembly mark is noted. The erection plan also includes details showing the anchor bolt locations in the foundation. Anchor bolts are used to secure columns to the foundation and are placed in the concrete as the foundation is poured, so their location is critical. Engineering design drawings are similar to erection drawings but have more design detail and, in some cases, can be used as erection drawings. Typically, they are small and crowded with details, so larger-scale erection drawings are usually provided with only the information needed to identify and locate each member.

Each steel member is identified by an erection or piece mark, usually a letter and number giving the part a unique identification. For example, "B2" means beam number 2, "C2" means column number 2, and "G2" means girder number two. All identical members use the

Construction **Methods**

Fasteners for Metal-Building Construction

In metal-building construction, two commonly used fasteners are designed to join metal-to-metal and metal-to-wood connections for sheet metal panels. Metal-to-metal fasteners include drilling fasteners and self-tapping fasteners. The drilling fasteners, shown in Figure A, have a tip that drills the correct-sized hole through the sheet metal followed by threads that tie the pieces together. The fastener must have a high surface hardness and a head that withstands the use of a power-impact driver. Drilling fasteners are available in carbon steel and several classifications of stainless steel and in many sizes and lengths. Manufacturers should be consulted for this information.

The self-tapping fasteners, shown in Figure B, are installed in holes that are drilled or punched in the sheet metal. The size of the hole must correspond to those specified by the manufacturer for each of the screws they produce. Self-tapping fasteners are

available in carbon steel and stainless steel. Consult a manufacturer for available sizes.

Typical sheet-metal-to-wood fasteners, available in carbon steel, are shown in Figure C. They are power driven through predrilled holes in the metal into the wood. They are available with protective coatings, such as a polymer film and zinc-and-chromate finish. Manufacturers will paint heads a desired color. These fasteners come in lengths up to 2 in. (50.8 mm).

An examination of head types shows that a hexagon head is used for fasteners driven with socket-type power impact tools. Smaller types have slotted heads and also are power driven. A variety of washer faces are available. Some have metal washers that provide metal-to-metal contact. Others have a water-sealing gasket that is typically polyvinyl chloride (PVC), which is a plastic, and ethylene-propylenediene (EPDM), which is a synthetic rubber. Other materials are used also.

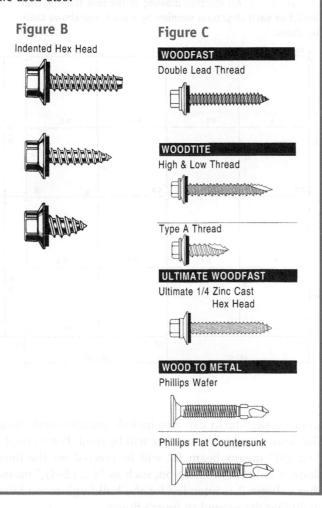

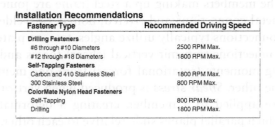

Figure 17.3 A typical shop drawing of a beam that has connectors shop welded that will be bolted on the site.

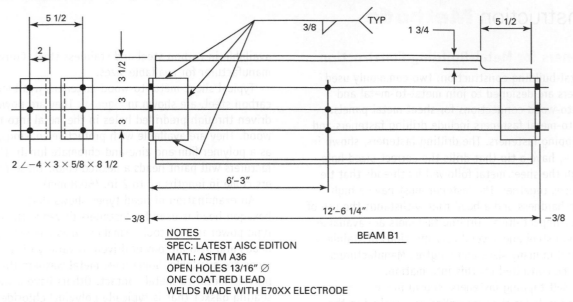

NOTES
SPEC: LATEST AISC EDITION
MATL: ASTM A36
OPEN HOLES 13/16" ⌀
ONE COAT RED LEAD
WELDS MADE WITH E70XX ELECTRODE

BEAM B1

Figure 17.4 An erection drawing of the roof framing that identifies each structural member by a mark and shows their locations.

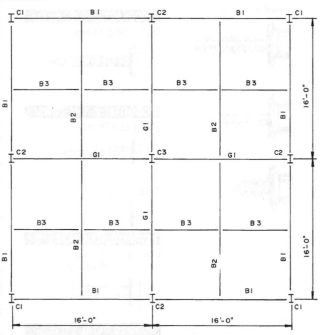

same mark. Marks can also include numbers indicating the floor on which a member will be used. For example, "B2 (3)" means beam B2 will be erected on the third floor. A column designation, such as "C2 (2–4)," means that column 2 is intended for the building's second tier, including the second to fourth floors.

Structural steel members are delivered on site with a primer coat and numbered as shown on the erection plan. Connectors are also primed and numbered. Steel members that will be encased in concrete are usually not primed. This allows better bonding between the concrete and the member. After the steel is erected, some field painting is required, for example, in areas where welding occurred. Welds are chipped free of any formed slag before they are painted. The selection of a paint system depends on the service life of the structure, the potential severity of environmental exposure, and the cost of the initial application and future maintenance.

THE STEEL FRAME

The members making up a steel frame are joined with metal connections and riveted, bolted, or welded. The connections typically utilize angles, tees, or plates. The connections transmit vertical (shear) forces and bending moments (rotational forces) from one member to the other. *Shear stress* is produced when vertical loads are applied to a member, creating a deformation in which parallel planes slide relative to each other. A *moment* is the property by which a force tends to cause a body to which it is applied to rotate about a point or line, as shown in Figure 17.5. *Bending moment* is the moment that produces bending at a section of a structural member. The various connections illustrated in the next section of this chapter are identified as shear

Figure 17.5 Structural members and connections are subject to shear forces and bending moments.

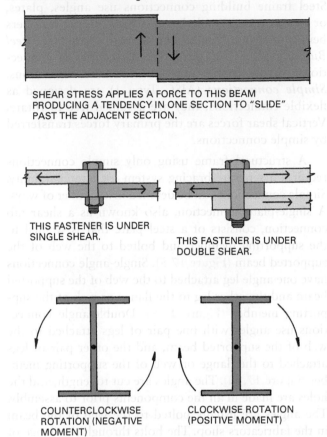

SHEAR STRESS APPLIES A FORCE TO THIS BEAM PRODUCING A TENDENCY IN ONE SECTION TO "SLIDE" PAST THE ADJACENT SECTION.

THIS FASTENER IS UNDER SINGLE SHEAR.

THIS FASTENER IS UNDER DOUBLE SHEAR.

COUNTERCLOCKWISE ROTATION (NEGATIVE MOMENT)

CLOCKWISE ROTATION (POSITIVE MOMENT)

connections, moment connections, or connections controlling both.

Frame Stability

A building using structural steel framing (or concrete or wood) must be designed to resist wind, live load, and earthquake forces acting upon it. A number of strategies exist to provide stability to frame structures (Figure 17.6). The rigid frame utilizes moment connections between members to resist lateral forces. These connections prohibit adjacent members from changing the geometry between them. Braced frames use diagonal members to provide triangulation that braces the framing from deformations. Used often in tall buildings, a rigid core acts as a vertical beam that is stabilized with a deep foundation. The core is frequently built of reinforced concrete and houses stairs, elevators, and service chases. Flat shear walls of concrete or steel, stiffened in both directions, can be used to stabilize a structural frame. For very tall, slender buildings, hollow tubes provide maximum efficiency in resisting wind forces. The bundled tube concept combines the inherent strength of individual tubes in a bundled configuration. Mega frames describe structures in which large members constitute the overriding lateral support of a building.

Figure 17.6 Methods for stabilizing structural steel frames.

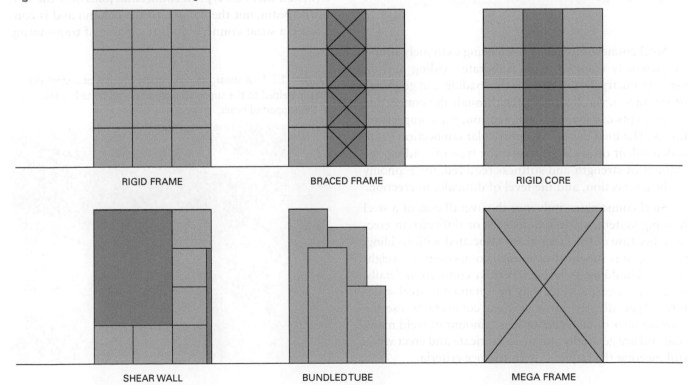

RIGID FRAME

BRACED FRAME

RIGID CORE

SHEAR WALL

BUNDLED TUBE

MEGA FRAME

STEEL FRAME CONNECTIONS

A connection detail entails the assembly of one or more joints that transmit forces between two or more members or connection elements. There are four basic connectors used in making structural steel connections: bolts, welds, pins, and rivets. In most cases, a combination of all of them is used. Bolts and welds are the most common connectors used in steel frame construction (Figure 17.7). Pins are used for connections at the ends of various support members that require freedom of movement to accommodate expansion and contraction of the structural frame.

Figure 17.7 Steel frame connections are engineered to transmit forces between two or more members or connection elements.

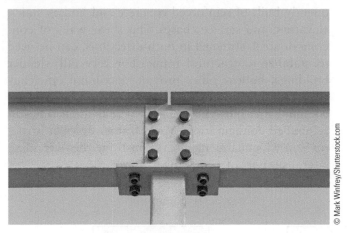

Steel connections range from being extremely simple to extremely complex with elaborate loading and diverse geometry. Depending on the loading and geometry of the union, the load transfer through the connecting components can involve shear, tension, and compression forces. The final design of a particular connection detail is dependent on several factors: the type of loading, the amount of strength and stiffness required, the economy of the connection, and the level of difficulty in erection.

Steel connections influence the overall cost of a steel framing system, as well as its ease or difficulty in erection. Because of the labor costs associated with welding, weld metal is the most expensive component by weight in a steel building. By using repetitive connection details, a significant cost savings may be obtained in steel structures. Typically, the most efficient connections use the least number of fasteners or least amount of weld material, and are generally simple to fabricate and erect while still meeting the required performance criteria.

Simple Connections

Steel frame building connections use angles, plates, or tees as transitional elements between the members being connected. The *Specification for Structural Steel Buildings* (AISC 2005) defines two types of connections: simple connections and moment connections. *Simple connections* (AISC Type 1) are designed as flexible connections; they are assumed free to rotate. Vertical shear forces are the primary forces transferred by simple connections.

A structural frame using only simple connections requires a separate bracing system for lateral stability. Simple connections can be detailed in a number of ways. A single-plate connection, also known as a shear tab connection, consists of a steel plate that is welded to the supporting member and bolted to the web of the supported beam (Figure 17.8). Single-angle connections have one angle leg attached to the web of the supported beam and the other leg to the flange or web of the supporting member (Figure 17.9). Double-angle connections use angles with one pair of legs attached to the web of the supported beam, and the other pair of legs attached to the flange or web of the supporting member (Figure 17.10). The angles are cut to length, and the holes are made in all the components prior to assembly. The angles are usually bolted to the web of the beam in the fabricators shop. The bolts through the flange of the column are added as the beam is erected on the construction site. This type of connection joins only the web of the beam, not the flanges, to the column and is considered a shear connection. It is capable of transmitting

Figure 17.8 A shear tab connection consists of a steel plate that is welded to the supporting member and bolted to the web of the supported beam.

Figure 17.9 Typical welded beam to column framed, seated and end plate connection details.

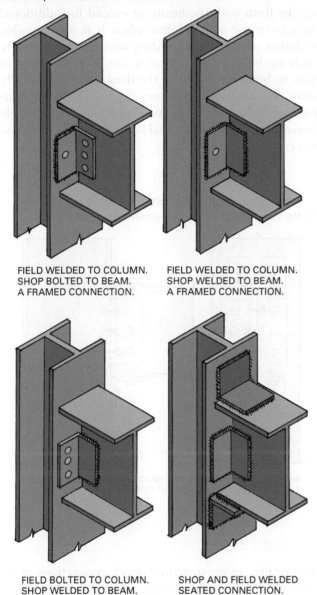

FIELD WELDED TO COLUMN.
SHOP BOLTED TO BEAM.
A FRAMED CONNECTION.

FIELD WELDED TO COLUMN.
SHOP WELDED TO BEAM.
A FRAMED CONNECTION.

FIELD BOLTED TO COLUMN.
SHOP WELDED TO BEAM.
A FRAMED CONNECTION.

SHOP AND FIELD WELDED
SEATED CONNECTION.

vertical, or shear, forces from the beam to the column. However, because it does not connect the beam flanges to the column, it is of no value in transmitting bending forces from one to the other. Shear end plate connections use a plate that is welded perpendicular to the end of the supported web and attached to the supporting member. A *seated connection* uses an angle that is mounted with one leg vertical against the supporting column, while the other leg provides a seat upon which the beam is mounted. A tee connection uses the stem of a structural T section connected to the supported member and the flange attached to the supporting member.

Figure 17.10 A simple bolted beam-to-column connection.

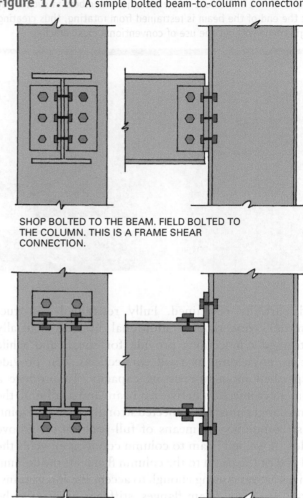

SHOP BOLTED TO THE BEAM. FIELD BOLTED TO
THE COLUMN. THIS IS A FRAME SHEAR
CONNECTION.

SHOP BOLTED TO THE BEAM. FIELD BOLTED TO
THE COLUMN. THIS IS A SEATED SHEAR
CONNECTION.

Moment Connections

Simple connections provide for shear only and require additional support to stabilize the structural frame, such as diagonal bracing or shear panels. Rigid or moment connections provide for shear and have sufficient rigidity to hold the intended geometry between connected members. A moment connection (AISC Type 2) in structural steel design is a connection joint between a beam and a column where the end of the beam is restrained from rotating, thus creating a rigid frame without the use of conventional cross bracing (Figure 17.11). Moment connections have flange stresses developed independently of the shear connections. This force transfer does not preclude the need for shear connections used to support the beam.

The *Specification for Structural Steel Buildings* defines two types of moment connections: fully restrained

Figure 17.11 Moment connections join members in such a way that the end of the beam is restrained from rotating, thus creating a rigid frame without the use of conventional cross-bracing.

and partially restrained. Fully restrained construction does not require additional bracing. Partially restrained connections provide for shear, and while not as unyielding as rigid connections, can provide a specified moment-resisting capacity. To produce a moment connection between a beam and a column, the beam flanges must be connected strongly across the joint, most commonly by means of full-penetration groove welds. A welded beam to column connection welds the full end of the beam to the column flange. If the column flanges are not strong enough to accept the forces transmitted from the beam flanges, stiffener plates may be installed inside the flanges of the column to help distribute forces into the body of the column (Figure 17.12).

Figure 17.12 Stiffener plates are installed inside the flanges of a beam to help distribute forces into the body of the column.

A bolted end plate connection uses a plate, shop welded to the beam end that is bolted to the column. The plate may be flush with the beam, or extend for additional strength or use as a column splice. It is also possible to design moment transmitting connections that rely solely on bolting. Metal angles can be used bolted in pairs to both the web and the flange, creating a fully restrained connection (Figure 17.13). Moment connections are considerably more expensive than simple connections due to additional labor and higher fabrication costs.

Figure 17.13 Metal angles bolted in pairs to both the web and the flange result in a fully restrained connection.

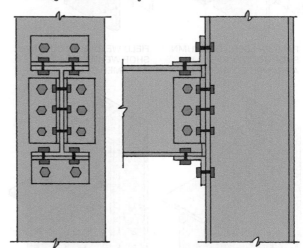

SHOP BOLTED TO THE BEAM. FIELD BOLTED TO THE COLUMN. THIS IS A MOMEMT–RESISTING CONNECTION.

Riveted Connections

Historically, most structural steel connections were either welded or riveted. Today, high-strength bolts have completely replaced structural steel rivets. A rivet is a mechanical fastener used to join steel members. Rivets consist of a cylindrical shaft with a head on one end. Rivets are installed in holes that are drilled or punched in framing members. They are heated to a white heat, inserted into the holes of the members to be joined, and a head is formed by a pneumatic hammer on the opposite side. When the rivet cools, it shrinks, shortens in length, and pulls the members together (Figure 17.14). Rivets are still used for some shop made connections. They are made from high-strength carbon and alloy steels according to ASTM dimensional and chemical specifications.

Figure 17.14 Structural steel connections were historically joined using rivets.

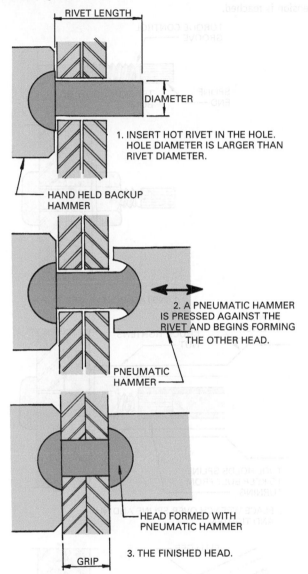

1. INSERT HOT RIVET IN THE HOLE. HOLE DIAMETER IS LARGER THAN RIVET DIAMETER.

HAND HELD BACKUP HAMMER

2. A PNEUMATIC HAMMER IS PRESSED AGAINST THE RIVET AND BEGINS FORMING THE OTHER HEAD.

PNEUMATIC HAMMER

HEAD FORMED WITH PNEUMATIC HAMMER

3. THE FINISHED HEAD.

Figure 17.15 High-strength steel bolts have the ASTM identification markings on the heads and nuts.

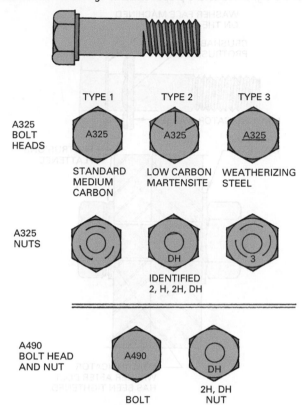

Bolted Connections

Bolts used to join steel frame members are either carbon steel, as specified by ASTM A449, ASTM A325, or A490 high-strength bolts. ASTM A449 bolts are used in bearing connections in which their lower strength is adequate. ASTM A325 and A490 high-strength bolts are used for friction connections. They have higher shear strength than do carbon A449 bolts and high tensile strength. Bolts are identified by head and nut markings (Figure 17.15).

Bolts used in friction connections must be tightened to a minimum of 70 percent of their ultimate tensile strength.

Bolts are tightened with calibrated torque wrenches, which develop the specified tension values in the connector. They have an automatic cut-off and a gauge that registers the tensile stress developed. Another way to determine tension is the turn-of-the-nut method. Here, the nut is turned until tight and then turned an additional, specified amount, such as one-third to one full turn. Tightening can also be accomplished with a direct tension indicator. *Load indicator washers*, metal washers with rounded protrusions on one face, are placed under the bolt head. As the nut tightens, they flatten (Figure 17.16). The degree of flattening is measured with a feeler gauge and correlated with the amount of tension developed. Still another method uses tension-control bolts (Figure 17.17). These bolts have a splined end that is inserted in a power wrench that grips both the nut and the splined end. When the specified torque is reached, the splined end twists off at the torque control groove.

Plain smooth hardened washers may be used to distribute the load of a bolt or nut over a larger area. The hardened washer is used to prevent galling, which can render the tension calibration ineffective. *Galling* is the wearing or abrading of one material against another

Figure 17.16 Load-indicator washers are used to verify that bolts are tightened to the proper tension.

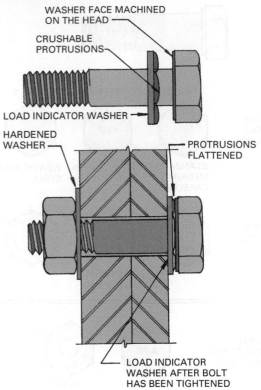

Figure 17.17 Connections can be tensioned using tension-control bolts, which have a spline that shears when the proper tension is reached.

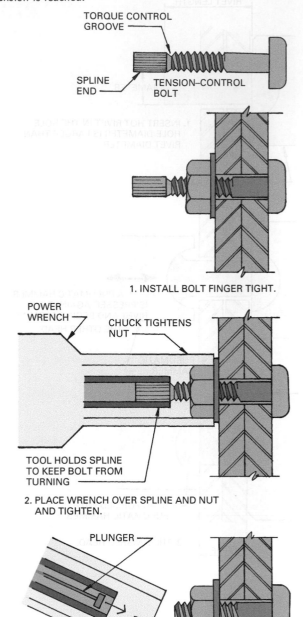

under extreme pressure. Boltholes are usually $1/16$ in. larger in diameter than the bolt, but some field conditions may require a larger allowance. The dimension of oversize holes must be approved by a structural engineer.

The amount of threads that fall within the inside of the hole in the connection influences the strength of the connection. High-strength bolts have shorter threads than other types, resulting in minimum encroachment of threads into the grip. However, the bolt length must also provide sufficient protrusion beyond the connection to hold a washer and the nut. To achieve higher strengths in bearing connections, the number of threads that fall within the grip must be less than the thickness of the outside metal piece. A typical beam-to-beam connection is shown in Figure 17.18. Notice the top of the beam butting the girder has a clearance cut to facilitate flushness. The notch (called a cope, block, or cut) allows for clearance around the flange.

Welded Connections

Welding joins members as if they were monolithic. Welded connections are stronger than the members they join and are able to transmit both shear and moment forces. The load capacity of a welded connection depends on

the connector, type of weld, base material, and strength of the electrode material. AISC provides tables indicating the sizes and capacities of welded connections based on the use of certain electrodes.

Figure 17.18 A typical beam-to-beam connection using a clearance cut to facilitate clearance around the flange.

Figure 17.19 Connecting plates can be riveted or bolted to a member in the fabrication shop and welded to a connecting member on site.

© Dwight Smith/Shutterstock.com

Welding entails heating metal on two adjacent plates and then adding additional molten metal between them to form a molten pool. When cooled and solid, the pool acts like a glue and a strong bond is made. Welded beam connections consist of the same general types used for bolted and riveted connections. Connecting plates may be riveted or bolted to a member in the fabrication shop and welded to a connecting member on site (Figure 17.19), or welded in the shop and bolted in the field. When field connections are welded, connecting bolts or clamps are used to hold the members in correct alignment. Welds in a fabrication shop are made by clamping a connector to a member and welding it in place. Figure 17.20 shows a fabrication drawing for a shop-welded, field-bolted connection.

Other common welds include single U and its dif-

Welding symbols are used on engineering drawings to indicate the size and type of weld (Figure 17.21). The symbols used in welding design are specified in international standards, such as ISO 2553 and ISO 4063. American standard symbols are outlined by the American National Standards Institute and the American Welding Society and are noted as ANSI/AWS. In engineering drawings, each weld is identified by an arrow that points to location of the weld. The arrow is annotated with letters, numbers and symbols to indicate the exact specification of the weld.

Although there are several types of welding processes, electric arc welding is generally used for structural steel connections. Electric arc welding involves passing an

Figure 17.20 A detailed drawing of a beam with connections shop-welded to the beam and prepared to be field-bolted to a column or girder.

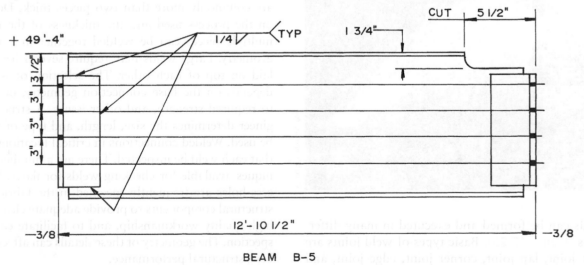

BEAM B-5

Figure 17.21 Standard symbols for welded connections.

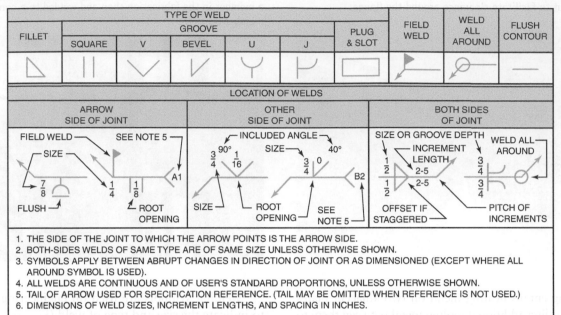

1. THE SIDE OF THE JOINT TO WHICH THE ARROW POINTS IS THE ARROW SIDE.
2. BOTH-SIDES WELDS OF SAME TYPE ARE OF SAME SIZE UNLESS OTHERWISE SHOWN.
3. SYMBOLS APPLY BETWEEN ABRUPT CHANGES IN DIRECTION OF JOINT OR AS DIMENSIONED (EXCEPT WHERE ALL AROUND SYMBOL IS USED).
4. ALL WELDS ARE CONTINUOUS AND OF USER'S STANDARD PROPORTIONS, UNLESS OTHERWISE SHOWN.
5. TAIL OF ARROW USED FOR SPECIFICATION REFERENCE. (TAIL MAY BE OMITTED WHEN REFERENCE IS NOT USED.)
6. DIMENSIONS OF WELD SIZES, INCREMENT LENGTHS, AND SPACING IN INCHES.

electric current through a metal electrode, known as the welding rod, to the piece to be welded (Figure 17.22). The electrode is kept slightly above the metal, which causes an arc to jump the gap. This produces considerable heat, which melts the electrode and a small area of the members being welded. As the electrode moves along the line of the weld, it leaves a continuous bead of metal that fuses the members together as it cools.

Figure 17.22 Electric arc welding involves passing an electric current through a metal electrode, known as the welding rod, to the piece to be welded.

© langdu/Shutterstock.com

Welds can be formed and executed in many different ways (Figure 17.23). Basic types of weld joints are the butt joint, lap joint, corner joint, edge joint, and T-joint. A common weld for lap, corner, and T-joints is the fillet weld. The weld made is made by running a weld bead in a tapered angle between two pieces of material, resulting in a triangular weld. A butt weld is made when two plates are end to end, with a gap between them, and the whole gap is filled with molten weld material, in one or more passes. Small rectangular backup bars are sometimes welded beneath the end of each beam flange to prevent the welding arc from burning through.

Other common welds include single-U and double-U joints. Instead of having straight edges like the single-V, they are curved, forming the shape of a U. Lap joints are commonly more than two pieces thick. Depending on the process used and the thickness of the material, multiple pieces can be welded together in a lap joint geometry. Large welds may require several beads to be laid on top of each other. The selection of weld type depends on the base connection geometry, in addition to required strength, and other issues. A structural engineer determines the size, length, and type of welds to be used. Welded connections in critical locations require that each weld be inspected. There are a number of techniques available for checking welds for flaws. Weld access holes are frequently specified in the fabrication of structural components to provide adequate clearance for high quality workmanship, and to facilitate ease of inspection. The geometry of these details can affect a member's structural performance.

Figure 17.23 Common types of full penetration welds for steel frame construction.

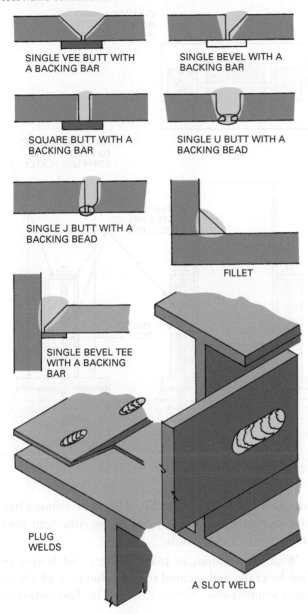

SINGLE VEE BUTT WITH A BACKING BAR

SINGLE BEVEL WITH A BACKING BAR

SQUARE BUTT WITH A BACKING BAR

SINGLE U BUTT WITH A BACKING BEAD

SINGLE J BUTT WITH A BACKING BEAD

FILLET

SINGLE BEVEL TEE WITH A BACKING BAR

PLUG WELDS

A SLOT WELD

THE ERECTION PROCESS

Steel erection is considered one of the most dangerous of all construction activities. In today's industry, upfront planning is necessary to ensure safety during steel erection. Site-specific erection planning is developed during pre-construction conferences attended by the erector, the general contractor, and the project engineer and fabricator. In this process, the sequence of erection activities is considered, including material deliveries, material staging and storage, and coordination with other trades and construction activities. Planning includes consideration

of the crane selection and placement procedures, including site preparation, paths for overhead loads, and other rigging supplies and equipment (Figure 17.24).

Figure 17.24 Finished members are delivered on site as near to their point of erection as possible in the order they will be needed.

© Richard Thornton/Shutterstock.com

Finished members are delivered on site in the order in which they are needed during erection. The erecting contractor is responsible for setting in place each member of the frame and securing it as required by the design drawings. As the steel arrives on site, members must be placed as near to their point of erection as possible in the order they will be needed.

After verifying the correct placement and elevation of the anchor bolts, an erection crew lifts the first level of columns with a ground-level crane and places them over the bolts cast in the foundation. Columns in multistory buildings generally bridge over two stories (Figure 17.25). The columns utilize steel base plates, welded during fabrication, that distribute the load on the column over a larger area of the foundation. There are several methods for setting columns. Smaller columns have a steel plate leveled over a bed of grout before erection (Figure 17.26). Larger columns have leveling nuts on the base plate that are adjusted to plumb the column. Then grout is worked below the base plate. A column with stiffener plates is shown in Figure 17.27. Very large plates required for large, heavy columns may have the base plate leveled and grouted before the column is welded to it. Large-diameter holes are sometimes specified near the plate's center to allow

Figure 17.25 Columns in multistory buildings generally span over two stories.

Figure 17.26 Typical base connections for steel columns.

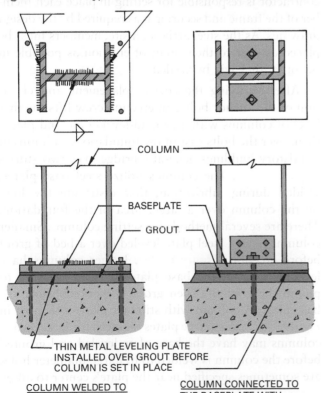

COLUMN

BASEPLATE

GROUT

THIN METAL LEVELING PLATE
INSTALLED OVER GROUT BEFORE
COLUMN IS SET IN PLACE

<u>COLUMN WELDED TO
THE BASEPLATE</u>

<u>COLUMN CONNECTED TO
THE BASEPLATE WITH
BOLTED ANGLES</u>

Figure 17.27 Stiffener angles can be welded to a column base for added strength.

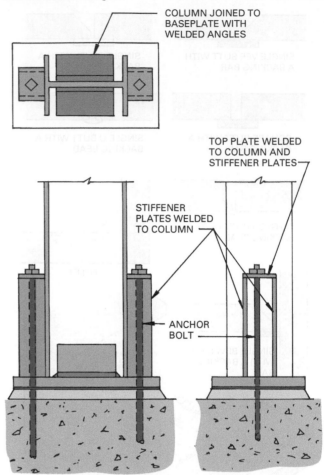

COLUMN JOINED TO
BASEPLATE WITH
WELDED ANGLES

TOP PLATE WELDED
TO COLUMN AND
STIFFENER PLATES

STIFFENER
PLATES WELDED
TO COLUMN

ANCHOR
BOLT

grout to be forced below it. For additional column base details, consult the publications of the American Institute of Steel Construction.

With the columns in place, girders and beams are lifted by crane and secured to the columns with the required connections (Figure 17.28). The first two-story

Figure 17.28 With the columns in place, girders and beams are lifted by crane and secured to the columns with the required connections.

tier of the structure is then straightened and plumbed with temporary diagonal cables and turnbuckles. With the structure in the correct alignment, all bolted connections are tightened, final welds completed, and any permanent diagonal bracing elements are installed. As the structure rises, temporary wood or steel decking is attached to the skeleton frame. The decking provides a working surface and protects workers below from potential falling debris.

Once the first level is complete, the second level of two-story columns is lifted with a crane and connected to the columns below with a splice plate (Figure 17.29). Column splices may be bolted or welded. A crane sets the second-level columns, beams, and girders. As the floors are framed, the floor decking is installed (Figure 17.30). Stringent safety regulations are observed that mandate the use of safety nets, railings, and the protection of openings.

Figure 17.29 Types of wide flange column connections.

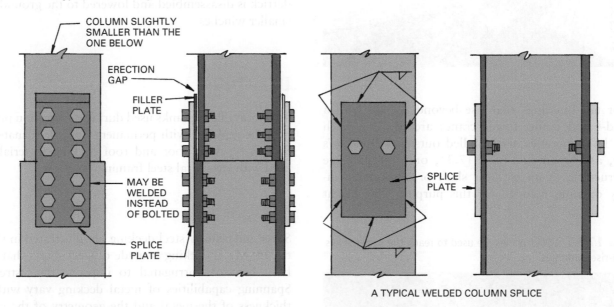

A TYPICAL WELDED COLUMN SPLICE

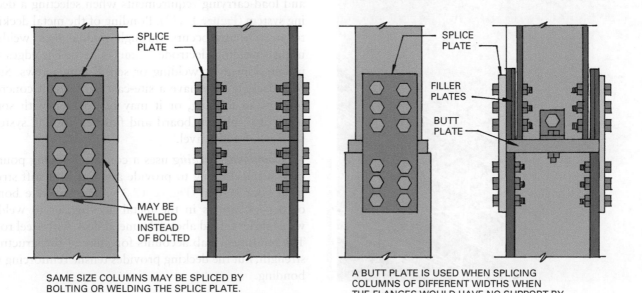

SAME SIZE COLUMNS MAY BE SPLICED BY BOLTING OR WELDING THE SPLICE PLATE.

A BUTT PLATE IS USED WHEN SPLICING COLUMNS OF DIFFERENT WIDTHS WHEN THE FLANGES WOULD HAVE NO SUPPORT BY THE COLUMN BELOW.

Figure 17.30 As the floors are framed, the floor decking is installed to provide a working surface and protect workers below from potential falling debris.

© Christian Delbert/Shutterstock.com

For tall buildings that rise beyond the reach of a ground-based crane, tower cranes are used to reach upper levels. Cranes are installed outside a building's frame, as shown in Figure 17.31, or erected inside the structure, in an elevator shaft or a special, temporary opening planned for this purpose. The height

Figure 17.31 Tower cranes are used to reach the upper levels of high-rise buildings.

© Labrador Photo Video/Shutterstock.com

of most cranes can be increased by adding additional sections.

Another lifting device for multistory construction is a guy derrick. It consists of a mast, a boom that pivots at the base of the mast, hoisting tackle, and supporting guy lines. The guy derrick can lift itself to floors constructed above it. The hoisting operation is controlled by a power winch located on grade. As the derrick rises to higher levels, the winch remains on the ground. A typical construction procedure is shown in Figure 17.32. Since the derrick is mounted on the structural frame, the loading must be considered as the design is developed. When the frame is completed, the derrick is disassembled and lowered to the ground with smaller winches.

DECKING

Temporary floor planks used during the erection process must be replaced with permanent floor deck materials. Several types of floor and roof decking materials are used with structural steel framing.

Metal Decking

Sheet and cellular steel decking are illustrated in Chapter 15. Metal decking is made of steel sheets that have been bent or corrugated to improve their strength. Spanning capabilities of metal decking vary with the thickness of the metal and the geometry of the corrugations and range from 6 ft. to 15 ft. (1.8 m to 4.5 m). Designers consider spans between supporting members and load-carrying requirements when selecting a decking system (Figure 17.33). Bonding of the metal decking to steel beams occurs through puddle tack welding using a welding electrode (Figure 17.34). The edges are usually joined by welding or self-drilling screws. Steel roof decking may have a site-cast lightweight concrete or gypsum topping, or it may be covered with some form of insulation board and finished roofing system, such as tar and gravel.

Composite decking uses a concrete topping poured onto metal decking to provide a strong and stiff structural floor system (Figure 17.35). The concrete bonds onto perforations in the metal decking, or to welded wire fabric welded above the metal deck with steel rods. The reinforced slab accounts for most of the structural strength, but the decking provides tensile reinforcing via bonding.

Figure 17.32 A guy derrick is positioned on the building's highest level and can be elevated as the height of the building increases.

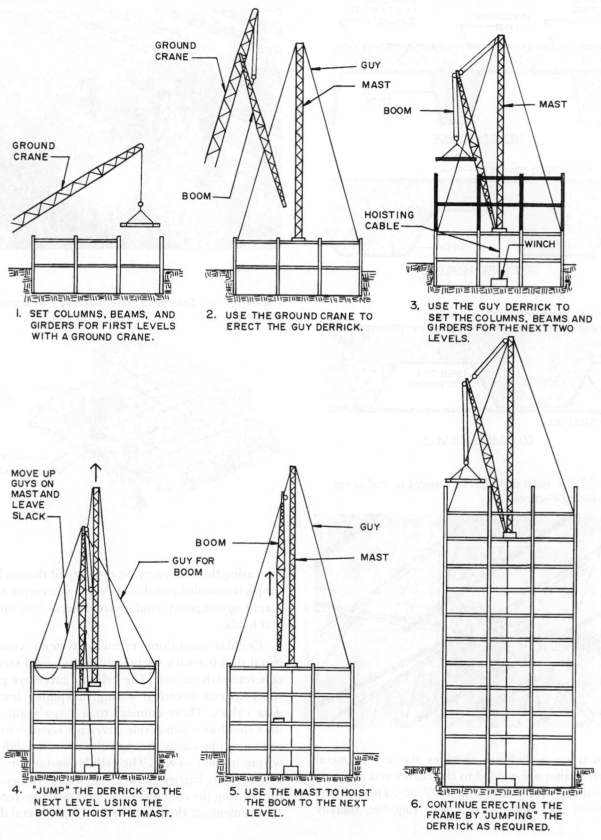

GROUND
CRANE

GROUND
CRANE

GUY

MAST

BOOM

1. SET COLUMNS, BEAMS, AND
GIRDERS FOR FIRST LEVELS
WITH A GROUND CRANE.

2. USE THE GROUND CRANE TO
ERECT THE GUY DERRICK.

BOOM

MAST

HOISTING
CABLE

WINCH

3. USE THE GUY DERRICK TO
SET THE COLUMNS, BEAMS AND
GIRDERS FOR THE NEXT TWO
LEVELS.

MOVE UP
GUYS ON
MAST AND
LEAVE
SLACK

GUY FOR
BOOM

4. "JUMP" THE DERRICK TO THE
NEXT LEVEL USING THE
BOOM TO HOIST THE MAST.

BOOM

GUY

MAST

5. USE THE MAST TO HOIST
THE BOOM TO THE NEXT
LEVEL.

6. CONTINUE ERECTING THE
FRAME BY "JUMPING" THE
DERRICK AS REQUIRED.

Figure 17.33 Typical types of metal decking.

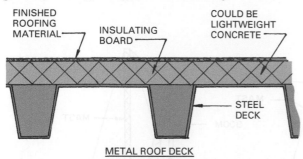

FINISHED
ROOFING
MATERIAL   INSULATING   COULD BE
            BOARD   LIGHTWEIGHT
                          CONCRETE

STEEL
DECK

METAL ROOF DECK

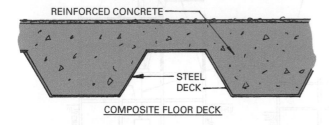

REINFORCED CONCRETE

STEEL
DECK

COMPOSITE FLOOR DECK

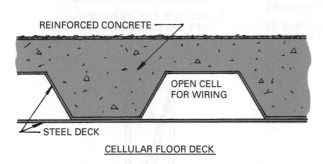

REINFORCED CONCRETE

OPEN CELL
FOR WIRING

STEEL DECK

CELLULAR FLOOR DECK

Figure 17.34 Metal decking is tack welded to steel beams using a welding electrode.

Copyright © Constructionphotographs.com

Sometimes a designer specifies the use of metal *shear studs* that are welded to the beams and protrude up into the concrete slab (Figure 17.36). These tie the steel beam and the concrete slab together, thereby

Figure 17.35 Composite decking serves as a form for a concrete floor or roof slab.

© Robert A. van het Hof/Shutterstock.com

Figure 17.36 Shear studs serve to tie the steel structure to the concrete slab.

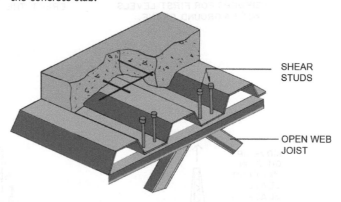

SHEAR
STUDS

OPEN WEB
JOIST

increasing the load-carrying capacity of the steel beam. Properly installed metal roof decking can serve as shear diaphragms against wind, seismic forces, and other lateral loads.

Cellular steel-floor raceway systems consist of metal decking with wiring raceways and a structural concrete slab on top. The cellular raceways provide space to run electrical wiring, telephone lines, and data cables. These connect to a large main header duct that has a removable cover for lay-in wiring. Inserts provide access through the concrete slab to the wiring in the raceway. The cellular system provides a fire-resistant barrier between floors, serves as a form for pouring the concrete deck, and provides tensile reinforcement for the concrete floor slab. Metal decking is generally plug-welded to the steel beams, girders, and joists.

Concrete Decking

Steel frame buildings may also use concrete floor and roof decks in lieu of metal decking. Cast-in-place concrete may be poured with the appropriate reinforcing over formwork erected in the steel frame. More common is the use of pre-cast concrete spanning elements to construct floor and roof assemblies. Detailed in Chapter 9, these include hollow-core units and pre-cast slabs and channels of various designs. Long-span pre-cast concrete roof and floor units are generally pre-stressed. Short-span and cast-in-place members use conventional reinforcement.

Pre-cast slabs and channels are supported on the structural steel framing to form decks and, in some cases, ceilings (Figure 17.37). They are secured to beams and joists by welding metal plates that are cast in them to the steel structure or by using clips, as shown in Figure 17.38.

Figure 17.37 Pre-cast flat slabs used to provide a structural deck.

© Stephen Rees/Shutterstock.com

Pre-cast hollow-core decking units, used for long spans, are placed on the steel framing with a crane. For better bonding with topping materials, their top surface is often left rough. The topping helps tie pre-cast units together and increases a roof's or floor's structural integrity. Reinforcing can be cast in the topping.

Figure 17.38 Pre-cast concrete decking is secured to steel framing with metal clips.

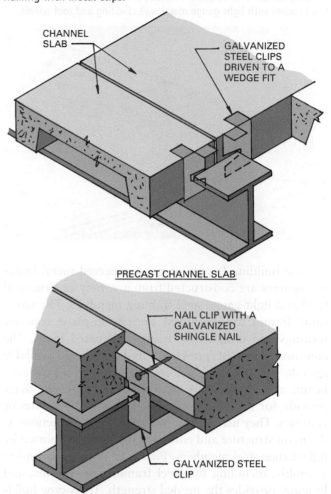

CHANNEL SLAB

GALVANIZED STEEL CLIPS DRIVEN TO A WEDGE FIT

PRECAST CHANNEL SLAB

NAIL CLIP WITH A GALVANIZED SHINGLE NAIL

GALVANIZED STEEL CLIP

PRECAST TONGUE–AND–GROOVE DECK PLANK

PRE-ENGINEERED METAL BUILDING SYSTEMS

A pre-engineered metal building is a complete building system that has been engineered and fabricated to specifications. Generally, they consist of steel purlins (horizontal members that connect trusses) spanning rigid steel frames with light gauge metal wall cladding and roof panels (Figure 17.39). The system provides a clear span and more-flexible structure compared to conventional steel-framed building. Pre-engineered metal buildings can accommodate many types of occupancies, from offices to large multipurpose facilities. Some systems achieve a clear span of up to 150 ft. (45 m) or larger with no columns required, making it a practical solution for storage facilities, religious, and sport facilities.

Figure 17.39 A pre-engineered metal building is a complete building system that consists of steel purlins spanning between rigid steel frames with light gauge metal wall cladding and roof panels.

The building frames of pre-engineered metal building systems are constructed from a variety of structural steel and light-gauge steel framing members. The structural frames are anchored on cast-in-place concrete footings and designed to resist anticipated loads. The common types of pre-engineered buildings available include the welded rigid frame, truss-type, post-and-beam, and sloped roof. Welded rigid frame buildings provide for the widest possible spans without interior columns. They use rigid frames of rolled steel sections as the main structure and enclosing materials supported by light-gauge steel members (Figure 17.40). The complete assembly, including the steel frame, girts, purlins, and bracing, provides the needed strength. Truss-type buildings use steel columns and open-web joists and girders to form a roof. Joist girders span the long distances

Figure 17.40 Welded rigid frame buildings provide use rigid frames of rolled steel sections as the main structure and enclosing materials supported by light-gauge steel members.

between steel columns. Smaller open-web joists serve as truss purlins between joist girders.

Manufacturers distribute the systems through a network of franchised dealers. The dealers work with owners and architects to establish a structural system that meets the requirements of the building program. The structure is designed by a certified engineer and fabricated by a pre-engineered manufacturer as a complete system.

OTHER STEEL CONSTRUCTION SYSTEMS

Pre-fabricated steel truss components are used for roof trusses, open-web joists, and space frames. A truss is a coplanar (forces operating in the same plane) assembly of structural members joined at their ends. The diagonals form triangles, producing a rigid framework. Triangulation is the concept behind the design of a truss. A triangle is the only multisided geometric figure that is rigid and will not move or rotate (unless a structural member bends or a connection fails). This principle is illustrated in the design of the bridge shown in Figure 17.41.

Figure 17.41 Trusses use triangulation to produce a rigid structure.

Roof trusses act like beams and support a roof and all imposed loads. A typical steel roof truss consists of top and bottom chords and a system of webs (Figure 17.42). The slope of the top chord can vary as desired, or it can be parallel to the bottom chord for flat roofs. Parallel-chord trusses are used for floors and flat roofs and as major structural beams and girders. A typical trussed-roof framing structure includes trusses, purlins, and a deck. As shown in large multifloor buildings, steel trusses can carry the loads of floors, roofs, and ceilings.

Figure 17.42 Open-web joists are end bearing and can span long distances in floor and roof constructions.

Open-web joists are a type of truss. They are lightweight structural members made from steel angles and bars. They, like other trusses, are end bearing and can span long distances in floor and roof constructions. Steel open-web joists typically require cross-bracing to prevent twisting.

Space frames are a three-dimensional truss system that can achieve very long spans. They provide a structural frame that spans in several directions and over large areas with a minimum number of vertical supports. Space frames are generally made from tubular members. However, wide flange beams and T members are also used. The attractiveness of the assembly is lost if it must be covered to meet fire codes. Some codes do not require fire protection if the space frame is more than 20 ft. (6.1 m) above the floor.

Tensile structures and cable-net structures are framed using high-strength cables of cold-drawn steel suspended between supporting members and secured with cable stays. These lightweight structures can span large distances in an almost unlimited number of shapes and curvatures. Cable-net structures may be covered in a variety of different materials to provide a weather-tight enclosure. Glass or polycarbonate panels can be fixed to cables with mechanical connections and sealed with rubber gaskets. Fabric coverings can be used to provide a lightweight and translucent enclosure for tensile steel structures.

Light-Gauge Steel Framing

Light-gauge structural steel shapes are formed from flat cold-rolled pieces of carbon steel. Gauge thicknesses range from No. 12 to No. 20. Some shapes are formed from a single steel sheet, but others have several forms shape-welded together. They are generally either galvanized or primed with zinc chromate. Load-carrying capacities and spans should be obtained from a manufacturer. Light-gauge metal is widely used in commercial, industrial, and residential construction (Figure 17.43).

Figure 17.43 Light gauge steel framing uses a construction system similar to wood frame construction.

Typical construction details for light-gauge steel structural members are shown in **Figure 17.44**. The floors can consist of steel decking with a concrete topping or precast concrete planks. Roofs can be any decking material available. Stud walls can support lightweight steel or open-web joists and various roof trusses or metal rafters. The finished exterior can be any material typically used.

Figure 17.44 Construction details used with light gauge steel construction.

FIRE PROTECTION OF THE STEEL FRAME

Structural steel is an incombustible material that will not melt even during a building fire. However, when subjected to sustained extreme heat, its properties, including strength, are affected. This can lead to steel column and beam failure during a prolonged fire. To protect a building's occupants and structural integrity, codes require coating certain steel frames with a fire-resistant material. These requirements depend on type of construction, building heights, floor area, occupancy, fire protection systems, and building location. Many materials are used to protect structural steel, including concrete, tile, brick, stone, gypsum board, gypsum blocks, fire-resistant plasters, sprayed-on mineral fibers, intumescent fire-retarding coatings, liquids, and flame shields.

Unprotected mild steel loses about half its room temperature strength at temperatures exceeding 1,000°F (542°C). Fire-resistance ratings are given as the number of hours a material can withstand fire exposure, as specified by standard test procedures. Most standard fire tests on structural steel members are conducted by the National Institute of Standards and Technology and the Underwriters Laboratories.

Insulating concrete can be used to protect steel members. The amount of protection depends on the thickness of the cover, the concrete mix, and the method of support. Lightweight concretes, those with aggregates, such as perlite, vermiculite, expanded slag, pumice, or sintered flyash, repel fire better than normal concretes due to greater resistance to heat transfer and a higher moisture content. Other insulating materials provide greater protection and are much lighter than concrete.

Masonry units, such as concrete block, brick, gypsum block, and hollow clay tile, are often used to enclose structural steel members. In addition to a material's protection, those with hollow cores can be filled with insulation (such as vermiculite) or mortar to increase their fire-protection qualities. Both concrete and masonry protection (Figure 17.45) is heavy, limiting its use.

Figure 17.45 Structural steel can be fire protected by concrete or masonry encasement.

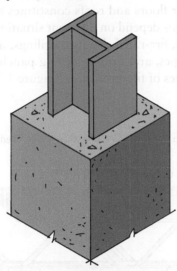

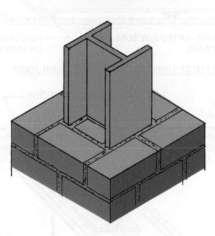

Lightweight plaster troweled over a metal lath is a lightweight fire-protection material. Lightweight plaster using perlite or vermiculite aggregates has good insulating qualities.

Fire-resistant sheet products are secured mechanically to structural steel members. Those with hard surfaces, such as gypsum board, can also be painted and serve as a finished interior surface (Figure 17.46).

Figure 17.46 Examples of gypsum board enclosures for fire protection.

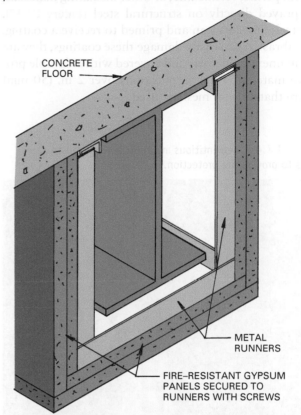

CONCRETE FLOOR

METAL RUNNERS

FIRE–RESISTANT GYPSUM PANELS SECURED TO RUNNERS WITH SCREWS

GYPSUM BOARD FIRE–RESISTANT BEAM CLADDING

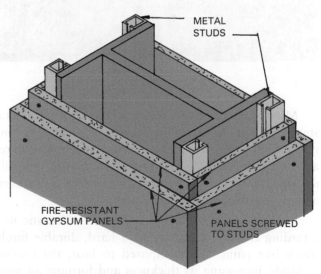

METAL STUDS

FIRE–RESISTANT GYPSUM PANELS

PANELS SCREWED TO STUDS

GYPSUM BOARD FIRE–RESISTANT COLUMN CLADDING

Sprayed cementitious materials, such as gypsum plaster with perlite, vermiculite, or other insulating materials, are sprayed directly on structural steel (Figure 17.47). The steel must be clean and primed to receive a coating. Since abrasion can easily damage these coatings, they are used in unexposed areas and covered with a durable protective material. Thick applications over 2 in. (50 mm) require that a metal mesh be used.

Figure 17.47 Cementitious materials are sprayed on steel frames to provide fire protection.

Mineral fiber slabs are also used to enclose structural steel members. Mineral fibers have excellent heat flow retardation properties and can withstand temperatures above 1,000°F (542°C). This type of coating is easily damaged and requires a protective covering if vulnerable to possible impacts or weather.

Intumescent coatings are a sprayed-on mastic fire-retarding coating. They dry to a hard, durable finish, much like paint. When exposed to heat, the coating expands, increasing its thickness and forming an insulation blanket. These coatings are available in various colors and can serve as a finished surface.

Liquid-filled columns are used to reduce steel's heat. Tube- or box-type columns are filled from a water main that replenishes water lost due to fire heat. Vent valves release the steam produced. Pumps are sometimes used to maintain circulation within the system. Antifreeze is added if the columns are exposed to freezing temperatures.

Flame shields are metal barriers that deflect flames and reflect heat away from exterior structural steel members. The installation of ceilings that provide needed fire protection for floors and roofs constitutes another procedure. Designs depend on different situations, but plastered ceilings, fire-rated dropped ceilings, acoustic tiles of various types, and drop-in ceiling panels all provide various degrees of fire protection (Figure 17.48).

Figure 17.48 Fire-rated ceiling assemblies used to protect steel framing.

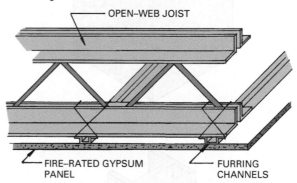

FIRE–RATED CEILING TIED TO AN OPEN–WEB JOIST

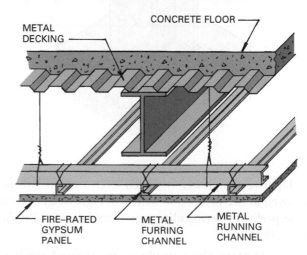

FIRE RATED SUSPENDED CEILING PROTECTING LONG SPAN STEEL BEAMS

Review Questions

1. What are the purposes of the various types of drawings used for the design and erection of steel-framed buildings?

2. How are structural steel members lifted for installation as the erection process proceeds?

3. What are the commonly used fastening methods for joining structural steel members?

4. How can specified tension values be ensured when bolting structural steel members?

5. What techniques can be used to provide for the stability of a structural steel frame?

6. What are the two types of beam connections?

7. What is the difference between bearing-type and friction-type riveted connections?

8. What techniques are used to provide fire protection for structural steel framing?

9. What types of decking are used with structural steel framing?

10. What is a truss?

11. What type of structural system is used to support tensile structures?

Key Terms

Bending Moment

Composite Decking

Design Drawing

Erection Plan

Galling

Load Indicator Washers

Moment

Seated Connection

Shear Stress

Shear Studs

Simple Connection

Activities

1. Contact an architect or general contractor and try to secure drawings of steel-framed buildings. Examine the framing and connection drawings. Note the system used to identify members and connections.

2. Select one method for fireproofing steel framing listed in your local building code and make a sketch showing details for beam and column fireproofing.

3. Visit a site where a steel-framed building is under construction. It can be a multistory or factory-manufactured metal building. Prepare a report with sketches to show types of connections used.

4. Examine local building codes and describe each type of construction in which structural steel framing may be used.

Additional Resources

Detailing for Steel Construction, American Institute of Steel Construction, Chicago, IL.

Design Guide for Anchored Brick Veneer Over Steel Studs, Western States Clay Products Association, Los Angeles, CA.

Manual of Steel Construction, American Institute of Steel Construction, Chicago, IL.

Standard Specifications, Load Tables and Weight Tables for Steel Joists and Girders, and other related publications, Steel Joist Institute, Myrtle Beach, SC.

Structural/Seismic Design Manual—Volume 3, International Code Council, Falls Church, VA.

Waite, T., and NAHB Research Center, *Steel-Frame House Construction*, National Association of Home Builders, Washington, DC.

Other resources include:

Extensive number of publications and free downloads at *www.aisc.org*, American Institute of Steel Construction, Chicago, IL.

Numerous technical Fact Sheets related to metal building construction, North American Insulation Manufacturers Association, Alexandria, VA.

See Appendix C for addresses of professional and trade organizations and other sources of technical information.

Wood, Plastics, and Composites
CSI MasterFormat™

Wood, Plastics, and Composites

Upon completion of this chapter, the student should be able to:

- Understand the structural composition of trees and identify species used in construction.
- Make decisions pertaining to the influence of defects in lumber on various applications.
- Know standard inch and metric lumber sizes.
- Use information about lumber grades to secure products that serve an intended purpose.

- Use data from the In-Grade Testing Program.
- Use information about properties of wood when selecting species.
- Apply the structural properties of wood when designing structural members.
- Take action to protect wooden structures from damage by insects and moisture.

Build Your Knowledge

For further study on these materials and methods, please refer to:

Chapter 19 Products Manufactured from Wood

Chapter 20 Wood and Metal Light-Frame Construction

Chapter 21 Heavy Timber Construction

Chapter 34 Flooring

Wood in the form of lumber and timbers is perhaps the most familiar of construction materials. From ancient pole structures and the half-timbered buildings of northern Europe, to today's ubiquitous wood homes, the knowledge of wood construction has developed and evolved over millennia. Wood is a natural organic material. It is unique among construction materials in that it is a renewable resource. Carefully managed timber farms and natural wild growth provide a continuing source of wood. Wood in the form of lumber and timbers is one of the most familiar construction materials. In addition, wood is used to produce a variety of reconstituted products, such as plywood, particleboard, and hardboard. Wood is also used in the manufacture of paper and cardboard. Wood fibers also provide a source of nitrocellulose for the manufacture of explosives.

Because there are many different species of trees, there is a wide variation in the properties of wood. As a result, wood is useful for many applications, both structural and decorative. It is important to understand the properties of a wood species before selecting it for a particular application.

TREE SPECIES

Several hundred species of wood grow throughout the world. Some are abundant and continuously farmed, while others are rare and endangered. Some species, because of their properties, availability, and abundance, are used commonly in construction. Woods are divided into two classes: hardwoods and softwoods (Figure 18.1). This division is based on botanical differences and not on their actual hardness or softness. The major forest areas providing construction lumber in the United States are shown in Figure 18.2.

Figure 18.1 Woods are divided into two classes, deciduous (hardwoods) and coniferous (softwoods).

Figure 18.2 Major forest regions in the United States.

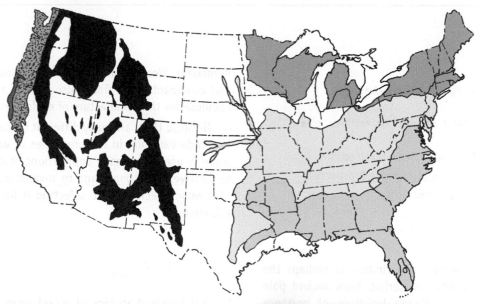

SOUTHERN FOREST
LOBLOLLY PINE, SHORTLEAF PINE,
SLASH PINE, LONGLEAF PINE, GUM,
HICKORY, CYPRESS, OAK, PECAN,
ASH.

CENTRAL FOREST
GUM, HICKORY, OAK, YELLOW
POPLAR, BLACK WALNUT, ELM,
MAPLE, RED CEDAR.

NORTHERN FOREST
EASTERN SPRUCE, NORTHERN
HEMLOCK, WHITE PINE, JACK
PINE, FIR, MAPLE, BEECH, BIRCH.

WESTERN FOREST
LODGEPOLE PINE, PONDEROSA
PINE, SUGAR PINE, IDAHO WHITE
PINE, DOUGLAS FIR, WHITE FIR,
RED CEDAR, LARCH, SPRUCE, ASPEN,
ELM, ASH, COTTONWOOD.

WEST COAST FOREST
SITKA SPRUCE, WESTERN RED
CEDAR, WEST COAST HEMLOCK,
DOUGLAS FIR, ALDER,
PONDEROSA PINE.

REDWOOD REGION
REDWOOD, DOUGLAS FIR.

Softwoods

Much of the production of wood for commercial use involves the softwoods class. *Softwoods* are coniferous, trees that bear cones, and, with a few exceptions, have needle-like leaves that stay green all year long. About 80 percent of the world's production of lumber is from softwoods, much of it from North America. Common softwoods used in construction include pine, fir, spruce, cedar, redwood, and hemlock. Softwoods are used for a variety of building components, including structural framing, sheathing, roofing, subflooring, siding, trim, and millwork.

Hardwoods

Hardwood, or deciduous, trees are broad-leaved and shed their leaves in the winter. Common North American hardwoods include ash, birch, cherry, hickory, maple, mahogany, oak, and walnut. Tropical hardwoods include ipe, garapa, tigerwood, teak, and zebrawood. Hardwoods are often more expensive than softwoods and are used in cabinetry, furniture making, paneling, interior trim, and flooring.

THE STRUCTURE OF WOOD

Wood uses energy from the sun, water, and extracted carbon dioxide through the process of photosynthesis. Trees are anchored deep into the ground by their root systems. The taproot draws minerals and water from the ground. Side roots grow out around the tree, and feeder roots grow off the side roots. They absorb soil water and minerals in the ground to provide nutrients for the tree. Water and minerals flow up into the sapwood (xylem). The leaves use sunlight and carbon dioxide to change water and minerals into food. This food flows down into the tree through the inner bark (phloem).

Each part of a tree serves a specific purpose in the tree's growth and development (Figure 18.3). At the center is a small core called the pith. During the early years of growth, the pith helps support the stem and feed the tree. As the tree matures, the pith ceases to function. Next to the pith is the *heartwood*, which is hard, mature wood that forms the largest part of the trunk. It strengthens the tree and displays the color associated with a particular species. The next layer, *sapwood*, is a living layer that carries water and food throughout the tree. It is soft, usually light in color, and contains

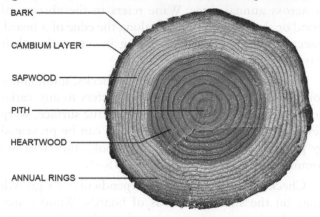

Figure 18.3 A cross-section of a tree, illustrating its structure.

BARK
CAMBIUM LAYER
SAPWOOD
PITH
HEARTWOOD
ANNUAL RINGS

more moisture than does heartwood. The heartwood and sapwood form growth rings that can be seen when a tree is cut down. The light, soft rings develop in the spring, when the tree grows rapidly, and are referred to as springwood. The dark, hard rings form in the summer, when growth is slow, and are called summerwood. When a tree is cut down, its approximate age can be determined by counting the hard summerwood annual rings.

Next to the sapwood is the cambium layer. It forms new cells that are either xylem or phloem. Sapwood and heartwood are formed with xylem cells. On the outside of the cambium layer are the phloem cells, which form the inner bark layer. Phloem cells in the inner bark carry food to the roots. As more phloem cells are formed, the outer layer of the inner bark changes into outer bark, a layer that protects the tree. Within the tree are medullary rays that run perpendicular to the growth rings. They carry food and water from the cambium layer to the interior of the tree.

WOOD DEFECTS

A piece of wood may have natural defects that occur while a tree grows or seasoning defects that are produced as the wood is dried for use. Natural defects include knots, shake, wane, insect holes, and pitch pockets. Knots occur where imbedded branches in tree trunks are cut. Knots weaken wood and are a factor lumber graders consider when judging wood. Pitch is the accumulation of sap or resin in pockets in a tree. It occurs in softwoods and usually does not cause major problems. Shake can result when a tree is racked or bent, as in a windstorm. Shake appears as small cracks running with or across annual rings. Wane refers to the absence of wood or the presence of bark along the edge of a board. Insect holes are caused by boring insects eating their way into wood.

Seasoning defects include warp, checks, stain, honey combing, and casehardening. *Warp* refers to any variation in a board's shape beyond a flat, true surface. Warp develops as wood loses moisture and can be prevented by proper seasoning. Cup, bow, crook, and twist are common forms of warping (Figure 18.4).

Checks are small cracks perpendicular to growth rings on the surface or ends of boards. Wood sometimes develops a stain on its surface after it is cut into lumber. Stains may be brown, green, or blue and generally do not influence the wood's strength. Honeycombing refers to cracks in a board's interior resulting from improper seasoning. When a board's outer surface is drier and subjected to more stress than its interior, casehardening can result. Again, this is caused by improper seasoning.

LUMBER MANUFACTURING

The production of all types of lumber begins in the forest where trees are felled (Figure 18.5). After branches are cut away, the log is moved to a loading site. Logs

Figure 18.5 A logger felling a mature tree.

© Mikhail Olykainen/Shutterstock.com

may be skidded to the site by a tractor or hauled down a steep hillside with a long cable called a choker line. Once they reach the loading site, they are generally loaded on trucks. The trucks operate on dirt roads built into the forest especially for the logging operation (Figure 18.6). At the mill, logs are stacked into large piles and often sprayed with water to keep them from drying out and splitting. Most mills use a mechanical peeler to grind off the bark. After peeling, a log is ready to be cut into lumber of various types.

Figure 18.4 Examples of common wood defects.

Twist

Cup

Heartwood Stain

Pith & check

Wane

Rot

Check

Figure 18.6 Southern pine logs being moved to a mill.

Logs are hauled onto the carriage of a large band saw machine called a heading. Small mills use circular saw blades 3 ft. to 4 ft. (.91 m to 1.22 m) in diameter. The carriage moves a log into the blade, which cuts off slabs of wood. The saw operator, called the sawyer, judges how to efficiently cut the most marketable wood from each log and adjusts the machinery to rotate and advance the log into the saw. Modern mills use computer controls to assist with this process (Figure 18.7). Slabs cut from the log fall on a conveyor and move to an edger that squares edges and cuts the slabs to desired widths. Pieces then move to a trim saw, which cuts the rough, square-edged wood to standard lengths. Hardwoods are not edged or cut to length.

Figure 18.7 A log being processed by a precision computer-controlled laser-guided system.

Lumber is the term used to describe wood that has been cut from the tree and formed to produce dressed boards and timber. Most lumber used for construction purposes is plain-sawed. This method produces the maximum yield and the widest stock from the log (Figure 18.8). Plain-sawed lumber produces boards

Figure 18.8 Cutting patterns for plain-sawed and quarter-sawed lumber.

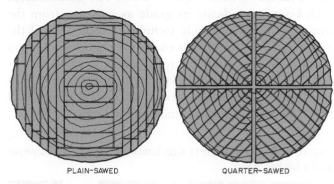

PLAIN-SAWED   QUARTER-SAWED

that have a broad grain running their length. Another common sawing method, *quarter-sawing*, produces boards with the edges of the annual rings showing on the face. Flooring is often quarter-sawed because the annual rings are hard and withstand wear better than do the wide areas of springwood exposed on plain-sawed boards. Examples of plain-sawed and quarter-sawed wood grains are shown in Figure 18.9.

Figure 18.9 Examples of plain-sawed and quarter-sawed wood profiles.

Once cut to size, the lumber moves on a conveyor past a *lumber grader* (Figure 18.10). The grader marks each board to indicate its grade as it passes from the trim saw. This is a highly technical job because of the extensive specifications for each of the many grades and types of lumber. Finally, green lumber goes to an air-drying area where it is stacked with sticks between each layer. After a number of weeks, the lumber is moved, still stacked, to a kiln for final moisture reduction.

Figure 18.10 Once cut to size, lumber is moved on a conveyor past a lumber grader.

© Lakeview Images/Shutterstock.com

After the lumber has been dried, it is dressed to finished size in a planing mill. All dimensional lumber (up to but not including 5 in. or 114 mm), if surfaced dry, must be marked S-Dry (surfaced dry). If it was surfaced green (moisture content more than 19 percent), it must be grade stamped S-GRN. If the lumber is surfaced at 15 percent, moisture content it is often stamped MC15. Hardwood lumber is generally shipped dried but unsurfaced to manufacturers that use it for furniture, cabinets, and other products.

SEASONING LUMBER

Wood is hygroscopic, expanding when it absorbs moisture and shrinking when it loses it. During its life, a tree contains considerable water. In this condition, wood is called green. To be useful in construction, furniture, and other products, it must be in a dry condition. *Seasoning* refers to the process of converting wood from a *green lumber* to a finished lumber state through drying, which results in a recommended and desired moisture content. Lumber-seasoning methods include air drying, kiln drying, dehumidification, and solar drying, with air and kiln drying being the most common. Seasoning requirements for softwoods, as specified by the Southern Pines Inspection Bureau, are given in Table 18.1.

Table 18.1 Seasoning Requirements for Softwoods[a]

Items (Nominal)	Moisture Content Limit	
	Maximum (Dry)	Kiln-Dried (KD or MC15)
D&btr grades		
1" & 1¼"	15%	12% on 90% of pieces 15% on remainder
1½", 1¾" & 2"	18%	15%
Over 2" not over 4"	19%	15%
Over 4"	20%	18%
Paneling		
1"		12%
Boards		
2" and less and dimension 2" to 4"	19%	15%
Decking		
2" thick	19%	15%
3" and 4" thick		15% on 90% of pieces 18% on remainder
Heavy dimension		
Over 2" not over 4"	19%	15%
Timbers		
5" and thicker	23%	20%

[a]As specified by the Southern Pine Inspection Bureau.

Air-Dried Lumber

Air drying involves stacking boards with stickers (wood strips) inserted between layers (Figure 18.11). This

Figure 18.11 Lumber is stickered and stacked for air drying.

© mironov/Shutterstock.com

permits air to circulate and aid in the drying process. Air drying is a slow process and does not provide control over drying duration. On hot, dry, windy days, wood will season too quickly, and checking and warping may occur. On cool days with high humidity, wood dries very little. It is difficult to regulate desired moisture content when air drying. Some dimension lumber and lower grades of softwood lumber are often air dried. Structural timbers are too large for quick air drying, so they are often shipped green.

Moisture content of lumber air dried for construction purposes should be in the 15 to 19 percent range in the United States. In some parts of the country, such as the Southwest, it can be slightly lower than 15 percent, while in the moist Pacific Northwest, it may be closer to 20 percent. Lumber grading rules for the various softwoods specify maximum moisture contents for each grade regardless of how they are dried.

Hardwoods are usually air dried for a period of time, then kiln dried. It is difficult to get moisture contents of hardwoods low enough for furniture making and cabinetry by air drying only. Normally, a moisture content of 6 to 8 percent is required for these purposes, necessitating kiln drying.

Kiln-Dried Lumber

Kiln drying involves stacking lumber in the same manner for air drying inside a kiln. In the kiln, temperature, humidity, and air circulation are controlled to reduce a wood's moisture content. Air temperatures in the kiln reach 180°F (82°C), with an equally high relative humidity (**Figure 18.12**). Since both temperature and humidity are controlled, lumber can be quickly dried to any desired moisture content. Most woods take less than two weeks to dry in a kiln. Kiln drying reduces defects and produces a product that will not expand or contract as much as air-dried wood.

Dehumidification and Solar Kilns

Dehumidification and solar kilns are relatively new. The dehumidification method uses electricity to dry the lumber. Solar kilns use the sun's energy to produce needed heat and are the most economical. Those in current use can handle only small amounts of wood.

Unseasoned Lumber

Most lumber more than 2 in. thick is air dried. The thicker the stock, the longer it takes to achieve the desired moisture content of 19 percent. Most stock in these thicknesses is used in a green condition (more than 19 percent moisture content). It continues to dry and shrink after use, which sometimes causes problems. Moisture content can be checked on-site with a battery-operated moisture meter.

LUMBER SIZES

After rough-sawed wood has been dried, it is taken to a planing mill for surfacing and shaping into useful products, such as boards, dimension stock, timbers, molding, and trim (**Figure 18.13**). Softwood lumber is produced in standard sizes. Hardwoods are often *dressed* to thickness but not to any specific width or

Figure 18.12 Lumber being loaded in a kiln to reduce moisture content.

Figure 18.13 After seasoning, lumber is taken to a planing mill for surfacing and shaping into useful products.

length. After the lumber is planed or shaped, it must be stored in weatherproof sheds so the moisture content will not increase.

SOFTWOOD LUMBER

Softwood lumber is sold by designating its nominal size, the size of the board after it has been rough cut at a sawmill. Inch and metric sizes are shown in Table 18.2, Table 18.3, and Table 18.4. The dressed or finished size is the actual size of the board after drying and surfacing. For example, a typical wall stud has a nominal size of 2 in. × 4 in. and a finished size of 1½ in. × 3½ in. A dry, dressed metric-size stud is 38.10 mm × 88.90 mm. If stock is surfaced green, it is made larger so that when it dries and shrinks it will be the same size as dried stock. Worked stock refers to boards that undergo some machining operation, such as moldings or tongue-and-groove flooring.

Table 18.2 Standard Sizes of Surfaced Dimension Lumber

Thickness of Stock					
Customary Units (in.)			Metric Units (mm)		
Nominal	Actual Dry	Actual Green	Nomen-clature[a]	Actual Dry	Actual Green
2	1½	1⁹/₁₆	38	38.10	39.69
2	2	2¹/₁₆	51	50.80	52.39
3	2½	2⁹/₁₆	64	63.50	65.09
3	3	3	76	76.20	77.79
4	3½	3⁹/₁₆	89	88.90	90.49
4	4	4¹/₁₆	102	101.60	103.19
Width of Stock					
2	1½	1⁹/₁₆	38	38.10	39.69
3	2½	2⁹/₁₆	64	63.50	65.09
4	3½	3⁹/₁₆	89	88.90	90.49
5	4½	4⁵/₈	114	114.30	114.47
6	5½	5⁵/₈	140	139.70	142.87
7	6½	6⁵/₈	165	165.10	168.28
8	7¼	7½	184	184.15	190.50
10	9¼	9½	235	234.95	241.30
12	11¼	11½	286	285.75	292.10
14	13¼	13½	337	336.55	342.90
16	15¼	15½	387	387.35	393.70

[a]Nomenclature means the size used to describe the members. Actual size may be slightly larger or smaller.

Source: Canadian Wood Council and the Southern Forest Products Association

Table 18.3 Standard Sizes of Surfaced Timbers

Thickness and Width of Stock					
Customary Units (in.)			Metric Units (mm)		
Nominal	Actual Dry	Actual Green	Nomen-clature[a]	Actual Dry	Actual Green
5	—	4	114	—	114.3
6	—	5	140	—	139.7
7	—	6	165	—	165.1
8	—	7	191	—	190.5
9	—	8	216	—	215.9
10	—	9	241	—	241.3
12	—	11	292	—	292.1
14	—	13	343	—	342.9
16	—	15	394	—	393.7
18	—	17	445	—	444.5
20	—	19	495	—	495.3

[a]Nomenclature means the size used to describe the member. Actual size may be slightly larger or smaller.

Source: Canadian Wood Council and the Southern Forest Products Association

Table 18.4 Metric Lengths for Softwood Lumber

Nominal Length (ft.)	Metric Length (m)
3	0.91
4	1.22
5	1.52
6	1.83
7	2.13
8	2.44
9	2.74
10	3.05
11	3.35
12	3.66
13	3.96
14	4.27
15	4.57
16	4.88
17	5.18
18	5.49
19	5.79
20	6.10
21	6.40
22	6.71
23	7.01
24	7.32

Source: Canadian Wood Council

As it seasons from a green to a dry condition, lumber shrinks in direct proportion to its loss of moisture. This makes it possible to establish separate sizes for green and dry-surfaced lumber used together in building to insure uniformity (size, strength, and stiffness) once the green lumber reaches its required moisture content. The dressed sizes of green and dry softwood lumber are specified in American Softwood Lumber Standard PS 20–15 and the Canadian National Lumber Grades Authority standard CSA 0141. American and Canadian standards are coordinated and accepted in both countries. They are based on dry lumber having a moisture content of 19 percent or less, and green lumber having more than 19 percent. The standard sizes for finish lumber, boards, siding, and other wood products are given in Table 18.5 and Table 18.6.

Softwood lumber is sold in standard lengths of 2-ft multiples ranging from 6 ft. to 24 ft. Some special lengths, such as studs pre-cut to standard ceiling heights, are cut to exact desired dimensions. Metric lumber lengths are specified in meters (m) and decimal parts of a meter, as shown in Table 18.4.

Hardwood Lumber

Hardwood lumber is sold in random widths and lengths and often is not surfaced to standard thicknesses. Since hardwoods are used in the manufacture of cabinets and furniture, boards are needed in many sizes and thicknesses. Cutting to standard widths and lengths would cause great waste. Standard rough thickness range from ⅜ to 1½ in. in ¼ in. increments. Hardwood lumber is also available in 2, 3, and 4 in. thicknesses.

Hardwood lumber in Canada is sized so metric units and customary units are within 1 percent of each other. The standard thicknesses of rough lumber are 15 mm to 60 mm in 5 mm increments, and 70 mm to 100 mm in 10 mm increments. Surfaced lumber thicknesses are 5 mm less than the rough size for green or air-dried lumber and 5 mm to 8 mm for kiln-dried lumber. Standard lengths are 1.2 m through 4.8 m in increments of 30 cm. Hardwood lumber is sold by volume in cubic meters (m³).

Buying Lumber

Most lumber in the United States is sold by the board foot. A *board foot* is equal to a piece of lumber with an actual size of 1 in. thick, 12 in. wide, and 1 ft. long. For example, a board 1 in. thick, 12 in. wide, and 8 ft. long contains 8 board feet. These are nominal sizes. The

Table 18.5 Standard Sizes for Softwood Products

	Thickness (in.)		Width (in.)		
	Nominal	Worked	Nominal	Face	Overall
Bevel siding	½	³/₁₆ × ⁷/₁₆	4	3½	3½
	⅝	⁷/₁₆ × ⁹/₁₆	5	4½	4½
	¾	³/₁₆ × ¹¹/₁₆	6	5½	5½
	1	³/₁₆ × ¾	8	7¼	7¼
Drop siding Rustic and drop siding (dressed and matched)	⅝ 1	⁹/₁₆ ²³/₃₂	4 5 6 8 10	3⅛ 4⅛ 5⅛ 6⅞ 8⅞	3⅜ 4⅜ 5⅜ 7⅛ 9⅛
Rustic and drop siding (shiplapped)	⅝ 1	⁹/₁₆	4 5 6 8 10 12	3 4 5 6⅝ 8⅝ 10⅝	3⅜ 4⅜ 5⅜ 7⅛ 9⅛ 11⅛
Flooring	⅜ ½ ⅝ 1 1 1	⁵/₁₆ ⁷/₁₆ ⁹/₁₆ ¾ 1 1¼	2 3 4 5 6	1⅛ 2⅛ 3⅛ 4⅛ 5⅛	1⅜ 2⅜ 3⅜ 4⅜ 5⅜
Ceiling	⅜ ½ ⅝ ¾	⁵/₁₆ ⁷/₁₆ ⁹/₁₆ ¹¹/₁₆	3 4 5 6	2⅛ 3⅛ 4⅛ 5⅛	2⅜ 3⅜ 4⅜ 5⅜
Partition	1	²³/₃₂	3 4 5 6	2⅛ 3⅛ 4⅛ 5⅛	2⅜ 3⅜ 4⅜ 5⅜
Paneling	1	²³/₃₂	3 4 5 6 8 10 12	2⅛ 3⅛ 4⅛ 5⅛ 6⅞ 8⅞ 10⅞	2⅜ 3⅜ 4⅜ 5⅜ 7⅛ 9⅛ 11⅛
Shiplap	1	¾	4 6 8 10 12	3⅛ 5⅛ 6⅞ 8⅞ 10⅞	3½ 5½ 7¼ 9¼ 11¼
Dressed and matched	1 1¼ 1½	¾ 1 1¼	4 5 6 8 10 12	3⅛ 4⅛ 5⅛ 6⅞ 8⅞ 10⅞	3⅜ 4⅜ 5⅜ 7⅛ 9⅛ 11⅛

Source: Southern Forest Products Association

actual size is ¾ in. × 11 in. ⅜ in. Usually, stock less than 1 in. thick is figured as 1 in. To calculate board feet, multiply thickness in inches by width in inches by length

Table 18.6 Metric Sizes for Surfaced Boards, Shiplap Siding, and Centre Match Siding

Actual Thickness (mm) Dry, Dressed	Actual Width (mm) Dry, Dressed
17	38
19	64
25	89
32	114
	140
	165
	184
	210
	235

Source: Canadian Wood Council

Figure 18.14 Examples of board foot and metric lumber calculations.

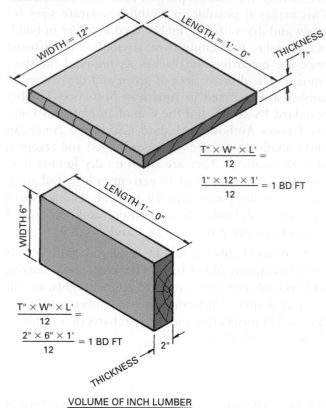

VOLUME OF INCH LUMBER

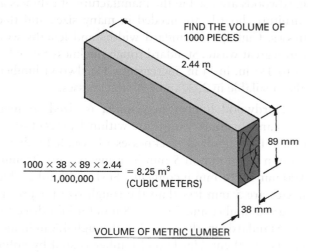

VOLUME OF METRIC LUMBER

in feet and divide by 12 (which converts the width to feet) (Figure 18.14).

Metric lumber is sold by the cubic meter (m³). The volume is based on the actual size. This gives the actual volume of the piece in cubic meters. It is computed by the formula:

Volume = thickness (mm) × width (mm) × length (m).

Thickness and width are given in millimeters (mm), and the length is in meters (m). An example is shown in Figure 18.14. The Canadian Wood Council provides tables for converting metric volume to board feet.

Other materials, such as molding and trim, are sold by the linear foot. Posts and pilings are sold by the piece. Wood shingles are sold by the square (a square covers 100 ft.²).

LUMBER GRADING AND TESTING

Lumber Grades

The grades of softwood lumber are based on *American Softwood Lumber Standard* PS 20–15, published by the U.S. Department of Commerce. Under the provisions of this standard, a National Grading Rule Committee (NGRC) was established to develop uniform nationwide grade requirements for dimensional lumber. The functions of the committee were to "establish, maintain, and make fully and fairly available grade strength ratios, nomenclature, and descriptions of grades of dimension lumber conforming to American Lumber Standards." The specifications developed are known as the *National Grading Rule for Dimensional Lumber* (NGRDL). They form a part of the grading rules of all wood association grading rules, such as those of the Western Wood Products Association, the Southern Pine Inspection Bureau, West Coast Lumber Inspection Bureau, and the Redwood Inspection Service. The grading rules of these individual associations pertain to those species produced in their geographic areas and contain the *National Grading Rules for Dimensional Lumber* specifications plus additional specs unique to that association (Table 18.7).

Table 18.7 U.S. and Canadian Softwood Grading Authorities by Region

United States			
Softwood Region	**Species**	**Lumber Association**	**Grading Authority**
Western wood region	Western red cedar	Western Wood Products Association	Western Wood Products Association
	Ponderosa pine		West Coast Lumber Inspection Bureau
	Douglas fir		
	White fir		
	Western hemlock		
	Englemann spruce		
	Western larch		
	Sitka spruce		
	Lodgepole pine		
	Idaho white pine		
	Sugar pine		
Southern pine region	Shortleaf pine Longleaf pine	Southern Forest Products Association	Southern Pine Inspection Bureau
Redwood region	Redwood Douglas fir	California Redwood Association	Redwood Inspection Service
Canada			
	Spruce	Canadian Wood Council and 12 lumber grading authorities	Canadian National Lumber Grades Authority
	Pine, several types		
	Fir, several types		
	Larch		
	Douglas fir		
	Hemlock		
	Western red cedar		
	Aspen		
	Poplar		

Grades of Canadian lumber are identical with those used in the United States (or the same as the requirements of *American Softwood Lumber Standard* PS 20–15). The grades are published by the Canadian National Lumber Grades Authority (NLGA) and appear in the publication *Standard Grading Rules for Canadian Lumber* (Table 18.7). The mills that manufacture lumber according to these standards place a grade stamp on each piece (Figure 18.15).

Softwood lumber grading standards cover the nomenclature and bending strength ratios for structural light framing, light framing, structural joists and planks, appearance framing, and stud grades. The bending stress ratios indicate the relationships between basic stresses of clear wood free of strength-reducing defects (knots, splits) and the stresses developed

Figure 18.15 Typical grade stamp for dimension lumber used by various grading authorities in the United States and Canada.

in a particular grade of lumber of the same species. Some examples using customary units are given in Table 18.8. Data for Canadian species are given in kilograms per square meter (kg/m²) and are available from the Canadian Wood Council. A description of the most commonly used lumber grades for softwoods, as

Table 18.8 Working Stresses for Selected Softwood Species

Species and Commercial Grade	Size Classification	Allowable Unit Stresses[a] in lb/in.²						
		Extreme Fiber in Bending "F_b"[b]		Tension Parallel to Grain "F_t"	Horizontal Shear "F_v"	Compression Perpendicular to Grain "F_c";	Compression Parallel to Grain "F_c//"	Modulus of Elasticity "E"
		Single-Member	Repetitive Member					
Douglas Fir-Larch (Surfaced dry or surfaced green. Used at 19% max. m.c.)								
Select Structural		2,100	2,400	1,200	95	385	625	1,800,000
No. 1		1,750	2,050	1,050	95	385	625	1,800,000
No. 2	2" to 4" thick	1,450	1,650	850	95	385	625	1,700,000
No. 3	2" to 4" wide	800	925	475	95	385	625	1,500,000
Appearance		1,750	2,050	1,050	95	385	625	1,800,000
Hem-Fir (Surfaced dry or surfaced green. Used at 19% max. m.c.)								
Select Structural		1,650	1,900	975	75	405	1,300	1,500,000
No. 1		1,400	1,600	825	75	405	1,050	1,500,000
No. 2	2" to 4" thick	1,150	1,350	675	75	405	825	1,400,000
No. 3	2" to 4" wide	650	725	375	75	405	500	1,200,000
Appearance		1,400	1,600	825	75	405	1,250	1,500,000
Southern Pine (Surfaced at 15% maximum moisture content, K.D. 15. Used at 15% max. m.c.)								
Select Structural		2,150	2,500	1,250	105	565	1,800	1,800,000
No. 1	2" to 4" thick	1,850	2,100	1,050	105	565	1,450	1,800,000
No. 2	2" to 4" wide	1,550	1,750	900	95	565	1,150	1,600,000
No. 3		850	975	500	95	565	675	1,500,000

[a]Values shown are for normal loading conditions at maximum 19% m.c., as in most covered structures.
[b]Values in bending for "repetitive member uses" are intended for design of members spaced not over 24" o.c., in groups of not less than three members and when joined by floor or roof sheathing elements

used by the Southern Pine Inspection Bureau, is shown in Table 18.9. Other U.S. regional and Canadian associations have established similar grades for species harvested in their regions. The higher the grade, the higher the allowable stresses. Lower grades of lumber are more economical.

Lumber Classifications as to Manufacture

Rough lumber has not been dressed and shows saw marks on all four surfaces. Blanked lumber is dressed to a size larger than standard dressed sizes but smaller than the nominal size. It may be surfaced on one surface (S1S), one edge (S1E), or surfaced on all four sides (S4S).

Dressed lumber has one or more surfaces smoothed in any combination of surfaces and edges, such as S1E, S2E, S1S1E, S1S2E, S4S, and so on. For example, S1E means surface one edge, S4S means surface four sides, and S1S1E means surface one side and one edge.

Worked lumber has been dressed and matched, shiplapped, or patterned. Matched lumber has a tongue on one edge and a groove on the other, providing a tongue-and-groove joint. Shiplapped lumber has been worked or rabbeted on both edges of each piece to provide a lapped joint when the pieces fit together. Patterned lumber has been worked to a shaped or molded form in addition to being dressed, matched, or shiplapped.

Re-sawn lumber is produced by re-sawing any thickness of lumber into thinner lumber. Ripped lumber is produced by sawing any width lumber in narrower pieces.

Size Classifications

The following size classifications, summarized in Table 18.10, are used by the Southern Pine Inspection Bureau.

Table 18.9 Southern Pine Softwood Lumber Grades

Product	Grade	Character of Grade and Typical Uses
Finish	B&B	Highest recognized grade of finish. Generally clear, although a limited number of pin knots permitted. Finest quality for natural or stain finish.
	C	Excellent for painted or natural finish where requirements are less exacting. Reasonably clear but permits limited number of surface checks and small tight knots.
	C&Btr	Combination of B&B and C grades; satisfies requirements for high-quality finish.
	D	Economical, serviceable grade for natural or painted finish.
Boards S4S	No. 1	High quality with good appearance characteristics. Generally sound and tight-knotted. Largest hole permitted is 1/16". A superior product suitable for wide range of uses, including shelving, form, and crating lumber.
	No. 2	High-quality sheathing material, characterized by tight knots. Generally free of holes.
	No. 3	Good, serviceable sheathing, usable for many applications without waste.
	No. 4	Admit pieces below No. 3 which can be used without waste or contain usable portions at least 24" in length.
Dimension Structural light framing 2" to 4" thick 2" to 4" wide	Select Structural Dense Select	High quality, relatively free of characteristics that impair strength or stiffness. Recommended for uses where high strength, stiffness, and good appearance are required.
	Structural	Provide high strength; recommended for general utility and construction purposes.
	No. 1	Good appearance, especially suitable where exposed because of the knot limitations.
	No. 1 Dense	Although less restricted than No. 1, suitable for all types of construction.
	No. 2	Tight knots.
	No. 2 Dense	Assigned design values meet wide range of design requirements. Recommended for general construction purposes where appearance is not a controlling factor. Many pieces included in this grade would qualify as No. 2 except for single limiting characteristic. Provides high-quality, low-cost construction.
	No. 3	
Studs 2" to 4" thick 2" to 6" wide 10[1] and shorter	Stud	Stringent requirements as to straightness, strength, and stiffness adapt this grade to all stud uses, including load-bearing walls. Crook restricted in 2" × 4"-8' to ¼", with wane restricted to ⅓ of thickness.
Structural joists and planks 2" to 4" thick 5" and wider	Select Structural Dense Select	High quality, relatively free of characteristics that impair strength or stiffness.
	Structural	Recommended for uses where high strength, stiffness, and good appearance are required.
	No. 1	Provide high strength; recommended for general utility and construction purposes.
	No. 1 Dense	Good appearance; especially suitable where exposed because of the knot limitations.
	No. 2	Although less restricted than No. 1, suitable for all types of construction. Tight knots.
	No. 2 Dense	
	No. 3 No. 3 Dense	Assigned stress values meet wide range of design requirements. Recommended for general construction purposes where appearance is not a controlling factor. Many pieces included in this grade would qualify as No. 2 except for single limiting characteristic. Provides high-quality, low-cost construction.
Light framing 2" to 4" thick 2" to 4" wide	Construction	Recommended for general framing purposes. Good appearance, strong, and serviceable.
	Standard	Recommended for same uses as Construction grade, but allows larger defects.
	Utility	Recommended where combination of strength and economy is desired. Excellent for blocking, plates, and bracing.
	Economy	Usable lengths suitable for bracing, blocking, bulkheading, and other utility purposes where strength and appearance not controlling factors.
Appearance framing 2" to 4" thick 2" and wider	Appearance	Designed for uses such as exposed-beam roof systems. Combines strength characteristics of No. 1 with appearance of C&Btr.
Timbers 5" × 5" and larger	No. 1 SR No. 1 Dense SR No. 2 SR No. 2 Dense SR	No. 1 and No. 2 are similar in appearance to corresponding grades of 20 dimension. Recommended for general construction uses. SR in grade name indicates Stress Rated.
Structural lumber	Dense Str. 86 Dense Str. 72 Dense Str. 65	Premier structural grades from 20 through and including timber sizes. Provides some of the highest design values in any softwood species with good appearance.

Table 18.10 Classifications, Sizes, and Grades of Softwood Lumber

	Standard Units		
Classification	**Thickness Nominal (in.)**	**Width Nominal (in.)**	**Grades**
Finish	3/8–4	2–16	B&B, C, C&Btr, D
Boards	1½	2–12+	No. 1, No. 2, No. 3, No. 4
Structural light framing	2–4	2–4	Select Structural, No. 1, No. 2, No. 3
Light framing	2–4	2–4	Construction, Standard, Utility
Studs	2–4	2–6	Stud
Structural joists and planks	2–4	5 and wider	Select Structural, No. 1, No. 2, No. 3
Appearance framing	2–4	2 and wider	Appearance
Timbers, nonstress	5 and larger	5 and larger	Square-edge and sound, No. 1, No. 2, No. 3
Timbers, stress-rated	5 and larger	5 and larger	Sel, Str, SR, DNS, Sel Str SR, No. 1 SR, No. 1 DNS SR, No. 2 SR, No. 2 DNS SR
	Metric Units		
Classification	**Thickness Nomenclature[a] (mm)**	**Width Nomenclature[a] (mm)**	**Grades**
Finish	8–64	38–286	B&B, C, C&Btr, D
Boards	17–32	38–387	No. 1, No. 2, No. 3, No. 4
Structural light framing	38–102	38–387	Sel Str, No. 1, No. 2, No. 3
Light framing	38–102	38–387	Construction, Standard, Utility
Studs	38–89	38–140	Stud
Structural joists and planks	38–89	114–387	Sel Str, No. 1, No. 2, No. 3
Appearance framing	38–89	38 and wider	Appearance
Timbers, nonstress	114 and larger	114 and larger	Square edge and sound, No. 1, No. 2, No. 3
Timbers, stress-rated	114 and larger	114 and larger	Sel Str SR, DNS Sel Str SR, No. 1 SR, No. 2 DNS SR, No. 3 DNS SR

[a]Nomenclature means the size used to describe the members. Actual size may be slightly larger or smaller.

Source: Southern Forest Products Association (standard) and Canadian Wood Council (metric)

Boards are less than 2 in. in nominal thickness and 1 in. or more in width. If they are less than 6 in. wide, they are classified as strips. **Dimension lumber** is from 2 in. up to but not including 5 in. thick and 2 in. or more in width. It is subdivided into five classes: structural light framing, light framing, structural joists and planks, appearance framing, and studs. **Timbers** are 5 in. or more in their smallest dimension and are subdivided into classes, such as beams, posts, and girders.

Stress-Rated Lumber

Each piece of **stress-rated lumber** is evaluated at a mill by mechanical stress-rating equipment that subjects each piece to bending stress. The modulus of elasticity (E), a measure of stiffness, is measured by instruments on the machine. The stress **grade** is electronically calculated and includes consideration of the effects of grain slope, knots, growth rate, moisture content, and

density. The machine then marks the grade stamp on the piece.

The grade stamp on machine stress-rated lumber indicates that the stress-rating system used to make the test meets certification requirements. The grade stamp shows the agency trademark, the mill name or number, the phrase "Machine Rated," the species, and the modulus of elasticity rating in millions of pounds per square inch (Figure 18.16).

Figure 18.16 Grade stamp for machine stress rated lumber.

The Southern Pine Inspection Bureau has established fifteen categories of stress-graded lumber 2 in. or less in thickness. These categories can be used for most structural purposes, such as for trussed rafters. There are five classes with lower allowable bending stresses in relation to the modulus of elasticity. The classes work when lower bending stress can be used, such as in floor joists.

The Southern Pine Inspection Bureau also has two grades of stress-rated timbers 5 in. × 5 in. and larger: Select Structural (SR) and Dense Select Structural (DSR). They are divided into No. 1 SR, No. 1 Dense SR, and No. 2 SR, No. 2 Dense SR. There are also a series of grades for stress-rated industrial lumber. Other regional associations in the United States and Canada have similar grading procedures.

In-Grade Testing Program

The In-Grade Testing Program is the result of a twelve-year research program conducted by the U.S. Department of Agriculture Forest Products Laboratory in cooperation with U.S. and Canadian lumber industry associations. The program was initiated to verify softwood lumber design values for visually graded lumber and to provide a scientific basis for wood engineering similar to that used for steel and concrete. Thousands of lumber specimens of many species, grades, and sizes were tested, and design values were applied.

The In-Grade Testing Program developed six new species groups for Western lumber species and Eastern lumber species (Table 18.11). The Canadian species groupings are shown in Table 18.12. Southern pine is a separate grouping for woods in the south and southeastern United States.

Table 18.11 Western Lumber Species Groups

Douglas Fir-Larch	Spruce-Pine-Fir (South)
Douglas fir	Engelmann spruce
Western larch	Sitka spruce
Douglas Fir-South	Lodgepole pine
Douglas fir grown in AZ, CO, NV, NM, & UT	Western Woods
	Ponderosa pine
Hem-Fir	Sugar pine
Western hemlock	Idaho white pine
Noble fir	Mountain hemlock
California red fir	Western Cedars
Grand fir	Incense cedar
Pacific silver fir	Western red cedar
White fir	Port orford cedar
	Alaska cedar

Table 18.12 Canadian Lumber Species Groups

Douglas Fir-Larch
Douglas fir
Western larch
Hem-Fir
Western hemlock
Amabilis fir
Spruce-Pine-Fir
White spruce
Red spruce
Black spruce
Engelmann spruce
Lodgepole pine
Jack pine
Alpine fir
Balsam fir
Northern Species
All species graded in accordance with the NLGA standard grading rules for Canadian lumber

Their design values are available in the Supplement to the National Design Specifications for Wood Construction. The various lumber associations also have publications pertaining to species in their areas.

Base Values The design data are presented in the form of base values for the various species groupings. Base values can be adjusted for a particular application in which the structural lumber is to be used. The Southern Forest Products Association refers to these as empirical values. Base values are assigned to six Basic Properties of wood. The six basic properties are: (1) extreme fiber stress in bending (Fb) (bending strength); (2) tension parallel to the grain (Ft); (3) horizontal shear (Fv); (4) compression parallel to the grain (Fc//); (5) compression perpendicular to the grain (Fc') (side-grain crushing); and (6) the modulus of elasticity (E or MOE) (stiffness) (Table 18.13).

Adjustment Factors Base values are adjusted for various Conditions of Use. Conditions of use and their application to base values are shown in Table 18.14.

The seven Conditions of Use are:

Size Factors (Cf)—Applied to dimension base values

Repetitive Member Factors (Cr)—Applied to size-adjusted Fb (bending stress)

Duration of Load Adjustment (Cd)—Applied to size-adjusted values

Horizontal Shear Adjustments (Ch)—Applied to Fv (horizontal shear) values

Table 18.13 Base Values for Western Dimension Lumber[a]

Species or Group	Grade	Extreme Fiber Stress in Bending "F_b" Single	Tension Parallel to Grain "F_t"	Horizontal Shear "F_v"	Compression Perpendicular "F_c"	Compression Parallel to Grain "$F_{c//}$"	Modulus of Elasticity "E"
Douglas Fir/ Larch	Select Structural	1,450	1,000	95	625	1,700	1,900,000
	No. 1 & Btr.	1,150	775	95	625	1,500	1,800,000
	No. 1	1,000	675	95	625	1,450	1,700,000
	No. 2	875	575	95	625	1,300	1,600,000
	No. 3	500	325	95	625	750	1,400,000
	Construction	1,000	650	95	625	1,600	1,500,000
	Standard	550	375	95	625	1,350	1,400,000
	Utility	275	175	95	625	875	1,300,000
	Stud	675	450	95	625	825	1,400,000

[a]Sizes: 2" to 4" thick by 2" and wider.

Source: Western Wood Products Association

Flat Use Factors (Cfu)—Applied to size-adjusted Fb (bending stress)

Adjustments for Compression Perpendicular to Grain (Cc') (compression perpendicular) values

Wet Use Factors (Cm)—Applied to size-adjusted values

How to Apply Adjustment Factors   How to find the adjusted bending stress, Fb, for a Select Structural 2 × 6 in. member from the Douglas Fir-Larch group: The base value for DF-L in SS grade is 1,450 psi. This is multiplied by the size value, which is 1.3—1,450 × 1.3 = 1,885 psi (size adjusted Fb). This can now be adjusted for other conditions of use, such as for a Repetitive Member Factor, which is 1.15—1,885 × 1.15 = 2,167 psi (adjusted Fb).

Various lumber associations have published new span tables for members, such as joists and rafters. These tables include some applications of conditions of use. Before using the span tables, it is necessary to note which conditions of use have been applied.

PHYSICAL AND CHEMICAL COMPOSITION OF WOOD

Since wood is a naturally occurring material, it has considerable variation in its physical properties, including color, density, weight, and strength. The physical and chemical composition of wood determines its properties and, therefore, its uses.

Porosity of Wood

Wood is a cellular material, as shown in Figure 18.17 and Figure 18.18. Softwood cellular structure contains large longitudinal cells called tracheids, and smaller radial cells called rays, both of which store and transfer nutrients. The annual rings are also cellular. The structure of hardwoods is more complex, having two different types of longitudinal cells, small-diameter fibers, and larger diameter vessels or pores, which transport the sap of the tree. They also have a higher percentage of rays than is found in softwoods.

The surface area of these cells is very large and gives wood several important properties. First, it makes it possible for wood when dry to absorb toxic chemicals needed to prevent decay and insect attack. Second, it can absorb moisture repellents to minimize moisture exchange and, therefore, control shrinkage. Third, the cells are air pockets that provide insulating qualities. Fourth, it enables wood to shrink and swell as moisture content varies. Fifth, it contributes to the ease of adherence of paint, adhesives, and other synthetic resins used on wood surfaces.

Composition of Wood

The cells are made of mainly cellulose and hemicellulose fibers that are bonded together with an organic substance called *lignin*. Cellulose and hemicellulose are complex glucose compounds. Glucose is a sugar useful to fungi and insects as a food. Although the exact amounts of cellulose, hemicellulose, and lignin vary with different species of wood, the general composition in kiln-dried woods is shown in Table 18.15. These elements are what

Table 18.14 Adjustment Factors for Base Values for Western Dimension Lumber

Size Factors (C_F) (Apply to Dimension Lumber Base Values)

		F_b				
Grades	Nominal Width (depth)	2" & 3" thick nominal	4" thick nominal	F_t	$F_{c\perp}$	Other Properties
Select Structural, No. 1 & Btr., No. 1, No. 2 & No. 3	2", 3" & 4"	1.5	1.5	1.5	1.0	1.0
		1.15	1.0			
	5"	1.4	1.4	1.4	1.1	1.0
	6"	1.3	1.3	1.3	1.1	1.0
	8"	1.2	1.3	1.2	1.05	1.0
	10"	1.1	1.2	1.1	1.0	1.0
	12"	1.0	1.1	1.0	1.0	1.0
	14" & wider	0.9	1.0	0.9	0.9	1.0
Construction & Standard	2", 3", & 4"	1.0	1.0	1.0	1.0	1.0
Utility	2" & 3"	0.4	—	0.4	0.6	1.0
	4"	1.0	1.0	1.0	1.0	1.0
Stud	2", 3", & 4"	1.1	1.1	1.1	1.05	1.0
	5" & 6"	1.0	1.0	1.0	1.0	1.0

Repetitive Member Factor (C_r) (Apply to Size-Adjusted F_b)

Where 2"- to 4"-thick lumber is used repetitively, such as for joists, studs, rafters, and decking, the pieces side by side share the load and the strength of the entire assembly is enhanced. Therefore, where three or more members are adjacent or are not more than 24" apart and are joined by floor, roof, or other load distributing elements, the F_b value can be increased 1.15 for repetitive member use.

Repetitive Member Use

$$F_b \times 1.15$$

Duration of Load Adjustment (C_d) (Apply to Size-Adjusted Values)

Wood has the property of carrying substantially greater maximum loads for short durations than for long durations of loading. Tabulated design values apply to normal load duration. (Factors do not apply to MOE or F_c).

Load Duration	Factor
Permanent	0.9
Ten years (normal load)	1.0
Two months (snow load)	1.15
Seven day	1.25
One day	1.33
Ten minutes (wind and earthquake loads)	1.6
Impact	2.0

Confirm load requirements with local codes. Refer to Model Building Codes or the National Design Specification for high-temperature or fire-retardant treated adjustment factors.

Flat Use Factors (C_{fu}) (Apply to Size-Adjusted F_b)

Nominal Width	Nominal Thickness	
	2" & 3"	4"
2" & 3"	1.00	—
4"	1.10	1.00
5"	1.10	1.05
6"	1.15	1.05
8"	1.15	1.05
10" & wider	1.20	1.10

Horizontal Shear Adjustment (C_H) (Apply to F_v Values)

Horizontal shear values are based upon the maximum degree of shake, check or split that might develop in a piece. When the actual size of these characteristics is known, the following adjustments may be taken.

2" Thick Lumber

For convenience, the table below may be used to determine horizontal shear values for any grade of 2"-thick lumber in any species when the length of split or check is known and any increase in them is not anticipated.

3" and Thicker Lumber

Horizontal shear values for 3" and thicker lumber also are established as if a piece were split full length. When specific lengths of splits are known and any increase in them is not anticipated, the following adjustments may be applied.

When Length of Split on Wide Face is:	Multiply Tabulated F_v Value by:	When Length of Split on Wide Face is:	Multiply Tabulated F_v Value by:
No split	2.00	No split	2.00
½ of wide face	1.67	½ of narrow face	1.67
¾ of wide face	1.50	1 of narrow face	1.33
1 of wide face	1.33	1½ of narrow or more	1.00
1½ of wide face or more	1.00		

Adjustments for Compression Perpendicular to Grain (C_c) (For Deformation Basis of 0.02" Apply to F_c Values)

Design values for compression perpendicular to grain (F_c) are established in accordance with the procedures set forth in ASTM Standards D 2555 and D 245. ASTM procedures consider deformation under bearing loads as a service ability limit state comparable to bending deflection because bearing loads rarely cause structural failures. Therefore, ASTM procedures for determining compression perpendicular to grain values are based on a deformation of 0.04" and are considered adequate for most classes of structures. Where more stringent measures need to be taken in design, the following formula permits the designer to adjust design values to a more conservative deformation basis of 0.02":

$$Y_{02} = 0.73\, Y_{04} + 5.60$$

Example:	Douglas Fir-Larch: $Y_{04} = 625$ psi
	$Y_{02} = 0.73\,(625) + 5.60 = 462$ psi

Wet Use Factors (C_M) (Apply to Size-Adjusted Values)

The design values shown in the accompanying tables are for routine construction applications where the moisture content of the wood does not exceed 19%. When use conditions are such that the moisture content of dimension lumber will exceed 19%, the Wet Use Adjustment Factors below are recommended:

	Property	Adjustment Factor
F_b	Extreme fiber stress in bending	0.85*
F_t	Tension parallel to grain	1.0
F_c	Compression parallel to grain	0.8**
F_v	Horizontal shear	0.97
F_c	Compression perpendicular to grain	0.67
E	Modulus of elasticity	0.9

*Fiber Stress in Bending Wet Use Factor 1.0 for size-adjusted F_b not exceeding 1,150 psi.
**Compression Parallel to Grain in Wet Use Factor 1.0 for size-adjusted F_c not exceeding 750 psi.

Figure 18.17 A greatly enlarged example of wood cellular structure. Tracheids (TR) are vertical cells that make up the major part of the structure. Rays (WR) are cells that run radially from the center to the outside of the tree. Annual rings (AR) are made up of small hard summerwood cells (SM) and larger, softer springwood cells (S). Resin is in horizontal resin ducts (HRD) and vertical resin ducts (VRD) centered in fusiform wood rays (FWR). Simple pits (SP) allow sap to pass back and forth between the ray cells and the longitudinal cells. Border pits (BP) transfer sap between longitudinal cells. Face RR indicates a radial cut through the wood and TG indicates a tangential cut.

Figure 18.18 Hardwood trees have a more complex structure. The rays (WR) make up the major mass and produce the grain features. Large vertical cells with pores (P) move the sap. Wood fibers (F) give the tree structural strength. Pits (K) transfer sap from one cell to another, and rays (WR) run radially from the center of the tree. A radial cut is indicated by RR and a tangential cut by TG. An annual ring is shown by AR.

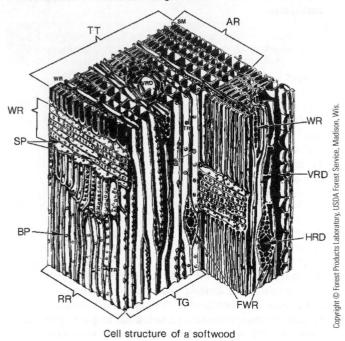

Cell structure of a softwood

Copyright © Forest Products Laboratory, USDA Forest Service, Madison, Wis.

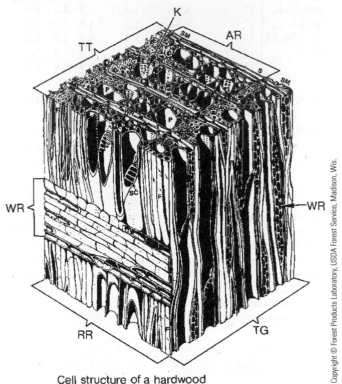

Cell structure of a hardwood

Copyright © Forest Products Laboratory, USDA Forest Service, Madison, Wis.

Table 18.15 Composition of Kiln-Dried Wood

Material	Softwoods Percent	Hardwoods Percent
Cellulose	40–50	40–50
Hemicellulose	20	15–35
Lignin	23–33	16–25
Extraneous materials	5–10	5–10

give wood its strength, susceptibility to decay, and hygroscopic properties. *Hygroscopic* refers to the property of wood that permits it to absorb and retain moisture. Cellulose provides strength in tension, toughness, and elasticity. Lignin, because it bonds the fibers together in fiber bundles, gives wood its compressive strength. The remaining substances in wood do not contribute to its structure but give each species its unique color, odor, taste, density, resistance to decay, and flammability.

Paper products are made using the cellulose and hemicellulose in wood. The wood is cut into chips and cooked to separate the fibers from the lignin. The fibers form the pulp used to make paper and cardboard products. Hemicellulose plays an important part in fiber-to-fiber bonding in the papermaking process.

Cellulose is also used in some textiles, plastics, and other products requiring cellulose derivatives. Extractive materials derived from wood include tannins, coloring matters, essential oils, fats, resins, waxes, gums, starch, and simple metabolic intermediates.

HYGROSCOPIC PROPERTIES OF WOOD

Wood is hygroscopic, expanding when it absorbs moisture and shrinking when it loses it. During its life, a tree contains considerable water. In this condition, wood is called green. The moisture content of wood must be reduced to a level acceptable for its intended purpose.

Moisture Content

The *moisture content* of wood is the weight of water in wood divided by the weight of the wood, when oven dry, expressed as a percentage. Green samples are weighed, oven dried, and weighed again to ascertain moisture content. For example, if a sample weighs 4 oz. green and 3 oz. dry, then it has a moisture content of 1 oz. of water (or 33 percent). On the job, moisture content is checked with a moisture meter. When a probe is placed in contact with the wood, this battery-operated unit shows the moisture content on a meter.

The moisture content of lumber from newly cut trees varies by species but ranges from 30 percent to as much as 200 percent. The moisture content of sapwood is higher than that of heartwood. For example, in the various species of pine, the heartwood moisture content ranges from 30 to 98 percent, and the sapwood ranges from 100 to 220 percent. In general, hardwoods have lower moisture contents when freshly cut than do softwoods.

Fiber Saturation Point

In a living tree or freshly cut wood (green), moisture is present in the cell fibers as absorbed water and in the cell cavities as free water. As the wood begins to dry, the water in the cell cavities begins to disappear. Once all free water in the cell cavities is gone, the cell fibers making up the cell walls contain any remaining water. This is the *fiber saturation point*. Most softwoods have a moisture content of about 30 percent at this point. As moisture is removed from the cell walls, they begin to shrink. Therefore, the fiber saturation point is the point at which shrinkage begins (Figure 18.19).

The total shrinkage in lumber occurs between the fiber saturation point, 30 percent, and the desired moisture content, such as the 15 percent required for most lumber. Shrinkage is approximately proportional to the amount of moisture loss. Wood shrinks about 1/30 of the total shrinkage with each 1 percent reduction in moisture below the fiber saturation point. The reduction mentioned above, 30 percent to 15 percent, is about half the possible shrinkage.

Equilibrium Moisture Content

Green lumber gives off moisture to the air and will continue to take on and give off moisture as the air's moisture content varies. When this point is reached, the wood has reached its *equilibrium moisture content*.

Since wood reacts to constantly varying changes in humidity and temperature, it is constantly seeking to reach equilibrium with the surrounding air. Variations tend to be seasonal, and there is less variation inside heated buildings than on the exterior.

As wood seeks equilibrium, it gradually shrinks or expands. Excessive expansion or contraction can cause problems in a building, such as sticking doors and windows. Cracks occur in the interior wall finish if studs and plates experience excessive shrinkage. To minimize the amount of expansion and contraction, wood should be installed at a moisture content as close as possible to the equilibrium moisture content it will have after installation. The recommended moisture content for interior wood, such as flooring, trim, cabinets, windows, and doors, is between 6 and 12 percent. This varies by region, from the dry Southwest to the more moisture-laden areas of the Northwest and Southeast (Figure 18.20). Since wood shipped to a job can have its moisture content increased or decreased during shipping or storage, it is recommended that wood elements be stored in the area where they will be used for several days so they can approach equilibrium before installation.

The moisture content for exterior wood products is usually 12 to 15 percent for most areas of the country. Products such as wood siding are exposed to greater variations in humidity and temperature and reach equilibrium at a higher moisture content. For example, surfaced framing lumber having a moisture content at maximum 15 percent is stamped MC15. If it is surfaced and has a moisture content at a maximum of 19 percent, it is stamped S-DRY (surfaced dry). If it has more than 19 percent moisture, it is stamped S-GRN (surfaced green).

To ensure that lumber does not exceed the moisture content stamped on it, mills sometimes dry it several percentages below the specified level. In dry regions, framing lumber should have a moisture content of not more than 15 percent when interior finishes are installed. In most of the country, a maximum moisture content of 19 percent is satisfactory.

Figure 18.19 As wood dries, the cells lose free water.

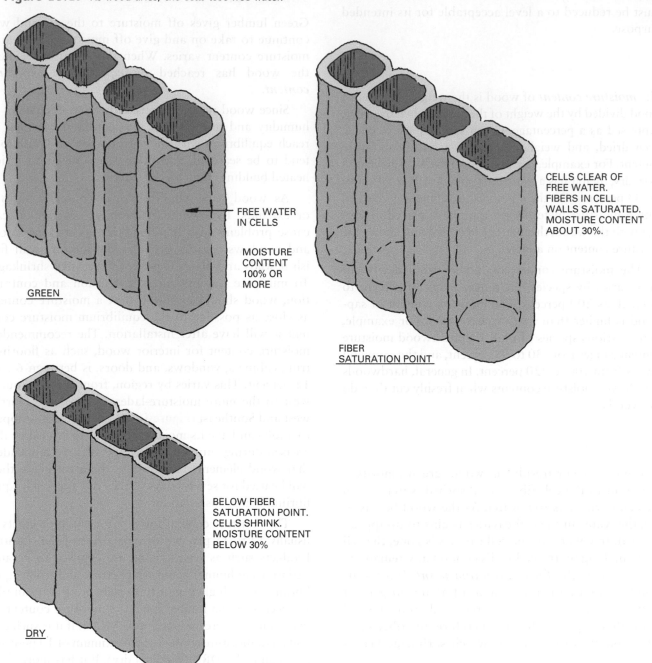

GREEN

FREE WATER
IN CELLS

MOISTURE
CONTENT
100% OR
MORE

FIBER
SATURATION POINT

CELLS CLEAR OF
FREE WATER.
FIBERS IN CELL
WALLS SATURATED.
MOISTURE CONTENT
ABOUT 30%.

BELOW FIBER
SATURATION POINT.
CELLS SHRINK.
MOISTURE CONTENT
BELOW 30%

DRY

How Moisture Affects Wood Properties

The moisture content of wood affects its size, dimensional stability, strength, stiffness, decay resistance, glue bonding, and paintability.

Size and dimensional stability are directly affected by the moisture content of wood. Shrinkage and swelling occur only after the moisture content falls below the fiber saturation point. Most woods shrink and swell very little parallel with the grain (longitudinally). This shrinkage has very little influence on construction uses. Wood shrinks and swells a great deal in thickness and width across the grain. Shrinkage is greatest in the direction parallel to the annual growth rings (tangential) and about twice as much as the shrinkage across the rings (radial).

Figure 18.20 The equilibrium moisture content varies in different sections of the country.

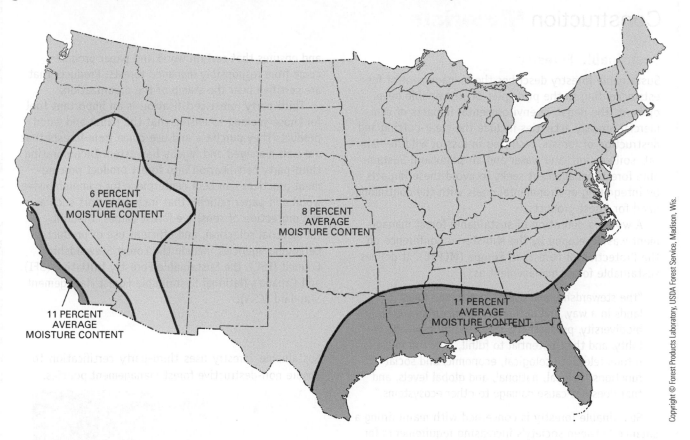

6 PERCENT
AVERAGE
MOISTURE CONTENT

8 PERCENT
AVERAGE
MOISTURE CONTENT

11 PERCENT
AVERAGE
MOISTURE CONTENT

11 PERCENT
AVERAGE
MOISTURE CONTENT

Figure 18.21 shows the combined effects of tangential and radial shrinkage as wood dries from its green condition. Notice that the shrinkage is affected by the direction of the growth rings. Tangential shrinkage is about twice as great as radial shrinkage. Tangential shrinkage is that which is tangent to the circumference

Figure 18.21 Typical shrinkage and distortion of various wood shapes in relation to the direction of growth rings.

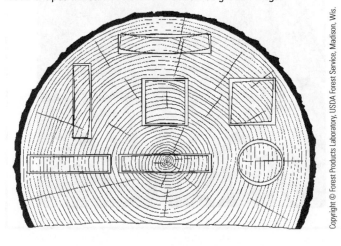

of the tree. Radial shrinkage is that which occurs parallel to a line passed through the center of the tree. This produces distortions in the lumber. The nature of the distortion depends on the location the piece of lumber occupied in the tree. These distortions due to shrinkage are greater in plain-sawed lumber.

Most hardwoods shrink more than softwoods, and heavier species shrink more than lighter species. Stock with a large cross-sectional area, such as 6 in. × 6 in. timbers, do not shrink as much proportionately because the inside does not dry at the same rate as the outside. The outer layers dry faster, become set, and keep the inner area from shrinking normally as it dries.

Softwood lumber shrinks about ¹⁄₃₂ in. per inch of face width while drying from a green condition to about 19 percent moisture content. This means a rough cut 2 in. × 10 in. board shrinks about ¼ in. in width as it dries to 19 percent moisture content and about ⁵⁄₁₆ to ³⁄₈ in. when it reaches 15 percent moisture content.

As the moisture content of wood decreases, its strength increases. This is caused by the stiffening and

Construction Materials

Sustainable Forestry

Sustainable forestry describes the management of forests according to the principles of sustainable development. The negative environmental impacts of commercial forestry activities include the clear-cutting and destruction of forests, resulting in loss of wildlife habitat, soil erosion, and water and air pollution. Sustainable forest management seeks to avoid these impacts by integrating environmental goals with the continuing need for forest products.

A working definition of sustainable forest management was developed by the Ministerial Conference on the Protection of Forests in Europe (MCPFE). It defines sustainable forest management as:

> "the stewardship and use of forests and forest lands in a way, and at a rate, that maintains their biodiversity, productivity, regeneration capacity, vitality, and their potential to fulfill, now and in the future, relevant ecological, economic, and social functions, at local, national, and global levels, and that does not cause damage to other ecosystems."

Sustainable forestry is concerned with maintaining a balance between society's increasing requirements for forest products and the preservation of forest health and biodiversity. The elements of sustainable forestry include sustainable harvesting techniques, preserving habitat for the biodiversity of plants and animals, minimizing the use of toxic chemicals, and protecting soil and water quality.

Forest managers must assess and integrate a wide array of sometimes-conflicting factors, including commercial values, environmental considerations, and community interests, to produce sound forest management plans. In most cases, forest managers develop their plans in consultation with citizens, businesses, government organizations, and other interested parties in and around the forest tract being managed.

Growing environmental awareness and consumer demand for more socially responsible businesses helped third-party forest certification emerge as a credible tool for communicating the environmental and social performance of forest operations. With forest certification, an independent organization develops standards of good forest management, and independent auditors issue certificates to forest operations that comply with those standards. This certification verifies that forests are well managed, as defined by a particular standard, and ensures that certain wood and paper products come from responsibly managed forests. Products that are certified bear the stamp of the organization.

Third-party forest certification is an important tool for those seeking to ensure that the paper and wood products they purchase and use come from forests that are well managed and legally harvested. Incorporating third-party certification into forest product procurement practices can be a centerpiece for comprehensive wood and paper policies that include factors such as the protection of sensitive forest values, thoughtful material selection, and efficient use of products. Certifying agencies include the Forest Stewardship Council (FSC), the Sustainable Forestry Initiative (SFI), and Canada's National Sustainable Forest Management Standard (CSA).

Sustainable forestry uses third-party certification to ensure non-destructive forest management policies.

© Stephen Rees/Shutterstock.com

strengthening of the fibers in the cell walls and the fact that as the wood shrinks it becomes a denser material. Crushing strength and bending strength increase a great deal in dry woods, while stiffness increases only moderately. Shock resistance depends on the pliability of the material and is lower for dry wood than for the original green stock. Green wood also bends farther before it breaks than dry wood.

Wood that maintains a moisture content below 20 percent will be free from decay. This can be accomplished in any of several ways, including painting or treating with wood-preserving chemicals. The recommended moisture content for wood that will be painted is 16 to 20 percent. Green wood should not be painted, nor should wood that has surface moisture, from rain, for example.

Reducing the moisture content of wood also improves the strength of glue bonds. Wood used for interior purposes is glued at 5 to 6 percent moisture content. Exterior wood can be glued satisfactorily at 10 to 12 percent moisture content. Veneers, such as those used to make plywood, usually have 2 to 6 percent moisture content. Woods glue satisfactorily up to about 15 percent moisture content, but they should be in equilibrium with the air in which they will be used. Interior wood at 12 to 15 percent moisture can be glued satisfactorily, but the glued members will check and warp as the wood loses moisture and seeks equilibrium with the air.

Specific Gravity

Specific gravity is a ratio of the weight of a certain volume of a material to the weight of an equal volume of water at 39.2°F (4°C). The specific gravity of wood is found by weighing a piece of wood of known volume, such as 6 in.^{3}, and dividing this by the weight of an equal volume of water. Water has a specific gravity of 1. If a material has a specific gravity less than this, it will float; if it is greater than 1, it will sink. Examples of specific gravity for selected woods are given in Table 18.16.

The specific gravity of wood fibers forming the cells is about 1.5, so the fibers would sink. However, most woods float, because much of the volume is made up of air-filled cells. The size of the cells and the thickness of the cell walls influence the specific gravity. Specific gravity is an approximate indicator of the density of wood and an indication of strength. In general, woods that are dense weigh more, have a higher specific gravity, and can handle higher stresses.

Table 18.16 Specific Gravity and Weight of Woods Commonly Used in Construction[a]

Softwoods	Specific Gravity	Density[b]
Cedar, western	0.32	20.0
Cypress	0.46	28.7
Douglas fir, coast	0.48	30.0
Larch, western	0.52	32.4
Pine, western white	0.38	23.7
Pine, lodgepole	0.41	25.5
Pine, ponderosa	0.40	25.0
Pine, shortleaf	0.51	31.6
Redwood	0.40	25.0
Spruce, Englemann	0.35	21.6
Spruce, Sitka	0.40	25.0
Hardwoods		
Birch, yellow	0.62	38.7
Cherry, black	0.50	31.2
Maple, sugar	0.63	39.3
Oak, red	0.61	37.9
Oak, white	0.64	39.9

[a]Oven-dry samples.
[b]Density in points per cubic foot based on specific gravity as shown and 0% moisture.

Source: Forest Products Laboratory, USDA Forest Service, Madison, Wis.

STRUCTURAL PROPERTIES OF WOOD

Wood is an orthotropic material. Orthotropic pertains to a mode of growth that is vertical. Wood has unique and independent mechanical properties in the directions of three mutually perpendicular axes: longitudinal, radial, and tangential. The longitudinal axis is parallel to the fibers (grain). The radial axis is normal to the growth rings (perpendicular to the grain in the radial direction). The tangential axis is perpendicular to the grain but tangent to the growth rings.

Wood is a fibrous material. The fibers are bonded together, with lignin forming the walls of the cells making up the material. The fibers in hardwoods are about 1/25 in. long and from 1/8 to 1/3 in. in softwoods. The strength of wood does not depend on the length of the fibers but on the thickness of the cell walls and the direction of the fibers in relation to applied loads. Most fibers are oriented with their lengths parallel with the vertical dimension of the tree. Strength of wood parallel with the fibers (parallel with the grain) is greater than the strength perpendicular to the fibers (perpendicular with the grain).

Structural members under an external load, such as wind, furniture, or people, produce internal forces called stresses in a member to resist these external forces.

Tensile stresses result when an external force tends to stretch a member. Compressive stresses occur when a member is under a squeezing force. When a member, such as a beam, is loaded so that the applied force is acting approximately perpendicular to the member, the load produces a bending stress. When a member is under bending stresses, it develops compressive stresses in the upper part and tensile stresses in the lower part.

The strength of wood under various stresses is found by testing samples in a laboratory. Tests include bending, shear, stiffness, tension, and compression. Samples free of any defects are tested to find the basic stresses for each species used in construction. Samples of the various grades in each species are tested to find realistic stresses

for each. These are lower than the basic stresses and are called working stresses. The working stress takes into account things that lower the load a member can carry, such as knots, pitch pockets, or checks.

The working stresses for several species of wood used in construction are shown in Table 18.17. These are for normal loading conditions, which include dead loads (weight of structural and finish materials) and live loads (weight of occupants or furniture). They are based on the normal duration of loading, which assumes a fully stressed member under the full maximum design load for 10 years and the application of 90 percent of this maximum normal load continuously through the remainder of the life of the structure.

Table 18.17 Working Stresses for Selected Species of Structural Light Framing Lumber[a]

Species	Grade	Extreme Fiber in Bending "F_b"	Tension Parallel to the Grain "F_t"	Compression Parallel to the Grain "F_c"	Horizontal Shear "F_v"	Compression Perpendicular to the Grain "F_c"	Modulus of Elasticity "E"
Southern Pine	Dense Select Structural	2,500	1,500	2,100	105	475	1,900,000
	Select Structural	2,150	1,250	1,800	105	405	1,800,000
	No. 1 Dense	2,150	1,250	1,700	105	475	1,900,000
	No. 1	1,850	1,050	1,450	105	405	1,800,000
	No. 2 Dense	1,800	1,050	1,350	95	475	1,700,000
	No. 2	1,550	900	1,150	95	405	1,600,000
	No. 3 Dense	1,000	575	800	95	475	1,500,000
	No. 3	850	500	675	95	405	1,500,000
	Stud	850	500	675	95	405	1,500,000
Hem-Fir	Select Structural	1,650	975	1,300	75	405	1,500,000
	No. 1/Appearance	1,400	825	1,050/1,250	75	405	1,500,000
	No. 2	1,150	675	825	75	405	1,400,000
	No. 3	650	375	500	75	405	1,200,000
Douglas Fir-Larch	Select Structural	2,100	1,200	1600	95	625	1,800,000
	No. 1/Appearance	1,750	1,050	1,250/1,500	95	625	1,800,000
	No. 2	1,450	850	1000	95	625	1,700,000
	No. 3	800	475	600	95	625	1,500,000

[a]Structural light framing 2″ to 4″ thick and 2″ to 4″ wide MC 15%

Working stresses are given in pounds per square inch and include working stresses for bending (Fb), horizontal shear (Fv), vertical shear (V), modulus of elasticity (E), compression parallel to the grain (Fc), fiber stress in tension (Ft), and compression perpendicular to the grain (Fc').

Bending

Wood beams under load deflect, producing bending stresses in the fibers. Wood has high fiber strengths in bending. However, if a beam is loaded enough to produce stresses greater than the fiber strength of the wood, the beam will fail. The stresses developed in the fibers are greater the farther they are from the central axis of the beam. Fibers located twice as far from the central axis of the beam as other fibers have twice the stress. Therefore, when a beam bends, the maximum stresses are developed in the outer fibers of the top and bottom of the beam. This is called the extreme fiber in bending, Fb (Figure 18.22).

Figure 18.22 Forces designated as extreme fiber stresses in bending occur along the faces of the beam.

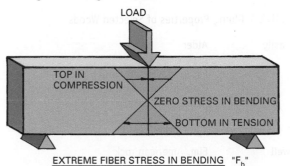

LOAD

TOP IN COMPRESSION

ZERO STRESS IN BENDING

BOTTOM IN TENSION

EXTREME FIBER STRESS IN BENDING "F_b"

LOADS APPLIED TO A BEAM CAUSE IT TO BEND, PRODUCING TENSION IN FIBERS ALONG THE FACE FARTHEST FROM THE LOAD AND COMPRESSION IN FIBERS ALONG THE FACE NEAREST THE LOAD. THESE INDUCED STRESSES ARE DESIGNATED AS "EXTREME FIBER STRESSES IN BENDING, F_b".

Shear

A beam is subject to vertical and horizontal shear (Figure 18.23 and Figure 18.24). **Vertical shear** refers to the tendency for one part of a beam to move vertically in relation to an adjacent part, allowing

Figure 18.23 Horizontal shear stress tends to cause fibers to slice horizontally, much the same as when you bend a deck of cards.

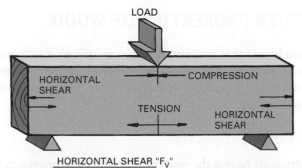

LOAD

HORIZONTAL SHEAR

COMPRESSION

TENSION

HORIZONTAL SHEAR

HORIZONTAL SHEAR "F_V"

HORIZONTAL SHEAR STRESSES TEND TO SLIDE FIBERS OVER EACH OTHER HORIZONTALLY. INCREASING THE BEAM CROSS SECTION DECREASES SHEAR STRESSES.

Figure 18.24 Vertical shear tends to cause fibers in one part of a beam to move vertically in relation to those adjacent them.

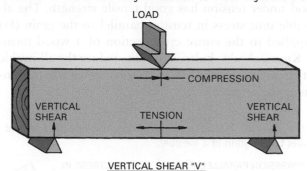

LOAD

COMPRESSION

VERTICAL SHEAR

TENSION

VERTICAL SHEAR

VERTICAL SHEAR "V"

VERTICAL SHEAR STRESSES TEND TO CAUSE ONE PART OF A MEMBER TO MOVE VERTICALLY IN RELATION TO THE ADJACENT PART.

the beam to slip down between supports. **Horizontal shear** refers to the tendency of the sapwood fibers to move horizontally in relation to the bottom fibers. Horizontal shear is more of a factor in beam failure than vertical shear. The fiber stress in horizontal shear must be kept below the working stress values indicated by the symbol Fv.

Modulus of Elasticity

The modulus of elasticity (E) is a measure of a beam's resistance to deflection, or its stiffness. Stiffness is important when selecting columns, beams, rafters, joists, and other structural members (Figure 18.25).

Figure 18.25 The modulus of elasticity is a measure of the stiffness of a beam.

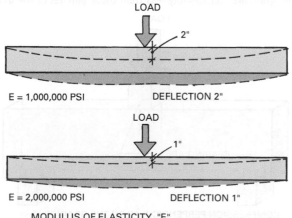

LOAD

2"

E = 1,000,000 PSI   DEFLECTION 2"

LOAD

1"

E = 2,000,000 PSI   DEFLECTION 1"

MODULUS OF ELASTICITY "E"

THE MODULUS OF ELASTICITY (E) IS A RATIO OF THE AMOUNT A MATERIAL WILL DEFLECT IN PROPORTION TO THE APPLIED LOAD.

Tension

Wood under tension has good tensile strength. The allowable unit stress in tension parallel to the grain (Ft) is applied to the entire cross-section of a wood member. Knots, checks, holes, and splits reduce the allowable unit stress in a member. Wood is weak in tension perpendicular to the grain (Figure 18.26).

Figure 18.26 Typical tension and compression stress applied parallel to the grain of a member.

COMPRESSION PARALLEL WITH THE GRAIN "F$_c$"

LOADS ARE SUPPORTED ON THE ENDS OF A PIECE. INTERNAL STRESS IS THE SAME ACROSS THE WHOLE CROSS SECTION AND THE FIBERS ARE UNIFORMLY STRESSED PARALLEL TO AND ALONG THE FULL LENGTH OF THE PIECE.

FIBER STRESS IN TENSION "F$_t$"

LOADS ARE PULLING ON THE ENDS OF A PIECE. INTERNAL STRESS IS THE SAME ACROSS THE WHOLE CROSS SECTION AND TENDS TO STRETCH THE PIECE. LENGTH DOES NOT AFFECT TENSILE STRENGTH.

TENSION PARALLEL WITH THE GRAIN

Compression

The unit of stress in compression parallel to the grain is several times greater than that perpendicular to the grain (Figure 18.27). Selected values for compressive strength perpendicular to grain (Fc) and parallel to grain (Fc) are given in Table 18.17.

Figure 18.27 The working compression stresses perpendicular to the grain are considerably less than those parallel to the grain.

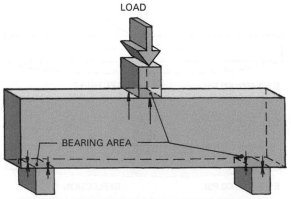

LOAD

BEARING AREA

COMPRESSION PERPENDICULAR TO THE GRAIN "F$_{c\perp}$"

WHERE A MEMBER BEARS ON SUPPORTS, THE LOADS TEND TO COMPRESS THE FIBERS. THE BEARING AREA MUST BE SUFFICIENT TO PREVENT SIDE GRAIN CRUSHING.

Gluing Properties

Glue bonding is improved as moisture content decreases, with 10 to 12 percent moisture effective for exterior products. In addition, some species of wood bond better than others (Table 18.18). A satisfactory glue joint is one that is as strong as the wood itself. Properly made joints are often stronger than the wood, with the wood failing before the glue joint does. The density and structure of the wood also influence the strength of the bond. In general, heartwood does not bond as well as sapwood, and hardwoods do not bond as well as softwoods.

Table 18.18 Gluing Properties of Selected Woods

Bond easily	Alder
	Aspen
	Fir: white, grand, noble, pacific
	Pine: eastern, white, western white
	Red cedar, western
	Redwood
	Spruce, Sitka
Bond well	Elm, American rock
	Maple, soft
	Sycamore
	Walnut, black
	Yellow poplar
	Douglas fir
	Larch, western
	Pine: sugar, ponderosa
	Red Cedar, eastern
	Mahogany: African, American
Bond satisfactorily	Ash, white
	Birch: sweet, yellow
	Cherry
	Hickory: pecan, true
	Oak: red, white
	Pine, southern
Bond with difficulty	Persimmon
	Teak
	Rosewood

Source: Forest Products Laboratory, USDA Forest Service, Madison, Wis.

OTHER PROPERTIES OF WOOD

Wood has other properties, some unique to it as a material, that must be considered. These include thermal properties, decorative features, decay resistance, and insect damage.

Thermal Properties

Softwoods have a thermal conductivity of approximately 1 BTU/in. of thickness. Therefore, wood is a good insulator but not as good as other materials used specifically for insulation. As the moisture content of wood increases, it becomes less efficient as an insulator, because the moisture increases thermal conductivity. Lighter (less dense) woods are better insulators because of their larger cell structure.

Wood also experiences thermal expansion. It expands when heated and contracts when cooled. This is not a factor for most construction purposes because the amount of change is so small. Expansion due to moisture increases is much greater and is the main factor to consider.

Decorative Features

Wood has unique decorative features that make it a valuable finish material. Various species have unique colors, including the mature heartwood and the new growth sapwood. Species vary in density, producing wood with smooth, closed grain to rough, very open grain, which can facilitate beautiful decorative appearances. Wood can also be stained or bleached to alter its color (Figure 18.28).

Figure 18.28 Wood can be easily stained or bleached to alter its color.

© somchai rakin/Shutterstock.com

Wood takes on decorative colors and texture when it is permitted to weather with no protective coating. Some woods, such as cedar and cypress, are decay resistant and weather to a silver-gray. Some species develop a rough, checked surface when weathered, while others do not. All woods tend to warp as they weather, some more than others. Decay resistant woods, such as redwood, cedar, and cypress, warp less than most. Narrow boards, such as 1 × 6 siding, warp less than wider boards.

Decay Resistance

Some species of wood have natural decay-resistant substances in their cell structures that enable the heartwood to resist attack by decay or insects, such as termites.

Decay-resistant woods include cedars, redwood, cypress, black walnut, and black locust. Insect-resistant woods include cypress, some cedars, and redwood. These species may be used for all above-ground uses that are exposed to weather. To be most effective, they should be 100 percent heartwood. Wood used in the ground as foundations for permanent structures should be pressure treated with a wood preservative.

Wood can be attacked by fungi (microscopic plants) that cause decay, molds, and stains. Fungi develop in wood when the moisture content is above 20 percent, temperatures are mild (40° to 100°F), there is sufficient oxygen, and the wood provides an adequate food supply. Wood in the ground is exposed to moisture, oxygen, and a uniform mild temperature, and therefore ideally situated for fungi attack (decay). Wood submerged in water lacks an adequate supply of oxygen.

Often parts of a building under renovation are dry but found to be heavily decayed. Although this condition is often called "dry rot," it occurs because the wood over the years has been intermittently wet (as from a leaking roof) and dry.

A harmless type of fungi called white pocket is found in living softwood trees. When the tree is cut into lumber, the fungi ceases to develop. It will not spread in the cut lumber or transfer to other lumber stored on it. It is an acceptable defect and permitted in most grades of lumber, excepting select grades.

Molds and stains are also caused by fungi but do not damage the wood. The wood may be discolored, but its strength is not impaired. If the wood is used where appearance is important, staining may be undesirable. Molds can often be removed by surface brushing. Stains appear as blue, blue-black, or brown specks, as streaks or as spots.

Insect Damage

Insects bore holes in living trees and cut lumber. Some holes are very small, called pinholes, and others are larger, called grub holes. Lumber graders watch for this damage because it can influence the strength of the stock. When structural integrity is important, insect-damaged lumber is usually rejected.

Completed buildings are also subject to attack by insects, particularly termites. Most parts of the country have termites, but they occur in larger numbers in milder climates. Most termites are subterranean, living in nests in the ground and building tunnels through the earth to get to material containing cellulose, which is their food. This type of termite must have a moist atmosphere to exist. When they attack wood in a building, they must build mud tunnels up the foundation to reach the floor joists. The tunnels protect them from the atmosphere and keep them moist.

Termites will also enter a building through wood that has direct contact with the earth or through very small cracks in a foundation or concrete floor slab. They digest the insides of wood members, leaving a thin outer layer to protect them from drying.

Following are things a builder can do to reduce termite attack:

1. Install a metal termite shield on top of the foundation. Termites have been known to build tunnels around these shields.

2. Remove all wood scraps, paper, and cardboard from the construction site. Any scraps buried, as when backfilling, will provide a food supply and attract a colony of termites.

3. Chemically treat the soil to form a barrier to repel termites. Many cities require this as part of the permitting process.

4. Use chemically treated, pressure impregnated wood for sills, posts, and other parts near the soil.

5. Use poured concrete foundations rather than concrete block. They are less likely to crack and have no open cells through which termites can build tunnels sight unseen inside the foundation.

6. If a concrete block foundation is used, it should be capped with a 4 in. solid concrete cap.

A few areas have a non-subterranean type of termite. They can live in damp or dry wood and do not require contact with the ground. They tend to leave fine wood particles (sawdust) in openings where they entered the wood.

Pressure-Treated Wood and Plywood

Pressure treatment is a process that forces preservatives deep into the cellular structure of wood. When properly treated, the wood resists attacks by insects and fungal decay for many years. The most commonly used approved wood preservatives for general use are Alkaline Copper Quat (ACQ), Copper Azole (CA), and Bardac 22C50. Detailed current information on these and other wood preservatives can be found at *www.epa.gov*.

Before specifying a type of pressure-treated wood, the local building code and a product's manufacturer should be consulted. The Environmental Protection Agency (EPA) regularly reviews requirements and sometimes approves the use of a particular treatment for a set number of years. The American Wood Preservers Association (AWPA) is involved in setting the standards for pressure-treated wood.

Metal fasteners and connections of pressure-treated components should be stainless steel or hot-dipped galvanized steel containing 1.85 oz. of zinc.

Safety Information

There are certain hazards when working with pressure-treated wood. Workers should always wear gloves, goggles, hats, and a high-quality dust mask or respirator. Pressure-treated wood should never be allowed to come into contact with drinking water. It must be properly disposed of and never burned.

Chromated Copper Arsenate (CCA)

Chromated Copper Arsenate (CCA) is no longer approved as the primary wood preservative used in residential, recreational, and general consumer construction. It has been replaced by arsenic-free alternatives such as Alkaline Copper Quat (ACQ), Copper Azole (CA), and Bardac®. CCA is approved for a variety of commercial and industrial applications, such as poles, marine piling, highway guardrails, highway noise barriers, and some agricultural uses.

Alkaline Copper Quat (ACQ)

Alkaline Copper Quat was developed as an alternative to CCA. It is an EPA-approved, non-arsenic, non-chromium, water-based wood preservative that provides a durable lumber and plywood product used for structural applications where protection from decay and insects is required (Figure 18.29). It is approved for full exposure to aboveground, ground-contact, and freshwater applications. Typical uses include building framing, playground equipment, posts, and decks. It is suitable for use in sensitive aquatic environments. It can be painted and stained and should be sealed with a water-repellant wood sealer.

Three formulations of ACQ are available. ACQ-B is used mainly to treat Western wood species, such as Douglas fir. ACQ-C is used in all parts of the United States. ACQ-D is used in all parts of the country except on the West Coast.

Figure 18.29 Alkaline Copper Quat-treated wood is widely used for posts and decks.

© Marcus P. Turner/Shutterstock.com

Bardac-Treated Wood

Bardac treatment can be applied to framing lumber and plywood using a pressure treatment process. It is used on wood framing that is continuously protected from the weather. It is not suitable for exposure to the ground. It protects against wood-destroying insects and decaying fungi. Framing members are sawed and assembled using fastener systems typically employed with untreated lumber.

Copper Azole (CA)

Copper Azole is a water-based wood preservative that prevents decay from fungi and insects. There are two types: CBA-A and CA-B. CBA-A is primarily used for treating softwood species, including Southern Pine and hemlock-fir. CA-B is primarily used to treat the wide range of softwood species.

Copper Azole–treated wood is approved for full exposure to above-ground, ground-contact, and fresh-water applications. Typical applications include millwork, shingles, shakes, siding, plywood, structural lumber, and wood used on above ground applications, such as decks and playground equipment. It should be sealed with water-repellant sealer and reapplied regularly to keep surfaces from cracking and splitting.

Pentachlorophenol

Pentachlorophenol was used for many years as a wood preservative. It is not EPA-registered for residential uses as a preservative.

Methyl Isothiocyanate

Methyl isothiocyanate is registered by the EPA as a wood preservative pesticide to control wood rot and decay-causing fungi. It is used on large structural timbers, such as utility poles, pilings, and bridge timbers. It is a restricted-use product for use only by specially trained certified applicators. It is reviewed regularly by the EPA.

Applying Preservatives

Wood products are treated by pressure and nonpressure processes.

Pressure Treatment Pressure treatment involves impregnating wood with a desired preservative by placing the wood in a closed vessel and raising the pressure considerably above atmospheric pressure. The pressure forces the preservative into the wood until the desired amount has been absorbed. Considerable preservative is absorbed with relatively deep penetration. There are several different pressure processes in use, but the basic principles are the same. Pressure processes provide a closer control over preservation retentions and penetrations and generally better protection than nonpressure methods.

Nonpressure Processes A wide range of nonpressure processes are used. These differ widely in penetration and retention; therefore, the degree of protection afforded varies. Application by brushing is the easiest but least effective. Dipping for a few minutes in a preservative gives greater assurance that all faces and checks have been coated. This is often used for assembled window sashes, frames, and other exterior millwork. A third method is cold-soaking. Well-seasoned wood is soaked for anywhere from several hours to several days in low-viscosity oil preservatives. Some woods, such as pine, are successfully treated this way and have a long life when in contact with the ground. The diffusion process is used with green and dry wood. It uses a water-borne preservative that diffuses out of the treating solution or paste into the wood. The double diffusion process is more effective. It involves steeping wood first in one chemical and then another. Other diffusion processes include injecting a preservative at the ground line of a pole, applying a paste to the surface, and pouring the chemical into holes bored in the pole at ground line.

Another nonpressure process is a thermal process. In a hot bath, the air in the wood expands and some air is forced out. In the cooling bath that follows, the air contracts to create a partial vacuum that forces liquid into the wood. It is used for fence posts and lumber. One last nonpressure process is the vacuum process. It utilizes a quick, low initial vacuum in a chamber, followed by a brief immersion in the preservative, followed by a final high vacuum. It is used to treat millwork with water-repellent preservatives and construction lumber with water-borne and water-repellent preservatives.

Effect of Treatment on Strength

Chemicals in water-borne salt preservatives, such as chromium, copper, arsenic, and ammonia, are reactive with wood. They are potentially damaging to mechanical properties and can cause corrosion in mechanical fasteners. At retention levels required for ground contact, mechanical properties are essentially unchanged, but maximum load in bending, impact bending, and toughness are reduced somewhat. Heavy salt loadings for marine use may reduce bending strength by 10 percent and working properties by 50 percent.

Quality Certification

The American Wood Preservers Association (AWPA) is a professional society responsible for establishing consensus standards for the wood preserving industry. The American Lumber Standards Committee certifies wood preservation inspection agencies to provide quality control services to individual wood preserving plants. The accrediting agencies permit the use of their quality stamp, which indicates to the consumer that a product meets established standards.

The National Wood Window and Door Association (NWWDA) has a testing, plant inspection, and certification program for non-pressure-treated products. They provide standards for water-repellent preservatives, wood window units, wood flush doors, wood sliding patio doors, wood skylight and roof windows, and wood swinging doors.

FIRE-RETARDANT TREATMENTS

In certain applications, wood must be treated with fire-retardant chemicals. The two general methods for applying fire-retardant chemicals are pressure-impregnating the wood with water-borne or organic solvent-borne chemicals, or applying fire-retardant chemical coatings to the wood's surface. Pressure-treating chemicals include inorganic salts and complex chemicals. Salts are most commonly used. They react to temperatures below the ignition point of wood, causing the combustible vapors generated in the wood to break down into nonflammable water and carbon dioxide. After treatment, the wood should be dried to its original required moisture content.

Coatings have low surface flammability, and when they are exposed to fire, they form an expanded low-density film. The film insulates the surface from high temperatures. Most fire-retardant chemicals do not resist exposure to weather, so it is necessary to use leach-resistant types for exterior use, such as on wood shakes. Fire-retardant treatment results in some slight reduction in the strength properties of wood, so design values for allowable stresses are reduced. Most chemicals cause fasteners to corrode. Designers must select a combination of chemicals and metal fasteners that can coexist without corrosion. Crystal salts in wood have an abrasive effect on cutting tools. Carbide-tipped cutting tools should be used when working with fire-retardant wood. Gluing is also a problem. Special resorcinol-resin adhesives have proven acceptable. Fire-retardant wood can be painted if the moisture content has been reduced sufficiently.

Review Questions

1. How can you tell a coniferous tree from a deciduous tree?

2. What natural defects are found in wood?

3. What are the two ways wood is seasoned?

4. What is the recommended moisture content for lumber used for framing?

5. Why is surfaced green softwood lumber produced more than softwood lumber that is surfaced when dry?

6. What standard is followed in the production of softwood lumber?

7. How many board feet are in a 2 × 10 × 12 ft piece of stock?

8. What is the volume of 250 pieces 38 mm × 140 mm × 3.66 m?

9. How does a contractor know lumber received on a job has been dried to a 15 percent moisture content?

10. How can you differentiate between dimension lumber and timbers?

11. What is stress-rated lumber?

12. What is meant by the equilibrium moisture content?

13. In what ways does moisture affect wood?

14. What will be the size of a 5 in. × 8 in. rough cut timber after it dries?

15. What is the recommended moisture content for wood to be painted?

16. Explain why wood floats even though the wood fibers have a specific gravity of 1.5.

17. What is a good moisture content for producing a satisfactory glue bond?

18. Which species of wood have natural resistance to decay?

19. What types of preservatives are used on wood?

20. What are the two processes used to apply fire retardant chemicals to wood?

Key Terms

Air Drying	Hardwood	Quarter-Sawing
Board Foot	Heartwood	Rough Lumber
Boards	Horizontal Shear	Sapwood
Dimension Lumber	Hygroscopic	Seasoning
Dressed	Kiln Drying	Softwood
Equilibrium Moisture Content	Lignin	Stress-Rated Lumber
Fiber Saturation Point	Lumber	Timber
Grade	Lumber Grader	Vertical Shear
Green Lumber	Moisture Content	Warp

Activities

1. Collect a variety of samples of wood species used in construction and conduct various tests, such as:

    a. Tension and compression tests.

    b. Moisture-content tests. Place samples outdoors for several weeks and test again. What happened to the moisture content?

    c. Drive common nails into the samples and devise a way to measure the pounds of pull required to remove them. Which woods had the best holding characteristics?

2. Prepare a visual aid showing actual samples of species used in construction. Label each and list its properties and characteristics. Try to find samples having commonly found defects and identify them.

3. Collect samples of freshly cut (green) woods used in construction. Machine them to a carefully selected thickness, width, and length. Measure the moisture content. In an oven, or by some other method (air drying), let the samples dry until they reach a moisture content of 15 percent, then measure the size of the samples. Record your findings in a written report.

4. Visit a local lumber yard and examine and make a list of the sizes and species available. Pay special attention to the type of treatment used on pressure-treated rot-resistant woods.

5. Invite a local exterminator to address the class and cite the chemicals and applications used to treat a building for termites and other insects. Ask questions about hazards and safety precautions taken.

Additional Resources

Timber Construction Manual, Herzog, Natterer, Schweizer, Volz, Winter, Birkhauser Edition Detail.

Timber Construction Manual, American Institute of Timber Construction, Englewood, CO.

Wood Building Technology, Canadian Wood Council, Ottawa, Ontario, Canada.

Wood Handbook: Wood as an Engineering Material, Forest Products Laboratory, Madison, WI.

Other resources include:

Many publications available from U.S. Department of Housing and Urban Development, HUD USER, Rockville, MD.

Numerous publications from the Southern Forest Products Association, Kenner, LA.

Pressure-treated wood information: Environmental Protection Agency; ASRC Aerospace Corp.; Ariel Rios Building, 1200 Pennsylvania Ave. NW, Washington, DC 20460; www.epa.gov/oppad001/reregistration/cca.

Forestry Stewardship Council, United States, www.fscus.org/green_building.

Products Manufactured from Wood

LEARNING OBJECTIVES

Upon completion of this chapter, the student should be able to:

- Use technical information on industrial plywood to make construction product decisions.
- Select the appropriate reconstituted wood-panel products for various applications.

- Identify and select hardwood plywood for construction applications.
- Identify and select manufactured wood structural components used for building construction.
- Be aware of the types of many products manufactured from wood.

Build Your Knowledge

For further study on wood products and materials and methods, please refer to:

Chapter 29 Doors, Windows, Entrances, and Storefronts

Chapter 33 Interior Walls, Partitions, and Ceilings
            Topic: Wood Product Wall Finishes

Chapter 34 Flooring

Until recently, building larger structures out of wood has not been possible because of the relative weakness of conventional wood construction methods. Today, the development of a new generation of engineered wood building products has expanded the structural capabilities of the material.

Engineered wood products are manufactured by bonding together wood strands, veneers, lumber, or other forms of wood fiber with glue to form larger, more efficient composite structural units. These products include wood panels (including plywood, oriented-strand board, and composite panels), glued-laminated timber, and other structural composite lumber (SCL) products.

Engineered wood products display highly predictable and reliable performance characteristics and provide enhanced design flexibility over solid lumber. These products are extremely resource efficient because they make use of

otherwise-undervalued species, and previously discarded waste products, such as wood chips and sawdust. Engineered wood provides a new material of high consistency with reduced swelling and shrinkage that allow for the use of smaller sections to achieve longer spans. Not only are these products more environmentally friendly, but they are also often less expensive than building materials such as steel or concrete. This chapter presents information on the variety of products and components made from wood.

PLYWOOD AND OTHER PANEL PRODUCTS

A range of veneered and bonded manufactured panels are made from wood. These include plywood, oriented-strand board, waferboard, particleboard, and hardboard, among others.

Plywood Panel Construction

Plywood, manufactured for nearly a century, is one of the most widely used wood products. The manufacturing process uses a lathe to peel thin veneer from water-saturated logs (Figure 19.1). **Plywood** panels are made by bonding together the thin layers, or plies, of wood. The grain in each ply layer is arranged perpendicular to

Figure 19.1 Plywood manufacturing uses a lathe to peel thin veneers from a log.

© BAY ISMOYO/AFP/Getty Images

the grain of the next layer. Panels usually have an odd number of plys, such as three, five, or seven. Each ply may be a single thickness veneer or two veneers glued together.

Plywood types commonly used in building construction are veneer core, lumber core, particleboard core, and medium-density fiberboard core. The veneer-core panel has from three to nine plies (Figure 19.2). The lumber-core panel consists of strips of solid wood glued together as the core and two veneers glued to each side. Particleboard-core and medium-density fiberboard-core panels contain a single sheet of one of these materials as a core and a single ply glued to each side. Each type has the same number of plies on each side of the core to ensure balanced construction.

Figure 19.2 Veneer-core panel plywood has from three to nine plies that are positioned with their grain perpendicular to each other.

© jocic/Shutterstock.com

Specifications for Plywood Panels

Some grades of veneered plywood panels are manufactured under specifications or performance testing standards of U.S. Product Standard PS 1-83, *Construction and Industrial Plywood*. This manufacturing specification was developed cooperatively by members of the plywood industry and the Office of Product Standards Policy of the National Bureau of Standards. Other veneered panels, including a number of performance-rated composite and non-veneered panels, are manufactured under provisions of APA–The Engineered Wood Association performance standards. APA-rated panels that meet PS 1-83 requirements have the designation "PS 1-83" in the APA trademark.

In Canada, there are three standards for softwood plywood, all of which are in metric terms: CSA 0121-M *Douglas Fir Plywood*, CSA 0151-M *Canadian Softwood Plywood*, and CSA 0153-M *Poplar Plywood*. Allowable stresses and section properties for Douglas fir plywood are laid out in CAN3-086-M.

Construction and Industrial Plywood

Construction and industrial plywood are manufactured according to the specifications in U.S. Product Standard PS 1-83. Canadian plywood is made following the standards just mentioned.

Grades   The outer veneers of construction and industrial plywood are classified in five appearance groups: N, A, B, C, and D. N is the best grade and D the poorest (Figure 19.3).

Species   Construction and industrial plywood is made using about seventy species of wood. These may be mixed within a panel and include hardwoods and softwoods. Species used are divided into five groups, depending on their strength and stiffness. Group 1 includes the strongest and stiffest species. Group 5 includes those with the lowest properties (Table 19.1). The inner plies in Groups 1, 2, 3, and 4 may be any species in Groups 1, 2, 3, and 4. Inner plies in Group 5 may be any of the species listed. Front and back plies are of the same species and must derive from that group number.

Sizes of Panels   In customary units, standard nominal thicknesses of sanded construction and industrial plywood panels are ⅛ through 1¼ in., in ⅛ in. increments. Unsanded panels are available in thicknesses of 5/16 to 1¼ in. Panel widths are 36, 48, and 60 in., with lengths available from 60 in. to 144 in., in 12 in. increments (Table 19.2). Soft converted metric thicknesses are also shown in Table 19.2. Panels sized 1200 × 2400 mm are

Figure 19.3 Plywood veneer grades.

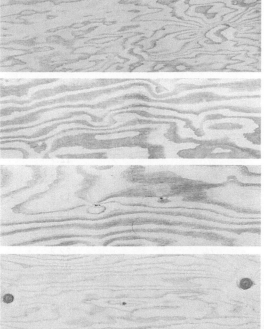

A Smooth, paintable. Not more than 18 neatly made repairs, boat, sled, or router type, and parallel to grain, permitted. Wood or synthetic repairs permitted. May be used for natural finish in less demanding applications.

B Solid surface. Shims, sled or router repairs, and tight knots to 1 inch across grain permitted. Wood or synthetic repairs permitted. Some minor splits permitted.

C Improved C veneer with splits limited to $\frac{1}{8}$ inch width and knotholes or other open defects limited to $\frac{1}{4} \times \frac{1}{2}$ inch. Wood or synthetic repairs permitted. Admits some broken grain.

Plugged

C Tight knots to $1\frac{1}{2}$ inches. Knotholes to 1 inch across grain and some to $1\frac{1}{2}$ inches if total width of knots and knotholes is within specified limits. Synthetic or wood repairs. Discoloration and sanding defects that do not impair strength permitted. Limited splits allowed. Stitching permitted.

D Knots and knotholes to $2\frac{1}{2}$ inch width across grain and $\frac{1}{2}$ inch larger within specified limits. Limited splits are permitted. Stitching permitted. Limited to Exposure 1.

Table 19.1 Plywood Outer Veneer Group Numbers

Group 1	Group 2		Group 3	Group 4	Group 5
Apitong	Cedar, Port Orford	Maple, black	Alder, red	Aspen	Basswood
Beech, American	Cypress	Mengkulang	Birch, paper	Bigtooth	Poplar, balsam
Birch	Douglas fir 2	Meranti, red	Cedar, Alaska	Quaking	
Sweet	Fir	Mersawa	Fir, subalpine	Cativo	
Yellow	California red	Pine	Hemlock, eastern	Cedar	
Douglas fir 1	Grand	Pond	Maple, bigleaf	Incense	
Kapur	Noble	Red	Pine	Western red	
Keruing	Pacific silver	Virginia	Jack	Cottonwood	
Larch, western	White	Western white	Lodgepole	Eastern	
Maple, sugar	Hemlock, western	Spruce	Ponderosa	Black (western	
Pine	Lauan	Red	Spruce	poplar)	
Caribbean	Almon	Sitka	Redwood	Pine	
Ocote	Bagtikan	Sweetgum	Spruce	Eastern white	
Pine, southern	Mayapis	Tamarack	Engelmann	Sugar	
Loblolly	Red	Yellow poplar	White		
Longleaf	Tangile				
Shortleaf	White				
Slash					
Tanoak					

Table 19.2 Plywood Panel Thicknesses and Dimensions

Nominal Thickness	
In.	**mm[a]**
1/4	6.4
5/16	7.9
11/32	8.7
3/8	9.5
7/16	11.1
15/32	11.9
1/2	12.7
19/32	15.1
5/8	15.9
23/32	18.3
3/4	19.1
7/8	22.2
1	25.4
1 3/32	27.8
1 1/8	28.6

Nominal Dimensions (width × length)		
ft.	**mm**	**m[a]**
4 × 8	1,219 × 2,438	1.22 × 2.44
4 × 9	1,219 × 2,743	1.22 × 2.74
4 × 10	1,219 × 3,048	1.22 × 3.05

[a]*Soft converted metric sizes*

Source: APA—The Engineered Wood Association

designed for use in hard metric-designed buildings, whereas panels 1219 × 2438 mm are soft conversions and can be used in buildings designed using customary units.

APA Performance-Rated Panels

APA performance-rated panels are manufactured to performance standards of APA–The Engineered Wood Association, which establishes performance criteria for designated construction applications. The panels specified by APA are plywood, oriented-strand board, composite panels (Com-Ply®), and APA-rated siding, sheathing, and underlayment (Figure 19.4).

Plywood panels are made using all veneer plies, producing the strongest type of panel. *Oriented-strand board (OSB)* is manufactured from strands or wafers arranged in one general direction. These layers of oriented strands are bonded together at right angles to each other. Usually three to five layers are bonded to form a panel. Most OSB panels are textured on one side to produce a non-slick surface. Oriented-strand board is rapidly replacing plywood as the most common sheathing material for walls, roofs, and floors. OSB's advantages include greater stiffness, uniform strength, and lower cost. Manufacturers have developed more-durable OSB sheathing and subflooring products that control the delamination and swelling associated with earlier versions of the

Figure 19.4 APA Performance-Rated Panels.

Plywood
All-veneer panels consisting of an odd number of cross-laminated layers, each layer consisting of one or more piles. Many such panels meet all of the prescriptive or performance provisions of U.S. Product Standard PS 1-83/ANSI A199.1 for Construction and Industrial Plywood.

APA Rated Siding
All-veneer panels constructed in the same manner as plywood panels. Available in a variety of panel sizes and thicknesses. Various surface textures and designs available.

Composite (Comply)®
Panels of reconstituted wood cores bonded between veneer face and back plies.

Oriented Strand Board
Panels of compressed strand-like particles arranged in layers (usually three to five) oriented at right angles to one another.

Copyright © APA-The Engineered Wood Association

material. The panels can be produced in sheets up to 8 ft. (2.43 m) wide and lengths of up to 16 ft. (4.87 m).

Composite or Com-Ply® panels have a core of reconstituted wood bonded between solid wood veneers. This produces a panel that allows for efficient use of wood materials and has a wood grain surface on the front and back.

APA-rated siding is made by bonding wood veneers in the same manner as plywood. It is available in a variety of surface textures and designs. Panels are available in 4 × 8, 4 × 9, and 4 × 10 ft. dimensions. A guide to APA performance-rated panels is given in Figure 19.5.

APA-rated sheathing is used for subfloors, walls, and roof sheathing where strength and stiffness are required. Nominal panel thicknesses range from 5/16 to 3/4 in. Structural I is a type of rated sheathing used where cross-panel strength and increased resistance to wracking are needed. It is used on structural diaphragms and panelized roofs. APA-rated Sturd-I-Floor® is a single-layer flooring for use under carpets. It may eliminate the need for installing additional underlayment. Panels are available with square and tongue-and-groove edges, and in thicknesses from 9/32 to 1 1/8 in.

Exposure Durability Classifications APA performance-rated panels are manufactured in four exposure durability classifications: Exterior, Exposure 1, Exposure 2, and Interior. These classifications signify how well the bonding agent used to join the veneers can resist exposure to moisture.

Exterior panels use a waterproof adhesive designed for applications subject to permanent exposure to moisture and weather. Exposure 1 panels use the same waterproof adhesive as exterior panels, but are designed to withstand exposure to moisture for longer periods before being permanently protected. They are also used where panels may be subjected to occasional moisture after they are in service. They differ from exterior panels in some compositional aspects and are not recommended for permanent exposure to the weather.

Exposure 2 panels are used in applications in which they will be briefly exposed to the elements during construction before being permanently protected. They are actually an interior panel with an intermediate adhesive.

Interior panels are manufactured with interior glue intended for indoor use only. They are identified by the abbreviation "INT-APA" and the omission of glueline information on their trademark. Veneer grades used on APA performance-rated panels are A, B, C, C plugged, and D (see Figure 19.2).

Product Identification APA trademarks on the types of plywood used in construction are shown in Figure 19.6. The sanded panels with B-grade or better veneer are used for various construction applications. The specialty grades include panels designed for a specific use, such as Plyform® for concrete forms and underlayment. Most of the information on the trademark is self explanatory, save for span rating. A span rating of 32/16 indicates the panel can be used as roof sheathing with rafters spaced up to 32 in. on-center and as subflooring with floor joists spaced up to 16 in. on-center.

Specialty Plywood

Many plywood manufacturers make various specialty plywood products that are not classified in the national standards for plywood manufacture (Figure 19.7). The designer must refer to information supplied by a manufacturer when selecting these products, because they meet no standard. Several of the most frequently used are *overlaid plywood*, siding panels, and interior paneling. The APA does have standards and trademarks for several specialty panels.

Overlaid plywood is a high-grade exterior-type panel that has a resin-impregnated fiber ply bonded to one or both sides. It is made in two types: high density and medium density. High-density overlay (HDO) panels have a hard, smooth chemically resistant surface, and are made in several colors that require no additional finish. Medium-density overlay (MDO) panels have a smooth, opaque, non-glossy surface that hides the grain of the veneers below it. They are used when a high-quality paint finish is needed.

Siding panels are used as finished exterior siding and are available in a variety of surface finishes, such as rough-sawed, V-grooved, and reverse-batten grooves.

Paneling is used on interior walls as a finish material. It is usually prefinished and only requires installation. It is available in a wide variety of wood species and surface features.

Hardwood Plywood

Hardwood plywood is made using various species of hardwood veneers for the face and back surfaces of the panel. They are bonded to plywood or particleboard panels that form the core. Hardwood plywood is manufactured following standards established by the American Society for Testing and Materials and the American National Standards Institute. The Hardwood Plywood and Veneer Association provides certification services.

Figure 19.5 APA–The Engineered Wood Association trademarks and uses of APA performance-rated panels.

GUIDE TO APA PERFORMANCE RATED PANELS(A)(B)
FOR APPLICATION RECOMMENDATIONS, SEE FOLLOWING PAGES.

APA RATED SHEATHING
Typical Trademark

APA
RATED SHEATHING
40/20
SIZED FOR SPACING
EXPOSURE 1
THICKNESS 0.578 IN.
000
PS 2-10 SHEATHING
PRP-108 HUD-UM-40
19/32 CATEGORY

APA
RATED SHEATHING
32/16
SIZED FOR SPACING
EXPOSURE 1
THICKNESS 0.451 IN.
000
PS 2-10 SHEATHING
PRP-108 HUD-UM-40
15/32 CATEGORY

Specially designed for subflooring and wall and roof sheathing. Also good for a broad range of other construction and industrial applications. Can be manufactured as OSB, plywood, or other wood-based panel. BOND CLASSIFICATIONS: Exterior, Exposure 1. COMMON PERFORMANCE CATEGORIES: 3/8, 7/16, 15/32, 1/2, 19/32, 5/8, 23/32, 3/4.

APA STRUCTURAL I RATED SHEATHING(C)
Typical Trademark

APA
RATED SHEATHING
STRUCTURAL I
32/16
SIZED FOR SPACING
EXPOSURE 1
THICKNESS 0.451 IN.
000
PS 1-09 C-D PRP-108
15/32 CATEGORY

APA
RATED SHEATHING
32/16
SIZED FOR SPACING
EXPOSURE 1
THICKNESS 0.451 IN.
000
STRUCTURAL I RATED
DIAPHRAGMS-SHEAR WALLS
PANELIZED ROOFS
P2 2-10 SHEATHING
PRP-108 HUD-UM-40
15/32 CATEGORY

Unsanded grade for use where shear and cross-panel strength properties are of maximum importance, such as panelized roofs and diaphragms. Can be manufactured as OSB, plywood, or other wood-based panel. BOND CLASSIFICATIONS: Exterior, Exposure 1. COMMON PERFORMANCE CATEGORIES: 3/8, 7/16, 15/32, 1/2, 19/32, 5/8, 23/32, 3/4.

APA RATED STURD-I-FLOOR
Typical Trademark

APA
RATED STURD-I-FLOOR
24oc
SIZED FOR SPACING
T&G NET WIDTH 47-1/2
EXPOSURE 1
THICKNESS 0.703 IN.
000
PS 2-10 SINGLE FLOOR
PRP-108 HUD-UM-40
23/32 CATEGORY

APA
RATED STURD-I-FLOOR
20oc
SIZED FOR SPACING
T&G NET WIDTH 47-1/2
EXPOSURE 1
THICKNESS 0.578 IN.
000
PS 1-09 UNDERLAYMENT
PRP-108
19/32 CATEGORY

Specially designed as combination subfloor-underlayment. Provides smooth surface for application of carpet and pad and possesses high concentrated and impact load resistance. Can be manufactured as OSB, plywood, or other wood-based panel. Available square edge or tongue-and-groove. BOND CLASSIFICATIONS: Exterior, Exposure 1. COMMON PERFORMANCE CATEGORIES: 19/32, 5/8, 23/32, 3/4, 1, 1-1/8.

APA RATED SIDING
Typical Trademark

APA
RATED SIDING
24oc
SIZED FOR SPACING
EXTERIOR
THICKNESS 0.578 IN.
000
PS 1-09
PRP-108 HUD-UM-40
19/32 CATEGORY

APA
RATED SIDING
303-18-S/W
16oc GROUP 1
SIZED FOR SPACING
EXTERIOR
THICKNESS 0.322 IN.
000
PS 1-09
PRP-108 HUD-UM-40
11/32 CATEGORY

For exterior siding, fencing, etc. Can be manufactured as plywood, as other wood-based panel or as an overlaid OSB. Both panel and lap siding available. Special surface treatment such as V-groove, channel groove, deep groove (such as APA Texture 1-11), brushed, rough sawn and overlaid (MDO) with smooth- or texture-embossed face. Span Rating (stud spacing for siding qualified for APA Sturd-I-Wall applications) and face grade classification (for veneer-faced siding) indicated in trademark. BOND CLASSIFICATION: Exterior. COMMON PERFORMANCE CATEGORIES: 11/32, 3/8, 7/16, 15/32, 1/2, 19/32, 5/8.

APA RATED SHEATHING – WALL
Typical Trademark

APA
RATED SHEATHING
WALL-24oc
SIZED FOR SPACING
EXPOSURE 1
THICKNESS 0.354 IN.
000
PS 2-10 SHEATHING
PRP-108 HUD-UM-40
3/8 CATEGORY

Specially designed for wall sheathing. Not intended for roof or floor sheathing. Can be manufactured as OSB, plywood, or other wood-based panel. BOND CLASSIFICATION: Exposure 1. COMMON PERFORMANCE CATEGORIES: 3/8, 7/16, 15/32.

(a) Specific grades, Performance Categories and bond classifications may be in limited supply in some areas. Check with your supplier before specifying.

(b) Specify Performance Rated Panels by Performance Category and Span Rating. Span Ratings are based on panel strength and stiffness. Since these properties are a function of panel composition and configuration as well as thickness, the same Span Rating may appear on panels of different Performance Categories. Conversely, panels of the same Performance Category may be marked with different Span Ratings.

(c) For some Structural I plywood panel constructions, the plies are special improved grades. Panels marked PS 1 are limited to Group 1 species. Other panels marked Structural I Rated qualify through special performance testing.

Figure 19.6 A trademark of the APA–The Engineered Wood Association.

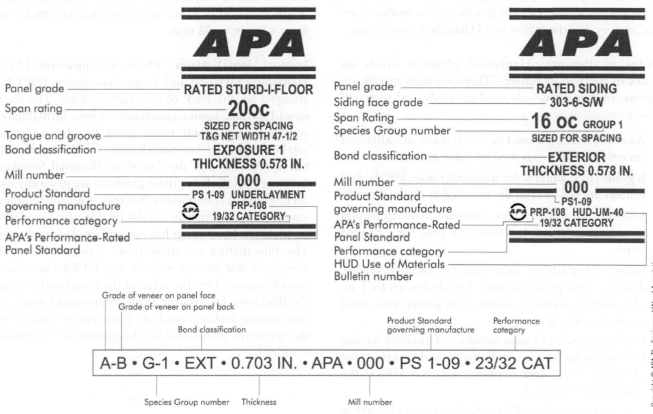

SANDED GRADES

Grade of veneer on panel back

Grade of veneer on panel face — **A-C** GROUP 1

Species Group number

Bond classification — EXTERIOR THICKNESS 0.328 IN.

Mill number — 000

Product Standard governing manufacture — PS1-09

Performance category — 11/32 CATEGORY

UNSANDED GRADES

Panel grade — RATED SHEATHING

Span Rating — 32/16

Bond classification — EXPOSURE 1 THICKNESS 0.451 IN.

Mill number — 000

Product Standard governing manufacture — PS 1-09 C-D PRP-108

Performance category — 15/32 CATEGORY

APA's Performance-Rated Panel Standard

STURD-I-FLOOR

Panel grade — RATED STURD-I-FLOOR

Span rating — 20oc

SIZED FOR SPACING T&G NET WIDTH 47-1/2

Tongue and groove —

Bond classification — EXPOSURE 1 THICKNESS 0.578 IN.

Mill number — 000

Product Standard governing manufacture — PS 1-09 UNDERLAYMENT PRP-108

Performance category — 19/32 CATEGORY

APA's Performance-Rated Panel Standard

SPECIALTY PANELS

Panel grade — RATED SIDING

Siding face grade — 303-6-S/W

Span Rating — 16 OC GROUP 1

Species Group number — SIZED FOR SPACING

Bond classification — EXTERIOR THICKNESS 0.578 IN.

Mill number — 000

Product Standard governing manufacture — PS1-09

APA's Performance-Rated Panel Standard — PRP-108 HUD-UM-40

Performance category — 19/32 CATEGORY

HUD Use of Materials Bulletin number

Grade of veneer on panel face

Grade of veneer on panel back

Bond classification

Product Standard governing manufacture

Performance category

A-B • G-1 • EXT • 0.703 IN. • APA • 000 • PS 1-09 • 23/32 CAT

Species Group number   Thickness   Mill number

Figure 19.7 Various specialty plywood products.

© Maksim Striganov/Shutterstock.com

Canadian hardwood plywoods are made according to metric standard CSA 0115-M, *Hardwood and Decorative Plywood.*

Species   Species used in hardwood plywood are divided into four categories. These categories reflect the modulus of elasticity (stiffness) of each species and its specific gravity. Category A is the stiffest and D has the lowest rating.

Grades of Veneers   Hardwood plywood offers six grades of hardwood veneers. There are also specific requirements for softwoods used in hardwood plywood panels:

A grade (A) is the best. The face is made of hardwood veneers carefully matched by color and grain.

B grade (B) is suitable for a natural finish, but the face veneers are not as carefully matched as on the A grade.

Sound grade (2) provides a smooth face. All defects have been repaired. It is used as a base for a painted finish.

Industrial-grade (3) face veneers can have surface defects. This grade permits knotholes up to 1 in. (25 mm) in diameter, small open joints, and small areas of rough grain.

Backing grade (4) uses unselected veneers having knotholes up to 3 in. in diameter and certain types of splits. No defect is permitted that affects the strength of the panel.

Specialty grade (SP) includes veneers having characteristics unlike any of those in the other grades. The characteristics are agreed upon between manufacturer and purchaser. For example, species such as wormy chestnut or bird's-eye maple are considered specialty grade.

Construction of Hardwood Plywood   The construction of hardwood plywood is described by identifying the core. The commonly available constructions are:

Hardwood and softwood veneer core: Have an odd number of plies, such as 3-ply, 5-ply, and so on.

Hardwood and softwood lumber core: Used in 3-ply, 5-ply, and 7-ply constructions.

Particleboard core: Used in 3-ply and 5-ply constructions.

Medium-density fiberboard core: Used in 3-ply construction.

Hardboard core: Used in 3-ply construction.

Special cores: Used in 3-ply construction. Special cores are made of materials not listed above.

Sizes and Thicknesses of Panels   Hardwood plywood is available in panels 48 in. wide and 96 and 120 in. long. Standard thicknesses are ¼ in., ⅜ in., ½ in. and ¾ in. Lumber-core panels and particleboard-core panels are available in only the ¾ in. thickness. Canadian hardwood plywood is available in incremental thicknesses from 4.8 mm to 32 mm.

Product Identification   The backstamp certified by the Hardwood Plywood and Veneer Association (HPVA) is stamped on the back of decorative panels. The flame-spread rating, smoke-generated rating, formaldehyde emissions, and bond types given on the stamp are established by the American Society for Testing and Materials (ASTM) and the American National Standards Institute (ANSI). Since these products are generally used on interiors, it is important that they meet the minimum standards. Stock panels usually do not carry the HPVA certification backstamp but are still graded by the mill. The information on these panel stamps is normally simple, in text format without any HPVA logo, and is usually stamped on the edge of the panel and contains the thickness, face-veneer grade, back-veneer grade, the face-cutting method (such as plain cut or rotary cut), the species of wood, and the designation of the product standard.

Reconstituted Wood Products

In addition to the APA performance-rated panels discussed, a number of reconstituted wood panels are used, primarily in cabinet and furniture construction. These include hardboard, particleboard, fiberboard, and waferboard (Figure 19.8).

Figure 19.8 Reconstituted wood panels include hardboard, MDF, particleboard, and oriented-strand board.

© 68/Steve Wisbauer/Ocean/Corbis

Hardboard is made from wood chips converted into fibers and bonded into panels under heat and pressure, and manufactured following standards developed by the American Hardboard Association. It is available in thicknesses from ¹⁄₁₂ up to 1⅛ in. The most frequent panel size is 4 × 8 ft., but other sizes are available through special order. Metric thicknesses for standard hardboard are 2.1 mm, 2.5 mm, 3.2 mm, 4.8 mm, 6.4 mm, 7.9 mm, and 9.5 mm. Panels are 1200 mm × 2400 mm.

Hardboard is available in five classes, which are defined as follows:

Class 1: Tempered class is impregnated with siccative material and stabilized by heat and special additives to impart substantially improved properties of stiffness, strength, hardness, and resistance to water and abrasion, as compared with the Standard Class.

Class 2: Standard class remains in the form in which it comes from the press and receives no further treatment. It has high strength and water resistance.

Class 3: Service-tempered class is impregnated with siccative material and stabilized by heat and additives. Its properties are substantially better than Service Grade.

Class 4: Service grade remains in the form in which it comes from a press but has less strength than Standard Class.

Class 5: Industrialite class is a medium-density hardboard that has moderate strength and lower unit weight than the other classes.

Hardboard is manufactured in the United States according to the following standards: ANSI/AHA A135.4-1988, *Basic Hardboard*; ANSI/AHA A135.5-1988, PF *Hardboard Paneling*; and ANSI/AHA A135.6-1990, *Hardboard Siding*. These are promoted by the American Hardboard Association (AHA). Panels manufactured to these standards have an AHA grade stamp.

Canadian hardboard products are manufactured following CGSB 11-GP-3M, *Hardboard*, and CGSB 11-GP-5M, *Hardboard for Exterior Cladding*.

Particleboard is made from wood chips, water, and a synthetic resin binder bonded with heat and pressure. It is manufactured to standards developed by the National Particleboard Association. It is mainly used for furniture and cabinet construction.

The general uses and grades are shown in Table 19.3. The grades are identified by a letter designation followed by a hyphen and a digit or letter. The second digit or letter designation indicates the grade identification within a particular density or product description. For example, "M-2" indicates a medium-density particleboard, Grade 2.

If there is a third designation it denotes a special characteristic, such as panel M-3 Exterior Glue. This is a medium-density panel, Grade 3, with exterior glue.

Table 19.3 Grades and Uses of Particleboard

Type	Grade	Use
High density	H-1, H-2, H-3	High-density industrial
	H-1, H-2, H-3 Exterior Glue	High-density exterior industrial
Medium density	M-1	Commercial
	M-2, M-3	Industrial
	M-1, M-2, M-3 Exterior Glue	Exterior construction
		Exterior industrial
Medium density—specialty grade	M-S	Commercial
Low density	LD-1, LD-2	Door core
Underlayment	PBU	Underlayment
Manufactured home decking	D-2, D-3	Flooring in manufactured homes

Source: National Particleboard Association

Grades 1, 2, and 3 relate to the relative magnitude of the mechanical properties. For example, Grade 3 has the highest modulus of elasticity, modulus of rupture, and hardness. Grade 1 has the lowest of these properties. Canadian particleboard is made following ANSI 208.1, *Particleboard Standards*.

Medium-density fiberboard (MDF) is manufactured in a similar way to hardboard but with fibers that are not as compressed. The use of refined fibers results in a highly homogenous board with an exceptionally fine surface. MDF is used as cabinetry, underlayment, and some interior wall finishes.

Waferboard is made by bonding large wood flakes 1½ in. or longer into panels having the same thicknesses and sizes as particleboard.

In the United States, fiberboard consists of two products, softboard and hardboard. In Canada, these are listed separately, and softboard panels are identified as fiberboard. Softboard panels are manufactured from loosely bound paper pulp and other types of fibers into panels with insulating properties. They are used as rigid insulation on walls and roofs.

In Canada, softboard panels have no standard thickness. This is established by a panel's manufacturer. They are cut into 1200 × 2400 mm and 4 × 8 ft. panels. Canadian fiberboard is manufactured according to CSA Standard A247-M, *Insulating Fiberboard*. In the United States, ¾ in., 1 in. 1½ in., and 2 in. thicknesses are available, and sheet sizes are 24 in. × 48 in., 48 in. × 48 in., and 4 ft. × 8 ft.

STRUCTURAL BUILDING COMPONENTS

The continuing development of engineered wood has increased its use as a structural component. *Structural composite lumber (SCL)* is lumber manufactured by parallel orientation of wood fibers of various geometries, bonded with a structural adhesive. Engineered wood trusses, and beams are now available in lengths that span large distances and have excellent load carrying capabilities, making them suitable for commercial applications.

Wood trusses are widely used for both floor and roof construction. A *truss* is a triangulated structural unit made by assembling structural wood members into a rigid frame. A wood truss consists of upper and lower chords that are connected by diagonal pieces called web members. For small spans, wood trusses are made from 2 × 4 in. and 2 × 6 in. lumber joined with wood or metal gusset plates. Gusset plates are sheets of steel that

are used to connect beams and girders to columns or to connect truss members. Trusses may also be made from considerable larger members, including glue-laminated timbers to achieve very long spans.

Wood trusses are carefully engineered and factory built under controlled conditions to produce a consistent and reliable quality. An advantages in the use of wood trusses is that they speed on-site erection time and can span the width of some buildings without interior load-bearing walls (Figure 19.9).

Figure 19.9 Wood trusses are made from 2 x stock joined with metal gusset plates.

© Mark Herreid/Shutterstock.com

Glued-laminated wood members, or glulams, are formed of solid sawn lumber that is finger-jointed end to end and then face bonded in laminations. The individual laminations, or lams, are selected and positioned based on their performance characteristics, and bonded together with durable, moisture-resistant adhesives. The strongest material is placed on the outer faces of beams, with the weaker material placed near neutral axis. Glulams are the only engineered wood product that can be produced in compound curved shapes.

Standard glulams used 2 x laminations are glued together under pressure to form simple beams or arches. The lams are arranged to provide maximum strength and minimum shrinkage. For more-complex shapes, ¾-inch-thick wood strips are used with laminations stepped to correspond to the curving shape. After being shaped and bonded, the beams are surface planed, sanded, and drilled to meet design requirements. They can also be pressure treated when required with fire retardant or wood preservative.

Glue-laminated lumber is available in depths from 6 to 72 in. or greater, and in lengths up to 100 ft. and longer (Figure 19.10). The strength and durability of glulam beams make them an ideal choice for large, open designs

Figure 19.10 Glued-laminated beams can span long distances without intermediate supports.

© David Papazian/Flame/Corbis

Figure 19.11 Laminated veneer lumber is made by bonding together layers of wood veneer, with the grain of all veneers running parallel to the long direction.

where long spans are required. As structural members, they also find use as vertical columns and portal frames.

Glulams bearing the American Plywood Association Engineered Wood Systems trademark are manufactured according to ANSI/AITC A190.1-2007, *Structural Glued Laminated Timber,* standards. They are available in three grades according to appearance. An industrial grade appearance is used where appearance is not a primary concern or members will be covered by other finish materials. Architectural appearance provides a good surface finish for exposed beams, while premium appearance is used for high-end applications. The structural capacity of the member is not influenced by its appearance grade.

Laminated-veneer lumber (LVL) is an engineered wood product produced by bonding thin wood veneers together into a large billet (**Figure 19.11**). Unlike plywood, where the grain of each layer is laid at right angles to the next, the grain of all veneers in an LVL run parallel to the long direction. Douglas fir and southern pine veneers are used for their strength and stiffness. The final product is considerably denser than plywood. The

billets are sawn to desired dimensions, depending on the construction application. LVL's are manufactured in thicknesses of 1-½ to 3-½ in., widths up to 18 in., and lengths up to 80 ft. Laminated-veneer lumber is the most widely used of the structural composite lumber products. The material displays almost no shrinking, checking, twisting, or splitting. Some of the product's many uses are in headers and beams, hip and valley rafters, scaffold planking, and the flange material for prefabricated wood I-joists.

Laminated-strand lumber (LSL) is a structural engineered wood that utilizes wood from species of wood that are not large or strong enough to be useful in solid lumber products. The product uses 12 in. long strands that are cut from logs, dried, and immersed in resin before being pressed into solid billets (**Figure 19.12**). The strands in the LSL are aligned parallel to each other to take advantage of the natural strength of the wood. LSL's are manufactured up to 5½ in. in thickness, 8 ft. in width and up to 35 ft. in length. Because the species of wood used lack inherent strength, the LSL is not used as a primary structural member. Common uses include small headers and millwork, and as components in door and window construction.

Figure 19.12 Laminated strand lumber uses wood strands aligned in parallel to take advantage of the natural strength of the wood.

Figure 19.13 A typical plywood-lumber beam (box beam) with plywood webs and solid wood flanges and stiffeners.

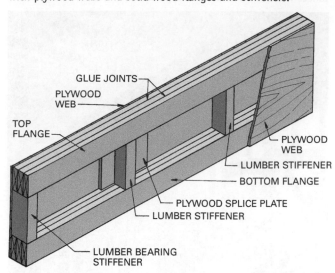

A comparison of the physical properties of glued-laminated lumber, laminated-veneer lumber, and solid lumber is shown in Table 19.4. Although actual data vary for species, grades, and manufacturing requirements, this comparison shows that glued manufactured wood structural members exhibit better properties than do those of most grades and species of solid woods.

Plywood-lumber beams (box beams) are made according to engineered specifications as to the size of members and nailing requirements. They are glued and built in factories under controlled conditions. One example is shown in Figure 19.13.

Stressed-skin panels are prefabricated panels using solid lumber stringers and headers and plywood skins. They can be produced in a variety of shapes for use in floors, walls, and roofs. They facilitate a building's rapid erection by covering large areas and spanning long distances.

Parallel-strand lumber (PSL) is a relatively new product that utilizes almost all the wood from solid logs. It is manufactured under the registered name Parallam® and made from Douglas fir or southern pine. The logs are peeled to produce a veneer that is dried and screened to remove strength-reducing defects. Then the sheets of veneer are clipped into strands up to 8 ft. in length and ¹⁄₁₀ and ³⁄₈ in. thick. Small defects are removed, and the strands are coated with a waterproof adhesive. The long-oriented strands are fed into a rotary belt press and cured under pressure using microwave energy. This produces a PSL billet that can be cut to standard sizes in lengths up to 66 ft. (Figure 19.14).

Table 19.4 Properties of Laminated Veneer, Glued Laminated Members, and Solid Lumber

Property	LVL[a]	Glued Laminated[b]	Solid Lumber[c]
Extreme fiber in bending (F_b) psi	3,000	1,600–2,400	760–2,710
Horizontal shear (H_v) psi	290	140–650	95–120
Compression perpendicular to grain ($F_{c\perp}$) psi	3,180	375–650	565–660
Compression parallel with grain (F_c) psi	—	1,350–2,300	625–1,100
Tension parallel with grain (F_t) psi	2,300	1,050–1,450	450–1,200
Modulus of elasticity (E) million psi	2.0	1.2–1.8	1.3–1.9

[a]Limited to one size member
[b]Data from one manufacturer
[c]Data for one species of lumber

Figure 19.14 Parallel strand lumber is formed with long oriented strands that are pressed and cured under pressure using microwave energy.

Figure 19.15 I-joists use flanges made of solid lumber and a composite wood web.

© David Papazian/Flame/Corbis

Parallam products have been tested for fire resistance, fastener-holding ability, moisture response, long-term loading, flexural strength, stiffness, and internal bond and have proved to exceed the performance of competing wood products.

Parallam structural products include beams, headers, posts, and columns. **Beam** and column sizes are given in Table 19.5. They are cut, drilled, and installed in the same manner as solid wood and laminated wood structural members.

Table 19.5 Stock Sizes of Parallam® Parallel Strand Lumber

Parallam Beams and Headers	
Thickness (in.)	**Depth (in.)**
1¾, 3½, 5¼, 7	7¼, 9¼, 9½, 11¼ 11½, 11⅞, 12, 12½ 14, 16, 18
Parallam Columns	
Dimensions (in.)	
3½ × 3½, 3½ × 5¼, 3½ × 7, 5¼ × 5¼, 5¼ × 7, 7 × 7	

Source: MacMillan Bloedel Limited

Another type of load-bearing structural beam is the wood *I-joist*, sometimes referred to as T-JI. Their characteristic "I" shape, commonly found in steel beams, provides high bending strength and stiffness characteristics. This allows them to span greater distances using substantially less material than would be required for solid lumber spanning the same distance. I-joists are made of softwood veneer-bonded top and bottom flanges connected by a thin composite panel web (Figure 19.15). The flanges are

commonly made of laminated veneer lumber and the web of either plywood or oriented strand board. Flanges are finger-jointed to make use of shorter wood scrap during manufacturing. The flanges are grooved to receive the veneer web, and assembled under pressure with waterproof glue. Wood I-joists are manufactured in various thicknesses with depths from 9¼ to 24 in., and lengths up to 80 ft. I-joists are used extensively in residential construction and are finding increased use in commercial applications as well. Their light weight makes them easy to handle and precludes the need for costly handling equipment.

Structural insulated panels (SIPs) are a composite building material consisting of an insulating layer of foam sandwiched between two layers of structural board (Figure 19.16). The board is usually oriented strand board and the foam expanded polystyrene foam, extruded polystyrene foam, or polyurethane foam. SIPs panels perform structurally similar to beams, with the rigid insulation core acting as the web, and the OSB sheathing as the flanges. SIPs replace several components

Figure 19.16 A structural insulated roof panel being readied for installation.

The Murus Company Inc.

Aldo Leopold Legacy Center

Architect: The Kubala Washatko Architects	Building type: Visitor and Interpretive Center	Completed April 2007
Contractor: The Bolt Company	Size: 11,900 sq. ft. (1,100 sq. m)	Rating: U.S. Green Building Council
Location: Baraboo, Wisconsin	Project scope: 3 1-story buildings	Level: Platinum (61 points)

Overview

Aldo Leopold, one of the founders of the modern conservation movement, advocated a holistic stewardship of the land, including careful management of natural resources and the protection of biodiversity. The headquarters of the Aldo Leopold Legacy Center consists of a grouping of buildings facing an exterior courtyard. In addition to meeting spaces, the center includes offices, the Leopold archive, and an interpretive hall. The building is located on the site of the Leopold Reserve, a property where the Leopold family planted thousands of trees as part of their restoration efforts during the 1930s and 1940s. The goal for the project was to provide a model of how sensitive design and construction techniques can integrate the built environment with natural systems. Part of the design concept was to restore and enhance the previously damaged site.

Design

The design process incorporated an integrated team, with architect, contractor, and environmental and other consultants working closely together from the onset of the project. The team sought to integrate simple, low-tech strategies with new, cutting-edge technologies to create ecologically sensitive buildings that fit well into their rural setting.

Working with the U.S. Fish and Wildlife Service, the project was designed to manage all stormwater on site. Native landscaping and pervious paving surfaces for parking and drives allows all water to naturally filter back into the ground.

Energy

The building utilizes high levels of thermal insulation to reduce heat loss across the exterior envelope. A heavy timber frame is clad in structural insulated panels, resulting in insulation values twice what is required by local building codes. Photovoltaic arrays on the rooftop produce more than the required energy for building operations, allowing surpluses to be sold back to the local utility. A ground-source heat pump provides radiant heating and cooling within the building floor slab. The resulting buildings use 70 percent less energy than a conventional project of the same size.

Materials

The design team chose to harvest trees planted by the Leopold family directly from the building site in the construction of the building. The existing 1,500-acre forest reserve had been poorly managed, resulting in overcrowding and a declining growth rate. Almost 100,000 board feet of lumber was harvested from the site and used for structural beams and trusses, siding, flooring, doors, windows, and other interior finishes. The lumber was debarked, air dried, and milled on-site, reducing both costs and transportation energy in securing materials. The surrounding forest health was improved through the careful thinning and harvesting of the timbers. The lumber used for the project was certified through the Forest Stewardship council. Other materials used were chosen based on their high recycled content and low emission potential, including composite wood products, native stone, recycled content cellulose insulation, and zero VOC interior paints.

The project incorporates other sustainable strategies, including natural daylighting to reduce electric consumption, passive solar design, waterless urinals, and water-efficient plumbing fixtures. Ongoing learning programs coupled with building energy and occupancy sensors utilize the buildings themselves in the education of visitors. The Aldo Leopold Legacy Center demonstrates how human activities, when carefully planned and executed, can work in harmony with ecological systems for the benefit of both.

The Aldo Leopold Center utilized wood for both the building structure and finishes that was harvested and processed directly on the project site.

of conventional building, such as studs and joists, insulation, vapor barrier, and air barrier. The panels are used for many different applications, such as exterior wall, roof, floor, and foundation construction.

In the United States, SIPs panels are available in sizes from 4 ft. (1.22 m) to 24 ft. (7.32 m) in width. Elsewhere, typical product dimensions are 300, 600, or 1,200 mm wide and 2.4, 2.7, and 3 m long, with roof panels up to 6 m long. Structures using SIPs panels have tighter building envelopes and higher insulation value, resulting in fewer drafts and lower operating costs for maintaining comfortable interior environments.

The standardization of SIPs also leads to reduced construction time and requires fewer trades for system integration. An example of SIPs construction used as a full building system is shown in Figure 19.17.

Figure 19.17 Structural insulated panels can be used for floors, walls, and roofs.

The Murus Company Inc.

Cross-laminated timber (CLT) is an engineered wood technology from Europe that is finding increased use in North America. CLT is formed in a similar fashion to glue-laminated wood beams, bonding layers of thick planks in a perpendicular arrangement with permanent adhesives. Imperfections like knots in the rough timber, are removed in the factory to reduce weakness and enhance structural performance. CLT differs from glulam in that it is formed into panels rather than beams, and the layers are bonded perpendicularly to one another, resulting in structural strength across two dimensions. Cross-laminated sections of kiln-dried soft woods are assembled into large solid panels that form walls, roofs, floors, and even lift shafts and stairs. Standard panel dimensions range from 2 in. up to 12 in. in thickness, though panels as

thick as 20 in. can also be produced. Panel length and width dimensions are mostly limited by what can be physically transported to a site and can typically be up to 16 ft. wide and 50 ft. long (Figure 19.18). Wall, floor, and roof elements can be pre-cut in the factory to any dimension and shape, including openings for doors, windows, stairs, etc.

Figure 19.18 Cross-laminated timber is formed by bonding layers of thick planks in a perpendicular arrangement into large panels.

© Science photo/Shutterstock.com

OTHER PRODUCTS MANUFACTURED FROM WOOD

Wood shingles and shakes are popular as a finished roof covering and as exterior siding on wood-framed buildings. Windows made of wood are available in a wide range of sizes, styles, and quality. A variety of types and styles of doors are made from both solid- and hollow-core wood construction. They are available in solid wood and with wood veneer, hardboard, and plastic laminate faces.

Cabinets are generally made from wood and wood products, although some steel cabinets exist. They come in a wide variety of styles and use a wide range of woods, such as birch, cherry, and walnut.

Hardwoods, softwoods, and engineered wood products are used for finish flooring. Wood flooring is popular because of the warmth and beauty of its grain and color. The most commonly used hardwoods are oak, beech, birch, pecan, and maple. Softwoods include southern pine, western larch, bald cypress, eastern Englemann, eastern spruce, red pine, ponderosa pine, eastern hemlock, and Douglas fir. More on the variety of wood products and their uses are described in subsequent chapters.

Review Questions

1. What are the differences between plywood, oriented-strand board, and composite panels?

2. What are the plywood exposure durability classifications? Give an example of where each could be used.

3. What are the differences between laminated-veneer lumber and glued-laminated members?

4. What appearance groups are established for the outer veneers of construction plywood?

5. How is the strength of the woods used in plywood indicated?

6. What is a performance-rated wood panel?

7. What are the grades of outer veneers for hardwood plywood?

8. What are the classes of hardboard?

Key Terms

APA Performance-Rated Panel

Cross-Laminated Timber (CLT)

Engineered Wood Product

Glue-Laminated Wood

Hardboard

Hardwood Plywood

I-Joist

Laminated-Veneer Lumber (LVL)

Oriented-Strand Board (OSB)

Overlaid Plywood

Parallel Strand Lumber (PSL)

Particleboard

Plywood

Structural Composite Lumber (SCL)

Structural Insulated Panels (SIP)

Truss

Waferboard

Activities

1. Collect samples of as many kinds of plywood products as you can find. Prepare a display. Identify each one with as much technical information as you can gather.

2. Immerse samples of interior and exterior plywood in water. Continue until delamination occurs. Keep a record of the length of time it takes each to fail. Did some types never fail?

3. Cut 2-inch-wide strips of ½ in. thick fir plywood 4 ft. long. Support a strip between two chairs with the flat side facing down. Add weights incrementally and note the amount of deflection until the strip breaks or slides off a chair. Then stand the strip on its edge and repeat the test. In which position did it carry the greatest load? Why?

4. Build scale models of the commonly used trusses.

5. Collect samples of the various types of wood flooring. Your building supply dealer may have manufacturers' samples for you to use. Label each.

Additional Resources

APA Engineered Wood Construction Guide, and numerous other related technical publications, APA—The Engineered Wood Association, Tacoma, WA.

Plywood Handbook and Plywood Design Fundamentals, CAN-PLY, Canadian Plywood Association, North Vancouver, BC, Canada.

Williamson, T. G., *APA Engineered Wood Handbook*, McGraw-Hill Publishing Co.

Wood Engineering Handbook, Forest Products Laboratory, Madison, WI 53705.

Wood Reference Handbook, Canadian Wood Council, Ottawa, Ontario, Canada.

See Appendix C for addresses of professional and trade organizations and other sources of technical information.

Wood and Metal Light Frame Construction

LEARNING OBJECTIVES

Upon completion of this chapter, the student should be able to:

- Develop an understanding of the methods used to construct light-frame buildings.

- Understand the differences in framing when using the various materials and products available.

- Be aware of the influence codes and ordinances have on the design of light-frame buildings.

Build Your Knowledge

For further study on these materials and methods, please refer to:

Chapter 22 Finishing the Exterior and Interior of Light Wood Frame Buildings

Chapter 33 Interior Walls, Partitions, and Ceilings
Topic: Wood Product Wall Finishes

Figure 20.1 A flexible construction system, light wood frame construction is capable of providing an almost unlimited range of building forms and designs.

© alexmisu/Shutterstock.com

Figure 20.2 A contemporary wood frame structure.

© photobank.ch/Shutterstock.com

Construction systems using wooden beams, columns, and frames have been utilized since the dawn of time. The first structures in the United States used simple rough-hewn logs, stacked to form a massive construction. The advent of steam-powered sawmills and the machine made nail in the late nineteenth century made possible a light wood frame construction system. Derived from the "balloon frame," today's ubiquitous platform frame is still used in most residential and light commercial construction.

A very flexible construction system, light wood frames are capable of providing an almost unlimited range of building forms and designs. Architects shape designs that are classic (Figure 20.1) or contemporary (Figure 20.2), simple or complex, low cost or expensive, that can be fitted with almost any electrical, heating, air conditioning, plumbing, and security system desired. Wood-frame structures are insulated, sealed, and waterproofed to facilitate long life and low maintenance. The system can be used in almost any climate and on any site that will accept an adequate foundation.

The system has evolved from employing only solid wood members to the use of a variety of reconstituted wood products, as discussed in Chapter 19. Many problems that developed with early wood frame construction—wood decay, swelling and shrinking of members, sticking doors and windows, and creaking floors—have been overcome by improved construction techniques and materials. Considerable effort has been made to utilize factory-assembled panels and modules, thereby reducing on-site labor costs associated with what is often referred to as "stick built" construction.

WOOD FRAMING AND THE BUILDING CODES

Building Codes

Building codes play an important role in the design and construction of wood and light-gauge metal frame construction. The IBC now provides increased opportunities for wood-frame construction as compared with that allowed under previous codes. Chapter 23 of the IBC governs materials, design, construction, and quality of wood members and their fasteners, covering wood in buildings of Construction Types III, IV and V. Type III construction uses fire retardant-treated wood framing within exterior wall assemblies with a two-hour rating or less. Type IV construction, also known as Heavy Timber, uses exterior walls that are of noncombustible materials and interior building elements that are of solid or laminated wood without concealed spaces (see Chapter 21). Type V construction permits the use of structural elements, exterior walls, and interior walls made from any of the materials permitted by the code. Platform frame construction typically falls under this classification.

Allowable design loads on wood building components are carefully detailed in the codes, and structural components must meet minimum deflection allowances. The International Building Code (IBC) allows wood-frame construction for up to five stories in building occupancies that range from business and mercantile to multi-family, senior, and student housing.

Fire-resistance ratings for various assemblies of buildings are specified. The fire code stipulates requirements for fire ratings and approved noncombustible materials. Typically, these requirements are met by using additional gypsum wallboard on interior walls and, in some cases, adding fire-resistant sheathing to exterior

walls. In some areas, sprinkler systems are required by local codes. Controls are placed on materials regulating flame-spread and smoke-developed ratings.

Construction in areas where basic wind speeds equal or exceed 110 miles per hour must meet special design criteria. Areas subject to earthquakes have seismic design requirements. Codes also regulate the design of interior spaces and such things as light, ventilation, glazing, and sanitation. Additional information on building codes is given in Chapter 2.

PREPARING THE SITE

Once the construction contracts are signed, bids accepted, and building permits issued, the site must be readied for construction. Unwanted brush and trees in the way of the building, sidewalks, or drives are removed. In some areas, tree removal is carefully regulated, and each tree must be marked and its removal approved before the site is cleared. When possible, mature trees and other plantings should be preserved.

The surveyor stakes the corners of the foundation according to dimensions given on the site plan. The builder sets up batter boards on each corner, as shown in Figure 20.3. Batter boards are usually 1×6 lumber held by 2×4 posts driven into the ground. A chalk line is pulled across the boards on each side of the proposed building based on the locations of the corner stakes. Once lines are located, they are tied to a nail driven into the batter board or run in a saw cut in the board. Figure 20.4 illustrates the typical corner layout. The corners are checked for square using the 3-4-5 right triangle method. When the base of a right triangle measures 3 ft. and the vertical side measures 4 ft., the hypotenuse will measure 5 ft. If it does not, the corner is not square. Some prefer to use longer measurements in multiples of three, such as 9, 12, and 15 ft.

Excavation

A plumb line is dropped from the batter board lines to establish the corner of the foundation and excavation (Figure 20.5). Excavations are then dug with a bulldozer or backhoe to the required width and depth. Footings must rest on undisturbed soil, so care must be exercised against over excavating. If soil crumbles on the sides of the footing excavation, wood formwork must be installed. Additional footings for columns, piers, and fireplaces are located and dug. Specified reinforcing is placed in the footing excavation and concrete poured.

Figure 20.3 Batter boards and chalk lines are used to locate the corners of the foundation.

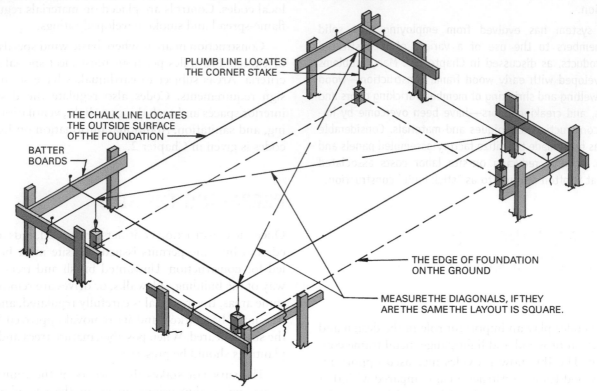

PLUMB LINE LOCATES
THE CORNER STAKE

THE CHALK LINE LOCATES
THE OUTSIDE SURFACE
OF THE FOUNDATION

BATTER
BOARDS

THE EDGE OF FOUNDATION
ON THE GROUND

MEASURE THE DIAGONALS, IF THEY
ARE THE SAME THE LAYOUT IS SQUARE.

Figure 20.4 The square of a corner marked by the chalk line can be checked by 3-4-5 method. Notice the batter boards are set back about 4 ft. from the corner stake.

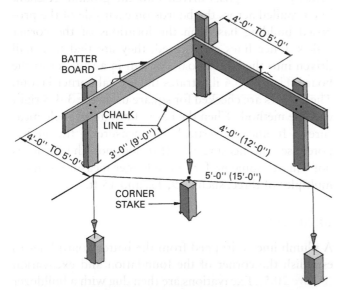

BATTER
BOARD

CHALK
LINE

4'-0" TO 5'-0"

4'-0" TO 5'-0"

3'-0" (9'-0")

4'-0" (12'-0")

5'-0" (15'-0")

CORNER
STAKE

FOUNDATIONS FOR LIGHT WOOD FRAME CONSTRUCTION

Footings for light frame construction are made of cast-in-place reinforced concrete. The foundation stem wall may be concrete block, brick, cast-in-place concrete, or pressure-treated wood. Detailed information on foundations is given in Chapter 6. The following discussion will review common foundations for wood light framed buildings, including basements, crawl spaces, and concrete slab on grade floors.

Basements are used predominantly in colder climate regions where the depth of the frost line requires deep excavations. They provide considerable space for a small cost, especially considering that a crawl space requires a footing and stem wall but provides no living space.

Basement walls can be cast-in-place reinforced concrete or concrete block. Two typical basement details can be found in Figure 20.6. Footings are poured first. When they have reached adequate strength, the concrete block wall can be laid, or the forms for a poured concrete foundation installed. In both cases, vertical steel reinforcing acts to tie the footing to the stem wall.

The foundation wall is usually topped with a metal flashing termite shield that projects to the interior and prevents termites from reaching the wood framing. A treated wood plate attached with anchor bolts caps the foundation walls. After the basement foundation has been built, the soil should not be backfilled until the floor platform is in place to provide protection against horizontal stresses created by the weight of the soil.

Figure 20.5 The basement excavation is dug to the specified depth and footings are dug. Footing forms are used if the sides of the footing excavation crumble.

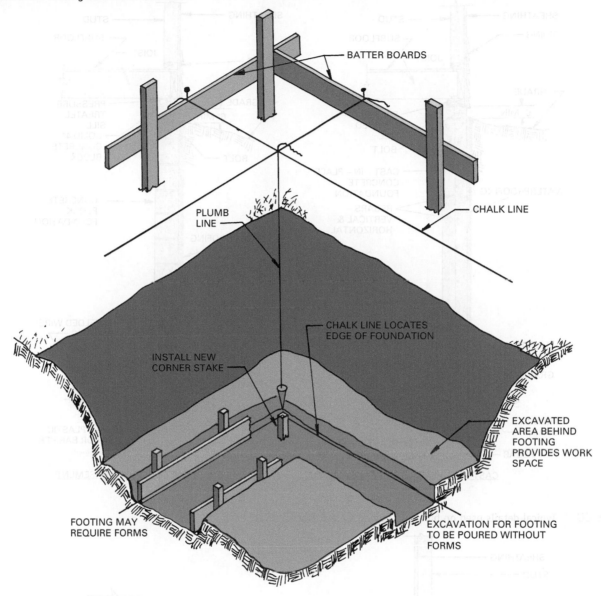

BATTER BOARDS

CHALK LINE

PLUMB LINE

CHALK LINE LOCATES EDGE OF FOUNDATION

INSTALL NEW CORNER STAKE

EXCAVATED AREA BEHIND FOOTING PROVIDES WORK SPACE

FOOTING MAY REQUIRE FORMS

EXCAVATION FOR FOOTING TO BE POURED WITHOUT FORMS

The foundations in Figure 20.7 are typical for a building with a crawl space. A brick veneer exterior requires that a foundation have a brick ledge that is usually near grade. Building codes limit the minimum height between the ground and the bottom of the floor joists. Since plumbing, heating, air-conditioning, and electrical service are often run in the space below the joists, this height is often set at well above the minimum. Anything below a 2 ft. (610 mm), high crawl space is claustrophobic and difficult to work in when servicing utilities.

Since it is important to keep the ground in the crawl space dry, the exterior of the foundation below grade should be waterproofed, and gravel and a French drain should be placed at the footing, draining to a drywell or daylight. The ground in the crawl space should be covered with plastic sheet material to retard moisture passage from soil to crawl space. The perimeter of the crawl space should be insulated in colder climate regions.

Several foundations for structures with concrete slab on grade floors are illustrated in Figure 20.8. A monolithic poured foundation and floor is used in areas where freezing does not occur. Stem walls are used where footings reach deep below grade. Anchor bolts for the connection of the bottom plate are set in concrete before it hardens.

Figure 20.6 Typical basement wall details.

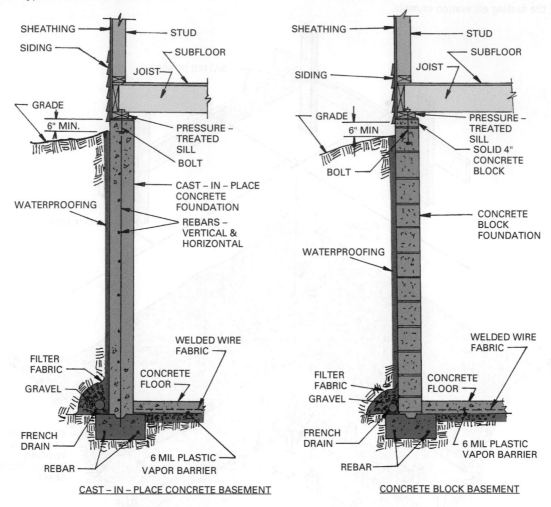

CAST – IN – PLACE CONCRETE BASEMENT

CONCRETE BLOCK BASEMENT

Figure 20.7 Typical details used on buildings having a crawl space.

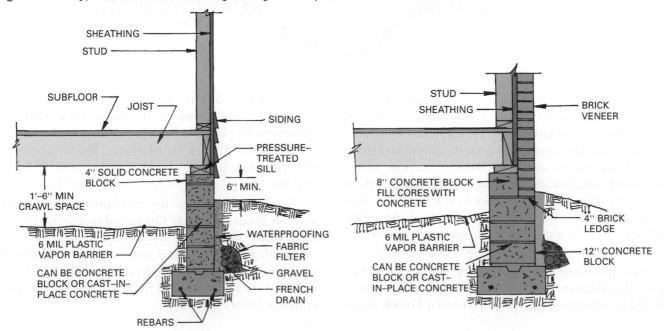

Figure 20.8 Types of concrete slab floor construction used with wood light-frame construction.

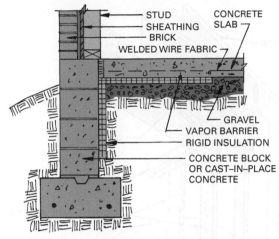

STUD
SHEATHING
BRICK
WELDED WIRE FABRIC
CONCRETE SLAB
GRAVEL
VAPOR BARRIER
RIGID INSULATION
CONCRETE BLOCK OR CAST–IN–PLACE CONCRETE

A GROUND SUPPORTED CONCRETE SLAB

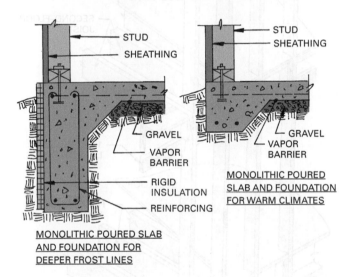

STUD
SHEATHING
GRAVEL
VAPOR BARRIER
RIGID INSULATION
REINFORCING

MONOLITHIC POURED SLAB AND FOUNDATION FOR DEEPER FROST LINES

STUD
SHEATHING
GRAVEL
VAPOR BARRIER

MONOLITHIC POURED SLAB AND FOUNDATION FOR WARM CLIMATES

Wood Foundations

Wood foundations are sometimes used for buildings with basements or crawl spaces. The materials are stress graded to insure they can withstand lateral soil and sub-surface water pressures. Wood materials must have the stamp of the American Wood Preservers Association (AWPA). Foundation panels are often shop-built and then shipped to site. The studs forming the panel face the inside of the building and are insulated in the same manner as an exterior above-grade wall. Panels must be carefully assembled using bronze, copper, silicon, or stainless steel nails that resist corrosion.

Interior Piers and Columns

In addition to perimeter foundation walls, footings for piers or columns to support beams that will carry

the floor joists and other point loads must also be considered. Concrete blocks compose a typical pier used in wood light-frame buildings with crawl spaces (Figure 20.9). The size of the pier depends upon the loads to be carried.

In buildings with basements, steel columns and beams can support floor loads at the interior. Figure 20.10

Figure 20.9 Concrete block piers or columns support interior beams that carry the floor structure.

© James R. Martin/Shutterstock.com

Figure 20.10 In buildings with basements, steel columns and beams can support floor loads at the building interior.

© Christina Richards/Shutterstock.com

shows an example. Steel baseplates are used to bolt the column into the footing or slab. A finished foundation, ready for carpenters to install the first floor platform, is shown in Figure 20.11.

Figure 20.12 The balloon frame used wall studs running the full two stories, with floors hung from inlet ledger boards.

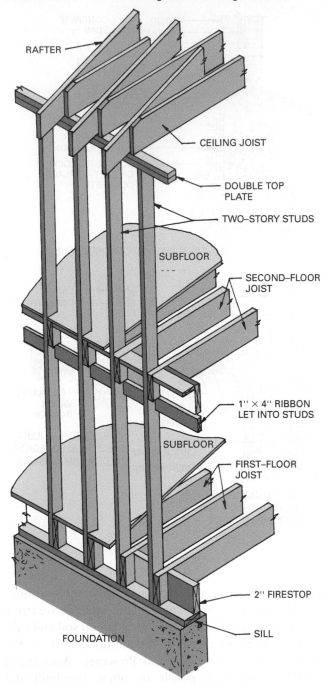

Figure 20.11 A carpenter lays out floor joists for installation.

© Wendy Kaveney Photography/Shutterstock.com

EVOLUTION OF WOOD LIGHT FRAME CONSTRUCTION

Early wood-frame construction in the U.S. colonies involved building a structural frame of heavy timbers that were fitted together with mortise and tenon joints. The spaces between frame members were infilled with smaller wood members or masonry. This construction required a large group of workers to hand cut, fit the joints, and lift the timbers into place. The infill walls below the horizontal timbers were non-load-bearing.

The forerunner of the current widely used *platform framing* system of wood light frame construction was developed in the early 1800s. It involved using smaller vertical members called studs to carry the roof and second floor, and eliminated the heavy timber frame. The system was called the *balloon frame* because it seemed light enough to float in comparison to heavy timber construction. The studs not only carried the loads but also served as the enclosure of the building (like the infill walls did on the heavy timber frame). Balloon frames used wall studs running the full two stories, with floors hung from inlet ledger boards. The detail in Figure 20.12 shows studs resting on the sill on top of the foundation and running continuously to the double top plate. The floor joists rest on a ribbon, let into and nailed to the stud. Balloon framing has been replaced by platform framing and is seldom used anymore.

PLATFORM FRAMING

Platform framing involves building the first floor deck on top of the foundation (Figure 20.13). Upon this, walls for the first floors are assembled, erected, and braced. Joists for the second floor are laid on top of the double plate of the exterior first-floor walls and supported by

Figure 20.13 The first floor frame provides a platform on top of the foundation on which the building can be assembled.

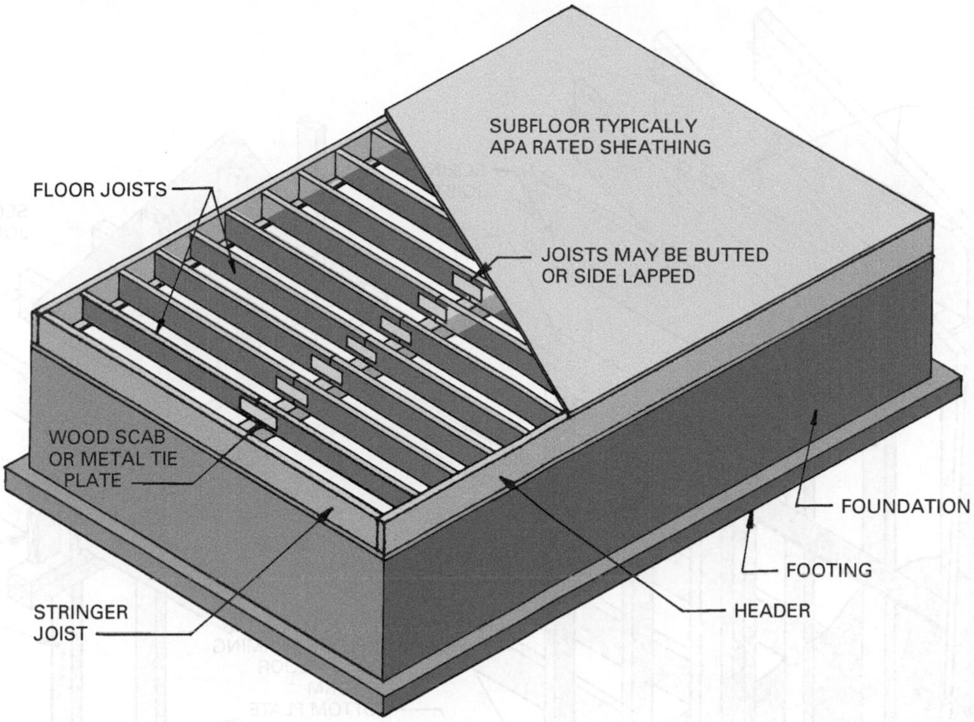

load-bearing interior walls. Second floor joists are covered with subflooring, forming the second floor platform, and walls for the second floor are assembled, erected, and braced. The ceiling joists are laid on the double plate of the second-floor exterior walls and supported by interior load-bearing walls. Finally, the rafters, which rest on the double plate of the second-floor walls, are erected (Figure 20.14). The walls and roof are sheathed, providing solid surfaces for siding and finished roofing materials. Sheathing greatly strengthens the roof framing and the rigidity of the entire structure.

If roof trusses are used, there is no need for ceiling joists, because the bottom chord of the truss serves as one. Trusses are erected bearing on the double top plate of the second-floor walls. Trusses are discussed in more detail later in this chapter.

Floor Framing

Floor joists are the horizontal members of a frame that transfer loads to sills and girders. When floor joists cannot span the entire length or width of a foundation, a beam is required to support the floor at the mid-span. The beam may be built of steel, wood, or a manufactured wood product. It is supported at intervals by piers or columns (Figure 20.15).

In Figure 20.16, the first floor has been built on the foundation using box-sill construction. Box-sill construction has floor joists resting on a treated lumber sill that is bolted to the foundation. The joists butt a band or *rim joist*, which, in effect, forms the floor framing into a box. The rim joist is end-nailed to the joist and the joists are toenailed to the sill. Sometimes metal connectors are used. Floor joists may consist of dimensional lumber or engineered wood I-joists and are usually spaced 16 or 24 in. on center (Figure 20.17).

After the first floor joists are in place, the subfloor is glued and nailed to them (Figure 20.18). Subfloors provide the structural floor upon which finish materials are installed. Usually tongue and groove 3/4 in. plywood is affixed with construction adhesive and then nailed in place.

Wall Framing

Figure 20.19 illustrates the framing for a typical exterior wall with an opening. Studs are typically 2 × 4 stock, but 2 × 6 studs provide space for additional insulation. The studs are generally spaced 16 in. (406 mm) on center, however, other spacing, such as 12 and 24 in. on center (305 and 610 mm) are also used. The load-carrying capacity of a wall and the size of its studs determine the

Figure 20.14 Typical platform framing details for a two-story building.

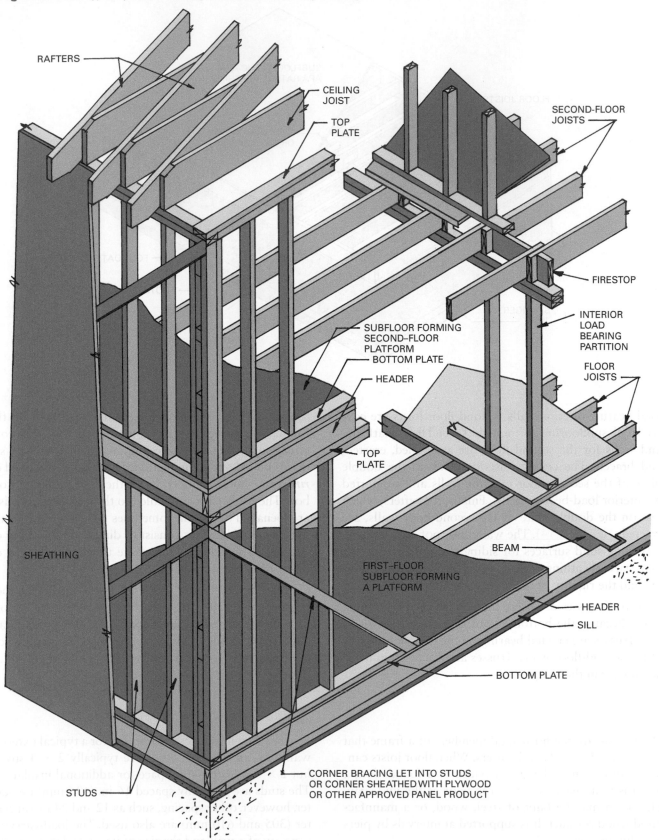

RAFTERS

CEILING JOIST

TOP PLATE

SECOND-FLOOR JOISTS

FIRESTOP

INTERIOR LOAD BEARING PARTITION

SUBFLOOR FORMING SECOND-FLOOR PLATFORM

BOTTOM PLATE

HEADER

FLOOR JOISTS

TOP PLATE

SHEATHING

BEAM

FIRST-FLOOR SUBFLOOR FORMING A PLATFORM

HEADER

SILL

BOTTOM PLATE

STUDS

CORNER BRACING LET INTO STUDS OR CORNER SHEATHED WITH PLYWOOD OR OTHER APPROVED PANEL PRODUCT

Figure 20.15 When floor joists cannot span the width of a foundation, a beam is required to support the floor at the mid-span.

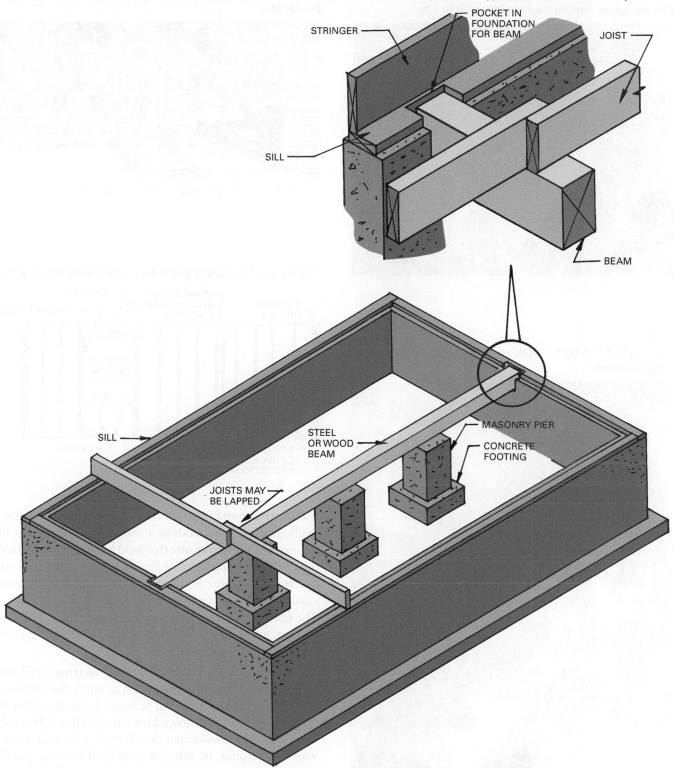

STRINGER

POCKET IN FOUNDATION FOR BEAM

JOIST

SILL

BEAM

SILL

STEEL OR WOOD BEAM

MASONRY PIER

CONCRETE FOOTING

JOISTS MAY BE LAPPED

spacing to be used. On-center spacing is always set in four foot modules to accommodate panels of sheathing that are 4 × 8 ft. (1.2 × 2.4 m) to ensure the panel edges will meet on a stud.

Wall construction begins with the careful layout in pencil of all wall components on the top and bottom plates. Walls use a single plate at the bottom and a double plate at the top. The double top plate locks intersecting

Figure 20.16 The rim joist provides a band into which floor joists are nailed, forming the floor framing into a box.

Figure 20.17 A light-weight, engineered I-joists being seated in a joist hanger.

Figure 20.18 Plywood subfloors are glued and nailed to floor joists.

Figure 20.19 Framing details for a typical wood framed wall.

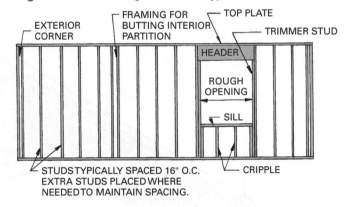

walls together and helps transfer ceiling loads more effectively. The layout of studs in a wall follows the layout of the joists beneath to ensure that load bearing is transferred directly to the foundation. All components of the wall are laid in place on the floor deck and then nailed together, either by hand or with pneumatic nailers (Figure 20.20). The wall is then lifted into position, straightened between corners, nailed to the floor framing, plumbed to vertical, and temporarily braced (Figure 20.21).

Framing an opening in a load-bearing wall requires that a header large enough to carry the imposed loads span the opening and transfer loads to adjacent studs. Figure 20.22 shows how an opening is framed. Figure 20.23 shows header details for 2 × 4 and 2 × 6 in. walls. Openings in interior non-load-bearing partitions can be framed with a single flat member. Framing details for a door opening in a load-bearing wall is shown in Figure 20.24. Note that the rough opening door size includes the size of the door, plus space for the doorframe, and a little room to allow a carpenter to plumb the doorframe.

Figure 20.20 Wall are laid out on the floor deck and nailed together, either by hand or with pneumatic nailers.

Figure 20.21 The completed wall is lifted into position, nailed to the floor framing, plumbed, and temporarily braced.

The corners of exterior walls must be built to allow for sidewalls to butt front or rear walls and leave a nailing surface for interior finish materials. Figure 20.25 shows the preferred framing method for framing corners to leave an opening for the insertion of insulation later in the construction process. Where interior partitions meet exterior walls, the top plate overlaps the lower plate of the double top plate, as shown in Figure 20.26. Where interior partitions butt an outside wall or another partition, they can be joined by installing blocking or extra studs, as shown in Figure 20.27.

When the wall framing is completed, and all walls have been straightened and braced, the exterior sheathing can be installed. Sheathing is an exterior covering placed on the studs that serves as a base for exterior cladding (Figure 20.28). Sheathing may be plywood,

Figure 20.22 This opening is framed with a solid wood header.

oriented-strand board, asphalt-impregnated fiber, or rigid-foam insulation sheets. When plywood or oriented-strand board are used, they provide the required lateral bracing for the wall. When rigid-foam plastic sheets are used, the wall is braced with let-in diagonal bracing (1 × 4 boards notched into studs diagonally and then nailed when the wall is plumb), or sheets of plywood or oriented-strand board are nailed at each corner and at intervals along the wall, as specified by an architect.

Ceiling Framing

Ceiling joists are supported on exterior walls and load-bearing interior partitions. They carry the finish ceiling materials and insulation. If they are subject to other loads, such as a second floor storage area, the architect will size them accordingly. When they become joists for the second floor of a two-story building, their framing is identical to the framing of the first floor (Figure 20.29). The first ceiling joist is set in from the outer edge of the plate on the gable end to permit construction of the framing for the gable-end wall. Ceiling joists can be trimmed, as shown in Figure 20.30, to follow the rafter shape.

Figure 20.23 Load-bearing header details.

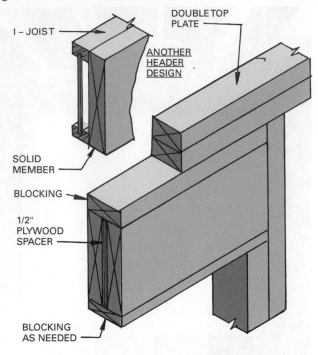

I – JOIST

DOUBLE TOP PLATE

ANOTHER HEADER DESIGN

SOLID MEMBER

BLOCKING

1/2" PLYWOOD SPACER

BLOCKING AS NEEDED

DOUBLE 2" THICK MEMBERS WITH PLYWOOD SPACER

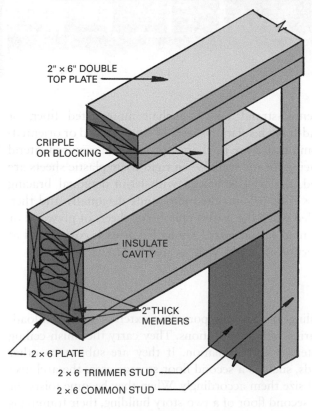

2" × 6" DOUBLE TOP PLATE

CRIPPLE OR BLOCKING

INSULATE CAVITY

2" THICK MEMBERS

2 × 6 PLATE

2 × 6 TRIMMER STUD

2 × 6 COMMON STUD

2 × 6 WALL WITH AN INSULATED HEADER

Figure 20.24 A section through the framing of a door opening in a wood-framed wall.

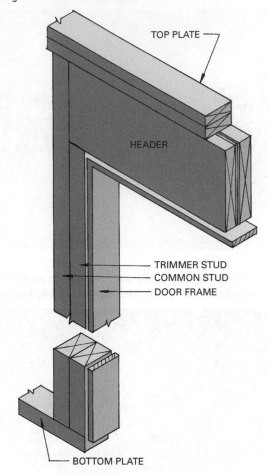

TOP PLATE

HEADER

TRIMMER STUD

COMMON STUD

DOOR FRAME

BOTTOM PLATE

Figure 20.25 An exterior wall corner framing detail.

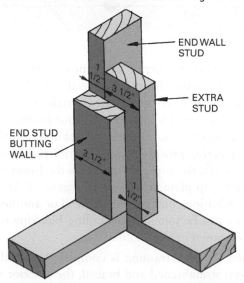

END WALL STUD

1 1/2"

3 1/2"

EXTRA STUD

END STUD BUTTING WALL

3 1/2"

1 1/2"

Construction Methods

Optimum Value Engineering

Optimum Value Engineering (OVE), also referred to as advanced or in-line framing, refers to a series of wood-framing techniques designed to utilize less lumber in the construction of a structure. Advanced framing construction detailing can reduce the amount of lumber used in a building, while still providing adequate structural strength. Many OVE strategies have the additional benefit of reducing the amount of labor required in the construction process. Individual wall lengths are usually dimensioned in multiples of 4 ft. in length, and the location of doors and windows must be carefully planned to align with the 24 in. on center framing.

The most commonly used framing procedure employed entails the use of 2 × 4 studs spaced at 16 inches (406 mm) on center. This practice has become the nationwide standard over the last 100 years. The on-center spacing can be enlarged to 24 inches (610 mm) with no other necessary changes or detrimental effects to structural integrity. In-line framing aligns studs with the floor and roof joists. For example, 2 × 6 studs align with 2 × 10 engineered joists above, transferring their load directly "in line" to the foundations. An enlargement of the on-center spacing of structural members provides benefits beyond the conservation of material resources. A lower percentage of wall studs has the additional benefit of increasing the area available for insulation and reducing the amount of thermal bridging inherent in wall wood studding.

Headers are the structural members over door and window openings that transfer ceiling and roof loads to vertical "trimmers" and the foundations. The standard header is built from two two-inch-thick framing members with a piece of ½ in. plywood sandwiched between them, which equal the required 3½ in. wall width of the standard 2 × 4 wall. In order to expedite the construction process, headers are commonly built of 2 × 12 (38 × 305 mm wide) material. A header of this size can be fastened directly to the top plate of a typical eight foot wall to provide the standard 6 ft. 8 in. window and door head height. Although full 2 × 12 headers are not required in interior non-load-bearing partitions, most builders put them in every opening to ensure safety. By carefully sizing headers to their anticipated load, tremendous material savings result.

Engineered lumber, such as I-joists and rafters, combines smaller dimensional lumber and particleboard and reduces the amount of cut off waste. Use of OVE techniques can result in reduction of cut-off waste from standard-sized building materials, reduction in the number of top plates needed, and reduction of the number of studs in building corners and exterior bearing walls. Higher energy efficiency resulting from decreased thermal bridging across structural members can have a measurable impact on heating and cooling costs.

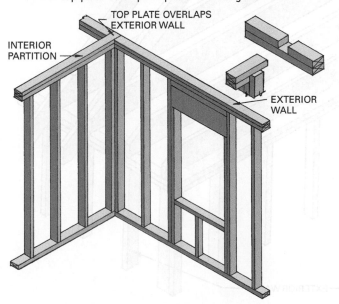

Figure 20.26 When exterior wall partitions butt exterior walls, the double top plate overlaps to provide a strong connection.

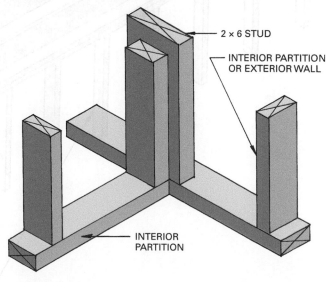

Figure 20.27 A connection detail between the exterior wall and an interior partition.

Figure 20.28 Sheathing serves as a base for exterior cladding and provides the required lateral bracing for the structure.

© TFoxFoto/Shutterstock.com

Figure 20.29 Ceiling joists are supported on exterior walls and load-bearing interior beams and partitions.

© Christian Delbert/Shutterstock.com

Figure 20.30 Ceiling joists commonly span from the exterior wall to a load-bearing interior partition.

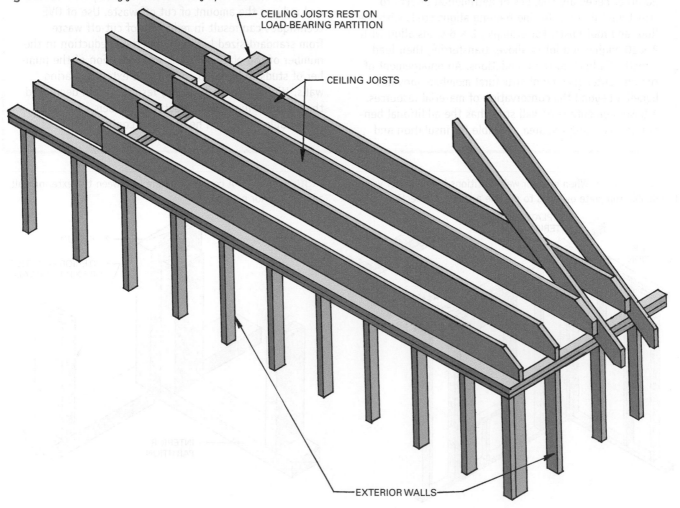

CEILING JOISTS REST ON
LOAD-BEARING PARTITION

CEILING JOISTS

EXTERIOR WALLS

Framing the Roof

Flat, gable, shed, and hip roofs forms are frequently used for wood construction. Roof framing is one of the most difficult parts of wood light frame construction and requires skilled carpentry. The following discussion is limited to the basic stick-built gable and hip roofs (Figure 20.31). The various components of roof framing are identified in Figure 20.32. Rafters rest on top of a wall's double plate and are nailed to it. In areas with high winds, metal straps are used to provide a stronger

Figure 20.31 Roof framing is one of the most difficult parts of wood light-frame construction and requires skilled carpenters.

connection to the plate (Figure 20.33). Overhangs on the gable end can be built using lookouts supported on the gable end wall and nailed to the nearest rafter beyond (Figure 20.34). The gable end can be site framed with 2 × 4 in. studs or built with a pre-manufactured gable end truss. Information on finishing cornices can be found in Chapter 22.

WOOD TRUSSES

Wood Roof Trusses

Wood roof trusses are widely used for framing wood light-frame buildings. A roof truss is a triangulated structural unit made by assembling structural wood members into a rigid frame. A typical example is shown in Figure 20.35.

The advantages of wood trusses are that they speed on-site erection time and can span the width of most buildings without interior load-bearing walls. Since they are precisely engineered and factory built, they have a consistent, reliable quality. One disadvantage is that they make it difficult to use an attic for storage. However, trusses are available that allow the center of a building to be open for storage or second-floor rooms. A few of the many types of wood trusses available are shown in Figure 20.36.

Figure 20.32 The nomenclature used to describe the various components of a "stick built" roof.

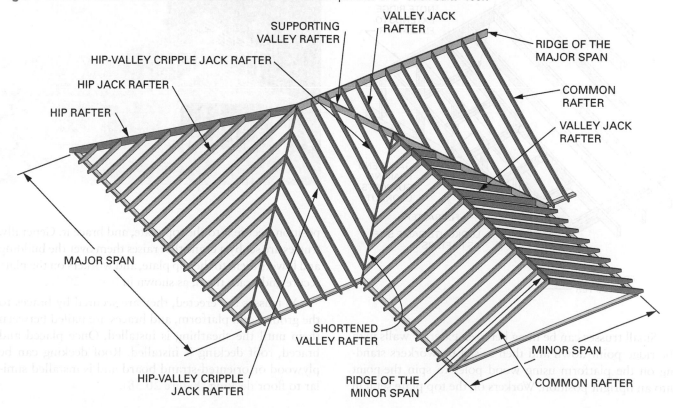

Figure 20.33 Rafters, trusses, and roof joists are toenailed to the top plate or fastened with metal connectors for increased strength.

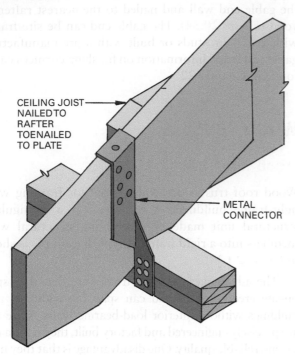

CEILING JOIST NAILED TO RAFTER TOENAILED TO PLATE

METAL CONNECTOR

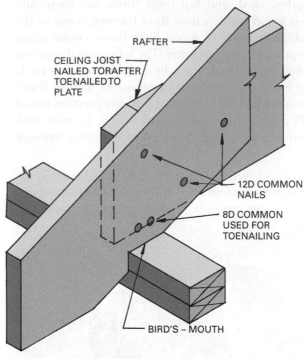

RAFTER

CEILING JOIST NAILED TO RAFTER TOENAILED TO PLATE

12D COMMON NAILS

8D COMMON USED FOR TOENAILING

BIRD'S – MOUTH

Figure 20.34 Gable-end roof overhangs are framed using lookouts that cantilever over the exterior wall to support the gable-end rafter.

Figure 20.35 Wood trusses speed the on-site erection time and can span the width of small buildings without interior load-bearing walls.

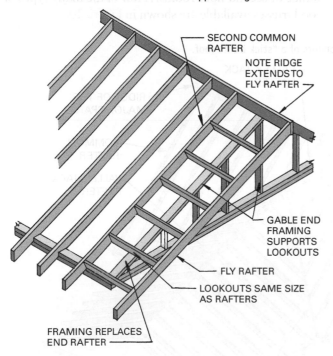

SECOND COMMON RAFTER

NOTE RIDGE EXTENDS TO FLY RAFTER

GABLE END FRAMING SUPPORTS LOOKOUTS

FLY RAFTER

LOOKOUTS SAME SIZE AS RAFTERS

FRAMING REPLACES END RAFTER

© B Brown/Shutterstock.com

Small trusses can be hung between exterior walls with the ridge point down and then raised by workers standing on the platform using wood poles to spin the point into an upright position. Workers on the top plate set it in

position, secure it to the top plate, and brace it. Generally, trusses are set by a crane that raises them over the building and lowers them to the top plate, and workers on the plate secure and brace them, as shown in Figure 20.37.

As trusses are erected, they are secured by braces to the ground and platform, and braces are nailed between them until the sheathing is installed. Once placed and braced, roof decking is installed. Roof decking can be plywood or oriented-strand board and is installed similar to floor decking (Figure 20.38).

Figure 20.36 Standard wood truss configurations.

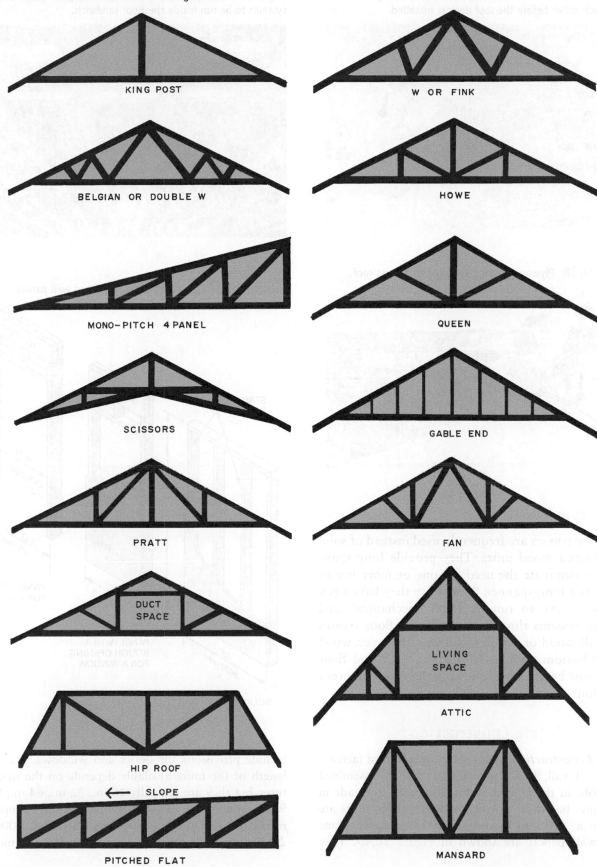

KING POST

W OR FINK

BELGIAN OR DOUBLE W

HOWE

MONO–PITCH 4 PANEL

QUEEN

SCISSORS

GABLE END

PRATT

FAN

DUCT SPACE

LIVING SPACE

ATTIC

HIP ROOF

SLOPE

PITCHED FLAT

MANSARD

Figure 20.37 Roof trusses are set and temporarily braced against each other before the roof deck is installed.

Figure 20.38 Plywood decking is nailed to a truss roof.

Figure 20.39 Wood floor trusses allow space for mechanical systems to be run inside the floor sandwich.

© arturasker/Shutterstock.com

Figure 20.40 Examples of prefabricated wall panels.

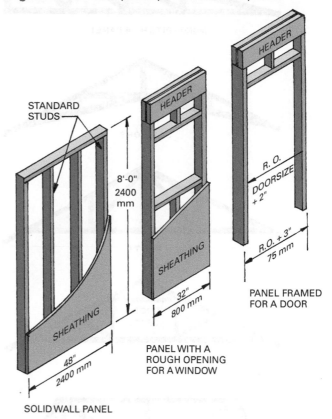

Wood Floor Trusses

Wood floor trusses are frequently used instead of solid or engineered wood joists. They provide long spans and often eliminate the need for one or more beams to support a long-spanned floor. Since they have open webs, it is easy to run electrical, mechanical, and plumbing systems through them. Wood floor trusses may be all wood or utilize metal webs between wood top and bottom chords (Figure 20.39). Wood floor trusses must be installed following the manufacturer's instructions.

Panelized Frame Construction

Panelized construction uses a series of standard factory-constructed wall, floor, and roof units that are assembled on the job. In the United States, the units are made in 16 in. units based on a 4 in. module. Metric units are based on a 100 mm module in 400 mm units. Some typical wall panels are shown in Figure 20.40. They include provisions for doors and windows. The actual length of the units available depends on the manufacturer, but they are typically 16 in., 32 in., 64 in., 80 in., 96 in., and 144 in. based on a 4 in. module. Comparable metric units are 400 mm, 800 mm, 1,600 mm, 2,000 mm, 2,400 mm, and 3,600 mm, based on a 100 mm module.

Both framed panels and structural insulated panels use similar installation procedures. Units are provided with a groove that is secured to a bottom plate bolted to the foundation. The panels are tied together with a top plate. Manufacturers install the sheathing, and sometimes windows, in the factory. The design of the panels must include ways to join exterior and interior corners. This is accomplished by adding extra studs and providing an overlapping flange of sheathing, as shown in Figure 20.41. Slots are provided for the connection of interior partitions to exterior panels.

Figure 20.41 Connection details for structural insulated panels.

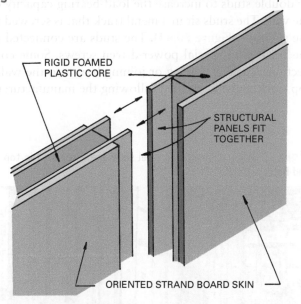

RIGID FOAMED PLASTIC CORE

STRUCTURAL PANELS FIT TOGETHER

ORIENTED STRAND BOARD SKIN

Roofs for panelized construction are often framed with factory-assembled gable ends and trusses. They are set in place with a crane and braced until sheathing is applied. The gable end comes sheathed and has the siding applied. The floor can be built with conventional joists and plywood subflooring. However, large factory-assembled floor panels can be used, and widely spaced beams can replace the floor joists.

FASTENERS FOR WOOD FRAME CONSTRUCTION

Nails, screws, bolts, and metal connectors are used to connect the members of wood-frame construction. A variety of fasteners are used for specific applications, and carpenters must be familiar with the selection of appropriate fastening devices. Most nails are cut from rolls of metal wires and are known as wire nails. Screws are used where greater holding power is required. Lag screws are very large wood screws with a hex head designed to be turned by a wrench for very strong connections.

Nails

Nails are made of steel, aluminum, brass, and other metals. Uncoated steel nails are referred to as bright nails. A number of coatings are applied to nails to increase holding capacity or prevent corrosion. Galvanized nails are coated in zinc to prevent rusting in moisture-laden environments.

The size of nails used for wood construction is designated by the penny (d) classification. The shortest nail in the penny system is 1 in. long and is referred to as 2d, while the longest (60d) is 6 in. long (Figure 20.42). Common nails have a smooth shank with a pointed end. Box nails are similar to common nails but thinner in diameter. Finish nails use a very small head that can be driven below the surface of the wood and filled with putty to hide the nail head. Casing nails are longer finish nails used to secure exterior trim materials.

Figure 20.42 Nails sizes in the penny system.

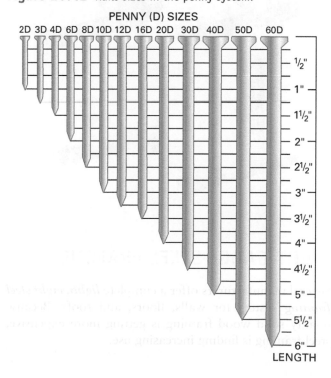

PENNY (D) SIZES

2D 3D 4D 6D 8D 10D 12D 16D 20D 30D 40D 50D 60D

½"
1"
1½"
2"
2½"
3"
3½"
4"
4½"
5"
5½"
6"

LENGTH

Wood Connectors

Metal connectors are widely used in wood construction. Consisting of metal pieces, they are used to connect wood to wood, wood to concrete, or wood to masonry. Wood-to-wood anchors, also known as framing anchors or hurricane ties, are used to provide extra connecting strength between members of a structure. Joist and beam hangers support horizontal framing members. Framing ties and anchors are manufactured in a range of unique shapes for particular applications (Figure 20.43).

Figure 20.43 Commonly available metal connectors for wood frame construction.

LIGHT-GAUGE STEEL FRAMING

Several manufacturers offer a complete *lightweight steel framing* system for walls, floors, and roofs. Because quality solid wood framing is getting more expensive, steel framing is finding increasing use.

Steel members are perforated to lighten them and permit the passage of utility lines, such as plumbing and electrical wiring. They are noncombustible and various types of exterior sheathing and siding can be applied; they are easy to insulate; and gypsum and other interior finish materials can be secured to them. Steel members can be preassembled into wall panels to speed erection. Steel framing is widely used for interior and exterior wall construction in commercial and, more recently, residential construction.

The lightweight steel-framing system includes studs, tracks, joists, bracing, hangers, and other accessories needed to assemble the unit. Bearing walls can be single or double studs to increase the load-bearing capacity of the wall. The studs sit in a metal track that is screwed to the subfloor (Figure 20.44). The studs are connected to the track with special power-driven screws. Some connections, such as bracing or joining beams to the wall's top track, must be welded following the manufacturer's

Figure 20.44 Light-gauge steel studs are assembled in top and bottom track.

specifications. Posts are made by assembling studs inside tracks. Headers over door and window openings are built up using joists secured inside track material.

Assembled walls require metal furring for strength that are screwed and spaced as specified (Figure 20.45). Furring strips are thin strips of wood or metal attached perpendicular to the studs in a horizontal fashion. Floors and ceilings are assembled using lightweight steel joists. Bridging may consist of straps or sections of joists forming solid bridging. In a multistory building, the second-floor joists rest directly above a first-floor stud that may be single or double. Short sections of joist are welded in on the end of each floor joist to stiffen it and help carry the load of the second-floor wall. Roofs are built using trusses assembled from metal channels. Refer to Chapter 17 for additional details.

Steel studs designed for nailing to wood top and bottom plates provide rapid assembly of walls and partitions. Walls are assembled by laying out studs on the subfloor and nailing them to the wood plates in the same manner as a wood stud. The wall is then lifted into place. Door and window openings are often

lined with wood to allow for easier fastening of finish materials (Figure 20.46). Additional lightweight steel-framing details for commercial construction can be found in Chapter 17.

Figure 20.46 Window and door openings may be cased in wood for easier fastening of finish materials.

Figure 20.45 Furring channels (hat track) are used in both ceiling and wall installations.

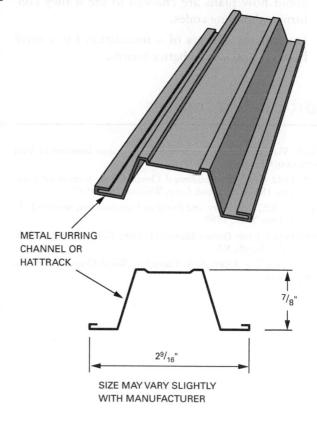

METAL FURRING
CHANNEL OR
HAT TRACK

$7/8"$

$2^9/_{16}"$

SIZE MAY VARY SLIGHTLY
WITH MANUFACTURER

Review Questions

1. What are the commonly used framing members used in light frame construction?

2. What is box-sill construction?

3. When is a beam required under floor joists?

4. How do the studs in balloon framing differ from those in platform framing?

5. How are second-floor joists supported when using balloon-frame construction?

6. What components make up an assembled truss unit used for truss-framed construction?

7. What is meant by panelized construction?

8. What are the advantages of using wood floor and roof joists?

9. What are the members used to construct a wall with lightweight steel members?

10. How are posts built using lightweight steel members?

Key Terms

Balloon Frame

Lightweight Steel Framing

Panelized Construction

Platform Framing

Rim Joist

Activities

1. Invite a local building official to speak to the class about how plans are checked to see if they conform to building codes.

2. Lay out the corners of a foundation for a small building and set the batter boards.

3. Visit as many wood frame construction sites as possible and observe the procedures used to lay out and construct light-frame buildings.

4. Construct a two-story scale model of a small building using platform framing.

Additional Resources

AISC Manual of Steel Construction, American Institute of Steel Construction, Chicago, IL 60601.

JLC Field Guide to Residential Construction, Journal of Light Construction, 186 Allen Brook Lane, Williston, VT 05495.

Spence, W.P., *Carpentry and Building Construction*, Sterling Publishing Co., New York, 1999.

Structural/Seismic Design Manual (Volume 2), International Code Council, Falls Church, VA.

Wood Building Technology, Canadian Wood Council, Ottawa, Ontario, Canada.

Wood Engineering and Construction Handbook, prepared by APA—The Engineered Wood Association, published by McGraw-Hill Corporation, New York.

Wood Frame House Construction, Metric Edition, Canada Mortgage and Housing Corp., Ottawa, Ontario, Canada.

Other resources include:

Publications of the Construction Institute of the American Society of Civil Engineers, Reston, VA.

See Appendix C for addresses of professional and trade organizations and other sources of technical information.

Heavy Timber Construction

Upon completion of this chapter, the student should be able to:

- Understand the requirements of the building code as it pertains to heavy timber construction.

- Become familiar with the materials and methods for framing heavy timber buildings and the types of connections used.

Build Your Knowledge

For further study on these materials and methods, please refer to:

Chapter 18 Wood, Plastics, and Composites

Chapter 19 Products Manufactured from Wood

Topic: Structural Building Components

Heavy timber construction is a method of creating framed structures using large exposed timbers joined together with wood joinery, pegs, or metal connectors. Diagonal bracing or panelized sheathing serves to strengthen the frame against racking. Timber-braced wall construction developed during the Middle Ages and was eventually brought to North America by British carpenters. To provide enclosure for these early structures, the spaces between the timbers were infilled with brick, rubble, or wattle and daub (Figure 21.1). Timber frame buildings tend toward an aesthetic of strength and craftsmanship. Since the frames require no interior load-bearing walls, interior spaces are open and flexible.

Wood products commonly used in heavy timber construction include solid timbers, glue-laminated members, parallel-strand lumber, and laminated-veneer lumber. The selection of material is governed by required load-bearing capacities, appearance, and availability of timbers. Modern complex structures and timber trusses often incorporate steel connections for both structural and architectural purposes. Sheathing and decking consists of thick wood members capable of spanning

Figure 21.1 Half-timbering utilized braced frames infilled with brickwork or wattle and daub.

© Torsten Lorenz/Shutterstock.com

distances between wood structural members. Heavy timbers can also be combined with conventional wood framing as in-fill panels (Figure 21.2). The structural frame can be enclosed in a variety of framing materials, normally located at the exterior of the frame to expose the timbers to the interior.

Figure 21.2 This heavy timber frame building uses tongue-and-groove decking on the roof and standard stud construction on the exterior wall areas between the columns.

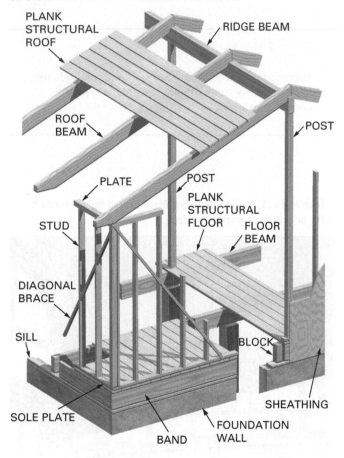

BUILDING CODES

Requirements for heavy timber construction are out-lined in the building codes. Heavy timber construction typically falls into Type IV and, in some cases, Type V. Under Type IV, exterior walls are constructed of ap-proved noncombustible materials and internal members may be of solid or laminated wood without concealed spaces or approved noncombustible materials. Under Type V, exterior walls, load-bearing walls, partitions, floors, and roofs may be any approved material that in-cludes heavy timber members. Type V construction uti-lizes heavy timber in conjunction with light wood frame walls for smaller buildings.

The code specifies the minimum size and type of per-mitted structural members for columns, floor framing, roof framing, floors, and roof decks. Approved types of connections are detailed, and the fire-resistance rat-ing of the structural elements is specified. Codes require seismic analysis and design in earthquake-prone areas. Diaphragms, shear panels, and selection and spacing of fasteners are specified. The building codes restrict the height of heavy timber buildings to three to five floors, depending on design and occupancy. Refer to Chapter 2 for more information on building codes.

FIRE RESISTANCE

Heavy timbers are able to absorb heat and flames much longer than other structural materials. By producing a char when exposed to fire, heavy timbers are able to maintain their structural capacity longer than would structural steel. The char produces a layer of fire pro-tection that shields the wood from sustained fire dam-age and prevents strength loss. Much of the structural strength is retained until the cross-sectional area be-comes so small that it will no longer carry the load.

TIMBER FRAME CONSTRUCTION

Solid timber construction is used for residential, com-mercial, educational, and religious buildings, among others. Timber-framed structures differ from light wood-framed buildings in a number of ways. Timber framing uses fewer, larger members, commonly with dimensions in the range of 6 in. to 12 in. (15 cm to 30 cm), as op-posed to light wood framing, which uses many more timbers with small cross-areas. The methods of fasten-ing frame members also differ: in conventional framing, members are joined with nails, while timber structures use complex joints that are usually fastened using only wooden pegs.

In the design of timber structures, the choice of material, the types of connections used, and the selec-tion of a structural system all play an important role. An architect and engineer must work together from the onset to define a project's objectives. The general layout of timbers is resolved based on the arrange-ment of functional areas. Once a species is selected, each member is sized individually along with connec-tion details. Commonly used species of wood include Douglas fir, pine, spruce, oak, and hemlock. Wood, as a hygroscopic material, can experience significant amounts of expansion and contraction with moisture content, especially perpendicular to the grain. Timber frames are engineered to manage this movement through careful detailing.

Timber frame members and components are normally prefabricated in mechanized shops and transported to the site for assembly. A timber frame structure is made up of walls, floors, roofs, and bents. Bents are full height frame constructions that run perpendicular to the walls and provide the primary structural posts of the frame. The bents are typically spaced in bays from 12 ft. to 16 ft. (3.65 m to 4.87 m) in width. Pre-cut and dressed timbers are laid out on the floor for assembly. The bents are constructed with mortise and tenon connections between column, beam, and bracing elements to produce a rigid, braced frame. The bents are raised in series and braced to one another with horizontal purlins. To provide structural stability, diagonal bracing is used to prevent racking, or movement of the structural frame. Braces are sized to at least half the length of the beam-to-beam span of the post. A wide range of materials can be used to serve as exterior sheathing between timber members. Figure 21.3 shows a traditional timber frame building using a series of bents as the structural frame.

Figure 21.3 This traditional heavy timber construction uses wood joinery to produce an exposed interior frame free from metal connectors.

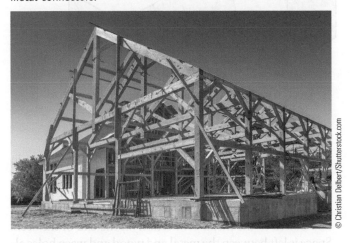

© Christian Delbert/Shutterstock.com

Timber Joinery

Timber joinery is the traditional method of connecting heavy timber structural members. It uses interlocking wood joints, such as tenons, dados, or halving joints, that are cut into the wood members. Once hand worked by experienced carpenters, these details are today produced by modern machinery. Some common wood-to-wood connection details are shown in Figure 21.4. Although these connections take longer to make, the finished joint demonstrates hand craftsmanship and the natural beauty of the wood. Experienced timber framers still use timber joinery, although metal connectors have largely replaced their use. The structural framing in Figure 21.5 shows typical beam and column construction and a heavy timber frame roof.

Timber frames may be supported on conventional concrete foundations or masonry walls. Figure 21.6 illustrates a detail for timber beam floors supported on a concrete wall. The beam is hung from a steel bracket anchored into the concrete bearing wall. A space is left between wood and masonry to prevent moisture migration through the wall from reaching the wood.

Beam-to-column connections may be made with metal connectors or wood-bearing blocks bolted to a column with *split-ring connectors* to control vertical forces (Figure 21.7). Beam-to-girder or beam or purlin-to-beam connections generally use some type of metal connector. However, wood ledgers with metal lateral ties are used in some cases. An engineered connection using metal connectors and wood angled supports bolted to the connector is shown in Figure 21.8. The work of a structural engineer is critical when designing structural systems and details to carry imposed loads.

ENGINEERED TIMBER FRAME CONSTRUCTION

The use of engineered wood allows designers to employ larger members with better structural properties than solid wood. Columns, beams, joists, rigid frames, arches, domes, and decking can be made from glued-laminated or other engineered wood products. Information about these products is available from the American Institute of Timber Construction (AITC) and in Chapter 19.

Glued-laminated members are engineered, stress-rated members made by laminating soft wood members with adhesives and resins. Standard glulam timber forms are shown in Figure 21.9. The actual sizes of commonly available rectangular glued-laminated members are shown in Table 21.1. Glulams are formed using adjustable presses that can produce a wide variety of forms, their length limited only by transportation restrictions. In addition to beams, rectangular, tapered, and varying-shaped columns are also manufactured.

Figure 21.4 Typical joints used in joining solid-wood framing members.

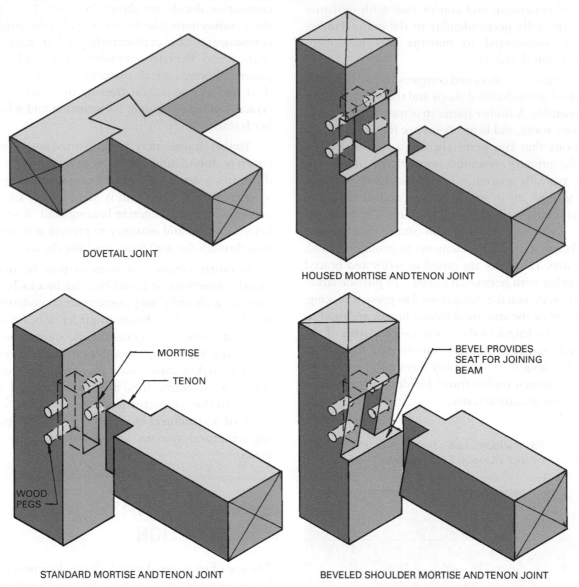

DOVETAIL JOINT

HOUSED MORTISE AND TENON JOINT

MORTISE

TENON

WOOD PEGS

STANDARD MORTISE AND TENON JOINT

BEVEL PROVIDES SEAT FOR JOINING BEAM

BEVELED SHOULDER MORTISE AND TENON JOINT

Arches and Domes

Glued-laminated arches and domes may be two-hinged, with hinges at each base, or three-hinged, with an additional hinge at the crown (Figure 21.10). A *hinge joint* is any joint that permits movement but in which there is no appreciable separation of adjacent members. These members produce considerable horizontal thrust at their base, which is controlled by the foundation and tie rods.

Laminated arches can span long distances, up to 70 ft. (21.4 m) or more. Span capabilities depend on the type of arch, imposed loads, roof pitch, and species of wood. Arches can be left exposed inside a building, forming an open ceiling (Figure 21.11).

Typical construction details for arches are shown in Figures 21.12 through 21.14. Figure 21.12 shows details for connecting parabolic arches and domes to a buttress foundation. The shoe plate is anchored to the foundation and bolted to the metal connection on the end of the arch. Space is left between the metal and wood and weep holes allow for the removal of water. A bridge pin connects the base shoe imbedded in concrete to the arch bracket. Several types of crown connections are shown in Figure 21.13. Long-span arches may require sections added on site because the assembled arch would be too large to transport.

Glued-laminated wood domes are designed as a radial arch or a triangulated system. The triangulated

Figure 21.5 A pegged wood joinery beam, column, and rafter connection.

© troy/Shutterstock.com

Figure 21.6 Wood beams are set in metal anchors that are tied to the foundation wall.

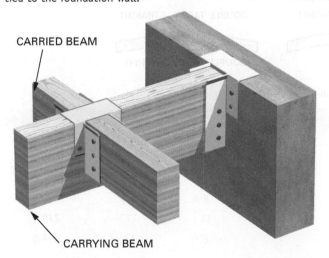

CARRIED BEAM

CARRYING BEAM

Figure 21.7 Ways to tie solid wood beams to columns.

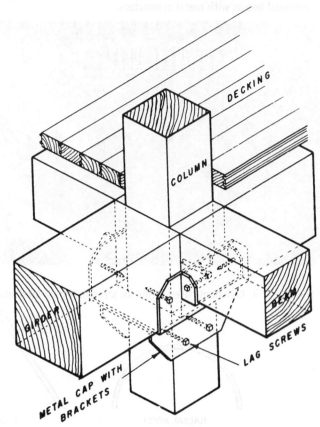

DECKING

COLUMN

GIRDER

BEAM

LAG SCREWS

METAL CAP WITH BRACKETS

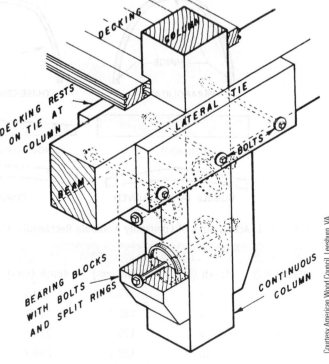

DECKING

COLUMN

DECKING RESTS ON TIE AT COLUMN

LATERAL TIE

BOLTS

BEAM

BEARING BLOCKS WITH BOLTS AND SPLIT RINGS

CONTINUOUS COLUMN

Courtesy American Wood Council, Leesburg, VA

Figure 21.8 This connection has angled wood members tied to overhead beams with metal connectors.

© Keith Hunter/Arcaid/Architecture/Corbis

arch can span greater distances than the radial arch. The arch shown in Figure 21.14 acts as both roof and exterior wall. Another type of dome construction, shown in Figure 21.15, produces a dramatic curved ceiling. Arched laminated wood structural members are also used for a variety of other projects, such as pedestrian bridges, towers, and exterior pavilions.

Columns, Beams, and Trusses

Several ways to secure both solid and glued-laminated beams to foundations are shown in Figure 21.16. Heavy steel clips and bolts resist both vertical and horizontal forces. It is recommended that beams rest on a metal plate. Column-to-foundation connections are shown in Figure 21.17. The column should rest on a metal plate and be at least 3 in. (76 mm) above grade or the finished

Figure 21.9 Commonly used forms for structural glue-laminated members.

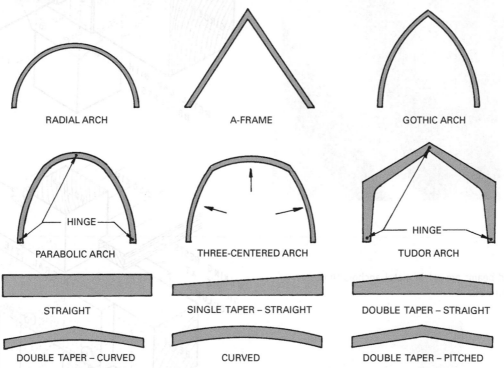

RADIAL ARCH   A-FRAME   GOTHIC ARCH

HINGE

PARABOLIC ARCH   THREE-CENTERED ARCH   TUDOR ARCH

HINGE

STRAIGHT   SINGLE TAPER – STRAIGHT   DOUBLE TAPER – STRAIGHT

DOUBLE TAPER – CURVED   CURVED   DOUBLE TAPER – PITCHED

Table 21.1 Actual Sizes of Commonly Available Rectangular Glued-Laminated Members

Laminations 1½ in. (38 mm) Thick				Laminations 1⅜ in. (35 mm) Thick			
Width (in.)	depth (in.)	width (mm)	depth (mm)	width (in.)	depth (in.)	width (mm)	depth (mm)
3⅛	7½	79.3	190.5	3	6⅞	76.2	174.6
5⅛	6	130.1	152.4	5	6⅞	127	174.6
5⅛	9	130.1	228.6	5	8¼	127	209.5
5⅛	10½	130.1	266.7	5	11	127	279.4
6¾	9	171.4	228.6	6	8¼	171.4	209.5

Figure 21.10 Arches and domes may be two hinged or three hinged. The horizontal thrust at the base is controlled with metal tie rods.

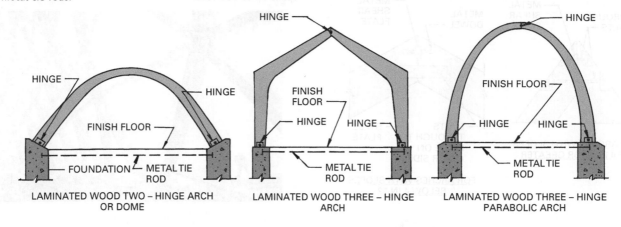

LAMINATED WOOD TWO – HINGE ARCH OR DOME

LAMINATED WOOD THREE – HINGE ARCH

LAMINATED WOOD THREE – HINGE PARABOLIC ARCH

Figure 21.11 Glue-laminated arches provide a lofty yet warm atmosphere to this commercial development.

© Ms Deborah Waters/Shutterstock.com

Figure 21.12 Dome-type arches are anchored to concrete foundations with steel anchor plates secured to them.

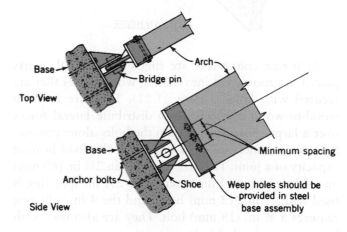

TRUE HINGE ANCHORAGE FOR ARCHES.

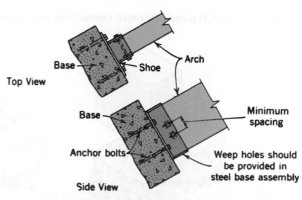

ARCH ANCHORAGE WHERE TRUE HINGE IS NOT REQUIRED.

© West Coast Lumber Inspection Bureau

floor. Large hold-down anchors are used to meet structural tie-down and tilt-over requirements specified by building codes. Steel connectors are used to secure beams and girders to the columns (Figure 21.18). The design engineer must indicate the type of fastener to be used and the size and number of bolts. Several beam-to-beam and beam-to-girder connections are shown in Figure 21.19.

Heavy timber trusses are used to span the width of a building and support purlins that carry the roof decking. A bolted, multimember truss is shown in Figure 21.20.

Connection Details

Two connectors used to provide additional strength to a bolted joint are split-ring connectors and shear plate connectors.

Figure 21.13 Typical connections for the crown of arches.

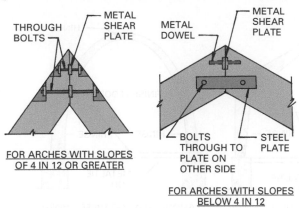

THROUGH BOLTS

METAL SHEAR PLATE

FOR ARCHES WITH SLOPES OF 4 IN 12 OR GREATER

METAL DOWEL

METAL SHEAR PLATE

BOLTS THROUGH TO PLATE ON OTHER SIDE

STEEL PLATE

FOR ARCHES WITH SLOPES BELOW 4 IN 12

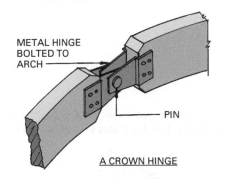

METAL HINGE BOLTED TO ARCH

PIN

A CROWN HINGE

Split-ring connectors are ring-shaped metal inserts placed in grooves cut into mating wood pieces that are secured with bolts (Figure 21.21). They are used on wood-to-wood connections to distribute lateral forces over a larger bearing area than the bolts alone provide. Split-ring connectors greatly increase the load-bearing capacity of a joint. They are available in 2½ in. (60 mm) and 4 in. (102 mm) diameters. The 2½ in. split ring is used with ½ in. (12 mm) bolts, and the 4 in. split ring requires a ¾ in. (19 mm) bolt. They are also used with lag screws instead of bolts.

Figure 21.14 Glue-laminated wooden domes can achieve spans up to 500 feet.

Figure 21.15 The glulam structure of this wooden roof becomes an architectural feature.

Figure 21.16 Typical glued-laminated connections used to secure a beam to the foundation.

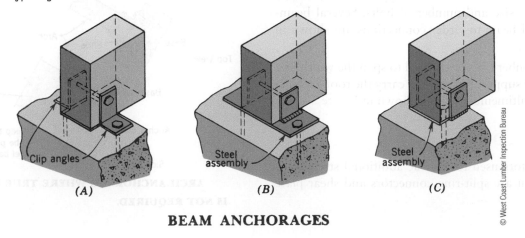

Clip angles

Steel assembly

Steel assembly

(A)

(B)

(C)

BEAM ANCHORAGES

Figure 21.17 Examples of glued-laminated column-to-foundation connections.

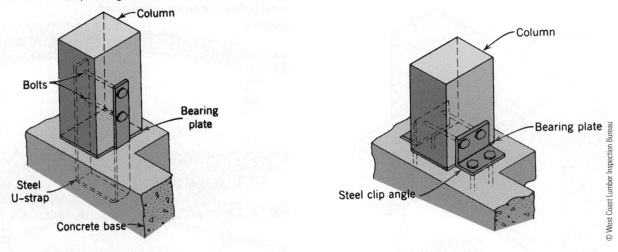

Figure 21.18 Typical beam-to-column and girder-to-column connections.

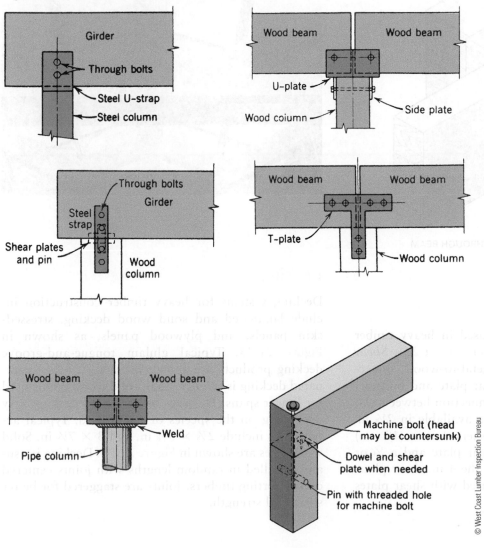

Figure 21.19 Typical beam-to-girder connections.

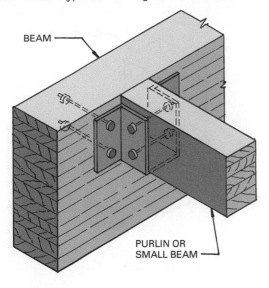

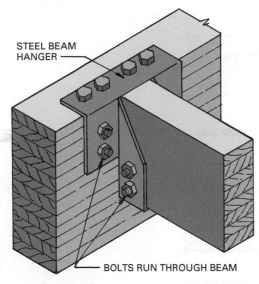

Figure 21.20 These multimember heavy wood trusses carry metal purlins that secure a glazed roof.

© Erick Margarita Images/Shutterstock.com

Another type of connector used in heavy timber construction is a shear plate (Figure 21.22). *Shear plate connectors* are used for metal-to-wood connections. The example shows a shear plate and bolts securing a steel plate forming a connection between two wood members. Shear plates are available in 2⅝ in. (52 mm) and 4 in. (102 mm) diameters. A ¾ in. (19 mm) bolt is used with the 2⅝ in. shear plate and a ¾ or ⅞ in. (22 mm) bolt is used with the 4 in. shear plate. Lag screws are also sometimes used with shear plates instead of bolts.

Decking

Decking systems for heavy timber construction include laminated and solid wood decking, stressed-skin panels, and plywood panels, as shown in Figure 21.23. Typical glulam tongue-and-groove decking products are shown in Figure 21.24. Laminated decking is stronger than solid wood and is used for longer spans. The sizes of laminated decking vary depending on the species of wood used. Typical actual sizes include 2⅞ × 5⅜ in. and 3 × 7⅛ in. Solid wood sizes are shown in Figure 21.25. Decking is usually installed in random lengths with joints centered on supporting timbers. Joints are staggered for better structural strength.

Figure 21.21 Split-ring connectors are inserted in recesses cut in the joining members.

Figure 21.22 Shear plates distribute lateral forces over a larger bearing area in metal-to-wood connections.

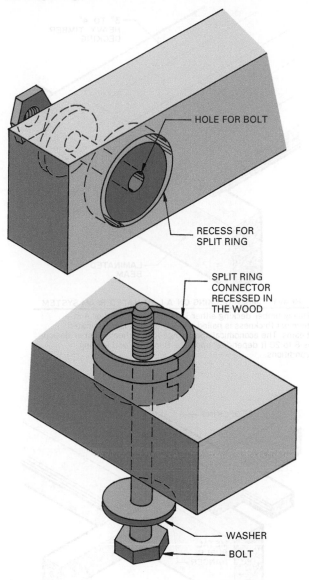

HOLE FOR BOLT

RECESS FOR SPLIT RING

SPLIT RING CONNECTOR RECESSED IN THE WOOD

WASHER

BOLT

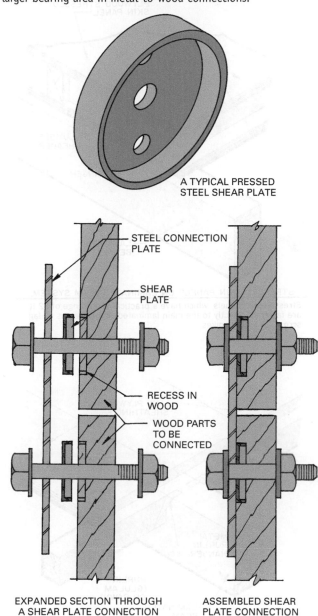

A TYPICAL PRESSED STEEL SHEAR PLATE

STEEL CONNECTION PLATE

SHEAR PLATE

RECESS IN WOOD

WOOD PARTS TO BE CONNECTED

EXPANDED SECTION THROUGH A SHEAR PLATE CONNECTION

ASSEMBLED SHEAR PLATE CONNECTION

Sheathing

The walls of timber frames may be sheathed in solid wood similar to the kind used for roof decking. Other panelized systems can also be utilized. Structural insulated panels consist of two rigid wood panel materials with a foamed insulating material secured between them, either by gluing billets, as in EPS (expanded polystyrene), or with polyurethane foamed and formed in place. These panels can be used as both sheathing and roof decking, providing a complete insulated enclosure at the exterior of the frame (Figure 21.26). An added advantage of panels is a decreased dependency on bracing and auxiliary members, since the panels can span considerable distances and serve to increase the stiffness of the timber frame. Using this method of enclosure ensures that timbers can only be seen from inside the building. The use of structural insulated panels results in a building with efficient heat insulation properties.

Figure 21.23 Types of decking used with glued-laminated framing.

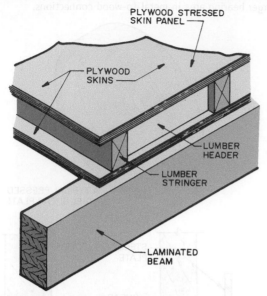

STRESSED SKIN PANELS ON LAMINATED BEAM SYSTEM.
Stressed skin panels, which have a practical span range of 32 ft, are fastened directly to the main laminated timber beams by lag screws or gutter spikes.

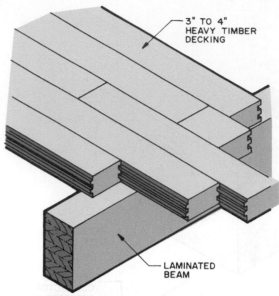

HEAVY TIMBER DECKING ON A LAMINATED BEAM SYSTEM.
Heavy timber decking either laminated or solid 3 or 4 in. nominal thickness is nailed directly to the main laminated beams. The economical span range for the heavy timber decking is 8 to 20 ft depending upon the thickness and loading conditions.

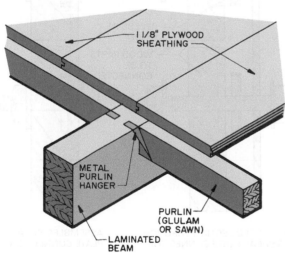

ONE AND ONE-EIGHTH-INCH PLYWOOD ON A LAMINATED BEAM AND PURLIN SYSTEM.

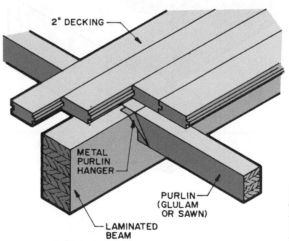

TWO-INCH DECKING ON A LAMINATED BEAM AND PURLIN SYSTEM.
Two-inch nominal thickness decking with an economical span range of 6 to 12 ft is nailed directly to glulam or sawed wood roof purlins, typically on 8 ft centers. Purlins are connected to the main laminated timber beams by metal purlin hangers.

Construction **Techniques**

Metal Fasteners

There are many types of metal fasteners used to join wood members. These range from nails and staples, used for light frame construction, to bolts, side plates, and other types of hardware. The effectiveness of metal fasteners depends on their being large enough to carry loads and transfer them over large areas so the wood fiber in contact with the fastener is not deformed. The spacing between metal fasteners and between the ends and edges of wood members and fasteners is critical to a successful union (Figure A).

Metal fasteners are subject to various types of loads, as shown in Figure B. Architects and engineers design connections so the calculated loads are sustained by the connectors and their placement in the wood members.

Figure A

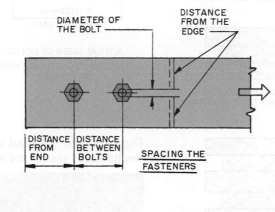

DIAMETER OF THE BOLT

DISTANCE FROM THE EDGE

DISTANCE FROM END

DISTANCE BETWEEN BOLTS

SPACING THE FASTENERS

Figure B

FASTENER LOADED (LATERALLY) IN SINGLE SHEAR

FASTENER LOADED (LATERALLY) IN DOUBLE SHEAR

FASTENER (NAIL) LOADED IN WITHDRAWAL. RELIES ON FRICTION FROM COMPRESSED WOOD FIBERS.

FASTENER (NAIL) LOADED LATERALLY WILL CARRY HIGHER LOADS.

Figure 21.24 Examples of the types of glued-laminated decking available.

SQUARE ENDS, SMOOTH
FACE, EQUAL THICKNESS

V-JOINT, SMOOTH FACE,
EQUAL THICKNESS

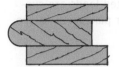

SQUARE ENDS, SMOOTH
FACE, CENTER THICKER

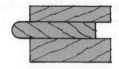

SQUARE ENDS, SMOOTH
FACE, BOTTOM THICKER

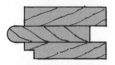

CHANNEL GROOVE, SMOOTH
FACE, EQUAL THICKNESS

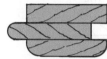

BULLNOSE, SMOOTH
FACE, EQUAL THICKNESS

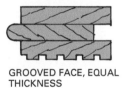

GROOVED FACE, EQUAL
THICKNESS

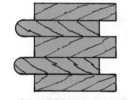

SQUARE ENDS, SMOOTH
FACE, DOUBLE TONGUE
AND GROOVE, EQUAL
THICKNESS

Figure 21.25 Sizes of glued-laminated wood decking made from Douglas fir, larch, and southern pine.

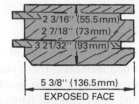

ACTUAL SIZES OF
DOUGLAS FIR, LARCH,
AND SOUTHERN PINE
GLUED LAMINATED
DECKING

2 3/16'' (55.5 mm)
2 7/18'' (73 mm)
3 21/32'' (93 mm)

5 3/8'' (136.5 mm)
EXPOSED FACE

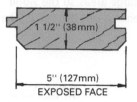

1 1/2'' (38 mm)

5'' (127 mm)
EXPOSED FACE

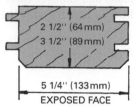

2 1/2'' (64 mm)
3 1/2'' (89 mm)

5 1/4'' (133 mm)
EXPOSED FACE

ACTUAL SIZES OF DOUGLAS FIR AND LARCH SOLID DECKING.

NOTE: EXPOSED FACE MAY BE SMOOTH, BRUSHED, GROOVED,
OR STRIATED.

© West Coast Lumber Inspection Bureau

Figure 21.26 Structural insulated panels can be used to simultaneously sheathe and insulate a heavy timber frame.

The Murus Company Inc.

Review Questions

1. What are the types of wood products used to frame heavy timber buildings?

2. What is the difference in reaction to fire between timber structural systems and steel systems?

3. What is meant by a hinged arch?

4. What two types of design are used for glued-laminated dome construction?

5. Why do codes require steel hold-downs?

6. What are the types of span details used with glued-laminated wood decking?

7. Why are split-ring connectors used?

8. When are shear plates used?

Key Terms

Heavy Timber Construction

Hinge Joint

Shear Plate Connector

Split-Ring Connector

Timber Joinery

Activities

1. Build a scale model of the structural frame of a heavy timber building.

2. Make sketches of the various types of metal connectors.

3. Copy from the local building code the requirements that apply specifically to heavy timber construction.

Additional Resources

Post-Frame Building Design Manual, National Frame Builders Association, Lawrence, KS.

Timber Construction Manual, American Institute of Timber Construction, Englewood, CO.

Wood Building Technology, Canadian Wood Council, Ottawa, Ontario, Canada.

Wood Handbook, Forest Products Laboratory, Madison, WI.

Wood Reference Handbook, Canadian Wood Council, Ottawa, Ontario, Canada.

See Appendix C for addresses of professional and trade organizations and other sources of technical information.

Finishing the Exterior and Interior of Light Wood Frame Buildings

Upon completion of this chapter, the student should be able to:

- Have an understanding of the many materials and design possibilities available for finishing the exterior of light wood frame buildings.

- Be familiar with the many materials and processes and some of the installation procedures required to finish light wood frame building interiors.

Build Your Knowledge

For further study on these materials and methods, please refer to:

Chapter 24 Thermal Insulation and Vapor Barriers

Chapter 31 Interior Finishes

Chapter 33 Interior Walls, Partitions, and Ceilings

Chapter 34 Flooring

Once the structural frame of a building is complete and sheathing, decking, and roof drainage plane are applied, exterior finish work can begin. Roof eaves and rakes must be framed and finished. Roofing contractors arrive at the site to install finished roofing and seal the building off from the weather. Doors, windows, and exterior siding are installed to complete the enclosure. Only then can interior finish work begin. While the interior finish work proceeds, final grading, drives, and landscape work can occur simultaneously. Detailed information on the various materials used to enclose light frame buildings is given in subsequent chapters. This chapter outlines the basic construction procedures in finishing light frame buildings in the order that they occur.

FINISHING THE EXTERIOR

Exterior finish work includes framing eaves and overhangs; installing roofing, siding, and gutters; installing windows and exterior doors; and doing the exterior painting or staining. With the installation of the exterior enclosure, the building is weather tight and ready for interior work.

Framing the Eaves and Rake

The eaves and rake are framed after the roof is sheathed. Eaves and rakes are designed to protect walls from excessive wear, provide shade, and complement the architectural style of the building. The *eaves* are the part of the roof edge that runs parallel with the roof ridge. The *rake* of a roof is perpendicular to the ridge and eaves. Common details include flush-roof edges and overhangs. Typical designs for sloped roofs are shown in Figure 22.1 and Figure 22.2. The finished board covering the ends of the rafters is called the *facia*. A rough fascia, usually 2 × material, provides spacing support for the ends of the rafters and a means of securing gutters and the finished fascia. The finished fascia, commonly 1 in. (38 mm) thick, can be painted composite

Figure 22.1 A closed cornice provides no overhang and uses a frieze board at the intersection of the roof and wall sheathing.

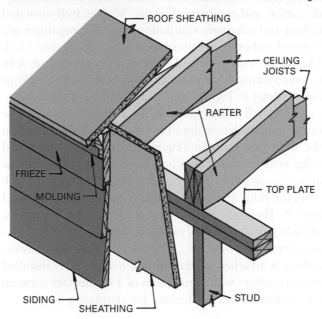

Figure 22.2 This wide-boxed cornice has a continuous attic vent strip.

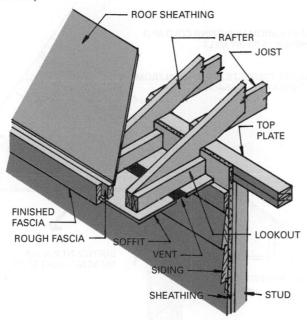

wood or wood covered with aluminum or vinyl. Rakes are finished in a fashion similar to eaves. Rake fascia may be set flush over siding or extend from a gable end. A *soffit*, the underside of an eave or rake, can be finished in vinyl, aluminum, or plywood, hardboard, or another

exterior reconstituted wood product. Vinyl and aluminum soffits are perforated, providing for air flow into the attic (Figure 22.3). Solid soffits material requires the addition of metal vent strips to provide attic ventilation (Figure 22.4).

Figure 22.3 This vinyl soffit is perforated to provide attic ventilation.

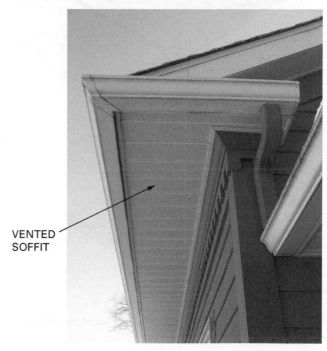

Ventilating the Attic

Ventilation is the exchange of air to allow for drying and improved air quality. A building envelope must control unwanted air leakage and ventilate unwanted moisture. With airtight construction detailing and proper ventilation, energy costs of a building are reduced. Attic ventilation is required to remove hot air in the summer and to prevent the accumulation of moisture in both summer and winter. In the winter, ventilation is necessary to keep an attic's air temperature congruent with the outside air. In an attic with insufficient insulation, heated interior air causes the air temperature in the attic to rise and snow on the roof to melt. An *ice dam* forms when the melted snow refreezes as it runs to an uninsulated eave (Figure 22.5). This snowmelt can back up under the shingles and possibly leak into the exterior wall and ceiling insulation. In cold climates, extra layers of builder's felt are laid from the eave up the roof to provide additional waterproofing.

Figure 22.4 Aluminum or plastic strip vents are nailed between soffit materials for continuous eave ventilation.

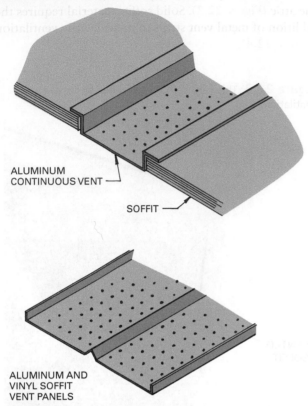

ALUMINUM
CONTINUOUS VENT

SOFFIT

ALUMINUM AND
VINYL SOFFIT
VENT PANELS

Figure 22.5 Ice dams are formed when melted snow flows over the cold eave and refreezes.

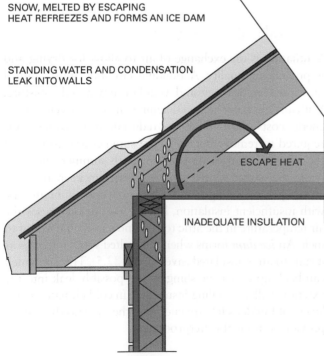

SNOW, MELTED BY ESCAPING
HEAT REFREEZES AND FORMS AN ICE DAM

STANDING WATER AND CONDENSATION
LEAK INTO WALLS

ESCAPE HEAT

INADEQUATE INSULATION

INADEQUATE INSULATION/UNVENTED

When heated air comes into contact with colder attic air, its relative humidity increases, which can cause condensation and moisture damage. With a well-insulated ceiling and adequate ventilation, attic temperatures are lowered and excessive moisture is removed (Figure 22.6). The use of both soffit and ridge vents is an effective way to vent an attic. Each rafter cavity is vented from soffit to ridge. (Refer to Figures 22.2, 22.3, and 22.4.) It is necessary to install some type of baffle at the exterior wall so the ceiling insulation does not block the flow of air from the soffit into the attic (Figure 22.6). Pre-manufactured ridge vents are used to allow the flow of air to exit the attic (Figure 22.7). The roof decking is cut back an inch from the ridge at each side and a vent assembly is nailed over it. There are many types of additional ventilators, including roof-mounted exhausts, power fans, and gable-end vents (Figure 22.8). On roofs where the finished ceiling is attached to the rafters, insulation is installed between rafters with a minimum of 1 in. airspace between the insulation and the decking for ventilation.

Figure 22.6 Sufficient insulation and soffit and roof vents prevent snow from melting.

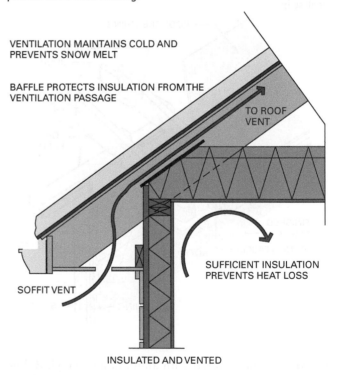

VENTILATION MAINTAINS COLD AND
PREVENTS SNOW MELT

BAFFLE PROTECTS INSULATION FROM THE
VENTILATION PASSAGE

TO ROOF
VENT

SUFFICIENT INSULATION
PREVENTS HEAT LOSS

SOFFIT VENT

INSULATED AND VENTED

The Finished Roof Material

After roof sheathing is covered with a water shield or builders felt, the finished roofing can be installed. Common roofing materials for light frame buildings include

Figure 22.7 Pre-manufactured ridge vents are widely used for attic ventilation.

Figure 22.8 Attic ventilation can be achieved with gable-end, dormer, cupola, or roof-mounted exhaust vents.

asphalt and wood shingles, clay, concrete or slate tiles, and sheet metal. Roofing materials are covered in detail in Chapters 26 and 27. The common residential roofing shingle is composed of a fiberglass base that is saturated with asphalt and has mineral granules embedded in the surface to form a protective coating (Figure 22.7). Shingles are usually applied by roofing contractors whose crews are trained to properly install both flashing and shingles according to manufacturer's recommendations. Clay and concrete tiles, wood shingles and shakes, and metal roofing materials are also used with light wood frame construction.

After the finished roofing is in place, gutters may be installed, although this is often delayed until the exterior is finished, to keep them from being damaged. Gutters are commonly made from either aluminum or vinyl. The finished roofing material extends about ¾ in. over the edge of the fascia and directs water into the gutter. The gutter also serves as a decorative finish on the fascia (Figure 22.9).

Figure 22.9 Gutters and downspouts are installed after all the exterior finish work is complete.

Installing the Weather Shield

Prior to the installation of windows and siding, the exterior sheathing is covered with asphalt-impregnated builder's paper or a watertight, vapor-permeable plastic sheet made from synthetic fibers. House wrap is a thin, plastic material commonly known by the names Typar or Tyvec (Figure 22.10). It provides a waterproof layer that permits water vapor in the wall cavity to pass through to the outside, thus preventing it from being trapped inside the wall cavity. A weathershield covers corners, window and door openings, plates and sills, and

Figure 22.10 A watertight, vapor-permeable plastic sheeting is installed under the exterior finished cladding.

is designed to survive prolonged periods of exposure to the weather. Some types of sheathing have a cover of perforated aluminum foil that serves the same purpose.

Installing Windows

Wood, steel, aluminum, and aluminum and vinyl-clad wood windows are commonly used in light frame construction. These are described in Chapter 29. Windows are usually fully assembled and ready for installation when delivered on site.

Installation occurs from the outside of a building after the sheathing and weather shield have been installed. Most windows have a metal or plastic nailing flange secured to the unit. The window is placed in the opening, checked for plumb, and nailed to the studs and header through the flange (Figure 22.11). The installation procedure begins by checking the opening to see that the sill is square and level. The house wrap

Figure 22.11 The installation of a manufactured window unit:

1. The house wrap is cut and pulled into the opening.

2. A bead of caulk is applied to adhere and seal the unit to the opening.

3. The unit is placed in the opening, checked for level and plumb, and nailed through the flange.

4. The window is sealed with an adhesive tape placed to ensure the drainage of moisture.

is carefully cut to ensure proper drainage around the window. A bead of caulk is applied around the opening to provide an airtight seal. The window is set into the opening, shimmed to level, and checked for square before being nailed securely through the nailing flange. The nailing flanges are covered with an asphalt adhesive tape for both air and waterproofing. In the second type of installation, the window unit is nailed to the studs and header through the casing (sometimes called the brick molding) (Figure 22.12).

Figure 22.12 Exterior window frames are plumbed with shims and nailed to the studs through the jamb and wedges.

© TroxFoto/Shutterstock.com

Exterior Door Installation

A variety of wood, metal, and fiberglass exterior doors and frames are available. These are described in Chapter 29. Like windows, most exterior doors come in an assembled, pre-hung unit. Typical exterior doorframe construction is shown in Figure 22.13. Careful installation of the unit is essential to ensure proper operation of the door and protection against the weather.

To install an exterior door, the doorframe is set in the opening and checked for plumb and level. The housewrap and caulking are treated as during window installation. Wedge-shaped shims between the studs and door jamb are used to position the door in the rough opening. When the door is properly positioned, the jamb is nailed through the shims to the framing. The door is checked and adjusted for proper operation before the exterior casings are installed. Units that come pre-hung with casings can be nailed through the casing first and then shimmed for final fastening.

Installing the Exterior Finish

Exterior siding may be wood, plywood, wood shingles, hardboard, plastic, vinyl, stucco, metal, brick, or stone. Details for installing common wood siding types are shown in Figure 22.14 and Figure 22.15. Wood siding is usually ¾ in. wide and face-nailed through the sheathing into the studs. When installing lap siding, the nails in the lower end of the board do not penetrate the siding board directly below. This allows each piece to expand and contract independently of each other. Concrete fiber siding, or hardboard, is available as lap siding and in panels. Hardboard products are manufactured with a wood grain finish that resembles conventional wood lap siding when installed (Figure 22. 16). The product is extremely durable and will not rot or deteriorate.

Hardboard panel siding is available in 4 × 8 ft. (1220 × 2440 mm) and 4 × 9 ft. (1220 × 2745 mm) sheets (Figure 22.17). It typically has batten strips nailed over it but may also be installed with a reveal (a slight offset between panels).

Aluminum and vinyl siding are available in a variety of sizes, colors, and surface textures (Figure 22.18). They must be nailed to solid sheathing or directly into the studs because they have little structural strength. A starter strip is nailed at the bottom of the wall (Figure 22.19). Specially designed channel posts are placed at internal and external corners and where the siding butts a surface, such as the wood frame of a door or window. Special nails are driven into premade slots, that allow for expansion and contraction of the materials. A typical installation detail is shown in Figure 22.20. Wood shingles used as siding are applied over nailable sheathing, such as plywood, and may be installed in a single or double course (Figure 22.21).

Stucco exterior surfaces may be constructed using a finish coat of a Portland cement, lime, and sand mixture (Figure 22.22). The stucco is troweled over a wire mesh

Figure 22.13 Exterior doorframes are plumbed with shims and nailed to studs through the jamb and shims.

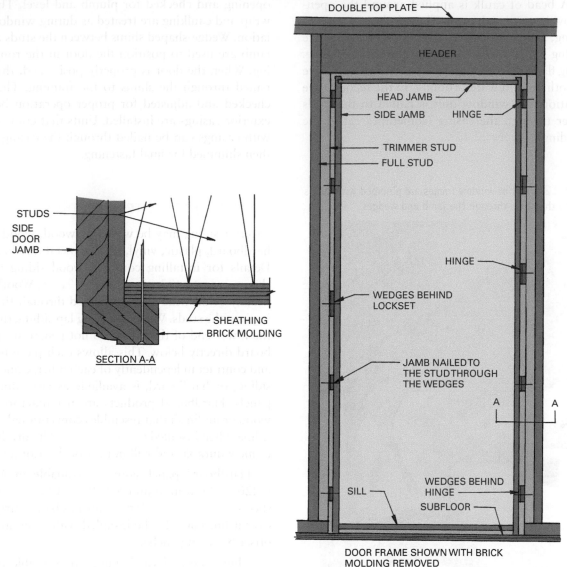

STUDS
SIDE
DOOR
JAMB

SHEATHING
BRICK MOLDING

SECTION A-A

DOUBLE TOP PLATE

HEADER

HEAD JAMB
SIDE JAMB — HINGE

TRIMMER STUD
FULL STUD

HINGE

WEDGES BEHIND
LOCKSET

JAMB NAILED TO
THE STUD THROUGH
THE WEDGES

A   A

WEDGES BEHIND
HINGE

SILL

SUBFLOOR

DOOR FRAME SHOWN WITH BRICK
MOLDING REMOVED

that is nailed to the sheathing (Figure 22.23). Another exterior finish similar to stucco that is finding increasing use is composed of a glass-fiber-reinforced Portland cement mixture containing special bonding agents. It is applied over a glass-fiber mesh bonded to the sheathing. This system has no means of draining water that may penetrate the wall assembly from around poorly flashed and caulked windows and other openings. Moisture such as this can cause the sheathing and wood wall framing to deteriorate.

Brick and stone veneers are also used as exterior siding on light frame construction. The masonry rests on a brick ledge constructed as part of the foundation. Metal ties are nailed to the studs and inserted in the mortar joints between masonry units. Wall ties are generally spaced every 16 in. (406 mm) vertically (Figure 22.24). Masonry and stone materials are covered in Chapters 11 through 14.

FINISHING THE INTERIOR

Once the building is weathertight, electricians, plumbers, and heating, ventilation, and air-conditioning (HVAC) contractors begin their rough-in work. Electricians drill holes in wood studs or use existing cavities to run electrical wiring and mount the required lighting and outlet boxes (Figure 22.25). Plumbers run water, waste disposal, and vent lines inside framed walls

Figure 22.14 Types of horizontal wood siding profiles and their nailing patterns.

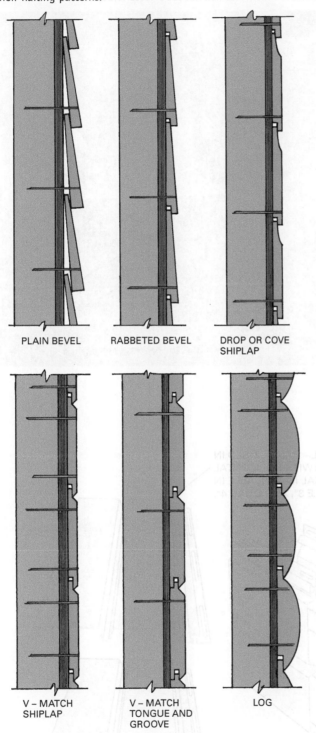

PLAIN BEVEL

RABBETED BEVEL

DROP OR COVE SHIPLAP

V – MATCH SHIPLAP

V – MATCH TONGUE AND GROOVE

LOG

Figure 22.15 Types of vertical wood siding with nailing patterns.

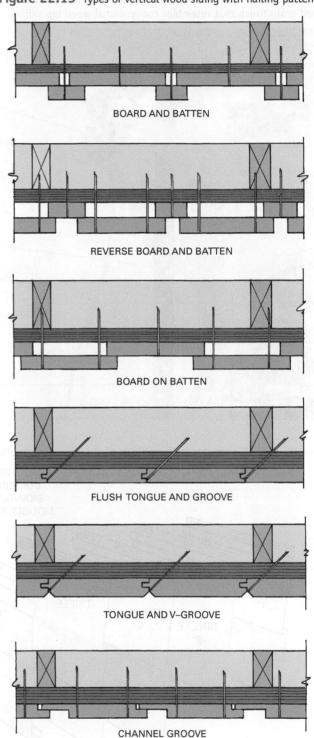

BOARD AND BATTEN

REVERSE BOARD AND BATTEN

BOARD ON BATTEN

FLUSH TONGUE AND GROOVE

TONGUE AND V–GROOVE

CHANNEL GROOVE

(Figure 22.26). Where pipes are close to the surface of wood studs, metal plates are used to prevent puncturing water lines during the installation of interior finish materials. The heating contractor installs sheet metal air ducts or runs hot water pipes to register locations.

Once all rough-in work is completed, the local building inspector checks for code compliance. The finish work and installation of fixtures, switches, outlets, registers, and other equipment will proceed after interior walls and ceilings are completed. The plumber usually

Figure 22.16 Concrete fiber hardboard is available with a wood grain finish that resembles conventional wood lap siding.

Figure 22.17 Hardboard panel siding provides an extremely durable finish that will not rot or deteriorate.

© Christina Richards/Shutterstock.com

Figure 22.18 Common configurations of vinyl siding and soffits.

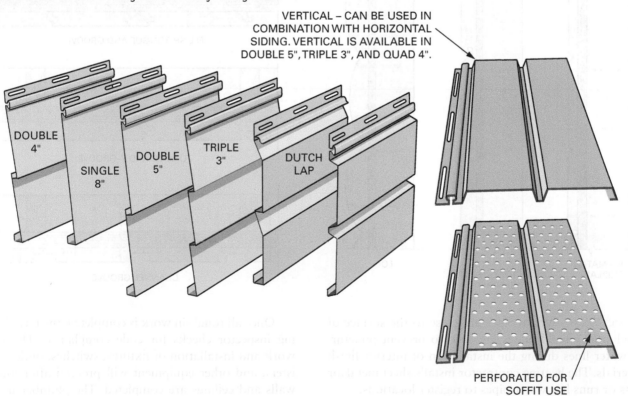

VERTICAL – CAN BE USED IN COMBINATION WITH HORIZONTAL SIDING. VERTICAL IS AVAILABLE IN DOUBLE 5", TRIPLE 3", AND QUAD 4".

DOUBLE 4"

SINGLE 8"

DOUBLE 5"

TRIPLE 3"

DUTCH LAP

PERFORATED FOR SOFFIT USE

Figure 22.19 A typical installation for vinyl and aluminum siding uses a starter strip and interlocking units that are nailed directly into the studs.

Figure 22.20 Vinyl siding is installed with special nails driven into premade slots that allow for expansion and contraction of the siding.

Figure 22.21 Wood shingles and shakes can be used as siding and are applied in a lapped pattern.

sets shower and tub fixtures to allow finish materials to be installed around them. The HVAC contractor temporarily covers duct openings to prevent construction dirt and dust from entering the system.

INSTALLING INSULATION

Once the interior is protected from the weather and the mechanical systems are installed and approved by a building inspector, insulation is installed in the walls, floors, and ceilings. Insulation materials create a space between the exterior and interior environments, breaking contact and preventing heat loss. Insulation is placed wherever the interior is exposed to exterior temperatures. A typical installation is shown in Figure 22.27. Among the materials used for insulating are glass fibers, mineral fibers (rock), organic fibers (paper, cotton), and plastics. Aluminum films are also used to reflect heat. These are described in Chapter 24.

Figure 22.22 Stucco exterior surfaces are constructed using a finish coat of a Portland cement, lime, and sand over a wire mesh that is nailed to the sheathing.

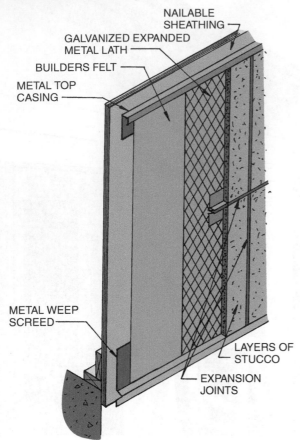

NAILABLE SHEATHING

GALVANIZED EXPANDED METAL LATH

BUILDERS FELT

METAL TOP CASING

METAL WEEP SCREED

LAYERS OF STUCCO

EXPANSION JOINTS

Figure 22.24 Metal ties nailed to studs are used to secure masonry veneers to wood frame walls.

Figure 22.25 Electrical wiring and service panels are installed after the building is weather tight.

© Christina Richards/Shutterstock.com

Figure 22.23 Portland cement stucco is troweled on a surface in consecutive layers.

© sima/Shutterstock.com

Flexible insulation is manufactured in rolls of either blanket or batt form. Blankets and batts may have an asphalt laminated paper covering that creates a vapor barrier when placed facing the warm side of a building. The paper provides tabs used to staple the insulation to the studs (Figure 22.28). Blankets should fit snugly against top and bottom plates and are cut to fit tightly around plumbing and electrical boxes.

Figure 22.26 Rough-in plumbing lines for water, waste disposal, and vent lines are installed inside of the framed walls.

© Lisa F. Young/Shutterstock.com

Figure 22.27 Fiberglass insulation is cut to fit around windows and other openings.

© Christina Richards/Shutterstock.com

Figure 22.28 Faced insulation has a water-resistant paper barrier that is stapled to the studs.

Figure 22.29 The insulation properties of a wall can be improved by using rigid plastic foam insulation at the exterior of the sheathing.

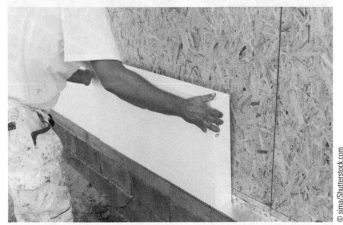

© sima/Shutterstock.com

Unfaced batts are held in place by friction until the interior finish is applied. Proper installation of batt insulation is essential to providing an uninterrupted insulation enclosure.

Various loose-fill and spray insulations are available today that overcome batt insulation installation problems. These are sprayed into a wall in successive layers and provide superior air-leakage control. For ceiling applications, plastic sheeting is used to contain the insulation before the finish material is installed.

The insulation properties of a wall can be improved by using rigid plastic foam insulated sheathing at the exterior of the sheathing (**Figure 22.29**). This can be installed to extend past the framing, down over

the foundation into the ground. It must be protected by a hard material where it is exposed to the weather. Some sheets are made with a protective coating already applied. Insulation may also be used to reduce the passage of sound through interior walls, floors, and ceilings. Various types of sound-control batts are available.

Interior Wall and Ceiling Finishes

Interior walls and ceilings in light frame construction are generally covered with gypsum board after the insulation has been installed. Other finishes include plaster, wood, and hardboard paneling. Gypsum and plaster products are discussed in Chapter 33 and wood in Chapter 19. Gypsum board and plaster finishes provide a hard, durable surface that can be covered with a wide range of decorative materials. In addition, they provide excellent fire ratings for the covered areas. Gypsum panels are usually installed by subcontractors specializing in this work (Figure 22.30). Plaster finishes require the services of highly skilled plasterers, who may not be available in some areas. Wood, plywood, and hardboard paneling are installed by finish carpenters.

Figure 22.30 Gypsum wallboard is a widely used material for finishing interior walls.

Usually, wood paneling is applied over a substrate of gypsum wallboard.

Ceilings are typically finished with gypsum board, plaster, some form of composition tile, or panels set in a suspended metal frame. See Chapter 33 for additional details.

INTERIOR FINISH CARPENTRY

Interior finish carpentry, sometimes referred to as *millwork*, involves the application of trim around doors and windows; the intersection of walls, floors and ceilings; and other interior moldings. Moldings are lengths of material, shaped in a variety of patterns, for use in a particular location. Wood is used to make most moldings, although plastic, composite wood products, and metal are sometimes used. Other millwork includes the installation of wainscoting, cabinets, built-in shelving, and stair finish. Finish carpentry requires the efforts of the most skilled craftspeople and cabinetmakers. Interior trim and millwork are among the final materials installed and care must be taken to not damage finishes during construction.

Installing Interior Doors

One task performed by a finish carpenter is the installation of interior doors and jambs. The inside frame of doors and windows are referred to as jambs. Interior doors in a variety of styles and finishes are manufactured in pre-hung units. Manufacturers supply detailed installation instructions with each unit. Additional information about doors is given in Chapter 29.

Most interior swinging doors arrive pre-hung on a solid wood door jamb. The carpenter places the jamb into the rough opening and plumbs the sides by driving wood shims between the frame and the wall stud. Once the door is set, it is nailed through the side jamb and wedges into the stud. A long level is used to check for plumb (Figure 22.31). The door is closed and the spacing between it and the jambs is checked for uniformity and to be certain the door does not stick on the jamb. Several types of split door jambs are also available. They are made with the casing attached to each half of a split jamb. The frames are slid into the rough opening from opposite sides and nailed to the studs.

Figure 22.31 An interior doorframe is set plumb using wood shims to position it in the rough opening.

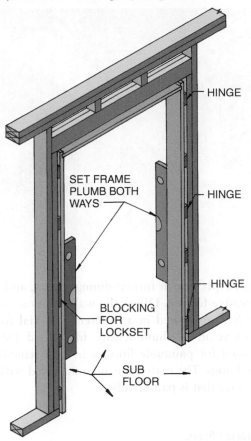

HINGE

SET FRAME
PLUMB BOTH
WAYS

HINGE

HINGE

BLOCKING
FOR
LOCKSET

SUB
FLOOR

Figure 22.32 Interior window openings are trimmed on both sides with a casing after the finish wall materials are installed.

© Christina Richards/Shutterstock.com

Installing Casings and Moldings

After door jambs are in place and gypsum wallboard installed, the finish carpenter can apply the casings and baseboards. *Casings* are the moldings used to trim around doors, windows, and other openings and are available in a wide range of sizes and patterns. An architect specifies the type to use. The installation of interior window casing is shown in Figure 22.32. The casing is usually set ¼ in. back from the edge of the jamb and nailed into the jamb and wall stud.

Traditional window casings and sills are commonly installed as shown in Figure 22.33. Here a wooden apron supports the extension of the windowsill, and the side casings die into the sill. Other designs utilize a wood windowsill only with a gypsum board return applied directly to the framing at the jambs and header, butting finished wallboard directly into the window frame.

Baseboard and molding is used where a wall meets the floor. It is usually mitered at outside corners and coped at inside corners. Coping refers to cutting the ends of the base molding to the shape of the piece it will butt. If the floor will be carpeted, the base is usually placed a little above the subfloor. If hardwood flooring will be used, the base is installed after the flooring is in place, and a small ¾ round molding covers the crack between the flooring and the baseboard (Figure 22.34). Flooring is discussed in Chapter 34.

Other types of molding are often used in high-end architectural applications. For example, a variety of crown moldings are used where walls and ceiling meet (Figure 22.35). Chair rails are used to protect walls from

Figure 22.33 A window with an apron supported sill and casings on three sides.

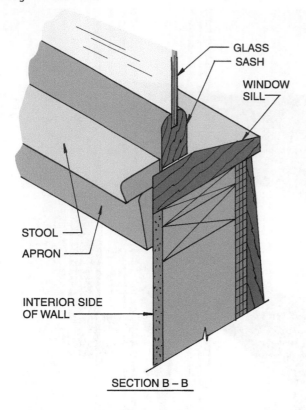

SECTION B – B

Figure 22.35 Crown moldings are used at the intersection of walls, beams, and ceilings.

© John Wollwerth/Shutterstock.com

damage, especially in formal dining rooms, and to add a decorative feature. Often, the wall below a chair rail is paneled or finished in a different material from the wall above it. Medium-density fiberboard (MDF) is often used for paintable finishes in wall paneling and cap moldings. The material is manufactured with a fine grit surface that is paint ready.

Figure 22.34 Baseboard and base shoe installation details for wood and carpeted floors.

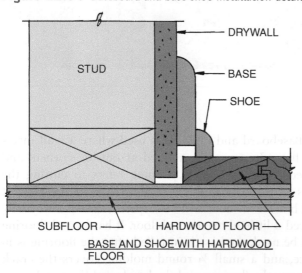

BASE AND SHOE WITH HARDWOOD FLOOR

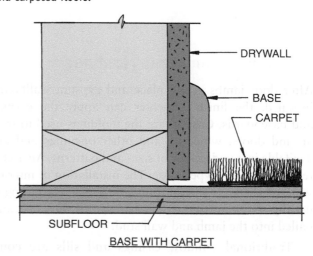

BASE WITH CARPET

Installing Cabinets

Manufactured kitchen and bath cabinets come in a wide variety of styles, materials, and finishes. Most cabinets used today are made in a factory and shipped to sites

ready for installation. Figure 22.36 shows standard modular sizes of both base and wall cabinet units. A finish carpenter installs cabinetry following the architectural drawings.

Figure 22.36 Manufactured kitchen cabinets are made in a factory and shipped to the site ready for installation.

© Kristin Smith/Shutterstock.com

Some installers prefer to install wall cabinets first because they can stand directly below them while working. Others set base cabinets first and use them as a support to hold the wall cabinets as they are installed.

Base cabinets are placed on the subfloor and leveled by shimming them at the floor. They are screwed through the back rail to the wall studs. Some cabinets have adjustable metal or plastic legs for leveling purposes. Walls must be checked for straightness. If there is a bow, the base cabinets must be shimmed out so they are straight.

Wall cabinets are installed a specified distance above the base cabinets, usually 18 in. (45.7 cm). They are often supported on the base cabinets and adjusted until they are level. Then they are shimmed to get them plumb and screwed through the back rails into the studs (Figure 22.37).

Countertop materials include plastic laminate, solid plastics, granite and other stones, sheet metal, and ceramic tile, among others (Figure 22.38). Thin materials utilize a particleboard underlayment for strength. A carpenter cuts openings needed for sinks and lavatories. A plumber installs these when the building nears completion.

Installing Stairs

Stair design is shown on architectural drawings; they specify the type and quality of the treads, handrail, newel posts, balusters, and other parts. Most wood stairs are built by carpenters on site. Stringers are constructed of solid or engineered wood, while railings and finish materials may be pre-manufactured. Completely manufactured stairs can be specified and assembled on site. In the

Figure 22.37 Wall cabinets are shimmed to plumb and screwed through the back rails into the wall framing.

© Lisa F. Young/Shutterstock.com

Figure 22.38 Wood subcounters provide a surface for the finish countertop material.

© Lisa F. Young/Shutterstock.com

design of stairs, an architect must observe the requirements of local building codes that regulate riser height, tread width, stair width, and handrail specifications.

The design of a typical wood-frame stair is illustrated in Figure 22.39. The load is carried by the carriage or stringer. Generally, a stair has three or more carriages. The rough framing for a stair with a landing is shown in Figure 22.40. A riser is a board covering the vertical distance between treads. Wood treads are commonly 1 in. thick. Some stairs have the treads and risers set in housed carriages. They are held in place with glue and wedges. There are many factory-manufactured stair components that provide interesting and attractive appearances (Figure 22.41).

Other Finish Items

Finish carpenters complete a number of additional tasks as a building nears completion. These include installing hardware, doorstops, bath accessories, and fireplace

Figure 22.40 Framing details for a stair with landing.

© David Papazian/Spirit/Corbis

Figure 22.39 A typical detail for a straight-run wood-frame stair.

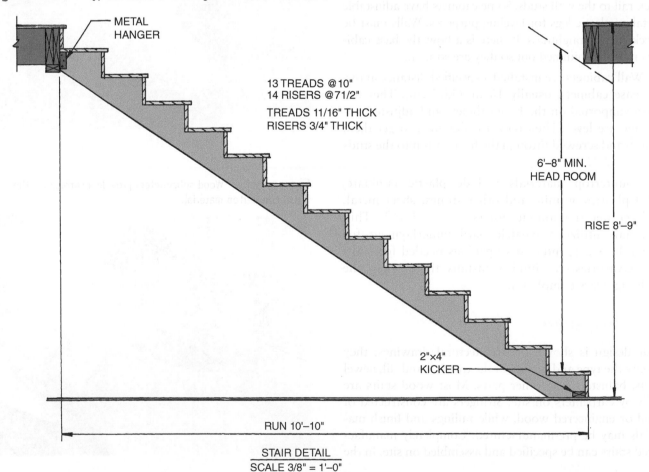

METAL
HANGER

13 TREADS @ 10"
14 RISERS @ 7 1/2"

TREADS 11/16" THICK
RISERS 3/4" THICK

6'–8" MIN.
HEAD ROOM

RISE 8'–9"

2"×4"
KICKER

RUN 10'–10"

STAIR DETAIL
SCALE 3/8" = 1'–0"

Figure 22.41 A staircase using commercially available components.

© Chad McDermott/Shutterstock.com

mantels. Some items, such as mirrors and closet shelving, may be installed by the subcontractor supplying the materials. Finally, the building is thoroughly cleaned, the subfloor scraped clean and sanded if necessary, and the carpet, composition floor covering, or tile, marble, or slate floor covering is installed (Figure 22.42). If prefinished hardwood floors are installed, they need to be cleaned and waxed. Unfinished floors are sanded and finished as specified.

Figure 22.42 The interior is finished after painting and the installation of the finished flooring material.

© Sergey Shlyaev/Shutterstock.com

Other trades have to phase in their work at the proper time. This includes painting and the installation of wallpaper and other wall coverings. Window shades and curtains are installed after carpet is down and all interior finishing is complete.

Review Questions

1. What materials can be used for the finished soffit and fascia?

2. Why is attic ventilation important?

3. What are some of the ways a roof can be vented?

4. What types of exterior siding are commonly used?

Key Terms

Casings

Eaves

Facia

Ice Dam

Millwork

Rake

Soffit

Activities

1. Visit a construction site regularly and follow the building phases, from finishing the exterior through the work of the trades involved in interior finishing. Keep a log indicating when each part started, thereby producing a construction schedule for the job.

2. Sketch several ways used for framing eaves and rakes.

3. Carefully examine the construction of a staircase in a building and see if it meets recommended construction standards. If it does not, prepare a report that includes sketches citing what you would do to improve the construction.

Additional Resources

Gypsum Construction Handbook, United States Gypsum Company, Chicago, IL.

Spence, W. P., *Installing and Finishing Flooring; Roofing Materials and Installation; Interior Trim-Making, Installing Finishing; Installing and Finishing Drywall; Carpentry* and *Building Construction;* and *Staircases, Balustrades and Landings*, Sterling Publishing Co., New York.

Wood Building Technology, Canadian Wood Council, Ottawa, Ontario, Canada.

See Appendix C for addresses of professional and trade organizations and other sources of technical information.

Plastics

Upon completion of this chapter, the student should be able to:

- Learn the properties and characteristics of plastic materials used in construction.

- Select the proper plastic material for various construction applications.

Build Your Knowledge

For further study on these materials and methods, please refer to:

Chapter 24 Thermal Insulation and Vapor Barriers

Chapter 29 Doors, Windows, Entrances, and Storefronts

Chapter 39 Plumbing Systems

Topics: Piping, Tubing, and Fittings

Plastic products have rapidly become a major construction material, finding use for both structural and nonstructural purposes (Figure 23.1). Because extensive research in plastic development has resulted in a variety of plastics that have properties not available in conventional materials, plastic is replacing materials such as wood and metal in many aspects of construction. Plastics resist corrosion and moisture; they are tough, lightweight, and easily formed into useful products. Since they are chemically derived, a wide range of plastics with special properties can be developed. This makes them particularly useful as a construction material because of the extensive range of applications that exist.

Vinyl siding is available in a variety of colors, surface textures, and sizes and provides a tough, maintenance-free exterior. Wood windows are clad in vinyl, eliminating the need for painting, and some windows are framed entirely from solid extruded plastic. Other types of plastics are replacing glass glazing, because the plastic is lighter and more resistant to shattering. Plastic lavatories, showers, and bathtubs have largely replaced those of ceramic-coated metal or cast-iron fixtures. A high percentage of components used in electrical systems, such as boxes and wiring insulation, are plastic. Plastic film is used for vapor barriers to reduce air infiltration, and plastic foams are used for insulation.

Since there are so many different kinds of plastics, each with varying properties, designers and constructors must be careful to use individual plastics for their designated purposes. For example, using plastic pipe designed to carry cold water for hot water lines will eventually cause problems. Many families of plastic materials exist, and within each, properties can vary widely. Plastics from several families may serve the same purpose, but others exhibit unique, special characteristics.

The term **plastic** is used today to describe manmade **polymers** that contain carbon atoms covalently bonded with other elements. Plastic is obtained by breaking down materials found in nature, such as petroleum, coal, and natural gas (Figure 23.2). Plastics are **synthetic materials**, resulting from chemical manipulations of natural materials. While plastics are composed of organic materials, they are manmade and not found in nature. When produced, they are soft and exhibit **plastic behavior**, meaning they can be formed into desired shapes. Most are made up of molecules built around a carbon atom. An exception to this is a plastic composed of a group of materials built around the silicon atom.

Figure 23.1 A variety of plastic products are widely used in the construction industry.

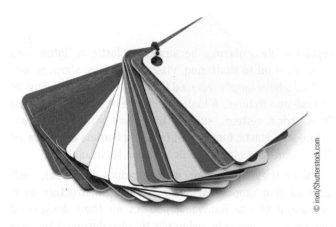

MOLECULAR STRUCTURE OF PLASTIC

The properties of plastics can be made to vary by manipulating the molecular structure of the material. Most materials classified in the plastics family are based on the carbon atom. The carbon atom bonds with other atoms through covalent bonding. *Covalent bonding* is a process in which small numbers of atoms are bound into molecules. A single molecule is known as a *monomer*. When several monomers link together to form a chain, they create a polymer. This chain formation is called polymerization. Plastic polymers are called macromolecules because they are composed of many smaller molecules ("macro" means large). Since the small molecules are joined together in a chainlike condition, they are often called chains. Plastics are made up of these polymers. Note that many plastics are described by the prefix "poly," such as polyvinyl and polystyrene. Plastic is a polymer. Polymers are moldable materials, sold in the form of granules, powder, flakes, liquids, or pellets for processing into useful products (Figure 23.3).

The molecular bonding process can be shown by the following illustration. The carbon atom has a *valence* of four, which means there are four points at which other atoms can attach themselves to the carbon atom to form covalent bonds. Other elements have different valences. For example, oxygen atoms have a valance of two. Two oxygen atoms can bond to the four carbon valence points creating carbon dioxide (CO_2) (one carbon atom and two oxygen atoms).

Classifications

Plastics can be divided into two basic classifications: thermoplastics and thermosetting materials. A new classification of bioplastics is finding wider use.

Thermoplastics

Thermoplastics are plastic materials that can be softened or re-melted by the application of heat and reformed. This quality allows for the recycling and reuse

Figure 23.2 The production of plastics and plastic products.

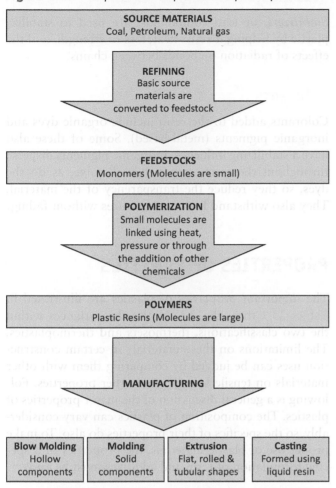

SOURCE MATERIALS
Coal, Petroleum, Natural gas

REFINING
Basic source materials are converted to feedstock

FEEDSTOCKS
Monomers (Molecules are small)

POLYMERIZATION
Small molecules are linked using heat, pressure or through the addition of other chemicals

POLYMERS
Plastic Resins (Molecules are large)

MANUFACTURING

Blow Molding	Molding	Extrusion	Casting
Hollow components	Solid components	Flat, rolled & tubular shapes	Formed using liquid resin

Figure 23.3 Polymers are moldable materials in the form of granules, powder, flakes, liquids, or pellets for processing into useful products.

© XXLPhoto/Shutterstock.com

of products made with thermoplastics, although some thermoplastic materials will experience contamination and chemical degradation if they are reheated frequently.

Thermoplastic materials are composed of long chain-like molecules that are unattached to each other. These molecules can slide past one another and change shape. At normal temperatures, 70°F (21°C), the material retains its shape because the movement of molecules is slight. However, when heat is applied, the bonding between molecules weakens, and the material expands and becomes soft. At high temperatures (which vary according to composition), the plastic softens and has sufficient flow to permit molding. As the molded part cools, the plastic holds the new shape.

Thermosetting Plastics

Thermosetting plastics, also called *thermosets*, are plastics that cannot be reheated and reformed once they have been softened, constituted, and cured. They are formed with a chemical process that produces a strong bond between molecules and prevents their sliding by each other. The final form of the material is irreversible.

Bioplastics

Bioplastics, or organic plastics, are derived from renewable biomass sources, such as vegetable oil or cornstarch, rather than from petroleum, as are fossil fuel plastics. The production and use of bioplastics is generally regarded as more sustainable when compared with petroplastics (plastic production from petroleum) because it relies less on fossil fuel carbon sources, resulting in fewer greenhouse emissions. Because the material biodegrades, it significantly reduces hazardous waste caused by oil-derived plastics, which remain solid for hundreds of years. Bioplastics represent a new era in manufacturing technology.

ADDITIVES

Most plastic resins in pure form do not have the properties needed for particular applications. Therefore, *additives* are mixed with the resin to modify their properties. Frequently, more than one additive is needed to obtain the desired properties. The commonly used additives include plasticizers, fillers, stabilizers, and colorants.

Plasticizers

Plasticizers are added to plastic resin to reduce brittleness, increase flexibility, resiliency, and moldability, and, in some cases, to improve impact resistance. A plasticizer weakens the bond between the chainlike molecules that make up a plastic. Since the molecules are able to slide past each other, the plastic can bend and flex. In doing this the strength, heat resistance, and dimensional stability are lessened (Figure 23.4).

Figure 23.4 Some plastic products, such as this vinyl floor covering, have plasticizers added to the resin to reduce brittleness, increase flexibility and resiliency, and improve impact resistance.

© Lilyana Vynogradova/Shutterstock.com

Fillers

Some *fillers* are used to reduce a plastic's cost by providing bulk. Others are used to improve a specific property. Fillers used to increase bulk or ease of molding include finely ground hardwood and nutshells. Hardness is improved by adding mineral oxides and mineral powders. Heat resistance is enhanced by adding inorganic fillers, such as clay, silica, ground limestone, or asbestos. Quartz or mica increase electrical resistance. Toughness is improved by adding fibers, such as hemp, cotton, sisal, rayon, polyester, or nylon. Mechanical properties are strengthened with the addition of glass or metal fibers.

Stabilizers

Stabilizers, or lead compounds, are used to stabilize plastic by helping it resist heat, loss of strength, and the effects of radiation on bonds between chains.

Colorants

Colorants added to the resin include organic dyes and inorganic pigments (metal-based). Some of these also have a stabilizing influence. Inorganic pigments disperse throughout the resin, rather than dissolve, as do the dyes, so they reduce the transparency of the material. They also withstand high temperatures without fading.

PROPERTIES OF PLASTICS

The important properties of plastics are illustrated in Tables 23.1 through 23.5. Notice the variances within the two classifications, thermosets and thermoplastics. The limitations on these materials in certain construction uses can be judged by comparing them with other materials on tensile strength and other properties. Following is a general discussion of the major properties of plastics. The composition of plastics can vary considerably, so the specifics of their properties do also. To make proper use of plastic materials, the properties of a specific formulation of resin and additives must be known.

Table 23.1 A Comparison of the Tensile Strengths of Plastics with Other Commonly Used Construction Materials

Material	Tensile Strength, psi
Thermoset plastics	175–35,000
Thermoplastics	1,000–20,000
Reinforced plastics	800–50,000
Plastic laminates	800–50,000
Glass (annealed)	3,000–6,000
Cast iron	20,000–100,000
Hot-rolled carbon steel	43,000–130,000
Southern pine	500–1,250
Wrought aluminum	5,000–82,000

Mechanical Properties

Mechanical properties to consider for plastics used in construction are tensile strength, stiffness, toughness (or impact strength), hardness, and creep. As shown in

Table 23.2 A Comparison of the Modulus of Elasticity of Plastics with Other Commonly Used Construction Materials

Material	Modulus of Elasticity (E), 10^6 psi
Thermoset plastics	0–2.4
Thermoplastics	0–1.8
Reinforced plastics	0–5.00
Cast iron	15–25
Wood	0.5–2.6
Aluminum alloys	9–11
Glass	9–11
Concrete	1–5
Steel, carbon, high strength, and stainless	29.0

Table 23.3 A Comparison of the Service Temperatures of Plastics with Other Commonly Used Construction Materials

Material	Service Temperature, °F
Plastics	80–500
Plastic foam (polystyrene)	140–150
Carbon steel	900–1,350
Wood	225–400
Glass	350–1,600
Concrete	450–1,350
Aluminum alloys	300–575

Table 23.4 A Comparison of the Coefficient of Thermal Expansion of Plastics with Other Commonly Used Construction Materials

Material	Coefficient of Thermal Expansion 10^{-6} in./in./°F
Thermoset plastics	7–167
Thermoplastics	19–195
Reinforced plastics	7–38
Cast iron	5.9–6.6
Carbon steel	6.5
Wood	7–38
Glass	1
Aluminum alloys	9
Concrete	7

Table 23.1, the tensile strength of most plastics is lower than steel and more closely resembles wood. Plastic laminates and fiberglass-reinforced plastics have tensile strengths approaching steel.

Table 23.5 A Comparison of the Thermal Conductivity of Plastics with Other Commonly Used Construction Materials

Material	Thermal Conductivity BTU/hr./ft.²/in./°F
Thermoset plastics	1–2
Thermoplastics	0.5–5.0
Plastic foams	0.15–0.26
Wood	0.5–1.5
Glass	5–6
Brick	3–6
Concrete	5.5–9.5
Steel alloys	300–500
Aluminum alloys	500–1,500
Copper	440–2,550

Stiffness is measured by the modulus of elasticity (E). The greater the modulus of elasticity, the stiffer the material. Overall, plastics are not as stiff as steel, aluminum alloys, and concrete (Table 23.2), so they deflect more under load. One advantage plastic structural members have is that they are lighter than conventional materials.

Impact strength (toughness) is the ability of a material to resist impact from an object striking it. In general, thermoplastics exhibit better impact strength than thermosets. Polyurethane is particularly resistant and is used for automobile bumpers. Polyethylene and polyvinyl chloride (PVC) have a wide range of impact strength and provide an array of uses, from bottles to piping. Compared to other construction materials, plastics are quite high in impact strength. This is why acrylic and polycarbonate sheet material is used for glazing in security situations.

Hardness is a measure of abrasion resistance. In general, plastics are not as good as glass or steel in resisting abrasion, as evidenced by visible scratches on plastic glazing resulting from improper cleaning.

Creep is the term used to describe the tendency of a material to flow in normal temperatures and to have a permanent change in size and shape when under continued stress. Although all materials creep, it is insignificant in materials such as steel. Creep is a significant factor for many plastics, even at room temperature. It is most apparent in thermoplastics, since they contain unattached molecules that can begin to slip past each other when under stress. When the duration of stress is short, the plastic may return to its original condition. Long term stress however may produce permanent deformation.

Laminates, fiberglass-reinforced plastics, and some thermosets resist creep better than thermoplastics, but they do not resist stress as effectively as steel and concrete.

Electrical Properties

Plastics have excellent electrical insulating properties that enabled the development of greatly improved electrical equipment. In addition, they have good heat resistance, a requirement for the use of a material in many electrical devices. Silicones and fluorocarbons are widely used for electrical components. Epoxies are employed to encapsulate electric components subjected to temperatures as high as 300°F (150°C), and phenolics are used as housings for electrical parts, switches, and receptacles. In other words, plastics have excellent *dielectric strength* (the maximum voltage a dielectric can withstand without rupture). A dielectric is a nonconductor of electric current. Since electrical components are subject to damage by arcing, arc resistance is important. *Arc resistance* is measured by the total elapsed time in seconds an electric current must arc to cause a part to fail. Plastic parts carbonize and become conductors, burn, develop thin lines between electrodes, or become glowing hot. Plastics used for applications in which arcing is a possibility have an arc resistance time from 100 to 300 seconds.

Thermal Properties

Two major factors to consider when examining thermal properties are the influence of temperature on strength and expansion and contraction of the material. Many plastics lose strength at comparatively low temperatures. The *service temperature* of plastics, the maximum temperature at which a plastic can be used without affecting its properties, is low when compared with other construction materials (Table 23.3). Some plastics soften below 200°F (94°C), so they become softer if placed in boiling water. In general, thermosets have higher service temperatures than thermoplastic materials. Polyamide thermosetting resins resist intermittent heat up to 930°F (500°C) under low stress and up to 480°F (250°C) under continuous stress.

All building materials expand and contract, a property that must be taken into consideration as a designer selects materials and prepares design details. Most plastics have a higher coefficient of thermal expansion than other construction materials (Table 23.4). In general, thermosets have a lower rate of expansion than do thermoplastics.

Plastics are poor conductors of heat. Unmodified plastics compare favorably with brick, concrete, and glass (Table 23.5). Notice that foamed plastics have the lowest thermal conductivity and are excellent insulators used for all types of insulation. When comparing plastics in general with other materials, it can be seen that plastics have the lowest coefficient of thermal conductivity.

Flammability is a measure of how easily something will ignite or burn. Since plastics are organic materials, they will burn. The range of flammability, however, is great. Cellulosics are highly flammable and are often banned by building codes. Polyurethane and polyvinyls give off toxic fumes when burning, and building codes specify protective measures for their use. Vinyl siding, for example, begins to burn at 698°F (370°C), whereas wood ignites at about 400°F (206°C). In general, thermoplastics are more flammable than are thermosets, though these properties can be changed with additives. Some plastics can develop a char layer that serves as a shield to deter burning. The National Aeronautics and Space Administration uses nylon and phenolic resins on the exterior of spacecraft that are exposed to the searing heat of reentry.

Although plastics are no more combustible than wood, they can produce toxic fumes, which are more likely than flames to cause death in a fire. Most building codes have a separate chapter devoted to the use of plastics in building construction.

Chemical Properties

While plastics do not corrode like metals, they can deteriorate and be damaged by chemical attack. When a metal corrodes, its surface is damaged to the extent that it loses weight. Deteriorated plastics gain weight because attacking chemicals combine with the plastic resin. This causes swelling, crazing, and discoloration. In addition, there is loss of impact, flexural, and tensile strengths. Usually the chemical resistance of a plastic decreases as the temperatures increase.

The *weatherability* of a plastic involves consideration of moisture, ultraviolet light, heat, and chemicals found in the air as ozone and hydrochloric acid. High-density polyethylene (HDPE) has great resistance to acids, water absorption, and weathering. It is used for electrical wire insulation. Acrylics also have great resistance to weathering and excellent optical qualities, so they are widely used for glazing.

Density

The density of plastic materials is in general lower than other commonly used construction materials. Glass-reinforced plastics (GRP) are lighter than steel and aluminum.

Specific Gravity

The specific gravity of plastics varies considerably, ranging from 0.06 for foams to 2.0 for fluorocarbons. Specific gravity is the ratio of the mass of a volume of material to the mass of an equal volume of water at a standard temperature. The specific gravity of water is 1, so some plastics will float (lighter than water) and others will sink. The specific gravity for softwoods is 0.5, about 2.7 for aluminum, and about 8.0 for steel.

Optical Properties

Some plastics have optical properties equal to that of glass. Acrylics are as transparent as fine optical glass and have a light transmission of 92 percent. Polystyrene, polypropylene, and polycarbonates also exhibit transmission qualities of 90 percent or better.

MANUFACTURING PROCESSES

Plastic resins in the form of liquid, powder, or beads are produced by chemical companies and sold to manufacturers of plastic products. A manufacturer combines these polymers with additives to modify the properties of the polymer to meet the needs of the product to be produced. The modified polymer is then processed by equipment of various types to produce the products.

Common manufacturing processes include blow molding, calendaring, compression molding, casting, extrusion, expandable bead molding, form molding, injection molding, laminating, rotational casting, transfer molding, and thermoforming.

Blow molding involves placing a heated, preformed plastic tube called a parison in a forming die. Air pressure is raised inside the plastic tube, forcing it to conform to the shape of the die. When it cools, the die is opened and the part is removed. Most bottles are formed this way.

Calendaring involves moving a plastic material in a liquid state through a series of rollers, which creates a thin plastic film as the material solidifies. Vapor barrier and floor covering materials are typical calendared products.

Compression molding involves placing plastic resin in powder form in a heated mold to which heat and pressure are applied. The resin melts and fills the mold.

Casting involves pouring plastic in a liquid state into a cavity in a mold. The liquid plastic fills the cavity and hardens. Extrusion is a process in which a semi-liquid plastic is forced under pressure through an opening in a die. The shape of the die opening determines the shape of the extruded member.

Expandable bead molding is a process in which small granules of resin, such as polystyrene, are mixed with an expanding agent and placed in a steam-heated rolling drum. When the granules have expanded, they are cooled and transferred to a mold in which they are heated until they fuse together.

Form molding requires that an expanding agent be mixed with plastic granules or powder and injected into a mold, where it is heated, melting the resin and forming a gas that expands the resin to fill the mold cavity.

Injection molding uses granules or powder resin. This is fed into the heated cylinder of the injection-molding machine and forced by a ram into a cold mold, where it solidifies (Figure 23.5).

Figure 23.5 Injection molding feeds heated plastic materials into a machine and forces it with a ram into a cold mold.

Laminating is a process in which several layers of material are bonded together to form a single sheet. An example is laminating a colored plastic veneer to plywood backing for use as wall paneling. The materials to be laminated are impregnated with the plastic resin, placed together, and bonded by the application of heat and pressure.

Rotational molding forms hollow one-piece items from polyethylene powders. The resin is placed inside the mold, which is heated as it rotates about two axes. The resin is distributed to the surfaces of the mold by centrifugal force and fused by the heat.

Transfer molding is a combination of compression and injection molding. The resin is made liquid in the transfer chamber outside the mold and then injected into the mold, where it fills the cavity and solidifies.

Thermoforming involves two commonly used procedures: vacuum forming and pressure forming. Vacuum forming involves placing a heated sheet of plastic over a mold cavity and pulling a vacuum below it. Atmospheric pressure forces the sheet to the shape of the mold. Pressure forming involves placing a heated sheet of plastic over a mold cavity and increasing the pressure behind the sheet, forcing it into the cavity of the mold.

Recycled Plastic Products

Discarded plastic and packaging make up a significant component of the waste stream, and the amount of plastics being discarded annually is not expected to decrease any time in the near future.

The implementation of an identification code by manufacturers of plastic products and packaging has greatly enhanced the recovery and sorting of the variety of plastics in the waste stream. Six polymer codes identified numerically as: 1-Polyethyleneterephthalate (PET), 2-High-Density Polyethylene (HDPE), 3-Polyvinyl Chloride (PVC), 4-Low-Density Polyethylene (LDPE), 5-Polypropylene (PP), and 6-Polystyrene (PS) constitute the majority of post-consumer plastics.

Post-consumer resins (PCRs) are used as a substitute for the raw materials in products made traditionally from other materials. One example of this reuse is recycled-plastic lumber (RPL) (Figure 23.6). The manufacture of RPL from post-consumer and post-industrial resins consumes large quantities of waste plastics that would otherwise be destined to landfills, and converts the waste into useful, long-lasting products. As a substitute for treated wood, RPL products offer the advantages of being resistant to insects, rot, moisture, and many chemicals. Plastic lumber materials are also benign

Figure 23.6 As a substitute for treated wood, recycled-plastic lumber (RPL) products are resistant to insects and moisture, and require no chemical treatments to maintain their properties.

to the environment and require no chemical treatments to achieve or maintain their properties.

Another example of PCRs used in construction is polyester/PET or polyethylene terephthalate carpet. The carpet is made by grinding soft drink bottles into chips that are heat-treated and forced through a die to produce thin fibers. PET carpet is non-allergenic, more stain resistant than other carpet materials, and highly resistant to mold and mildew.

PLASTIC CONSTRUCTION MATERIALS

The following discussion gives a brief description of the plastics most commonly used in construction and some of their major uses. They are divided into thermoplastics and thermosets.

Since plastic materials are frequently identified in technical literature and professional magazines by approved abbreviations, designers and constructors must be familiar with them. A list of these abbreviations, as compiled by the American Society for Testing and Materials, is provided in Table 23.6.

Thermoplastics

The majority of products used in construction are made from thermoplastic materials rather than thermosets. Following are commonly used thermoplastic materials and products using them.

Acrylonitrile-Butadiene-Styrene Acrylonitrile-butadiene-styrene (ABS) plastics are a combination of high-impact polystyrene, which is very tough, and acrylonitrile, which improves rigidity, tensile strength, and chemical resistance. ABS plastics are widely used for pipes and fittings for water lines with water up to 180°F (83°C) (non-pressure), gas supply lines, waste, drain, and sewage vent systems (Figure 23.7). They are also used for hardware, such as handles and knobs.

Acrylics The most widely used acrylic is polymethyl methacrylate (PMMA). It has excellent optical clarity and finds use in glazing. Typical uses include door and window lights and roof domes and skylights. It is also used to make lighting fixtures, but will soften at 200°F (94°C). Other uses include outdoor signs, corrugated roofing, and molded hardware. Acrylics are also dispersed as fine particles in a liquid to producing latex used to make paints.

Cellulosics Two common plastics based on the cellulose molecule are cellulose acetate (CA) and cellulose acetate-butyrate (CAB). Cellulose acetate is not used for

Table 23.6 ASTM[a] Abbreviations for Plastics

Thermoplastics	
Type	**Abbreviation**
Acrylonitrile-butadiene-styrene	ABS
Acrylic:	
Polymethyl methacrylate	PMMA
Cellulosics:	
Cellulose acetate	CA
Cellulose acetate-butyrate	CAB
Cellulose acetate-propionate	CAP
Cellulose nitrate	CN
Ethyl cellulose	EC
Fluorocarbons:	
Polytetrafluoroethylene	PTFE
Nylons:	
Polyamide	PA
Polyethylene	PE
Polypropylene	PP
Polycarbonates	PC
Styrene:	
Polystyrene	PS
Styrene-acrylonitrile	SAN
Styrene-butadiene plastics	SBP
Vinyl:	
Polyvinyl acetate	PVAc
Polyvinyl butyral	PVB
Polyvinyl chloride	PVC
Thermoset Plastics	
Type	**Abbreviation**
Epoxy, epoxide	EP
Fiberglass reinforced plastics	FRP
Melamine-formaldehyde	MF
Phenolic:	
Phenol-formaldehyde	PF
Polyester	—
Polyurethane:	
Urethane plastics	UP
Silicone plastics	SI
Urea-formaldehyde	UF

[a]*American Society for Testing and Materials*

construction purposes, but CAB can be made resistant to weathering and is used in piping for gas and chemicals. Cellulosics are also used in coating compounds and adhesives.

Figure 23.7 ABS plastics are used to make plumbing pipe and fittings.

© Ayakovlev/Shutterstock.com

Fluorocarbons The most widely used fluorocarbon is polytetrafluoroethylene (PTFE). It has high resistance to chemical degradation and a service temperature from 2,450°F (2,234°C) to 1,500°F (262°C). PTFE has a low coefficient of friction and is marketed under the trade name Teflon. It is used for pipe that must handle corrosive liquids at high temperatures and for parts that require easy sliding surfaces.

Ethylene tetrafluoroethylene (ETFE) is a fluorocarbon-based polymer designed to have high corrosion resistance and strength over a wide temperature range. In addition, it has a high melting temperature and does not emit toxic fumes when ignited. It is also recyclable. Compared to glass, ETFE film is 1 percent the weight, transmits more light, and costs considerably less to install. Both PTFE and ETFE are finding wider use in the construction industry as a cladding material (**Figure 23.8**).

Nylons Polyamides (PA), also known as nylon, are tough, high in strength, and have good chemical resistance. Since they resist abrasion well, they are used for molded parts, such as locks, rollers, gears, and cams. They do not weather well.

Figure 23.8 The Alliance Arena is one of the world's largest structures made of ETFE laminate.

Figure 23.9 Polycarbonate glazing panels have good heat-resistance and light-transmission qualities.

Polycarbonates Polycarbonates (PC) have high-impact strength and good heat resistance. They are dimensionally stable and transparent. Polycarbonates find use in light fixtures, molded parts, and signs. They are also used in place of glass in areas where damage is likely, such as in skylights.

Polycarbonate plastic sheets, such as Lucite and Lexan, have high impact strength, reaching up to 250 times the strength of glass and 30 times that of acrylic. They are suitable for glazing openings in areas where high security is needed and can be laminated to produce bullet-resistant panels. Light transmission varies with thickness but reaches 75 percent for 0.50 in. (12 mm) panels to 86 percent for 0.125 in. (3 mm) panels. Some manufacturers produce sheets in transparent solar tints, translucent white, and a variety of other colors. Double-skinned units with internal ribs and dead air spaces are available for use on vertical and sloped glazing (Figure 23.9).

Polyethylene Polyethylene (PE) is light, strong, and flexible, even at low temperatures. It has good water resistance and low vapor transmission. Its major use in construction is as a vapor barrier on walls, floors, and ceilings. It is also used on basement walls as part of a waterproofing application. The wide use of polyethylene makes for an important environmental issue. Polyethylene is not biodegradable and takes several centuries to efficiently degrade, though it can be recycled.

Polypropylene Polypropylene (PP) is much like polyethylene but is more heat resistant and stiffer. It is used for hot water pipes, waste disposal systems, and strong fibers for carpeting.

Polystyrene Polystyrene (PS) is a water-resistant, dimensionally stable, transparent plastic that maintains its properties at low temperatures but begins to soften around 200°F (94°C). It is brittle and has poor weathering qualities. When produced in a foamed condition, it is widely used as insulation. Polystyrene is used in rigid insulation sheets, as cores for insulated doors and sandwich panels and for roof insulation (Figure 23.10).

Figure 23.10 Polystyrene is a water-resistant, dimensionally stable plastic used frequently in insulation products.

© Dieter Heinemann/Westend61/Alloy/Corbis

Vinyls The term vinyl describes a large group of plastics developed from the ethylene molecule. Those used in construction are polyvinyl chloride (PVC) and polyvinyl butyral (PVB).

Polyvinyl chloride is the most widely used vinyl in the manufacture of products used in construction. It is dimensionally stable and has high impact resistance, high abrasion resistance, and good aging qualities. A wide variety of products are made from PVC, including siding, gutters, floor tile, pipe, and window frames (Figure 23.11). It can be bonded to other materials, such as plywood panels, to provide a protective skin. PVC is available as a rigid or flexible foam and used as a core material in panel construction. It is copolymerized with other plastics to produce a binder for terrazzo floors and a number of adhesives.

Figure 23.11 Polyvinyl chloride is used to produce plastic piping and exterior products such as siding, gutters, soffits, and fasciae.

© harper kt/Shutterstock.com

Polyvinyl butyral (PVB) is used as an inner layer in safety glass and as a protective coating on fabrics. Another type of vinyl, polyvinyl acetate (PVAc), is used in mortars, paints, and adhesives.

Thermosets

Thermoset materials are used for products requiring greater heat resistance and stiffness than that afforded by thermoplastics. Thermosets find limited use because they are brittle, harder to form, and cannot be reformed.

Epoxies Epoxies, mainly used because of their excellent adhesive qualities, have good chemical and moisture resistance. They are widely employed in coating compounds and as adhesives. Epoxies bond to almost any material and are used in the assembly of panels, the bonding of veneers and overlays, and as protective coatings. They are used as mortars for bonding concrete block and in patching material for damaged concrete. Epoxies are also used to produce some types of fiberglass-reinforced plastic.

Formaldehyde Formaldehyde plastics are incapable of plastic deformation and are hard, strong, heat resistant, and brittle. Three types find some use in construction.

Phenol-formaldehyde (PF), generally referred to as phenolics, has fillers such as glass fibers added to improve impact resistance and strength. It has good electrical and thermal properties. Phenolic plastics are the most widely used of the thermosets. They are molded to form hardware and electric parts, such as switches, boxes, and circuit breakers. An important use involves the coating of kraft paper that forms the base of high-pressure plastic laminates used for countertops and other surfaces. The cardboard interior structure of hollow-core doors and panels is impregnated with phenolic resin. It is also used to make some types of adhesives, protective coating materials, and foamed insulation.

Melamine-formaldehyde (MF), generally referred to as melamine, are hard, scratch-resistant plastics that withstand chemical attack. They are used in the production of high-pressure plastic laminates, such as those used for countertops. They are used as an adhesive in the production of plywood and to mold hardware and electrical fixtures.

Urea-formaldehyde (UF) is used for similar application as melamine-formaldehyde. Urea-formaldehyde is not as hard and does not have the same heat-resisting properties as melamine-formaldehyde.

Formaldehyde can be toxic, allergenic, and carcinogenic, and its use in many construction materials makes it one of the more common indoor air pollutants. At concentrations above 0.1 ppm (parts per million) in air, formaldehyde can irritate the eyes and mucous membranes. Formaldehyde inhaled at this concentration may cause headaches, a burning sensation in the throat, and difficulty breathing, as well as triggering or aggravating asthma symptoms.

Polyesters A large number of plastics fall under the generic name polyester. Polyesters include Mylar, from which drafting film is made; alkyds, used for paints and enamels; and fiberglass-reinforced plastics. Reinforced plastic interior moldings are available made from fiberglass-reinforced polyester. These are easy to cut, install, and come primed and ready for painting. They are available in small and very large, complex-profile single pieces, which makes installation easy. Complex wood moldings require that several individual pieces of molding be cut and fitted together. Other plastic moldings and trim for interior and exterior use are available made from cellular vinyl, polyvinyl chloride (PVC), and polymers. Saturated and unsaturated polyesters are produced. The most important group is the unsaturated polyesters. They can be linked to a monomer, such as styrene, forming a strong, hard thermosetting plastic. Their major use is in the production of reinforced plastics to which glass fibers are added. This produces a stable, tough, impact-resistant, and high-strength material. Major uses include molded bathtubs, showers, sheets for roofing and partitions, curtain wall exterior laminates, and window frames and sashes. Reinforced polyesters have ultimate strengths up to 50,000 psi (344,500 kPa) and E values as high as 3×10^6 psi. The strength of the plastic is proportional to the amount of glass reinforcement used, because the glass fiber provides a tensile strength approaching 150,000 psi (1,033,500 kPa).

Polyurethanes (UP) Polyurethanes are used to produce low-density foams that can be varied from soft, open cell types that are flexible to a tough, closed-cell, rigid material. These foams have very low thermal conductivity and make excellent insulation. They have good chemical and heat resistance and good tensile strength. Polyurethane foams are used to fill wall, ceiling, and floor spaces by spraying them on surfaces or in cavities, after which they solidify. They are also used to insulate pipes, ducts, and wall panels. Rigid insulation sheets are widely used for many purposes, such as insulating flat roofs (over which a hot tar and gravel roof is applied). Polyurethanes also find use in the manufacture of

elastomers (synthetic rubbers), caulking, adhesives, and glazing materials.

Silicones (SI) Silicone plastics are not based on the carbon atom, as all those previously discussed, but on the silicon atom. They have good corrosion resistance and are efficient electrical insulators. Heat-resistance properties enable them to have a service temperature ranging from 280°F (227°C) to 500°F (260°C). An important property involves their ability to withstand exposure to the elements.

Since silicons are very water repellent, they are applied in liquid form to exterior wall and masonry materials to provide a water-resistant coating (Figure 23.12). This stops water from penetrating a wall, such as a brick exterior wall, yet has the permeability to allow internally developed moisture to pass through as a vapor. Silicone rubber is soft, heat resistant, and does not harden at low temperatures. It is used in sealants and gaskets where watertightness is required.

Figure 23.12 Silicone sealants and caulks are in common use to seal gaps, joints, and crevices in buildings.

© CSImagemakers/Shutterstock.com

Review Questions

1. What two atoms form the nucleus for the derivation of the material plastics?

2. What is the process called that enables carbon atoms to bond with other atoms?

3. What is a single molecule called?

4. What is formed when several monomers link together?

5. What is the valence of the carbon atom?

6. What are the two major classifications of plastics?

7. Why is it possible to heat and reform thermoplastic materials?

8. Why is it that thermoset plastic cannot be reheated and reformed into new shapes?

9. What are the major groups of additives used in formulating plastics?

10. How do plasticizers change the properties of plastics?

11. What is a common way to reduce the cost of plastics?

12. What is added to plastic to increase its strength?

13. What colorants are used to produce desired colors?

14. What are the most important mechanical properties of plastics used in construction?

15. What are the two major considerations when considering the thermal properties of plastics?

16. Which plastics have the lowest thermal conductivity?

17. What is the major danger produced by burning plastics?

18. When plastics deteriorate due to chemical action, what type of damage occurs?

19. Which plastics have good optical properties?

20. Identify the following plastics by their standard abbreviation: ABS, PMMA, PA, PE, PVC.

21. What is the major use for ABS plastics?

22. What type of plastic would you choose for door and window glazing?

23. What type of plastic is used in the manufacture of latex paints?

24. The vapor barriers used in walls and floors most likely are made from what type of plastic?

25. What type of plastic weathers well and is used for exterior purposes, such as siding, gutters, and window frames?

26. What thermoset plastic is known for its use as a strong adhesive?

27. What plastics are used in the manufacture of high-pressure plastic laminates?

28. What type of plastic is used in the manufacture of showers and bathtubs?

29. What thermoset plastic is widely used to produce foamed insulation commonly sprayed on roofs and in wall cavities?

30. What uses are made of silicones in building construction?

Key Terms

Additive	Flammability	Stabilizer
Arc Resistance	Hardness	Synthetic Material
Bioplastics	Monomer	Thermoplastic
Covalent Bonding	Plastic	Thermoset
Creep	Plastic Behavior	Valence
Dielectric Strength	Plasticizer	Weatherability
Elastomers	Polymers	
Filler	Service Temperature	

Activities

1. Collect samples of various plastic products used in construction. Try to identify the type of plastic.

2. Develop a series of tests you can perform with the facilities available and test plastics samples. For example, see which will ignite the quickest when exposed to a continuous flame and which will burn after the flame is removed. Run tensile tests, expose samples to acids, hot water, etc., and record what happens to each.

Additional Resources

Plastic Pipe and Building Products, Volume 08.04, American Society for Testing and Materials, West Conshohocken, PA.

Structural Plastic Selection Manual, American Society of Civil Engineers, New York.

See Appendix C for addresses of professional and trade organizations and other sources of technical information.

Thermal and Moisture Protection

CSI MasterFormat™

CHAPTER 24

Thermal Insulation and Vapor Barriers

LEARNING OBJECTIVES

Upon completion of this chapter, the student should be able to:

- Select appropriate types of insulation for various applications in the design of a building.

- Understand how insulation is used to control heat transfer.

- Be aware of the problems caused by moisture penetrating the insulation and learn how to design an assembly of materials to reduce moisture transmission.

> **Build Your Knowledge**
>
> For further study on these materials and methods, please refer to:
>
> Chapter 23 Plastics

Structures are exposed to a variety of weather extremes, including temperature swings, driving rain and wind, and below ground water. Energy conservation and moisture control are crucial in protecting a building's structural integrity while ensuring the comfort of its occupants. Insulation is used to reduce unwanted heat losses and gains that increase demands on heating and cooling systems. Thermal insulation in buildings is an important factor in achieving both energy efficiency and interior thermal comfort.

Moisture-related problems are among the largest set of problems buildings experience within the United States. Moisture management in building construction entails preventing water, vapor, and air transport, while promoting drying and ventilation. Our understanding of water management is expanding as new techniques and products for controlling moisture both inside and outside of buildings are developed.

The building envelope, its walls, windows and roof, constitutes the interface between the controlled indoor environment and outdoor climate. While the indoor environment is designed to maintain comfort conditions, assumed at 70°F–72°F (21°C–22°C), the outdoor climate can experience sharp temperature differences, depending on location. The building envelope must resist the transfer of heat in addition to controlling both air and water infiltration under a variety of constantly changing conditions. All building surfaces that separate interior conditioned (heated or cooled) spaces from the exterior environment must be insulated (Figure 24.1). This crucial layer between the interior and exterior environment is known as the **thermal envelope**. In addition to establishing the required durability, strength and rigidity, the thermal envelope must control the flow of heat, airflow and vapor flow. It must resist the penetration of rain, control solar radiation, reduce sound transmission, and ensure fire resistance.

HEAT TRANSFER

One key to energy efficient design is an understanding of the fundamental physics of heat flow and transfer. Heat is transferred by radiation, convection, or conduction, or in most cases a combination of these. To avoid heat loss in winter and heat gains in summer, insulation is used to break the transfer of heat. The flow of heat involves the transfer of energy through air and solids at the molecular level. Groups of molecules exhibit movement according to the amount of energy contained within them. A warm region indicates that the molecules contain a higher level of energy, while cold regions

Figure 24.1 Any building surface that separates conditioned air from outside air is insulated to form the thermal envelope.

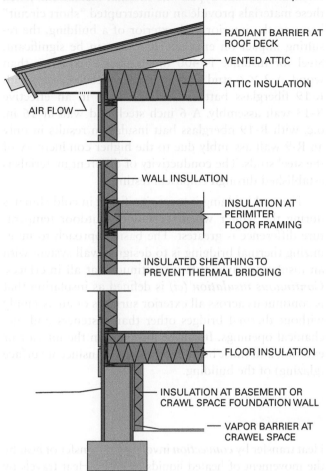

RADIANT BARRIER AT ROOF DECK

VENTED ATTIC

ATTIC INSULATION

AIR FLOW

WALL INSULATION

INSULATION AT PERIMETER FLOOR FRAMING

INSULATED SHEATHING TO PREVENT THERMAL BRIDGING

FLOOR INSULATION

INSULATION AT BASEMENT OR CRAWL SPACE FOUNDATION WALL

VAPOR BARRIER AT CRAWEL SPACE

contain less energy. Heat transfer always moves from a region of higher energy to one of lower energy. Heat transfer, then, is the movement of thermal energy from a hotter material to a cooler one. If there is no difference in temperature, there can be no heat transfer. When an object or fluid resides at a different temperature than its surroundings or another object, transfer of thermal energy, also known as heat exchange, occurs in such a way that the body and its surroundings reach thermal equilibrium.

Heat quantity is measured in *British thermal units (BTUs)*. One BTU is the amount of heat needed to raise the temperature of 1 lb. of water by 1°F. A BTU is approximately equal to the heat generated by a wooden match. In the metric system, heat quantity is measured in calories. One calorie is the heat required to raise the temperature of 1 gram of water 1°C. Calories can be converted to *joules (J)* by multiplying them by 4.18. A joule is a derived metric unit used to describe work, energy, and quantity of heat.

The transfer of heat occurs across the floors, walls, and roofs of buildings in all climate conditions. In designing the materials of a multilayered construction, the insulating values of the individual layers are combined to resist the flow of heat across the envelope. The amount of transferable heat depends on the characteristics of a material, its thickness, and the quality of construction detailing. Porous and fibrous materials that contain air pockets, such as wood, transfer less heat than do solid materials, such as concrete. A thick section of material provides better protection than a thin section of the same material.

Conduction

Conduction is the transfer of thermal energy from a region of higher temperature to a region of lower temperature through direct molecular movement within a material or between materials in direct physical contact. For example, when solar radiation heats a masonry wall, the heat is transferred through to the other side of the wall by conduction. How easily a material will conduct heat depends on its molecular structure. Generally, the denser a material is the better it will conduct heat. For this reason, dense metals like steel and aluminum are excellent conductors. Most gases, including air, are poor conductors and, therefore, good insulators. Conductive barriers often incorporate a layer or pockets of air to reduce heat transfer. Porous materials like fiberglass that contain many air spaces are used to reduce the transfer of heat.

The rate of conduction through an assembly is proportional to the thickness of the material, the cross-sectional area over which it travels, the temperature gradients between its surfaces, and the thermal conductivity of the material. Thermal conductivity (k) is a measure used to indicate the amount of heat that will be conducted through a square foot of area of a material per a specified unit of thickness. A lower k value indicates better insulating qualities. In the United States' customary system of measurement, thermal conductivity is indicated as BTU per in./hr ft^2 °F (or BTU per inch of thickness per square foot per hour per degree Fahrenheit).

Thermal conductance (C) is a measure used to indicate the amount of heat that will pass through a specified thickness of material. It is the reciprocal of thermal resistance: C = 1/R. Thermal conductance is indicated as C = BTU/hr ft^2 °F (or BTU per hour per square foot per degree Fahrenheit).

Thermal resistance (R) is a measure used to indicate the ability of a material to resist the flow of heat through it. The higher the R-value, the greater the insulating effectiveness. Thermal resistance is indicated in hours per square foot per degree Fahrenheit per (hr · ft.² °F)/ BTU. In measuring insulating capacity, R-values are commonly used rather than the conductivity k, because R-values can be added. To calculate the thermal resistance of a layered construction assembly, the individual values of each layer in the system is added. R-total is the overall thermal resistance for a construction assembly, or the sum of individual R-values of component materials. To illustrate how to determine the total resistance of an assembly, consider a typical wood frame wall. Moving from the inside out, there is a layer of still air along the wall that contributes to the resistance (R-0.68); wallboard (R-0.45); an air space (R-1.01); sheathing over the studs (R-1.32); wood siding (R- 0.81); and the outside air that is moving, thus contributing very little. Ignoring the contribution of outside air, which varies with weather conditions, the R-value for the building wall is:

$$R\text{-total} = 0.68 + 0.45 + 1.01 + 1.32 + 0.81 = 4.27$$

Thermal transmittance (U) is a measure of the amount of heat that would pass through an assembly of various materials, such as the exterior wall described above in BTU/(hr.ft² °F). The U-value is the reciprocal of the total resistance (R) of the assembly.

When a material or assembly is uninsulated from exterior to interior, it creates a *thermal bridge*, allowing for conductive heat transfer from the warm side to the cold (Figure 24.2). Thermal bridging occurs in building envelopes when relatively high thermal conductivity

Figure 24.2 A thermographic image of a building during winter demonstrates the location of thermal bridges in dark red.

materials, such as steel and concrete, create pathways for heat loss that bypass the thermal insulation. When these materials provide an uninterrupted "short circuit" between the interior and exterior of a building, the resulting impact on envelope R-value can be significant. Steel for instance is 400 times more conductive than wood. A 2 × 6 stud wall at 16 in. on center (o.c.) with R-19 fiberglass batt insulation results in an effective R-13 wall assembly. A 6 inch steel stud wall at 16 in. o.c. with R-19 fiberglass batt insulation results in only an R-9 wall assembly due to the higher conductivity of the steel studs. The conductivity of different materials is established through empirical testing.

Thermal bridging is most significant in cold climates during the winter when the indoor–outdoor temperature difference is greatest. The basic approach to minimizing thermal bridging is to design a wall system with an insulation layer that is continuous at all interfaces. *Continuous insulation (ci)* is defined as insulation that is continuous across all exterior surfaces of an assembly without thermal bridges other than fasteners and mechanical openings. It can be installed on the interior or exterior, and must be integral to any translucent surface (glazing) of the building.

Convection

Heat transfer by *convection* involves the transfer of heat by the movement of heated liquids or gases. Heat travels by natural or forced (mechanical) currents of air that absorb the heat brought to a space. As warm air passes over a cool surface, it transfers some of its heat to the cool material. A hot water heating convector, for example, transfers heat to the air by convection. This convection current is a means for heat transfer that occurs in every building.

The transfer of thermal energy from air to a surface is known as surface conductance, sometimes referred to as the film conductance or coefficient. Convection can also be caused by stack effect, driven by buoyancy as warm air rises.

Infiltration is the unwanted transfer of conditioned indoor air with outside air through small cracks or leaks within the exterior building envelope. Air leakage or convection across the thermal envelope is driven by air pressure differences, wind, and mechanical ventilation. Infiltration or unintended leakage of air in buildings has serious consequences. The U.S. Department of Energy reports that up to 40 percent of the energy used by buildings for heating and cooling is lost due to infiltration. The building enclosure must be designed to limit airflow through the assembly by selecting one layer of the envelope and sealing it using durable tapes, adhesive sheet products, or liquid-applied materials.

When the transfer or air is intentional, through windows or fans, it is referred to as ventilation. The magnitude of convective heat exchange can be calculated by multiplying the amount of building air exchanged in cubic feet per minute (CFM) by its heat capacity and the temperature difference between the interior and exterior air.

Radiation

Radiation involves the transmission of heat by electromagnetic waves. Heat is transferred by radiation when surfaces exchange electromagnetic waves, such as light, infrared radiation, or UV radiation. Radiation exchange occurs between two surfaces when one is warmer than the other, and they are in "view" of each other. We experience radiant heat exchange when we are exposed to sunlight or an open fireplace.

The rate of radiant heat transfer depends on the temperature differential between the surfaces, the distance between the surfaces, and the thermal absorptivity of the surfaces. Shiny materials tend to reflect radiant energy while dark materials will absorb it. A white metal roof, for instance, will reflect more radiant energy than a brown or black roof, thus increasing the energy efficiency of a building by reducing heat gain.

Occupant thermal comfort in a building is a function of the influence of both air and surface temperatures. The *mean radiant temperature (MRT)* is the average temperature experienced from the combination of all the surface temperatures in a room, including walls, floors, ceilings, furniture, and people. Occupants in a room at 72°F air temperature may still feel uncomfortably cold if the walls and ceiling are at 50°F. Conversely, they may feel uncomfortably warm if the walls are 85°F. While the air temperature is the same in both cases, the radiative cooling or warming of occupants relative to the walls and ceiling will affect their comfort level.

Radiation is a significant component of heat transfer in buildings and is especially important for building surfaces exposed to the sun and large temperature differences. Radiant barriers work in conjunction with an air space to reduce radiant heat transfer across the air space. Radiant barriers are typically used in reducing downward heat flow in attics because upward heat flow tends to be dominated by convection. Radiant barriers utilize thin sheet materials that have at least one reflective surface, usually a coating of aluminum.

INSULATION MATERIALS

A variety of materials are used to manufacture thermal insulation. Organic fibrous insulation materials include cotton, wood, cellulose, wool, and synthetic fibers. Organic cellular materials include polyurethane, polystyrene, cork, cotton, recycled materials, and foamed rubber. Mineral fibrous materials include rock, glass, slag, and asbestos melted and spun into fibers. The type and amount of insulation for a building is determined by the building function, local climate conditions, building code requirements, its effectiveness in resisting heat transfer, and its cost relative to the savings expected from improving the energy efficiency of the building. Typical R-values of insulation materials are given in Table 24.1.

Table 24.1 R-Values of Insulation Materials

Batts and Blankets	
Fiberglass	R-3.3–4.3/inch
Mineral wool	R-3.3–3.7/inch
Cotton/wool	R-3.5–3.8/inch
Plastic fiber	R-3.8–4.3/inch
Loose Fill	
Cellulose	R-3.7/inch
Fiberglass	R-3.3–4.3/inch
Perlite	R-2.7–2.9/inch
Vermiculite	R-2.1–2.3/inch
Rigid	
Extruded polystyrene	R-5/inch
Expanded polystyrene	R-5/inch
Polyisocyanurate	R-6–6.5/inch
Foamed/Sprayed	
Open cell polyurethane	R-3.6–3.8/inch
Closed cell polyurethane	R-5.6–6.8/inch
Icynene	R-3.6–4/inch
Soy	R-6.9/inch
Cementitious	R-3.9/inch
Insulating Masonry	
Insulated Concrete Forms (ICF)	12" wall can reach an R-value of R-24
Insulated blocks	R-9 (8" block w/vermiculite)
Aerated Autoclaved Concrete (AAC)	R-1.1/inch
Other	
Aerogel	up to R-10/inch
Gas-filled panels	R-5–11/inch
Strawbale	R-2.4–3/inch

Batts and Blankets

Batts and blankets are flexible insulation mats made from a variety of materials, including fiberglass, mineral wool, cotton and wood fibers, and recycled plastic materials. Batts are available in standard sizes usually 48 in. (1.2 m) lengths, and blankets in rolls up to 8 ft. (2.4 m). Some thinner blankets are available in longer rolls. Common thicknesses are 3½ in., 6½ in., 9½ in., and 12 in. Common widths include 15 in. and 23 in. sized to correspond to a standard 16 in. or 24 in. on-center framing layout. Special types are available for particular applications, such as extra low-flame-spread types and water-resistant batts for furred-out masonry walls. Special batts are also available for use over panels in suspended ceilings, and as sound transmission barriers for walls, floors, or ceilings.

Batts and blankets are available unfaced (Figure 24.3), faced on one side (with moisture-resistant kraft paper that forms a vapor barrier), and faced with aluminum foil (that forms a fire-resistant surface). Some batts have facing on both sides and are used in wall, floor, and ceiling applications. In wood construction, batt insulations are stapled to the face of a stud. In metal studs, the insulation is designed to be held by friction between studs or joists (Figure 24.4).

Figure 24.3 Fibrous insulation works by entrapping air within the fibers of batts.

© Madlen/Shutterstock.com

Both batts and blankets can be cut to size with a knife. The proper installation of batt and blanket insulation is critical. Insulation that is improperly installed can lower the material's effectiveness. The thermal insulation layer must be continuous and attention must be paid to sealing around outlets, pipes, and other obstructions.

Figure 24.4 Fiberglass insulation is friction fit in metal stud walls.

© pryzmat/Shutterstock.com

Spaces left around door and window frames are often filled with spray foam insulation.

Fiberglass   Fiberglass insulation is one of the most commonly used insulation materials in the United States. It is made by blowing a jet of air through molten glass, slag, or rock, and works by entrapping air within the fibers of batts. The material often contains some recycled glass content. Phenol formaldehyde (a known carcinogen) and acrylic binders are used and the fibers are an irritant during installation. Installers of fiberglass insulation should always wear protective clothing and dust masks. New medium- and high-density batts with higher R-values are now available. While more expensive than low-density batts, they are ideal for limited cavity spaces.

Mineral Wool   Mineral wool insulation comes in two forms, rock wool made from virgin basalt, and slag wool made from steel-mill slag (Figure 24.5). Though the individual fibers in mineral wool conduct heat well, when pressed into rolls and sheets, their ability to entrap air makes them excellent insulators. Mineral wool is stiff and brittle, but a new softer, mineral wool batt product is now available. The new, highly compressible material expands to continuously press against framing members. Mineral wool is fire resistant and can aid in soundproofing. The cost of soft mineral wool batt insulation is about 15 percent higher than fiberglass.

Cotton   Cotton insulation is a post-consumer recycled material made from 85 percent denim, polyester, and nylon recycled fibers (Figure 24.6). Ammonium sulfate or borate are used as fire retarder and mold inhibitor. It

Figure 24.5 Mineral wool insulation is highly compressible, fire resistant, and can aid in soundproofing.

Figure 24.6 Insulation made from recycled denim and cotton fibers.

is available as batt or loose fill in standard widths and thickness with a kraft paper facing on one side. Cotton insulation meets the same Class I standards for fire resistance as fiberglass insulation and imparts no skin irritation during installation.

Wool Made from natural wool, this insulation has a very low embodied energy. The material uses no dangerous chemicals and is 100 percent biodegradable. Unlike manufactured materials, when wool absorbs moisture it actually emits energy that warms the wool and prevents condensation. Wool is able to absorb, retain, and release moisture without affecting its thermal properties or performance. The material does not support dust or mold, and is naturally fire retardant; it only chars when exposed to flames.

Loose-Fill Insulation

Loose-fill insulation is usually composed of materials in bulk form. It can be installed by blowing or pouring the fill into attics, finished wall cavities, and hard-to-reach areas (Figure 24.7). It is an ideal insulator because it conforms to spaces, thoroughly filling nooks and crannies in the construction assembly. It can also be sprayed in place with water-based adhesives. Some loose-fill insulations are made of recycled materials and are relatively inexpensive. Loose-fill insulation is available as a granular or loose fibrous material. Granular insulation is generally poured, whereas the fibrous variety is machine-blown into areas to be insulated. Granular insulation includes perlite (expanded volcanic rock), vermiculite (expanded mica), cork, and expanded polystyrene. These are available in different densities and R-values.

Figure 24.7 Loose-fill insulation is machine blown into the areas to be insulated.

Granular insulation is used when filling vertical cavities, such as cores in concrete block and wall cavities.

Fibrous loose insulation is made by blowing a jet of air through molten glass, slag, or rock to form thin fibers that compose a wool-like substance when gathered. Fibrous loose insulation is difficult to work into tight spaces. It is widely used in large areas, such as insulating ceilings, where a sprayed uniform layer can be applied.

Cellulose Cellulose insulation is made from recycled newsprint and other paper stock that has been treated for fire and pest resistance (Figure 24.8). A high flashpoint mineral oil is added to reduce dust during installation and enhance bonding of the product. A small amount of water is inserted to activate the starches within the material to provide the required adhesion. Cellulose is increasingly regarded as a superior insulating material, for three main reasons. First, it provides higher levels of acoustical insulation and fire protection. Second, the material is less subject to a decline in performance over time. Finally, cellulose insulation is very effective in creating an airtight building envelope, thereby protecting against unwanted infiltration heat losses. Cotton, wool, hemp, corncobs, straw dust, and other harvested natural materials are also beginning to find increased use as loose-fill insulation.

Figure 24.8 Cellulose insulation, made from recycled newsprint, is sprayed into wall cavities, and blown into the attic over a plastic vapor barrier.

Rigid Insulation

Rigid insulation board is made using organic fibers, such as wood, mineral wool, glass, corkboard, and several forms of expanded plastics, such as expanded and extruded polystyrene and polyisocyanurate foam. This form of insulation products are used in all parts of a building, including roof, wall, and floor insulation. Care must be taken when choosing a rigid insulation material, as some products are not specified for contact with moisture and should not be installed near water. Sheets of rigid insulation are attached to wall systems with mechanical fasteners or adhesives, or friction fit between framing members. Fasteners used incorporate large plastic washers to hold the foam in place. Certain adhesives may dissolve some types of foam boards; adhesives must be specified according to manufacturer recommendations. Rigid insulation panels come in a variety of sizes in widths up to 4 ft. and lengths up to 12 ft. (Figure 24.9).

Figure 24.9 Rigid insulation sheets are used in all parts of a building, including roof, wall, and floor insulation.

Expanded Polystyrene (EPS)   EPS is created by reacting benzene with ethylene with a pentane (non-ozone depleting) blowing agent. If expanded polystyrene comes in contact with water, it will absorb moisture, which lessens its insulating value.

Extruded Polystyrene Extruded polystyrene is created by reacting benzene with ethylene. The result is a closed-cell foam that resists moisture absorption. It can be installed in wet locations and is often used to insulate foundations and roof decks (Figure 24.10). While the product can contain recycled content, it uses an HCFC blowing agent, the only remaining ozone-depleting blowing agent for plastic foams.

Figure 24.10 Tongue-and-groove extruded polystyrene panels being installed on a sloped roof.

© Ralph Orlowski/Getty Images

Figure 24.11 Blown insulations generally utilize polyurethane- or phenol-based compounds that combine for high insulation levels.

© Joe Belanger/Shutterstock.com

Polyisocyanurate Polyisocyanurate is made by combining isocyanate with polyol. Normally faced with a foil facing, the material sometimes contains up to 10 percent recycled content and uses a non-ozone-depleting blowing agent. This insulation material may deteriorate if exposed to UV radiation (sunlight) and the performance may decline as the product continues to off-gas.

Wood Fiber Wood and cane fiberboard are typically asphalt impregnated and are commonly used for exterior sheathing and roof insulation. Rigid insulation sheets made from granulated cork are used for roof, wall, and floor insulation and are available in thicknesses from 2 in. to 12 in. and sheet sizes of 24 in. × 36 in.

Blown Insulation

Blown insulations generally utilize polyurethane- or phenol-based compounds that combine for high insulation levels. Most, including polyurethane and isocyanine insulation, contain hazardous chemicals, such as benzene and toluene. Environmental concerns exist during raw material production, transport, manufacture, and installation. Although CFCs are no longer used as blowing agents, many spray insulations use HCFCs or HFCs. Both are potent greenhouse gases, and HCFCs have some ozone-depletion potential.

During installation, the ingredients are carefully measured and mixed by special equipment that maintains isocyanate and polyol levels at a one-to-one ratio. These are pumped as separate materials into a proportioning unit, which heats each and pumps them into separate heated hoses. Once mixed, they are pumped through hoses into building cavities or sprayed in layers on flat and sloping surfaces (Figure 24.11). Since the material expands as it hardens, the amount injected into a cavity must be

carefully measured. If a wall cavity is overfilled, the pressure could damage wall finishes. After the sprayed polyurethane foam insulation is in place, it must be protected from fire, exposure to moisture, and ultraviolet radiation.

Open-Cell Spray Polyurethane Foam While open-cell spray polyurethane uses water as the blowing agent, it does emit VOCs during and after installation. The material provides a good air barrier, and is vapor permeable and moisture resistant with a good fill of the cavity.

Closed-Cell Spray Polyurethane Foam Closed-cell polyurethane uses three to four times more material use than open-cell. The product also emits VOCs during and after installation, so flushing with air after installation is required. The material is roughly three times as expensive as traditional polyurethane insulations.

Icynene An insulation and air barrier material that reduces the need for petroleum-based polyol and uses 100 percent water-blown foam. Applied as a liquid, the material expands to 100 times its volume to fill cracks while remaining flexible so that the integrity of the building envelope seal remains intact over time. Like other blown insulations, the material minimizes air leakage and seals the building envelope for optimal air tightness.

Aerogel and Other Innovations

Aerogel Aerogel is a low-density material derived from gel in which the liquid component of the gel has been replaced with gas (Figure 24.12). The result is an extremely low-density solid with several remarkable thermal insulation properties. Since it is 99.8 percent air, it appears semi-transparent, making it ideally suitable as a thermal insulation material for windows and skylights. The material feels like expanded polystyrene (styrofoam)

Figure 24.12 Aerogel is a material in which the liquid component of a gel has been replaced with gas, resulting in an extremely low-density solid with remarkable thermal insulation properties.

© Roger Ressmeyer/Documentary/Corbis

to the touch. Insulation blankets are now being made of synthetic aerogel impregnated onto a fabric material.

Gas-Filled Panels (GFP)   Gas-filled panel insulation materials consist of pockets of low-conductivity gas in a honeycomb foil substrate (Figure 24.13). GFP insulation has inner and outer skins comprised of an aluminum foil-polymer laminate and five internal alumized low-emissivity (low-e) films that expand with gas to form a 1½ in. thick flexible honeycomb panel. The panels are available with air, argon, krypton, or xenon as the gas-fill, providing insulating values of R-5.0 for air, R-6.4 for argon, R-7.6 for krypton, and R-11 for xenon.

Figure 24.13 Gas-filled panel insulation materials consist of pockets of low-conductivity gas in a honeycomb foil substrate.

© Bplanet/Shutterstock.com

Structural Insulating Panels (SIPs)   Structural insulated panels consist of two rigid wood panel materials with a foamed insulating material secured between them, created either by gluing billets, as in EPS (expanded

polystyrene) or with polyurethane foamed and formed in place. These panels can be used as bearing walls or sheathing and roof decking, over a structural frame. An added advantage of SIPs is a decreased dependency on bracing and auxiliary members, since the panels can span considerable distances and serve to increase the stiffness of the structural frame.

Reflective Barriers

Radiant, or reflective, barriers inhibit heat transfer by thermal radiation. Since thermal energy may also be transferred via conduction or convection, radiant barriers are typically used in addition to other insulation. On a sunny summer day, solar energy is absorbed by a roof, heating the roof sheathing and causing the underside of the sheathing and the roof framing to radiate heat downward toward the interior. When a radiant barrier is placed directly underneath the roofing material incorporating an air gap, much of the heat radiated from the hot roof is reflected back, with little radiant heat emitted downward.

Reflective barriers are usually made from aluminum or copper foil in sheets or rolls (Figure 24.14). Rolls

Figure 24.14 Radiant barriers are placed directly underneath the roofing material to reflect heat radiated from the hot roof.

© Ozgur Coskun/Shutterstock.com

Construction Materials

Superinsulation

Superinsulation (SI) is an approach to building design that provides higher-than-normal levels of thermal insulation. Superinsulated buildings are intended to be heated predominantly by internal sources, including waste heat generated by lighting, appliances, and the body temperature of occupants. They typically provide no conventional heating system, instead relying on just a small backup heater. Superinsulated homes have been demonstrated to perform well even in very cold climates, but their construction requires close attention to air tightness and insulation.

There is no established definition for superinsulation, but superinsulated buildings typically include the following aspects: they utilize very thick insulation with R-values typically in the area of R-40 for walls and R-60 for roofs; careful attention is paid to detailing insulation where walls meet roofs, foundations, and other walls; airtight construction is achieved through a thorough sealing of all penetrations, especially around doors and windows; windows are generally smaller and of high quality, often with a triple layer of glazing to prevent heat loss. Because of the hermetic nature of the envelope, superinsulated construction must incorporate a ventilation system to control air quality, sometimes in the form of a heat-recovery ventilator that provides fresh air.

A variety of insulation types are used in tandem to produce superinsulated buildings. Cavity walls may be insulated with loose-fill batts, then covered both inside and out with rigid insulation sheets, and finally finished with a radiant barrier before exterior finishes are applied. Straw-bale construction is a building method that uses straw bales as structural elements, insulation, or both. It is, by nature, "superinsulation," and easier to air seal, particularly in conjunction with a slab on grade and plastered exterior surfaces.

Superinsulated buildings conserve energy without impacting an occupant's lifestyle. They tend to produce more comfortable interiors, with no drafts, cold spots, or temperature stratification. Because of thicker walls and better windows, less outside noise penetrates their interiors. Superinsulated construction typically costs 5 to 7 percent more compared to conventional construction. Most of the cost accrues from better windows and doors, more insulation, and the additional labor involved with air sealing. While initial costs are higher, the return on the investment is quick, since SI buildings typically save 75 percent of heating and cooling costs.

SI building practices are included in Canadian building codes and are standard practice in many parts of Europe. In the United States, changes to the *Model Energy Code* have been implemented by the Department of Energy and the Department of Housing and Urban Development. Monetary incentives, such as Energy-Rated Mortgages, which increase the borrowing power of buyers of energy-efficient homes, are now available. Programs like the ENERGY STAR Homes Program, sponsored by the Environmental Protection Agency and the Department of Energy, encourage builders to erect SI homes and provide a rating system to evaluate energy performance.

---

are typically 24 in. and 48 in. (610 mm and 1,220 mm) wide and 500 ft. (152.5 m) long. They are available in single thickness layers or in a multilayer batt with dead air spaces between the layers. The foil utilizes reflective properties to reject the passage of heat and increase the effectiveness of the dead air spaces. Some fiber insulation batts have reflective foil bonded to one side that serves as a vapor barrier. Reflective insulation is used in residential and commercial construction on walls, floors, ceilings, and roofs. It can decrease heat flow by as much as 25 percent, reducing the amount of conventional insulation needed and, in some cases, reducing the size of air-conditioning units required.

WATER AND VAPOR CONSIDERATIONS

All buildings during their lifetime will have some form of water or moisture intrusion. Appropriate management of these water problems is crucial in protecting a building's structural integrity and ensuring the health of both the building and its occupants. Water-damaged building materials and finishes can become significant sources of microbiological growth if not properly detailed. Water-borne contamination in building environments can lead to potential health problems for occupants, ranging from simple irritation and allergic responses to acute sensitivity diseases.

As our understanding of both the physics of moisture management and the potential health problems associated with indoor pollution increases, the importance of controlling moisture infiltration into buildings is becoming easier. Moisture management in buildings strives to balance the entry and removal of water within the building envelope. A variety of waterproofing membranes, coatings, and sealers are used to effectively manage moisture protection.

Rainwater Protection

Rainwater is the most severe form of moisture intrusion to the building envelope. Moisture as rainwater can be drawn through construction assemblies by a variety of natural forces. Water concentrates around window and door penetrations, the roofline and construction joints, and the base of exterior walls. The building envelope acts as the interface between the interior and exterior of buildings, and must resist rainwater intrusion. The following forces are the most mechanisms by which rainwater can enter a construction assembly.

The force of water entering by gravity is greatest on penetrations in vertical surfaces and sloped horizontal surfaces. These exterior assemblies must remove water from envelope surfaces through appropriate slopes, drainage planes, and proper flashing. Capillary action is a natural wicking force that draws water from the exterior into the envelope cavity through porous materials or small cracks. This occurs primarily at the base of exterior walls but can happen at any opening in an assembly.

Surface tension enables water to adhere to and travel along the underside of building components, such as joints in masonry walls or window heads. This water is further pulled into the building by gravity or unequal air pressures. Wind-driven rain during heavy rainstorms can force water inside the building if the envelope is not resistant to these forces. For example, door sealants and gaskets that are not properly designed to adjust with the window can create air gaps that can allow water into the interior. To avoid moisture problems in the extreme weather conditions, building envelope design must control water from all of these factors. Waterproofing involves applying a material on the surface of an assembly of materials, such as a foundation, to make it impervious to water.

Water Vapor Transfer

Depending on temperature conditions, water can exist in liquid, vapor, or solid states. Water is always present as vapor in both indoor and outdoor air. Moisture-related problems arise from the presence of excessive moisture or changes of state, such as condensation or freezing within a wall. The potential for vapor flow across a building assembly is a function of the vapor pressure differential across the building envelope.

Diffusive vapor flow describes the transfer of moisture in its gaseous state through the various layers of an exterior wall system or assembly due to vapor pressure differential. Vapor pressure results from high moisture content in the air. Hot, humid air has a higher pressure than cooler, dry air. A high amount of water vapor can create significant force, in some instances enough to bubble and peel paint on exterior finishes. Vapor diffuses through walls at a rate proportional to the vapor pressure difference. When one side of a wall is much drier than the other side, the vapor will diffuse faster.

The usual water vapor migration direction is from higher temperatures toward lower temperatures, so in cold climates, where buildings are heating load dominated, the primary vapor drive is from the interior toward the exterior. In warmer climates, buildings are primarily air-conditioned and the primary vapor drive across the wall is from the exterior toward the interior. Vapor diffusion issues tend to be greatest in cold climates, where even small amounts of internally generated moisture can condense inside cold wall cavities during the cooling season. The fact that the interior of a building is at a different temperature than exterior conditions means that there are temperature gradients across the wall from the interior to the exterior. Within this gradient, the location of the dew point temperature, or where migrating moisture might condense, can be determined.

Wall dew point analysis is used to calculate what the temperature profile in a wall will be and where the dew point will occur. The procedure is used to determine how far moisture is allowed to penetrate the wall and where a vapor retarder should be located. Once the location of the dew point is known, some components within the wall assembly can be made more resistant to moisture, and possible damage to the wall will be minimized or eliminated. The detailed method is outlined in the ASHRAE *Handbook of Fundamentals*.

The procedure for calculating water vapor diffusion involves evaluating each layer of the wall system, including its thickness, permeance to vapor transmission, and thermal resistance (R-value).

Vapor Retarders

A vapor retarder, or barrier, is a thin sheet material that is used to provide a barrier against the flow of air-borne moisture (Figure 24.15). Vapor retarders typically consist

Figure 24.15 Vapor barriers prevent moisture generated inside a building from penetrating the insulation.

© pryzmat/Shutterstock.com

Table 24.2 Perm Values of Vapor Control Materials

Vapor Control Material	Permeance
Thermal Insulation	
Extruded polystyrene (1 in.)	1.0–1.2
Expanded polystyrene (1 in.)	2.0–5.8
Polyisocyanurate 1 (foil face)	2.0
Polyisocyanurate 1 (no foil)	1.6
Plastic and Metal Film	
Aluminum foil (1 mil)	0
Polyethylene (4 mil)	0.08
Polyethylene (6 mil)	0.06
Polyethylene cross laminated (4 mil)	0.02
Building Paper, Felt and Roofing Paper	
Saturated and coated rolled roofing	0.05
Kraft paper and asphalt laminated	0.03
Kraft thermal insulation facing	1.0
Foil thermal insulation facing	0.5
15 lb. asphalt felt	1.0
Paints	
Primer Sealer	0.9
Semi-gloss vinyl acrylic enamel	6.6
Vinyl acrylic primer	8.26

of treated paper, thin plastic sheeting, or low-permeance paints that resist the diffusion of moisture through the wall, ceiling, or floor assemblies of a building. A vapor retarder is also used to prevent water vapor generated inside a building, such as by cooking and bathing, from penetrating walls and condensing as moisture on insulation.

The permeability of vapor retarders is rated in perms, a measure of the rate of transfer of water vapor through a material. Vapor retarders are rated as either impermeable, semi-permeable, or permeable. Materials with a perm rate of less than .1 are considered impermeable, and are classified as class 1 vapor retarders; class 2 vapor retarders have ratings of greater than .1 perm and less than 1.0, whereas class 3 retarders have a perm rating of greater than 1.0 and less than 10 (Table 24.2). Class 1 vapor retarders include polyethylene film, aluminum foil, sheet metal, oil-based paints, vinyl finishes, and foil-faced insulated sheathing. Class 2 retarders include materials like brick, extruded and expanded polystyrene, oriented-strand board, and asphalt-backed kraft paper insulation facing. Materials like plywood, fiberglass and cellulose insulation, and latex paint constitute class 3 vapor retarders.

The location of vapor retarder within an assembly depends on the exterior and interior climate and the type of wall assembly. While a detailed dew point analysis is required in most applications, some general rules of thumb exist. Vapor retarders should be installed on the warm side of an enclosure. In cold climates, moisture from the warm interior flows toward the cold exterior. This wetting from the interior is retarded by an interior vapor control layer. To allow any moisture left in the assembly to escape, permeable exterior sheathing materials should be specified. In hot, humid climates, moisture from warm exterior flows toward the cooler interior. An exterior vapor retarder and permeable interior finishes will allow drying to the interior. In both cases, one side of the assembly facilitates the removal of moisture. Under no circumstances should a vapor retarder be used on both sides of an exterior envelope.

A vapor retarder is not required in all situations. The building envelope may perform as an adequate barrier to vapor diffusion. The installation of vapor barriers such as polyethylene on the interior of air-conditioned assemblies has been linked with moldy buildings and should be avoided.

Vapor barriers are also used to control moisture and humidity resulting from soil water evaporation in the crawl space of wood frame structures. The vapor barrier is installed up the walls of the crawl space and over the exposed earth floor.

Control of Air Leakage

All buildings, no matter how tightly detailed, will experience some degree of air leakage due to small openings in the envelope construction. This leakage of air carries a certain amount of moisture with it that is transferred into, or out of, the building. Most of the incoming air can be counteracted through positive pressurization of the HVAC system. A tightly sealed exterior envelope will minimize any remaining air leakage and reduce the amount of pressurization required of the HVAC system. Moisture contributed by air leakage is a significant source and should be a serious concern in the design of enclosure systems. The design of the building envelope for minimizing air leakage may be more critical than the design of the vapor barrier.

Air infiltration from the exterior of a building is controlled by applying an air barrier on the exterior over the sheathing, or just behind the exterior finish (Figure 24.16). The simplest approach to air tightening a wall is to select one of the layers and air tighten it using durable tapes, adhesive sheet products, or liquid-applied materials. Air barrier systems (ABS) must be durable enough to take wind gust, and must be continuous across the exterior envelope. A commonly used material is a spun-bound olefin formed into a sheet of very fine high-density polyethylene fibers. It resists tearing, punctures, and will not rot. It has a perm rating of 94, so any moisture vapor that gets into the wall from inside the building can pass through to the exterior. In all cases,

Figure 24.16 Air infiltration from the exterior of a building is controlled by applying a plastic barrier over the exterior sheathing just behind the exterior finish.

vapor barriers and air infiltration wrap must be carefully installed, overlapped at joints, and sealed to reduce penetration.

Walls constructed out of very permeable materials, such as concrete block, can be air tightened using applied elastomeric (flexible) coatings, specially formulated paints, air barrier sheet products, and liquid-applied spray or trowel-on material. Transition peel-and-stick membranes are most commonly used at window and door perimeters.

Review Questions

1. What types of organic fibers are used for insulation?
2. What kinds of organic cellular materials are used for insulation?
3. What kinds of mineral cellular materials are used for insulation?
4. What kinds of mineral fibrous materials are used for insulation?

5. In what forms is thermal insulation available?
6. In what three ways can heat be transferred?
7. What are the four terms used to describe thermal properties of a material?
8. How are insulation materials rated on their ability to resist heat flow?
9. What materials are used for vapor barriers?

Key Terms

British thermal unit (BTU)

Conduction

Continuous Insulation (ci)

Convection

Infiltration

Joule (J)

Mean Radiant Temperature (MRT)

Radiation

Thermal Bridge

Thermal Conductance (C)

Thermal Envelope

Thermal Resistance (R)

Thermal Transmittance (U)

Activities

1. Try to set up an experiment to get a rough evaluation of the relative effectiveness of various types of insulation. For example, you might build a box from ¾ in. plywood, leaving the top open. Place a heat source in the box, such as a number of high-wattage incandescent lamps. Cover the open top with an insulation product and measure the temperature of the exterior surface after a set number of hours. Repeat with various types and thicknesses of insulation. Be alert for the possibility of fire and do not leave the experiment unattended.

2. Collect samples of various insulation products and label them, identifying their composition and properties.

Additional Resources

Insulation Manual, National Association of Home Builders, Washington, DC.

Moisture Control Handbook, U.S. Department of Energy, Oak Ridge, TN.

Thermal Design Guide for Exterior Walls, American Iron and Steel Institute, Washington, DC.

Wood Building Technology, Canadian Wood Council, Ottawa, Ontario, Canada.

Other resources include:

Publications in whole-building design and energy-simulation software, Sustainable Buildings Industry Council, Washington, DC.

See Appendix C for addresses of professional and trade organizations and other sources of technical information.

Bonding Agents, Sealers, and Sealants

Upon completion of this chapter, the student shall be able to:

- Select the proper bonding agents for various construction assemblies.
- Understand how the various bonding agents cause adherence.

- Select appropriate sealers for use on exterior applications.
- Decide the best way to protect joints in exterior construction.
- Select waterproofing materials best suited for various applications.

Build Your Knowledge

For further study on these materials and methods, please refer to:

Chapter 23 Plastics

               Topics: Silicones, Thermoplastic Resins, Thermoset Resins

Chapter 26 Bituminous Materials

Bonding agents, sealers, and sealants made from natural and manmade substances are used widely for both interior and exterior purposes. Bitumens and various synthetic resins are used to produce many adhesive products.

BONDING AGENTS

A *bonding agent* is a compound that joins materials together by bonding to their contacting surfaces. In many building assemblies and products, bonding agents are used to permanently fasten both structural and decorative elements. Bonding agents used in the fabrication of construction products and for on-site applications typically join materials by mechanical action or by specific adhesion. In *mechanical action*, the bonding agent enters the pores of a porous material, hardens, and forms a mechanical link. Bonding agents that join by *specific*

adhesion are used to bond dense materials without pores, such as glass and metal. The bonding is caused by the attraction of unlike electrical charges. The positive and negative charges in the bonding agent are attracted by the electrical charges on the surface of the material to be bonded. This molecular attraction provides a strong holding force.

Bonding agents can have their properties varied for specific conditions and materials. Some are a combination of two or more types, such as a phenol and resorcinol resin combination or a urea resin blended with a melamine resin. They are available as powders, solids, liquids, and pastes. Some require the addition of a catalyst.

A summary of commonly used bonding agents, their uses, and the materials they join is given in Table 25.1.

Curing of Bonding Agents

The curing of bonding agents is accomplished by loss of solvents, anaerobic environments, catalysts in two-part mixtures, and cooling of hot melts. Loss-of-solvent bonding agents cure by the loss of volatile liquids, water, or organic solvents used to dissolve the base material. The solvents evaporate or soak into porous materials to be bonded. Sometimes a solvent may damage the materials on which it is applied, so manufacturer's instructions should always be followed. *Anaerobic bonding agents* maintain a fluid condition when exposed to oxygen but set hard when oxygen is omitted, as occurs when two

Table 25.1 Commonly Used Adhesives and Their Applications

Adhesive	Bonded Material	Typical Uses
Acrylic	Plastics to metal, plastics to plastics, rubber to metal	Curtain walls
Casein	Wood to paper, wood to wood	All interior wood-joining needs
Cyanoacrylate (anaerobic)	Acrylics, phenolic, rubber, glass, polycarbonates, ceramics, steel, copper, aluminum	Any use (known as "super glue"); electronic and electrical devices
Epoxy	Almost any material except a few plastics and silicones	Interior and exterior uses, panels, glass to metal, curtain walls
Melamine formaldehyde	Paper, textiles, hardwood, interior plywood	Interior uses, plywood manufacturing
Natural rubber	Leather, paper, cork, foam rubber	Pressure-sensitive tape
Neoprene rubber (contact cement)	Many plastics, ceramics, aluminum	Plastic laminates, other interior uses
Nitrile rubber	Many plastics, ceramics, glass, aluminum	General uses
Phenol	Wood, cardboard, cork	Exterior plywood, any exterior use
Polyvinyl acetate	Porous materials (wood)	Various interior and exterior applications
Resorcinol	Rubber, paper, cork, asbestos, wood	Furniture, wood beams, columns
Silicone	Glass, ceramics, aluminum, polyester, acrylics, phenolic, rubber, steel, textiles	Sealant, gasket material
Polyurethane	Many plastics, glass, copper, aluminum, ceramics	Bonding dissimilar materials (e.g., on steel and glass sun roofs)
Urea	Many plastics, glass, copper, aluminum, ceramics	Particleboard, furniture, cabinets
Vinyl butyral	Glass	Laminating glass

components are clamped together. Two-part mixtures supply the resin and catalyst in separate containers. When mixed, the catalyst causes cross-linking of the resin. *Hot-melt adhesives* come in solid form, become liquid when heated, and set rapidly when heat is removed.

Types of Bonding Agents

Bonding agents may be divided into three major classes: adhesives, glues, and cements. Following are some of the more commonly used types.

Adhesives *Adhesive bonding agents* are made from synthetic materials. They fall into two types: thermoplastic and thermosetting.

Thermoplastic Adhesives Thermoplastic polymer adhesives generally have less resistance to heat and moisture and less long-term resistance to loads than do thermosetting polymers. They are moisture resistant but not used where they will be exposed to moist conditions.

Aliphatic resins are a type of polyvinyl resin and are stronger and more heat-resistant than other polyvinyls. They are generally yellow in color, but some polyvinyls are white. Aliphatic resins are used for furniture and general carpentry work **(Figure 25.1)**.

Figure 25.1 Polyvinyl acetate (white glue) and aliphatic resin types of polyvinyl resin (yellow or carpenter's glue) are used for furniture and general carpentry.

© Kristian Septimius Krogh/Getty Images

Alpha-cyanoacrylate, often called "superglue," is used to bond metals, plastics, and other dense materials. It is not recommended for use with porous materials, such as wood.

Hot-melts are a mixture of polymers sold in solid form, such as rods, pellets, ribbons, or films. They are placed in an electric hot-melt applicator that melts the adhesive as it is applied to a surface **(Figure 25.2)**. There

Figure 25.2 Hot melts are applied with a handheld electric hot melt gun.

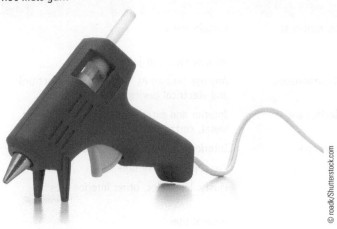

© roadk/Shutterstock.com

Figure 25.3 Epoxies are two-part bonding agents, with the resin and catalyst in separate tubes.

© 1125089601/Shutterstock.com

are various types for bonding plastics, particleboard, softwoods, and hardwoods. Hot-melts set up fast but are not very strong and are used primarily for interior purposes where little stress is expected.

Polyvinyl acetate adhesives, generally referred to as "white glue," have moderate moisture-resistance but high dry-strength and are recommended for interior use only. They are used to join paper, wood, and vinyl plastics (sometimes with metal) and are one of the most widely used adhesives. Typical applications include furniture assembly, bonding plastic laminates, and flush doors. A waterproof-type for bonding exterior wood is also available.

Thermoset Adhesives Thermosetting polymers are widely used as a structural adhesive because they undergo an irreversible chemical change and, if subjected to heat after use, will not soften and flow. They resist moisture and many chemicals and are very strong. Thermosetting adhesives are available in a variety of forms. Epoxy adhesives are produced in a two-part liquid form. An epoxy resin and an epoxy hardener are mixed immediately before use (Figure 25.3). Epoxy resins will bond to almost any material and produce a strong joint. Epoxies are used in the construction of curtain walls and for bonding steel and concrete in applications such as bridge construction. Epoxies are being used for more structural applications and are available in a variety of types for special uses.

Melamine adhesives are produced as a powder with a catalyst to be added when used. They are cured by applying heat to about 300°F (150°C) and provide a good bond to paper and wood. Melamine adhesives are used as fortifiers in urea resins for hardwood plywood, gluing lumber, and scarf joining softwood plywood.

Phenol resin adhesives have good bonding qualities to paper and wood, good shear strength, and are resistant to moisture and temperature. They are cured in a hot press at about 300°F (150°C). Phenol resins are widely used in the manufacture of plywood and particleboard.

Resorcinol resin adhesives are used when a waterproof bond in wood products is required. The two-part adhesive consists of a resin and a catalyst. The combination must be used within eight hours of mixing. One important use is in the manufacture of glued-laminated wood structural members. Some adhesives are a mixture of resorcinol and phenol resins.

Urea resin adhesives are available in powder form and are mixed to proper consistency with water. They are moisture resistant but not waterproof. They bond well to paper and wood and have good resistance to heat and cold. Normal clamp time is about sixteen hours but can be reduced to minutes or seconds using a radio frequency glue-drying machine. Typical uses include interior hardwood plywood, interior particleboard, flush doors, and furniture.

Polyurethane adhesives are available as low-viscosity liquid or high-viscosity mastic. Both provide superior strength and stiffness to an adhesive bonded assembly for wood, stone, metal, ceramics, plastics, and other materials. Sold under the brand name Liquid Nails, among others, they are capable of withstanding exposure to moisture and some types are considered waterproof.

Glues *Glues* are bonding agents made from animal and vegetable products, such as bones, blood albumin, hides, fish, and milk.

Animal glue is produced as flakes or dry powder that is mixed with water and heated. The heated mixture is applied to wood surfaces. It is primarily used for furniture manufacture. Animal glue has excellent bonding and shear strength but has been largely replaced by newer products. Another form, liquid hide glue, a ready-mixed form of animal glue, is more widely used. It is not moisture-resistant and is used only on interior products, such as furniture, paper, and textiles.

Blood albumin glue is a form of animal glue used to bond paper products. It finds limited use on some forms of interior plywood. It has moderate bonding power with wood, poor resistance to moisture, and only fair resistance to heat and cold. Fish glues are used for sealing cardboard boxes and on packing tape.

Casein glue is made from dried milk curds in powdered form mixed with water. It is water resistant and used for interior wood and on exterior products that will be sheltered and not be directly subjected to the weather, such as laminated timbers. Used on wood and paper, it provides a strong joint, resists heat well, but has only moderate resistance to cold.

Vegetable glues are made from soybeans, starch, and dextrin and are used for bonding interior plywood, paper, and wallpaper.

Cements *Cements* are made from synthetic rubber, such as neoprene, nitrile, and polysulfide, suspended in a liquid.

Cellulose cements, such as cellulose acetate, cellulose nitrate, and ethyl cellulose, are used for interior purposes, for bonding plastics and glass, and for porous materials, such as wood and paper. Cellulose nitrate is more commonly known as a general household cement and is sold in tubes for ease of application. Cellulose cements have moderate resistance to temperature change but good moisture resistance.

Contact cements are neoprene cements that stick immediately upon contact, requiring no clamping time. They are used to bond plastic laminates on countertops, plastics to plastics, plastics to wood, and wood to wood. Although they have very high bonding strength and good resistance to moisture, they are generally used for interior purposes only.

Mastic is elastomeric construction cement. Sometimes referred to as liquid nails, it is sold in tubes and applied with a caulking gun. Typical uses include bonding plywood subfloors to joists; bonding wall paneling to studs; laminating gypsum wall-board, styrene, and similar materials; and assembling wall, floor, and roof panels. Mastic is water resistant and develops full strength after several weeks (**Figure 25.4**).

Figure 25.4 Mastic is an elastomeric construction cement applied with a caulking gun.

Buna-N resins are a form of acrylonitrile butadiene rubber used for general-purpose interior cements. They are liquid cements that have good moisture resistance and strength. A summary of some of the properties of these bonding agents is given in Table 25.2.

SEALERS FOR EXTERIOR MATERIALS

Sealers are applied to the surface of a material to protect against penetration by water and moisture. Brick walls are often sealed to keep moisture from infiltrating mortar joints and leaking into the inside of the wall. Sealers have adhesive properties and are closely related to bonding agents. In addition to bonding to surfaces, they form an unbroken film over them and fill any minute pores, cracks, or other openings. Large cracks or defects must first be packed with some form of sealant before a sealer is applied, because sealers will not span most cracks.

Acrylic sealer is a high-solid-content clear acrylic sealer. It maintains the original appearance of masonry or concrete while protecting it from moisture, airborne dirt, and other pollutants. It is used on products such as exposed aggregate panels, brick, and stone. Acrylic sealer is virtually unaffected by prolonged exposure to moisture, common acids, ultraviolet rays, oils, and aliphatic solvents. It reduces damage from the freeze–thaw cycle, prevents efflorescence, and stains. It is available in a number of variations for different applications, such as methyl methacrylate acrylic polymer.

Table 25.2 Characteristics of Bonding Agents

	Form	Moisture Resistance	Application Temperature	Clamp Time
Thermoplastic Adhesives				
Aliphatic	Liquid	Low	45°F (7°C)	1 hr.
Alpha-cyanoacrylate	Liquid	Low	75°F (21°C)	Minutes
Hot melts	Solid	Low	Electronic welder 300°F (150°C)	None
Polyvinyl acetate	Liquid	Low	60°F (16°C)	1 hr.
Thermosetting Adhesives				
Epoxy	Liquid	Water resistant	60°F (16°C)	None
Melamine resin	Liquid	Water resistant	Hot press 250°F (122°C)	Minutes
Phenol resin	Liquid and powder	Waterproof	300°F (150°C)	Minutes
Resorcinol resin	Liquid	Waterproof	70°F (21°C)	16 hr.
Urea resin	Powder and liquid	Water resistant	70°F (21°C)	1 to 3 hr., or seconds with high-frequency glue curing machine
Glues				
Animal	Flakes and powder	Low	70°F (21°C)	2 to 3 hr.
Liquid hide	Liquid	Low	70°F (21°C)	2 to 3 hr.
Blood albumin	Liquid	Moderate	70°F (21°C)	2 hr.
Vegetable or fish	Liquid	Low	70°F (21°C)	2 hr.
Casein	Powder	Water resistant	32°F (0°C)	2 hr.
Cements				
Contact cement	Liquid to thick	Water resistant	70°F (21°C)	None
Mastic	Paste in tubes	Water resistant	70°F (21°C)	15 min.
Cellulose	Liquid	Low	70°F (21°C)	Minutes
Buna N	Liquid	Water resistant	70°F (21°C)	Minutes

Applied by brush or roller, asphalt driveway sealer is a thin, quick-drying sealer that produces a black protective coating. Coal tar coatings are used to seal surfaces and provide protection against corrosive conditions encountered in industrial plants, water and sewage systems, and other such industries. They resist moisture and most acids, alkalies, corrosive vapors, and atmospheric corrosion on metal, concrete, and masonry surfaces.

Silicone sealers come in liquid form for use on brick, concrete block, stucco, cement plaster, and concrete. They produce a water-repellent coating that protects surfaces but does not prevent water vapor from escaping. They seal surfaces, minimize efflorescence, and protect from absorbing staining materials, such as dirt and soot. Since protection against moisture

penetration is provided, damage due to freeze–thaw cycles is also reduced.

Epoxies are used to provide a waterproof coating. They are often employed to seal floors in areas where moisture or chemicals exist, such as in food processing and industrial applications. Other uses include floors around swimming pools, decks, showers, restrooms, loading docks, and on stair treads. When mixed with an aggregate, such as emery, epoxies provide a textured, slip-resistant surface available in a variety of formulations.

Polyurethane sealers are used to minimize concrete problems, such as scaling, spalling, chloride penetration, and damage due to freeze–thaw cycles. Some sealers provide ultraviolet light stabilization. They do not inhibit the transmission of vapor out of the concrete.

Wood Sealers

Sealers on wood products are used to keep the various layers of finish, such as stains and fillers, from bleeding through the final topcoat. A sealer also provides the required base for the finished topcoat. The sealer used depends on the recommendations of the manufacturer of the topcoat. If the wood is to be stained and varnished, shellac is often used as a sealer over the stain. Then wood filler can be applied and another sealer, often more shellac, follows. Then the varnish topcoat is applied. Synthetic resin sealers are also used. Lacquer topcoats require a lacquer sanding sealer. After careful sanding, the lacquer topcoats go on. Other finishes, such as polyesters and polyurethanes, may or may not require a sealer.

For exterior and interior paints, a manufacturer will have specific recommendations for sealers. Following is an example for one manufacturer:

Drywall sealer – latex primer

Plaster sealer – wall and wood primer

Wood sealer – alkyd enamel undercoat

Concrete block sealer – block filler

Masonry sealer – latex primer

Ferrous metal sealer – water-based acrylic paint

Walls to receive wallpaper or fabric coverings also must have sealers; usually some form of casein glue is used.

SEALANTS

A *sealant* is a material used to seal joints between construction members and to protect materials against the penetration of moisture, air, corrosive substances, and foreign objects. Examples include expansion joints in large masonry walls, spacers between glass and frames, and openings between exterior siding and door and window units. The sealing of joints between these and other parts of a building is essential to ensure the integrity of the entire structure.

Joints in exterior walls are needed to allow materials to expand and contract. These are called expansion joints, and they must allow for this movement. If a foreign object, such as a rock, gets into a joint, contraction may be blocked, and the object may cause the materials on each side to spall or crack (Figure 25.5).

The two basic methods for protecting joints involve the use of sealants and prefabricated covers. A sealant is a flexible adhesive material that is worked into a joint,

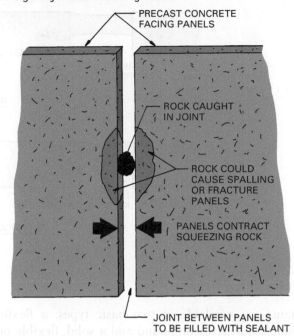

Figure 25.5 Joints between masonry materials must be free of foreign objects before being sealed.

PRECAST CONCRETE FACING PANELS

ROCK CAUGHT IN JOINT

ROCK COULD CAUSE SPALLING OR FRACTURE PANELS

PANELS CONTRACT SQUEEZING ROCK

JOINT BETWEEN PANELS TO BE FILLED WITH SEALANT

bonds to the sides, and sets into a firm but rubbery plastic material. Prefabricated covers are typically metal and made to allow for anticipated movement. The sealant filling a joint is a substitute for the material that was removed. As such, it must meet the performance requirements of the adjoining material. It must maintain the integrity of the assembly of materials while still allowing for movement between them.

Moldable sealants may be deformable or elastic/elastomeric. The deformed sealant is installed in its natural stretched shape. It stretches as the joint widens and deforms as the joint width is reduced. The elastic/elastomeric sealant stretches as the joint widens and shrinks back to its normal size as the joint width contracts (Figure 25.6).

Sealant Performance Considerations

A key factor in sealant design is the percent of elongation the sealant can safely stretch and still give expected protection. High-performance sealants have a 50 percent elongation, while intermediate types are usually rated up to 25 percent. A sealant must have excellent adhesion to the material with which it is expected to bond. It must be flexible and have minimum internal shrinkage through years of use, resist staining the material around it, and have a tough non-tacky elastic surface skin such that dirt and solid objects do not stick to it.

Figure 25.6 Elastic/elastomeric and deformable sealants allow for expansion and contraction at construction joints.

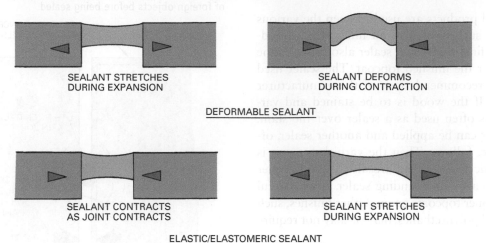

Types of Sealants

Sealants are available in two basic types: a flexible, moldable adhesive compound and a solid, flexible preformed shape. Preformed sealants are available in a variety of materials and shapes (Figure 25.7).

Moldable sealants are manufactured in three performance levels. Low-performance sealants last for four to seven years and are used where limited joint movement is expected. Intermediate-performance sealants are more expensive but last between seven to fourteen years. They are used in joints with the greatest movement. High-performance sealants have a life expectancy of twenty to thirty years and are the most expensive. They are used in joints where the anticipated movement is the greatest. Manufacturers' recommendations on joint size and the allowable percent of elongation must be observed.

Moldable bulk sealants are available in pourable form, in knife grade (such as glazing compounds), gunable form for manual or pneumatic caulking guns, and as preformed tapes that may or may not be cured.

The actual makeup of the following commonly used sealants can vary considerably and many special-purpose sealants are available.

Polyurethane sealants are general purpose sealants used for areas such as precast concrete and masonry joints, glazing and sealing around door and window openings and swimming pools. Some are designed for exterior concrete joints in roads and sidewalks. This type is resistant to damage from fuels and oils. Some are formulated for application by gun, and others are liquid and poured.

Epoxy crack filler is available for use on very small fractures that occur in concrete, concrete block, and brick walls.

Figure 25.7 Preformed sealants are designed for specific applications.

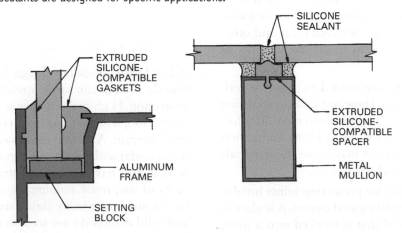

Silicone rubber sealants are available in a variety of compounds. Some are for interior use on nonporous surfaces where high humidity and temperature extremes exist, such as around bathtubs. Another type is used to seal exterior building joints and will bond to nonporous materials, such as glass, ceramics, and most metals and plastics. It can be gunned in below-freezing temperatures, is available in a variety of colors, and has a life expectancy of up to thirty years.

Latex sealant adheres to wood, metal, concrete, masonry, marble, porcelain, ceramic tile, glass, and many types of plastics. Some types permit joint movement up to 25 percent.

Polysulfide polymer sealants are elastomeric and bond to all masonry, concrete, wood, glass, and metal surfaces. Available in a variety of formulations, including those with one and two components, they are widely used in applications such as sealing joints between curtain wall panels, pre-cast concrete, and various window assemblies. Polysulfide polymer sealants cure in about twenty-four hours to a rubber-like material with excellent stretch capabilities.

A urethane bitumen sealant is a two-part catalyzed 100 percent solid polyurethane-coal tar elastomeric compound. It bonds to concrete, stone, brick, glass, wood, metal, cement asbestos, concrete block, and most plastics, with the exception of polyethylene. It offers excellent resistance to acids and commonly used solvents.

Sealants and Joint Design

The width of a joint between two members, such as pre-cast concrete facing panels, must be determined according to the anticipated structural and thermal movement. For example, one type of sealant available tolerates a joint movement of +100 to −50 percent. The manufacturer requires the joint width be two times the expected joint movement and at least ¼ in. (6 mm) wide. A backup rod is used to control the depth of the sealant (Figure 25.8). Other types of sealer joints are shown in Figure 25.9.

Figure 25.8 A backer rod is used to control the depth of the sealant.

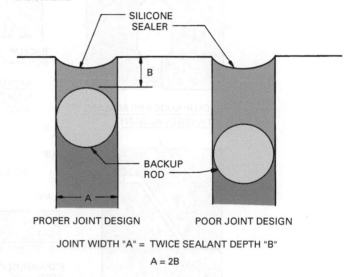

PROPER JOINT DESIGN   POOR JOINT DESIGN

JOINT WIDTH "A" = TWICE SEALANT DEPTH "B"

A = 2B

Figure 25.9 Examples of joints requiring backup rods or the use of bond breaker tape.

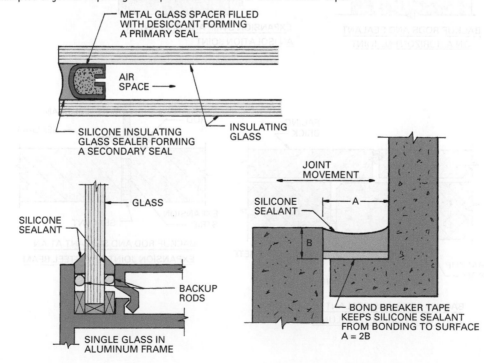

Backup Materials

Proper installation of sealants depends on the use of proper backup materials installed to control the depth of the sealant, as shown in Figure 25.10. The backup material not only limits the depth of the sealant but also serves as a bond breaker to keep the sealant from

bonding to the back of the joint. Some backers are installed in the factory as part of a unit, as in a manufactured window. Others are applied to joints in the field, such as in control joints in a concrete floor slab.

Backup materials vary depending on location and use. Rods or tubing made from polyethylene, butyl,

Figure 25.10 Typical applications of the use of backup rods and expansion strips.

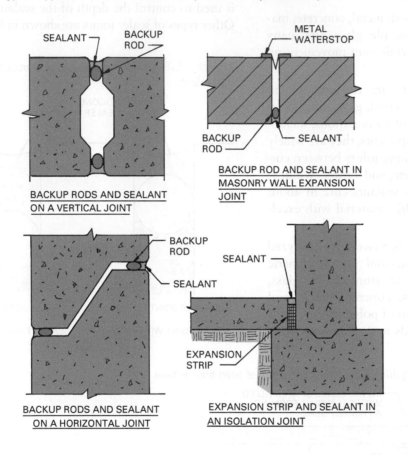

BACKUP RODS AND SEALANT ON A VERTICAL JOINT

BACKUP ROD AND SEALANT IN MASONRY WALL EXPANSION JOINT

BACKUP RODS AND SEALANT ON A HORIZONTAL JOINT

EXPANSION STRIP AND SEALANT IN AN ISOLATION JOINT

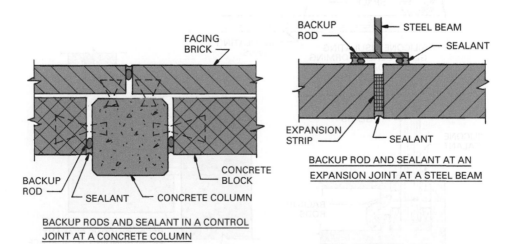

BACKUP RODS AND SEALANT IN A CONTROL JOINT AT A CONCRETE COLUMN

BACKUP ROD AND SEALANT AT AN EXPANSION JOINT AT A STEEL BEAM

urethane, and neoprene that control the depth of a sealant are most frequently used. Other rod types form a primary water seal to keep out moisture during construction until the sealant can be installed. Typical uses for rods and tubes are in joints between pre-cast curtain wall panels and expansion joints in long masonry walls.

Tapes and polyethylene film are another form of backup material. They act as a bond breaker, keeping a sealant from adhering to joint surfaces where bonding is not desired.

Other materials, such as resin-impregnated fiberboard, corkboard, and dense plastic foamed strips, are used in places such as isolation and control joints in concrete and masonry construction. Rods and tubes can also be used for these purposes. Some examples are shown in Figure 25.10.

Caulking and Glazing

Caulking is a procedure for sealing joints, cracks, or other small openings with caulking compound (sealer) (Figure 25.11). *Caulking compound* is a resilient mastic material. Glazing compounds are a form of sealer used to set glass in place in frames. They serve to seal out water and air and form a cushion allowing for expansion and contraction of the glass and frames.

Figure 25.11 Caulking compound is a resilient mastic material used to seal joints, cracks, and other small openings.

© Vadim Ratnikov/Shutterstock.com

Caulking and glazing compounds are commonly made from silicone, acrylic, butyl, polysulfide, or polyurethane. These five types are considered gunable, meaning their consistency allows for application with a caulk gun. Other types are referred to as "knife grade," because they are applied with a putty knife.

WATERPROOFING MEMBRANES AND COATINGS

Waterproofing involves applying a material on the surface of an assembly of materials, such as a foundation, to make it impervious to water. Many parts of a building require waterproofing, including foundation walls, roofs, exterior wood or masonry walls, and exposed structural components, including steel. Waterproofing must be able to resist the forces that tend to force water through the assembly of materials. These forces include:

Gravity, which forces water through horizontal areas, such as a roof or deck.

Hydrostatic pressure on one side of a horizontal or vertical assembly (most commonly due to subsoil water).

A difference in air pressure on one side of a horizontal or vertical assembly.

For surfaces not subjected to hydrostatic, gravity, or air pressure differences, waterproofing can be a lightduty coating, such as silicone or coal tar pitch. These provide a moisture-resistant membrane that dampproofs the assembly. Surfaces under pressure require a heavy-duty membrane, such as tar and felt, or a synthetic membrane. Waterproofing is most effective if applied on the surface directly facing the source of moisture.

Waterproofing can be accomplished by:

- Applying a built-up bituminous membrane of felt and hot or cold tar pitch.
- Applying a heavy coating, such as Portland cement plaster or a trowelable asphalt.
- Bonding an elastomeric membrane to a wall.
- Applying a thin film or coating to the exterior of the wall, such as liquid silicone or coal tar pitch.
- Adding waterproofing admixtures to concrete as it is mixed.
- Applying a dry coating that emulsifies in place, such as bentonite clay.

A summary of the most frequently used types of waterproofing is given in Table 25.3.

Waterproofing coatings and membranes are not self supporting but must be bonded to the surface to be treated. In addition, waterproofing must be able to adjust to the stresses caused by movements of the assembly and any cracks or deterioration without losing its waterproofing capabilities. It should be noted that waterproofing coatings and membranes can be damaged

Table 25.3 Types of Waterproofing

| | | Built-up Membranes | | | |
Sheet membranes	Composite membranes	Hot applied	Cold applied	Liquid membrane	Applied coating
Butyl	Elastomeric, backed	Asphalt, type I, II, III	Bitumen emulsion	Butyl	Acrylic, silicone
Ethyene propylene	Polyethylene and rubberized bitumen	Coal tar pitch, type B	Bitumen, fiberated cement	Urethane	Asphalt emulsions, cut backs
Neoprene	Polyvinyl chloride backed	Felts, saturated and coated	Felts, coated	Polychlorene (neoprene)	Cementitious with admixtures
Polyethylene	Saturated felts and bitumen coated		Bentonite clay	Polyurethane, coal tar	Epoxy, bitumen
Polyvinyl chloride			Fabric, saturated		Urethane, bitumen
			Glass fiber mesh, saturated		Bitumen, rubberized
			Cementitious membrane		

Construction **Techniques**

Waterproofing Tips

There are a number of ways to waterproof a foundation wall. The manufacturer of a system usually requires the contractor to employ a certified applicator if the manufacturer's guarantee is to be valid. Common systems include liquid membranes, sheet membranes, cementitious coating, built-up systems, and bentonite.

Liquid membranes are applied with a roller, trowel, or spray. The liquid solidifies into a rubbery coating. Different materials are available, such as polymer-modified asphalt and various polyurethane liquid membranes.

Sheet membranes tend to be self-adhering rubberized asphalt coverings, typically an assembly of multiple layers of bitumen and reinforcing materials. Some companies manufacture PVC and rubber butyl sheet membranes.

Cementitious products are mixed on site and applied with a brush. Some have an acrylic additive available that improves bonding and makes the cementitious coating more durable. One disadvantage is that these coatings will not stretch if the foundation cracks, thus opening leakage possibilities.

The widely used hot tar and felt membrane is an example of a built-up system. Alternate layers of hot tar and at least three layers of felt are bonded to a foundation. Bentonite is a clay material that expands when wet. It is available in sheets that are adhered to foundations. As groundwater penetrates the clay, it swells many times its original volume, providing a permanent seal against water penetration.

Surface Preparation   Regardless of the type of waterproofing system used, it is important to prepare a surface before application. This includes (1) drying the wall and footings; (2) removing the concrete form ties, making certain they break out inside the foundation to prevent penetration of the waterproof membrane; (3) clearing the wall of all dirt or other loose material; and (4) sweeping the wall free of dust and mud film residue. The residue left when wet mud is wiped off and left to dry on a foundation can inhibit bonding. Finally, any openings around pipes or other items that penetrate the wall must be grouted.

Safety   Waterproofing presents some hazards that must be controlled. First, possible cave-ins of soil around excavations could bury workers. Normal shoring procedures should be observed. Many materials used are flammable and solvent based, presenting a potential fire hazard. Workers should not smoke or use tools that might cause ignition. Solvent fumes can be very harmful so workers must wear respirators. Fumes are usually heavier than air and settle around a foundation in the excavated area. The solvents, asphalt, and other materials used may cause skin problems, so protective clothing, including gloves and eye protection, is required with many products. Manufacturer recommendations should be followed closely.

during installation and construction. Roof membranes can be punctured by construction traffic. Protective pads are available for installation atop roof membranes to protect areas where workers walk. Damage frequently occurs during backfilling of foundation walls. A protective material can be placed over the waterproofing to keep rocks from piercing it.

Bituminous Coatings

Hot-applied and emulsified coal tar pitch, hot-applied and cold-applied asphalt, and emulsified asphalt can be administered to a foundation by brush, roller, or spray. These are effective only for situations in which hydrostatic pressure is not a factor.

A waterproofing system that will resist hydrostatic pressure consists of alternate layers of hot-mopped coal tar pitch, asphalt, or cold-mopped emulsified asphalt over layers of mineral or glass fiber felts (applied in much the same way as for laying a built-up roof). The number of layers of felt and asphalt depends on the hydrostatic conditions. Manufacturers of these systems have established specifications for various conditions.

Another form of light-duty waterproofing is a bituminous binder with asbestos fibers forming a trowelable mix. It is hand-troweled on concrete foundations and often used to bond foamed plastic insulation boards to a foundation. Bituminous binders are only good for damp-proofing a wall not subject to hydrostatic pressure. A thinner version can be sprayed using a mastic pump. Refer to Chapter 26 for additional information.

There are a variety of solvent-based asphalt damp-proofing compounds. They are available in thick, semi-solid, and spray mastics and give damp-proofing properties to interior and exterior above- and below-grade surfaces. They are also used on metal to prevent corrosion.

Liquid Coatings

An acrylic copolymer waterproof coating is available in a variety of colors. It may be obtained with fillers and texturing aggregates that are fused onto a concrete or masonry surface. This offers waterproofing protection and an attractive finish coating. Important uses are on above-grade exterior concrete walls, columns, and spandrels. The coatings are also used on Portland cement plaster and stucco walls, giving a textured, sand-like finish in color that minimizes surface defects. They are applied by brush, roller, or spray.

Clear silicone damp-proof liquids are widely used on exterior concrete, masonry, and wood surfaces. They are not a surface coating but penetrate a material, carrying solids into its pores. They do not color a wall but do reduce efflorescence. They can be applied with a brush, roller, or spray (**Figure 25.12**).

Figure 25.12 Clear silicone dampproof liquids penetrate a material, carrying solids into its pores.

© Gary Gladstone/Getty Images

A liquid waterproofing material that is self-curing is available in a polyurethane rubber with a coal tar additive. It is applied to decks, roof slabs, and floors with a brush, roller, or squeegee. It will cover small cracks and is flexible enough to cover surfaces of irregular shapes. It can be covered with a concrete slab, hardboard, or mineral surface roll roofing (**Figure 25.13**).

Figure 25.13 Self-curing polyurethane rubber forms a seamless waterproof membrane.

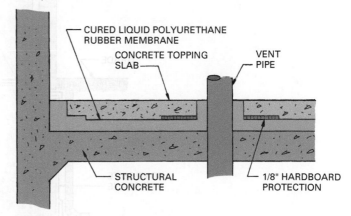

CURED LIQUID POLYURETHANE RUBBER MEMBRANE

CONCRETE TOPPING SLAB

VENT PIPE

STRUCTURAL CONCRETE

1/8" HARDBOARD PROTECTION

Another form of liquid waterproofing uses a silicate liquid gel that reacts with soluble calciums in concrete to form a glass gel in the microscopic pores. The gel will penetrate 1.5 in. to 2.0 in. (38 mm to 50 mm), providing a strong moisture-proof barrier that also reduces efflorescence.

Synthetic Sheet Membranes

Synthetic membranes are available made from neoprene, polyvinyl chloride, polyethylene, butyl, and ethylene propylene. These sheet materials are bonded to foundation walls using adhesives recommended by the system manufacturer. Since they are flexible membranes, they tend to adjust to settling and compaction of the soil and are not likely to rupture. They are available as single materials or composite membranes. Following are examples of a few of these products.

Composite Membranes   Composite membranes are sheet products made by laminating two or more waterproofing materials. One frequently used composite membrane is made of a rubberized asphalt layer with a polyethylene film bonded to the outer surface (Figure 25.14). It remains stable below ground and below water. When the sheets are lapped, the rubberized asphalt back bonds to the polyethylene face of the sheet beside it. Under most conditions it will bridge gaps up to ¼ in. (6 mm). It is available in rolls 48 in. (121 mm) wide and 60 ft. (18.3 m) long. Since the back surface is very tacky, it is covered with a peel-off paper covering (Figure 25.15).

Another composite membrane is made by laminating a chlorinated polyethylene (CPE) film to a nonwoven polyester fabric. This composite is bonded with water-based acrylic adhesives, avoiding problems that

Figure 25.14 Composite membranes are sheet products made by laminating two or more waterproofing materials.

© Michael N. Paras/Getty Images

occur from fumes of solvent-based adhesives. It can be installed on surfaces that are damp at the time of installation.

Sheet Membranes   Sheet membranes are waterproofing products composed basically of one major waterproofing material. One such product is made from polyvinyl chloride alloyed with high-density polymer resins. It is not affected by aging, mildew, or corrosion and remains flexible at low temperatures with good abrasion and tear resistance. It is bonded, and laps are sealed with a special adhesive. Some forms of PVC membrane can be

Figure 25.15 Composite membranes provide a high-quality waterproofing.

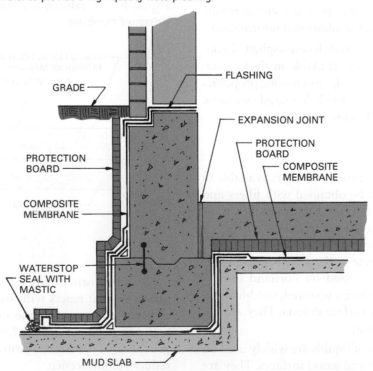

GRADE

FLASHING

EXPANSION JOINT

PROTECTION BOARD

PROTECTION BOARD

COMPOSITE MEMBRANE

COMPOSITE MEMBRANE

WATERSTOP SEAL WITH MASTIC

MUD SLAB

formulated to be resistant to gasoline and oil. The seams can be welded with hot air or bonded with an adhesive specified for that purpose.

Another type of sheet membrane is a chlorinated polyethylene (CPE) product. It is available in a range of thicknesses with and without integral reinforcement. It is suitable for use above and below grade and on horizontal and vertical surfaces. Laps can be joined by chemical or thermal fusion. It will bridge cracks up to ¼ in. (6 mm).

Cement-Based Waterproofing

A number of cement-based heavy-duty waterproof coatings are available. These have carefully graded aggregates that produce a high-density and high-strength waterproof coating. Some types are applied with a trowel, while others require special procedures. They are used to waterproof concrete masonry and stone in interior and exterior locations. They are employed on water reservoirs, swimming pools, basements, parapet walls, and other heavily exposed surfaces.

Lead Waterproofing

Lead waterproofing sheets are frequently used for projects in which waterproofing must be of uncompromised security. Examples include areas of grass, fountains, and reflecting pools that may be built over underground facilities, such as parking or retail, or on the roof of a building. The gauge of the lead sheets varies for different purposes but must be at least 6 lb. lead (³⁄₃₂ in. or 2.3 mm) thick to allow for the burning (welding) necessary to join sheets. If lead comes in contact with cementitious materials, it must be protected with a bituminous coating. Lead wool is also used to waterproof joints in above-ground installations.

Bentonite Clay Waterproofing

Bentonite is a clay formed from decomposed volcanic ash, with a high content of the mineral montmorillonite. It can absorb large amounts of water, causing it to swell many times its original volume and form a waterproof barrier. The dried, finely ground particles are usually applied as a waterproofing membrane in three ways.

Bentonite Panels Panels, usually 4 × 4 ft. (1,200 × 1,200 mm), consist of a biodegradable paper covering over bentonite clay particles. One type is ³⁄₁₆ in. (2.5 mm) thick with a corrugated kraft board core. It is used on vertical walls and under structural slabs. Another type is ⅝ in. (16 mm) thick, composed of layers of corrugated

kraft board with the center layer holding the bentonite clay. The hollow outer layers allow space for the expansion of the clay and reduce upward pressure against a thin nonstructural slab.

The panels are ready for installation when received on site and can be applied at all temperatures and over moist substrates. They can be nailed to green concrete walls. The panels are overlapped 1 in. (38 mm). A hydrated sodium bentonite gel is used to fill gaps around pipes and fittings. If the backfill contains rocks that may pierce the panels, they should be covered with a protective material (Figure 25.16).

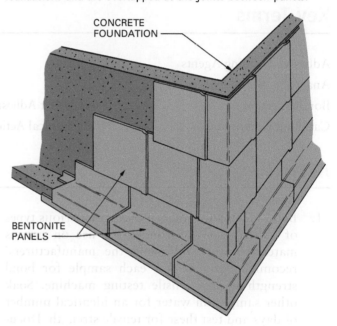

Figure 25.16 Bentonite panels can be nailed to concrete foundations and are overlapped at the joints between panels.

Sprayed Bentonite Bentonite clay mixed with a modified asphalt that serves as an adhesive to bond the clay to a surface is applied by spraying. A ⅜ in. (10 mm) membrane is built up for normal applications, but thicker layers can be used for situations in which hydrostatic pressure is severe.

Bentonite–Sand Mixture Bentonite clay mixed with sand is used to produce a waterproof barrier under concrete slabs. The mix is carefully measured and spread over the area to be covered with concrete. It is then covered with a polyethylene sheet to protect it from the moisture in the concrete. The reinforcing is placed on top and the slab is poured. If too much bentonite is used, the forces of expansion can crack the slab.

Review Questions

1. What are the two ways bonding agents hold materials together?
2. In what ways do various bonding agents cure?
3. What are the three major classifications of bonding agents?
4. What are the two classes of adhesives?
5. What is the source of materials for making glues?
6. From what materials are cements made?
7. What is the difference between a sealer and a sealant?
8. What is a working joint?
9. What are the two basic methods of protecting exterior joints?
10. What are the two types of sealants?
11. What are the performance levels of sealants?
12. What purpose does a sealant backup rod serve?
13. Why are tapes sometimes used with sealants in joints?
14. What forces tend to force water through an assembly of materials?
15. What is bentonite?

Key Terms

Adhesive Bonding Agents

Anaerobic Bonding Agents

Bonding Agent

Caulking Compound

Cement

Glue

Hot-Melt Adhesives

Mechanical Action

Sealant

Sealers

Specific Adhesion

Activities

1. Bond identical wood samples with various types of adhesives recommended for use on porous materials. Cure following the manufacturers' recommendations. Test each sample for bond strength with a tensile testing machine. Soak other samples in water for an identical number of days and test these for tensile strength. Document your findings.

2. Bond identical metal samples with various types of adhesives. After the recommended curing time, test for strength of bond with a tensile testing machine.

3. Visit several commercial buildings and note how various expansion joints and other areas needing to be sealed are protected. Observe any evidence of sealant failure. What would you recommend be done to repair any failures?

Additional Resources

Joint Sealers; Structural Sealant Glazing Systems and Fenestration Sealants Guide Manual, American Architectural Manufacturers Association, Schaumburg, IL.

The NRCA Roofing and Waterproofing Manual, National Roofing Contractors Association, Rosemont, IL.

A Professional's Handbook on Grouting, Concrete Repair, and Waterproofing, available from Five Star Products, 425 Stillson Road, Fairfield, CT 06324.

Specifications for Weatherstrips and Sealants, American Architectural Manufacturers Association, Schaumburg, IL.

Other resources include:

Chemprobe Coating Systems, TSE Inc., PO Box 23559, Columbia, SC 29224.

Publications from the Gorilla Group, Santa Barbara, CA 93103, 800-966-3458.

Publications from Architectural Products Outerwear, LLC, PO Box 347, Wood-Ridge, NJ 07075.

See Appendix C for addresses of professional and trade organizations and other sources of technical information.

Bituminous Materials

Upon completion of this chapter, the student should be able to:

- Describe the properties of bitumen and how these properties influence decisions as to how bituminous materials might be used.

- Interpret the results of laboratory tests and use these for decision making.
- Be aware of the array of bituminous products available and their applications.

Build Your Knowledge

For further study on these materials and methods, please refer to:

Chapter 4   The Building Site

                Topic: Paving

Chapter 25   Bonding Agents, Sealers, and Sealants

                Topics: Waterproofing, Membranes, and Coatings

Chapter 27   Roofing Systems

Bitumen is a mixture of organic liquids made of complex hydrocarbons that occur naturally or are heat-produced from materials like coal and wood (Figure 26.1). Highly viscous, black, and sticky bitumens exist in gaseous, liquid, semisolid, or solid states.

Bitumen can be made from non-petroleum-based renewable resources, such as sugar, molasses and rice, corn and potato starches. Bitumen can also be made from waste material through fractional distillation of used motor oils, which are otherwise disposed of by burning or dumping into landfills. Non-petroleum-based bitumen binders can be made lighter in color. Roads made with lighter-colored pitch absorb less heat from solar radiation and stay cooler than darker surfaces, reducing their contribution to the urban heat island effect.

Asphalt, tar, and coal tar pitch are the most commonly used bituminous materials in construction. Asphalt is found

Figure 26.1 Natural deposits of bitumen are formed from the remains of ancient, microscopic algae and other once-living organisms.

in natural deposits or produced from petroleum. Tar is produced by the distillation of wood and coal. Fractional distillation of tar produces coal tar pitch.

PROPERTIES OF BITUMENS

Perhaps the most significant property of bitumens is their excellent water resistance. A variety of different bitumen products are produced as waterproof materials. Bitumens bond well to dry solid surfaces because they exist in the semi-fluid state needed by adhesives to bond. They will not bond to wet surfaces.

Bitumens are flammable and will ignite when heated to their flashpoint. This influences the temperatures used to heat bituminous materials for various applications.

Another important property of bitumens is their softening point. Although this varies with the composition of a product, it is crucial to consider in relation to temperatures conditions in various applications. For example, the softening point for roofing asphalts ranges from 200 to about 220°F (95 to 104°C), while coating-grade asphalts such as those used for waterproofing have a softening point of 50°F to 55°F (11°C to 13°C).

Bitumens exhibit cold-flow properties; they tend to flow, spread, or lose their shape. This is more pronounced at higher temperatures and offers the advantage of causing a check or crack on a roof membrane to seal when the sun heats up the asphalt.

The *viscosity* of asphalt is an important property in classifying bitumens for different applications. Viscosity describes the ability of a material to stay in place when subjected to heat. Asphalts have good viscosity properties up to 135°F (58°C), with some able to withstand temperatures up to 275°F (136°C).

Asphalts rate high in their ductility properties. Their molecules remain constant even when they are deformed by heat and pressure. Asphalts can expand and contract while remaining bonded to materials upon which they are placed.

ASPHALT

Asphalt is a dark-brown to black cementitious material in semisolid or solid form, consisting of bitumen found in deposits of natural asphalt. A manufactured asphalt is produced from residues from the distillation of petroleum. Petroleum provides the raw material for the manufacturing of most asphalt used today (Figure 26.2).

Asphalt is used for many construction components, including roofing, siding, and paving materials. It is added to paints, acid- and alkali-resistant coatings, and damp and waterproofing. Asphalt is employed to coat organic fiber, fibrous plastic panels, and papers of various types. It can be found in adhesives, cementitious materials, and in the manufacture of drain and sewer pipe. Some types, such as waterproof coatings, are applied on site, while others, such as coating for fibrous panels, are applied in a factory. Table 26.1 lists the major types of bitumens and their uses.

Table 26.2 and Table 26.3 detail the types of asphalt used for coatings, cements, paints, and adhesives.

A grade classification is used to describe the liquefaction of the asphalt. Grade 0 is the thinnest, and grade 5, which is like a paste, has the thickest consistency. Steam-refined or oxidized asphalt having a petroleum solvent is classified as slow curing (SC), medium curing (MC), and rapid curing (RC) (Table 26.4). Asphalt emulsified in chemically treated water is graded as slow setting (SS), medium setting (MS), and rapid setting (RS).

Roofing asphalt is available in four types:

Type I – Dead level

Type II – Flat

Type III – Steep

Type IV – Special steep

Physical characteristics for these four types of roofing asphalts are given in Table 26.5.

Asphalt cements are binders used to produce high-quality asphalt pavements. They are highly viscous and made in several grades based on consistency. Asphalt cements are tested for viscosity at 140°F (60°C) and are semisolid at normal ambient temperatures.

Asphalt cements are graded from the softest, AR 1000, to the hardest, AR 16000. AR 4000 is a general-purpose grade, and AR 8000 is used in hot climates. The grade chosen depends on climate and the quality of the aggregate.

Asphalt cements are blended with aggregates graded into a range of sizes. Typical aggregates include crushed stone, gravel, and sand. Aggregates compose about 90 percent of a paving mix's weight. The aggregates give the mix its strength, and the asphalt acts as a binder. In some cases, tars are added because they increase resistance to damage from gasoline spilled on paved surfaces. Asphalt cement paving is used for paving roads, drives, and parking lots (Figure 26.3).

Cutback asphalt is a broad classification of residual asphalt materials left after petroleum has been processed to produce gasoline, kerosene, diesel oil, and lubricating oils. The residual asphalt is then blended with various solvents to produce cutback asphalt in three classifications: rapid curing (RC), medium curing (MC), and slow curing (SC). Cutback asphalt is often mixed with granular soil to stabilize a roadbed before paving or to serve as a binder for a finished road surface for light traffic.

An emulsion group of asphalts consists of emulsified asphalt cement mixed with water and is used in road construction. The emulsion is made by adding heated asphaltic cement (in liquid form) to water mixed with an emulsifying agent, such as soap or bentonite clay.

Figure 26.2 The process for producing asphalt products from crude oils.

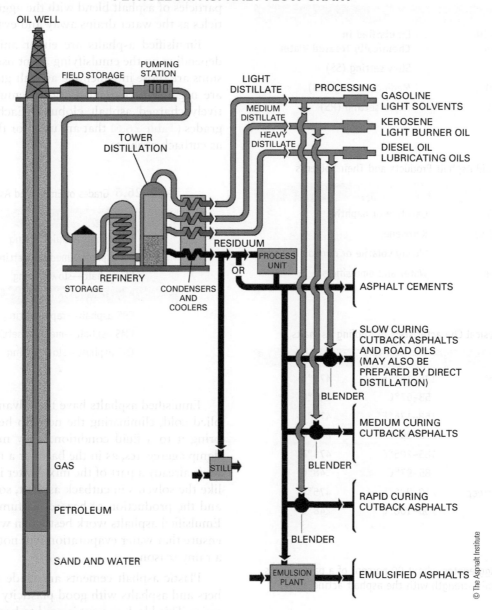

PETROLEUM ASPHALT FLOW CHART

© The Asphalt Institute

Table 26.1 Major Classes and Uses of Bituminous Mixtures

Class	Use
Liquid	Alleviate dust (spraying)
	Waterproofing
	Impregnation or saturation of materials
Medium consistency	Sealing compound
	Road surface binder
	Adhesive compound for roofing
	Expansion joint caulking
Solid	Electrical insulation compound
	Molded products

Table 26.2 Grades of Asphalt Used for Paints, Coatings, Adhesives, and Cements

Grades of Liquification
0 (Thinnest)
1
2
3
4
5 (Thickest)

Table 26.3 Types of Asphalt Used for Paints, Coatings, Adhesives, and Cements

Type	
Steam Refined with Petroleum Solvent	**Emulsified in Chemically Treated Water**
Slow curing (SC)	Slow setting (SS)
Medium curing (MC)	Medium setting (MS)
Rapid curing (RC)	Rapid setting (RS)

Table 26.4 Liquid Asphalt Products and Their Solvents

Classification	Solvent
Rapid curing (RC)	Gasoline or naphtha
Medium curing (MC)	Kerosene
Slow curing (SC)	Slowly volatile or nonvolatile oils
Asphalt emulsions	Water and emulsifiers

Table 26.5 Physical Characteristics of Roofing Asphalts

Type	Softening Point	Flashpoint
Type I, Dead level	135–151°F	475°F
	58–67°C	248°C
Type II, Flat	158–176°F	475°F
	70–80°C	248°C
Type III, Steep	185–205°F	475°F
	85–97°C	248°C
Type IV, Special steep	210–225°F	475°F
	100–108°C	248°C

Figure 26.3 Aggregates, about 90 percent of a paving mix's weight, give the mix strength with the asphalt acting as a binder.

© bogdanhoda/Shutterstock.com

A protein stabilization agent is also added to prevent particles from blending within the mix. The suspended particles of asphalt blend with the aggregate or soil particles as the water drains away and evaporates.

Emulsified asphalts are either anionic or cationic, depending on the emulsifying agent used. Anionic emulsions are those in which the asphalt globules in the mix are negatively charged. Cationic emulsions have positively charged asphalt globules. Each comes in three grades (Table 26.6) that are used for the same purposes as cutback asphalts.

Table 26.6 Grades of Emulsified Asphalts

Anionic
RS asphalt—rapid setting
MS asphalt—medium setting
SS asphalt—slow setting
Cationic
CRS asphalt—rapid setting
CMS asphalt—medium setting
CSS asphalt—slow setting

Emulsified asphalts have the advantage of being applied cold, eliminating the need to heat the asphalt to bring it to a fluid condition. They may be applied to damp aggregates, as in the base for a road, because water is already a part of the mix. Water is not volatile, unlike the solvents in cutback asphalt, so danger from fire and the production of hazardous fumes is not present. Emulsified asphalts work best when weather conditions ensure that water evaporation will not occur, such as in a rainy season.

Plastic asphalt cements are made using asbestos fibers and asphalts with good plasticity and elastic properties. This black cement is used to bond flashing and to make roof repairs. It does not flow at high temperatures or become brittle at cold temperatures, allowing for expansion and contraction of materials without breaking.

Quick-setting asphalt cement is much like plastic asphalt cement but has greater adhesive properties and sets up rapidly. It is used to cement the free tabs on shingles and the laps between layers of roll roofing.

Asphalt roofing tape is a porous fabric strip saturated with asphalt. It is available in rolls 4 to 36 in. (100 to 914 mm) wide and 50 yd. (45.5 m) long. It is used with plastic asphalt cements to patch holes in roofs, to patch seams, and to seal flashing.

Asphalt Paving

Asphalt is widely used for paving because it is a cementitious substance and will bind other ingredients, such as aggregates. It withstands exposure to hot and cold weather, enjoys resistance to salt, acids, and alkalis, and experiences little cracking due to its flexible, plastic, and ductile nature.

A bituminous binder is used to hold and protect the mineral aggregate used on paving surfaces. Asphalt cements are used as binders for quality paving. Cutback asphalt is used to stabilize a roadbed before paving and as a binder on the surfaces of light duty roads. Emulsion asphalts may be applied cold to damp aggregates. Asphalt paving mix designs are tested by a variety of methods. The purpose of these tests is to produce a blend of asphalt and aggregates of various sizes that will withstand traffic and still be easy to place.

Laboratory Tests of Asphalt

Since asphalts are used where temperatures may vary considerably, standard tests ascertain their flow properties. Various testing procedures have been developed by the Asphalt Institute, the American Association of State Highway and Transportation Officials, and the American Society for Testing and Materials.

Viscosity Test   Viscosity tests determine the flow properties of asphalts at application and service temperatures. One testing method uses a capillary tube viscometer. The viscometer is mounted in a constant temperature bath, and preheated asphalt is poured into the large side of the viscometer until it reaches a fill line. The viscometer is left in the bath for a prescribed length of time. A partial vacuum is pulled on the small tube, and the time it takes the asphalt to move between two timing marks is recorded. This is an indication of flow or viscosity.

Flashpoint Test   A flashpoint test determines the temperature to which asphalt may be heated without the danger of an instantaneous flash occurring when exposed to an open flame. The flashpoint temperature is well below the temperature at which the asphalt will burn. The burn temperature is called the "fire point."

The test is made by placing a sample in a brass cup and heating it at a prescribed rate. A small flame is played over the surface periodically. The temperature at which the vapor produces a flash is the flashpoint temperature (Figure 26.4).

Figure 26.4 This apparatus is used to find the flashpoint of asphalt samples.

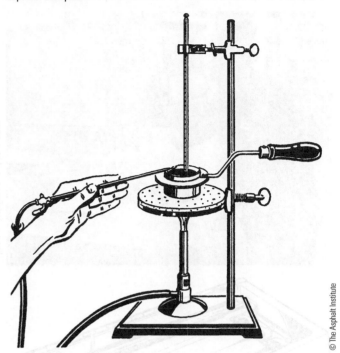

© The Asphalt Institute

Thin Film Oven Test   This test subjects a sample of asphalt to hardening conditions similar to those that occur in a hot-mix plant operation. A 50 ml asphalt sample is placed in a cylindrical flat bottom pan 5.5 in. (140 mm) in diameter and ⅜ in. (9.5 mm) deep. An asphalt layer about in. (3 mm) deep is placed in the pan. This is heated at 325°F (163°C) in an oven with a revolving shelf for five hours. The sample is then subjected to a penetration test.

Ductility Test   The *ductility* test is used to determine how much an asphalt sample will stretch at various temperatures below its softening point. A standard briquette is molded, brought to the specified test temperature, and pulled at a specified rate until the sample completely separates. The elongation in centimeters, at which the final thread of asphalt breaks, is used to indicate ductility (Figure 26.5).

Solubility Test   This test is used to ascertain the purity of asphalt cement. A predetermined amount of asphalt cement is dissolved in trichloroethylene. The soluble portion represents the active cementing constituents. Inert matter is filtered out and measured. The result is given as a percent of the soluble content.

Specific Gravity Test   The specific gravity of a material is the ratio of the weight of a given volume of the material to the weight of an equal volume of water. Water has a specific gravity of 1. The specific gravity of a material will

Figure 26.5 The ductility test is made to find out how much an asphalt sample will stretch at various temperatures.

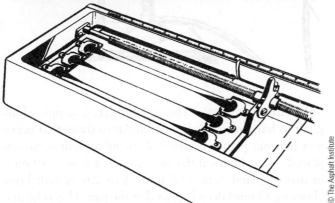

vary with the temperature because the volume of a material changes with temperature fluctuations. Asphalts with a specific gravity of 1.1 are 1.1 times as heavy as water.

Softening Point Test   Various grades of asphalt soften at different temperatures. The softening point is found by the ring and ball test. The heated asphalt is poured into a brass ring. The sample is suspended in a water bath, and a steel ball of specified weight and diameter is placed in the center of the sample. The bath is heated at a controlled rate until the ball reaches the bottom of the glass bath container. The temperature of the water at this point is the softening point.

Distillation Test   The distillation test is used to find the proportions of asphalt and diluent present in a sample. It is also used to find how much diluent distills off the sample at various temperatures. This information is useful to ascertain an asphalt's evaporation characteristics, which influence the rate at which road construction asphalt will cure after it is applied.

OTHER PRODUCTS MADE WITH BITUMINOUS MATERIALS

In addition to asphalt, a wide range of products are manufactured using bituminous materials as their major components. Most common among these are coal tar pitch, felts, stabilizers, and surfacing materials.

Coal Tar Pitch

Coal tar pitch is a dark brown to black hydrocarbon obtained through the distillation of coke-oven tar. It is available in several grades and is used as the basis for a number of paints, roofing products, and waterproofing materials. It has a softening point near 150°F (65°C).

Coal tar enamel is made from coal tar pitch with added mineral fillers. It is used to protect pipe in pipeline work. Cold-applied coal tar products have a solvent added to liquefy them. Hot-applied coal tar coatings give better protection than cold coal tar coatings.

Felts

Felt is a sheet material made from the cellulose fibers of organic materials, such as wood, paper, rags, glass fibers, and asbestos.

Saturated felts, sometimes called tar paper, are made with an organic mat saturated with coal tar pitch or asphalt and coated with a layer of thin asphalt (Figure 26.6). Tar paper is used as an underlayment for shingles, as

Figure 26.6 Tar paper is used as an underlayment for shingles, sheathing paper, and as laminations in built-up roof construction.

sheathing paper, and as laminations in built-up roof construction. Tar paper is also used to produce roll roofing and shingles.

Saturated felt is available in three weights: Types 15, 20, and 30, according to the weight per square of felt. For example, Type 15 weighs 15 pounds per square (100 sq. ft.). Saturated felt is available in rolls 36 in. (914 mm) wide and up to 144 ft. (44 m) long.

Ice and Water Shield

Ice and water shield is a roofing membrane composed of two waterproofing materials bonded into one layer. Comprised of a rubberized asphalt adhesive backed by a layer of polyethylene, it comes in 36 in. (914 mm) × 75 ft. (22.8 m) rolls. The rubberized asphalt surface is backed by a release paper to protect the sticky side. The material is used as waterproofing in cavity walls, and for trouble spots on roofs, such as along eaves, in valleys, and in other areas where leaks are more likely.

Fiberglass Sheet Material

Fiberglass mats can be impregnated with asphalt but are not "saturated" because the glass fibers will not absorb the asphalt. The asphalt forms a coating on the surface and fills the spaces between the fibers. Fiberglass mats are available in rolls 36 in. (914 mm) wide and 108 ft. (11 m) long.

Fireproofing Paper

Fireproofing paper is made using asbestos fibers either in a pressed mat-like felt or in woven sheets. The various sheet products are used as underlayment for finished roofing materials, as vapor barriers in walls and floors, and for other similar applications. They should not be exposed to the weather because coal tar pitch oxidizes rapidly when subjected to the sun's ultraviolet rays.

Waterproof Coatings

Asphalt waterproofing is used on masonry walls above and below grade. Below grade, it is used to resist the pressure of subsurface water and prevent it from passing through the foundation. As it is not subjected to high temperatures below grade, asphalts with lower softening points can be used. Above grade, it resists passage of water through walls or roof decking (**Figure 26.7**). Where it will be exposed to sunlight, asphalts with a higher softening point should be utilized.

Figure 26.7 Bituminous waterproofing coatings are applied in one or more hot or cold mopped coats.

© Vadim Ratnikov/Shutterstock.com

A bituminous waterproofing coating is applied in one or more coats mopped on either hot or cold. Cold-applied coats can be reinforced by the addition of glass, plastic, or asbestos fibers. Cutbacks and emulsions are used extensively for this purpose. They may be covered with a plastic or felt membrane.

A damp-proofing board product is made with an asphalt core covered on both sides by layers of asphalt-impregnated paper or felt treated with a weather-resistant coating. Additional information on waterproof coatings can be found in Chapter 25.

BITUMINOUS ROOF COVERINGS

The following topics discuss the various types of bituminous roof-covering materials in common use. It should be noted that coal tar pitch and asphalt are not compatible and should not be used where they will come in contact with each other.

Roll Roofing

Roll roofing uses either organic felt or fiberglass mats as a base material. A viscous bituminous coating is applied to this base, forming the exposed surface. Roll roofing is made in four types: smooth surfaced, mineral surfaced, mineral surfaced selvage edged, and pattern edged. Smooth-surfaced roll roofing has both sides covered with a fine talc or mica to keep the surfaces from sticking as it is made into rolls.

Mineral-surfaced roll roofing has mineral granules in a wide range of colors rolled into the surface, producing a surface that is attractive and protects the bitumen from the sun's ultraviolet rays. The minerals also increase the fire resistance of the product. Mineral-surfaced roll roofing is available in rolls 36 in. (915 mm) wide and from 36 ft. to 72 ft. (8 m to 22 m) long. Smooth-surfaced rolls come in weights of 50 lb. to 65 lb. per square, mineral-surfaced in rolls of 90 lb. per square, and fiberglass-reinforced mineral fiber in rolls of 75 lb. per square.

Mineral-surfaced selvage-edge roofing is of the same construction as mineral-surfaced roll roofing except that only 17 in. of the 36 in. wide surface is covered with granules. The 19 in. selvage edge is used for lapping with an adjoining layer, forming a two-ply covering.

Pattern-edged roll roofing is a mineral-surfaced product that has a 4 in. uncoated band in its center. The roll is semi-cut along this strip to form two 18 in. wide patterned roofing strips that are thin lapped 2 in. over the layer below. Roll roofing can be installed in a single or double thickness. The double thickness provides increased protection over a longer period of time.

Hot Bitumen Built-Up Roof Membranes

A built-up roof consists of alternate plies of organic or fiberglass roofing felt with a hot bitumen coating mopped over each layer (Figure 26.8). The design of the roof varies by situation but generally consists of three or more layers of felt with a bitumen layer over each and a bitumen topcoat with aggregate rolled on top.

Felts provide the needed reinforcement to keep bitumens in each layer from alligatoring. "Alligatoring" refers to surface cracking caused by oxidation and shrinkage stresses, which can result in a repetitive mounding of the asphalt surface similar to an alligator's hide.

The aggregate surface forming the top of the roof is typically derived from gravel, marble chips, or slag. Gravel and marble chips are usually applied at 400 pounds per 100 sq. ft. and slag at 300 pounds per 100 sq. ft. Conventional built-up roofs can be designed for slopes up to 3 in. (76 mm) per 12 in. (305 mm). Slopes above in. (12 mm) per 12 in. (305 mm) usually require the use of steep asphalt.

Modified Asphalt Roofing Systems

Modified asphalt roll roofing is composed of polymer-modified bitumen reinforced with one or more plies of fabric, such as polyester glass fiber. These membranes are of uniform thickness and have consistent physical properties throughout the membrane area.

A variety of modifiers and types of reinforcing plies are designed for use on almost every type of construction assembly, including new roofing, re-roofing, domes, and

Figure 26.8 Hot bitumen built-up roof membranes consist of alternate plies of roofing felt with a hot bitumen coating mopped over each layer.

© mikeledray/Shutterstock.com

spires. Modified membranes are also used below grade for waterproofing canals, water reservoirs, and landfills.

Most modified bitumen membranes are made using either styrene-butadiene-styrene (SBS) or atactic poly-propylene (APP). APP-modified membranes are generally applied using a propane torch to heat and soften the underside of the membrane. This surface becomes a molten adhesive that is placed on the substrate, rolled for adhesion, and bonds when it cools.

SBS modifies the bitumen by forming a polymer lattice within the bitumen. When this polymer lattice cools, the membrane acts like rubber. SBS membranes are more flexible than APP membranes and are used where flexibility is needed, such as when the substrate may be subject to movement or deflection. SBS-modified membranes are either mopped in hot asphalt, self-adhered, or adhered with cold-process adhesives. Some types can have their joints heat-welded. Some have factory-applied mineral aggregate surfaces to protect them from ultraviolet damage. Those without this covering need some form of ultraviolet protective coating.

Cold-Applied Asphalt Roofing Systems

Cold-applied systems use some form of coated base sheet, fabric, or similar reinforcement over which the principal waterproofing agent, which is a liquid, is applied at ambient temperatures. The selection of the cold-process application depends on the levels of maintenance, repair, and service expected, and on the compatibility between the cold-applied materials and any existing substrate.

The reinforcing felts and fabrics include organic-coated base sheets, organic mineral-surfaced cap sheets, organic felt sheets, fiberglass ply sheets, fiberglass base sheets, fiberglass mineral-surfaced cap sheets, single-ply smooth surfaced sheets, single-ply mineral-surfaced sheets, fiberglass fabrics, polyester fabrics, cotton fabric, and jute burlap fabric. The base sheet, cap sheet, and single-ply roofing are bonded with cold-applied adhesives and cold-applied surface coatings to form a roof membrane.

The coatings and adhesives are designed to be brushed or sprayed at normal room temperatures. These include filled and non-filled asphalt cutbacks, asphalt emulsions, coal tar coatings, and aluminum pigmented asphalt. Toppings include gravel and aluminum chips placed in the topcoat while it is still wet. They block ultraviolet light, can reflect heat, and are decorative.

Review Questions

1. What will prevent bitumen from adhering to a concrete wall?

2. What is the point at which bitumen will burn?

3. What property of bitumen allows it to spread after it is in place?

4. What binder is used with aggregate to produce quality asphalt pavements?

5. From what source are most asphalts obtained?

6. How do cutback asphalts differ from asphalt emulsions?

7. What are the two major uses for cutback asphalts?

8. What can happen that will cause asphalt emulsions on a roadbed not to harden?

9. What asphalt product is used to make roof repairs?

10. What are the big differences between saturated felts and fiberglass felts?

11. Why are fiberglass felt mats not saturated with asphalt?

12. What types of fiber are used in making fireproofing paper?

13. What are the fire rating classes used to rate asphalt roofing materials?

14. Which fire rating class indicates the highest fire protection?

15. How does a built-up roof differ from one covered with roll roofing?

16. What are the various materials used to make fiberglass shingles?

Key Terms

Asphalt	Coal Tar Pitch	Felt
Bitumen	Ductility	Viscosity

Activities

1. Collect samples of as many bituminous products as possible.

2. Visit construction sites when the finished roofing is being applied and observe the installation techniques and safety protection used.

3. Visit a road paving project and observe the procedures for preparing, delivering, and laying the asphalt final coat. Try to find out what was done to prepare the base. If possible, find the local or state requirements for street and parking area paving.

Additional Resources

CertainTeed Shingle Applications Manual, CertainTeed Corporation, Roofing Products Group, PO Box 860, Valley Forge, PA 19482.

The NRCA Roofing and Waterproofing Manual, National Roofing Contractors Association, Rosemont, IL 60018.

Residential Asphalt Roofing Manual, Asphalt Roofing Manufacturers Association, Washington, DC 20005.

Spence, W. P., *Roofing Materials and Installation*, Sterling Publishing Co., 387 Park Avenue South, New York, NY 10016-8810.

Other resources include:

CD-ROMs: *Installation Tips & Techniques and Avoiding Common Roof Installation Mistakes*, GAF Materials Corporation, 1361 Alps Road, Wayne, NJ 07470.

See Appendix C for addresses of professional and trade organizations and other sources of technical information

Roofing Systems

Upon completion of this chapter, the student should be able to:

- Understand the vast array of roofing systems available.

- Select the proper roofing system for various applications.
- Understand the proper installation of roofing systems.

Build Your Knowledge

For further study on these materials and methods, please refer to:

Chapter 11 Clay Brick and Tile

Chapter 16 Nonferrous Metals

Chapter 19 Products Manufactured from Wood

Topic: Wood Shingles and Shakes

Chapter 24 Thermal Insulation and Vapor Barriers

Chapter 26 Bituminous Materials

Topic: Roof Coverings

Proper roof design and construction is critical to a building's ability to provide a safe and comfortable enclosure. The roof establishes the first line of defense against the onslaught of rain, snow, and sun. It must insulate the interior from temperature extremes, and protect against water vapor condensation. Because the roof surface is continually exposed, the roof experiences the largest temperature differential over the course of a day. Roofing materials are particularly susceptible to expansion and contraction caused by direct sunlight. In cold climates, where snow and ice accumulate, roofing systems must also be able to accommodate cycles of freezing and thawing. Atmospheric conditions can negatively affect certain roofing materials. Strong winds can damage or dislodge shingles. Asphalt roofing may be damaged under high temperatures. Metal roofing other than those that are corrosion resistant, like copper, can be corroded by salt or industrial exhausts.

The roof may be part of a new building design, or may involve the replacement or re-cover of an existing system. Because roofing is subject to decay over time, approximately 75 percent of roofing construction is in reroofing. Manufacturers of a variety of deck, insulation, and finished roofing materials provide technical information and installation instructions.

Roofing materials are tabulated by the square. A square is the amount of roofing required to cover 100 square feet of roof area. A primary factor in the type of roofing system chosen is the weight of a material. The heavier a material, the greater its dead load, and the larger will be the structure required to support it. Copper roofing, for instance, weighs about 1 pound per foot, whereas clay tile may weigh over 10 pounds per foot.

Roof systems are generally divided into two categories: low-slope and steep-slope. Low-slope roofs are nearly flat and provide slow rainwater runoff. Steep-slope roofs permit the rapid runoff of rain, reducing the likelihood of penetration through their water-resistant surfaces. This chapter outlines the details, components, and materials commonly used for residential and commercial construction.

BUILDING CODES

Building codes govern the materials, design, construction, and quality of a roof structure and covering. This includes the ability to resist rain and wind, durability specifications, compatibility of materials, physical characteristics, and fire protection. Because of its large surface area, roofing plays an important role in fire protection. Roofing systems are rated as entire systems, including the roof deck, method of attachment to the deck (fasteners, hot bitumen, cold adhesives), vapor retarder, thermal insulation, roof membrane, and surfacing. ASTM Standard *E108 Standard Test Methods for Fire Tests of Roof Coverings* specifies four classifications of roof covering.

Class A roof coverings are effective against severe fire exposure. Roof coverings of this class are not readily flammable and do not spread fire, afford a fairly high degree of fire protection to the roof deck, and possess no flying brand hazard. They include clay tile, concrete, slate, some fiberglass asphalt shingles, and other materials that have been certified by an approved testing agency (Figure 27.1). Class A coverings may be used on buildings of all types.

Figure 27.1 Slate shingles are an example of a Class A fire-resistant roof covering.

© Chris Bradshaw/Shutterstock.com

Class B roof coverings are effective against moderate fire exposure. Class B coverings afford the same protection as Class A, but may require occasional repairs in order to maintain their fire-resistant properties. They include metal sheets and shingles, some composition shingles, and other materials certified by an approved testing agency.

Class C roof coverings are effective against light fire exposure. While not readily flammable, they may require sporadic repairs or renewals in order to maintain their fire-resistant properties. They include materials certified as Type C by an approved testing agency. Non-classified roof coverings are not permitted on any

buildings covered by most codes. In some cases, these may be approved on some types of storage buildings.

Local weather conditions are also addressed in the building codes as strong winds can damage or dislodge shingles and sheet materials. Underwriters Laboratories lists wind-rated systems in their *Roofing Materials and Systems Directory* as Class 30, 60, or 90. Factory Mutual Research Corporation lists additional wind-rated roofing systems in their *Approval Guide* with ratings ranging from 60 to 210 psf.

ROOFING COMPONENTS

The following discussion lists some of the principal considerations in the design and detailing of a roof system.

Roof Decks

A properly functioning roof depends on a structurally sound deck that is compatible with the roofing system. The roof deck as well as the structure below must carry the dead loads of the structure, roofing, and any ballast materials. Additional live loads are generated by wind causing uplift, snow loads, and rainwater. The deck must be able to support potential ponding of water in the case of primary drains becoming clogged. Large roof decks must be provided with construction joints that allow for the expansion and contraction of the roof membrane.

A variety of materials are frequently used as roof decking. Principle deck materials for membrane roofing include steel, cast-in-place concrete, pre-cast concrete, gypsum, wood, plywood, OSB, and structural wood fiber. Steel is the most widely used, followed by concrete and wood products (Figure 27.2).

Figure 27.2 Steel is widely used for commercial construction roof decking.

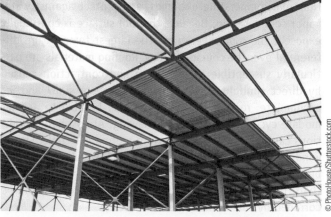

© PhotoHouse/Shutterstock.com

Corrugated steel decking requires a covering layer, typically mechanically fastened rigid insulation or site-cast lightweight concrete. Concrete decks are made with either cast-in-place forms or pre-cast concrete structural members. Pre-cast concrete planks must be topped with a concrete fill or grouted between planks. A lightweight insulating concrete can also be cast on top of pre-cast concrete decking to achieve a smooth surface.

Cement-wood-fiber roof decking is made by bonding treated wood fibers with Portland cement or other binders, and compressing them into structural panels. Gypsum concrete decks are produced by mixing gypsum, wood fibers, or mineral aggregates with water and casting it on form boards. Pre-cast gypsum planks are also used.

Plywood and oriented strand board are the two commonly used wood materials used for roof decks. Wood planks of solid wood decking or glued-laminated members made by bonding solid dimensional lumber into a structural decking member are also used. Wood decks must be smooth and free of large gaps or knotholes.

Regardless of the type of deck material selected, no water can be on the deck when roofing is installed to prevent water vapor from being trapped beneath the membrane.

Vapor Retarders

The flow of water vapor from inside a building into a roofing system must be carefully controlled. Water vapor can cause decking to deteriorate and potentially damage some types of finished roofing surfaces. Changes in temperature and humidity create a difference in water vapor pressure between the inside and outside of a building, causing water vapor to migrate into insulation. Vapor retarders should be applied to the warm side of the insulation in roof decks in cold climates (Figure 27.3). In

Figure 27.3 Vapor retarders are placed on the warm side of the roof assembly.

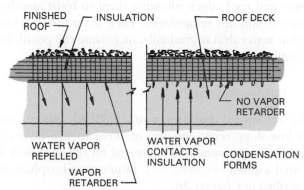

warm, humid climates, winter condensation in the roof assembly is less of a problem, and vapor retarders are often omitted. When they are used, they are placed at the exterior of the roof assembly, as air conditioning in warm climates may actually cause a reverse vapor migration. Vapor retarders must be used on top of concrete and poured gypsum decks, and any tears or holes in the material must be sealed.

Insulation

Resistance to heat flow through a roof structure should be as high as is practicable and cost effective. In general, an R factor of greater than 20 is recommended. (Note: $U = 1/R$, therefore the U factor should be less than 0.05 BTU/hr-ft^2 °F.) Three possible configurations exist for the location of thermal insulation in low-slope roof assemblies.

The insulation can be installed below the deck. Batt insulation is placed above the vapor retarder and beneath the deck. While the installation is relatively easy, this leaves the deck and membrane exposed to the weather and more prone to wear. Since the insulation increases the roof temperature on hot days, the roof materials age faster and the expansion and contraction stresses on the membrane are greater.

A second method places the insulation between the deck and the membrane (Figure 27.4). Rigid insulation panels or lightweight insulating concrete is placed on the deck, with the vapor retarder beneath. The deck is protected, but water can be trapped in the insulation. Topside vents should be provided to allow water vapor to escape from within the roof structure. Finally, the insulation can be installed above the membrane on the exterior of the roof. The type of insulation used must retain its insulating value when wet and not decay or disintegrate. Sheets of rigid insulation boards are laid loose or adhered to the membrane with a coat of hot asphalt. The entire roof assembly is thus held down and protected from sunlight by a layer of ballast. When insulation is placed above the membrane, the membrane also serves as the vapor retarder.

Rigid roof insulation materials include cellular glass, composite boards, glass fiber, perlitic boards, polyisocyanurate foam boards, polystyrene boards, polyurethane foam boards, and wood fiber boards. Roof insulation may be placed as a single or double layer. The double layer is installed with offset joints, eliminating any leakage of heating or cooling energy that may occur between joints in the first layer. A base layer can be mechanically joined to the deck when steel decking is used, with the

Figure 27.4 Insulation for low-slope roofs can be installed below the deck, between the deck and the membrane, or above the membrane.

© Don Cline/Shutterstock.com

second layer bonded to it with hot bitumen or adhesive. On concrete decks, the first layer is often bonded directly to the deck with hot bitumen.

The required slope for low-slope roofing systems can be created with the use of tapered insulation. Standard sizes and slopes are available or can be custom designed for a roof layout to provide the slope required for positive drainage.

Flashing

Flashing is made of thin materials that are impervious to water and are used to prevent water penetration and provide for water drainage. Typical flashing locations in roof construction include the roof perimeter, joints between roofs and walls, at expansion joints, and around objects, such as pipes, that penetrate the roof surface. Materials used for flashing include copper, galvanized steel, lead, aluminum, stainless steel, bituminous sheet material, and plastics. In some cases, a combination of materials is used, such as galvanized steel covered with bitumen, which prevents corrosion of the steel when in contact with mortar. Flashing materials often have different thermal expansion and contraction characteristics from those of the roofing membrane. These differences can cause cracking and tears in a roof material unless allowance is made for this movement.

Finished Roofing

A range of finished roofing materials are available providing the designer with considerable choices. Table 27.1 describes many of these materials. The specific properties of products provided by various manufacturers must be known and the manufacturer's installation recommendations must be followed.

LOW-SLOPE ROOF ASSEMBLIES

A *low-slope roof* is defined as a roof having a pitch of less than 3:12. "Low slope" denotes a roof that is nearly flat and from which water drains slowly. The actual slope permitted varies with the type of roofing system used and the manufacturer's recommendations. Typical slopes fall in a range of ¼ in. rise per 12 in. run (6.4 mm per 305 mm) to 2 in. rise per 12 in. run (50.8 mm per 305 mm).

Low-slope roofing is often used on commercial projects where the minimum slopes required by steep roofing materials make their use impractical for larger buildings. A low-slope roof can efficiently cover a building of any horizontal dimension with a more economical structure underlying the roof than is required for steep-roof systems. Low-slope membrane systems are tightly sealed at joints and roof edges, allowing them to resist standing water conditions temporarily. Their main disadvantage is that water drains gradually, increasing the possibility of water damage.

Built-Up Roofing

The traditional *built-up roofing (BUR)* system used on low-slope roofs consists of three to five layers of bitumen-impregnated felts, layered on the roof deck in heated asphalt (Figure 27.5). Bitumen and asphalt are described in Chapter 26.

Table 27.1 Materials Used for Finished Roofing

Material	Type of Roof	Descriptive Factors	Weight per Square (lb./100 ft.²)	Weight per Square Meter (kg/m²)
Aluminum (sheet, shingles)	Steep-slope	Fire resistant, long life, range of colors	5–90	2.44–4.39
Asphalt (built-up)	Steep-slope, low-slope	Granular topping applied influences fire class, life 20–30 years	100–600	4.88–29.3
Asphalt shingles (fiberglass, asphalt)	Steep-slope	Fire resistance varies with product, range of colors, life 20–30 years	235–325	11.47–15.86
Cement-fiber tile	Steep-slope	Fire resistance, long life, heavy, use in warm climates	950	46.4
Clay tile	Steep-slope	Fire resistant, long life, heavy	800–1,600	39–78
Copper (sheet)	Steep-slope, low-slope	Fire resistant, long life, can be soldered	0.019" thick 160 0.040" thick 320	7.8 15.6
Lead, copper coated	Steep-slope, low-slope	Fire resistant, long life	1/32" thick 200 1/16" thick 400	9.76 19.52
Monel (Ni-Cu)	Steep-slope	Fire resistant, long life	22 gauge 1,424 26 gauge 827	69.5 40.4
Perlite-portland cement	Steep-slope	High fire rating, lightweight, long life	900–1,000	43.9–48.8
Plastic (single-ply membrane)	Low-slope	Long life, requires careful installation, limited fire classification, several types available	Loose laid Ballasted, 1,000–1,200 Fully adhered, 30–55	48.8–58.6 1.5–2.7
Plastic (liquid applied)	Low-slope	Limited fire classification follow manufacturer's directions	20–50	0.98–2.4
Slate	Steep-slope	Fire resistant, heavy, long life	3/8" thick 800 1/4" thick 900 3/8" thick 1,100 1/2" thick 1,700 3/4" thick 2,600	39.0 43.9 53.7 83.0 126.9
Stainless steel, terne coated	Steep-slope	Fire resistant, long life	90	3.89
Steel (sheet, shingles)	Steep-slope, low-slope	Fire resistant, long life, durable colors	Copper coated 130 Galvanized 130	6.3 6.3
Wood (shingles, shakes)	Steep-slope	No fire resistance unless treated, limited life	200–450	9.8–22.0
Zinc	Steep-slope	Fire resistant, long life, can be painted	9 gauge 670 12 gauge 1050	32.7 51.2
Terneplate copper-bearing sheet	Steep-slope	Fire resistant, long life	30 gauge 540 26 gauge 780	26.4 38.1
Mineral-surfaced cap sheet	Low-slope	Limited fire resistance, limited life	55–60	2.68–2.9
Modified bitumen	Low-slope	Fire resistance, 10 years or more	100	4.9

Figure 27.5 Overlapping sheets of asphalt-saturated felt are bonded to the rigid insulation with hot roofing asphalt.

SECOND LAYER OF RIGID INSULATION

ROOFING ASPHALT

ASPHALT-SATURATED FELT

GRAVEL   ROOFING ASPHALT

Figure 27.6 Roofing asphalt is heated in an asphalt kettle and pumped to the roof.

© mikeledray/Shutterstock.com

Figure 27.7 Heated asphalt is applied hot in layers over felts.

© mikeledray/Shutterstock.com

In built-up roofing membranes, bitumen is usually applied hot over felts, and finished with a top surface, such as gravel or slag aggregate or a cap sheet. Saturated felts, sometimes called tar paper, are made with an organic mat saturated with coal tar pitch or asphalt and coated with a layer of thin asphalt. Tar paper is used as an underlayment for shingles, other roofing materials, and as laminations in built-up roof construction. Saturated felt is manufactured in three weights: Types 15, 20, and 30, according to the weight per square of felt. For example, Type 15 weighs 15 pounds per square (100 sq. ft.). The felt is sold in rolls 36 in. wide and up to 144 ft. long. Felts provide the needed reinforcement to keep bitumens in each layer of a built-up roof from alligatoring. "Alligatoring" refers to surface cracking caused by oxidation and shrinkage stresses, which can result in a repetitive mounding of the asphalt surface similar to an alligator's hide.

Typical built-up roof construction may be laid on insulated or uninsulated roof decks. In the installation of a built-up roof assembly, the deck is initially covered with one ply of base coat, usually of 40 lb. felt, which acts as a vapor retarder. Roofing asphalt is brought to the site in a tank truck and heated in an asphalt kettle (Figure 27.6). The heated asphalt is pumped to a tank on the roof and moved to needed areas. The top of the base coat is hot mopped with coal tar or asphalt (Figure 27.7). A 12 in.

wide layer of 15 lb. felt is applied at the lowest roof edge. Secondary layers of 24 in. and 36 in. wide felt is laid above and hot mopped between each layer. Next, three to five layers of asphalt-coated felt are applied, bonded with coatings of hot-mopped bitumen. Finally, the entire surface is flood coated and allowed to cool a bit (Figure 27.8).

Figure 27.8 The roof is flood coated and allowed to cool before being covered with an aggregate ballast.

© Per Mikaelsson/age fotostock/SuperStock

The flood coating is covered with an aggregate wear surface that protects the bitumen from solar radiation and prevents uplift of the thin roofing materials. The aggregate surface forming the top surface is typically derived from gravel, marble chips, or slag. Gravel and marble chips are usually applied at 400 pounds per 100 sq. ft., and slag at 300 pounds per 100 sq. ft.

Conventional built-up roofs can be designed for slopes up to 3 in. per foot. Slopes above ½ in. per foot require the use of steep asphalt. Another type of built-up roofing is an assembly using an asphalt glass-fiber roof membrane covered with a mineral-surfaced inorganic cap sheet. A mineral-surfaced roof material consists of a base felt coated on one or both sides with asphalt and surfaced with mineral granules. An asphalt glass-fiber membrane is fastened to the nailable deck, with additional layers bonded with hot asphalt (Figure 27.9). The mineral-surfaced inorganic cap sheet is bonded to the asphalt glass-fiber base with hot asphalt. It is recommended that this system be used on roofs with a slope of ¼ in. per foot or greater.

Figure 27.9 One installation detail for using mineral-surfaced cap sheets.

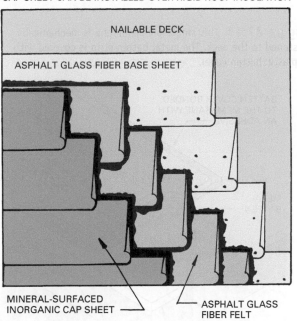

CAP SHEET CAN BE INSTALLED OVER RIGID ROOF INSULATION

NAILABLE DECK

ASPHALT GLASS FIBER BASE SHEET

MINERAL-SURFACED INORGANIC CAP SHEET

ASPHALT GLASS FIBER FELT

Modified Bitumen Membranes

Modified bitumen membranes consist of a bituminous material to which polymeric compounds have been added to increase cohesion, toughness, and resistance to flow. Two common membranes available are styrene-butadiene-styrene (SPS) and atactic-polypropylene (APP). SBS sheets have a reinforcement mat coated with an elastomeric blend of asphalt and SBS rubber. APP membranes use a reinforcement mat coated with a blend of asphalt and APP plastic. While fiberglass reinforcement is most frequently used, membranes are available from various manufacturers with different reinforcements. The major difference between SBS and APP products is the blended asphalt used. The blend creates a product that has greater elongation, strength, and flexibility than traditional roofing asphalts. Recovery after elongation, flexibility, and cold-weather performance of SBS membranes are superior to those of APP membranes.

Modified bitumen membranes can be either self-adhered, loose laid, or laid in hot asphalt (Figure 27.10). The seams between sheets are sealed by torching or applying a hot asphaltic adhesive. SBS products are generally installed using hot asphalt as the bonding material. They are applied as cap sheets over a base of hot asphalt and roofing felts. The cap sheet SBS membrane may have a ceramic granule surfacing to protect it from ultraviolet light, or it may be left unsurfaced. The unsurfaced type must be coated with asphalt and gravel to give it ultraviolet protection.

Figure 27.10 A typical assembly for a partially adhered single-ply modified-bitumen roof system.

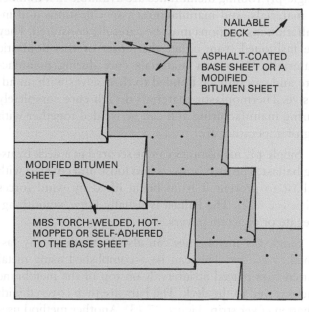

NAILABLE DECK

ASPHALT-COATED BASE SHEET OR A MODIFIED BITUMEN SHEET

MODIFIED BITUMEN SHEET

MBS TORCH-WELDED, HOT-MOPPED OR SELF-ADHERED TO THE BASE SHEET

APA products are applied by a method called torching, made possible by the unique properties of the modified bitumen. The back coating of modified asphalt is heated with a propane torch to the point at which the sheet can be bonded to the substrate (Figure 27.11). APA products cannot be installed with hot-mopped asphalt.

Figure 27.11 APA products are applied by a method called torching.

© Dmitry Kalinovsky/Shutterstock.com

Single-Ply Membranes

Single-ply roofing membranes can be applied over almost any type of roof deck, as well as over existing asphalt or built-up roofs when re-roofing a building. Roof insulation is required over a roof deck unless it is composed of very smooth concrete, plywood, or splinterfree solid wood decking. The roof must drain freely and have sufficient slope and drains to carry away water. Single-ply roofing membranes are available in a number of materials, and manufacturer's specifications and installation instructions must be carefully consulted. They are fashioned from either thermoset or thermoplastic materials. Thermoset materials cure during manufacture and can only be bonded to themselves with an adhesive. Thermoplastic materials do not cure completely during manufacturing and can be welded together with high-temperature air.

Single-ply membranes can be secured to a deck by using ballast. The membrane is laid loose and covered with ballast to prevent it from being lifted by wind forces (Figure 27.12). The ballast is usually large, smooth aggregate or concrete pavers.

Single-ply membranes can also be mechanically fastened to a deck. This can be accomplished using metal batten bars placed at intervals on top of the membrane and screwed to the deck. The bars are then covered with a batten cover strip (Figure 27.13). Another method uses screws and metal discs placed over the edge of a layer of membrane and screwed into the deck (Figure 27.14). The adjoining membrane is lapped over this attachment. A third method involves placing large-diameter discs on top of the membrane, spaced as required, and screwing them to the deck. A waterproof cover is placed over the discs.

Figure 27.12 A single-ply protected membrane is covered with a layer of rigid insulation and ballast.

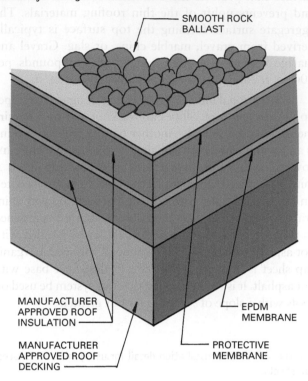

SMOOTH ROCK BALLAST

MANUFACTURER APPROVED ROOF INSULATION

MANUFACTURER APPROVED ROOF DECKING

EPDM MEMBRANE

PROTECTIVE MEMBRANE

Figure 27.13 This single-ply membrane is mechanically fastened to the deck. The metal batten strip is covered with a plastic batten cover.

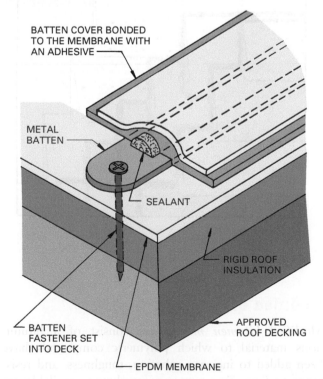

BATTEN COVER BONDED TO THE MEMBRANE WITH AN ADHESIVE

METAL BATTEN

SEALANT

BATTEN FASTENER SET INTO DECK

RIGID ROOF INSULATION

APPROVED ROOF DECKING

EPDM MEMBRANE

Figure 27.14 Some membranes are fastened to the deck with metal discs and sealed with a heat-welded lap joint.

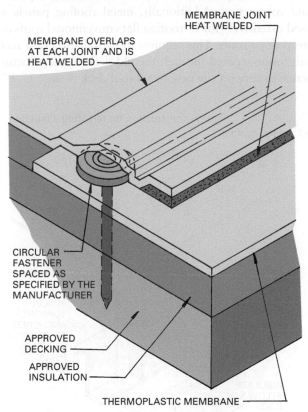

MEMBRANE JOINT HEAT WELDED

MEMBRANE OVERLAPS AT EACH JOINT AND IS HEAT WELDED

CIRCULAR FASTENER SPACED AS SPECIFIED BY THE MANUFACTURER

APPROVED DECKING

APPROVED INSULATION

THERMOPLASTIC MEMBRANE

Figure 27.15 A polyvinyl chloride (PVC) thermoplastic membrane is adhered to the roof deck.

© Mark E. Gibson/Encyclopedia/Corbis

A number of single-ply membrane products are commonly used in commercial low-slope roofing. Chlorosulfonated polyethylene (CSPE) is a thermoset membrane that completes its cure after installation. It has excellent weathering qualities and is resistant to ozone, sunlight, and most chemicals. Ethylene propylene diene monomer (EPDM) is a thermoset elastomeric compound produced from propylene, ethylene, and diene monomer. It has good resistance to weathering, ultraviolet rays, and abrasion. EPDM membranes can be installed ballasted, fully adhered, or mechanically fastened.

Polyvinyl chloride (PVC) is a thermoplastic membrane produced by the polymerization of vinyl chloride monomer, stabilizers, and plasticizers. It may be installed either fully adhered or mechanically fastened (Figure 27.15). PVC membranes are resistant to weather and chemical atmospheres, have good fire resistance, and are easy to bond. They do not work well with bituminous materials. Thermoplastic Olefin membrane (TPO) is a blend of polypropylene and ethylene-propylene polymers. Colorant, flame retardants, UV absorbers, and other proprietary substances can be blended with the TPO to achieve the necessary physical properties.

Spray- and Liquid-Applied Membranes

Spray- and liquid-applied roof coatings are installed in liquid form with a roller or spray gun, usually in several coats. They are used primarily where roofs have complex shapes, such as domes and shells. Both can be applied hot or cold, on horizontal and vertical surfaces. Spray-applied roof coatings consist of a polyurethane foam insulation layer applied to a deck and then topped with a protective coating. The protective coating is usually acrylic, polyurethane, or silicone. Sprayed polyurethane foam (SPF) provides a seamless, monolithic waterproof surface that adheres directly to the substrate. Sometimes, mineral granules or aggregate are applied to the wet top coating and used to provide additional protection.

Liquid-applied coverings are available from various manufacturers as one- or two-component elastomeric materials. They are applied by spraying, brushing, or rolling over the roof decking, which is typically plywood or concrete. Any joints require special attention as recommended by a manufacturer.

Low-Slope Metal Roofing

Various manufacturers have metal roofing systems designed for low-slope roof assemblies. Numerous panel profiles and seaming methods have been developed to produce a watertight roof. Panels are available in a wide range of colors and may be galvanized or Galvalume steel, aluminum, copper, or terne-coated stainless steel. Galvalume is a trade name for a patented steel sheet coated with a corrosion-resistant aluminum-zinc alloy applied by a continuous hot-dip process. Terne-coated stainless-steel panels are made from nickel-chrome stainless steel coated on both sides with an alloy of 80 percent lead and 20 percent tin. They are recommended for use in severe chemical and marine environments. Metal roofing panels are available embossed to create a stucco-like finish. Various types of other coatings are available, such as siliconized polyester and epoxy-based coatings.

Low-slope standing-seam metal roofing systems require a minimum slope of ¼:12 for positive drainage. This type of roofing is generally attached to the roof structure with metal clips. The clips are joined to the bar joist or Z-purlin of the structure with metal fasteners (Figure 27.16). The hold-down clip hooks onto the leading edge of the panel. The clips and edges of the butting panels are interlocked during the seaming process. The standing seam is formed on the roof with a portable seaming machine that runs along the seam and folds the members into a

Figure 27.16 This metal roofing is held to the structural frame with a metal clip that is rolled into a standing seam.

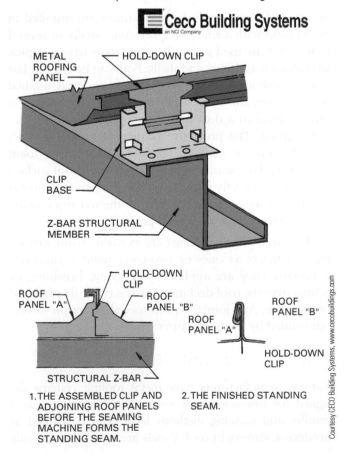

Courtesy CECO Building Systems, www.cecobuildings.com

watertight joint. The clip fits into a slot in the clip base that permits the clip to move with the roof during expansion and contraction. Additionally, metal roofing panels are used extensively for re-roofing flat conventional roofs and steep-slope roofs. This requires the installation of a metal sub-frame secured through the old deck to the structural frame to support the new metal roof deck (Figure 27.17).

Figure 27.17 Typical construction for reroofing a building with metal roofing.

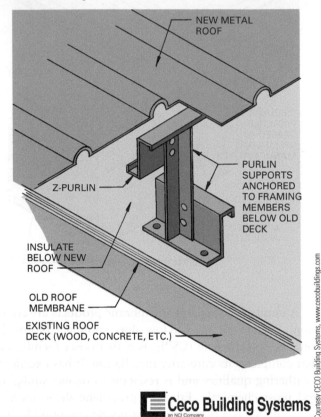

Courtesy CECO Building Systems, www.cecobuildings.com

Low-Slope Roof Flashing

Wood nailers are used at roof edges and where roof insulation terminates. They provide a surface to which the roof membrane and flashing can be nailed. Wood nailers must be securely fastened to the deck to prevent wind damage to flashing. The wood used is treated with a wood preservative. Care must be taken that the preservative does not damage roofing materials.

When a low-slope roof abuts a parapet wall, a wood ledger is used to protect the edge of the insulation. A wood header butts the vertical wall and has a layer of insulation between it and the wall. The corner is sloped with a wood cant strip. Metal flashing is set into the wall and overlaps the felt laid up over the cant strip (Figure 27.18). Roof

Figure 27.18 This detail shows a roof butting a masonry wall. A wood ledger butts the vertical wall and has a layer of insulation between it and the wall. The corner is sloped with a wood cant strip. Metal flashing is set into a mortar joint and overlaps the felt laid up over the cant strip. This assembly allows movement between the roof and the masonry wall.

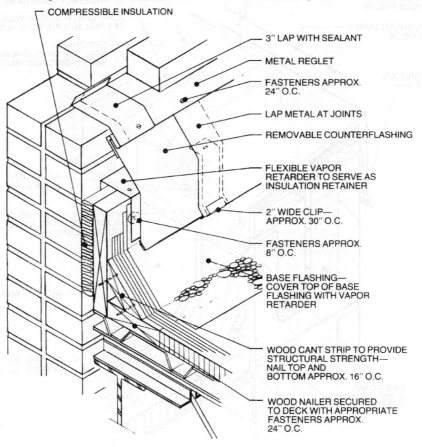

COMPRESSIBLE INSULATION

3" LAP WITH SEALANT

METAL REGLET

FASTENERS APPROX. 24" O.C.

LAP METAL AT JOINTS

REMOVABLE COUNTERFLASHING

FLEXIBLE VAPOR RETARDER TO SERVE AS INSULATION RETAINER

2" WIDE CLIP— APPROX. 30" O.C.

FASTENERS APPROX. 8" O.C.

BASE FLASHING— COVER TOP OF BASE FLASHING WITH VAPOR RETARDER

WOOD CANT STRIP TO PROVIDE STRUCTURAL STRENGTH— NAIL TOP AND BOTTOM APPROX. 16" O.C.

WOOD NAILER SECURED TO DECK WITH APPROPRIATE FASTENERS APPROX. 24" O.C.

edges are often finished with one-piece metal parapet caps (Figure 27.19). The roofing membrane is wrapped up the edge over a cant strip. Base and counter flashing are used over the membrane to assure water tightness. Base flashing is provided by upturned edges of the watertight membrane on a roof. Counter flashing is secured on or into a wall, curb, pipe or other surface, to cover and protect the upper edge of a base flashing and its associated fasteners.

As the size of a roof increases, the danger of tears from expansion and contraction of the membrane also increases. Expansion joints are used to provide allowances for movement (Figure 27.20). Also included in the design are area dividers located between expansion joints (Figure 27.21). They must not restrict the flow of water. Spacing varies, depending on the climate, but they are generally located every 150 to 200 ft. (45–61m).

Low-Slope Roof Drainage

Positive drainage of water on a low-slope roof is a priority that impacts the effectiveness and durability of the system throughout its service life. Most low-slope roof problems are the result of ponding, water accumulation lasting too long in one place and not being drained or evaporated. Positive drainage is created by routing water around roof penetrations to lower points on the roof via tapered insulation or a sloped decking structure.

Tapered roof insulation systems are a key component of both BUR and single-membrane systems. Historically, tapered insulation systems were created on site by cutting and shaping rigid roof insulation components. Today, many manufacturers offer pre-cut tapered insulation sheets that can be installed to achieve tapered positive drainage designs. Low-slope roof assemblies must always comply with a minimum slope of ¼ in. /ft.

Interior drains and scuppers are common drainage components used on commercial buildings with large roof surfaces. Interior drains use a drain pipe that runs through the building and connects to a storm drain below. Interior drain systems must have an absolutely watertight connection at the point where the drain penetrates the roof membrane. A means of preventing debris at the roof

Figure 27.19 This light-metal parapet cap uses a wood nailer and cant strip. The base flashing is overlapped with a metal parapet cap. This construction is used only when the deck is supported by the wall.

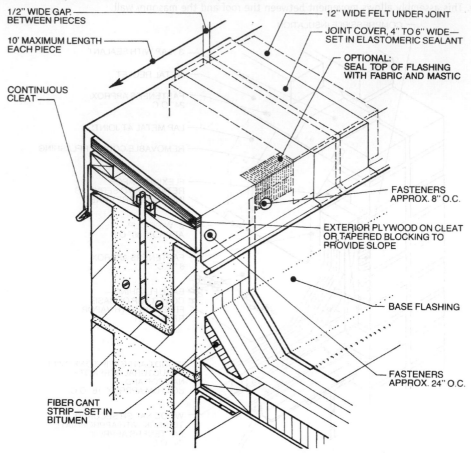

surface from entering the drain system must be provided. Many drains on flat roofs use a birdcage-type cover to keep the roof drain from clogging (Figure 27.22).

The use of scuppers as either primary or secondary drains is common on low-slope commercial roofs. A scupper is a penetration through a parapet wall or drip edge that allows water to drain off the roof (Figure 27.23). A parapet wall is a wall that extends above the roofline at its perimeter. A scupper is usually connected to a downspout that runs down the outside of the building. If scuppers are primary drains, they are installed around the perimeter of the roof at low points. If a scupper is an emergency or secondary drain, it is installed about two inches above the roof membrane to address extreme ponding conditions.

STEEP-SLOPE ROOF ASSEMBLIES

A steep roof is generally defined as a roof with a slope of 3:12 or greater. *Steep-slope roof* assemblies have sufficient slope to permit water to drain quickly from its surface. Gutters and downspouts are used to capture

and guide the water flow. A variety of roofing materials and systems are available for use on steep slopes. Their selection is governed by the slope, structural requirements, and the desired architectural appearance.

"Slope" is a common term used to express the steepness of a roof in terms of unit run and unit rise. The unit run is the number used as a base for the roof angle and specified as 12 in. The unit rise is the number of inches the roof will rise vertically for every unit of run. For example, if the unit rise is 6 in., then the roof will rise 6 in. for every 12 in. it covers horizontally, and the slope will be expressed as 6:12. Minimum slopes for steep-slope roofing materials are given in Table 27.2.

Shingle Roofing

Shingles are small units applied to the roof in overlapping layers. These elements are typically flat rectangular shapes laid in rows or courses from the bottom edge of the roof up, with each successive higher row overlapping the joints in the row below. Shingle roofs minimize the effects of movement through the ability of roofing

Figure 27.20 Expansion joints allow for expansion and contraction in the roof.

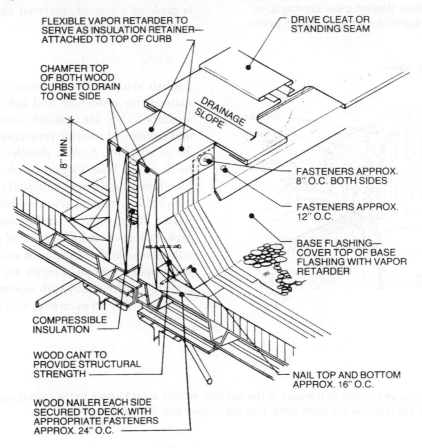

FLEXIBLE VAPOR RETARDER TO
SERVE AS INSULATION RETAINER—
ATTACHED TO TOP OF CURB

CHAMFER TOP
OF BOTH WOOD
CURBS TO DRAIN
TO ONE SIDE

DRIVE CLEAT OR
STANDING SEAM

DRAINAGE
SLOPE

8" MIN.

FASTENERS APPROX.
8" O.C. BOTH SIDES

FASTENERS APPROX.
12" O.C.

BASE FLASHING—
COVER TOP OF BASE
FLASHING WITH VAPOR
RETARDER

COMPRESSIBLE
INSULATION

WOOD CANT TO
PROVIDE STRUCTURAL
STRENGTH

NAIL TOP AND BOTTOM
APPROX. 16" O.C.

WOOD NAILER EACH SIDE
SECURED TO DECK, WITH
APPROPRIATE FASTENERS
APPROX. 24" O.C.

Figure 27.21 An area divider is a raised wood member attached to the deck and flashed.

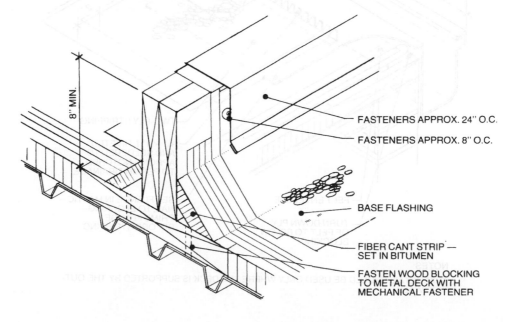

8" MIN.

FASTENERS APPROX. 24" O.C.

FASTENERS APPROX. 8" O.C.

BASE FLASHING

FIBER CANT STRIP—
SET IN BITUMEN

FASTEN WOOD BLOCKING
TO METAL DECK WITH
MECHANICAL FASTENER

Figure 27.22 Roof drains are located in various parts of the roof and carry away water through pipes running down inside the building. Installation details are provided by the manufacturer.

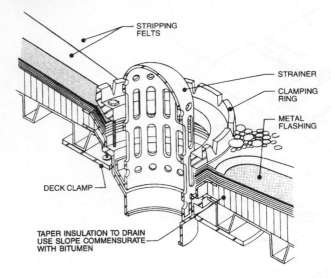

STRIPPING FELTS

STRAINER

CLAMPING RING

METAL FLASHING

DECK CLAMP

TAPER INSULATION TO DRAIN
USE SLOPE COMMENSURATE
WITH BITUMEN

units to move with respect to one another. A roofing tile is made of a ceramic material and uses similar installation methods to shingles.

Asphalt Shingles

Asphalt shingles are the most commonly used roofing material for residential and light commercial roof construction. They are available in a wide variety of styles and colors and provide protection from the weather for 20 to 30 years. Asphalt shingles have either an organic or fiberglass base. Organic asphalt shingles consist of a wood fiber base that is saturated with asphalt and coated with colored mineral granules. Fiberglass asphalt shingles have a fiberglass mat that has top and bottom layers of asphalt, also coated with mineral granules. A self-healing adhesive seals the shingles to one another and prevents wind uplift. Asphalt shingles are die cut from the mat, typically in three tabs with square edges. Most common are strips 12×36 in. in size, with the exposed surface cut

Figure 27.23 A scupper is an opening in the edge of the roof that permits water to drain to the ground, usually through a downspout. Notice the layers of felt laid over the sheet metal from the scupper that is secured to the deck.

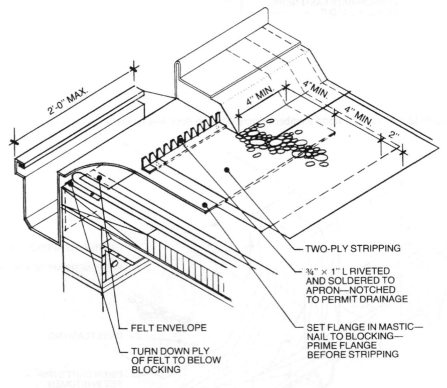

2'-0" MAX.

4" MIN.

4" MIN.

4" MIN.

2"

TWO-PLY STRIPPING

¾" × 1" L RIVETED
AND SOLDERED TO
APRON—NOTCHED
TO PERMIT DRAINAGE

SET FLANGE IN MASTIC—
NAIL TO BLOCKING—
PRIME FLANGE
BEFORE STRIPPING

FELT ENVELOPE

TURN DOWN PLY
OF FELT TO BELOW
BLOCKING

NOTES:
THIS DETAIL SHOULD BE USED ONLY WHERE THE DECK IS SUPPORTED BY THE OUT-
SIDE WALL.

Table 27.2 Typical Minimum Slopes for Roofing on Steep-Slope Roofs[a]

Material	Allowable Slope
Asphalt shingles	4:12 if one layer of asphalt-saturated felt underlayment is used. 2:12 if two layers of asphalt-saturated felt underlayment are used.
Clay and concrete tile	4:12 if interlocking tile are used with one layer of 4 lb. cap sheet underlayment. Less than 3:12 when noninterlocking tile are used with two layers of No. 40 asphalt-saturated felt underlayment.
Slate	4:12 with one layer of 30 lb. asphalt-saturated felt underlayment. 2:12 with double underlayment.
Wood shingles and shakes	3:12 for shingles and 4:12 for shakes with 30 lb. felt underlayment and interlayment with shakes or shingles.
Metal shingles and shakes	3:12 and 4:12 depending on the material and style. Requires 30 lb. asphalt-saturated felt underlayment.

[a]*Consult local building codes and observe the manufacturer's recommendations on acceptable slope and installation details.*

to resemble three smaller shingles. They are available in weights of 100 to 300 lb. per square.

A typical installation starts with the underlayment. Underlayment protects the roof deck from moisture before the roofing is applied. The material is rolled out on the deck starting at the bottom of the roof and is overlapped in successive layers. The underlayment is doubled at the eave, as shown in Figure 27.24. In cold climates, this double layer should extend over the roof until it is 24 in. inside the exterior wall. A corrosion-resistant metal drip edge is placed along the edge of the roof at the eave and rake to facilitate water dripping away from the roof edge. The underlayment goes below the drip edge on the rake and on top at the eave.

On most roofs, shingles are applied starting from the rake. A starter course is laid with tabs facing up to fill spaces between tabs on the first regular course of shingles. Additional courses are overlaid in a lapped configuration (Figure 27.25). The maximum exposure of the shingles depends on weather conditions and ranges between 4 and 6 in. (10–15 cm).

Figure 27.25 Asphalt shingles are spaced so that slots between tabs do not align.

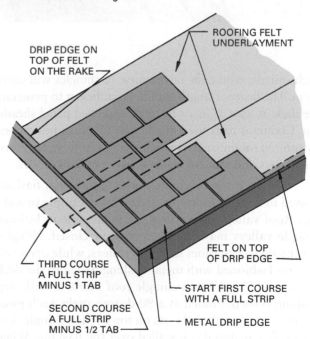

A number of visual shingle patterns are possible, depending on the shingle tab arrangement. Asphalt and fiberglass shingles are fastened using nails or staples (Figure 27.26). The fasteners are placed at the midpoint of

Figure 27.24 A double-layer asphalt-saturated roofing felt underlayment with drip-edge flashing.

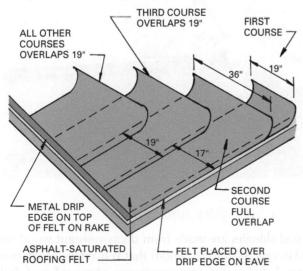

Figure 27.26 Asphalt and fiberglass shingles are fastened using nails or staples placed at the midpoint of each shingle, where the next course of shingles will cover them.

© Christina Richards/Shutterstock.com

Figure 27.27 Metal valley flashing is placed over asphalt felt that is bonded to the roof with plastic asphalt cement.

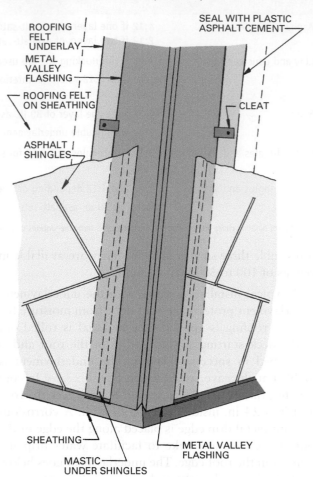

each shingle, where the next course of shingles will cover them. The fastener length should be sufficient to penetrate the deck at least ⅜ in. or through approved panel sheathing. Common nails used are 11- or 12-gauge hot-dipped galvanized roofing nails with shanks to 1 in. long. Required staple sizes and spacing are specified in building codes.

To prevent water from entering a building, flashing is used in various locations on roofs susceptible to leakage. Roof valleys are particularly vulnerable to leaking. Shingle valleys may be either open or closed. Shingles cover valley centerlines in closed valleys, while open valleys are fashioned with metal flashing nailed to the deck (Figure 27.27). When a shingle roof meets a wall, step flashing is used. Folded at a 90-degree angle, each piece is nailed to the wall and rests on top of each shingle. The wall siding material is installed over the flashing. When the roof meets a brick wall, flashing is set into the mortar joint. Finally, hips and ridges are capped with single tabs overlapping each other (Figure 27.28). Special vented ridge components are used to ensure adequate ventilation is provided.

Figure 27.28 Cap shingles being applied to a vented ridge.

Wood Shingles and Shakes

Wood shingles are made from decay-resistant wood (red cedar, cypress, and redwood) that is machine sawn, overlapped, and nailed on a continuous or spaced roof deck.

Wood shakes are split rather than sawn pieces of wood and have a much rougher face texture than do wood shingles. Both are tapered and installed with their thicker end facing toward the bottom of the slope. Treated shingles are pressure impregnated with fire-retardant chemicals that saturate into the wood cells. Building codes in many areas require that shingles and shakes be treated to achieve a class B rating. Shingles require a minimum slope of 3:12, and shakes a 4:12 slope to assure proper drainage.

Usually, wood shingles are applied over 1 × 4 in. or 1 × 6 in. wood boards called purlins, which are spaced between the base and ridge of the roof. A purlin is a horizontal roof member used to support roofing materials. The size and spacing used depends on the amount of shingle to be exposed to the weather. Solid decking is used at the eaves and runs upward until it is 24 in. inside exterior walls (Figure 27.29). Wood shakes are placed over 1 × 6 in. spaced wood decking. The deck is covered with roofing felt. Shingles are doubled or tripled at eaves, while shakes are usually doubled. A layer of felt lays over the top edge of each row of shakes. Hips and ridges are finished with cap shingles that may be hand-cut or purchased ready made. Valleys use metal flashing placed over roofing. The nails used should be of hot dipped zinc or aluminum with their heads driven flush (Figure 27.30).

Figure 27.30 Wood shingles and shakes are fastened with corrosion resistant nails or staples.

Slate Shingles

Slate shingles are available in a variety of colors originating out of the rock from which it was quarried. They are split to thickness and trimmed to size from larger slabs of slate. Holes for fasteners are either drilled or punched. Slate is a heavy material, weighing up to 2,600 lbs. per square; hence, the roof structure must be designed to carry the weight. Slate shingles are fire resistant and have a long life but are more costly than most roofing materials.

Slate roofing is divided into two classifications: textural (or random) and standard commercial. Textural slate is delivered to the job site in a range of thicknesses and sizes and must be sorted by roofers. Standard commercial slate is graded at the quarry by thickness, length, and width, with thicknesses of usually ¼ in.

ASTM C406 *Standard Specification for Slate Roofing* addresses material characteristics, physical requirements, and sampling procedures for the selection of roofing slate. Roofing slate is available in three grades: grade S1 has an expected life of 75 years; S2 40 to 75 years; and S3 20 to 40 years.

Slate tile is usually applied over solid sheathing covered with an asphalt-saturated roofing felt underlayment. Codes require double underlayment for roofs with less than 4:12 slopes. The material is fastened with large-head copper nails, with each course of slate covering the joints in the course below it. Plastic cement is applied over joints to be covered. Each slate must have at least two fasteners. Slates are laid in the same manner as other shingles. The exposed ends may be in even rows or laid in a random edge. Hips and

Figure 27.29 Typical installation details for wood shakes.

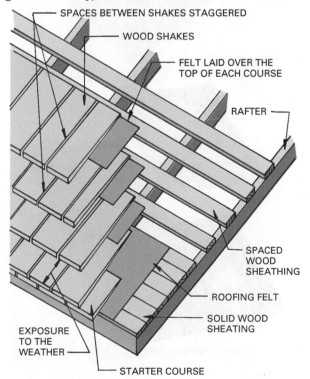

SPACES BETWEEN SHAKES STAGGERED

WOOD SHAKES

FELT LAID OVER THE TOP OF EACH COURSE

RAFTER

SPACED WOOD SHEATHING

ROOFING FELT

SOLID WOOD SHEATING

EXPOSURE TO THE WEATHER

STARTER COURSE

ridges cannot be nailed and are held in place with a waterproof elastic slater's cement that is colored to match the slate.

Clay and Concrete Roof Tiles

Tile roofing is made from concrete or clay units that overlap and interlock to create a durable, fire-resistant, and heavy roof. Some forms of concrete tile are made lighter using lightweight aggregates. Clay roofing tile may be unglazed or glazed. Unglazed tile ranges from an orange-yellow to a dark red, depending on the clay. Glazed tile is available in a wide range styles and colors (Figure 27.31).

Interlocking tiles are used on roofs with slopes of 3:12 or more. Shingle tiles, often called Norma tiles, require a 5:12 slope. Flat interlocking clay tile is installed on solid sheathing that has been covered with asphalt-saturated roofing felt (Figure 27.32). A flat under-eave tile or an under-eave tile with an apron is used along the eaves. The gable rake is covered with special right and left rake tiles. Ridges and hips are covered with V-shaped tiles set in a cement mortar and nailed.

Flat clay tile are installed much like slate shingles (Figure 27.33). Spaced sheathing can be used for roofs with over 4:12 slopes, with roofing felt laid between layers. A number of curved profile tiles are available. These include gable rake, curved ridge, and hip tiles. In addition to

Figure 27.31 Common shapes of clay roofing.

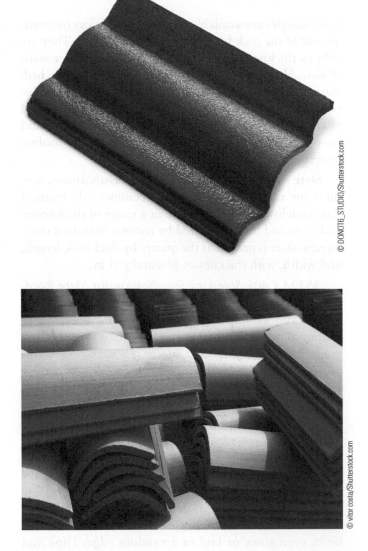

Figure 27.32 A typical installation of clay roofing tiles.

Figure 27.33 Eave, rake, and ridge details for clay tile roofing.

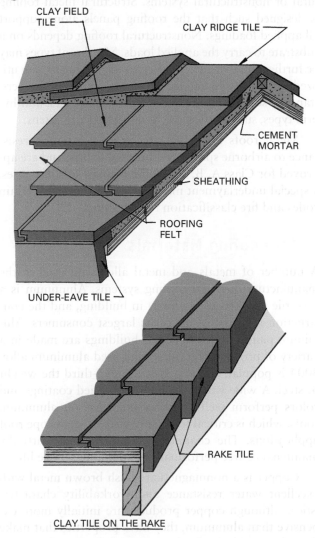

CLAY FIELD TILE

CLAY RIDGE TILE

CEMENT MORTAR

SHEATHING

ROOFING FELT

UNDER-EAVE TILE

RAKE TILE

CLAY TILE ON THE RAKE

being nailed to sheathing, they are set in a cement mortar. Valleys are flashed as described for other roofing products.

Cement-fiber shingles are made by combining Portland cement and various fibrous materials. One common product is a composition of Portland cement, organic and inorganic fibers, perlite, and iron-oxide pigments for color. Their installation parallels that of slate roof shingles (Figure 27.34).

Figure 27.34 Cement-fiber shingles are installed much like slate shingles.

Photovoltaic Roofing

The photovoltaic (PV) effect causes certain semiconductor materials (silicon, gallium, arsenide, copper indium dieseline, and cadmium telluride) to convert light energy into electrical energy at the atomic level. Photovoltaic cells are encapsulated between a transparent glass front and a backing material, and mounted in an aluminum frame to form photovoltaic modules that can be installed on a roof surface. Photovoltaic modules are typically mounted on a stable, durable metal rack that can support the array and withstand wind, rain, hail, and other adverse conditions.

Unlike conventional, rack-mounted solar panels, manufacturers are now producing solar cell-embedded tiles, shingles, and other low-profile photovoltaic installations that try to imitate the appearances of conventional roofing materials (Figure 27.35). Solar shingles, also called photovoltaic shingles, are solar cells designed to look like conventional asphalt shingles. There are several varieties of solar shingles including shingle-sized solid panels that take the place of conventional shingles in a strip, semirigid design containing several silicon solar cells that are sized more like conventional shingles, and newer systems using various thin film solar cell technologies that match conventional shingles in both size and flexibility.

Figure 27.35 Solar shingles are solar cells designed to integrate with conventional asphalt shingles.

Figure 27.36 While expensive in initial cost, metal roof systems provide lightweight and long-lasting weather protection.

Solar electric roofing tiles combine individual solar cells onto a fiber-cement slate or clay standardized roofing material for pitched-roof buildings. Also available are panel roofing products that combine structural metal roofing sheets with embedded solar cells.

For flat roof applications, PV modules made from thin, flexible, and lightweight amorphous silicon solar cells laminated in advanced polymers are being manufactured as the top layer of flat roof membranes as well. The membranes are adhered with a peel-and-stick adhesive backing or are mechanically fastened.

STEEP-SLOPE METAL ROOFING ASSEMBLIES

Metal roofing consists of cold-formed, corrugated, fluted, or ribbed metal sheets that serve as the exterior covering of the structure (Figure 27.36). Metal roofing for steep-slope roof systems is usually installed in sheets that are fastened together to maintain water tightness. Steep-slope metal roofs accelerate the rate of water run-off due to their smooth surface characteristics. They also reduce the likelihood of debris build-up, and the formation of mold and mildew. While expensive in initial cost, metal roof systems provide lightweight, attractive, and long-lasting weather protection. The durability and reliability of steep-slope systems have led metal roof manufacturers to warranty their products for up to 40 years. Steep-slope metal roof systems typically weigh from 40 to 135 pounds per 100 square feet, making them among the lightest roofing materials. A waterproof membrane is installed under metal roofing to guard against leaks.

Steep-slope metal roofing is available in either structural or nonstructural systems. Structural metal roofing is designed such that the roofing panels alone support all applied loadings. Nonstructural roofing depends on a substrate to carry the applied loads. Metal roof types may be further subdivided according to the joining system utilized, encompassing lapped seams and exposed fasteners, standing seams and hidden metal clip fasteners, and hybrid types, such as those with snap seams or battens.

Metal roofs are widely recognized for their resistance to airborne sparks and burning debris, and are approved for Class A, B, and C fire ratings. In some cases, a special underlayment may be required to meet certain codes and fire classification requirements.

Metal Roofing Materials

A number of metals and metal alloys are used in the manufacture of metal roofing systems. Aluminum is a versatile material used widely in building, and the construction industry is one of its largest consumers. Aluminum panels for commercial buildings are made in a variety of profiles. The commonly used aluminum alloy 3004 is popular, as the panels are one-third the weight of steel. A wide variety of factory-applied coatings and colors perform well and stay colorfast on aluminum roofs, which is critical for highly visible steep-slope roof applications. The coatings also help ensure virtually maintenance-free performance and a long service life.

Copper is a nonmagnetic reddish brown metal with excellent water resistance and workability characteristics. Although copper products are initially more expensive than aluminum, they have properties that make them less costly in the long run, such as resistance to corrosion and other damaging conditions. Copper roofs can last for decades, developing their distinctive greenish patina. Copper and copper alloys provide an excellent roofing choice and are often used for other exterior uses, including siding, flashing, and guttering.

Many styles of steel roof and siding systems are available. A variety of coatings are used to protect the steel surfaces, including galvanizing, zinc-aluminum alloy with a covering paint, siliconized polyester in a variety of colors, fluoro-polymer paint finish, and weathering copper coating. Terne is produced by coating metals such as carbon steel and stainless steel with a specially formulated alloy containing zinc and tin to dramatically increase corrosion resistance. Historically, terne roofing products contained roughly 80 percent lead and 20 percent tin. However, in the latter half of the twentieth century, as lead was found to have potentially harmful effects on health, the lead/tin alloy was replaced with a new zinc/tin alloy.

Zinc is a bluish-white metal that is brittle and has low strength. It is often referred to as a white metal and is widely used as a protective coating over steel to prevent corrosion. Zinc is a natural material that never fades and retains its look over its entire life span. It is also a noncorrosive, environmentally friendly product with a clear water runoff.

Standing Seam Roofing

The great advantage of standing seam metal roofs is hinted at in the name. Seams, the weakest aspect in any roof and a potential entry point for moisture, are raised above the level of the roofing panel. Panels run from the ridge of the roof all the way down to the eaves. Because the metal panels run unobstructed from the top to the bottom of the roof, the installation has no horizontal seams and a far fewer total number of seams. To address thermal expansion of the metal roofing, a system of floating attachment is the convention for standing seam metal roofing. Hold-down clips use two components, one fixed to the structure and a second locking into the seam and moving with the panel material (Figure 27.37). Differential movement is thus taken up between the two components of the attachment clip.

Roof panels for standing seam roofs are available either pre-formed or site-formed. Site-formed panels are created from rolls of metal that are run through mobile forming machines that crimp the metal into rigid panels. Two types of metal roof systems are most commonly used commercially today: Architectural Standing Seam Metal Roofs (ASSMR) and Structural Standing Seam Metal Roofs (SSSMR).

Figure 27.37 A standing seam roof is formed by rolling the edges of each metal roof panel with the cleat that is secured to the deck.

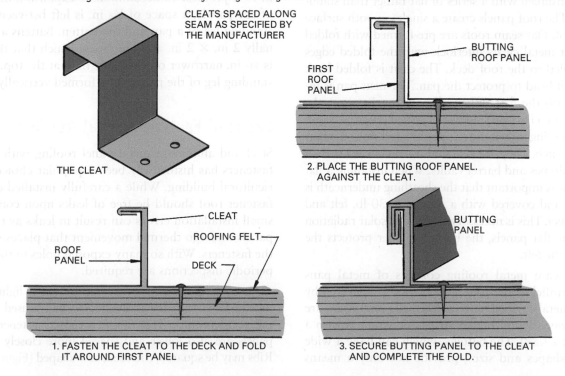

CLEATS SPACED ALONG SEAM AS SPECIFIED BY THE MANUFACTURER

THE CLEAT

CLEAT

ROOFING FELT

ROOF PANEL

DECK

1. FASTEN THE CLEAT TO THE DECK AND FOLD IT AROUND FIRST PANEL

BUTTING ROOF PANEL

FIRST ROOF PANEL

2. PLACE THE BUTTING ROOF PANEL. AGAINST THE CLEAT.

BUTTING PANEL

3. SECURE BUTTING PANEL TO THE CLEAT AND COMPLETE THE FOLD.

Significant differences exist in the supporting structures, materials, profiles, insulation, installation, and details of the systems.

An SSSMR is a system of metal roof panels supported and attached by clips fastened directly to the building structure. Clips are used to elevate the panels above the structural framing system, essentially allowing the roof to float. These panels are capable of spanning between the structural supports and can resist snow, dead, live, concentrated, and wind loads without benefit of any substrate material.

ASSMRs attach panels by concealed anchor clips to a continuous solid support substrate. The panels carry no structural loads and are not capable of spanning between structural supports without substrate materials, such as wood, metal, or concrete decks.

The installation of a standing seam roof begins with the application of underlayment over the roof decking. A preformed ridge cap is placed over the roof peak and secured per manufacturer recommendations. Every subsequent vertical panel is attached to the ridge and the underlying deck with clips. A crimping tool is used to crimp the raised edges of the panel ends together. This creates the raised seam appearance and provides a watertight seal to the finished roof. The standing seam can be anywhere from ½ to 1½ in. high.

Flat and Batten Seam Roofing

A flat seam metal roof consists of a series of metal roofing panels installed with a series of flat rather than standing seams. The roof panels create a single, smooth surface across a roof. Flat seam roofs are pre-formed with folded edges. Sheet metal cleats interlock with the folded edges and are nailed to the roof deck. The cleat is folded back over the nail head to protect the pan. The next pan is interlocked with the first. When all pans are in place, the edges are beaten flat and soldered or sealed (Figure 27.38). Flat seam roofing systems are typically used on roofs that have a low pitch. They are also find use on curved surfaces, such as domes and barrel vaults. When installing a flat seam roof, it is important that the sheathing underneath is continuous and covered with a minimum 30 lb. felt and building paper. This is necessary because as solar radiation heats up the flat panels, the building paper protects the integrity of the felt.

Batten seam metal roofing consists of metal pans that run parallel to the roof slope and are separated by wood- or metal-formed battens. Vertical panel edges are placed between wood batten strips and covered with a cap (Figure 27.39). The battens, which can have a wide variety of shapes and sizes, provide not only a means

Figure 27.38 A flat seam metal roof consists of a series of metal roofing panels installed with a series of flat rather than standing seams.

© ppl/Shutterstock.com

of securing the roofing, but also permit a wide variety of roof profiles. To accommodate expansion movement of the panels, a space of ¹⁄₁₆ in. is left between the upstanding leg of a pan and the batten. Battens are nominally 2 in. × 2 in. and are tapered such that their base is ¹⁄₁₆ in, narrower on each side than at the top. The upstanding leg of the pan is then formed vertically.

Exposed Fastener Metal Roofing

Steel and aluminum metal panel roofing with exposed fasteners has historically been a popular choice for agricultural building. While a carefully installed exposed-fastener roof should be free of leaks upon completion, small installation errors can result in leaks as the metal panels undergo thermal movement that places stress on the fasteners. With so many exposed holes in the panels, periodic inspections are required.

Exposed-fastener panels typically use a lighter-gauge metal when compared to the heavier gauge used in standing seam roofing. The ribs in exposed-fastener roofing panels are also lower and spaced more closely together. Ribs may be squared, rounded, or v-shaped (Figure 27.40).

Figure 27.39 A batten seam is constructed by snapping a metal batten cap over the cleats that secure the edge of the metal roofing to the deck.

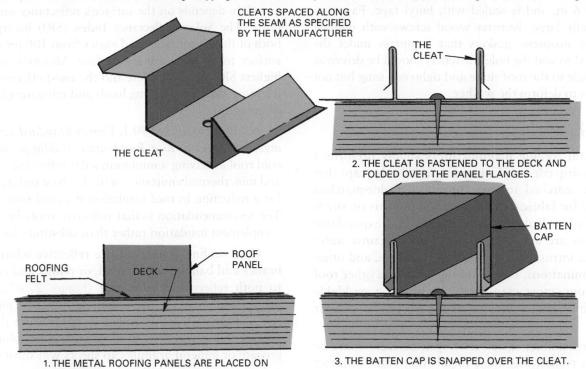

CLEATS SPACED ALONG
THE SEAM AS SPECIFIED
BY THE MANUFACTURER

THE CLEAT

THE
CLEAT

2. THE CLEAT IS FASTENED TO THE DECK AND
FOLDED OVER THE PANEL FLANGES.

ROOFING
FELT

DECK

ROOF
PANEL

BATTEN
CAP

1. THE METAL ROOFING PANELS ARE PLACED ON
THE DECK.

3. THE BATTEN CAP IS SNAPPED OVER THE CLEAT.

Figure 27.40 The ribs in exposed fastener roofing panels may be squared, rounded, or v-shaped.

© PhotoHouse/Shutterstock.com

Figure 27.41 Exposed fastener roofing is secured by large-diameter wood screws with integral EPDM or neoprene gaskets that compress under the screw head to seal the hole.

© Steven Frame/Shutterstock.com

The panels are available in standard sizes or can be factory cut to exact lengths. Ordering panels cut to exact lengths simplifies installation and reduces potential corrosion problems at field cuts. Excessive thermal movement can be a problem for steel panels longer than 40 feet. While traditionally installed over battens, most panels in commercial applications are now installed over solid plywood or OSB decking, with a minimum 30 felt lb. felt underlayment.

The installation of an exposed-fastener roof begins by anchoring a drip edge along the roof perimeter. A first panel is fit along the downwind end of the roof, such that the edge of each overlapping panel faces away from prevailing winds. After the first panel is screwed down, the next panel is set in place, overlapping the first. Side laps are typically sealed with butyl tape and held together with gasketed sheet-metal screws **(Figure 27.41)**. When horizontal seams are required, the upper panel laps the lower by 6 in. and is sealed with butyl tape. Fasteners are typically large diameter wood screws with integral EPDM or neoprene gaskets that compress under the screw head to seal the hole. Fasteners should be driven at a right angle to the roof slope and tightened snug but not so tight as to deform the washer.

Steep-Roof Flashing and Drainage

Most metal roofing manufacturers supply preformed flashings, drip edges, rake moldings, and ridge caps that are color matched to their products. Color-matched coil stock for fabricating custom components on site is also available. Rubber closure strips and expandable foam tapes are used to seal panel ends against water and insect intrusion at eaves, valleys, ridges, and other panel terminations. For plumbing vents and other roof penetrations, most manufacturers supply a moldable base flashing that conforms to the rib profile of their roofing.

When detailing metal roof systems, care must be taken not to mix dissimilar metals that come in direct contact with one another. The galvanic scale ranks a metal's tendency to react in contact with another metal in the presence of an electrolyte, such as water or even moisture from the air. In the presence of an electrolyte, dissimilar metals in contact will corrode more rapidly than similar metals in contact. For example, an aluminum gutter secured with copper nails produces galvanic action at the point of contact. When it is not possible to avoid contact between dissimilar metals, they can be given a coat of unleaded paint or separated with plastic or other non-conducting material. A joint can also be caulked or otherwise sealed to keep out moisture, thus eliminating the electrolyte.

Drainage design of steep-slope metal roof systems must address the sizing and locations of gutters and downspouts. The location and size of a gutter must be designed for the slope of the roof so that the gutter is not overshot or overloaded in individual locations. Gutter systems are sized according to the size of the roof, likely rainfall intensity for the project location, and the size and number of downspouts.

COOL ROOFING

A *cool roof* is defined as a roof surface that stays relatively cool when compared to the surrounding temperature. Materials with highly reflective, light-color surfaces help roofs to absorb less heat and stay considerably cooler

than conventional roofing materials. The temperature of a surface depends on the surface's reflectance and emittance. The Solar Reflectance Index (SRI) incorporates both of these properties and varies from 100 for a white surface to zero for a black surface. Materials with the highest SRI are the coolest and the most effective in reducing a building's cooling loads and mitigating the heat island effect.

ASHRAE Standard 90.1, *Energy Standard for Buildings Except Low-Rise Residential Buildings,* defines a cool roof as having a minimum solar reflectance of 0.70 and min. thermal emittance of 0.75. The standard allows for a reduction in roof insulation if a cool roof is used. The recommendation is that reflective roofs be used to complement insulation rather than substitute for it.

Cool roofing systems utilize reflective white membranes and ballast, white metal, or paints and coatings to both reflect and emit heat (Figure 27.42). Coatings include organic and complex inorganic pigments (CICPs) that are either rolled or sprayed on the roof surface. For membrane roof systems, light-colored aggregate or mineral granule cap sheets will substantially reduce maximum surface temperatures over black membranes.

Figure 27.42 Highly reflective, light-color surfaces help roofs to absorb less heat and stay considerably cooler than conventional roofing materials.

© pedrosala/Shutterstock.com

Construction Methods

Green Roofs

With increased interest in sustainable building technologies, the practice of applying vegetation directly to a roofing system is on the rise. Green roofs are classified as either intensive or extensive. An intensive green roof is installed on an adequately supported roof structure and creates actual roof gardens encompassing trees, bushes, and terraced surfaces. Extensive green roofs are used for smaller applications in both low-slope and pitched roofs. They are characterized by a shallow-growing medium and utilize low maintenance sedums, grasses, and mosses.

The basic components of a green roof include four essential layers:

A waterproofing layer may consist of a liquid-applied membrane, a specially designed single-ply roofing, or multiple layers. Correct and meticulous application of the waterproofing membrane is essential to insure the water tightness of a green roof. A thorough water flood test is usually conducted to check for membrane leaks before other layers are applied.

Green roofs have a drain layer that prevents water from accumulating on the roof or in the substrate. Pre-manufactured drainage systems consisting of a high-strength plastic core with two layers of geo-synthetic fabric attached to top and bottom are available. The geo-synthetic fabric acts as a root barrier that prevents plant roots from affecting the membrane and drainage systems.

The third layer consists of the growing medium or soil. Because natural soils are heavy, particularly when wet, green roofs often involve the use of lightweight, engineered soil mixes. Soil cannot contain any silt that might clog the filter fabric and should be water permeable and resistant to rot, heat, frost, and shrinkage.

Vegetation is the final and most visible layer of the green roof. Plants add aesthetic value and ultimately determine the success or failure of the roof. Compatibility of plant selection with the local environment is essential. Characteristics of plants typically used on green roofs include a shallow root system, good regenerative qualities, resistance to direct sun, drought, frost, and wind tolerance, and compatibility with local ranges of temperature, humidity, rainfall, and sun.

Green roofs exhibit a number of positive attributes when compared to conventional roofing systems. The soil layers and plants protect the roofing membrane from ultraviolet radiation damage, which extends the life of the roof. Roofs can reduce and slow the amount of storm-water runoff. Many municipalities now offer tax breaks and other incentives to builders in an effort to encourage the use of green roofs.

© 18042011/Shutterstock.com

Review Questions

1. List the four fire classifications for roof coverings.
2. What types of roof decking are used for low-slope roofs?
3. Why are vapor retarders an important part of a roof assembly?
4. What materials are used for vapor retarders?
5. Describe the process for installing a built-up roof covering.
6. What is a modified bitumen?
7. How are single-ply roofing membranes secured to a roof deck?
8. What types of single-ply roofing membranes are available?
9. Describe the makeup of a spray-applied roof coating.
10. How are standing-seam metal roof panels joined to form a watertight joint?
11. How are standing-seam metal roof panels secured to a deck?
12. Why are rooftop walkways needed?
13. List several locations where flashing is required.
14. What roof covering materials are used on steep-slope roofs?

Key Terms

Built-Up Roofing   Low-Slope Roof   Steep-Slope Roofs
Cool Roof   Modified Bitumen Membrane
Flashing   Single-Ply Roofing Membrane

Activities

1. Collect samples of as many roofing products as possible. Label each, listing its properties, fire rating, and advantages and disadvantages.

2. Set up a partial roof deck outside the classroom and install as many types of roofing systems as possible.

3. Ask a roofing contractor to visit the classroom and discuss bidding procedures, relative costs of various systems, and safety requirements that must be met during installation.

4. Visit various construction sites and observe the procedures being used for the installation of roofing systems.

Additional Resources

Schunk, Oster, Barthel, Kiessl, *Roof Construction Manual*, *Birkhauser Edition Detail, Munich*.

Concrete and Clay Roof Installation Manual, Roof Tile Institute, Eugene, OR.

NRCA four-volume publication, *Roofing and Waterproofing Manual*, National Roofing Contractors Association, Rosemont IL.

Residential Asphalt Roof Manual, the Asphalt Roofing Manufacturers Association, Calverton, MD.

Spence, W. P., *Roofing Materials and Installation*, Sterling Publishing Co., 387 Park Avenue South, New York 10016.

Steep-Slope Roofing Materials Guide, National Roofing Contractors Association, Rosemont, IL 20018.

Other resources include:

CD-ROMs: *Installation Tips & Techniques and Avoiding Common Roof Installation Mistakes*, GAF Materials Corporation, 1361 Alps Road, Wayne, NJ 07470.

See Appendix C for addresses of professional and trade organizations and other sources of technical information.

08

Openings
CSI MasterFormat™

Courtesy Eva Kultermann

Glass

LEARNING OBJECTIVES

Upon completion of this chapter, the student should be able to:

- Recognize the various types of glass products used in building construction.

- Select appropriate glass products for various applications.

Build Your Knowledge

For further study on these materials and methods, please refer to:

Chapter 29 Doors, Windows, Entrances, and Storefronts

Chapter 30 Cladding Systems

Glass has always been present in nature as the result of high temperature processes involving silica rocks or sands. Glasses are formed when crystalline materials melt at very high temperatures. Naturally occurring glass is produced when lightning strikes sandy beaches or is formed in volcanic lava. Glass is an inorganic mixture that has been fused at a high temperature and cooled without crystallization. It has an unusual internal structure because mechanically it is rigid and has the characteristics of a solid, yet the atoms in glass are arranged in a random order similar to those of a liquid. Glass is technically a super-cooled liquid. Glass has great intrinsic strength being weakened only by surface imperfections that account for its reputation to be fragile. Special tempering can minimize surface flaws. Tens of thousands of workable glass compositions exist.

The selection of glazing systems for a building is a process of ever-growing complexity. In the recent past, the range of glazing products available to designers was much smaller than what is available today. Windows and glazed walls have historically been regarded as net energy drains on a building. Recent developments and the introduction of insulated windows, glass tinting, and selective film coatings are beginning to change this perception. With the current concern with energy efficiency, choosing windows that will enhance the energy performance of a building becomes more crucial (Figure 28.1).

Figure 28.1 Glazing can provide daylight and view while enhancing the energy performance of a building.

Only by gaining a comprehensive understanding of the key physical parameters used to characterize glazing systems can an appropriate choice be made.

Glass is a major component in bringing daylight into a building. As lighting is often the single-largest consumer of electricity in buildings, all efforts should be made to integrate the use of daylight in building design. The challenge is to find a balance between the utilization of useful daylight and minimizing the solar heat gains that accompany it. By utilizing daylight, the need for electrical lighting will be reduced, as will the heat gains produced by artificial lighting, thereby reducing cooling loads. Shading and diffusing devices can be used to offset the possibility of glare, which accompanies natural light.

TYPES OF GLASS

Glass consists of three basic components, formers, and fluxes and stabilizers. The former, a coarse sand called silica oxide, makes up 75 percent of the mix. Fluxes that help the former melt at lower, more practical temperatures include soda ash, potash, and lithium carbonate. Additional lime and alumina stabilizers provide water resistance and stability.

The six basic types of glass are classified by their ability to resist heat (Table 28.1). Soda-lime-silica is the type commonly used for door and window glazing, as well as consumer bottles. About 90 percent of all glass produced is of this type. Its general composition includes 74 percent silica, 15 percent soda, 10 percent lime, and 1 percent alumina. Soda-lime-silica glass is easy to form and cut, has fair chemical resistance, and does not resist high temperatures or rapid thermal changes. It is easily shattered into small, sharp pieces.

Table 28.1 Thermal Properties of Commonly Used Glass

Categories of Glass	Thermal Expansion	Heat Resistance
Soda-lime-silica	High	Low
Lead-alkali-silica	High	Low
Fused silica	Low	High
96% silica	Low	High
Borosilicate (Pyrex)	Medium	Medium
Aluminosilicate	Medium	Medium

Lead-alkali-silica glass has similar properties to soda-lime but is more expensive.

Fused silica glass is composed of about 99 percent silicon dioxide and is the most expensive. It has the highest resistance to heat, managing temperatures ranging from 1,650 to 2,190°F (900 to 1,200°C). It also has the highest corrosion resistance and affords excellent transmission of ultraviolet rays.

Ninety-six percent silica is used when high thermal hardness and the ability to withstand the thermal shock of going rapidly from hot to cold is required. Borosilicate glass is a thermally hard glass that has been used for many years. It is best recognized by its Corning trade name, Pyrex®. Aluminosilicate glass is more costly than borosilicate but can withstand high service temperatures and is similar in its ability to withstand thermal shock.

This chapter is largely devoted to the uses of soda-lime-silica glass, because it is the most widely used glass for construction applications.

THE MANUFACTURE OF GLASS

Although several glassmaking processes have been used for many years, most glass today is produced worldwide by the *float glass process*. The first production of glass by this method was in 1959 by the English firm Pilkington Brothers, Ltd.

Float Glass

The float glass process involves producing molten glass in a furnace from which it is conveyed to a float bath. Here the molten glass floats across a bath of molten tin (Figure 28.2). The molten tin provides a very flat surface that supports the glass as it is polished by the application of heat from above. The heat melts out any irregularities. The ribbon of glass moves to a cooling zone, which permits the glass to solidify enough to be conveyed on to the annealing lehr (furnace). After the glass has been annealed, it is cut into lengths, inspected, and packed. The sheets of glass produced by this method have parallel surfaces; a smooth, clear finish; and high optical clarity.

Float glass is a flat glass available as regular float glass or heavy float glass. Thicknesses range from 3/32 to ½ in. (2.5 to 12 mm). Regular float glass is made in three forms with specific qualities: silvering, which is used in selected high-quality pieces for optical uses and mirrors; mirror glazing, for general-purpose mirrors; and glazing for door and window glazing. Float glass provides clarity and visual transparency with a minimum of distortion. Float glass is used for many products, such as reflective glass, mirrors, tinted glass, laminated glass, and insulating glass.

Figure 28.2 The float glass manufacturing process.

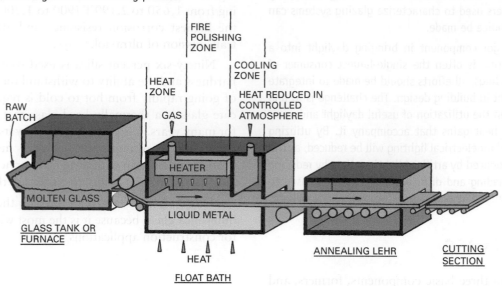

Sheet Glass

Sheet glass is a type of flat glass that is more economical than float glass. It is made using older methods that involve drawing a ribbon of molten glass along a series of rollers where its thickness is established and it is annealed, cooled, and cut to size (Figure 28.3). Sheet glass has more distortion than float glass and is not as widely used. It is available in single strength, ³⁄₃₂ in. (2.3 mm) thick; double strength, ⅛ in. (3.1 mm) thick; and heavy sheet ³⁄₁₆ in. (4.7 mm) and ⁷⁄₃₂ in. (5.6 mm) thick. Picture glass is a thinner version, ³⁄₆₄, ¹⁄₁₆, and ⁵⁄₆₄ in. (1.2, 1.6, and 2.0 mm) thick that is used for covering pictures and for other purposes where strength is not a factor. Sheet glass is available in three grades: AA (best), A (good), and B (general glazed), and as clear, tinted, reflective, tempered, or heat-treated products.

Figure 28.3 Sheet glass after being cooled and cut to size.

PROPERTIES OF SODA-LIME-SILICA GLASS

The following discussion presents the general properties of soda-lime-silica glass. Glass properties can be varied by altering the composition of its ingredients.

Mechanical Properties

When deciding on a glass for a specific application, its ability to withstand breakage is a major consideration. Glass is brittle yet remains elastic up to its ultimate tensile strength. This means that glass can be bent up to a breaking point and, when released, will return to its original position. Glass breaks by bending or stretching; therefore, tensile strength is a major mechanical property to be considered.

Glass does not have a clearly defined tensile strength because its actual strength depends on the surface condition of the glass. A small scratch or nick is sufficient to cause tensile stresses to concentrate at that weaker point. Therefore, actual tensile strength is less than the theoretical tensile strength. Mechanical strength can be increased by chemical and heat treatments.

Thermal Properties

Heat can be transferred by conduction, convection, and radiation. The thermal conductivity of glass is high, but lower than most metals. The U-value for a single-strength sheet of glass is about 1.04 to 1.10. Double glazing with a ½ in. (12 mm) air space reduces the U-value to about 0.49 to 0.56. Typical thermal properties for selected flat glass products are given in Table 28.2.

Table 28.2 Thermal Properties of Flat Glass

Glass	Thickness (in.)	Winter Nighttime U-value/R-value	Summer Daytime U-value/R-value
Clear single	¼	1.3/0.78	1.04/0.96
Clear double	⅛	0.49/2.04	0.52/1.92
Clear double	¼	0.49/2.04	0.56/1.79
Clear double with Low-E film	¼	0.31/3.32	0.32/3.03
Light brown single	¼	1.13/0.88	1.10/0.91
Dark brown single	¼	0.89/0.88	0.89/1.88
Double glass light brown/clear	¼	0.31/2.04	0.33/1.75
Double glass dark brown/clear	¼	0.41/2.04	0.47/1.72
Light green single	¼	1.13/0.88	1.10/0.91
Dark green single	¼	0.95/1.14	0.98/1.12
Double light green/clear	¼	0.50/2.04	0.59/1.75
Double dark green/clear	¼	0.42/2.50	0.50/2.13
Clear insulating glass with suspended Low-E film	¼	0.23/4.30	0.37/2.70

For single-glazed glass, the majority of thermal resistance is borne at the exterior and interior surfaces. Indoors, about two-thirds of heat flows by radiation to room surfaces and one-third flows by convection. Heat transfer of the interior surface of glazing can be reduced a great deal by adding low-emissivity metallic films (referred to as Low-E) to the glass. This also reduces the U-value for glazing units with air spaces. A typical double-glazed opening with suspended Low-E film has a U-value of 0.31 to 0.32. Other glass products, such as tinted glass, reflective glass, and insulating glass, also reduce heat gain and loss.

The thermal expansion of glass must be considered as glazing units designed. The coefficient of expansion of soda-lime glass is about 4.5×10^{-6}, while the coefficient for aluminum is about 13×10^{-6}. An aluminum window frame contracting under low temperatures imposes stress on glass edges unless sufficient clearance is allowed for the difference in thermal movement.

Chemical Properties

Glass used in typical applications in building construction is very durable and more resistant to corrosion than are many other materials. Since it is not porous, it will not absorb moisture or chemical elements in the ground or atmosphere. A few exceptions may occur in isolated industrial applications. Hot concentrated alkali solutions and superheated water can cause soda-lime-silica glass to dissolve, while hydrofluoric acids will cause corrosion.

Electrical Properties

Glass is a good electrical insulator and is widely used for applications where this and other properties make it useful. Light fixtures often consist of glass, as do light bulb envelopes, because it is strong, heat resistant, and an electrical insulator.

Optical Properties

Glazing products are not completely transparent to incoming radiant energy. A fraction of solar radiation is reflected, another absorbed, and a third transmitted. Clear sheet glass permits the passage of about 86 to 89 percent of visible light. Double-glazed lights (window panes) permit about 80 to 82 percent light passage. Typical daylight transmittance figures are shown in Table 28.3.

When glass is treated to reflect or absorb light, the percent of light transmitted is greatly affected. Tinted glass frequently transmits less than 50 percent of the solar radiation striking it. Glass with reflective coatings transmits 6 to 50 percent of visible light. Optical quality

Table 28.3 Daylight Transmittance for Selected Glass Units

Glass	Percent Daylight Transmittance
Clear single	86–89
Clear double	80–82
Clear double with Low-E film	72
Clear triple	72–74
Clear insulating with Low-E film	37–69
Light brown single	52
Light green single	75
Light gray single	41

is also influenced by lack of distortion. If the front and back surfaces of a sheet of glass are not parallel, images viewed through the sheet will form some angles, appearing wavy or distorted. With the increased use of float glass, distortion has been reduced and optical properties improved.

HEAT-TREATING GLASS

The properties of glass can be improved by various heat-treating methods. These include annealing, tempering, and heat strengthening.

Annealing

Annealing is part of the normal production of glass. After a glass ribbon has been formed, it is passed through an annealing lehr, where temperatures are carefully controlled. The temperature of the glass is raised high enough to relieve strains developed during forming in the float bath. Then the temperature is slowly lowered, allowing all parts of the glass to cool uniformly.

Tempering

Tempering increases strength. Tempering involves raising the temperature of glass just shy of its softening point and then blowing jets of cold air on both sides to cool it quickly. Surfaces harden and shrink while the interior is still fluid. As the interior cools and shrinks, the exterior remains unchanged in size, resulting in compression forces along the surfaces and edges and

tension in the interior. The opposing compressive and tension forces balance each other, resulting in a stronger glass. Tempered glass is three to five times as resistant to damage as annealed glass. When the thin tempered skin on the glass is broken, the entire sheet disintegrates into small pebble-like particles instead of sharp slivers (Figure 28.4).

Figure 28.4 When the skin on tempered glass is broken, the entire sheet disintegrates into small particles instead of sharp slivers.

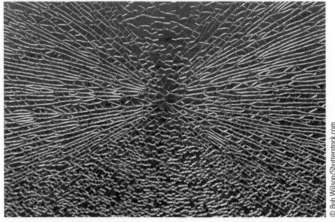

© Rob Wilson/Shutterstock.com

Heat Strengthening

Heat-strengthened glass is heated and cooled much like tempered glass. Heat-strengthened glass has a compression range of 3,000 to 10,000 lb. per sq. in. (surface) or 5,500 to 9,700 lb. per sq. in. (edge), making it about twice as strong as annealed glass.

Chemical Strengthening

Glass can be *chemically strengthened* via immersion in a molten salt bath, causing larger potassium ions in the salt to replace the smaller sodium ions already in the glass. This crowds the surface, causing compressive stresses.

FINISHES

A variety of finishes are available for glass, depending on its end use. Although most glass produced is clear and transparent, finishes make it useful for other applications (Figure 28.5).

Figure 28.5 Finishes available for glass include etched (a), patterned (b), fritted (c), and tinted (d) among others.

© Simon Bratt/Shutterstock.com

© Irina_QQQ/Shutterstock.com

© Markus Pfaff/Shutterstock.com

Etching

Etching glass produces a surface that can range from almost opaque to smooth but translucent. The degree of opacity depends on the length of time the glass is exposed to hydrofluoric acid or other etching compounds. The glass may be dipped or sprayed with the etching fluid. The finished surface is opaque but remains smooth to the touch.

Sandblasting

Glass is sandblasted by blasting its surface with coarse-grained sand particles blown by compressed air. This produces a translucent finish that is generally rougher than etched glass. Etching and sandblasting abrade the surface, greatly lowering the strength of the sheet.

Patterned Finish

Patterned glass has a texture or pattern rolled into the surface as the glass is drawn through the furnace. It may be colorless or tinted and is available in a wide variety of patterns. It may be relatively transparent from one side and translucent from the other.

Silvering

Mirrors are produced by spraying silver nitrate and tin chloride on the surface of glass as it moves on a conveyor. For additional protection, this layer can be covered with shellac, varnish, or paint or a layer of copper can be electroplated over it.

Considered the highest quality, float glass mirrors are available in silvering, mirror glazing, and glazing quality, with silvering being the best. Sheet glass used for window glazing can be used for low-quality mirrors. These come in two grades: A and B.

Ceramic Frit

Glass surfaces can be coated with a ceramic frit that is fired at high temperatures. Ceramic frit is composed of small powdered particles produced by quenching a molten glossy material. This produces a colored surface layer that is highly weather resistant. Both translucent and opaque colored layers are available. These are often used for finished exterior surfaces on wall spandrels. A special type of fritted glass adds the frit in visible patterns of dots or lines, to control solar heat gain. The pattern is created by opaque or transparent glass fused to a substrate glass material under high temperatures.

Tinting

Glass can be infused with color by tinting. One function of tinted glass is to minimize the amount of solar gains entering a building. Tinted glass is also known as heat-absorbing glass, since the glass itself absorbs much of the incoming energy. Tinting is achieved by adding color-producing ingredients to the glass at the beginning of its production process. The color is technically not a surface finish but rather is infused in the glass.

Tinted glass is often used to reduce glare. Blue-green glass achieves a moderate reduction in glare and brightness. Bronze tints are used primarily to reduce heat gains. Gray tints give a wide range of light transmittance values. Translucent glass reduces the entry of direct sunlight yet permits considerable transmission of diffused light.

PHYSICAL PROPERTIES OF GLAZING SYSTEMS

A number of physical characteristics are used to describe the performance of glazing systems. Manufacturers supply these performance parameters directly with commercially available glazing units. Requirements for the energy performance of glazing systems are given in the *International Energy Conservation Code*.

Solar Heat Gain Factor (SHGF)

The total solar energy transmittance of a glazing unit is its solar heat gain factor (SHGF). It is the energy entering the interior through a glazing unit directly, either as direct or diffuse beam radiation, with a smaller amount absorbed by the glass being transferred inside via infrared radiation or convection. The concept is important in determining the amount of heat gain that will enter a space through a glazing unit. Values for SHGF are given on a scale from zero to one, with "one" representing 100 percent of solar heat being transmitted and "zero" indicating no heat being transmitted. The lower a window's solar heat gain coefficient, the less solar heat it transmits.

U-Value (U)

The U-value is a measure of the overall heat transfer of an assembly, or the rate of conductive, convective, and radiative heat that occurs due to the difference in temperature between the inside and outside. This can be either heat loss or heat gain, depending on exterior climate conditions. For a window unit, the U-value represents the sum of heat transfer from the center of the glass, the edge of glass, and the frame. The smaller the U-value, the better the thermal performance of the window. Heat transfer can be minimized with coatings that reduce radiative heat losses and specialized frames that limit conductive losses.

Visible Light Transmittance (Vt)

Visible light transmittance (Vt) is that portion of incoming solar radiation that occurs in visible wavelengths, or electromagnetic radiation to which the human eye is sensitive: a small amount from 380 to 760 nanometers. Only this visible transmittance is useful for daylighting purposes. The visible component is about one-half of the total incoming radiation; the rest is mostly in the infrared, with small amounts in the ultra-violet. In general, a high value (.6 or above) of Vt is desirable, as this will reduce the need for artificial lighting.

Shading Coefficient (Sc)

The solar heat gain of glass is related to the total energy transmittance with a shading coefficient. The *shading coefficient (Sc)* is defined as the ratio of solar heat gain of a glazing unit to the solar heat gain of clear float glass. The lower the shading coefficient, the more efficient the glazing unit is at reducing solar gains.

Emissivity (e)

Another important characteristic of glazing materials is their emissivity, which is the ability of an object to radiate energy. A material that radiates energy perfectly will have an emissivity of 1, while one that radiates no heat will have a value of 0. Materials with smooth and reflective surfaces will have a lower emissivity than those with rougher, more textured surfaces. Clear float glass has an emissivity of 1. The emissivity of a glazing unit can be altered by the introduction of coatings on its glass surfaces. Low-emissivity (Low-E) coatings reduce the radiative heat transfer of a window unit. The positioning of the coating within the unit can produce either enhanced solar control or solar gains.

GLASS PRODUCTS

An extensive number of glass products are available for specific purposes. Specifications vary by manufacturer.

Tempered Glass

Tempered glass is used when strength beyond that of standard annealed glass is required. Tempered glass is made in accordance with Federal Specification DD-G-1403B, which requires minimum compression values of 10,000 lb. per sq. in. (surface) or 9,700 lb. per sq. in. (edge). Fully tempered glass can meet safety standards required by ANSI Z97.1-1975 and Federal Standard CPSC 16 CFR 1201 *Safety Standard for Architectural Glazing Materials*. Building codes specify that tempered

glass be used in interior and exterior doors and sidelights, any glass within 36 inches (horizontal) of walking surfaces, glass sheets with area greater than 9 sq. ft., and glass in guards and railings.

Laminated Glass

Laminated glass is used in areas where the glazing is subject to possible breakage or security of glazed openings is required. Typical applications from a safety standpoint include residential and commercial door glazing and windows in high-impact areas, such as gymnasiums. Special security applications may be required in detention centers and banks.

Laminated glass is made by bonding layers of float glass with intermediate layers of plasticized polyvinyl butyral (PVB) resin or polycarbonate (PC) resin. The glass is chemically strengthened by immersing it in a molten salt bath. Laminated glass is made to meet ANSI Z97.1-1975 and Consumer Products Safety Commission Regulation 16 CFR1201 Categories 1 and 11.

Many variations and thicknesses are available from various manufacturers. Laminated glass is available with ultraviolet filtering laminates and wire glass laminates. The widths of the various products vary by manufacturer, but thicknesses from ¼ to ⅜ in. (6 to 75 mm) are common. Typical lamination designs are shown in Figure 28.6.

Laminated glass for use in high-traffic security areas is composed of three sheets of annealed glass and two intermediate layers. It will withstand impact by heavy wood and other small, lightweight objects. When more security is needed, a five-glass panel with four plastic intermediate layers is used to provide protection against attack by bricks, bottles, and chairs. A laminate for high-risk areas has a polycarbonate core from ¼ to ⅜ in. (6 to 9 mm) thick, with special intermediate-layer materials plus chemically strengthened glass. Made with multiple cores for greater strength, laminated glass can withstand pounding by sledgehammers, and is used where maximum security is needed. Bullet-resistant laminated glass has a similar construction as those just mentioned but with additional layers of glass and PVB intermediate layers. It is available in thicknesses ranging from 1³⁄₁₆ to 2½ in. (30 to 63 mm). Greater thicknesses are available on special order.

Another type of laminated glass is *acoustical glass*. It is manufactured with a sound-absorbing plastic sandwiched between two or more pieces of glass. The sound reduction is achieved because this soft interlayer allows

Figure 28.6 Several types of laminated glass used to increase the security of glazed openings.

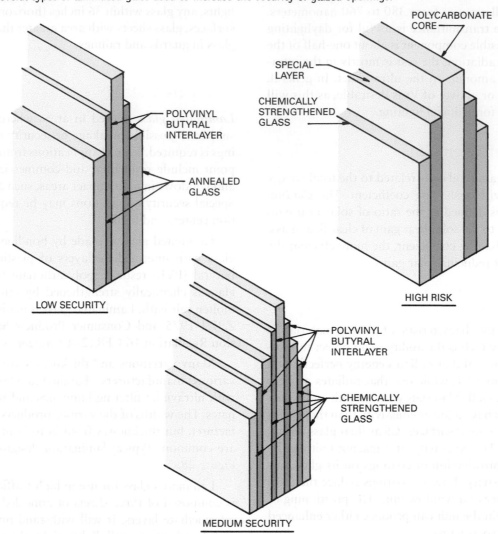

POLYVINYL
BUTYRAL
INTERLAYER

ANNEALED
GLASS

LOW SECURITY

POLYCARBONATE
CORE

SPECIAL
LAYER

CHEMICALLY
STRENGTHENED
GLASS

HIGH RISK

POLYVINYL
BUTYRAL
INTERLAYER

CHEMICALLY
STRENGTHENED
GLASS

MEDIUM SECURITY

the glass unit to bend in response to pressure from sound waves. It is also available in insulated constructions, thus reducing heat loss and gain as well as sound transmission. Acoustical glass comes in thicknesses of ¼, ½, and ¾ in. (6, 12, and 18 mm), with sound class ratings of 36, 40, and 41. It is commonly used for office partitions and in radio, television, and recording studios.

Reflective Glass

Adhered reflective metal coatings reduce the heat build-up of the glazing associated with tinted glass. Here the majority of incoming radiation is simply reflected directly from the exterior surface. To understand how reflective glass reduces the amount of solar energy transmitted through a glass opening, consider the following discussion.

When solar energy strikes a glass surface, it can be (1) transmitted through the glass, (2) reflected, or

(3) absorbed in the glass. Of that portion that is absorbed in the glass, part of the energy will be re-radiated and convected (transmitted) inward and part of it will be re-radiated and convected outward. The solar heat gain consists of that portion that is transmitted through the glass plus the portion of the absorbed energy that is re-radiated and convected to the inside. The solar heat rejected is that portion of the original solar incidence that is returned to the exterior. This includes the energy reflected from the glass surface plus any absorbed energy that is re-radiated and convected outward.

Another source of energy transfer through a glass window results from convection. Convection is the transfer of heat by currents of air resulting from differences in air temperature and density in a heated space. The total heat gain of a window is the sum of solar heat gain and convective heat gain (Figure 28.7).

Figure 28.7 Compared to clear glass, reflective glass reduces the amount of solar energy transmitted through a glazed opening.

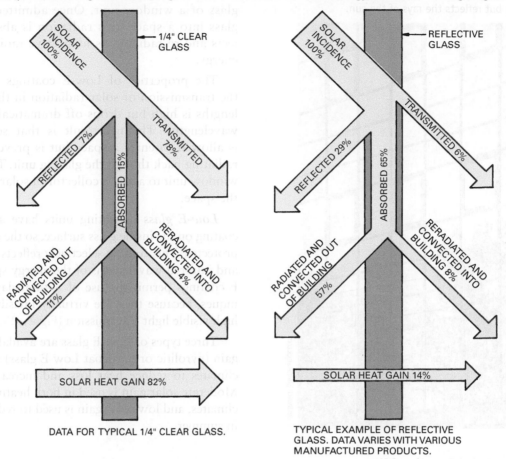

DATA FOR TYPICAL 1/4" CLEAR GLASS.

TYPICAL EXAMPLE OF REFLECTIVE GLASS. DATA VARIES WITH VARIOUS MANUFACTURED PRODUCTS.

It is these factors designers consider when choosing a type of glazed opening.

Reflective glass has one surface covered with thin, transparent layers of metallic film. Several metals and mineral oxides are used, producing a variety of colors and heat reflection ratings. On single-sheet glazing, the metallic film is applied to the surface facing the inside of a building. On insulating units, the film is on the outside face of the glass facing the inside of the building (Figure 28.8). Reflective glass is available in colors such as silver, green, blue, bronze, copper, and gold. Variations of these colors are available from different manufacturers (Figure 28.9).

Visible light transmittance varies by coating but ranges in general from 8 to 45 percent. Coatings are available on clear or tinted heat-absorbing glass. Most reflective glasses are ¼ in. (6 mm) thick, but thicker sheets are available. A summary of the solar-optical properties of clear, tinted, and reflective float glass is given in Table 28.4.

Figure 28.8 Reflective glass has one surface covered with thin, transparent layers of metallic film.

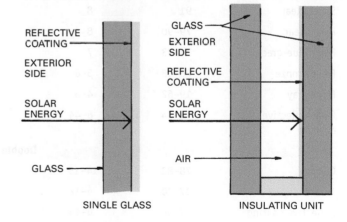

SINGLE GLASS

INSULATING UNIT

Low-E Glass

Low-E glazing units have special coatings applied to the glass panes that are designed to reduce a window unit's heat transfer. Solar radiation in the form of

Figure 28.9 High-performance reflective glass lets daylight enter the building but reflects the rays of the sun.

© tandemich/Shutterstock.com

short wavelengths has the ability to pass through the glass of a window unit. Once admitted through the glass into a space, this radiation is absorbed by objects and re-radiated as long-wave radiation, or heat energy.

The properties of Low-E coatings are such that the transmission of solar radiation in the short wavelengths is high but drops off dramatically in the long wavelengths. The net result is that solar radiation is allowed to enter a space but is prevented from re-radiating back though the glazing unit. This allows the window unit to act as a collector of solar energy within the space.

Low-E glass insulating units have a thin metallic coating on the inside glass surface, so the coating is fully protected. This coating selectively reflects the ultraviolet and infrared wavelengths of the energy spectrum. Low-E coatings permit the use of natural daylighting techniques, because they are virtually invisible and have a high visible light transmission (Figure 28.10).

Three types of Low-E glass are available. High solar gain (pyrolitic or hardcoat Low-E glass) is used in cold climates to reduce heat loss and increase solar gains. Moderate solar gain is used in both heating and cooling climates, and low solar gain is used to reduce heat gains in summer.

Table 28.4 Solar Optical Properties of Glass

	Light (%)		Solar Energy (%)		U-value Btu/hr./ft.²/°F	
	Transmitted	Reflected	Transmitted	Reflected	Winter	Summer
Single Glazed						
Sheet clear	91	8	86–89	7	1.13	1.02–1.03
Float clear	79–90	8	58–86	7	1.00–1.13	.98–1.03
Float blue-green	74–83	7	48–64	6	1.10–1.13	1.08–1.09
Float bronze	26–68	5–6	24–65	6	1.08–1.13	1.08–1.09
Float gray	19–62	4–6	22–63	5	1.06–1.13	1.08–1.09
Float reflective coating	8–34	6–44	6–37	14–35	0.90–1.11	0.89–1.12
Double Glazed						
Clear	78–82	14–15	60–71	14	0.49–0.57	0.55–0.62
Tinted	37–70	6–12	34–55	6–11	0.49–0.57	0.57–0.64
Reflective coating	7–30	6–44	5–29	14–35	0.41–0.49	0.47–0.58
Triple Glazed						
Clear	70–73	20	46–60	19–20	0.31–0.38	0.40–0.46
Tinted	11–56	8–17	25–46	8–16	0.31–0.38	0.40–0.46
Reflective coating	16–22	8–44	5–22	8–44	0.28–0.31	0.34–0.41

Figure 28.10 Low-E glass uses special low-emissivity coatings to reduce heat loss and regulate solar gains.

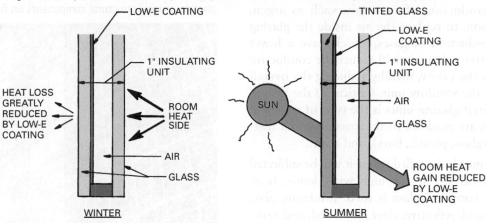

Insulating Glass

Insulating glass is a manufactured glazing unit composed of two layers of glass with an airtight, dehydrated air space between them. The insulated, or double-pane, window utilizes two panes of glass separated by a spacer with a hermetically sealed air gap between. New triple-pane units are also available from selected manufacturers. A wide range of multiple pane units are designed for use on commercial and residential buildings. Examples of these are shown in **Figure 28.11**. Insulating glass lowers heating and cooling costs by reducing air-to-air heat transfer.

Figure 28.11 Typical frame details for insulated glass units.

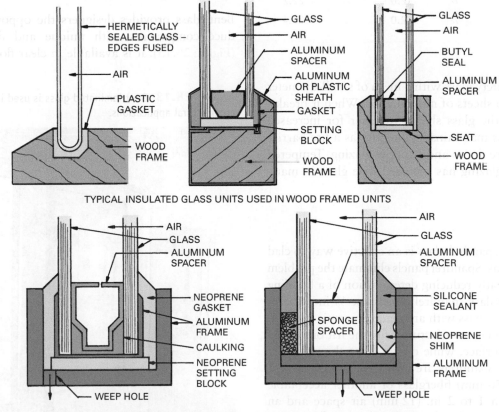

TYPICAL INSULATED GLASS UNITS USED IN WOOD FRAMED UNITS

TYPICAL INSULATED GLASS UNITS USED FOR LARGE COMMERCIAL WINDOWS

A better rating of thermal resistance can be achieved through the introduction of materials, such as argon, krypton, or xenon, to replace the air inside the glazing cavity. The introduction of gases, which have a lower thermal conductivity than air, will reduce the conductive heat loss within the cavity, thereby reducing the overall heat transfer of the window unit. Critical to the performance of insulated glazing units is the type of edge seal used. Edge seals are made of metal spacers with a thermal break, fiberglass, plastic, butyl, and foam.

When glass must be installed where it will be subjected to unusual thermal stresses or high wind loads, heat-strengthened or tempered glass is used. Insulating glass units also use tinted, reflective, clear laminated, and Low-E glass. The sizes of typical units are shown in Table 28.5.

Table 28.5 Sizes of Insulating (Float) Glass

| Maximum size | 90 × 144 in. (2286 × 36576 mm) | | | | |
| Minimum size | 15 × 20 in. (380 × 508 mm) | | | | |

| Glass Thicknesses | | Air Space | | Unit Thickness | |
in.	mm	in.	mm	in.	mm
3/32	2.5	1/4	6.0	7/16	11.0
1/8	3.0	1/4	6.0	1/2	12.7
3/32	4.0	1/4	6.0	9/16	14.2
3/16	5.0	1/2	12.0	7/8	22.2
1/4	6.0	1/2	12.0	1	25.4

Wired Glass

Wired glass is safety glass with a mesh of small-diameter wires rolled into sheets of molten glass. When it breaks, the wires hold the glass shards together for increased safety. Wire glass maintains its integrity as a fire barrier and is used in fire doors and window glazing. Tempered and laminated glazing has replaced wire glass in many applications.

Spandrel Glass

Ceramic-coated spandrel glass is an effective way to clad exterior wall areas. Spandrel panels eliminate the problem of corrosion, greatly reducing deterioration of a building (Figure 28.12). Glass spandrel panels use 1/4 in. (6 mm) heat-strengthened glass with an opaque ceramic frit fired on its surface. Some manufacturers fire the frit on the exposed exterior surface, while others favor the protected interior surface. Spandrels are insulated, usually with 1 to 2 in. (25 to 50 mm) fiberglass or another acceptable insulation with a 1 to 2 in. (12 mm) air space and an aluminum foil vapor barrier facing the panel's interior

Figure 28.12 Glass spandrel panels are used between vision glass to conceal structural components on fully glazed facades.

© Roman Sigaev/Shutterstock.com

side. Panel sizes vary, but typical sizes available include minimum panels of 18 × 33 in. (457 × 838 mm) and maximum panels of 84 × 144 in. (2,130 × 3,660 mm).

Bent Glass

Bent glass provides designers the opportunity to produce components with unique and dramatic looks (Figure 28.13). It is available in clear float, tinted float,

Figure 28.13 Bent laminated glass is used in variety of architectural applications.

© iofoto/Shutterstock.com

obscure, wire, Pyrex, patterned, and other types. Some forms of reflective glass can also be bent. Laminated safety glass and double-glazed thermal insulating units can be fabricated. Thickness and size specifications must be developed in consultation with the manufacturer.

Ceramic Glass

Ceramic glass shares many properties with traditional crystalline ceramics. Ceramic glass is the standard for many fire-rated glazing applications. Depending on the product, it can offer fire-ratings for up to 90 minutes, high-impact safety ratings, and sound reduction (Figure 28.14).

Figure 28.14 Ceramic glass can be used for fire-rated glazing applications.

Architectural Beveled Glass

Architectural beveled glass provides a unique transition from outside to inside and is an outstanding architectural detail. The glass is made in a wide range of standard sizes and designs as well as custom designs. The pieces of beveled glass are assembled into panels with lead caming.

Channel Glass

Channel glass, sometimes referred to under the trade name Profilit, consists of self-supporting U-shaped cast-glass channels (Figure 28.15). The glass is held in an extruded metal perimeter frame. It can be installed vertically or horizontally, in single line or double assembly. The glass is manufactured in a variety of colors, patterns, and translucencies. The U-shaped channels can extend uninterrupted up to 23 ft. vertically, and nearly unlimited horizontally. To reduce thermal bridging, insulating aerogel can be incorporated into the overlapping channels.

Figure 28.15 Channel glass consists of self-supporting U-shaped cast glass channels.

Glass Block

Glass blocks are made as solid and cavity units and are laid in a mortar bed similar to masonry units. Various designs permit light to be diffused, reduced, or reflected. The translucence or transparency varies by manufacturer (Figure 28.16).

Figure 28.16 Glass blocks are laid in a mortar bed similar to masonry units.

© Souchon Yves/Shutterstock.com

Glass blocks are made by fusing two halves of pressed glass together, creating a space with a partial vacuum. This produces an insulating value equal to a 12 in. (305 mm) thick concrete wall (U-value 0.51, R-value 1.96). Glass blocks are very strong and provide excellent security.

Glass blocks serve well as solar glazing in walled areas facing in a southerly direction. Blocks are available with a variety of surface patterns that help provide privacy, brightness, and control light transmission. Blocks also can have a highly reflective thermally bonded oxide surface coating that reduces heat gain and transmitted light. Selected properties of glass block are given in Table 28.6.

Smart Glazing

The past decades have seen the development of a number of new, dynamic glazing products that have the ability to respond to changing environmental conditions.

Chromogenic glazing materials display an ability to change their optical properties in response to fluctuations in solar radiation intensity, changes in temperature, or through the introduction of electric charges (photochromic, thermochromic, electrochromic). The transmittance, reflectance, and absorption characteristics of the glazing are altered according to changing environmental conditions. Originally developed for sunglasses, these products offer a new way to regulate solar gains over the range of daily and yearly environmental conditions.

Studies conducted to test control strategies find that electrochromic windows have excellent optical clarity; no coating aberrations; uniform density of color across their entire surface during and after switching; smooth, gradual transitions when switched; and excellent synchronization (or color-matching) between groups of windows during and after switching.

Photovoltaic Glazing

Research is now concentrating on the development of building integrated photovoltaic (BIPV) systems. This new technology will enable glazing to not only provide

Table 28.6 Properties of Single Cavity Glass Block

	Nominal Sizes		Compressive Strength (psi)	Light Transmission (%)	U-value
	3" thick	4" thick			
Clear	3 × 6 6 × 6 4 × 8 8 × 8 4 × 12 6 × 8 12 × 12	6 × 6 8 × 8 12 × 12	400–600	75	.52
Clear with reflective coating		8 × 8 12 × 12	400–600	5–21	.56
Light-diffusing pattern		8 × 8 12 × 12	400–600	39	.60
Decorative pattern	3 × 6 6 × 6 4 × 8 4 × 12 8 × 8 6 × 8 12 × 12		400–600	20–75	.65

enclosure and lighting for a building but also to simultaneously generate electricity.

A typical PV glazing unit uses two panes of clear or tinted glazing with the photovoltaic modules adhered to the front of the window's inside pane (Figure 28.17). Most PV glazing is available only by special order. The PV cells can be spaced in a variety of patterns and densities to aid in blocking sunlight from interior spaces.

Self-Cleaning Glass

Self-cleaning glass is manufactured with the same advanced pyrolytic technology used to produce glass panels for electronic and photovoltaic solar cell applications. A pyrolitic surface is applied to clear float glass via a chemical vapor deposition. The surface is an integral part of the glass, not just a coating. The glass uses outdoor daylight to clean windows in a two-step process. The surface uses energy from the sun to actually break down, loosen, and destroy dirt

Figure 28.17 Photovoltaic glazing provides enclosure and partially shaded light for a building while simultaneously generating electricity.

© Anthony West/Eureka/Corbis

and other organic matter. The water sheets off the surface when it rains so windows dry without spots and streaks.

Review Questions

1. What are the basic types of glass?
2. What type of glass is used in most construction applications?
3. What are the two types of float glass?
4. What are the grades of sheet glass?
5. How does a small scratch or nick in glass reduce the tensile strength of a sheet?
6. How does the thermal conductivity of glass compare with that of metal?
7. Why is glass resistant to corrosion?
8. How does glass rate on optical clarity?
9. Which types of soda-lime-silica glass are annealed?
10. List the types of heat-treated glass from weakest to strongest.
11. In addition to heat-treating glass, how can it be strengthened?
12. List the types of surface finishes commonly used on glass.
13. What type of glass helps reduce heat transfer into a building in the summer and heat loss to the outside during winter?
14. How does tinting affect the performance of glass?
15. What are the main reasons for using laminated glass?
16. How does insulating glass differ in construction from laminated glass?
17. What three things occur to produce solar heat gain when solar energy strikes a glass surface?
18. How much visible light is transmitted through reflective glass?
19. What are the two important features of wired glass?
20. What type of finish is used on spandrel glass?

Key Terms

Acoustical Glass

Annealing

Chemical Strengthening

Chromogenic Glazing

Float Glass Process

Insulating Glass

Laminated Glass

Low-E Glass

Reflective Glass

Shading Coefficient (Sc)

Tempered Glass

Activity

1. Collect samples of glass products used for glazing, identify each, and describe their properties. Glass

manufacturers often have small sample pieces available.

Additional Resources

Schittich et al., *Glass Construction Manual*, Birkhauser Edition Detail.

Glass Block Installation Manual, Pittsburgh Corning Corp., 800 Presque Isle Drive, Pittsburgh, PA 15239.

Merritt, F. and Rickets, J., *Building Design and Construction Handbook*, McGraw-Hill Publishing Co., New York.

Spence, W. P., *Encyclopedia of Construction Methods and Materials*, Sterling Publishing Co., 387 Park Avenue South, New York 10016.

Spence, W. P., *Windows & Skylights*, Sterling Publishing Co., 387 Park Avenue South, New York 10016.

See Appendix C for addresses of professional and trade organizations and other sources of technical information.

Doors, Windows, Entrances, and Storefronts

LEARNING OBJECTIVES

Upon completion of this chapter, the student should be able to:

- Know the types of doors, windows, entrances, and storefronts available.

- Apply this information to the decision-making processes as these units are selected for a building.

- Understand the properties of various units as they relate to fire, security, privacy, and operation.

Build Your Knowledge

For further study on these materials and methods, please refer to:

Chapter 19 Products Manufactured from Wood
Topic: Wood Doors and Windows
Chapter 28 Glass
Topic: Glass Products

DOORS

Doors are one of the most essential parts of any building. They provide the necessary access between the interior and exterior, and facilitate passage and privacy between interior spaces. In addition, they afford security, privacy, and sound control, as well as thermal and fire protection. The choice of a door depends on the volume of traffic, security requirements, and desired appearance. In residential buildings, interior doors provide privacy and sound control, whereas exterior doors provide security, protection from the weather, and an important part of the exterior appearance. In commercial buildings, doors must permit ingress and egress as required by building codes. They are subject to considerable use and must be durable, with hardware able to withstand continued operation.

Doors can be classified into three basic operating types: swinging, sliding, and folding. Specialty doors used frequently in commercial construction include overhead and revolving doors. Most doors are factory built and manufacturers produce standard sizes of various door types. The type, size, and location of doors must be carefully designed and coordinated, with appropriate rough opening sizes and lintels in the structural frame. Doors are available manufactured from wood, hardboard, plastic, metal, and glass. Typical residential and commercial door types and operation are shown in Figure 29.1 and Figure 29.2.

DOOR OPERATING AND MATERIAL TYPES

Before introducing door types and materials, it is important to understand the terms given to both the members of the door itself and the frame that supports it (Figure 29.3). The frame that holds a door in place is called the *jamb*. The vertical members of the jamb are known as side jambs, while the top is called the head jamb. In wood doors, the jamb may be composed of square-edged wood to which a stop is applied. The stop is the part of the jamb against which the door closes. Wood jambs may also be rabbeted, or cut out of a single piece of wood to provide an integrated stop. Most metal door jambs are manufactured with an integral doorstop. A pre-hung door is a manufactured unit that arrives on the job site already fitted and hinged with one side of the casing trim applied. Casing refers to the molding used to trim around doors, windows, and other wall openings.

Figure 29.1 Common door types and styles.

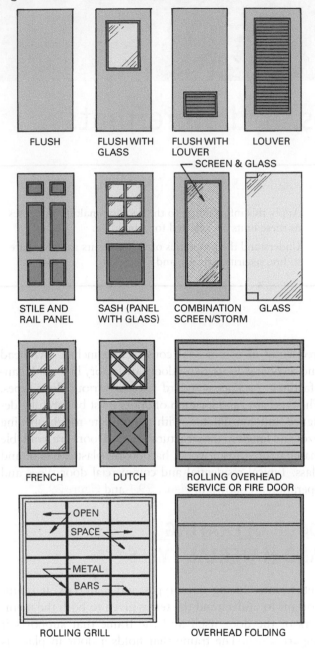

Figure 29.2 Common door operating types.

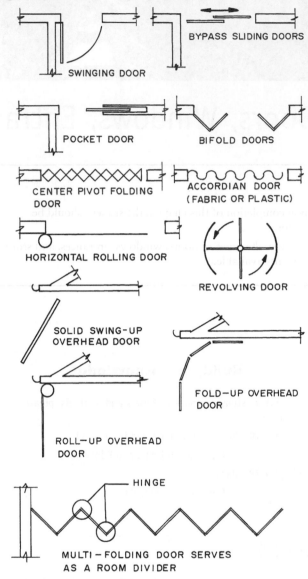

The bottom member of the door jamb is referred to as the sill or threshold. Thresholds are often constructed of wood or aluminum and commonly come with adjustable weather-strip inserts to protect against air and water infiltration in exterior doors. The jamb fits inside of what is called the rough opening, the opening in the wall before any door framing material has been applied. The rough opening is always slightly larger than total size of the door and frame together, to prevent structural loads from interfering with the operation of the door. A lintel or header at the top of the rough opening transfers any structural loads to the adjacent wall.

The sides of the doors itself are named according to their location. The hinge stile is the side from which the door swings, while the lock stile receives the door handle and locking hardware. The top and bottom pieces that make up the horizontal frame members are called rails. Panel doors may have two or more intermediate horizontal rails. Flush doors have a smooth flat surface of wood, metal, or plastic. Panel doors make use of inset panels in a variety of styles. When glass is used in doors, it is referred to as a light. Sidelights, although not part of the door itself, are glass openings that are fixed to the doorframe on one or both sides. Transoms are similar

Figure 29.3 Nomenclature for doors and their frames.

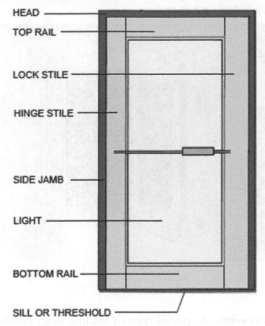

HEAD
TOP RAIL
LOCK STILE
HINGE STILE
SIDE JAMB
LIGHT
BOTTOM RAIL
SILL OR THRESHOLD

to sidelights but are installed directly above the door. When an opening comprises more than one door, the individual door units are referred to as leaves.

Swinging Doors

Swinging doors are perhaps the most commonly used operating type. They are hinged along one side to allow the door to swing away from the doorframe in one direction only. The axis of rotation is usually vertical, although in some cases, like hinged overhead doors, the axis may be horizontal above the doorframe. While they require clearance for the door to swing, swinging doors provide the most convenient operation for access and passage. Swinging doors are also the most effective operating type for weather resistance and acoustical privacy.

For most applications, a swinging door opens to one side only, and closes against the door stop. Double swinging doors have special hinges that allow them to open either outward or inward. They are usually supplied with closing hardware to keep the doors shut when not in use.

French doors are derived from a historic French design called a casement door. They are composed of double doors, generally with multiple glass panels in each door leaf, that can swing both out as well as in.

The direction in which a door swings is referred to as door handing. A door is designated having either a left-hand (LH) or right-hand (RH) swing. The type of swing is determined by standing on the side of the door that swings away from the viewer. A right-hand door has the hinges on the viewer's right side, while a left-hand door has the hinges on the viewer's left side. Door handing is used when ordering doors from material suppliers.

Swinging doors are manufactured in a variety of sizes. Standard sizes for commercial exterior doors are from 3 ft. to 4 ft. (.91–1.21 m) in width and from 6 ft. 8 in. to 8 ft. (2–2.4 m) in height. Door manufacturers publish detailed catalogues outlining available sizes and styles. Local building codes should be consulted when specifying doors for particular occupancies.

Bypass and Folding Doors

A sliding or *bypass door* is a door unit that has two or more sections (Figure 29.4). The doors slide in either direction past each other along one axis on parallel overhead tracks. The doors in a bypass unit will overlap slightly when closed, in order to provide complete enclosure, and not have a visible gap between them. Sliding doors find use in applications with limited space where a swinging door may interfere with the movement in a corridor for instance. They are also commonly used for storage closets to allow full access to one side at a time. Exterior sliding doors present security concerns and should be provided with heavy-duty locks, top and bottom lever bolts, and a reinforced center-meeting stile. The door panels should be designed to prevent their being lifted out from the bottom track when in the locked position.

Figure 29.4 Bypass doors slide in either direction past each other along one axis on parallel overhead tracks.

© photobank.ch/Shutterstock.com

A *pocket door* is a sliding door that disappears into a compartment, or pocket, in the adjacent wall when fully open. Pocket doors are used when there is no room for the swing of a hinged door. They usually move on rollers suspended from an overhead track, although some also feature tracks or guides along the floor. Both single- and double-door versions are used, depending on how wide an opening is desired. Pre-manufactured pocket door-frames are installed during the rough framing of wall assemblies (Figure 29.5).

Figure 29.5 A pre-assembled pocket doorframe during installation.

A folding door is a type of door that uses multiple hinged leaves that open by folding back in sections (Figure 29.6). They may be bi-fold or multifold doors that fold up to one or both sides of an opening. Larger folding door units are able to provide an efficient way to divide larger spaces, or provide large adjustable exterior

Figure 29.6 Folding doors use multiple hinged leaves that open by folding back in sections.

© Dainis/Shutterstock.com

wall openings. A special type of folding door is referred to as an accordion door. Accordion doors use multiple smaller leaves of wood or fabric that are suspended from an overhead track. They are frequently used as retractable partitions to provide quick visual and acoustic division between adjacent building spaces. Acoustical models include multi-ply sweep strips at the top and bottom on both sides of the door for enhanced sound isolation.

Revolving Doors

Revolving doors are used frequently to provide access to larger commercial buildings. A revolving door comprises three or four doors that connect to a central shaft and rotate on a vertical axis within a circular enclosure (Figure 29.7). Revolving doors are efficient entryways as they allow greater numbers of people to pass in and out of a building. Sizes of revolving doors can be as large as 10 ft. wide, allowing for the entry of individuals with strollers or luggage. Revolving doors are more energy efficient than swinging or sliding doors because they prevent air exchange between the interior and exterior of a building enclosure. The leaves of revolving doors usually incorporate glass, to allow people to see one another as they are passing through the entry. Many revolving doors incorporate circular glass walls of annealed glass set in stainless-steel or aluminum frames. The codes vary in crediting legal exiting requirements for revolving doors and often mandate that they be paired with hinged doors located directly adjacent to them.

Figure 29.7 Revolving doors are efficient entryways that allow greater numbers of people to pass in and out of a building.

© Christopher Dodge/Shutterstock.com

Figure 29.8 Overhead doors travel along a jamb-mounted guide track and open by rolling into a position above the door opening.

© Jow Gough/Shutterstock.com

Revolving doors may be manually or automatically operated. Manual revolving doors rotate through push-bars that rotate all doors simultaneously. Automatic revolving doors are powered from the central shaft, or along the perimeter. Most are equipped with safety sensors that control the speed and prevent the doors from turning too rapidly.

Specialty Doors

Additional door types are manufactured to meet special needs other than providing occupant access within a building. There are many variations of rolling doors, such as those used on aircraft hangers and openings in manufacturing plants. Many are power activated. Various types of rolling grille doors are used to provide security for stores opening onto a public mall.

Overhead doors are used for large openings required at garages, loading docks, and commercial storefronts (Figure 29.8). The door travels along a jamb-mounted guide track and opens by rolling into a position above the door opening. Required clearance issues need to be addressed to allow for motors and door and roller space. Overhead doors are generally constructed of steel, aluminum, wood, or fiberglass panels or slats.

Wood and Plastic Doors

A variety of types and styles of doors are made from both solid and hollow-core wood construction. They are available in solid wood and with wood veneer, hardboard, or plastic laminate faces. Other doors have wood stiles and rails covered with a fiberglass skin. Wood and plastic doors can easily incorporate glass or louvered openings. Many companies manufacture doors to the specifications of the National Wood Window and Door Association (NWWDA). These standards establish minimum requirements for material, design, construction, and pressure treatment of components. Doors meeting these standards are recognized by the NWWDA seal.

Wood doors are typically used only for residential and interior commercial applications, since wood tends to fail when exposed to moisture and exterior weather conditions. While not as durable as other door materials, wood is often chosen for its warm appearance.

Flush doors have wood veneers or hardboard face panels bonded to a core of solid wood strips (called a solid-core door) or a core made of wood or kraft paper forming a hollow, honeycomb interior (called a hollow-core door) (Figure 29.9). Stile and rail doors have solid wood vertical and horizontal members enclosing panels made of wood, glass, or louvered sections. Some are made from pressed wood fibers, forming a paneled hardboard door. Fire-rated wood doors use a mineral composition core. Acoustical doors use sound-dampening cores or sheets of lead over the core. Solid wood-core doors reduce sound transmission and retard fire better than do hollow-core doors.

Figure 29.9 The construction of a hollow-core flush door.

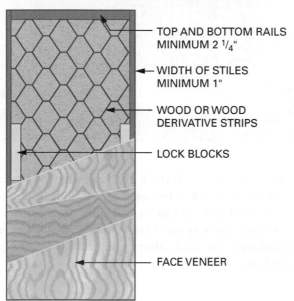

TOP AND BOTTOM RAILS
MINIMUM 2 1/4"

WIDTH OF STILES
MINIMUM 1"

WOOD OR WOOD
DERIVATIVE STRIPS

LOCK BLOCKS

FACE VENEER

PVC is used for both doors and rot-resistant frames in single- or double-wide configurations. The material is impact resistant, durable, easily cleaned, and rust, rot, and moisture proof. Plastic doors find use in food processing facilities, agricultural buildings, clean rooms, water treatment facilities, and retail occupancies.

Fiberglass Doors

Fiberglass doors provide a number of advantages over wood doors. They provide durability, resisting denting, splitting, cracking, and warping. These doors are manufactured with an insulating foam core, offering more superior insulating properties than do solid wood doors. Fiberglass doors are available in both smooth and textured wood-grain finish and can be painted or stained.

Hollow-Core Metal Doors

Hollow-core metal doors are available in steel and aluminum. They are used on interior and exterior openings and are available in a wide range of designs. Insulated metal core doors, pre-hung in a wood jamb, are used frequently for exterior doors in residential applications (Figure 29.10). Exterior surfaces may be flush or panelized, integrate windows or louvers, and finished with a baked enamel finish on a smooth or textured surface. Decorative doors may be covered with plastic veneers of wood grain or a variety of colors and patterns.

Figure 29.10 A typical insulated core metal door in a wood jamb, frequently used in residential applications.

© Beth Van Trees/Shutterstock.com

The exact construction of hollow metal doors varies by manufacturer, but two types of designs are common. One uses a tubular framework over which the face panels are welded. Insulation and sound-deadening material are placed in the hollow core. Another type uses a core material, such as kraft paper honeycomb, polyurethane, polystyrene, a steel grid, mineral fiberboard, or vertical steel stiffeners, over which metal panels are bonded. The door may be seamless, full-flush, flush-panel, or industrial-tube construction (Figure 29.11). Seamless doors have the edge seams hidden by welding and grinding them smooth. Full-flush doors have visible seams on the edges but no seams on their faces. Flush-panel doors are of either a stile-and-panel or a stile-and-rail construction. Face panels may be slightly recessed. Industrial tube doors are made using tubular steel stiles and rails with recessed panels. Metal doors are classified by grades and models, as shown in Table 29.1. The higher grades utilize a thicker-gauge metal. Hollow steel door types are designated by letter symbols, such as F, for flush; L, for louvered; VL, for vision lite; and FG, for full glass.

Hollow Metal Doorframes

Hollow metal doorframes are available in steel or aluminum and are supplied knocked down or preassembled. A wide range of designs can accommodate many types of interior and exterior wall constructions (Figure 29.12). Hollow metal frames are anchored to masonry walls with a metal strap or wire loop (Figure 29.13). Three anchors are typically required on each jamb. Several strap designs are used to secure hollow metal doorframes to wood or metal studs. Brackets secure door bases to floors (Figure 29.14).

The frame and the door are prepared to receive a lockset and strike plate. The holes, recesses, and screw holes are precut into the frame. Pre-hung doorframes have doors installed before a unit ships to a job. The required steel gauge for doorframes is related in standards to the door grade and model. The higher the door grade, the thicker the steel used for the doorframe.

Metal Fire Doors and Frames

Fire doors are an assembly of a door and frame manufactured and installed to give protection against the passage

Figure 29.11 Standard steel door types.

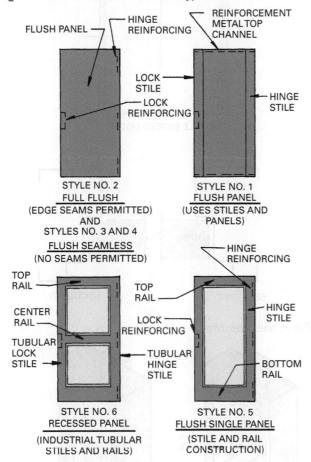

Table 29.1 Grades and Models of Standard Steel Doors

Grade	Model
Grade I—Standard duty, 1⅜" and 1¾" thick	Model 1—Full flush design, hollow metal and composite
	Model 2—Seamless design, hollow metal and composite
Grade II—Heavy duty, 1¾" thick	Model 1—Full flush design, hollow metal and composite
	Model 2—Seamless design, hollow metal and composite
Grade III—Extra heavy duty, 1¾" thick	Model 1 and 1A[a]—Full flush design, hollow metal and composite
	Model 2 and 2A[a]—Full flush design, hollow metal and composite
	Model 3—Stile and rail, flush panel

[a]1A and 2A are made with a heavier gauge metal than models 1 and 2

Figure 29.12 Typical metal doorframe details.

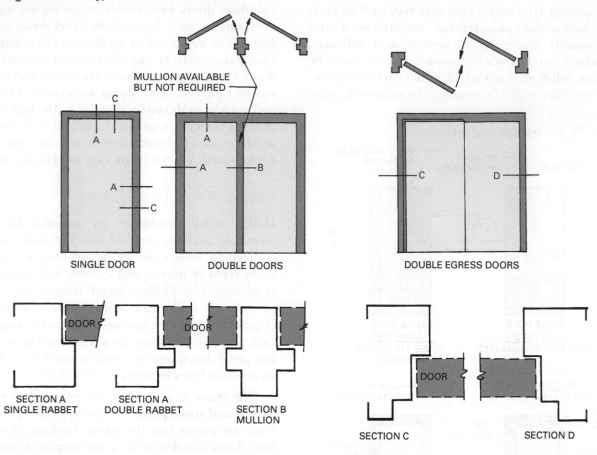

MULLION AVAILABLE
BUT NOT REQUIRED

SINGLE DOOR   DOUBLE DOORS   DOUBLE EGRESS DOORS

DOOR   DOOR   DOOR

SECTION A   SECTION A   SECTION B   SECTION C   SECTION D
SINGLE RABBET   DOUBLE RABBET   MULLION

Figure 29.13 Anchors used to secure steel doorframes to different wall assemblies.

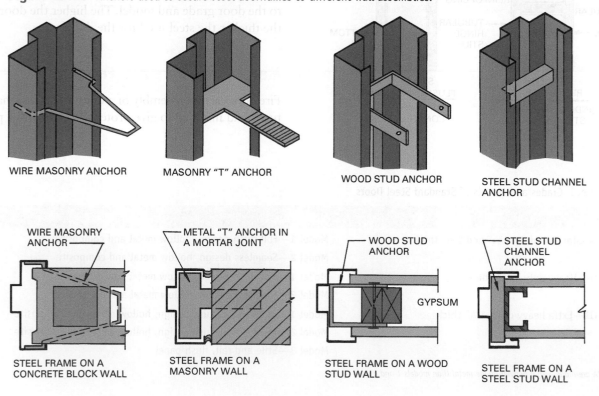

WIRE MASONRY ANCHOR   MASONRY "T" ANCHOR   WOOD STUD ANCHOR   STEEL STUD CHANNEL
                                                                            ANCHOR

WIRE MASONRY   METAL "T" ANCHOR IN   WOOD STUD   STEEL STUD
ANCHOR   A MORTAR JOINT   ANCHOR   CHANNEL
                                                                                       ANCHOR

                                                                            GYPSUM

STEEL FRAME ON A   STEEL FRAME ON A   STEEL FRAME ON A WOOD   STEEL FRAME ON A
CONCRETE BLOCK WALL   MASONRY WALL   STUD WALL   STEEL STUD WALL

Figure 29.14 Metal doors are secured to the floor with metal base anchors.

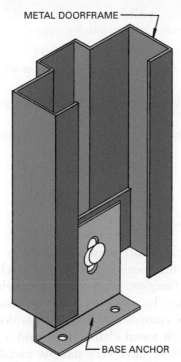

METAL DOORFRAME

BASE ANCHOR

Figure 29.15 A metal fire door with panic hardware.

© FocusDzign/Shutterstock.com

of fire through an opening in a wall. The design, location, and installation are controlled by national standards and local building codes. Approved fire door assemblies are required to meet the test requirements specified in ASTM E152, *Methods of Fire Tests of Door Assemblies*. Standards for fire doors and windows are outlined in ANSI/NFPA 80, a publication prepared by a committee of the National Fire Protection Association and approved by the American National Standards Institute (ANSI).

All fire door assemblies must bear the label of an approved agency and include the name of the manufacturer, the fire-protection rating, and the maximum transmitted temperature. Some assemblies are automatic, keeping the door open and closing it automatically in case of elevated temperature or exposure to smoke. Doors must be fully operable without the use of keys or special effort. Panic hardware consists of bars that open the door when a person pushes against it. It was developed after fire incidents showed that people tend to congregate at doors that may jam under pressure. Panic hardware is required for educational buildings and assembly occupancies of 100 persons or more (Figure 29.15).

Fire doors are classified by an hourly rating designation, an alphabetical letter designation, or a combination of hourly and letter designations. Horizontal access doors use a special listing indicating the fire-rated floor, floor-ceiling, or roof-ceiling assembly for which the door will be used.

The hourly designation indicates the duration of the fire test exposure in hours and is called the fire-protection rating. The alphabetical letter designation in use follows:

Class A—Openings in firewalls and in walls dividing a single building into fire areas.

Class B—Openings in enclosures of vertical communications through buildings and in two-hour-rated partitions providing horizontal fire separations.

Class C—Openings in walls or partitions between rooms and corridors having a fire-resistance rating of one hour or less.

Class D—Openings in exterior walls subject to severe fire exposure from outside a building.

Class E—Openings in exterior walls subject to moderate or light fire exposure from outside a building.

Special listings are tested in accordance with NFPA 251, *Standard Methods of Fire Tests of Building Construction and Materials*. It indicates the fire-rated assembly and its hourly rating. Fire door rating details are shown in Table 29.2. Labels on doors

Table 29.2 Fire Door Ratings

Class	Hour	Glazing	Location
A	3	No glazing	In fire wall and walls that divide a building into fire areas
B	1½	100 sq. in. of glazing	In enclosures of vertical communication through buildings
C	¾	1,296 sq. in. of glazing per light	Openings in walls requiring a fire resistance rating of 1 hr. or less
D	1½	No glazing	Openings in exterior walls subject to fire exposure from the outside of the building
E	¾	720 sq. in. glazing	Openings in exterior walls subject to moderate or light fire exposure from outside the building

specify the class of door, which indicates the time interval the door will meet. For example, a Class A door has a 3 hr. time interval. Metal doorframes have fire ratings of ¾, 1½, and 3 hours. The hardware used must be fire rated to maintain the fire rating of the door and frame.

Typical types of materials used for fire doors include:

1. Wood-core fire doors have wood, hardboard, or plastic laminate faces bonded to a solid wood core or a wood particleboard core.

2. Hollow metal fire doors may be of flush or panel design with a steel face of 20-gauge or thicker.

3. Metal-clad fire doors have wood cores or stiles and rails with insulated panels. The face is 24-gauge steel or lighter.

4. Sheet-metal fire doors are made using 22-gauge or lighter steel.

5. Tin-clad fire doors have a solid wood core covered with a 24- to 30-gauge terneplate or galvanized steel facing.

6. Composite fire doors consist of some combination of wood, steel, or plastic laminate bonded to a solid core material.

7. Rolling steel fire doors are steel doors that move on an overhead barrel enclosed in a hood. They may have an automatic closing mechanism.

8. Curtain-type fire doors have interlocking steel blades forming a steel curtain in a steel frame.

Special-purpose fire doors include a fire door and frame assembly. Acoustical fire doors, security fire doors, armored-attack-resistant fire doors, radiation-shielding fire doors, and pressure resistant fire doors are examples. A variety of automatic fire vents are also manufactured.

Glass Doors

Glass entry doors are used frequently in commercial applications that benefit from having a clear view of the interior function. Doorways and their surrounding walls may be pre-assembled or constructed on site from the component pieces. Glass doors may be frameless, with metal channels to hold a pivot hinging apparatus or have very narrow metal frames on all sides (Figure 29.16). A wider frame that provides a stronger door is used in areas where heavy traffic is expected. The glass must meet building code safety requirements, which typically require that safety glass be used in glass door construction.

Tempered safety glass is designed to prevent injuries in the event that the glass is broken. Rather than breaking into dangerous shards, tempered glass crumbles into tiny pieces that pose minimal threat to those nearby. Local codes should always be consulted to determine safety requirements. Safety glazing must be used in all unframed swinging doors, regardless of how large the area of the glass is. It is also required on framed doors with individual glass panes larger than 9 sq. ft., (.836 sq. m.) or doors with glass located less than 18 in. (45.7 cm) above the floor.

DOOR HARDWARE

Countless types of hardware, including locks and handles, hinges, kick plates, and doorstops, are used in conjunction with doors. Door hardware ranges from simple knob key sets to sophisticated automated access systems. When specifying door hardware, local building codes, related industry standards and their requirements for door openings including fire-ratings, means of egress, and accessible opening requirements must always be consulted.

Figure 29.16 Glass door types used in commercial construction.

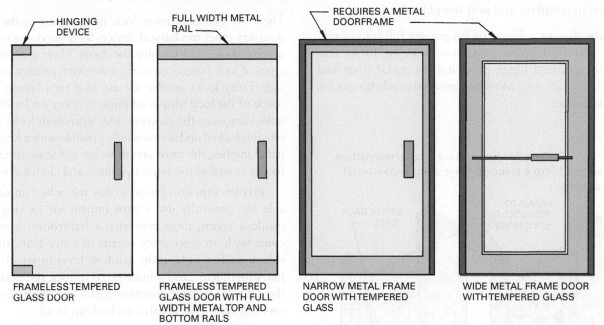

FRAMELESS TEMPERED GLASS DOOR

FRAMELESS TEMPERED GLASS DOOR WITH FULL WIDTH METAL TOP AND BOTTOM RAILS

NARROW METAL FRAME DOOR WITH TEMPERED GLASS

WIDE METAL FRAME DOOR WITH TEMPERED GLASS

Hinges

A hinge consists of two leaves, one of which is fastened to the door and the other to the doorframe (Figure 29.17). The bearings keep the hinge aligned and help the hinge last longer by reducing wear. The pin holds the leaves together and provides the axis on which the door turns. The knuckle is a loop of metal through which the pin passes. Hinge pins are either removable or fixed, and are sometimes decorative. Non-removable pins (NRP) are used in applications where security is required for an exterior an out-swinging door.

Figure 29.17 Components of a standard hinge.

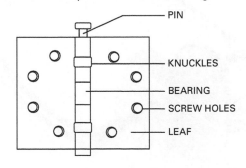

PIN

KNUCKLES

BEARING

SCREW HOLES

LEAF

A minimum of three hinges are required for doors 7 ft. 6 in. (2.3 m) high and taller. A further hinge should be added for every additional 30 in. (76 cm) in height. Design criteria for specifying hinges include the type weight and dimensions of the door, frequency of use, and the type of pinheads and screws. There is an enormous variety of door hinges available in today's commercial hardware market.

Full-Mortise Hinges Full-mortise hinges have both leaves mortised into the door and frame. The majority of commercial full-mortise hinges use 1¾ in. wide leaves, and measure 4½ in. by 4½ in. when laid flat.

Half-Mortise Hinges Half-mortise hinges are hinges that have one leaf mounted to the visible front of the frame and the other leaf mounted in a mortise, on the edge of the door.

Full-Surface Hinges Full-surface hinges have one leaf fastened to the surface of the frame and the other to the surface of the door, leaving both leaves visible when the door is closed.

Swing-Clear Hinges Swing-clear hinges are designed such that when the door is open to 90 degrees, the door itself is outside of the opening. When fully open, clear-swing hinges permit the door to open 180 degrees, allowing for a clear, full opening. Swing-clear hinges can be used to meet ADA requirements for door openings in remodeling projects where the addition of a larger door is not possible.

Spring Hinges Spring hinges are spring loaded in order to shut a door when left in the open position. They are available in full-mortise, with or without radius corners, and in template and non-template versions.

Template Hinges   Template hinges are full hinges that have a standard screw pattern and sizing to fit into an ANSI standard hinge prep, hollow metal door and frame (Figure 29.18). Most commercial grade hinges are template hinges.

Figure 29.18 Template hinges have a set screw pattern and sizing to fit into a standard hinge prep, hollow metal door, and frame.

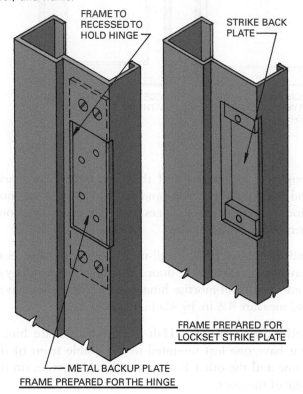

Continuous Hinges   Continuous hinges extend the full height of the door. They are widely used on aluminum storefronts and hollow metal door applications. Continuous hinges are appropriate for high-frequency-of-use doors, where added durability is necessary. Continuous hinges are made in aluminum, steel, and stainless steel, and are available in different types to accommodate different applications.

Pivot Hinges   Pivot hinges are used on very heavy doors in high-traffic applications and on many aluminum storefront doors. Since the weight of the door rests on the bottom pivot, the door does not hang as it does with other types of hinges; therefore, there is less risk of the door sagging over time.

Locking Mechanisms

There are probably more lock mechanisms in the world than any other mechanical device. Securing doors typically involves locking or latching the doors. There are four basic types of lock functions: entry, storeroom, privacy, and passage. Entry locks involve the use of a turn button on the inside of the lock trim in addition to a key for locking and unlocking from the exterior side. Storeroom locks always remains locked and access is only possible with a key. As the name implies, the most common use is for securing storerooms as well as mechanical, janitor, and electrical rooms.

Privacy function locks do not use a key on the outside but generally use a turn button for locking when inside a room, most typically a bathroom. Most will come with an emergency means of entry from the outside, usually a hole in the knob or lever in which a pin-like instrument can be inserted to unlock the door from the outside. Passage function locks use knobs or levers on both sides, but involve no locking at all.

There are two basic types of locks: bored and mortise. *Bored locks* are installed into a hole drilled through the face of a door (Figure 29.19). They interact with a bolt installed in the edge of the door, called a latch bolt. A bored lock is simply a two-piece handle and spindle that is attached by threaded screws through the hole in the face of the door. The latch bolt extends into the metal piece in the frame, which is called a strike. Bored locks are relatively inexpensive and are used predominantly on interior and residential applications.

Figure 29.19 Bored locks are installed into a hole drilled through the face of a door and interact with a latch bolt installed in the edge of the door.

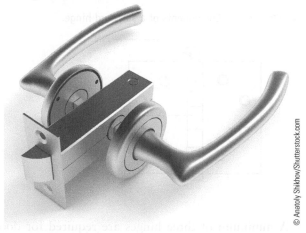

A *mortise lock* uses a rectangular lock body installed in an opening that is mortised out of the edge of the door (Figure 29.20). Mortise locks are more expensive,

Figure 29.20 Mortise locks use a rectangular lock body installed in an opening that is mortised out of the edge of the door.

© nikitabuida/Shutterstock.com

but they provide more durability and a longer life. The handle or knob with plate is secured with threaded bolts that extend through the lock body and a spindle that connects the inner and outer handles.

WINDOWS

Various windows are manufactured primarily for use in commercial buildings. Windows are available made from solid wood, wood clad with plastic or aluminum, solid plastic, steel, stainless steel, aluminum, bronze, and composite materials.

A major consideration in the selection of windows is energy savings. Various types of energy-efficient window glazings are available, including Low-E glass, double glazing, and glass with louver blinds set between the panes of glass for shading. Some double-glazed units have a gas inserted in the space between the panes. See Chapter 28 for additional information on glass and glazing. Windows must have airtight unions between both fixed and moving parts to reduce air infiltration. Metal units transmit heat and cold rapidly through their material, so designs must provide insulation, called thermal breaks, between adjacent interior and exterior parts.

Quality windows are tested and certified against ASTM standards for air leakage, water infiltration, uniform wind-load structural requirements, and uniform load deflection.

The energy performance of *fenestration* (windows and other exterior openings) products is rated by the National Fenestration Rating Council (NFRC). The NFRC combines U-value, solar heat-gain factors, optical properties, air infiltration, condensation resistance, and other characteristics into a uniform rating system to reflect annual energy performance. Products are certified and labeled, providing a means for comparing products manufactured by various companies. The NFRC certification label is shown in **Figure 29.21**.

Figure 29.21 A generic example of the National Fenestration Rating Council certification label indicating the energy performance attributes of a window unit.

© National Fenestration Rating Council

A wide range of locking and operating devices are available and should be considered as window units are selected. These include manual and electric operating devices, hinges and other hardware, and safety devices, such as a keyed lock or safety bar that limits how much a window can be opened.

Local fire codes also influence the choice of windows. These codes typically control the sizes of windows, height of sills above floors, and the ability to open a window from the inside without the necessity of special equipment. Windows used in firewalls must be fire rated and labeled to restrict the spread of fire and smoke. Security windows are tested to meet the standards required to provide resistance to forced entry. Finally, the material and finish must fit with the surrounding area and be able to withstand any corrosive elements to which it will be exposed.

Window Terms and Types

Most windows are shipped to the construction site as complete units that will require only the installation of trim on site. There are a number of basic terms used to describe the parts of a window.

The *sash* is the moving segment of a window, although sash is sometimes referred to as fixed sash. As with doors, the glass in a window is called a light or pane. The head is the horizontal member at the top of the window frame, while the jamb refers to the vertical members at the sides. Drip cap flashing is used at the exterior head to throw water away from the window. The horizontal member at the bottom of the window is called the sill. The sill normally extends beyond the wall below with a drip, to make sure that water that accumulated on the window does not penetrate the wall under the window. "Mullion" is the term given to internal window divisions, located between individual glass panels.

A number of different window operating types are commonly used. A fixed window is an inoperable unit used to provide view and light only. A double-hung window consists of an upper and lower sash, with both panels able to slide vertically in the plane of the window. A single-hung window is similar except that only one sash is operable. Most units use metal sash slides provided with springs, sash balances, or compression weather stripping to hold the windows in any position.

A casement window is side-hinged, and swings either outward or inward. *Casement windows* provide the largest opening area of any window type and are often used to provide natural ventilation. An *awning window* has a sash that is top-hinged, and swings outward by means of a crank or lever. A *hopper window* is similar but bottom-hinged. Sliding windows are essentially a double-hung turned on its side. A special type of window known as the European tilt/turn has the ability to open like a casement and a hopper.

Window types for a building project are called out on the floor plans of architectural drawings by numbers or letters. The number or letter designation is keyed to a window schedule that gives the window's style of operation, its manufacturer, size, and rough opening size.

Wood Windows

Wood frame windows are available with fixed and operable sashes (Figure 29.22). They may be clad with plastic or aluminum or left natural and painted. Wood windows provide better insulation than do metal and plastic windows. However, they swell and shrink as the moisture content changes. The cladding of wood window exteriors has reduced this problem considerably.

Wood windows can be installed in wood frame walls by nailing into a metal or plastic flange, through the sheathing, and into the wood studs and headers in the

Figure 29.22 Wood frame windows may be clad with plastic or aluminum, left natural or painted.

© Jacek_Kadaj/Shutterstock.com

rough opening (Figure 29.23). Other windows are designed with a wood molding that is nailed to the sides of the wood frame rough opening. An example of another installation method involves the use of metal clips to secure a window in a wall framed with metal studs as shown in Figure 29.24. Windows come with various types of glazing and insect screens.

Plastic Windows

Plastic windows are available in the same basic types as described for wood windows. Structural and casement frames are made from polyvinyl chloride (PVC) and glass fibers and a polyester resin. The material will not rust, swell, peel, or corrode and never requires painting. Plastic windows have good thermal properties, and frames are made with dead air pockets that increase the insulation value (Figure 29.25). PVC also has excellent sound-insulation value, and members can be produced with the accuracy needed to provide an airtight fit.

PVC windows are available in a range of colors. Some color the exterior members and leave the interior

Figure 29.23 An installation detail for a clad casement window secured in a wood frame wall by nailing through the plastic flange.

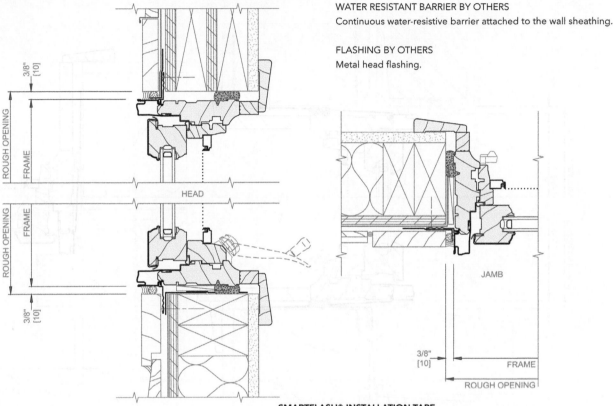

FIN INSTALLATION DETAILS
Aluminum-Clad Wood Exterior
Casement with Wood Trim / Siding

WATER RESISTANT BARRIER BY OTHERS
Continuous water-resistive barrier attached to the wall sheathing.

FLASHING BY OTHERS
Metal head flashing.

SMARTFLASH® INSTALLATION TAPE
Integrate the window and door flashing tape into the wall system.

SHIM AND PLUMB UNITS AS REQUIRED BY OTHERS.

INSTALLATION FIN #42G7
Attach unit to wall construction with nails or screws at each pre-punched hole in the installation fin.

PELLA WINDOW AND DOOR INSTALLATION SEALANT AND WATER RESISTANT BACKER ROD.
Apply continuous sealant according to label directions and ASTM C1193.

LOW EXPANSION, LOW PRESSURE POLYURETHANE INSULATING WINDOW AND DOOR FOAM SEALANT - (DO NOT USE HIGH PRESSURE OR LATEX FOAMS).
Apply a continuous 1" bead of insulating foam to provide a full interior seal.

JAMB EXTENSION #283G (OPTIONAL)

PELLA 2-1/2" RANCH-1 TRIM (#30AY) (OPTIONAL)

Scale 3" = 1' 0"
- Items in bold are available from Pella.
- These details are for typical single punch openings.
- See page W-ID-3 for typical sealant details.
- Refer to the appropriate Pella® Installation Instruction for step-by-step instructions.

Figure 29.24 An installation detail for a clad double-hung window installed in a steel stud frame wall with clips that are secured to the metal studs.

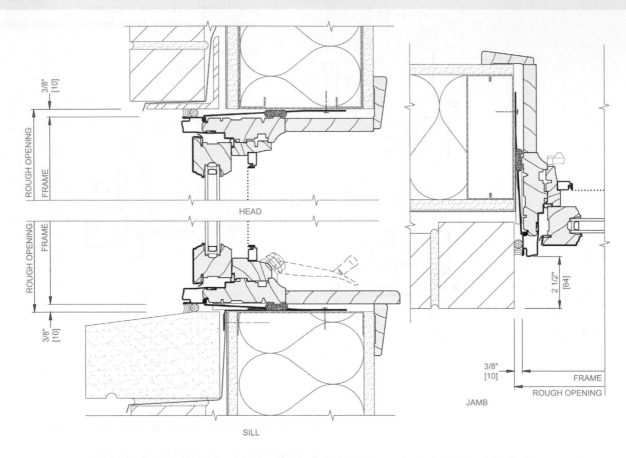

WATER-RESISTANT BARRIER BY OTHERS.
Continuous water-resistive barrier attached to the wall sheathing.

THROUGH-WALL CAVITY FLASHING WITH WEEPS AT HEAD AND SILL BY OTHERS.

SMARTFLASH® INSTALLATION TAPE
Integrate the window and door flashing tape into the wall system.

SHIM AND PLUMB UNITS AS REQUIRED BY OTHERS.

INSTALLATION CLIPS #5071
Place into fin groove and anchor to wall construction. Ensure that the interior bead of insulating foam continues around clips.

PELLA WINDOW AND DOOR INSTALLATION SEALANT AND WATER RESISTANT BACKER ROD.

Apply continuous sealant according to label directions and ASTM C1193.

PELLA 2-1/2" RANCH-1 TRIM (#30AY) (OPTIONAL)

LOW EXPANSION, LOW PRESSURE POLYURETHANE INSULATING WINDOW AND DOOR FOAM SEALANT - (DO NOT USE HIGH PRESSURE OR LATEX FOAMS)

Apply a continuous 1" bead of insulating foam to provide a full interior seal.

Add blocking at jamb to close cavity and provide backing for backer rod and sealant. See page F-9 for alternate detail. Seal wall cavity around head and jamb of window's rough opening.

Scale 3" = 1' 0"

- Items in bold are available from Pella.
- These details are for typical single punch openings.
- See page W-ID-3 for typical sealant details.
- Refer to the appropriate Pella® Installation Instruction for step-by-step instructions.

Figure 29.25 Plastic window frames are made with dead air pockets to increase their insulation value.

© alterfalter/Shutterstock.com

Figure 29.26 Steel window units are made from structural-grade steel, resulting in narrower frame members.

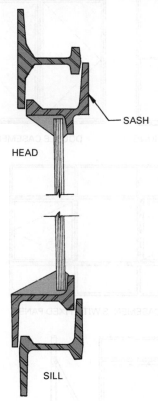

SASH

HEAD

SILL

exposed surfaces an off-white. Windows are fastened to walls using steel wall anchors with zinc-coated screws applied through slots in the frame as specified by a manufacturer.

Steel Windows

Steel window units are rigid and strong and utilize narrower frame members than do other materials, producing a smaller sight line (Figure 29.26). Made from structural-grade steel, they are available primed or galvanized and ready for painting. Some types have a polyvinyl chloride or urethane plastic coating to provide permanent protection. The units are weather stripped and available with single and double glazing. Manufacturers provide a range of stock sizes but will produce custom-designed units of any size. Steel windows are typically casement type, but other designs are available (Figure 29.27).

Residential-grade windows are made from lighter-gauge steel than are windows designed for use in industrial and commercial buildings, which are classified as intermediate grade. Heavy-duty intermediate grade windows are used where the size of a window requires additional strength or must meet other conditions, such as resisting high winds. Manufacturers' catalogs list the gauges and recommended uses for their products.

Stainless-steel windows are usually more expensive than steel or aluminum windows but provide long life and may save money on replacement costs. They are made from various types of stainless steel formed into structural members that are welded to form a unit. The sizes and types of window vary by manufacturer and many are custom built to an architect's design. Typical types available include folding, awning, casement, and various hinged types.

Aluminum Windows

Aluminum windows are made from extruded aluminum structural members. Since aluminum is easy to work, complex unit designs are readily produced. Aluminum has good structural properties and resists

Figure 29.27 Typical steel window operating types.

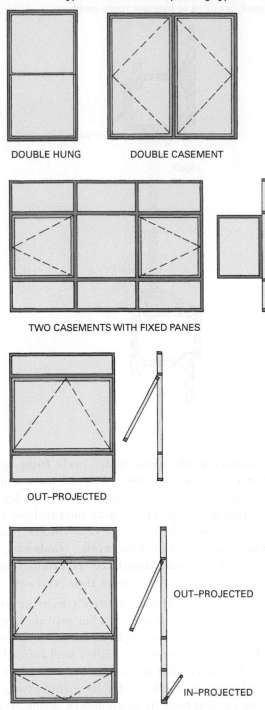

DOUBLE HUNG

DOUBLE CASEMENT

TWO CASEMENTS WITH FIXED PANES

OUT–PROJECTED

OUT–PROJECTED

IN–PROJECTED

ARCHITECTURAL PROJECTED

corrosion and damage from weather. Various organic protective coatings can be applied in areas where damage may occur. Aluminum can also be colored using anodized finishes. Additional information on aluminum is given in Chapter 16.

The frames of aluminum windows are provided with a *thermal break*, an insulating material in the sash members that reduces the flow of heat either inward or outward. Aluminum windows are assembled using both mechanical connections and welding. Any connectors, fasteners, hardware, or anchors must be aluminum or a material (such as stainless steel) that is compatible with aluminum so corrosion does not occur from galvanic action.

Several aluminum alloys are used in the manufacture of aluminum windows. These affect the properties of a product and vary with the purpose for which a window will be used. Standards for the quality of aluminum windows are established by the Architectural Aluminum Manufacturers Association (AAMA), and vary according to the type of window and its intended use. These windows are of three types: residential, commercial, and monumental. They are available in the same operating types as described for plastic and steel windows. A section through a typical aluminum window is shown in Figure 29.28.

Figure 29.28 A section through an aluminum window showing the thermal break material designed to slow conductive heat transfer through the metal.

Composite Materials

Increasing use is being made of composite materials in construction. Composites are formed by combining several different materials. Composite windows are made entirely of composites (a cellular PVC material) and wood by-products, producing a material that looks like wood but will not split, warp, or rot. It has twice the insulation value of wood and is available in several colors.

Special Windows

A number of special windows are manufactured for special applications, such as skylights, security, pass, and storm windows. Skylights and roof windows perform the same basic function as other windows, but they are located in a roof. Skylights are glazed openings in a roof that admit light only. Roof windows are operable like other windows, but are located in a roof. Skylights can have fixed or operable glazing, or the sash on roof windows can be opened to provide ventilation. They may be installed as single units, as is common in residential construction, or in strings or clusters, admitting considerable light (Figure 29.29). Skylights are glazed

Figure 29.29 Skylights may be ganged together to admit more evenly distributed daylight.

with glass or plastic lights. Some operable units open manually with a crank, and other types are electrically opened. They may be designed to blow out the glazing if an explosion occurs or to open automatically if the temperature increases to serve as a fire vent.

Most skylights used in commercial, industrial, and monumental work come with aluminum frames. The small units used in residential applications have wood frames. Small-domed and barrel skylights may have single or double glazing.

Skylights are available for enclosing large spaces, such as malls, building entries, or pool areas (Figure 29.30). They also find use in enclosed walkways, protecting interiors from the weather and providing abundant light. They are glazed with glass and plastic materials.

Figure 29.30 Commercial and public buildings make extensive use of roof windows.

Security windows (also called detention windows) are tested to classify their resistance to forced entry. This includes testing the locking device, the impact resistance of the sash and frame, the resistance of security bars, and the resistance of the glazing. The degree of security is specified as one of four classes:

Class 1:   Minimal
Class 2:   Moderate
Class 3:   Medium
Class 4:   Relatively high

In some occupancies, window security is specified in building codes that require a high-grade window in certain areas of a wall. If windows are protected by devices such as bars, screens, or shutters, provisions for meeting exit requirements in the fire codes must be met. Manufacturers indicate the ANSI/ASTM security standards their windows meet. In addition to a secure window, the window frame, locking mechanism, hinges, and glazing must also be secure. Laminated glass used on security windows is discussed in Chapter 28.

ENTRANCES AND STOREFRONTS

A *storefront* is a non-residential system of entrance doors and fixed windows mulled as a composite structure. Typically designed for high use/abuse and strength, it is field assembled and field glazed (Figure 29.31). The storefront system is usually installed between floor and ceiling on low-rise buildings.

Entrances and storefronts are typically designed with extruded aluminum members. The aluminum frame is in-filled with flat architectural glass products, such as tempered and heat-strengthened glass, laminated glass, or insulating glass.

Typically, tempered glass doors are available with door closers, top and bottom pivots, locks, and push/pulls. Units may include one or more doors, sliding glass panels, and transoms that can be manually or automatically operated. The metal frame, typically aluminum, is usually anodized, and is available in a variety of colors. Stainless-steel frames are also used. Glazing is available in clear, obscure, and many colors.

Entrances may be custom-built for a particular opening using stock door sizes and custom-built frames (Figure 29.32). The dimensions of a storefront system must be verified on site prior to shop fabrication. The frames are mechanically joined and then welded. Components must be designed to allow for expansion and contraction without causing buckling, excessive opening of joints, or overstressing of welds and fasteners. The units must be set on an appropriately prepared surface. The units are then set level and plumb, and operating hardware is adjusted to ensure proper operation. Anchorage devices and fasteners include threaded fasteners for concrete and masonry inserts, toggle bolts, through-bolts, lag bolts, wood screws, and other connectors. Mounting brackets must be securely mounted to the building structure in a including sufficient reinforcements to ensure a strong and rigid installation.

Figure 29.31 A storefront is a non-residential system of entrance doors and fixed windows that is field-assembled and field-glazed.

Figure 29.32 Entrances and storefronts are normally designed with extruded aluminum members and flat architectural glass products.

Entrances and storefronts are subject to a variety of requirements in local building codes. Among these are considerations of fire resistance; wind and snow loads; hazards, such as human-impact loads; the means for supporting the doors and related glazing, hardware, locking arrangements, requirements for power-operated doors, and the size and number of doors required.

Review Questions

1. Describe the construction of various types of hollow-core metal doors.

2. Explain how hollow metal doorframes are secured to walls and floors.

3. Identify each of the letter designations used on metal fire doors.

4. List some of the specialty doors available.

5. What are some of the types of energy-efficient glazing available?

6. What association has established energy performance rating for fenestration products?

7. What two methods are commonly used to secure wood windows to wall framing?

8. What are the grades of steel windows?

9. How do windows made of composite materials compare with solid wood windows?

10. What are the classes of security windows?

11. What are typical building code requirements placed upon entrances and storefronts?

Key Terms

Awning Window	Fenestration	Pocket Door
Bored Lock	Hopper Window	Sash
Bypass Door	Jamb	Storefront
Casement Window	Mortise Lock	Thermal Break

Activity

1. Walk through a commercial development and list the various doors, windows, entrances, and storefronts used. Record the materials and glazing used.

Check your local codes to see if those you observed meet the requirements specified.

Additional Resources

How to Install Interior and Exterior Door Frames, Wood Molding & Millwork Products Association, Woodland, CA.

GANA Glazing Manual, GANA Fully Tempered Heavy Glass Doors and Entrance Systems Design Guide, Glass Association of North America, Topeka, KS.

Performance Specifications and Standards for Doors and Windows, American Architectural Manufacturers Association, Schaumburg, IL.

Spence, W. P., *Doors & Entryways*, Sterling Publishing Co., 387 Park Avenue South, New York 10016.

Spence, W. P., *Encyclopedia of Construction Methods and Materials*, Sterling Publishing Co., 387 Park Avenue South, New York 10016.

Spence, W. P., *Windows & Skylights*, Sterling Publishing Co., 387 Park Avenue South, New York 10016.

See Appendix C for addresses of professional and trade organizations and other sources of technical information.

Cladding Systems

Upon completion of this chapter, the student should be able to:

- Understand the forces that must be considered when designing cladding systems.

- Know the building code requirements pertaining to cladding design and installation.
- Distinguish between various types of cladding systems.

Build Your Knowledge

For further study on these materials and methods, please refer to:

Chapter 9   Pre-cast Concrete

Chapter 12   Concrete Masonry

Chapter 25   Bonding Agents, Sealers, and Sealants
             Topic: Sealants

Chapter 28 Glass

Historically, buildings were constructed using exterior bearing-wall construction systems. Bearing walls, typically of stone or masonry, provided both the primary structural system and served to enclose the interior. The development and increased use of modern materials such as structural steel and reinforced concrete allowed for the use of relatively small columns and beams to support all of the building loads. As the enclosing walls of buildings were no longer required for structural support, the exterior skin could be non-load bearing, much lighter and more transparent. *Cladding* is defined as the non-load-bearing exterior facing that encloses a building. Cladding may be constructed of wood, stone, brick, a variety of metals, plastic, glass, or other suitable materials (Figure 30.1). This chapter introduces the various types of cladding systems that find common use in commercial buildings.

CLADDING DESIGN CONSIDERATIONS

Exterior cladding is exposed to the full gamut of exterior weather conditions, and must resist the forces generated by wind, water, and movement due to temperature changes or seismic events. The design and construction details of cladding systems require careful engineering analysis and the consideration of a wide range of factors. The various elements and the final assembled materials must be tested to ensure they will meet the design requirements of the building codes. Following are some of the critical factors to be considered when selecting and designing cladding systems.

Structural Performance

Because structural performance relates directly to life safety in a building, it is one of the primary concerns of cladding design that is codified in detail. The structural design of cladding must first consider the primary loads generated by gravity, or self-weight of the envelope, followed by wind forces. Secondary loads include the actions generated by seismic forces, thermal and freeze/thaw actions, and maintenance factors such as loads due to window washing.

Structural requirements include the ability of cladding to support its own weight, span distances between

Figure 30.1 Some commonly used cladding systems and materials.

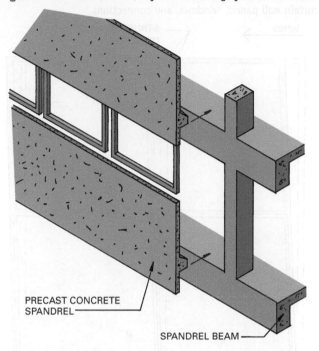

PRECAST CONCRETE
SPANDREL

SPANDREL BEAM

PRECAST CONCRETE CURTAIN WALL

RIBBON
WINDOWS

BRICK OR STONE
SPANDREL

ANGLE IRON

SPANDREL BEAM

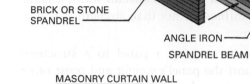

MASONRY CURTAIN WALL

METAL MULLIONS SECURED TO
BUILDING STRUCTURE

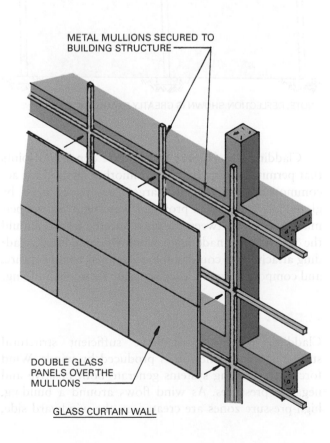

DOUBLE GLASS
PANELS OVER THE
MULLIONS

GLASS CURTAIN WALL

HORIZONTAL METAL MULLIONS

VERTICAL METAL MULLIONS SECURED
TO THE BUILDING STRUCTURE

METAL OR GLASS
SPANDREL SET IN
THE MULLIONS

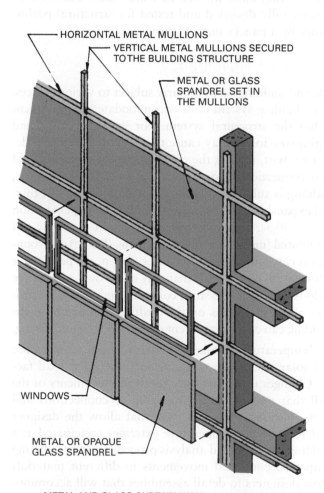

WINDOWS

METAL OR OPAQUE
GLASS SPANDREL

METAL AND GLASS CURTAIN WALL

connectors, accommodate forces of cyclical expansion and contraction, resist wind loads, and absorb seismic movement. The method of connecting cladding to a structural frame is also subject to analysis. One consideration is whether a designer should rely on local and national wind-load standards for the structural analysis, or conduct wind tunnel testing for a specific design. Some jurisdictions require a detailed wind analysis for buildings above a specified height.

Transportation to a site, lifting into place, and an ability to support its own weight are factors that must be accounted for in a panel's structural design. Since cladding is non-load bearing, it should never be required to carry additional loads. Cladding stiffness requirements are a major structural factor in the resistance of wind loads. In the structural design of curtain walls, it is the requirements of stiffness rather than strength that usually govern design.

The connections securing a panel to a structural frame must support the panel's weight and resist other forces acting upon it, such as wind loads. Connections are generally designed and tested for structural performance by a panel's manufacturer.

Control of Movement

Since all buildings are regularly subject to various forces, their cladding system must accommodate the movement within the structural system. For example, wind and earthquake forces may cause the structural frame to deflect or twist, putting the cladding panels, windows, and their connections under stress (Figure 30.2). In addition, cladding is subject to the forces created by gravity, thermal expansion and contraction and moisture penetration (Figure 30.3). The weight of construction materials and differential (uneven) settling or heaving (lifting) of a foundation can cause beams to deflect. Creep, the permanent deformation of materials that may occur over a long time periods, can trigger beams, columns, and girders to produce additional stress on the cladding. Moisture on the cladding can cause movement due to swelling and drying.

Temperature fluctuations, due to both temperature and solar effects, are a significant movement load factor. Changes in temperature result in movements of the wall that must be accommodated at connections and joints. Various references exist that allow the designer to determine the temperature extremes anticipated at a building site. Detailed analysis procedures for predicting temperature-induced movements in different materials allow designers to detail assemblies that will accommodate the imposed loads.

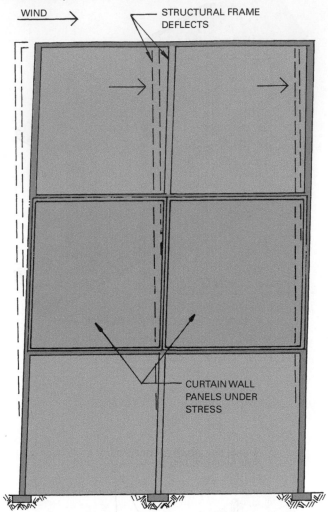

Figure 30.2 Both wind and earthquakes put stress on the curtain wall panels, windows, and connections.

WIND

STRUCTURAL FRAME DEFLECTS

CURTAIN WALL PANELS UNDER STRESS

NOTE: DEFLECTION SHOWN IS GREATLY EXAGGERATED.

Cladding panels must incorporate expansion joints that permit a sliding overlap or another method for accommodating movement. Large glass panes must be glazed into frames that provide a watertight seal yet permit movement between the glass and the frame. Should the design prove inadequate, windows could break, cladding attachments could pull loose, panels could rupture, and components could even separate from the building.

Control of Wind Forces

Cladding systems must have sufficient structural strength to resist the forces produced by winds. Wind forces on cladding systems generate both positive and negative pressures. As wind flows around a building, high-pressure zones are created on the windward side,

Figure 30.3 Many forces acting on the structural frame can put excessive stress on the curtain wall, leading to possible failure.

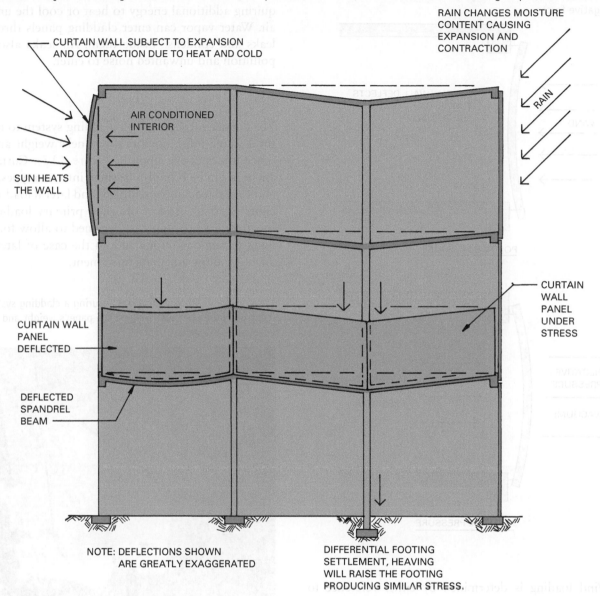

CURTAIN WALL SUBJECT TO EXPANSION
AND CONTRACTION DUE TO HEAT AND COLD

RAIN CHANGES MOISTURE
CONTENT CAUSING
EXPANSION AND
CONTRACTION

RAIN

AIR CONDITIONED
INTERIOR

SUN HEATS
THE WALL

CURTAIN
WALL
PANEL
UNDER
STRESS

CURTAIN WALL
PANEL
DEFLECTED

DEFLECTED
SPANDREL
BEAM

NOTE: DEFLECTIONS SHOWN
ARE GREATLY EXAGGERATED

DIFFERENTIAL FOOTING
SETTLEMENT, HEAVING
WILL RAISE THE FOOTING
PRODUCING SIMILAR STRESS.

and low-pressure zones occur along the sides parallel to the wind and on the leeward side. High pressure on the windward side creates a pushing force on cladding materials that causes an inward deflection. Wind blowing around a building is likely to create a negative pressure. Negative pressure (suction) tends to pull the cladding from the structure. On high-rise buildings, these negative pressures are usually greatest near the corners of the building.

Indoor air pressure, due to mechanical ventilation systems, may be higher than outdoor air pressure,

creating additional stress on the curtain wall from inside the building. If a cladding panel lacks adequate stiffness or connections, it may be pulled from the structural frame (Figure 30.4). Wind becomes stronger at greater heights from the ground, so the higher the façade of a high-rise building, the greater the wind velocity and forces it is subjected to. In an area with surrounding multistory buildings, wind pressures are influenced by the direction of prevailing winds, the shape and orientation of the building and those around it, and the topography of the surrounding area.

Figure 30.4 Wind forces subject the curtain wall to positive and negative forces.

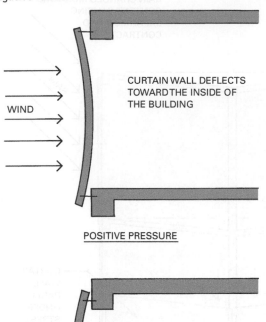

WIND

CURTAIN WALL DEFLECTS TOWARD THE INSIDE OF THE BUILDING

POSITIVE PRESSURE

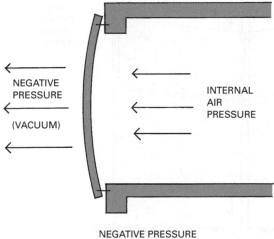

NEGATIVE PRESSURE

(VACUUM)

INTERNAL AIR PRESSURE

NEGATIVE PRESSURE

leaks permit unconditioned air to enter a building, requiring additional energy to heat or cool the unwanted air. Water vapor can enter cladding panels through air leaks and condense inside them. Air leaks also permit pollution and unwanted noise to enter.

Connections

The connections securing a cladding system to its structural frame must support the panel's weight and resist other forces acting upon it (Figure 30.5). Curtain wall anchors can be broadly classified into two types, gravity and lateral load anchors (fixed) and lateral load-only anchors (slotted). Aside from their primary load-carrying function, anchors must be designed to allow for adjustment to site conditions and, in the case of lateral load anchors, allow adequate movement.

Figure 30.5 The connections securing a cladding system to its structural frame must support the panel's weight and allow for adequate movement.

Wind loading is determined by either reference to relevant building codes or through specific modeling by wind tunnel testing. Procedures are provided by the codes to calculate external and internal pressures, with special factors included for the design of cladding elements. Wind load charts published by the American Society of Civil Engineers (ASCE) allow designers to determine the structural resistance requirements needed for an assembly. The use of *boundary layer wind tunnel (BLWT)* has increased steadily since the 1950s and is frequently used for large buildings and those of unusual height or shape. BLWT takes into account not only the effects of prevailing wind speed and direction, but also those generated by the local topography and adjacent buildings.

The methods for controlling air infiltration closely resemble those for containing water penetration. Air

Curtain wall anchors come in countless sizes, shapes, and configurations. Common anchor types have an embedded part that is cast in concrete, or mechanically fastened to the floor slab edge, and a mounting assembly that fixes the cladding to the embedded anchor. Connections are generally designed and tested for structural requirements by a panel's manufacturer. Examples of various connection details are shown later in this chapter.

Solar Control

Cladding must prevent the conduction of heat into and out of a building through the exterior walls. Thermal conduction is the process of heat transfer through a material. Cladding systems use insulation and thermal breaks in materials that resist the transfer of heat through an assembly. A thermal break is an insulating material or gap placed between thermally conductive materials in contact to retard the passage of heat or cold. The performance parameters of glass and the use of multiple glazing and energy efficient glass are discussed in Chapter 28.

Fire, Noise, and Pollution Control

Cladding is subject to the requirements of building and fire codes. Fire resistance ratings for cladding are determined by testing assemblies following ASTM procedures. These evaluate the combustibility of materials in the cladding, the fire-resistance ratings of the assembled panels and spandrels, and the proximity to other buildings. The required fire resistance of non-load-bearing cladding varies with a building's occupancy and fire-separation distance. Fire separation specifies the distance between adjacent buildings. It is used to assess the risk of fire spreading from one structure to another. Distances specified in the code are usually given in feet and measured between a building and the closest lot line from the centerline of the street, or from a halfway distance between two buildings on the same site.

Cladding connectors must hold cladding in place during a fire for the time specified by the fire-resistance rating of the assembly. Exterior trim and other wall finish materials are subject to established fire-resistance ratings. In addition, codes contain requirements for fire protection of openings in walls.

Firestopping uses approved materials to prevent the movement of flame and gases through small vertical and horizontal concealed openings in building components (Figure 30.6). Materials commonly approved for use as firestops include masonry set in mortar; concrete, mortar, or plaster on metal lath; plasterboard; sheet metal of

Figure 30.6 One example of firestopping a curtain wall at the spandrel beam.

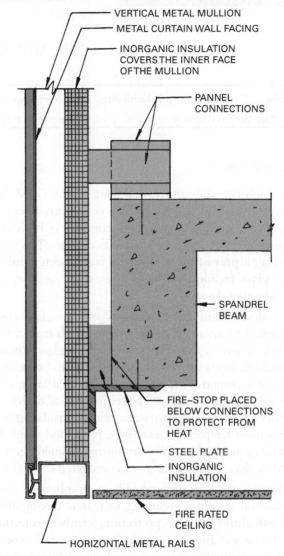

- VERTICAL METAL MULLION
- METAL CURTAIN WALL FACING
- INORGANIC INSULATION COVERS THE INNER FACE OF THE MULLION
- PANNEL CONNECTIONS
- SPANDREL BEAM
- FIRE–STOP PLACED BELOW CONNECTIONS TO PROTECT FROM HEAT
- STEEL PLATE
- INORGANIC INSULATION
- FIRE RATED CEILING
- HORIZONTAL METAL RAILS

approved gauge; and asbestos cement board; mineral, slag, or rock wool, if solidly compacted into a confined space. These materials, commonly referred to as *safing*, must be permanently fastened in place.

The importance of controlling the influx of exterior noise and air pollution into a building varies with the type of occupancy. For example, a hospital requires greater control than other occupancies. In other cases, it may be necessary to keep noise or pollution inside a building, such as in the case of manufacturing or chemical processing operations. Cladding systems for these occupancies must be airtight, well insulated, and, in many cases, have sound-deadening material as part of their assembly.

GSW Administration Building

Architect: Sauerbruch Hutton Architekten	Building type: Commercial office
Location: Berlin, Germany	Completed 1999

Overview

The high-rise slab and lower wing of the GSW Administration building houses the corporate offices of a large real estate company. The complex of buildings incorporates an existing 1961 office tower. The most striking aspect of the project is a west-facing double-skin glass façade that incorporates red and orange sun-shading devices.

A double-skin façade consists of two glass skins separated by an air space ranging in width from a few inches to several feet. The main layer of glass, usually insulating, serves as part of the conventional structural wall or a curtain wall, while the additional layer, usually single glazing, is placed either in front of or behind the main glazing. The air space acts as an insulating barrier against temperature extremes, noise, and wind. As demonstrated in the GSW Administration building, sun-shading devices can be integrated between the two skins.

Double-skin facades provide a number of environmental benefits, including increased day-lighting, acoustical and pollution protection, weather protection of the facade and shading devices, and wind protection for manually operated windows in high-rise applications, allowing for natural ventilation that helps eliminate sick building syndrome (SBS). The façade also provides solar pre-heating of ventilation air, which serves to reduce heating and cooling demands.

Day Light and Ventilation

The narrow floor plate and fully glazed west wall of the GSW building provides ample daylight for all offices. The building further supports natural airflow through added ventilation elements integrated into the design of the east façade (Figure A). Fresh air enters from the east side of the tower through specially designed louvered vents. The air is drawn through the building to the western façade where it is exhausted by convection through operable glazing units in the

outer layer of the façade. The natural thermal lift that occurs in the gap between the two facades draws the air through the narrow floorplate.

Sun Shading

In order to prevent excessive heat gains from the fully glazed west façade, adjustable shading devices are integrated into the double layer of the façade. Vertical louvers can be automatically operated by thermostatically controlled sensors or manually adjusted by occupants (Figure B).

Energy Conservation

The double-skin construction of both the east and west facades provides an insulated air space, minimizing heat loss by conduction. The concrete structure of the floor slabs act to moderate temperature extremes in both summer and winter. In summer, the thermal mass of the slabs serves as a heat sink, absorbing heat during the day that is flushed by ventilation at night. In winter, the slabs absorb solar radiation in the day, and re-radiate the heat at night.

The operations of the ventilation and thermal storage systems allow the complete omission of mechanical air conditioning. Central mechanical heating is used only as a backup system for extreme cold weather. Double-skin facades are expensive, with additional costs generated by the engineering of these systems and the amount of additional glass required.

Maintenance Concerns

Cladding should be selected to last over the expected life span of a building. The material components used, including connections, seals, and joints, should have a proven record of durability. Maintenance requirements over the life of a building will be greater if materials of reduced durability are selected. The design team must choose between more costly, durable materials with higher initial costs or inexpensive, less-durable materials that incur higher maintenance costs over time. Generally, masonry coverings require little maintenance and tend to show weather streaking less than do other impervious skins.

Perhaps the most common maintenance that is regularly performed on a cladding system is washing of the exterior vision glass. On most commercial office buildings, window washing is conducted at least twice per year to maintain a clear view through the glass from within the building. The effects of atmospheric pollution on exterior metal components can lead to deterioration and staining of the finish. It is recommended that all exterior trim and cap components be included in regular exterior maintenance activities.

The exterior of cladding systems are cleaned using stationary scaffolds, swing stages, or cradles hanging from gantry arms on a trolley. A *swing stage* is a scaffold that is suspended by ropes or cables from a block-and-tackle system attached by roof anchors that can be raised or lowered along a building façade (Figure 30.7). Smaller buildings are normally cleaned by a single worker suspended from a cradle chair, whereas larger buildings will employ either permanent or rental swing stages.

Figure 30.7 A swing stage is a scaffold that is suspended by ropes or cables from roof anchors that can be raised or lowered along a building façade for maintenance and repair.

© hxdbzxy/Shutterstock.com

WATER MANAGEMENT FOR CLADDING SYSTEMS

A typical cladding system is an assembly of similar and dissimilar materials that are connected with a diversity of joints designed to withstand the forces acting on them. A variety of cladding assemblies have been developed that can effectively prevent moisture degradation.

Methods of Water Penetration

Cladding must resist the penetration of rain, snow, and ice into a building's interior. Water penetration is caused by wind driven rain and other forces (Figure 30.8). In

Figure 30.8 Forces that can cause water penetration in joints and other openings in exterior walls.

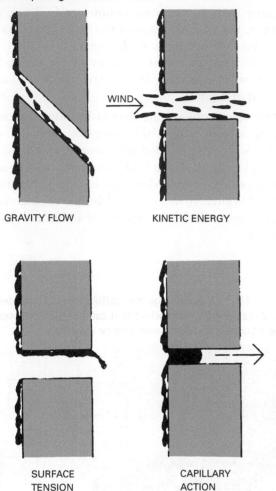

GRAVITY FLOW   KINETIC ENERGY

SURFACE   CAPILLARY
TENSION   ACTION

multistory structures, large amounts of wind-driven rain tend to cascade down a building's face. Gravity works constantly to force water downward into improperly designed joints. Wind-driven rain possesses sufficient kinetic energy to pass through openings in joints and wall surfaces. *Kinetic energy* is the energy in a body caused by motion. Covering joints with some type of batten or internal baffle can mitigate the penetration of kinetic energy driven water.

Water also can penetrate poorly designed joints through surface tension that allows it to adhere to and travel along the underside of building components such as soffits and window heads. This can be prevented by adding an outer edge drip. A drip is a channel cut on the underside of a sill or other component that breaks the surface tension and causes the water to drip. Another force to consider when designing joints is *capillary action*. This

can occur when the connecting surfaces of a joint provide a very small opening along which water can cling and travel. One way to control this is to provide an air gap in the joint, which breaks the capillary path.

These four forces are generally controlled by sealing every joint and opening in the building with gaskets and sealants in an effort to close every hole and gap (Figure 30.9). Sealant systems are not flawless, and may be improperly designed or installed. Because the sealants are exposed to the destructive impacts of rain, wind, and temperature extremes, they may be pulled loose by excessive movement or fail due to weathering. To deal with these shortcomings, cladding systems often make use of secondary internal drainage systems that anticipate the penetration of water and provide a means to conduct it out of the assembly. In some types of panels, such as aluminum-faced, provision is made for internal drainage of water that may penetrate a joint. These provisions can consist of flashing, drainage channels, and weep holes.

Figure 30.9 Techniques for controlling water penetration at joints.

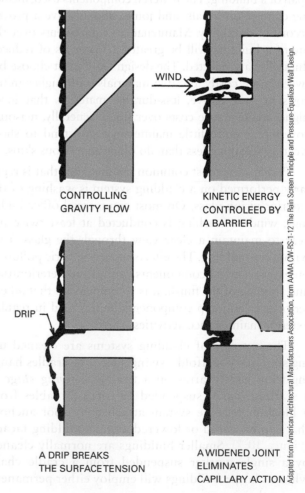

CONTROLLING   KINETIC ENERGY
GRAVITY FLOW   CONTROLEED BY
                                A BARRIER

DRIP

A DRIP BREAKS   A WIDENED JOINT
THE SURFACE TENSION   ELIMINATES
                                CAPILLARY ACTION

The Rainscreen Principle

Wind forces can cause a difference in pressure between a cladding panel's outside face and its interior. When air pressure is greater at the exterior of a panel, water can be forced through a joint into the panel. This type of penetration can be controlled by the rainscreen principle.

The *rainscreen principle* is a pressure equalization method for cladding systems. A rainscreen is composed of a wall's exposed outer surface backed by an air space. Joints are designed to prevent water from penetrating but are not sealed. This permits air pressure on the panel's face to equalize with the air pressure inside the panel. Water droplets are not driven through the panel joints or openings because the rainscreen principle means that wind pressure acting on the outer face of the panel is equalized in the cavity. During extreme weather, a minimal amount of water may penetrate the outer cladding. Drainage elements in the cavity remove any water that penetrates between panel joints. Figure 30.10 illustrates the rainscreen principle in a metal cladding system.

Figure 30.10 This aluminum curtain wall panel has divided air spaces that are vented with protected openings to the exterior. This equalizes the air pressure in the panel with the outside air. Techniques for controlling water penetration at joints.

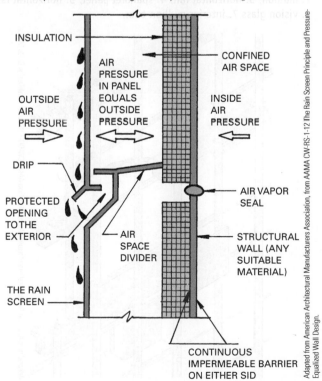

The size of the air space behind the rainscreen is limited by compartmentalizing it into relatively small subdivisions, with each area having openings to the exterior. A waterproof air barrier must be provided on the interior side of the panel. This is necessary to prevent leakage to the interior of the building. Pressure equalization techniques can also be used with glazing details (Figure 30.11). Glazing units may have a water-resistant seal, a pressure equalization chamber along the edge of the glass, and an inner vapor seal.

Figure 30.11 A typical detail for using the rainscreen principle with glazed units. Techniques for controlling water penetration at joints.

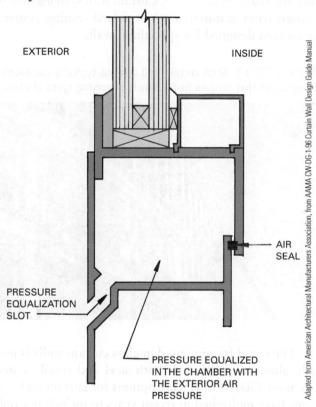

The rainscreen principle is also used to resist water penetration by air pressure differences in traditional wood and masonry walls. These assemblies form a double-wall construction that uses an outer layer to keep out the rain and an inner layer to provide thermal insulation, prevent excessive air leakage and provide a drainage system to the exterior. The structural frame of the building and wall insulation is kept dry, as water never reaches it. When properly detailed and built, a rainscreen wall is an effective envelope solution to external moisture penetration.

METAL AND GLASS CURTAIN WALL SYSTEMS

Glass curtain wall systems typically use aluminum frames that contain in-fill panels of varying types of glass. The framing is supported by the building structure and carries no gravity loads other than its own and that of the glass. Lateral loads such as wind are transferred to the building structure at floor and column lines. Different means of affixing the glass are chosen depending on the final aesthetic desired. Glass curtain walls are widely used because they provide a lightweight exterior wall system of exceptional transparency that can be rapidly erected (Figure 30.12). Curtain wall systems may be chosen from manufacturer's standard catalog systems, or custom designed for specialized walls.

Figure 30.12 Glass curtain wall systems typically use aluminum frames that contain in-fill panels of varying types of glass.

© Stephen B. Goodwin/Shutterstock.com

The metal framing used in glass curtain walls is usually aluminum, although both steel and stainless steel are used. Glass and glazing options for curtain wall systems have multiplied in recent years to include not only standard float glass, but also tempered, laminated, heat-reflective, insulated, and tinted glass, among others. Opaque spandrel glass is used for panels that conceal structural building elements such as floors, columns, floors, HVAC components, electrical wiring, and plumbing. Spandrel panels are typically located between the vision glass on each floor of a building. Spandrel glass is heat-strengthened with a ceramic frit, a molten glossy material, fired onto its surface. It is recommended that rigid insulation or some other rigid backing material be bonded to the back of the glass spandrel panel to hold the glass in place in case of fracturing due to high internal temperatures or storm damage.

The American Architectural Manufacturers Association (AAMA) is the primary source for industry standards regarding typical curtain wall types. Fabrication and installation details can be found in the *Aluminum Curtain Wall Design Guide Manual* published by the AAMA. Curtain walls can be classified by their method of fabrication and installation into the following basic types.

Stick System

The *stick system* was the first curtain wall type developed by manufacturers, and remains the most common. In the stick system, vertical mullions are installed, followed by horizontal rails into which glazing and spandrel panels are placed (Figure 30.13). First, the mullions in one or two-story lengths are attached to mechanical anchors on the structural frame (Figure 30.14). Next, horizontal rails equal in length to the width of the infill panels are mounted between them. The glass and spandrel panels can be installed from either the interior or exterior. In exterior glazed systems, glass and opaque panels are installed from the outside of the curtain wall frame. Exterior glazed systems require cranes, scaffolding, or swing stages to access to the exterior of the wall

Figure 30.13 The stick curtain wall system: 1. anchors, 2. mullion, 3. horizontal rail, 4. spandrel panel, 5. horizontal rail, 6. vision glass, 7. interior mullion trim.

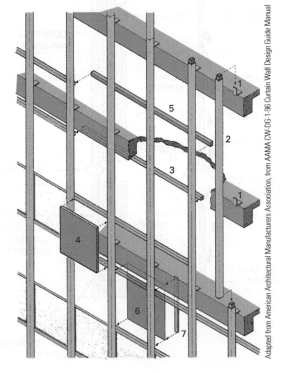

Adapted from American Architectural Manufacturers Association, from AAMA CW-DG-1-96 Curtain Wall Design Guide Manual

Figure 30.14 The metal framing used in glass curtain walls is attached to mechanical anchors on the structural frame.

© Dmitry Kalinovsky/Shutterstock.com

Figure 30.15 The unit curtain wall system: 1. anchor, 2. preassembled framed curtain wall unit.

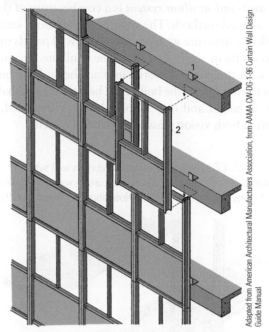

Adapted from American Architectural Manufacturers Association, from AAMA CW-DG-1-96 Curtain Wall Design Guide Manual

for installation as well as repair or replacement activities. Interior glazed systems install the glass or opaque panels into the curtain wall frame from the inside of the building. Interior installation is often preferred for high-rise construction because of the ease of access both during construction and in replacing glass.

The stick system is relatively economical in material costs, with minimum requirements for assembly facilities and shipping. However, since each component of a wall is installed piece by piece in the field, it is more labor intensive than other systems. The stick system is most commonly used in low-rise construction and in smaller buildings.

Unitized System

The *unitized system* uses large preassembled wall panels. The vertical edges of the units are joined, serving as a mullion, and the bottom of one panel is joined with top of the one below it, forming a horizontal rail (Figure 30.15). Large factory-assembled framed units can be made, complete with spandrel panels and vision glass installed, or the panels may be fabricated as frames with the vision glass fixed on site. The panels are designed for sequential installation with interlocking split-vertical mullions and nesting horizontal rails at expansion joints. With significant fabrication and shipping requirements, the unitized system is more shop labor dependent, with less fieldwork required than the stick system. Unitized systems are most common in large high-rise buildings, although they find application on low-rise buildings as well. The panels may be one, two or three floors tall, with a width of infill panels from 5 ft. to 6 ft. (1.52–1.82m) (Figure 30.16).

Figure 30.16 A preassembled curtain wall panel is hoisted into place.

© anyaivanova/Shutterstock.com

Unit and Mullion System

The *unit-and-mullion system* is a combination of the stick and unitized methods. The system uses mullions secured to a building structure with preassembled wall panels installed between them (Figure 30.17). The mullions are installed first with the panels lowered in place behind the mullions from the interior of the building. The panels are usually one story in height and may be pre-glazed or unglazed, incorporating both vision glass and spandrel panels.

Figure 30.17 The unit-and-mullion curtain wall system: 1. anchors, 2. mullion, 3. preassembled curtain wall unit lowered into place behind the mullion from the floor above, 4. interior mullion trim.

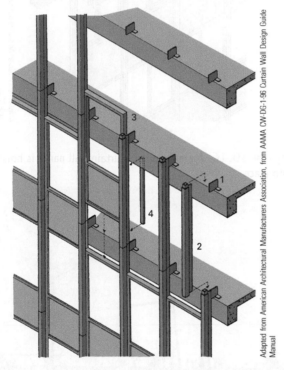

Adapted from American Architectural Manufacturers Association, from AAMA CW-DG-1-96 Curtain Wall Design Guide Manual

Column-Cover-and-Spandrel System

The *column-cover-and-spandrel system* uses long spandrel panels that span column covers. The glazing may be installed above the spandrel, and columns are enclosed with a hollow cover (Figure 30.18). While column cover and spandrel systems are similar to unit and mullion systems, they differ in that the building frame is emphasized by the vertical column covers. These systems are relatively new and are finding increasing use because they offer more design options. Whole units can be pre- or site-assembled with infill vision glass and spandrel panels between the columns, and with the column covers in place. Since these systems are customized, framing construction tolerances are more critical as the units are manufactured to fit precisely within existing column bays.

Figure 30.18 The column-cover-and-spandrel curtain wall system: 1. column cover, 2. spandrel panel, 3. glazing unit.

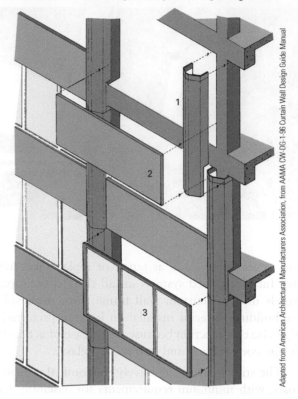

Adapted from American Architectural Manufacturers Association, from AAMA CW-DG-1-96 Curtain Wall Design Guide Manual

Design Examples Construction details for a glass curtain wall using structural silicone sealant to attach panels to metal mullions hidden behind the glass are shown in Figure 30.19. This design clads the entire wall with glass curtain wall panels that admit natural light and permit occupants to see out. They are referred to as vision-type glass ribbon windows. The glass may be any of several types, such as reflecting or heat resisting. In Figure 30.20, a fully glazed construction is used, with opaque glass spandrel panels covering interior structural members and other construction components, such as ceiling panels, heating, plumbing, and electrical runs.

Glass-Fixing Methods

Glazing is the system or process used to support the glass in the frame and seal the dissimilar mating surfaces from the elements. There are several basic options for affixing curtain wall glass to a structural frame. Relatively common in smaller commercial and residential applications, *wet glazing* uses preformed tape or a gunable liquid sealant to set the glass in the frame.

Dry glazing utilizes a manufactured compression gasket of rubber or vinyl that supports the glass (Figure 30.21). Dry glazing is common for larger, commercial installations.

Figure 30.19 Details for a double-glazed curtain wall assembly.

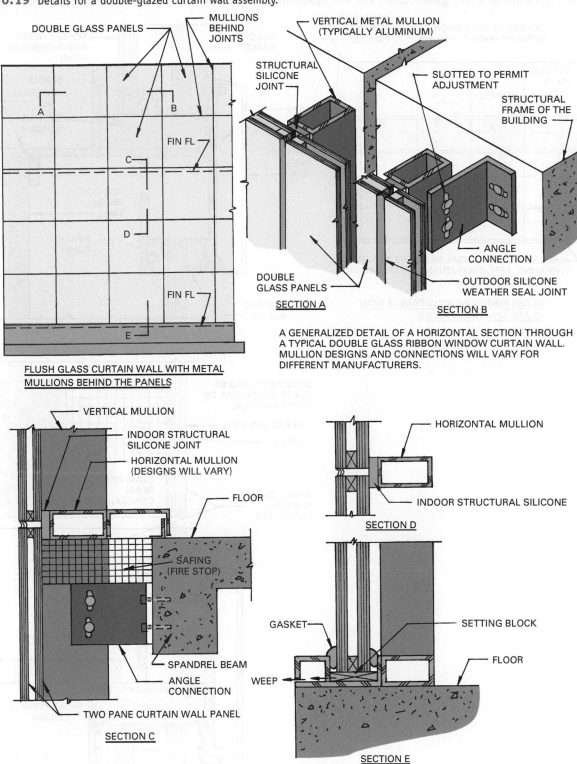

DOUBLE GLASS PANELS

MULLIONS BEHIND JOINTS

A   B

FIN FL

C

D

FIN FL

E

FLUSH GLASS CURTAIN WALL WITH METAL MULLIONS BEHIND THE PANELS

VERTICAL METAL MULLION (TYPICALLY ALUMINUM)

STRUCTURAL SILICONE JOINT

SLOTTED TO PERMIT ADJUSTMENT

STRUCTURAL FRAME OF THE BUILDING

ANGLE CONNECTION

DOUBLE GLASS PANELS

OUTDOOR SILICONE WEATHER SEAL JOINT

SECTION A   SECTION B

A GENERALIZED DETAIL OF A HORIZONTAL SECTION THROUGH A TYPICAL DOUBLE GLASS RIBBON WINDOW CURTAIN WALL. MULLION DESIGNS AND CONNECTIONS WILL VARY FOR DIFFERENT MANUFACTURERS.

VERTICAL MULLION

INDOOR STRUCTURAL SILICONE JOINT

HORIZONTAL MULLION (DESIGNS WILL VARY)

FLOOR

SAFING (FIRE STOP)

SPANDREL BEAM

ANGLE CONNECTION

TWO PANE CURTAIN WALL PANEL

SECTION C

HORIZONTAL MULLION

INDOOR STRUCTURAL SILICONE

SECTION D

GASKET

WEEP

SETTING BLOCK

FLOOR

SECTION E

Dry glazing generally minimizes quality-control issues because it is pre-manufactured, unlike wet glazing where assemblies are glazed on site. Glazing beads or bars provide combined sealing and retaining through external aluminum, steel, wood, or other materials that exert a contact pressure along the full length of the glass and supporting construction. Permanently elastic pads of silicone (EPDM) ensure a good seal and adequate elasticity.

Figure 30.20 Details for a fully glazed curtain wall with opaque glass spandrel panels covering interior structural members.

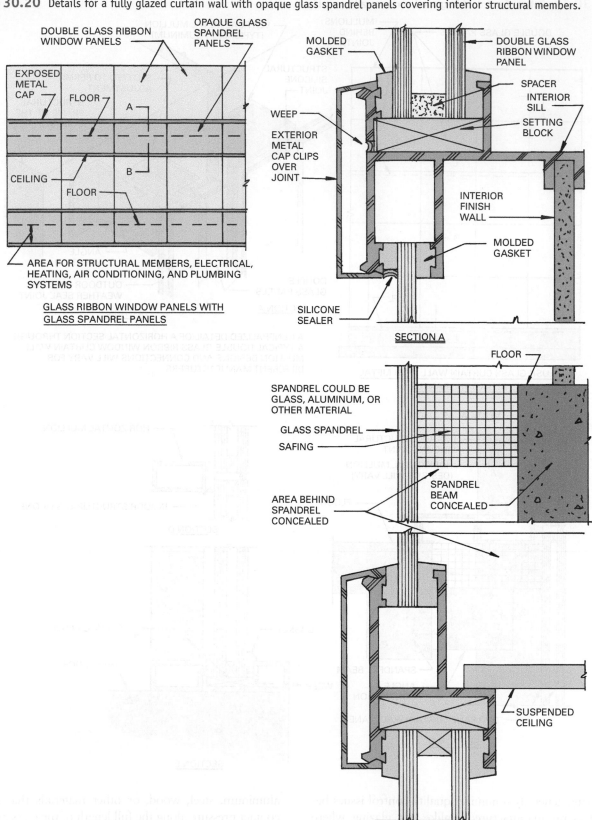

DOUBLE GLASS RIBBON WINDOW PANELS

OPAQUE GLASS SPANDREL PANELS

EXPOSED METAL CAP

FLOOR

A

CEILING

B

FLOOR

AREA FOR STRUCTURAL MEMBERS, ELECTRICAL, HEATING, AIR CONDITIONING, AND PLUMBING SYSTEMS

GLASS RIBBON WINDOW PANELS WITH GLASS SPANDREL PANELS

MOLDED GASKET

WEEP

EXTERIOR METAL CAP CLIPS OVER JOINT

SILICONE SEALER

DOUBLE GLASS RIBBON WINDOW PANEL

SPACER INTERIOR SILL

SETTING BLOCK

INTERIOR FINISH WALL

MOLDED GASKET

SECTION A

SPANDREL COULD BE GLASS, ALUMINUM, OR OTHER MATERIAL

GLASS SPANDREL

SAFING

AREA BEHIND SPANDREL CONCEALED

FLOOR

SPANDREL BEAM CONCEALED

SUSPENDED CEILING

SECTION B

Chapter 30 Cladding Systems

Figure 30.21 Dry glazing utilizes a manufactured compression gasket of rubber or vinyl to support the glass.

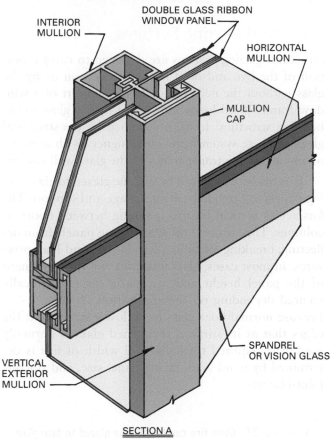

INTERIOR MULLION

DOUBLE GLASS RIBBON WINDOW PANEL

HORIZONTAL MULLION

MULLION CAP

SPANDREL OR VISION GLASS

VERTICAL EXTERIOR MULLION

SECTION A

Continuously fixed linear support systems use adhesion and/or compressed based on fixing mechanisms. Pressure-plate systems use linear plates that are bolted to the frame to compress the glass along its edges (Figure 30.22). Loads are transferred through gaskets or foam tapes that acts as a pad between the plate and the glass. A cover cap is used to cover the bolts and provide a finished appearance (Figure 30.23). The system is installed on site and is labor intensive.

Figure 30.22 Continuously fixed linear glazing support systems use adhesion and/or pressure-plate systems.

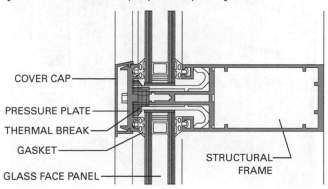

COVER CAP

PRESSURE PLATE

THERMAL BREAK

GASKET

GLASS FACE PANEL

STRUCTURAL FRAME

Figure 30.23 The installation of a cover cap used to conceal the mechanical anchoring and provide a finished appearance.

Structural sealant systems use structural adhesives like silicone to affix glass panels on either two or four glass sides. Structural adhesives have been in use for 30 years and provide a way to set glass without mechanical fixing. Double-sided structural-bonding tape, modified epoxy adhesives, and silicone all afford adhesion, stability, and movement capability. The linear edge gluing can also act as a thermal break and increase the soundproofing of the assembly. If both sides of a double-glazed unit are glued to a frame, one side must be softer than the other to allow for thermal movement. Combined plate-and-sealant systems use both a pressure plate and structural sealants. Commonly the pressure plate is used for horizontal joints and the sealants for vertical ones.

Point-Supported Glazing

Point-supported glazing systems utilize mechanical anchors at discrete locations near the glass edge in place of continuous edge supports. Point supports are typically anchored to the primary structural system through base plates or other fabricated mechanical anchors.

Point-supported glazing requires minimal interior frame-work and uses no exterior pressure plates. Holes drilled into the glass by the manufacturer allow the glass to be supported by a variety of point-supported components.

Simple systems use an exterior plate and interior counter plate that clamp the glass and are connected by a bolt. The plates are fitted with neoprene washers and work to disperse loads to a larger area of the glass. Spring plates can be added to accommodate thermal movement. The use of a single interior plate with a counter-sunk bolt fitting produces a flush and uninterrupted glass wall. The bolts can be face mounted or counter-sunk in the glass for a completely smooth surface.

Structural standoff systems use stainless-steel stand-offs or spiders that are fixed to building support structures such as concrete columns, steel mullions and posts or metal/cable truss systems (Figure 30.24). Spider fix-ings are made from high-quality stainless steel and used in high wind load areas where conventional fixings are not sufficient to cope with loadings. The system can al-low for an unlimited height in glazing when supplied with intermediary structural supports at given spans.

Figure 30.24 Structural standoff systems use stainless-steel standoffs or spiders fixed to the building frame.

Point-fixing of glass can also be achieved without penetrating the glass pane. Economical options exist for clamping glazing without the need to drill. In this method, it is important that the glass not be restrained, as the structure moves under load. Clamping plates are often used to fix glass to cable net structures that use cable truss systems to support the glazing.

Structural Glazing Systems

Structural glazing systems are designed to carry a por-tion of the live and dead loads imposed on or by the glass without the need of additional support of a win-dow frame. Wind loads transferred by the glass to the building structural frame are carried by the structural glazing. These systems provide designers with a means of amplifying the transparency of the glass wall system.

Fins made of glass can be silicone glazed to glass fac-ing panels to provide both anchorage and support. The fins act as vertical beams, spanning between floors or columns. They prevent the glass facing panels from de-flecting, breaking, or falling out due to wind-load pres-sures. In most cases, glass fins must run the full length of the panel height and are glazed or mechanically secured depending on the application (Figure 30.25). Because normal float glass has a lower strength at the edges that at its surface, toughened glass is normally used. The required thickness and width of fins is de-termined by wind load, facing panel size, and silicone joint-bite size.

Figure 30.25 Glass fins can be silicone glazed to face glass panels to provide both anchorage and support.

While still rare, the use of glass mullions or fins is finding increased use in monumental spaces, such as convention centers or airport terminals. These systems have very long lead times and are more proprietary in nature. The involvement of the manufacturer early in the design process is crucial in selecting system components and determining their structural capabilities. Product and installation costs for structural glazing systems are higher than for conventional frame-supported curtain wall systems.

MASONRY, STONE, EIFS AND METAL CLADDING SYSTEMS

In addition to glass, curtain walls can be faced at the exterior with any construction material that is noncombustible and can withstand exterior weather conditions.

There are countless types and combinations of cladding materials used, from lightweight assemblies of plaster and metal, to more substantial facings of masonry, stone, or concrete. Most require the use of a backup wall or frame to anchor the facing to the building's structural frame.

Masonry Curtain Walls

Masonry curtain walls are usually supported by steel shelf angles secured to the structural frame of a building. The shelf may be anchored to a concrete spandrel beam, bolted, or welded to a structural-steel spandrel beam (Figure 30.26). Walls can be laid up a brick at a time, as is done with low-rise masonry construction. Provisions are made for horizontal and vertical expansion joints to permit panels to expand and contract as temperatures vary. Masonry curtain wall panels can also be prefabricated on the ground and lifted into place with a crane.

Figure 30.26 Typical details for a brick masonry curtain wall placed over concrete and steel structural systems.

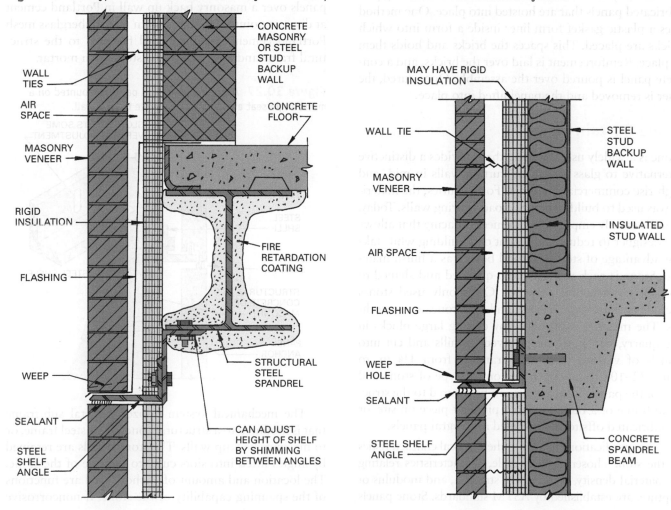

WALL TIES

AIR SPACE

MASONRY VENEER

RIGID INSULATION

FLASHING

WEEP

SEALANT

STEEL SHELF ANGLE

CONCRETE MASONRY OR STEEL STUD BACKUP WALL

CONCRETE FLOOR

FIRE RETARDATION COATING

STRUCTURAL STEEL SPANDREL

CAN ADJUST HEIGHT OF SHELF BY SHIMMING BETWEEN ANGLES

MAY HAVE RIGID INSULATION

WALL TIE

MASONRY VENEER

AIR SPACE

FLASHING

WEEP HOLE

SEALANT

STEEL SHELF ANGLE

STEEL STUD BACKUP WALL

INSULATED STUD WALL

CONCRETE SPANDREL BEAM

In either case, to secure them, high-bond mortars are used for their excellent bonding characteristics and high compressive and tensile strength.

Masonry curtain walls typically have a backup wall, often of concrete masonry units or steel studs. The backup wall is non-load bearing and supports the sheathing, insulation, and air and vapor barriers, as well as the interior finish materials. The curtain wall is affixed to the backup wall with masonry ties. Ties are stiff but can flex enough to accommodate differential movements between the veneer and the backup wall. Most single-width masonry curtain walls experience some penetration by driving rain. This moisture is handled by the rainscreen principle, which provides a cavity behind the masonry and weep holes to drain the moisture. The cavity acts as a pressure equalization chamber and provides some degree of pressure equalization between the exterior and interior surfaces of the masonry. Some architectural designs permit the structural system to be visible on the exterior of the finished building, with cladding systems filling the spaces between.

Masonry curtain walls are also constructed using prefabricated panels that are hoisted into place. One method uses a plastic gasket form liner inside a form into which bricks are placed. This spaces the bricks and holds them in place. Reinforcement is laid over the bricks, and a concrete panel is poured over the assembly. Once cured, the liner is removed and the panel lifted into place.

Stone Curtain Walls

Stone is a widely used cladding that provides a distinctive alternative to glass and metal curtain walls in mid- and high-rise commercial buildings. For centuries, the material was used to build structural, load-bearing walls. Today, stone is mainly employed as a veneer or facing that allows the designer to reduce the weight of a building while taking advantage of stone's natural beauty as a finish material. Stone is rock that has been quarried and shaped to size for construction purposes. Commonly used stones include granite, slate, marble, and limestone, among others. The material is produced by cutting large blocks in the quarry, which are then moved to mills and cut into panels of various thicknesses ranging from 1¼ in. to 4 in. (32–102 mm), depending on the type of stone and size of the panel. Stone panels can be secured to the structural frame of a building either piece by piece on site, or be fabricated offsite and installed as integral panels.

The specifications related to the physical characteristics of the stone chosen, for example, characteristics relating to material density, compressive strength, and modulus of rupture, are established by ASTM standards. Stone panels

are available in a variety of finishes, ranging from polished, honed, rubbed, flame or thermal finished, to a natural sawn appearance. The texture of the stone has a great influence on the finished appearance. Fine-grained stones can have a smoother, polished surface, while coarse-grained stones present a more porous face. Porosity pertains to the ability of stone to resist moisture penetration. Porous rock tends to permit some minerals to dissolve and stain the exposed face. It also is not durable and can be damaged by freeze–thaw cycles. Stone exposed to harsh conditions must have sufficient durability to withstand freeze–thaw cycles as well as corrosive conditions.

There are two primary methods used to attach stone veneer to a building's exterior. One is installed in individual panels on site using either cement or mechanical means, while the other is pre-fabricated offsite and installed in integral panels. In the conventional piece-by-piece stone facade system, the stone is installed by using steel angles, strap anchors, dowels, and mortar, or by other mechanical systems (Figure 30.27 and Figure 30.28). One method of cladding a soffit is shown in Figure 30.29. A typical set method applies the stone panels over a masonry back up wall in Portland cement at discreet points. Another system uses a fiberglass mesh Portland cement backing board fastened to the structural frame and adhering to the stone with mortar.

Figure 30.27 A stone curtain wall panel is mounted on a metal angle seat and tied to a concrete backup wall.

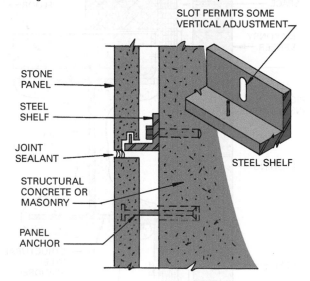

The mechanical system utilizes a metal sub-frame that is mounted to a structural concrete or steel frame, or to masonry backup walls. The stone panels are retained by angles fitted into slots cut into the side of the stone. The location and amount of anchors used are functions of the spanning capability of the stone. If noncorrosive

Figure 30.28 A stone curtain wall panel mounted over a steel structural system and steel studs.

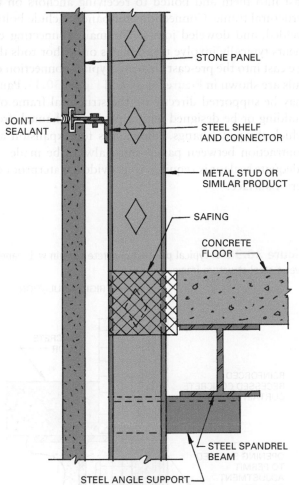

STEEL STUD WALL WILL CONTAIN INSULATION

STONE PANEL

JOINT SEALANT

STEEL SHELF AND CONNECTOR

METAL STUD OR SIMILAR PRODUCT

SAFING

CONCRETE FLOOR

STEEL SPANDREL BEAM

STEEL ANGLE SUPPORT

Figure 30.29 A typical stone soffit detail as used at window returns.

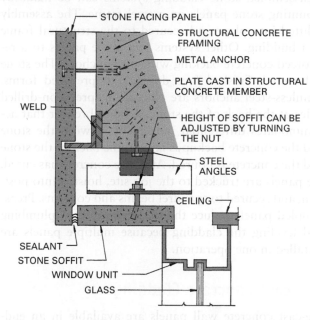

STONE FACING PANEL

STRUCTURAL CONCRETE FRAMING

METAL ANCHOR

PLATE CAST IN STRUCTURAL CONCRETE MEMBER

WELD

HEIGHT OF SOFFIT CAN BE ADJUSTED BY TURNING THE NUT

STEEL ANGLES

CEILING

SEALANT

STONE SOFFIT

WINDOW UNIT

GLASS

metals are used, they must be protected from electrolytic reaction. Joints between panels are typically ¼ in. (.635 cm) in width but vary widely by product. Manufacturers often supply or recommend engineering approved connections and installation details.

A stone facade system that is finding more application in high-rise buildings consists of an aluminum curtain wall system into which stone veneer is installed instead of glass panels (Figure 30.30). The precut stone veneer is simply slipped into what would otherwise be aluminum window frame. In the prefabricated system,

Figure 30.30 Typical types of steel curtain wall sub-frames used with stone panels.

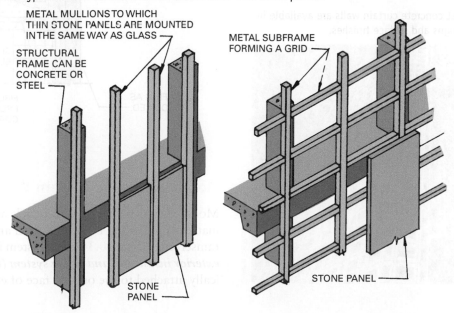

METAL MULLIONS TO WHICH THIN STONE PANELS ARE MOUNTED IN THE SAME WAY AS GLASS

STRUCTURAL FRAME CAN BE CONCRETE OR STEEL

METAL SUBFRAME FORMING A GRID

STONE PANEL

STONE PANEL

panels are constructed offsite by specialty contractors. Preassembled stone cladding systems can be made by mounting stone panels on a steel frame. The assembly is lifted into place and secured to the structural frame of a building. Other systems join stone panels to a re-inforced concrete backing with steel anchors. The stone is lowered, finished face down, into prepared forms. Stainless-steel anchors are placed into precision-drilled holes at the back of the stone. A bond breaker that ac-commodates differential movement between the stone and the concrete backup is laid on the back of the stone and the concrete is poured. After the concrete has cured, the panels are trucked to the job site, hoisted into posi-tion, and secured to spandrel beams and columns. Preas-sembled panels reduce the time required for plumbing and leveling the cladding because multiple panels are installed in one operation.

Pre-cast Concrete Cladding

Pre-cast concrete wall panels are available in an end-less variety of designs, surface finishes, and thicknesses. For cladding applications, they may be pre-stressed or conventionally reinforced non-load-bearing units (Fig-ure 30.31). They may be hollow, solid, or of an insu-lated sandwich construction with a flat, ribbed, or other surface configuration. Flat panels are often up to two stories high, and ribbed panels can be cast up to four stories high. Solid panels are typically from 4 to 10 in. (10.1–25.4 cm) thick, while sandwich and hollow-core panels vary from 5½ to 12 in. (14–30 cm) thick. The use of glass-fiber reinforcing in addition to steel enables the production of lighter and thinner panels.

Figure 30.31 Pre-cast concrete curtain walls are available in an endless variety of designs and surface finishes.

© Aaron Roe/Shutterstock.com

Pre-cast concrete panels can be used in connection with a pre-cast or cast-in-place concrete, or steel framing systems. The panels are installed by means of anchors cast into them and bolted to receiving anchors on the structural frame. Connections for panels include bolted, welded, and doweled joints. The main connecting ele-ments typically involve metal plates or anchor rods that are cast into the pre-cast element. Typical connection de-tails are shown in Figure 30.32 and Figure 30.33. Panels may be supported directly on the structural frame of a building or be designed with a projecting haunch from which the panel hangs. Allowances for expansion and contraction between panels must always be made. An adequate sealant is required to provide a waterproof ex-terior surface.

Figure 30.32 A typical pre-cast concrete curtain wall panel over a steel structure frame.

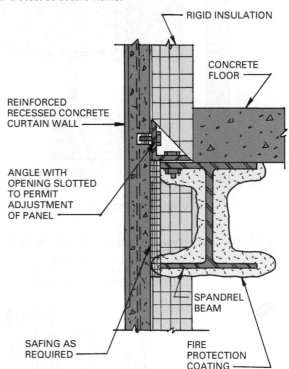

Modified Stucco Curtain Walls

Modified stucco curtain wall panels are an assembly of materials providing both insulation and a weather-tight rainscreen (Figure 30.34). The system is referred to as an *exterior insulation and finish system (EIFS)*. EIFS is typ-ically attached to the outside face of exterior walls with

Figure 30.33 Several ways pre-cast concrete curtain walls are secured to concrete spandrel beams.

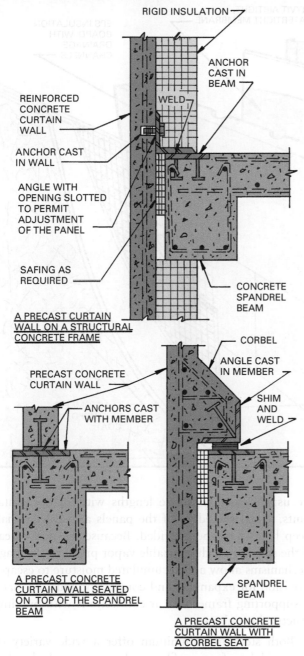

RIGID INSULATION

ANCHOR CAST IN BEAM

WELD

REINFORCED CONCRETE CURTAIN WALL

ANCHOR CAST IN WALL

ANGLE WITH OPENING SLOTTED TO PERMIT ADJUSTMENT OF THE PANEL

SAFING AS REQUIRED

CONCRETE SPANDREL BEAM

A PRECAST CURTAIN WALL ON A STRUCTURAL CONCRETE FRAME

CORBEL

ANGLE CAST IN MEMBER

PRECAST CONCRETE CURTAIN WALL

SHIM AND WELD

ANCHORS CAST WITH MEMBER

A PRECAST CONCRETE CURTAIN WALL SEATED ON TOP OF THE SPANDREL BEAM

SPANDREL BEAM

A PRECAST CONCRETE CURTAIN WALL WITH A CORBEL SEAT

Figure 30.34 The continuous exterior insulation of EIFS systems encloses a building in a protective blanket of rigid insulating.

© Marcin Krzyzak/Shutterstock.com

into it. A final synthetic plaster finish coat is troweled over the base coat. Some EIFS systems include only the insulation and EIFS coatings. Other types also include plastic edge trim, water-resistive barriers, a drainage cavity, and other accessories.

Curtain wall panels of one or more stories can be constructed of steel studs covered with EIFS materials. The assembled panels are lifted into place and bolted or welded to connectors on the structural frame of a building in the same manner as described for other types of curtain wall construction. The continuous exterior insulation of EIFS systems encloses a building in a protective blanket of rigid insulating. Boards with thicknesses of 1 to 12 in. (2.5–30 cm) can provide insulating high R-values.

A number of similar panel systems are manufactured for lightweight exterior wall cladding. For example, a composite panel similar to the one just described is sometimes used, but with a hard polymer exterior finish. Another type provides a stucco, exposed aggregate, or simulated brick or stone exterior surface. EIFS systems provide architects endless design opportunities in terms of form, details, color, and texture. Features such as soffits, curves, accent bands, and sculptured details are easy to fabricate and install in the insulation material.

Because EIFS systems rely on a perfect seal at the exterior surfaces, they are susceptible to entrapment of moisture inside the wall system. The EIFS Industry Members Association (EIMA) publishes a set of installation details that addresses correct installation details for EIFS systems. For wood-framed buildings, a secondary weather barrier should be placed over the sheathing before the exterior cladding is installed. This barrier

adhesive or mechanical fasteners. Most systems begin with a sheathing material over which expanded polystyrene (EPS) insulation board is secured with special mechanical fasteners. The substrate often uses a board with inorganic glass mat facings; a water-resistant, silicone-treated gypsum core; and an alkali-resistant surface coating (Figure 30.35). A base coat is troweled over the insulation board; then a reinforcing mesh is embedded

Figure 30.35 The Dryvit® Infinity EIFS provides a water-tight membrane and an insulation board with drainage channels to capture, control, and discharge any incident moisture that may enter the system.

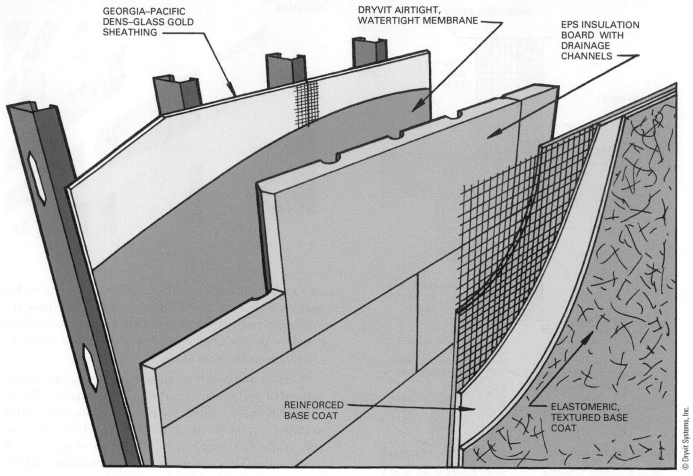

GEORGIA–PACIFIC DENS–GLASS GOLD SHEATHING

DRYVIT AIRTIGHT, WATERTIGHT MEMBRANE

EPS INSULATION BOARD WITH DRAINAGE CHANNELS

REINFORCED BASE COAT

ELASTOMERIC, TEXTURED BASE COAT

© Dryvit Systems, Inc.

protects the structure from water intrusion and allows moisture to exit by draining on top of the barrier.

Metal Cladding Systems

A number of metal exterior cladding systems are frequently used for commercial construction (Figure 30.36). Curtain walls may be built of prefabricated metal panels that are capable of meeting requirements for strength, durability, insulation, and appearance. Metal panels generally incorporate an insulation core sandwiched between a thin lightweight facing and backing. Some metal panels are made of single thickness with ribs formed in them for strength and decorative purposes. The panels can be fastened in place in a light frame, attached to secondary framing members, anchored to brackets at each floor level, or connected directly to the structural frame of the building (Figure 30.37).

Flat and profiled metal sheets are supported to framing members at a right-angle to their profile. They

are usually used in single lengths without horizontal joints. When the edges of the panels are sealed, small weep holes must be provided. Because the direct heat of the sun can produce sizable vapor pressure, drainage mechanisms allow any accumulated moisture to escape. Provision for expansion and contraction must be made in supporting frames and at connections with building structural members.

Both steel and aluminum offer a wide variety of trapezoidal surface profiles, and can be applied either horizontally or vertically. Manufacturers provide specialized bending processes to produce pre-formed corner panels and junctions. Metal panels should be shaped so that changes in surface appearance will not be noticeable as the metal expands and contracts. Light-gauge metal panels are often given decorative pattern to help to stiffen the sheets while also hiding movements due to temperature variations. Flat sheets may be formed with a slight curvature and stiffened on the rear side with ribs, so that thermal or structural movement will

Figure 30.36 Curtain walls may be built of prefabricated metal panels that are capable of meeting requirements for strength, durability, insulation, and appearance.

© Oledjio/Shutterstock.com

Figure 30.37 Metal cladding panels can be attached to secondary framing members, or connected directly to the structural frame of the building.

© Dmitry Kalinovsky/Shutterstock.com

only change the curvature a little and not reverse it. Flat sheets may also be laminated to one or more flat stiffening sheets, like mineral fiber panels or a second light-gauge metal sheet, to prevent distortions or oil canning. The term "oil canning" describes distortions in thin-gauge metal panels that are fastened in a way that restricts their expansion and contraction under normal thermal conditions.

SEALING JOINTS FOR CLADDING SYSTEMS

The detailing of joints and proper installation of seals and flashings is a key factor in controlling water penetration. Sealing materials must bond to joint surfaces, repel rain and air infiltration, and resist the stresses caused by movement.

Sealant Joints

Most cladding systems rely on sealants to maintain weather tightness between components and panels. A sealant is a material used to seal joints between construction members and protect materials against the penetration of moisture, air, corrosive substances, and foreign objects. The sealing of joints between these and other parts of a building is essential to ensure the integrity of the entire structure. If sealants are not properly specified or poorly installed, a leaky building is unavoidable.

The choice of joint sealant depends on the materials to be sealed, the function of the joint, and possible changes that may occur in the joints after the sealant has been installed. The sealant filling a joint is a substitute for the material that it joins. As such, it must meet the performance requirements of the adjoining materials. It must maintain the integrity of the assembly of materials while still allowing for movement. A sealant must also demonstrate excellent adhesion to the material with which it is expected to bond. It must be flexible and have minimum internal shrinkage through years of use, resist staining the material around it, and have a tough non-tacky elastic skin, such that dirt and solid objects do not stick to it.

Sealants are composed of flexible compounds or solid materials. Working joints are designed to expand and contract with movement. Non-working joints are joined by mechanical fasteners and do not accommodate movement (Figure 30.38).

Working joints are made from flexible adhesive sealants that are worked into a joint, bond to the sides, and set into a firm but rubbery plastic substance (Figure 30.39). A key factor in the design of working joints is the percent of elongation the sealant can safely stretch to and still give expected protection. Elastic/elastomeric sealants

Figure 30.38 Common details for working and non-working sealant joints.

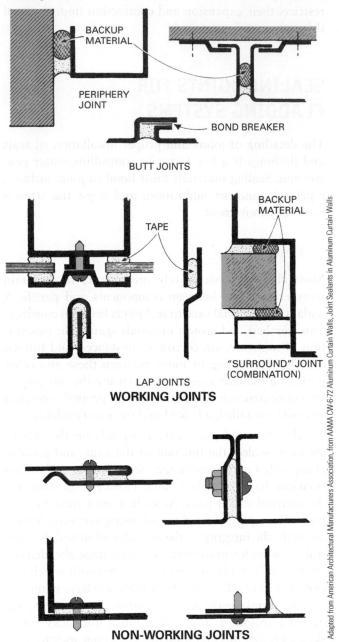

WORKING JOINTS

NON-WORKING JOINTS

Adapted from American Architectural Manufacturers Association, from AAMA CW-6-72 Aluminum Curtain Walls, Joint Sealants in Aluminum Curtain Walls

stretch as a joint widens, and shrink back to their normal size as the width contracts. High-performance sealants have a 50 percent elongation, while intermediate types are usually rated up to 25 percent. Sealant joint width is typically sized to withstand about four times the anticipated movement. Sealant manufacturers publish recommendations of the proper joint width-to-depth ratios for their materials. Very small joints, of less than ¾ in. are not recommended, due to difficulties with applying sealants effectively.

Figure 30.39 These pre-cast concrete curtain wall panels have their joints filled with a flowable sealant.

Types of Sealant Materials

Liquid sealant materials have adhesive qualities and are applied with either a sealant gun or a knife. They adhere to the surfaces on which they are applied and cure into a rubbery material that creates a water- and air-tight joint. Flowable sealants allow for expansion and contraction in a joint because they can stretch and contract without fracturing. Flowable sealant materials are grouped into three classes, determined by the amount of change in joint size they can accommodate.

Low-performance sealants have minimum movement capacity and are used in stable joints. Typical materials include oil-and-resin-based compounds, bituminous-based caulks and mastics, and polybutene compounds.

Medium-performance sealants can accommodate elongations in the range of 65 to 612.5 percent. Typical materials in this group include acrylics, butyls, and neoprene. These materials are available on the market, but are generally considered older technology and are not recommended for the perimeter sealing of high-performance curtain walls. High-performance sealants are used when cyclic movement is expected to be between 612.5 to 625 percent. Typical materials in this group include polymercaptans, polysulfides, polyurethanes, silicones, and some solvents that release acrylics.

Urethanes are an effective, all-purpose sealant with relatively low initial cost. They bond well with masonry, precast concrete, and stone, and have a relatively slow cure rate, allowing for longer working time. Urethanes require a primer coat for good performance finishes on metal, and service-life limitations may require more maintenance. Urethanes are available in a wide variety of colors.

Silicones provide a high-performance sealant for curtain wall construction, offering a wide selection of curing mechanisms and times, and adhesion characteristics. They are capable of carrying the structural loads imposed on an assembly, and can seal both glazing and perimeter components in one application. Silicones derive from the same chemical family as glass, which explains their excellent adherence to it. Initial cost and compatibility with masonry and concrete must be considered when selecting silicone sealants, and color selection is more limited than with urethane.

Proper installation of sealants depends on the use of backup materials installed behind the joint to control the depth of the sealant. Joint filler are inserted into joints that are deeper than required for the sealant installation. A backup material, sometimes called backer rod, not only limits the depth of the sealant joint, but also serves as a bond breaker to keep the sealant from bonding to elements behind the joint. Sealant backing also provides the necessary resistance to sealant installation pressure and helps to attain a proper substrate as the sealant is tooled. For moving joints, both the sealant and joint filler must be compressible. Some backing materials are installed in the factory as part of a unit, as in a manufactured window. Others are installed in the field directly before the sealant application.

Backup materials vary depending on location and use. Rods or tubing made from polyethylene, butyl, urethane, and neoprene are the most frequently used. Closed-cell sealant backing is made primarily from polyethylene foam, while open-cell sealant backing uses urethane foam. Bi-cellular sealant backing is composed of both open- and closed-cell polyethylene or polyolefin foam. Closed-cell sealant backing should be no more than 25 percent to 30 percent larger than the joint width, so that it remains in compression and in place during sealant installation. Open-cell and bi-cellular sealant backing must be at least 25 percent larger. Care should be taken not to poke holes in closed-cell sealant backing

material during installation that can cause air bubbles to form in the sealant and compromise its ability to seal properly. Properly installed sealant backing prevents sealant leakage and creates a proper sealant profile.

Other rod types form a primary water seal that help to keep out moisture during construction before the sealant can be installed. Typical uses for rods and tubes are in joints between pre-cast curtain wall panels and expansion joints in long masonry walls. Tapes and polyethylene film are another form of backup material. They act as a bond breaker, keeping a sealant from adhering to joint surfaces where bonding is not desired.

Solid Sealant Materials

Solid sealant materials include tapes and gaskets. Preformed solid tapes are made from polybutene or polyisobutylene and have adhesive on one or both sides. They are available in rectangular, square, circular, and wedge shapes (Figure 30.40). These materials are completely cured when manufactured. They seal by being compressed, and are used predominantly on lap joints.

Figure 30.40 Solid sealant materials include tapes and gaskets that are available in a variety of shapes.

© Peter Sobolev/Shutterstock.com

Pre-formed cellular tapes are made from polyurethane, neoprene, vinyl nitrile, and polyvinyl chloride. They contain a chemical solvent and are delivered in a tightly compressed condition. When installed, they expand, filling the joint and bonding to contacting surfaces as they cure.

Gaskets are pre-formed solid elastomeric materials designed to fit the configuration of components and panels to be sealed. They are compressed and forced into a joint, forming a watertight seal. Dense and cellular rubber products are generally formulated from

EPDM, neoprene, silicone, and thermoplastic polymers. Extruded and molded gaskets are fabricated primarily from silicone formulations, although some polyurethane formulations are also available. They are manufactured as extruded strips in various standard colors and widths from 1 to 6 in., and can serve as custom-molded components for corners and other transition areas.

Compression sealant systems use pre-compressed products usually fabricated of a polyurethane, acrylic-impregnated cellular component with a silicone rubber exterior face. When inserted into a joint opening, the foam expands to be compressed in the joint and the rubber face is sealed to the substrates. Typical products are capable of up to 25 percent movement.

Review Questions

1. What types of cladding are typically used for multistory buildings?

2. What are the major considerations when designing a cladding system?

3. What problems do wind forces cause for cladding systems?

4. What forces can cause movement within a cladding system?

5. What is the rainscreen principle?

6. What forces may cause water to penetrate the joints in cladding?

7. What materials are used as firestopping?

8. Define a curtain wall.

9. Describe each of the systems used for metal and glass curtain wall design.

10. Define the different ways that glass can be affixed to framing systems.

11. What is the difference between working joints and non-working joints?

12. Describe the classes of flowable sealants.

13. Describe the types of solid sealant materials used in curtain wall construction.

Key Terms

Boundary Layer Wind Tunnel (BLWT)

Capillary Action

Cladding

Column-Cover-and-Spandrel System

Dry Glazing

Exterior Insulation and Finish System (EIFS)

Glazing

Kinetic Energy

Point-Supported Glazing

Rainscreen Principle

Safing

Stick System

Swing Stage

Unit-and-Mullion System

Unitized System

Wet Glazing

Activities

1. Try to find buildings with various cladding materials and examine the finished installation. Prepare a report detailing how the architect handled the various factors, such as meeting fire resistance requirements, controlling condensation, and resisting water penetration.

2. Examine manufacturers' brochures (as in Sweet's Catalogs). Sketch the designs proposed and indicate the materials to be used, such as glazing materials, sealants, and thermal breaks. Show and label construction details, such as how units are installed, insulated, and vented, and how condensation drainage is provided.

3. Invite a local architect to address your class on how decisions are made regarding various cladding systems and relate personal experiences with the actual performance of systems.

Additional Resources

Aluminum Curtain Wall Design Guide Manual, Architectural Manufacturers Association, Palatine, IL.

Exterior Wall Construction in High-Rise Buildings, Canada Mortgage and Housing Corp., Ottawa, Ontario.

Howard, P. J., Editor, *Precast Concrete Cladding*, Halsted Press, New York.

The Metal Curtain Wall Manual, American Architectural Manufacturers Association, Palatine, IL.

Reed, R. D., Editor, *The Building Systems Integration Handbook*, the American Institute of Architects, published by Butterworth Heinman, 80 Montvale Avenue, Stoneham, MA 02180.

See Appendix C for addresses of professional and trade organizations and other sources of technical information.

Finishes, Paints, and Coatings
CSI MasterFormat™

Interior Finishes, Paints, and Coatings

LEARNING OBJECTIVES

Upon completion of this chapter, the student should be able to:

- Understand the integration of various interior finishing processes and other aspects of a building's construction.
- Select interior finish materials based on knowledge of interior finish processes.

- Recognize the range of protective and decorative coatings available and the properties of each.
- Select appropriate protective and decorative coatings for interior and exterior work.

Build Your Knowledge

For further study on these materials and methods, please refer to:

Chapter 22 Finishing the Exterior and Interior of Light Frame Buildings

Topic: Interior Finish carpentry

Chapter 32 Acoustical Materials

Topic: Sound Control

We spend the majority of our lives in interior environments. The quality and character of interior spaces and finishes provide the backdrop that will fulfill functional requirements. Interior finishes refers to wall, ceiling, and floor finishes; materials applied to them, such as paneling and wainscoting; and materials used for acoustical treatment, decoration, and other features (Figure 31.1). Interior finish installation is carefully regulated by building codes, which vary depending on the proposed occupancy of a building.

Interior finishes include a wide range of materials, as shown by the MasterFormat outline for Division 9. Details on these systems and materials are described in other chapters.

In addition to meeting code requirements, the selection of interior finish materials also depends on factors such as cost, appearance, durability, health issues, acoustical

Figure 31.1 Interior finishes refer to a wide range of surface materials that are applied to walls, ceilings, and floors.

© romakoma/Shutterstock.com

considerations, and fire requirements. An architect and owner work to select materials that meet the requirements of both design intent and building codes.

Interior finishing begins after a building has been enclosed, protecting the interior from the weather. The roof must be finished, cladding installed, and doors and windows set. The electrical and mechanical trades can then begin installing electrical, communications, telephone, and computer systems, as well as waste and potable water lines; automatic sprinkler systems; and heating, cooling, and ventilation systems. Installation of transportation systems, such as elevators, escalators, and moving walks and ramps, can also start.

A registered interior designer is often brought on to the project team to lead interior space planning and the selection of materials and furnishings. Renderings and material samples are used to evaluate finish options based on aesthetics and functional requirements. Once chosen, finish material selections are outlined in finish schedules. The schedule consists of a matrix indicating which finishes are used on specific interior surfaces, with a list identifying each finish by pattern, color, number, brand, and manufacturer (Figure 31.2).

Figure 31.2 Finish schedules list the interior surface materials selected by room and component.

FINISH SCHEDULE

ROOM NAME	FLOOR			BASE		CEILING		WALLS			REMARKS
	CER. TILE	WOOD	EXIST.	WOOD	CER. TILE	GYP. BD.	EXPO. STR.	GYP. BD.	EXPO. BRK	CER. TILE	
101 BASEMENT			●				●		●		
201 ENTRY	●			◗		●		◗	◗		
202 F.F. RETAIL		●		◗		◗	◗	◗	◗		
203 F.F. BATH	●				●	●		◗		◗	
204 F.F. KIT.		●				●		◗	◗		
301 S.F. RETAIL		●		◗		◗	◗	◗	◗		
202 S.F. BATH	●				●	●		◗		◗	
303 S.F. KIT.		●				●		◗	◗		
STAIRWELL		●				●		◗	◗		

PERFORMACE CRITERIA FOR INTERIOR FINISHES

Finish materials are selected based on the expected occupancy and types of operations in a building, the interrelationship of people and functions, and any special equipment or functional needs. Their performance criteria range from appearance, cost, durability, and maintenance

characteristics, to code requirements and environmental concerns. Because of their large influence on interior noise levels, the acoustical properties of interior finishes are also considered.

Environmental Concerns

The environmental characteristics of products must be considered when selecting interior finish materials. Studies show that we spend 80 percent of our time indoors, in an environment where hundreds of pollutants have been identified in building finish materials. Environmentally preferred products are those that display one or more environmental properties when compared to similar products used for the same purpose. Examples include materials made from rapidly renewable natural resources, those containing post-consumer or post-industrial recycled content, and materials that are themselves readily recycled (Figure 31.3).

Figure 31.3 Examples of environmentally friendly interior finishes: composite countertop, recycled-glass tile, and bamboo flooring.

Products that off-gas volatile organic compounds (VOCs) and other toxins that reduce indoor air quality should be avoided. More than 500 pollutants have been identified in building materials, and the focus on tighter envelope construction for energy efficiency results in less ventilation of harmful vapors. Sources of VOCs include oil-based paints, plastics, adhesives and glues, carpet backing, and some cleaning agents. Designers and contractors must observe that installed VOC ratings of materials and coatings meet with recommended minimums. Manufacturers list VOC ratings of products in specifications and sales literature. Generally speaking, natural products not made from synthetic or petroleum-based resources create the healthiest interior environment.

With the increased interest in green building rating systems, finish materials are now available that are certified by independent organizations to meet stringent indoor air-quality standards. The GREENGUARD Environmental Institute (GEI) is an example of a not-for-profit organization that oversees the GREENGUARD Certification Program. GEI's mission is to improve public health and quality of life through programs that improve indoor air. As an ANSI-Accredited Standards Developer, GEI establishes acceptable indoor air standards for indoor products, environments, and buildings.

Appearance

Interior finishes are chosen according to their appearance to set the mood and atmosphere of a space. Finish systems generally hide the structural, mechanical, and electrical systems and provide quality interior surfaces. In certain commercial applications, portions of the structural system are left exposed intentionally to become part of the interior design. Materials used are directly influenced by the desired appearances of interior spaces and their proposed use. A hotel lobby, for instance, may have darker textured walls, carpeted floors and stairs, and elaborate wall hangings. The lighting system can provide soft general illumination and spotlight features such as paneled walls and paintings. The interior of a health and fitness center will be in sharp contrast to this by utilizing light, durable wall surfaces, acoustical ceilings, considerable natural light, and floor coverings to suit the activity in a given area. The halls and lobby may have durable tile wainscoting; the aerobics room a soft, resilient wall covering; and the basketball area a composition floor.

The color of an interior finish establishes the mood and occupant reactions to an area. Warm colors, such as those ranging from reds through orange and yellow, create a feeling of friendliness and warmth. Cool colors, ranging from greens through blues and purples, create a more distant feeling. Lighter colors increase illumination levels, while darker colors absorb light, reducing the amount of light reflected within a space. Generally, interior finishes use a range of both warm and soft colors (Figure 31.4). Textures of finished walls and ceiling surfaces influence not only appearance, but also other qualities, such as acoustical properties and light reflection.

Figure 31.4 Interior finishes and colors are chosen according to their appearance to set the mood and atmosphere of a space.

© Jerry Portelli/Shutterstock.com

Cost

Like all other construction materials, selection of interior finishes involves consideration of material costs and the labor to install them, including a contractor's overhead and profit. Other cost factors include the expected life-cycle of materials before replacement is required, regular maintenance, minor repairs, and the possibility of increased replacement costs years later because of inflation and increased labor costs. Using life-cycle costing, calculations can reveal if it would be less costly over the long run to use higher-quality, more-expensive materials. This is frequently the case if an owner plans to retain title to a building over a long time.

Code Requirements

Life safety, fire codes, and local regulations must be consulted when selecting interior finish materials. The design team and material specifiers should identify all applicable codes early in the project development to assure code compliance and minimize the need for reselections. All materials must also comply with the Americans with Disabilities Act.

Interior finish and trim materials are subject to a wide range of fire code requirements. Interior finishes must be protected from damage by fire in order to prevent a structural failure. The combustibility of an interior finish material is rated by testing the flame spread of the surface of the material. "Flame spread" is the term used to describe the rate at which combustion will move across a material. The *flame-spread rating* is a single number that designates the ability of a material to resist flaming combustion over its surface. The rate of flame travel across various materials is measured under ASTM testing procedures. Noncombustible cement-asbestos board has a rating of 0. Untreated specified species of wood have a designated rating of 100. The lower the rating, the slower flame will spread across the surface of

a material. Acceptable flame-spread ratings for different types of occupancies and building areas are specified by building codes (Table 31.1).

Flame-spread ratings are classified as:

Class A flame spread, 0–25

Class B flame spread, 26–75

Class C flame spread, 76–200

Smoke-developed ratings classify materials by the amount of smoke they will give off as they burn. Most codes prohibit materials with a rating of 450 or more for use inside a building. Materials are tested using ASTM test procedures. Class A, Class B, and Class C ratings permit smoke amounts to range from 0 to 450.

Table 31.1 Maximum Flame-Spread Class *(Courtesy International Code Council)*

INTERIOR WALL AND CEILING FINISH REQUIREMENTS BY OCCUPANCY[k]

GROUP	SPRINKLERED[l]			NONSPRINKLERED		
	Vertical exits and exit passageways[a, b]	Exit access corridors and other exitways	Rooms and enclosed spaces[c]	Vertical exits and exit passageways[a, b]	Exit access corridors and other exitways	Rooms and enclosed spaces[c]
A-1 & A-2	B	B	C	A	A[d]	B[e]
A-3[f], A-4, A-5	B	B	C	A	A[d]	C
B, E, M, R-1, R-4	B	C	C	A	B	C
F	C	C	C	B	C	C
H	B	B	C[g]	A	A	B
I-1	B	C	C	A	B	B
I-2	B	B	B[h, i]	A	A	B
I-3	A	A[j]	C	A	A	B
I-4	B	B	B[h, i]	A	A	B
R-2	C	C	C	B	B	C
R-3	C	C	C	C	C	C
S	C	C	C	B	B	C
U	No restrictions			No restrictions		

For SI: 1 inch = 25.4 mm, 1 square foot = 0.0929 m².

a. Class C interior finish materials shall be permitted for wainscotting or paneling of not more than 1,000 square feet of applied surface area in the grade lobby where applied directly to a noncombustible base or over furring strips applied to a noncombustible base and fireblocked as required by Section 803.4.1.

b. In vertical exits of buildings less than three stories in height of other than Group I-3, Class B interior finish for nonsprinklered buildings and Class C interior finish for sprinklered buildings shall be permitted.

c. Requirements for rooms and enclosed spaces shall be based upon spaces enclosed by partitions. Where a fire-resistance rating is required for structural elements, the enclosing partitions shall extend from the floor to the ceiling. Partitions that do not comply with this shall be considered enclosing spaces and the rooms or spaces on both sides shall be considered one. In determining the applicable requirements for rooms and enclosed spaces, the specific occupancy there of shall be the governing factor regardless of the group classification of the building or structure.

d. Lobby areas in Group A-1, A-2 and A-3 occupancies shall not be less than Class B materials.

e. Class C interior finish materials shall be permitted in places of assembly with an occupant load of 300 persons or less.

f. For churches and places of worship, wood used for ornamental purposes, trusses, paneling or chancel furnishing shall be permitted.

g. Class B material is required where the building exceeds two stories.

h. Class C interior finish materials shall be permitted in administrative spaces.

i. Class C interior finish materials shall be permitted in rooms with a capacity of four persons or less.

j. Class B materials shall be permitted as wainscotting extending not more than 48 inches above the finished floor in exit access corridors.

k. Finish materials as provided for in other sections of this code.

l. Applies when the vertical exits, exit passageways, exit access corridors or exitways, or rooms and spaces are protected by a sprinkler system installed in accordance with Section 903.3.1.1z or 903.3.1.2.

The *fuel-contribution rating* is a measure of the amount of combustible substances in a given material. The *fire-resistance rating* indicates a material's capacity to withstand fire for a specified time and under conditions of standard intensity such that it will not fail structurally or permit the non-fire side to become hotter than a stipulated temperature in a specified number of hours.

Additional code requirements are set by the International Building Code (IBC) to address maintenance, cleanliness, and health issues. Toilet rooms in all occupancies other than dwelling units are required to utilize materials that are not adversely affected by moisture. Walls directly adjacent toilets and sinks must have a smooth, hard, and non-absorbent surface to at least 4 ft. above the floor.

Durability and Maintenance

Durability of interior finishes includes the ability of exposed material (and any protective or decorative coating) to resist damage from abrasion or dirt. In areas where wear and tear is expected, hard-surfaced materials provide considerable protection. In some spaces, such as commercial kitchens and restrooms, the ability of a material to withstand water and high humidity is important. Floor coverings are especially vulnerable to wear and damage. A lobby of a busy building will require a very durable floor covering, while other areas may be able to use something less robust. Materials and products appropriate to the function and level of use should be selected. Extra consideration must be given to finishes specified in heavy-use areas and specific functional areas. Finally, the expected life of a product and replacement costs must factor in when determining degrees of durability.

The ability to easily clean materials with minimum damage to them is another factor (Figure 31.5). The use of easily maintained finishes is critical. While certain finishes may provide excellent durability, the designer must consider maintenance and the effort required to maintain the appeal of certain products.

Acoustical Considerations

The acoustical properties of finish materials on walls, ceilings, and floors are another major consideration. The type of material, texture of a surface, and any coatings influence the acoustical properties. Under consideration is the ability of a finish to control sound created within an area, as well as restrict its flow through walls, ceiling, and floor to adjoining areas. Sound transmitted through

Figure 31.5 Maintenance requirements and the effort required to preserve the appeal of finishes is considered during materials selection.

© hxdbzxy/Shutterstock.com

materials may occur because of vibrations caused by the actions of occupants or by machinery operating in a building. Assemblies can be tested to ascertain their impact-noise rating. Impact noise can be reduced by insulating floors and covering them with soft materials, such as underlayment, carpet padding, and carpet. This must be considered as a finish is planned. Acoustical design and materials are discussed in Chapter 32.

ELECTRICAL AND MECHANICAL SYSTEMS INTEGRATION

The installation of electrical and mechanical systems is discussed in Chapters 39 through 41. Building must be designed to facilitate the installation of these systems. Horizontal runs of electrical, plumbing, and ventilation ducts are often placed below the floor and concealed with a suspended ceiling. Some buildings feature cellular floor construction using metal or pre-cast hollow-core concrete decking through which wires and plumbing can be run (Figure 31.6). A design developed for office occupancies use a raised floor system that provides access below the finished floor for flexibility. Restrooms usually have a plumbing wall installed so the plumbing can be placed out of sight yet be accessible for servicing

Figure 31.6 Electrical and mechanical systems may be run horizontally below structural decks or through cellular metal decking or hollow-core concrete slabs.

Courtesy of constructionphotographs.com

Figure 31.7 The services required for multistory buildings are usually distributed vertically using a central service core.

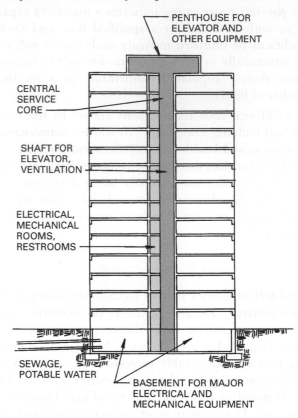

when needed. The restrooms may be part of a central *service core*, which stacks plumbing walls on each floor directly above the one below. This facilitates the vertical run of plumbing lines.

In multistory buildings, areas to house the major electrical service entrance; sewer connections to the central sewage system; potable water; elevator operating equipment; and heating, cooling, and ventilation units are generally placed in a basement or subbasement. These systems are run up through the building to each floor through a vertical central service core (Figure 31.7) from which they are distributed as required to each floor. The service core may also house stairs, elevators, restrooms, electrical and mechanical chases, fire-protection equipment, and ventilation airshafts. Each floor will have an area allotted to the services provided by the central service core. A typical example is shown in Figure 31.8. Areas such as electrical and mechanical rooms are

enclosed on each floor with a floor and ceiling. Areas such as air and elevator shafts are often open for the entire height of the building. The enclosed rooms must meet fire codes and be designed to contain noise generated within them. Mechanical rooms may contain air-handling fans and pumps capable of transmitting sound vibrations through the floor to occupied spaces outside the core. Adjacent interior finishes must control these and other disturbances.

Interior finishes on walls of air and elevator shafts are carefully controlled by codes. These shafts can serve as a passage for flames, smoke, and gases from a fire on one floor to all floors above. An interior finish is critical to the design of the service core.

It should be noted that in high multistory buildings, major electrical and mechanical equipment is usually located on a number of intermediate floors. For example, major equipment on the tenth floor could service the core areas on floors ten through nineteen, and another major equipment area on floor twenty could service core areas on floors twenty through twenty-nine. Similarly, banks of elevators typically serve only the first ten floors, while a second bank might serve floors eleven through twenty.

Figure 31.8 The central service core provides space for services, utilities, and vertical transportation on each floor.

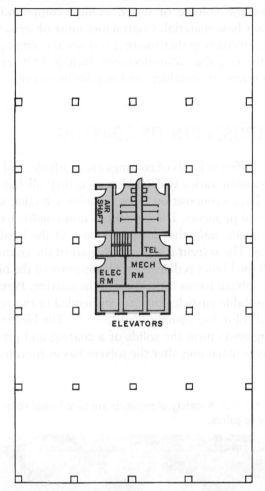

PAINTS AND COATINGS

Paints and coatings are layers of materials in liquid form applied to surfaces to decorate, preserve, protect, and seal them. When a liquid coat solidifies, it leaves a thin layer over the *substrate* to which it is applied. *Coatings* play a crucial role in protecting materials from dirt, solar radiation, moisture, chemicals, and corrosion, as well as providing *abrasion resistance*. They also provide decorative surfaces. Many coatings are applied in three layers: a primer coat, an intermediate coat, and a finish coat, also referred to as a topcoat.

The *primer coat* (sometimes called an undercoat) is applied directly to the substrate material. It must have good *adhesion* to the surface, appropriate flexibility, permit the intermediate coat to bond to it, retard corrosion of the substrate, and resist weathering long enough to permit the application of the intermediate and finish coats.

The *intermediate coat* is applied over the primer coat. It must provide an adequate film thickness and structural strength, bond to the primer coat, permit the finish coat to bond to it, and create a barrier to chemicals and other environmental contaminates.

The *topcoat* (finish coat) is applied over the intermediate coat. It provides the final finish surface, the desired aesthetic, and may serve other functions, such as providing a nonskid surface or resisting mildew.

The total coating system must be one that is compatible with the substrate, and each coating must be compatible with every other. If this does not occur, the coating will deteriorate and have to be completely removed before the substrate can be recoated.

Coatings may be clear, semitransparent, or opaque. *Clear coatings* let most of the natural color and texture of the substrate show through. They are used when the natural qualities of materials are desired, such as the color and grain of wood or the color and texture of colored concrete.

Semitransparent coatings allow some color and texture of an original surface to show but obscure much of it. They are used when the appearance of a substrate needs augmenting, for example by adding some color yet permitting the existing material to show.

Opaque coatings completely obscure the color and much of the texture of a substrate. They are used when a solid, uniformly colored surface is desired. The material composing the substrate is often not identifiable.

FEDERAL REGULATIONS

Federal regulations that affect the use of paint and other coatings include the 1992 Residential Lead-Based Paint Hazard Reduction Act, which provides guidelines for the evaluation, reduction, and notification of any known lead-based paint hazards in residential occupancies. The 1970 Clean Air Act is legislation that enforces regulations to protect the public from exposure to airborne contaminants that are known to be hazardous to human health, including paints and coatings.

Lead in Coatings

Lead is now prohibited from use in paints. An ingestion of loose paint dust or chips can cause lead poisoning. Typical hazards include harm to workers on remodeling projects where paint containing lead was used in previous years and harm to children living in older houses and apartments who may eat loose paint chips pulled from the trim and cabinets. See the "Construction Techniques" section of this chapter for additional details.

Air-Quality Regulations

Information related to air-quality regulations can be found in the *Code of Federal Regulations*. The regulation for finishing materials is called the *National Volatile Organic Compound (VOC) Emissions Standards for Architectural Coatings*, citation i40CFE 59.400. The standard regulates emissions from many sources, including products used as finishing materials in construction that emit *volatile organic compounds (VOCs)*.

Emissions from solvents in coatings are one of the largest sources of air pollution. We are surrounded by paints, stains, and varnish, both at home and in other buildings. Although paints are becoming less toxic, many are made from petrochemicals, particularly those for exterior use.

VOCs include any solvent, propellant, or other substance, except water, that evaporates from a coating as it dries. VOC is calculated as the combined weight of all solvents in a given volume of coating, excluding certain exempted materials. VOC is specified in grams per liter (g/l), which can be converted to pounds per gallon (lb./gal.) by dividing g/l by 119.8.

Architectural coatings are limited to about 250 to 700 grams per liter (2.1 to 5.8 pounds per gallon). A few examples are given in Table 31.2. These limits are the

amount of VOC per liter or gallon of coating thinned to the manufacturer's maximum recommendations, excluding the volume of water, exempt compounds, or colorant base material. Contractors must observe VOC ratings of coatings they use or determine if exempt products are available. Manufacturers include VOC ratings of products on containers and in sales literature.

COMPOSITION OF COATINGS

Many different kinds of coatings are available, and their composition varies widely. However, they all have the same basic composition. This includes a binder, a solvent, and pigments. The *binder* is a non-volatile natural or synthetic *resin* that forms the base of the hardened coating. The *solvent* is the volatile part of the coating in which the binder is dispersed. The mixture of the binder in the solvent forms the *vehicle* of the coating. *Pigments* are insoluble particles that are suspended in the vehicle to add color and opacity (Figure 31.9). The binder plus the pigments form the solids of a coating and produce the layer remaining after the solvent has evaporated.

Figure 31.9 A variety of pigments are used to add color and opacity to paints.

Table 31.2 Limits on VOC Emissions for Selected Finishing Materials

Coating Category	Grams VOC per liter	Pounds VOC per gallon
Concrete protective coating	400	3.3
Clear fire-retardant coatings	850	7.1
Flat exterior coatings	250	2.1
Flat interior coatings	250	2.1
Lacquers	680	5.7
Nonflat exterior coatings	380	3.2
Nonflat interior coatings	380	3.2
Enamels	450	3.8
Rust-preventative coatings	400	3.3
Interior sealers	400	3.3
Clear shellac	730	6.1
Opaque stains	350	2.9
Varnishes	450	3.8
Clear wood sealer	550	4.6

For more information, see the National Volatile Organic Compound Emission -Standards for Architectural Coatings, Code of Federal Regulations, Environmental Potection Agency.

Clear coatings are made of binder and solvent but contain no pigments. Semitransparent coatings have a small amount of pigment to provide some opacity. Opaque coatings contain considerable pigment and totally obscure the face of a substrate. The binder bonds itself to the substrate. It also must have the properties needed to meet the requirements of the coating. For example, it must have plasticizers to make it flexible, stabilizers to enable it to resist solar radiation or sources of heat, and driers to control the rate of curing.

Coatings are either solvent based or water based. Solvent-based coatings are volatile, while **water-based coatings** have the binders and pigments dissolved in water. Solvent-based coatings require the use of a solvent, such as paint thinner, to thin the mix and clean brushes and spray guns. The solvent used depends on the composition of a coating. Water-based coatings are thinned, and the tools used to apply them are cleaned with water.

Coatings are specified for specific uses. Manufacturer's recommendations clearly indicate whether a coating can be used for exterior or interior purposes, or both. They specify the type of substrate to which a coating can be applied and whether it resists attack by chemicals, heat, and ultraviolet rays. In addition, the final appearance is indicated as clear, semitransparent, or opaque and high gloss, satin gloss, or flat (no gloss). Coating specifications may include wet film thickness and/or dry film thickness.

Wet Film Thickness

To ensure optimum performance, the proper thickness of paint must be applied. This is specified by the minimum wet film (MWF) thickness and stated as "applied at the rate of 33 mils per coat minimum wet film thickness (MWF)." A mil is a unit of measure equal to 0.001 in. (0.025 mm).

Wet film thicknesses vary depending on **spreading rate** per gallon. The wet film thickness of a coating can be determined by measuring the wet film thickness with special gauges designed for that purpose or by using an industry-developed formula that relies on the spreading rate. Table 31.3 shows some typical wet film thicknesses

Table 31.3 Wet Film Thickness Based on the Spreading Rate

Spreading Rate (sq. ft./gal.)	Wet Film Thickness (mils)
1,600	1.0
1,000	1.6
700	2.3
400	4.0
200	8.0
160	10.0

found using the formula. The MWF thickness for various coatings can also be found by noting the recommended spreading rates indicated by the manufacturer of a coating material.

Dry Film Thickness

Specifications state the minimum dry film (MDF) thickness per coat for a coating material. Dry thickness determines the amount of protection a coating will provide. Sufficient wet mils must be applied to get a specified dry film thickness.

The dry film thickness is found by multiplying the wet film thickness in mils by the percent of solids in the paint. This gives a theoretical dry film thickness. The percent of solids in a coating is determined by the manufacturer of a material.

FACTORS AFFECTING COATINGS

After a coating has been applied to a substrate and has hardened, a number of external factors must be considered in selecting a coating. Water resistance is one of the most common factors affecting coatings. Water vapor may penetrate a coating from the outside, resulting in damage to the substrate that causes the coating to peel and crack. Water vapor can also penetrate the substrate from behind the coating, again causing blisters and peeling.

Solar radiation affects coatings with ultraviolet radiation, which can cause pigment fading and other chemical reactions in the solvents and binders, leading to coating deterioration. This permits ultraviolet radiation to reach the substrate, possibly causing further damage. Solar radiation can increase the temperature of both the coating and substrate, causing both to expand. A coating must be flexible enough to expand and contract without cracking. This includes resisting expansion of a substrate that may exceed the normal expansion of a coating at the same temperature. Likewise, freezing temperatures cause damage, especially if moisture has penetrated a coating.

Coatings are subject to the effects of dust and dirt. The extent of possible damage depends on their location. A ceiling, for example, usually has minimum exposure, while a baseboard or door has greater exposure. Location also dictates exposure to abrasion. Impact due to natural causes (such as hail), normal wear (as on doors), and vandalism must be considered.

In some situations, coatings are affected by chemical fumes, solutions, and reactions. Chemical fumes are generated by many sources, including power plants and automobile emissions. Seawater, oils, solvents, and other chemical solutions heavily impact coatings. *Alkalies* are substances like lye, soda, or lime that can be destructive to coatings. Soluble alkaline salts in mortar and concrete can dissolve and crystallize on the surface, damaging the coating. This is called *efflorescence*. Sealants may react unfavorably, and rust may streak a coating surface. Wood knots heavy with resin may bleed through a coating. These and other possible reactions must be carefully considered before choosing a coating material.

The absorption characteristics of the surface of a substrate will affect a coating. Some surfaces are hard and smooth, and may absorb little of an applied coating. Such surfaces require surface treatment by sanding, sandblasting, or etching for better adhesion. Various porous surfaces have different rates of absorption, providing different levels of adhesion, which may cause some cracking of a coating.

New coatings applied over existing ones must be compatible. The surface of the old coating must be stable and have good adhesion to the substrate or the new coating will fail. Old coatings chalk as they age. If this is the only deterioration, a new coating can be applied over it. If there are cracks and loose peeling sections, the old coating must be removed before recoating.

Following is a discussion of commonly used field-applied coatings. There are many other products finished in the factory that use automated application and drying techniques.

CLEAR COATINGS

In addition to traditional clear coatings, such as shellac, varnishes from natural resins, and lacquer, a number of products using synthetic resins are available. Clear coatings used in exterior locations in general do not have the durability of opaque coatings, because they lack the pigments that protect against ultraviolet damage from the sun. They let the natural color and texture of a substrate show as well as protect it from moisture, abrasion, and other forms of damage.

Natural-Resin Varnishes

Varnishes are one of the oldest finishes used to coat wood surfaces. Several types are available, and their properties and composition vary considerably. Varnishes fall into two broad classifications: natural-resin varnishes and synthetic varnishes.

Natural varnishes are made from either fossil resins or resins obtained from a variety of trees in tropical countries. There are three basic types of natural-resin varnish: linseed oil, tung oil, and spirit. The vehicle is some form of drying oil into which the resin is dissolved. The oil-to-resin ratio determines the classification of a varnish, expressed as the number of gallons of oil that are mixed per 100 lb. (45 kg) of resin. Varnishes made with natural resins and oils are called oleoresinous varnishes.

Turpentine is a common solvent for varnish, although mineral spirits, naphtha, and paint thinner are also used. The solvent evaporates, causing the varnish to harden. The drying oils cure by oxidation and polymerization following the loss of the solvent. This is why varnish is a slow-drying coating.

Linseed oil varnishes are available in three types: long-oil, medium-oil, and short-oil. Long-oil varnish is sold under the name "spar varnish." It is used on exterior surfaces exposed to intermittent wetting. Synthetic resin varnishes are better suited for very moist conditions. Medium-oil varnishes dry faster than long-oil types and have a harder film but are not as water resistant. Synthetic binders, alkyd and phenolic, are used to produce a modified spar varnish. Short-oil varnishes contain the least oil, dry rapidly, are brittle, and do not resist abrasion very well.

Tung oil varnishes are used in areas where heavy use is expected, such as on school furniture. They are usually factory applied.

Spirit varnish, also known as shellac, uses a resin obtained from the exudation of the lac insect, which is found in Southeast Asia and India. The resin is dissolved in denatured alcohol and is orange in color (referred to as orange shellac). It can be bleached, producing white shellac (Figure 31.10).

Shellac is available in various grades, depending on the amount of resin dissolved in a gallon of denatured alcohol. The grades are referred to as "cuts," such as a 4 lb. (1.8 kg) cut, which denotes 4 lb. of resin dissolved in one gallon of denatured alcohol. Shellac dries rapidly but does not resist moisture. It is mainly used in construction to seal knots and other resinous places in wood over which water-resistant coatings are applied. Shellac will not withstand exposure to sunlight.

Synthetic-Resin Varnishes

Synthetic resins utilize plastic materials suspended in a solvent. The most commonly used resins are alkyds,

Figure 31.10 Shellac is a natural finish material made from a sticky substance given off by the lac insect, which thrives in India and Burma.

© Thomas M Perkins/Shutterstock.com

polyurethane, silicone, epoxy, acrylics, and phenolics. The vehicle may consist of the same drying oils used for natural resins, although synthetic materials have been developed. A listing of some clear synthetic resin varnishes and their solvents are shown in Table 31.4.

Table 31.4 Clear Finishes

Binder	Base	Uses
Acrylic	Solvent or water	Waterproofing, sealing surface against dirt Used on concrete, masonry, stucco
Alkyd (spar varnish)	Solvent	Used on interior and exterior protected surfaces
Phenolic (spar varnish)	Solvent	Exterior wood exposed to moisture, marine applications
Silcone	Solvent	Waterproofing, sealing surface against dirt Used on concrete, masonry, stucco, wood
Polyurethane	Solvent	Resists chemical attack, abrasion, heavy foot traffic

Acrylic resin varnishes produce a thermoplastic film that is resistant to age-induced yellowing, ultraviolet rays, and oxidation. They have good gloss retention, are almost colorless, and are used on metal, wood, plastics, textiles, and paper. Acrylic resin coatings use either a solvent base or a water base. The solvent-based formulation has a high gloss and is impermeable to water vapor. The water-based formulation has a semigloss finish and is permeable to water.

Alkyd resin varnishes are made by a chemical reaction between an alcohol and an organic acid. They are used for waterproofing and for reducing dirt retention on concrete, masonry, and stucco walls. Some types are used on exterior metal surfaces.

Phenolic resin varnishes include phenol formaldehyde and modified phenolic types. Phenol formaldehyde varnishes are excellent for materials exposed to weather and are used on wood marine products. They are resistant to caustic substances and acids but tend to yellow with age and lose gloss with exposure to the sun. Modified phenolic varnishes do not resist weathering as well but have an abrasion-resistant quality that makes them useful for interior applications, such as furniture and floors.

Epoxy varnishes have excellent resistance to caustic materials, adhesion qualities, and resistance to abrasion. Epoxy coatings come in a variety of forms, each with special characteristics. Some are highly flexible. They can be used on a variety of materials, such as concrete, metal, and wood. Epoxies are often used as primers. Some are formulated to resist solvents, heat, chemicals, and saltwater.

Polyurethane resin varnishes protect against abrasion resistance better than natural varnishes and offer resistance to solvents, chemicals, and oxidation. Polyurethane coatings have a wide range of applications on wood products and wood floors (**Figure 31.11**) and are used as penetrating *sealers* to control dust on concrete floors. They provide a fast drying, durable, mar-resistant surface. Three types are one-component coatings, and two types have two components. Type 1 cure by exposure to oxygen in the air. Type 2 requires 30 percent humidity in the air to cure. Types 1 and 2 can be applied on site. Type 3 cures with heat. Types 3, 4, and 5 are factory applied.

Silicone varnishes have excellent heat resistance, water resistance, and resistance to corrosive atmospheric conditions. A silicone-acrylic coating has a high gloss and high resistance to blistering and crazing. Silicones are also used to modify alkyds by improving their

Figure 31.11 A fast-drying polyurethane topcoat is used on interior floor and millwork surfaces.

© David Papazian/David Papazian/Corbis

durability. Silicone-polyester coatings will resist damage by heat up to 550°F (288°C) if the mix contains 75 percent silicone.

Silicones bond well to wood, but steel must be cleaned and bonderized to achieve proper adhesion. Silicone solutions are formulated for application to concrete and masonry materials; they provide a water-repellent coating by penetrating pores, leaving no film on a surface.

Lacquer

Lacquer is made from synthetic materials and is quick drying due to the rapid evaporation of the solvent. Lacquer has a nitrocellulose base used in combination with various resins and plasticizers. Drying oils are added to improve adhesion and elasticity. A variety of natural and synthetic resins are used to improve adhesion and hardness and to give a desired gloss. Those used depend on the end-use of the lacquer. Commonly used resins are alkyds, epoxy, acrylic resins, and cellulose acetate.

Solvents commonly used in lacquer include diethylene glycol, acetone, and amyl, ethyl, butyl, and isopropyl acetate. Formulating lacquer requires several different solvents to dissolve the synthetic and natural materials used. This blending of solvents affects the gloss, ease of flow, setting time, and amount of bubbling experienced when applied.

Thinners are sometimes added to lacquer before application. This is especially important when using sprayed lacquers. Thinners adjust consistency and rate of drying. In general, lacquers dry to the touch in five to ten minutes and form a firm film in thirty minutes to three or four hours, depending on their formulation. Some sprayed lacquers can have a second coat applied just fifteen to twenty minutes after the first. Commonly used lacquer thinners are toluol, xylol, benzol, and ethyl, amyl, butyl, and isopropyl alcohol. Lacquer thinner spilled on a hardened lacquer finish will dissolve the finish.

Most lacquers are sprayed on. A brushing lacquer can be applied with a pure bristle brush but dries more slowly. Wood surfaces require a wash coat of shellac or lacquer sealer before a finish coat is applied. Several finish coats are required because lacquer film is very thin. Lacquer is used mainly on interior products, such as cabinets and furniture. Pigments are added to provide color and opacity.

OPAQUE COATINGS

Opaque coatings obscure the natural color of a surface, provide protection, and add a decorative feature through a wide range of colors. The type of coating material apllied depends on the surface to be coated and its intended use. Exposure to weather, soil, chemical fumes, saltwater, and other conditions must be considered when selecting a type of coating. Some of the frequently used coatings for various materials are listed in Table 31.5.

Before a surface is coated, it must first be cleaned and prepared to receive a primer, if required. After priming, the requisite number of finish coats are applied as needed for coverage.

Table 31.5 Commonly Used Primers and Topcoats

Material	Primer	Topcoat	Remarks
Aluminum	Vinyl red lead	Vinyl	Exposure to weather
	Zinc chromate	Chlorinated rubber	Exposure to rain, saltwater spray
	Zinc chromate	Alkyd or acrylic	Used on trim, flashing
	Self-priming	Epoxy ester	Exposure to fumes
Ferrous metal	Self-priming	Phenolic	Exposure to weather, high humidity
	Zinc silicate	Silicate, alkyd	Exposure to weather
	Zinc silicate	Silicone, aluminum pigmented	Exposure to weather
	Self-priming	Vinyl	Exposure to rain, saltwater spray
	Red lead	Vinyl	Will not resist abrasion
	Self-priming	Acrylic	Exposure to corrosion, chemicals, abrasion
	Self-priming	Urethane	Exposure to acids, alkalis, chemicals
	Self-priming	Epoxy	Apply hot to metal to be below ground
	Chlorinated rubber with red lead or zinc chromate	Coal tar	Exposure to rain, saltwater spray, chemicals
		Chlorinated rubber	Normal exterior conditions
	Red lead	Oil-based paints	Not abrasion resistant, exposure to severe weather
	Red lead and alkyd-based red lead	Alkyd	
	Zinc-polystyrene	Polystyrene	Chemical fumes, exposure to fresh and saltwater
Ferrous metal (galvanized)	Zinc dust or zinc chromate-zinc dust	Alkyd	Does not require topcoat
	Zinc dust or zinc oxide	Chlorinated rubber	Exposure to rain, saltwater spray
Ferrous metal in ground	Self-priming	Coal-tar-epoxy	Used on pipelines, buried structural steel
Gypsum wallboard	Vinyl	Alkyd	Light duty
	Self-priming	Acrylic	Heavy duty
Gypsum plaster	Self-priming	Acrylic	Plaster must be dry
Concrete and concrete masonry (dry), brick masonry	Self-priming	Acrylic	Interior locations, scrubbable
	Self-priming	Vinyl	Dry locations
	Self-priming	Epoxy esters	Exterior use, resists fumes, scrubbing
	Self-priming	Polychloroprene	Resists water, solvents, impact, exterior uses
	Self-priming	Urethane	Washable, interior locations
Concrete floors, no moisture exposure	Self-priming	Urethane	Light-to-moderate traffic
	Self-priming	Epoxy	Moderate-to-high traffic
Concrete, heavy moisture	Self-priming	Alkali-resistant chlorinated rubber	Water reservoirs, swimming pools
Portland cement plaster	Self-priming	Vinyl	Dry locations
	Styrene-butadiene	Alkyd	Dry locations
Wood, interior	Self-priming	Vinyl	Walls and floors
	Self-priming	Alkyd	Doors, paneling, trim, light-duty floors
	Self-priming	Urethane	Surfaces subject to impact, scrubbing, heavy-duty floors
	Self-priming	Acrylic	Surfaces subject to impact, scrubbing
Wood, exterior	Self-priming	Urethane	Porch decking, exterior stairs
	Self-priming	Alkyd	Siding, plywood, cedar shakes, trim
	Self-priming	Acrylic	Siding, plywood
	Self-priming	Phenolic	Siding, plywood, trim
	Oil-based primer	Oil-based vehicle	Wood siding, exterior trim, plywood siding
	Self-priming	Epoxy	Any exterior wood

Primers for Opaque Coatings

A primer is a coating applied to a surface to seal and prepare it for a finish topcoat. Primers help hide discoloration and stains and seal areas that may bleed through a finish topcoat. Resin in wood presents a common bleeding problem. Primers are also used to smooth porous surfaces, such as those on concrete blocks. The use of the proper primer is vital to the successful application of a finish topcoat, and manufacturer's recommendations must always be followed. Table 31.5 gives examples of some commonly recommended primers and associated topcoats.

Cleaning Metal Surfaces for Priming The preparation of steel surfaces depends on which primer and finish topcoat will be used. In some cases, cleaning with a detergent or solvent is adequate. Wire brushes, sandpaper, and sandblasting are also used to clean surfaces and roughen them. Galvanized metal can be cleaned with a diluted acetic acid, steel with phosphoric acid. After a surface is washed and dried, the proper primer can be applied. Some metals are allowed to weather for six to nine months prior to the application of a primer.

Aluminum painted on site should be allowed to weather for at least a month. It can then be wiped with mineral spirits or other solutions as recommended by the coating manufacturer. Most aluminum products arrive on site with a factory-applied finish.

Copper, bronze, and their alloys can be cleaned by wiping them with mineral spirits or a dilute solution of hydrochloric or acetic acid, which must then be completely washed away. Corrosion and material stuck on a surface can be removed by brushing or light sanding.

An etching primer is available for cleaning ferrous and nonferrous metals and some alloys. It cleans and chemically etches a surface, leaving a thin protective film. Generally, a standard prime coat is applied over this material.

Preparing Wood Surfaces for Priming Exterior wood to be painted should have a moisture content of 12 to 15 percent. Interior wood should have a moisture content of about 6 percent. Wood with knots that might bleed should have a coat of shellac or other sealer applied before priming. Mold, fungus, and other stains must be removed.

Preparing Concrete Surfaces for Priming It is advisable to let concrete completely cure before priming. Concrete aged less than thirty days is generally considered not suitable for painting. Paint manufacturers' recommendations on curing time should be observed.

Concrete surfaces should be free of dirt, oil, and other substances. Loose material can be removed by brushing or sanding.

Priming Coats A primer is a first coat applied to a substrate. It seals and fills the pores of a surface, inhibits rust in ferrous metals, and improves the adhesion of subsequent coats of paint (Figure 31.12). The following discussion covers some of the frequently used primers.

Figure 31.12 A primer coat is applied directly to the substrate and must resist wear long enough to permit the application of the intermediate and finish coats.

© aurenar/Shutterstock.com

Gypsum plaster can be primed with latex (acrylic), alkyd, or oil-based primer after a thirty-day curing period. Latex flat interior finish may be used as a primer if the topcoat is a gloss or semigloss coating. Damp new plaster requires an alkali-resistant primer.

Gypsum wallboard and other paper-faced products are primed with polyvinyl acetate if an alkyd topcoating is used. Latex (acrylic) topcoatings are self-priming. Portland cement plaster often is painted with a vinyl topcoating, which requires no primer.

A wide range of primers and topcoatings are used on concrete, concrete masonry, and brick masonry. Most of these topcoats are self-priming. Surfaces of concrete and masonry are usually rough and porous. They can be coated with a latex–Portland cement grout or latex block filler to produce a smoother surface. This results in a better, watertight coverage because the topcoat will not have to bridge pores in the surface. A thinned coat of catalyzed-epoxy coating material can also be used. Concrete walls can be primed with a latex primer-sealer, after which a topcoat of latex or alkyd paint can be applied.

Aluminum is often primed with zinc chromate. Bare ferrous metals use a variety of primers, depending on which topcoat will be used. Since ferrous metals rust rapidly, they must be coated with a rust-inhibiting primer. Red lead is the oldest of the primers and is almost exclusively used as a corrosion-inhibiting metal primer, typically on bridges and structural steel. Zinc silicate and zinc chromate are used when there will be considerable exposure to the weather. Coatings designed for special purposes often have a primer developed specifically for use with that product. Etching primers that clean and seal surfaces also are effective.

Galvanized metal can be chemically etched before painting to provide needed adhesion. Etching can be omitted if a latex metal primer designed for galvanized metal is used. Another primer is a varnish-based material pigmented with zinc dust, zinc oxide, or Portland cement.

Types of Opaque Coatings

Opaque coatings are used to hide the color of previous coatings or substrate and obscure much of the grain or texture of a substrate (Figure 31.13). Commonly used opaque coatings are listed in Table 31.5.

Figure 31.13 Opaque coatings obscure the natural color of a surface, provide protection, and add a decorative feature through a wide range of colors.

Alkyd Coatings *Alkyd* coatings have been a major type of organic coating. However, because of requirements relating to the amount of volatile organic compounds (VOCs) that alkyd coatings release to the environment, water-based products are largely replacing them.

Alkyd emulsions are formulated using a variety of synthetic alkyd resins in different coating formulations. Alkyd coatings have only mild alkali resistance but provide excellent water resistance and weather well. This makes them useful for exterior applications as enamels for exterior stairs and porches, for example. Alkyd coatings retain their color well and permit the formulation of a range of light colors. With some reformulation, alkyd emulsions are used in making baking enamels, like those used on kitchen appliances.

Alkyd resins are added to other coatings to improve both adhesion and durability. For example, when modified with phenolic resins, water resistance and alkali resistance are improved and a coating will penetrate rusted surfaces. When alkyd resins are formulated with rust-inhibiting pigments, such as red lead, iron oxide, or zinc chromate, they are used as rust-inhibiting primers. A gloss enamel resistant to chalking is produced by adding vinyl chloride acetate to alkyd resins. Silicone aids in color and gloss retention.

Chlorinated Rubber Chlorinated rubber-based coatings are solvent thinned and have excellent resistance to alkalis and acids and some resistance to saltwater and salt-air exposure. They exhibit superior water and water vapor resistance and are used around swimming pools and on basement walls. They have good abrasion resistance and can be used on masonry, plastic, concrete, and metal surfaces. Chlorinated rubber coatings are not recommended for use on wood because they are not permeable and blistering can occur, but they do resist attack by microorganisms. However, they do bond to metal, and, through separating dissimilar metals, they are used to stop corrosion caused by galvanic action.

Enamel Coatings *Enamels* are a form of pigmented coating that use varnish as a vehicle. They form a gloss or semigloss film that is hard and durable. Oil-based and resin-based paints form similar coatings and can be classified as enamels.

Baked enamels are formulated for application in a factory by spraying. They are generally thermosetting materials that cure to a hard coating at 200°F to 300°F (93°C to 149°C). The coating becomes insoluble in the solvent used in its formulation. Enamels are available in a wide range of colors, are hard and washable, and resist alkalis and acids.

Epoxy Epoxy coatings are available in a variety of formulations. Epoxy-ester is an epoxy resin reacted with a drying oil. It has properties similar to phenolic varnishes and alkyd resins but offers better resistance to chemical fumes and exposure to water. Epoxy-polyamide offers excellent resistance to chemical fumes, oils, atmospheric acids, and alkalies. It resists abrasion; adheres to concrete, metal, or wood; and will cure even when wet. Epoxy-bitumen may be formulated with coal tar or asphalt. They are used on items buried in soil, such as tanks and piping. Epoxy-polyester coatings are heavy-bodied two-part systems used to protect masonry and concrete. They have high-solids vinyl filler that is applied to a concrete surface followed by a high-solids epoxy-polyester pigmented topcoat.

Epoxy resins are chemically setting and solvent less. They harden with the application of heat rather than by the evaporation of solvent. When exposed to weather, they may fade and chalk, but the film remains undamaged.

Latex Coatings *Latex* is a term applied to emulsions containing synthetic resins that are thinned with water. Since they contain no flammable solvent, they present no fire hazard when stored or applied. Latex emulsion coatings dry rapidly and release only minimum amounts of volatile organic compounds into the atmosphere. The common types available are acrylic, polyvinyl acetate, and styrene-butadiene. They are used for coating interior and exterior vertical surfaces.

Acrylic Coatings Acrylic coatings are thermoplastic resins that have a range of properties varying from hard coatings to softer finishes. They are water based and available in clear and pigmented coatings. They offer excellent protection to concrete against weathering. They are also used as factory-applied coatings on aluminum and steel wall panels because they have good durability and resistance to salt spray and chemicals. Acrylic coatings remain flexible and do not lose their color as they age. Some acrylic coatings are solvent based and are, therefore, not latex emulsions, using solvents such as xylol or toluol. One-part acrylic emulsions (latex) are used on interior and exterior vertical surfaces, such as wood, masonry, gypsum wallboard, plaster, and metal. They are permeable to water vapor.

Two-part epoxy modified acrylic coatings are water based and used for interior and exterior vertical surfaces. They are tough coatings that resist stains and will withstand scrubbing.

Styrene-Butadiene Coatings These water-based coatings are formulated using styrene and butadiene, producing a rubber-like film. They resist alkali well and are permeable to water vapor. Styrene-butadiene coatings are used as exterior fillers over porous concrete surfaces. Some formulations are used on interior masonry, plaster, and gypsum wallboard, producing a washable film that resists abrasion. They generally are not used on wood.

Vinyl Coatings Vinyl and polyvinyl acetate emulsions are available as water-based coatings. One formulation is used for interior application and another for exterior use. Polyvinyl chloride copolymerized with polyvinyl acetate is an opaque solvent-based coating that has poor adhering properties and requires a special primer. It has good durability and resistance to oils, alkalies, acids, and saltwater.

Oil-Based Coatings Oil-based coatings are formulated by combining a body, pigment if needed, and a vehicle (drying oil, a thinner, and a drier). The paint body is a solid, fine material that provides superior *hiding power*. White lead, lithopone, titanium white, and zinc oxide are used for bodies of white oil-based paint. Both natural and synthetic pigments are added to give the coating color. Natural pigments are obtained from minerals, as well as animal and vegetable products. For example, red lead is a typical red pigment. Synthetic pigments are generally derived from coal tar.

Oil-based paints use a nonvolatile fluid vehicle that suspends the body particles in solution. It is composed primarily of drying oil with small amounts of thinner and dryer. Thinners are volatile and evaporate. They are used to regulate the flow of the paint. Turpentine, a product formed by distilling gum from pine trees, is a high-quality thinner. Naphtha and benzene are used in some formulations. Driers accelerate the oxidation and hardening processes. Organic salts of iron, zinc, cobalt, and manganese are commonly used driers.

Oil-based paints are permeable to water vapor and, thus, minimize blistering over porous surfaces, such as wood. They should not be used in corrosive or alkaline conditions. Since they have excellent wetting properties, they are widely used as primers.

Construction Techniques

Testing for Lead Paint

Lead paint found in older buildings can poison construction workers doing remodeling or additions. Common symptoms of lead poisoning include stomach cramps, nausea, loss of appetite, headache, and joint and muscle aches. The use of lead pigments in paint was banned in 1978, but the practice continued until 1980. Most houses built before 1980 contain some lead paint.

The Residential Lead-Based Paint Hazard Reduction Act of 1992 gives detailed instructions on what is required of a contractor before beginning work on projects where lead paint exists, and how workers and occupants in a building must be protected. For example, workers must wear respiratory filter masks and protective clothes, and work-site facilities for their cleanup and clothes changing must be provided. Those who deal regularly with lead paints are required to undergo periodic blood testing. OSHA follows meticulous regulations when inspecting work where lead pigmented paints are known to exist.

Before accepting work where lead paints may be encountered, a contractor should have tests conducted on painted surfaces. Three tests are used: x-ray fluorescence, laboratory analysis, and chemical spot tests.

A portable x-ray fluorescence analyzer measures the amount of lead on a painted surface (expressed in milligrams per square centimeter). The analyzer can read multiple layers of paint but cannot indicate which of the layers contain lead pigment.

Laboratory analysis is the most accurate test method for lead. Atomic absorption spectrophotometry gives results as a percentage of lead by weight. Paint samples are removed from a surface, making certain all layers of paint down to the substrate in each chip will be tested. No portion of the substrate should be adhered to the backs of the samples. The chips are then sent to laboratories for testing.

Spot tests for lead use a chemical kit containing a swab or dropper used to apply a chemical reagent that reacts to lead by changing color if paint contains 0.5 percent or more lead by weight. These kits are sold by many retail paint stores. Since the test kit only checks top layers of paint, it is necessary to scrape down and test each layer. Generally, this test is used for preliminary analysis because it is inexpensive and fast. If it appears to indicate the presence of lead, paint chip samples should then be sent to a laboratory for a more accurate determination.

Phenolic Coatings

Phenolic coatings are made by polymerization of phenol and a formaldehyde reactant. They are solvent based and dry by the evaporation of the solvent, leaving a strong, flexible coating. Phenolic coatings work on exterior concrete, plaster, wood, metal, and gypsum wallboard. They are used where resistance to acids, alkalies, and some solvents is required, as well as immersion in hot distilled water. Two-part phenolic coatings have a catalyst added on site and harden via a chemical reaction. This coating protects against harsh conditions, including chemical fumes.

Urethane or Polyurethane Coatings

Urethane resin coatings are available as one-part and two-part formulations. One-part coatings are moisture cured (in reaction to atmospheric moisture) and are generally clear (no pigment). They have better abrasion resistance than alkyd enamels. An oil-modified one-part urethane coating is available. Although it has better gloss retention, it has lower resistance to chemical attack.

Two-part formulations range from hard to rubber-like surface films with good resistance to abrasion, water, and solvents. Adhesion to steel and concrete is poor, and surfaces must be carefully prepared. Both one- and two-part types are used for heavy-duty wall coatings and surfaces subjected to heavy traffic, such as a gymnasium floor. They resist scrubbing, abrasion, and impact and have better abrasion resistance than regular varnishes.

SPECIAL-PURPOSE COATINGS

Special-purpose coatings are formulated to meet a specific need and are, therefore, not suitable for most coating situations. Several widely used types include bituminous, asphalt, reflective, and fire-retardant coatings. Bituminous and asphalt coatings are described in Chapter 26.

Reflective Coatings

Reflective coatings absorb the ultraviolet band of solar radiation and reflect it as visible light. The life expectancy of reflective coatings exposed to sunlight is about one year.

Pigmented Fire-Retardant Coatings

Varieties of pigmented fire-retardant coatings are available from various manufacturers. Products are rated for surface-burning characteristics on combustible and noncombustible substrates. Class 1 flame spread (0–25) is required by codes for numerous applications. Many pigmented fire-retardant paints meet this requirement. Although these coatings retard the spread of flame on a surface, they do not protect a substrate from fire or heat. If substrate protection is required, an intumescent coating must be used.

Intumescent Fire-Retardant Coatings

When exposed to flame, *intumescent coatings* develop a thick, rigid foam protective layer that insulates a substrate and prevents the spread of fire. They are applied to wood, hardboard, cellulose board, and other wood-based products. They are noncombustible and produce no toxic fumes. A common type uses a urea-formaldehyde resin with an intumescent agent. As a water-based material, it dries rapidly via water evaporation. When exposed to fire or heat of about 350°F (178°C), it expands and develops a thick, insulating mat hundreds of times thicker than the original paint film. Usually, one or two coats are sufficient. The important factor is not the number of coats but the amount of coating applied per square foot. The insulating layer delays contact between fire and the combustible material below it. This impedes flame spread on a surface, holds down smoke, and, since it retards heat transfer, delays the ignition of a substrate, giving occupants additional time to evacuate a building Table 31.6. These coatings should not be applied over other covering materials and are generally not used for exterior applications, due to their water sensitivity.

STAINS

Stains are used on exterior wood to provide color and weather protection. Those designed for interior wood provide color, and protection is accomplished with a transparent topcoat, such as varnish or lacquer.

Exterior Stains

Exterior stains are blends of oil, driers, resins, a coloring pigment, a wood preservative (such as creosote or pentachlorophenol), a *mildewcide*, and a water repellent. Low-pigmented stains are penetrating types made to soak into wood. Heavily pigmented stains have the same formulation but contain more pigment. Heavily pigmented stains are frequently used for staining cedar shakes and shingles. They usually do not have the durability of paints. Both types are used on a variety of exterior wood applications, including siding, plywood, fencing, decks, and trim.

Exterior wood stains are available in solid and semitransparent formulations. Solid stains hide wood color but allow its texture to show. Semitransparent stains retain the natural wood color (Figure 31.14).

Stains that are oil- and water-based are available for application on wood. Oil-based alkyd stains are solvent thinned and available in opaque and semitransparent types. Acrylic latex stains are water soluble.

Table 31.6 Fire-Retardant Coatings[a] Surface-Burning Characteristics (Based on 100 for Untreated Red Oak)

Coating Type[b]	DS - Clear	PR - Clear	PR - White		"DS 11" - Clear	
Surface	Douglas Fir	Douglas Fir	Douglas Fir	Cellulose Board	Douglas Fir	Cellulose Board
Flame spread	5	5	5	5	10	10
Smoke development	0	0	0	0	30	20
Number of preliminary coats	None	None	None	None	None	None
Number of fire-retardant coats	2	2	2	2	2	2
Rate per coat (sq. ft./gal.)	200	200	200	200	200	200
Number of overcoats	None	None	None	None	None	None

[a]Tests conducted in accordance with ASTM E/84 (UL 723 and ULC-S-102).
[b]Coatings tested are Exolit Fire Retardant Coatings manufactured by American Vamag Company, Inc.

Figure 31.14 Semi-transparent exterior stains and sealers provide excellent water-repellent properties.

© Jeffrey Sheldon/Shutterstock.com

Interior Stains

Interior stains are used on doors, trim, cabinets, and other wood products that require a furniture-quality finish. Many types are available (Figure 31.15). A summary of them is shown in Table 31.7. Note that a number of vehicles are used and that some are penetrating stains, whereas others are pigmented stains. Penetrating stains are made with dyes and do not obscure wood grain. Pigments in pigmented stains stay on the surface of the wood and partially obscure the grain. Some stains are penetrating compounds that contain aliphatic hydrocarbons. When applied, their color penetrates the wood and a protective coat forms over it. They are combustible and good ventilation is needed or a National Institute of Occupational Health and Safety (NIOSH)-approved respirator should be worn as it is applied.

Figure 31.15 Stains used for interior wood provide color, with protection provided by a transparent topcoat, such as varnish or lacquer.

© Fotokostic/Shutterstock.com

Table 31.7 Wood Stains Commonly Used for Interior Application

Type of Stain	Vehicle Solvent	Staining Action	Remarks
Alcohol	Alcohol	Penetrating	Dries quickly but fades easily Sold in powder form Will raise grain
Gelled wood stain	Mineral spirits or turpentine	Pigmenting	Slow drying and does not fade Sold in gelled form
Latex stain	Water	Pigmenting	Slow drying and does not fade Sold in liquid form
Non-grain-raising stain	Alcohol Glycol	Penetrating	Dries quickly and fades easily but does not raise the grain Sold in liquid form
Oil stain (penetrating)	Mineral spirits or turpentine	Penetrating	Dries quickly and fades easily but does not raise the grain Sold in liquid form
Oil stain (pigmenting)	Mineral spirits or turpentine	Pigmenting	Slow drying and does not fade Sold in liquid form
Penetrating resin stain	Mineral spirits	Penetrating	Sold in liquid form Contains stain and protective coating in one coating
Water	Water	Penetrating	Dries quickly but fades easily Sold in powder form Will raise the grain

Fillers

Woods fall into two general groups: close-grained, such as maple, and open-grained, such as oak and walnut. When finishing these woods, filler is applied to achieve a smoother final surface. Close-grained woods are smooth and use a liquid filler to seal surface pores. Some varnishes act as liquid fillers. Open-grained woods have visible open pores that must be filled with a paste wood filler to achieve a smooth surface. Paste wood fillers contain a vehicle such as linseed oil, a solvent such as mineral spirits, silex (ground quartz), and a drier. Silex mixed with the linseed oil makes a paste. It is forced into pores and the excess is wiped off before it hardens. Silex is a neutral-colored material, and color pigment can be added so the filled pores match the wood.

Review Questions

1. What materials are included in Division 9 of the MasterFormat?

2. What factors are considered as an interior finish is planned?

3. How are electrical and mechanical services distributed in multistory buildings?

4. How do flame-spread, smoke-developed, and fuel-contributed considerations influence interior finish?

5. What acoustical measurements are considered when selecting interior finish materials?

6. What purpose do coatings serve on interior finishes?

7. What is the purpose of regulations related to volatile organic compounds in coatings?

8. What are the VOC limits for architectural and industrial maintenance coatings?

9. What are the major components of a coating?

10. What are the two types of coating bases?

11. What do MWF and MDF mean when used in coating specifications?

12. What are the major elements that affect the life of coatings?

13. What are major natural-resin varnishes?

14. What are major synthetic-resin varnishes?

15. What is a primer?

16. What methods are used to clean metal surfaces for painting?

17. What is the recommended moisture content for wood to be painted?

18. What are the major types of opaque coatings?

19. What are some special-purpose coatings?

20. What is meant by a flame-spread rating?

21. What are flame-spread classifications?

22. What does a smoke-developed rating indicate?

23. What does a fuel-contributed rating indicate?

Key Terms

Abrasion Resistance

Adhesion

Alkali

Alkyd

Binder

Clear Coating

Coating

Durability

Efflorescence

Enamel

Epoxy Varnish

Fire-Resistance Rating

Flame-Spread Rating

Fuel-Contribution Rating

Hiding Power

Intermediate Coat

Intumescent Coating

Lacquer

Latex

Mildewcide

Oil-Based Paint

Opaque Coating

Pigment

Primer Coat

Resin

Sealer

Semitransparent Coating

Service Core

Smoke-Developed Rating

Solvent

Spreading Rate

Stain

Substrate

Topcoat

Varnish

Vehicle

Volatile Organic Compounds (VOCs)

Water-Based Coating

Activities

1. Visit a paint store and collect color charts and samples. Prepare a display citing the properties and some suggested uses for each type of material collected.

2. Secure several lead sampling kits and test finishes in several old local residences and campus buildings for lead. Write up your findings and submit a report to the class.

Additional Resources

The GREENGUARD Environmental Institute, www.greenguard .org.

Lead Paint Safety, U.S. Department of Housing and Urban Development National Lead Information Center, 1-800-424-5323.

Merritt, F., and Rickets, J. *Building Design and Construction Handbook (5th ed.)*, McGraw-Hill Publishing Co., New York.

Paint Tests for Chemical, Physical, and Optical Properties: Appearance, Volume 06.01, American Society for Testing and Materials, West Conshohocken, PA.

Performance Specifications and Standards for Coatings and Finishes, American Architectural Manufacturers Association, Schaumburg, IL.

Small Entity Compliance Guide: National Volatile Organic Compound Emissions Standards for Architectural Coatings, U.S. Environmental Protection Agency, Research Triangle Park, NC 27711.

Williams, R., and Knaebe, M., *Finishes for Exterior Wood*, U.S. Department of Agriculture, Forest Service, Forest Products Laboratory, Madison, WI.

Other resources include:

Many publications available from U.S. Department of Housing and Urban Development, HUD USER, Rockville, MD.

See Appendix C for addresses of professional and trade organizations and other sources of technical information.

Acoustical Materials

Upon completion of this chapter, the student should be able to:

- Understand the factors involved when considering the specifications and recommendations for the acoustical treatment of a building.

- Select appropriate materials to provide acoustical control of various interior spaces.

Build Your Knowledge

For further information on these materials and methods, please refer to:

Chapter 12   Concrete Masonry

Chapter 28   Glass
              Topic: Glass Products

Acoustics is the science of sound, including the generation, transmission, and effects of sound waves. Acoustical materials are used to reduce the levels of sound within an area by absorption and to control sound transmission between adjacent areas caused by sound vibrations in a building structure.

SOUND

Sound is the sensation produced by human organs sensing vibrations transmitted through the air. This includes vibrations traveling in the air at a speed of about 1,130 ft./sec. (345 m/sec.) at sea level, and mechanical vibrations transmitted through an elastic medium, such as steel. Sound is the movement of air molecules traveling in a wavelike motion. A sound wave produces changes in the atmospheric pressure above and below the existing static pressure. This deviation in atmospheric pressure is called sound pressure.

Sound Waves

Sound waves move in a spherical direction from a source in all three dimensions (Figure 32.1). There is a delay between the time a sound is created and the time it is heard. Since it takes audio sound about a second to travel 1,130 ft. (345 m) in the air, a person 2,260 ft. (690 m) away will hear the sound two seconds after it is generated. Thunder is not heard for several seconds when lightning strikes in the distance due to the different speeds of sound and light. Sound propagates at different speeds in various materials. In wood, it travels about 11,000 ft./sec. (3,355 m/sec.) and in steel about 16,000 ft./sec. (4,880 m/sec.).

Figure 32.1 Sound travels in all directions simultaneously.

©iStock.com/studioaraminta

Frequency of Sound Waves

The *frequency* of sound is the number of cycles of like waveforms per second. Sound travels in sine waves, as shown in Figure 32.2. Wavelength equals the velocity of a sound (1,130 ft./sec. in air), divided by the frequency (number of cycles). For example, a sound having a frequency of 15 cps would have a wavelength of 1,130/15, or 75 ft. The frequency, cycles per second, is expressed in units of hertz (Hz). A hertz is one cycle per second (cps), so, in the above example, the frequency is expressed as 15 hertz (15 cps = 15 Hz).

The frequency of sound determines the pitch. Low, deep sounds have low frequencies, and high-pitched sounds have high frequencies. The human ear can receive sounds ranging from a low of about 20 Hz to a high of about 20,000 Hz (or 20 KHz).

The frequency of sound to be considered varies, often rapidly and constantly. For example, the sound produced by music from a radio often ranges from low to high frequencies rapidly and often. Frequency from other sources, such as a dishwasher, is more limited in range.

Figure 32.2 The frequency of sound is the number of cycles of like waveforms per second.

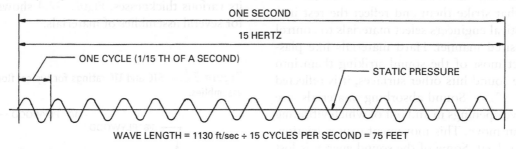

Sound Intensity

Sound intensity depends on the strength of the force that sets off the sound vibrations. Sound intensity is measured in *decibels (dB)*. The decibel scale for normal applications ranges from 0 dB, which is just below the lowest audible sound (about 20 Hz), to 120 dB, which can produce a feeling of vibration in the ear. Intensities above this can cause actual damage to human hearing and structural damage to buildings. Representative sound levels in decibels for selected situations are given in Table 32.1. The intensity of sound varies in much the same way as described for frequency. Music can produce very high decibel levels, but a dishwasher produces lower, steady decibel intensity.

Table 32.1 Sound Pressure Levels of Selected Sounds

Sound Levels (decibels)	Source of Sound	Sensation
140	Near a jet aircraft	Deafening
130	Artillery fire	Threshold of pain
120	Elevated train, rock band, siren	Threshold of feeling
110	Riveting, air-hammering	Just below threshold of feeling
80–100	Power mower, thunder close by, symphony orchestra, noisy industrial plant	Very loud
60–80	Noisy office, average radio or TV, loud conversation	Loud
40–60	Conversation, quiet radio or TV, average office	Moderately loud
20–40	Average residence, private office, quiet conversation	Quiet
0–20	Whisper, normal breathing, threshold of audibility	Very faint

SOUND CONTROL

A common acoustic issue in virtually any type of occupancy is sound transmission. Sound transmission can be both airborne and caused by structure-borne vibration. Airborne sound travels through the air and moves through materials, assemblies, and partitions. Sound can also pass under doorways, through ventilation shafts, and around obstructions. Sound transmission can cause noise, confidentiality, and privacy issues. Sound generated within a space is controlled by acoustical materials with sound-absorbing properties and by materials that reduce sound transmission through assemblies of materials, such as a wall or floor.

Sound-absorbing materials take in some of the sound waves that strike them and reflect the rest into an area. Acoustical engineers select materials to control sound in a desired manner. Hard materials like plaster walls reflect most of the sound striking them into a room. As the sound hits other surfaces, it is reflected again (Figure 32.3). Sound-absorbing materials are porous and have openings in and out of which vibrating air particles can move. This movement causes friction, which generates heat. Some of the sound energy is lost as heat, which may be reflected into a room or transmitted through the material.

Figure 32.3 Sound is reflected when it strikes hard, non-absorptive surfaces.

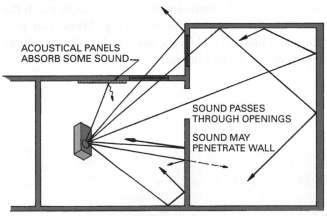

Sound Transmission Class

Sound transmission class (STC) is a single-number rating that indicates the effectiveness of a material or an assembly of materials in reducing the transmission of airborne sound. The larger the STC number, the more effective the material is as a sound transmission barrier.

Materials are tested following the specifications in ASTM E90-70 *Standard Test Method for Laboratory Measurement of Airborne Sound Transmission Loss of Building Partitions*. Typical building code STC requirements for residences, apartments, and hotels are as follows:

1. All separating walls and floor-to-ceiling assemblies must provide an STC of 50.

2. All penetrations in assemblies, such as openings for piping and electrical wiring, must be sealed to maintain the STC 50 rating.

3. Entrance doors and their seals must have an STC rating of 26 or more.

Manufacturers of materials used to reduce sound transmission indicate STC ratings for each material and its various thicknesses. Figure 32.4 shows STC ratings for several assemblies of materials.

Figure 32.4 STC and IIC ratings for typical floor-ceiling assemblies.

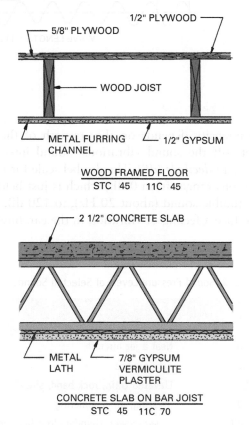

Impact Isolation Class

Impact isolation class (IIC) is a single number giving an approximate measure of the effectiveness of floor/ceiling assembly to provide isolation against sound transmission from impacts or structure-borne sound. Impacts include walking, skidding, or dropping items on the floor, all of which set up vibrations that radiate to areas below. Improved performance for floors with low IIC ratings can be realized by applying carpet or other sound-absorbing materials to them. Several design suggestions are shown in Figure 32.5. IIC ratings of 45 to 65 are common for floor-ceiling assemblies in multifamily dwellings. Building codes typically require that floor-ceiling assemblies have an IIC rating of 50.

Figure 32.5 STC and IIC ratings for concrete-floor construction.

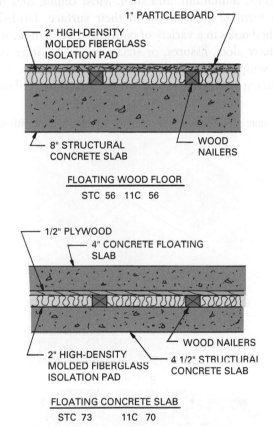

FLOATING WOOD FLOOR
STC 56   IIC 56

FLOATING CONCRETE SLAB
STC 73   IIC 70

Noise-Reduction Coefficient

The *noise-reduction coefficient (NRC)* is an indication of the amount of airborne sound energy absorbed by a material. The single-number rating is the average of the sound absorption coefficients of an acoustical material at frequencies of 250, 500, 1,000, and 2,000 Hz. The larger the NRC number, the greater the efficiency of a material to absorb sound. Some typical examples are given in Table 32.2.

Table 32.2 Noise-Reduction Coefficients for Selected Materials

Material	NRC
Unpainted brick wall	0.02–0.05
Painted brick wall	0.01–0.02
Glazed clay tile	0.01–0.02
Concrete wall	0.01–0.02
Lightweight concrete block	0.45
Heavyweight concrete block	0.27
Standard plaster wall	0.01–0.04
Gypsum wall	0.01–0.04
Acoustical plaster wall	0.21–0.75
Glass	0.02–0.03
Fiberglass	0.50–0.95
Wood panel	0.10–0.25
Mineral wool	0.45–0.85
Acoustical tile	0.55–0.85
Carpeting	0.45–0.75
Vinyl floor covering	0.01–0.05

ACOUSTICAL MATERIALS

The control of sound transmission and the absorption of sound can be accomplished using a wide variety of materials. Some standard construction materials, such as brick or concrete block, are used to control sound transmission. Other materials are especially designed for acoustical applications. Many of these materials are discussed in detail in other chapters.

Floor Coverings

The installation of carpet and vinyl composition floor covering reduce impact noise and dampen airborne noise. Carpeting is much more effective than vinyl floor covering and its effectiveness is increased when installed over a cushion. See Chapter 34 for additional information on carpets. Special sound-reduction mats use small elevated nodes, resulting in only 5 percent of the surface area making contact with the subfloor below (Figure 32.6). The low-profile mats can be used under most finish flooring materials to increase both STC and IIC ratings.

Acoustical Plaster

Acoustical plaster is composed of a plaster made with perlite or vermiculite aggregate. It may be applied with a brush but is usually sprayed on surfaces. It has the advantage of uniformly covering curved and irregular shapes. The plaster is applied in several layers, with a

Figure 32.6 Special sound-reduction mats are used below finished flooring materials to reduce impact noise.

© s74/Shutterstock.com

finished thickness of about ½ in. (12 mm). It has an NRC of about 0.21 to 0.75. See Chapter 33 for more information about plaster.

Another form (though not a plaster) of spray-applied acoustical wall and ceiling covering is composed of cellulosic fibers in a bonding agent. It will bond to almost any surface and provides acoustical control and thermal insulation.

Acoustical Ceiling Tile and Wall Panels

A variety of acoustical ceiling tiles and wall panels are available. They are made from various materials, such as wood fibers, sugarcane fibers, mineral wool, gypsum, fiberglass, aluminum, and steel. Most ceiling tiles have some form of perforation in their surface. Drilled or punched holes in a variety of patterns are common. Some tiles have slots, fissures, or striations, while other types use a sculptured, irregular molded surface (Figure 32.7). Following are brief descriptions of some tiles available.

Figure 32.7 Acoustical surface treatments used on ceiling tile: (a) fissured, (b) aluminum, (c) wood fiber, and (d) textured surface with holes.

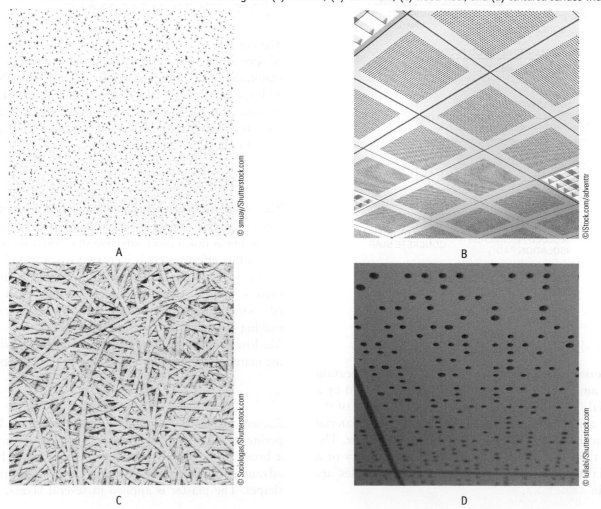

Wood fiber and sugarcane fiber ceiling tiles are available in thicknesses ranging from ⅝ in. (16 mm) to 1½ in. (38 mm). Tile sizes commonly available include 12 × 12 in. (305 × 305 mm), 12 × 24 in. (305 × 610 mm), 24 × 24 in. (610 × 610 mm), and 24 × 48 in. (610 × 1220 mm).

Ceiling tiles and wall panels are also made from molded mineral fibers and are cast with a wide choice of surface textures. Some manufacturers use mineral fiber or fiberglass sound-deadening panels covered with fabric. This provides an attractive finished ceiling or wall panel. This type has STC ratings of 35 to 40, NRC ratings of 0.60 to 0.80, and flame-spread ratings of 0 to 25.

Some ceiling tile and wall panels can resist damage from moisture and impacts. A common type is a ceramic ceiling tile made from mineral fibers in a ceramic bond. It is fire resistant and used in areas of high humidity with STC ratings of 40 to 44, NRC ratings of 0.50 to 0.60, and flame-spread rating 0. Another type uses perforated aluminum or steel tiles. Still others apply a vinyl coating over a perforated aluminum tile that is backed by a mineral fiber substrate. The sound passes through the holes and is absorbed by the mineral fiber. Metal tiles have STC ratings of 35 to 45, NRC ratings of 0.60 to 0.70, and flame-spread rating 0. Another type that resists damage is made by bonding wood fibers into a porous panel resembling particleboard that resists moisture and heavy blows.

Lightweight panels are made utilizing a vinyl covering over a laminate of high-density molded glass fiber bonded to a core of 1 in. or 2 in. (25 mm to 50 mm) sound-absorbing fiberglass. Wall panels made from these materials are applied in hallways, restaurants, gymnasiums, offices, and other places where sound control is necessary (Figure 32.8). Some use a perforated zinc-coated steel or aluminum panel bonded to a 2 in. (50 mm) thick fiberglass panel that is mounted several inches off a wall. New decorative, natural wood panels with perforations use a backing of acoustical fleece and fiberglass to achieve sound control (Figure 32.9).

Another type of acoustical treatment involves the use of ceiling baffles. Baffles are acoustical panels hung from a ceiling to reduce airborne sound in large spaces, such as factories, restaurants, or auditoriums (Figure 32.10). Vinyl-covered fiberglass baffles use a lightweight panels available 1 in. (25 mm) thick, 10 in. (254 mm) high, and from 2 to 4 ft. (610 to 1220 mm) long. Flame-resistant polyethylene is bonded over a 1 in. (38 mm) fiberglass core. It is flexible, like a blanket, and hung from a ceiling (Figure 32.11). Another type of baffle uses rigid wood-fiber sound-deadening panels. These baffles can be joined to form an attractive grid on a ceiling.

Figure 32.8 Acoustical wall panels of molded glass fiber are used to control sound within a space.

© View Pictures Ltd/SuperStock

Figure 32.9 Perforated wood acoustical panels provide both a patterned finish and sound control properties.

© ASchindl/Shutterstock.com

Figure 32.10 Acoustical ceiling baffles are used to reduce airborne sound in large spaces.

© Jesse Kunerth/Shutterstock.com

Figure 32.11 Fabric baffles at the ceiling used to control noise in a passageway.

© J. Lang/Shutterstock.com

Sculptured acoustical wall units made from a high-density molded fiberglass layer bonded to a sound-absorbing glass fiber blanket are used to attenuate sound and provide a decorative feature (Figure 32.12). They can also be covered with a wide range of fabrics, typically in round, octagonal, and triangular shapes.

Figure 32.12 Smooth-surface mineral-composite panels with a foam backing are used to attenuate sound and provide a decorative wall surface.

© Ron Dale/Shutterstock.com

Sound Barriers

A wide range of materials can be used to block the transmission of sound through an assembly, such as a wall. Many of these, such as brick, concrete, concrete block, and gypsum, are discussed in earlier chapters. These materials control sound by providing mass, which blocks its transmission. For example, a 4 in. (100 mm) brick or concrete masonry unit has an STC of about 40. A 4 in. (100 mm) concrete floor has an STC of 44.

Lead sheet material is used in commercial buildings to block sound transmission. For example, a lead sheet can be laid on top of a suspended ceiling or hung from the bottom of the floor above, blocking the transmission of sound that penetrates the ceiling. Lead is available in sheets and foils as thin as 0.0005 in. (0.013 mm). It is used in walls, doors, and other areas where a thin but effective sound barrier is needed.

Acoustical sealant is available as a pumpable material and is applied to all openings through which sound may penetrate. For example, spaces around electrical boxes and pipes that pierce a wall or floor must be sealed. Openings, even very small ones, reduce the effectiveness of an otherwise efficient sound barrier.

CONSTRUCTION TECHNIQUES

The following discussion outlines common ways to construct walls, floors, and ceilings to reduce sound transmission.

Frequently used methods for increasing the STC rating of wood frame walls are shown in Figure 32.13.

Figure 32.13 Assembly details for increasing the STC rating of wood frame walls.

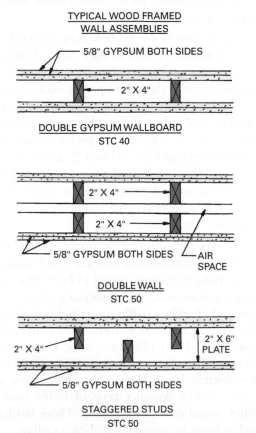

TYPICAL WOOD FRAMED
WALL ASSEMBLIES

5/8" GYPSUM BOTH SIDES

2" X 4"

DOUBLE GYPSUM WALLBOARD
STC 40

2" X 4"

2" X 4"

5/8" GYPSUM BOTH SIDES   AIR SPACE

DOUBLE WALL
STC 50

2" X 4"   2" X 6" PLATE

5/8" GYPSUM BOTH SIDES

STAGGERED STUDS
STC 50

Adding layers of gypsum board can improve the STC rating of an assembly. Staggered or double-stud framed walls achieve higher ratings than do single-stud walls. Each of these assemblies could have sound attenuation batts (SABs) installed to increase the STC value. Fiberglass insulation reduces noise transfer through building assemblies and improves room sound quality. Depending on the construction method used, SABs can improve STC ratings by 4 to 10 dBs. Examples of sound-control methods for steel framed walls incorporate layers of gypsum board and plaster (Figure 32.14). The use of resilient channels screwed to the face of the framing results in higher STC ratings (Figure 32.15).

Figure 32.15 A sound-deadening wall using metal furring and a sound-attenuation blanket.

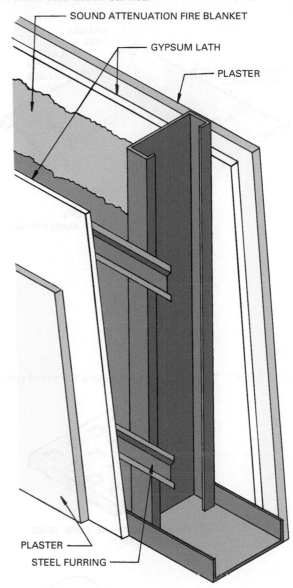

Figure 32.14 Assembly details for increasing the STC rating of steel frame walls.

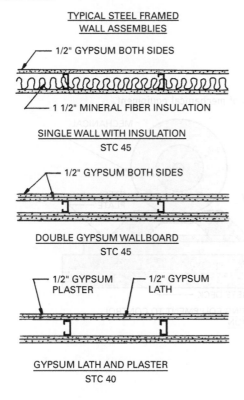

Floor-ceiling assemblies for wood and steel joist assemblies appear in Figure 32.4. The floating-floor technique shown in Figure 32.5 is used in commercial construction. The same type of construction can be used with residential wood frame floors.

Acoustical ceiling tiles can be glued to ceiling substrates, while others can be nailed or stapled to a ceiling. Tile installed with bonding agents has nut-size daubs of adhesive placed on its back. The tile is pressed against a ceiling substrate and slid into place.

If ceiling tiles are to be installed to the bottom of floor joists without a substrate, wood 1 × 3 in. (25 × 75 mm) strips are nailed perpendicular to the joists at 12 in. (305 mm) on center spacing. The ceiling tiles are then nailed or stapled through their tongues to the wood strips. Suspended-ceiling systems consist of a grid of metal runners, hung from overhead on wires, into which acoustical ceiling panels fit. Light fixtures are also designed to fit into this grid. A typical system is shown in Figure 32.16.

Figure 32.16 A suspended ceiling is hung from overhead wires that carry metal runners into which the acoustical tile is laid.

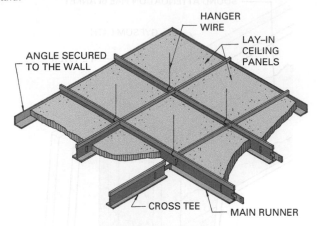

CONTROLLING SOUND FROM VIBRATIONS

Mechanical equipment mounted on the structural frames of buildings produce vibration sounds that can travel for great distances to other parts of a building. Vibrating water and sewer pipes also create sound that must be controlled. Much of the sound can be dampened by mounting the mechanical equipment on various types of isolator pads. Rubber, neoprene, cork, and fiberglass pads are used for small pieces of equipment or larger pieces located in basement areas (Figure 32.17). For heavier-duty units mounted within a building or on its roof, a steel-spring mounting is used. The base of the spring pad must itself be isolated from the structure with an isolator pad to prevent the transmission of audible high-frequency vibration through the spring to the structure.

Figure 32.17 Types of pads used to damp the sound produced by vibrations of mechanical equipment.

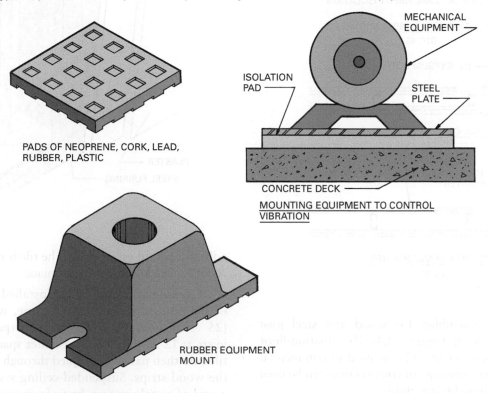

Review Questions

1. In what two ways do acoustical materials control sound?

2. What is sound pressure?

3. In what directions do sound waves move?

4. How long will it take a sound to travel one mile from its source?

5. What is the wavelength of a sound having a frequency of 25 cps?

6. What is the frequency in hertz of a sound having a frequency of 35 cps?

7. How does sound frequency influence the pitch of a sound?

8. What is the normal range of sound that can be heard by the human ear?

9. How many decibels of sound will damage a person's hearing?

10. What is the difference between sound-absorbing materials and sound-reflective materials?

11. What are typical STC and IIC requirements for apartment construction?

12. How does acoustical plaster differ from standard gypsum plaster?

13. What acoustic product is used to reduce airborne sound at the ceiling in a large area, such as an auditorium?

14. What are the three types of grids used on suspended ceilings?

15. How is sound transmission of vibrations from mechanical machinery controlled?

Key Terms

Acoustics

Decibel (dB)

Frequency

Impact Isolation Class (IIC)

Noise-Reduction Coefficient (NRC)

Sound

Sound Transmission Class (STC)

Activities

1. Visit rooms used for various types of activities, such as a classroom, music room, auditorium, and mechanical room in your school. List the materials on the ceiling, walls, and floor, plus any special acoustical treatments. Write a report explaining what might be done to improve the acoustics of each space.

2. Ask the director of the physical plant of your school to show you the devices used on mechanical equipment that reduce the vibration transmissions into the structure. Write a report and make sketches of the various devices.

Additional Resources

See Appendix C for addresses of professional and trade organizations and other sources of technical information.

CHAPTER 33

Interior Walls, Partitions, and Ceilings

LEARNING OBJECTIVES

Upon completion of this chapter, the student should be able to:

• Be aware of the code requirements for various interior walls, partitions, and ceilings.

• Select appropriate materials and assemblies for interior walls, partitions, and ceilings.

• Understand the installation of the various types of finish walls and ceilings.

Build Your Knowledge

For further study on these materials and methods, please refer to:

Chapter 19   Products Manufactured from Wood
             Topic: Plywood and Other Panel Products
Chapter 31   Interior Finishes

Interior walls and ceilings are constructed in a variety of ways. Some finishes, like an exposed masonry wall, are a direct result of the means of construction. Others use further layers of materials that are applied to the structural frame.

Interior walls and partitions may be load bearing or non-load bearing and serve to divide open space into separate and rentable areas. They also serve to enclose openings such as stairs, shafts, and exit pathways. Constructed with wood studs, steel studs or masonry, various types of finish material can be applied over a partition structure. Walls and partitions afford space for electrical and mechanical systems that travel inside wall cavities (Figure 33.1). They also provide privacy, security, acoustical control, insulation, and fire and smoke protection.

Ceilings are a major component of the architecture of a space. Ceilings make available space for electrical and mechanical systems while simultaneously providing sound attenuation and fire protection. Interior finish materials used

Figure 33.1 Interior walls and ceilings provide space for electrical and mechanical systems that travel inside construction assemblies.

© 68/Ocean/Corbis

on partitions and ceilings include gypsum board products, plaster, wood, hardboard panels, fiberboard, and a variety of metals. Detailed information of the characteristics of these materials is given in other chapters.

FIRE-RESISTANT INTERIOR ASSEMBLIES

Passive fire protection involves the application of fire-resistance-rated construction assemblies that are used to control the spread of fire and smoke, and provide sufficient time for evacuation and firefighting. The fire-resistance property of a material or assembly is based on the amount of time, either minutes or hours, that an assembly can withstand a fire before failure. An assembly that can resist a fire for an hour is designated as a one-hour assembly. The fire codes define one-, two-, three-, and four-hour assemblies. Building codes apply minimum fire-resistance ratings to building elements based on their construction type.

Materials and assemblies are tested to determine their endurance according to test methods like ASTM E 119, *Standard Test Methods for Fire Tests of Building Construction and Materials*. Similar test procedures are available from the National Fire Protection Association (NFPA 251 *Standard Methods of Tests of Fire Endurance of Building Construction and Materials*) and Underwriters Laboratories (UL 263 *Standard for Fire Tests of Building Construction and Materials*).

To protect a building's occupants and structural integrity, the International Building Code (IBC) mandates fire-rated construction assembly requirements for structural systems and interior partitions. These requirements depend on the type of construction, building height, floor area and location, occupancy type, and the type of fire protection system used. Fire-resistance ratings are given as the number of hours a material can withstand fire exposure, as specified by standard test procedures. Table 33.1 lists minimum fire-resistance requirements for various building elements, as specified in the International Building Code.

To define the quality of fire-resistive construction, the IBC has established six basic fire-resistive assemblies. Fire-separation walls are used to separate a single building into two or more separate areas. These walls must have minimum fire-resistance ratings of 2 to 4 hours, depending on the occupancy group (Table 33.2). Fire-separation

Table 33.2 Fire-Resistance-Rating Requirements for Fire-Barrier Assemblies Between Fire Areas

Occupancy Group	Fire-Resistance Rating (hours)
High Hazard Groups 1 and 2	4
Factory Group 1	3
High Hazard Group 3	3
Storage Group 1	3
Assembly	2
Business	2
Educational	2
Factory Group 2	2
High Hazard Groups 4 and 5	2
Institutional	2
Mercantile	2
Residential	2
Storage Group 2	2
Utility and Miscellaneous	1

For additional information, see Chapter 7 in the International Building Code published by the International Code Council, Falls Church, VA.

Source: International Code Council

Table 33.1 Fire-Resistance-Rating Requirements for Building Elements (Hours)

Building Element	Type I A	Type I B	Type II A	Type II B	Type III A	Type III B	Type IV HT	Type V Aᵈ	Type V B
Structural frameª Including columns, girders, trusses	3	2	1	0	1	0	HT	1	0
Bearing walls									
Exteriorᶠ	3	2	1	0	2	2	2	1	0
Interior	3	2	1	0	1	0	1/HT	1	0
Non-bearing walls and partitions Exterior					See Table 602				
Non-bearing walls and partitions Interiorᵉ	0	0	0	0	0	0	See Section 602.4.6	0	0
Floor construction Including supporting beams and joists	2	2	1	0	1	0	HT	1	0
Roof construction Including supporting beams and joists	1	1	1	0	1	0	HT	1	0

For additional information, see Chapter 7 in the International Building Code published by the International Code Council.

Source: International Code Council

walls are used to control the spread of fire between areas on a floor, such that if a fire breaks out in one area, it will not immediately spread. Fire-separation walls must extend from the top of a fire-resistant floor to the bottom of a fire-resistant ceiling (or roof slab above), and be securely fastened to them. Opening sizes in fire-separation walls are restricted by code, and openings must be protected by fire doors, windows, or shutters.

Fire barriers are used to separate shafts from other building areas, in horizontal exits, and passageways, and in the separation of occupancies. Fire barriers have fire-resistance ratings of 1 to 4 hours, depending on the application or occupancy.

Shaft enclosures involve additional requirements for fire barriers used in shaft construction. Shafts include elevator hoist-ways, mechanical chases, and laundry chutes. Shafts require a fire-resistance rating of 1 hour if the shaft connects fewer than four stories, and 2 hours if the shaft connects more than four stories.

Fire-rated partitions are used to separate dwelling units in the same building, sleeping units in hotels and care facilities, and tenant spaces in covered malls, in corridors, and in elevator–lobby separation walls. Fire partitions have a fire-resistance rating of at least 1 hour. A typical assembly for a fire-rated partition designed to offer effective sound control and a high fire rating uses steel studs, gypsum lath on steel furring, a gypsum finish coat, and a sound-attenuation blanket in the wall cavity. Fire-resistant drywall partitions are used to enclose shafts in multistory buildings, including those for elevators, mechanical equipment, stairwells, and air returns.

Smoke barriers are fire-resistant membranes used to restrict the movement of smoke. Smoke barriers divide areas of a building into separate smoke compartments. They form a continuous membrane from one outside wall to another and from floor slab to roof (or floor slab above). Doors in smoke barriers must meet special code requirements and have automatic closers that activate when detectors sense smoke. They typically have a 1-hour minimum fire-resistance requirement. A similar assembly is the smoke partition, which is not required to have any fire-resistance rating.

Horizontal assemblies apply to floor and roof construction. All floor and roof assemblies must be made of materials allowed by the type of building construction, and the fire-resistance rating of floor and roof assemblies cannot be less than that required by the building type of construction. Fire-resistance ratings for horizontal assemblies range from 1 to 2 hours. In addition to these assemblies, building codes also

establish fire-resistance requirements for opening protection (doors and windows), duct and other through-penetrations, and joint systems.

INTERIOR GYPSUM WALL PARTITIONS

Metal-Stud Non-Bearing Interior Walls

Partition walls in commercial buildings are frequently framed using wood or light-gauge metal framing. Wood framing must adhere to the fire-resistance requirements of the local code, which sometimes requires the use of fire-treated wood. Metal framing studs and tracks are formed from light-gauge steel in a range of standard sizes. Wall-framing procedures follow closely the conventions of wood framed walls. Metal studs for non-load-bearing interior walls and partitions are typically made from 18- to 20-gauge steel sheets, with either solid or perforated webs (Figure 33.2). Perforations in the web of metal studs lighten their weight and facilitate the installation of plumbing, electrical, and communications lines. For exterior and load-bearing walls, heavier-gauge steel

Figure 33.2 Metal studs for non-load-bearing interior walls and partitions are typically made with either solid or perforated webs.

is used. Metal studs using a round rod bent on a diagonal truss design between vertical cords are used primarily for walls covered with gypsum lath or metal lath. Metal studs are typically manufactured in lengths ranging from 8 ft.-0 in. to 24 ft.-0 in. and tracks in 10 ft.-0 in. lengths.

Studs are secured to structural floor and ceiling assemblies using runner tracks. When the structural system consists of concrete walls and floors, the tracks are secured with power-driven fasteners. Guns use a gunpowder cartridge to drive a steel fastener through a track and into a concrete floor. The metal studs fit snuggly in the tracks and are secured with self-drilling, self-tapping screws (Figure 33.3).

Figure 33.3 The metal bottom track is fastened to concrete floors with steel fasteners driven by a gun that uses a small charge of gunpowder.

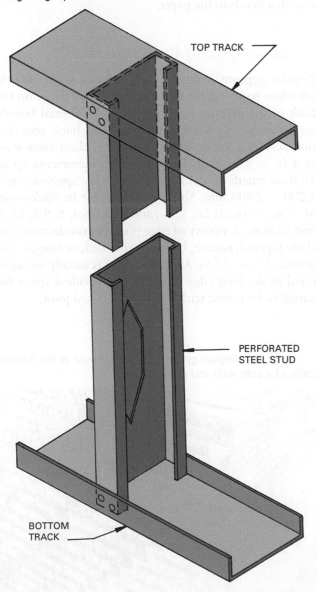

The screws are made of hardened steel and driven with an electric screwdriver. Metal-framed walls are noncombustible, and easy to insulate; and gypsum and other interior finish materials can be easily secured to them.

The actual assembly of a metal-framed wall depends on desired sound transmission levels, thermal insulation, and building code fire-resistance specifications. An example of a typical assembly for a cavity wall that can serve as a *fire separation wall* or party wall is shown in Figure 33.4. This is a continuous, vertical non-load-bearing wall assembly with metal studs and furring, sound attenuation, fire blankets, gypsum liner panels, and water- and fire-resistant gypsum facing panels. The assembly uses C-H steel studs with steel E-studs at wall ends. A wall with this construction can achieve a 2-hour fire rating. A typical assembly for a non-load-bearing fire-resistant partition for enclosing shafts, air ducts, and stairwells is shown in Figure 33.5. It uses a C-H stud, a gypsum panel liner, and a single layer of fire-resistant gypsum for a 2-hour fire rating. Fire-resistance ratings of metal partition walls can be increased by adding layers of gypsum board. Figure 33.6 illustrates several steel-frame partition types and their fire ratings.

Figure 33.4 A cavity-type separation wall utilizing a gypsum liner panel and a sound-attenuation fire blanket.

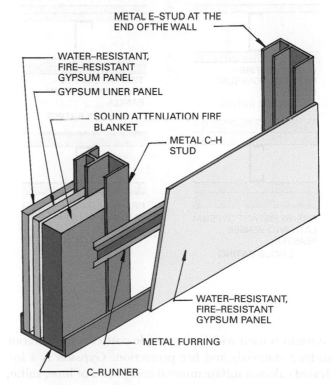

Figure 33.5 Typical construction of a single-layer fire-resistant non-load-bearing cavity shaft wall as used on elevator and mechanical shafts, air ducts, and stairwells.

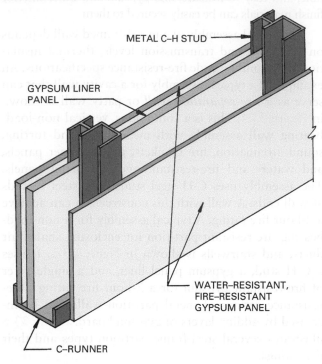

METAL C–H STUD

GYPSUM LINER PANEL

WATER–RESISTANT, FIRE–RESISTANT GYPSUM PANEL

C–RUNNER

Figure 33.6 Fire ratings for different types of metal-framed gypsum wallboard and plaster partitions.

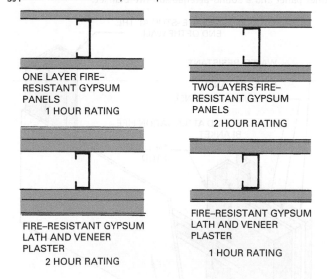

ONE LAYER FIRE–RESISTANT GYPSUM PANELS
1 HOUR RATING

TWO LAYERS FIRE–RESISTANT GYPSUM PANELS
2 HOUR RATING

FIRE–RESISTANT GYPSUM LATH AND VENEER PLASTER
2 HOUR RATING

FIRE–RESISTANT GYPSUM LATH AND VENEER PLASTER
1 HOUR RATING

Gypsum

Gypsum is used widely to provide cost-effective interior surface materials and fire protection. *Gypsum* is a hydrated calcium sulfate mineral compound of lime, sulfur, and water found in natural rock formations. The mined rock is crushed, ground, and calcined (heated). Calcining evaporates most of the chemically combined water. The *calcined gypsum*, called plaster of Paris, is then ground into a fine powder that is used to produce products such as wallboard and gypsum plaster. When calcined gypsum is mixed with water, the plaster absorbs the water and solidifies. This bonds the gypsum and any additives into a solid, hardened mass.

Gypsum board, often called drywall or wallboard, is the generic name for a series of panel products having a noncombustible core of calcined gypsum with paper surfacing on faces, backs, and long edges. Gypsum board products are made by mixing calcined gypsum with water that is fed between continuous layers of paper on a board machine. As the board moves down a conveyer, the calcium sulfate re-crystallizes or re-hydrates, forming a solid core that bonds to the paper.

Types of Gypsum Board Products

Regular gypsum wallboard is produced with smooth, off-white paper on the finish side and gray paper on the back of the gypsum core (Figure 33.7). Special boards are available with an aluminium foil back covering that acts as a vapor barrier. Standard sheet sizes start at 4 ft. × 8 ft., with lengths in 2-ft. increments up to 16 ft. in length available. Metric sizes are approximately 1,200 × 2,400 mm. Sheets are available in thicknesses of ¼ in., 5⁄16 in., 3⁄8 in., ½ in., and 5⁄8 in. (6.4, 8, 9.5, 12.7, and 16 mm). A variety of edge profiles manufactured include tapered, square, bevelled, rounded, or tongue and groove (Figure 33.8). A tapered seam is usually incorporated at the long edge of sheets to provide a space for panels to be joined with a smooth finished joint.

Figure 33.7 Regular gypsum wallboard is used as the finished surface for both walls and ceilings.

Figure 33.8 Gypsum board is manufactured with a variety of edge profiles.

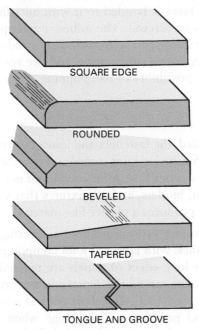

SQUARE EDGE

ROUNDED

BEVELED

TAPERED

TONGUE AND GROOVE

Fire-resistant gypsum board (Type X) has improved fire-resistance qualities due to the addition of special core materials and is used in assemblies that must meet fire code ratings. Pre-decorated gypsum board has finish surfaces covered with a printed, textured, painted, coated, or vinyl film. It provides a finished surface that requires no further treatment after installation.

Water-resistant gypsum board has a water-resistant core faced with a green water-repelling paper. It is used as a base for the installation of ceramic tile and plastic panels in wet areas. Available with regular or fire-resistant core, it comes in ½ in. and ⅝ in. thicknesses. It is not recommended for use on ceilings with joist spacing greater than 12 in. which could exceed panel strength.

Backerboard is used as the base ply or plies in assemblies that require more than one layer of gypsum board. It is also used as a backing for acoustical tile, plywood panelling, and other decorative wall panelling. It is available in regular and fire-resistant panels. Gypsum core board is a 1 in. thick panel used in shaft walls and laminated gypsum panels.

Gypsum linerboard has a special fire-resistant core enclosed in a moisture-resistant paper. It is used as a liner panel on shaft walls, stairwells, chaseways, corridor ceilings, and area-separation walls.

Gypsum lath is used as a base to receive layers of hand or machine-laid plaster. It is available in 16 in. (406 mm) and 24 in. (610 mm) panels that are 48 in. (1,220 mm) long, and in 16 in. (9 mm) and 24 in. (12 mm) thicknesses.

Installing Gypsum Wallboard

Gypsum wallboard is cut by scoring the front face with a utility knife along the edge of a metal T-square (Figure 33.9). The board is then bent to break the core, and the paper is scored at the back to separate the panel parts. Panels are fastened to wood studs and ceiling joists with special nails or power-driven screws and to metal studs and metal furring with self-drilling, self-tapping screws (Figure 33.10). Panels

Figure 33.9 Gypsum wallboard is cut by scoring the front face with a utility knife, bending the panel to break the core, and scoring the paper at the back to separate the panel parts.

©iStock.com/finabelle

Figure 33.10 Gypsum wallboard is fastened to wood and metal studs and ceiling joists with power-driven screws.

© Alex Wong/Getty Image News/Getty Images

can be installed with the long edge perpendicular to or parallel to the studs. Perpendicular application is usually used because it reduces the amount of joints that need taping. Perpendicular application also places the strongest dimension of the panel across the studs.

Ceiling panels are installed first and run perpendicular or parallel to joists, depending on which method produces the fewest joints. Panels may be single or double nailed. Nails on single-nailed ceilings are usually spaced 7 in. (178 mm) on center and 8 in. (203 mm) on center on sidewalls (Figure 33.11). Screws are spaced 12 in. (305 mm) on a ceiling and 16 in. (406 mm) on sidewalls. To reduce the possibility of nail popping after a wall is finished, panels can be double nailed as shown in Figure 33.12. Fasteners are driven until their heads are set in shallow dimples without breaking the paper covering. Nails are driven with a drywall hammer with the required shape and diameter.

Figure 33.11 Suggested single-nailing pattern for gypsum wallboard panels.

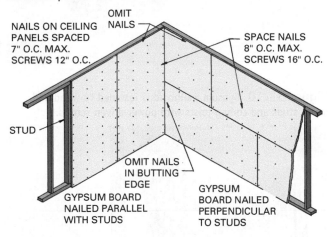

Figure 33.12 Suggested double-nailing pattern for gypsum wallboard panels.

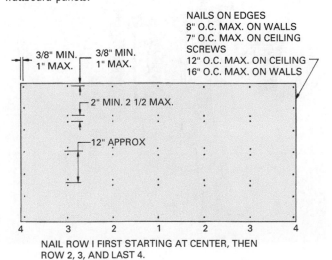

NAIL ROW I FIRST STARTING AT CENTER, THEN ROW 2, 3, AND LAST 4.

Gypsum panels are often installed in a double layer. The first layer is nailed or screwed to the framing, and the second layer is bonded to it with an adhesive (plus a few nails or screws). The adhesive is applied to the backs of panels with a trowel. This produces a wall with greater sound attenuation and fire resistance and reduces the possibility of nail popping.

Finishing Gypsum Wallboard

Once installed, the fasteners and joints between gypsum panels are finished in joint compound to create a smooth and continuous surface. *Joint compound* is a mixture of marble dust, binders, and admixtures that when mixed with water produces a plaster-like material.

The finishing process begins by covering the joints between panels with layers of joint compound and tape. Because the long edges of panels are tapered, the compound and tape can fill the depression and create a flush surface (Figure 33.13). The short ends of panels are not tapered, and produce a slight bulge when taped. For this reason, end joints are avoided during installation

Figure 33.13 The joints between panels are covered with layers of joint compound and tape.

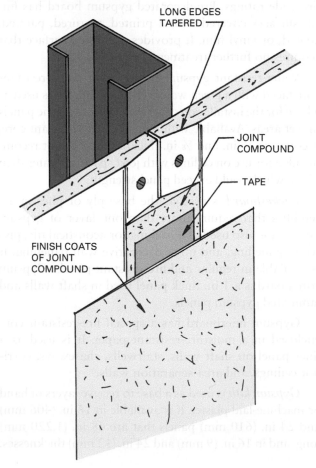

whenever possible. Both types of joints are covered with a thin layer of compound into which a fiber-reinforced tape is pressed with a finishing knife or taping tool (Figure 33.14). A thin skim coat is applied over the tape and the compound is allowed to dry overnight. Second and third layers of compound are applied over the base layer, allowed to dry, and sanded. Each layer is wider than the one before and is feathered out with progressively wider taping knives (Figure 33.15). Both the tape and joint compound can be applied to wall and ceiling joints manually or mechanically with a variety of machines.

Figure 33.14 A tool called a banjo is used to apply the first layer of tape and joint compound.

© George Peters/E+/Getty Images

Figure 33.15 The final layer of joint compound is applied with wide taping knives.

© Bloomberg/Bloomberg/Getty Images

Figure 33.16 External corners are covered with a metal or plastic corner bead-and-joint compound.

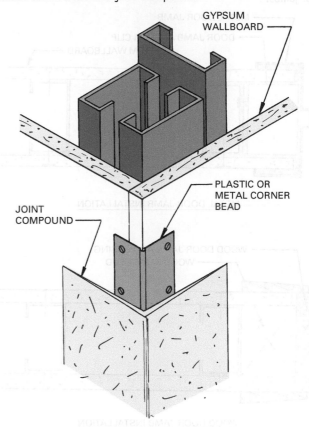

GYPSUM WALLBOARD

PLASTIC OR METAL CORNER BEAD

JOINT COMPOUND

The compound may be hand sanded, but a pole-sanding unit with a vacuum greatly reduces dust. Nail and screw heads use no paper and are covered with three layers of joint compound. Internal corners are finished with a folded paper or fiberglass tape bonded in joint compound and finished in the same manner described for joints. External corners are finished with metal or plastic corners nailed or screwed to the studs and covered with joint compound (Figure 33.16). Other finishing accessories include edge trim, control joints, and trim for round corners. Gypsum wallboard, like other building materials, is subject to

movement induced by changes in moisture, temperature, or both. To relieve the stresses that occur because of such movement, control joints are required in both partitions and ceilings. In long expanses of partitions, control joints should be installed at 30 ft. (9.1 m) intervals, from floor to ceiling. Additionally, control joints should be included in ceilings to limit dimensions in either direction to 50 ft. (15.2 m). Thin sheets of gypsum wallboard can be installed on curved framing with special curved corner trim (Figure 33.17). Doorframes on walls with metal or wood studs are installed as shown in Figure 33.18.

Figure 33.17 Thin sheets of gypsum wallboard can be installed on rounded framing and finished with special curved corner trim.

©iStock.com/George Peters

Figure 33.18 Typical installation details for metal and wood door jambs.

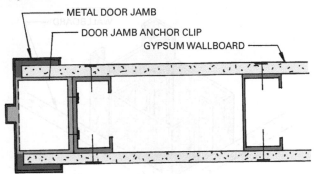

METAL DOOR JAMB
DOOR JAMB ANCHOR CLIP
GYPSUM WALLBOARD

METAL DOOR JAMB INSTALLATION

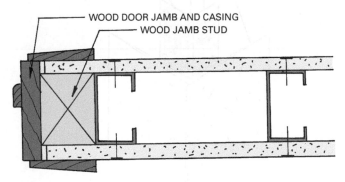

WOOD DOOR JAMB AND CASING
WOOD JAMB STUD

WOOD DOOR JAMB INSTALLATION

PLASTER FINISHES

A plaster finish consists of a supporting base, either gypsum or metal lath, and one or more coats of plaster. The main ingredient in *plaster* is usually gypsum, but some types use Portland cement. Gypsum plasters meet the requirements of ASTM C28, *Standard Specification for Gypsum Plasters*, or ASTM C587, *Standard Specification for Gypsum Veneer Plaster*.

The following plaster products are used for interior finish applications. *Plaster of Paris* is calcined gypsum mixed with water to form a thick paste. Used in construction for ornamental plastering and repairing plaster walls, it sets in fifteen to twenty minutes. If mixed with lime putty, it can be used as a finish coat on plaster walls. Plaster of Paris hardens rapidly and experiences little shrinkage.

Regular gypsum plaster is used with sand aggregate and applied by hand. Lightweight plaster is used with lightweight aggregates and is manually applied. Machine-applied gypsum is used with either sand or lightweight aggregates. Gypsum neat plaster is a calcined gypsum plaster mixed at a mill with other ingredients to control its working quality and setting time. Sand, perlite, and vermiculite are used as aggregates. Gypsum plaster with wood fiber is manufactured in powder form for lighter weight and increased fire resistance.

Plaster may be applied to a solid substrate or on expanded metal lath. Gypsum lath is a board product made of hardened plaster to which a paper facing is applied to which the plaster is towelled. Expanded *metal lath* is made from thin steel sheets that are cut and stretched to create an open mesh into which the plaster is pressed. Plaster on metal lath is applied either manually or by machine in a series of layers. The three-step process includes an initial scratch coat, followed by a brown coat and additional levelling coats (Figure 33.19). Plaster applied over gypsum lath is usually a two-coat system, but three coats may be used (Figure 33.20).

Figure 33.19 A three-coat plaster wall finish over metal lath.

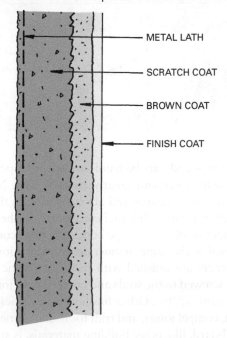

METAL LATH
SCRATCH COAT
BROWN COAT
FINISH COAT

Figure 33.20 A two-coat plaster wall finish over gypsum lath.

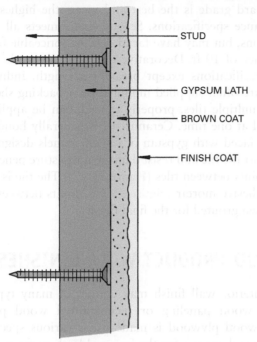

STUD

GYPSUM LATH

BROWN COAT

FINISH COAT

The first coat is applied with pressure so it bonds to the lath and provides a level coat with a rough surface. The second coat goes over the first coat and is finished to the desired texture. Various accessories are used with plaster finishes, much as is described for drywall. Typical among these are casing beads, external corner beads, flexible corner beads, and expansion joints.

Plaster over Masonry and Concrete Walls

Plaster may be applied directly to masonry, concrete, and clay tile walls (Figure 33.21). Concrete block walls are porous and provide a satisfactory base. Clay tile and brick walls may be used as a base for plaster walls, but they must be sufficiently porous to provide suction for plaster and also scored to increase mechanical bonding. Smooth-surface glazed or semi-glazed tile cannot be

Figure 33.21 Plaster may be applied directly to masonry, and concrete walls that are not in direct contact with the exterior.

© Andy Sotiriou/The Image Bank/Getty Images

covered. These require that the wall be furred and a metal or gypsum plaster base be installed. Monolithic concrete walls are porous but usually require an application of an agent to produce the adhesive bond necessary for direct application of gypsum plasters. Generally, masonry walls in contact with the exterior are not plastered directly to the masonry because of the likelihood of water seepage and condensation that can saturate the plaster.

Masonry concrete and clay tile walls can have gypsum lath installed by securing metal or wood furring to their surface, or building a frame wall to hold the gypsum wallboard (Figure 33.22). On exterior walls, this provides an air space, keeping the *plaster base* and

Figure 33.22 Masonry walls can be covered directly with a plaster finish or use furred out walls.

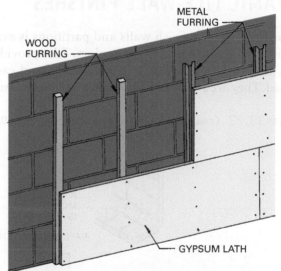

METAL FURRING

WOOD FURRING

GYPSUM LATH

FUR THE WALL WITH WOOD OR METAL FURRING

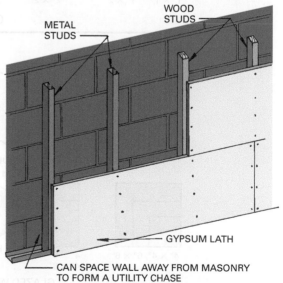

METAL STUDS

WOOD STUDS

GYPSUM LATH

CAN SPACE WALL AWAY FROM MASONRY TO FORM A UTILITY CHASE

BUILD A WALL OVER THE MASONRY WALL

plaster away from the masonry wall. Furring can be shimmed to help true a wall with minor surface irregularities. Furring may be placed horizontally or vertically.

STRUCTURAL CLAY TILE PARTITIONS

Structural clay tile partitions are made with hollow clay masonry units that are glazed on one or both sides. The glazes provide a smooth surface that is impervious to penetration by water and is available in a wide range of colors. The tiles resist abrasion, wear, and mildew and can be routinely scrubbed and sanitized. Information about tile sizes and wall construction can be found in Chapter 11.

CERAMIC TILE-WALL FINISHES

Ceramic tile used to finish walls and partitions is available in a wide range of sizes and colors. Tiles provide a hard, water-resistant facing and are durable and easily cleaned. They are available in a variety of shapes and sizes

(Figure 33.23). There are three grades of ceramic tile. Standard grade is the best and meets the highest performance specifications. Second grade meets all specifications, but may have facial defects noticeable from a distance of 10 ft. Decorative thin wall tile grade meets all specifications except breaking strength. Individual tiles are often supplied mounted on a backing-sheet so that multiple tiles, properly spaced, can be applied to a wall at one time. Ceramic tile is generally bonded to walls faced with gypsum or cement panels designed to support the tile and resist damage if moisture penetrates the joints between tiles (Figure 33.24). The tile is set in an adhesive mortar (Figure 33.25). Joints between tiles are then grouted for the final finish.

WOOD-PRODUCT WALL FINISHES

An interior wall finish may consist of many types of solid wood paneling or reconstituted wood panels. Hardwood plywood is made using various species of hardwood veneers for the face and back surfaces of the

Figure 33.23 Commonly available ceramic and glazed wall tiles.

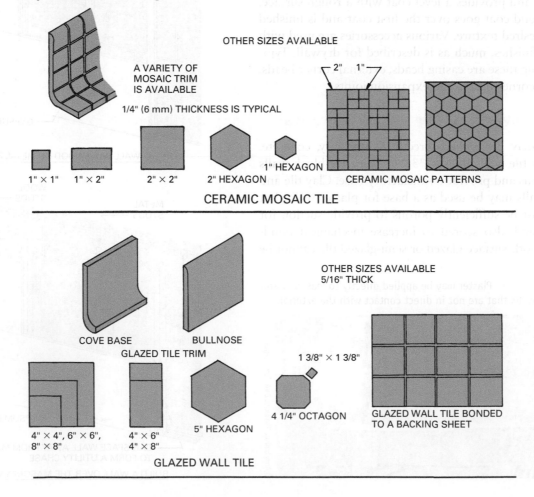

Figure 33.24 Substrate details for walls with ceramic tile facing.

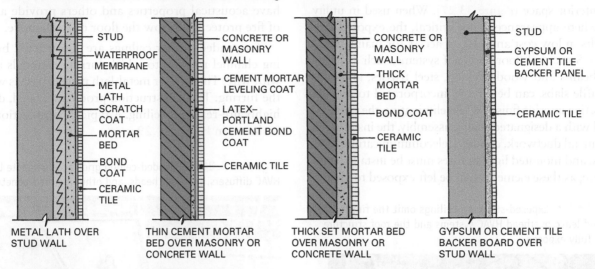

METAL LATH OVER STUD WALL

- STUD
- WATERPROOF MEMBRANE
- METAL LATH
- SCRATCH COAT
- MORTAR BED
- BOND COAT
- CERAMIC TILE

THIN CEMENT MORTAR BED OVER MASONRY OR CONCRETE WALL

- CONCRETE OR MASONRY WALL
- CEMENT MORTAR LEVELING COAT
- LATEX–PORTLAND CEMENT BOND COAT
- CERAMIC TILE

THICK SET MORTAR BED OVER MASONRY OR CONCRETE WALL

- CONCRETE OR MASONRY WALL
- THICK MORTAR BED
- BOND COAT
- CERAMIC TILE

GYPSUM OR CEMENT TILE BACKER BOARD OVER STUD WALL

- STUD
- GYPSUM OR CEMENT TILE BACKER PANEL
- CERAMIC TILE

Figure 33.25 Ceramic tiles are often supplied mounted on a backing sheet so multiple tiles can be applied to a wall at one time.

© Jupiterimages/Stockbyte/Getty Images

Figure 33.26 Book-matching is used on high-quality wood veneer installations.

© pics721/Shutterstock.com

panel. High-quality wood veneers, including mahogany, maple, and cherry wood, are commonly used. They are bonded to plywood or particleboard panels that form the core. Panels are available prefinished, including some with vinyl wall coverings. Details about these products are given in Chapter 19.

Wood panels may be applied in a process called book-matching. With book-matching, every other piece of veneer, referred to as a leaf, is flipped so that adjacent leaves are mirrored, like the pages of a book (Figure 33.26). Book-matching is used on more upscale and high-quality wood veneer installations. Because of their flammability, the use of wood interior wall finishes may be severely limited or prohibited by code regulations for certain building types and occupancies.

CEILING CONSTRUCTION

In addition to providing an important role in shaping interior spaces by providing a finished surface to overhead areas, ceilings incorporate a variety of important functions.

Ceiling assemblies enclose space below the floor structure in which electrical, mechanical, and fire-suppression systems can be accommodated out of view. Most ceilings support lights and HVAC diffusers, and provide a measure of fire protection to the floor or roof above. They also contribute to the acoustical treatment of a room, and provide a light-reflecting surface that can enhance illumination.

Ceiling finish materials may be plastic, metal, plaster, gypsum panels, wood, or one of many fibrous panels available. In some occupancies, an architectural design referred to as exposed-structure ceilings, or open plenum ceilings, is becoming increasingly popular. Exposed-structure ceilings omit the finished ceiling and leave overhead HVAC systems and the roof or floor deck fully exposed as an integral part of the interior space (Figure 33.27). When used in utility spaces where appearance is not critical, the exposed ceiling results in low cost and ease of access for maintenance activities. Structural floor and roof systems, such as heavy timber beams and wood decking, steel trusses, and concrete waffle slabs, can be directly incorporated to provide an attractive interior finish. While eliminating the costs associated with a designated ceiling assembly, the installation of rigid metal ductwork, conduit, telecommunications connections, and mounted light fixtures must be installed with more care, as these elements will be left exposed to view.

Figure 33.27 Exposed-structure ceilings omit the finished ceiling and leave overhead HVAC systems and the roof structure and deck fully exposed.

Ceilings systems are constructed by attaching finish materials to the bottom of structural floor and roof assemblies. The materials may be attached through direct contact, they may be furred directly below the structure, or be suspended from it. Ceiling height minimums for certain components are regulated by the building codes. For instance, most codes require that exit egress paths have a ceiling height of no lower than 7 ft. 6 in. Ceiling heights for commercial spaces vary depending on the local jurisdiction and vary between 13 and 15 ft. in height.

Suspended Ceilings

Suspended ceilings are widely used in commercial construction. They offer considerable advantages by allowing electrical and mechanical systems to be hidden, while allowing for a variety of finished-surface materials.

Suspended ceilings are hung from a floor or roof structure by a series of wires that support a metal grid that holds the finished ceiling. Wire lengths can vary, so a flat horizontal ceiling can be installed below a structural floor or roof that has members of various sizes. Typically, a lighting system is set in the grid and is flush

with the ceiling (Figure 33.28). Some types of panels have acoustical properties and others provide a degree of fire protection below the floor or roof above.

Suspended-plaster ceilings are constructed by hanging channel runners to which furring channels are tied with wire. The wire or metal-lath plaster base is wired to the furring. This construction produces a hard, durable, highly fire-resistant ceiling. A typical construction detail is shown in Figure 33.29.

Figure 33.28 Suspended-ceiling grids accommodate lighting, HVAC diffusers, sprinkler heads, and other required penetrations.

Figure 33.29 A suspended plaster ceiling has the main channel hung from overhead by wires, and metal furring channels wired to the main channel. The plaster coats are as described for finished walls.

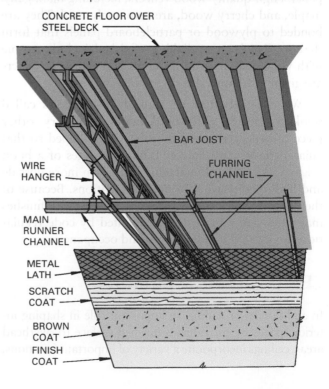

Various types of suspended acoustical systems use a suspended metal grid into which lightweight panels and lights are placed. Grids are assembled with some form of locking connection (Figure 33.30). Drop-in panels may have an exposed grid or a concealed grid (Figure 33.31). These systems also accommodate heat ducts, sprinkler heads, and other required penetrations. Some grid systems have recessed panels, and a large number of surface textures are available. Ceiling panels are made with materials such as aluminum, fiberglass-reinforced polymer gypsum cement, gypsum panels faced with vinyl and fabrics, fiberglass, and wood fibers.

Another suspended-ceiling system uses a metal grid with gypsum panels secured to it with screws. In this system, the main runners are spaced 48 in. (1219 mm) on center and the furring cross-channels either 16 in. (406 mm) or 24 in. (610 mm) on center. The gypsum drywall panels are attached to the grid with conventional drywall screws (Figure 33.32). The ceiling grid is secured to walls with a wall track, and partitions are secured to the ceiling by screwing them to a drywall furring cross-channel or main runner.

Figure 33.30 An exploded view of a suspended metal grid with acoustical tile resting on the grid flanges.

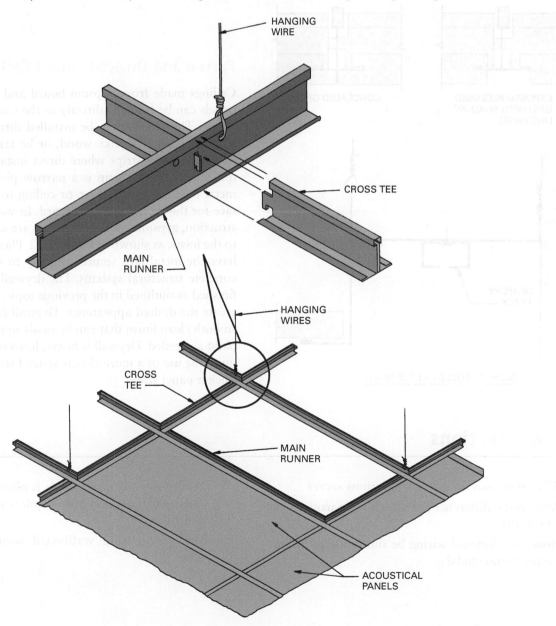

Figure 33.31 Drop-in ceiling panels may have an exposed, recessed, or concealed grid.

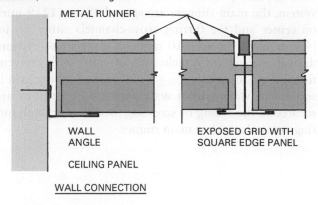

METAL RUNNER

WALL
ANGLE

CEILING PANEL

EXPOSED GRID WITH
SQUARE EDGE PANEL

WALL CONNECTION

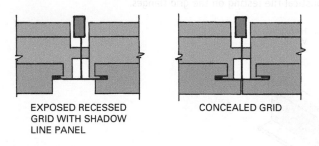

EXPOSED RECESSED
GRID WITH SHADOW
LINE PANEL

CONCEALED GRID

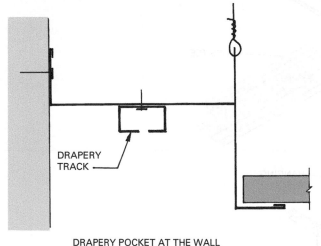

DRAPERY
TRACK

DRAPERY POCKET AT THE WALL

Figure 33.32 A suspended-ceiling system using a metal grid with gypsum panels secured to it with screws.

© Pavel L Photo and Video/Shutterstock.com

Furred and Direct-Contact Ceilings

Ceilings made from gypsum board and other material panels can be fastened directly to the roof or floor deck above. The panels may be installed directly to certain structural materials like wood, or be fastened to intermediate furring strips where direct installation may be difficult. A furring strip is a narrow piece of wood or metal fixed to a wall, floor, or ceiling to provide a surface for the fixing of plasterboard. In wood-frame construction, gypsum wallboard panels are screwed directly to the joists, as shown in Chapter 22. Plaster ceilings can have the metal lath secured directly to steel, wood, or concrete structural systems. The drywall is sanded and finished as outlined in the previous topic, and painted to create the desired appearance. Drywall ceilings create a smooth clean finish that can be easily updated with new paint as needed. Drywall is heavy, however, and may require the use of a more closely spaced structural grid or thicker panel sizes.

Review Questions

1. What purposes do interior partitions serve?

2. What is the difference between a fire partition and a firewall?

3. How can electrical wiring be run inside partitions having metal studs?

4. What assembly of materials is often used in a fire-rated partition that also requires effective sound control?

5. How is gypsum wallboard secured to metal studs?

6. What can be done to reduce nail popping in gypsum wallboard installation?

7. How are external corners on gypsum wallboard finished?

8. What base materials are used when constructing a plaster interior finish?

9. What types of plaster coats are applied over metal lath?

10. Why can concrete block walls have plaster applied directly to them?

11. How can you plaster over a glazed tile wall?

12. Describe the construction of typical solid gypsum partitions.

13. What are the merits of using glazed hollow clay tiles for partition construction?

14. What is a suspended ceiling?

Key Terms

Backerboard	Gypsum	Plaster
Calcined Gypsum	Gypsum Lath	Plaster Base
Fire-Rated Partition	Joint Compound	Plaster of Paris
Fire Separation Wall	Metal Lath	Smoke Barriers

Activities

1. Examine catalogs of companies that manufacture interior walls, partitions, and ceiling systems. For each, list materials used, fire rating, and sound transmission class, and sketch a section of the assembly.

2. Tour various buildings on your campus and record the location of different types of interior wall assemblies, partitions, and ceilings. Do those in older buildings meet the current codes? What should be done to upgrade each system found lacking?

Additional Resources

Gorman, J. R., Jaffe, S., Pruter, W. F., and Rose, J. J., *Plaster and Drywall Systems*, BNI Building News, 1612 Clementine Street, Anaheim, CA 92802.

Gypsum Construction Guide, National Gypsum Company, Gold Bond Building Products, 2001 Rexford Road, Charlotte, NC 28211.

Gypsum Construction Handbook, United States Gypsum Company, 125 S. Franklin Street, Chicago, IL 60606.

Handbook for Ceramic Tile Installation, Tile Council of North America, Anderson, SC.

Melander, J., and Isberner, A., "Portland Cement Plaster," *Journal of Light Construction Publishers*, Richmond, VT.

Remodelers Guide to Suspended Ceilings, Chicago Metallic Corp., 4849 S. Austin Avenue, Chicago, IL 60638.

Spence, W., *Installing and Finishing Drywall*, Sterling Publishing Co., 387 Park Avenue South, New York, 10016.

Other resources include:

Many publications from the Gypsum Association, 810 First Street NE, Washington, DC 20002.

See Appendix C for addresses of professional and trade organizations and other sources of technical information.

CHAPTER 34

Flooring

Upon completion of this chapter, the student should be able to:

- Understand the importance of building codes in the selection of flooring.
- Select the appropriate flooring materials for various applications.

- Understand the types of flooring available and their sizes and properties.
- Understand various types of carpet construction.
- Choose carpet for various applications.
- Understand installation procedures for carpeting.

Build Your Knowledge

For further information on these materials and methods, please refer to:

Chapter 8   Cast-in-Place Concrete

Chapter 11   Clay Brick and Tile

Chapter 13   Stone

Figure 34.1 Finish flooring forms a large part of the first impression occupants receive as they enter and move about a facility.

© Rade Kovac/Shutterstock.com

In many commercial environments, flooring is the surface that experiences the most use, and facilitates the ease of movement of people and equipment. Flooring sets the stage for all commercial and institutional activities. It forms a large part of the first impression occupants receive as they enter and move about a facility, shaping their opinions about the organization's ability to provide a safe and well-organized space (Figure 34.1). In the industry today, there are three broad categories of floor coverings: hard, soft, and resilient. Significant variations exist in the characteristics of the products found within each. Selecting finished floor materials depends on the flooring function and can vary extensively between areas like bathrooms that are prone to slips, versus more sedentary activity areas such as offices, or high-traffic areas such as corridors and walkways.

Finished flooring consists of the underlying sub-floor that provides support, and the floor covering and surface finish that create the walking surface. In selecting flooring systems, the designer must consider the characteristics of the substrate, expected traffic loads on a floor, its maintenance requirements, fire resistance, building code requirements, and appearance. The finish flooring must meet all of the required conditions at an acceptable cost. Some applications may require that special conditions be met, such as a need for water resistance, enhanced hygiene, or other unique situations like those found in hospitals or laboratories.

PERFORMANCE CRITERIA FOR FLOORING

The prevention of slips, trips, and falls is a primary performance criterion for any finish flooring material (Figure 34.2). Slips occur when there is insufficient friction or traction between the occupant and the walking surface. Trips also occur when the foot strikes an object, like a flooring transition, and the momentum causes a fall. Material on the sole of the footwear can also catch on the flooring surface and cause the foot to be abruptly halted, which could impair balance.

Figure 34.2 The prevention of slips, trips, and falls is a primary performance criterion for any finish flooring material.

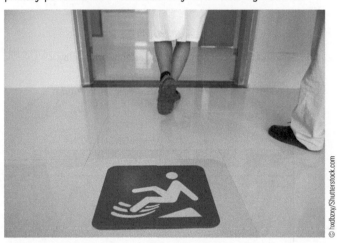

© hxdbzxy/Shutterstock.com

National standards for slip and slide resistance of floorings in different types of buildings and work areas have recently been established. The American National Standards Institute (ANSI) and National Floor Safety Institute (NFSI) have developed test methods for flooring materials. According to ANSI/NFSI B101.1-2009 and ANSI/NFSI B101.3-2012, walkway slip resistance can be measured and categorized in one of three traction ratings: high, moderate, or low. Floors categorized as high-traction offer a low risk of a slip and fall, while moderate- and low-traction floors present a higher risk.

In contrast to slips, which are affected by the surface properties of the floor, trips are a design issue. Planning for the thickness of different floor coverings and how transitions are addressed between different materials or thresholds is a key consideration. The Americans with Disabilities Act (ADA) sets applicable standards for transitions and recommends that flooring thresholds should be less than ¼ in. vertical or between ¼ in. and ½ in. beveled. Ideally, there should be few changes to flooring types, no thresholds, and careful transitions from carpet to resilient flooring.

According to the Environmental Protection Agency (EPA), building materials in general, and flooring in particular, can have a large impact on air quality, which, in turn, can affect the occupants. Floor covering should have minimum emissions of volatile organic compounds (VOCs). All surface floorings should minimize the need for surface coating. Carpet and carpet cushion should meet the Carpet and Rug Institute's (CRI) Green Label Plus.

Finished floors are classified into Class 1 and Class 2 finishes according to their flame resistance by NFPA 253 and ASTM E648. Once flooring is installed, there can be many changes in due to variation in maintenance protocols, different surface finishing products, and the wear and tear over time. Cleaning products specified should meet Green Seal GS-37.

HARD-SURFACE FLOORING

Hard-surface flooring is made of inflexible materials that provide little or no give. A variety of types and styles are used frequently in commercial settings, including concrete, clay tile, stone, brick, wood products, and terrazzo. Most hard-surface flooring is expensive, often costing several times more than the common resilient flooring. However, because of their easy maintenance and very long durability, they have the lowest life-cycle cost of any flooring material. A hard floor is much easier to clean, generally traps less dirt and supports good indoor air quality.

Concrete Flooring

Finished concrete floors are widely used in commercial construction to provide a durable and economic flooring solution. A concrete floor surface may be finished in several ways. A slab can be poured, leveled, and screeded, and its surface smoothed with a wood float. A smoother surface can be produced by smoothing the wood float surface with a steel trowel before the concrete firmly sets. Various textured surfaces can be achieved by techniques such as broom brushing a surface after troweling but before it sets fully.

Stamped concrete flooring uses concrete that is patterned or embossed to resemble brick, slate, stone, tile, or wood, and other patterns and textures. Installation involves pouring wet concrete and then pressing patterns onto the concrete before it is fully dry. Manufacturers provide a variety of imprinting tools to produce patterns. The tools provide a rigid working surface, are equipped with handles for ease of placing, and typically include grout or joint lines in the pattern. A release agent is used to allow the form tools to be easily removed once the concrete has set.

An exposed aggregate finish is made by leveling a slab, embedding aggregates with a float to get a level surface. After the concrete begins to set, it is sprayed with water, leaving the aggregate exposed and forming the finished surface. A high-quality smooth-finished surface can be produced by covering the base concrete with a 1 in. thick specially formulated concrete layer before the base concrete hardens.

Colored surfaces are produced by adding color pigments, usually metallic oxides, to a concrete mix as a batch is prepared. Another technique involves pouring and finishing a floor, then spreading a dry shake coloring material prepared especially for it. The dry shake is spread evenly over the surface before it hardens and floated to smooth the surface and even out the color. Other finishes include painting, staining, applying an abrasive aggregate to a surface to produce a nonslip finish, adding metallic aggregate to a surface to produce a more wear-resistant finish, and the use of various chemical coatings to harden a surface.

Heavy-duty concrete floors are used in many industrial applications. They are usually constructed as a two-layer floor. The reinforced floor base is poured and topped with a durable concrete layer made with special abrasion-resistant aggregate, such as emery or iron filings. A technique called scoring gives a concrete slab the appearance of tile with a much more durable surface. Shallow cuts are made into the surface of the concrete and can be made to resemble grout lines (Figure 34.3).

Figure 34.3 Concrete slabs can be scored to give the appearance of tile, with a much more durable surface.

© photobank.ch/Shutterstock.com

Brick and Stone Flooring

Brick flooring and brick pavers are used to produce durable flooring. Brick pavers are thinner than standard brick, ranging from ½ in. to 2½ in. (12 mm to 57 mm) thick, and are preferred to standard brick for flooring applications. Pavers are also available in a variety of glazed brick, concrete, and asphalt units. Brick flooring can be utilized for both interior and exterior floors. Hard-backed types are suitable for heavy use on industrial floors.

Bricks can be laid with the large flat face exposed, or on edge exposing the narrow face. Many techniques and patterns are used to lay brick pavers (Figure 34.4). Exterior brick patios are often laid on

Figure 34.4 Typical floor details for clay brick, clay, or concrete pavers.

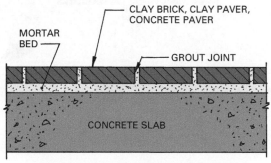

BRICKS AND PAVERS SET ON A MORTAR BED

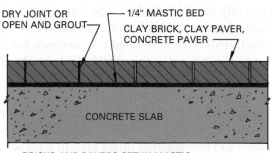

BRICKS AND PAVERS SET IN MASTIC

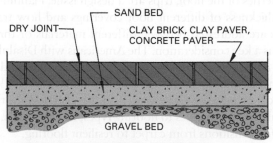

BRICK AND PAVERS SET ON A SAND BED

a bed of sand with no mortar between units. When a floor will be subject to heavy loads, pavers are laid over a concrete slab.

Stone finish flooring is typically slate, marble, or granite. When used on a building's interior, stones are laid over a wood substrate or concrete slab in the same manner as brick pavers. Sand-laid stone is usually used for exterior work, and the joints are left open. Interior stone floors may have joints filled with grout or be set tight without it. Most stone flooring is finished with applications of a clear sealing compound, and must be waxed periodically to maintain their sheen and luster (Figure 34.5).

Figure 34.5 A variety of stone products, including slate, marble, and granite, is used for finish flooring.

Flagstone is thin, flat stone from ½ in. to 4 in. in thickness. Flagstones laid over a concrete base are usually ¾ in. to 1 in. thick. If laid over a sand or loam base, pieces 1¼ in. to 1½ in. thick are required. Its surface may be left rough or polished. Random flagstones, with the exception of minor shaping, are left in their natural shape. Trimmed flagstones are random-shaped pieces with several edges sawed straight. Trimmed rectangular flagstones are pieces with four sides sawed, forming square and rectangular pieces.

Another stone floor covering material is cast stone tile flooring, formed by impregnating and encapsulating stone aggregate with high-technology resins that create a very dense surface that meets Underwriters Laboratory requirements for slip resistance.

Clay Tile Flooring

Clay tile products are used for flooring include quarry tile, paver tile, and ceramic tile. Quarry tile is unglazed and made from shales and fire clays. It is durable, cleans easily, resists damage from freezing, thawing, and abrasion, and is used on floors subject to heavy and extra-heavy use. Paver tiles may be glazed or unglazed and are like ceramic tiles but larger (Figure 34.6). Typical sizes are 4 in. × 4 in. (102 mm × 102 mm), 4 in. × 8 in. (102 mm × 203 mm), and 8 in. (203 mm) hexagons. They are weather and abrasion resistant. Typical uses include residential and moderate duty commercial floors, such as in shopping

Figure 34.6 Paver tiles are weather and abrasion resistant.

centers and restaurants, and heavy-duty floors, such as those found in a commercial kitchen. A variety of trim and edging materials are available.

Ceramic mosaic tile is available glazed or unglazed in square, rectangular, and hexagonal shapes. It comes in a wide range of sizes, from small, 1 in. or 2 in. square or hexagonal shapes, to 12 in. and 16 in. squares for large areas. Ceramic tile is used for interior floors, countertops, walls, swimming pools, and exterior floors and walls with special installation procedures. Tile with a variety of slip-resistant surfaces is available as base and bullnose, and many other trim tiles. Glazed ceramic floor tile is available with smooth and textured surfaces. Textured surfaces provide greater slip resistance. Tiles are available in 4 in. and 8 in. squares, 8 in. hexagons, and 4 in. × 8 in. rectangles. They are used for interior moderate-duty commercial floors, such as in restaurants and shopping malls. Some types are made with an abrasive grain surface to provide slip resistance. Although various manufacturers produce a variety of sizes and shapes, square, rectangular, and hexagonal tiles are most common. Various trim units, such as stair-nosing and cove base units, are also available.

Porcelain tile is also made from clay but uses denser clays than those used for ceramic tiles. Porcelain tiles are baked at very high temperatures for longer periods of time so that almost all the water is eliminated. This longer drying time makes porcelain tile harder, denser, and more impervious to water than standard ceramic tiles. Porcelain generally costs more than ceramic tile but can be used for exterior and high-traffic areas.

Mortars and Adhesives for Tile Floors

Clay floor tile is installed using a variety of bonding agents. Portland cement mortar is recommended for a thick-setting bed. Epoxies and furans that do not contain Portland cement can be used but are more expensive. Portland cement mortar is a mixture of Portland cement, hydrated lime, sand, and water. It can be used to provide a leveling bed up to $1/14$ in. (32 mm) thick. Tiles are set into this bed while it is still plastic (Figure 34.7). After curing, tiles may be bonded to the leveling bed with a thin- or dry-set coat or latex Portland cement mortar.

Dry-set mortar is a mixture of Portland cement, resinous materials, sand, and water. It is a thin-set mortar (about $3/32$ in. or 2.4 mm thick). It may be applied over the hardened thick-set Portland cement mortar bed and used to bond tiles in place.

Epoxy mortar contains no Portland cement but is a mix of epoxy resin and a hardener. It is a thin-coat

Figure 34.7 Cement mortars for tile products are applied with a notched trowel. Temporary plastic spacers maintain consistent grout widths between tiles.

© Titi Matei/Shutterstock.com

adhesive with high bond strength and is resistance to impact and chemical attack.

Latex Portland cement mortar is a mixture of Portland cement, a latex additive, and sand. It is a thin-coat application, usually about $1/8$ in. (3 mm) thick. Organic adhesives harden by evaporation. They are applied with a notched trowel, leaving a thickness of about $1/16$ in. (1.6 mm), and are not used for exterior applications and interior uses exposed to considerable water.

Furan mortar is a mixture of furan resin and a hardener. It has high resistance to chemicals.

Tile Grout

Grout is a mortar used for filling the gaps between floor tiles (Figure 34.8). Grouts typically are a mixture

Figure 34.8 Grout is spread with a rubber trowel. After setting for 15–30 minutes, the excess grout is wiped up with a sponge and water.

© Lisa S./Shutterstock.com

of Portland cement, fine sand, and lime. They generally come premixed, requiring the addition of water. Latex, furan, epoxy, or silicone rubber can be added to influence properties. A manufacturer should be consulted to ensure choosing the most desirable grout for a situation. The grout color chosen has a large influence on the appearance of the finished floor.

Terrazzo Flooring

Terrazzo provides an attractive, hard, long-lasting, and easily cleaned surface. The material is widely used in major public spaces such as shopping centers, public transport facilities, and similar applications.

Terrazzo is a matrix consisting of marble or granite chips, Portland cement, and water or a synthetic resin. It is placed as a top layer over a concrete substrate, steel decking, or concrete slab. Portland cement terrazzo is divided into sections by metal or plastic dividing strips, which reduce the possibility of cracking. The three methods for casting Portland cement terrazzo are monolithic, bonded, and sand cushion (Figure 34.9). Monolithic terrazzo uses a ⅝ in. (16 mm) thick layer placed as a topping on a green concrete slab. Bonded terrazzo has a topping 1¾ in. (44.5 mm) deep or thicker. The cured concrete slab is cleaned and coated with a neat Portland cement, and a concrete under bed of about 1 in. is laid. Divider strips are placed on the underbed and pushed into it. The terrazzo topping (25 mm) is leveled over the underbed.

Sand-cushion terrazzo has a 12 in. (12.7 mm) bed of dry sand laid over a concrete slab. The sand is covered with a waterproof membrane, and a concrete underbed is laid over it. Dividing strips are placed on the underbed, and the terrazzo topping is poured and leveled. The sand cushion isolates the slab from the floor slab, protecting it from damage caused by movement of the floor slab.

After the terrazzo topping has hardened, it is ground smooth and polished (Figure 34.10). The polishing produces a gloss and brings out the color of the chips. Sometimes, a less-glossy nonslip surface is ground. A protective sealer is applied over the finished floor. When used in public areas, terrazzo should have the required wet-slip resistance. Terrazzo pre-cast units are available for a variety of specialty applications. Synthetic resin terrazzo is cast in square and rectangular shapes in thicknesses from ⅛ to ¼ in. (3 mm to 6 mm). Pre-cast stair treads, risers, window stools, wall base, and other shapes are available precast in Portland cement terrazzo.

Figure 34.9 Terrazzo floor installation details over different subfloor materials.

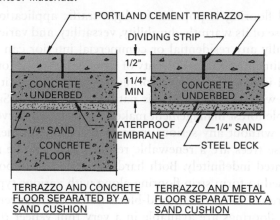

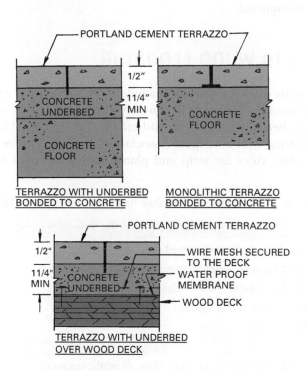

Figure 34.10 Once hardened, terrazzo is ground smooth and polished to produce a gloss and bring out the color of the aggregate.

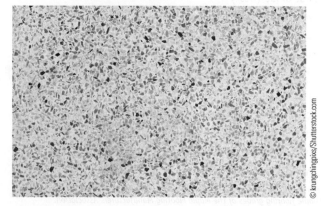

© krungchingpixs/Shutterstock.com

WOOD FLOORING

Wood flooring is popular for light-traffic applications because of its warmth, durability, versatility, and variety. Virtually any residential or commercial interior can be complimented with a diversity of wood flooring options. Wood is unique among flooring materials in that it is a renewable resource that is natural and organic. Sustainable forest management makes it possible to harvest wood without any serious impact on the environment, because trees are a renewable resource that can be regenerated indefinitely. Both hardwoods and softwoods are used to form strip flooring along with various types of parquet and heavy wood-block flooring. Engineered wood floorings are available in a very thin veneer that is prefinished.

SOLID WOOD FLOORING

Woods used for flooring include both hardwoods and softwoods. The most commonly used hardwoods are oak, beech, birch, pecan, and maple. The national Oak Flooring Manufacturers Association publishes uniform grading rules for strip and plank flooring. Grades for

unfinished oak flooring from best to economy are clear, select, no. 1 common, and no. 2 common. Softwoods include southern pine, western larch, bald cypress, eastern Engelmann, eastern spruce, red pine, ponderosa pine, eastern hemlock, and Douglas fir.

Grading rules for wood flooring are shown in Table 34.1. They specify requirements for kiln drying, grading, and control of moisture and establish standard sizes. Many manufacturers produce additional sizes for special applications, such as very thick flooring for industrial use.

Wood flooring is produced in four basic types: strips, planks, parquet (thin wood blocks), and solid end-grain blocks. Oak flooring is made from both plain-sawed and quarter-sawed wood. Maple, beech, and birch flooring are plain sawed because these woods are hard, dense, and close-grained. Southern pine flooring is graded under the rules of the Southern Pine Inspection Bureau. It is available in both flat grain and edge grain. Douglas fir flooring is available in both flat grain and edge grain.

Strip Flooring

Strip flooring is available in soft and hard woods. It is manufactured with square edges, side matched (tongue and

Table 34.1 Grades of Unfinished Hardwood and Softwood Flooring

Unfinished Oak[a]	Prefinished Oak[a]
Clear Plain or Clear Quartered (best appearance)	Prime (excellent)
Select Plain or Select Quartered (excellent appearance)	Standard and Better (mix of Standard and Prime)
Select and Better (mix of Clear and Select)	Standard (variegated)
No. 1 Common (variegated)	Tavern and Better (mix of Prime, Standard, and Tavern)
No. 2 Common (rustic)	Tavern (rustic)
Beech, Birch, and Hard Maple[a]	**Pecan[a]**
First Grade White Hard Maple (face all bright sapwood)	First Grade Red (face all heartwood)
First Grade Red Beech (face all red heartwood)	First Grade White (face all bright sapwood)
First Grade (best)	First Grade (excellent)
Second and Better (excellent)	Second Grade Red (face all heartwood)
Second (variegated)	Second Grade (variegated)
Third and Better (mix of First, Second, and Third)	Third Grade (rustic)
Maple[b]	**Softwood[c]**
First (highest quality)	B and Btr (best quality)
Second and Better (mix of First and Second)	C (good quality, some defects)
Second (good quality, some imperfections)	C and Btr (mix of B & Btr and C)
Third and Better (mix of First, Second, and Third)	D (good economy flooring)
Third (good economy flooring)	No. 2 (defects but serviceable)

[a]*Hardwood flooring grades of the National Oak Flooring Manufacturers Association.*
[b]*Maple flooring grades of the Maple Flooring Manufacturers Association.*
[c]*Softwood flooring grades of the Southern Pine Inspection Bureau.*

groove), and side and end matched (tongue and groove on sides and ends). Standard installation details are shown in **Figure 34.11**. It usually has a small hollow relief area cut on the bottom and is available in solid wood and laminated

Figure 34.11 Installation details for solid wood flooring.

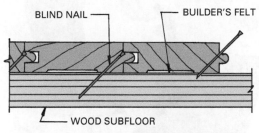

BLIND NAIL — — BUILDER'S FELT

— WOOD SUBFLOOR

BLIND NAIL TO WOOD SUBFLOOR OVER JOISTS

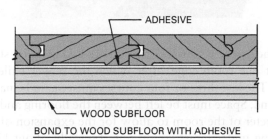

ADHESIVE

WOOD SUBFLOOR

BOND TO WOOD SUBFLOOR WITH ADHESIVE

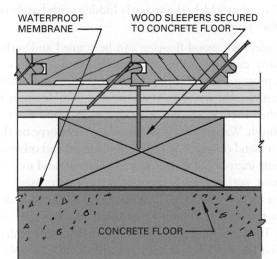

WATERPROOF MEMBRANE — — WOOD SLEEPERS SECURED TO CONCRETE FLOOR

CONCRETE FLOOR

WOOD SUBFLOOR NAILED TO WOOD SLEEPERS SECURED TO CONCRETE FLOOR. IN SOME CASES THE SUBFLOOR CAN BE OMITTED.

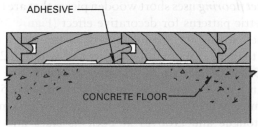

ADHESIVE

CONCRETE FLOOR

IN SOME SITUATIONS A WOOD FLOOR CAN BE BONDED TO A CONCRETE FLOOR WITH AN ADHESIVE.

wood strips. Strip flooring is supplied either unfinished or prefinished. Unfinished flooring usually has square, tight-fitting edge joints, which produces a smooth floor. Prefinished strip flooring usually has a V-joint on the edges and ends of each strip. A wide range of woods and finishes are available. Sizes of strip flooring are listed in **Table 34.2**.

Table 34.2 Stock Sizes for Wood-Strip Flooring

Southern Pine Flooring[a]				
	Actual Thickness		**Actual Width**[b]	
Species	**in.**	**mm**	**in.**	**mm**
Longleaf pine, slash pine, shortleaf pine, loblolly pine	5/16	8	1 3/8	35
	7/16	11	2 3/8	60
	9/16	14	3 3/8	86
	3/4	76	4 3/8	101
	1	101	5 3/8	136
	1 1/4	107		

[a]Source: Southern Pine Inspection Bureau
[b]Widths available in all thicknesses.

Hardwood Strip Tongue-and-Groove Flooring[c]				
	Actual Thickness		**Actual Width**	
Species	**in.**	**mm**	**in.**	**mm**
Oak, beech, birch, hard maple, hickory	3/4	19	1 1/2, 2	38, 51, 57, 82.5
	11/32	8.7	2 1/4, 3 1/4	38, 51
	15/32	12	1 1/2, 2	38, 51
	33/32	26	2, 2 1/4, 3 1/4	51, 57, 82.5

[c]Source: National Oak Flooring Manufacturers Association

Maple Strip Tongue-and-Groove Flooring			
Actual Thickness		**Actual Width**	
in.	**mm**	**in.**	**mm**
25/32, 33/32	20, 26	1 1/2, 2 1/4, 3 1/4	38, 57, 82
41/32	32.5	2 1/4, 3 1/4	57, 82

Planks

Planks may be solid wood or laminated. Solid wood planks are usually 3/4 in. (18 mm) thick and from 3 1/2 in. to 8 in. (87 mm to 204 mm) wide. Laminated planks are usually thinner, ranging from 7/16 in. to 3/8 in. thick. They are used to produce a floor representative of the wide flooring used in the early days of this country. Some are secured with screws covered with wood plugs.

Wood Strip and Plank Floor Installation

Strip and plank flooring have similar installation details. The flooring is typically installed with its length parallel to the longest room dimension (Figure 34.12). Wood flooring can be laid directly on solid wood or plywood sub-flooring. On concrete subfloors, a moisture barrier or treated plywood is used as a substrate. A typical installation uses sleepers to lift the flooring off the subfloor. 2 × 4 lumber is laid at 12 in. (304.8 mm) on center and is covered with a vapor barrier prior to the installation of the flooring. Tongue-and-groove flooring is attached with nails through the tongue of each board, referred to as *blind nailing*. Because the nail is driven through the groove, it is not visible on the face. Specially designed nail guns are specifically designed for wood flooring. They use a spring-loaded button that is hit with a special mallet. The spring drives a nail at the correct angle to both secure the board to the floor and fix it tightly against the adjacent board (Figure 34. 13). Pneumatic-power nailers assure a perfect installation and are less labor intensive than nailing flooring boards by hand.

Figure 34.13 Pneumatic power-nailers assure a perfect installation and are less labor intensive than nailing flooring boards by hand.

© Christina Richards/Shutterstock.com

The first few courses of wood are laid by hand starting at one room edge, after which the power nailer is used. The last three or four courses must again be nailed by hand. Space must be left between the flooring and the perimeter of the room to allow for the expansion of the flooring under moist conditions. Typically, about ½ in. (12.7 mm) in width, the space is hidden with baseboards or other trim components.

Unfinished wood flooring can be stained and surfaced in many different colors and types of finishes. Floor finishes protect wood flooring from wear, dirt, and moisture while giving the wood an attractive and durable sheen. Oil-modified urethane is the most common wood floor surface finish. Water-based urethane is a water-borne urethane with a blend of synthetic resins, plasticizers, and other film-forming ingredients that produces a durable and moisture-resistant surface. While generally more expensive, they have a milder odor than do oil-modified finishes and dry in about two to three hours. Penetrating oil sealers are made from Tung or linseed oil, with additives to improve drying and hardness. While still in use, varnish, lacquer, and shellac have been largely superseded by the products listed above.

Parquet Flooring

Parquet flooring uses short wooden pieces that are laid in geometric patterns for decorative effect (Figure 34.14). Parquet strips are manufactured in standard short strips that are exact multiples of its face width. A 2¼ in. (57.15 mm) wide strip for example is available in 9 in., 11¼ in., 13½ in., and 15¾ in. (228.6 mm, 285.75 mm, 343 mm, and 400 mm) lengths. Each strip is milled with tongue and grooves at both its sides and ends.

Figure 34.12 Strip-and-plank flooring are typically installed with their length parallel to the longest room dimension.

© cappi thompson/Shutterstock.com

Figure 34.14 A parquet-strip floor being laid in a herring-bone pattern.

© carmakoma/Shutterstock.com

Common installation patterns include herringbone and basket-weave patterns. Parquet blocks are pre-manufactured flooring units consisting of square or rectangular units. Wood that contrasts in color and grain, such as oak, walnut, and cherry, are sometimes used to provide color differences. In high-end applications, richly patterned mahogany and other tropical hardwoods are also used. Three types of blocks are available: unit, laminated, and slat block. Blocks are available in thicknesses of $5/16$ in., $3/8$ in., and $3/4$ in. (7.93 mm, 9.52 mm, and 19.05 mm) with 9 in. × 9 in. and 12 in. × 12 in. (228.6 mm × 228.6 mm and 304 mm × 304 mm) sizes.

Solid-block flooring for use in areas where heavy traffic occurs, as in a factory, and where easy maintenance is required, is cut so the face exposes the end grain. Canadian hardwood strip flooring is manufactured in the customary sizes compatible with those used in the United States and in metric sizes. Strip and parquet flooring are measured by surface area in square meters (m²). The two Canadian industry standards for strip and hardwood flooring are published by the Canadian Lumbermen's Association.

Engineered and Laminate Wood Flooring

Engineered and laminate flooring are manufactured products that use a multi-ply construction to create a stable product that contracts less than many other flooring materials. Designed to look like wood flooring, laminate is economical; does not need to be nailed, sanded, stained or finished; and is resistant to scratches, fading, and stains. The factory-applied finishes are moisture resistant and look like natural wood without a finish. Plastic finishes such as urethane and acrylic are applied to the wood fibers to make them hard and durable.

Engineered flooring is typically a laminate of three layers of solid wood with tongue-and-groove edges and ends. The pre-finished top veneer is visible and available in an endless variety of wood species. *Laminate flooring* is a rigid floor covering that has a top layer consisting of one or more thin sheets of fibrous material, often a paper product impregnated with amino-plastic thermosetting resins, such as a melamine. The increasing popularity of this flooring is in its precise milling. The exactness of the manufacturing process allows pieces to be joined without any unevenness of the finished surface.

There are two types of laminate flooring: direct pressure and high pressure. Direct-pressure laminates fuse the top wear layer to a core material using pressure between 400 and 500 pounds per square inch. High-pressure laminates fuse the multiple layers at around 1,000 pounds per square inch. The laminate and backer layer are bonded to a water-resistant, high-density fiberboard core with urea-based adhesive.

Most laminated flooring is installed over a thin foam underlayment that is applied over the subfloor. Instead of being glued or fastened to the underlayment, the boards are either loose laid or glued to each other along the ends and edges **(Figure 34.15)**. A first course is laid

Figure 34.15 Engineered flooring is a laminate material with tongue-and-groove edges and ends.

© Nagy-Bagoly Arpad/Shutterstock.com

with the groove end facing the wall. Glue is applied to each subsequent course, and the pieces are tightened to each other with a hammer and tapping block.

Bamboo Flooring

While not technically a wood, bamboo is also a popular material for modern floors (Figure 34.16). Bamboo is one of the fastest-growing, inexpensive, and most sustainable flooring materials available. Bamboo is an attractive alternative for flooring because its physical characteristics are similar to those of hardwoods. Bamboo is a rapidly renewable material that is harvested about every five to ten years without the need to replant. Bamboo flooring is inexpensive and provides a unique aesthetic. In addition to being moisture and insect resistant, bamboo is also much harder than typical hardwood flooring, including oak. Pre-manufactured bamboo floors are typically available in planks with either vertical or horizontal grain orientation. The pieces are milled with tongue and groove ends and side and are laid similar to other engineered-wood flooring products.

Figure 34.16 Bamboo is one of the fastest growing, inexpensive, and most sustainable flooring materials available.

© Mr. Green/Shutterstock.com

RESILIENT FLOORING

Resilient floor coverings are made of materials that have more elasticity and pliability than do hard-surface floor coverings. The broad category encompasses many different manufactured products, including linoleum, sheet vinyl, vinyl composition tile, cork, and rubber.

Resilient flooring is defined as any non-textile organic floor surfacing material made in sheet or tile form or formed in place as a seamless material. It includes but

is not limited to surfaces made of asphalt, cork, linoleum, rubber, vinyl, vinyl composition, and polymeric-poured floors. These materials are not only extremely sturdy and stain resistant; they also possess a natural resilience or springiness that is much more comfortable to stand on. Resilient flooring is used in a variety of applications where properties like durability, traction, comfort, and water resistance are primary considerations. Depending on the project type, this can include retail areas, corridors, patient rooms, laboratories, and work and storage areas.

Linoleum and Sheet Vinyl

Linoleum is a natural material made of linseed oil, wood products, ground cork, mineral fillers, and pigments that are bonded to a jute or synthetic backing. The product is available in backed or un-backed sheets as well as tile form. Thicknesses range from .08 in. to .125 in. (2 mm to 3.17 mm), with sheet widths of around 6 ft. 5 in. (1,841.5 mm), and lengths up to 90 ft. (27.4 m). Although linoleum was once a printed product, manufacturers today produce homogeneous, marbleized linoleum. Most products have a factory-applied low-maintenance surface finish that can be maintained during the life of the product using sealers and polishes. Linoleum is suitable for dry areas subject to occasional wetting but is not recommended in wet applications where enhanced slip resistance is required. Vinyl sheet with the appropriate wet-slip-resistance classification should be used in these applications. Linoleum floors are also more costly and sensitive to chemicals and abrasives than vinyl floors.

Since its development, polyvinylchloride (PVC) or sheet vinyl has been one of the principal flooring materials used in commercial facilities. A durable, resilient, and impervious sheet material, it provides an economical, low-maintenance surface. PVC is a polymeric thermoplastic material with vinyl chloride polymers and copolymers acting as a binder to which plasticizers, stabilizers, fillers, and pigments added to form a stable compound. Sheet vinyl falls into two categories. Homogeneous materials use a single-ply sheet with the wear layer extending throughout the full sheet thickness. Heterogeneous materials use a multiple-ply sheet consisting of separate layers of different materials that are laminated together. The sheet usually consists of a wear layer bonded to a backing layer. There can be considerable variation in the composition and thickness of layers between products.

Both linoleum and sheet vinyl sheets can be cut to the shape of the floor using a straightedge and a flooring knife (Figure 34.17). The material is attached to

Figure 34.17 Sheet vinyl is a durable, resilient, and impervious sheet material that provides an economical, low-maintenance surface.

© Ernest R. Prim/Shutterstock.com

the substrate with adhesives, such as floor epoxies. The sub-base must have a smooth, flat surface free from imperfections that may show through the floor covering. Raised surfaces will wear rapidly and could damage the installation and constitute a safety hazard. Joints in sheet vinyl installations are formed by heat welding using a vinyl-welding rod, cold welding, or the use of sealants.

Vinyl Composition Tile (VCT)

Vinyl composition tile (VCT) is a type of resilient flooring that is used in a variety of commercial applications. In comparison with sheet vinyl, VCT is a harder, less-flexible material with a higher proportion of fillers. It is widely used in retail and other commercial installations because of its durability, ease of maintenance, and low installed and life cycle cost. It is also more brittle and subject to impact damage. Sub-bases should be sealed and water-resistant adhesives used where regular water damage is expected.

The word "vinyl" describes a large group of plastics developed from the ethylene molecule. Vinyl composition tiles (VCT) are made by compressing and heating vinyl resins, plasticizers, stabilizers, fibers, and pigments into a uniform thickness (**Figure 34.18**). Standard tiles are available in 9 in., 12 in., 18 in., and 36 in. (22.8 mm, 30 mm, 45.7 mm, and 91.4 mm) squares and some rectangular shapes. Standard thicknesses are ⅛ in. and ³⁄₃₂ in. (2.4 mm and 3.2 mm), although thicker tiles are available for commercial applications subject to heavy wear. The modular tiles are available in a multitude of colors, offering endless design flexibility and tile-patterning possibilities.

Figure 34.18 Vinyl composition tile is widely used in retail and other commercial installations because of its durability, ease of maintenance, and low installed and life-cycle cost.

© Claudio Divizia/Shutterstock.com

The installation of VCT is similar to many other tile materials. For square patterns, two guidelines are established: parallel to the walls and at right angles to each other. The lines are located such that the tiles are either centered in the space, or leave an equal tile cut on either side. Along these lines, a first course of tiles is set in adhesive applied with a notched trowel. Once the tiles have set, subsequent tiles can be laid in a stair-step fashion, simplifying the installation (**Figure 34.19**). Vinyl composition tile is cut with a utility knife. The tile is scored at the desired location and then snapped along the score line.

Figure 34.19 The installation of VCT is similar to other tile materials.

Courtesy of constructionphotographs.com

While many vinyl composition tiles are made with a factory finish, most require the application of sealers and polishes. A coat of sealer under the polish is recommended for areas of high traffic, or where staining potential is high.

Rubber Flooring

Rubber has recently re-emerged as a viable alternative to PVC. Institutional and commercial facilities frequently use rubber tile, sheet, and stair treads because they are strong, durable, and slip resistant. The surface texture can vary from a smooth or marbleized finish to a chip design. Most types have some form of gridded surface that aids traction and a roughened or recessed back grid. Embossed designs, including circular, square, hammered, or diamond-plate patterns, are also available.

Rubber flooring can be manufactured using natural, synthetic, or recycled rubber, and is available in sheet or tile form (Figure 34.20). Most rubber flooring is made from synthetic rubbers such as SBRs (synthetic butyl rubber), EPDM (ethylene propylene diene), and new synthetic compounds. A number of new flooring products are available that utilize rubber from recycled automobile tires. Products made from natural rubber are usually more expensive. The composition of rubber sheets can include inert fillers such as clay, chalk, and non-volatile plasticizers. The properties of rubber flooring products can vary significantly with manufacturer and product. 12 in. and 36 in. (305 mm and 915 mm) square tiles are common, with sheets ranging from 36 in. to 50 in.

Figure 34.20 Rubber flooring made of natural, synthetic, or recycled rubber can be used where a slip-proof resilient surface is required.

© K. Miri Photography/Shutterstock.com

(915 mm to 1,270 mm) in width, in ⅛ in. and ³⁄₁₆ in. (3 mm and 4.7 mm) thicknesses. Rubber flooring is comfortable for walking, wears well, and resists damage from oils, solvents, alkalis, acids, and other chemicals.

Installation techniques vary with application, and manufacturer's recommendations should always be observed. Typically, rubber flooring is bonded to concrete or wood subfloors with an approved adhesive. On-grade concrete slabs require the use of a moisture-resistant adhesive. Manufacturers also produce rubber flooring products for use as stair treads, entrance mats recessed in the floor, and long runners that are not bonded to the subfloor.

Cork Flooring

Similar to rubber and linoleum, cork is a historic material that is finding renewed popularity. Cork, like wood flooring, provides a warm and natural aesthetic but with the added advantage of resilience (Figure 34.21). According to the Cork Institute of America, the material encapsulates millions of tiny air bubbles that contribute to its elasticity and resilience. Cork flooring is also one of the most sustainable flooring products available. The material is harvested in ten-year cycles from the bark of the cork oak tree. A protective inner layer of bark remains after harvesting, allowing the tree to continue to grow and regenerate new bark.

Figure 34.21 Cork, like wood flooring, provides a warm and natural aesthetic but with the added advantage of resilience.

© Eugene Sergeev/Shutterstock.com

The high resiliency of cork flooring makes it a good choice for occupancies that require prolonged standing. It also provides excellent sound-deafening properties. Cork is available in tile and sheet form. The raw material is boiled, then ground, and compressed using adhesive resins. Cork flooring planks are often laminated on a high-density fiberboard base. Tiles are available with either prefinished or unfinished surfaces. The unfinished form must be sanded and sealed after installation. Common finishes include polyurethane, wax, and oil. Because cork is a porous material, care must be taken not to wet mop the flooring during maintenance operations.

Tile cork flooring is usually adhered directly to the subfloor with adhesive. Cork flooring strips have tongue-and-groove edges similar to hard wood-strip flooring. Most forms are simply snapped together rather than being nailed through the tongue. Space must be left at the floor edges as the material experiences some expansion and contraction with changing humidity levels.

Seamless Flooring Materials

Seamless flooring materials are made from plastics and resins that chemically cure to form a rigid, highly moisture- and abrasion-resistant surface for high-performance interior floor finishes. These are usually applied in areas that require superior water resistance, such as lavatories, medical facilities, or food processing plants. Various types of poured seamless floor coverings are available. Some are formed from plastic, such as acrylics, latex and epoxy resins, polyester, and urethanes. Poured epoxy flooring can be from ⅛ in. to ¼ in. (3.17 mm to 6.35 mm) in thickness, and provides excellent adhesion to the substrate, low odor and VOC levels, and low flammability. Polyester and vinyl-ester resins can be installed over a sloping slab to support the flow of liquids to drains. Sloping is recommended in all wet areas since flooring that is immersed under ponding water tends to deteriorate more quickly.

Seamless flooring materials are mixed according to manufacturers recommendations, poured, and spread with rollers over a concrete or wood subfloor to provide a seamless surface (Figure 34.22). Some products are applied in a single coating and then allowed to cure under specified

Figure 34.22 Seamless flooring materials are poured and spread over a subfloor to provide a completely seamless surface.

conditions. Other systems utilize a base coat, grout coat, and optional topcoat. Floor finishes range from a standard gloss to matte, low-gloss surfaces. For skid-resistant finishes, granular or rubberized particles can be added to the mixed product. Color chips can be also added for aesthetic effect, usually while the topcoat is still wet, immediately after application. A wide range of seamless flooring for athletic use is also available. This includes coatings for showers and saunas; basketball, tennis, and volleyball courts; tracks; and locker rooms. Materials used include elastomers, rubbers, polymers, and urethane products.

CARPETING

Carpet is widely used as floor covering in commercial construction. It competes well with other finish floor products in terms of cost, durability, and maintenance. The material is available in a wide range of fibers, textures, and colors (Figure 34.23). It is manufactured in widths from 6 ft. to 25 ft. (1.8 m to 7.62 m) and greater, resulting in a finished surface that produces few seams and can efficiently cover large floor areas. Although carpeting is often selected for its warmth and comfort, it also provides excellent sound-deadening properties by reducing both impact noise and sound reflection within a space.

Figure 34.23 Carpet competes well with other finish floor products in terms of cost, durability, and maintenance.

There are several elements to consider in selecting the right carpet. Room location and use will determine the required durability. The best-quality carpet is recommended for heavy-traffic areas due to increased toughness. The selected color, texture, and pattern are another important issue. Different color, texture, and pattern will give each room a different feel and look. Finally, the cost of carpet, carpet cushion materials, and their installation must be considered.

Introduction to Carpet Systems

Carpet is made in various qualities suitable for different applications. Building codes specify flammability requirements for carpets in various locations and occupancies. The flame resistance of carpets is most commonly ascertained by the Flooring Radiant Panel Test. A carpet specimen is mounted on the floor of a test chamber and exposed to intense radiant heat from above, and the rate of flame spread is measured.

The use of carpet affects indoor air quality in a variety of ways. All carpet types trap dust, moisture, and pollutants to varying degrees, depending on the depth of the pile, the carpet density, and the type of carpet. *Synthetic fibers* are traditionally made from petroleum and can off-gas. Seam sealants, carpet padding, and carpet treatments often contain volatile organic compounds, formaldehyde, and other pollutants. In response to these issues, the Carpet and Rug Institute (CRI) has developed the "Green Label" indoor air-quality (IAQ) testing and labeling program for carpet and the adhesives used with them. CRI tests each carpet line four times a year for four categories of emissions. Those that contain no hazards to building occupants are certified by the Green Label.

More than 3 million tons of carpeting is produced every year and about 2 million tons are discarded. Carpet accounts for a considerable amount of municipal solid waste. The carpet industry has begun voluntarily to take responsibility for its products once their useful life is over. Discarded carpet is collected by manufacturers and reused in the production of new carpet and other products.

Although the construction of carpets varies, Figure 34.24 illustrates the terms commonly used to describe their makeup. Pile yarns form the top surface that is exposed to wear and abrasion. Pile yarns are made from wool or one or more synthetic fibers. Backing yarns consist of weft yarns (running the width of the carpet) and warp yarns (running the length of the carpet). Stuffer yarns run lengthwise and give strength and stability to the carpet. The warp yarns pass over the weft yarns, pulling the pile yarns into place.

Carpet Fibers

Wool, a *natural fiber*, has been a major component in carpet construction for many years. The development of synthetic fibers, however, has enabled carpet manufacturers to produce products with a wider range of characteristics. Carpet piles are often a blend of natural and synthetic fibers. For example, a wool carpet may have a blend of 10 to 30 percent nylon fibers to increase its toughness. Acrylic, modacrylic, nylon, polyester, and polypropylene (olefin) are commonly used synthetic fibers.

Wool Wool used in carpets is a natural fiber imported into the United States from wool-producing countries, such as Australia, New Zealand, and Scotland. Different breeds of sheep make for a widely varied product. Wool fibers range in size from 3½ in. to 7 in. (89 mm to 178 mm) long. The wool pile is made by spinning the fibers into yarn. Wool has good resistance to abrasion and aging, resists damage from mildew and sunlight, and is relatively unaffected by weak alkalis and organic solvents. Generally, wool is more expensive than synthetic fibers.

Acrylic Synthetic modacrylic fibers are of the acrylic group, containing at least 35 percent (but less than 85 percent) acrylonitrile. They are soft, resilient, quick drying, abrasion and flame resistant, and resist damage from acids, alkalis, sunlight, and mildew.

Nylon Nylon is a synthetic petrochemical product produced as a continuous filament but often cut into short lengths, varying from 1½ in. to 6 in. (38 mm to 152 mm). The short fibers are spun into yarn much the same as wool fibers. Nylon is a strong fiber that has anti-stain properties and is easily dyed. It is resilient, low in moisture absorbency, and resistant to mildew, aging, and abrasion.

Polyester Polyester fibers are synthetic fibers that have high tensile strength, good resistance to mineral acids, and excellent mildew, aging, and abrasion resistance. Prolonged exposure to sunlight may cause some loss of strength. They are not as durable as nylon fibers.

Polypropylene (olefin) Polypropylene is a synthetic fiber that is a class of olefin consisting of 85 percent

Figure 34.24 Terms used to identify carpet components.

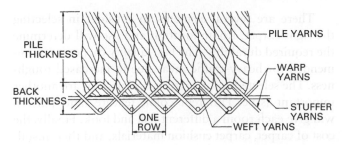

propylene by weight. It has the lowest moisture-absorption rate of all carpet fibers in use. It resists mildew, abrasion, aging, sunlight, and common solvents. While not as resilient as nylon, it is less expensive. Olefin is very lightweight, strong, and soil resistant.

CARPET CONSTRUCTION

Woven Carpets

Woven carpets are produced on a loom, similar to woven fabrics. Woven carpets include Axminster, loomed, velvet, and Wilton construction.

Axminster carpet is woven on a loom that inserts each tuft of pile yarn individually, allowing the color of each tuft to vary. This provides great flexibility for design patterns. Axminster is usually a cut pile, even in height. The backing is heavily ridged and sized, making it so stiff that it must be rolled lengthwise (Figure 34.25).

Figure 34.25 Construction details of Axminster carpet.

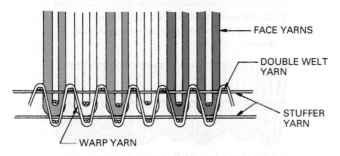

Loomed carpet has the pile yarn bonded to a 3⅙ in. (5 mm) thick rubber cushion in a low-loop single-level pile. The pile back has a waterproof coating to which a rubber cushion is bonded. In a *velvet weave carpet*, the pile, stuffer, and weft yarns are held together with double-warp yarns (Figure 34.26). The setting of the

Figure 34.26 Velvet weave construction has double-warp yarns.

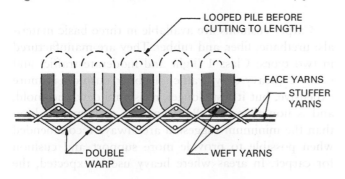

loom establishes the height of the piles, which begin as loops before being cut by a knife-edged wire to a desired height. A textured surface can be produced by setting wires to form and cut loops of different heights. Patterned designs cannot be produced by this loom, but tweed effects can be created by varying yarn colors.

Wilton carpet is produced on a loom similar to the one used for velvet weaves, but a Wilton loom has a mechanism that feeds yarns of various colors. Different color yarns are selected by a series of punched cards. The cards regulate which color yarn is looped over the loop-forming wires. These become the visible pile tufts, and the other colored yarns are buried in the body of the carpet, producing a stiff-backed, multicolored product (Figure 34.27). Wilton looms can produce sculptured and embossed textures by varying the pile height and effecting a combination of cut and uncut loops. Uncut loop Wilton weaves are also available.

Figure 34.27 Wilton construction is used to produce carpets containing multicolored yarns.

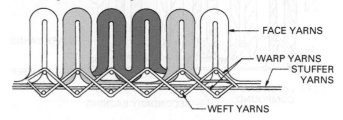

Tufted, Knitted, Flocked, and Bonded Carpet

Tufted carpet construction involves stitching pile yarn through a backing material, much like sewing on a power sewing machine. The back is coated with a latex-bonding agent that holds the tufts to the base and bonds a second backing over it. The first backing is usually made of cotton, polypropylene olefin, or jute. The second backing is usually a coarse jute fabric (Figure 34.28). Tufted carpets are available in level loop, multilevel, cut, uncut, and a mixture of cut and uncut piles. Carved and textured surfaces can also be produced.

Knitted carpet is made by looping together the pile yarn, stitching, and backing in a single operation. A coat of latex spread on the back bonds and stiffens the carpet (Figure 34.29). Knitted carpets are available in solid colors and tweeds and in single or multilevel uncut pile.

Figure 34.28 Tufted carpet is made by stitching pile yarns through a backing material and bonding two layers of backing over it.

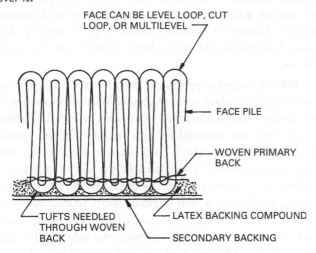

FACE CAN BE LEVEL LOOP, CUT LOOP, OR MULTILEVEL

FACE PILE

WOVEN PRIMARY BACK

TUFTS NEEDLED THROUGH WOVEN BACK

LATEX BACKING COMPOUND

SECONDARY BACKING

Figure 34.29 Knitted carpet construction involves looping the pile yarns, stitching, and backing in a single operation.

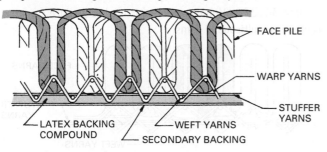

FACE PILE

WARP YARNS

STUFFER YARNS

LATEX BACKING COMPOUND

WEFT YARNS

SECONDARY BACKING

Flocked carpet construction involves electrostatically spraying short strands of pile yarn onto an adhesive-coated backing sheet. The pile strands become vertically imbedded in the adhesive, a second backing is laminated to the first, and the unit is cured.

Fusion-bonded carpet is made by bonding pile yarn between two parallel sheets of backing. These are coated with vinyl adhesive, making a sandwich product with the pile yarn in the center. After the adhesive hardens, the pile yarn is cut in the middle to form two pieces of cut-pile carpet. A secondary backing is often bonded to the first backing sheet (Figure 34.30).

Carpet Cushions

Carpet is usually installed over a *cushion* because this increases its life, absorbs considerable traffic noise, has insulation value, and provides a soft, resilient floor covering. There are two common types of carpet cushion installations: carpet can be installed over a separate

Figure 34.30 Fusion-bonded carpet construction involves bonding pile yarn to backing sheets coated with a vinyl adhesive.

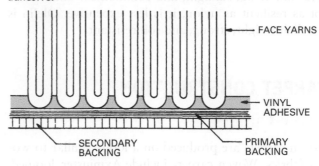

FACE YARNS

VINYL ADHESIVE

SECONDARY BACKING

PRIMARY BACKING

cushion, or carpet can be made with a foam cushion attached to its back (Figure 34.31). Carpet can also be laid without a cushion.

Figure 34.31 Types of carpet cushions.

1. Carpet cemented directly to the floor without any cushion is called direct glue down.

2. Carpet can be installed over separate cushion by stretching the carpet in over tack strip or by gluing the cushion to the floor and the carpet to the cushion.

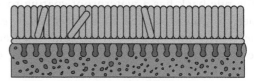

3. Carpet with cushion attached.

Carpet cushions are available in three basic materials: urethane, fiber, and rubber. They are manufactured in two types: Class 1 (light and moderate traffic) and Class 2 (heavy-duty traffic). Rubber cushion is more expensive, but it lasts longer, resists mildew and mold, and is non-allergenic. Better grades of carpet cushion than the minimum suggested are always recommended when possible to provide more support and cushion for carpet. In areas where heavy use is expected, the

Carpet Cushion Council suggests using firmer grades of cushion. These areas include stairways, halls, and areas where heavy furnishings are used.

Recent development of sustainable carpet cushion materials, such as soybean plant backing, will enhance indoor air quality when used with certified carpet.

Carpet Installation

Carpet is installed over concrete, plywood, particleboard, and other firm, smooth substrates in several ways. The most frequently used method uses tack strips. These are thin metal or wood strips with tack points sticking through them. They are nailed to a particleboard, plywood, or concrete subfloor (Figure 34.32). The carpet is rolled out over the cushion, and any seams are secured.

Figure 34.32 Wood tack strips at the room perimeter are used to hold carpet in place.

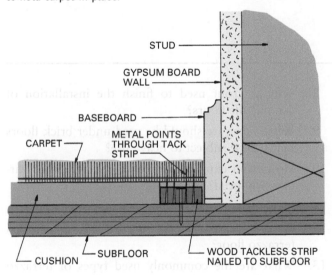

Seams between sections of carpet are bonded with a hot-melt tape applied beneath the seam. An electric seaming iron is placed below the carpet on one edge of the tape to melt the adhesive. This edge of the carpet is pressed into the hot adhesive. Then the other edge of the tape is heated and the butting carpet pressed against the tape. The seam is then pressed with a roller, closing the gap and firming the bond.

After the seam adhesive hardens, the carpet is stretched against one wall. A knee kicker or stretcher tool is used to stretch the carpet into place, after which it is pressed on to the tack strip (Figure 34.33). Then the carpet is stretched toward the other side of the

Figure 34.33 A stretcher tool stretches the carpet to the opposite wall, where it is pinned into the tack strip.

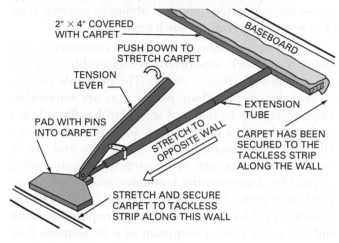

room. At the wall edge, the carpet is pressed into the space between the baseboard and the tack strip with a stair tool.

Carpet can be also be glued directly to a wood or concrete floor without a cushion. If placed in a frequently used area, uncushioned carpet will wear out faster. When placing carpet over concrete, the base must be dry and not sweat with changes in temperature. The application of a coat of sealer applied to the concrete is recommended. A plywood subfloor can be installed directly over concrete or on wood sleepers. Sleepers provide an air space that can be ventilated along the walls.

Carpet tiles, also known as carpet squares, are an alternative to the standard rolls of carpet. They are typically used in commercial applications, such as in schools and airports. The tiles do not require any padding or glue to install them. Individual carpet tiles can be removed for cleaning, and replaced when dry. If one area of the floor becomes damaged, the tile can be replaced without replacing the entire floor.

RAISED-ACCESS FLOORING

Raised-access flooring provides an elevated floor surface above the structural floor system to provide space for the passage of mechanical and electrical services. They are widely used in data technology centers and computer rooms where the large amount of equipment requires the routing of electrical supply and communications wiring. They also find use in commercial office spaces where broad areas have minimal partition walls and are subject to frequent changes in furniture or cubicle layout. Raised-access floors provide unlimited ability in meeting complete rewiring operations when tenants change.

Raised-access floors are typically constructed of a metal framework supported by adjustable supports called pedestals (Figure 34.34). The pedestals provide support for surface floor panels that are removable. The panels may be made of steel-clad hardboard or a steel panel with a cementitious solid or hollow core. The panels are usually 2 ft. × 2 ft. in size and can be covered with a variety of flooring finishes, including carpet tiles, high-pressure laminates, marble, or stone. The height of the floor is designed to accommodate the volume of cables and other services required beneath and can range from a minimum of 6 in. to more than 4 ft. Electrical and communications outlets can be installed anywhere within a panel system. Special handles allow workers to access the system when repair is necessary.

Figure 34.34 Raised-access flooring provides an elevated surface above the structural floor system to provide space for the passage of mechanical and electrical services.

© Mika/Comet/Corbis

Review Questions

1. What types of finish flooring are in common use?

2. What type of finish flooring presents the greatest hazard in a fire?

3. What techniques are used to secure wood flooring to a substrate?

4. How can a designer find the fire-spread and smoke-production data for various finish flooring products?

5. What type of floor tile is used for areas subject to heavy use?

6. How is the surface of rubber flooring designed to provide slip resistance?

7. What types of mortars and adhesives are used to install tile floors?

8. Why is grout used to finish the installation of ceramic tile floors?

9. What substrate should be used under brick floors that will be subject to heavy use?

10. How is an exposed aggregate concrete floor surface produced?

11. How are concrete floors given a colored surface?

12. What substrates can be used when constructing a terrazzo floor?

13. What are the commonly used types of terrazzo flooring?

14. Where are precast terrazzo products used?

15. What materials are used to form seamless floors?

Key Terms

Axminster Carpet	Engineered Flooring	Grout
Blind Nailing	Epoxy Mortar	Knitted Carpet
Carpet Tiles	Flocked Carpet	Laminate Flooring
Cushion	Furan Mortar	Latex Portland Cement Mortar
Dry-Set Mortar	Fusion-Bonded Carpet	Linoleum

Loomed Carpet

Natural Fiber

Parquet Flooring

Raised-Access Flooring

Resilient Flooring

Synthetic Fibers

Terrazzo

Tufted Carpet

Velvet Weave Carpet

Vinyl Composition Tiles (VCT)

Wilton Carpet

Activities

1. Visit floor covering dealers to obtain samples of as many different products as possible. Arrange the samples in a display and label each, citing the brand, material makeup, and places where its use would be appropriate.

2. Visit buildings under construction when floor finishing crews are at work. Observe the tools and procedures used.

3. Have the managers of various flooring supply dealers visit class and explain how they help a client select a covering material.

4. Collect carpet samples from manufacturers and local carpet retail outlets. Arrange a display and label the samples, detailing the fibers, methods of construction, and suitable applications.

5. Review the local building code pertaining to carpeting and prepare a brief summary.

Additional Resources

Guide for Flooring Covering Installers, Mannington, Mills, Inc., P.O. Box 30, Salem, NJ 08079.

Installing Hardwood Flooring and other publications, National Oak Flooring Manufacturers Association.

Sheet Flooring Installation Guide, Armstrong Floor Products, P.O. Box 3001, Lancaster, PA 17604.

Spence, W. P., *Finish Carpentry*, Sterling Publishing Co., 387 Park Avenue South, New York 10016.

Spence, W. P., *Installing and Finishing Flooring*, Sterling Publishing Co., 387 Park Avenue South, New York 10016.

Standard for Installation of Commercial Carpet, and many other publications, the Carpet and Rug Institute, Dalton, GA. Carpet and Rug Institute (www.carpet-rug.com)

Other resources include:

Publications of the Carpet Cushion Council, Riverside, CT 06878.

See Appendix C for addresses of professional and trade organizations and other sources of technical information.

Specialties
CSI MasterFormat™

Equipment
CSI MasterFormat™

Furnishings
CSI MasterFormat™

© Roger Bruce/Shutterstock.com

Specialties, Equipment, and Furnishings

Upon completion of this chapter, the student should be able to:

- Identify specialty products as classified by the Construction Specifications Institute.
- Make informed decisions when deciding on and specifying specialty items.

- Understand the extensive array of residential and commercial equipment a designer and contractor must be able to specify and install.
- Understand both the range of furnishings used in residential and commercial buildings and the places that various art forms are used.

Commercial and industrial buildings have widely varying requirements when it comes to items needed to prepare a facility for use by occupants. Products listed as specialties, equipment, and furnishings encompass a vast spectrum of components used to provide a limited special function.

The architect provides the location and layout of all specialty items according to the functions and uses required for a project. Submittals for specialties, equipment, and furnishings are required as outlined in the contract documents. The contract outlines the contractor's responsibilities relative to procurement, receiving, storage, and installation of all items in these divisions. The contractor must plan and coordinate all blocking, support, and services for installation of the products in these divisions. This chapter outlines some of the materials, products, and equipment found in Divisions 10, 11, and 12.

Visual Display Boards

Chalkboards are available with the exposed surface made from slate, porcelain enamel on steel, aluminum, or glass (Figure 35.1). The glass units use tempered float or plate glass coated with a colored, vitreous glaze that contains a fine abrasive. Chalkboards may be permanently mounted on a wall or they can be portable. Some have panels that slide horizontally or vertically. They are available in sizes up to 6 ft. × 12 ft. (1.8 m × 3.6 m). Natural slate chalkboards are typically from ¼ in. to

Figure 35.1 A chalkboard is one of the more common visual display devices.

© Ronny Lambotte/Shutterstock.com

⅜ in. (6 mm to 10 mm) thick, in widths up to 4 ft. (1.2 m), and in random lengths up to 6 ft. (1.8 m). Other chalkboards are available in widths up to 4 ft. (1.2 m) and lengths up to 10 ft. (3.05 m).

Whiteboards may have a vinyl sheet surface or one of porcelain enamel on steel. The substrate can be particleboard, fiberboard, sheetrock, hardboard, plywood, or a plastic honeycomb. Special ink markers are used to write on a board's surface. Some are a form of watercolor and can be wiped off the marker board with a wet cloth. Another type has semi-permanent ink and requires a special liquid cleaner.

Pegboard panels are usually ¼ in. (6 mm) thick hardboard with holes spaced 1 in. (25 mm) on center. They are typically manufactured with an aluminum frame and special wire hanging clips.

Bulletin boards, or *tackboards*, typically have a cork surface laminated to a fiberboard substrate. Some have a woven fabric facing and can hold paper with pins or hook-and-loop fasteners that attach to the fabric. Various types of manual and motorized projection screens are available. They may be integrated into a larger unit with chalkboards, tackboards, and markerboards.

Compartments and Cubicles

These products include enclosures, such as metal, plastic laminate, and stone toilet compartments; shower and dressing *compartments*; and various *cubicles*. Also included are dividers, screens, and curtains for showers, hospital cubicles, toilets, and urinals. Toilet enclosures may be supported by a floor, ceiling, and wall (Figure 35.2). Toilet enclosures hung from a ceiling make floor cleaning easier. Ceiling-hung units require a structural steel ceiling support.

Figure 35.2 Toilet enclosures may be supported by the ceiling, wall, and floor.

© Konstantin L/Shutterstock.com

Metal toilet compartments may be stainless steel or steel with a baked-enamel finish. Plastic laminate compartments have a ⅛ in. (1.6 mm) plastic laminate sheet adhered to particleboard or high-pressure plastic laminate core. Stone compartments are typically made of marble.

Cubicle enclosures are hung from an overhead track whose track path can be varied to suit the conditions. It can be mounted directly to a ceiling or suspended.

Louvers, Grilles, Vents, and Screens

A variety of *grilles*, screens, louvers, and vents are considered specialty items. These include louvers and other devices for ventilation that are not an integral part of a building's mechanical system, as well as operable, stationary, and motorized metal wall louvers; louvered equipment enclosures; door louvers; and various types of wall and soffit vents. For example, a curtain wall louver is a specialty product. Interior and exterior grilles and screens of any material are used for air distribution, sun screening, and other functions in addition to ventilation purposes (Figure 35.3).

Figure 35.3 These wooden louvers are used to provide sun shading .

© Arunkumar. T.D./Shutterstock.com

Grilles are open gratings or barriers used to cover, conceal, protect, or decorate an opening. They can be steel, aluminum, cast iron, or wood (Figure 35.4). Various types of grating are used to cover trench frames that

Figure 35.4 These cast-metal grilles provide security and are a major architectural feature.

Figure 35.6 A cast-iron tree guard protects the tree roots from surface damage yet permits rain to reach them.

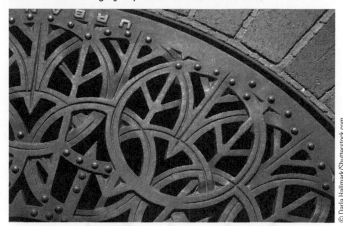

control surface water flow. These are typically cast iron (Figure 35.5). Street and parking lot construction must be designed to carry away surface water from rain and melting snow. Curb inlets are used as intakes for this purpose. Tree grates are another form of grating used both inside and outside buildings. They provide protection while permitting a tree to receive water (Figure 35.6).

Figure 35.5 This cast-iron trench drain is used to control surface water.

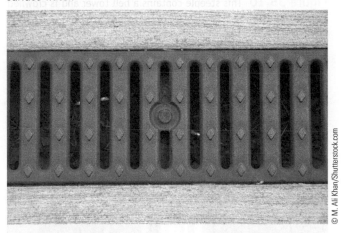

Figure 35.7 Drinking fountains are a part of the service wall system.

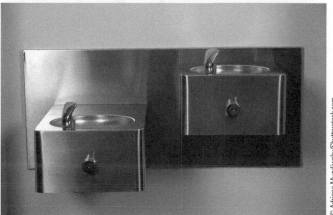

Service Wall Systems

Service wall systems incorporate services such as clocks, fire hoses, drinking fountains, fire extinguisher cabinets, telephones, and waste receptacles (Figure 35.7).

Walls and Corner Guards

Wall and corner guards are protective devices made from metal, plastic, rubber, and other materials that can withstand impacts and, for exterior use, exposure to the weather. Figure 35.8 shows metal corner guards used to protect the corners of columns in a warehouse. Rubber bumpers are secured to walls, at a truck loading dock, for example, to protect them and trucks from damage. Other types of guards, typically constructed of molded rubber or plastic, are used to protect wall corners that might be damaged by moving carts, forklifts, or other passing traffic.

Access Flooring

Raised access flooring is freestanding flooring made in modular units and raised above a basic floor system to provide space to run mechanical and electrical services. The systems commercially available have some type of adjustable metal pedestals on which the modular floor panels rest. The panels typically are covered with vinyl composition floor covering, although carpet can be used.

Figure 35.8 Metal corner guards are used to protect building elements in high traffic areas.

© LovePHY/Shutterstock.com

Figure 35.9 A manufactured direct-vent gas fireplace is quickly installed and has a high heat output.

© Nicola Gavin/Shutterstock.com

Pest Control

Various mechanical, chemical, and electrical pest-repellent systems and protective devices are available. These include devices that control rodents, birds, and insects.

Fireplaces and Stoves

Manufactured fireplaces and stoves, including dampers, metal chimneys, and fireplace screens and doors, are specialties. Manufactured fireplaces typically are metal-framed firebox and vent units designed to accommodate gas-fired logs. Some may be vented vertically through a metal pipe or horizontally through a wall, while others do not require venting. They are delivered completely assembled. A wide range of gas and wood-burning stoves are available. Some are used not only as a heat source but also as part of a room's decor (Figure 35.9).

Manufactured Exterior Specialties

These items include stock and custom designed steeples, spires, cupolas, and weather vanes. A *steeple* is an ornamental construction usually ending in a spire (Figure 35.10). Usually erected on a roof or tower, a steeple can hold a clock or bells. A *spire* is a tall, pointed pyramidal roof built on a tower or steeple. A *cupola* is a light-roofed structure mounted on a roof. Cupolas

Figure 35.10 This steeple contains a bell tower and is crowned with a tall spire.

© Ethan boisvert/Shutterstock.com

typically are circular or polygonal. In addition to adding architectural beauty, a cupola often serves as a roof vent (Figure 35.11).

Figure 35.11 A cupola with louvered sides used to provide attic ventilation.

© Mark Winfrey/Shutterstock.com

Flagpoles

This specialty area includes complete flagpole assemblies, including required hardware and accessories. Flagpoles are typically made from wood, metal, or fiberglass. They may be anchored in the ground or wall mounted.

Identifying Devices

Identifying devices include directories, direction signs, bulletin boards, and various letters, signs, and plaques used to communicate or identify. These may be fixed devices, or they may be lighted, computerized, or some other type of electronic device.

Directories usually have replaceable letters and a locked glass door. They are available as wall-mounted or freestanding units. Bulletin boards may have a cork face or a cork face covered with nylon or vinyl fabric. The substrate is typically a fiberboard panel.

Various types of accessibility signs have become standardized for use across the country. Many building codes require accessible elements to be identified by the international symbol of accessibility (Figure 35.12).

Figure 35.12 The International Symbol of Accessibility.

The international symbol of accessibility is posted to indicate all elements and spaces of accessible facilities. This includes facilities such as parking spaces reserved for individuals with disabilities, accessible passenger-loading zones, accessible entrances, and accessible toilet and bathing facilities. In addition, signs are used to identify volume-control telephones, text telephones, and assistive listening systems (Figure 35.13).

Figure 35.13 Symbols used to identify telephones and assistive listening devices for individuals with hearing impairments.

Examples of code-regulated signs include exit identifiers and signs indicating the direction of a means of egress. Illuminated exit signs are used to denote stairs, elevators, or doors exiting directly from a building to the outside (Figure 35.14). Signs also are used to identify areas of refuge and the direction to move to reach

Figure 35.14 Lighted exit signs are one of the more important identifying devices in a building.

© fmua/Shutterstock.com

them. An area of refuge is a designated space that provides protection from smoke and fire. The maximum occupant capacity for assembly rooms and other areas where numbers of people gather is indicated by posted signs. Control of parking and identification of facilities for the handicapped are other major parts of any sign system related to a building. These and many other signs are required by building codes.

Other symbol signs are widely used to present a graphic message and to help communicate with people using various languages. These are found extensively in public buildings, such as airports, hotels, and hospitals, and are used as highway traffic control signs (Figure 35.15).

Pedestrian Control Devices

Products used for pedestrian control include portable posts and railings, rotary gates, turnstiles, electronic detection and counting systems, and items to limit access (Figure 35.16 and Figure 35.17). These strong, durable products can withstand exposure to the weather.

Lockers and Shelving

Lockers provide temporary storage in areas such as airports and athletic facilities (Figure 35.18). All types of open manufactured metal and wood shelving used for general storage are specialty items. This includes open shelving, various manufactured shelving systems, and mobile storage systems.

Fire Protection Specialties

This specialty includes all portable firefighting devices and their storage facilities but excludes items directly

Figure 35.16 A non-penetrable turnstile provides total passageway security.

© Manfred Steinbach/Shutterstock.com

connected to an installed fire protection system. Typical items include fire extinguishers, extinguisher cabinets, fire blankets and their storage cabinets, and any fire-extinguishing units on wheels.

Figure 35.15 A variety of symbols is used to communicate graphically.

Figure 35.17 This self-closing turnstile permits electronic card access.

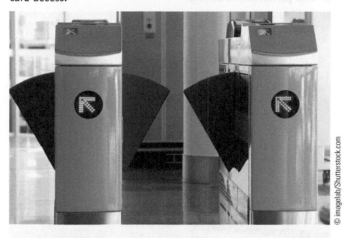

Figure 35.18 Double-tier lockers are widely used in schools and athletic facilities.

Protective Covers

Various types of awnings, canopies, marquees, covered walkways, and sheltered bus stop and car stop structures are available as standard manufactured units or as custom designed and built structures. They are made from any of the commonly used materials that withstand weather exposure.

Postal Specialties

Postal specialty products include all types of postal service facilities and devices, such as view windows, letter slots, letter boxes, collection boxes, and chutes. In large commercial buildings, these can become an extensive system used to speed mail collection (Figure 35.19).

Figure 35.19 Post office boxes typically used in multi-unit residential buildings.

Demountable Partitions

Demountable (movable) partitions include various room dividers, screens, and enclosures. They may be freestanding post-and-panel units that hang from a wall or ceiling, or they can be secured to a floor and ceiling. Some types are solid, offering privacy and a measure of sound control (Figure 35.20). Others are made of open mesh or screen panels and give an area security while still permitting visual oversight. Certain types incorporate folding gates or doors (Figure 35.21). They are made of a variety of materials, including wood, fabrics, plastic laminates, and metal.

Operable Partitions

Operable partitions are manually or power operated, allowing a moving partition to enclose or divide a larger area. They are suspended from a ceiling track and supported by a steel floor track. Some are top-hung and have a rubber seal at the floor. They include folding

Figure 35.20 Low demountable partitions provide privacy while using the general illumination and heating and air-conditioning systems.

Figure 35.21 These floor-to-ceiling prefabricated wall panels are used to enclose an area and provide privacy and security, yet allow visual oversight when required.

panel partitions, accordion folding partitions, and sliding and coiling partitions. Folding and accordion doors are outside of this classification.

Exterior Protection Devices for Openings

An extensive range of protection devices are manufactured, including products such as sun-control screens, shutters, louvers, screens for security, and panels that provide insulation and protection against storms. These may be fixed or controlled manually or electrically (Figure 35.22).

Telephone Specialties

Telephone specialties include manufactured telephone enclosures that may be completely enclosed, supported

Figure 35.22 These perforated metal louvers are used to provide partial sun screening.

on a floor, or wall hung (Figure 35.23). Telephone directory and shelving units are usually wall hung; they are available in many sizes. Outdoor telephone enclosures are weather tight and equipped with integral lighting for night use (Figure 35.24). Some companies will manufacture enclosures to an architect's design.

Figure 35.23 A row of individual floor mounted telephone kiosks.

Toilet and Bath Accessories

Toilet and bath accessories include all items manufactured for use in connection with bathroom facilities. These encompass products used in most commercial restrooms (Figure 35.25), plus special units designed for use in hospitals and areas where accessibility and security are important. Items in this specialty area include

Figure 35.24 Exterior telephone units are equipped with integral lighting.

© Sheri Armstrong/Shutterstock.com

Figure 35.25 Toilet and bath accessories include all items manufactured for use in lavatory facilities.

© Carolyn Brule/Shutterstock.com

mirrors, air fresheners, electric hand dryers, grab bars, waste receptacles, and dispensers for soap, toilet tissue, towels, razor blades, and paper cups. Hospital products include special cabinets and bedpan storage units. Security items include mirrors that resist theft and breakage and specially designed soap and tissue dispensers.

Scales

Any type of weighing device is considered a specialty item.

Wardrobe and Closet Specialties

These specialties are units used to store clothing, such as hat and coat racks. Among them are wardrobes that provide hanging space as well as several drawers. Other units typically used in hospitals and dormitories combine a wardrobe with a vanity area and lavatory. Wardrobes often are constructed with a plastic laminate over a substrate, such as particleboard. Racks are usually steel.

A wide variety of racks, wall mounted and floor supported, are used to store coats, hats, and packages. Automatic coat- and hat-checking systems are used in restaurants, hotels, theaters, and other places where large numbers of people gather. They are activated by hand and foot remote controls or by automatic selection devices using a customer's check number to locate and retrieve garments.

RESIDENTIAL AND COMMERCIAL EQUIPMENT

Division 11 of the CSI MasterFormat covers equipment related to residential and commercial construction. The items included in this section are shown on the introductory page to this chapter. A few of these are discussed in the following pages.

Maintenance Equipment

Maintenance equipment consists of freestanding and built-in equipment used to provide maintenance for a building. This includes devices such as window-washing systems, automatic vacuum systems, floor- and wall-cleaning equipment, and various types of housekeeping carts. Central-vacuum systems use a centrally powered vacuum unit connected to pipes run through walls to sealed outlets in various rooms. The flexible plastic hoses with various end attachments are connected to an outlet when an area needs vacuuming.

Window-washing systems for high-rise buildings have hoists mounted on a building's roof. The exterior washing equipment is controlled by an apparatus that usually rides on tracks on the roof (Figure 35.26). Some systems lower scaffolding from a roof, providing a platform from which workers can wash windows.

Figure 35.26 Window-washing equipment for high-rise buildings sometimes use hoists mounted on the roof from which scaffolding is lowered.

© oksana.perkins/Shutterstock.com

Security and Ecclesiastical Equipment

Security and vault equipment is designed to store money or other valuables and includes vault doors, day gates, security and emergency systems, safes, and safe deposit boxes.

Teller and service equipment is designed for use where money transactions occur. This includes service and teller windows, package transfer units, automatic banking systems, and teller equipment systems.

Mercantile Equipment

This includes equipment used in retail and service stores, such as display cases (Figure 35.27), cash registers, and checking and food processing machinery. Mercantile equipment also includes items in specialty establishments, such as barber and beauty shops.

Figure 35.27 Retail display fixtures are designed to exhibit a wide variety of merchandise of different sizes, weights, and values.

© Losevsky Pavel/Shutterstock.com

Ecclesiastical equipment includes items related to churches, such as baptisteries and various chancel items. These articles vary according to the denomination and desires of a congregation. Some items are manufactured stock, and others are custom built and can utilize all types of materials, including stone, wood, and metal.

Solid-Waste-Handling Equipment

Waste-handling equipment includes units that collect, shred, compact, and incinerate solid waste. The systems manufactured include package incinerators, compactors, storage bins, pulping machines, pneumatic waste transfer systems, and various chutes and storage collectors.

Office Equipment

Office equipment includes products such as copiers, computers, printers, fax machines, modems, and computer servers (Figure 35.28).

Office equipment is manufactured by many companies. Since some of these products must communicate with each other, care should be exercised when different brands are combined.

Construction Materials

Security of Valuables

Security and vault equipment is a major item for banks, wholesale and retail stores, hotels, and many other business establishments. Available are both built-in and freestanding equipment especially designed for the storage of money and other valuables, such as artwork, jewelry, and stock certificates.

Figure A

© Sebastian Kaulitzki/Shutterstock.com

Large vaults are available for storing sizable items. Vaults require ventilation systems and specialized security networks.

Vault doors have fire ratings up to at least six hours (Figure A). The controls include a time lock, day guard lock, alarm activator, lighting, emergency ventilator, and a connection for a telephone handset. Manufacturers offer centralized twenty-four-hour alarm systems for installations within the United States and maintain fleets of trucks manned by trained service technicians.

Safe deposit boxes come in stackable, interlocking sections to fit various vault sizes (Figure B). Various types of key systems are available, which greatly reduces the ability to improperly duplicate a key.

Figure B

© Andriy Rovenko/Shutterstock.com

Residential Equipment and Unit Kitchens

Residential equipment includes all appliances that are either built-in or freestanding, such as stoves, washers, dryers, freezers, microwaves, dishwashers, trash compactors, refrigerators, surface cooking units, ovens, range hoods, and indoor barbecue units. When combining units built by different manufacturers, it is important to consider possible differences in appearance (Figure 35.29).

Unit kitchens are manufactured assemblies that combine a refrigerator, cooking element, sink, and cabinets into one unit. They are delivered totally assembled for rapid installation (Figure 35.30).

Fluid Waste Treatment and Disposal

This section includes a wide range of equipment used to treat and dispose of fluid waste. A few of the products

Figure 35.28 Equipment for offices includes items such as computer network servers.

Figure 35.30 Unit kitchens are integrated single units that fit along one wall.

Medical Equipment

The array of medical equipment used for human and animal healthcare is extensive. It includes sterilizing equipment, examination and treatment equipment, optical and dental equipment (Figure 35.31), and items used in operating rooms and radiology facilities (Figure 35.32). Technical changes and improvements occur rapidly in this arena, and designers planning these facilities must be constantly alert for new equipment. These items must be of the highest quality.

Figure 35.29 Kitchen appliances include a wide range of electric and gas units in a variety of sizes and colors.

Figure 35.31 Typical equipment in a dental operatory.

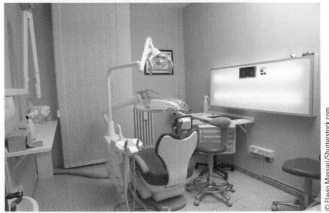

include an oil/water separator, fluid pumping stations, sewage and sludge pumps, scum removal equipment, aeration equipment, and package sewer treatment plants. Unit selection comprises a major part of the engineering design of a waste treatment system. The units must be durable and dependable.

The many items used in laboratories for testing, research, and developing new products may be preassembled by a manufacturer or designed for assembly on site. Typical items include fume hoods, incubators, sterilizers, refrigerators, and emergency safety appliances designed specifically for laboratory use, along with a range of special service fittings and the accessories to use them.

Figure 35.32 Medical facilities utilize a large number of testing, treatment, and examination equipment.

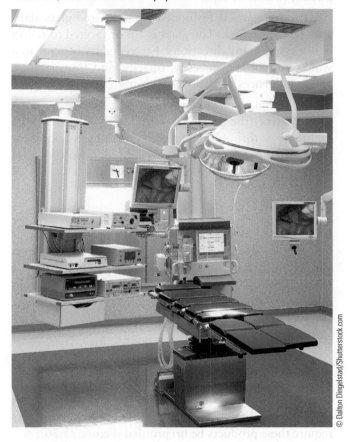

© Dalton Dingelstad/Shutterstock.com

FURNISHINGS

Furnishings include products that are decorative in nature, such as artwork, rugs, and window treatments, and other items, such as casework, furniture for commercial buildings, and multiple seating. An interior decorator plays a major role in the selection of these items and sometimes becomes involved in the design of special furnishings. The architect is also an important player in the overall interior design process.

ARTWORK

Murals and Wall Hangings

A popular form of interior artwork is photo murals. They are available in a wide range of sizes and scenes. Murals painted on walls are much more expensive but frequently become an important art item.

Mosaic murals are formed by bonding pieces of stone—such as marble, clay, or glass—on heavy paper to form a desired design. These sheets are then installed

against a mortar setting bed on a wall. When the setting bed hardens, the paper is soaked off the face, leaving the mosaic mural exposed (Figure 35.33).

Figure 35.33 Mosaics are formed by bonding pieces of stone, clay, or glass into patterns.

© vuttichai chaiya/Shutterstock.com

Many types of wall decorations are used, including paintings, prints, fabric wall hangings, and *tapestries* of various types. High-quality woven rugs are also used as wall hangings.

Sculpture

Sculpture is another form of artwork used in both interior and exterior locations. Some types are carved from stone, and others are cast in bronze, lead, or aluminum (Figure 35.34). Sculpture made by welding various shapes of metal is called reconstructed sculpture. Additional materials, such as stone, may also be integrated into a reconstructed piece. Relief artwork is usually a panel-type piece with images raised above its surface, producing a three-dimensional appearance.

Stained Glass

Stained-glass windows are mosaics of translucent glass shards cut to a desired shape (Figure 35.35). Sometimes the glass requires additional detailing, shading, or texturing. This is accomplished by painting it with special mineral pigments and firing it in a kiln. The two major types of stained glass are leaded and faceted.

Leaded glass windows have glass pieces joined with H-shaped lead strips called cames. Large windows require round bracing bars be wired to the leaded glass. This provides support while allowing for thermal

Figure 35.34 Cast-metal sculpture is used frequently to produce striking art for public spaces.

© Kushch Dmitry/Shutterstock.com

movement. Exterior leaded glass is pressed into a bed of glazing sealant or tape.

Faceted glass is fashioned from colored glass slabs 1 in. (25 mm) thick. The glass is cut or faceted to reflect light rays in many directions. Faceted glass is made in a wide range of translucency, opacity, and transparency, which influences the color and degree of reflected light.

Faceted glass is set in a reinforced concrete or epoxy resin matrix (the process is similar to setting brick). Light passes through the faceted glass pieces bound together by the matrix. The panels are mounted in wood or metal frames or in a masonry wall.

Ecclesiastical Artwork

Ecclesiastical artwork includes pieces of religious significance to specific denominations. They can take a variety of forms, some of which are discussed in this chapter. Specific types can include altar pieces and religious symbols.

FABRICS

This section relates to fabrics, leathers, and furs for upholstery, curtains, draperies, and wall hangings; this includes the various fabric treatments and fillers. Fire codes may require these products be fireproofed (Figure 35.36).

Figure 35.35 Leaded glass is formed by joining glass pieces cut in various sizes and shapes with lead cames.

© Chris Howey/Shutterstock.com

Figure 35.36 Fabrics are used extensively in interior design for upholstery, curtains, draperies, and wall hangings.

© Dana Hoff/Beateworks/Corbis

MANUFACTURED CASEWORK

Manufactured casework includes stock cabinets fashioned from wood, steel, and plastic laminates. It includes countertops, sinks, and any accessories or fixtures mounted on countertops. For residential applications it includes both kitchen and bath casework (Figure 35.37).

Figure 35.37 Residential casework includes kitchen and bath cabinets.

Medical applications include casework designed for dental, hospital, and optical uses, for nurses stations (Figure 35.38), and for veterinary uses. All types of laboratory casework used in medical, research, and other laboratories are part of this section.

Various manufacturers produce specialized casework with a limited market. Included is casework for banks, dormitories, ecclesiastical uses, hotels, motels, schools, restaurants, and various display units.

WINDOW TREATMENT

Interior window treatment can be decorative, and it can be practical, blocking the intrusion of sunlight and providing privacy. Blinds, often called Venetian blinds, are adjustable narrow wood or metal slats that can be easily positioned to block sun and provide privacy or permit a view of the outside (Figure 35.39). They can be raised to a window's head so they are completely out of view.

Figure 35.38 Casework for medical applications includes such items as cabinetry for nurses stations.

Figure 35.39 Venetian blinds have horizontal wood or metal slats that can be adjusted to block the view and sun, or opened to permit the entrance of light.

Vertical blinds are made with fabric slats hanging from the top of a window (Figure 35.40). Shutters made of wood or plastic are also used on window interiors. Some have adjustable louvers that let in light but prohibit much of an exterior view.

Figure 35.40 Vertical blinds are widely used in residential and commercial construction.

A number of shades are manufactured, including insulating, lightproof, translucent, and woven wood or plastic types. They can be pulled down for privacy or rolled up for viewing. Accordion types with fiberglass fabric are also available.

Curtains of various styles can serve the same purpose as shades. Some drapes have metal tracks across the top that permit opening or closing over windows. Other units remain on each side or above a window, serving a decorative purpose. Fire codes must be observed when selecting shades and curtains.

FURNITURE AND ACCESSORIES

Furniture under CSI MasterFormat Division 12 includes all types of freestanding units for commercial and residential use. The listing of open office furniture includes office partitions, storage, work surfaces, shelving, and light fixtures. General furniture includes all types from residential through commercial areas, such as classroom, hotel, library (Figure 35.41), medical, office, and restaurant

Figure 35.41 Libraries use large shelving units to store their inventory of publications.

furniture. Furniture accessories, such as clocks, ashtrays, lamps, and waste receptacles, are also included.

RUGS AND MATS

This section includes loose rugs and mats, gratings, and foot grilles (carpet covering floor areas is included in Division 9). Typical products include chair pads, floor mats, entrance tiles, and floor runners. These are made from a variety of materials, including rubber, carpet, vinyl, aluminum, and stainless steel.

MULTIPLE SEATING

Multiple seating includes seating for areas in which audiences gather, such as in schools, restaurants, theatres, churches, stadiums, and auditoriums (Figure 35.42).

Figure 35.42 Multiple seating includes products used in any area where large numbers of people gather.

It includes fixed, portable, and telescoping seating. Booths and various table and seat modules, as found in restaurants, also are included.

INTERIOR PLANTS AND PLANTERS

Extensive use of live and artificial plants is common in all types of commercial buildings. This section includes plants, plant holders, and any landscaping accessories and maintenance materials required (Figure 35.43).

Figure 35.43 An extensive variety of live and artificial plants are used for decoration.

© Jill Lang/Shutterstock.com

Review Questions

1. What finished surfaces are available when selecting chalkboards?
2. What is a whiteboard?
3. What are the typical ways toilet enclosures are hung?
4. What materials are used for panels on toilet enclosures?
5. What purposes do grilles serve?
6. What services are available with various service wall systems?
7. Why are access floors used?
8. How are gas-fired fireplaces vented?
9. What is the difference between a steeple, a spire, and a cupola?
10. What types of identifying devices are commonly used?
11. How do building codes govern the design of identifying devices?
12. What are some frequently used pedestrian control devices?
13. Where are protective covers used?
14. What are the ways various partition systems may be installed?
15. List some exterior protection devices.
16. What types of equipment are used for building maintenance?
17. What types of ecclesiastical equipment can you name?
18. What operations are performed by solid-waste-handling equipment?
19. How does a unit kitchen differ from a typical residential kitchen?
20. What types of units are used for fluid waste treatment and disposal?
21. What types of window blinds are commonly available?
22. How are mosaic murals made?
23. What types of sculpture are commonly available?
24. How are leaded stained glass pieces joined?
25. What are the differences between leaded and faceted glass?
26. What types of units are used for window treatment?
27. Where is multiple seating commonly found?

Key Terms

Chalkboard	Grille	Steeple
Compartment	Leaded Glass Window	Tackboard
Cubicle	Mosaic	Tapestries
Cupola	Raised Access Flooring	Whiteboard
Faceted Glass	Spire	

Activities

1. Take a walk around your school or a major building and record the location and use of any specialty items you observe. Decide if the proper choice was made and suggest replacements for those you consider improperly used.

2. Visit a hospital, a large restaurant, and a school and record the types of equipment that have been installed. Discuss the merits of each with the person responsible for operation and maintenance and record that person's comments on each item. Does your interviewee feel that any items are not adequate for their assignment?

3. Visit various commercial buildings and shopping malls and record the types of furnishings used both inside and out.

Additional Resources

Sweets Catalog File, Architects, Engineers, and Construction Edition, Section 10, McGraw-Hill Construction, New York.

See Appendix C for addresses of professional and trade organizations and other sources of technical information.

DeChiara, J., Panero, J., and Zelnik, M., *Interior Design and Space Planning*, McGraw-Hill, New York.

Special Construction
CSI MasterFormat™

36

Special Construction

Upon completion of this chapter, the student should be able to:

- Understand the range of special construction facilities available and how they function.

Division 13 *Special Construction* concentrates on construction components designed for a very specific function, such as a swimming pool, which require detailed construction specifications and methods customized to that unique building task. It also includes the installation of individual subsystems, such as a bank vault, within otherwise conventionally constructed buildings. Division 13 construction projects usually require professional and construction experience, making them more expensive to design and build than standard projects. Division 13 structures and systems often offer less flexibility for future space reconfiguration due to the unique nature of their function.

Special construction encompasses an extensive array of structures, systems, and assemblies, each designed to serve a specific purpose. Construction ranges from complex facilities and large structures to individual components of a small scale, such as special-purpose rooms or an ornamental fountain.

SPECIAL FACILITY COMPONENTS

The CSI MasterFormat lists special facility components for water-related constructions, including swimming pools, fountains, aquariums, and ice rinks, among others.

Pools and Spas

Swimming pools are constructed in a variety of sizes, from large public pools to residential applications and specialty spas. Pools are designed to withstand all anticipated loadings, both in an empty and full state. Small pools may be constructed of gunite, a mixture of pea stone, sand, cement, flyash, and water, sprayed on a reinforcing framework (Figure 36.1). Surfaces can be finished by painting or with a vinyl membrane lining. Other pools and tubs are made of a prefabricated fiberglass (Figure 36.2). Larger pools use heavy reinforced concrete construction clad with tile or a variety of other finish materials.

Figure 36.1 Some pools are constructed of gunite sprayed on a reinforcing framework.

© Jeff R. Clow/Shutterstock.com

All swimming pools must be equipped with a filtration system to clarify the water. Systems normally consist of one or more filter units containing sand, diatomaceous earth, or a cartridge-type filter. A disinfectant feeder keeps the microbiological, chemical, and physical characteristics of pool water within prescribed limits. Pool equipment rooms house pumps, filters, water treatment feeders, and heaters.

Figure 36.2 Pools and tubs can be made of a prefabricated fiberglass structure.

Saunas and steam rooms require insulated enclosures with integral mechanical systems to supply heat and steam. Saunas are usually clad in cedar or similar woods that perform well in extreme environments (Figure 36.3). Saunas can be divided into two basic types: conventional saunas that warm the air, or infrared saunas that warm objects. Infrared saunas use various heating agents, such as charcoal or active carbon fibers.

Figure 36.3 Saunas are often finished in water- and moisture-resistant materials such as cedar or redwood.

Heat sources include wood, electricity, natural gas, and other methods, such as solar power. There are wet saunas, dry saunas, steam saunas, and those that work with infrared waves. There are two main types of stoves: continuous heating and heat storage. Continuously heating stoves have a small heat capacity and are heated quickly, while heat storage stoves have a large heat (stone) capacity and can take much longer to achieve high temperatures.

SPECIAL-PURPOSE ROOMS

Special-purpose rooms are designed and constructed to meet specific performance requirements, such as fire protection, thermal control, or sound control, among others. The CSI MasterFormat lists planetariums, athletic rooms, soundproof rooms, clean rooms, cold-storage rooms, insulated rooms, shelters, booths, saunas, steam rooms, and vaults under the category "special-purpose rooms."

An example of one such room is an anechoic chamber, a shielded room designed to attenuate sound or electromagnetic energy (Figure 36.4). Anechoic chambers were originally used to study acoustics caused by internal reflections of a room, but more recently they have also been used to provide shielded environments for radio frequency and microwaves.

Figure 36.4 An anechoic chamber is a shielded room designed to attenuate sound or electromagnetic energy.

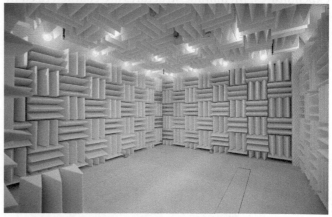

Anechoic chambers range from small compartments to chambers as large as aircraft hangars. The size of an anechoic chamber depends on the size of the objects to be tested and the frequency range of the radio or microwave signals used. Radio frequency interference (RFI) is the unwanted reception of radio signals. Radio frequency interference sources include lightning, electrical equipment, fluorescent lighting, cell phones, and transmitting equipment from radio stations.

Special Structures

"Special structural systems" refers to innovative structures that are used to enclose a variety of activities. While the concept is difficult to define, special structures listed in CSI MasterFormat Division 13 include space frames or grids; air-supported or air-inflated buildings; geodesic domes; and fabricated engineered structures. Considering the lack of standards, codes, and direct training for special structures in most engineering curriculums, their design and construction would have not been feasible without the originality of a number of talented engineers who used their deep-seated engineering knowledge to create innovative structures around the globe. Special structural systems generally find use for buildings that require large, unobstructed interior areas.

FABRIC AND MEMBRANE STRUCTURES

Membrane structures are architecturally innovative forms constructed of fibers or films. They are capable of providing a variety of free-form building designs for both permanent and temporary applications. The benefits of membrane structures are many. Less lighting is required, as most fabrics used are translucent, allowing for the entry of daylight and resulting in reduced energy expenditures. Membrane structures are characterized by having a very small mass relative to the applied load, reducing foundation expenditures (Figure 36.5). They

Figure 36.5 This fabric roof structure uses a PTFE membrane to provide diffuse light, shade, and weather resistance.

cost about half of what a traditional structure costs. Membrane structures use a variety of structural support systems including frame supported, grid-shell and cable net, and air-supported.

Frame-supported membrane structures utilize a frame or series of frames that form a load-bearing structure without the aid of any fabric. The membrane may however contribute toward the stability of the structure. Frames are typically made of structural steel elements and often incorporate truss forms.

A grid shell is a curvilinear surface (synclastic or anticlastic) that is composed of linear elements configured to form squares, triangles, and/or parallelograms (Figure 36.6). The grid can be of any material but most often wood or steel is used. They may be single or double-layered, sometimes employing in-plane cables for stability and shear resistance. Large-span timber grid shells are constructed by laying out the main members in a flat lattice and then deforming the grid into the desired form. A lamella structure is a type of grid shell that is made from smaller elements that brace against each other to create a spanning element. Lamella structures rely on all members to be held together in tension in order to gain strength as a system.

Figure 36.6 A grid shell is a curvilinear surface that is composed of linear elements configured to form squares, triangles, and/or parallelograms.

Cable nets use single-layered anticlastic (opposing curvatures) surface made of two sets of closely spaced cables that are orthogonal (or nearly so) to one another. The net supports the fabric or mesh material. Cable nets are tensile structures. A tensile structure is characterized by the tensioning of a fabric or pliable material, typically with wire or cable, to provide the key structural support to the structure. There is no compression or bending within the elements, only tension. The basic supporting elements for cable net structures consist of masts and cables. A cable-and-strut system uses a planar or curvilinear structure composed of short discontinuous compression elements (struts) connected by tensile cables to form a structure (Figure 36.7). Such structures are often referred to as "tensegrity" structures, those that have elements both in tension and in compression.

Figure 36.7 A cable-and-strut structure uses cables and masts to form a net that supports a membrane.

Fabric Membranes

Most membrane structures are constructed of fabric, meshes, or films engineered and fabricated to meet structural, flame-retardant, and weather-resistant specifications. The type of fabric chosen depends on the following material properties. Tensile strength is fundamental for architectural fabrics that function primarily in tension, and acts as an indicator of relative strength. The initial strength of a fabric is determined by the size and strength of the threads and the number of threads per linear inch.

Tear strength is important in that if a fabric ruptures, it generally will do so by tearing. Tearing can occur when a local stress concentration or damage results in the failure of one fiber, which thereby increases the stress on the remaining fibers. Adhesion strength is a measure of the strength of the bond between the base material and coating or film laminate that protects it. Adhesion strength is used to evaluate the strength of bonded joints in connecting strips of fabric into a single assembly. Both flame-retardant and flame-proof fabrics are available.

Materials for membrane structures include polyester, woven fiberglass, PTFE, and ETFE. The most common fabric used is polyester, due to its inherent properties of strength, durability, cost, and stretch. The polyester is either coated or laminated with layers of PVC film or other liquid polymer. Woven fiberglass is coated with a PTFE, either Teflon or silicone. PTFE (poly tetra fluoro ethylene) is the most chemically resistant plastic known. It has excellent thermal and electrical insulation properties. ETFE (ethylene tetra fluoro ethylene) has a much higher melting point than other plastics, in addition to excellent chemical and high energy radiation-resistance properties. PTFE allows maximum 40 percent light transmission, while ETFE allows 90–95 percent light transmission.

Typically, membranes are coated with synthetic materials for increased strength, durability, and weather resistance. An additional form of topcoating is also applied to the exterior to facilitate easy cleaning. Topcoating provides a hard surface on the outside of the material, forming a barrier that aids in preventing dirt from sticking to the material, while allowing the fabric to be cleaned with water.

The hardware used in membrane structures is taken from yacht- and bridge-building technology, including shackles, turnbuckles, toggles. Perimeter attachments often use steel tubes to which the membrane is mechanically clamped.

Air-Supported Structures

Air-supported (pneumatic) structures derive their structural integrity by using pressurized air to inflate a flexible, structural fabric, making air the main support of the structure (Figure 36.8). To maintain structural integrity, the structure is pressurized to produce internal pressure that is equal to or exceeds external pressures, such as wind forces. Air-supported structures use either single or multiple wall enclosures made of a flexible membrane material. A single skin utilizes one outer layer, with the air pressure exerting directly against that skin.

Denver International Airport

Architect: Fentress Bradburn Architects	Building type: Airport	Completed 1995
Contractor: PCL Construction Services and BL Harbert	Size: 189,000. sq. ft. (17,558. sq. m.)	
Location: Denver, Colorado	Project scope: Airport Terminal Building	

Overview

The planning for the new airport began in 1985 when the city of Denver decided to establish a new airport facility on land 20 miles outside of Denver. By 1989, after years of poor management and repeated design changes, the City and County of Denver commissioned Fentress Bradburn Architects to re-design the new Passenger Terminal Complex at DIA, today known as the Elrey B. Jeppesen Terminal. To address the problems of critical cost and scheduling overruns, Fentress Bradburn Architects decided to employ a membrane roof structure for its accelerated installation time, reduced construction costs, and increased energy efficiency.

Design

The roof's unique shape, material, and color, stemmed from a desire to create an iconic building appearance. The architects established a fabric and cable form for the terminal roof that would echo the snow capped Rocky Mountains in the background (Figure A). The

Figure A

© Ambient Ideas/Shutterstock.com

distinctive tensile membrane structure resulted in cost savings to fit the terminal's budget, and reduced construction time by using prefabrication methods.

The roof structure of the terminal hall is composed of 17 folded plate modules positioned 150 ft. (45.72 m) apart and supported by steel masts located 60 ft. (18.28 m) on center. From the terminal floor to the top of the masts, the roof elevation varies from 106 ft. to 126 ft. (32.30 m to 38.4 m) in height. Ridge cables suspended from the top of the masts bear the weight of the structure, resist forces due to snow loads, and serve to stabilize the masts. Cables in the valleys of the folded plates are anchored at the perimeter of the structure to hold the membrane in place, resist wind uplift, and provide positive roof drainage. The light-weight membrane roof allows for much greater spans than traditional roofing systems, decreasing construction material costs.

Membrane Materials

The double-layer roof membrane uses a Teflon-coated PTFE fiberglass fabric that weighs less than five ounces per square foot. The membrane's substrate consists of woven glass filaments that provide both the mechanical strength to resist punishing wind loads and the required non-combustibility. Covering the substrate is polytetra fluoroethylene (PTFE), a UV-resistant fluoro-polymer that provides long-standing weather protection. PTFE fiberglass is known for its durability, translucency, high reflectivity, and non-combustibility. The use of the lightweight and flexible membrane drastically reduced the weight of the steel and concrete structure considered in the earlier design schemes. Unlike most conventional roofing systems that generally have a lifecycle of 20 to 25 years,

PTFE fiberglass roofs can last in excess of 30 years. Additional maintenance savings are realized due to the material's self-cleaning properties.

Natural Daylighting

Because of the translucent nature of the PTFE membrane roof, the terminal hall is supplied with diffused sunlight throughout the daytime hours. The abundance of available daylight in the Denver climate means that the hall requires little artificial lighting during much of the year (Figure B). The roof's transparent membrane drastically reduces the energy consumption spent in powering artificial lighting, while

Figure B

© Joseph Sohm/Shutterstock.com

Figure 36.8 Air-supported structures use pressurized air to inflate a flexible, structural fabric.

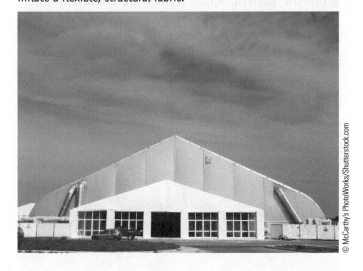

© McCarthy's PhotoWorks/Shutterstock.com

A double skin is similar to a single skin, but with an attached liner that is separated from the outer skin and provides an airspace which serves for insulation, acoustic, aesthetic, or similar purposes.

simultaneously reducing the cooling loads that would result from heat generated by the lights. Additionally, the white colored membrane reflects about 75 percent of incident solar radiation landing on its surface thereby reducing solar heat gains.

Construction Process

The roof installation began once the terminal floor structure had been erected. The steel masts were delivered to site prefabricated with hardware and rigging components attached and raised by conventional boom cranes. The oval shaped truss rings at the top of the masts were welded around the mast bases and fitted with caps, glass skylights, mechanical equipment, and lighting systems. With the truss rings in place, the valley cables were strung between the masts. The substrate fabric was laid out on the floor below with two halves of a bay resting together, one on top of the other, prior to being hoisted into place by the ridge cables. Once the ridge cable was fastened in place, each fabric bay was unfolded and seamed at the valley cables to the neighboring bays. With all bays seamed together and tensioned, the weather tight PTFE membrane was heat welded into place.

An air-inflated structure uses air-pressurized membrane beams, arches, or other elements to enclose space. Occupants of such a structure do not inhabit the pressurized area used to support the structure. Some air-supported structures use unreinforced membranes, and the membrane itself is the primary structural element. Another type uses a membrane that is reinforced by a network of cables or webbing. The webbing forms the primary structural system, and the membrane spans the area between the cables or webbing. The lightweight structures are secured by ground anchors or are attached to a foundation.

The inflation system for air-supported structures typically consists of one or more blowers and includes provisions for automated control to maintain the required inflation pressures (Figure 36.9). Inflation systems must be designed to prevent over-pressurization of the structure. In addition to the primary inflation system, in buildings exceeding 1,500 square feet in area, an auxiliary inflation system must be provided with sufficient capacity to maintain the inflation of the structure in case of mechanical failure. The auxiliary inflation

Figure 36.9 The exterior skin of the Allianz Arena uses 2,784 pressurized air cushions made of ETFE.

Figure 36.10 Geodesic domes provide a structurally strong, light-filled space free of internal supports.

steel or wood members. Domes can range in size from small enclosures to spaces as large as 150 ft. (50 m.) in diameter. A geodesic dome provides a highly aerodynamic structure that is able to withstand considerable wind loading. Numerous manufactures provide frame materials and engineering for geodesic domes.

Space Frames

Space frames were separately developed by Alexander Graham Bell around 1900 and by Buckminster Fuller in the 1950s. A *space frame* is a truss-like, lightweight rigid structure constructed by interlocking struts in a geometric pattern. Space frames are essentially three-dimensional trusses that can accomplish very long spans with minimal internal supports (Figure 36.11). They derive their strength from the inherent rigidity of

system should operate automatically when there is a loss of internal pressure, or when the primary blower system becomes inoperative. All access to the air-supported interior must be equipped with two sets of doors or a revolving door airlock. Air locks are used as entrances, and emergency exits are counter-pressure balanced, and self-closing to prevent de-pressurization.

Geodesic Structures

A *geodesic dome* is a spherical shell structure based on a network of triangular elements that provide triangulated rigidity and distribute stresses across the entire structure.

The architect Buckminster Fuller coined the term "geodesic dome," receiving a U.S. patent in the late 1940s. Geodesic domes are very strong and provide an enclosed space free of structural supports. The infill material most widely used is glass, although other opaque materials are also utilized (Figure 36.10). The basic structure can be erected quickly and efficiently using small, lightweight

Figure 36.11 Space frames derive their strength from the inherent rigidity of a triangular frame.

a triangular frame, with tension and compression loads transmitted along the length of each strut. The simplest form of a space frame is a horizontal slab of interlocking square pyramids and tetrahedra built from aluminum or tubular steel struts. A tetrahedron is a polyhedron composed of four triangular faces, three of which meet at each vertex.

All elements in a space frame structure carry loads, resulting in structural redundancy. Concentrated loads are distributed individually and over the entire system. Space frames are generally supported by a regular column grid. Materials used to construct space frames include steel tubes, hollow sections, and I-beams. Space frames are an increasingly common architectural structure, especially for large roof spans in modern commercial and industrial buildings.

INTEGRATED ASSEMBLIES

Integrated ceilings are a type of integrated assembly that is pre-engineered and prefabricated for assembly on site (Figure 36.12). A wide range of materials and products is available, as discussed in Chapter 35. Some integrated systems include lighting, air supply registers, ducts, air-return grilles, fire-sprinkler systems, and communication linkage. These systems are coordinated with CSI MasterFormat Divisions 15 and 16.

Figure 36.12 Integrated ceilings incorporate lighting, mechanical, communications, and sprinkler systems.

© yxm2008/Shutterstock.com

PRE-ENGINEERED STRUCTURES

All types of pre-engineered prefabricated buildings and structures are classified as special construction, even if they are erected on temporary foundations. Some of the structures included are metal building systems, glazed structures (such as a greenhouse), portable buildings, grandstands and bleachers, cable-supported, fabric, and log structures.

WASTE AND UTILITIES SECTIONS

A number of special constructions occur in the area of waste treatment and systems involved with utilities. Filters, underdrains, and media include the piping and filters used in water and fluid waste treatment. Filter media include anthracite, charcoal, diatomaceous earth, mixed, and sand media. Digester covers and appurtenances include special tank covers and assemblies used on digestion tanks.

Oxygenation systems include site-assembled piping systems and related equipment for the dissolution and mixing of gaseous oxygen in liquid waste. This includes oxygen generators, oxygen storage, and dissolution systems.

Utility control systems include the operating and monitoring systems for water supply, wastewater, and electrical power generation plants. They include metering devices, display panels, control panels, and sensing and communicating equipment.

MEASUREMENT AND CONTROL INSTRUMENTATION

Measurement and control instrumentation are covered in a section of Division 13. The types of controls vary by industry and can include mechanical, electrical, fluid, pneumatic, and computer-controlling devices. These systems may be totally automated or semi-automated. They can monitor phenomena such as temperatures or pressures, regulate a system so it operates at the required speed and quality, and control safety within a manufacturing, processing, or assembly plant.

Devices listed as recording instruments are used to measure and record various incidents such as seismic information, stresses in structures, and meteorological information, solar and wind energy. Transportation control instrumentation describes systems used to monitor and control the various aspects in transportation systems, such as airport control, railway control, subway control, and transit vehicle control.

Review Questions

1. What types of covering materials are used for air-supported structures?

2. How are entrances to air-supported structures treated so air pressure within a structure is maintained as people enter and leave?

3. What is meant by special-purpose rooms?

4. What types of structures are available as pre-engineered units?

5. What kinds of materials are used to construct space frames?

6. What is the basic structural element of the geodesic dome?

Key Terms

Air-Supported (Pneumatic) Structure　　Geodesic Dome

Space Frame

Activity

1. Locate any of the special construction areas mentioned and arrange a visit. Ask the person in charge to conduct a tour and explain how the system operates.

Additional Resources

Sweets Construction Catalog, Architects, Engineers, and Contractors, Edition, Division 13, McGraw-Hill Construction, Two Penn Plaza, New York 10121.

See Appendix C for addresses of professional and trade organizations and other sources of technical information.

Conveying Equipment
CSI MasterFormat™

Conveying Systems

Upon completion of this chapter, the student should be able to:

- Understand the codes related to the design and installation of conveying systems.

- Understand the various operating mechanisms used on elevators.

- Be familiar with additional conveying systems used in buildings.

Conveying systems, each designed for a special purpose, find common use in both commercial and industrial applications. Some facilitate the movement of people, while others transport materials and equipment. They range from a simple dumbwaiter used to carry mail between floors in a building, to large, complex elevator systems that make vertical circulation possible in high-rise buildings.

The modern passenger elevator was developed by Elisha Otis, founder of the Otis Elevator Company. Otis developed a safety device that prevents elevators from falling if the hoisting cable fails. For multistory buildings, no design decision is more important than the design of the vertical transportation equipment. Passenger service, freight elevators, and escalators represent a major building expense, and their quality of service is critical to occupant satisfaction.

ELEVATOR CODE STANDARDS

Elevators, escalators, moving walks, and dumbwaiters are designed and installed following the code requirements of the *American National Standard Safety Code for Elevators and Escalators,* ANSI/ASME A17.1, and local building codes. Standard elevator sizes and shapes have been developed by National Elevator Industries, Inc. (NEII). Standards such as *Elevator Engineering Standard Layouts and Suggested Minimum Passenger Elevator Requirements for the Handicapped* are also available. These standards establish basic rules that require specific compliance and should be coordinated with the design criteria of the complete elevator installation. Following are some examples.

Hoistways, the enclosed shaft in which elevators move, must be enclosed for their entire height with fire-resistant materials, such as masonry, drywall, or concrete. Pits must also be constructed of noncombustible material and waterproofed to prevent groundwater entry. Hoistway tops must be enclosed by concrete or metal floors. Hoistway assembly design must prevent the accumulation of gases and smoke in case of fire. No windows are permitted, and the hoistway and machinery space must be free of any pipes or ducts.

The elevator machine room must have fire-resistant enclosures and doors. The only machinery allowed in the room is that needed for the operation of the elevator. Since regular maintenance will be necessary, permanent and easy access to the machine room is required.

Electrical equipment and wiring must conform to the *National Electrical Code,* ANSI/NFRA 70. All main electrical feeds are installed outside the hoistway. The only electrical equipment allowed in the hoistway are those devices directly connected with the elevator. The machine room should have permanent lighting and natural or mechanical ventilation.

Buffers are required in the pit. Buffers are energy-absorbing units located at the bottom of the hoistway that absorb any impact from a car that descends below the normal lowest level. Code requirements also include stipulations for required tests, emergency-condition operation, venting, opening protection, signals, and signs.

The Americans with Disabilities Act specifies the following for elevator requirements in new construction. One passenger elevator should serve each level, including mezzanines, in all multistory buildings. An exception is that elevators are not required in facilities fewer than three stories unless the building is a shopping center, shopping mall, or the office of a health care provider.

Elevators are not considered part of the system of egress for code purposes because they might not operate as required during an emergency, such as a fire or an earthquake, and people could be stranded between floors. If power fails, an elevator will automatically return to the lowest landing and the doors will open, allowing passengers to exit. In the event of a fire, the elevator fire service will be activated by the building's smoke alarm or by the fire service key switch located in a hallway. When this happens, all car calls are canceled, cars return to the main floor, and the doors open. Since elevators are vital for use by firefighters and other emergency personnel, they may be reactivated for emergency use.

ELEVATOR COMPONENTS

An elevator consists of a hoisting mechanism connected to a car or platform that slides vertically on guides on the sides of a fire-resistant hoistway (Figure 37.1). It is used to move passengers and materials between floors of multistory buildings. Passenger elevators are designed to transport people between floors. Freight elevators carry

Figure 37.1 Elevator hoistways are constructed with fire-resistant walls, top, and pit.

© trekandshoot/Shutterstock.com

materials from a loading dock to upper floors. Safety regulations permit only the operator or others needed to handle the materials being transported. Hospital elevators have cars large enough to transport patients on stretchers or beds along with their attendants. They can also serve as passenger elevators.

Elevator Hoistways

An elevator *hoistway* is a vertical, fire-resistant enclosed shaft in which an elevator moves. It has a *pit* at the bottom and openings at each floor. The openings are protected by automated doors controlled by an operating system. Traction elevators require a penthouse on the roof above the shaft. Codes typically require that elevator shafts have a 2-hour fire rating and that the opening be protected with doors having a 1-hour fire rating. A metal or concrete deck is required on the top of the hoistway.

Hoistway Doors

Building codes specify permissible door types and their opening requirements. Passenger elevators generally employ horizontal sliding doors, although swinging doors may also be used. Vertical sliding doors are used on freight elevators.

Elevator doors are closed automatically before the car leaves the landing zone. The *landing zone* is an area 18 in. (457 mm) above or below the landing floor. The elevator car will not move unless all doors on all levels are closed and locked. Doors have locks that prohibit them from being opened by someone on the landing side. Emergency access is available for maintenance and rescue personnel.

Machine Rooms

Machine rooms are part of the hoistway and provide a fire-resistant enclosure for the required hoisting machinery, controls, pumps, and hydraulic oil storage. Since control systems are computerized, the area should be equipped with temperature control. Only equipment involved with the operation of the elevator is permitted in the machine room. The size of the machine room will vary depending on the type of elevator.

Machine rooms for electric elevators are generally situated in a penthouse on the roof over the hoistway (Figure 37.2). The structural system must be designed to carry the weight of the machinery plus loads required for lifting and lowering the car. In some cases, machine

Figure 37.2 A traction electric elevator generally has the hoisting machine at the top of the hoistway.

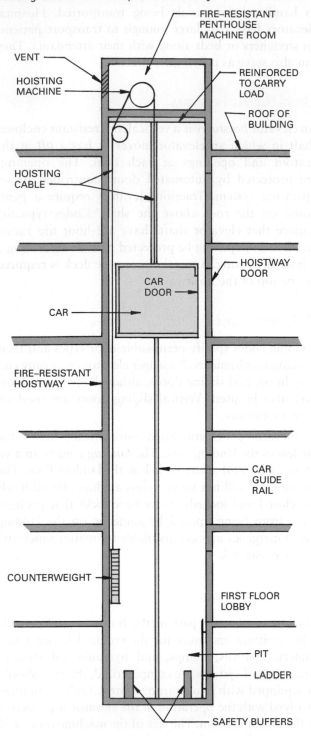

Figure 37.3 A hydraulic elevator locates machine rooms on the side of the hoistway.

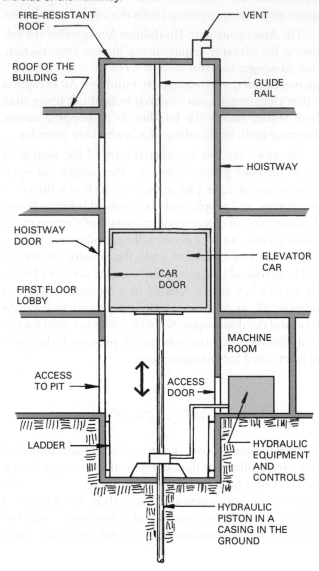

Venting Hoistways

During a fire, the elevator hoistway can function as a vertical flue for gases and smoke and serve to carry away these noxious materials via venting at the top. Some building designs use other means of containing and venting smoke and gases from a fire site to prevent them from moving to unaffected floors. Codes may require the hoistway to be vented, and other equipment for controlling smoke is considered as the hoistway is designed.

Hoistway Sizes

Hoistways are sized according to car shapes, sizes, and door sizes, with consideration given to space

rooms for traction elevators can be located in the pit at the bottom of the shaft. Hydraulic elevator machine rooms containing the hydraulic equipment and controls are usually located to the side of the hoistway (Figure 37.3).

requirements for guide rails and brackets, counter-weight systems, running clearances, and ancillary equipment. Sufficient air space must be provided around cars and elevator counterweights to minimize buffeting and airborne noise during operation. The clear inside dimensions for hoistways housing standard-size elevators are specified in the *National Elevator Industry Standard: Elevator Engineering Standard Layouts*. Maximum and minimum clearances between the hoistway and cars, moving weights, and other equipment must be observed. For specially designed elevators, the hoistway may accommodate larger or smaller car and moving equipment, and the required clearances become critical.

ELEVATOR CARS

The *elevator car*, a platform enclosed by walls and a roof, is designed to carry passengers, freight, and other specialty loads, such as hospital beds and stretchers. The car is guided in its vertical travel by means of guide rails mounted in the hoistway. Each car contains lighting, controls, venting equipment, handrails, telephones, and various types of wall, ceiling, and floor-finish materials. Passenger elevators are often customized to complement building architecture (Figure 37.4). Freight elevator cars must have durable, abrasion-resistant interior finishes. Typically, they have steel floors and walls.

The car in a hydraulic elevator is mounted on top of a piston. Cars on electric elevators utilize a steel frame to which wire hoisting ropes are connected. As passengers step on and off an elevator, the load constantly changes, making it difficult to keep the platform level with the floor. A self-leveling device called the Micro-drive automatically levels the car floor. The design of elevator cars is detailed in ANSI/ASME A17.1 *Safety Code for Elevators and Escalators*.

Car doors are automatically operated and cannot be opened while a car is moving or outside a landing zone. A car will not move if the door is open. Separate inner and outer doors are an essential part of the modern elevator. Both are opened and closed by an electric motor on the car, equipped with an electronic sensor that reverses the door if it encounters an obstruction. Both single and double entrance elevator cars are available. Used on hospital and freight elevators, double-entrance cars open to the front and rear (Figure 37.5). Elevator car doors may be center-opening, two-speed sliding doors, and single sliding doors (Figure 37.6). For high-speed systems, center-opening doors provide

Figure 37.4 Passenger elevators are often custom finished to integrate with the building architecture.

© Kutlayev Dmitry/Shutterstock.com

Figure 37.5 Elevator doors may open in one or more directions.

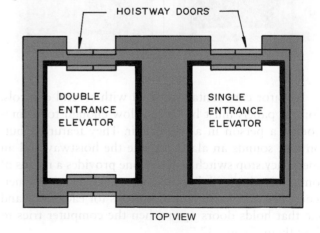

HOISTWAY DOORS

DOUBLE ENTRANCE ELEVATOR

SINGLE ENTRANCE ELEVATOR

TOP VIEW

the quickest access and egress. Two-speed sliding doors provide the widest possible opening but operate slower than center-opening doors. Single sliding doors are the most economical and the slowest. They move right or left, depending on the car design, and the opening width is limited by the width of the car.

Figure 37.6 Typical types of elevator car door operation.

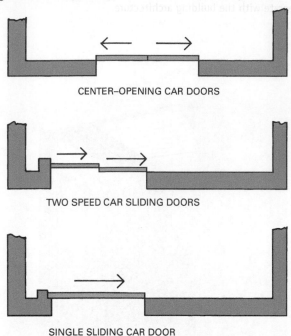

CENTER–OPENING CAR DOORS

TWO SPEED CAR SLIDING DOORS

SINGLE SLIDING CAR DOOR

© Adam Radosavljevic/Shutterstock.com

Figure 37.7 Elevator control panels must be located low enough to accommodate a person in a wheelchair.

© Vladimir Mucibabic/Shutterstock.com

Elevator cars contain a panel with several controls. Control panels must be located low enough to accommodate a person in a wheelchair. They feature a button that sounds an alarm outside the hoistway and an emergency stop switch. A telephone provides a means of communication with building personnel in case of emergency. The panel also houses buttons for each floor and one that holds doors open when the computer tries to close them (Figure 37.7).

The capacity of a car is determined by the maximum number of people it can safely carry. The capacity is the total load in pounds divided by 150 (the allotted weight per person). For example, an elevator rated at 1,500 lbs. would be marked to carry a maximum of 10 people.

Car sizes vary with the different manufacturers, but typical sizes for passenger elevators range from

6 ft. × 4 ft. (1.8 m × 1.2 m) to 8 ft. × 6 ft. (2.4 m × 1.8 m). Doors are typically 3 ft. to 4 ft. (0.9 m to 1.2 m) wide and 7 ft. to 9 ft. (2.1 m to 2.7 m) high. Hospital elevators are generally 9 ft. × 7 ft. (2.7 m × 2.1 m). Freight elevators range from 5 ft. × 7 ft. (1.5 m × 2.1 m) to 12 ft. × 16 ft. (3.7 to 4.9 m). They have clear door widths from 5 ft. to 12 ft. (1.5 m to 3.7 m).

ELEVATOR TYPES

Elevators are either electric or hydraulic. The electric elevator uses an electric motor to supply power. The mechanism includes an electric motor, a brake, a driving sheave or drum, gearing, belts, and wire rope. Electric elevators are used on low-, medium-, and high-rise buildings and are faster than hydraulic elevators.

ELECTRIC PASSENGER ELEVATORS

Electric passenger elevators suspend cars from wires and use weights to counterbalance them. A car is guided by vertical guide rails. The electrically driven hoisting mechanism may be located at the top or bottom of the shaft. The wire rope runs over traction sheaves. The two types in common use are geared and gearless traction mechanisms.

Traction Driving Mechanisms

Electric elevators are powered by traction machines consisting of an electric motor connected to a driving sheave. This connection can be direct or operated through a series of gears. A wire rope runs through grooves in the face of the sheave, or traction is provided by friction.

Gear-driven traction machines provide slower rising speeds and are used when slower speeds are desired. Gearing may utilize a helical gearbox or a worm gear. Gearless direct drive machines have the traction sheave connected directly to the motor. They provide high speeds and are usually used on high-rise buildings.

Geared Traction Elevators

A geared electric passenger elevator is shown in Figure 37.8. This hoisting mechanism uses a worm gear to drive a large spur gear connected to the traction sheave. A *sheave* is a pulley with a grooved rim for retaining a wire rope used to transmit force to the rope. The wire rope runs over the traction sheave and moves as the sheave rotates. This type of drive is used when low speeds and high lifting capacity are required. Speeds can be changed by varying the size of the spur gear, which changes the gear ratio. Typical car rise speeds for geared-traction passenger elevators range from 350 ft./min. to 500 ft./min. (106 m/min. to 152 m/min.). The range of lifting capacity is typically from 2,000 lb. to 4,500 lb. (900 kg to 2,025 kg). The hoisting mechanism is usually housed in a penthouse on the roof. Geared traction freight elevators typically have a rise from 50 ft./min. to 200 ft./min. (15 m/min. to 60 m/min.) and carry loads up to about 20,000 lb. (9,072 kg).

Gearless Traction Elevators

A *gearless traction elevator* is shown in Figure 37.9. The traction sheave and brake are mounted directly on the motor shaft (Figure 37.10). The speed of rotation of the traction sheave (which contains the wire rope) is the same as the speed of the motor. The speed can be varied by using a DC electric motor built to run at the speed required. Car rise speeds can range from 500 ft./min. to 1,200 ft./min. (152 m/min. to 366 m/min.). Gearless elevators can also be operated using a motor generator drive. The range of lifting capacity is typically from

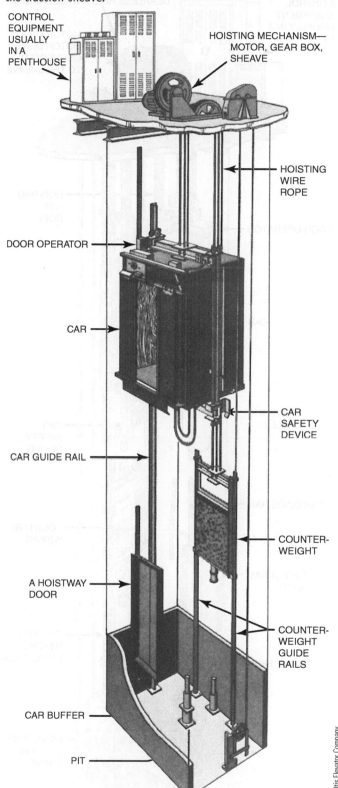

Figure 37.8 A geared electric passenger elevator uses an electric motor to drive a gearbox that regulates the speed of the traction sheave.

CONTROL EQUIPMENT USUALLY IN A PENTHOUSE

HOISTING MECHANISM—MOTOR, GEAR BOX, SHEAVE

HOISTING WIRE ROPE

DOOR OPERATOR

CAR

CAR SAFETY DEVICE

CAR GUIDE RAIL

COUNTER-WEIGHT

A HOISTWAY DOOR

COUNTER-WEIGHT GUIDE RAILS

CAR BUFFER

PIT

© Otis Elevator Company

Figure 37.9 A gearless electric passenger elevator has the traction sheave and brake connected directly to the motor shaft.

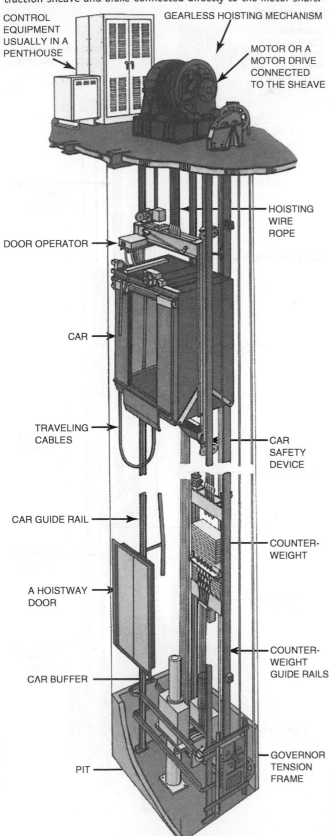

© Otis Elevator Company

Figure 37.10 Gearless traction elevators have the traction sheave and brake mounted directly on the motor shaft.

© momente/Shutterstock.com

2,000 lb. to 7,000 lb. (900 kg to 3,150 kg). The hoisting mechanism is normally housed in a penthouse on the roof.

Electric Elevator Control

The elevator control system regulates the starting, stopping, safety devices, speed of movement, and direction of movement, acceleration, and deceleration of the car. Two types of controls are in general use: multi-voltage and variable-voltage variable-frequency control (VVVF).

Multi-voltage control is used with machines having DC motors. It controls speed by varying the voltage supplied to the motor armature. Multi-voltage control provides a smooth regulation of speed and is widely used on passenger elevators.

VVVF control is used to control AC motors and produces a smooth overall operation of the car. It is more efficient than multi-voltage control and DC motor operation, and it is gaining popularity over DC controls and motors.

Platform Safeties

Elevators lifted by hoisting cables are required to have *platform safeties,* devices designed to clamp onto the steel guide rails upon activation, quickly braking the elevator to a halt. A device called a governor monitors speed and activates a car safety if the car exceeds safe velocities. A *car safety* applies brake shoes against the guide rails, stopping the car. The governor also switches off the electrical power to the motor. In addition, some types activate a brake shoe to the motor drive shaft.

Cables (Roping)

Cables connected to the top of the car crossbeam carry the weight of the car and its live load. Traction-type machine cars must be suspended from at least three hoisting ropes (ANSI 17.1). The wire rope consists of steel strands laid helically around a hemp core, and wrapped again around a steel wire core.

The arrangement of the roping of traction-type elevators greatly affects the speed of the car and the loads on the hoisting wires. Typical roping systems are shown in Figure 37.11.

Figure 37.11 Different roping arrangements used with electric traction elevators.

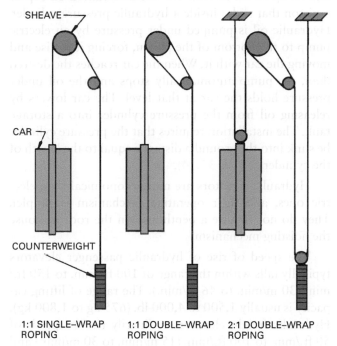

SHEAVE

CAR

COUNTERWEIGHT

| 1:1 SINGLE–WRAP ROPING | 1:1 DOUBLE–WRAP ROPING | 2:1 DOUBLE–WRAP ROPING |

Single-wrap roping has the wire pass over the driving sheave once before it moves on to the counterweight. The deflection sheave moves the wire clear of the moving car. This provides a 1:1 ratio, which means it will move the car 1 ft. (305 mm) for every 1 ft. (305 mm) of travel along the circumference of the sheave.

Double-wrap roping has the wire wrap twice around the driving sheave and a secondary sheave and then on to the counterweight. Double-wrap roping can be used to provide a 1:1 or 2:1 ratio. The 1:1 double-wrap ratio causes less wear on the wire but increases the load on the sheave. The 2:1 double-wrap roping is used for heavily loaded elevators. This system moves the car 1 ft. (305 mm) for every 2 ft. (610 mm) of rope travel along the circumference of the sheave. Other roping systems are also used.

Winding Drum Machines

A winding drum hoisting mechanism has a gear-driven drum. The hoisting wire is attached to the drum and winds or unwinds around it as it rotates. This type is used with dumbwaiters and some small residential elevators.

Counterweights

Counterweights are stacked steel plates that slide on vertical steel guides (Figure 37.12). As an elevator rises, they lower, and as the elevator lowers, they rise. Counterweights generally equal the weight of the unloaded car and hoisting wires, plus a percentage of the load capacity of the elevator. Since they rely on gravity, they help raise the car, thus reducing the overall power required.

Figure 37.12 Counterweights equal to the weight of the car, hoisting wires, and load capacity of the elevator move on vertical steel guides within the hoistway.

© Dmitry Kalinovsky/Shutterstock.com

Operating Systems

Early elevators were operated by elevator operators using a motor controller.

Typical operator-controlled systems include the car-switch and signal systems. In the car-switch system, the operator controls the direction of travel and initiates car movement. In the signal system, the operator presses buttons indicating the floor stops and then presses a start button. The car automatically stops at each floor for which a button was pressed.

Automatic operating systems do not require an operator. Signals are initiated by the passengers or by an automatic operating device. Typical automatic operating systems include the single, selective collective, and group systems.

Single automatic systems start when a passenger pushes a button to indicate the floor desired. The car moves automatically to that floor, stops, and the doors open. By pressing a button on the wall next to the elevator doors, anyone can call a car to any floor.

The selective collective automatic system has two call buttons, up and down. When the up button in a car or on a hall wall is pressed, the car moves up, stopping at floors where the hall wall up button has been pressed. When it reaches the top up floor, it descends, stopping as indicated by passengers using the elevator's control panel and on floors where a hall wall down button has been pressed (Figure 37.13).

Figure 37.13 Call buttons on the wall bring the elevator to the floor. The overhead direction arrows indicate the direction the arriving elevator is moving.

Group automatic systems control the operation of several cars that serve the same floors using a supervisory control system. The control system automatically decides which cars to send to various floors and coordinates the flow of traffic. The cars are dispatched from the first floor at predetermined intervals, and whichever car is nearest a floor where a passenger is waiting stops to make the pickup. This reduces waiting time and increases the number of passengers that may be carried.

The supervisory control system is able to make adjustments in car flow and direction to handle times when requirements vary, such as late afternoon in a high-rise office building when many workers are trying to leave at the same time. The use of computers to make adjustments and control car movement has increased the efficiency of group car operations in heavily loaded situations.

Computerized systems vary, but many use a traffic information database to decide which pattern of car utilization will produce the shortest waiting time. For example, in peak up traffic times, the cars may make fewer stops per round trip, returning to the lobby more often. A car may be assigned to serve only one group of floors and return quickly to the lobby. Another car can serve another group of floors. This reduces the stops for each car and facilitates quicker lobby return for more trips. The reverse can occur when the down trips become heavily loaded. The control system display in the lobby indicates which floors each elevator will serve.

HYDRAULIC ELEVATORS

Hydraulic elevators are used mainly for low-rise installations up to six stories. The car is mounted on top of a piston that slides inside a hydraulic pressure cylinder. Hydraulic oil is pumped under pressure by an electric pump to the bottom of the piston, forcing it to rise and moving the car with it. When the car reaches the desired floor, the pump automatically stops and the oil under pressure holds the car at that level. The car lowers by releasing oil from the pressure cylinder into a storage tank. The installation requires that the pressure cylinder be sunk into the ground a distance equal to the length of the cylinder (Figure 37.14).

Hydraulic elevators are more economical than electric ones, and their operating mechanism is simpler. They do not require a penthouse on the roof to house the hoisting mechanism.

The speed of rise of hydraulic passenger elevators typically falls within the range of 100 ft./min. to 150 ft./min. (30 m/min. to 46 m/min.). The range of lifting capacity is usually 1,500 to 4,000 lb. (675 kg to 1,800 kg). Hydraulic freight elevators ordinarily rise at speeds of 50 ft./min. to 100 ft./min. (15 m/min. to 30 m/min.) and carry loads from 3,000 lb. to 12,000 lb. (1,350 kg. to 5,400 kg). Since gravity affects the system, these elevators lower faster than they rise.

FREIGHT ELEVATORS

Freight elevators are designed to carry general freight, motor vehicles, and other heavy concentrated loads. Hydraulic systems can be used for freight elevators in low-rise buildings, while electric systems are necessary in mid- and high-rise buildings. Three classes of freight elevators are specified in the *American National Standard Safety Code for Elevator and Escalators* (ANSI/ASME A17.1). Class A is limited to carrying a maximum of one-quarter of the rated load. It is for

Figure 37.14 A hydraulic elevator moves the car with a long piston controlled by a hydraulic pump and valves.

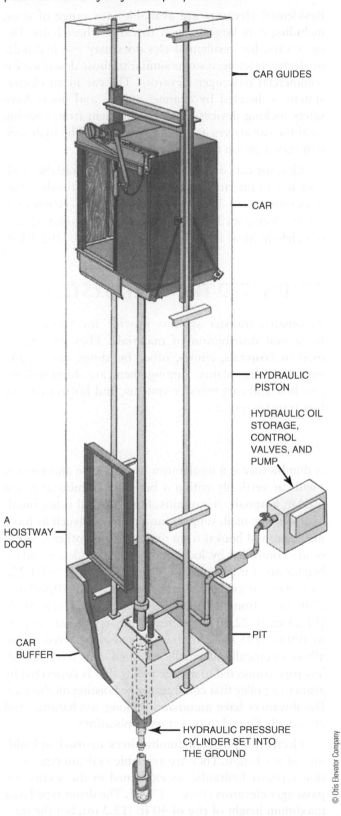

CAR GUIDES

CAR

HYDRAULIC PISTON

HYDRAULIC OIL STORAGE, CONTROL VALVES, AND PUMP

A HOISTWAY DOOR

CAR BUFFER

PIT

HYDRAULIC PRESSURE CYLINDER SET INTO THE GROUND

© Otis Elevator Company

general freight that is loaded and unloaded by hand (often with a lightweight hand truck). Class B handles only motor vehicles. Class C elevators are designed to handle heavy concentrated loads. A Class C1 elevator can carry materials plus industrial trucks or forklifts. Class C2 elevators are used when materials are loaded by industrial trucks but the trucks are not carried with the materials. Class C3 elevators carry heavy concentrated loads other than trucks.

Pre-manufactured freight elevators are available, but many are custom designed. Size of car and loading capacity vary widely with industrial requirements. Platforms are engineered to carry anticipated loads and withstand extra weight and movement when being loaded. For example, a truck loading materials can cause sideways movement and shock impacts if a load is not slowly lowered to the floor. A load placed off-center in a car produces eccentric loading, the possibility of which must be considered as the system is designed.

The doors on freight elevator cars are often made of an open steel mesh and operate vertically. The door to the hoistway is typically a solid, vertical bi-parting door with a locking system that will not allow it to open until the car arrives at the floor (Figure 37.15).

Freight elevators are generally automatically operated (in the same manner as described for passenger elevators). Some operator-controlled systems are available.

Figure 37.15 Freight elevators are designed to carry the anticipated loads and resist actions from forklifts and other devices used to load them.

© Thomas Northcut/Getty Images

OBSERVATION ELEVATORS

Observation elevators move up a hoistway secured to the exterior of a building. Cars are designed with glass walls to provide panoramic views of surrounding areas as the elevator rises (Figure 37.16). Elevators may be geared, gearless, or hydraulic, and the hoisting machinery is hidden. The lower end of the hoistway (at the ground) must be enclosed with a safety barrier. As with all circulation systems, elevators doors and cars must be large enough to accommodate wheelchair use (Figure 37.17).

Figure 37.16 Observation elevators are glass enclosed and travel outside of a hoistway or in a hoistway open on one side.

© iurii/Shutterstock.com

Figure 37.17 Elevators in public occupancies must be large enough to accommodate wheelchair users.

© Image Source/Getty Images

RESIDENTIAL ELEVATORS

Residential elevators are available in a range of sizes, including cars large enough to carry wheelchairs. Designs vary, but residential elevators may use hydraulic or electric traction systems similar to those described for commercial passenger elevators. The car in an electric system is directed by channel guides, and doors have safety-locking devices that prevent them from opening until the car arrives at a designated floor. The hydraulic unit uses a piston to raise the car.

Elevator cars are steel-reinforced and available with a variety of interior wall, floor, and ceiling finishes. Car sizes range from 36 in. (914 mm) to 42 in. (1066 mm) square. A custom 36 in. × 48 in. car can accommodate a wheelchair. Most handle loads up to 450 lb. (202.5 kg).

AUTOMATED TRANSFER SYSTEMS

Automated transfer systems provide for vertical and horizontal distribution of materials. They are widely used in hospitals, clinics, office buildings, hotels, and manufacturing plants. Among them are dumbwaiters, tote box and cart transfer systems, and horizontal tote box transfer systems.

Dumbwaiters

A dumbwaiter is a mechanism used to raise and lower a small car vertically within a building. Dumbwaiters are used in hospitals, restaurants, libraries, and office buildings to move mail, supplies, and materials (such as food, medicine, and books) from one floor to another. Car sizes are controlled by local and national codes. Standard heights are 3 ft., 3½ ft., and 4 ft. (915, 1,067, and 1,220 mm), with a maximum platform size of 9 sq. ft. (0.837 m²). Units range from light-duty lifts that carry 25 lb. to 50 lb. (11.25 kg. to 22.50 kg) to heavy-duty types that carry up to 500 lb. (225 kg). Dumbwaiter can be powered manually or electrically. Manually operated units have an endless rope connected to a large pulley that is connected by gears to a pulley that connects to the hoisting mechanism. Dumbwaiters have automatic braking mechanisms and are usually limited to two-story applications.

Electrically powered dumbwaiters are used on buildings of any height. They are available as drum type, traction type, or hydraulic, as explained in the section on passenger elevators (Figure 37.18). The drum type has a maximum height of rise of 40 ft. (12.2 m), but the traction type height is unlimited.

Figure 37.18 Dumbwaiter details for floor and counter-level loading.

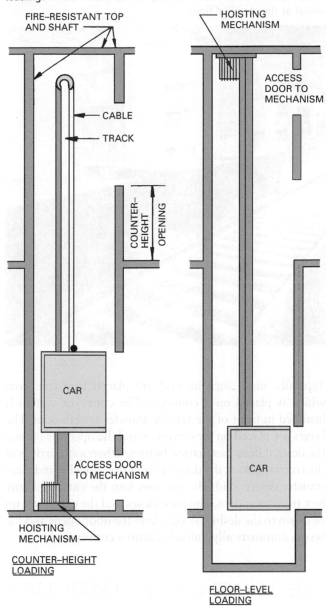

COUNTER–HEIGHT LOADING

FLOOR–LEVEL LOADING

Figure 37.19 Modern dumbwaiters are available in a variety of styles and door arrangements.

© Goodshoot/Corbis

dumbwaiters ride on vertical rails secured to each floor with brackets, with the lifting mechanism typically located on the top of the shaft (Figure 37.20).

Cart and Tote Box Transfer Systems

Cart and tote box transfer systems use high-performance dumbwaiter and elevator lift equipment that may work in connection with a horizontal transfer system. The systems available will lift up to 1,000 lb. (450 kg). Carts come in various sizes but may be up to 30 in. (762 mm) wide, 55 in. (1,397 mm) deep, and 65 in. (1,651 mm) high. Tote boxes are about 15 in. (381 mm) wide, 20 in. (508 mm) long, and 10 in. (254 mm) deep. They can be carried on standard dumbwaiters. Small carts with weights not exceeding the capacity of a standard dumbwaiter can be carried by them. Tote box systems use counter-high doors.

Horizontal transfer systems move tote boxes and other items into a vertical transfer system (Figure 37.21).

Electric dumbwaiters move 50 ft./min. to 150 ft./min. (15.25 m/min. to 45.75 m/min.). The higher speeds are typically used in buildings more than 50 ft. (15 m) high. Some types permit the standard speed to be reduced to as low as 25 ft./min. (7.6 m/min.) for carrying fragile items. Cars can open from the front, front and rear, or front and side (Figure 37.19). Standard doors are power or manually operated with bi-parting, slide-up, slide-down, and swinging door configurations.

Dumbwaiters may be counter loading or floor loading. Floor-loading types enable carts to load directly onto the dumbwaiter. Electric traction and drum

Figure 37.20 A traction electric-powered dumbwaiter.

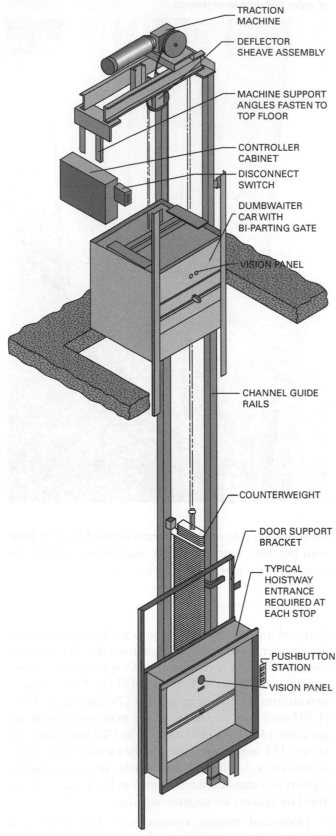

TRACTION
MACHINE

DEFLECTOR
SHEAVE ASSEMBLY

MACHINE SUPPORT
ANGLES FASTEN TO
TOP FLOOR

CONTROLLER
CABINET

DISCONNECT
SWITCH

DUMBWAITER
CAR WITH
BI-PARTING GATE

VISION PANEL

CHANNEL GUIDE
RAILS

COUNTERWEIGHT

DOOR SUPPORT
BRACKET

TYPICAL
HOISTWAY
ENTRANCE
REQUIRED AT
EACH STOP

PUSHBUTTON
STATION

VISION PANEL

TRACTION (TR)

© Matot, Inc.

Figure 37.21 Horizontal transfer systems redistribute items in tote boxes that are automatically moved onto a conveyor upon arrival at the specified floor.

© FK Photo/FK PHOTO/Corbis

Typically, items such as mail are placed in a tote box, which is placed on a conveyor. The conveyor system is installed in front of the vertical transfer system door. The boxes are placed on the conveyor and the operator presses the desired floor destination button. When a car arrives at the receiving floor, the door opens automatically, and a car transfer device loads the tote box into the car of the tote box transfer system. The door closes and the car moves up or down to the desired floor, where the door opens and the box is automatically unloaded onto a conveyor.

WHEELCHAIR LIFTS AND STAIR LIFTS

Wheelchair lifts are used for both interior and exterior locations. They are constructed of a steel platform with steel sides and a front gate that lowers to form a ramp that helps load the wheelchair. The ramp has a rubber skid-proof surface. Wheelchair lifts are operated by an electric motor and have an automatic stop switch that activates whenever a user releases the mechanism that controls movement. Lifts will not operate until all entry and exit doors are closed.

Typical maximum lifting capacity falls in the range of 400 lb. to 500 lb. (180 kg. to 225 kg), and the speed of rise is generally about 8 ft./min. (2.8 m/min.). Various lifting heights are available, with 3 ft. to 9 ft. (.914 m to 2.74 m) being common. The platform area is generally about 12 sq. ft. (1.1 m²).

Another type of wheelchair lift moves the passenger and wheelchair up a stair to another level. This system uses a platform with side enclosures to move a wheelchair up a stair on a rail system fastened to the wall or stair treads. It can travel up a multilevel straight stair. Two lifts are used when turning a 90° or 180° corner is necessary (Figure 37.22).

Figure 37.22 This stair-climbing wheelchair moves around corners and across stair landings by using two lifts.

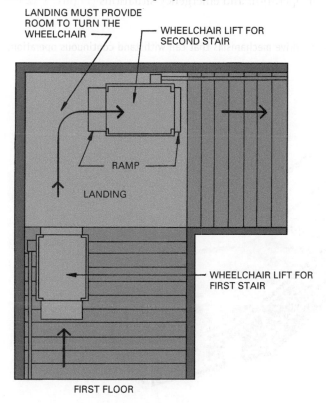

LANDING MUST PROVIDE ROOM TO TURN THE WHEELCHAIR

WHEELCHAIR LIFT FOR SECOND STAIR

RAMP

LANDING

WHEELCHAIR LIFT FOR FIRST STAIR

FIRST FLOOR

Stair lifts are used to move people who do not use a wheelchair but have difficulty climbing stairs. The lift consists of a chair that runs along a track that may be mounted on the stair or along the wall of the stairwell. The chair can move around corners and across stair landing. The rail can continue past the top of the stair, permitting a passenger to disembark a safe distance from the stairway. Lifts are electrically powered, and a series of gears and shafts provide the motion (Figure 37.23).

ESCALATORS

Escalators are inclined, continuous, power-driven stairways used to move passengers up or down between floors. Because escalators are an integral part of both horizontal and vertical circulation, they are generally placed directly

Figure 37.23 A chair lift moves a person up a stair while comfortably seated.

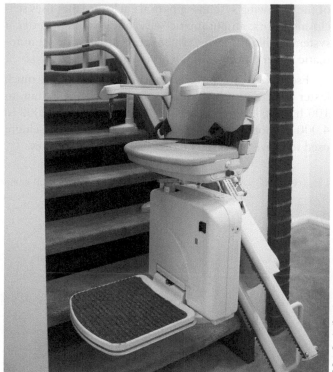

© Peter Dazeley/Getty Images

in the main line of traffic. Adequate space must be allowed at each landing, because people will be concentrated in these areas (Figure 37.24). An alternate method for moving between floors, such as stairs parallel with the escalator, is necessary when codes require it. This provides access should the escalator fail or fall under repair. Escalators may be used as a required means of exiting if they meet all requirements for an emergency egress stairway. This includes having floor openings enclosed to provide fire and smoke protection and the provision of an approved

Figure 37.24 Escalators require a large open area at each landing to provide sufficient space as people crowd on and off.

© Christophe Testi/Shutterstock.com

sprinkler system. Escalators not serving as a required exit should have floor openings enclosed or protected by one of the following systems: a partial enclosure that uses self-closing doors, an automatic self-closing rolling shutter, a system of high-velocity water-spray nozzles, or an automatic water curtain with an air exhaust system.

Escalators can move large numbers of people much faster than elevators. Typical speeds of up 90 ft./min. to 100 ft./min. (27 m/min. to 30 m/min.) can move 2,000 to 4,000 or more people per hour. However, they are seldom used to move passengers more than five or six stories.

Escalator Components

An escalator has a welded steel-truss structural frame. The stair is a series of moving steps that are cast metal and grooved to provide safe footing. The treads and risers are secured to a continuous chain that is moved by an electric-geared driving unit. Each side of the stair has a solid balustrade that covers the ends of the stair and supports a handrail. The handrail moves at the same speed as the stair. The escalator has electronic control devices for operation and emergency situations (Figure 37.25).

Figure 37.25 Escalators are built using heavy structural components and drive mechanisms that can withstand continuous operation.

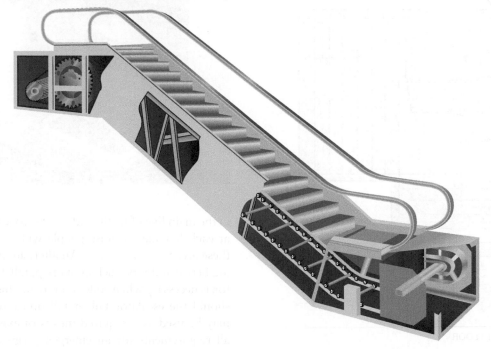

Standards and Safety

Escalator standards are available in the publication *American National Standard Safety Code for Elevators and Escalators*, ANSI/ASME A17.1, and the *Life Safety Code of the National Fire Protection Association*. In Canada, the requirements are given in Standard CAN 3B-44. The safety features covered by the codes include emergency stop buttons, broken step and drive chain switches, an electronically released brake that activates during power failures, a switch to prevent the exceeding of design speeds, step lights, switches to stop the escalator if a handrail breaks, switches to control handrail speed, a switch to shut down operation if a step breaks, smooth balustrades that protect sides of steps, and landing plates that protect against items getting caught as the steps flow under the floor.

Installation Examples

To provide smooth movement of passengers, escalators often are installed in pairs with one moving up and one heading down. They may be located in a parallel arrangement (Figure 37.26) or in a crisscross pattern (Figure 37.27).

Escalator Sizes

The stair sizes established in ANSI/ASME A17.1 require a maximum rise of 8 in. and a max depth of 15 in. (203 mm and 381 mm). The widths of escalator treads available are typically 24 in., 32 in., and 40 in. (0.6 m, 0.8 m, and 1.0 m). They are built on an angle of 30° and have various maximum rises, depending on design. Rises of 20 ft. to 30 ft. (6.1 m to 9.2 m) are common for a single unit. Figure 37.28 shows design details for one escalator installation.

Figure 37.26 Escalators may be installed in a parallel arrangement.

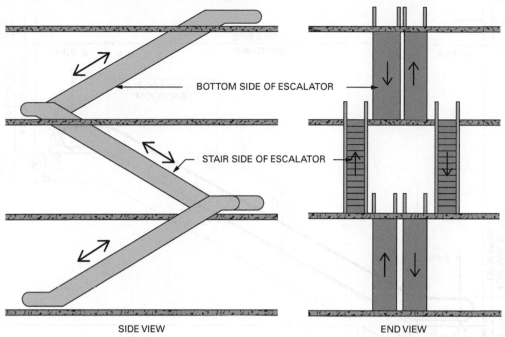

BOTTOM SIDE OF ESCALATOR

STAIR SIDE OF ESCALATOR

SIDE VIEW   END VIEW

Figure 37.27 A characteristic crisscross escalator installation.

© zhu difeng/Shutterstock.com

MOVING WALKS AND RAMPS

Moving walks are horizontal conveyor belts designed to move people (Figure 37.29). Usually used on level walks, they may have a slight rise or fall, seldom exceeding 5°. Moving ramps convey people up or down inclines (with a maximum slope of 12°) and connect with moving walks. They are used where large numbers of people need to be transported over long distances, such as in an airport terminal.

Typical widths of the moving flexible rubber-covered belt are 24 in., 32 in., and 40 in. (0.6 m, 0.8 m, and 1 m). The 24 in. belt can accommodate one adult, the 32 in. belt provides room for an adult and a child, and the 40 in. accommodates two adults or one adult with luggage.

The sides of the walks and ramps are enclosed with solid balustrades covered with an endless moving rubber handrail. The steel truss structural system is set in a concrete pit (Figure 37.30). The electrically driven apparatus has a mechanical and electrical safety system and controls similar to those used on escalators.

OTHER CONVEYING EQUIPMENT

In addition to the equipment already discussed in this chapter, there is a wide variety of other less-widely used conveying equipment for both people and materials.

Figure 37.28 Typical escalator installation details.

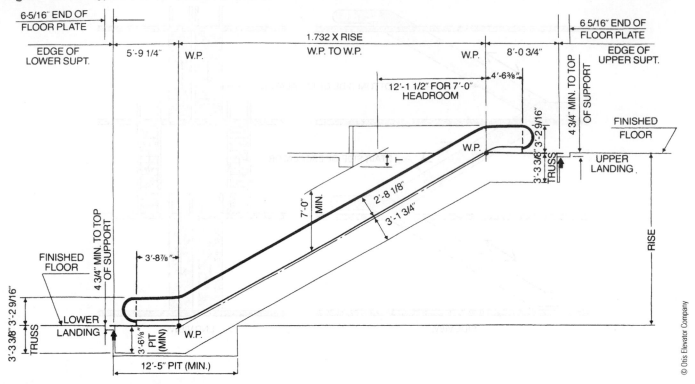

© Otis Elevator Company

Figure 37.29 A moving walk is used for long-distance horizontal travel within buildings.

© ArtisticPhoto/Shutterstock.com

These include shuttle transit, conveyors, pneumatic tube systems, material conveyors, cranes, hoists, and special-purpose material-moving systems.

Shuttle Transit

Shuttle transit systems provide horizontal transportation for distances beyond the practical limits of moving walks. They have many applications, including linking office or retail areas to remote parking facilities or moving passengers between terminals in large airports. Business and industrial parks sometimes use them to provide rapid internal transportation on large campuses (Figure 37.31).

Figure 37.30 A generalized illustration showing structural details through a moving walk.

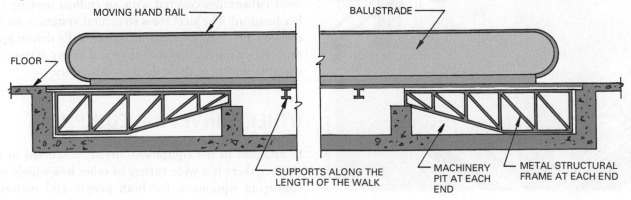

Figure 37.31 People are moved rapidly and comfortably between buildings by monorail systems.

Shuttle transit systems function much like elevators. The cars operate on automatic controls and are dispatched on a regular schedule or on call. Each boarding station has signs to indicate scheduled arrival times. The system uses standard elevator gearless traction machine drives and cable equipment. A steel cable is attached along the side of the shuttle. The steel guide rails and power rails are located adjacent the running surface. Vertical loads are supported by a cushion of air developed between pads on the bottom of the car and the guideway running surface. The cars may be operated singly or several can be joined to increase passenger capacity.

CONVEYERS AND PNEUMATIC TUBE SYSTEMS

Various types of conveyor and pneumatic tube systems are used to move materials within a building. An electric track vehicle conveyor system consists of self-propelled electric cars that travel over a network of tracks connecting stations within a building. Cars typically carry up to 50 lb. (22.5 kg) and move vertically and horizontally at speeds up to 120 ft./min. (36.6 m/min.). They are used to transport paperwork, tools, test samples, and other small items (Figure 37.32). Systems can be installed in a number of ways. The simplest is a single-track system in which the car moves back and forth on one track. A dual

Figure 37.32 This electric track vehicle conveyor system moves containers within the building following a single or dual track.

1. Container
2. Track
3. Horizontal 90-degree bend
4. Vertical 90-degree bend
5. Transfer unit
6. Power supply unit
7. CRT and control center
8. Reversing station
9. Through station
10. Fire door

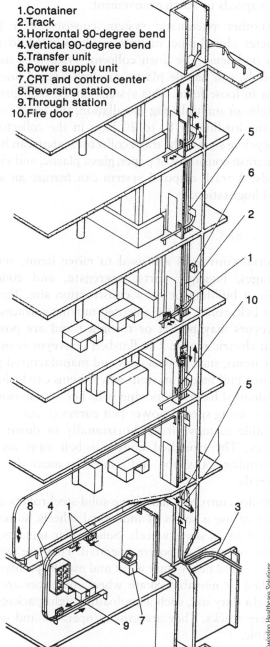

track permits cars to go both ways at the same time, providing faster service. The loop system stores cars and sends them forward when called, after which they return to storage until called again.

Pneumatic Tube Systems

Computerized pneumatic tube systems serve a variety of purposes. The system shown in Figure 37.33 transmits small items placed in tube-like carriers. They are conducted through a system of transmission piping and managed by a computerized control center. The system can be a simple single-zone route or have multiple zones, which speeds up carrier movement.

Another pneumatic system consists of a large-diameter closed-pipe vacuum network used to move trash or linens. The linen collector system, as used in hospitals, can handle plastic or cloth-bagged linens or linens in loose form. This system can be configured as a single- or multiple-bag installation, the latter permitting the loading of 10 to 20 bags in the collector before cycling starts. The trash collector system can handle loose trash consisting of paper, glass, plastic, and viscous liquids. A trash disposal system can feature an added shredding station.

Material Conveyors

Material conveyors are used to move items, such as packages, luggage, parts, aggregate, and concrete, within a building or on a construction site. They include belt, roller, and segmented moving surfaces. Belt conveyors may be flat or troughed and are powered by an electric motor. The flat-belt conveyor is used to move items, such as packages and manufactured parts, within a building to stations where items can be loaded or unloaded from the moving belt. The system can turn corners using special power belt curves (Figure 37.34). It is able to move items horizontally or down small inclines. The troughed conveyor belt runs on rollers forming a U-shape and is used to move dry loose materials.

Roller conveyors may have solid-steel rollers across a unit or use a series of individual wheels, sometimes referred to as skate wheels. Solid roller units may be gravity operated or power operated. They are used for medium- and heavy-duty work and may be permanently installed or movable. Skate wheel conveyors are used for light duty use, such as unloading light packages off delivery trucks. They are gravity operated and usually portable.

Figure 37.33 A computerized pneumatic tube conveying system transmits items stored in tube-shaped containers through tube transmission piping.

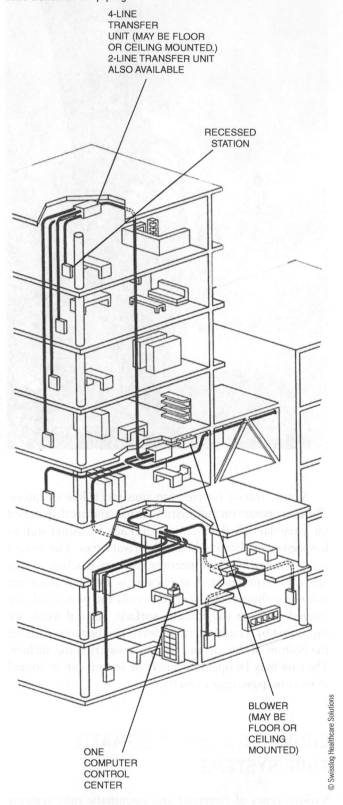

4-LINE TRANSFER UNIT (MAY BE FLOOR OR CEILING MOUNTED.) 2-LINE TRANSFER UNIT ALSO AVAILABLE

RECESSED STATION

BLOWER (MAY BE FLOOR OR CEILING MOUNTED)

ONE COMPUTER CONTROL CENTER

Figure 37.34 Items are moved directly on roller conveyors on solid roller or skate wheels.

Cranes and Hoists

Cranes and hoists are used to move materials and heavy freight at outdoor locations and within buildings. The choice of system depends on the applications required and is an important design aspect of a structure. Overhead, monorail, and under-hung cranes run on rails generally supported by the structural frame of a building, so loads imposed by the crane must be carefully calculated. Cranes can also be installed with a secondary structural system independent of the building structure.

Overhead cranes move along the top of fixed overhead rails. They are often used for specific tasks, such as moving steel members from storage to fabrication areas and moving finished members to storage or shipping areas (Figure 37.36). The trolley moves horizontally along a steel track with the lifting action produced by a hoist that is part of the trolley. A monorail has the hoist slung below a single steel track in a similar arrangement. The under-slung crane is much like the overhead crane except it travels along the bottom flange of a track. These cranes are powered by electric motors and are generally controlled by an operator on the floor. A control cable extends from hoist to floor. The operator can manipulate the hoist to lift or lower loads and the trolley to move along the track.

A segmented conveyor has a moving surface made up of flat sections joined with hinge-like connectors (Figure 37.35). Airport luggage conveyors of this type can handle heavy loads and are power driven.

Figure 37.35 Power-driven segmented conveyors use a moving surface made up of flat sections joined with hinge-like connectors to move materials around corners.

Figure 37.36 An overhead crane moves a trolley along a steel track with lifting action produced by a hoist.

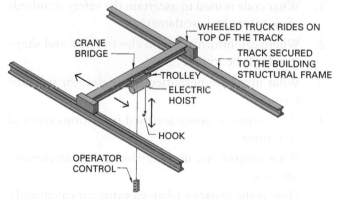

A gantry crane has an overhead track and a trolley that runs on a horizontal track supported by legs that run on rails secured to the floor. Gantry cranes are used by industries that must move very heavy loads. They may operate inside or outside a building and do not require support from its structural system. On large cranes, the operator sits in a cab located below the bridge.

A diesel-powered mobile gantry is shown in Figure 37.37. It moves on wide-base radial tires, so it is more versatile than the track-bound gantry. It is available in a wide range of structural configurations. The cab is located to give the operator excellent visibility and is reinforced with steel guards. It also has a heater, defroster, windshield wipers, tinted glass, and a dome light. The control panel includes a load weight indicator. It is available with two-wheel and four-wheel drive.

Figure 37.37 A diesel-powered mobile gantry moves on wide-base radial tires for increased maneuverability.

Review Questions

1. What code is used to ascertain the safety standards of elevators and escalators?

2. Where can information on elevator size and shape standards be found?

3. What items are permitted in an elevator machine room?

4. What types of doors are used on various types of elevators?

5. What controls are normally located on an elevator car panel?

6. How is the capacity of an elevator car calculated?

7. What happens to elevator cars in case of a power failure?

8. What types of traction machines are used for hoisting elevators?

9. What are the major differences between gear-driven and gearless traction machines?

10. How is the speed of gearless traction elevator drive mechanisms varied?

11. What are the two types of elevator control systems?

12. What types of roping systems are used with traction elevators?

13. What types of elevator operating systems are available?

14. What mechanism is used for moving hydraulic elevators?

15. What are the classes of freight elevators?

16. What purposes do automated transfer systems serve?

17. What types of lifts are used to move the physically handicapped from floor to floor?

18. What escalator installation configurations are generally used?

19. What emergency features are included in an escalator system?

20. What is a moving walk?

21. Where is a shuttle transit system likely to be used?

22. What are the commonly used material conveyors?

Key Terms

Buffer	Gearless Traction Elevator	Moving Walk
Car Safety	Hoistway	Pit
Elevator Car	Hydraulic Elevator	Platform Safety
Escalator	Landing Zone	Sheave

Activity

1. Arrange to visit buildings having a variety of conveying systems. Inspect the mechanical rooms and other behind-the-scenes aspects of the mechanism. Ask about maintenance activities and schedule a visit for each system. Detail these in a written report.

Additional Resources

American National Standard Safety Code for Elevators, Dumbwaiters, Escalators, and Moving Walks, American National Standards Institute, New York.

Other resources include:

Publications from the National Association of Elevator Contractors, Conyers, GA.

Publications from the National Elevator Industry, Teaneck, NJ.

See Appendix C for addresses of professional and trade organizations and other sources of technical information.

Fire Suppression
CSI MasterFormat™

© mrfiza/Shutterstock.com

Fire-Suppression Systems

Upon completion of this chapter, the student should be able to:

• Have an understanding of the fire codes and the agencies that maintain them.

• Develop a working knowledge of the various types of fire-suppression systems.

Build Your Knowledge

For further study on these materials and methods, please refer to:

Chapter 35 Specialties, Equipment, and Furnishings
Topic: Fire-Suppression Specialties

Fires affect thousands of buildings each year, resulting in injury and damage to property. Fire-resistant design and construction is of utmost concern during both building design and operations. New construction must be designed to incorporate automatic fire-protection systems that are effective in detecting, controlling, and extinguishing a building fire in its early stages. Fire-protection engineers should be involved in all phases of project development in order to ensure a high degree of protection of human life from fire and the products of combustion as well as to reduce the potential loss from fire to property.

Building fire protection is typically achieved through two basic strategies: active protection and passive protection. **Passive fire protection** involves the application of fire-resistance-rated construction assemblies that are used to compartmentalize parts of a building in order to control the spread of fire and smoke and provide sufficient time for evacuation and firefighting. Many fire-resistant materials and assemblies have been outlined in previous chapters.

Active fire protection is characterized by devices and systems that set in motion a physical reaction to fight a fire. Fires can be controlled or extinguished either manually or automatically. Manual approaches include the use of a fire extinguisher or a standpipe system. Automatic fire-suppression systems include various types of sprinkler systems, foam and fog extinguishing systems, gas and chemical systems.

FIRE CODES

The International Code Council and the National Fire Protection Association administer model codes that are adopted by state or local jurisdictions and enforced by fire prevention officers within municipal fire departments. The fire codes compile a series of regulations mandating minimum requirements to prevent fire hazards that may occur from building operations, the storage or use of hazardous materials, or from other situations (Figure 38.1). While concentrated chiefly on preventing fires, the code addresses ensuring that the necessary training and equipment is provided, and that the layout and design of a building supports fire prevention and containment goals. The fire codes also direct the inspection and maintenance procedures for various fire protection systems in order to maintain optimal active and passive fire-protection measures. In addition, the Underwriters Laboratory, Inc. (UL) tests and approves fire-protection equipment and issues reports in its Fire Protection Equipment List.

Figure 38.1 The Fire Codes address a range of related issues, from emergency planning and preparedness to automatic sprinkler and alarm systems.

FIRE AND SMOKE MANAGEMENT

Fire and smoke management are an important part of a buildings life-safety system and are often dictated by code for different building types, sizes, and configurations. They are employed to protect building content and the area surrounding a fire from spreading, and provide a clear path for fire fighters to extinguish a fire quickly.

Fire Management

Fire management is typically provided by compartmentalizing a building into smaller areas (Figure 38.2). The

Figure 38.2 Fire management is provided by compartmentalizing a building into smaller areas separated by fire-resistant assemblies.

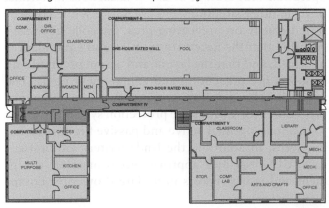

purpose of compartmentalizing is to confine the fire, heat, smoke, and toxic gases to their area of origin until the fire is extinguished, or burns itself out. In addition, compartmentalization protects the designated egress ways and areas of refuge for both occupants and fire-fighters. A fire compartment or area, by definition, is surrounded by exterior walls, fire barriers, fire-rated walls, and horizontal assemblies. Building areas are separated using fire-rated assemblies like floors, walls, and ceilings to contain the fire in a given area for some defined interval of time (see Chapter 33 for more details on these assemblies). The compartments use self-closing doors, fire dampers, and other self-closing devices to trap smoke and create a smoke reservoir.

HVAC and mechanical trades must interface with this compartmentalization any time their systems penetrate them. All penetrations must be sealed or otherwise protected to maintain the integrity of the fire separation. Typically, this is accomplished by fire-stopping around the penetrating system and, in the case of ducts, by providing fire dampers that will contain the fire on one side of the separation. A *fire damper* is a device that prevents the spread of fire through rated walls and partitions where they are penetrated by air ducts (Figure 38.3). Fire dampers are required by all building codes to maintain the required fire-resistance ratings of construction assemblies when they are breached by air ducts or other HVAC components. A fire damper uses a heat-detection device known as a fusible link. A *fusible link* is an apparatus consisting of two strips of metal soldered together with a fusible alloy that is designed to melt at a specific temperature. When temperatures reach above 165°F. (73.8°C), the link melts and automatically closes the

Figure 38.3 Fire dampers are used to maintain the required fire-resistance ratings of construction assemblies when they are penetrated by air ducts or other HVAC components.

fire damper. Fusible links are also used as the triggering method in fire-sprinkler systems and automatic door-release mechanisms that close fire doors.

Smoke Control Systems

Smoke control systems are intended to limit and control the movement of smoke during a fire to allow for the evacuation of occupants in a nontoxic environment. These systems respond to a fire not by acting as a barrier, but by activating equipment that exhausts smoke by natural or mechanical ventilation.

Smoke control design uses dedicated smoke exhaust fans or reversible HVAC supply fans, coupled with numerous detectors and dampers to manipulate the movement of smoke. The most common approach involves pressurizing areas on either side of two fire compartments to create a pressurized sandwich of air that will move smoke away from the protected areas. Most high-rise buildings employ a stairwell pressurization system to keep the stairs clear of smoke. Stairwells are frequently the principal escape routes for the occupants of a building and the primary access routes for the fire fighting teams responding to the emergency. As such, it is critical that stairwells remain clear of smoke for as long as possible to safely evacuate the building and bring the fire under control.

Many smoke and fire management and control systems use the HVAC air handling equipment to provide the make-up and exhaust air capacity required for the smoke control and management functions. In most systems, all air-handling units will shut down and smoke dampers will close when an alarm is activated. Operation of individual fans and dampers is then controlled manually by fire personnel or the building operator of the smoke control management system.

A major component of smoke control involves the mechanical venting of smoke, or smoke exhaust. The method uses the principle of exhausting smoke through exhaust fans or other openings that are located high in the building (Figure 38.4). Fresh, or make-up air, is introduced through other openings that are lower than the exhaust points. Systems used for smoke exhaust systems include power ventilator fans, fusible link automated smoke vents, automatic louver vents, or melt-out vents that open during extreme heat. The sizes and types of venting devices and controls depend on the systems design specifications. Large building areas may require numerous vents spread over equal areas of the roof.

Figure 38.4 Smoke-removal systems use exhaust fans or other openings that are located high in the building to release smoke.

FIRE-SUPPRESSION SYSTEMS

Fire-suppression systems include various types of sprinkler systems, foam and fog extinguishing systems, gas systems, chemical systems, and manual water extinguishing.

Automated Water Sprinkler Systems

Water is the most common substance used to extinguish building fires. It is readily available, economical, and serves to smother and cool a fire. As water interacts with fire, it turns to vapor, which increases its volume and helps to disperse the oxygen the fire needs to continue growing.

Automatic fire-suppression sprinkler systems use the water supply from a municipal water system. Tall buildings sometimes require a backup supply that can be accommodated in a storage tank on the roof (Figure 38.5). The water inflow to the system is controlled by a valve. When water flow is detected, an alarm valve activates. The riser has a *Siamese connection*

Figure 38.5 A simplified illustration of an automatic water fire-suppression sprinkler system.

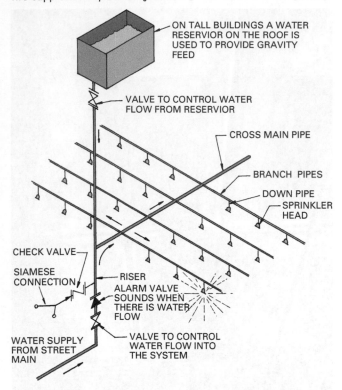

located outside the building for use by the fire department to pump additional water when needed from an outside source, such as a secondary water supply or a street hydrant (Figure 38.6). Each area on each floor protected by sprinklers has a similar system. In large buildings, separate risers to larger portions of the structures are used. Automatic water sprinkler systems include wet pipe, dry pipe, deluge, pre-action, and various

Figure 38.6 A Siamese connection is used by the fire department to connect a supplemental source of water to the building's fire-suppression system.

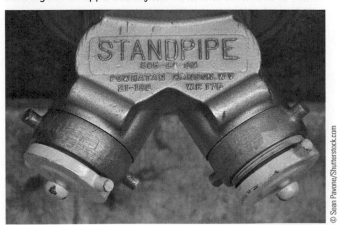

types of water fog and liquid foams, each of which use some type of sprinkler head or nozzle.

The most common type of water fire suppression is the wet-pipe system. Pipes are filled with water under pressure in the system of pipes at all times. When sprinkler heads are activated by the heat from a fire, water is released immediately (Figure 38.7). In order to limit water damage, only sprinkler heads over the area where fire is located activate. In areas where pipes may freeze, the system is filled with an antifreeze and water mixture.

Figure 38.7 A wet-pipe sprinkler system maintains water under pressure in the pipes.

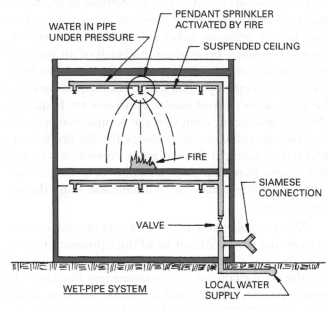

Dry-pipe systems are maintained under pressure with either air or nitrogen. When a sprinkler head opens due to heat from a fire, air pressure is released, causing the dry-pipe to open (Figure 38.8). The pipes fill quickly with water, which moves through the open sprinkler heads. Since the pipes are dry, this system is widely used in areas subject to freezing temperatures. Normally, upright sprinkler heads are used because pendant heads may hold water and freeze. Dry-pipe systems require additional equipment designed to maintain the necessary air pressure in the pipes.

Deluge systems are designed to deliver as much water as quickly as possible. Sprinkler heads or spray nozzles are kept open at all times and the pipes are dry. The system activates by a sensitive fire-detection valve that opens the deluge valve rather than activating each sprinkler head (Figure 38.9). The deluge system delivers water to an entire area because all sprinklers are open.

Figure 38.8 A dry-pipe system maintains the sprinkler pipes dry and under air pressure.

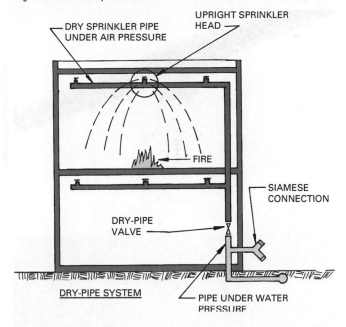

Figure 38.9 A deluge-pipe system activates all the sprinkler heads in an area at the same time.

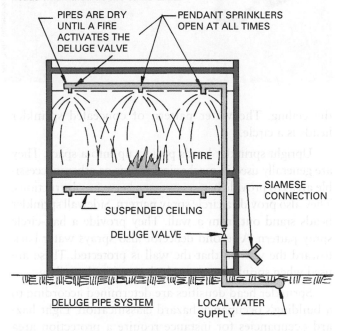

As the name implies, these systems are able to deliver vast quantities of water simultaneously. Deluge systems are often used in buildings like flammable liquid storage facilities where a rapid spread of fire is possible.

Pre-action systems operate with dry pipes that may or may not be pressurized. They use a fire-detection system that is more sensitive than typical. This system opens the pre-action valve, allowing water to flow

only to sprinkler heads opened by heat from the fire (**Figure 38.10**). Careful attention must be paid to the design of buildings using pre-action systems, as there is a delay in the activation of the sprinklers, which could allow a fire to spread. The amount of water delivered by the system must be large enough to offset this delay. Pre-action systems find use in buildings where the contents are particularly sensitive to water damage such as server rooms or museums.

Figure 38.10 A pre-action pipe system maintains a dry pipe until the sprinkler valve is opened by a sensitive fire detector. Water is sprayed only by sprinklers activated by the fire.

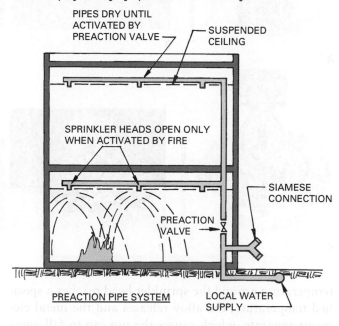

Water-fog systems use a standard sprinkler piping system and have spray heads or nozzles instead of sprinkler heads. They are used in areas where highly flammable materials are stored. The fog tends to cool materials, keeping them below ignition temperature.

Sprinkler Heads

Automatic sprinkler heads activate when exposed to excessive temperatures (usually about 165°F for the particular area being protected). When activated, a valve opens, releasing water stored in the attached piping. Sprinkler heads of various designs are available, as shown in **Figure 38.11**. Except for deluge systems, heads activate individually, preventing water from being released in areas that are not threatened by fire. A fusible-link sprinkler head has a two-part metal element that is fused by a heat-sensitive alloy. The link holds the pip cap, or plug, in place. Once the ambient

Figure 38.11 Typical sprinkler heads: (a) an upright head that is on the top of the exposed branch pipe and below the ceiling, (b) a recessed pendant that extends below the line of the ceiling, (c) an adjustable concealed head with a round plate that provides a decorative appearance, and (d) an adjustable pendant used in residential applications and standard commercial systems.

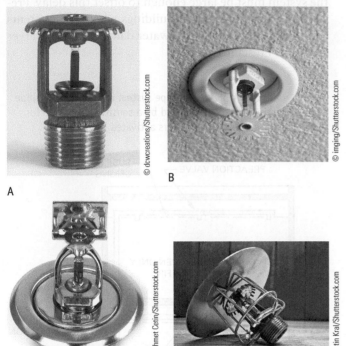

A

B

C

D

Figure 38.12 Pendant sprinkler heads hang down from the ceiling, whereas upright sprinkler heads project up into a space.

temperature around the sprinkler head reaches a specified temperature, the alloy releases and the metal elements separate, which causes the pip cap to fall away and release water.

Glass-bulb sprinkler heads have a small glass reservoir that holds a heat-sensitive liquid. This glass bulb holds the pip cap in place. When the ambient temperature of the liquid reaches a certain level, the liquid expands, causing the glass bulb to break, which allows the pip cap to fall away, releasing water. As with the fusible-link heads, water is released only where the ambient temperature reaches a certain level, which helps limit water damage.

Sprinkler heads may be upright or pendant types (Figure 38.12). Pendant sprinkler heads hang down from the ceiling and spray water in a circle pattern. Concealed pendant sprinkler heads are recessed in a ceiling and are covered with a decorative cap. The cap will fall away about 20°F (6.66°C) prior to the activation temperature of the sprinkler. Once the sprinkler reaches its rated activation temperature, the head will drop below the ceiling. The water pattern of concealed sprinkler heads is a circle.

Upright sprinkler heads project up into a space. They are generally used in mechanical rooms or other inaccessible areas to provide better coverage between obstructions. They also provide a circle spray pattern. Sidewall sprinkler heads stand out from a wall. They provide a half-circle spray pattern. A second deflector also sprays water back toward the wall so that the wall is protected. These are used when sprinklers cannot be located in the ceiling.

Sprinkler head densities are determined according to a building's occupancy hazard classification. Light hazard occupancies for instance require a protection area per sprinkler head of 200 ft.^{2}, while in extra-hazard occupancies, the area covered per sprinkler head may not exceed 90 ft^2.

Foam Fire-Suppressant Systems

Foam fire-suppressant systems are used on fires involving gasoline or chemicals for which water systems are ineffective. Foams are masses of air-filled or gas-filled bubbles formed by mechanical or chemical techniques. Chemical foam forms by a reaction between water and several chemicals, resulting in bubble-filled foam produced by carbon dioxide. The foams are lighter than water and tend to float on the surface of burning liquids, smothering them and sealing in gases (Figure 38.13). It also blocks airflow, thereby restricting the supply of oxygen to a fire. Foam agents are conductive in nature and cannot be used on electrical fires.

Figure 38.13 Foam systems produce large amounts of air-filled or gas-filled bubbles that fill a compartment with foam from large-diameter ducts.

© Markus Scholz/Corbis Wire/Corbis

Foams tend to expand when activated and are classified by their expansion ration. Low-expansion foams have an expansion ration of 20:1 and are used primarily in extinguishing flammable liquids. These foams can be distributed through a water sprinkler piping system. High-expansion foam forms by passing air through a screen constantly wetted with a chemical solution and a small amount of water, and has an expansion ration

of 200:1. It is moved to fire areas in large ducts and generally fills the entire compartment on fire. Various foam extinguishing agents and their properties can be found in the NFPA *Fire Protection Handbook*.

Gas Fire-Suppression Systems

Gas fire-suppression systems have the advantage of being able to flood areas, suppressing fire with little harm to contents. This is especially important in buildings with sensitive equipment, such as computer rooms, commercial aircraft, telephone exchanges, and libraries. Gas used for fire suppression is usually carbon dioxide or Halon 1301.

Carbon dioxide covers fire with a blanket of heavy gas that reduces the oxygen content in the surrounding air so much that combustion is extinguished. It is used in compartmentalized areas, such as electrical cabinets, but never in areas where people may be present. Carbon dioxide is stored in liquid form under high pressure and is released as a gas that cools and smothers a fire. Fire-sensing systems activate control valves that release gas through a series of pipes with nozzles directed at the point of a potential fire.

Halon 1301 extinguishes fire by interfering with the combustion process, thus preventing it from occurring. Low concentrations of Halon 1301 extinguish flames rapidly, and higher concentrations completely diffuse into the atmosphere, preventing further burning or explosion. Halon 1301 is stored as a gas in cylinders under pressure (Figure 38.14). It vaporizes as it enters a fire area through discharge nozzles. The vapor diffuses into the surrounding atmosphere, leaving no residue.

While Halon 1301 is an effective fire-suppression agent, it is an ozone-depleting gas. Since the early 1990s, manufacturers have successfully developed safe and effective Halon alternatives. Generally, Halon replacement agents available today fall into two broad categories: in-kind (gaseous extinguishing agents) and not-in-kind (alternative technologies). In-kind gaseous agents generally include Halocarbons and Inert Gases.

DRY CHEMICAL SYSTEMS

Dry chemical systems do not penetrate burning materials but remain on the surface and smother fire. Chemicals frequently used include sodium bicarbonate, potassium chloride, and monoammonium phosphate. They are effective on flammable liquids, electrical, and ordinary

Figure 38.14 Cylinder gas canisters store gases used to extinguish fires without damaging equipment or other building contents.

© Noppasin/Shutterstock.com

Figure 38.15 Manual fire-suppression systems use hose stations connected to the water supply system on each floor of a building.

© Angelo Giladelli/Shutterstock.com

combustible materials. They have a high tolerance for extreme weather conditions and temperatures.

Automatic systems pipe chemicals directly to areas in which fire might occur. These tend to be limited areas, such as a piece of equipment that is composed of materials that make it a constant fire hazard. Dry chemicals are also widely used in handheld fire extinguishers.

MANUAL FIRE-SUPPRESSION SYSTEMS

Manual fire-suppression systems have fire hose stations on each floor of a building that are connected to the water piping system (Figure 38.15). When water pressure is insufficient to supply stations on the upper floors of a building (or will not supply water in adequate quantity), standpipes are added to the system.

Standpipe systems consist of a series of pipes, which connect a water supply to hose connections that allow for manual fire fighting within a building (Figure 38.16). They are designed to provide a pre-piped water system for building occupants or the fire department, not unlike a fire hydrant within a building. Older buildings were equipped with standpipe systems as the main fire-suppression method. Modern commercial buildings typically use a combination system, which supplies the fire sprinkler and standpipe system. When water pressure is insufficient to supply hose stations on the upper floors of a building (or will not supply water in adequate quantity), standpipes can be added to the system. Standpipes provide an extra supply when normal water pressures fail. Standpipe systems can include pumps to increase water pressure. A stand pipe can also provide a reserve of potable water for normal building use. Additional information on fire detection devices and alarm control systems is provided in Chapter 42.

Figure 38.16 A simplified schematic of a manual fire-suppression system utilizing standpipes.

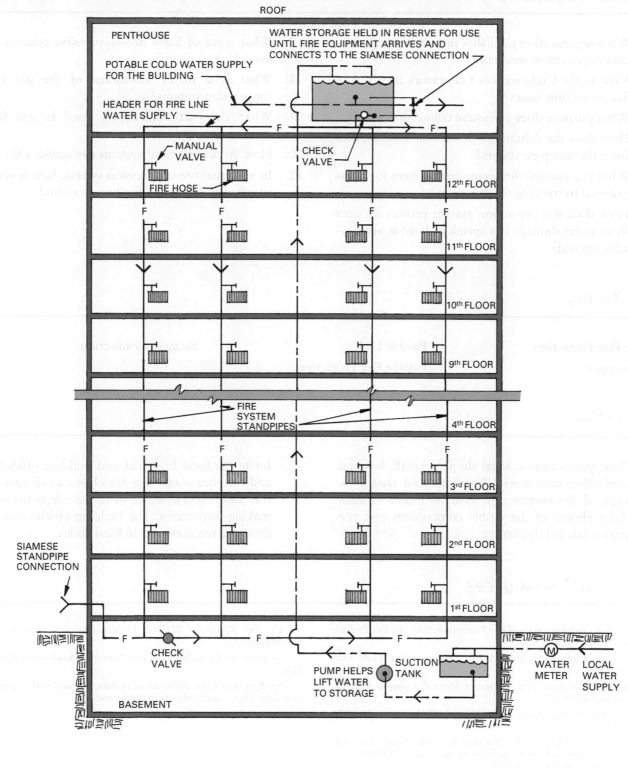

Review Questions

1. What organization publishes the major fire protection requirement manuals?

2. How is the Underwriters Laboratory involved in fire protection issues?

3. What purpose does a Siamese connection serve?

4. How does the deluge fire protection system differ from the wet-pipe system?

5. What is a suitable fire protection system for areas exposed to freezing temperatures?

6. How does the pre-action system protect an area from water damage if a sprinkler head is accidentally opened?

7. What types of foam fire-suppression systems are available?

8. What is a major advantage of the gas fire-suppression systems?

9. What gases are commonly used in gas fire-suppression systems?

10. How do dry chemical systems extinguish a fire?

11. In a manual fire-suppression system, how is water supplied to individual fire hose stations?

Key Terms

Active Fire Protection

Fire Damper

Fusible Link

Passive Fire Protection

Siamese Connection

Activities

1. Tour your campus, local shopping mall, hospital, and other commercial buildings and record the type of fire-suppression systems found in each. Take photos of the visible components and prepare a labeled display.

2. Invite the local fire chief and building official to address your class. The fire chief can discuss the role and responsibilities of the fire department in making inspections. The building official can address fire requirements in local codes.

Additional Resources

Cote, A.E., *Fundamentals of Fire Protection*, National Fire Protection Association, Quincy, MA.

Fire and Safety Inspection Manual, National Fire Protection Association, Quincy, MA.

Gypsum Association Fire Resistance Manual, Gypsum Association, Washington, DC.

International Fire Code, International Code Council, Falls Church, VA.

National Fire Alarm Code; *National Fuel Gas Code*; *National L-P Gas Code*, National Fire Code Subscription Service, National Fire Protection Association, Quincy, MA.

UL Fire Resistance Manual, Underwriters Laboratories, Inc., Northbrook, IL.

Uniform Fire Code, National Fire Protection Association, Quincy, MA.

See Appendix C for addresses of professional and trade organizations and other sources of technical information.

Plumbing
CSI MasterFormat™

Plumbing Systems

Upon completion of this chapter, the student should be able to:

- Relate plumbing codes to plumbing system design.
- Understand the types of piping, tubing, and fittings used on plumbing systems.

- Be knowledgeable about the sources of potable water and its distribution systems.
- Understand the components of sanitary piping systems.

Build Your Knowledge

For further study on these materials and methods, please refer to:

Chapter 15 Ferrous Metals

Topic: Steel Products

Chapter 16 Non-Ferrous Metals

Chapter 23 Plastics

A basic need for a successful building is a well-functioning plumbing system. Plumbing is the system of pipes, drains, valves, and other devices installed in a building for the distribution of water for drinking, heating, washing, and the removal of waterborne wastes, and the skilled trade of working with pipes, tubing, and plumbing fixtures in such systems. The plumbing system in any building is composed of two separate subsystems. One system brings freshwater in, and the other takes wastewater out. A typical building plumbing system includes hot and cold *potable (drinkable) water* distribution systems and a sanitary disposal system. Additional plumbing systems are designed to suit a special need, such as compressed air systems, fuel oil systems, natural gas systems, and oxygen gas systems.

PLUMBING CODES

Materials, plumbing system design, and installation procedures are strictly regulated by local plumbing codes. Local governments adopt the *International Plumbing Code*, sponsored by the International Code Council. The plumbing codes sponsored for many years by various code organizations have been cooperatively joined to form the *International Plumbing Code*.

POTABLE WATER SUPPLY

Of the approximately 370 billion gallons of water on earth, almost 98 percent resides in the oceans; less than 1 percent is fresh water in the atmosphere, rivers, lakes and groundwater; with the remaining 2 to 3 percent contained in glaciers and ice caps.

Potable (drinkable) water for public and privately owned central water systems may be obtained from rivers, lakes, ponds, surface runoff, and wells. Surface-collected water tends to contain a number of contaminates that must be removed to produce potable water. Quality problems include hardness caused by calcium and magnesium salts, discoloration caused by manganese or iron, corrosion caused by acidity in the water, pollution by sewage and organic matter, odor and taste problems caused by organic matter, and turbidity caused by silt and other suspended matter. Water treatment plants process water to correct these problems (**Figure 39.1**).

Figure 39.1 Water-treatment-plant processes include settling, screening, sedimentation and coagulation, filtration, disinfecting, aeration, and softening.

Water treatment includes processes such as screening water at intake, sedimentation (allowing particles to drop in a settling basin), coagulation (removing suspended matter with a chemical, such as hydrated aluminum sulfate, in a settling basin), filtration (removing suspended particles and some bacteria through a filter, such as sand or diatomaceous earth, or by chlorination filtration), disinfection (removing harmful organisms using bromine, iodine, ozone, or heat treatment), softening (removing calcium and magnesium), and aeration (exposing water to air, by spraying, for example, to improve taste and color).

In areas without a central supply, water can be obtained from wells. These are usually drilled, driven, or jetted. A drilled well involves placing pipe in a deep hole in the earth dug with a steel auger. A driven well is formed by placing a steel point on the end of sections of pipe

and driving them into the earth. A jetted well has a well point from which a high-pressure stream of water pumps, opening a hole in the earth for the well pipe to move into.

Well water, as a rule, does not require the extensive treatments needed for surface water. However, it must be tested for purity and treated as required to reduce hardness, iron, or magnesium.

Once potable water is delivered to a building, the quality must be maintained as the water is stored and distributed through the building piping system. Potable water systems must be kept isolated from all other plumbing systems.

POTABLE WATER DISTRIBUTION SYSTEMS

Codes specify that all fixtures in a building must be supplied with potable water at required pressures. Potable water in residential buildings generally uses water pressure from the water system main, or pumps, on a well to supply the fixtures (**Figure 39.2**). Water enters a building through a meter and a shut-off valve (**Figure 39.3**). In cold climates, the meter is located inside the building and the water service line below the frost line. The depth of the frost line varies in different geographical areas. The water continues to a water softener if required, then on to a water heater and the horizontal circulation lines. Water moves to fixtures in the upper stories through *risers* and to individual fixtures through small-diameter pipe connectors. Hot water runs from the water heater through a similar piping distribution system. An upfeed system is one in which the water moves up from the source into the building and to the fixtures. The piping used must consist of the proper size to carry the amount of water needed by each fixture. Circulation pipes and risers usually feed more than one fixture, so total expected flow must be calculated.

Low multistory buildings may also use an upfeed system if water demand is not excessive (**Figure 39.4**). A series of basement pumps provide the needed pressure to raise the water to desired heights and maintain pressure as fixtures are used.

Upfeed systems are useful on multistory buildings that are not high enough to warrant the expense of a rooftop water storage system but do have height limitations. A system may require two or more pumps. As water demand increases, a second or third pump will come online to increase flow and maintain pressure. It

Figure 39.2 Components of a typical residential potable water-distribution system.

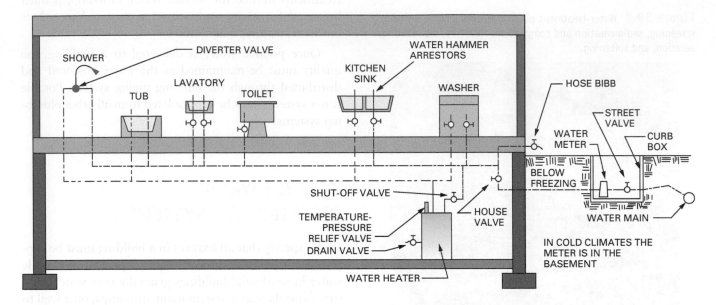

Figure 39.3 Water enters a building through a meter and a shut-off valve.

should be noted that the supply available from the central water system main should be large enough so that demands of the building can be met without reducing water service to neighboring buildings. Notice that this system does not have a reserve water supply.

Tall multistory buildings utilize a downfeed potable hot and cold water distribution system. Water from the central system is pumped from a street main or water storage suction tank to a roof storage tank with one or more pumps. This storage tank also holds a reserve for

the fire-suppression system (see Chapter 38) and a supply of potable water for fixtures. The water flows from a header pipe to downfeed risers from which branch water lines run to fixtures on each floor (Figure 39.5). Very tall buildings usually are divided into several zones, each having its own pumps and storage tanks (Figure 39.6).

GAS DISTRIBUTION SYSTEMS

Gas is piped to a meter outside a building by a utility company. The building's gas piping is run inside to individual fixtures that require fuel, such as a gas-fired furnace. Codes regulate materials and methods of installation. Generally, a building's gas inlet must be above grade. Gas piping within the building is black steel pipe meeting the requirements of ASTM A53 *Standard Specification for Pipe,* or ASTM A106 *Standard Specification for Seamless Carbon Steel Pipe for High-Temperature Service.* It uses steel or malleable iron fittings. Some codes permit the use of copper or brass pipe if the gas type will not cause them to corrode. Some types of plastic pipe are permitted for underground exterior installation. They must receive a protective coating to help resist corrosion from soil elements.

Figure 39.4 A simplified schematic of an upfeed potable water distribution system for a low-rise building.

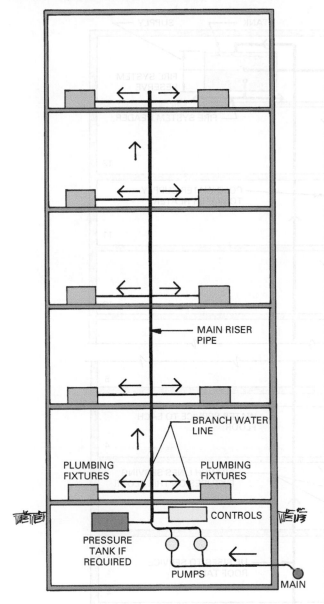

MAIN RISER PIPE

BRANCH WATER LINE

PLUMBING FIXTURES

PLUMBING FIXTURES

CONTROLS

PRESSURE TANK IF REQUIRED

PUMPS

MAIN

PIPING, TUBING, AND FITTINGS

Water supply is piped through buildings to a number of fixtures for various purposes. Potable water is delivered to sinks, lavatories, and water heaters. Typically, it is also run to toilets (water closets) and exterior hose bibs. Pipes for most applications are hidden inside the walls, floors, or ceilings of a building structure (Figure 39.7). Pipes can be exposed in areas where their visibility is not important, but must always be protected from freezing. Plumbing systems use valves to control the flow of water

into a building, within parts of large buildings, and to individual fixtures.

The types of pipe used for potable water systems are detailed in Table 39.1. Decisions on which to use are based on cost, length of service expected, and the required quality of the water. Some water contains large amounts of minerals and is very corrosive. Pipes that carries hot water must withstand temperatures of 180°F (82°C) or higher. In special applications, the effect of the liquid or gas to be carried must be considered. Gasoline, liquified petroleum gas, and similar materials have a substantial effect on pipes.

Plastic, copper, and steel pipe and fittings are manufactured with a variety of diameters and wall thicknesses (Figure 39.8). The thicker the pipe wall, the greater the pressure the pipe can carry.

Galvanized steel pipe is strong, long lasting, and can be used for hot and cold water. Piping and fittings are joined with threaded connections. The pipe is available in three grades: standard weight, extra strong, and double extra strong.

Welded steel pipe is made by rolling a flat hot steel strip into a circular shape and butt-welding the edges as they are pressed together. It is available in the same grades as galvanized steel pipe. The outside diameter is the same for all three grades and the difference in the thickness of the wall is taken up on the inside of the pipe. Welded steel pipe is joined by threaded fittings.

Red brass pipe is used for water lines, especially if the water contains corrosive elements. It is manufactured with threaded fittings and plain ends that are brazed to socket-type fittings.

Copper water tubing is an excellent hot and cold water distribution material. It is resistant to corrosion and does not rust. Flexible copper tubing is easy to bend into various shapes, reducing the number of fittings needed. The joints fit together in a socket arrangement and are secured by soldering (Figure 39.9). The solder used cannot contain lead because lead will contaminate the water, causing potential health problems.

Copper water tubing has a high coefficient of expansion. When carrying heated liquids, it can expand enough in length to cause damage to the pipe unless the expansion is allowed for. Typically, a section of pipe is bent into a loop, allowing it to expand and contract.

Different types of copper tubing are shown in Table 39.2. Copper water tubing is available in types K, L, and M. Type K has the heaviest wall and is marked

Figure 39.5 A simplified schematic of a downfeed potable hot and cold water-distribution system typically used in high-rise buildings.

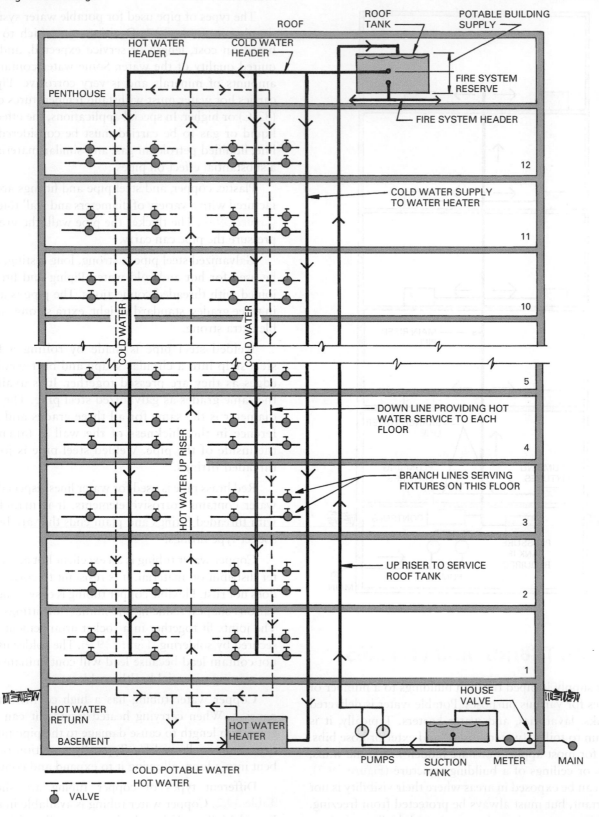

Figure 39.6 A simplified schematic of a two-zone downfeed hot and cold water system.

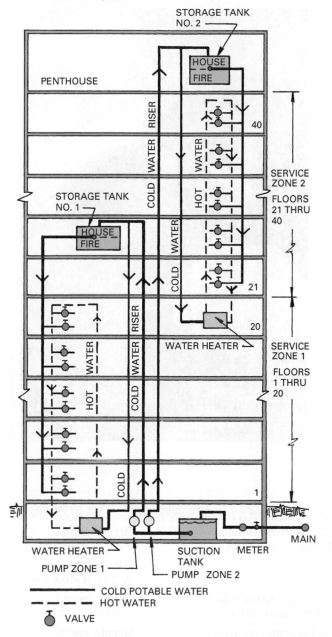

PENTHOUSE

STORAGE TANK NO. 2

HOUSE FIRE

COLD WATER RISER

STORAGE TANK NO. 1

HOUSE FIRE

WATER

HOT

COLD WATER

COLD WATER

40

SERVICE ZONE 2
FLOORS 21 THRU 40

21

20

WATER HEATER

SERVICE ZONE 1
FLOORS 1 THRU 20

COLD WATER RISER

HOT WATER

COLD

WATER

COLD

1

WATER HEATER

PUMP ZONE 1

SUCTION TANK

PUMP ZONE 2

METER

MAIN

——— COLD POTABLE WATER
— ∙ — ∙ — HOT WATER
◯ VALVE

Figure 39.7 Pipes for most applications are hidden inside the walls, floors, or ceilings of a building structure.

© David Papazian/Shutterstock.com

by a green stripe along its length. Because of its strength, it is widely used for underground runs. Type L has a medium wall and a blue stripe, while type M has a light wall and a red stripe. All three are used for potable water, heating and air-conditioning, and fuel and fuel oil systems. Type ACR is used for air-conditioning and refrigeration systems; DWV (yellow stripe) is used for

Table 39.1 Pipe and Tubing Used for Water Supply Systems

Type	Connections	Diameters	Special Qualities
Galvanized steel	Threaded	⅛–4 in. (3–102 mm)	Strong, long life
Welded steel	Threaded	⅛–4 in. (3–102 mm)	Strong, long life
Copper tube	Soldered, brazed	¼–6 in. (6–152 mm)	Corrosion resistant
Red brass	Threaded, brazed	⅛–6 in. (3–152 mm)	Corrosion resistant
Plastic	Solvent joined, heat fusion, serrated inserts	¼–6 in. (6–152 mm)	Lightweight, corrosion resistant

Figure 39.8 Potable-water piping utilizes plastic, copper, and steel pipe and fittings.

Figure 39.9 Copper pipes and fittings join in a socket arrangement and are secured by soldering.

drain, waste, and ventilation systems; and several other types identified as OXY and MED are used for medical gas applications.

Copper tubing is manufactured in hard and soft tempers. Soft-tempered pipe bends easily, reducing the need for many connections. The hard-tempered pipe requires soldered connectors to turn corners. Copper tube is available in diameters from ¼ in. to 8 in. (6 mm to 203 mm).

Plastic pipe is manufactured in several synthetic resins. The type of resin used greatly influences the strength and use of the pipe. Table 39.3 lists the

Table 39.2 Copper Tubing: Types, Uses, and Sizes

Type	Color	Uses	Nominal Diameter
K	Green	Potable water, fire protection, solar, fuel/fuel oil, HVAC, snow melting	Straight lengths ¼–12 in. (6–305 mm) Coils ¼–2 in. (6–50.8 mm)
L	Blue	Potable water, fire protection, solar, fuel/fuel oil, HVAC, snow melting	Straight lengths ¼–12 in. (6–305 mm)
M	Red	Potable water, fire protection, solar, fuel/fuel oil, HVAC, snow melting	Straight lengths ¼–12 in. (6–305 mm)
DWV	Yellow	Drain, waste, vent, HVAC, solar	Straight lengths 1¼–8 in. (32–203 mm)
ACR	Blue Soft ACR not marked	Air-conditioning, refrigeration, natural gas, liquified petroleum gas	Straight lengths ¾–4⅛ in. (9.5–105 mm)
OXY, MED, OXY/MED, OXY/MED OXY/ACR,	K-Green L-Blue	Medical gas	Straight lengths ¼–8 in. (6–203 mm)
G	Yellow	Natural gas, liquified petroleum gas	Straight lengths ⅜–1⅛ in. (9.5–28.5 mm) Coils ⅜–⅞ in. (9.5–22 mm)

Table 39.3 Major Types and Uses of Plastic Pipe

| Type | Condition | Connections | Maximum Operating Temperature | | Typical Uses |
			°F	°C	
Acrylonitrile butadiene styrene (ABS)	Rigid	Threaded, serrated fittings, solvent	100 pressure 180 nonpressure	38 82.9	Cold water, waste, vent, sewer, drain, conduit, gas
Chlorinated polyvinyl chloride (CPVC)	Rigid	Threaded, couplings, serrated fittings, solvent	180 at 100 psig (type 11)	82.9	Cold and hot water, chemical piping
Polyethylene (PE)	Flexible	Serrated fittings, fusion in socket, butt fusion	100 pressure 180 nonpressure	38 82.9	Cold water, gas, waste, chemicals
Polypropylene (PP)	Rigid	Mechanical couplings, butt fusion, socket fusion	100 pressure 180 nonpressure	38 82.9	Chemical piping, chemical drainage
Polyvinyl chloride (PVC)	Rigid	Solvent, threading, mechanical couplings, serrated fittings	100 pressure 180 nonpressure	38 82.9	Cold water, gas, waste, vents, drains, sewers, conduit
Styrene rubber plastic (SRP)	Rigid	Solvent, serrated fittings, elastomer seal	150 nonpressure (not used under pressure)	66	Sewage disposal field, storm drainage, soil drainage

commonly used resins for producing plastic pipe. Notice that most are not acceptable for hot water piping. Plastic pipe is lightweight, flexible, and available in long lengths. Various types are joined with solvent cement, elastomeric seals for bell-end piping and fittings, serrated insert fittings secured with stainless steel clamps, heat fusion used on plastics for which there is no solvent, and threaded fittings that can connect to threaded metal pipe.

Plastic piping and sanitary piping are manufactured in diameters from ½ in. (12.7 mm) to 48 in. (1,219 mm) and in lengths to 20 ft. (6 m). The wall thicknesses are identified by a schedule number, such as Schedule 40, 80, or 120, with the larger numbers indicating a thicker pipe wall.

Pipes made from glass, nickel, silver, and chrome are corrosion resistant and used for special applications. Most metal pipes have threaded connections.

PEX water tubing, made from high-density polyethylene (HDPE) is flexible plastic tubing used for both water supply and hydronic radiant heating systems. The use of PEX has increased because of its ability to bend and accommodate turns without using elbow joints (Figure 39.10). PEX is particularly useful in installations prone to freezing temperatures because it can expand to three times its size without bursting.

Figure 39.10 PEX water supply tubing can accommodate bends and turns without the use of elbows.

Pipe Connections

The most common methods for connecting pipe are shown in Figure 39.11. Water pipe uses butt-welded, socket, and threaded fittings. The bell and spigot connection is used on sewer lines, while the flanged connection is used in applications such as petrochemical and power-generation piping. No hub pipes are joined using gaskets and stainless steel straps (Figure 39.12).

Figure 39.11 Welded, socket, and threaded pipe and fitting connections.

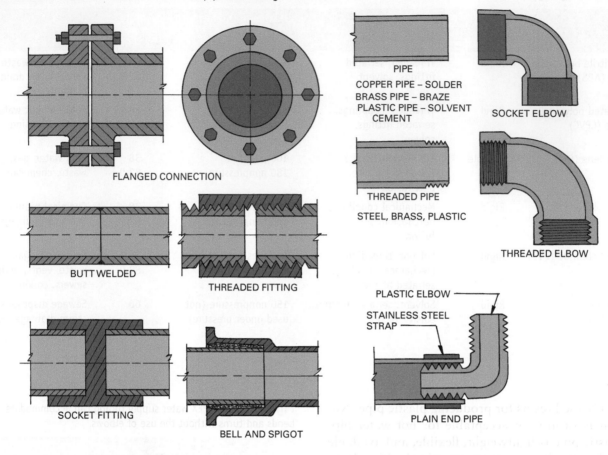

FLANGED CONNECTION

BUTT WELDED

THREADED FITTING

SOCKET FITTING

BELL AND SPIGOT

PIPE

COPPER PIPE – SOLDER
BRASS PIPE – BRAZE
PLASTIC PIPE – SOLVENT
CEMENT

SOCKET ELBOW

THREADED PIPE

STEEL, BRASS, PLASTIC

THREADED ELBOW

PLASTIC ELBOW

STAINLESS STEEL
STRAP

PLAIN END PIPE

Figure 39.12 No-hub pipes are joined with gaskets and steel straps.

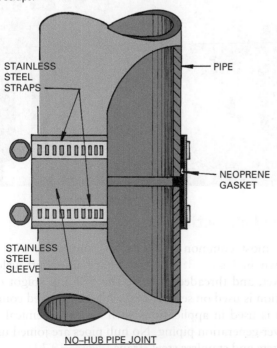

STAINLESS
STEEL
STRAPS

PIPE

NEOPRENE
GASKET

STAINLESS
STEEL
SLEEVE

NO–HUB PIPE JOINT

Water Pipe Fittings

Figure 39.13 illustrates the most commonly used water pipe fittings. Couplings are used to connect two pipes end to end. Elbows change the direction of the pipe. A tee permits one pipe to intersect another. The end of a pipe can be closed by installing a coupling and screwing a plug into it with a cap that screws over the end of the pipe.

VALVES

Valves are used to control the flow of water, oil, gas, and chemicals in pipe distribution systems. Some of the frequently used types are shown in Figure 39.14. A pressure regulator valve limits water pressure to acceptable levels, preventing damage to piping and equipment. A water hammer arrestor has a hydraulic piston that absorbs shock waves produced by sudden changes in water flow; this reduces banging in the pipes.

Backflow preventers keep water from backing up in a system, and expansion tanks mitigate excess pressure by absorbing extra water volume created when water is heated

Figure 39.13 Commonly used steel pipe fittings.

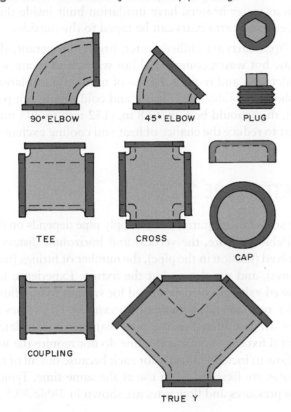

90° ELBOW   45° ELBOW   PLUG

TEE   CROSS   CAP

COUPLING   TRUE Y

and expands. Expansion tanks utilize a rubber bladder that flexes against the water pressure. A "T and P" valve is a safety relief valve that senses a buildup of temperature or pressure and opens to release the excess pressure and allow cold water to enter and prevent an explosion.

Float valves are used to control water levels in tanks. As the water level rises, the float rises and shuts off the supply when a set capacity has been reached. Strainers typically have a 20-mesh screen that collects dirt and debris. They may be on the main line or on lines to a particular piece of equipment. Saddle valves are installed on piping that carries water under pressure. They clamp on a pipe and then pierce it.

Gate-style stop-and-waste valves were once commonly used to shut off water to a building on the owner's side of the meter. Now the more dependable ball valve is used for shutoffs. Globe valves are used to control the flow of hot and cold water, oil, and gas.

PIPE INSULATION

Condensation forms on the exterior of cold water pipes when they contact warm humid air. Condensation can drip from a pipe into a wall cavity, ceiling,

Figure 39.14 Typical valves used in water distribution systems: (a) ball valve, (b) pressure-reducing valve, (c) stop and waste valve, (d) swing check valve, (e) strainer, (f) relief valve.

a   c   e

b   f

d

or floor, wetting insulation and penetrating drywall or other wall finishes.

To prevent damage from condensation, pipes are covered with preformed fiberglass or foam insulation that is usually ½ in. to 1 in. (12 mm to 25 mm) thick (Figure 39.15). The insulation is fitted around the pipe and taped. Hot water pipes are insulated to reduce heat

Figure 39.15 Hot and cold water pipes are insulated to maintain water temperature, prevent freezing, and stop condensation from dripping from the pipes.

© iStock.com/nsj-images

loss to the cooler atmosphere. Most plumbing units, such as water heaters, have insulation built inside their jackets, and extra batts can be taped to the outside.

Pipes carrying chilled water, brine, refrigerant, domestic hot water, commercial hot water, and steam will condensate, and require the use of minimum insulation, as shown in Table 39.4. If hot and cold piping run parallel, they should be 6 in. to 8 in. (152 mm to 203 mm) apart to reduce the chance of heat and cooling exchange.

WATER PIPE SIZING

The size of the required water supply pipe depends on the available pressure, the vertical and horizontal distances involved (friction in the pipe), the number of fittings (tees, elbows), and use demand at the fixture. Experience has allowed values to be determined for various fixture flows under recommended minimum pressures. The values of these are considered as the pipe diameter is calculated. Several fixtures on one water line do not require the water flow to increase directly for each because not all of the fixtures are likely to be in use at the same time. Typical flow pressures and flow rates are shown in Table 39.5.

Table 39.4 Minimum Insulation for Pipes (Inches)

| Fluid Design Operating Temperature Range (°F) | Insulation Conductivity | | Nominal Pipe Diameter (in.) | | | | | |
	Conductivity Range [BTU-in./(h-ft³-°F)]	Mean Rating Temperature (°F)	Runouts[a] up to 2	1 and less	1–1¼ to 2	2½ to 4	5 and 6	8 and up
Heating Systems (Steam, Steam Condensate, and Hot Water)								
Above 350	0.32–0.34	250	1.5	2.5	2.5	3.0	3.5	3.5
251–350	0.29–0.31	200	1.5	2.0	2.5	2.5	3.5	3.5
201–250	0.27–0.30	150	1.0	1.5	1.5	2.0	2.0	3.5
141–200	0.25–0.29	125	0.5	1.5	1.5	1.5	1.5	1.5
105–140	0.24–0.28	100	0.5	1.0	1.0	1.0	1.5	1.5
Domestic and Service Hot Water Systems[b]								
105 and greater	0.24–0.28	100	0.5	1.0	1.0	1.5	1.5	1.5
Cooling Systems (Chilled Water, Brine, and Refrigerant)[c]								
40–55	0.23–0.27	75	0.5	0.5	0.75	1.0	1.0	1.0
Below 40	0.23–0.27	75	1.0	1.0	1.5	1.5	1.5	1.5

Source: ASHRAE/IES Standard 90.1-1989, Energy Efficient Design of New Buildings Except Low-Rise Residential Buildings, © 1989 by the American Society of Heating, Refrigerating, and Air-Conditioning Engineers, Inc., Atlanta, Ga.
[a]Runouts to individual terminal units not exceeding 12 ft. in length.
[b]Applies to recirculating sections of service or domestic hot water systems and first 8 ft. from storage tank for non-recirculating systems.
[c] The required minimum thicknesses do not consider water vapor transmission and condensation. Additional insulation, vapor retarders, or both, may be required to limit water vapor transmission and condensation

Table 39.5 Minimum Flow and Pressure Required by Typical Plumbing Fixtures

Fixture	Flow Pressure (psi)	(kPa)	Flow Rate (gpm)	(L/s)
Ordinary basin faucet	8	55	2.0	0.13
Self-closing basin faucet	8	55	2.5	0.16
Sink faucet, ³⁄₈ in. (9.5 mm)	8	55	4.5	0.28
Sink faucet, ¹⁄₂ in. (12.7 mm)	8	55	4.5	0.28
Bathtub faucet	8	55	6.0	0.38
Laundry tub faucet, ¹⁄₂ in. (12.7 mm)	8	55	5.0	0.32
Shower	8	55	5.0	0.32
Ball-cock for closet	8	55	3.0	0.19
Flush valve for closet	15	103	15–40	0.95–2.52
Flushometer valve for urinal	15	103	15.0	0.95
Garden hose (50 ft., ³⁄₄-in. sill cock) (15 m, 19 mm)	30	207	5.0	0.32
Garden hose (50 ft., ⁵⁄₈ in. outlet) (15 m, 16 mm)	15	103	3.33	0.21
Drinking fountains	15	103	0.75	0.05
Fire hose 1¹⁄₂ in. (38 mm), ¹⁄₂ in. nozzle (12.7 mm)	30	207	40.0	2.52

Source: Manual of Individual Water Supply Systems, U.S. Environmental Protection Agency

SANITARY PIPING

Piping used for sanitary systems may be cast iron, copper, plastic, lead, glass, and clay. These are used for various *drainage, waste, and vent installations (DWV)*.

Cast iron soil pipe and fittings are gray iron castings suitable for installation and service for storm drain, sanitary, waste, and vent piping. Hub-type cast iron soil pipe and fitting specifications are detailed in ASTM A74, *Cast Iron Soil Pipe and Fittings.* They are available in extra heavy and service classifications in diameters, shown in Table 39.6. The pipes and fittings must be coated with a material to protect their surfaces. Hub-less cast-iron soil pipe and fittings are specified in ASTM A888-90,

Table 39.6 Cast-Iron Soil Pipe Uses and Sizes

Type	Uses	Nominal Diameter
Hubless soil pipe, standard and extra heavy	Sanitary and storm drains, waste, vents	2–15 in. (50.8–381 mm)
Hub type soil pipe, service and extra heavy	Sanitary and storm drains, waste, vents	2–15 in. (50.8–381 mm)

Standard Specification for Hubless Cast Iron Soil Pipe and Fittings for Sanitary and Storm Drain, Waste, and Vent Piping Applications. Neither type is intended to be used on pressurized systems. The selection of the proper design size ensures space for the free air needed for gravity drainage. Both types are available in a wide range of fittings. A few are shown in Figure 39.16 and Figure 39.17. The hub-type joints may be sealed with lead and *oakum* or with a neoprene compression gasket (Figure 39.18). Hub-less connections use a gasket and a stainless-steel casing and retaining clamps.

Copper tubing used for sanitary waste systems is classified as DWV (drainage, waste, vent). The available diameters are given in Table 39.2. Copper tubing is used in residential, low-rise, and high-rise buildings for all parts of drainage plumbing, including soil and vent stacks and soil, waste, and vent branches. In high-rise buildings, expansion must be considered as the system is designed. Changes of 1°F can cause DWV pipe to change up to 0.001 in. (0.025 mm) for a 10 ft. section. One way to control thermal movement is to anchor the pipe to the floor, as shown in Figure 39.19. The number of anchors is a design decision, but a rule of thumb is to anchor at every eight floors if the temperature rise is anticipated to be up to 50°F (10°C).

Figure 39.16 Hub-type cast iron soil pipe fittings.

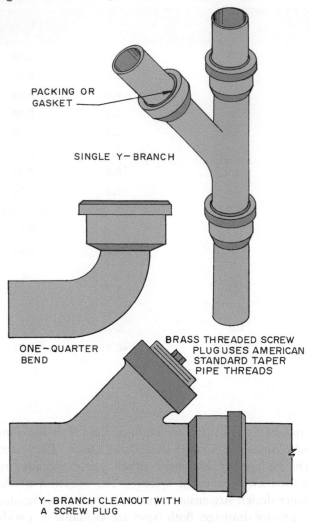

PACKING OR GASKET

SINGLE Y-BRANCH

ONE-QUARTER BEND

BRASS THREADED SCREW PLUG USES AMERICAN STANDARD TAPER PIPE THREADS

Y-BRANCH CLEANOUT WITH A SCREW PLUG

Figure 39.17 Hubless-type cast iron soil pipe fittings.

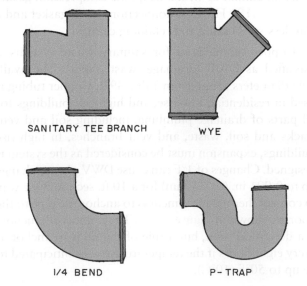

SANITARY TEE BRANCH

WYE

I/4 BEND

P-TRAP

Figure 39.18 Joints used to join hub type and hubless cast iron soil pipe.

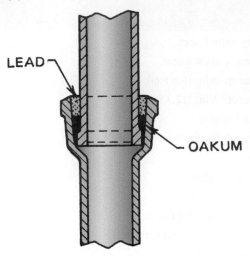

LEAD

OAKUM

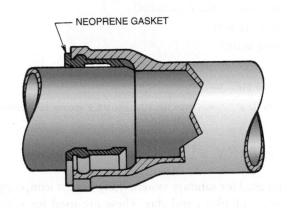

NEOPRENE GASKET

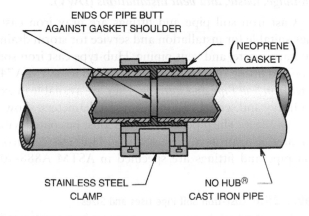

ENDS OF PIPE BUTT AGAINST GASKET SHOULDER

(NEOPRENE GASKET)

STAINLESS STEEL CLAMP

NO HUB® CAST IRON PIPE

Joints in copper DWV tubing are made by properly cleaning and fluxing the joining parts and allowing the solder to flow between them by capillary action. The space between the parts is sized to allow molten solder to flow into the joint. Brazed connections are used if greater strength is needed.

Figure 39.19 Thermal movement in long vertical runs of copper pipe can be controlled by anchoring the pipe to intermediate floors.

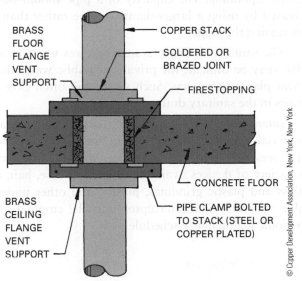

BRASS FLOOR FLANGE VENT SUPPORT

COPPER STACK

SOLDERED OR BRAZED JOINT

FIRESTOPPING

CONCRETE FLOOR

BRASS CEILING FLANGE VENT SUPPORT

PIPE CLAMP BOLTED TO STACK (STEEL OR COPPER PLATED)

© Copper Development Association, New York, New York

Figure 39.21 Adjoining surfaces of the plastic pipe are cleaned and primed before joining with solvent cement.

© Christina Richards/Shutterstock.com

Plastic pipe suitable for DWV systems include acrylonitrile butadiene styrene (ABS), styrene rubber plastic (SRP), and polyvinyl chloride (PVC). All three are used for sewer systems, and ABS and PVC can be used for drain, waste, and vent systems. Local plumbing codes should be checked to verify approved uses.

Plastic pipe and fittings have identification symbols on each piece, as shown in Figure 39.20. They are joined with solvent cement. Some plumbers prefer to clean pipe ends fitting insides by brushing on a coat of priming solvent. After about fifteen seconds, a thick coat of solvent cement is applied to the end of the pipe and the inner surface of the fitting (Figure 39.21). The pipe is inserted and twisted a quarter turn to spread the cement. At least three minutes must pass before starting the next joint.

Figure 39.20 Identification used on plastic piping.

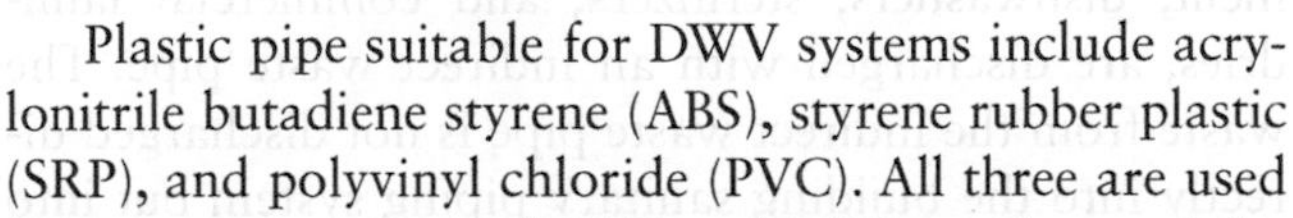

SIDE VIEW

SDR 11 CPVC 4120 400 PSI @ 73°F ASTM D-2846 NSF-PW DRINKING WATER 100 PSI @ 180°F

Lead and glass drain pipe is used for the disposal of special liquids, such as in a chemical plant or research laboratory. Lead and glass resist attack by many chemicals. Lead pipe is linked by welding the joints. Glass pipe is connected with a gasket and stainless steel clamp, as shown for hub-less pipe in Figure 39.11.

Clay pipe is used for waste disposal lines outside a building. Its use is carefully controlled by building codes. The pipe is made from clay and burned in a kiln (similar to clay brick). It is available in diameters from 4 in. to 36 in. (102 mm to 914 mm) and has a variety of fittings. It uses a hub joint filled with a packing compound. Clay pipe is impervious to acids and alkalines and does not deteriorate with age. It is not used where it might be subjected to shock or loads.

Calculating Waste Pipe Sizes

The diameter of pipes used in a sanitary piping system depends on the amount of waste they are to carry. Plumbing codes typically establish the number of drainage fixture units running through each pipe size (Table 39.7). A *fixture unit* is a measure of the probable discharge into the drainage system by the various plumbing fixtures. It is expressed in units of cubic volume per minute. The value for a particular fixture depends on the volume rate of discharge, the time duration of a single discharge, and the average time between successive discharges. A fixture unit is generally equal to 7½ gallons of flow.

Table 39.7 Drainage Fixture Units for Selected Plumbing Fixtures

Plumbing Fixture	Drainage Fixture Unit Value
Automatic clothes dryer	3
Bathtub with or without overhead shower	2
Clinic sink	6
Dental unit	1
Drinking fountain	½
Dishwasher, domestic	2
Floor drain, 2 in. waste pipe	3
Kitchen sink, 1½ in. trap	2
Kitchen sink with dishwasher	3
Lavatory, 1¼ in. waste pipe	1
Shower stall, residential	2
Urinal, stall, washout	4
Water closet, tank operated	4
Water closet, valve operated	6

Source: National Standard Plumbing Code, National Association of Plumbing, Heating, and Cooling Contractors

Sanitary Piping Systems

A *sanitary piping* system removes the water and other waste materials discharged at the various fixtures. The system carries the material through the building to a *building drain* from which it is discharged to the *building sewer* and on to the public *sanitary sewer* system or septic tank. The waste material flows from the building at a level below the lowest fixture and moves by gravity flow to the building sewer. If the public sewer is not below the sloped building sewer, waste will not drain and pumps must be used to lift the waste and move it to the public sewer.

Pipe diameters for sanitary piping systems are sized to carry the anticipated waste material flow rapidly without clogging pipes or creating annoying water noises. Pipe diameters also are selected to produce minimum pressure variations where fixtures connect to *waste pipes*, where waste pipes connect to *branch soil pipes*, and where branch soil pipes connect to the *soil stack*.

The system relies on gravity flow, so it is not under pressure. The effluence increases in speed as it proceeds down the piping system. Friction within the pipes limits speed and must be considered when sizing pipe.

Plumbing codes specify the required slope for horizontal piping. The requirement varies with pipe diameter but is typically from to ⅛ to ½ in. (3 mm to 12.7 mm) per foot of run. Flow velocity increases with the amount of slope, and high velocities in horizontal pipes can increase siphoning. The capacity of a pipe should be increased by using a larger-diameter pipe rather than by increasing the slope.

The sanitary drainage system receives waste matter that may be difficult for private or public waste treatment plants to process. Such waste will often clog the pipes in the sanitary drainage system.

Interceptors are used to catch large items of waste material near the point of origin, intercepting it before it moves very far into the system. Manufacturers have a variety of devices available to catch grease, hair, oil, glass and plastic grindings, plaster, and other undesirable materials. The interceptor must be emptied on a periodic maintenance schedule.

Indirect Wastes

Indirect waste, such as that from food-handling equipment, dishwashers, sterilizers, and commercial laundries, are discharged with an indirect waste pipe. The waste from the indirect waste pipe is not discharged directly into the building sanitary piping system but into a fixture connected to the building sanitary system. This provides an air gap between the two systems. An *air gap* is used to prevent one system from accidentally backing up into another. Toxic and corrosive waste discharges must be automatically diluted with water or chemically neutralized before being introduced into the building sanitary system.

RESIDENTIAL SANITARY PIPING

A typical simple sanitary piping system for a residence is shown in Figure 39.22. The waste from the fixtures drains through waste pipes or branch soil pipes into a soil stack. Each waste pipe is sized to carry the flow from a fixture. The branch soil pipe is sized to carry the flow from all the fixtures flowing into it. The soil stock extends below the building and connects to the building drain and sewer, which in turn connects to the central sewer system or a septic tank. Each fixture has a vent pipe connected to a *vent stack* running through the roof. The *vent pipe* keeps the water in the *trap* of the fixture under atmospheric pressure, thus eliminating the chance of its being siphoned out when another fixture is used (Figure 39.23). For example, if the waste system were a closed installation, a trap in a lavatory could have the water siphoned out when a connecting

Case Study

Bronx Zoo Eco-Restrooms

Architect: Edelman Sulton Knox Wood
Architects LLT

Contractor: Summit Construction
Services Group

Location: Bronx, New York

Building type: Service Building
Completed: 2006

Overview

Founded in 1895, the Wildlife Conservation Society (WCS) was one of the first conservation organizations in the U.S. The Bronx Zoo opened in 1899 with a clear mandate: to advance wildlife conservation and promote the study of zoology. An existing bathroom facility at one of the zoo gateway entrances had fallen into disrepair. Zoo management decided to inaugurate a new building that would showcase strategies to responsibly deal with human waste while simultaneously preventing pollution of local waterways. The aged facility had operated on a septic system that over time had leached into the adjacent Bronx River. In addition to the use of natural daylighting and other energy saving features, the new building incorporates a number of cutting edge water conservation technologies (Figure A).

Plumbing Fixtures

The Eco-Restroom's plumbing fixtures include fourteen foam-flush toilets, four waterless urinals,

and ten composting chambers. Composting foam-flush fixtures use between three to six ounces of water mixed with a drop of soap that flushes waste down to the composters below. In the composter, solid and liquid wastes are separated and treated by natural biological decomposition into a topsoil-like compost and strong liquid fertilizer (Figure B). Even when compared to new efficient 1.6 gpf toilets, the composting system has the potential of saving more than 100,000 gal. of water per year.

Graywater System

Water from the restroom's non-toilet fixtures is collected via conventional plumbing conduits in a tank in the basement. The graywater is distributed through buried pipes into the root zone of a small adjacent garden where plants use the water and nutrients contained in it. Because graywater is used for plant irrigation and toilet wastes are contained in the composters, there is

Figure A The exterior of the Bronx Zoo bathroom facility.

© Clivus Multrum

Figure B Some of the composting reactors.

© Clivus Multrum

no waste water discharge like there is in conventional buildings attached to sewers or septic systems.

Rainwater Harvesting

In addition to the composting toilets and graywater system, the new restroom also harvests rainwater to prevent stormwater runoff from entering the municipal sewer system. Rainwater is collected from the roof in a series of barrels, where it is stored for later use in landscape irrigation. Rainwater harvesting

is especially useful in urban areas where it can increase soil moisture levels for urban greenery and supplement the groundwater table through artificial recharge

The various green building technologies incorporated in the Eco-Restroom are clearly explained through signage designed to educate visitors about environmental issues. The project was recently named Eco Project of the Year by New York Construction, and plans are under way for two additional green buildings in the zoo.

Figure 39.22 A simplified schematic of a residential sanitary piping system.

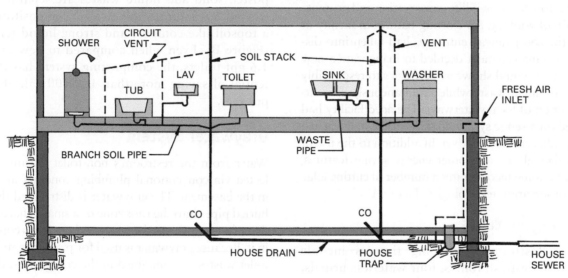

toilet is flushed. Some fixtures, such as toilets, have a trap built into the unit. The house drain will have one or more cleanouts. A *cleanout* is a pipe fitting with a removable plug through which an auger may be run to dislodge obstructions in the pipe.

Various types of traps are used (Figure 39.24). Lavatories typically have a p-trap, while tubs use a drum trap. To keep plumbing costs low, designers attempt to place fixtures on the same wall. This reduces the piping required and eliminates long horizontal runs to reach out-of-the-way fixtures (Figure 39.25). In wood frame construction, the wall containing the plumbing uses at least 2 in. × 6 in. studs.

MULTISTORY BUILDING SANITARY PIPING SYSTEMS

In multistory buildings, it is common to design *plumbing chases* on each floor directly above the other to run water and waste disposal systems the height of the building (Figure 39.26). Heating, air-conditioning, and fire-suppressing systems use the same chases for the efficient distribution of all systems. This consolidates the plumbing, reduces costs, and makes planning of the uses of the space on each floor more flexible. Plumbing fixtures are typically located next to the sides of the vertical chase. Horizontal runs to locations within a floor

Figure 39.23 Plumbing fixtures are vented to prevent water from being siphoned when a nearby fixture discharges water into the system.

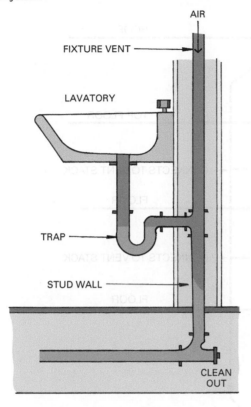

Figure 39.25 Plumbing costs can be reduced by aggregating plumbing fixtures.

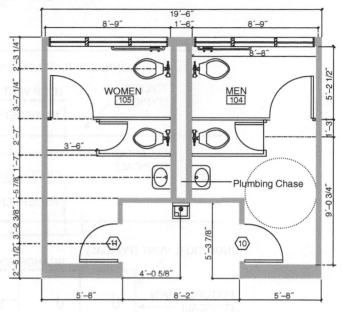

Figure 39.24 Commonly used trap configurations.

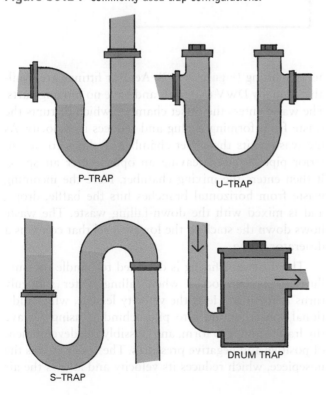

area can also be made from the chase. In other cases, pipe risers can be located inside the finished enclosure of a structural column. This enables placement of fixtures in various locations within the building.

A schematic of a small multistory building sanitary piping system is shown in **Figure 39.27**. The fixtures for the restrooms with their related piping are detailed on the second floor, with notes indicating that they are identical on all other floors. The soil stack, vent stack, and building drain are shown for the entire building. Notice the venting of multiple fixtures to horizontal circuit vents that connect to the vent stack. In some cases, the fixture waste pipes connect directly to the soil stack, but in other situations they feed into a horizontal branch soil pipe that connects to the vertical soil stack. The circuit vents, waste pipes, and branch soil pipes slope toward the soil stack, providing gravity flow.

The system in Figure 39.27 requires two sets of pipes, a vent system, and a waste disposal system. This two-pipe system is most commonly used in the United States. The venting controls the possibility of siphoning water from the traps, which would allow sewer gas to enter the building. Another system, the Sovent system, is used in Europe and Africa and has been approved for use in the United States by some national plumbing codes.

Figure 39.26 Plumbing chases stack fixtures on each floor for increased efficiency.

SIDE VIEW

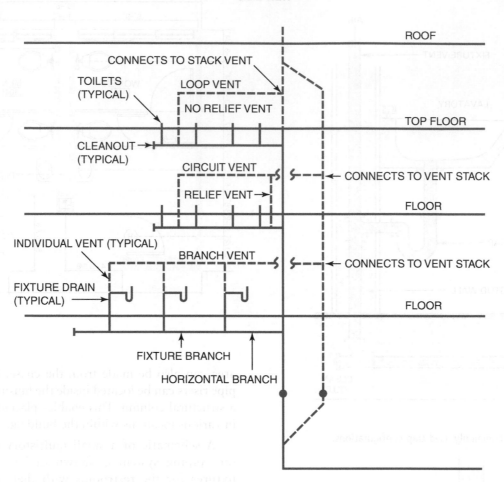

The Sovent System

The Sovent system is a vertical cast iron drainage and waste system that conveys waste from the upper levels of a building to the base of each Sovent stack. The stack begins just above the lowest deaerator fitting and continues up to just above the highest fixture connection (Figure 39.28). This includes horizontal stack offsets located at intermediate levels. The stack uses traditional fittings and pipe made from approved drain, waste, vent (DWV) materials. The Sovent stack penetrates the roof to the atmosphere much like the vent stack in traditional systems.

Waste flowing in a vertical pipe clings to the interior wall and moves down the pipe in a swirling motion. This leaves an open airway in the center needed for flow to continue. As the falling waste gathers speed, it meets air resistance, flattens out, and may form a complete blockage of the pipe. This blockage is eliminated by the Sovent

aerator fitting (Figure 39.29). Aerator fittings are available in many DWV materials and have no moving parts. The waste enters the offset chamber, which disrupts the waste from forming a plug and reduces its velocity. As the waste exits the offset chamber, it clings to the interior pipe surfaces, leaving an open center air space. It then enters the mixing chamber. Here the incoming waste from horizontal branches hits the baffle, drops, and is mixed with the down-falling waste. The waste flows down the stack to the lowest level that contains a deaerator fitting.

The deaerator fitting is designed to handle pressure fluctuations that occur when falling water suddenly turns horizontal. Here, the velocity lessens, while additional waste continues to pile behind, causing a wave (hydraulic jump) to form, and possibly the development of positive and negative pressures. The water strikes the nosepiece, which reduces its velocity and allows the air

Figure 39.27 A simplified schematic of a sanitary piping system for a low-rise multistory building. Underground tanks can collect rain water from the roof for distribution to the non-potable water system used for toilets and landscaping irrigation.

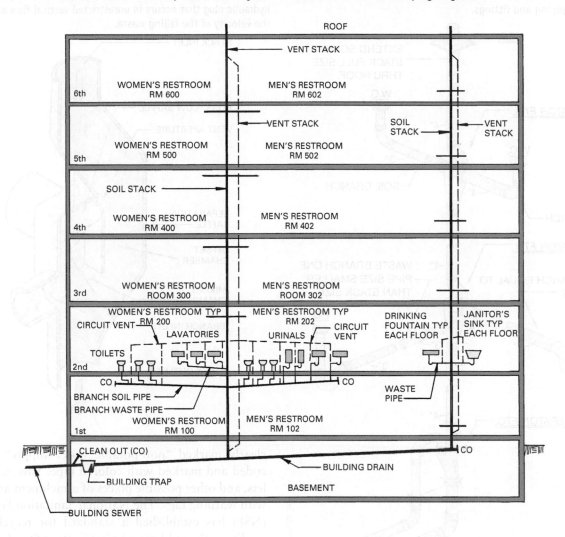

and waste to separate and continue to flow down the horizontal pipe. The relief line provides an outlet for pressure and eliminates any chance of the system's developing a vacuum, which would pull the water out of the traps at each fixture.

NON-POTABLE WATER SYSTEMS

Increasing awareness of the need for water conservation has produced new technologies and alternate methods of handling water and wastewater. In buildings where large amounts of water are used, a considerable flow of wastewater is produced. Recycling this flow helps conserve the water supply and reduces the load on wastewater disposal facilities.

Graywater is non-industrial wastewater generated from domestic processes, such as dish washing, laundry, and bathing. It comprises wastewater generated from all of the house's sanitation equipment except for the septic tank (water from toilets being "blackwater"). Graywater makes up between 50 to 80 percent of residential wastewater. By designing plumbing systems to separate graywater from blackwater, graywater can be recycled for flushing toilets, landscape irrigation, and exterior washing.

A schematic for a wastewater treatment and recycling facility is shown in **Figure 39.30**. Potable water is delivered to drinking fountains, lavatories, and other fixtures where it is needed. The potable supply is distributed in a completely separate system from the *non-potable water* system. The wastewater from the

Figure 39.28 A vertical waste stack using cast iron Sovent aerators and deaerators and standard code-approved drain, waste, vent (DWV) piping and fittings.

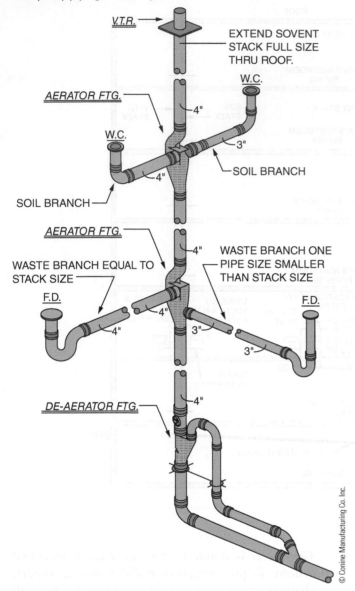

Figure 39.29 The cast iron Sovent aerator offsets the waste flow around the horizontal branch inlets and breaks up the hydraulic plug that occurs in unrestricted vertical flow and reduces the velocity of the falling waste.

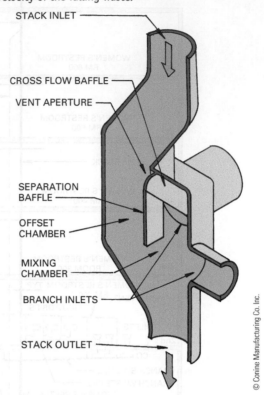

clearly marked "non-potable water supply" or color-coded and marked with colored tape. Valves, wall outlets, and other possible places of attachment are marked with warning tags. The National Sanitation Foundation (NSF) has established a standard for recycled water quality in the publication *Certification Standard No. 1.*

ROOF DRAINAGE

Areas such as flat roofs or balconies on multistory buildings collect water during storms and require a drainage system. This system is separate from the sanitary sewer and must drain into the public storm sewer system.

Flat roofs and other such surfaces usually slope toward interior roof drains. The roof drains are raised cast iron, plastic, or aluminum strainers set in the center of a sloped area. Water drains through downpipes, which may be hidden internally or mounted on the exterior of a building. The downpipes connect to a storm drain below the building, which connects to the storm sewer outside the building. Another technique is to slope the roof toward *scuppers*, which provide outlets in the

potable supply enters a pretreatment trash trap, and a sump provides temporary storage in case of a system mechanical failure. Usually this will hold several days' supply of wastewater. The wastewater proceeds automatically through the various processes, including biological treatment, filtering, color removal, disinfection, and ozone treatment. This produces treated water, which is kept in a storage reservoir. The recycled water is moved under pressure to the piping serving the toilets.

It is vital that the potable and non-potable systems are designed to be kept separate. All piping must be

Figure 39.30 A schematic for an in-building wastewater treatment and recycling facility.

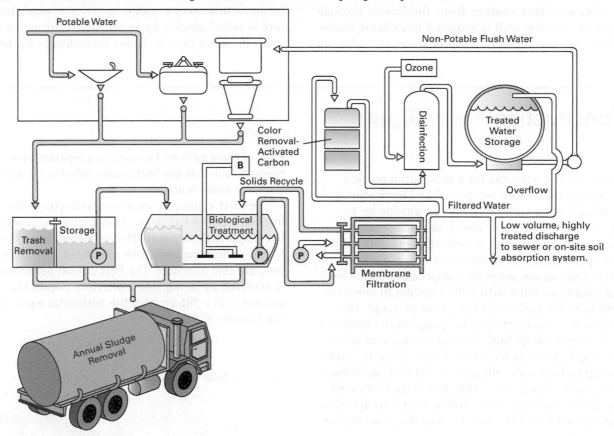

parapet wall and drain the surface water through them into downpipes and on to the storm sewer. See Chapter 27 for roof drain and scupper details.

Rainwater is a valuable natural resource that can be collected for use in landscape irrigation, cooling tower make-up water, and water to flush toilets and urinals. The ideal roofing for a water catchment surface is smooth, dense and non-toxic, such as corrugated aluminium and neoprene-based membranes. The basic components of a rainwater harvesting system include the catchment surface, leaf screens and debris diverters, and roof drains and piping that convey the water to storage tanks or cistern delivery systems use gravity flow or pumps to deliver the water to its end use. Treatment and purification can be added when used in a potable water system.

PLUMBING EQUIPMENT

Buildings use a variety of plumbing equipment that connects to distribution and drainage systems.

WATER HEATING

There are many ways to provide domestic hot water (DHW) needs for a building. Most water heaters utilize a storage tank that is heated by electricity, natural gas, heating fuel, or solar thermal energy. A thermostat monitors the water temperature and controls the heating mechanism. When the water is heated to a desired temperature, the heat source automatically shuts off. When the thermostat senses that the water is below the desired temperature, the heating cycle resumes. Direct-heating systems have the water in direct contact with a heated surface. Indirect systems use coils submerged in a in a tank or boiler. Service temperatures range from 105°F (40 °C) for lavatories, to 180 °F (82 °C) for commercial dishwashing equipment.

Storage gas water heaters are typically used for residential and small commercial applications. The fixtures are rated in terms of the gallons of hot water they can produce. They are available in a variety of heights, diameters, and sizes, with tank capacity ranging from a 40 gal. to 125 gal. (151.4 L to 473 L) and higher. The exhaust fumes from gas water heaters contain carbon monoxide, so direct venting from the heater through the roof or exterior wall is required to exhaust fumes. Tankless, or instantaneous water heaters, take up considerably less space by eliminating the storage tank. In an *instantaneous water heater*, the water temperature is raised quickly by being forced through a heated small diameter pipe and sent immediately to the point

Construction Techniques

Solar Hot Water

Water heating accounts for a substantial portion of energy use in both residential and institutional buildings. Solar water heating can provide up to 85 percent of hot water needs, depending on climate, without fuel cost or pollution and with minimal operational and maintenance expenses. Solar water systems are generally composed of solar thermal collectors filled with a fluid system to move the heat from the collector to its point of usage. The system may use electricity for pumping the fluid and have a reservoir or tank for heat storage and subsequent use. Systems are either direct, using the sun's energy to heat water directly, or indirect, utilizing a fluid, such as antifreeze, that indirectly heats water through a heat exchanger. Indirect systems are more widely used in cold climates where freeze protection is required.

The simplest of all solar water-heating systems is a batch system. A metal water tank painted with a heat absorbing black coating is placed in an insulating container with a glass cover that admits sunlight. Unheated water enters the solar collector and remains there until it is heated to the desired temperature. The heat is prevented from escaping by the glazing. The heated water is then stored for use as needed, with a conventional water heater providing any additional heating that might be necessary. Batch systems are inexpensive and require little maintenance.

A passive thermosiphon uses a flat plate collector and an integrated storage tank that is located higher than the collector. The storage tank receives heated water coming from the top of the collector by natural convection while colder water from the bottom of the storage tank is drawn into the solar collector to replace the heated water (Figure A).

Active systems use electrically powered pumps, valves, and other equipment to help circulate water or a heat-transfer fluid. Heated water from the collector circulates via the pump, which is activated by temperature sensors of the system controller. Figure B shows a schematic illustration of a solar water-heating system with typical control, safety, and operating features.

Many of these systems incorporate a separate solar pre-heat tank that pre-heats water entering the conventional water heater.

The most commonly used solar collector is the insulated glazed flat panel. Less-expensive panels, like polypropylene panels (for swimming pools), or higher-performing ones, like evacuated tube collectors, are also available. The best annual performance is achieved by facing solar collectors toward the equator with a tilt up from the horizontal equal to the latitude of the site.

Figure A Solar water heating.

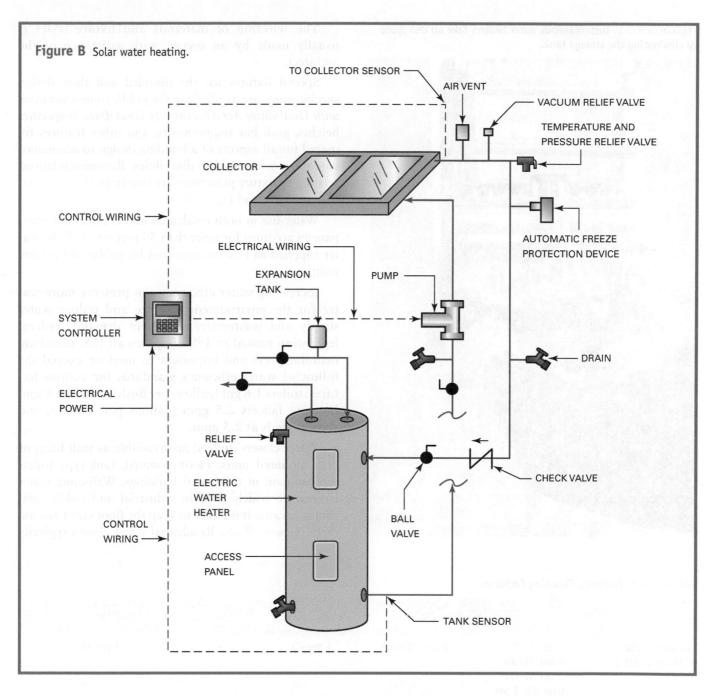

Figure B Solar water heating.

TO COLLECTOR SENSOR

AIR VENT

VACUUM RELIEF VALVE

TEMPERATURE AND
PRESSURE RELIEF VALVE

COLLECTOR

CONTROL WIRING

AUTOMATIC FREEZE
PROTECTION DEVICE

ELECTRICAL WIRING

EXPANSION
TANK

PUMP

SYSTEM
CONTROLLER

DRAIN

ELECTRICAL
POWER

CHECK VALVE

RELIEF
VALVE

ELECTRIC
WATER
HEATER

BALL
VALVE

CONTROL
WIRING

ACCESS
PANEL

TANK SENSOR

of use (Figure 39.31). Tankless water heaters require a stainless-steel vent pipe due to excessive heat build up. To offset this additional cost, many tankless heaters are installed on the exterior.

PLUMBING FIXTURES

Minimum requirements for the number of plumbing fixtures are specified in the various plumbing codes according to occupancy. An example of some selected types of

building occupancy requirements as specified by the *Uniform Plumbing Code* is given in Table 39.8. These are minimum recommendations and known conditions may warrant the use of additional fixtures.

Plumbing fixtures require a steady supply of clean water to assist with the discharge of waste materials. They are under constant wear from water, bacteria, and other harmful elements and must be made from durable materials having a smooth, nonporous surface. Typical materials include stainless steel, copper, brass, enameled cast iron, vitreous china, molded plastics, gel-coated fiberglass, and acrylic-faced fiberglass.

Figure 39.31 Instantaneous water heaters take up less space by eliminating the storage tank.

© sergeevspb/Shutterstock.com

The selection of materials and fixture styles is usually made by an owner, with assistance of the architect.

Special fixtures for the disabled and their design specifications are available in the publication *Americans with Disabilities Act Accessibility Guidelines*. It specifies heights, grab bar requirements, and other features required for all aspects of a building design to accommodate those with physical disabilities. Recommendations relating to fixture placement are shown in Figure 39.32, Figure 39.33, and Figure 39.34.

Water use in both residential and commercial occupancies accounts for more than 50 percent of all the water supplied to U.S. communities by public and private utilities.

Increasing water efficiency can preserve more water for the environment at large and reduce water supply and wastewater treatment demand. Federal legislation passed in 1992 requires all U.S. plumbing manufacturers and importers to meet or exceed the following water efficiency standards for various fixtures: toilets 1.6 gpf (gallons per flush), urinals 1.0 gpf, lavatory faucets 2.5 gpm (gallons per minute), and showerheads at 2.5 gpm.

Water closets (toilets) are available as wall-hung or floor-mounted units. Floor-mounted, tank-type toilets are common in residential buildings. Wall-hung water closets are widely used in industrial and public restrooms because it is easier to keep the floor clean around them (Figure 39.35). Residential water closets typically

Table 39.8 Minimum Plumbing Facilities

Type of Building or Occupancy	Water Closets	Urinals	Lavatories	Bathtubs or Showers	Drinking Fountains
Assembly places (theaters, etc.)	1 per 1–15 2 per 16–35 3 per 36–55 Over 55, 1 per additional 40	1 per 50 males	1 per 40	—	1 per 75
Hospitals Individual room Ward room	1 per room 1 per 8 patients	— —	1 per room 1 per 10 patients	1 per room 1 per 20 patients	— —
Restaurant	1 per 1–50 2 per 51–150 3 per 151–300	1 per 1–150 males	1 per 1–150 2 per 151–200 3 per 201–400	—	—
Worship place, assembly area	1 per 300 males 1 per 150 females	1 per 300 males	1 per toilet room	—	1 per 75
Worship place, educational and activities	1 per 250 males 1 per 125 females	1 per 250 males	1 per toilet room	—	1 per 75

Source: The Uniform Plumbing Code™, International Association of Plumbing and Mechanical Officials

Figure 39.32 Lavatories must be mounted with the top surface not more than 34 in. (865 mm) above the floor and positioned so a wheelchair user will have the required knee room. Any exposed pipes must be covered with protective insulation.

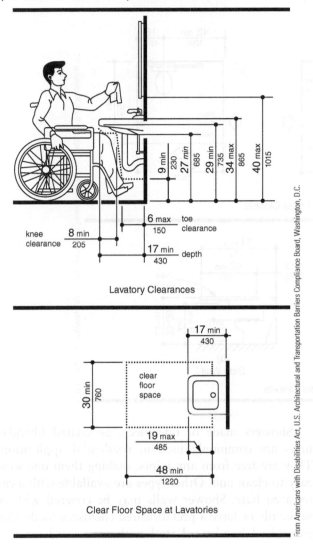

Lavatory Clearances

Clear Floor Space at Lavatories

From Americans with Disabilities Act, U.S. Architectural and Transportation Barriers Compliance Board, Washington, D.C.

use a flush tank, while commercial establishments use a high-pressure flushing system. Water closets have various methods for flushing. New low-flush models use as little as 1.0 gpf to 1.6 gpf and can cut annual water use by 20 to 25 percent, with a corresponding reduction in wastewater flow. Dual-flush toilets allow the user to choose either a full 1.6 gal. flush for solids or a 0.8 gal. flush for liquids.

A *composting toilet* is any system that converts human waste into an organic compost and usable soil through the natural breakdown of organic matter into its essential minerals. Microorganisms accomplish this over time, working through various stages of oxidation and anaerobic breakdown. Self-contained composting toilets complete the composting within the unit, while central ones flush waste to a remote composting unit in a utility space below the toilet. A composting toilet with a separate composting room for the compartment is called a Clivus Multrum. These units incorporate a composting reactor that is connected to one or more toilets and an exhaust system, often fan forced, to remove odors, carbon dioxide, water vapor, and the by-products of aerobic decomposition and provide oxygen (aeration) for the aerobic organisms in the composter. A means for draining and managing excess liquid and an access door for removal of the end product complete the installation.

Urinals are wall-hung units used in men's restrooms in public facilities to reduce the number of required water closets. They may have a manual high-pressure flush valve or be connected to an automatic high-pressure flushing system. Sometimes one is set lower than others for use by children. A more recent innovation is urinals that do not use water at all. Waterless urinals work completely without water or flush valves (Figure 39.36). They install to conventional waste lines and are equipped with a special cartridge that acts as a funnel, allowing liquid to flow through a sealant liquid that prevents any odors from escaping. The cartridge filters sediment, allowing the remaining liquid to pass freely down the drain. Waterless urinals can save up to 45,000 gallons of water per year per urinal.

Lavatories (sinks) in residences are typically mounted in a base cabinet. The lavatory may be set into a top covered with plastic laminate or ceramic tile. Others have a molded plastic top with the lavatory bowl and top as one integral piece. This reduces problems that occur around the stainless-steel edge of those set into the top. Another popular lavatory is a pedestal type. The lavatory is actually wall hung and the pedestal covers up the plumbing below. Wall-hung residential lavatories often have decorative metal legs instead of a pedestal.

Lavatories in public facilities are typically wall-hung units, although some are set into a wall-hung countertop without a cabinet base below. These usually have self-closing faucets that prevent anyone from letting the water run after they leave. One type of faucet has an infrared control that turns on the water when hands are placed below the faucet and turns it off when hands are removed. This saves on water use and reduces the cost of hot water.

Wash fountains are used in restrooms and dressing rooms in which a large number of people need to wash up at the same time, such as in an industrial plant. Wash fountains are typically half-round wall-hung

Figure 39.33 Water closets not in stalls require a minimum of 48 in. (1,220 mm) clearance from other items. The water closet must be 17 in. to 19 in. (430 mm and 485 mm) high. Grab bars to the side and rear are required.

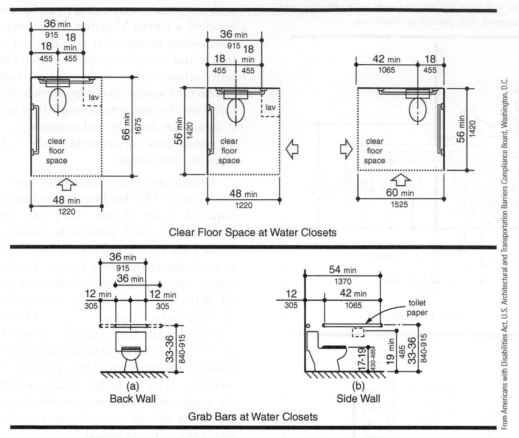

Clear Floor Space at Water Closets

Grab Bars at Water Closets

From Americans with Disabilities Act, U.S. Architectural and Transportation Barriers Compliance Board, Washington, D.C.

units or round freestanding units (Figure 39.37). Large units can accommodate up to eight people at one time.

Kitchen sinks are available in a range of types, with the most typical being some form of two-bowl unit. However, single and triple-bowl units are available. One bowl serves for general use, while the second contains the garbage disposal. Sinks are installed in plastic laminate, stone, and a variety of other countertop materials.

Service sinks are used by janitors to clean mops and for other cleaning activities. They are usually wall hung and deep. Mop service basins are floor-mounted units about 1 ft. (305 mm) high.

Bathtubs are available in a wide range of sizes and designs. Some have two closed sides and fit in a corner. Others have one closed side and fit between end walls. Whirlpool tubs have pumps that circulate the water. Some bathtubs made from gel-coated fiberglass have the tub and wall enclosure formed as a single unit, leaving no joints to form mold, leaving the unit easy to clean (Figure 39.38).

Showers made of one piece, gel-coated fiberglass units are commonly used in residential applications. They are free from any joints, making them one solid, easy-to-clean unit. Other types are available with a cast terrazzo base. Shower walls may be covered with ceramic tile or have a prefabricated enclosure made from galvanized bonderized steel with an enamel finish or molded plastic panels. Shower stall are often installed with tempered glass doors (Figure 39.39).

Various types of drinking fountains are available. Some are wall hung and protrude into the room, while others are recessed or semi-recessed into a wall. Freestanding and some recessed drinking fountains have a water cooling system in their base that produces temperature-controlled water. Units that provide access for the handicapped are wall mounted and have an electrically activated push-button valve on the front that requires only a light touch. Some pedestal types are designed for outdoor use and may have a foot-operated valve that is frost-proof. Stainless steel is the major material used on the surfaces exposed to water. Drinking fountains require a sanitary drain as well as a source of potable water.

Figure 39.34 Drinking fountains must project from a wall so a wheelchair can move below it, and have spouts no higher than 36 in. (915 mm) above the floor.

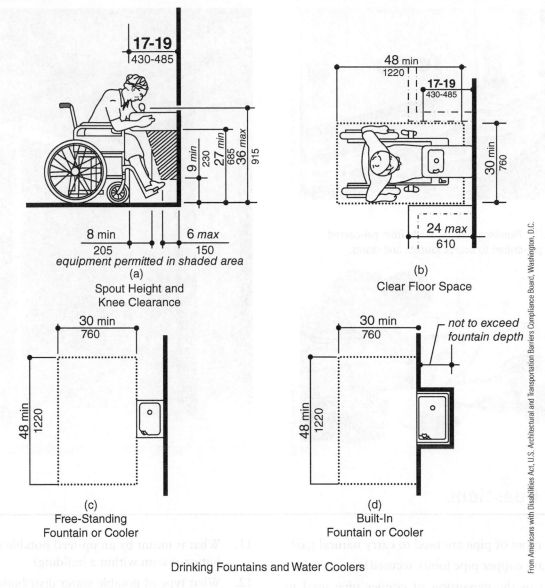

17-19
430-485

9 min 230

27 min 685

36 max 915

8 min 205

6 max 150

equipment permitted in shaded area
(a)
Spout Height and
Knee Clearance

48 min 1220

17-19
430-485

30 min 760

24 *max* 610

(b)
Clear Floor Space

30 min 760

48 min 1220

(c)
Free-Standing
Fountain or Cooler

30 min 760

not to exceed fountain depth

48 min 1220

(d)
Built-In
Fountain or Cooler

Drinking Fountains and Water Coolers

From Americans with Disabilities Act, U.S. Architectural and Transportation Barriers Compliance Board, Washington, D.C.

Figure 39.35 Public restrooms utilize high-pressure wall-mounted toilets.

© iStock.com/benkrut/Essentials Collection

Figure 39.36 Waterless urinals install to conventional waste lines and work completely without water or flush valves.

© iStock.com/jazz42

Figure 39.37 Wash fountains allow large numbers of occupants to wash up at the same time.

Figure 39.38 Plumbing fixtures made from gel-coated fiberglass are pre-drilled to accept faucets and drains.

Figure 39.39 Shower enclosures are often fitted with glass doors.

Review Questions

1. What types of pipe are used to carry natural gas?

2. How are copper pipe joints secured?

3. How can the expansion of copper pipe used to carry hot water be handled to prevent breakage?

4. How are the various types of copper water tubing identified?

5. What resins are used to produce plastic pipe?

6. What methods are used to join various types of plastic pipe?

7. Where would glass pipe be used?

8. Where would you use a relief valve?

9. Why are pipes insulated?

10. What processes are used at a water purification plant to produce potable water?

11. What is meant by an up-feed potable water distribution system within a building?

12. What type of potable water distribution system is used in large multistory buildings?

13. What materials are used for pipes in sanitary piping?

14. How does a plumbing contractor know what slope is required on sanitary piping installations?

15. What are indirect wastes?

16. How does the Sovent drainage and waste system operate without the venting piping used in traditional systems?

17. What uses are made of non-potable water within a building?

Key Terms

Air Gap

Branch Soil Pipe

Building Drain

Building Sewer

Cleanout

Composting Toilet

Drainage, Waste, and
    Vent Installations (DWV)

Fixture Unit

Graywater

Instantaneous Water Heater

Interceptor

Non-Potable Water

Oakum

Plumbing Chase

Potable Water

Riser

Sanitary Piping

Sanitary Sewer

Scupper

Soil Stack

Trap

Valve

Vent Pipe

Vent Stack

Waste Pipe

Activities

1. Arrange a tour of a multistory building in which the plumbing system can be inspected. Attempt to view the hidden aspects, such as plumbing chases and vertical distribution systems.

2. Visit the local water treatment plant and sewage disposal plant. Request the process be explained in the order treatment occurs. Obtain information about regulations within which each facility must operate.

Additional Resources

Americans with Disabilities Act Accessibility Guidelines, U.S. Architectural and Transportation Barriers Compliance Board, 1331 F Street, NW, Suite 1000, Washington, DC 20004.

Applications of Water Pipe Sizing, International Code Council, Falls Church, VA.

International Plumbing Code, International Code Council, Falls Church, VA.

International Private Sewage Disposal Code, International Code Council, Falls Church, VA.

LP-Gas Code; *National Fire Alarm Code*, National Fire Protection Association, Quincy, MA.

See Appendix C for addresses of professional and trade organizations and other sources of technical information.

Heating, Ventilating, and Air-Conditioning CSI MasterFormat™

Heating, Air-Conditioning, Ventilation, and Refrigeration

Upon completion of this chapter, the student should be able to:

- Understand the many factors to be considered when designing heating and air-conditioning systems.
- Make decisions concerning needs and types of ventilation for acceptable indoor air quality.
- Recognize the fuels used by heating, air-conditioning, and ventilation systems.

- Select appropriate air heating and cooling systems.
- Select appropriate designs for steam and hot-water heating systems.
- Discuss the systems and equipment available for cooling systems in large buildings.

Heating, ventilating, and air-conditioning are based on the principles of thermodynamics, fluid mechanics, and heat transfer. Heating, venting, and air-conditioning (HVAC) equipment is used to accomplish climate control in buildings, producing comfortable humidity and temperature levels that are closely regulated to maintain safe and healthy conditions. The aim is to provide thermal comfort, acceptable indoor air quality, and reasonable installation, operation, and maintenance costs. A solution for a residential building may be insufficient for a medium to large office building or a skyscraper. Solutions vary depending on climate, scale, occupancy, and make-up of the exterior envelope. For example, the design of an office building is more concerned with human comfort, while manufacturing plant design addresses a range of other considerations.

Design factors include the type and amount of glazing, insulation and air infiltration levels; heating and cooling loads generated by machinery; industrial processes; solar load; and the materials used for walls, ceilings, floors, and roofs and their coefficients of thermal conductivity. Space must be provided within a building to house mechanical equipment and to access various parts of the structure containing ducts, pipes, and other parts of a system. Other factors, such as a need to control humidity; removal of chemicals, noxious gases, or dust; availability and cost of fuel; cost of the various

systems; expected maintenance expenses; possibility of down time; and climatic factors, illustrate additional factors HVAC engineers consider as a system is designed.

HVAC CODES

The installation, control, and operation of heating, ventilating, and air-conditioning systems involve a wide range of equipment and fuels, and a host of safety considerations. These are regulated by the various HVAC codes and supporting manuals available from professional organizations. Some of the codes from the National Fire Protection Association include the *Liquefied Petroleum Gas Code*, the *National Fuel Gas Code*, and the *Boiler and Combustion Systems Hazards Code*. The International Code Council offers code publications such as the *International Mechanical Code* and the *International Fuel Gas Code*.

HEAT BALANCE

HVAC design is concerned with establishing indoor temperature and humidity at comfortable levels and maintaining them. In hot weather, a system must remove excess heat from a building's interior at the same rate that it gains heat. In cold weather, it must add interior heat at the same rate it loses heat to the cold outside environment. This provides a balance that achieves comfortable conditions without large fluctuations in temperature.

HEAT TRANSFER

Heat transfer is the movement of thermal energy from a heated item to a cooler item. When an object or fluid resides at a different temperature than its surroundings, transfer of thermal energy, also known as heat exchange, occurs in such a way that the solid or liquid body and the surroundings reach thermal equilibrium. Heat transfer always flows from a hot object to a cold one, a result of the second law of thermodynamics. This flow occurs via radiation, conduction, or convection.

Thermal radiation transfers heat energy emitted by a source, such as waves through space, which is absorbed by a material it touches. For example, the sun radiates heat energy through space to the earth, where it warms anything it strikes. Radiant solar energy can be harnessed by solar collectors and transferred through a system that moves it to a building's interior. The sun provides *radiant heat* through windows, which can help warm a room or make it uncomfortably hot. Likewise, a cold exterior room wall absorbs heat from occupants' bodies, leaving them cold (Figure 40.1). In the summer, a hot exterior wall radiates heat, potentially making occupants uncomfortable.

Thermal conduction involves the transfer of heat energy through a material by transmitting kinetic energy from one molecule to the next. The movement is caused by a temperature change in the material. For example, when a match is held on one end of a short piece of copper wire, the other end will quickly get hot. Heat gain or loss due to conduction occurs when your body has physical contact with a material. If you touch a metal object (a good thermal conductor) at room temperature, it will absorb body heat and feel cool to the touch. If you touch a wooden object (a poor thermal conductor), it will not absorb as much heat and feel warm to the touch.

Thermal convection involves the transfer of heat by the circulation of heated air, gas, or liquid. In natural convection, air surrounding a heat source receives heat, becomes less dense, and rises. The surrounding cooler air then moves to replace it. This cooler air then heats, and the process continues, forming a convection current. Heat transfers to the ceiling, walls, and other building

Figure 40.1 The HVAC engineer must consider various sources of radiant heat transfer energy and their influence on human comfort.

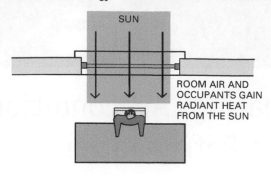

ROOM AIR AND OCCUPANTS GAIN RADIANT HEAT FROM THE SUN

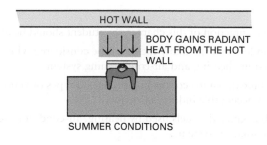

BODY GAINS RADIANT HEAT FROM THE HOT WALL

SUMMER CONDITIONS

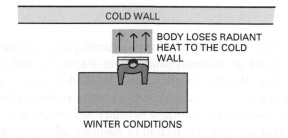

BODY LOSES RADIANT HEAT TO THE COLD WALL

WINTER CONDITIONS

elements. Convection continues until thermal equilibrium is reached. Hot-air furnaces supply forced heated air through ducts and use convection to heat a room. Hot-water- and steam-heating systems also use convection by heating air with radiators.

If air temperature is lower than body temperature, your body loses heat to the air, and your body cools. If air temperature is higher than body temperature, your body absorbs heat from the air and becomes warmer.

SENSIBLE AND LATENT HEAT

Sensible heat is heat that can be detected by touch when heat energy is added to or removed from a material. It is associated with a change in temperature.

Latent heat involves forces that change the state of a substance, such as when water turns to steam. This is an important factor to consider in the design of HVAC systems. For example, by adding heat to a unit of water until it reaches the boiling point, 212°F (100°C), an engineer can calculate the BTUs of heat required to reach this point. A *British thermal unit (BTU)* is the amount of heat required to raise the temperature of 1 lb. of water 1°F. If heat continues to be applied, the temperature of the water will not increase, but it will begin to boil and undergo a physical change into water vapor, also known as *steam*.

HUMAN COMFORT

HVAC systems are designed to maintain recommended room temperatures. These vary by season, with acceptable winter temperatures sitting a few degrees below acceptable summer conditions. For example, residences and similar occupancies, such as offices and classrooms, are typically most comfortable if kept in the range of 70°F to 75°F (21°C to 24°C) in the summer and 68°F to 74°F (20°C to 23.5°C) in the winter. The air temperature for most other occupancies, such as retail shops, medical facilities, and restaurants, is also kept within this range. Some areas, such as showers, will have a higher design temperature. Despite this range of recommended temperatures, people feel comfortable at different temperatures. Differing levels of physical activity, manual labor versus working at a desk for example, also necessitate different temperature requirements.

A space in which one area is cold while another is warm produces discomfort. Experiencing a rapid temperature change, such as exiting an air-conditioned building on a very hot day, may cause immediate discomfort. Some control can be exercised by keeping inside temperatures close to those outside, so long as the interior environment is comfortable.

Research by the American Society of Heating, Refrigerating, and Air-Conditioning Engineers (ASHRAE) has established tolerances for high ambient temperature (Table 40.1). (Ambient temperature is that of the surrounding air.) When a person is exposed to high temperatures beyond these limits, they are subject to heat stress, which can be dangerous. Likewise, exposure to very cold situations without adequate protection loses body heat faster than it can be replaced, which results in discomfort or worse. If uncorrected, this creates a life-threatening situation.

ASHRAE Standard 55–2013, *Thermal Environmental Conditions for Human Occupancy*, defines the range of indoor thermal environmental conditions acceptable to a majority of building occupants The factors involved include clothing, air speed, temperature, thermal radiation, and an occupant's activity level.

VENTILATION

Ventilation is used to control heat, condensation, humidity, odors, and contaminants in the air. Ventilating involves the process of replacing air in a space to control temperature or remove moisture, odors, smoke, heat, dust, and airborne bacteria. Ventilation includes both the exchange of air to the outside as well as circulation of air within a building. It is one of the most important factors for maintaining acceptable indoor air quality in buildings. Ventilation strategies include the use of natural flow, driven by wind and temperature, and mechanical flow, fan driven through ducts.

Air typically contains minute particles, such as dust raised by the wind and various pollens. The air where large numbers of people assemble, such as in a theater, becomes contaminated by carbon dioxide given off by occupants. This carbon dioxide must be removed and replaced with fresh air. In some situations, mist, fog, and various gases and vapors exist, as well as a variety of bacteria and other microorganisms. These can cause serious infections in humans and are a particular problem in medical facilities. Some business and industrial plants develop odors that must be removed. Unpleasant odors can cause physical discomfort, such as headaches and nausea. Others are an indication of airborne toxic substances that could cause physical harm to occupants.

ASHRAE outdoor air ventilation specifications for various indoor occupancies can be found in ASHRAE Standard 62.1–2004, *Ventilation for Acceptable Indoor Air Quality*. Selected examples from this publication for health care facilities are listed in Table 40.2. The standard lists outdoor air requirements for commercial and institutional spaces in proportion to the number of persons occupying the space. For example, hospital room

Table 40.1 High Ambient Temperature Tolerance Times

Air DB Temperature	Tolerance Time
180°F (82°C)	Almost 50 minutes
200°F (93°C)	33 minutes
220°F (104°C)	26 minutes
239°F (115°C)	24 minutes

Source: ASHRAE Handbook of Fundamentals, American Society of Heating, Refrigerating and Air-Conditioning

Table 40.2 Outdoor Air Requirements for Ventilation of Health Care Facilities (Hospitals, Nursing, and Convalescent Homes)

Application	Estimated Maximum** Occupancy P/1,000 ft² or 100 m²	Outdoor Air Requirements				Comments
		cfm/ person	L/s* person	cfm/ft²	L/s · m²	
Patient rooms	10	25	13			Special requirements or codes and pressure
Medical procedure	20	15	8			relationships may determine minimum
Operating rooms	20	30	15			ventilation rates and filter efficiency.
Recovery and ICU	20	15	8			Procedures generating contaminants may require higher rates.
Autopsy rooms	20			0.50	2.50	Air shall not be recirculated into other spaces.
Physical therapy	20	15	8			

*Table prescribes supply rates of acceptable outdoor air required for acceptable indoor air quality. These values have been chosen to dilute human bioeffluents and other -contaminants with an adequate margin of safety and to account for health variations among people and varied activity levels.
**Net occupiable space.
Source: ASHRAE Standard 62.1-2004, Ventilation for Acceptable Indoor Air Quality, American Society of Heating, Refrigerating and Air-Conditioning Engineers, Inc.

ventilation requirements are based on the expectation that the maximum occupancy will be ten occupants per 1,000 sq. ft. (100 m²). Physical therapy room ventilation requirements are based on occupancy of twenty persons per room. For residential spaces, the required

outdoor air supply is specified in air changes per hour or cubic feet per minute. ASHRAE 62.1–2004 recommendations for residential facilities are listed in Table 40.3.

Ventilated air is brought from outside and filtered before entering an interior air distribution system. In

Table 40.3 Outdoor Air Requirements for Ventilation of Residential Facilities (Private Dwellings, Single, Multiple)

Applications	Outdoor Requirements	Comments
Living areas	0.35 air changes per hour but not less than 15 cfm (7.5 L/s) per person	For calculating the air changes per hour, the volume of the living spaces shall include all areas within the conditioned space. The ventilation is normally satisfied by infiltration and natural ventilation. Dwellings with tight enclosures may require supplemental ventilation supply for fuel-burning appliances, including fireplaces and mechanically exhausted appliances. Occupant loading shall be based on the number of bedrooms as follows: first bedroom, two persons; each additional bedroom, one person. Where higher occupant loadings are known, they shall be used.
Kitchens[b]	100 cfm (50 L/s) intermittent or 25 cfm (12 L/s) continuous or openable windows	Installed mechanical exhaust capacity.[c] Climatic conditions may affect choice of the ventilation system.
Baths, Toilets[b]	50 cfm (25 L/s) intermittent or 20 cfm (10 L/s) continuous or openable windows	Installed mechanical exhaust capacity[c]
Garages: Separate for each dwelling unit	100 cfm (50 L/s) per car	Normally satisfied by infiltration or natural ventilation
Common for several units	1.5 cfm/ft² (7.5 L/s m²)	See "Parking garages"

[a]In using this table, the outdoor air is assumed to be acceptable.
[b]Climatic conditions may affect choice of ventilation option chosen.
[c]The air exhausted from kitchens, bath, and toilet rooms may utilize air supplied through adjacent living areas to compensate for the air exhausted. The air supplied shall meet the requirements of exhaust systems as described and be of sufficient quantities to meet the requirements of this table.
Source: ASHRAE Standard 62.1–2004, Ventilation for Acceptable Indoor Air Quality, American Society of Heating, Refrigerating and Air-Conditioning Engineers, Inc.

some cases, the interior air is recycled, having contaminants removed and odors reduced. It is not always permissible to release air from a building directly outside without some form of treatment. Exchanging outside air with inside air can also increase heating and cooling costs, but heat exchanger units are available to reduce them. Heat exchangers use the heat in the outgoing air stream to warm the cooler incoming fresh air.

Methods of Ventilation

Ventilation may be natural or mechanical. Natural ventilation utilizes outside air without using fans or other mechanical assistance. It can be achieved with operable windows, skylights, roof ventilators, and other openings. In more complex systems, warm air in a building can be allowed to rise and exit through upper openings (stack effect), thereby drawing in cool air naturally through openings in lower areas.

Mechanical, or forced, ventilation is used to regulate indoor air quality. Excess humidity, odors, and contaminants are controlled via dilution or replacement with outside air. Mechanical ventilation uses fans or blowers to pull air into a building through louvered exterior openings; it then directs the air to needed areas via ducts (Figure 40.2). Mechanical ventilation is more predictable than natural ventilation because the fans, openings, and ducts can be sized to meet specific needs, and mechanical systems operate continuously to provide the specified ventilation. Natural ventilation performance varies, due to uncontrollable fluctuation in temperature and exterior weather conditions.

Figure 40.2 A fixed-bar louvered grille is used with fans or blowers to pull exterior air into a building.

© Ken Schulze/Shutterstock.com

Some ventilation systems remove smoke, gases, and heat during fires. These are designed following the National Fire Protection Association Standard NFPA 90-A, *Installation of Air-Conditioning and Ventilating Systems*. Other ventilation systems, such as those in restaurant kitchens, are specifically designed to remove fumes and heat. The air contains greasy materials that coat ducts and fans and become a fire hazard. In addition to routine maintenance, these systems must have an automatic fire-control system. This typically includes fire dampers that close ducts when a fusible link breaks and activates some fire-smothering material, such as foam.

In certain occupancies, it is not permissible to recirculate ventilated air. Areas such as restrooms, biology and chemistry laboratories, hospital areas (including operating rooms), and industrial facilities in which various flammable vapor or hazardous fumes are developed, must exhaust the air, and in some cases treat it before release.

Types of Air Filters

The most commonly used air cleaners include fibrous panel filters, renewable filters, electronic filters, and combination air cleaners.

Viscous impingement, fibrous panel filters are made of porous, coarse fibers coated with an adhesive substance that captures particles in the air. They are inexpensive and good for lint but do little to control pollen or atmospheric dust. They are available in thickness from ½ in. to 4 in. (12.7 mm to 102 mm) and in a variety of standard sizes. A common application is in hot-air residential furnaces and air-conditioning systems. Filter materials include coated animal hair, glass, vegetable and synthetic fibers, metallic wool, expanded metals and foils, crimped screens, random matted wire, and synthetic open-cell foams. Some types, such as the glass-fiber residential filter, are disposable (Figure 40.3). Metal filters can be cleaned and reused. They are typically washed with water or steam and may be recoated with adhesive (Figure 40.4).

Dry-type extended surface air filters are made of random fiber mats or blankets using bonded glass-fiber, cellulose fibers, synthetics, and wool felt supported by a wire frame that forms pockets or pleats. Some types require replacement of the entire filter, while others permit the replacing of mats inside the wire frame. Generally, dry-type filters are more efficient and have higher dust-holding capacities than do viscous impingement fiber panel filters.

Figure 40.3 A disposable fiber air-filter panel.

© iStock.com/BanksPhotos

Figure 40.4 This metal and plastic fiber mesh air filter can be washed and reused.

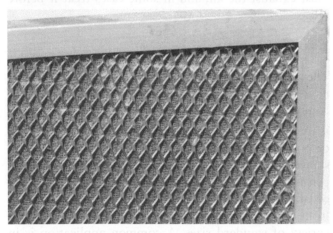

Figure 40.5 A cross-section through an ionizing electronic air cleaner.

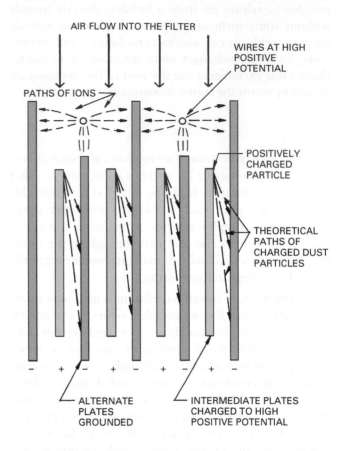

Renewable filters are viscous-impingement-type filters in roll form. They are either moved by hand or automatically unrolled across the filter area. As the clean filter material unrolls, the dirty portion rolls onto a roller at the bottom. When the entire roll is used up, it is removed and replaced. Random dry fiber filter material is also used.

Electronic air cleaners use electrostatic precipitation to collect dust, pollen, and smoke. The filters have ionization and collection plate sections. The ionization section has small diameter wires with a positive direct current potential of 6 to 25kV DC. They are suspended between grounded plates, as shown in Figure 40.5. The wires create an ionizing field through which particles in the air pass and receive a positive charge. The collecting plate section has a series of equally spaced parallel plates. Alternate plates have a 4 to 10kV DC positive charge. The negative plates that are not charged are

grounded. The charged particles in the air pass into the collecting section and are forced by their electrical fields onto the negative grounded plates.

Electronic air cleaner cells are cleaned with a detergent and hot water. Large commercial installations may include an automatic washing system that cleans the cells in place. Small units, as used in residential systems, generally are removed and cleaned manually.

Combination air cleaners are an assembly of the systems just discussed. Typically, a dry-type filter might be used in line after an electronic filter has processed the air. The dry filter can catch any particles that may escape the collector plates before they pass through the system.

FUEL AND ENERGY USED FOR HVAC

The major fuels used for mechanical systems include coal, fuel oil, natural gas, propane, butane, solid waste, and wood. All but natural gas require some form of storage. The energy sources include electricity, wind, geothermal, ground heat, and solar. Solar energy can be used directly with a means for storing heat. The others sources depend on an outside source of energy supply.

Coal

Coal is a combustible fossil fuel composed of elemental carbon, hydrocarbons, complex organic compounds, and various inorganic materials. The types commonly used for commercial and industrial heating and the production of electricity are anthracite, bituminous, lignite, semi-anthracite, and subbituminous. Coal is pulverized for burning in commercial furnaces. It is converted into liquid and gaseous products that may be used as fuels or made into coke, which is used to produce steel. Systems that use coal as a fuel must have emissions controls to reduce air pollution and a means for disposing of the ash. Anthracite coal has a heat content of about 14,600 BTU/lb or 33,980 kJ/kg (kilojoules per kilogram).

Fuel Oil

Fuel oil is a fossil fuel derived from petroleum refining. It is a mixture of liquid hydrocarbons containing about 85 percent carbon, 12 percent hydrogen, and 3 percent other elements. The American Petroleum Institute (API) has a system for grading fuel oil. Kerosene is Grade 1, characterizing it as a very light and easily vaporized liquid. As the grade numbers increase, so does the viscosity. Viscosity is a measure of the flow quality of a liquid. Grades 1 and 2 fuel oils are frequently used in residential and small commercial furnaces (Grade 3 is sometimes used for these purposes). They flow easily and can be vaporized and mixed with air as required for burning. Grades 4, 5, and 6 are heavier and used in large industrial and commercial burners. Grades 5 and 6 are very heavy and must be preheated before they can be pumped and burned. No. 2 fuel oil has a heat content of about 141,000 BTU/gal. or 39,300 kJ/L (kilojoules per liter).

Natural Gas

Natural gas is a combustible hydrocarbon gas consisting primarily of methane with small amounts of ethane and traces of butane, propane, pentane, and hexane. It is odorless in its natural form but an added odorant acts as a safety measure so leaks can be detected by smell. Natural gas has a heat content of approximately 1,050 BTU/ft.^{3} (39,100 kJ/m^3). It is obtained from wells drilled in earth strata that contain pockets of gas. Natural gas is supplied to buildings through underground utility pipelines, so on-site storage is not required. A utility company is responsible for maintaining a continuous supply.

Propane and Butane

Propane and butane are liquid petroleum products and very volatile. They are colorless, flammable gases derived from petroleum and natural gas. Supplied in liquid form, they are transported in pressurized cylinders. Propane and butane have a heat content of about 2,500 BTU/ft.^{3} (93,150 kJ/m^3).

Solid Waste

The collection of burnable solid waste is an ever-increasing activity, and its use as a fuel is becoming more popular. It takes specialized incinerator equipment to produce a clean burn resulting in the desired heat energy. Typically, biomass facilities are built near large urban areas to ensure a steady supply of waste or near a commercial or industrial plant that has contracted to use the energy. This has the advantage of helping reduce the growing problem of disposing of large amounts of waste materials while producing a usable product.

In some systems, the waste is shredded, and electromagnets draw out ferrous materials in the recycling process. These plants burn this refuse-derived fuel (RDF), with coal as a backup fuel. The fuel produced from the waste is the main fuel used to burn the incoming waste. About 50 percent of waste materials entering the plant drive turbine generators that produce electricity. In this case, the electricity is used by the city to operate sewer and water treatment plants, streetlights, and to serve other public and private customers.

Another waste-to-energy system is illustrated in Figure 40.6. It accepts mixed waste and sorts it for recycling and composting before turning the remainder into RDF. In all cases, major efforts are required to make waste-to-energy systems economical to operate. It requires cooperation from the local trash collectors and local and state governments.

Wood

Wood is a renewable resource, and the harvesting of trees is accompanied by a program of replanting. Wood has become a minor source of heat energy, typically used only to heat small residences. Overall, wood has a heat content of about 7,000 BTU/lb. (16,290 kJ/kg).

A variety of wood-burning stoves are available, most of which operate more efficiently than fireplaces. Many

Figure 40.6 This waste-to-energy system accepts unsorted waste, sorts, recycles, and composts it to produce refuse-derived fuel (RDF).

newer-designed stoves meet the Environmental Protection Agency's Phase II emission standards. These have the EPA approval label, consume less wood, and burn 80 to 90 percent cleaner than older wood-burning stoves.

The EPA Phase II stoves are available in two types: catalytic and non-catalytic. The catalytic stove has a cored combustor in the smoke path just past the main combustion chamber. The combustor has a catalytic coating that causes combustion gases to burn completely, thus reducing emissions. After the fire in the main combustion chamber reaches 500°F (262°C), a bypass damper opens, directing gases through the catalyst rather than straight up the chimney. The combustor does deteriorate with use and must be replaced occasionally.

The non-catalytic stove can also achieve low emissions, but it does this by keeping high temperatures in the main combustion chamber and adding turbulent secondary air at the top of the chamber where hot gases leave the stove. This air provides additional oxygen so combustion gases burn more completely.

Pellet stoves are designed to burn a manufactured product made from wood and wood by-products. The pellets are small, about the size of large feed grain, and are sold in waterproof bags. The concentrated material burns hot and clean, producing low emissions. Pellet fireplace inserts are also available. They use a motor-driven auger to feed pellets from a fuel storage hopper. The typical hopper holds enough fuel to last several days.

Some types of fireplace inserts function much like stoves. The insert is a metal liner with pyro-ceramic glass doors that permit generated heat to radiate into a room. They are airtight and burn hot, producing low emissions. This type of fireplace insert has emission qualities equal to the EPA Phase II woodstove. They can be installed in a wood-burning fireplace or have an approved brick or wood surrounding and are vented with a metal chimney. Some have an electric blower to increase heat distribution into a room.

Electricity

Electricity is a major energy source for operating HVAC systems. It powers blowers, pumps, and controls and supplies heat energy through various types of resistance heating units. Electric power usually has the lowest installation cost and is easy to install anywhere in a building. It does not, however, compete well with other fuels on a cost basis. Electricity requires no on-site storage, but large installations require space for transformers and switchgear. Power is usually provided by a utility company that guarantees a steady supply, although some on-site power generation is used on large installations (as discussed in Chapter 41).

Geothermal

Geothermal heat energy derives from the earth's internal heat. The typical example is a geyser that send up jets of hot water and steam from within the earth. This source of heat is limited to a geyser basin (an area containing a group of geysers). The system can be tapped to supply heat to an entire town from this central source.

Ground Heat

Ground temperatures fluctuate by season, but remain stable as depth increases. Exact temperatures vary with the geographic location. For example, in Minnesota's cold climate, the temperature is a relatively constant 50°F (10°C) at a depth of 16 ft. to 26 ft. (5m to 8m). Earth-sheltered houses use this stable temperature to provide a steady source of heat through exterior walls. If a 50°F (10°C) interior air temperature is maintained, very little supplemental heating is required. Ground-source heat pumps provide another important use of ground heat. Some systems use an earth coil that involves burying plastic pipe either in drilled holes or in a continuous ground loop. The water in the pipe absorbs ground heat and is pumped to a water source heat pump that extracts the heat and recirculates the water. Tapping well water also works. A heat pump removes water from a well, extracts the heat, and disposes of the water in a second well, putting it back into the aquifer. Groundwater temperatures vary by geographic region, but groundwater heat pumps in the range of 55°F to 60°F (13°C to 16°C) are commonly used.

Solar

Solar energy derived from the sun is a major source of heat energy. Solar energy can be tapped by various collectors, stored, and used to heat and cool buildings or produce electricity. It is also a source of natural light. A variety of solar heating systems are available. Active solar systems utilize mechanical devices to gather, store, and distribute heat energy as needed. In passive solar-heating systems, the thermal energy flow is by radiation, conduction, or natural convection.

AIR HEATING EQUIPMENT

Warm-air heating systems incorporate a heat-generating device (furnace), controls, and a distribution system of ducts. A furnace may be fired by oil, gas, electricity, and some solar systems that use warm-air distribution. Dual systems include an air-conditioning unit to both heat and cool buildings.

Warm-Air Furnaces

Warm-air furnaces use blowers to move heated air through ducts. These include upflow, downflow, and horizontal furnaces. An upflow warm-air furnace moves air out the top of the unit into a duct system. It is used in basements when ducts are run under a floor or on the floor when ducts are run in an attic (Figure 40.7).

Figure 40.7 Typical configurations of upflow warm-air heating systems.

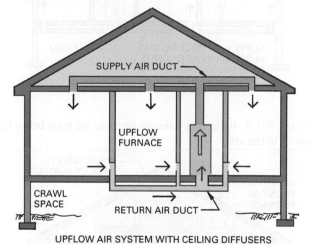

UPFLOW AIR SYSTEM WITH CEILING DIFFUSERS

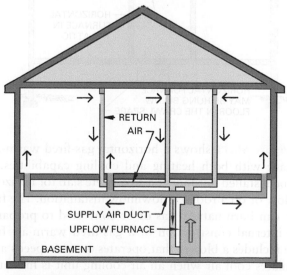

UPFLOW AIR SYSTEM WITH THE FURNACE IN THE BASEMENT

A downflow warm-air furnace moves air out the bottom of a unit. This type is typically used when ducts are run in a concrete slab floor or below the floor in a crawl space (Figure 40.8). Horizontal warm-air furnaces are mounted in an attic or hung below floor joists, as shown in Figure 40.9. Each of these may have an air-cooling unit that uses the furnace blower and ducts to provide cool air to building spaces (Figure 40.10).

Figure 40.8 Downflow warm-air heating systems are used in buildings with a crawl space or concrete floor.

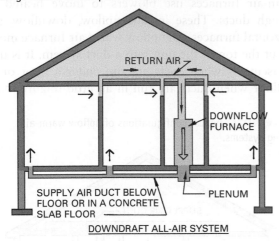

Figure 40.9 Horizontal warm-air furnaces are hung below the floor or in the attic.

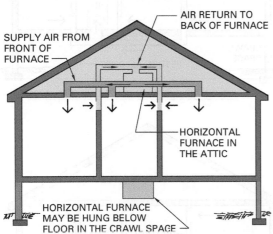

Figure 40.11 shows a horizontal gas-fired warm-air furnace with both heating and cooling capabilities. It can be installed outdoors on a concrete slab for horizontal flow or on a roof for downflow installation. The furnace can burn natural gas or be converted to propane. The internal construction of a gas-fired warm-air furnace includes a blower that operates on two speeds and circulates cool air when an air-cooling unit is mounted on top of the unit. These furnaces are available in a wide

Figure 40.10 This attic-mounted furnace works with an air-cooling unit that uses the furnace blower and ducts.

© iStock.com/perrygerenday

Figure 40.11 A rooftop single-package gas-fired warm-air furnace with an electric cooling coil to provide summer cooling in addition to heating.

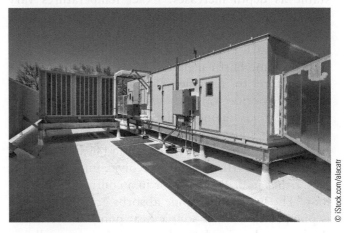

© iStock.com/alacatr

range of sizes and output ratings (BTUs per hour). Gas-fired furnaces can be switched on site from an upflow/horizontal configuration to a downflow setup.

Figure 40.12 shows a high-efficiency, gas-fired warm-air furnace that is typically used in residential applications. The variable-speed blower motor adjusts airflow for better humidity control and comfort. Warm-air furnaces may also be equipped with electric resistance coils to heat the air, which is then distributed through ducts by a blower. These all-electric furnaces are similar to the gas furnace shown; however, the high cost of electricity in some areas makes them more costly to operate. The efficiency of these all-electric systems is significantly increased when they are matched with an outdoor heat pump unit.

Oil-fired furnaces require a large exterior storage tank from which to pump fuel oil to the furnace burner.

Figure 40.12 This is a high-efficiency gas-fired furnace that can be installed in either the upflow or horizontal air discharge position. The furnace includes a hot surface ignitor that eliminates the pilot light. It is a direct vent furnace, which requires outdoor air for combustion. This unit is equipped with a variable-speed indoor blower motor.

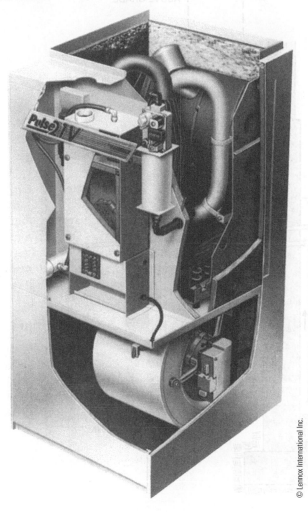

© Lennox International Inc.

A typical installation is shown in Figure 40.13. The location and method of placement of the unit is regulated by codes. Oil tank construction must meet Underwriters Laboratory (UL) specifications. Tanks may be placed above ground or underground below the frost line. Some codes permit oil storage tanks inside a building. An oil-fired furnace is much like gas-fired models, but it has an oil burner unit with a nozzle that breaks the oil into a fine spray. A high-voltage ignition system using spark electrodes ignites this vapor.

Heat Pumps

A heat pump is a machine that can heat or cool a building using forced air through ducts. It takes heat from one source, such as outside air, and transfers it to another, such as the air inside a building. The commonly available types include air-to-air, water-to-water, and air-to-water. Use of ground-source heat pumps is increasing.

The most frequently used heat pump is an air-to-air unit. It has a *compressor* similar to those used in refrigerators. A *refrigerant* (a gas) circulates in coils. In heating mode, the coils in the exterior unit enable the refrigerant to absorb heat from outdoor air. The refrigerant is then transferred to a compressor, where compression raises its temperature. It then transfers to indoor coils, a blower moves air over it, which captures the heat and distributes it through a building via ducts (Figure 40.14). In cooling mode, the reverse happens. Heat is absorbed by the inside coils and dispersed to outside air by the outside coils.

Figure 40.15 shows an air-to-air heat pump compressor unit. The assembled outdoor unit rests on a concrete slab, and the coils are fully exposed to the air. The fan on top pulls air through the coils.

A groundwater heat pump (GWHP) is a water-to-air system. A water-to-air system pumps water from a well to the heat pump. The well water typically maintains a constant temperature, which varies with the geographic area but is typically between 55°F and 65°F (13°C to 18°C). The refrigerant in the heat pump coils absorbs heat in the water and discharges it into a building through an air-handling unit and ducts. The unit can reverse the process and remove heat from inside a building and discharge it to the water, which is drained in a disposal well.

Water-to-water heat pumps can be used to simultaneously heat and cool water, if necessary. For example, the pump can remove heat from a liquid source, such as a fluid in a manufacturing operation that requires chilling, and move the heat to another liquid, such as water for cleaning operations.

Air-to-water heat pumps can be used to cool air in a room and transfer captured heat to water or another liquid that needs heating. An apartment complex could cool air in living spaces and use the resultant heat to keep water in a swimming pool at a desired temperature.

A ground-loop heat pump is used in areas where groundwater is not readily available, too costly to acquire, or of poor quality for heat pump use. The ground loop avoids the maintenance issues caused by the mineral content of water in the groundwater systems. The system may use an earth coil in a horizontal loop (Figure 40.16), or be installed vertically in drilled holes (Figure 40.17). The earth-coil system transfers heat to or from a water/antifreeze solution circulated through plastic pipes buried in the earth. The heat from the earth is transferred to

Figure 40.13 A typical installation for an oil-fired warm-air furnace.

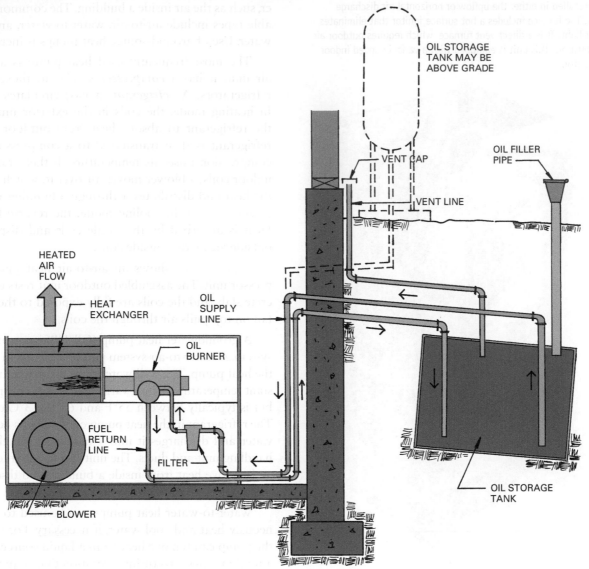

the refrigerant in a heat exchanger and compressed, and distributed by a blower through ducts. In cooling mode, the reverse occurs, with indoor heat being transferred to the water/antifreeze solution, which disperses it to the cooler ground.

The choice of a horizontal or vertical ground loop depends on the results of a study that analyzes surface and subsurface conditions, including soil type, moisture content, rock strata, and ground temperatures. Horizontal loops require sufficient open space free from paved surfaces, underground utilities, and trees and shrubs (Figure 40.18).

Efficient pumps designed to heat and cool one room with no ducts are manufactured as a single unit. The coil

and fan are housed in a decorative interior case. The compressor and exterior coil extend through an opening in the wall and are exposed to the outside air (Figure 40.19). Similar units are available with total electric heating and cooling capacities.

A natural gas heating and cooling unit uses natural gas as the primary energy source for both heating and cooling, operating much like the all-electric heat pump. The compressor is driven by a quiet natural gas engine, and its function is the same way as described for electric heat pumps. The unit has a recuperator that recovers engine heat, thus increasing efficiency. An auxiliary gas heater provides supplemental heat during extremely cold weather. The system is microprocessor-controlled

Figure 40.14 The heating cycle of an air-to-air electric heat pump.

OUTDOOR COIL

OUTSIDE AIR

REVERSING VALVE

COOL AIR TO THE OUTSIDE

COMPRESSOR

DUCT SYSTEM

INDOOR COIL

EXPANSION DEVICE

HEAT PUMP ON OUTDOOR SIDE

EXPANSION DEVICE

WARM AIR TO INSIDE

SYSTEM IS SHOWN IN THE HEATING MODE. THE CYCLE REVERSED WHEN COOLING.

AIR HANDLER ON INDOOR SIDE

Figure 40.15 An air-to-air heat pump.

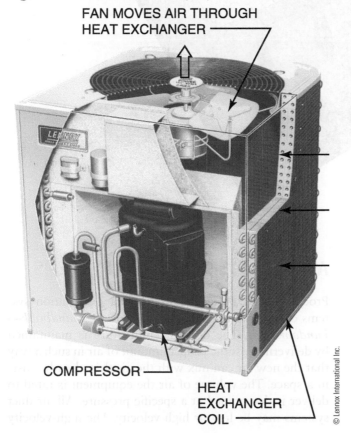

FAN MOVES AIR THROUGH HEAT EXCHANGER

COMPRESSOR

HEAT EXCHANGER COIL

to provide continuous comfort. A gas engine uses pressure formed by combustion of expanding gas to produce mechanical energy by supplying power to the motor shaft, which connects to the compressor.

AIR-CONDITIONING WITH DUCT SYSTEMS

Duct systems are also used to air-condition buildings. The heat pump previously discussed provides both heating and cooling modes. When oil, gas, or electric warm-air furnaces are used, a cooling coil unit is placed on the furnace and connected to an outside air-conditioner, as shown in Figure 40.20. They are connected by pipes that carry refrigerant from the compressor in the outdoor unit to the indoor coil. The furnace blower moves air over the cold coils and through the ducts into the building.

Residential external electric air-conditioning units are shown in Figure 40.21. The refrigerant absorbs heat from air blown over the indoor coil. It then moves to the outdoor air-conditioning compressor, where the absorbed heat is transferred to the atmosphere. The refrigerant then moves back to the interior coil to repeat the process.

Figure 40.16 A horizontal ground-loop heat pump installation. The ground loop is made of plastic pipe carrying a water/antifreeze mixture that is circulated through the pipe to the water-to-refrigerant heat exchanger in a heat pump.

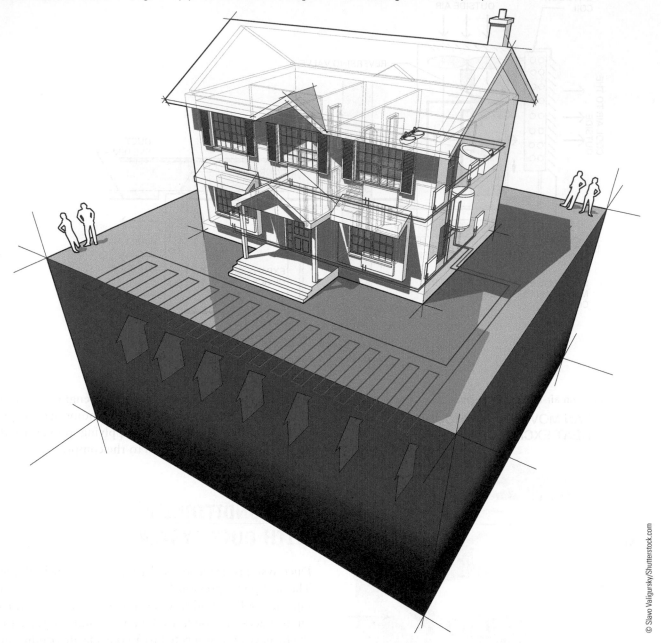

If a building utilizes hot water or steam heat, a similar system can be used to air-condition it. A duct system is installed for cooling, and the indoor unit is an air handler with a blower and coil. The outdoor air-conditioning unit is the same as the one just described.

Figure 40.22 shows a roof-mounted heating/cooling unit used on light commercial construction. When combined with a duct system, as shown in Figure 40.23, they can heat and cool discreet zones within a building. Each zone has its own *thermostat* that operates a zone damper.

ALL-AIR DISTRIBUTION SYSTEMS

Procedures for designing all-air duct distribution systems can be found in the 2013 ASHRAE *Handbook—Fundamentals*. The desired temperature is maintained by delivering just the proper amount of air in such a way that the new air can mix with the air that already exists in a space. The volume of air the equipment is rated to deliver is determined at a specific pressure. All-air duct systems may be low or high velocity. The high-velocity

Figure 40.17 A vertical ground-loop heat pump installation. In areas with low soil moisture vertical loops are recommended if the soil permits drilling deep enough to get adequate lengths on the vertical loops.

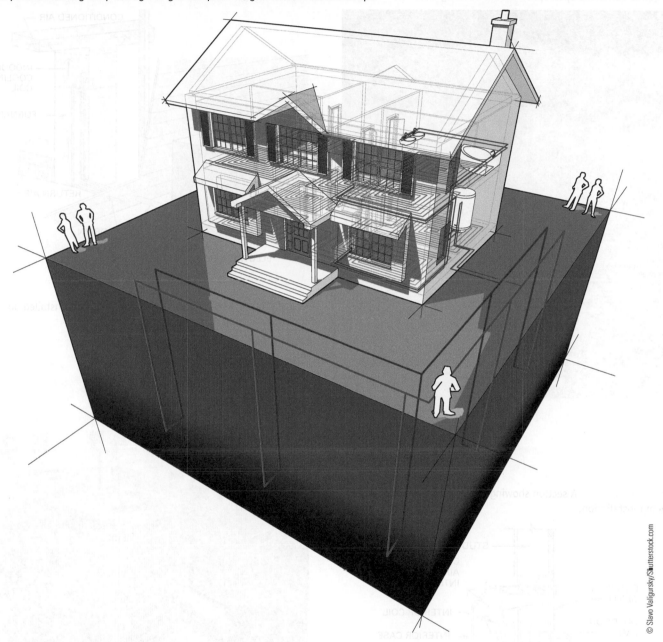

design requires higher air pressure but results in smaller-diameter ducts. Every obstruction and bend in the ductwork used adds to the pressure loss within the system. The designer must consider how duct size influences performance, supply air temperatures, airflow, air changes, zoning humidity control, heat gain and heat loss of the space to be conditioned, control of noise in the system, possible use of heat recovery devices, and many special factors, like those in hospitals, manufacturing plants, and laboratories.

Systems in common use include single zone, multi-zone, reheat, variable air volume, and dual duct. Various combinations of these can be used to meet design requirements.

Single-Zone Systems

Single-zone systems use one air-handling unit (AHU) to supply an entire building or a portion of a building that is considered a single zone. The furnace and air-handling unit can be installed outside of or within the space to be

Figure 40.18 The installation of a horizontal ground loop requires sufficient open space on the building site.

© Christine Langer-Pueschel/Shutterstock.com

Figure 40.20 A unit air conditioner can supply cooled refrigerant to cooling coils in the air handler.

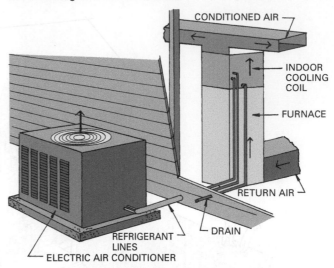

CONDITIONED AIR

INDOOR COOLING COIL

FURNACE

RETURN AIR

DRAIN

REFRIGERANT LINES

ELECTRIC AIR CONDITIONER

Figure 40.21 Exterior unit air conditioners are installed on concrete pads adjacent to interior utility rooms.

© iStock.com/WendellandCarolyn

Figure 40.19 A section showing a typical through-wall heat pump installation.

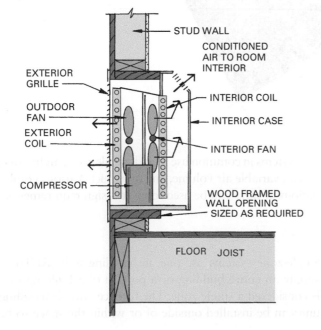

STUD WALL

CONDITIONED AIR TO ROOM INTERIOR

EXTERIOR GRILLE

INTERIOR COIL

OUTDOOR FAN

INTERIOR CASE

EXTERIOR COIL

INTERIOR FAN

COMPRESSOR

WOOD FRAMED WALL OPENING SIZED AS REQUIRED

FLOOR JOIST

conditioned. A *return-air* duct system is used to extract used air from a space and recycle it through the system to re-condition it by cooling or heating. A small residence may be conditioned with a single-zone system. Larger residences may use two or more heating and cooling sources, with two or more single-zone systems. A simple schematic is shown in Figure 40.24. It should be noted that this system supplies air at a constant rate. Therefore, room temperature is varied by changing the air temperature.

Figure 40.22 These roof-mounted units, designed for commercial buildings, are available as an electric-cooling/gas-heating unit or an electric-cooling/electric-heating unit.

© iStock.com/buchachon

Figure 40.23 A typical duct system for a roof-mounted heat pump or electric/gas-fired unit. The controls permit it to condition the air in in discreet zones.

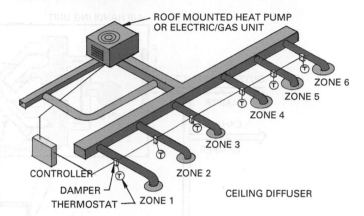

Figure 40.24 A single-zone heating/cooling duct system for a small building that has a single source of heating and cooling.

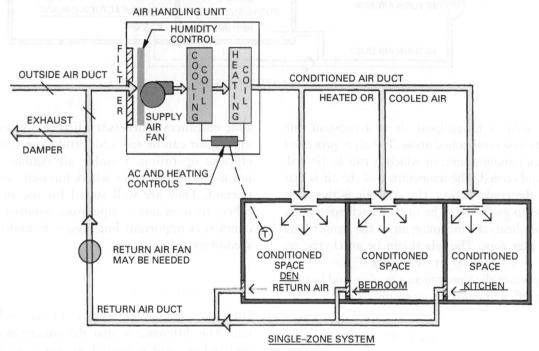

Multi-zone Systems

Multi-zone systems produce heated and cooled air in an air-handling unit. The air for each zone is blended at the air-handling unit into a single temperature and fed to the zone by a single duct (Figure 40.25). This enables a zone requiring the lowest air temperature to receive it or, if needed, the temperature can be raised to a higher level. Likewise, areas requiring higher temperatures can be accommodated by blending little or no cool air with the heated air sent to an area.

The flow of air in the ducts is regulated by motorized dampers controlled by thermostats. Since this system uses a constant volume of air, it passes by the heating and cooling coils even if conditioning is not needed. Because of this, the system is not cost efficient nor is it widely used.

Reheat Systems

A reheat system is a variation of the single-duct single-zone and the multi-zone systems. It supplies a single source

Figure 40.25 A multi-zone system designed to blend cool and heated air to produce different temperatures required for the various zones.

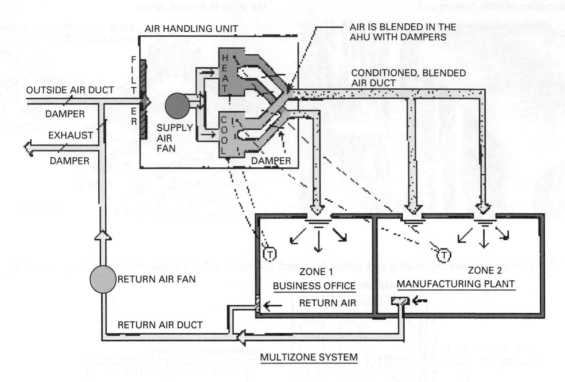

of preconditioned or recirculated air at a constant rate through ducts to several zoned areas. The air is processed through an air-handling unit, in which it can be filtered, humidified, and cooled. The temperature of the air is that required for the coolest zone. The cool air is then sent through ducts to each zone. The duct to each zone has a reheat coil that heats the incoming air to the temperature required for that zone. The reheat can be an electric resistance unit or a hot water or steam coil (Figure 40.26). This system permits the simultaneous cooling and heating of areas with different temperature requirements.

Variable Air Volume Systems

Variable air volume (VAV) systems control a space's environment by supplying air at a constant temperature and varying the quantity supplied rather than changing the temperature of the air (as is done in the single-duct, reheat, and multiple-zone systems). The air from the handling unit moves through single ducts to each zone, where a variable air volume terminal is located. This terminal varies the air supply to the space as needed, but the air temperature is held constant (Figure 40.27).

Variable air volume systems may use a common or separate fan system, and a reheat coil can be added at

zone entrances if necessary. If reheat is used, the volume of air can be reduced, producing a more energy-efficient operation. Variable air volume systems are not useful in situations where humidity control is important. They are well suited for use in offices and other areas where temperature control for human comfort is important but precise humidity control is secondary.

Dual-Duct Systems

Dual-duct systems are similar to the multi-zone system. The difference is that the conditioned air in the air-handling unit is moved to spaces by two parallel ducts. One duct carries cold air and the other warm air (Figure 40.28). In each space, mixing valves combine the warm and cold air in the proportion needed to meet the air temperature requirements of the space. These may be constant volume or variable volume air systems. Constant volume systems may use a reheat. Variable air volume systems may use a single fan or dual-supply fans. The dual-duct system is not cost efficient because it uses energy to both heat and cool air and then mix them to get desired temperatures. Additional details on these systems can be found in the 2012 ASHRAE Handbook, *HVAC Systems and Equipment*.

Figure 40.26 A reheat warm-air system supplies conditioned air at the lowest required temperature and uses reheat units to raise the temperature at zones requiring higher air temperatures.

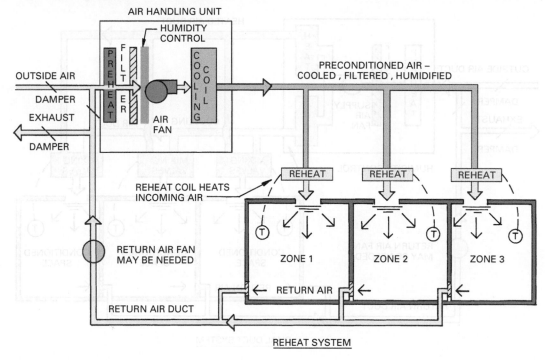

Figure 40.27 A variable-air-volume warm-air system controls air temperature in each zone by varying the amount of air flowing into that zone.

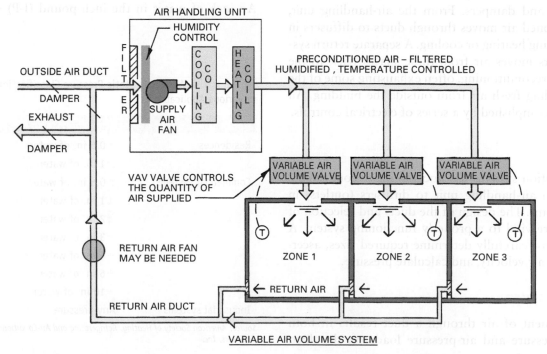

Figure 40.28 Dual-duct warm-air systems use separate warm and cool air ducts that meet at a mixing valve to produce air at the temperature required.

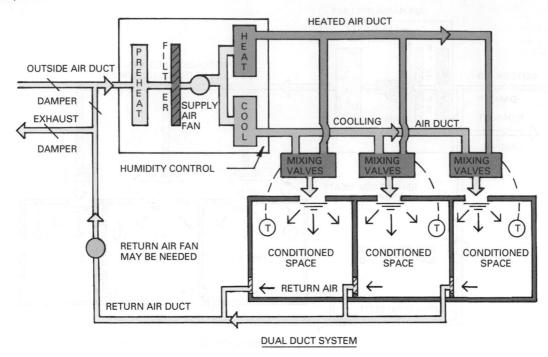

DUAL DUCT SYSTEM

TYPES OF DUCT SYSTEMS

Air distribution systems receive heated or cooled air from HVAC equipment. The air moves into an air-handling unit, where it is filtered, humidified, or dehumidified. The air-handling unit has a fan, filters, humidifiers, coils, and dampers. From the air-handling unit, the conditioned air moves through ducts to diffusers in rooms needing heating or cooling. A separate return system of ducts moves air from these spaces back to the furnace for reconditioning, often exhausting some of the air and adding fresh air from outside the building. All of this is accomplished by a series of electrical controls.

Ducts

Air distribution systems use ducts to move conditioned air from an air-handling unit to diffusers (outlets) in various rooms. The design of the ducts and selection of materials are vital to a properly functioning system. A designer must carefully determine required sizes, ascertain proper air velocity, and calculate pressure.

Duct Pressure

The movement of air through a duct results in both velocity pressure and air-pressure loads on the duct.

The duct design must calculate the actual static pressure on each section of a duct and specifies the pressure classification. The pressure classifications for residential, commercial, and industrial ducts are shown in Table 40.4. The pressure classifications for residential, commercial, and industrial ducts are specified in inches of water. An inch of water in the inch-pound (I-P) system is a

Table 40.4 Allowable Static Pressure Classifications for Ducts in Various Applications

Application	Allowable Static Pressure
Residences	±0.5 in. of water
	±1 in. of water
Commercial systems	±0.5 in. of water
	±1 in. of water
	±2 in. of water
	±3 in. of water
	+4 in. of water
	+6 in. of water
	+10 in. of water
Industrial systems	Any pressure

Source: American Society of Heating, Refrigerating and Air-Conditioning Engineers, Inc.

unit of head equal to a column of liquid water 1 inch high at 39.2°F. (Head is the energy per unit mass of fluid divided by gravitational acceleration.) The I-P system uses customary units (inches and pounds) rather than metric units.

A standard for duct design is published by the Air Conditioning Contractors of America (ACCA) Manual D, *Duct Design for Residential Winter and Summer Air Conditioning*. The Sheet Metal and Air Conditioning Contractors National Association (SMACNA) also has publications relating to duct design and installation.

Duct Classification

Duct systems are regulated by various laws, building codes, local ordinances, and standards. These must be considered by an engineer as a duct system is designed. Projects built for the federal government will have standards issued by various agencies, such as the General Services Administration and the Federal Construction Council.

Duct construction is classified in terms of pressure and use. Commercial duct systems include HVAC systems for applications in educational, business, general factory, and mercantile structures. Industrial duct systems include those used for industrial exhaust and air pollution control.

Model project specifications for the construction of ducts include Masterspec, produced by the American Institute of Architects (AIA), and Spectext, available from the Construction Specifications Institute (CSI). Residential ducts are specified by local building codes. A source for multifamily dwellings is National Fire Protection Association (NFPA) Standard 90A. Supply ducts may be galvanized steel, aluminum, or other materials rated by Underwriters Laboratory (UL) Standard 181. Rigid and flexible fiberglass supply ducts must meet the *Fibrous Glass Duct Construction Standards* of the SMACNA.

Commercial ducts are also usually regulated by NFPA Standard 90A and UL181. This classifies ducts into two groups:

Class 0. Zero flame spread, zero smoke spread

Class 1. 25 flame spread, 50 smoke developed

Class 0 ducts are made from iron, steel, aluminum, concrete, masonry, or clay tile. Class 1 ducts include many of the flexible and rigid fiberglass ducts manufactured. Industrial ducts are specified by NFPA Standard 91. These are used for duct systems that might convey flammable vapors or air containing various particles. Particle conveying ducts are available in four classifications:

Class 1. Non-particulate applications (*makeup air*, fresh air brought in to replace inside air; general ventilation; and gaseous emission control)

Class 2. Moderately abrasive particles in the air (sanding or buffing)

Class 3. Highly abrasive material in low concentration (handling sand or abrasive cleaning)

Class 4. Highly abrasive particles in high concentration

Abrasive ratings are specified in *Round Industrial Duct Construction Standards* by the SMACNA. Industrial ducts are generally made from galvanized steel, uncoated carbon steel, or aluminum. Aluminum is not used if the air will contain abrasive particles, and ducts carrying corrosive vapors must have appropriate protective coatings.

Duct standards include many other specifications, such as insulation, type and spacing of hangers, welding when necessary, seismic requirements, ducts exposed to outdoor atmospheric conditions, and ducts below ground.

Ducts are available in round, rectangular, and flat-oval shapes. Specifications for these vary, depending on their use. They are often insulated to control heat loss or gain and to control condensation that occurs when moist air strikes a duct.

FORCED-AIR DUCT SYSTEMS

Duct systems used in residential and small commercial buildings include the perimeter loop, perimeter radial, and extended plenum systems. The perimeter-loop system is typically used with concrete slab floors and downflow furnaces or in a building with a basement or crawl space (Figure 40.29). The perimeter duct is placed in the thickened edge of a slab. It is essential that the edge and bottom of the slab be insulated to reduce heat looses from the ducts. Registers are placed along the perimeter duct as needed. The return ducts in this system are located in the attic.

The perimeter radial system is also used in concrete slab construction but can be used in basements and crawl spaces, too. It uses a downflow furnace (Figure 40.30). A variation of this system uses an upflow furnace with the radial ducts in the attic (Figure 40.31).

Figure 40.29 A perimeter-loop warm-air duct system circulates air through a continuous duct system fed by several horizontal ducts.

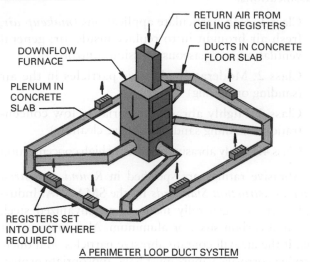

A PERIMETER LOOP DUCT SYSTEM

Figure 40.30 A perimeter-radial duct system extends individual ducts to each space to be heated and cooled.

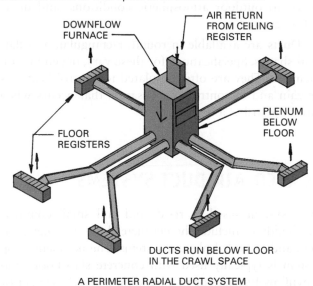

A PERIMETER RADIAL DUCT SYSTEM

Figure 40.31 A perimeter radial duct system can place the ducts in the attic and use ceiling diffusers.

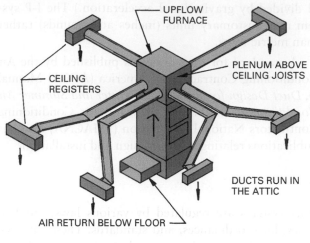

A PERIMETER RADIAL DUCT SYSTEM

Figure 40.32 The extended plenum runs from the furnace with individual ducts extending from it to the rooms to be heated and cooled.

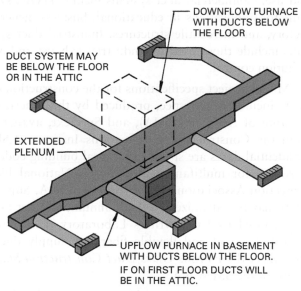

EXTENDED PLENUM DUCT SYSTEM

The extended-plenum system may have the furnace in the basement or on the first floor. It can also be used with horizontal furnaces in a crawl space or attic. The plenum is extended to provide needed airflow to feed ducts that run from it to outlets.

If the furnace is on the first floor, the plenum can be in the attic, with ducts running over ceiling joists and diffusers running through the ceiling into the room (Figure 40.32).

Horizontal warm-air furnaces are commonly placed below the floor or in the attic.

The ducts in residential and many small commercial systems are hidden from view, but in some large commercial and industrial buildings, they are left exposed. Typically these systems are used where large open areas exist, such as in an airport or a sports complex (Figure 40.33).

Figure 40.33 Exposed duct systems are used in buildings that have large open areas.

© iStock.com/claudio.arnese

SUPPLY-AIR OUTLETS AND RETURN-AIR INLETS FOR WARM-AIR HEATING SYSTEMS

The design and selection of *supply-air* outlets and return-air inlets is critical to providing adequate air-conditioning in a room. Poorly located or poorly chosen outlet and inlet grilles and diffusers can reduce the effectiveness of even the best heating and cooling units. An engineer has a considerable number of factors to consider, and a decision is often one of compromise. Following are examples of some of the things to be considered.

Air velocity and air temperature in a supply duct must be greater than that permitted within a room. This air must, therefore, be emitted into the room so that air velocity is not offensive and people are not exposed to areas of high temperature. This requires the system to mix the air supply with room air to control convection currents and uneven temperatures.

Air must be supplied to a room with a termination device (a grille or diffuser) that provides the required air dispersion pattern needed for human comfort. The termination device must throw the air across the room far enough to enable it to blend with the room air, thus reducing velocity and temperature before it drops into the room's lower zone. Engineers know that cool air is heavy and settles to the floor, but warm air is lighter and rises toward the ceiling. The location of windows must also be considered in a design because they can introduce a source of cold air that can produce occupant discomfort and cool circulating warm air.

Surface effect, another thing to be considered, is caused when the airstream from a duct outlet moves across a ceiling or contacts a wall. This creates a low-pressure

area along a surface, keeping the air in contact with it for most of the *throw*. Circular ceiling diffusers project an airstream across portions of a ceiling and create surface effect if they are long enough to cover the ceiling area. Grilles may also cause some ceiling effect (Figure 40.34).

Figure 40.34 Air-surface effect keeps the incoming air from a ceiling diffuser moving across the ceiling.

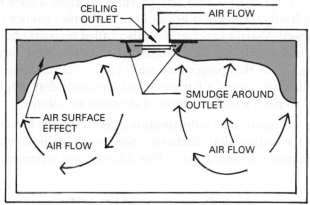

AIR SURFACE EFFECT WITH A CEILING DIFFUSER.

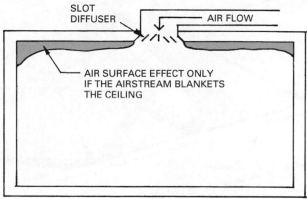

AIR SURFACE EFFECT WITH A SLOT DIFFUSER.

Smudging occurs with both slot diffusers and ceiling diffusers. Smudging is a band of discoloration that eventually appears on ceiling material around the edge of a diffuser. It is caused by dirt particles in the discharged airstream that are held in turbulence around a diffuser and cause smudging around the outlet. This can be alleviated by placing anti-smudge rings around outlets.

Another design consideration is an outlet's noise level. Outlets will transmit sound caused by mechanical equipment and high air velocities through a duct. High-pitched noise is frequently caused by air passing through an outlet. Recommended air velocities and decibel levels are found in publications of the American Society of Heating, Refrigerating, and Air-Conditioning Engineers.

Types of Supply Outlets

The basic types of supply outlets are grilles, slot *diffusers*, ceiling diffusers, and air distributing ceilings (plenum).

A grille is a rectangular opening with vanes that direct the airstream as it leaves an outlet. Adjustable grilles have adjustable vertical or horizontal vanes that deflect airstream in a horizontal or vertical direction. Some types have two sets of vanes at right angles to each other. Fixed-bar grilles have vanes that cannot be moved. Variable-area grilles (also called registers) have dampers that can vary the size of an outlet, thus varying the air that passes through (Figure 40.35). Stamped grilles are simply a flat sheet with openings that permit air to pass, with no control of direction or volume.

Diffusers are air-distribution outlets, usually located in the ceiling, that discharge conditioned air in several directions within a space. Slot diffusers are rectangular

Figure 40.36 Air-distribution ceilings or plenums use the space between the floor and ceiling as a plenum. Perforated ceiling panels feed air into the room below.

Figure 40.35 Grilles may have fixed bars or adjustable vanes that vary the direction and amount of airflow from the duct.

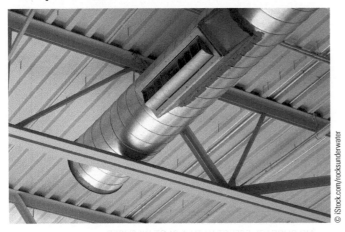

outlets with one or several slots, often installed in long lengths. Perpendicular slot diffusers discharge air perpendicular to or at a slight angle to the face of a diffuser. Parallel slot diffusers produce an airstream parallel with the face of a diffuser. Diffusers are often located at the sills of large windows and in the floor below glass doors or floor-length glass windows. They create an upflow of heated air over cold glass, producing increased comfort for occupants.

Air-distribution ceilings, or plenums, use the space between the bottom of the floor above and the top of a suspended ceiling as a plenum. The space is fed heated or cooled air by ducts located around its perimeter. The ceiling panels have perforations through which air enters the room below (Figure 40.36).

Ceiling diffusers may be round or square and contain a series of louvers that form air passages. They distribute the air supply in a uniform pattern around a diffuser (Figure 40.37). Flush ceiling diffusers have concentric louvers that extend the same distance from the core. Stepped-down diffusers have louvers that project beyond the shell of a unit. Perforated-face ceiling

Figure 40.37 This diffuser fits into the grid of the suspended ceiling and is fed from a supply duct above.

diffusers have a flat perforated face that is flush with and blends in with the surface of a suspended ceiling. They may have deflection vanes behind their exposed faces to direct airstream discharge. Variable ceiling diffusers have dampers that can be adjusted to control the amount of airflow from the diffuser.

Another type of ceiling air distribution diffuser is shown in Figure 40.38. The supply duct runs above a suspended ceiling and supplies conditioned air to the air diffuser. The diffuser is a rectangular unit that rests upon the horizontal metal members of a suspended ceiling. Below this, parallel beams with spaces between them are used to distribute air into the room. Airflow direction is controlled in two directions by vanes. Diffusers are available in a variety of several sizes.

Figure 40.38 This rectangular diffuser uses parallel beams with spaces between them to distribute air into the room.

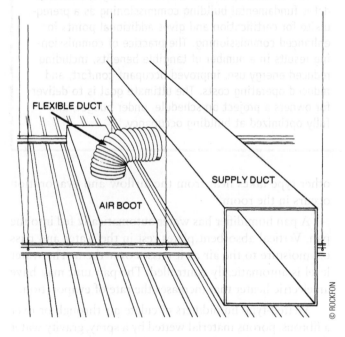

Diffuser Applications

Following are some general applications for the location of outlet and inlet diffusers. Outlet diffusers bring air into a space, and inlet diffusers are used with air returns or exhaust.

In areas where air-cooling needs predominate, outlets will be most effective in the ceiling, with air-returns located on walls at the floor or in the floor. This works because cool air sinks as it mixes with existing air in a room. In areas where the heating needs are predominant,

outlets are best located in floors or at the baseboard on walls. The return air can be high on the wall.

Outlet ducts are generally located on exterior walls and under windows and return-air ducts on interior walls. This permits a curtain of conditioned air to flow up the outside wall, warming or cooling its surface as needed to produce a more comfortable interior situation. A cold outside wall will pull body heat from an occupant even though the air temperature is adequate. Likewise, a hot wall will radiate heat to an occupant.

Return-Air Inlets and Exhaust-Air Inlets

Return-air inlets connect to ducts that return room air to a furnace for reconditioning. Exhaust-air inlets move air from inside a building through ducts and discharge it outside. They typically do not require vanes to deflect or redirect the airstream. Some types do have dampers to control airflow amounts. Fixed-bar grilles are commonly used. Bars may be situated on a slight angle or perpendicular to a grille's face (Figure 40.39). Fixed-bar louvered grilles used on the exterior of large commercial buildings for exhaust and fresh air intake can be from 30 ft. to 60 ft. in width and length.

Figure 40.39 A return-air grille installed in the ceiling.

HUMIDITY CONTROL

Control of a space's relative humidity is an important factor in the overall conditioning of its environment. Requirements vary depending on occupancy. *Humidity* describes the water vapor within a space. *Relative humidity* is a ratio of water vapor weight actually in the air to the maximum possible water vapor weight the air could contain at a given temperature; this ratio is expressed as a percentage. For example, if relative

Construction **Methods**

Building Commissioning

Building commissioning is the process of verifying that all HVAC, plumbing, electrical, and building fire and security systems are installed, calibrated, and operating as intended (by a building's owner) and as designed (by a building's architects and engineers). The process involves procuring the services of a qualified Commissioning Authority (CA) to conduct and verify commissioning activities.

The basic procedure for building commissioning encompasses a comprehensive pre-construction review of design documents for compliance with Owner's Project Requirements (OPR), periodic site observations during the construction phase, and systems performance testing at project completion. The commissioning team begins by documenting the design requirements for systems to be commissioned prior to approval of contractor submittals for equipment. Commissioning requirements are then incorporated into the construction documents. These include detailed specifications for submittal review, construction verification, and functional performance testing procedures.

During the construction process, the commissioning authority inspects system installations to ensure all components are properly set up and any problems discovered are corrected prior to performance testing. Functional performance testing begins once all systems are in place, powered, and programmed. Testing procedures include a thorough review of all operations, including start-up and shut-down procedures, systems balancing, and emergency and failure modes of the various systems. A variety of methods are used to simulate operations and evaluate whether or not all systems perform as specified in construction documents. Any deficiencies discovered are reported to the owner in a summary commissioning report, and a plan of action is formulated to correct them.

While the practice of building commissioning is still fairly new in the construction industry, it has quickly become common practice. The LEED rating system mandates fundamental building commissioning as a prerequisite for certification and gives additional points for enhanced commissioning. The practice of commissioning results in a number of tangible benefits, including reduced energy use, improved occupant comfort, and reduced operating costs. The ultimate goal is to deliver for owners a project on schedule, under budget, and fully optimized at building occupancy.

---

humidity is 100 percent, the air can hold no more water vapor, and an increase will cause moisture to condense and form water drops.

Human comfort depends a great deal on the relative humidity of air. Typically, indoor relative humidity should be kept between 30 to 60 percent. Low humidity causes drying of nose, throat, skin, and hair membranes. Furniture, cabinets, interior trim, and other wood products can shrink and check if the relative humidity is too low. Likewise, high relative humidity can cause doors and drawers to swell and stick. Heating air during winter removes moisture, so a humidifier is used to increase the relative humidity. In the summer, air in many geographic areas has a high relative humidity. An air-conditioning system must then dehumidify (remove water vapor from) the air.

Humidification Equipment

Humidifiers increase the amount of water vapor in air. They operate in two different ways. One type adds heat: as air flows through the humidifier's wet section, it picks up moisture, and the air is distributed to the room. The other type takes heat from the airflow and evaporation occurs in the room.

A pan humidifier has water automatically fed into the pan. Vertical absorbent plates rest in the water and pass off moisture to the air as it passes over them. The water level is automatically controlled. The pan unit may have an electric heater that increases the rate of evaporation.

Wetted-type humidifiers circulate air through or over a fibrous, porous material wetted by a spray, gravity water flow over it, or revolving through a water pan. One type uses a fan that moves air from the furnace plenum and blows it through the pad or rotating drum and back into the plenum.

A bypass humidifier is shown in Figure 40.40. The humidifier mounts on the furnace air supply and uses air moved by the furnace blower. Small humidifiers are available that can be mounted in ducts.

A spinning-disc atomizing humidifier injects small particles of water into the airflow where it is absorbed, raising the humidity Figure 40.41. Another type uses a spray nozzle to inject a mist into the airflow.

Figure 40.40 A bypass humidifier is installed on the main furnace branch duct.

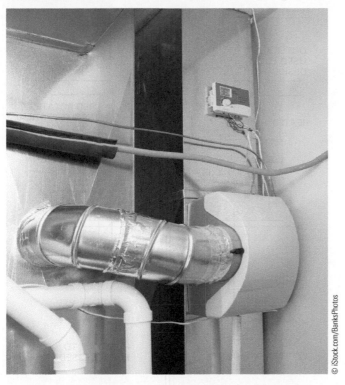

© iStock.com/BanksPhotos

Figure 40.41 A spinning-disc atomizing humidifier injects small particles of water into the air-flow where it is absorbed, raising the humidity.

© terekhov igor/Shutterstock.com

Various industrial humidifiers are also available. A pan humidifier can be heated by hot water, steam coils, or electric element. It may be installed on a duct or operated in a remote location with hot water vapor carried to the main duct system by a connecting duct. Other types use an enclosed grid or inject steam directly into the duct. A jacketed steam humidifier operates off a constant steam supply flowing into a steam trap in which any condensation drains off. The flow into the dispersing tube in the duct is controlled by a steam valve.

A self-contained steam humidifier uses electric heaters to convert water to steam and then injects it into the airflow in the duct. Various types of atomizing humidifiers are also used in industrial applications.

Desiccant Dehumidification

Dehumidification involves the removal of water vapor from air, gases, or other fluids by some mechanical or chemical means. A *dehumidifier* is a device used to remove moisture from the air, thus reducing the relative humidity.

Desiccation is the use of a desiccant to remove moisture from a material. A *desiccant* is any absorbent liquid or solid used to remove water or water vapor from a material. Dehumidification equipment uses both solid and liquid desiccant materials. *Absorbent* materials will extract substances from a liquid or gas medium with which they are in contact. *Adsorbent* materials have the ability to capture molecules of gases, liquids, or solids on their surfaces without changing the adsorbent material chemically or physically.

Dehumidification is typically accomplished in residences by the cooling of air during air-conditioning mode. For commercial and industrial applications, other methods are available.

Dehumidification is required for many industrial applications. They include maintaining a dry atmosphere in a warehouse, producing dry air to aid in the drying of a material in an industrial process, drying natural and liquified gas, and lowering the relative humidity in a plant manufacturing products using hygroscopic materials, such as wood.

Liquid-desiccant dehumidification systems remove moisture by passing air through a mist-like spray of a liquid desiccant, such as a glycol solution. The desiccant spray absorbs moisture from the air and passes into a regeneration chamber, where it is heated and gives off the moisture. The moisture is discharged by outside air flow (Figure 40.42).

Figure 40.42 Liquid-desiccant dehumidification uses a mist-like spray of the liquid desiccant to absorb moisture in the air.

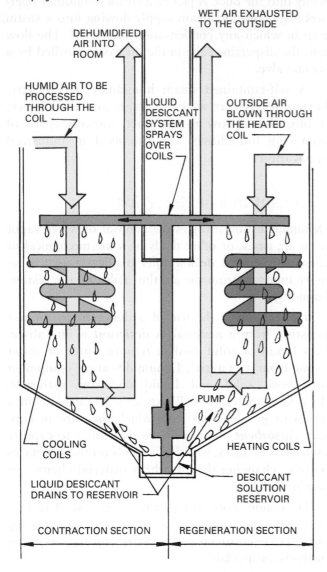

Figure 40.43 Solid-sorption dehumidification flows the humid air through a granular desiccant.

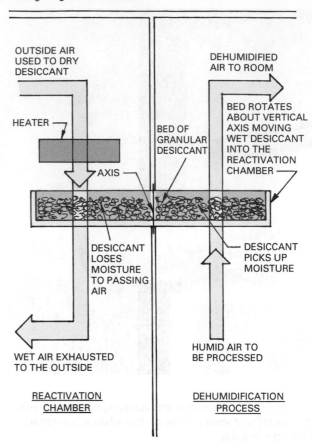

Solid-sorption dehumidification systems remove moisture by flowing air through a granular desiccant, such as silica gel or hygroscopic salts. "Sorption" is a general term used to include both absorption and adsorption. The granular desiccant has a vapor pressure below that of the vapor in the air. This difference in pressure drives water vapor in the air into the desiccant. When the desiccant becomes saturated, it is heated and dried in a reactivation chamber and can be reused (Figure 40.43).

HYDRONIC (HOT WATER) HEATING SYSTEMS

Hydronic heating systems are used in residential and commercial installations. *Hydronics* refers to the use of water as the heat-transfer medium in heating and cooling systems. A system consists of a boiler fueled by oil or natural gas, and a system of pipes, radiators, pumps, and controls. Three types of systems are typically used in residential and small commercial buildings. These are one-pipe, two-pipe direct-return, and two-pipe reverse-return systems.

One-Pipe Systems

One-pipe systems use a single loop of pipe to circulate water at 180°F (82.9°C) to radiators and return cooler water to a boiler for reheating and recirculating, as shown in Figure 40.44. Notice the water flows directly from the boiler to radiator 1, from which it flows into radiator 2. The water temperature at radiator 2 will be

Figure 40.44 A single-pipe hydronic system carries the water from the boiler through each terminal and back to the boiler.

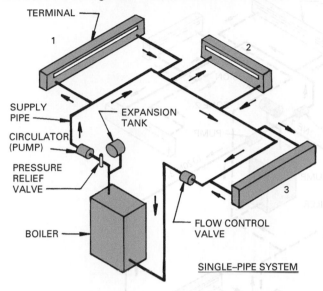

SINGLE–PIPE SYSTEM

Figure 40.45 The two-pipe hydronic direct-return hot water system has a separate return line for the cooled water, which flows in a direction opposite the supply line.

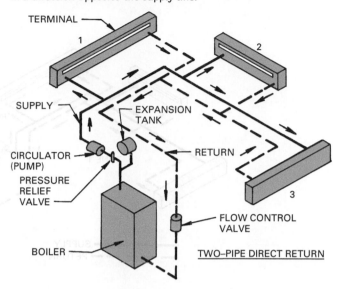

TWO–PIPE DIRECT RETURN

lower than at radiator 1. This loss can be compensated for by having a larger No. 2 radiator, but the loss continues through the remainder of the radiators on the loop. This system is difficult to keep in balance.

Two-Pipe Direct-Return System

Each of the radiators in this system receives hot water directly from a boiler through the hot water supply line. The cool water from each radiator returns directly to the boiler through a separate loop of pipe. The last radiator tends to receive less water because of higher pipe resistance to water flow. It has the longest hot water supply and cool water return lines. This is overcome by increasing the size of the hot water supply line and sizing the circulation pump to provide needed flow at the radiators on the end of the circuit (Figure 40.45). Balancing devices are used to regulate hot water flow through each radiator. In general, this system is not widely used because of the difficulty in getting a balanced distribution.

Two-Pipe Reverse-Return System

This system is similar to the two-pipe direct-return, except the return water flows in the same direction as the hot water. This provides a system having nearly the same pipe resistance on all radiators. For example, radiator 1 has the shortest hot water supply line but the longest return line. The reverse is true for radiator 3. This system is typically used in larger buildings with longer pipe runs (Figure 40.46).

Figure 40.46 The two-pipe reverse-return hydronic hot water system has the return water flowing in the same direction as the hot supply water, providing about the same pipe resistance to water flow for each terminal.

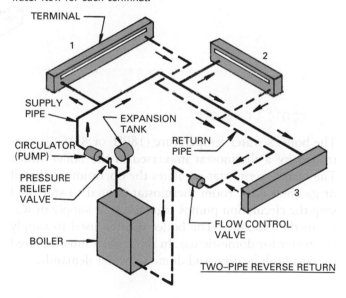

TWO–PIPE REVERSE RETURN

Multi-zone Two-Pipe Systems

A multi-zone system enables the temperatures in various parts of a building to be controlled separately. Each zone has a complete two-pipe system fed from a central boiler. The flow of hot water is individually controlled to each zone. This permits some zones to be kept at lower temperatures when not in use, resulting in a saving of energy costs. Large multistory and multi-use buildings will use multi-zone two-pipe hydronic systems (Figure 40.47).

Figure 40.47 A multi-zone two-pipe hydronic system is used to provide different temperatures in each zone.

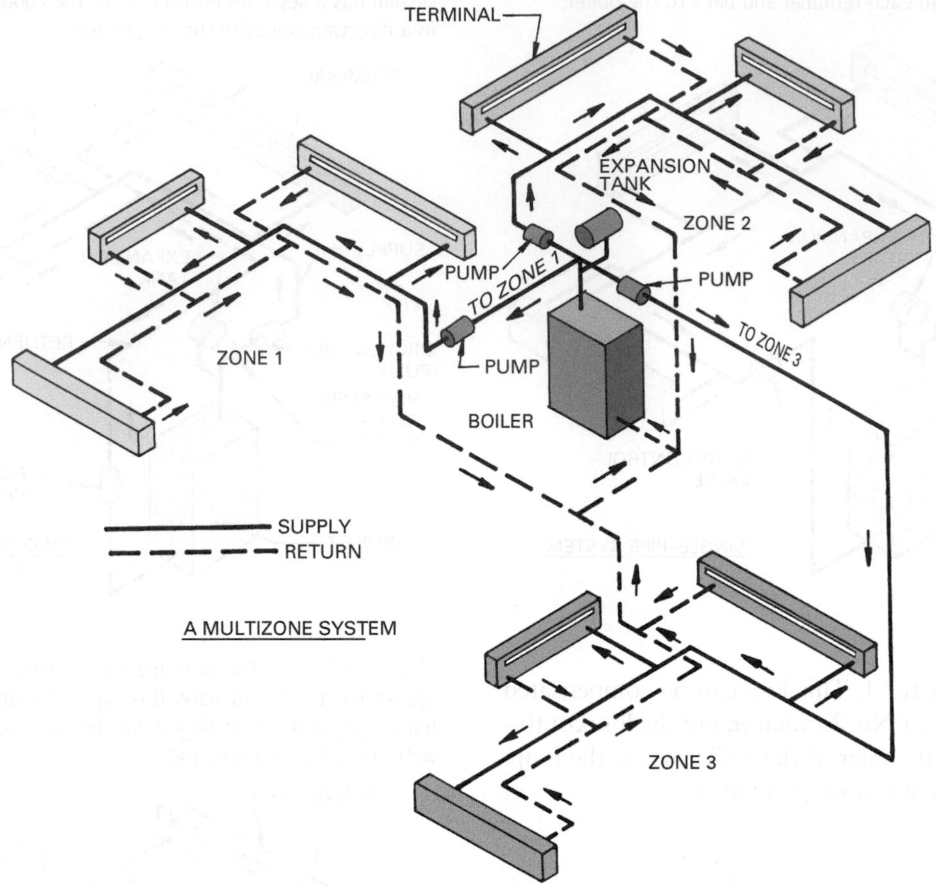

TERMINAL

EXPANSION
TANK

ZONE 2

PUMP

TO ZONE 1

PUMP

TO ZONE 3

PUMP

ZONE 1

BOILER

——————— SUPPLY
— — — — — RETURN

A MULTIZONE SYSTEM

ZONE 3

Hydronic Controls

The boiler's water temperature (180°F or 83°C) is controlled by a thermostat immersed in the boiler water. The water thermostat regulates the operation of the oil or gas burner. A room thermostat is used to start and stop the circulation pumps regulating the supply of water to the radiators. The boiler is often used to supply hot water for domestic use. In this case, it must be sized to meet both heating and domestic water demands.

Expansion Chambers

Hydronic heating systems require some form of expansion chamber (tank) (Figure 40.48). This serves as a space into which uncompressible water can flow as it expands within the system due to temperature increases. A closed-expansion system has a set volume of air that can be compressed and water the tank will hold as the air is compressed. The air pressure changes as the volume of water increases or decreases. The only time water can flow into or out of the

Figure 40.48 Types of expansion chambers used with hydronic heating systems.

tank is when the water expands or shrinks. The tank is sized to handle expansion due to the maximum temperature of the water according to guidelines available in the ASHRAE publication, *HVAC Systems and Equipment*.

The diaphragm-expansion system has a flexible membrane in the tank to separate air and water. As the expansive water enters the tank from the top, the diaphragm moves down, compressing the air. This system is widely used. Figure 40.49 shows a simplified illustration of a multistory hot-water heating system showing both closed and diaphragm-type expansion systems. Each terminal unit has a vent that opens automatically to relieve excessive pressure.

Figure 40.49 A simplified schematic of a multistory hot-water-heating system showing the diaphragm and closed-expansion systems.

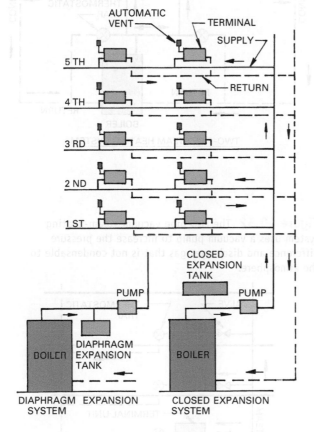

PIPE SYSTEMS FOR WATER HEATING AND COOLING

Water distribution systems can be used for heating with hot water and cooling with chilled water. The systems are usually closed and use circulators (pumps) to move water through the system and the terminal units. Hot water is supplied by a *boiler* and the cold water by a *chiller*. A chiller is a refrigerating machine used to remove heat from water that is circulated for cooling air in a building. Three-pipe and four-pipe distributions are used.

Two-Pipe Systems

Two-pipe systems provide heating and cooling supply to terminal units by running a supply pipe to each terminal. The second pipe holds the return in which the hot and chilled water are mixed and returned to the boiler and chiller, which results in warmed chilled water going into the chiller and cooled warm water going into the boiler. This results in increased costs to reheat and re-chill the water before it is recirculated.

Four-Pipe Systems

Four-pipe systems are used when a system provides both heating and cooling modes. In this system, each terminal receives separate hot water and chilled water supply and return lines, providing required heating and cooling any time either is needed. Systems that use the same coil for heating and cooling control the flow with two valves on each of the heating and cooling supplies at each terminal. Others put separate heating and cooling coils in the terminal unit (Figure 40.50).

Figure 40.50 A four-pipe hydronic system providing both heating and cooling modes.

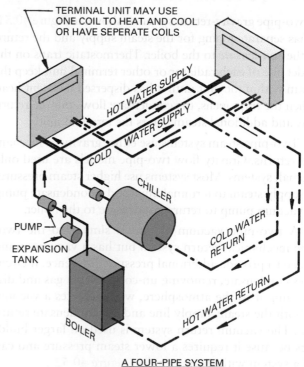

A FOUR-PIPE SYSTEM

STEAM HEATING SYSTEMS

A steam heating system has a boiler or other steam-generating device, a piping system, radiators or **convectors**, and controls. The boiler is usually oil or gas fired, but coal, wood, waste products, solar, nuclear, electrical energy, or cogeneration sources can also be used. The two types of steam heating systems are one-pipe and two-pipe.

Steam delivers considerably more heat per pound (0.45 k) than a pound (0.45 k) of water, but when it becomes vapor, steam expands much more than hot water. Therefore, a steam system requires larger diameter pipes than do hot water systems. Steam produces high pressures that force it through a piping system without the pumps used in hot water systems. Pipes must also be sized to allow gravity flow of the condensed water back to the boiler without interfering with the steam flow. Engineers use the design data for pipe sizes available in the *ASHRAE Handbook*. Steam space-heating systems are usually classified as low pressure.

It is more difficult to control the temperature of steam than hot water, so hot water is more widely used for space-heating systems. If a building requires steam for an industrial process, the steam supply is generally also used for space heating. This is often high-pressure steam requires the use of pressure-reducing valves in series to achieve acceptable pressures for space heating.

Two-Pipe Steam Heating Systems

A two-pipe gravity-return system is shown in Figure 40.51. It has separate piping for the steam supply and the return of the **condensate** to the boiler. Thermostatic traps on the outlet line of each radiator or other terminal unit keep the steam within the unit until it has dispersed its latent heat. When the trap opens, the condensate flows out the return line, and additional steam enters the terminal unit.

Two-pipe steam systems use either gravity or mechanical returns. Gravity flow two-pipe models are used only in small systems. Most systems use higher steam pressures to supply steam to terminal units and a condensate pump or vacuum pump to return condensate to the boiler.

A two-pipe vacuum system is similar to the two-pipe mechanical-return system but has a vacuum pump added to provide additional pressure difference. It circulates condensate, removing un-condensable gas and discharging it to the atmosphere, which creates a vacuum in both the steam supply line and the condensate return line. The vacuum return system is used on larger buildings because it requires a lower steam pressure and can fill a system with steam rapidly (Figure 40.52).

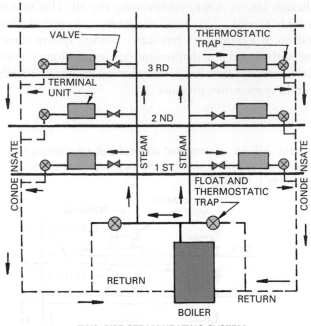

Figure 40.51 Two-pipe steam-heating systems have separate piping lines for the steam supply and the return condensate.

TWO–PIPE STEAM HEATING SYSTEM

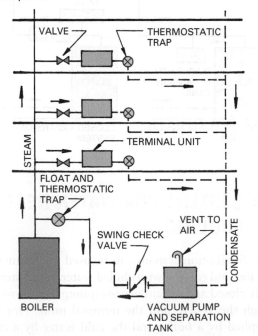

Figure 40.52 The two-pipe vacuum steam-heating system uses a vacuum pump to increase the pressure difference and discharges gas that is not condensable to the atmosphere.

TWO–PIPE VACUUM RETURN SYSTEM

BOILERS

Boilers are used to transfer heat from a fuel source to a fluid, such as water. The liquid is contained in a cast iron, steel, or copper pressure vessel that transfers heat to water to produce hot water or steam (Figure 40.53). Boilers are constructed according to the American Society of Mechanical Engineers (ASME) *Boiler and Pressure Vessel Code*. The Hydronics Institute publishes the *Testing and Rating Standard for Heating Boilers*.

Figure 40.53 Boilers are closed vessels that transfer heat from a fuel source to produce hot water or steam.

© iStock.com/alexandrumagurean

Boilers are classified by working pressure, temperature, fuel, size, and whether they are steam or water boilers. Steam boilers are used for space heating and auxiliary uses, such as in commercial laundries or industrial processes in which steam is required. Steam is typically used in large commercial and industrial buildings. Hot-water boilers are used for space heating and domestic hot water supply. They are typically used in residential and small commercial buildings.

Boilers are classified as high pressure or low pressure. High-pressure boilers are referred to as power boilers. They produce steam pressures above 15 psi and hot water pressures and temperatures above 160 psi and 250°F (122°C). They are typically of steel construction and use firetube or watertube designs.

Low-pressure boilers, referred to as heating boilers, are limited to a maximum steam pressure of 15 psi, a maximum hot water pressure of 160 psi, and a temperature of 250°F (122°C). They are made from cast iron, steel, or copper.

Firetube boilers have hot combustion gases pass through tubes surrounded by water. They are used in both low and high-pressure boilers. Watertube boilers have water inside tubes, and hot combustion gases pass over the tubes. Some boilers have the tubes in horizontal or nearly horizontal position, while some situate tubes vertically. Small, low-pressure boilers used for residential units may have helical coils. Watertube boilers are available ranging from small low-pressure units to large high-pressure steam units.

Boilers are available as condensing and noncondensing units. For many years, boilers operated so they did not condense flue gases in the boiler. This was done to prevent cast iron and steel parts from corroding. Non-condensing boilers are operated with a high flue gas return temperature to prevent moisture in the flue gas from condensing on cast iron and steel parts.

Condensing boilers salvage latent heat from the products of combustion by condensing the flue gas, increasing the efficiency of the boiler. This means lower flue gas temperatures can be used and the water vapor in the flue gas can condense and drain. The condensing section of the boilers must be made of materials that will withstand the required temperatures and resist corrosion. Certain types of stainless steel are used.

A pair of high-efficiency cast iron wet-base boilers are shown in Figure 40.54. They operate on oil or natural/propane gas, and models are available for hot-water and steam-heating systems. Some types may have a combination gas/oil-fired burner/boiler unit. These are typically used in schools, apartments, and commercial buildings, where a switch in fuels may be essential for continued heating service.

Cast-iron boilers are made by assembling individual cast-iron boiler sections, called watertubes, in which water for heating flows. The size and rating of a boiler depends on the number of sections assembled. Tubes may be vertical or horizontal. The boiler may be dry base, which means the firebox is beneath the fluid-backed sections, or wet base, which has the firebox surrounded by fluid-backed sections on all sides. Wet-leg boilers have the firebox surrounded on the top and sides by fluid-filled boiler sections.

Figure 40.54 High-efficiency cast iron wet-base boilers that operate on oil or gas are used in schools, apartments, and commercial buildings, where a switch in fuels may be essential for continued heating service.

Steel boilers are made by welding the parts forming the water chambers. One type of heat-exchange surface may be slanted, horizontal, or vertical firetubes. They may be dry-base, wet-leg, or wet-base design.

Copper boilers use finned copper tube coils run from headers or serpentine copper tubing coil. These are usually fired by natural gas. Electric boilers consume no fuel and produce no exhaust gas, so no flue is required. The electric electrodes are immersed in the water.

Another type of boiler used more in recent times is the gas-fired pulse-combustion boiler. Pulse combustion provides efficient combustion and high heat transfer rates. These boilers reach operating temperature much faster than conventional boilers; they burn cleaner than conventional gas boilers and much cleaner than coal or oil.

The pulse combustion process has no power burner and no moving parts and the combustion takes place in a sealed environment. The process begins with a charge of air and gas entering the burner/heat exchanger through metering valves. The charge is initially ignited by a spark plug. Once combustion starts, the spark plug shuts off. The positive pressure from combustion closes the metering valves and forces the hot gases created out the tailpipes into the exhaust decoupler. As this happens, the air and gas metering valves are sucked open, admitting a fresh charge of air and gas. This cycle repeats at a natural frequency of about thirty times a second.

The combustion process is so clean that the boiler does not require a chimney but vents through the roof or sidewall with small-diameter tubing. Pulse-combustion boilers may be condensing or non-condensing.

Large commercial hot-water and steam-heating systems use large boilers often installed in series, as shown in Figure 40.55. Typical boiler connections for multiple boiler installations are controlled by local codes.

Figure 40.55 Gas-fired boilers installed in series.

TERMINAL EQUIPMENT FOR HOT-WATER AND STEAM-HEATING SYSTEMS

Terminal units used on hot-water and steam-heating systems include natural convection units and forced convection units. Those used on high-pressure steam systems are more heavily constructed than those used for hot-water and low-pressure steam. These include cast-iron and steel radiators, convectors, finned tube units, and baseboards.

Natural Convection Terminal Units

Natural convection heating devices include various types of terminal units that may be wall hung, recessed, or in the form of a baseboard. These devices distribute heat by using natural air circulation. This occurs because heated air rises and cool air settles, producing a natural circulation. Cool air enters a unit below the finned heating tube and rises as it heats, exiting through a grille or other opening at the top of the unit. These devices are available in various styles and sizes.

Finned tube units have metal fins secured to a metal tube and are a convection-type heater. However, some radiant heat is produced. If placed where human contact may be possible, they have an enclosing cover, otherwise they need not be covered. The fins are usually aluminum or copper and secured to copper tubing that conducts steam or hot water from the boiler. The fins become hot and air heats as it flows by. Several examples are shown in Figure 40.56.

Baseboard units are located along a wall where it meets the floor. The heating element may be a finned tube, cast-iron, or aluminum unit. This unit is enclosed in metal, with openings at the floor and near the top of the enclosure, providing for natural air circulation.

Steel radiators are another device that heats by convection and radiation. These come in the form of panels typically mounted on a wall or ceiling, although they can be freestanding. They are heated with hot water circulating through tubes that feed flat hollow panels or a series of tubes (Figure 40.57). Steel radiators work with low-temperature hydronic

Figure 40.56 Typical finned tube convection terminal units.

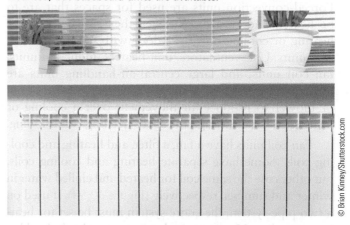

Figure 40.57 These steel hot water radiators are wall mounted, but baseboard units are available.

© Brian Kinney/Shutterstock.com

systems. Although they are not as hot as conventional hot water radiators, steel radiators are still somewhat hot to the touch. The panels are made of heavy-gauge steel or cast iron.

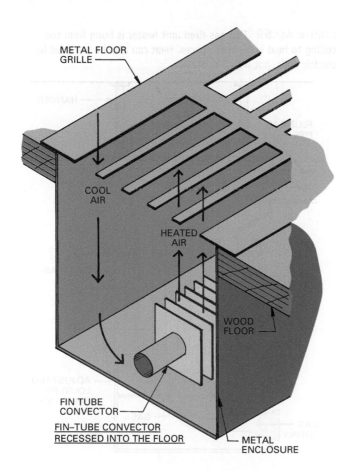

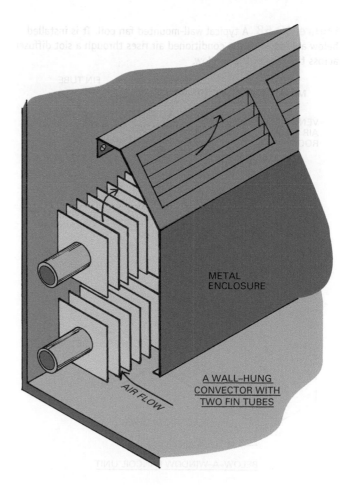

Forced-Convection Heating Units

Forced-convection units are used for spaces that are to be heated and cooled. However, they can be used for heating or cooling only as is done with natural convection units. Unit ventilators, unit heaters, induction units, fan coil units, and large central-air-handling units are available. Forced-convection units use some form of fan or blower to produce air movement over a cooling or heating coil and through the space needing conditioning.

Fan coil units have a fan, a filter, and heating and cooling coils. Some have separate heating and cooling coils, but others use the same coil for heated and chilled water in winter and summer, respectively (Figure 40.58). If used on a two-pipe system, the entire system must be set for heating or cooling. Hot water in the pipes must be drained before chilled water is introduced. In a large building this can take several days. If used with three or four-pipe systems, each fan coil unit has separate heating and cooling coils and can provide heating or cooling whenever required.

Fan coil units may be wall-mounted, ceiling-mounted, or high-rise vertical units. The ceiling-mounted units may sit below or above the ceiling. If above, short ducts are used for the air return and supply. High-rise units are usually placed in room corners, but this is not mandatory.

Unit ventilators are similar to fan coil units except they have an opening through an outside wall that allows outside air to be brought in. The opening is covered with a decorative louvered grille. The system regulates dampers to control the influx of outside air. Unit ventilators are typically used in large-occupancy rooms where frequent air changes may be necessary.

Induction units are much like fan coil units, but air movement in the room occurs via high-pressure air piped through a nozzle behind the coil. This causes room air to circulate through the coil and into the room. The nozzle and pressurized air replace the fan in the fan coil unit. They are usually mounted on an outside wall below a window.

Unit heaters have a fan that circulates air over some type of heat exchange surface, such as a hot water coil. They are enclosed in a metal case and usually suspended from the ceiling. Most common uses involve large open industrial plants or in businesses such as auto repair shops. They can use electric heating elements, natural gas or propane firing, or be part of a steam or hot water heating system. A typical gas-fired unit is shown in Figure 40.59.

Figure 40.58 A typical wall-mounted fan coil. It is installed below a window so the conditioned air rises through a slot diffuser across the face of the window.

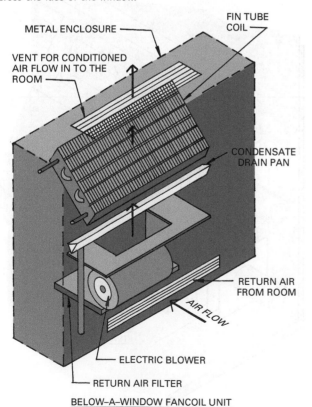

Figure 40.59 This gas-fired unit heater is hung from the ceiling to heat large open spaces. Heat can also be provided by electric coils, hot water, or steam.

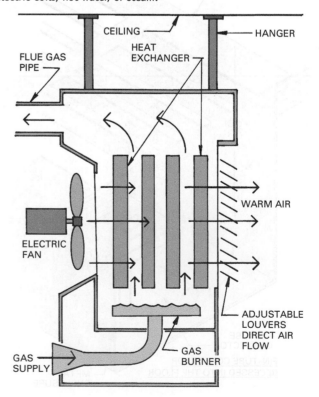

Unit heaters provide a rapid flow of heated air. They are also used to temper cold outside air introduced into a building through heavy-use openings, such as doors. They fill such areas with warm air, protecting occupants from sudden temperature changes. Unit heaters are controlled by thermostats and can be used to provide zoned heat by allowing one thermostat to control only the heaters in one part of a heated area.

Central air-handling units have a fan and a hot water or steam heating coil, which is used for cooling with a chilled water supply—a compressor is used if the water supply is not chilled. Heated and cooled air is distributed to various rooms through a duct system. Each space has terminal equipment, such as reheat coils, mixing boxes, and variable volume controls, to regulate air temperature. Reheat coils add additional heat to the air at the terminal device. Mixing boxes (also called blending boxes) are compartments in which two air supplies combine and then discharge into a room. Variable volume controls regulate airflow to control air temperature.

CHILLED-WATER COOLING SYSTEMS FOR LARGE BUILDINGS

The most commonly used cooling system uses chilled water to remove heat from the air in a building. This involves the use of mechanical equipment, including a means of refrigeration, cooling coils, and heat exchangers. The chilled water system is a closed circuit of piping in which water recirculates through a chiller and other equipment used to remove heat. It then moves through terminal units in the cooling space and back to the chiller. The three types of liquid chillers are centrifugal, absorption, and positive displacement.

CENTRIFUGAL CHILLERS

Centrifugal chillers are driven by an electric motor, a steam or gas turbine, or an internal combustion engine, which may be diesel or natural gas fueled. They use a vapor compression refrigeration system. The major parts of a mechanical vapor-compression chilled water system include a compressor, a *condenser*, an evaporator, and an expansion valve. An air-cooled refrigerant condenser is a device in which heat removal from the refrigerant is accomplished entirely by heat absorption via air flowing over the condensing surfaces. A *compressor* is a machine that mechanically compresses the refrigerant. An evaporator is that part of a refrigerating system in which the refrigerant is evaporated to absorb heat from the contacting heat source. An expansion valve reduces the temperature and pressure of the refrigerant as it passes through the valve.

The vapor compression refrigeration cycle used by centrifugal chillers is illustrated in Figure 40.60. This is

Figure 40.60 A simplified illustration of an air-cooled direct expansion (DX) vapor-compression refrigeration cycle used in centrifugal chillers.

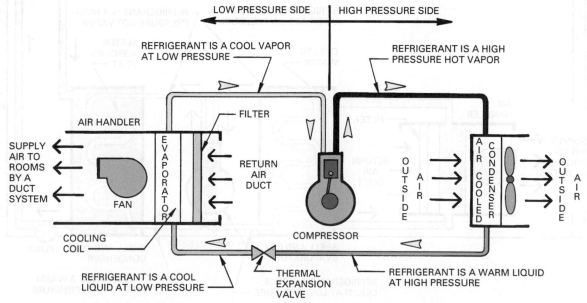

a direct-expansion (DX) cycle. The refrigerant in vapor form is compressed by the cylinder in the compressor. This causes it to become hot at a high pressure. The refrigerant moves to the condenser where a fan blows outside air through the condenser coil, removing much of the heat in the refrigerant and causing it to condense into a warm liquid, still at a high pressure. The pressure pushes the liquid refrigerant toward the evaporator. On the way, it passes through a thermal expansion valve that imposes a pressure drop, causing it to expand, which reduces the temperature. The cool liquid refrigerant passes through the cooling coil of the evaporator where it absorbs heat from air passing through the coil. When the refrigerant leaves the evaporator, it is a vapor. The vapor moves to the compressor, and the cycle repeats.

The centrifugal chiller system can also use a shell-and-coil heat exchanger and a cooling tower to dispose of heat from the refrigerant in the shell-and-coil condenser. A shell-and-coil heat exchanger has a coil of pipes within a shell or container. The pipes carry refrigerant through a second fluid held in the container. The shell-and-coil evaporator chills water in the shell or container and passes it through a cooling coil in the air handler. The air handler passes return air from the room and fresh outside air as required through the coil, conditioning the air supply to the rooms. The air flows through a system of ducts (Figure 40.61).

The shell-and-coil condenser water absorbs heat from the refrigerant and moves it to a cooling tower on the roof. The hot condenser water drips or is sprayed in fine droplets. A fan pulls outside air through the spray, removing much of the heat. Other cooling tower designs are available.

Figure 40.61 A schematic of the chilled-water vapor-compression refrigeration cycle using shell-and-coil heat exchangers and a cooling tower.

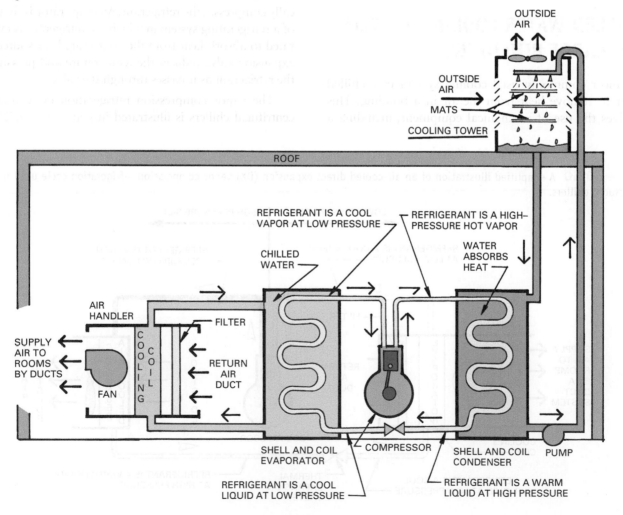

THE ABSORPTION CHILLER REFRIGERATION SYSTEM

An absorption chiller can be described as a refrigerating machine that uses heat energy and absorption input to generate chilled water. Absorption cooling uses an evaporated refrigerant (often water). Rather than the mechanical compression stage used in centrifugal chillers, it involves a process using an absorbent, such as a lithium bromide solution. This solution pulls vapor off evaporator coils, creating a cooling reaction. A source of heat, such as a gas burner, low-pressure steam, or hot water, is used to regenerate the absorbent solution by separating it from the absorbed vapor. Since a source of heat is used to operate an absorption chiller, it is especially economical to operate when a building has waste steam or other heat sources that can be used to provide a major amount of the heat required.

The absorption chiller operates quietly with few moving parts, generates minor vibration and is lightweight compared with compression type chillers. An absorption chiller is shown in Figure 40.62.

Figure 40.62 A direct-fired double-effect absorption chiller/heater.

© Lee Torrens/Shutterstock.com

Single-Effect Absorption Chillers

The basic absorption chiller refrigeration process for a water-cooled chiller is shown in Figure 40.63. Following is a description of this single-effect water-cooled absorption process.

The absorbent material, lithium bromide, has a great affinity for the refrigerant, water. The absorbent material, which contains some refrigerant, is heated in the generator. The refrigerant vapor is then driven to the condenser where the process converts it to a liquid (water). The concentrated absorbent moves from the generator to the absorber.

The liquid refrigerant flows from the condenser through an expansion valve to a low-pressure evaporator where it flashes to a vapor, providing cooling for the chilled water system refrigerant circulated to cooling coils within the building. This is a refrigerant in a system separate from the chiller system refrigerant. The expansion valve reduces the refrigerant pressure as it enters the evaporator, which is on the low-pressure side of the cycle.

Next, the chiller refrigerant moves to the absorber, where it is reabsorbed by the concentrated absorbent solution. This results in reduced pressure in the system beyond the expansion valve. The absorber is cooled to increase the rate of absorption of refrigerant into the concentrated absorption solution. This solution is pumped back into the generator, where the cycle repeats. The generator and condenser side operate at high pressure, and the evaporator and absorber operate at low pressure.

Double-Effect Absorption Chillers

Gas-fired double-effect absorption chillers/heaters are utilized in commercial applications with a central air-conditioning system using chilled water for cooling and hot water for heating. The condenser is water-cooled during the cooling period, and the heat is rejected through a cooling tower. The double-effect absorption cooling cycle has two generators. One is heated by natural gas and the other by the hot, semi-concentrated refrigerant vapor.

Cooling Cycle In the cooling cycle of a double-effect chiller/heater the gas burner heats a dilute absorbent solution (lithium bromide) in the high-temperature generator. The boiling process drives the refrigerant vapor and droplets of semi-concentrated solution to the primary separator. The semi-concentrated solution flows through a heat exchanger, where it is pre-cooled before flowing to the low-temperature generator.

In the low-temperature generator, the hot refrigerant vapor flowing from the primary separator heats the semi-concentrated solution. This liberates refrigerant vapor, which flows to the condenser. The concentrated absorbent solution is pre-cooled in the condenser and flows to the absorber.

In the condenser, refrigerant vapor is condensed on the surfaces of the cooling coils, and the latent heat is moved by the cooling water to a cooling tower, where it is expelled to the outside air. Refrigerant liquid accumulates in the condenser and passes through an orifice into the evaporator.

Figure 40.63 A simplified illustration showing the refrigeration process for a single-effect lithium-bromide water-cycle absorbent chiller.

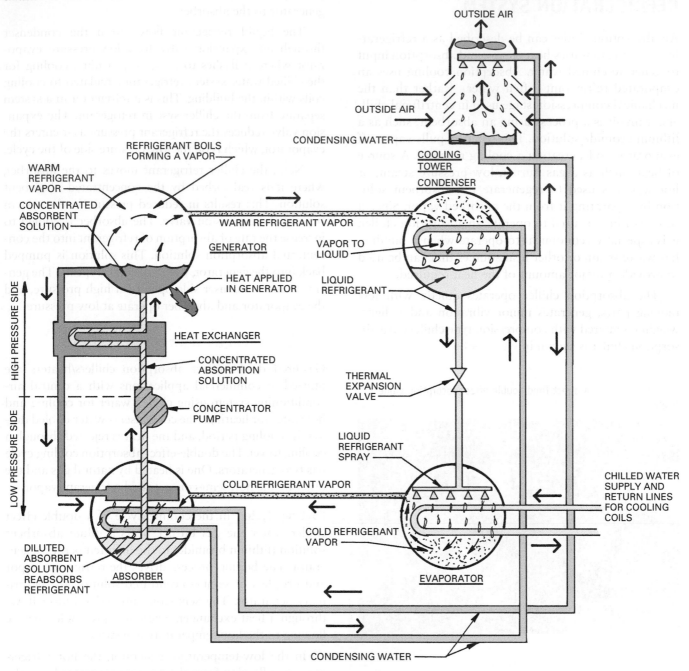

The pressure in the evaporator is lower than that in the condenser. As the refrigerant liquid flows into the evaporator, it boils on the surface of the chilled/hot water coil. Heat, equivalent to the latent heat of the refrigerant, is removed from the recirculating water to the cooling coils, which chills the water to about 44°F (6.7°C). The refrigerant vapor flows to the absorber.

In the absorber, the concentrated lithium-bromide solution (the absorbent) absorbs the refrigerant vapor

as it flows across the absorber coil. The cooling water in the coil removes heat from the solution, after which the diluted lithium bromide flows through the heat exchanger, is preheated, and moves to the high-temperature generator to start through the cycle again.

Heating Cycle In the heating cycle for a double-effect absorption chiller/heater, the lithium bromide absorbent solution is brought to a boil in the high-temperature generator and the vapor with concentrated lithium bromide solution is lifted to the primary separator. The hot refrigerant vapor and droplets of concentrated absorbent solution flow through the open changeover valve into the evaporator and the absorber. The heat is transferred to recirculating water coils in the evaporator. This heated water flows to fan coil units within a building, providing required space heating. The refrigerant (water) vapor mixes with the concentrated absorbent and returns to the generator, where the cycle is repeated.

Absorption chillers may be water cooled, as with a cooling tower, or air cooled. Air-cooled absorption chillers generally use ammonia as the refrigerant and water as the absorbent.

POSITIVE-DISPLACEMENT CHILLERS

Positive-displacement chillers use scroll, reciprocating, and rotary screw compressors, typically with electric motor drives. A scroll compressor is a positive-displacement compressor in which the reduction of the internal volume of the compression chamber is produced by a rotating scroll within a stationary scroll. A scroll is an involute spiral. A reciprocating compressor is a positive-displacement compressor in which the change in compression chamber volume is produced by the reciprocating movement of a piston. Positive-displacement compressors increase the pressure of the refrigerant vapor by reducing the volume of the compression chamber in which it has been injected using a mechanical means, such as a piston. A screw compressor uses a rotary motion to drive two intermeshing helical rotors to produce compression.

The capacity of the system is varied by having several compressors and turning them on and off as capacity requirements vary. Some reciprocating chiller compressors have an inlet suction valve. Reciprocating chillers use a compressive refrigeration cycle, usually driven by electric motors. They are smaller than centrifugal chillers and usually release condenser heat into the air rather than through a cooling tower.

EVAPORATIVE AIR COOLERS

Evaporative cooling is sensible cooling obtained by the exchange of heat produced by water jets or sprays injected into an airstream. As the air passes through water vapor, the vapor retains some of its heat, thus cooling it. This cooling process is effective in hot and dry areas. There are two basic types, direct and indirect.

Direct Evaporative Air Coolers

Direct evaporative air coolers have evaporative pads, usually of a wood fiber, a water-circulating pump, and a large fan (Figure 40.64). The fan pulls air through the pads, which cool the air, and moves it to spaces within a building. The water that does not evaporate collects in a pan, and a pump lifts this water back through the cycle.

Figure 40.64 A direct-evaporative air cooler pulls the air through wet-evaporative pads.

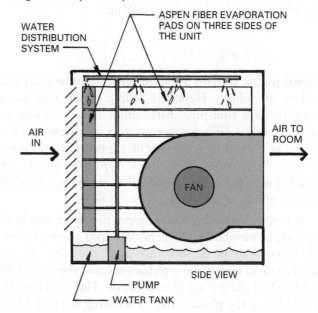

Indirect Evaporative Air Coolers

A variety of systems in use fall under this type. The following describes a package indirect evaporative air cooler. The unit contains a heat exchanger, a system to provide water vapor, a secondary fan, and a secondary air inlet. The heat exchanger is typically constructed with tubes, allowing one airstream to flow inside them and another over their exterior surface. Air filters are used to keep the system from becoming clogged with dust. The circulated water is usually treated to remove minerals, thus reducing corrosion in the heat exchanger.

The system is much like the direct evaporative system except the conditioned air is kept separate from the secondary air (which causes the water evaporation). Therefore, the conditioned air does not pick up moisture during the cooling process. This is why it is called an indirect system.

RADIANT HEATING AND COOLING

Radiant panel systems use panels in walls, ceilings, or floors whose surface temperatures can be controlled. The panels are usually operated by electrical resistance units or by circulating water or air. Radiant energy is transmitted through the air in straight lines. It does not heat the air, but it does raise the temperature of solid objects upon which it falls. The objects obtain the heat by absorption, and the heat can be reflected off a surface.

There are a variety of piping systems used for radiant heating and cooling. Some use bronze or copper tubing, while others utilize plastic tubing.

Piping in Ceilings

Several piping systems are used for radiant heating and cooling in ceilings. Hydronic ceiling panels can use a two-pipe or a four-pipe distribution system similar to those discussed for hot water systems. The design of the system is critical to a successful application and must be done by a qualified engineer. Typically, panels are installed with the pipes embedded above or in a finished plaster ceiling, as shown in Figure 40.65.

A suspended ceiling has the pipes tied to an overhead supporting member, with metal lath and plaster below. When wood or metal joists are used, the pipes are secured to them with metal pipe hangers, and the metal lath and finished plaster are placed below. The coils may be embedded in the plaster coat by securing the metal lath above the pipes and plastering over them. Other types of finished ceilings can be used, but plaster is most common. Similar installations can be utilized on wall surfaces.

Another type of ceiling panel consists of tubing bonded to flat metal panels that act as the exposed finished ceiling. The individual panels are hung between the channels of a suspended ceiling. This system is used to heat and cool interior air.

Piping in Floors

Radiant-heating piping may be embedded in concrete floors or placed above or below a wood subfloor. For concrete slab heating, ferrous and nonferrous pipe and PEX tubing may be used. The piping may be arranged in continuous coils or have header coils (Figure 40.66). Usually, 1 in. to 4 in. (25 mm to 101 mm) of concrete covers the pipe. The edges of the slab must be fully insulated with rigid insulation. Sometimes, insulating concrete is used. Piping may be placed on top of a wood

Figure 40.65 Typical radiant-heat ceiling panel piping installations.

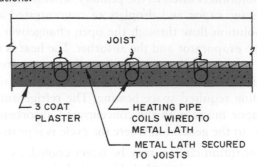

COILS IN PLASTER BELOW METAL LATH

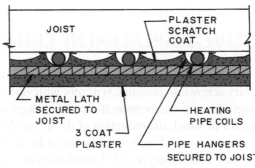

COILS IN PLASTER ABOVE METAL LATH

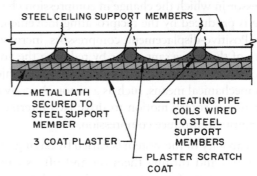

COILS IN SUSPENDED PLASTER CEILING

subfloor and covered with 1 in. to 2 in. (25 mm to 50 mm) of concrete or gypsum underlayment. Gypsum products designed specifically for floor heating are available, and concrete of structural quality should be used.

Piping under a subfloor is attached to the subfloor. Metal heat emission plates are used to improve heat transfer (Figure 40.67).

Piping may also be intertwined with a subfloor consisting of spaced strips, which allows the piping to be between them and above the floor joists. Metal heat-transfer plates are used. This is covered with wood subflooring and the finished floor covering.

Figure 40.66 Radiant heating pipes can be installed over a subfloor and then embedded in a gypsum or concrete underlayment.

Figure 40.67 This assembly installs the radiant-heating pipes on metal heat-emission plates below the subfloor.

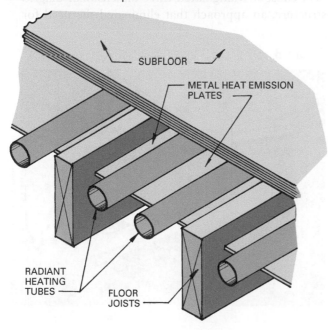

ELECTRIC HEATING SYSTEMS

Types of electrically heated systems include factory-assembled panels that mount on walls or ceiling, fabrics and wall-covering material containing resistance heating wires, and various types of electric resistance cables that may be embedded in concrete or laminated in drywall ceilings or in plaster.

One type of ceiling panel is made to fit into the grid formed by the structural members of suspended ceilings. The panels are available with various constructions, such as conductors embedded in a panel (maybe a gypsum panel) or some form of laminated panel.

Another ceiling heating system uses electric heating cable stapled to a ceiling covering material, such as gypsum board, plaster lath, or another fire-resistant material. The wires are covered by the various coats of plaster. If metal lath is used, it must first be covered with a brown coat of plaster to provide a nonelectrical conducting surface.

Electrical heating cable can also be laid in concrete slabs. The floor is laid in two pours. The first pour is 3 in. (76 mm) or more of insulating concrete. The cable lies on this slab and is fastened in place by stapling or special nail anchors. These hold the cables so the spacing is maintained when pouring the second layer of concrete. The second layer is usually 1 in. to 2 in. (25 mm to 50mm) thick and must not be insulating concrete. A finish floor can be laid over this slab.

Pre-fabricated infrared flooring systems are installed under a variety of finish flooring materials (**Figure 40.68**). Electrical current is passed through carbon strips to generate far infrared rays that distribute evenly within a space.

Figure 40.68 Infrared flooring systems warm the floor and the room with infrared waves that transfer heat like the sun.

Hearst Tower

Architect: Foster and Partners	Building type: Commercial Office	Completed 2006
Contractor: Turner Construction	Size: 860,000 sq.ft. (49,070 sq.m)	Rating: U.S. Green Building Council LEED NC Gold
Location: New York, New York	Project scope: Offices, auditorium, dining facilities.	LEED for Existing Buildings: O&M Platinum

Overview

In 1927, William Randolph Hearst set out to build a new headquarters building to house his growing publishing empire. The original 40,000 sq. ft. limestone building, designed by architect Joseph Urban, achieved historical landmark status in 1988. In 2000, the Hearst Corporation decided to carry out plans for an addition to the office building to house more than 2,000 employees and create a new icon for the New York City skyline (Figure A).

Figure A

© UniversalImages Group/Editorial/Getty Images

Design

An important design requirement was the preservation of the existing landmark façade and its integration with the new tower design. The new Hearst Headquarters is a combination of a tall structure and a re-modeling of the historic base. The original building façade was retained, its interior gutted and new structural supports incorporated. The new interior creates a six-story atrium with a waterfall, a restaurant, and communal areas for meetings and receptions (Figure B). The atrium features a three-story water sculpture, built with thousands of glass panels, which houses the main escalators and helps to cool and humidify the lobby space. Rain collected on the roof is stored in a tank in the basement for use in the cooling system, to irrigate plants and for the water sculpture in the main lobby.

In developing the tower addition, Foster and Partners chose a triangulated three-dimensional diagrid structure, an approach that eliminated the need for

Figure B

© UniversalImages Group/Editorial/Getty Images

corner columns and created a dynamic facade. The diagrid system creates a series of four-story triangles of stainless steel to provide lateral stiffness and strength that is clad in high performance glazing. The innovative structure required 20 percent less steel than would a conventional perimeter frame, and utilized over 90 percent recycled steel material. Around 86 percent of the removed building materials were also recycled for future use.

Natural Daylighting

The design of the tower attempted to maximize the use of natural daylight. The six-story lobby receives abundant natural light from skylights of ceramic fritted glass and 27 ft. (9 m) high clerestory windows that reach to the tenth story. Most of the office floors are designed without interior partitions, using only glass walls to enclose private offices to allow daylight to reach deep into the floor plates. Daylight sensors are used to activate artificial lighting and occupancy sensors reduce energy use by turning off lights when a room is unoccupied.

Mechanical Systems

The tower uses a low temperature air distributed central variable air volume (VAV) system. Four large

cooling towers serve a chiller plant, housing a series of centrifugal chillers. Heating is provided to the building via a steam to hot water heat exchanger that distributes hot water to individual fan powered VAV boxes located on each floor. The tower is served by four large air handling units that are capable of supplying 100 percent outdoor air.

Radiant flooring was integrated into the lobby atrium to supplement the forced-air cooling system. Radiant floors work especially well in spaces with high ceilings where the area being occupied need only to be conditioned to a height of 5 ft. to 6 ft. (1.5 m to 1.8 m) above the floor. The floor of the atrium uses polyethylene tubing embedded in concrete under a lime stone finish. The radiant tubing is filled with circulating water for cooling in the summer and heating in the winter.

Operations and Maintenance

Overall, the building was designed to use considerable less energy than the minimum code requirements, and earned a gold designation from the United States Green Building Council's LEED certification program. Because of stringent operational and upkeep policies established by the Hearst Corporation the tower went on to receive LEED Platinum status in 2012 in the LEED Operations and Maintenance category.

Review Questions

1. What is the difference between thermal radiation, thermal conduction, and thermal convection?

2. What is a major source of information about thermal conditions necessary for human comfort?

3. What types of ventilation are used to assist with the conditioning of inside air?

4. What types of air cleaners are currently available?

5. What are the three basic directions of heat flow from warm-air furnaces?

6. What fuels are used to fire warm-air furnaces?

7. How does an air-to-air heat pump produce heat?

8. What is meant by a ground-loop heat pump?

9. How can a gas-fired warm-air furnace be used to cool a building?

10. What are the types of all-air distribution systems?

11. What factors must a designer consider as an all-air duct distribution system is being designed?

12. How does a reheat warm-air distribution system differ from a variable air volume system?

13. How does a dual-duct warm-air system regulate air temperature?

14. What are the classes of commercial warm-air ducts?

15. What is the difference between the perimeter-loop and perimeter-radial duct systems?

16. What is meant by the surface effect from a ceiling warm-air diffuser?

17. What is a comfortable relative humidity for indoor air?

18. What types of residential and industrial humidifiers are available?

19. Why is the control of humidity in the air important?

20. How does a liquid-desiccant dehumidification system remove moisture from the air?

21. What types of hydronic heating systems are in use?

22. What is the difference between a two-pipe direct-return and a two-pipe reverse-return hydronic system?

23. Why do hydronic heating systems need an expansion chamber?

24. What is the advantage of a four-pipe system for hydronic heating and cooling?

25. How does the condensate in a one-pipe steam heating system return to the boiler?

26. What purpose do thermostatic traps on each radiator of a steam heating system serve?

27. Why is a two-pipe vacuum steam-heating system used in large buildings?

28. What is the major difference in the way water is heated in firetube and watertube boilers?

29. What is the difference in the operation of condensing and non-condensing boilers?

30. What is the unique feature identifying a wet-base boiler?

31. What are the commonly used natural convection terminal units?

32. What are the major components of a fan coil unit?

33. What types of refrigeration units are used to chill water for air-conditioning purposes?

34. How does a shell-and-coil condenser remove heat from the refrigerant?

35. How does a single-effect water-cooled absorption chiller cool the water for the air-conditioning system?

Key Terms

Absorbent	Desiccation	Return Air
Adsorbent	Diffuser	Sensible Heat
Boiler	Humidifier	Steam
British Thermal Unit (BTU)	Humidity	Supply Air
Chiller	Hydronics	Surface Effect
Compressor	Hydronic Heating System	Thermal Conduction
Condensate	Latent Heat	Thermal Convection
Condenser	Makeup Air	Thermal Radiation
Convector	Radiant Heat	Thermostat
Dehumidifier	Refrigerant	Throw
Desiccant	Relative Humidity	

Activities

1. Arrange visits to large multistory buildings and ask to see the equipment used for heating and cooling. Prepare a report describing what you saw.

2. Invite an engineer who designs multistory heating and cooling systems to address the class and review the steps of the design process and how decisions are reached.

3. Review the local building code and cite the general requirements relating to heating and air-conditioning systems for residential and commercial buildings. Report what the local building official checks as the building is under construction.

Additional Resources

ASHRAE Handbook—HVAC Applications; *ASHRAE -Handbook*—Refrigeration; *ASHRAE Handbook*—HVAC Systems and Equipment; *ASHRAE Handbook*—Fundamentals, American Society of Heating, Refrigerating, and Air-Conditioning Engineers, Inc., Atlanta, GA.

International Mechanical Code, The, Plumbing and HVAC Collection, International Code Council, Falls Church, VA.

National Fuel Gas Code; Flammable and Combustible Liquids Code; Liquefied Petroleum Gas Code, National Fire Protection Association, Quincy, MA.

Pitchers, N., *Combined Heating, Cooling, and Power Handbook*, Taylor and Francis Books, Atlanta, GA.

Standard for the Installation of Oil-Burning Equipment, National Fire Protection Association, Quincy, MA.

Stein, B. and J. S. Reynolds, *Mechanical and Electrical Equipment for Buildings*, John Wiley and Sons, New York, NY.

See Appendix C for addresses of professional and trade organizations and other sources of technical information.

Sheet Metal and Air Conditioning Contractors' National Association: www.smacna.org.

DIVISION

26

Electrical
CSI MasterFormat™

Electrical Equipment and Systems

Upon completion of this chapter, the student should be able to:

- Understand the various sources of electrical power and how it is distributed from a generating source.

- Describe an electrical current and the difference between AC and DC current.

- Specify the types of electrical conductors for various applications.

- Identify and discuss the various units used to measure, control, and distribute electrical power within a building.

- Specify lighting installations for meeting needs in a wide range of situations.

The modern building depends heavily on electricity to make it functional and habitable. Electricity provides power for lighting; runs motors for heating, ventilating, and air-conditioning; powers elevators, escalators, and other conveying systems; supplies the power to operate the communications, fire, and security systems; and is used to operate a vast array of electrical devices. In case of power interruption, emergency systems can be utilized to provide a temporary source of electricity for critical operations.

escalators, moving walks, conveyors) requires electric power. Internal communications and controls are electrically operated, and many building systems would not function without these systems. There are hundreds of special equipment items such as those found in hospitals, computer centers, and radio and television studios. The determination of electrical loads and internal systems is a major part of an adequately designed building.

ELECTRICAL LOADS

The loads put on an electrical system vary widely, depending on a building's occupancy. All buildings have extensive lighting requirements. Many functions, such as air-conditioning and heating, require the use of electric motors. Other systems, such as refrigeration and ventilation, utilize motors, compressors, and other electricity-consuming devices. The extensive range of appliances and other electrical devices produces varying loads as periods of demand fluctuate.

Industrial plants have heavy electrical demands to operate machinery that performs manufacturing operations, such as melting, fusing, and otherwise processing materials. Internal transportation (elevators,

BASICS OF ELECTRICITY

Electric current can be defined as the flow of electrons along a conductor, such as a copper wire. It is produced by a generator or battery that forces electrons to flow through the conductor to a consuming device—such as a light—and back to the producing source.

The flow of electric current in this continuing circuit resembles the flow of fluid in a hydraulic circuit (Figure 41.1). In a hydraulic system, a pump puts the fluid under pressure (pounds per square inch). The fluid flow is measured in a quantity, such as gallons. The fluid meets some resistance as it enters a fixture, and the flow is controlled by a valve. An electric circuit has a battery or generator to produce electricity

Figure 41.1 A comparison of electrical and hydraulic circuits reveals similarities. Electric switches and hydraulic valves control flow; electric current and hydraulic fluid flow in the circuit. Power is supplied by an electric device (a battery and a hydraulic pump); electric wire and hydraulic piping form the circuit; and resistance to flow occurs in both circuits.

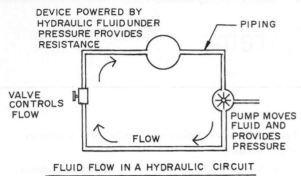

FLUID FLOW IN A HYDRAULIC CIRCUIT

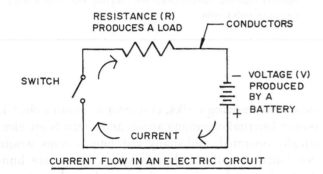

CURRENT FLOW IN AN ELECTRIC CIRCUIT

and the electromotive force (volts) to move it along the conductor in quantities measured in amperes. The flow of electricity finds *resistance* (ohms) when it enters a fixture, and the flow is controlled by a switch.

Units used to identify factors related to the flow of electric current are:

Ampere (A).   A unit of the rate of flow of electric current. An electromotive force of 1 ohm results in a current flow of 1 ampere.

Volt (E).   The unit of electromotive force (pressure) that causes electric current to flow along a conductor.

Ohm (Ω).   The unit of electrical resistance of a conductor. The symbol for ohm is the Greek capital letter omega (Ω).

Ohm's Law

Amperes, volts, and ohms are related to each other, and a change in one will affect the others. This relationship is identified as Ohm's Law. The relationship for each measure is shown by the following formula, in which I is the electric current or intensity of electron flow (measured in amperes), R is resistance (measured in ohms), and E is the electromotive force (measured in volts).

$$I = E/R \qquad \text{amperes} = \text{volts/ohms}$$
$$R = E/I \qquad \text{ohms} = \text{volts/amperes}$$
$$E = I \bullet R \qquad \text{volts} = \text{amperes} \bullet \text{ohms}$$

Conductors

Electric current flows along a material called a *conductor*. Good conductors such as copper, have a low resistance to the flow of electricity. Materials with a high resistance to electrical flow get hot because of the friction generated by the flow of electrons. These materials, such as nichrome, are used in applications like electric heaters, in which the production of heat is desired. Other materials, such as glass, ceramics, and plastics, do not conduct electricity and are called *insulators*. They are used on electrical devices to provide protection from the electricity. The switch lever on a light switch, for example, is made of a nonconducting material.

For electricity to do work, it must flow through a circuit. It flows from the positive connection to the negative connection, as shown in **Figure 41.2**. The negative electrons move to the positive pole, through the circuit to the consuming device, and back to the negative pole. A switch placed in the circuit can be opened to interrupt the flow of electrons and closed to complete the circuit.

Figure 41.2 Electricity flows through a circuit from the positive connection to the negative connection.

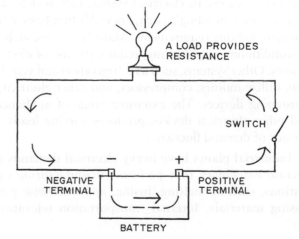

AC AND DC CURRENT

The two types of electric current are direct current (DC) and alternating current (AC). Direct current has a constant flow of electrons from negative toward positive to complete the circuit. Alternating current varies periodically in value and direction, first flowing in one way in the circuit and then flowing in the opposite direction. Each complete repetition is called a cycle. The number of repetitions per second is called the frequency and is measured in hertz (Hz). In the United States, the frequency for alternating current is 60 cycles per second or 60 hertz.

POWER AND ENERGY

"Energy" is the term used to express work. Energy is expressed in units of kilowatt-hours, foot-pounds, BTUs, joules, or calories. *Power* is the rate at which energy is used. Power is expressed in terms of watts, kilowatts, and other units shown in Table 41.1. Since power is the rate at which energy is used, time is a factor. The relationship between power and energy is shown by the following equation:

Power = energy/time

Energy = power • time

Table 41.1 Units of Power and Energy

English System	Metric System
Units of Power[a]	
Horsepower (hp)	Joule per second (J/sec)
BTU per second (BTU/sec)	Calorie per second (cal/sec)
Watt (W)	Watt (W)
Kilowatt (kW)	Kilowatt (kW)
Units of Energy[b]	
BTU	Calorie (cal)
Foot-pound (ft. • lb)	Joule (J)
Kilowatt-hour (kWh)	Kilowatt-hour (kWh)

[a]The rate at which work is done.
[b]The amount of work done.

The unit of electric power in electric circuits is expressed in watts (W) or kilowatts (kW). One kilowatt equals 1000 watts. One watt-hour of energy represents one watt of power used for one hour. A *watt* is one ampere flowing under an electromotive force of one volt. The power, W (watts), flowing into an electrical device having a resistance of R (ohms), in which the current is I (amperes), is found by the equation:

$$W = I^2 R$$

watts = amperes^2 × ohms

It can be seen, therefore, that power is related to current (amperes), electromotive force (voltage), and resistance (ohms).

ELECTRICAL CODES

Both the design and installation of electrical systems are carefully regulated by electrical codes. The National Fire Protection Association sponsors the model code, the *National Electrical Code Requirements for One- and Two-Family Dwellings*, as well as a series of standards publications related to electrical safety and equipment. The International Code Council published the *International Code Council Electrical Code*, which details information related to the administration and enforcement of the NFPA National Electrical Code. Underwriters Laboratories, Inc. (UL) tests and certifies electrical devices. Approved electrical devices carry the UL label.

ELECTRIC POWER SOURCES

Most electrical current is produced by some type of generator. These include hydroelectric, fossil fuel, renewable energy, and nuclear powered electrical generators.

Alternating current is produced by an AC generator, also called an alternator. The alternator is powered by any of the four sources of energy mentioned above.

Hydroelectric Power Generation

Electric power is generated by hydroelectric generation plants using the energy of falling water to turn a generator and produce electricity. The water turns the turbine, which drives the generator that produces the power. Transformers step up voltage to requirements needed to transmit it over electric lines to various destinations. A typical installation has numerous turbines, generators, and transformers (Figure 41.3).

Figure 41.3 Hydropower uses water turbines to drive electric generators.

© Ralf Broskvar/Shutterstock.com

Fossil Fuel-Powered Generators

Fossil fuels used to produce electricity include coal, oil, and natural gas. Fossil fuel power plants use rotating machinery to convert the heat energy of combustion into mechanical energy, which operates an electrical generator. The process may utilize a steam turbine, gas turbine, or, in small isolated plants, a reciprocating internal combustion engine.

Generated waste heat must be released to the atmosphere, often using a cooling tower or river or lake water as a cooling medium. Fossil-fueled power stations are major emitters of greenhouse gases. Flue gases containing carbon dioxide and other substances, such as nitrogen, nitrous-oxides, sulfur oxides, fly ash, and mercury are discharged into the air. Since fossil fuels are nonrenewable resources—and are becoming increasingly expensive—alternate sources, such as solar, wind, and nuclear fission, are finding increasing use. Experiments aimed at generating electricity by burning waste materials are also underway.

Nuclear-Powered Generators

Nuclear-powered electric-generation plants are similar to fossil-fueled plants in that they produce steam to run a steam turbine that drives a generator. The energy is produced in a nuclear reactor by fission. Fission is a process in which the centers (or nucleuses) of certain atoms are split when struck by subatomic particles called neutrons.

The products of fission fly apart at high speed and generate heat as they collide with surrounding matter. The fission reaction is controlled in the nuclear core of a reactor, which consists of fuel rods in a chemical form of plutonium or uranium and thorium. Heat energy is produced by the fission reaction of the nuclear fuel. The heat is removed by a coolant and used to produce steam to drive a steam turbine, which drives an electrical generator (Figure 41.4).

Spent fuel from the nuclear fission process is highly radioactive and must be handled and stored following strict guidelines. Spent fuel rods are stored in shielded basins of water usually located on site. The water provides both cooling for the continuously decaying fission products and shielding from radioactivity. To ensure safety in a nuclear plant, the concept of "defense in depth" is employed. There are several layers of protection, each independent of the others, so if one failed others will continue protecting the plant, the workers, and the general public. Work is now underway to find methods of reprocessing spent fuel that could potentially recover up to 95 percent of remaining uranium and plutonium.

ON-SITE POWER GENERATION

On-site power generation is used when a utility system is not available or is unable to provide reliable service. Some systems provide additional service during peak periods. Certain facilities, such as hospitals, have on-site backup power-generation systems that engage when the public utility system fails.

The methods used to generate electricity on site include wind turbines, solar photovoltaic cells, thermal sources, gas- or steam-powered turbines, and cogeneration.

Thermal sources use an engine or turbine coupled to the shaft of a generator to produce electricity. Internal combustion engines run on fuels such as diesel fuel, methane gas, or natural gas. These are typically used as emergency backup sources of electricity for a vast range of commercial facilities, such as hospitals, banks, computer centers, schools, and waste-water treatment plants. Backup systems include the generator set, transfer switches, and paralleling switchgear.

Turbines are either gas or steam powered. Gas turbines burn various types of gaseous and liquid fuels, such as natural gas or fuel oil. Steam turbines are driven by a source that produces a large quantity of high-pressure steam. The revolving turbine drives the generator, producing electricity. The steam is produced by a boiler fired with a fuel, such as coal, solid waste, natural gas, or oil.

Figure 41.4 A simplified illustration showing how a nuclear electric generating plant operates.

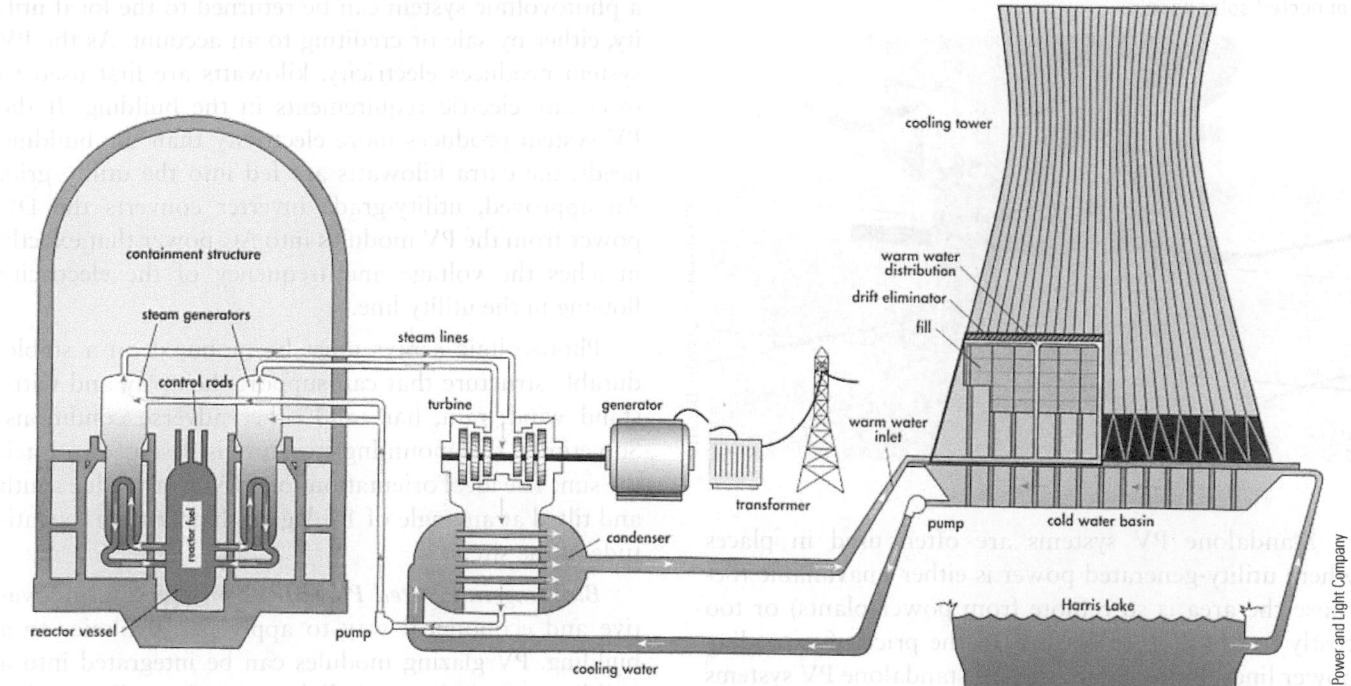

containment structure

steam generators

control rods

steam lines

turbine

generator

transformer

reactor fuel

condenser

reactor vessel

pump

cooling water

cooling tower

warm water distribution

drift eliminator

fill

warm water inlet

pump

cold water basin

Harris Lake

Solar Energy

Active solar systems use mechanical assistance to utilize solar energy with manufactured components that convert this energy to thermal energy and electrical power.

Photovoltaic (PV) systems harness solar radiation and convert it to electricity. Photovoltaic semiconductor materials, including silicon, gallium, arsenide, copper indium diselenide, and cadmium telluride, exhibit a property that causes them to absorb photons of light and release electrons. When these free electrons are captured, an electric current that can be used as electricity is generated. Silicon is the most popular photovoltaic material in use today.

Components of a Photovoltaic System The fundamental building block of a photovoltaic system is the solar cell, a thin wafer consisting of an ultra-thin layer of phosphorus silicon on top of a thicker layer of boron silicon (Figure 41.5). When sunlight strikes the surface of the cell, an electric field is generated between these two materials and provides momentum and direction to light-stimulated electrons, resulting in a flow of current. A typical silicon PV cell produces about 0.5–0.6 volt DC under open-circuit, no-load conditions.

Solar cells are connected in series or parallel circuits to produce higher voltages, currents, and power levels. Photovoltaic modules consist of PV cells sealed in

Figure 41.5 Solar cells consist of an ultra-thin layer of phosphorus silicon on top of a thicker layer of boron silicon.

a protective enclosure. Photovoltaic panels include one or more PV modules assembled as a pre-wired, field-installable unit. A photovoltaic array is the complete power-generating unit, consisting of any number of PV modules and panels (Figure 41.6).

Additional components are required to make up an entire PV system. Mounting components support the panels and point them toward the sun. Inverters take the direct-current electricity produced by the modules and "condition" that electricity, usually by converting it to alternate-current electricity. Standalone systems use batteries to store electricity for later use. All these items are referred to as "balance of system" (BOS) components.

Figure 41.6 A complete photovoltaic array is composed of connected solar panels.

Standalone PV systems are often used in places where utility-generated power is either unavailable (because the area is so remote from power plants) or too costly to hook up to because of the price of extending power lines. Batteries are used in standalone PV systems to store energy produced during the day and supply it to electrical loads as needed during nighttime and periods of cloudy weather.

PV grid-connected systems are similar in components to standalone systems except a building is also connected to utility grid power (Figure 41.7). The PV array can provide a portion of household energy demand, while grid power supplies the remainder. When the batteries' voltage reaches a preset low point, an automatic transfer switch connects the grid to electrical loads.

Figure 41.7 A schematic of a grid-connected photovoltaic power system.

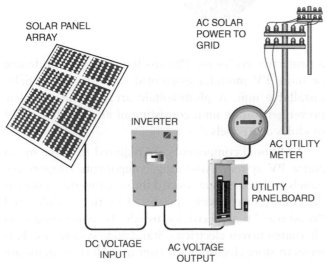

With *net metering*, excess electricity produced from a photovoltaic system can be returned to the local utility, either by sale or crediting to an account. As the PV system produces electricity, kilowatts are first used to meet any electric requirements in the building. If the PV system produces more electricity than the building needs, the extra kilowatts are fed into the utility grid. An approved, utility-grade inverter converts the DC power from the PV modules into AC power that exactly matches the voltage and frequency of the electricity flowing in the utility line.

Photovoltaic arrays must be mounted on a stable, durable structure that can support the array and withstand wind, rain, hail, and other adverse conditions. Sometimes, this mounting structure is designed to track the sun. The ideal orientation for a PV array is due south and tilted at an angle of 15 degrees higher than the latitude of the site.

Building-integrated PV (BIPV) provides an innovative and economical way to apply photovoltaics to a building. PV glazing modules can be integrated into a building as windows, skylights, curtain walls, roofing shingles and tiles, or shading elements (Figure 41.8). Because solar products replace conventional materials while simultaneously producing electrical power from sunlight, material savings can be realized.

Figure 41.8 Building-integrated photovoltaics used to provide shading while simultaneously generating electricity.

Wind Energy Systems

Another source of electrical energy is derived from pre-fabricated systems using wind to drive turbines. Wind turbines are available for a variety of applications, from residential standalone or grid-connected to farm or community-scale systems. Full-scale utility projects typically interconnect to existing local power distribution lines.

Horizontal axis turbines consist of four basic parts. A concrete or steel foundation is used to support the structure. The tower, commonly of lattice or tubular steel, contains the electrical conduits and supports the nacelle, which houses a generator and gearbox. The spinning blades are attached to the generator through a series of gears that increase the rotational speed of the blades and produce electricity. The most common tower is comprised of a white steel cylinder from 150 ft. to 200 ft. in height and 10 ft. in diameter (Figure 41.9). All towers require a ladder for access and a hoist for tools and equipment.

Figure 41.9 The most commonly used wind turbine is comprised of a white steel cylinder ranging from 150 ft. to 200 ft. in height.

New designs for vertical axis turbines, including egg-beater and helical forms, are under development, although they are generally less efficient than horizontal-axis systems. Similar to photovoltaic systems, wind turbines are integrated with an inverter to convert generated power to usable voltages. Battery backup is used if the system is not grid connected. Modern turbine technology is allowing architects and engineers to incorporate wind energy systems directly into buildings (Figure 41.10).

Figure 41.10 The Bahrain World Trade Center tower integrates wind turbines directly in the architecture.

Cogeneration

Cogeneration, also called combined heat and power (CHP), is the use of a heat engine or power station to simultaneously generate both electricity and useful heat. Conventional power plants emit heat, created as a by-product of electricity generation, into the environment through cooling towers or flue gases. CHP captures the waste heat and utilizes it for domestic or industrial heating purposes (Figure 41.11). For example, if an internal combustion engine or a turbine is used to produce on-site electricity, the heat produced can be reclaimed and used to heat water or produce steam, which can be used to heat the building or provide cooling using an absorption chiller. An auxiliary conventionally fired boiler is used in the system to provide extra hot water or steam if needed.

Figure 41.11 Cogeneration utilizes heat that normally would be wasted to produce on-site electricity and to heat or cool a building.

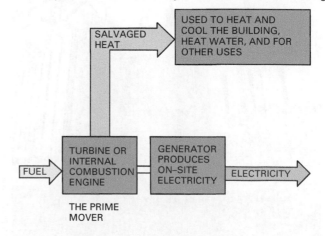

Salvaged heat may also be used to produce steam used to drive a turbine, providing a supplemental on-site supply of electricity. This is especially effective in areas where high-pressure steam is required for an industrial process—such as food processing or pulp and paper manufacturing—that produces a steady, high-temperature source of wasted heat.

ELECTRICAL SUPPLY

A public utility uses some form of energy or fuel to operate generators at their power plant that produce electricity for transmission to consumers. The electricity passes to a step-up transformer station, which increases voltage so the power can be transmitted with less current. This reduces line power losses and permits the use of smaller conductors. High-voltage transmission lines move the power to area transformer stations.

Area transformer stations step down the voltage of the power (Figure 41.12), and it is transmitted with overhead and underground lines to different areas. Industries using large amounts of electricity often receive this power into their own transformer stations for distribution and voltage regulation within their plants. For most commercial and private consumers, the power is sent to a local transformer where it is stepped down to single-phase 120V or 120/240V power (Figure 41.13). The 120/240V line is used to operate residential appliances

Figure 41.12 An area transformer station that receives power at 115,000V and steps it down to 23,000V three-phase for distribution to commercial and industrial uses.

© DSBfoto/Shutterstock.com

Figure 41.13 Pole-mounted and pad-mounted transformers are connected to high-voltage transmission lines. They step down the voltage for use by the consumer.

© Dmitry Matrosov/Shutterstock.com

(electric dryers, ranges, water heaters) and small industrial equipment requiring 240V. Larger commercial buildings may require either single-phase or three-phase at 120/208V, 120/240V, or 277/480V.

Power connects to a commercial building through a service entrance consisting of a disconnect switch, a meter, and distribution switchgear. Service entrances for residential and small commercial buildings consist of a meter and a panelboard.

ELECTRICAL DISTRIBUTION SYSTEMS IN BUILDINGS

Electrical distribution systems inside a building supply electrical power as needed to various sections and transmit information through an internal communications system. Standards for the design and installation of electrical distribution systems are maintained by the *National Electrical Code* published by the National Fire Protection Association. Underwriters Laboratory certifies electrical equipment and materials through the use of testing specifications and procedures.

The electrical power system of a commercial or industrial building includes the service entrance equipment, the interior distribution system, and the end equipment that uses the electricity. The service entrance incorporates equipment such as meters, switches, panelboards, switchgear, switchboards, fuses, and circuit breakers. The distribution system includes trays, wiring, raceways, wireways, conduit, and busways. The final part of the interior electrical system consists of the devices that use the electricity, including lighting, motors, heaters, and communications and industrial equipment. These constitute the loads that must be considered when designing a system.

The electrical distribution system consists of branch circuits that are sized according to the anticipated electrical load. Single-outlet circuits are used to feed a specific appliance or other piece of equipment only. Multiple-appliance-outlet circuits feed a number of small appliances, while multiple-outlet general-purpose circuits supply power for receptacles and lighting.

The actual design of a system varies considerably, depending on the requirements of the building and the

equipment to be installed. The design of distribution systems requires the services of an experienced electrical engineer.

The Service Entrance

The *National Electrical Code* defines a ***service entrance*** as the conductors and equipment for delivering electricity from a utility to the wiring system of the building served. The meter is generally mounted on the exterior of the building, with service supplied either overhead or underground. This enables the utility to read the meter and install or remove it without entering the building. Underground lines, run in conduit for access and protection from the elements, are common in dense urban areas. A typical service entrance for a residential or small commercial building is shown in **Figure 41.14**. From the

Figure 41.14 Typical above- and below-ground service entrances for residential and small commercial buildings.

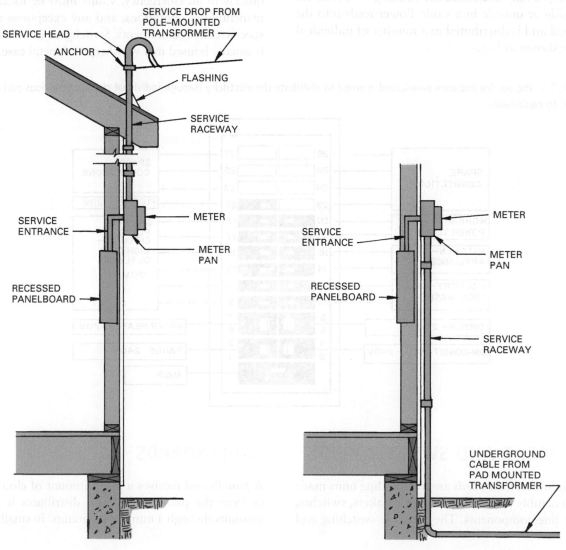

service entrance, panelboard branch circuits run to various parts of the building for lights, outlets, appliances, furnace, and other required services. Each circuit has a circuit breaker overload device in the panelboard. Service can be run to a subpanel in an area some distance from the panelboard where branch circuits can be taken from it (Figure 41.15). If the building has multiple occupancies, individual meters are set up for each unit.

A service entrance for commercial buildings can take various forms, depending on occupancy needs. It requires a service switch or circuit breaker, which disconnects power to the entire building. This is typically located outside the building, but exceptions are made. The power enters through a meter installation that includes a meter pan, a meter cabinet, and a current transformer in a cabinet or vault. The service switch and metering equipment may be assembled in one unit. The metering transformer steps down incoming electric current from the primary feeder to a voltage required for use in the building (120/240V, for example). It may sit outside the building on a concrete pad or inside or outside in a vault. Power feeds into the switchboard and is distributed to a number of individual circuits, as shown in Figure 41.16.

feeder protection to circuits connected to a main power source. The switchboard distributes the incoming electrical power in smaller amounts within a building. Switchboards are manufactured with all components enclosed within a metal cabinet and controlled by insulated handles and push buttons in a front panel (Figure 41.17).

Switchgear contains the same components and serves the same function as switchboards but handle higher voltages, usually above 600V. The term "switchgear" is also used to identify individual switching units not assembled in a panel.

Switchboards and switchgear used in commercial and industrial buildings are commonly placed in a basement vault designed specifically to house the switchgear. It must be adequately ventilated and meet requirements of the *National Electrical Code*. Typical space requirements are given in Table 41.2. The design of the vault should include adequate entrances and exits so equipment can be installed and removed, and workers in the room can exit quickly in an emergency. Vaults must be located in permanently dry conditions, and any exceptions must meet specific code requirements. Switchgear installed outdoors is usually housed in a weatherproof metal case.

Figure 41.15 The service entrance panelboard is wired to distribute the electricity through individual circuits to various parts of the building and to equipment.

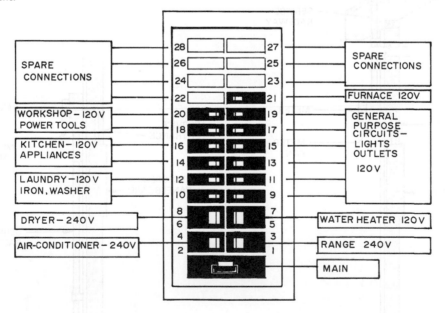

SWITCHGEAR AND SWITCHBOARDS

Switchgear and *switchboards* are freestanding units made up of an assembly of fuses and circuit breakers, switches, and other line components. They provide switching and

PANELBOARDS

A *panelboard* receives a large amount of electrical power from the public utility and distributes it in smaller amounts through a number of circuits. In small buildings,

Figure 41.16 A generalized power riser diagram feeding from a switchboard to panelboards on the various floors and to equipment on the roof and in the basement.

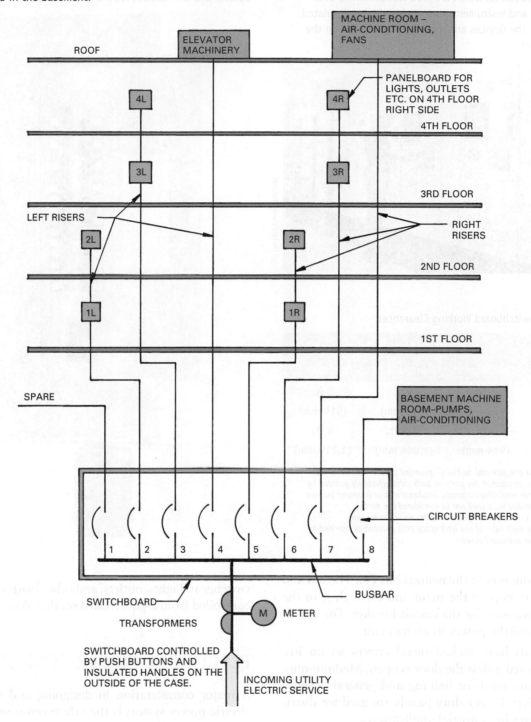

it serves the same basic purpose as a switchboard but on a smaller scale. A panelboard is typically used in residential and small commercial construction to distribute power to each of the circuits within a building after receiving input from the public utility line (**Figure 41.18**).

A typical panelboard using circuit breakers is illustrated in **Figure 41.19**. The incoming service entrance cable connects to the neutral bus bar and the main circuit breaker. The main controls the flow of power from the utility into the service panel. The white wire is

Figure 41.17 A switchboard is a single electric control panel or assembly of panels on which are mounted switches, over-current devices, and instrumentation enclosed in an insulated metal structure. The devices are controlled by handles on the front of the panel.

Table 41.2 Switchboard Working Clearances

Nominal-Voltage to Ground	Minimum Clear Distance		
	Condition 1[a]	Condition 2[b]	Condition 3[c]
0–150	3 ft.	3 ft.	3 ft.
	(914 mm)	(914 mm)	(914 mm)
151–600	3 ft.	3 1/2 ft.	4 ft.
	(914 mm)	(1,066 mm)	(1,219 mm)

[a]Exposed live parts on one side and no live or grounded parts on the other side of the working space, or exposed live parts on both sides effectively guarded by suitable wood or other insulating materials. Insulated wire or insulated busbars operating at not over 300 volts shall not be considered live parts.
[b]Exposed live parts on one side and grounded parts on the other side.
[c]Exposed live parts on both sides of the work space (not guarded as provided in Condition 1) with the operator between.

Figure 41.18 This panelboard has the main breaker at the bottom and two vertical rows of circuit breakers.

neutral and connects to the neutral bus bar. The red and black wires connect to the main and through it to the snap-on connections for the circuit breaker. The circuit breakers control the power to each circuit.

Panelboards have locked metal covers so no live parts are exposed unless the door is open. Medium-duty panelboards are used for lighting and general-purpose electrical outlets. Heavy-duty panels are used for distribution of power in industrial applications.

The panel may be surface mounted, semi-recessed, or recessed into a wall. Each circuit is numbered, with a description identifying its location given on a chart, such as No. 6—Main Office wall outlets.

In large construction, panelboards are located on the various floors and fed from the service switches and switchgear. Lights, outlets, and other loads on a floor are controlled from a panelboard on that floor.

ELECTRICAL POWER CONDUCTORS

A major consideration in designing and installing an electric power system is the safe transmission of power to its end use. Electric power systems have the potential to cause fires, property damage, and human injury and death. Electrical codes are strict, and inspection during construction must be thorough. A crucial consideration is whether or not to isolate electrical conductors as they pass through a building until they reach their point of use, such as an electric light.

Figure 41.19 Assembly details for a typical panelboard with circuit breakers.

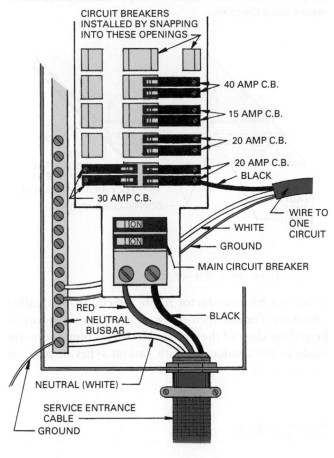

Figure 41.20 Commonly used types of electrical cables.

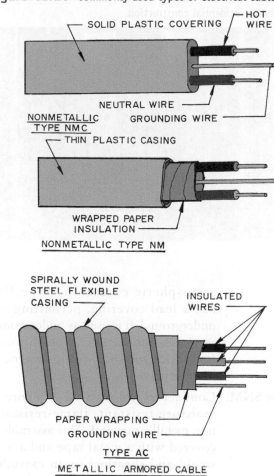

Cables

Electric cable design and use is strictly regulated by codes such as the *National Electrical Code* published by the National Fire Protection Association. The types of conductors (Figure 41.20) rated by this code are:

Type AC.   Insulated conductors are wrapped in paper and enclosed in a flexible, spiral-wrapped, metal covering. An internal copper bonding strip in contact with the metal covering provides a means of grounding. Referred to as **BX cable**, it is used only in dry locations.

Type ACL.   This type has the same insulation and covering as Type AC, but it uses lead-covered conductors that make it useful in wet applications.

Type ACT.   The individual copper conductors have a moisture-resistant fibrous covering and run inside a spiral metal sheath.

Type MC.   Insulated copper conductors are sheathed in a flexible metal casing. If it has a lead sheath, it can be used in wet locations. It is a heavy-duty industrial feeder cable.

Type MI.   Conductors are mineral insulated and sheathed in a gas-tight and watertight metal tube. This type can be fire-rated and is used in hazardous locations and underground.

Type NM or NMC.   Known as **Romex**, is non-metallic-sheathed cable used in protected areas (Figure 41.21). NM has a flame-retardant and moisture-resistant outer casing and is restricted to interior use. NMC is also fungus resistant and corrosion resistant and can be used on exterior applications.

Type SE or USE.   Moisture-resistant, fire-resistant insulated cable with a braid of armor providing protection against

Figure 41.21 Romex is non-metallic-sheathed cable used frequently for residential construction.

© David Papazian/Shutterstock.com

Figure 41.22 A typical under-carpet wiring system. It uses a flat power cable that runs under the carpet. Outlets are placed as needed along the cable.

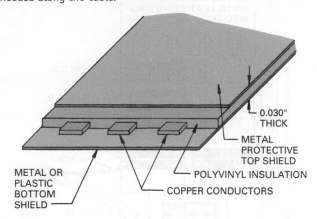

0.030" THICK

METAL PROTECTIVE TOP SHIELD

POLYVINYL INSULATION

COPPER CONDUCTORS

METAL OR PLASTIC BOTTOM SHIELD

atmospheric corrosion. Type USE has a lead covering, permitting use underground. It is used as underground service entrance cable. Type SE works for service entrance wiring or general interior use.

Type SNM. Conductors are situated in a core of moisture-resistant, flame-resistant, non-metallic material. This assembly is covered with a metal tape and a wire shield and sheathed in an extruded non-metallic material impervious to oil, moisture, fire, sunlight, corrosion, and fungus. It can be used for hazardous applications.

Type UF. Conductors are enclosed in a sheath resistant to corrosion, fire, fungus, and moisture, and can be directly buried in the earth.

Flat Conductor Cables

Flat conductor cables consist of copper cables formed flat and embedded in a plastic sheathing. A typical cable is 0.030 in. (0.78 mm) thick and around 2 in. (63.5 mm) wide (Figure 41.22). The conductor is placed on the subfloor, a shielding material is placed over it, and finished flooring is installed. These cables are used for standard 120V electric power and in communication and data systems. Outlets are installed as required by making connections through the carpet (Figure 41.23). Codes generally require that flat cable be covered with carpet squares to make it easily accessible.

Cable Bus and Busways

A *bus* is a bare conductor run in a metal trough, called a *busway*. The conductors are mounted on insulators to keep them clear of the trough. Electrical connections are made to the conductors with various types of plug-ins (Figure 41.24).

RACEWAYS

Raceways are used to support, enclose, and protect electrical wires.

Cable Trays

Open raceways, or *cable trays*, are open-faced metal channels used to provide support for electric wires that have adequate insulation and do not require extra protection. Cable trays only support the wires. Wires in this system are open to inspection and modifications (Figure 41.25).

Conduit

Conduit is a form of closed raceway. It supports insulated electric wire and provides protection. Conduit does not have conductors inside when installed. The conduit is installed first and the wires run through later (Figure 41.26). Conduit can be run inside walls and ceilings, below and through floors, and in concrete slabs. Codes regulate the use and locations of the various types. Metal and non-metal conduit can be left exposed to view when appearance is not important.

Figure 41.23 Flat-cable electric power, communications systems, and electronic systems are placed below the finished floor and have outlets mounted on top.

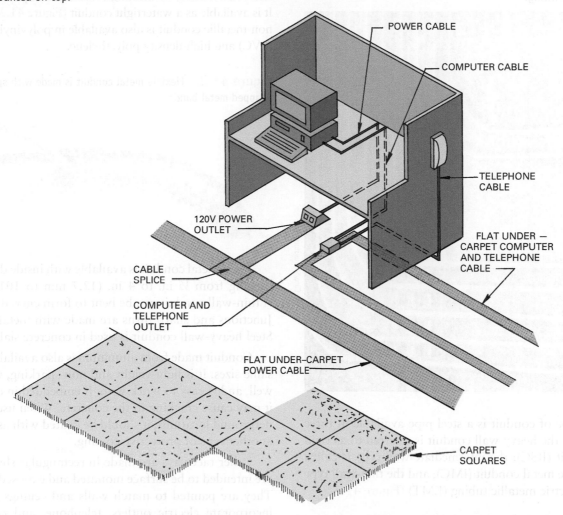

POWER CABLE

COMPUTER CABLE

TELEPHONE CABLE

120V POWER OUTLET

FLAT UNDER—CARPET COMPUTER AND TELEPHONE CABLE

CABLE SPLICE

COMPUTER AND TELEPHONE OUTLET

FLAT UNDER–CARPET POWER CABLE

CARPET SQUARES

Figure 41.24 These lights are mounted on a busway secured to the ceiling. The lights can be placed anywhere along the busway.

© newphotoservice/Shutterstock.com

Figure 41.25 Cable trays are open-faced metal channels that provide support for electric wires and access for inspection and modifications.

© Bork/Shutterstock.com

Figure 41.26 Surface-mounted steel conduit is used to protect insulated wiring.

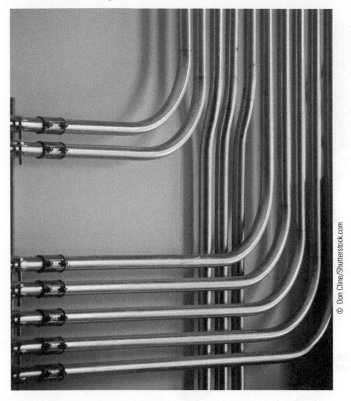

One type of conduit is a steel pipe available in three thicknesses: the heavy-wall conduit is referred to as rigid steel conduit (RSC); the intermediate-wall type is called intermediate metal conduit (IMC); and the thin-wall type is called electric metallic tubing (EMT) (Figure 41.27).

Figure 41.27 Common types of rigid metal conduit.

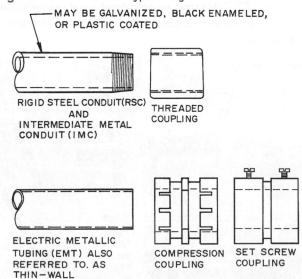

MAY BE GALVANIZED, BLACK ENAMELED, OR PLASTIC COATED

RIGID STEEL CONDUIT (RSC) AND INTERMEDIATE METAL CONDUIT (IMC)

THREADED COUPLING

ELECTRIC METALLIC TUBING (EMT) ALSO REFERRED TO. AS THIN-WALL

COMPRESSION COUPLING

SET SCREW COUPLING

EMT CONDUIT WALL THICKNESS TOO THIN TO PERMIT THREADING.

Flexible metal conduit, called Greenfield, is used for short runs, such as connecting a furnace to its power source. It is available as a watertight conduit (Figure 41.28). Rigid non-metallic conduit is also available in polyvinyl chloride (PVC) and high-density polyethylene.

Figure 41.28 Flexible metal conduit is made with spiral-wrapped metal bands.

Rigid metal conduit is available with inside diameters ranging from ½ in. to 4 in. (12.7 mm to 101.6 mm). A thin-wall conduit can be bent to form curved corners. Junctions and sharp turns are made with metal fittings. Steel heavy-wall conduit is used in concrete slabs.

Conduit made from aluminum is also available in the same sizes. It is lightweight and non-sparking, weathers well, and is easy to work with. If embedded in concrete, it may cause cracking in the concrete. When used in underground locations, it should be coated with asphalt or another type of protective coating.

Other raceways are made in rectangular shapes and are intended to be surface mounted and exposed to view. They are painted to match walls and ceilings and can incorporate electric outlets, telephone, and computer and communications wiring connections (Figure 41.29).

Figure 41.29 Rectangular raceways are used for surface-mounted wiring installations.

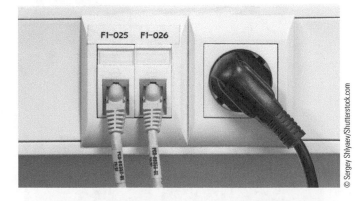

Metal raceways can be incorporated in a cellular steel floor deck. Metal decking serves as a substrate to support a cast-in-place concrete floor. Cells in the decking

are used to carry electrical and communications wires. Perpendicular to the cells in the steel decking are metal ducts spaced as required with outlets through which wires may be pulled to provide power for an area. These are typically spaced on a grid, permitting electric power to be available in a number of places in the floor.

METERS

The amount of electric power used is measured in watt-hours (Wh). Since the amount increases rapidly, it is reported in kilowatt-hours (kWh). One kWh is equal to 1,000 Wh. The amount used is measured by a kilowatt-hour *meter* (Figure 41.30). Three-wire meters are typically used for residential applications. Commercial and industrial applications having heavy demands and higher voltage requirements typically use four-wire meters.

Figure 41.30 The amount of electric power used is measured by a kilowatt-hour meter.

An electric utility supplies meters, which are usually placed outside buildings to provide ready access to utility staff. Meters can be installed inside a building if easy access is provided. The owner of a building provides meter pans and any current transformers needed to step down voltage within the building.

MOTOR CONTROL CENTERS

Motor control centers are used to start and stop motors and protect them from potential overloads. They have a disconnect switch that must be located within sight of

the controller and the motor. The overload protection is typically a heat-operated relay that opens the circuit when line temperatures rise. The motor circuit is closed after the overload situation has been corrected by pressing a reset button (Figure 41.31).

Figure 41.31 A motor control center utilizing electronic and programmable devices in addition to electromechanical controls. Motor controls start and stop motors and provide overload protection.

TRANSFORMERS

Transformers are used to change (transform) alternating current from one voltage to another. The voltage coming into a transformer is the primary voltage, and voltage leaving the transformer is the secondary voltage. For example, a transformer is used to step down a primary 4,160V current received from a utility distribution system to a secondary voltage, such as 480V, as it enters a building. Another transformer in the building's vault (or closet) could step this down to the 120V required for use within the building. The generally available primary voltages include 2,400V, 4,160V, 7,200V, 12,470V and 13,200V. Secondary voltages include 480V, 277V, 240V, 208V, and 120V.

Transformers generate heat, which must be removed to prevent overheating. They may be dry (air cooled) or liquid cooled. Dry transformers remove heat by circulating air through spaces in the transformer. Liquid-cooled transformers circulate oil through coolers that absorb the heat and transfer it to the outside air. Mineral oil is the least expensive liquid, but it is flammable and cannot be used in all locations. A number of other liquid coolants, such as silicone liquids, have low flammability and are widely used. Liquid transformers are used on large installations.

Transformers are specified by the voltages in and out, the amount of power they can handle in kilovolt-amps (kVA) (thousands of volts times the rated maximum amperage), also called insulation class, and sound rating. Insulation specification pertains to electrical insulation temperature ratings in degrees centigrade. Sound ratings have to do with the sound vibrations produced by the different types and sizes of transformers. Transformers are available in single-phase and three-phase construction. The term "phase" refers to the number of circuits, voltages, and currents in an alternating current system. Single-phase has only one circuit or path for the flow of current; three-phase has three paths of flow.

Transformers may be located indoors or outdoors. Dry-type outdoor installations must have weatherproof enclosures, and larger sizes must be kept away from combustible materials. Liquid-type transformers on or adjacent to buildings in which combustion is possible must have a means for protecting against fires caused by excess heat or oil leaks. This can be provided with fire-resistant enclosures, various types of barriers, or a sprinkler system.

SWITCHES

Switches are used to open electrical circuits and interrupt the flow of current. A typical switch used on low-power applications, such as lighting, has some means of physically opening and closing the circuit—manually moving a lever or pushing a button using an electric coil, a motor, or a spring. This action moves metal contacts that separate to open the circuit or close to complete the circuit. Solid-state switches also interrupt a circuit, but they do so by electronically creating a conducting or non-conducting condition with no moving parts. Switches are classified by the National Electric Manufacturers Association (NEMA) and Underwriters Laboratories (UL).

Different actions enable switches to perform specific functions. These are illustrated by the simple knife switches shown in Figure 41.32. A single-pole, single-throw (SPST) switch is moved to make or break a connection between two contact points. When the points are in contact, the circuit is complete, and when they separate or open, the circuit is broken.

A single-pole, double-throw (SPDT) action is used to control a unit, such as a light, from two locations. This requires an SPDT switch at both locations.

A double-pole, single-throw (DPST) action is similar to the SPST action. However, it opens both wires in the circuit. DPST switches are used when the wires are not

Figure 41.32 Different actions enable switches to perform specific functions.

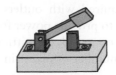

SINGLE–POLE, SINGLE–THROW
KNIFE SWITCH

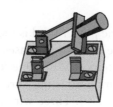

DOUBLE–POLE, SINGLE–THROW
KNIFE SWITCH

grounded, for example, switches used on 240V motors and appliances. They can also be used to control two circuits at the same time. Both circuits may be open or closed simultaneously.

A double-pole, double-throw (DPDT) action can be used to control more than one circuit at a time. It can be used to reverse the direction of rotation of a DC-motor by reversing the polarity.

Switches that control circuits with loads up to 30A, such as lighting, are shown in Figure 41.33. They are rated to carry various loads that typically run 15A, 20A, or 30A at 120V. They may be toggle, push, keyed, rocker, touch, tap-plate, or rotary type. They are available as SPST, DPST, DPDT, and SPDT and three-way and four-way switches. Three-way switches are used to control a light from two different locations. Four-way switches control a light from three different locations. Other switches can operate as timers that allow a unit, such as a fan, to operate for a set time after which it automatically shuts off. This type uses a spring-wound timer. A keyed switch provides security because a key is needed to operate it. Programmable switches are solid-state switches that can be programmed to switch a circuit on and off at preset times.

The switches described above do not contain fuses. However, fusible switches are available (Figure 41.34). The fuses and switches are enclosed in a metal box that can be closed and locked. The switch is manually operated with a handle on the outside of the box. Another type of switch using a solid-state rectifier enables incandescent lights to have high, low, and off controls. A rectifier is an apparatus in which electric current flows more readily in one direction than in the reverse direction for the purpose of changing alternating current into direct current.

Figure 41.33 Examples of switches typically used to control circuits with loads up to 30A.

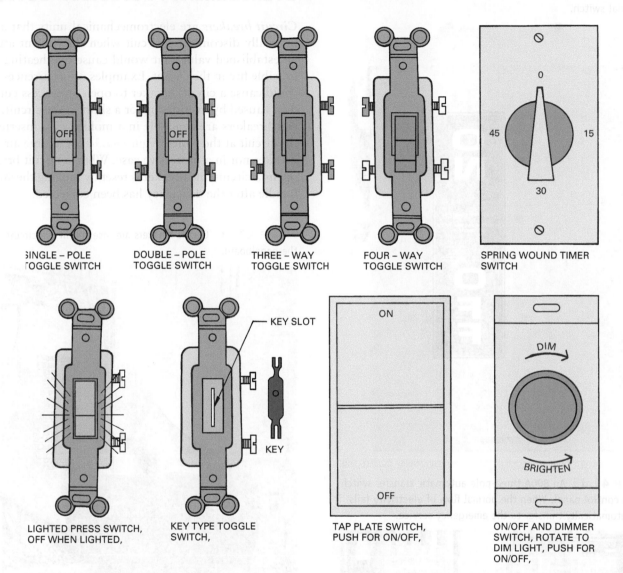

SINGLE – POLE TOGGLE SWITCH

DOUBLE – POLE TOGGLE SWITCH

THREE – WAY TOGGLE SWITCH

FOUR – WAY TOGGLE SWITCH

SPRING WOUND TIMER SWITCH

LIGHTED PRESS SWITCH, OFF WHEN LIGHTED,

KEY SLOT

KEY

KEY TYPE TOGGLE SWITCH,

TAP PLATE SWITCH, PUSH FOR ON/OFF,

ON/OFF AND DIMMER SWITCH, ROTATE TO DIM LIGHT, PUSH FOR ON/OFF,

Other Switches

There are many other switches designed to serve special purposes. A service disconnect, disconnects all electric power to a building except for specific emergency equipment requirements. Generally, these are located outside a building. They may be part of a switchboard.

A contactor is a form of switch that uses an electromagnet to close the contact blocks. It is operated from a remote location and can be activated by various devices, such as pushbutton or thermostat. A remote-control switch is a form of contactor that remains latched until its electric circuit is energized. In this way, the coil energizes only when the circuit opens. It is used on applications where circuits must be kept closed (a light installation, for example) for long periods of time.

Time-controlled switches use some type of timer, such as an electronic timer or a miniaturized low-speed motor that rotates a disc that contacts the switch to open and close the circuit.

An automatic transfer switch is a double-throw unit that switches to a source of emergency power when normal electrical service is interrupted (Figure 41.35). A typical application would be in the power supply for a hospital operating room.

Isolation switches are opened only after current flow in a circuit has been interrupted by another regular-use switch. They are not used to interrupt the flow of current in a circuit.

Figure 41.34 This is a heavy-duty fused three-pole three-wire industrial switch.

© SARAWUTK/Shutterstock.com

Figure 41.35 An 800A three-pole automatic transfer switch with a control panel. When the normal flow of electricity fails, it will automatically transfer to the emergency service.

© Alexander Tihonov/Shutterstock.com

OVERCURRENT PROTECTION DEVICES

Circuit breakers are electromechanical units that automatically disconnect a circuit when the current attains an established value that would cause overheating or a possible fire in the circuit. Examples of occurrences that could cause a circuit breaker to open are excess current flow caused by overloading or a short in the circuit. Circuit breakers are available in a molded case inserted in the circuit at the panel (Figure 41.36) or a large air-type breaker not in a protective case. When a circuit breaker opens a circuit, the breaker is reset by moving the switch handle after the deficiency has been corrected.

Figure 41.36 Circuit breakers are inserted in the circuit at the panelboard.

© Lisa F. Young/Shutterstock.com

A *fuse* is also used to protect circuits from overloads and shorts. Fuses have an internal metal link that melts when overheated, breaking the circuit. The two types of fuses are cartridge and plug (Figure 41.37).

Cartridge fuses are either knife-blade or ferrule. The blades on the first type slip into metal clips on a panelboard, and the ferrule type's copper rings on each end fit into clips. This connects the incoming power at the service entrance to the bus bars in the panelboard. Most cartridge fuses are discarded after they blow, but some types have replaceable links.

Ferrule-type cartridge fuses are rated from 10A to 60A and are generally used to protect currents for individual appliances, such as an electric stove. Knife-blade cartridge fuses work for service over 60A and are used

Figure 41.37 Commonly used plug and cartridge fuses.

Reprinted with permission, Bussmann Division, Cooper Industries

in the service entrance between the incoming power line and the circuits in the panelboard.

Plug fuses are available in 15A, 20A, 25A, and 30A sizes. They are used to protect individual circuits requiring small current requirements, such as a series of lights. They are installed in a fuse box or panelboard and serve the same function as a circuit breaker. However, when plug fuses are blown, they are discarded and replaced with new fuses after the deficiency has been corrected.

There are several varieties of plug fuses. The standard plug fuse will blow when the fusible link overheats. The time delay plug fuse has a fusible link that will melt immediately only when a short circuit occurs. If there is an overload, the link softens but does not break. If the overload is quickly removed, such as when the momentary load that starts a large electric motor ceases, it will not break. However, if the overload continues, the link will melt.

A ground fault circuit interrupter (GFCI) is an inexpensive electrical device that can be installed in an electrical system to protect from severe electrical shocks. Integrated into electrical receptacles (outlets), a GFCI constantly monitors current flowing through a circuit. If the current flow in the circuit differs even by a small amount from the returning current, the GFCI interrupts power. GFCIs are used in vulnerable areas that are close to water sources. Circuits that require GFCI protection are designated by the *National Electrical Code* and include receptacles in bathrooms, kitchen receptacles, and exterior plugs.

LIGHTING

Sufficient light is essential for the proper functioning of a building. Standards for lighting design are written and maintained by the Illuminating Engineering Society of North America (IESNA). Most spaces integrate natural and artificial light to work together and provide the type and degree of illumination required. Natural lighting can be controlled by various shading devices and the type of glazing used. Review Chapter 28 for information on glazing.

Lighting is a major source of energy use within buildings. In addition to the power used directly by light fixtures, many artificial light sources generate a substantial amount of heat within a building. This excess heat must be removed by the HVAC system, resulting in still-higher energy consumption. The use of daylighting, energy efficient fixtures, and motion sensors to activate lighting should all be considered in an effort to reduce the energy consumption associated with artificial lighting.

Illumination Design Considerations

Lighting systems are designed to provide the specific lighting needs each building area requires. The purpose of a lighting system is to impart a pleasant atmosphere for occupants and to provide sufficient visibility to perform required tasks. All of this must be accomplished using as little electrical power as possible.

When designing a lighting system, the ceiling height; the reflectance values of the ceiling, walls, floors, and furniture; and footcandle requirements all relate to one another. Standards and codes dictate the illuminance level, the degree of glare limitation, the luminous color and color rendering. The choice of the type of *luminaire* (light fixture) and its spacing, mounting, and height above the area needing lighting must be considered.

LIGHTING-RELATED MEASUREMENTS

Levels of Illumination

The illumination recommendations for residential and commercial buildings are detailed in Chapter 10 of the IESNA *Lighting Handbook* produced by the Illuminating Engineering Society of North America. The text outlines procedures for determining the quality and quantity recommendations for lighting design for various occupancies. Recommended illumination levels are given in **Table 41.3**.

Table 41.3 Recommended Illumination Levels

Type of Space	Guideline (footcandles)[a]
Commercial and Institutional Interiors	
Art galleries	30–100
Auditoriums/assembly spaces	15–30
Banks	
Lobby	50
Customer areas	70
Teller stations and accounting areas	150
Hospitals	
Corridors, toilets, waiting rooms	20
Patient rooms	
General	20
Supplementary for reading	30
Supplementary for examination	100
Recovery rooms	30
Lab, exam, treatment rooms	
General lighting	50
Close work and examining table	100
Autopsy	
General lighting	100
Supplementary lighting	1,000
Emergency rooms	
General lighting	100
Supplementary lighting	2,000
Surgery	
General lighting	200
Supplementary on table	2,500
Hotels (rooms and lobbies)	10–50
Labs	50–100
Libraries	
Stacks	30
Reading rooms, carrels, book repair and binding, check-out	70
Catalogs, card files	100
Offices	
Corridors, stairways, washrooms	10–20[b]
Filing cabinets, bookshelves, conference tables	30
Secretarial desks (with task lighting as needed)	50–70
Routine work (reading, transcribing, filing, mail sorting, etc.)	100
Accounting, auditing, bookkeeping	150
Drafting	200
Post Offices	
Storage, corridors, stairways	20
Lobby	30
Sorting, mailing	100

Table 41.3 Recommended Illumination Levels *(Continued)*

Type of Space	Guideline (footcandles)[a]
Commercial and Institutional Interiors	
Restaurants	50
Schools	
Auditoriums	15–30
Reading or writing, libraries	70
Lecture halls	70–150
Drafting labs, shops	100
Sewing rooms	150
Stores	
Circulation areas, stock rooms	30
Merchandise areas	
Serviced	100
Self-service areas	200
Showcases and wall cases	
Serviced	200
Self-service	500
Feature displays	
Serviced	500
Self-service	1,000
Industrial Interiors (Manufacturing Areas)	
Storage areas	
Inactive	5
Active (rough, bulky)	10
Active (medium)	20
Active (fine)	50
Loading, stairways, washrooms	20
Ordinary tasks (rough bench and machine work, light inspection, packing/wrapping)	50
Difficult tasks (medium bench and machine work, medium inspection)	100
Very difficult tasks (fine bench and machine work, difficult inspection)	200–500
Extremely difficult tasks	500–1,000
Garages	
Active traffic areas	20
Service and repair	100
Exteriors	
Building security	1–5
Parking	
Self-parking	1
Attendant parking	2
Shopping centers (to attract customers)	5
Floodlighting	5–50
Bulletins and poster panels	20–100

[a]To convert to lux, multiply value by 10.76.
[b]Must be at least 20 percent of the level of the adjacent work space.

Source: Illuminating Engineering Society of North America

Luminous power of a light source is measured in either footcandles or candelas, the metric unit. The *candela (cd)* was adopted in 1948 as the international standard of luminous intensity. The unit used to measure luminous power at a predetermined distance from a light source is the lumen. The light output of a lamp is measured in a physical quantity called lumens, (provided in product data). The *lumen (LU)* is the luminous power on an area of one square foot at a distance of 1 ft. from a one-footcandle light source—or on an area of 1 sq. m at a distance of 1 m from a one-candela light source. *Lux* is the metric unit used to describe the illumination produced by the luminous flux of one lumen falling perpendicular on a surface of one square meter. Lux is equivalent to 0.0929 footcandles. Lighting unit conversion factors are shown in Table 41.4. A related term, the *footlambert*, is a unit for measuring brightness or luminance. It is equal to 1 lumen per square foot when brightness is measured from a surface.

Table 41.4 Lighting Unit Conversion Factors

Unit	To convert from ...	Multiply by the correction factor ...	To convert to ...
Luminous Intensity (I)	Candlepower	1.00	Candela
Illuminance (E)	Footcandle	10.76	Lux
	Lux	0.09	Footcandle
Luminance (L)	Candela/m²	0.29	Footlambert
	Candela/in.²	1550.00	Candela/m²
	Candela/ft.²	10.76	Candela/m²
	Footlambert	3.43	Candela/m²

Measuring Illuminance Levels

Illuminance (E) is the total amount of visible light incident upon a point on a surface from all directions above the surface. It is a measure of light intensity striking a surface. One lumen of luminous flux uniformly imposed on 1 sq. ft. of area results in an illuminance of one *footcandle*.

The level of illuminance is commonly measured with a *luminance meter*, often referred to as a light meter. It has a photoelectric panel connected to a microammeter and electronic control circuitry and is calibrated to read in footcandles or lux. Some types have a remote sensor. The meter is held so the photoelectric panel is parallel to the plane of the area being tested.

Measuring Luminance

To evaluate available illumination, it is necessary to get a measure of luminance, or the quality of illumination. *Luminance (L)* is a measure of brightness and brightness contrasts caused by photometric luminance. It is what is seen rather than illuminance (the quantity in footcandles), and it is an important part of the evaluation of the total illumination available. Luminance is measured by a luminance meter. Current luminance recommendations are developed following IESNA procedures.

Glare

"Glare" refers to light that is intense and uncomfortable to a viewer. Glare can be generated by luminaires that shine directly into a viewer's eyes. Glare can be reduced by lowering the brightness of luminaires or relocating and shielding them, if possible. Reflected glare is found more frequently. It occurs when light reflects off a surface. It can be controlled by selecting surfaces with low reflectance coefficients, reducing the brightness of the luminaire, or changing the angle of a reflecting surface.

Reflection

The *reflectance coefficient* is a measure of the percentage of incident light reflected from a surface. The *luminous intensity* (force) of a luminaire can be reduced or enhanced by the reflective potential of surrounding ceilings, walls, and floors. Dark colors absorb light, and light colors reflect light. Therefore, more light produced by a luminaire is available if light-colored ceilings and walls are used. Bright colors, such as red or yellow, are used infrequently because they tend to produce glare.

Manufacturers of wall and ceiling materials and paints generally list the reflectance coefficients of their products. For example, if half the incident light on a surface is reflected, the reflectance coefficient is 50 percent or 0.50. The other 50 percent is absorbed by or transmitted through the material. The amount transmitted is called *luminous transmittance*. The rate of flow of light energy through a surface is called *luminous flux*.

The reflectance value of a ceiling used with a direct lighting system has little influence on the light projected, because direct lighting projects from a luminaire directly down into a room. Ceiling reflectance

Construction Materials

Choosing Light and Color

Light quality is measured by color temperature and the color-rendering index. Color temperature gives a measurement of the visual warmth or coolness of a light source, expressed in degrees Kelvin (°K). This measurement describes the quality of the actual light emitted by a lamp. The higher the color temperature, the cooler the light, and the lower the color temperature, the warmer the light. For example, typical inexpensive cool white fluorescent tube has a color temperature of about 4,100°K. Typical incandescent lamps have a color temperature of about 2,700°K. Light from an incandescent lamp is warmer, containing reds and yellows that contribute to the tones of surrounding materials, which people find warm and pleasing. And when high-temperature fluorescent tubes replace incandescent lamps, they cast a hot bluish to greenish light that gives surrounding materials a gray and flat appearance.

The color-rendering index (CRI) provides a measure of light quality that indicates how natural an object looks when under a light source. This index is given as a percentage. The closer the CRI is to 100 percent, the more natural things appear. When a natural appearance is critical, as in an art museum, seldom is a lamp with less than 70 percent used. Incandescent lamps have CRI ratings close to 100 percent, but fluorescent tubes vary from below 50 to about 90 percent.

Following are some rules of thumb to help make color selection easier.

1. All space colors (wall and floor coverings, furniture, drapes, accents, etc.) should be chosen under the lamp color specified for installation. Experience suggests that warm sources should be used at low lighting levels, cool sources at high levels. However, the choice may be influenced by space colors and by degree of luminaire brightness control.

2. Warm color schemes may appear overpowering if lighted with a warm source to relatively high levels—a cooler source should be used.

3. Cool color schemes may need warm sources, particularly at low lighting levels.

4. Well-shielded lighting systems, such as those using wedge louvers, or low brightness lenses will "cool off" a room. A slightly warmer lamp will counteract the effect.

5. Where color rendition is highly critical, high-CRI continuous-spectrum sources, such as Chroma 50 (C50) or Chroma 75 (C75), should be used.

6. Where both color and lighting level are important, the use of SP30, SP35, or SP41 GE Specification Series Color lamps, with three-peak rare-earth phosphors, will provide good color rendering and high efficiency.

7. The Specification Series Colors tend to make spaces look more colorful because the three-peak phosphors compress all colors into the blue, green, and red-orange bands. This increases the contrast between colors. Three-peak lamps do not, however, increase the contrast of black-on-white tasks. Claims that less light is needed for typical office and industrial tasks when three-peak lamps are used are scientifically unfounded.

8. The color of a light source does not affect the visual performance of people doing black-on-white visual tasks. Vision and productivity studies, however, indicate that productivity may be affected by the color contrast and appearance of a visual environment and that color can contribute strongly to appearance.

does greatly influence usable light if indirect lighting is used. Indirect lights project light up to a ceiling where it is reflected down into the room. The reflectance value of walls also influences reflected light. However, walls typically have doors, windows, cabinets, and other items that reduce the actual wall surface available. Floors have little influence on lighting; however, light-colored floors do reflect some light and provide visual comfort.

Room Cavity Ratio

The size of a space to be lighted affects the way light is distributed throughout the space. High ceilings reduce the efficiency of lighting. Open spaces can be more efficiently lighted than small rooms or partitioned space. The relation between the space proportions, room height, room perimeter, and the height of work surfaces needing illumination divided by the floor area is called the *room cavity ratio*.

LIGHTING SYSTEMS

Lighting systems fall into the categories of direct, indirect, semi-indirect, general-diffuse, and semi-direct (Figure 41.38). The exact classification depends on the amount of light from the luminaire directed up and down.

Direct lighting is the most efficient system, projecting 90–100 percent of the light down to a floor or work surface. A light-color reflective floor will reflect some light toward the ceiling, reducing ceiling darkness somewhat. Indirect lighting systems project most of the light up to the ceiling and some onto walls. The light is then reflected to the floor and work surfaces. This system requires that ceilings and walls have high reflectance coefficients. It produces an illumination that is uniform, free of glare, and generally diffuse. Indirect lighting uses baffles and requires that luminaires have more powerful lamps than direct lighting.

Semi-indirect lighting systems allow more light (10–40 percent) to project down than indirect systems do. They do this by using a translucent diffuser that permits some light to project down but reflects most of it up to the ceiling to produce a diffuse low-glare illumination.

Figure 41.38 The distribution of light can be varied depending on the type of fixture used.

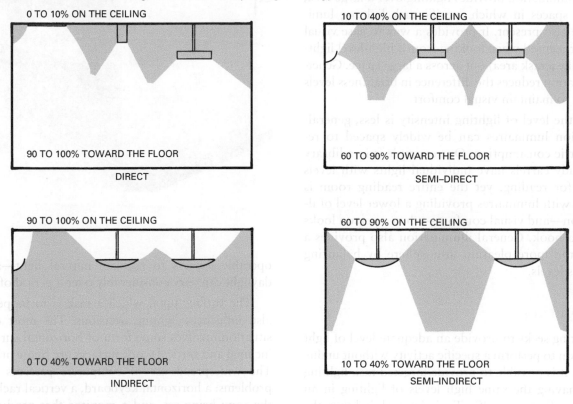

0 TO 10% ON THE CEILING

90 TO 100% TOWARD THE FLOOR

DIRECT

10 TO 40% ON THE CEILING

60 TO 90% TOWARD THE FLOOR

SEMI–DIRECT

90 TO 100% ON THE CEILING

0 TO 40% TOWARD THE FLOOR

INDIRECT

60 TO 90% ON THE CEILING

10 TO 40% TOWARD THE FLOOR

SEMI–INDIRECT

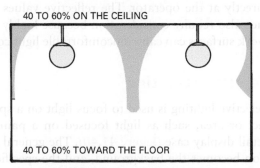

40 TO 60% ON THE CEILING

40 TO 60% TOWARD THE FLOOR

GENERAL DIFFUSE OR DIRECT–INDIRECT

General-diffuse systems (also called direct-indirect) project about equal amounts of light up and down. They produce a light ceiling and upper wall area but also project light down to the floor and work surfaces. This system typically uses a globe diffuser, which allows light to project in all directions.

Semi-direct lighting systems project most of their light down, with only a very small percentage projecting upward to light the ceiling. This produces efficient direct lighting while somewhat brightening dark ceiling areas (at least more so than when direct lighting is used).

General Illumination

General illumination provides lighting over a large area, including spaces in which specific task lighting luminaires are also present. It provides a way to ease visual discomfort caused when looking from a high-level lighted area (like a task area) out across a large space. General illumination reduces the difference in brightness levels and helps to maintain visual comfort.

Since the level of lighting intensity is less, general-illumination luminaires can be widely spaced to reduce electric consumption. One example is in a library where study carrels have individual lights with levels required for reading, yet the entire reading room is equipped with luminaires providing a lower level of illumination—and visual comfort—when a person looks up from a book. General illumination also provides a warmer and more pleasant atmosphere by balancing brightness levels.

Task Lighting

Task lighting seeks to provide an adequate level of light for a person to perform a specific activity without undue eyestrain. It is possible to provide targeted task lighting without having the same high levels of lighting in an entire space (Figure 41.39). To design task lighting, the activities to be performed, their location within a building, as well as the number of people performing tasks, their nearness to each other, and any patterns of movement must be specified. In many types of businesses, change in activities is normal, so the system must allow for easy revision, including relocating luminaires or redirecting light from existing luminaires. Since the level of light required may vary from time to time, luminaires must be individually switched so only those needed can be turned on or dimmer switches should be used. Windows near task activities must have blinds, shutters, or

Figure 41.39 This reception desk uses task lighting to provide a high level of illumination at the work surface.

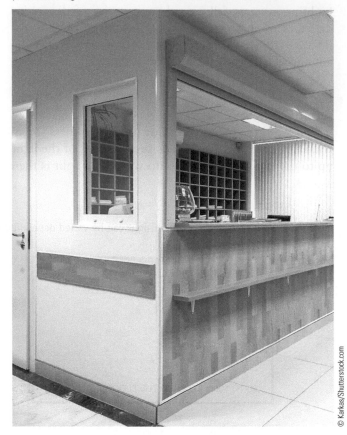

operable louvers to regulate natural light—available daylight can vary considerably over a period of hours.

The surface upon which a task is to be performed also influences lighting decisions. The most common situation involves some form of horizontal surface, but inclined and vertical work surfaces are frequently found. The widespread use of computers presents multiple problems: a horizontal keyboard, a vertical rack to hold the copy being set, and a monitor that produces light directly at the operator. The reflective values of a surface also require consideration because highly reflective work surfaces can cause uncomfortable light conditions.

Selective Lighting

Selective lighting is used to focus light on a specific object or area, such as light focused on a painting or a retail display case (Figure 41.40). The general illumination becomes the background, and the specific lighting focuses visual attention on an object.

Figure 41.40 Retail spaces make use of selective lighting to focus on specific merchandise.

© Karkas/Shutterstock.com

Daylight Integration

Daylighting is the use of light from the sun and sky to complement or replace electric light. Daylight reduces the electric energy needed to power lights as well as the heat generated from lights, thereby also reducing cooling requirements (Figure 41.41). Lighting and its associated cooling energy use constitute 30 to 40 percent of a commercial building's total energy expenditure.

Figure 41.41 The use of daylight reduces the electric energy needed to power lights.

© Galyna Andrushko/Shutterstock.com

Studies have shown that people are happier and more productive under daylight. Daylighting has been found to result in a reduced number of sick days in commercial offices and increased sales in retail spaces. Students learn better and day-lit schools have been proven to assist in higher levels of achievement.

The diffuse light of an overcast sky is similar in all orientations—it is soft and cool in both temperature and color. Every orientation provides useful daylight. North exposures give high-quality, consistent light with minimal heat gain, and minimal shading requirements. South orientations provide good access to strong illumination that varies throughout the day. Shading must be provided to prevent seasonal overheating. East and West orientations must also incorporate proper shading to prevent glare and heat gain.

LIGHTING CONTROL SYSTEMS

A lighting control system consists of a device that controls electric lighting and devices, alone or as part of a daylight harvesting system. Timers, dimmers, and occupancy sensors can provide large savings by turning lights down when they are not needed at full light output levels, and turning them off when they are not required at all.

The simplest form of lighting control is a simple on-off wall switch. Dimming can be performed in various ways, depending on the lamp type and the control strategy. Sophisticated digital timers offer the possibility to program their luminaire function according to a predefined schedule. Photoelectric sensors use cadmium sulfide as the active element for light detection to turn lights on in the evening and off in the morning according to measured lighting levels.

Occupancy sensors are switching devices that respond to the presence and absence of people in the sensor's field of view. Occupancy sensors automatically turn lights on when a person enters the room; keep lights on while the controlled space is occupied; and turn lights off within a predefined time interval after if the room is unoccupied. Newer occupancy sensors operate in dual mode with both PIR (passive infrared radiation) sensing, and sound detecting ultrasonic technology.

TYPES OF LAMPS

A variety of *lamps* are in use, each with particular advantages and disadvantages. The three major types of artificial lamps are incandescent, fluorescent, and high-intensity discharge. Manufacturer's catalogs list lamp size, input wattages, lumen output, and lamp life for their products.

The lamp input is in terms of electrical power and has the units of watts. Efficacy, the performance measure for electric lighting, is measured in units of lumens per watt (Lu/W) and varies with type and size of lamps. For example, a 100-watt incandescent lamp has about 17 lumens per watt, while a modern T8 fluorescent lamp with electronic ballast has 100 lumens per watt. The higher the lumen per watt rating of a lamp, the better the efficiency, meaning greater light output for a fixed-wattage input.

Incandescent Lamps

An *incandescent lamp* is a glass bulb containing a filament joined to a metal base with copper lead-in wires (Figure 41.42). The filament is a tungsten wire that resists the flow of electricity. In doing so, the wire gets hot and glows, producing light. A lot of the energy used to create the heat that lights an incandescent bulb is waste, making them less energy efficient than other types of lamps. The metal base screws into a socket that is part of a luminaire, or lighting fixture. Most incandescent lamps are filled with argon and nitrogen, which makes the use of a high filament temperature possible. The melting point of tungsten is 3,655K (6,170°F). Lamps filled with krypton gas have a longer life than argon and nitrogen lamps and cost more.

Figure 41.42 The working components of an incandescent lamp.

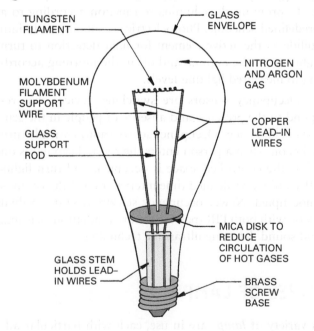

TUNGSTEN FILAMENT

GLASS ENVELOPE

MOLYBDENUM FILAMENT SUPPORT WIRE

NITROGEN AND ARGON GAS

GLASS SUPPORT ROD

COPPER LEAD–IN WIRES

MICA DISK TO REDUCE CIRCULATION OF HOT GASES

GLASS STEM HOLDS LEAD–IN WIRES

BRASS SCREW BASE

Incandescent lamps are made with several size bases and rated in watts, as shown in Table 41.5. The larger bases are used on lamps with higher wattages. The voltage at which the lamp is used influences its life. If a 120V lamp is burned at a voltage higher than 120, its life will

Table 41.5 Summary of Lamp Characteristics[a]

Type	Light Production	Typical Wattage Range
High-Intensity Lamps		
Mercury	Argon gas and mercury vapor	50–1,000
Metal-halide	Argon gas, mercury, and halide salts	175–1,500
High-pressures sodium	Sodium, mercury amalgam, and xenon gas	100–1,000
Incandescent		
Standard incandescent	Tungsten filament in a gas	6–2000
	Tungsten filament in a vacuum	5,000 and 10,000
Tungsten-halogen	Tungsten filament with a halogen, such as iodine or bromine	12–2,000
Fluorescent		
Preheat	Mercury vapor in an inert gas, such as krypton, argon, or neon, with the tube interior coated with a fluorescent material	15–90
Instant heat		20–75
Rapid start, high output		30–100
Rapid start, very high output		100–215

[a]Check manufacturers' catalogs for specific light production and wattage data.

be shortened, but it will produce more lumens per watt. If operated at a voltage below this, it will have a prolonged life but produce fewer lumens per watt. Some of the commonly available incandescent lamps are shown in Figure 41.43.

Tungsten-Halogen Lamps

Tungsten-halogen lamps (quartz-iodine) are a type of incandescent lamp. They have a tungsten filament and are filled with a halogen such as iodine or bromine, and an inert gas that reduces the evaporation of the filament. The bulb is made from quartz because it will withstand higher temperatures than glass. The halogen additive in the lamp reacts chemically with any tungsten deposited on the bulb and re-deposits it on the tungsten filament, improving the efficiency of the lamp. Tungsten halogen lamps are available in a wide range of wattages, as shown in Table 41.5. Examples of some typical lamps

Figure 41.43 Some of the types of incandescent lamps available.

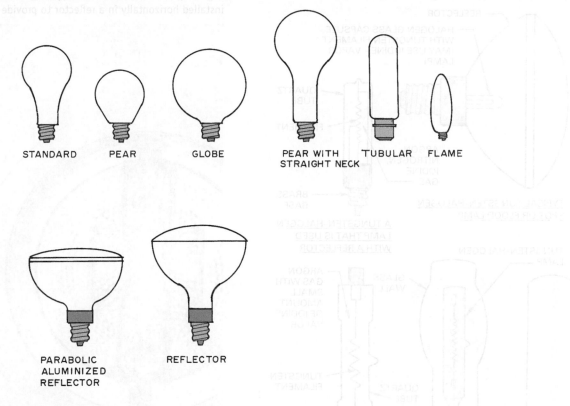

STANDARD   PEAR   GLOBE   PEAR WITH   TUBULAR   FLAME
                                              STRAIGHT NECK

PARABOLIC   REFLECTOR
ALUMINIZED
REFLECTOR

are shown in Figure 41.44. Two of the lamps shown have quartz tubes. One has a screw base and is installed vertically. The double-end lamp is installed horizontally. Both require reflectors, as shown in Figure 41.45.

Tungsten-halogen lamps with parabolic reflectors are available from 15W to 1,500W and in a number of shapes. The slanted backside of the bulb is a reflector. Both flood and spot lamps are available. Tungsten-halogen lamps operate at high temperatures, and even though the quartz envelope can withstand high temperatures, the lamps can explode. Therefore, manufacturers recommend some type of shielding or protective screen be used to contain potential flying fragments.

Fluorescent Lamps

Fluorescent lamps have a long glass tube sealed at each end. The tube is filled with a mixture of an inert gas, such as argon, and low-pressure mercury vapor. A cathode is placed in each end. The cathode produces electrons that start and maintain operation of a mercury arc, which produces an ultraviolet arc. The ultraviolet arc is absorbed by phosphors that coat the inside of the tube and cause it to fluoresce (radiate) light (Figure 41.46). Because fluorescent bulbs don't use heat to create light, they are far more energy efficient than regular incandescent bulbs.

Fluorescent lamps will not operate directly off 120V alternating current because it will not cause the arc discharge. Each luminaire incorporates a ballast that provides the starting and operating voltages. Standard lamps have a starter that preheats the cathodes, producing the high-voltage arc needed to start the lamp. Rapid-start lamps have the same basic construction as standard lamps, but a circuit keeps the lamp electrodes constantly preheated by means of low-voltage windings that are part of the ballast.

High-output (HO) and very-high-output (VHO) lamps require special ballasts. They are used where high output is required in a limited area, such as a merchandise display or an outdoor sign. They generate considerable heat that must be considered, but they function in cold situations where standard lamps will not light.

Instant-start fluorescent lamps use a high-voltage transformer to generate an arc, lighting the lamp without the preheating delay of standard lamps. They have a single pin at each end. The high-voltage start greatly reduces the life of the lamp, but it will start in lower temperatures than rapid-start lamps.

Fluorescent lamps are available in straight tubes, circles, and U-shapes. Small lamps are available that fit

Figure 41.44 Typical tungsten-halogen lamps.

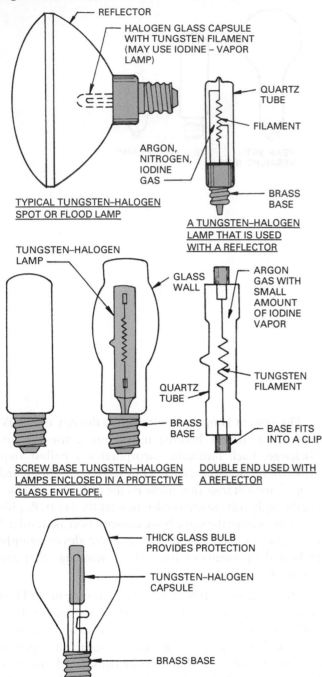

REFLECTOR

HALOGEN GLASS CAPSULE
WITH TUNGSTEN FILAMENT
(MAY USE IODINE – VAPOR
LAMP)

QUARTZ
TUBE

FILAMENT

ARGON,
NITROGEN,
IODINE
GAS

BRASS
BASE

TYPICAL TUNGSTEN–HALOGEN
SPOT OR FLOOD LAMP

A TUNGSTEN–HALOGEN
LAMP THAT IS USED
WITH A REFLECTOR

TUNGSTEN–HALOGEN
LAMP

GLASS
WALL

ARGON
GAS WITH
SMALL
AMOUNT
OF IODINE
VAPOR

QUARTZ
TUBE

TUNGSTEN
FILAMENT

BRASS
BASE

BASE FITS
INTO A CLIP

SCREW BASE TUNGSTEN–HALOGEN
LAMPS ENCLOSED IN A PROTECTIVE
GLASS ENVELOPE.

DOUBLE END USED WITH
A REFLECTOR

THICK GLASS BULB
PROVIDES PROTECTION

TUNGSTEN–HALOGEN
CAPSULE

BRASS BASE

GENERAL SERVICE
TUNGSTEN–HALOGEN LAMP

Figure 41.45 A tungsten-halogen quartz-iodine lamp
installed horizontally in a reflector to provide indirect lighting.

in table lamp sockets. Several different bases and lamp
holders are used, depending on type of lamp. Since fluorescent lamps are much more efficient than incandescent
lamps, lower wattages can be used. Typical wattage ratings are shown in Table 41.5.

Standard fluorescent lamps containing mercury are
classified as hazardous waste by the Environmental Protection Agency. A new long-mercury fluorescent lamp is now
available. It uses a chemical buffer to slow the absorption
of mercury by the phosphor crystals that coat the inside of
the tube, so less mercury is required in the lamp.

Each lamp receives a pre-measured amount of mercury encapsulated in a tiny glass bubble. As the lamp
leaves the factory, a radio-frequency beam heats a wire
wrapped around the bubble, which breaks, releasing the
mercury in the tube.

Compact Fluorescents A *compact fluorescent light
(CFL)* is a type of fluorescent lamp (Figure 41.47) made

Figure 41.46 A generalized illustration showing the construction of a fluorescent lamp with bi-pin bases.

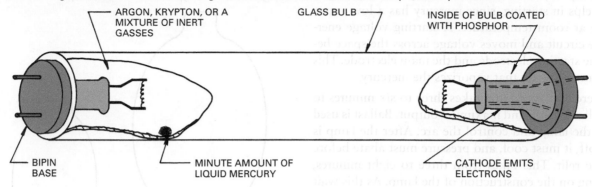

ARGON, KRYPTON, OR A MIXTURE OF INERT GASSES

GLASS BULB

INSIDE OF BULB COATED WITH PHOSPHOR

BIPIN BASE

MINUTE AMOUNT OF LIQUID MERCURY

CATHODE EMITS ELECTRONS

Figure 41.47 The characteristic shape of a compact fluorescent light.

© Hurst Photo/Shutterstock.com

in special shapes to fit standard light sockets. Unlike fluorescent tubes, which require a separate ballast independent of the bulb, CFLs incorporate integral ballast built directly into the light bulb. Compared to incandescent lamps that provide the same amount of visible light, CFLs generally use less power, and have a longer rated life. Compared to an incandescent lamp, a CFL can save in electricity costs over its lifetime. It can also save 2,000 times its own weight in greenhouse gases. Like all fluorescent lamps, CFLs contain mercury, which complicates their disposal.

High-Intensity Discharge Lamps

High-intensity discharge lamps (HID) produce light by passing an electric arc through a metallic vapor confined in a sealed quartz or ceramic tube. The high-intensity

discharge lamps available include mercury vapor, metal-halide, and high- and low-pressure sodium. With appropriate color correction, they can be used in many indoor and outdoor situations.

Mercury Vapor Lamps Mercury vapor lamps (M-V) are available in clear, white, white-deluxe, and color-corrected. The clear lamp produces a blue-green light that causes distortion of other colors. Color correction is accomplished by coating the outer bulb with phosphors to make the lamp useful for some indoor applications (Figure 41.48).

Figure 41.48 A simplified schematic of a typical mercury vapor lamp.

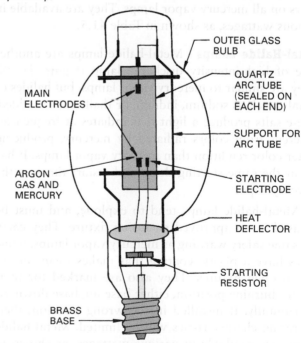

OUTER GLASS BULB

QUARTZ ARC TUBE (SEALED ON EACH END)

MAIN ELECTRODES

SUPPORT FOR ARC TUBE

ARGON GAS AND MERCURY

STARTING ELECTRODE

HEAT DEFLECTOR

STARTING RESISTOR

BRASS BASE

Mercury vapor lamps use mercury and argon gas, which helps in starting, since mercury has a low vapor pressure at room temperature. A starting voltage energizes the circuit and moves voltage across the space between the starting electrode and the main electrode. This creates an argon arc that vaporizes the mercury.

A mercury vapor lamp takes three to six minutes to reach full warm-up and maximum output. Ballast is used to start the lamp and control the arc. After the lamp is turned off, it must cool, and pressure must abate before it can be relit. This takes from three to eight minutes, depending on the construction of the lamp. As this wait time would leave an area in total darkness if a power failure of even a few seconds occurred, some form of emergency backup lighting is usually required. This could be a system using incandescent lamps connected to the electric system or a battery-powered emergency lighting system.

Mercury vapor lamps are not as efficient as fluorescent lamps, but they are more efficient than incandescent lamps. They are used on applications in which they remain in use for long periods, such as overnight in a parking lot. They should not be used if frequent switching on and off is required.

The mercury vapor lamp spectrum is in the ultraviolet (UV) range, but it is not harmful because the outer glass absorbs much of the UV. However, if the glass breaks, the lamp continues to function, and UV exposure could be hazardous; therefore, a safety warning appears on all mercury vapor lamps. They are available in various wattages, as shown in Table 41.5.

Metal-Halide Lamps Metal-halide lamps are another type of high-intensity discharge lamp (Figure 41.49). They are similar to mercury vapor lamps, but halides of metals, such as sodium, indium, or thallium, are added. These salts produce a light that radiates at frequencies different from colors radiated by mercury, producing better color rendition than mercury vapor lamps. It has about the same starting delay and restart times as the mercury vapor.

Metal-halide lamps tend to explode, and must be encased in an approved enclosing fixture. They carry the same safety warning as mercury vapor lamps. Some types have a plastic coating that makes them safe to use in open fixtures. They also are marked for their proper burning position, either base up, base down, or horizontally. If installed in the wrong position, their operating characteristics will be limited. Metal-halide lamps are available in various wattages, as shown in Table 41.5.

Figure 41.49 Typical metal-halide lamps.

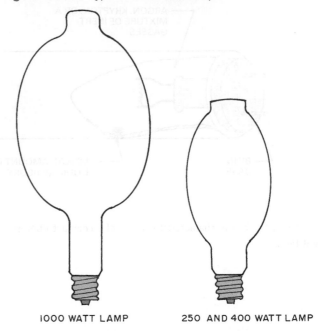

1000 WATT LAMP   250 AND 400 WATT LAMP

High- and Low-Pressure Sodium Lamps High-pressure sodium (HPS) lamps are arc discharge lamps containing sodium under high pressure in a glass arc tube. This produces a light with a yellow tint. They are highly efficient lamps, superior to all other types, and in areas where color is not important they can replace mercury vapor and metal-halide lamps (Figure 41.50). These lamps are available in various wattages, as shown in Table 41.5.

Figure 41.50 A simplified schematic of a high-pressure sodium lamp.

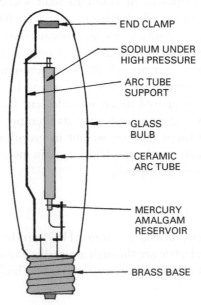

END CLAMP

SODIUM UNDER HIGH PRESSURE

ARC TUBE SUPPORT

GLASS BULB

CERAMIC ARC TUBE

MERCURY AMALGAM RESERVOIR

BRASS BASE

Low-pressure sodium lamps, referred to as SOX, produce a deep yellow light that is not acceptable for general interior lighting. They are widely used for street, highway, and parking lighting. They have a long life and are the most economical type of lighting.

Light-Emitting Diodes

Light-emitting diodes (LEDs) are small light bulbs based on semiconductor technology. The electrons of a semiconductor diode are brought from a state of high energy to a state of low energy, with the energy difference being emitted in the form of light. Until recently, LEDs were used only in single-bulb applications, such as in instrument panels and electronics. Today, LED bulbs are made by clustering many small bulbs encased in diffuser lenses that spread light in wider beams (Figure 41.51). Bulbs are available with standard bases to fit common household light fixtures. Despite their high expense, LEDs

Figure 41.51 Light-Emitting Diodes (LED) are light bulbs based on semiconductor technology.

© tarczas/Shutterstock.com

provide a number of advantages over traditional light sources, including low energy consumption, a long lamp life, and small size.

Review Questions

1. What is the rate of flow of electric current if it is available at 120V and has a resistance of 15 ohms?

2. What is the difference between alternating current and direct current?

3. What are the various ways electric power can be produced?

4. What fuels are used to power fossil-fuel electric-generating plants?

5. What power source is used to drive emergency generators?

6. How is electricity produced by utilizing normally wasted heat?

7. What are the types of electrical conductors recognized by codes?

8. What type of electrical conductor is used for underground applications?

9. What voltage current can be carried by flat conductor cables?

10. What types of conduit are in general use?

11. One kilowatt equals how many watt-hours?

12. What purpose do motor control centers serve?

13. What is the difference between the primary transformer voltage and the secondary voltage?

14. What are the commonly available switch actions?

15. What ways are various types of switches activated?

16. What devices are used to protect a circuit from overloads?

17. How does a type S plug fuse keep someone from putting the wrong size fuse in a panelboard?

18. What are the types of fuses in common use?

Key Terms

Building-Integrated PV (BIPV)

Bus

Busway

BX Cable

Cable Tray

Candela (cd)

Circuit Breaker

Cogeneration

Compact Fluorescent Light (CFL)

Conductor

Conduit

Daylighting

Electric Current

Footcandle (fc)

Footlambert

Fuse

Illuminance (E)

Incandescent Lamp

Insulator

Lamp

Light-Emitting Diode (LED)

Lumen (LU)

Luminaire

Luminance (L)

Luminance Meter

Luminous Flux

Luminous Intensity

Luminous Transmittance

Lux

Meter

Motor Control Center

Net Metering

Panelboard

Photovoltaic (PV)

Power

Raceway

Reflectance Coefficient

Resistance

Romex

Room Cavity Ratio

Service Entrance

Switch

Switchboard

Transformer

Watt

Activities

1. Invite a local building official to speak to the class about electrical code requirements.

2. Collect various electrical control and metering devices. Prepare a display and label each, giving the manufacturer's specifications.

3. Erect a section of a stud wall and install boxes for fuses, outlets, switches, lights, et cetera, and wire these. Bring power to the beginning end of the wiring and activate the system.

4. Walk through various buildings and note the types of lighting used. Prepare a report citing the location for each type found.

5. Arrange a visit to the electrical equipment room of a major building and have the building engineer explain the function of each unit. Describe the flow of power from the service entrance to the various floors of the building. View distribution equipment used on one of these floors.

Additional Resources

IESNA Lighting Handbook, and many other books on installation, standards, education, and design, Illuminating Engineering Society of America, New York, NY.

International Code Council Electrical Code, International Code Council, Falls Church, VA.

Kaufman, J. E., *IES Lighting Handbook Application*, Illuminating Engineering Society of America, New York, NY.

National Electrical Code®, National Fire Protection Association, Quincy, MA.

Other resources include:

Many electrical design, safety, and installation manuals, National Fire Protection Association, Quincy, MA.

See Appendix C for addresses of professional and trade organizations and other sources of technical information.

Electronic
Safety and Security
CSI MasterFormat™

© Nathan Griffith/Solus/Corbis

Electronic Signal and Security Systems

Upon completion of this chapter, the student should be able to:

- Understand the signal systems used to provide electronic safety and security.
- Identify various fire detection and alarm strategies in common use.

- Describe the various devices used in the design of security systems.
- Understand the principles of intrusion protection.
- Understand the functions of a building automation system.

Build Your Knowledge

For further information on these materials and methods, please refer to:

Chapter 38 Fire Suppression Systems

Electronic signal and security involve the installation and operation of various control, alarm, and communications systems. Topics listed in division 28 include fire detection and alarm, communications, access control, intrusion detection, and electronic monitoring and control systems. The signal systems used require a source, equipment that processes and transmits a signal, and a means of registering the signal through visual or auditory devices. Signals are typically conveyed via low-voltage wiring through a variety of sensors. Modern buildings are increasingly using an integrated-systems design approach that incorporates security with other operational functions into a building automation system (BAS). These systems require monitoring and control to function reliably.

AUTOMATIC FIRE DETECTION AND ALARM SYSTEMS

The primary purpose of a fire alarm system is to notify emergency responders and initiate the proper reaction from them. The second is to initiate fire safety functions,

which are building and fire control procedures that are intended to increase the level of safety for occupants or to control the spread of the harmful effects of fire.

The *National Fire Alarm Code* (NFPA 72) is a standard published by the National Fire Protection Association that regulates the location, performance, inspection, testing, and maintenance of fire alarm systems. NFPA 72 does not mandate the use of fire alarm systems. This is regulated by the *Life Safety Code* (NFPA 101) and other codes that determine whether a fire alarm system is required in a given occupancy.

Fire alarm systems are classified by the *National Fire Alarm Code* as household fire warning systems, protected premises, or off-premises systems. Household warning systems incorporate smoke and fire detectors to activate an auditory alarm. Protected premises alarms sound a warning only in the area of immediate danger and response is activated locally. Off-premises systems are alarms that are directly connected to the municipal fire department. Large multi-building campuses may incorporate a central supervisory station that receives signals from a number of buildings. Fire alarm systems should never be coupled with other systems, such as building automation, or energy management. Down time for any of these non-life safety systems could also take the fire alarm system out of service. Fire alarm zones should be coordinated with smoke compartments and sprinkler zones.

A variety of signal sources act as alarm-initiating systems by sensing the products of combustion at different stages of a fire's duration. Ionization detectors can detect microscopic particles emitted during the early stages of a fire (Figure 42.1). Photoelectric detectors use beams of light to respond to smoke particles visible to the naked eye during development of a fire. Heat detectors sense the actual fire to immediately activate alarm and fire suppression systems. *Fire flame detectors* are used in occupancies where combustible materials are present. They detect the presence of either infrared or ultraviolet radiation.

Figure 42.1 A common fire detector that senses smoke and sounds an alarm.

© magicoven/Shutterstock.com

Another type of automatic alarm uses normally locked doors that act as emergency exits. The doors are fitted with panic hardware that automatically signals an alarm when opened in an emergency. Fires may also be detected by building occupants with alarms activated via manually operated pull stations (Figure 42.2). Pull stations are generally located along designated exit paths. Various types of fire signal alarms are used as auditory indicators. The most common incorporate horns, bells, and flashing lights (Figure 42.3). These are often combined into one unit, providing sound and light signals.

The types and location of detection devices, alarms, and manual pull stations is specified by code for different occupancies.

Figure 42.2 Fire extinguishers are often placed in close proximity to manually activated fire alarm stations.

© Jennie Book/Shutterstock.com

Figure 42.3 Horns and bells are used to provide a loud warning when the fire-detection system senses a fire.

© Joe Gough/Shutterstock.com

Fire-suppression systems include automatic sprinklers, water-fog generators used on highly flammable solids or liquids, and a liquid foaming agent introduced into the water in a sprinkler system. Most systems incorporate a flow switch that senses that a sprinkler has been activated and sounds an alarm. Additional information on fire suppression systems can be found in Chapter 38.

SECURITY SYSTEMS

Electronic security systems provide surveillance and control for buildings to detect and prevent the occurrence of crime and vandalism. Systems range in size from small residential alarms to large central-control centers for multistory buildings. Large control centers can monitor different surveillance equipment, activate automatic response systems, and voice communication with emergency personnel.

Surveillance Systems

Visual surveillance systems use CCTV cameras mounted throughout a building to monitor problematic activity (Figure 42.4). Typical locations for surveillance cameras include building entrances, hallways, elevators, and exterior walks, as well as parking lots and garages. The cameras can be remote controlled to pan, tilt, and zoom for larger coverage areas. Small systems record information for review in case of an occurrence. Multiple cameras may be monitored continuously from a central command post, where information is both recorded and

Figure 42.4 Security cameras provide a wide coverage and record the activities within their area of surveillance.

stored. Emergency voice communications equipment is often incorporated to allow emergency personnel to remain in contact with the central control center.

Access Control Systems

Access control systems play a vital part in the security of many buildings by restricting access where required and monitoring the movement of staff and visitors. The signal systems used require a source, equipment that processes and transmits a signal, and a means of activating the signal through motive devices. Signals are typically conveyed via low voltage wiring through a variety of sensors and may be activated by keypad or electronic access cards. Modern buildings are increasingly using an integrated-systems design approach that incorporates security with other operational functions into a building automation system.

A number of means are used to provide access control to a building. Locks may be operated manually or controlled electronically and programmed to permit or deny access on a timed schedule. Multi-family apartment buildings incorporate access control within the doorbell system. A two-way intercom provides communication from the access point to the apartment with a pushbutton activating the lock. In very large occupancies, the doorbell is replaced with an electronic tenant list and push-button telephone. For office, hotel, or high-risk occupancies, electronic coded key cards are used to gain access and monitor building occupancy (Figure 42.5). The card transmits a signal that either opens a physical barrier or closes one in case of unauthorized access.

Intrusion detectors trigger an alarm when someone enters an unauthorized area. When an alarm is triggered, it sounds an auditory alarm, activates lights, and transmits information to the central control system where an operator can alert security staff. The most common detectors use magnetic devices mounted on doors and windows that transmit a signal when opened.

Electric through-wire hinges use conductors that are threaded through the hinge and direct electricity from the doorframe to power electric locks or enable closure from monitor switches in the door or locking hardware. Concealed magnetic contact hinges conceal a magnetic contact switch in the leaves of a full-mortise hinge. When the door is opened, the leaves are spread apart, breaking or making the contact. Electrified hinges allow for door monitoring, indicating whether a door is open or closed. New door systems include an identification and locking system for digital access control that uses transponders. A transponder is a wireless communications, monitoring, and controlling a device that picks up and automatically responds to an incoming signal.

Figure 42.5 Electronic coded key cards are used to gain access and monitor building occupancy.

© Grandpa/Shutterstock.com

Photoelectric devices use either light or infrared beams sent from a source to a receiver. When the beam of light is interrupted, the receiver signals the alarm to activate. Motion detectors use ultrasonic frequencies and microwaves to detect movement. Based on the Doppler Effect, the device detects changes in the frequency of waves reflected from a moving object. *Audio sensors* use controls that activate an alarm when a sound is detected.

One frequently used security system is found in banks and other businesses where valuable materials are present. It uses audio sensors, waterproof outlets, and an alarm control cabinet, which is usually located in the vault. *Alarm actuators*, controls that sense a particular activity and signal an alarm, are located at the teller and drive-in windows, and at the employees' desks, and surveillance cameras are placed in strategic locations.

ELECTRONIC MONITORING AND CONTROL

Large modern buildings typically incorporate a central control facility that combines the supervision and control of various building functions. Known as *building automation systems (BAS)*, these computerized controls are able to process data from remote systems and calculate operational responses to optimize building performance. Fire protection, HVAC and electrical and lighting controls, elevator functions, communications, and security systems can all be monitored and regulated from this central command. Energy management control systems that optimize HVAC functions that shut off systems when they are not needed, and cycle electrical loads are often integrated. The control center is staffed by trained personnel and includes automatic controls that respond electronically to preset signals (**Figure 42.6**).

Figure 42.6 A building automation control center must be staffed by trained personnel.

© ERproductions Ltd/Blend/Corbis

The core missions of a BAS system includes keeping the building climate within a specified range, providing lighting based on established schedules, and monitoring system performance and device failures. Most of the automation system is installed as hardware devices mounted to equipment or hidden in walls, floors, or ceilings. Various types of sensors provide input data that is sent to controllers that manage equipment in the network. The data is processed and email or text notifications are transmitted to building engineering staff.

Lighting can be turned on and off with a building automation system based on time of day or occupancy sensors and timers. The monitoring system adjusts according to conditions and reduces building energy and maintenance costs when compared to a non-controlled building. A building controlled by a BAS is often referred to as an intelligent building system.

Building automation systems are increasingly being powered by direct digital control (DDC) devices. The BACnet protocol, developed by ASHREA, provides mechanisms for computerized building automation devices to exchange information, regardless of the building service they perform.

Construction Techniques

Home Security Systems

The use of home security systems is increasing rapidly. A basic system uses sensors on doors and windows to activate an alarm. The system is armed and disarmed by a keypad located near the door that is most frequently used for entry by the occupant (Figure A). When the system is armed and an intruder opens a door or window, the alarm is triggered. This may simply be an alarm inside and outside the building, or the system may be connected to a central monitoring station that verifies if it is a true entry and notifies local police. Another system is wireless with a central control unit. Each door or window has a battery-powered transmitter that sends a signal to the base station, which sounds the alarm. The system has a keypad that can be carried about the building.

Many types of protection sensors are available. Exterior sensors provide protection around the perimeter of a building. These can include pressure detectors that sense vehicles entering the driveway, and photoelectric beams and motion detectors that can sense a person in the yard. Within the building, a system can have alarms on window screens that sound when the screen is cut, glass breakage detectors, and sensors on doors and windows that activate when they are opened, as well as interior motion sensors (Figure B).

Figure B A passive infrared (PIR) motion detector.

Figure A A close up of a security key pad.

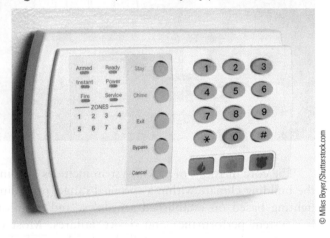

TELECOMUNICATIONS AND DATA SYSTEMS

Like electrical service, telecommunications and data systems involve a building service entry, and a distribution network that is run within cables that are surface mounted or hidden in wall, floor, and ceiling assemblies (Figure 42.7). Historically, telephone and data systems were installed by the providing utility. Today, these systems are run by the owner during the construction process. The dividing line that determines who is responsible for installation and maintenance of wiring and equipment is known as the demarcation point. In the United States and Canada, the demarcation point is the network interface device, normally mounted at the building exterior for easy access by

Figure 42.7 Telecommunications and data networks are run within cables that are surface mounted or hidden in wall, floor and ceiling assemblies.

the service provider. The incoming service is controlled at a panel or telecommunications closet. In residential occupancies, communication lines typically follow the route of the electric service, entering the building overhead or underground. In large buildings, telecommunications rooms are required to provide space for cable termination hardware and transport electronics.

The system of raceways, boxes, and outlets dedicated to communications systems of all sorts, is known as premise wiring. *Premise wiring* is the cabling, connectors, and accessories that are used to connect LAN (Local Area Network) and phone equipment within a commercial building. Premise wiring is made up of vertical and horizontal cable run from a central location throughout buildings to individual outlets. The transmission lines can be metallic, usually copper, or optical fiber. Most premise wiring is surface mounted to allow for frequent access.

Review Questions

1. What type of security units are used in areas requiring a very secure situation?

2. Where are fire flame detectors used?

3. What are the most commonly used type of fire alarms?

4. How are occupants alerted to the presence of carbon monoxide and natural and propane gas?

5. How can people entering the property near a building be detected?

6. Why are flashing light warning units used?

Key Terms

Alarm Actuators

Audio Sensors

Building Automation System (BAS)

Fire Flame Detector

Premise Wiring

Activities

1. Arrange a visit to a local business that has a security system. Ask to have security personnel demonstrate different system operations.

2. If there is a security firm in the area, try to arrange to have a sales representative visit the class and explain the goals of various systems and show examples of the detection and alarms available.

Additional Resources

Sweets Construction Catalog, Architects, Engineers, and Contractors Edition, Division 13, McGraw-Hill Construction, Two Penn Plaza, New York, NY 10121.

See Appendix C for addresses of professional and trade organizations and other sources of technical information.

The detailed divisions and level-two titles in the 2014 edition of *MasterFormat*®, published jointly by the Construction Specifications Institute (CSI), Alexandria, Virginia, sections of the 2014 *MasterFormat*® are shown below.

LEVEL 2 NUMBERS AND TITLES

This listing is intended to provide a useful map to the arrangement of the *MasterFormat* Numbers and Titles. A complete listing of all Numbers and Titles, with explanations and additional information, immediately follows this abbreviated listing.

PROCUREMENT AND CONTRACTING REQUIREMENTS GROUP

DIVISION 00—PROCUREMENT AND CONTRACTING REQUIREMENTS

INTRODUCTORY INFORMATION

PROCUREMENT REQUIREMENTS

00 10 00	**Solicitation**
00 11 00	Advertisements and Invitations
00 20 00	**Instructions for Procurement**
00 21 00	Instructions
00 22 00	Supplementary Instructions
00 23 00	Procurement Definitions
00 24 00	Procurement Scopes
00 25 00	Procurement Meetings
00 26 00	Procurement Substitution Procedures
00 30 00	**Available Information**
00 31 00	Available Project Information
00 40 00	**Procurement Forms and Supplements**
00 41 00	Bid Forms
00 42 00	Proposal Forms
00 43 00	Procurement Form Supplements
00 45 00	Representations and Certifications

CONTRACTING REQUIREMENTS

00 50 00	**Contracting Forms and Supplements**
00 51 00	Notice of Award
00 52 00	Agreement Forms
00 54 00	Agreement Form Supplements
00 55 00	Notice to Proceed
00 60 00	**Project Forms**
00 61 00	Bond Forms
00 62 00	Certificates and Other Forms
00 63 00	Clarification and Modification Forms
00 65 00	Closeout Forms
00 70 00	**Conditions of the Contract**
00 71 00	Contracting Definitions
00 72 00	General Conditions
00 73 00	Supplementary Conditions
00 80 00	*Unassigned*
00 90 00	**Revisions, Clarifications, and Modifications**
00 91 00	Precontract Revisions
00 93 00	Record Clarifications and Proposals
00 94 00	Record Modifications

SPECIFICATIONS GROUP

GENERAL REQUIREMENTS SUBGROUP

DIVISION 01—GENERAL REQUIREMENTS

FACILITY CONSTRUCTION SUBGROUP

DIVISION 02—EXISTING CONDITIONS

DIVISION 03—CONCRETE

03 70 00   **Mass Concrete**

03 71 00   Mass Concrete for Raft Foundations

03 72 00   Mass Concrete for Dams

03 80 00   **Concrete Cutting and Boring**

03 81 00   Concrete Cutting

03 82 00   Concrete Boring

03 90 00   *Unassigned*

DIVISION 04—MASONRY

04 01 00   Maintenance of Masonry

04 03 00   Conservation Treatment for Period Masonry

04 05 00   Common Work Results for Masonry

04 06 00   Schedules for Masonry

04 08 00   Commissioning of Masonry

04 10 00   *Unassigned*

04 20 00   **Unit Masonry**

04 21 00   Clay Unit Masonry

04 22 00   Concrete Unit Masonry

04 23 00   Glass Unit Masonry

04 24 00   Adobe Unit Masonry

04 25 00   Unit Masonry Panels

04 26 00   Single-Wythe Unit Masonry

04 27 00   Multiple-Wythe Unit Masonry

04 28 00   Concrete Form Masonry Units

04 29 00   Engineered Unit Masonry

04 30 00   *Unassigned*

04 40 00   **Stone Assemblies**

04 41 00   Dry-Placed Stone

04 42 00   Exterior Stone Cladding

04 43 00   Stone Masonry

04 50 00   **Refractory Masonry**

04 51 00   Flue Liner Masonry

04 52 00   Combustion Chamber Masonry

04 53 00   Castable Refractory Masonry

04 54 00   Refractory Brick Masonry

04 57 00   Masonry Fireplaces

04 60 00   **Corrosion-Resistant Masonry**

04 61 00   Chemical-Resistant Brick Masonry

04 62 00   Vitrified Clay Liner Plate

04 70 00   **Manufactured Masonry**

04 71 00   Manufactured Brick Masonry

04 72 00   Cast Stone Masonry

04 73 00   Manufactured Stone Masonry

04 80 00 to 04 90 00   *Unassigned*

DIVISION 05—METALS

05 01 00   Maintenance of Metals

05 03 00   Conservation Treatment for Period Metals

05 05 00   Common Work Results for Metals

05 06 00   Schedules for Metals

05 08 00   Commissioning of Metals

05 10 00   **Structural Metal Framing**

05 12 00   Structural Steel Framing

05 13 00   Structural Stainless-Steel Framing

05 14 00   Structural Aluminum Framing

05 15 00   Wire Rope Assemblies

05 16 00   Structural Cabling

05 17 00   Structural Rod Assemblies

05 19 00   Tension Rod and Cable Truss Assemblies

05 20 00   **Metal Joists**

05 21 00   Steel Joist Framing

05 25 00   Aluminum Joist Framing

05 30 00   **Metal Decking**

05 31 00   Steel Decking

05 33 00   Aluminum Decking

05 34 00   Acoustical Metal Decking

05 35 00   Raceway Decking Assemblies

05 36 00   Composite Metal Decking

05 40 00   **Cold-Formed Metal Framing**

05 41 00   Structural Metal Stud Framing

05 42 00   Cold-Formed Metal Joist Framing

05 43 00   Slotted Channel Framing

05 44 00   Cold-Formed Metal Trusses

05 45 00   Metal Support Assemblies

05 50 00   **Metal Fabrications**

05 51 00   Metal Stairs

05 52 00   Metal Railings

05 53 00   Metal Gratings

05 54 00   Metal Floor Plates

05 55 00   Metal Stair Treads and Nosings

05 56 00	Metal Castings		06 43 00	Wood Stairs and Railings
05 58 00	Formed Metal Fabrications		06 44 00	Ornamental Woodwork
05 59 00	Metal Specialties		06 46 00	Wood Trim
05 60 00	*Unassigned*		06 48 00	Wood Frames
05 70 00	**Decorative Metal**		06 49 00	Wood Screens and Exterior Wood Shutters
05 71 00	Decorative Metal Stairs		**06 50 00**	**Structural Plastics**
05 73 00	Decorative Metal Railings		06 51 00	Structural Plastic Shapes and Plates
05 74 00	Decorative Metal Castings		06 52 00	Plastic Structural Assemblies
05 75 00	Decorative Formed Metal		06 53 00	Plastic Decking
05 76 00	Decorative Forged Metal		**06 60 00**	**Plastic Fabrications**
05 77 00	Decorative Extruded Metal		06 61 00	Cast Polymer Fabrications
05 80 00 to 05 90 00	*Unassigned*		06 63 00	Plastic Railings
			06 64 00	Plastic Paneling
			06 65 00	Plastic Trim
			06 66 00	Custom Ornamental Simulated Woodwork

DIVISION 06—WOOD, PLASTICS, AND COMPOSITES

06 01 00	Maintenance of Wood, Plastics, and Composites		**06 70 00**	**Structural Composites**
06 03 00	Conservation Treatment for Period Wood		06 71 00	Structural Composite Shapes and Plates
06 05 00	Common Work Results for Wood, Plastics, and Composites		06 72 00	Composite Structural Assemblies
			06 73 00	Composite Decking
06 06 00	Schedules for Wood, Plastics, and Composites		06 74 00	Composite Gratings
			06 80 00	**Composite Fabrications**
06 08 00	Commissioning of Wood, Plastics, and Composites		06 81 00	Composite Railings
			06 82 00	Composite Trim
06 10 00	**Rough Carpentry**		06 83 00	Composite Paneling
06 11 00	Wood Framing		*06 90 00*	*Unassigned*
06 12 00	Structural Panels			
06 13 00	Heavy Timber Construction			
06 14 00	Treated Wood Foundations			
06 15 00	Wood Decking			
06 16 00	Sheathing			
06 17 00	Shop-Fabricated Structural Wood			
06 18 00	Glued-Laminated Construction			

DIVISION 07—THERMAL AND MOISTURE PROTECTION

06 20 00	**Finish Carpentry**		07 01 00	Operation and Maintenance of Thermal and Moisture Protection
06 22 00	Millwork			
06 25 00	Prefinished Paneling		07 03 00	Conservation Treatment for Period Roofing
06 26 00	Board Paneling			
06 30 00	*Unassigned*		07 05 00	Common Work Results for Thermal and Moisture Protection
06 40 00	**Architectural Woodwork**			
06 41 00	Architectural Wood Casework		07 06 00	Schedules for Thermal and Moisture Protection
06 42 00	Wood Paneling			
			07 08 00	Commissioning of Thermal and Moisture Protection
			07 10 00	**Dampproofing and Waterproofing**
			07 11 00	Dampproofing

07 12 00	Built-Up Bituminous Waterproofing
07 13 00	Sheet Waterproofing
07 14 00	Fluid-Applied Waterproofing
07 15 00	Sheet Metal Waterproofing
07 16 00	Cementitious and Reactive Waterproofing
07 17 00	Bentonite Waterproofing
07 18 00	Traffic Coatings
07 19 00	Water Repellents
07 20 00	**Thermal Protection**
07 21 00	Thermal Insulation
07 22 00	Roof and Deck Insulation
07 24 00	Exterior Insulation and Finish Systems
07 25 00	**Weather Barriers**
07 26 00	Vapor Retarders
07 27 00	Air Barriers
07 30 00	**Steep Slope Roofing**
07 31 00	Shingles and Shakes
07 32 00	Roof Tiles
07 33 00	Natural Roof Coverings
07 40 00	**Roofing and Siding Panels**
07 41 00	Roof Panels
07 42 00	Wall Panels
07 44 00	Faced Panels
07 46 00	Siding
07 50 00	**Membrane Roofing**
07 51 00	Built-Up Bituminous Roofing
07 52 00	Modified Bituminous Membrane Roofing
07 53 00	Elastomeric Membrane Roofing
07 54 00	Thermoplastic Membrane Roofing
07 55 00	Protected Membrane Roofing
07 56 00	Fluid-Applied Roofing
07 57 00	Coated Foamed Roofing
07 58 00	Roll Roofing
07 60 00	**Flashing and Sheet Metal**
07 61 00	Sheet Metal Roofing
07 62 00	Sheet Metal Flashing and Trim
07 63 00	Sheet Metal Roofing Specialties
07 64 00	Sheet Metal Wall Cladding
07 65 00	Flexible Flashing
07 70 00	**Roof and Wall Specialties and Accessories**
07 71 00	Roof Specialties

07 72 00	Roof Accessories
07 76 00	Roof Pavers
07 77 00	Wall Specialties
07 80 00	**Fire and Smoke Protection**
07 81 00	Applied Fireproofing
07 82 00	Board Fireproofing
07 84 00	Firestopping
07 86 00	Smoke Seals
07 87 00	Smoke Containment Barriers
07 90 00	**Joint Protection**
07 91 00	Preformed Joint Seals
07 92 00	Joint Sealants
07 95 00	Expansion Control

DIVISION 08—OPENINGS

08 01 00	Operation and Maintenance of Openings
08 03 00	Conservation Treatment for Period Openings
08 05 00	Common Work Results for Openings
08 06 00	Schedules for Openings
08 08 00	Commissioning of Openings
08 10 00	**Doors and Frames**
08 11 00	Metal Doors and Frames
08 12 00	Metal Frames
08 13 00	Metal Doors
08 14 00	Wood Doors
08 15 00	Plastic Doors
08 16 00	Composite Doors
08 17 00	Integrated Door Opening Assemblies
08 20 00	*Unassigned*
08 30 00	**Specialty Doors and Frames**
08 31 00	Access Doors and Panels
08 32 00	Sliding Glass Doors
08 33 00	Coiling Doors and Grilles
08 34 00	Special Function Doors
08 35 00	Folding Doors and Grilles
08 36 00	Panel Doors
08 38 00	Traffic Doors
08 39 00	Pressure-Resistant Doors
08 40 00	**Entrances, Storefronts, and Curtain Walls**
08 41 00	Entrances and Storefronts

DIVISION 10—SPECIALTIES

DIVISION 11—EQUIPMENT

11 32 00	Unit Kitchens
11 40 00	**Foodservice Equipment**
11 41 00	Foodservice Storage Equipment
11 42 00	Food Preparation Equipment
11 43 00	Food Delivery Carts and Conveyors
11 44 00	Food Cooking Equipment
11 46 00	Food Dispensing Equipment
11 47 00	Ice Machines
11 48 00	Foodservice Cleaning and Disposal Equipment
11 50 00	**Educational and Scientific Equipment**
11 51 00	Library Equipment
11 52 00	Audio-Visual Equipment
11 53 00	Laboratory Equipment
11 55 00	Planetarium Equipment
11 56 00	Observatory Equipment
11 57 00	Vocational Shop Equipment
11 59 00	Exhibit Equipment
11 60 00	**Entertainment and Recreation Equipment**
11 61 00	Broadcast, Theater, and Stage Equipment
11 62 00	Musical Equipment
11 66 00	Athletic Equipment
11 67 00	Recreational Equipment
11 68 00	Play Field Equipment and Structures
11 70 00	**Healthcare Equipment**
11 71 00	Medical Sterilizing Equipment
11 72 00	Examination and Treatment Equipment
11 73 00	Patient Care Equipment
11 74 00	Dental Equipment
11 75 00	Optical Equipment
11 76 00	Operating Room Equipment
11 77 00	Radiology Equipment
11 78 00	Mortuary Equipment
11 79 00	Therapy Equipment
11 80 00	**Facility Maintenance and Operation Equipment**
11 81 00	Facility Maintenance Equipment
11 82 00	Facility Solid Waste Handling Equipment
11 90 00	**Other Equipment**
11 91 00	Religious Equipment

11 92 00	Agricultural Equipment
11 93 00	Horticultural Equipment
11 94 00	Veterinary Equipment
11 95 00	Arts and Crafts Equipment
11 97 00	Security Equipment
11 98 00	Detention Equipment

DIVISION 12—FURNISHINGS

12 01 00	Operation and Maintenance of Furnishings
12 05 00	Common Work Results for Furnishings
12 06 00	Schedules for Furnishings
12 08 00	Commissioning of Furnishings
12 10 00	**Art**
12 11 00	Murals
12 12 00	Wall Decorations
12 14 00	Sculptures
12 17 00	Art Glass
12 19 00	Religious Art
12 20 00	**Window Treatments**
12 21 00	Window Blinds
12 22 00	Curtains and Drapes
12 23 00	Interior Shutters
12 24 00	Window Shades
12 25 00	Window Treatment Operating Hardware
12 26 00	Interior Daylighting Devices
12 30 00	**Casework**
12 31 00	Manufactured Metal Casework
12 32 00	Manufactured Wood Casework
12 34 00	Manufactured Plastic Casework
12 35 00	Specialty Casework
12 36 00	Countertops
12 40 00	**Furnishings and Accessories**
12 41 00	Office Accessories
12 42 00	Table Accessories
12 43 00	Portable Lamps
12 44 00	Bath Furnishings
12 45 00	Bedroom Furnishings
12 46 00	Furnishing Accessories

DIVISION 13—SPECIAL CONSTRUCTION

DIVISION 14—CONVEYING EQUIPMENT

FACILITY SERVICES SUBGROUP

DIVISION 21—FIRE SUPPRESSION

DIVISION 22—PLUMBING

DIVISION 23—HEATING, VENTILATING, AND AIR CONDITIONING (HVAC)

DIVISION 25—INTEGRATED AUTOMATION

DIVISION 28—ELECTRICONIC SAFETY AND SECURITY

SITE AND INFRASTRUCTURE SUBGROUP

Division 30 Reserved for Future Expansion

DIVISION 31—EARTHWORK

DIVISION 32—EXTERIOR IMPROVEMENTS

DIVISION 33—UTILITIES

DIVISION 34—TRANSPORTATION

DIVISION 35—WATERWAY AND MARINE CONSTRUCTION

Divisions 36 to 39 Reserved for Future Expansion

PROCESS EQUIPMENT SUBGROUP

DIVISION 40—PROCESS INTERCONNECTIONS

DIVISION 42—PROCESSING HEATING, COOLING, AND DRYING EQUIPMENT

DIVISION 43—PROCESS GAS AND LIQUID HANDLING, PURIFICATION, AND STORAGE EQUIPMENT

DIVISION 44—POLLUTION AND WASTE CONTROL EQUIPMENT

44 20 00	**Noise Pollution Control**	45 21 00	Leather and Allied Product Manufacturing Equipment
44 21 00	Noise Pollution Control Equipment	45 23 00	Wood Product Manufacturing Equipment
44 30 00	**Odor Control**	45 25 00	Paper Manufacturing Equipment
44 31 00	Odor Treatment Equipment	45 27 00	Printing and Related Manufacturing Equipment
44 32 00	Odor Dispersing and Masking/Counteracting Equipment	45 29 00	Petroleum and Coal Products Manufacturing Equipment
44 40 00	**Water Pollution Control Equipment**	45 31 00	Chemical Manufacturing Equipment
44 41 00	Water Pollution Containment and Cleanup Equipment	45 33 00	Plastics and Rubber Manufacturing Equipment
44 50 00	**Solid Waste Control and Reuse**	45 35 00	Nonmetallic Mineral Product Manufacturing Equipment
44 51 00	Solid Waste Collection, Transfer, and Hauling Equipment	45 37 00	Primary Metal Manufacturing Equipment
44 53 00	Solid Waste Processing Equipment	45 39 00	Fabricated Metal Product Manufacturing Equipment
44 55 00	Composting Equipment	45 41 00	Machinery Manufacturing Equipment
44 60 00	**Waste Thermal Processing Equipment**	45 43 00	Computer and Electronic Product Manufacturing Equipment
44 61 00	Waste-to-Energy Plants	45 45 00	Electrical Equipment, Appliance, and Component Manufacturing Equipment
44 62 00	Fluidized Bed Combustion Equipment	45 47 00	Transportation Manufacturing Equipment
44 63 00	Rotary Kiln Incinerators		
44 64 00	Gasification Equipment	45 49 00	Furniture and Related Product Manufacturing Equipment
44 65 00	Pyrolysis Equipment	45 51 00	Other Manufacturing Equipment
44 66 00	Hazardous Waste and Medical Waste Incinerators	*45 60 00 to 45 90 00 Unassigned*	
44 67 00	Heat Recovery Equipment for Waste Thermal Processing		
44 68 00	Synthesis Gas Cleanup and Handling Equipment		

44 70 00 to 44 90 00   Unassigned

DIVISION 45—INDUSTRY-SPECIFIC MANUFACTURING EQUIPMENT

45 05 00	Common Work Results for Industry-Specific Manufacturing Equipment
45 08 00	Commissioning of Industry-Specific Manufacturing Equipment
45 11 00	Oil and Gas Extraction Equipment
45 13 00	Mining Machinery and Equipment
45 15 00	Food Manufacturing Equipment
45 17 00	Beverage and Tobacco Manufacturing Equipment
45 19 00	Textiles and Apparel Manufacturing Equipment

DIVISION 46—INDUSTRY-SPECIFIC MANUFACTURING EQUIPMENT

46 01 00	Operation and Maintenance of Water and Wastewater Equipment
46 05 00	Common Work Results for Water and Wastewater Equipment
46 06 00	Schedules for Water and Wastewater Equipment
46 07 00	Packaged Water and Wastewater Treatment Equipment
46 08 00	Commissioning of Water and Wastewater Equipment
46 10 00	*Unassigned*

DIVISION 48—ELECTRICAL POWER GENERATION

The *MasterFormat®* Groups, Subgroups, and Divisions and cover used in this book are from *MasterFormat®*, published by CSI and Construction Specifications Canada (CSC), and are used with permission from CSI. For those interested in a more in-depth explanation of *MasterFormat®* and its use in the construction industry visit www .csinet.org/masterformat or contact:

CSI

110 South Union Street, Suite 100

Alexandria, VA 22314

800-689-2900: 703-684-0300

www.csinet.org

MASTERFORMAT

The detailed divisions and level-two titles in the 2014 edition of MasterFormat™, published jointly by the Construction Specifications Institute (CSI), Alexandria, Virginia, sections of the 2014 MasterFormat™ are shown below.

DIVISION 1

GENERAL REQUIREMENTS

Following is descriptive information about the contents under each level-two title in General Requirements.

SUMMARY

The first level-two title is Summary, which includes Summary of Work, Multiple Contract Summary, Work Restrictions, and Project Utility Sources.

- Summary of Work identifies the work to be covered in a single contract and describes provisions for future construction. It may include work by an owner, work covered in the contract documents, and specific things, such as products ordered in advance, owner-furnished and installed products, and the use of salvaged material and products.

- Multiple Contract Summary describes the construction delivered under *more* than one construction contract, such as a construction management contract and a multiple-prime contract. It includes the sequence of construction, contract interface, any construction by the owner, and a summary of the contracts.

- Work Restrictions pertains to restrictions that affect construction operations, such as the location of construction, occupancy requirements, and the use of the building premises and site during construction. If the construction involves a building already occupied, procedures for coordination with the occupants are detailed.

- Project Utility Sources includes the identity of utility companies that are to provide permanent services to the project.

PRICE AND PAYMENT PROCEDURES

The second level-two title is Price and Payment Procedures. It details the allowance-adjusting procedures for cash and quantity allowances for products, installation, testing, and contingencies.

- Allowances includes the adjusting procedures for cash and quantity allowances for products, installation, inspection, and testing contingencies.

- Alternates provides for the submission and acceptance procedures for alternate bids, including, when desired, a list and description of each alternate.

- Value Analysis includes the procedures and submittal requirements for value analysis, value engineering, application for consideration, and consideration of proposals.

- Contraction Modification Procedures details the procedures for making clarifications and proposals for change and for changing the contract.

- Unit Prices establishes the procedures associated with unit prices and measurement and payment. It may include a list and descriptions of the actual unit price items.

- Payment Procedures describes the procedures for submitting schedules of values and applications for payment.

ADMINISTRATIVE REQUIREMENTS

The third level-two title, Administrative Requirements, encompasses the areas of Project Management and Coordination, Construction Progress Documentation, Submittal Procedures, and Special Procedures.

- Project Management and Coordination includes the administration of subcontractors and coordination with other contractors and the owner. This includes project meeting and on-site administration.

- Construction Progress Documentation covers the requirements for scheduling, recording, and reporting progress. This includes such things as construction photographs, progress reports, site observation, purchase order tracking, scheduling of construction, and survey and layout data.

- Submittal Procedures includes the general procedures and requirements for submittals during the course of the construction. This can include items such as certificates, design data, field test reports, shop drawings, product data and samples, and source quality control reports.

- Special Procedures includes any special procedures that any of the project situations require, such as historic restoration, renovation or alteration; preservation; or hazardous material abatement.

QUALITY REQUIREMENTS

The fourth level-two title of Division 1 is Quality Requirements. This covers four areas: Regulatory Requirements, References, Quality Assurance, and Quality Control.

- Regulatory Requirements provides information necessary for conformance to requirements such as building codes, mechanical codes, electrical codes, and other regulations and to fee payments applicable to a project.

- References include lists of reference standards cited in the contract documents and the organizations whose standards are cited. This includes items such as symbols used, abbreviations, and acronyms.

- Quality Assurance references are used to provide for procedures to ensure the quality of construction. This includes field observations and tests performed by manufacturers' representatives during installation. Quality assurance is provided by fabricators, installers, manufacturers, suppliers, and testing agencies.

- Quality Control establishes procedures to measure and report the quality and performance of construction. It may require field samples and mockups assembled at a site. The contractor provides quality control plans. Control can be provided by inspections, inspection services, and testing laboratories.

TEMPORARY FACILITIES AND CONTROLS

The fifth level-two title, Temporary Facilities and Controls, provides requirements for installation, maintenance, and removal of temporary utilities, controls, facilities, and construction aids during construction. This includes Temporary Utilities, Construction Facilities, Temporary Construction, Construction Aids, Vehicular Access and Parking, Temporary Barriers and Enclosures, Temporary Controls, and Project Identification.

- Temporary Utilities includes all such services used during construction, such as electrical, water, lighting, gas, telephone, fire protection, fuel oil, gasoline, and diesel fuel.

- Construction Facilities includes any temporary facilities built on-site for use during construction. Typical examples include field offices, storage buildings, sanitary facilities, and first aid stations.

- Temporary Construction includes those facilities built to provide access to various parts of a site and project so as to facilitate the construction process or to accommodate the needs of the owner and occupants. Typical examples include temporary bridges, ramps, overpasses, decking, and turnarounds.

- Construction Aids include all requirements and procedures related to tools and equipment used during construction, such as scaffolding, cranes, hoists, and construction elevators.

- Vehicular Access and Parking provides requirements for and procedures related to access to the site and parking facilities to meet needs of the construction and owner operations. Typically this can include access roads, haul routes, parking areas, temporary roads, and control of traffic.

- Temporary Barriers and Enclosures includes all facilities and procedures for the protection of the occupants or existing spaces during construction. This can include air barriers, barricades, dust barriers, fences, noise barriers, pollution control, security, and the protection of trees and other vegetation.

- Temporary Controls includes site or environment controls required to allow construction to proceed, including such things as erosion and sediment control and pest control.

- Project Identification includes any signs used to identify the construction site and areas on the site.

PRODUCT REQUIREMENTS

The sixth level-two title, Product Requirements, includes a comprehensive series of requirements.

- Basic Product Requirements includes the basic requirements for new, salvaged, and reused products used in construction.

- Product Options includes the basic requirements for options the contractor may have in selecting products and how the determination is made for equal products.

- Product Substitution Procedures includes the basic requirements and procedures when proposals for the substitution of a product are made.

- Owner-Furnished Products includes the basic requirements for products that are to be furnished by the owner. This also involves scheduling, coordinating, handling, and storing owner-furnished products.

- Product Delivery Requirements specifies the basic requirements for packing, shipping, delivery, and acceptance of products at the site.

- Product Storage and Handling Requirements sets forth the basic requirements for storing and handling products on-site.

EXECUTION AND CLOSEOUT REQUIREMENTS

The seventh level-two title, Execution and Closeout Requirements, has eight sections.

- Examination involves the acceptance of conditions, existing conditions, and the basic requirements for determining acceptable conditions for installation.

- Preparation details the requirements for preparing to install, erect, or apply products, including activities such as field engineering, protection of adjacent construction, surveying, and construction layout.

- Execution includes the basic requirements for installing, applying, or erecting products that are new, prepurchased, salvaged, or owner furnished.

- Cleaning sets the requirements for maintaining the site in a neat condition during construction and the final cleaning in preparation for turning the project over to the owner.

- Starting and Adjusting involves establishing the initial checkout and startup procedures and any adjustments needed to ensure safe operation during the acceptance testing and commissioning.

- Protecting Installed Construction provides the requirements and procedures for protecting installed construction.

- Closeout Procedures includes the administrative procedures for substantial completion and final completion of the work.

- Closeout Submittals sets the procedures for closeout submittals, revised project documents, and delivery and distribution of spare parts and maintenance materials.

PERFORMANCE REQUIREMENTS

The eighth level-two title, Performance Requirements, has to do with the final requirements for preparing the facility for decommissioning.

- Commissioning requirements include commissioning summary, system performance evaluation, testing, adjusting, and balancing procedures.

- Demonstration and Training includes the requirements and procedures for the demonstration of the products and systems within the facility, including training the owner's operating and maintenance personnel.

- Operation and Maintenance sets forth the requirements and procedures for operating the facility after commissioning.

- Reconstruction involves any renovation or reconstruction of the existing facilities that may be required.

Life-Cycle Activities

The last and ninth level-two title, Life-Cycle Activities, includes the basic requirements for deactivating a facility or a portion of it from operation. This includes activities such as facility demolition and removal; hazardous materials abatement, removal, and disposal; and protection of deactivated facilities.

Appendix C

U.S. AND CANADIAN PROFESSIONAL AND TECHNICAL ORGANIZATIONS

DIVISION 3—CONCRETE

American Concrete Institute
PO Box 9094
Farmington Hills, MI 48333-9094

Concrete Reinforcing Steel Institute
933 North Plum Grove Road
Schaumberg, IL 60173-4758
www.crsi.org

American Society of Concrete Contractors
38800 Country Club Drive
Farmington Hills, MI 48331-3411
www.ascconline.org

Architectural Precast Association, Inc.
6710 Winkler Road, Suite 8
Fort Meyers, FL 33919
www.archprecast.org

Cement Association of Canada
Headquarters at 1500-60 Queen Street
Ottawa, Ontario
Canada K1P 5Y7
www.cement.ca
Regional Offices in Bedford, NB; Montreal, QC; Toronto, ON; Vancouver, BC; and Calgary, AB

National Concrete Masonry Association
13750 Sunrise Valley Drive
Herndon, VA 20171-4662
www.ncma.org

National Precast Concrete Association
10333 N. Meridian Street, Suite 272
Indianapolis, IN 46290
www.precast.org

National Ready Mixed Concrete Association
900 Spring Street
Silver Spring, MD 20910
www.nrmca.org

Portland Cement Association
5420 Old Orchard Road
Skokie, IL 60077-1083
www.cement.org

Post-Tensioning Institute
1717 W. Northern Avenue, Suite 114
Phoenix, AZ 85071-5471
www.post-tensioning.org

Tilt-Up Concrete Association
PO Box 204
Mt. Vernon, IA 52314
www.tilt-up.org

DIVISION 4—MASONRY

Brick Industry Association
11490 Commerce Park Drive, Suite 300
Reston, VA 20191-1525
www.gobrick.com

Building Stone Institute
PO Box 507
Purdys, NY 10578
www.buildingstoneinstitute.org

Cast Stone Institute
10 W. Kimball Street
Winder, GA 30680
www.caststone.org

Indiana Limestone Institute of America
Stone City Bank Building, Suite 400
1502J L Street
Bedford, IN 47421
www.iliai.com

International Masonry Institute
53 W. Jackson Boulevard, Suite 308
Chicago, IL 60604

The Masonry Society
3970 Broadway, Suite 201-D
Boulder, CO 80304-1135
www.masonrysociety.org

Marble Institute of America
28901 Clemens Road, Suite 100
Cleveland, OH 44145
www.marble-institute.com

National Concrete Masonry Association
13750 Sunrise Valley Drive
Herndon, VA 20171
www.ncma.org

National Lime Association
200 N. Glebe Road, Suite 806
Arlington, VA 22203
www.lime.org

Tile Council of North America, Inc.
100 Clemson Research Boulevard
Anderson, SC 29625
www.tileusa.com

Western States Clay Products Association
2550 Beverly Boulevard
Los Angeles, CA 90057-1019
www.brick-wscpa.org

DIVISION 5—METALS

The Aluminum Association
1525 Wilson Road, Suite 600
Arlington, VA 22209
www.aluminum.org

Aluminum Anodizers Council
1000 N. Eand, Suite 214
Wauconda, IL 60084
www.anodizing.org

American Galvanizers Association
6881 S. Holly Circle, No. 108
Englewood, CO 80112
www.galvanizeit.org

Aluminum Extruders Council
1000 N. Rand Road, Suite 214
Wauconda, IL 60084
www.aec.org

American Institute of Steel Construction
1 East Wacker Drive, Suite 3100
Chicago, IL 60601
www.aisc.org

Association for Iron & Steel Technology
186 Thorn Hill Road
Warrendale, PA 15086-7528
www.aist.org

Copper Development Association
260 Madison Avenue, 16th Floor
New York, NY 10016
www.copper.org

Metal Building Manufacturers Association
1300 Summer Avenue
Cleveland, OH 44115-2851
www.mbma.com

Metal Building Contractors and Erectors Association
242 East Main Street
Louisville, KY 40202
www.mbcea.org

Metal Construction Association
4700 W. Lake Avenue
Glenview, IL 60025
www.metalconstruction.org

National Association of Architectural Metal
Manufacturers
8 S. Michigan Avenue, Suite 1000
Chicago, IL 60603
www.naamm.org

National Ornamental & Miscellaneous Metals
Association
532 Forest Parkway, Suite A
Forest Park, GA 30297
www.nomma.org

Sheet Metal and Air-Conditioning Contractors
National Association
4201 Lafayette Center Drive
Chantilly, VA 30151
www.smacna.org

Specialty Steel Industry of North America
3050 K Street NW, Suite 400
Washington, DC 20007
www.ssina.com

The Structural Engineering Institute of the American
Society of Civil Engineers
1801 Alexander Bell Drive
Reston, VA 20191-4400
www.constructioninst.org

Steel Framing Alliance
1201 15th Street NW, Suite 320
Washington, DC 20005-2842
www.steelframing.org

Steel Erectors Association of America
PO Box 4891
Chapel Hill, NC 27515-4891
www.seaa.net

Steel Stud Manufacturers Association
8 S. Michigan Avenue, Suite 1000
Chicago, IL 60603
www.ssma.com

Steel Joist Institute
3127 Mr. Joe White Avenue
Myrtle Beach, SC 28577-6760
www.steeljoist.org

Wire Reinforcement Institute
942 Main Street, Suite 300
Hartford, CT 06103
www.wirereinforcementinstitute.org

Wire Association International
PO Box 578
Guilford, CT 06437
www.wirenet.org

Wire Fabricators Association
710 E. Ogden Avenue 600
Naperville, IL 66583

DIVISION 6—WOOD, PLASTICS AND COMPOSITES

American Fiberboard Association
853 N. Quentin Road
Palatine, IL 60067

American Forest and Paper Association
1111 Nineteenth Street NW, Suite 800
Washington, DC 20036
www.afandpa.org

American Institute of Timber Construction
7012 South Revere Parkway, Suite 140
Englewood, CO 80112
www.aitc-glulam.org

American Lumber Standards Committee
PO Box 210
Germantown, MD 20875-0210
www.alsc.org

American Wood Council
1111 Nineteenth Street NW, Suite 800
Washington, DC 20036
www.awc.org

Architectural Woodwork Institute
46179 Westlake
Potomac Falls, VA 20165
www.awinet.org

American Wood Preservers Association
PO Box 388
Selma, AL 36702
www.awpa.com

APA—The Engineered Wood Association
7011 South 19th Street
Tacoma, WA 91411-0700
www.apawood.org

Building Products Information
320-979 de Bourgogne
Ste-Foy, Quebec
Canada G1W 2L4

California Redwood Association
405 Enfrente Drive, Suite 200
Novato, CA 94940
www.calredwood.org

Canadian Wood Council
1400 Blair Place, Suite 210
Ottawa, Ontario
Canada K1P 6B9
www.cwc.ca

Canadian Plywood Association
735 West 15th Street
North Vancouver, British Columbia
Canada V7M 1T2
www.canply.org

Canadian Particleboard Association
27 Goulburn Avenue
Ottawa, Ontario
Canada K1N 8C7

Canadian Hardwood Plywood Association
27 Goulburn Avenue
Ottawa, Ontario
Canada K1N 8C7

COFI Plywood Technical Centre
735 West 15th Street
North Vancouver, British Columbia
Canada V7M 1T2

Composite Panel Association
18928 Premiere Court
Gaithersburg, MD 20879-1569
www.pbmdf.com

Composite Wood Council
18922 Premiere Court
Gaithersburg, MD 20879-1574
www.pbmdf.com

The Construction Institute of the American Society of
Civil Engineers
1810 Alexander Bell Drive
Reston, VA 20191-4400
www.constructioninst.org

Cedar Shake and Shingle Bureau
PO Box 1176
Sumas, WA 98295-1178
www.cedarbureau.org

Forest Products Laboratory
One Gifford Pinchot Drive
Madison, WI 53726
www.fpl.fs.fed.us

Forest Products Society
2801 Marshall Court
Madison, WI 53705
www.forestprod.org

Forest Stewardship Council – U.S. (FSC-US)
212 Third Avenue North, Suite 280
Minneapolis, MN 55401
www.fscus.org

Hardwood Council
PO Box 525
Oakmont, PA 15139
www.americanhardwoods.org

Hardwood Manufacturers Association
400 Penn Center Boulevard, Suite 530
Pittsburgh, PA 15235
www.hardwoodinfo.com

Hardwood Plywood and Veneer Association
Laboratory and Testing Service
PO Box 2789
Reston, VA 20195-0789
www.hpva.org

Laminated Timber Institute of Canada
c/o Western Archrib Structures
PO Box 5648
Edmonton, Alberta
Canada T6C 4G1

Maple Flooring Manufacturers Association
60 Revere Drive, Suite 500
Northbrook, IL 60062
www.maplefloor.org

Modular Building Institute
413 Park Avenue
Charlottesville, VA 22902
www.mbinet.org

National Lumber Grades Authority
103-4400 Dominion Street
Burnaby, British Columbia
Canada K1N 8C7
www.nlga.org

National Hardwood Lumber Association
PO Box 34518
Memphis, TN 38184-0518
www.nhla.com

National Frame Building Association
4840 Bob Billings Parkway
Lawrence, KS 66049
www.nfba.org

Quebec Wood Export Bureau
979 de Bourgagne, Bureau 450
Ste-Foy, Quebec
Canada G1W 2L4
www.quebecwoodexport.com

Structural Board Association
412-45 Sheppard Avenue East
Willowdale, Ontario
Canada M2N 5W9

Southern Forest Products Association
PO Box 641700
Kenner, LA 70064-1700
www.sfpa.org

Southern Pine Association
PO Box 641700
Kenner, LA 70064-1700
www.southernpine.com

Southern Pine Council
PO Box 641700
Kenner, LA 70064-1700

Structural Board Association
454 Sheppard Avenue E, Suite 412
Willowdale, Ontario
Canada M2N 5W9
www.osbguide.com

Structural Insulated Panel Association
PO Box 1699
Gig Harbor, WA 98335
www.sips.org

Truss Plate Institute
218 North Lee Street, Suite 312
Alexandria, VA 22314
www.tpinst.org

Truss Plate Institute of Canada
c/o M. Tek Canada, Inc.
100 Industrial Road
Bradford, Ontario
Canada L3Z 3G7
www.tpic.ca

The Wood Flooring Manufacturers Association
PO Box 3009
Memphis, TN 38173-0009
www.nofma.org

Wood Moulding & Millwork Producers Association
507 First Street
Woodland, CA 95695-4025
www.wmmpa.com

Western Wood Products Association
522 SW Fifth Avenue, Suite 600
Portland, OR 97204-2127
www.wwpa.org

Wood Truss Council of America
6300 Enterprise Lane
Madison, WI 53719
www.woodtruss.com

PLASTICS ORGANIZATIONS

Association of Postconsumer Plastic Recyclers
2000 L Street NW, Suite 835
Washington, DC 20006
www.plasticsrecycling.org

Extruded Polystyrene Foam Association
1223 Dale Boulevard
Woodbridge, VA 22193
www.XPSA.com

Plastics Institute of America
333 Aiken Street
Lowell, MA 01854
www.plasticsinstitute.org

Society of Plastics Engineers
PO Box 403
Brookfield, CT 06804-0403
www.4spe.org

The Society of Plastics Industry, Inc.
1667 K Street NW, Suite 1000
Washington, DC 20006-1620
www.plasticsindustry.org

Vinyl Siding Institute
1801 K Street NW, Suite 600K
Washington, DC 20006
www.vinylsiding.org

DIVISION 7—THERMAL AND MOISTURE PROTECTION

Adhesives Manufacturers Association
401 N. Michigan Avenue, Suite 2400
Chicago, IL 60611-4267
www.adhesives.org/ama

Asphalt Institute
2696 Research Park Drive
Lexington, KY 40511-8480
www.asphaltinstitute.org

Asphalt Roofing Manufacturers Association
1156 15th Street NW
Washington, DC 20005
www.asphaltroofing.com

Cellulose Insulation Manufacturers Association
1368 Keowee Street
Dayton, OH 45402
www.cellulose.org

Extruded Polystyrene Foam Association
4223 Dale Boulevard
Woodbridge, VA 22193
www.xpsa.com

Insulation Contractors of America
1321 Duke Street, Suite 303
Alexandria, VA 22314
www.insulate.org

Midwest Roofing Contractors Association
3840 Bob Billings Parkway
Lawrence, KS 66049
www.mrca.org

North American Insulation Manufacturers
Association
44 Canal Center Plaza, Suite 310
Alexandria, VA 22314
www.naima.org

National Roofing Contractors Association
10255 W. Higgins Road, Suite 600
Rosemont, IL 60018
www.nrca.net

Roof Consultants Institute
7424 Chapel Hill Road
Raleigh, NC 27607
www.rci-online.org

Roof Tile Institute
PO Box 40337
Eugene, OR 97404-0049
www.rooftile.org

Sealant, Waterproofing, and Restoration Institute
14 W. Third Street, Suite 200
Kansas City, MO 64105
www.swrionline.org

Tile Roofing Institute
230 East Ohio, Suite 400
Chicago, IL 60611
www.tileroofing.org

Tile Roofing Institute Technical Office
PO Box 40337
Eugene, OR 97404-0048
www.tileroofing.org

Western States Roofing Contractors Association
1400 Marsten Road, Suite N
Burlingame, CA 94010-2422
www.wsrca.com

DIVISION 8—OPENINGS

American Architectural Manufacturers Association
1827 Walden Office Square, Suite 550
Schaumburg, IL 60173
www.aamanet.org

Builders Hardware Manufacturers Association
355 Lexington Avenue, 17th Floor
New York, NY 10017
www.buildershardware.com

Efficient Windows Collaborative
1850 M Street, NW, Suite 600
Washington, DC 20036
www.efficientwindows.org

Glass Association of North America
2945 SW Wanamaker Drive, Suite A
Topeka, KS 66614-5321
www.glasswebsite.com

Glass Technical Institute
12653 Portada Place
San Diego, CA 29130

Glazing Contractors Association
43636 Woodward Avenue
Bloomfield Hills, MI 48302

National Fenestration Rating Council
8484 Georgia Avenue, Suite 320
Silver Spring, MD 20910
www.nfrc.org

National Wood Window and Door Association
1400 E. Touhy Avenue, Suite 470
Des Plains, IL 60018-3337
www.wdma.com

Steel Window Institute
1300 Summer Avenue
Cleveland, OH 44145
www.steelwindows.com

DIVISION 9—FINISHES

Carpet and Rug Institute
PO Box 2048
Dalton, GA 30722
www.carpet-rug.com

Carpet Cushion Council
PO Box 546
Riverside, CT 06878
www.carpetcushion.org

Ceilings and Interior Systems Construction
Association
1500 Lincoln Highway, Suite 202
St. Charles, IL 60174
www.cisca.org

Foundation of the Wall and Ceiling Industry
803 W. Broad Street, Suite 600
Falls Church, VA 22046
www.awci.org

Green Seal
1001 Connecticut Avenue, NW Suite 827
Washington, DC
20036-5525 USA
www.greenseal.org

Gypsum Association
810 First Street NE, Suite 510
Washington, DC 20002
www.gypsum.org

International Institute for Lath and Plaster
3127 Los Fotir Boulevard
Los Angeles, CA 90039

National Oak Flooring Manufacturing Association
PO Box 3009
Memphis, TN 38103
www.nofma.org

National Paint and Coatings Association
1500 Rhode Island Avenue, NW
Washington, DC 20005
www.paint.org

National Terrazzo & Mosaic Association, Inc.
201 North Maple Avenue, Suite 208
Purcellville, VA 20132
www.ntma.com

National Tile Contractors Association
PO Box 13629
Jackson, MS 39236
www.tile-assn.com

National Wood Flooring Association
16388 Westwoods Business Plaza
Ellisville, MO 63021

Painting and Decorating Contractors of America
3913 Old Lee Highway
Fairfax, VA 22030-2401
www.pdca.com

Resilient Floor Covering Institute
401 E. Jefferson Street, Suite 102
Rockville, MD 20850

Rubber Manufacturers Association
1400 K Street NW, Suite 900
Washington, DC 20005
www.rma.org

Tile Council of America, Inc.
100 Clemson Research Boulevard
Anderson, SC 29625
www.tileusa.com

Wallcoverings Association
401 N. Michigan Avenue, Suite 2200
Chicago, IL 50611-4267
www.wallcoverings.org

THERE ARE NO REFERENCES FOR -DIVISION 10—SPECIALTIES AND DIVISION 11—EQUIPMENT

DIVISION 12—FURNISHINGS

American Society of Furniture Designers
144 Woodland Drive
New London, NC 28127
www.asfd.com

Business and Institutional Furniture Manufacturers
Association
2335 Burton Street NE
Grand Rapids, MI 49506

Contract Furnishings Council
1190 Merchandise Mart
Chicago, IL 60654

THERE ARE NO REFERENCES FOR DIVISION 13—SPECIAL CONSTRUCTION

DIVISION 14—CONVEYING EQUIPMENT

Conveyor Equipment Manufacturers Association
6724 Lone Oak Boulevard
Naples, FL 34109
www.cemanet.org

National Association of Elevator Contractors
1298 Wellbrook Circle NE, Suite A
Conyers, GA 30207-2872
www.naec.org

National Elevator Industry, Inc.
400 Frank W. Burr Boulevard
Teaneck, NJ 07666

THERE ARE NO DIVISIONS 15 THROUGH 20

DIVISION 21—FIRE SUPPRESSION

Building and Fire Research Laboratory
National Institute of Standards and Technology
100 Bureau Drive, MS 8600
Gaithersburg, MD 20899-8600
www.bfrl.nist.gov

Fire Suppression Systems Association
5024R Campbell Boulevard
Baltimore, MD 21236-5974
www.fssa.net

Fire Research Program
Institute for Research in Construction
Montreal Road Campus, M59
Ottawa, Ontario
Canada K1A 0R6
www.irc.nrc-cnrc.ga.ca/fr

Fire Equipment Manufacturers Association
1300 Summer Avenue
Cleveland, OH 44115-2851
www.yourfirstdefense.com

International Fire Code Institute
5360 S. Workman Mill Road
Whittier, CA 90501-22988
www.ifci.com

National Fire Protection Association
1 Batterymarch Park
Quincy, MA 02269-9101
www.nfpa.org

National Fire Laboratory Institute for Research in Construction
Montreal Road Campus
Ottawa, Ontario
Canada K1A 0R6

National Fire Sprinkler Association
40 Van Barrett Road
Patterson, NY 12563-1000
www.nfsa.org

Society of Fire Protection Engineers
7315 Wisconsin Avenue, Suite 620E
Bethesda, MD 20914
www.sfpe.org

DIVISION 22—PLUMBING

American Society of Mechanical Engineers
3 Park Avenue
New York, NY 10016-5990
www.asme.org

American Society of Plumbing Engineers
3617 Thousand Oaks Boulevard, Suite 210
Westlake Village, CA 91362-3649
www.aspe.org

American Society of Sanitary Engineers
29901 Clemens Road 100
West Lake, OH 44145
www.asse-plumbing.org

American Water Works
6666 W. Quincy A
Denver, CO 80235
www.awwa.org

Cast Iron Soil Pipe Institute
5959 Shallowford Road, Suite 419
Chattanooga, TN 37412
www.cispi.org

International Association of Plumbing and
Mechanical Officials
2001 Walnut Drive S
Walnut, CA 91789-7825

National Association of Plumbing, Heating, and
Cooling Contractors
PO Box 6808
Falls Church, VA 22040
www.phccweb.org

National Fire Sprinkler Association
40 Van Barrett Road
Patterson, NY 12563-1000
www.nfsa.org

Plastic Pipe Institute
1801 K Street NW, Suite 600K
Washington, DC 20006-1301
www.plasticpipe.org

Plumbing Manufacturers Institute
1340 Remington Road, Suite A
Schaumburg, IL 60173
www.pmihome.org

Plastic Pipe and Fittings Association
800 Roosvelt Road, Building C, Suite 312
Glen Ellyn, IL 60137
www.ppfahome.org

Waste Material Management Division
U.S. Department of Energy
1000 Independence Avenue SW
Washington, DC 20585

Water Quality Association
4151 Naperville Road
Lisle, IL 60532
www.wqa.org

DIVISION 23—HEATING, VENTILATING, AND AIR-CONDITIONING

Air-Conditioning and Refrigeration Institute
430 N. Fairfax Drive, Suite 425
Arlington, VA 22203
www.ari.org

Air Conditioning Contractors of America
2800 Shirlington Road, Suite 300
Arlington, VA 22206
www.acca.org

Air Movement and Control Association International,
Inc.
30 W. University Drive
Arlington Heights, IL 60004
www.amca.org

American Boiler Manufacturers Association
950 N. Glebe Road, Suite 160
Arlington, VA 22203-1824
www.abma.com

American Society of Heating and Air-Conditioning
Engineers
1791 Tullie Circle NE
Atlanta, GA 30329-2305
www.ashrae.org

American Solar Energy Society
2400 Central Avenue, Suite A
Boulder, CO 80301
www.ases.org

American Gas Association
400 North Capitol Street NW, 4th Floor
Washington, DC 20001
www.aga.org

Air Diffusion Council
1901 N. Roselle Road, Suite 800
Schaumburg, IL 60195

Building Commissioning Association
1400 SW 5th Ave, Suite 700
Portland, OR 97201
www.bcxa.org

Cooling Tower Institute
530 Wells Fargo Drive 218
Houston, TX 77090
www.cti.org

Florida Solar Energy Center
1679 Clearlake Road
Cocoa, FL 32922-5703
www.fsec.ucf.edu

Heat Exchange Institute
1300 Summer Avenue
Cleveland, OH 44115-2851
www.heatexchange.org

Heating, Refrigeration, and Air Conditioning Institute of Canada
5045 Orbitor Drive, Building 11, Ste. 300
Mississauga, Ontario
Canada L4W 4Y4
www.hrai.ca

The Hydronics Institute, Inc.
PO Box 218
Berkeley Heights, NJ 07922-0218
www.ahrinet.org

International Society for Indoor Air Quality and Climate
Box 22038, Sub 32
Ottawa, Ontario
Canada K1V 0W2

Institute of Heating and Air Conditioning Industries
454 W. Broadway
Glendale, CA 91204
www.ihaci.org

National Renewable Energy Laboratory
1617 Cole Boulevard
Golden, CO 80401
www.nrel.gov

Northeast Sustainable Energy Association
50 Miles Street, Suite 3
Greenfield, MA 01301

Plumbing, Heating, and Cooling Contractors
National Association
180 S. Washington Street
Falls Church, VA 22046
www.phccweb.org

Sheet Metal and Air Conditioning Contractors
National Association
4201 Lafayette Center Drive
Chantilly, VA 20151
www.smacna.org

U.S. Department of Energy
19901 Germantown Road
Germantown, MD 20874-1290
www.doe.gov

DIVISION 26—ELECTRICAL

American Lighting Association
PO Box 420288
Dallas, TX 75342
www.americanlightingassoc.com

American Public Power Association
2301 M Street NW, Suite 300
Washington, DC 20037
www.appanet.org

Edison Electric Institute
701 Pennsylvania Avenue, NW
Washington, DC 20004-2696
www.eei.org

Electric Power Research Institute
3412 Hillview Avenue
Palo Alto, CA 94034
www.epri.org

Electric Power Supply Association
1401 H Street NW, Suite 760
Washington, DC 20005
www.epsa.org

Illuminating Engineering Society of North America
120 Wall Street, 17th Floor
New York, NY 10005-4001
www.iesna.org

International Association of Lighting Designers
The Merchandise Mart, Ste. 11-114A
World Trade Center
Chicago, IL 60654
www.iald.org

Lighting Research Institute
120 Wall Street, 17th Floor
New York, NY 10005

Lighting Research Center
Watervliet Facility
Rensselaer Polytechnic Institute
110 8th Street
Troy, NY 12180
www.lrc.rpi.edu

National Lighting Bureau
8811 Colesville Road, Suite G106
Silver Spring, MD 20910
www.nlb.org

National Electric Manufacturers Association
1300 N 17th Street, Suite 1867
Rosslyn, VA 22209
www.nema.org

Underwriters Laboratories, Inc.
333 Pfingsten Road
Northbrook, IL 60062-2096
www.ul.com

DIVISION 28—ELECTRONIC SAFETY AND SECURITY

American Society for Industrial Security
1625 Prince Street
Alexandria, VA 22314
www.asis.online.org

National Burglar and Fire Alarm Association
7101 Wisconsin Avenue, Suite 901
Bethesda, MD 20814
www.alarm.org

National Crime Prevention Council
1700 K Street NW, 2nd Floor
Washington, DC 20006-3817
www.weprevent.org

National Crime Prevention Institute
University of Louisville
Burhans Hall
Louisville, KY 40292-0001
www.louisville.edu/a-s/ja

National Fire Protection Association
1 Batterymarch Park
Quincy, MA 02169-7471
www.nfpa.org

Security Hardware Distributors Association
100 N. 20th Street, 4th Floor
Philadelphia, PA 19103-1443
www.shda.org

Security Industry Association
635 Slaters Lane, Suite 110
Alexandria, VA 22314
www.siaonline.org

DIVISION 31—EARTHWORK

American Nursery and Landscape Association
1000 Vermont Avenue NW, Suite 300
Washington, DC 20005
www.anla.org

American Society of Civil Engineers
1801 Alexander Bell Drive
Reston, VA 20191-4400
www.asce.org

American Society of Landscape Architects
636 Eye Street NW
Washington, DC 20001
www.asla.org

ASFE: Professional Firms Practicing in the Gensciences
8811 Colesvill Road
Silver Spring, MD 20910
www.asfe.org

American Association of State Highway and
Transportation Officials Systems
444 Capital Street NW, Suite 249
Washington, DC 20001
www.aashto.org

American Water Works Association
6660 W. Quincy Avenue
Denver, CO 80235
www.awwa.org

Canadian Society of Landscape Architects
1339 Fifteenth Avenue NW, Apartment 310
Calgary, Alberta
Canada T3C 3V3

Environmental Information Association
6935 Wisconsin Avenue, Suite 306
Chevy Chase, MD 20815-6112
www.eia-usa.org

The Geo-Institute of the American Society of Civil
Engineers
1801 Alexander Bell Drive
Reston, VA 20191-4400
www.geoinstitute.org

Information Sources for Recycled and Energy-Efficient Materials

Athena Sustainable Materials Institute – Head Office
629 St. Lawrence St.
PO Box 189
Merrickville, Ontario, Canada, K0G 1N0
www.athenasmi.org

Alliance to Save Energy
1200 18th Street NW, Suite 800
Washington, DC 20036
www.ase.org

American Solar Energy Society
2400 Central Avenue, Suite G-1
Boulder, CO 80301
www.ases.org

American Wind Energy Association
122 C Street NW, Suite 400
Washington, DC 20001-2109
www.awea.org

Association of Energy Engineers
4025 Pleasantdale Road, Suite 420
Atlanta, GA 30340
www.aeecenter.org

American Council for an Energy-Efficient Economy
1001 Connecticut Avenue NW, Suite 801
Washington, DC 20036-5525
www.aceee.org

Environmental Building News – BuildingGreen, LLC
122 Birge Street, Suite 30
Brattleboro, VT 05301
www.buildinggreen.com

Canada Green Building Council
47 Clarence Street, Suite 202
Ottawa, ON K1N 9K1
www.cagbc.org

Center for Energy Policy and Research
New York Institute of Technology
Old Westbury, NY 11568

Center for Resourceful Building Technology
PO Box 100
Missoula, MT 59806
www.mountains.com/crbt

Construction Materials Recycling Association
P.O. Box 122
Eola, IL 60519
www.cdrecycling.org

Energy Efficiency and Renewable Energy Clearing House
PO Box 3048
Merrifield, VA 22136
www.eere.energy.gov

Energy Efficient Building Association
PO Box 22307
Eagan, MN 55122-0307

The Green Building Initiative
2104 SE Morrison,
Portland, Oregon 97214
www.greenglobes.com

The Geo-Institute of the American Society of Civil Engineers
1801 Alexander Bell Drive
Reston, VA 20191-4400
www.geoinstitute.org

NAHB Research Center
400 Prince George's Boulevard
Upper Marlboro, MD 20774-8731
www.nahbrc.org

Northeast Sustainable Energy Association
50 Miles Street
Greenfield, MA 01301
www.nesea.org

National Energy Management Institute
601 N. Fairfax Street, Suite 250
Alexandria, VA 22314
www.nemionline.com

National Renewable Energy Laboratory
1617 Cole Boulevard
Golden, CO 80401-3393
www.nrel.gov

National Research Council Canada and the Institute for Research in Construction
Montreal Road Campus, Building M-59
Ottawa, Ontario
Canada K1A 0R6
www.nrc.ca

National Energy Foundation
5225 Wiley Post Way, Suite 170
Salt Lake City, UT 84116
www.nef1.org

Sustainable Buildings Industry Council
1112 16th Street NW, Suite 240
Washington, DC 20036
www.SBICouncil.org

U.S. Green Building Council
2101 L Street, NW Suite 500
Washington, DC 20037
www.usgbc.org

Building Code Agencies

International Code Council
5203 Leesburg Pike, Suite 708
Falls Church, VA 22041-3401
www.iccsafe.org

National Fire Protection Association
1 Batterymarch Park
Quincy, MA 02169-7471
www.nfpa.org

International Association of Plumbing and
Mechanical Officials
20001 Walnut Drive South
Walnut, CA 91789-2825
www.iapmo.org

American Society for Testing and Materials
100 Barr Harbor Drive
West Conshohocken, PA 19428
www.astm.org

American Society of Mechanical Engineers
Three Park Avenue
New York, NY 10016-5990

American Concrete Institute
PO Box 9094
Farmington Hills, MI 48333
www.concrete.org

National Conference of States on Building Codes and
Standards
505 Hunter Park Drive, Suite 210
Herndon, VA 20170
www.ncsbcs.org

International Fire Code Institute
5360 S. Workman Mill Road
Whittier, CA 90501-2298
www.ifci.com

Other Organizations

American Water Works Association
6666 West Works Avenue
Denver, CO 80235-3098
www.awwa.org

American Architectural Manufacturers Association
1827 Walden Office Square, Suite 550
Schaumburg, IL 60173
www.aamanet.org

American Institute of Architects
1735 New York Avenue NW
Washington, DC 20006
www.aia.org

American National Metric Council
1735 N. Lynn Street, Suite 950
Arlington, VA 22209-2022

American National Standards Institute
25 West 42nd Street, 4th Floor
New York, NY 10036
www.ansi.org

ASTM International
100 Barr Harbor Drive
W. Conshohocken, PA 19428-2959
www.astm.org

American Society of Mechanical Engineers
3 Park Avenue
New York, NY 10016-5990
www.asme.org

American Society of Civil Engineers
World Headquarters
1801 Alexander Bell Drive
Reston, VA 20191-4400
www.asce.org

American Council for Construction Education
1717 N. Loop, 1604 East, Suite 320
San Antonio, TX 78232-1570
www.acce-hq.org

Association of Construction Inspectors
1224 N. Nokomis NE
Alexandria, MN 56308
www.iami.org

Building Research Council
University of Illinois
1 E. St. Mary's Road
Champaign, IL 61820

Canadian Housing Information Center
700 Chemin Montreal
Ottawa, Ontario
Canada K1A 0P7
www.cmhc-schl.gc.ca

Canadian Mortgage and Housing Corporation
700 Montreal Road
Ottawa, Ontario
Canada K1A 0P7
www.cmhc-schl.gc.ca

The Canadian Standards Association
5060 Spectrum Way
Mississauga, Ontario
Canada L4W 5N6
www.csa.ca

Canadian Construction Materials Centre
Institute for Research in Construction
National Research Council Canada
Building M-24, 1500 Montreal Road
Ottawa, Ontario
Canada K1A 0R6

Canadian Home Builders Association
150 Laurier Avenue West, Suite 200
Ottawa, Ontario
Canada K1P 5J4
www.chba.ca

The Construction Specifications Institute
99 Canal Center Plaza, Suite 300
Alexandria, VA 22314-1588
www.csinet.org

Construction Specifications Canada
100 Lombard Street, Suite 200
Toronto, Ontario
Canada M5C 1M3
www.csc-dcc.ca

The Coasts, Oceans, Ports, and Rivers Institute of the
American Society of Civil Engineers
1801 Alexander Bell Drive
Reston, VA 20191-4400
www.coprinstitute.org

Environmental Hazards Management Institute
PO Box 932
Durham, NH 03824

The Environmental & Water Resources Institute of
the American Society of Civil Engineers
1801 Alexander Bell Drive
Reston, VA 20191-4400
www.ewrinstitute.org

Federal Specifications
General Services Administration, Specifications
Activity Office
Building 197
Washington Navy Yard
2nd and M Street SE
Washington, DC 20407

Home Builders Institute
1090 Vermont Avenue NW
Washington, DC 20005
www.hbi.org

Institute for Research in Construction
Montreal Road Campus, M59
Ottawa, Ontario
Canada K1A 0R6
www.irc.nrc-chrc.gc.ca

National Association of Home Builders
1201 15th Street NW
Washington, DC 20006
www.nahb.org

National Association of Home Builders Remodelers
Council
1201 15th Street NW
Washington, DC 20005
www.nahb.org/remodelers

National Fire Protection Association
1 Batterymarch Park
Quincy, MA 02169-7471
www.nfpa.org

National Institute of Building Sciences
1090 Vermont Avenue NW, Suite 700
Washington, DC 20005-4905
www.nibs.org

National Association of the Remodeling Industry
780 Lee Street, Suite 200
Des Plaines, IL 60016
www.nari.org

National Institute of Science and Technology
U.S. Department of Commerce
Building 820, Room 306
Gaithersburg, MD 20899
www.nist.gov/metric

National Research Council of Canada
NRC Corporate Communications
1200 Montreal Road, Building M-58
Ottawa, Ontario
Canada K1A 0R6

National Institute of Standards and Technology
U.S. Department of Commerce
100 Bureau Drive MS2150
Gaithersburg, MD 20899-2150
www.nist.gov/srm

Occupational Safety and Health Administration
200 Constitution Avenue NW
Washington, DC 20210
www.osha.gov

Rubber Manufacturers Association
1400 K Street NW, Suite 900
Washington, DC 20005
www.rma.org

Sweets Construction Catalog
Architects, Engineers, and Contractors
McGraw-Hill Construction
Two Penn Plaza
New York, NY 10121
www.construction.com

The Transportation & Development Institute of the
American Society of Civil Engineers
1801 Alexander Bell Drive
Reston, VA 20191-4400
www.tanddi.org

U.S. Metric Association
10245 Andasol Avenue
Northridge, CA 91325-1504
www.metric.org

Underwriters Laboratories, Inc.
333 Pfingsten Road
Northbrook, IL 60062-2096
www.ul.com

Underwriters Laboratories of Canada
7 Underwriters Road
Scarborough, Ontario
Canada M1R 3A9
www.ulc.ca

U.S. Environmental Protection Agency
Ariel Rios Building
1200 Pennsylvania Avenue NW
Washington, DC 20460
www.epa.gov

U.S. Department of Commerce
Building 820, Room 306
Gaithersburg, MD 20899
www.commerce.gov

U.S. Department of Housing and Urban Development
HUD USER
PO Box 23268
Washington, DC 20026-3268
www.huduser.org

U.S. Department of Energy
Energy Efficiency & Renewable Energy Center
1000 Independence Avenue
Washington, DC 20585 0121
www.eere.energy.gov

U.S. Environmental Protection Agency
Atmospheric Pollution Prevention Division
401 M Street, Mail Code 6202JSW
Washington, DC 20460

Air and Radiation Docket and Information Center
U.S. Environmental Protection Agency
401 M Street SW
Washington, DC 20460

U.S. EPA Main Library
Environmental Protection Agency
109 Alexander Drive
Research Triangle Park, NC 27711
www.epa.gov

U.S. Department of Energy
Office of Scientific Information
PO Box 62
Oak Ridge, TN 37831

The level-two numbers and titles used in this publication are from MasterFormat™ 2014 Edition, published by the Construction Specifications Institute (CSI) and Construction Specifications Canada (CS), and used with permission from CSI 2014.

Appendix D

METRIC INFORMATION

Tables giving metric equivalents for common fractions, two-place decimal inches, and millimeter-to-decimal-inches are shown below.

Base SI Units		
Quantity	Unit	Symbol
Length	Meter	m
Mass	Kilogram	kg
Time	Second	s
Electric current	Ampere	A
Thermodynamic temperature	Kelvin	K
Amount of substance	Mole	mo
Luminous intensity	Candela	cd

Supplementary SI Units		
Quantity	Unit	Symbol
Plane angle	Radian	rad
Solid angle	Steradian	sr

Derived Metric Units with Compound Names		
Physical Quantity	Unit	Symbol
Area	Square meter	m^2
Volume	Cubic meter	m^3
Density	Kilogram per cubic meter	kg/m^3
Velocity	Meter per second	m/s
Angular velocity	Radian per second	rad/s
Acceleration	Meter per second squared	m/s^2
Angular acceleration	Radian per second squared	rad/s^2
Volume rate of flow	Cubic meter per second	m^3/s
Moment of inertia	Kilogram meter squared	$kg \bullet m^2$
Moment of force	Newton meter	N•m
Intensity of heat flow	Watt per square meter	W/m^2
Thermal conductivity	Watt per meter Kelvin	W/m•K
Luminance	Candela per square meter	cd/m^2

SI Prefixes		
Multiplication Factor	Prefix	Symbol
$1\ 000\ 000\ 000\ 000\ 000\ 000 = 10^{18}$	exa	E
$1\ 000\ 000\ 000\ 000\ 000 = 10^{15}$	peta	P
$1\ 000\ 000\ 000\ 000 = 10^{12}$	tera	T
$1\ 000\ 000\ 000 = 10^9$	giga	G
$1\ 000\ 000 = 10^6$	mega	M
$1\ 000 = 10^3$	kilo	k
$100 = 10^2$	hecto	h
$10 = 10^1$	deka	da
$0.1 = 10^{21}$	deci	d
$0.01 = 10^{22}$	centi	c
$0.001 = 10^{23}$	milli	m
$0.000\ 001 = 10^{26}$	micro	m
$0.000\ 000\ 001 = 10^{29}$	nano	n
$0.000\ 000\ 000\ 001 = 10^{212}$	pico	p
$0.000\ 000\ 000\ 000\ 001 = 10^{215}$	femto	f
$0.000\ 000\ 000\ 000\ 000\ 001 = 10^{218}$	atto	a

Metric Unit to Imperial Unit Conversion Factors[a]		
Metric Units		Imperial Equivalents
Length		
1 millimeter (mm)	=	0.0393701 inch
1 meter (m)	=	39.3701 inches
	=	3.28084 feet
	=	1.09361 yards
1 kilometer (km)	=	0.621371 mile
Length/Time		
1 meter per second (m/s)	=	3.28084 feet per second
1 kilometer per hour (km/h)	=	0.621371 mile per hour
Area		
1 square millimeter (mm^2)	=	0.001550 square inch
1 square meter (m^2)	=	10.7639 square feet
1 hectare (ha)	=	2.47105 acres
1 square kilometer (km^2)	=	0.386102 square mile
1 cubic millimeter (mm^3)	=	0.0000610237 cubic inch
1 cubic meter (m^3)	=	35.3147 cubic feet
	=	1.30795 cubic yards

Metric Unit to Imperial Unit Conversion Factors[a] (continued)

Metric Units		Imperial Equivalents
Area		
1 milliliter (mL)	=	0.0351951 fluid ounce
1 liter (L)	=	0.219969 gallon
Mass		
1 gram (g)	=	0.0352740 ounce
1 kilogram (kg)	=	2.20462 pounds
1 tonne (t) (= 1,000 kg)	=	1.10231 tons (2,000 lb.)
	=	2204.62 pounds
Mass/Volume		
1 kilogram per cubic meter (kg/m³)	=	0.0622480 pound per cubic foot
Force		
1 newton (N)	=	0.224809 pound-force
Stress		
1 megapascal (MPa) (51 N/mm²)	=	145.038 pounds-force per sq. in.
Loading		
1 kilonewton per sq. meter (kN/m²)	=	20.8854 pounds-force per sq. ft.
1 kilonewton per meter (kN/m) (51 N/mm)	=	68.5218 pounds-force per ft.
Moment		
1 kilonewtonmeter (kNm)	=	737.562 pound-force ft.
Miscellaneous		
1 joule (J)	=	0.00094781 Btu
1 joule (J)	=	1 watt-second
1 watt (W)	=	0.00134048 electric horsepower
1 degree Celsius (°C)	=	32 1 1.8 (°C) degrees Fahrenheit

Notes: 1. 1.0 newton = 1.0 kilogram − 9.80665 m/s² (International Standard Gravity Value)

2. 1.0 pascal = 1.0 newton per square meter

[a]Multiply the number of metric units by the imperial (English) equivalent to convert a measurement from metric units to imperial (English) units.

Brick and Block Masonry lb./ft.² kg/m²
Imperial Unit to Metric Conversion Factors[b]

Imperial Units		Metric Equivalents
Length		
1 inch	=	25.4 mm
	=	0.0254 m
1 foot	=	0.3048 m
1 yard	=	0.9144 m
1 mile	=	1.60934 km

Brick and Block Masonry lb./ft.² kg/m² (continued)

Length/Time		
1 foot per second	=	0.3048 m/s
1 mile per hour	=	1.60934 km/h
Area		
1 square inch	=	645.16 mm²
1 square foot	=	0.0929030 m²
1 acre	=	0.404686 ha
1 square mile	=	2.58999 km²
Volume		
1 cubic inch	=	16387.1 mm³
1 cubic foot	=	0.0283168 m³
1 cubic yard	=	0.764555 m³
1 fluid ounce	=	28.4131 mL
1 gallon	=	4.54609 L
Mass		
1 ounce	=	28.3495 g
1 pound	=	0.453592 kg
1 ton (2,000 lb.)	=	0.907185 t
1 pound	=	0.0004539 t
Mass/Volume		
1 pcf	=	16.1085 kg/m³
Force		
1 pound	=	4.44822 N
Stress		
1 psi	=	0.00689476 MPa
Loading		
1 psf	=	0.0478803 kN/m²
1 plf	=	0.0145939 kN/m
Moment		
1 pound-force ft.	=	0.00135582 kNm
Miscellaneous		
1 Btu	=	1055.06 J
1 watt-second	=	1 J
1 horsepower	=	746 W
1 degree Fahrenheit	=	(°F 2 32)/1.8 °C

Notes: 1. 1.0 newton = 1.0 kilogram − 9.80665 m/s² (International Standard Gravity Value)

2. 1.0 pascal = 1.0 newton per square meter

[b]Multiply the number of imperial (English) units by the metric equivalent to convert a measurement from imperial (English) units to metric units.

WEIGHTS OF BUILDING MATERIALS

Brick and Block Masonry	lb./ft.³	kg/m³
4" brickwall	40	196
4" concrete brick, stone or gravel	46	225
4" concrete brick, lightweight	33	161
4" concrete block, stone or gravel	34	167
4" concrete block, lightweight	22	108
6" concrete, stone or gravel	50	245
6" concrete block, lightweight	31	152
8" concrete block, stone or gravel	55	270
8" concrete block, lightweight	35	172
12" concrete block, stone or gravel	85	417
12" concrete block, lightweight	55	270

Concrete	lb./ft.³	kg/m³
Plain, slag	132	2155
Plain, stone	144	2307
Reinforced, slag	138	2211
Reinforced, stone	150	2403

Lightweight Concrete	lb./ft.³	kg/m³
Concrete, perlite	35–50	561–801
Concrete, pumice	60–90	961–1442
Concrete, vermiculite	25–60	400–961

Wall, Ceiling, and Floor	lb./ft.²	kg/m²
Acoustical tile, $\frac{1}{2}$"	0.8	3.9
Gypsum wallboard, "	2	9.8
Plaster, 2" partition	20	98
Plaster, 4" partition	32	157
Plaster, $\frac{1}{2}$"	4.5	22
Plaster on lath	10	49
Tile, glazed, $\frac{3}{8}$"	3	14.7
Tile, quarry, $\frac{1}{2}$"	5.8	28.4
Terrazzo, 1"	25	122.5
Vinyl composition floor tile	1.4	69
Hardwood flooring, $\frac{25}{32}$"	4	19.6
Flexicore 6", lightweight concrete	30	14.7
Flexicore 6", stone concrete	40	196

Wall, Ceiling, and Floor	lb./ft.²	kg/m²
Plank, cinder concrete, 2"	15	73.5
Plank, gypsum, 2"	12	58.8
Concrete reinforced, stone, 1"	12.5	61.3
Concrete reinforced, lightweight, 1"	6–10	29.4–49
Concrete plain, stone, 1"	12	58.8
Concrete plain, lightweight, 1"	3–9	14.7–44.1

Partitions	lb./ft.²	kg/m²
2–4 wood studs, gypsum wallboard 2 sides	8	39.2
4" metal stud, gypsum wallboard 2 sides	6	29.4
6" concrete block, gypsum wallboard 2 sides	35	171.5

Roofing	lb./ft.²	kg/m²
Built-up	6.5	31.9
Concrete roof tile	9.5	46.6
Copper	1.5–2.5	7.4–12.3
Steel deck alone	2.5	12.3
Shingles, asphalt	1.7–2.8	8.3–13.7
Shingles, wood	2–3	9.8–14.7
Slate, "	14–18	68.6–88.2
Tile, clay	8–16	39.2–78.4

Stone Veneer	lb./ft.³	kg/m³
2" granite, $\frac{1}{2}$" parging	30	481
4" limestone, $\frac{1}{2}$" parging	36	577
4" sandstone, $\frac{1}{2}$" parging	49	785
1" marble	13	208

Structural Clay Tile	lb./ft.²	kg/m²
4" hollow	23	368
6" hollow	38	609
8" hollow	45	721

Structural Facing Tile	lb./ft.²	kg/m³
2" facing tile	14	68.6
4" facing tile	24	118
6" facing tile	34	167

(Continued)

Wood	lb./ft.²	kg/m²
Ash, white	40.5	198
Birch	44	202
Cedar	22	108
Cypress	33	162
Douglas fir	32	157
White pine	27	132
Pine, southern yellow	26	127
Redwood	26	127
Plywood, ½"	1.5	7.4

Residential Assemblies	lb./ft.²	kg/m²
Wood framed floor	10	49
Ceiling	10	49
Frame exterior wall, 4" studs	10	49
Frame exterior wall, 6" studs	13	64
Brick veneer of 4" frame	50	245
Brick veneer over 4" concrete block	74	363
Interior partitions with gypsum both sides (allowance per sq. ft. of floor area—not weight of material)	20	320

Suspended Ceilings	lb./ft.²	kg/m²
Acoustic plaster on gypsum lath	10–11	49–54
Mineral fiberboard	1.4	6.9

Metals	lb./ft.³	kg/m³
Aluminum	165	2643
Copper	556	8907
Iron, cast	450	7209
Steel	490	7850
Steel, stainless	490–510	7850–8170

Glass	lb./ft.²	kg/m²
¼" (6.3 mm) plate or float	3.3	16.2
½" (12.7 mm) plate or float	6.6	32.3
1/32" (0.79 mm) sheet	2.8	13.7
¼" (6.3 mm) sheet	3.5	17.2
1/8" (3.2 mm) double strength	1.6	7.8
7/32" (5.6 mm) sheet	2.85	14.0
¼" (6.3 mm) laminated	3.30	16.2
½" (12.7 mm) laminated	6.35	31.1
2" (50.2 mm) bullet resistant	26.2	128.4
1" (25.4 mm) insulating, ½", (6.3 mm) air space	6.54	32.0
¼" (6.3 mm) wired	3.5	17.1
3/8" (9.5 mm) wired	5.0	24.4
37/8" = 53/4" square, (98.0 – 146 mm) glass block	16.0	78.4

NAMES AND ATOMIC SYMBOLS
OF SELECTED CHEMICAL ELEMENTS

Name	Atomic Symbol	Name	Atomic Symbol
Aluminum	Al	Iron	Fe
Antimony	Sb	Krypton	Kr
Argon	Ar	Lanthanum	La
Arsenic	As	Lead	Pb
Barium	Ba	Lithium	Li
Beryllium	Be	Lutetium	Lu
Bismuth	Bi	Magnesium	Mg
Boron	B	Manganese	Mn
Bromine	Br	Masurium	Ma
Cadmium	Cd	Mercury	Hg
Calcium	Ca	Molybdenum	Mo
Carbon	C	Neodymium	Nd
Cerium	Ce	Neon	Ne
Cesium	Cs	Nickel	Ni
Chlorine	Cl	Niobium	Nb
Chromium	Cr	Nitrogen	N
Cobalt	Co	Osmium	Os
Copper	Cu	Oxygen	O
Dysprosium	Dy	Palladium	Pd
Erbium	Er	Phosphorus	P
Europium	Eu	Platinum	Pt
Fluorine	F	Polonium	Po
Gadolinium	Gd	Potassium	K
Gallium	Ga	Praseodymium	Pr
Germanium	Ge	Protactinium	Pa
Gold	Au	Radium	Ra
Hafnium	Hf	Radon	Rn
Helium	He	Rhenium	Re
Holmium	Ho	Rhodium	Rh
Hydrogen	H	Rubidium	Rb
Illinium	Il	Ruthenium	Ru
Indium	In	Samarium	Sm
Iodine	I	Scandium	Sc
Iridium	Ir	Selenium	Se

(Continued)

Name	Atomic Symbol
Silicon	Si
Silver	Ag
Sodium	Na
Strontium	Sr
Tantalum	Ta
Tellurium	Te
Terbium	Tb
Thallium	Tl
Thorium	Th
Thulium	Tm
Tin	Sn
Titanium	Ti
Tungsten	W
Uranium	U
Vanadium	V
Xenon	Xe
Ytterbium	Yb
Yttrium	Y
Zinc	Zn
Zirconium	Zr

COEFFICIENTS OF THERMAL EXPANSION[a] FOR SELECTED CONSTRUCTION MATERIALS

Material	Multiply by 10^{26} in./in./°F	Multiply by 10^{26} mm/mm/°C
Concrete		
Normal weight concrete	5.5	9.8
Gypsum		
Gypsum panels	9.0	16.2
Gypsum plaster	7.0	12.6
Wood fiber plaster	8.0	14.4
Masonry		
Brick (varies some)	3.0	5.6
Concrete masonry units	5.2	9.4
Marble	7.3	13.1
Granite	4.7	8.5
Limestone	4.4	7.9

Material	Multiply by 10^{26} in./in./°F	Multiply by 10^{26} mm/mm/°C
Metal		
Iron, gray cast	5.7	10.5
Iron, malleable	5.6	10.5
Steel, carbon (ASTM A285)	5.6	10.5
Steel, high strength (ASTM A141)	6.4	11.7
Steel, stainless (type 201)	8.7	15.7
Steel, stainless (type 405)	6.0	10.8
Nickel (211)	7.4	13.3
Copper (CA110)	9.4	16.5
Bronze, commercial	10.2	19.3
Brass, red	10.4	19.9
Aluminum, wrought	12.8	23.0

Material	Multiply by 10^{26} in./in./°F	Multiply by 10^{26} mm/mm/°C
Polymer, Thermosetting		
Phenolics	45.0	81.0
Urea-melamine	20.0	36.0
Polyesters	45.0	75.5
Epoxies	33.0	72.0

Material	Multiply by 10^{26} in./in./°F	Multiply by 10^{26} mm/mm/°C
Polymer, Thermoplastic		
Polyethylene, high density	70.0	120.0
Polypropylene	50.0	90.0
Polystyrene	38.0	68.5
Polyvinyl chloride (PVC)	30.0	54.0
Acrylonitrile- butadiene- styrene (ABS)	50.0	90.0
Acrylics	40.0	72.0
Wood		

For most hardwoods and softwoods the *parallel-to-grain* thermal expansion values range from 1.7×10^{26} to 2.5×10^{26} in./in./°F *or* 3.1×10^{26} to 4.5×10^{26} mm/mm/°C.

Linear expansion coefficients across the grain are proportional to wood density. They range from 5 to 10 times greater than the parallel-to-grain coefficients and, therefore, are of more concern.

Material	Multiply by 10^{26} in./in./°F	Multiply by 10^{26} mm/mm/°C
Foam Insulation		
Polystyrene	3.5	6.3
Polyurethane	2.7	4.9

Material	Multiply by 10^{26} in./in./°F	Multiply by 10^{26} mm/mm/°C
Glass		
Glass, soda lime window sheet	47.0	85.0
Glass, soda lime plate	48.0	87.0

[a]The change in dimension of a material per unit of dimension per degree change in temperature.

CONVERSION TABLES

Millimeters to Decimal Inches									
mm	**In.**	**mm**	**In.**	**mm**	**In.**	**mm**	**In.**	**mm**	**In.**
1	0.0394	21	0.8268	41	1.6142	61	2.4016	81	3.1890
2	0.0787	22	0.8662	42	1.6536	62	2.4410	82	3.2284
3	0.1181	23	0.9055	43	1.6929	63	2.4804	83	3.2678
4	0.1575	24	0.9449	44	1.7323	64	2.5197	84	3.3071
5	0.1969	25	0.9843	45	1.7717	65	2.5591	85	3.3465
6	0.2362	26	1.0236	46	1.8111	66	2.5985	86	3.3859
7	0.2756	27	1.0630	47	1.8504	67	2.6378	87	3.4253
8	0.3150	28	1.1024	48	1.8898	68	2.6772	88	3.4646
9	0.3543	29	1.1418	49	1.9292	69	2.7166	89	3.5040
10	0.3937	30	1.1811	50	1.9685	70	2.7560	90	3.5434
11	0.4331	31	1.2205	51	2.0079	71	2.7953	91	3.5827
12	0.4724	32	1.2599	52	2.0473	72	2.8247	92	3.6221
13	0.5118	33	1.2992	53	2.0867	73	2.8741	93	3.6615
14	0.5512	34	1.3386	54	2.1260	74	2.9134	94	3.7009
15	0.5906	35	1.3780	55	2.1654	75	2.9528	95	3.7402
16	0.6299	36	1.4173	56	2.2048	76	2.9922	96	3.7796
17	0.6693	37	1.4567	57	2.2441	77	3.0316	97	3.8190
18	0.7087	38	1.4961	58	2.2835	78	3.0709	98	3.8583
19	0.7480	39	1.5355	59	2.3229	79	3.1103	99	3.8977
20	0.7874	40	1.5748	60	2.3622	80	3.1497	100	3.9371

Fractional Inches to Millimeters							
mm	**In.**	**mm**	**In.**	**mm**	**In.**	**mm**	**In.**
1/64	0.397	17/64	6.747	33/64	13.097	49/64	19.447
1/32	0.794	9/32	7.144	17/32	13.494	25/32	19.844
3/64	1.191	19/64	7.541	35/64	13.890	51/64	20.240
1/16	1.587	5/16	7.937	9/16	14.287	13/16	20.637
5/64	1.984	21/64	8.334	37/64	14.684	53/64	21.034
3/32	2.381	11/32	8.731	19/32	15.081	27/32	21.431
7/64	2.778	23/64	9.128	39/64	15.478	55/64	21.828
1/8	3.175	3/8	9.525	5/8	15.875	7/8	22.225
9/64	3.572	25/64	9.922	41/64	16.272	57/64	22.622
5/32	3.969	13/32	10.319	21/32	16.669	29/32	23.019
11/64	4.366	27/64	10.716	43/64	17.065	59/64	23.415
3/16	4.762	7/16	11.113	11/16	17.462	15/16	23.812
13/64	5.159	29/64	11.509	45/64	17.859	61/64	24.209
7/32	5.556	15/32	11.906	23/32	18.256	31/32	24.606
15/64	5.953	31/64	12.303	47/64	18.653	63/64	25.003
1/4	6.350	1/2	12.700	3/4	19.050	1	25.400

Decimal Equivalents of Common Fractions

4ths	8ths	16ths	32nds	64ths	To 4 Places	To 3 Places	To 2 Places	4ths	8ths	16ths	32nds	64ths	To 4 Places	To 3 Places	To 2 Places
				1/64	.0156	.016	.02					33/64	.5156	.516	.52
			1/32		.0312	.031	.03				17/32		.5312	.531	.53
				3/64	.0469	.047	.05					35/64	.5469	.547	.55
		1/16			.0625	.062	.06			9/16			.5625	.562	.56
				5/64	.0781	.078	.08					37/64	.5781	.578	.58
			3/32		.0938	.094	.09				19/32		.5938	.594	.59
				7/64	.1094	.109	.11					39/64	.6094	.609	.61
	1/8				.1250	.125	.12		5/8				.6250	.625	.62
				9/64	.1406	.141	.14					41/64	.6406	.641	.64
			5/32		.1562	.156	.16				21/32		.6562	.656	.66
				11/64	.1719	.172	.17					43/64	.6719	.672	.67
		3/16			.1875	.188	.19			11/16			.6875	.688	.69
				13/64	.2031	.203	.20					45/64	.7031	.703	.70
			7/32		.2188	.219	.22				23/32		.7188	.719	.72
				15/64	.2344	.234	.23					47/64	.7344	.734	.73
1/4					.2500	.250	.25	3/4					.7500	.750	.75
				17/64	.2656	.266	.27					49/64	.7656	.766	.77
			9/32		.2812	.281	.28				25/32		.7812	.781	.78
				19/64	.2969	.297	.30					51/64	.7969	.797	.80
		5/16			.3125	.312	.31			13/16			.8125	.812	.81
				21/64	.3281	.328	.33					53/64	.8281	.828	.83
			11/32		.3438	.344	.34				27/32		.8438	.844	.84
				23/64	.3594	.359	.36					55/64	.8594	.859	.86

(Continued)

879

Design Services Of carlisle Publishing Services (continued)

Decimal Equivalents of Common Fractions

4ths	8ths	16ths	32nds	64ths	To 4 Places	To 3 Places	To 2 Places	4ths	8ths	16ths	32nds	64ths	To 4 Places	To 3 Places	To 2 Places
	3/8				.3750	.375	.38		7/8				.8750	.875	.88
				25/64	.3906	.391	.39					57/64	.8906	.891	.89
			13/32		.4062	.406	.41				29/32		.9062	.906	.91
				27/64	.4219	.422	.42					59/64	.9219	.922	.92
		7/16			.4375	.438	.44			15/16			.9375	.938	.94
				29/64	.4531	.453	.45					61/64	.9531	.953	.95
			15/32		.4688	.469	.47				31/32		.9688	.969	.97
				31/64	.4844	.484	.48					63/64	.9844	.984	.98
					.5000	.500	.50						1.0000	1.000	1.00

Metric Equivalents of Two-Place Decimal Inches

Inch	Millimeter Equivalent	Inch	Millimeter Equivalent
.02	.508	.52	13.208
.03	.762	.53	13.462
.04	1.016	.54	13.716
.05	1.270	.55	13.970
.06	1.524	.56	14.224
.08	2.032	.58	14.732
.09	2.286	.59	14.986
.10	2.540	.60	15.240
.12	3.048	.62	15.748
.14	3.556	.64	16.256
.15	3.810	.65	16.510
.16	4.064	.66	16.764
.18	4.572	.68	17.272
.19	4.826	.69	17.526
.20	5.080	.70	17.780
.22	5.588	.72	18.288
.24	6.096	.74	18.796
.25	6.350	.75	19.050
.26	6.604	.76	19.304
.28	7.112	.78	19.812
.30	7.620	.80	20.320
.31	7.874	.81	20.574
.32	8.128	.82	20.828
.34	8.636	.84	21.336
.35	8.890	.85	21.590
.36	9.144	.86	21.844
.37	9.398	.87	22.098
.38	9.652	.88	22.352
.40	10.160	.90	22.860
.41	10.414	.91	23.114
.42	10.668	.92	23.368
.44	11.176	.94	23.876
.45	11.430	.95	24.130
.46	11.684	.96	24.384
.47	11.938	.97	24.638
.48	12.192	.98	24.892
.50	12.700	1.00	25.400

Millimeters to Decimal Inches

mm	In.	mm	In.	mm	In.	mm	In.	mm	In.
1 =	0.0394	21 =	0.8268	41 =	1.6142	61 =	2.4016	81 =	3.1890
2 =	0.0787	22 =	0.8662	42 =	1.6536	62 =	2.4410	82 =	3.2284
3 =	0.1181	23 =	0.9055	43 =	1.6929	63 =	2.4804	83 =	3.2678
4 =	0.1575	24 =	0.9449	44 =	1.7323	64 =	2.5197	84 =	3.3071
5 =	0.1969	25 =	0.9843	45 =	1.7717	65 =	2.5591	85 =	3.3465
6 =	0.2362	26 =	1.0236	46 =	1.8111	66 =	2.5985	86 =	3.3859
7 =	0.2756	27 =	1.0630	47 =	1.8504	67 =	2.6378	87 =	3.4253
8 =	0.3150	28 =	1.1024	48 =	1.8898	68 =	2.6772	88 =	3.4646
9 =	0.3543	29 =	1.1418	49 =	1.9292	69 =	2.7166	89 =	3.5040
10 =	0.3937	30 =	1.1811	50 =	1.9685	70 =	2.7560	90 =	3.5434
11 =	0.4331	31 =	1.2205	51 =	2.0079	71 =	2.7953	91 =	3.5827
12 =	0.4724	32 =	1.2599	52 =	2.0473	72 =	2.8247	92 =	3.6221
13 =	0.5118	33 =	1.2992	53 =	2.0867	73 =	2.8741	93 =	3.6615
14 =	0.5512	34 =	1.3386	54 =	2.1260	74 =	2.9134	94 =	3.7009
15 =	0.5906	35 =	1.3780	55 =	2.1654	75 =	2.9528	95 =	3.7402
16 =	0.6299	36 =	1.4173	56 =	2.2048	76 =	2.9922	96 =	3.7796
17 =	0.6693	37 =	1.4567	57 =	2.2441	77 =	3.0316	97 =	3.8190
18 =	0.7087	38 =	1.4961	58 =	2.2835	78 =	3.0709	98 =	3.8583
19 =	0.7480	39 =	1.5355	59 =	2.3229	79 =	3.1103	99 =	3.8977
20 =	0.7874	40 =	1.5748	60 =	2.3622	80 =	3.1497	100 =	3.9371

Fractional Inches to Millimeters

In.	mm	In.	mm	In.	mm	In.	mm
1/64 =	0.397	17/64 =	6.747	33/64 =	13.097	49/64 =	19.447
1/32 =	0.794	9/32 =	7.144	17/32 =	13.494	25/32 =	19.844
3/64 =	1.191	19/64 =	7.541	35/64 =	13.890	51/64 =	20.240
1/16 =	1.587	5/16 =	7.937	9/16 =	14.287	13/16 =	20.637
5/64 =	1.984	21/64 =	8.334	37/64 =	14.684	53/64 =	21.034
3/32 =	2.381	11/32 =	8.731	19/32 =	15.081	27/32 =	21.431
7/64 =	2.778	23/64 =	9.128	39/64 =	15.478	55/64 =	21.828
1/8 =	3.175	3/8 =	9.525	5/8 =	15.875	7/8 =	22.225
9/64 =	3.572	25/64 =	9.922	41/64 =	16.272	57/64 =	22.622
5/32 =	3.969	13/32 =	10.319	21/32 =	16.669	29/32 =	23.019
11/64 =	4.366	27/64 =	10.716	43/64 =	17.065	59/64 =	23.415
3/16 =	4.762	7/16 =	11.113	11/16 =	17.462	15/16 =	23.812
13/64 =	5.159	29/64 =	11.509	45/64 =	17.859	61/64 =	24.209
7/32 =	5.556	15/32 =	11.906	23/32 =	18.256	31/32 =	24.606
15/64 =	5.953	31/64 =	12.303	47/64 =	18.653	63/64 =	25.003
1/4 =	6.350	1/2 =	12.700	3/4 =	19.050	1 =	25.400

abrasion resistance Resistance to being worn away by rubbing or friction.

absorbent A material that extracts a substance for which it has an affinity from a liquid or gas and changes physically and chemically during the process.

accelerator An admixture used to speed up the setting of concrete.

acceptable indoor air quality Air containing no known contaminants at harmful concentrations.

access floor A freestanding floor raised above the structural floor to provide access to mechanical and electrical services.

acoustical glass A glazing unit used to reduce the transmission of sound through the glazed opening by bonding a soft interlayer between the layers of glass.

acoustical plaster A plaster used to control sound made from calcined gypsum mixed with lightweight aggregates.

acoustics The science of sound generation, sound transmission, and the effects of sound waves.

active fire protection Characterized by devices and systems that set in motion a physical reaction to fight a fire.

additive Materials mixed with the basic plastic resin to alter its properties.

adhesion The ability of a coating to stick to a surface.

adhesive A substance used to hold materials together by surface attachment.

adhesive bonding agents Bonding agents made from synthetic materials.

admixture A material other than Portland cement, aggregate, and water that is added to concrete to alter its properties.

adsorbent A material that extracts a substance for which it has an affinity from a liquid or gas and that changes physically and chemically during the process.

agglomeration A process that bonds ground iron ore particles into pellets to facilitate handling.

aggregate Inert granules such as crushed stone, gravel, and expanded minerals mixed with Portland cement and sand to form concrete.

air drying Drying wood by exposing it to the air.

air-entrained Portland cement A Portland cement with an admixture that causes a controlled quantity of stable, microscopic air bubbles to form in the concrete.

air gap An unobstructed vertical distance between the lowest opening of pipe that supplies a plumbing fixture and the level at which the fixture will overflow.

air-supported (pneumatic) structure A structure that derives its structural integrity from the use of internal pressurized air to inflate a pliable material.

alarm actuators Controls which sense a particular activity and signal an alarm or a pre-set process to control the situation.

alkali A substance such as lye, soda, or lime that can be destructive to paint films.

alkyd Synthetic resin modified with oil for good adhesion, gloss, color retention, and flexibility.

alloying elements Any substance added to a molten metal to change its mechanical or physical properties.

alumina A hydrated form of aluminum oxide from which aluminum is made.

anaerobic bonding agents Bonding agents that set hard when not exposed to oxygen.

angle of repose The steepest angle of a surface at which loose material will remain in place rather than sliding.

annealing Heating a metal or glass to a high temperature followed by controlled cooling to relieve internal stresses.

anodizing An electrolytic process that forms a permanent, protective oxide coating on aluminum.

APA performance-rated panels Plywood manufactured to the structural specifications and standards of APA—The Engineered Wood Association.

arc resistance The total elapsed time in seconds an electric current must arc to cause a part to fail.

arch A curved construction spanning an opening and supported at the sides or ends.

architectural terra-cotta Hard fired clay masonry units made with a textured or sculptured face.

ashlar masonry Stone cut into rectangular shapes.

asphalt Dark brown to black hydrocarbon solids or semisolids having bituminous constituents that gradually liquefy when heated.

audio sensors Controls which activate a security or safety system when a sound is detected.

Authority Having Jurisdiction The governmental agency that regulates the construction process.

autoclave A high-pressure steam room that rapidly cures green concrete units.

awning window Window that has a sash that is top-hinged, and swings outward by means of a crank or lever.

Axminster carpet Carpet formed by weaving on a loom that inserts each tuft of pile individually into the backing.

backerboard Used as the base ply or plies in assemblies that require more than one layer of gypsum board.

balloon framing A system of framing a wood-framed building in which all vertical structural members (studs) of the exterior bearing walls and partitions extend the full height of the frame, from the bottom plate to the top plate, and support the floor joists and roof.

bank measure The volume of soil in situ in cubic yards.

batch The amount of concrete mixed at one time.

bauxite Ore containing high percentages of aluminum oxide.

beam A straight horizontal structural member whose main purpose is to carry transverse loads.

bending moment The moment that produces bending on a beam or other structural member.

beneficiation A process of grinding and concentration that removes unwanted elements from iron ore before the ore is used to produce steel.

883

binder Film-forming ingredient in paint that binds the suspended pigment particles together.

bioplastics Derived from renewable biomass sources, such as vegetable oil or cornstarch, rather than from petroleum.

bio-remediation The use of bacteria in site remediation that are able to digest and remove toxic chemicals.

bioswale A landscape element consisting of a linear drainage ditch with sloped sides covered in vegetation designed to remove silt and pollution from surface runoff water.

bitumen A generic term describing a material that is a mixture of predominantly hydrocarbons in solid or viscous form derived from coal and petroleum.

bituminous coatings Coatings formulated by dissolving natural bitumens in an organic solvent.

bleed Excess water that rises to the surface of concrete shortly after it has been poured.

blind nailing When flooring is attached with nails through the tongue of each board; because the nail is driven through the groove, it is not visible on the face.

board foot The measure of lumber having a volume of 144 in.³.

boards Lumber less than 2 in. (50.8 mm) thick and 1 in. (25.4 mm) or more wide.

boiler A closed vessel used to produce hot water or steam.

bond pattern Describes the arrangement of masonry units in a masonry wall.

bonding agent A compound that holds materials together by bonding joining surfaces.

bored lock Installed into a hole drilled through the face of a door.

boundary layer wind tunnel (BLWT) Takes into account the effects of prevailing wind speed and direction, and also those generated by the local topography and adjacent buildings; frequently used for large buildings and those of unusual height or shape.

branch soil pipe A pipe in a plumbing system that discharges into a main or submain and into which no other branch pipes discharge.

brick A solid or hollow masonry unit made from clay or shale, molded into a rectangular shape, and fired in a kiln to produce a hard, strong unit.

brick bond The pattern of masonry units laid to form a multi-wythe wall.

brick pavers Solid clay units whose selection is based on weather resistance, abrasion resistance, and appearance relating to bond patterns, size, and tolerances.

brick veneer A facing of brick that is laid against a wall but not structurally bonded to it.

British thermal unit (BTU) The amount of heat required to raise the temperature of 1 lb. of water 1°F.

brown coat The second coat of plaster in a three-coat plaster finish.

brownfield site A property, the expansion, redevelopment, or reuse of which may be complicated by the presence or potential presence of a hazardous substance, pollutant, or contaminant.

buffer, elevator Energy-absorbing units placed in an elevator pit.

building automation system A central control system that combines the supervision and control of various building functions.

building code A governmental ordinance that regulates the design, construction, and maintenance of buildings by enforcing minimum standards to safeguard the life, health, property, and general welfare of the public.

building construction The addition of a building to a piece of real estate.

building drain The lowest horizontal piping of a plumbing drainage system that receives the discharge from soil, waste, and other drainage pipes within a building and carries the waste to the building sewer.

building information modeling A software program that administers a process for developing and managing all information of a building project throughout its life cycle.

building permit A certificate issued by the authority having jurisdiction authorizing the construction of a project after review of the construction documents.

building section A drawing that provides a representation of a building after a vertical plane has been cut through it and the front portion removed.

building sewer Horizontal piping that carries the waste discharge from the building drain to the public sewer or septic tank.

building-integrated PV (BIPV) photovoltaic materials used to replace conventional building materials in parts of the building envelope, such as the roof, skylights, or facades.

built-up roofing A continuous, semi-flexible roof membrane consisting of plies of saturated felts, coated felts, fabrics, or mats that have surface coats of bitumens; the last ply is covered with mineral aggregates, bituminous materials, or a granular-surface roofing sheet.

bull float A tool used to level concrete in a form after it has been screeded.

burning Curing bricks by placing them in a kiln and subjecting them to a high temperature.

bus A rigid electric conductor enclosed in a protective busway.

busway A rigid conduit used to protect a bus running through it.

BX cable A cable sheathed with spirally wrapped metal strip identified as Type AC.

bypass door A door unit that has two or more sections.

cable tray A ladder-like metal frame, open on the top, used to support insulated electrical cables.

caisson A watertight structure within which work can be carried out below the surface of water.

caisson foundation Foundations formed by drilling holes in the earth and filling them with concrete and reinforcing steel.

calcareous clays Clays containing at least 15 percent calcium carbonate.

calcined gypsum Ground gypsum that has been heated to drive off the water content.

camber A slightly arched surface in a structural beam designed to resist applied loadings.

candela (cd) A metric unit of luminous intensity that closely approximates candlepower.

candlepower (cp) A term used to express the luminous intensity of a light source. It is the same magnitude as a candela.

capillary action The movement of a liquid through small openings of fibrous material by the adhesive force between the liquid and the material.

car, elevator The load-carrying unit of an elevator, consisting of a platform, walls, ceiling, door, and a structural frame.

car safeties, elevator Devices used to stop a car and hold it in position should it travel at excessive speed or free fall.

carbohydrates Organic compounds that form the supporting tissue of plants.

carpet tiles An alternative to the standard rolls of carpet that are typically used in commercial applications since they do not require padding or glue and the tiles can be individually removed for cleaning or replacement.

casement window A single-hinged window that swings either outward or inward.

casing The trim moldings used around doors, windows, and other openings.

cast alloys Products made by pouring molten metal into either sand or permanent molds.

cast iron A hard, brittle metal made of iron that contains a high percentage of carbon.

cast-in-place concrete Concrete members formed and poured on the building site.

cast steel products Steel items produced by casting.

caulking compound A resilient material used to seal cracks and prevent leakage of water.

cavity wall A masonry wall made up of two wythes of masonry units separated by an air space.

cellulose A complex carbohydrate that constitutes the chief part of the cell walls of plants, wood, and paper, and yields fiber for many products.

cement board A panel product made with an aggregate Portland cement core reinforced with polymer-coated glass-fiber mesh on each side.

cementitious materials Materials that have cementing properties.

cements 1. A powder with adhesive and cohesive properties that sets into a hard, solid mass when mixed with water.
2. A bonding agent made from synthetic rubber suspended in a liquid.

certificate of occupancy A certificate issued by the authority having jurisdiction certifying that a completed building is in compliance with all locally adopted building regulations and is in proper condition to be occupied.

chalkboard A surface often made of slate for writing on with chalk.

change order A written directive to the contractor issued after the execution of an agreement that authorizes an addition, deletion, or revision to the project, along with the related adjustments in contract budget and time.

chase A recessed area in a wall for holding pipes and conduit that passes vertically between floors.

chemical strengthening A process for strengthening glass that involves immersing it in a molten salt bath.

chiller A refrigerating machine composed of a compressor, a condenser, and an evaporator, used to transfer heat from one fluid to another.

chromogenic glazing Glazing materials that can change their optical properties in response to changes in solar radiation intensity, temperature, or through the introduction of an electric charge (photo chromic, thermo chromic, electro chromic).

circuit breaker An electrical device used to open and close a circuit by non-automatic means or to open a circuit by automatic means at a predetermined overcurrent without damage to itself.

cladding The external finish covering the structural frame of a building.

clay A cohesive material made up of microscopic particles (less than 0.00008 in. or 0.002 mm).

clay tile A unit made from fired and sometimes glazed clay and used as a finish surface on floors and walls.

cleanout An opening in waste water piping systems that permits cleaning obstructions from the pipe.

clear coating A transparent protective or decorative film.

coal tar pitch A dark-brown-to-almost-black hydrocarbon solid or semisolid material derived by distilling coke-oven tar.

coating A paint, varnish, lacquer, or other finish used to create a protective or decorative layer.

coefficient of thermal expansion The total amount of heat, expressed in BTU, that passes by conduction through a 1 in. thickness of a homogeneous material per hr., per ft.², per °F, which is measured as the temperature difference between the two surfaces of the material.

cofferdam A temporary watertight enclosure around an area of water-bearing soil or an area of water from which the liquid is removed (via pumping), allowing construction to take place in a water-free area.

cogeneration The utilization of normally wasted heat energy to produce electricity or to heat or cool a building.

column-cover-and-spandrel system Uses long spandrel panels that span column covers.

compact fluorescent light (CFL) A type of fluorescent lamp made in special shapes to fit in standard household light sockets.

compaction The act of compressing soil to increase its density.

compartment A small area within a larger area enclosed by partitions.

competitive bidding A method of selecting a general contractor by inviting bids on a competitive basis.

composite decking Uses a concrete topping poured onto metal decking to provide a strong and stiff structural floor system.

composite panels Panels having a reconstituted wood core bonded between layers of solid veneer.

composting toilet A system that converts human waste into organic compost and usable soil through the natural breakdown of organic matter into its essential minerals.

compressive stress Stresses created when forces push on a member and tend to shorten it.

compressor A mechanical device for increasing the pressure of a gas.

computer-aided drafting (CAD) Software used by architects, engineers, drafters, and others to create precision drawings or technical illustrations.

concentrated load Any load that acts on a very small area of a structure.

concrete A solid, hard material produced by combining Portland cement, aggregates, sand, and water (and sometimes admixtures).

concrete masonry units Factory-manufactured concrete units, such as concrete brick or block.

concrete pump A pump that moves concrete through hoses to the area where it is to be placed.

condensate A liquid formed by the condensation of a vapor.

condenser A heat exchange unit in which a vapor has some heat removed, causing it to form a liquid.

conduction The process of heat transfer through a material to another part of that material or to a material touching it.

conductivity, electric A measure of the ability of a material to conduct electric current.

conductor, electric Wire through which electric current flows.

conduit A steel or plastic tube through which electrical wires are run.

console A decorative bracket used to support a cornice, window or other object.

consolidation The process of compacting freshly placed concrete in a form.

construction and demolition debris (C&D) Waste materials generated by building and demolition activities, such as scrap, damaged or temporary materials, packaging, and other wastes.

construction bid A decorative bracket used to support a cornice, window or other object.

construction change directive A document that describes a change to the original contract documents and specifies an adjustment in construction time and cost.

construction document (CD) The final phase of the design process that consists of two interdependent components: the drawings and the specifications.

construction drawings Drawings included in construction documents that graphically describe the dimensional relationships, form, sizes, and quantities of all building elements.

construction management A project delivery method in which an owner hires a construction manager to act as his or her agent by overseeing both design and construction activities.

construction management at risk Method in which the management firm acts as general contractor and assumes all the liability and responsibility of a general contractor.

construction management for fee Method in which the manager does not guarantee a budget or schedule, and will not perform any work on site, but instead acts purely as an advisor.

construction mock-up A full-size model of a proposed construction system built to judge the appearance of an assembly, examine its construction details, and test for performance under actual site conditions.

construction phase Phase that involves the physical realization of the finished building.

continuous flight auger Like a large drill bit that can form a bored pile ranging from 20 to 48 inches (20 to 121 cm) in diameter.

continuous insulation (ci) Insulation that is continuous across all exterior surfaces of an assembly without thermal bridges other than fasteners and mechanical openings.

contract documents The drawings and specifications that designate the exact requirements for the construction of a project.

control joint A groove formed in concrete or masonry structures to allow a place where cracking can occur, thus reducing the development of high stresses.

convection The process of carrying heat from one spot to another by movement of a liquid or gas. The heated liquid or gas expands and becomes lighter, causing it to rise, while the cooler, heavier dense liquid or air settles.

convector A unit designed to transfer heat from hot water or steam to the air by convection.

cool roof A roof surface that stays relatively cool when compared to the surrounding temperature.

core boring A costly but most reliable method of subsurface soil testing that involves penetration through all materials and to great depths.

corona The overhanging vertical member of a cornice.

corrosion The deterioration of metal or concrete by chemical or electrochemical reaction caused by weather exposure.

covalent bonding A process in which small numbers of atoms are bonded into molecules.

creep Permanent dimensional deformation occurring over a period of time in a material subjected to constant stress at elevated temperatures.

critical path method (CPM) Most widely used scheduling technique in which the minimum time required for a certain task is calculated, along with the possible start and finish times for the project activities.

cross-laminated timber (CLT) An engineered wood product formed by bonding layers of thick planks in a perpendicular arrangement with permanent adhesives into panels.

cubicle A small enclosed space often used in modular office partitions.

cupola A small roofed structure built on a roof, usually to vent the area below the roof.

curing Protecting concrete after placing so that proper hydration occurs.

current, electric The flow of electrons along a conductor.

cushion A layer of resilient material applied to a floor over which a carpet is to be laid.

daylighting The use of natural daylight to offset electric lighting use.

dead load The structural load on a building resulting from the weight of the actual materials of construction, such as walls, floors, roofs, and finishes.

decibel (dB) A unit for measuring sound energy or power.

dehumidifier A cooling, absorption, or adsorption device used for removing moisture from the air.

desiccant Any absorbent, adsorbent, liquid, or solid that removes water or water vapor from a material.

desiccation The process of evaporating or removing water vapor from a material.

design-bid-build The traditional three step project delivery method in which the architect and contractor each sign separate contracts with the owner.

design-build A project delivery method in which an owner signs a single contract with one entity to perform both design and construction services.

design drawings Structural drawings prepared by the structural engineer that provide the necessary information for the detailing and fabrication of steel structural members.

design for disassembly (DFD) An emerging subfield of sustainable construction that promotes disassembly over demolition in an effort to maximize material recycling.

design intent A statement that defines the anticipated aesthetic, functional, and performance characteristics of a construction project.

design phase Phase that determines the actual geometry, materials, and performance characteristics of the finished structure.

dewatering Pumping subsurface water from an excavation to maintain dry and stable working conditions.

dielectric strength The maximum voltage a dielectric (nonconductor) can withstand without fracture.

diffuser A circular, square, or rectangular air distributing outlet, usually in the ceiling, that has members to discharge supply air in several directions, mixing the supply air with the secondary air in a room.

dimension lumber Lumber cut and dressed to standard sizes.

dimension stone Large squared blocks.

direct reduction Method of producing pure pellets that contain more than 90 percent iron from ore that has very high iron content.

displacement piles A foundation system that dislodges soils radially and vertically as a slender wood, steel, or pre-cast concrete pile shaft is driven or mechanically vibrated into the ground.

double Ts T-shaped precast floor and roof units that span long distances unsupported.

dressed The actual size of the board after drying and surfacing.

dressed lumber Lumber having one or more sides planed smooth.

dry-press process The process used to make bricks when the clay contains 10 percent or less moisture.

dry-set mortar A mixture of Portland cement, resinous materials, sand, and water that is a thin-set mortar and can be used to bond tiles in place.

dry glazing Utilizes a manufactured compression gasket of rubber or vinyl that supports the glass.

drypack A stiff granular grout.

ductility A measure of the ability of a material to be stretched or deformed without breaking.

durability (of a coating) The ability of a coating to hold up against destructive agents, such as weathering and sunlight.

DWV Drain, waste, vent pipe installation.

dynamic load Any load that acts on a very small area of a structure.

eave The part of a roof that extends beyond the side wall.

efflorescence A white soluble salt deposit on the surface of concrete and masonry, usually caused by free alkalies leached from mortar by moisture moving through it.

elastic limit The highest stress that can be imposed on a material without permanent deformation.

elastomers A macromolecular material that returns to its approximate initial

dimensions and shape after being subjected to substantial deformation.

elevation A drawing that provides a representation of the exterior face of a building, delineating geometries and the materials of construction.

elevator A hoisting and lowering mechanism equipped with an enclosed car that moves in a hoistway between floors in a multi-story building.

embodied energy The energy consumed by all processes associated with the production of a material or building.

enamel A classification of paints that dry to a hard, flat semigloss or gloss finish.

engineered flooring Typically consists of a laminate of three layers of solid wood with tongue-and-groove edges and ends.

engineered wood product Manufactured by bonding together wood strands, veneers, lumber, or other forms of wood fiber with glue to form larger, more efficient composite structural units.

environmental site assessment (ESA) A report prepared for a building site that identifies potential or existing environmental contamination.

environmentally preferable products (EPPs) Materials that have a lesser or reduced effect on human health and the environment when compared to competing products that serve the same purpose.

epoxy mortar Contains no Portland cement, but is instead a mix of epoxy resin and a hardener that is a thin-coat adhesive with high bond strength and resistant to impact and chemical attack.

epoxy varnish A clear finish having excellent adhesion qualities, abrasion and chemical resistance, and water resistance.

equilibrium moisture content The moisture content at which wood neither gains nor loses moisture when surrounded by air at a specified relative humidity and temperature.

erection plan An assembly drawing showing where each structural steel member is located on a building frame.

escalator A continuously moving stair used to move people between the floors of a multi-story building.

expansion joint A joint used to separate two parts of a building to allow expansion and contraction movement of the parts.

existing conditions survey A survey of all utilities, infrastructure, built structures, and significant natural conditions existing above or below grade.

exterior insulation and finish system (EIFS) System of modified stucco curtain wall panels that are an assembly of materials providing both insulation and a weather-tight exterior finish.

faceted glass window A window made by bonding glass pieces with an epoxy resin matrix or reinforced concrete.

facia The finished board covering the ends of the rafters.

falsework Any additional temporary bracing that serves to support the formwork and permanent structure until it is able to support its own weight.

fast-track project A construction delivery method wherein design and construction activities occur simultaneously producing shorter project schedules.

fatigue strength A measure of the ability of a material or structural member to carry a load without failure when the loading is applied repeatedly.

felt A sheet material made using a fiber mat that has been saturated and topped with asphalt.

fenestration The windows and openings in a building enclosure.

ferrous metals Iron-based metallic materials.

fiber saturation point The moisture content of wood at which the cell walls are saturated but there is no water in the cell cavities.

fibered gypsum A neat gypsum with cattle hair or organic fillers added.

fillers Inert material added to a plastic resin to alter its strength and working properties and to lower its cost.

finish coat The third or final coat of gypsum plaster.

finish plaster The topcoat of plaster on a wall or ceiling.

fire-rated partition A partition assembly that has been tested and given a rating indicating the length of time it will resist a fire in hours.

fire-resistance rating A partition assembly that has been tested and given a rating indicating the length of time it will resist a fire in hours.

fire-resistant The capacity of a material or assembly of materials to withstand fire or provide protection from it.

fire-resistant gypsum A gypsum product that has increased fire-resistance properties due to the addition of fire-resistant materials in the gypsum core.

fireclays Deep mined clays that are able to withstand heat.

fire damper A device that prevents the spread of fire through rated walls and partitions where they are penetrated by air ducts.

fire flame detector A control that detects a flame that does not produce smoke when a fire first starts.

fire separation wall Used to control the spread of fire between areas on a floor, such that if a fire breaks out in one area, it will not immediately spread.

fire wall A construction of noncombustible materials that subdivides a building or separates adjoining buildings to retard the spread of fire.

fixture unit A measure of the potable discharge into a waste disposal system from its various plumbing fixtures expressed in cubic volume per minute.

flame-spread rate The rate at which flames will spread across the surface of a material.

flame-spread rating A numerical designation given to a material to indicate its comparative ability to restrict flame combustion over its surface.

flame-retardant Having resistance to the propagation of flame and therefore not readily ignitable. Indicates a lower resistance than fire-resistant.

flammability A measure of the extent to which a material is flammable.

flashing A thin impervious material used to prevent water from penetrating the joints between building elements.

float glass process A glass manufacturing process in which the molten glass ribbon flows through a furnace supported on a bed of molten metal.

flocked carpet Carpet formed by electrostatically spraying short strands onto an adhesive-coated backing material.

floor plan A drawing that provides a representation of a building after a horizontal plane has been cut through it and the top portion removed.

flux A mineral added to molten iron to cause impurities to separate into a layer of molten slag on top of the iron.

footcandle (fc) The unit of illumination equal to 1 lumen per square foot.

footing The base of the foundation that transmits building loads directly to the soil.

footlambert A unit for measuring brightness or luminance. It is equal to 1 lumen per square foot when brightness is measured from a surface.

formwork Either temporary or permanent molds into which concrete is poured.

foundation The element that supports structural loads superimposed on it through transmitting elements into solid bearing below.

framed connections Connections joining structural steel members with a metal, such as an angle, that is secured to the web of the beam.

framing plan A drawing showing the location, size, and spacing of structural members.

freeze cycle day A day when the temperature of the air rises above or falls below 32°F or 0°C.

frequency The number of cycles per second of current or voltage in alternating current, of a sound wave, or of a vibrating solid expressed in hertz.

fuel-contributed rating A rating of the amount of combustible material in a coating.

furan mortar A mixture of furan resin and a hardener that is highly resistant to chemicals.

fuse An overcurrent protection device that opens an electric circuit when a fusible element breaks from heat due to overcurrent passing through it.

fusible link An apparatus consisting of two strips of metal soldered together with a fusible alloy that is designed to melt at a specific temperature.

fusion-bonded carpet Carpet formed by bonding pile yarn between two sheets of backing material and cutting the pile yarn in the center, forming two pieces of carpet.

galling The wearing or abrading of one material against another under extreme pressure.

galvanic corrosion Corrosion that develops by galvanic action when two dissimilar metals are in contact in the atmosphere.

galvanize To coat steel or iron with zinc to prevent rust.

gearless traction elevator An elevator with the traction sheave connected to a spur gear that is driven by a worm gear connected to the shaft of the electric motor.

general contractor The primary contractor who is responsible for all the work on a construction site and oversees the work of subcontractors.

geodesic dome A dome made of prefabricated short, straight, triangular sections interconnected to give stability in all directions.

glaze A ceramic coating fused to the surface of bricks at high temperatures.

glazing The system or process used to support the glass in the frame and seal the dissimilar mating surfaces from the elements.

glued-laminated wood A structural wood member made by bonding laminas of dimension lumber.

glues Bonding agents made from animal and vegetable products.

good indoor air quality Indoor air in which there are no known contaminants at harmful concentrations.

grade (1) Related to soil, the elevation or slope of the ground. (2) In relation to lumber, a means of classifying lumber or other wood products based on specified quality characteristics.

grade beam A ground-level reinforced structural member that supports the exterior wall of a structure and bears directly upon columns or piers.

grading The act of remodeling an existing land form to provide a level area for a structure and create circulation paths and drainage and landscape features.

gravel Hard rock material in particles larger than an in. (6.4 mm) in diameter but smaller than 3 in. (76 mm).

graywater Non-industrial wastewater generated from domestic processes such as dish washing, laundry, and bathing.

green building A building that provides the specified building performance requirements while minimizing disturbance to and improving the functioning of local, regional, and global ecosystems both during and after its construction and specified service life.

GreenFormat A newly released version of the MasterFormat that incorporates new environmentally friendly products and construction procedures that were previously not included.

green lumber Lumber having a moisture content greater than 19 percent.

Greenhouse Effect A process by which the earth's atmosphere traps solar radiation by the existence of gases— such as carbon dioxide (CO_2), water vapor, and methane—that allow incoming sunlight to pass through them but absorb the heat radiated back from the earth's surface.

grille An open grate used to cover or conceal, protect, or decorate an opening.

groundwater Water that exists below the surface of the earth and passes through the subsoil.

groundwater level The level of free-flowing water below the ground surface.

grout A viscous mixture of Portland cement, water, and aggregate used to fill cavities in concrete. Also refers to a specially formulated mortar used under the baseplates of steel columns and in connections in precast concrete.

gypsum Hydrous calcium sulfate.

gypsum board A gypsum panel used for interior wall and ceiling surfaces. It contains a gypsum core and surfaces covered with paper.

gypsum lath A panel having a gypsum core and a paper covering providing a bonding surface for plaster.

gypsum neat plaster A gypsum plaster with no aggregates or fillers added. Sometimes called unfibered gypsum.

hardboard A general term used to describe a panel made from interfelted ligno-cellulosic fibers consolidated under heat and pressure.

hardness A measure of the ability of a material to resist indentation or surface scratching.

hardwood A botanical group of trees that have broad leaves that are shed in the winter.

hardwood plywood Plywood with various species of hardwoods used on the outer veneers.

haunch A projection used to support a member, such as a beam.

heartwood The wood extending from the pith to the sapwood.

heat capacity The ability of a material to store and release heat

heat treating Heating and cooling a solid metal to produce changes in physical and mechanical properties.

heat-treatable alloys Aluminum alloys whose strength characteristics can be improved by heat treating.

heavy construction The construction of large infrastructure projects, such as highways or bridges.

heavy timber construction A type of wood-frame construction using heavy timbers for columns, beams, joists, and rafters.

hiding power The ability of a paint to hide a previous color or substrate.

hinge joint A joint that permits some bending action (similar to that of a hinge) and in which there is no appreciable separation of the joining members.

hoistway, elevator A fire-resistant vertical shaft in which an elevator moves.

hollow clay masonry A unit whose core area is 25 to 40 percent of the gross cross-sectional area of the unit.

hollow core door A door with face veneers on the outer surfaces, wood stiles around the edges, and a hollow interior supported by a honeycomb grid.

hopper window A window that has a sash that is bottom-hinged, and swings inward by means of a crank or lever.

horizontal shear The tendency of top wood fibers to move horizontally in relationship to bottom fibers.

hot-melt adhesives Adhesives that bond when heated to a liquid form.

humidifier A device used to add moisture to the air.

humidity The amount of water vapor within a given space.

hydrate The capacity of lime to soak up water several times its weight.

hydration A chemical reaction between water and cement that produces heat and causes the cement to cure or harden.

hydraulic elevator An elevator having a car mounted on top of a hydraulic piston that moves by the action of hydraulic oil under pressure.

hydronic heating system A system that circulates hot water through a series of pipes and convectors to heat a building.

hydronics The science of cooling and heating water.

hydroxide of lime The product produced by the chemical reaction during the slaking or hydrating of lime.

hygroscopic The ability of a material to readily absorb and retain moisture from the air.

I-joist A wood joist made of an assembly of laminated veneer wood top and bottom flanges and a web of plywood or oriented strandboard.

ice dam Ice that forms on an overhang, causing water buildup behind it to penetrate under roofing material.

igneous rock Rock formed by the solidification of molten material.

illuminance (E) The density of luminous power in lumens per a specified area.

impact isolation class (IIC) An index of the extent to which a floor assembly transmits impact noise from a room above to a room below.

impact strength The ability of a material to resist fracture when struck with a rapidly applied load.

incandescence Emitting visible light as a result of being heated.

incandescent lamp A glass bulb containing a filament joined to a metal base with copper lead-in wires.

independent footing Footings supporting a single structural element, such as a column.

indoor air quality (IAQ) Acceptable indoor air quality is air in which there are no known contaminants at harmful concentrations.

industrial construction The construction of large manufacturing or utility infrastructure projects.

infiltration The unwanted transfer of conditioned indoor air with outside air through small cracks or leaks within the exterior building envelope.

instantaneous water heater A water heating system without a storage tank that heats water instantaneously as taps are activated.

insulated concrete form (ICF) Type of form used for residential and light commercial foundation construction with integral stay-in-place insulation.

insulating glass A glazing unit used to reduce the transfer of heat through a glazed opening by leaving an air- or gas-filled space between layers of glass.

insulator, electric A material that is a poor conductor of electricity used on electrical devices to provide protection from electricity.

integrated design process A project delivery approach that incorporates all stakeholders throughout the design and construction sequence. Used especially for sustainable building projects.

interceptor A trapping device designed to collect materials a sewage treatment plant cannot handle, such as grease, glass or metal chips, and hair.

intermediate coat A layer of coating between a primer coat and a topcoat whose main purpose is to provide structural strength.

intumescent coatings Coatings that expand to form a thick fire-retardant coating when exposed to flame.

invitational bidding A method of preselecting a general contractor based on experience and qualifications rather than cost.

jamb The frame that holds a door or window in place.

joint compound A plastic gypsum mixture similar to plaster used to seal joints and fasteners in gypsum wallboard installations.

jointing Forming control, expansion, or construction joints in concrete construction.

joule A meter-kilogram-second unit of work or energy.

jute A coarse fiber, obtained from two East Indian plants, used in paper and carpet products.

kiln (1) A chamber with controlled humidity, temperature, and airflow in which lumber is dried. (2) A low-pressure steam room in which green concrete units are cured.

kiln-dried Wood products dried in a kiln.

kinetic energy The energy possessed by a body because of its motion.

knitted carpet Carpet formed by looping pile yarn, stitching, and backing together.

lacquer A fast-drying clear or pigmented coating that dries by solvent evaporation.

laminate flooring A rigid floor covering that has a top layer consisting of one or more thin sheets of fibrous material, often a paper product impregnated with amino-plastic thermosetting resins, such as a melamine.

laminated glass Glass panels that have outer layers of glass laminated to an inner layer of transparent plastic.

laminated-veneer lumber A structural lumber manufactured from wood veneers so that the grain of all veneers runs parallel to the axis of the member.

lamp A general term used to describe a source of artificial light. Often called a bulb or tube.

landing zone, elevator The area 18 in. (5490 mm) above or below a landing floor.

latent heat Heat involved with the action of changing the state of a substance, such as changing water to steam.

lateral load Loads moving in a horizontal direction, such as those caused by wind.

latex A water-based coating, such as styrene, butadiene, acrylic, and polyvinyl acetate.

latex Portland cement mortar A mixture of Portland cement, a latex additive, and sand that is a thin-coat application.

leaded glass window A window made of small glass pieces joined with lead cames.

LEED The Leadership in Energy and Environmental Design is a third-party certification program that provides a nationally accepted standard for the design, construction, and maintenance of high-performance buildings.

life cycle analysis A procedure for compiling and analyzing the inputs and outputs of resources and energy and their associated environmental impacts directly attributable to the functioning of a material or service system throughout its life cycle.

light-emitting diode (LED) Small light bulbs based on semiconductor technology; the electrons of a semiconductor diode are brought from a state of high energy to a state of low energy, with the energy difference being emitted in the form of light.

lightweight steel framing Structural steel framing members made from cold-rolled lightweight sheet steel.

lignin An amorphous substance that penetrates and surrounds the cellulose strands in wood, binding them together.

lime A white-to-gray powder produced by burning limestone, marble, coral, or shells.

linoleum A natural flooring material made of linseed oil, wood products, ground cork, mineral fillers, and

pigments that are bonded to a jute or synthetic backing.

lintel block A precast concrete structural member used to span an opening in a wall and carry the weight of the masonry units above it.

liquid limit (LL) Related to soils, the water content expressed as a percentage of dry weight at which the soil will start to flow when tested by the shaking method.

live load The structural load on a building resulting from occupancy and use—consists of people, furnishings, and environmental forces, such as wind, snow, and seismic activity.

load indicator washers Metal washers with rounded protrusions on one face that ensure the accurate tension for bolts.

loomed carpet Carpet formed by bonding pile yarn to a rubber cushion.

low-E glass Low emissivity glass with a thin metallic coating that selectively reflects ultraviolet and infrared wavelengths.

low-slope roofs Roofs that are nearly flat.

lumber A product produced by harvesting, sawing, drying, and processing wood.

lumber grader Persons employed by sawmills and wood processing plants who inspect and grade lumber according to industry standards.

lumen (lm) A unit for measuring the flow of light energy. See luminous flux.

luminaire A complete lighting unit consisting of one or more lamps plus elements needed to distribute light, hold and protect the lamps, and connect power to the lamps. Also called a lighting fixture.

luminance The luminous intensity of a surface of a given area viewed from a given direction.

luminance meter A photoelectric instrument used to measure luminance. Also called a light meter.

luminescence The emission of light not directly caused by incandescence.

luminous flux The rate of flow of light energy through a surface, expressed in lumens.

luminous intensity The force that generates visible light expressed by candela, lumens per steradian, or candlepower.

luminous transmittance A measure of the capacity of a material to transmit incident light in relation to the total incident light striking it.

lux A unit of illumination equal to 1 lumen per square.

makeup air Air brought into a building from outside to replace interior exhausted air.

malleability The characteristic of a material that allows plastic deformation in compression without rupture.

manufactured stone An architectural building unit manufactured to simulate natural cut stone used in masonry applications.

marker-board A surface that can be written on with a water-based or semi-permanent ink that can be removed.

masonry veneer Consists of a single wythe of masonry that is anchored to a load-bearing backup construction with an air space between them.

MasterFormat A standard developed by the Construction Specifications Institute for writing specifications using a system of descriptive titles and numbers to organize construction activities, products, and requirements.

mat foundation (also called spread foundation) A large, single concrete footing equal in area to the area covered by the footprint of the building.

mean radiant temperature (MRT) The average temperature experienced from the combination of all the surface temperatures in a room, including walls, floors, ceilings, furniture, and people.

means of egress A continuous and unobstructed path of vertical and horizontal egress from any occupied portion of a building or structure to a public way.

mechanical action The bonding of materials by adhesives that enter pores and harden, forming a mechanical link.

melting temperature The temperature at which a material turns from a solid to a liquid.

metal lath Perforated sheets of thin metal secured to studs that serve as the base for a finished plaster wall.

metals Materials refined from ores that have been extracted from the earth.

metamorphic rock Rock formed by the action of pressure or heat on sedimentary soil or rock.

meter, electric A device measuring and recording the amount of electricity passing through it in kilowatt-hours.

mildewcide An agent that helps prevent the growth of mold and mildew on painted surfaces.

millwork Any wood products that have been manufactured or preassembled, such as moldings, doors, windows, and built-in units.

mixers Furnaces where molten pig iron is moved to be blended to equalize the chemical composition and kept in a liquid state until used.

mock-up A full scale model of a construction component built on its construction site to judge appearance and test performance characteristics.

modified bitumen membrane A roofing membrane composed of a polyester or fiberglass mat saturated with a polymer-modified asphalt.

modillion A stone scroll supporting a corona.

modular bricks Bricks whose actual size plus a mortar joint can be assembled on a 4-inch standard module.

modulus of elasticity (E) The property of a material that indicates its resistance to bending. It is the ratio of unit stress to unit strain.

moisture content The amount of water contained in wood, expressed as a percentage of the weight of the wet wood to the weight of an oven-dry sample.

moment A force that acts at a distance from a point and that tends to cause a body to rotate about that point.

monomer An organic molecule that can be converted into a polymer by chemical reaction with similar molecules or organic molecules.

mortar A plastic mixture of cementitious materials, water, and a fine aggregate.

mortar flow A measure of the consistency of freshly mixed mortar related to the diameter of a molded truncated cone specimen after the sample has been vibrated a specified number of times.

mortise lock Uses a rectangular lock body installed in an opening that is mortised out of the edge of the door.

mosaic A decoration made up of small pieces of inlaid stone, glass, or tile.

motor control center Controllers used to start and stop electric motors and protect them from overloads.

moving walk A conveyor belt system operating at floor level used to move people in a horizontal direction.

multiple prime contract When a number of specialty contractors are engaged to complete different parts of the work; often used on large and complex projects.

natural fibers Fibers found in nature, such as wool and cotton.

negotiated contract A project delivery process in which the construction contractor is selected without the process of competitive bidding.

net metering A process wherein excess electricity generated by a photovoltaic system is sold back to a utility grid.

noise-reduction coefficient (NRC) A single number indicated by the amount of airborne sound energy absorbed into a material.

noncalcareous clays Clays containing silicate of alumina, feldspar, and iron oxide.

non-displacement piles End-bearing units that are drilled using large-diameter drilling rigs.

nonferrous metals Metallic materials in which iron is not a principal element.

non-heat-treatable alloys Alloys that do not increase in strength when they are heat treated.

non-metallic inorganic materials Materials extracted from soil and refined to produce a variety of products. Typical inorganic materials include sand, limestone, glass, brick, cement, gypsum, mortar, and mineral wool insulation.

non-potable water Recycled water that can be used by plumbing fixtures, such as toilets, when there is no possibility of it becoming intermingled with potable water used for human consumption.

non-pre-stressed units Concrete structural members in which the reinforcing steel is not subject to pre-stressing or post-tensioning.

notice to proceed A written directive from the owner to the contractor that sets the date that the contractor can begin the work under the conditions of the contract.

oakum A caulking material made from hemp fibers treated with tar.

off-gassing A chemical process that occurs when solid but chemically unstable materials evaporate at room temperature and slowly release contaminants.

oil-based paints Paint composed of resins requiring solvent for reduction purposes.

opaque coatings Coatings that completely obscure the color and much of the texture of a substrate.

organic matter (or material) A class of compounds comprising only those existing in plants and animals.

oriented-strand board A panel made from wood strands that has the strand face oriented in the long direction of the panel.

overlaid plywood Plywood panels whose exterior surfaces are covered with a resin-impregnated fiber ply.

oxidation A chemical reaction between a material and oxygen in the atmosphere.

oxide layer In aluminum, a very thin protective layer formed naturally on aluminum due to its reaction to oxygen.

panelboard A panel that includes fuses or circuit breakers used to protect the circuits in a building from overloads.

panelized construction Construction that uses preassembled panels for walls, floors, and roof.

parallel strand lumber Lumber made from lengths of wood veneer bonded to produce a solid member.

parquet flooring Uses short wooden pieces that are laid in geometric patterns for decorative effect.

particleboard A sheet product manufactured from wood particles and a synthetic resin or other binder.

passive fire protection Involves the application of fire-resistance-rated construction assemblies that are used to compartmentalize parts of a building in order to control the spread of fire and smoke and provide sufficient time for evacuation and firefighting.

permeability The rate at which water flows through a material.

permeable paving Paving materials, typically concrete, stone, or plastic, that promote the absorption of rain and snowmelt.

photovoltaic (PV) The use of semiconductor materials to convert solar radiation to electric current that can be used as electricity.

pig Another word for ingots.

pig iron A high-carbon-content iron produced by a blast furnace and used to manufacture cast iron and steel.

pigments Paint ingredients mainly used to provide color and hiding power.

pilaster A vertical projection from a masonry or concrete wall providing increased stiffening.

pile cap Distributes loads equally among the piles below the cap.

pile foundation A foundation that uses long wood, concrete, or steel piles driven into the earth.

pile hammer A machine for delivering blows to the top of a pile, driving it into the earth.

pit, elevator The part of a hoistway that extends below the floor of the lowest landing to the floor at the bottom of the hoistway.

pitch Related to carpets, the number of tufts in a 27 in. width of carpet.

plaster Typically made up of mostly gypsum, but some types use Portland cement.

plaster base Any material suitable for the application of plaster.

plaster of Paris A calcined gypsum mixed with water to form a thick, paste-like mixture.

plastic An organic material that is solid in its finished state but is capable of being molded.

plastic behavior The ability of a material to become soft and formed into desired shapes.

plastic limit (PL) Related to soils, the percent moisture content at which soil begins to crumble when rolled into a thread $\frac{1}{8}$ in. (3 mm) in diameter.

plasticity index (PI) The difference between the liquid limit and the plastic limit.

plasticizers Liquid material added to some plastics to reduce their hardness and increase pliability. Also, an additive to concrete and mortar to increase plasticity.

platform framing A wood structural frame for light construction with the studs extending only one floor high upon which the second floor is constructed.

platform safety A device used in elevators designed to clamp onto the steel guide rails upon activation, quickly braking the elevator to a halt.

plumbing chase A hollow wall area accommodating piping used for drain waste or vent in plumbing systems.

plywood All-veneer panels consisting of an odd number of cross-laminated layers, each layer consisting of one or more piles. Many such panels meet

all of the prescriptive or performance provisions of U.S. Product Standard PS 1-83 / ANSI A199.1 for Construction and Industrial Plywood.

pocket door A sliding door that disappears into a compartment, or pocket, in the adjacent wall when fully open.

pointing When raked space is filled with mortar.

point-supported glazing Systems that utilize mechanical anchors at discrete locations near the glass edge in place of continuous edge supports.

polymer A chemical compound formed by the union of simple molecules to form more complex molecules.

Portland cement A cementitious binder used in concrete, mortar, and stucco, obtained from pulverizing clinker consisting of hydraulic calcium silicates.

post-consumer recycled material Material that has served its intended use, been collected from the consumer, and reprocessed.

post-industrial recycled content Materials created using an industrial waste or by-product, diverting it from landfill, and integrating it into new projects.

post-tensioned A method used to place concrete under tension in which steel tenons are tensioned after the concrete has been poured and hardens.

potable water Water that is safe to drink and meets standards of a local health authority.

power The rate at which work is performed, expressed in watts or horsepower.

pozzolan A siliceous or siliceous and aluminus material blended with Portland cement that chemically reacts with calcium hydroxide to form compounds possessing cementitious properties.

pre-design phase The initial part of the design process in which a project team determines the goals and objectives of a project through design sketches and feasibility studies.

premise wiring The cabling, connectors, and accessories that are used to connect Local Area Network (LAN) and phone equipment within a commercial building.

preservatives Chemicals forced into wood to provide protection from decay, insects, and other damaging conditions.

pre-stressed units Structural concrete units cast in molds in a factory that have stresses introduced as they are being cast (or post casting).

pre-tensioned A method used to place a concrete member under tension by pouring concrete over steel tendons that are under tension before the concrete is poured.

primer coat A base coat in a paint system. It is applied before the finish coats.

programming The research and decision-making process that determines the specific set of needs that a building is expected to fulfill, and identifies the scope of work to be designed.

project manual The completed specifications for a project.

punch list A listing of remaining items needing installation or repair near the completion of a construction project.

quarter-rsawing Sawing lumber so the hard annual rings are nearly perpendicular to the surface.

raceway An enclosed channel designed to carry wires and cables.

radiant heat Heat transferred by radiation.

radiation The transmission of heat by electromagnetic waves.

raft foundation Reinforced concrete slabs several feet in thickness that cover the total footprint of a building.

rainscreen principle The principle that states that wall cladding can be made watertight by placing wind-pressurized air chambers behind the joints, which reduces the air pressure differentials between the inside and outside that could cause water to move through the joints.

raised-access flooring Provides an elevated floor surface above the structural floor system to provide space for the passage and easy access of mechanical and electrical services.

rake The sloping portions at the gable end of a building.

rammed earth A soil-cement mixture that is pneumatically rammed into forms to create walls generally 18 to 24 inches thick.

rapidly renewable materials Materials that are able to regenerate themselves in ten years or less.

reduction (1) A process in which iron is separated from oxygen with which it is chemically mixed by smelting the ore in a blast furnace. (2) With regard to aluminum, the electrolytic process used to separate molten aluminum from the alumina.

reflectance coefficient A measure stated as a percentage of the amount of light reflected off a surface.

reflective glass Glass having a thin layer of metal or metal oxide deposited on its surface to reflect heat and light.

refrigerant A medium of heat transfer that absorbs heat by evaporating at low temperatures and pressure and releases heat when it condenses at a higher temperature and pressure.

reinforced masonry Masonry construction that has steel reinforcing bars inserted to provide tensile strength.

relative humidity The ratio of the amount of water vapor present in air to that which the air would hold if saturated at the same temperature.

renewable materials Materials derived from a living plant, animal, or ecosystem that has the ability to regenerate itself.

request for information (RFI) Used by the contractor to obtain clarification on specific components and assemblies that are not fully detailed or understood through a comprehensive review of the drawings and specifications.

resilient flooring Finished flooring made from a resilient material, such as polyvinyl chloride or rubber.

resin A natural or synthetic material that is the main ingredient of paint and binds the ingredients together.

resistance, electric The physical property of a conductor or electric-consuming device to resist the flow of electricity, reducing power and generating heat.

retarder An admixture used to slow the setting of concrete.

return air Air removed from a space and vented or reconditioned by a furnace, air conditioner, or other apparatus.

rigging Hoisting equipment used to lift materials above grade.

rim joist A type of sill used in frame construction in which the floor joists butt and are nailed to a rim joist and rest on the sill.

riser, plumbing A water supply running vertically through a building to

provide water to the various branches and fixtures on each floor.

rock A solid mineral material found naturally in large masses.

romex A non-metallic-sheathed electric cable of type NM or NMC.

room cavity ratio A relationship between the height of a room, its perimeter, and the height of the work surface above the floor divided by the floor area.

rough lumber Lumber that has not been dressed and shows saw marks on all four surfaces.

rows Related to carpets, the number of tufts per inch.

rubble Rough stones in their natural irregular shapes and sizes.

safing Fire-stopping material placed in spaces between a floor and curtain walls in high-rise construction.

sand Fine rock particles from 0.002 in. (0.05 mm) in diameter to less than 0.25 in. (6.4 mm) in diameter.

sanitary piping Pipe that carries liquid and water-borne waste from plumbing fixtures into which storm, surface, and groundwater are not admitted.

sanitary sewer A sewer that receives sanitary sewage without the infusion of other water such as rain, surface water, or other clear water drainage.

sapwood The wood near the outside of a log just under the bark.

sash The moving segment of a window.

scratch coat The first coat of gypsum plaster that is applied to the lath.

screeding The process of striking off the surface of freshly poured concrete with a screed so it is flush with the top of a form.

scupper An outlet in a parapet wall for the drainage of overflow water from the roof to the outside of a building.

sealants A mastic used to seal joints and seams.

sealer A material used to seal the surface of a material against moisture.

seasoning Removing moisture from green wood.

seated connections Connections that join structural steel members with metal connectors, such as angles, upon which one member, such as a beam, rests.

sedimentary rock Rock formed from the deposit of sedimentary materials on the bottom of a body of water or on the surface of the earth.

segregation The tendency of large aggregate to separate from the sand-cement mortar in a concrete mix.

seismic area A geographic area where earthquake activity may occur.

semitransparent coatings Coatings that allow some of the texture and color of a substrate to show through.

sensible heat Heat that causes a detectable change in temperature.

service core A fire-resistant vertical shaft through a multistory building used to route electrical, mechanical, and transportation systems.

service entrance The point at which power is supplied to a building and where electrical equipment such as the service switch, meter, overcurrent devices, and raceways are located.

service temperature The maximum temperature at which a plastic can be used without altering its properties.

shading coefficient (Sc) The ratio of the solar heat gain of a window to the solar heat gain of clear float glass.

shales Clays that have been subjected to high pressures, causing them to become relatively hard.

shear plate connector A circular metal connector recessed into a wood member that is to be bolted to a steel member.

shear stress The result of forces acting parallel to an area but in opposite directions, causing one portion of the material to "slide" past another.

shear studs Metal studs welded to a steel frame that protrude up into a cast-in-place concrete deck.

sheave A pulley over which an elevator wire hoisting rope runs.

sheeting Wood, metal, or concrete members used to hold up the face of an excavation.

sherardize To coat steel with a thin cladding of zinc by heating it in a mixture of sand and powdered zinc.

shop drawing A drawing prepared by a contractor or material supplier that gives precise directives for the fabrication of a construction component.

shrinkage limit (SL) Related to soils, the water content at which a soil volume is at its minimum.

Siamese connection An outside connection to which firefighters connect an alternate source of water to boost the water used by a building's fire suppression system.

sick building syndrome A medical condition caused by exposure to contaminants within a building resulting in headaches, dizziness, nausea, and respiratory and skin problems.

silt Fine sand with particles smaller than 0.002 in. (0.05 mm) and larger than 0.00008 in. (0.002 mm).

simple connection Steel frame connections that are designed as flexible connections and are assumed free to rotate.

single-ply roofing membrane A roofing membrane composed of a sheet of waterproof material secured to the roof deck.

sintering A process that fuses iron ore dust with coke and fluxes into a clinker.

site plan A drawing of a construction site, showing the proposed locations of buildings, site contours, and other improvements.

slab on grade A reinforced concrete foundation that is placed directly on the ground to provide the structural support for a building.

slag A molten mass composed of fluxes and impurities removed from iron ore in a furnace.

slake The process of adding water to quicklime, hydrating it and forming lime putty.

slump A measure of the consistency of freshly mixed concrete, mortar, or stucco.

slump test A test made on site to ensure that the required consistency of a concrete mix has been achieved.

slurry A liquid mixture of water and bentonite clay or Portland cement.

slurry wall A wall built of a slurry used to hold up the sides of an area to be excavated.

smelting A process in which iron ore is heated, separating the iron from the impurities.

smoke barriers Continuous membranes used to resist the passage of smoke.

smoke-developed rating A relative numerical classification of the fumes developed by a burning material.

soffit The underside of a cornice or other overhanging assembly.

soft-mud process A process used to make bricks when the clay contains moisture in excess of 15 percent.

softwood A botanical group of trees that have needles and are evergreen.

soil anchors Metal shafts grouted into holes drilled into the sides of an excavation to stabilize it.

soil load test Test that uses incremental loads that are applied to a platform erected on an area of undisturbed soil and measured to determine the settlement characteristics of the soil.

soil stack A vertical plumbing pipe into which waste water flows through waste pipes from plumbing fixtures.

solid masonry A unit whose core does not exceed 15 percent of the gross cross-sectional area of the unit.

solvent Liquids used in paint and other finishing materials that give the coating workability and that evaporate, permitting the finish material to harden.

sound A sensation produced by the stimulation of the organs of hearing by vibrations transmitted through the air. It is the movement of air molecules in a wavelike motion.

sound transmission class (STC) A single number rating of the resistance of a construction to the passage of airborne sound.

space frame A three-dimensional truss that forms a rigid, stable structure able to span long distances.

special units Concrete masonry units that are designed and made for special applications.

specific adhesion The bonding of dense materials through the attraction of unlike electrical charges.

specifications Part of the construction documents that describes in writing the procedures, materials, methods, and performance requirements of a construction project.

spire A tall pyramidal roof built upon a tower or steeple.

split-ring connector A ring-shaped metal insert placed in circular recesses cut in joining wood members that are held together with a bolt or lag screw.

spread footing Provides a stable and continuous foundation around the full perimeter of a structure; most common shallow foundation type.

spread foundation A foundation that distributes the load over a large area.

spreading rate The area over which a paint can be spread expressed in square feet per gallon.

stabilizers Additives used to stabilize plastic by helping it resist heat, loss of strength, and the effect of radiation on the bonds between the molecular chains.

stain A solution of coloring matter in a vehicle used to enhance the grain of wood during the finishing operation.

standard penetration test (SPT) Test that uses a pipe-like weight of around 140 lbs placed on top of a drive rod and is repeatedly raised and dropped. The number of hits or blows required to drive the weight every 6 inches allows for a determination of bearing capacity.

static load Any load that does not change in magnitude or position with time.

steam Water in a vapor state.

steeple A tower-like ornamental construction, usually square or hexagonal, placed on the roof of a building and topped with a spire.

steep-slope roofs Roofs with sufficient slope to permit rapid runoff of rain.

stick system System where vertical mullions are installed, followed by horizontal rails into which glazing and spandrel panels are placed; the first curtain wall type developed by manufacturers and remains the most common.

stiff-mud process A process used to make bricks from clay that has 12 to 15 percent moisture.

stone Rock selected or processed by shaping to size for building or other use.

storefront A non-residential system of entrance doors and fixed windows mulled as a composite structure.

storm water pollution prevention plan (SWPPP) A plan that must be implemented by the construction manager and conform to best management practices of the Environmental Protection Association's Storm Water Management for Construction Activities or local standards and codes.

story pole A rod divided into equal parts with each part equal to the height of one course of masonry units.

strain The deformation of a body or material when it is subjected to an external force.

strain hardening Increasing the strength of a metal by cold-rolling.

strand-casting machine A machine that casts molten steel into a continuous strand of metal that hardens and is cut into required lengths.

stress The intensity of internally distributed forces that resist a change in the form of a body due to an applied force. It is measured in pounds per square inch (psi), mega-Newtons per square meter (MN/m²) or mega-Pascals (MPa).

stress-rated lumber Lumber that has its modulus of elasticity determined by actual tests.

stretcher A masonry unit that is positioned with its bottom side in mortar and its face in the plane of the wall; the most common masonry unit position.

structural clay tile Rectangular masonry units made from clay and fired to harden them for use in load-bearing and non-load-bearing walls and partitions.

structural composite lumber (SCL) Lumber manufactured by parallel orientation of wood fibers of various geometries, bonded with a structural adhesive.

structural insulated panels (SIP) A structural panel consisting of two high-strength facing materials bonded to a rigid insulation core.

subcontractor An entity that is contractually obligated to a general contractor to complete a portion of work on-site.

substrate A subsurface to which another material is bonded.

suction The rate at which clay masonry units absorb moisture.

suction rate The weight of water absorbed when a brick is partially immersed in water for one minute—expressed in grams per minute or ounces per minute.

superintendent A general contractor's on-site representative responsible for continuous field supervision, coordination, and completion of the work.

supplementary cementing materials (SCMs) Mineral admixtures consisting of natural or by-products of other manufacturing processes that are used to reduce the amount of Portland cement in a concrete mix.

supply air Conditioned air entering a space from an air conditioning, heating, or ventilating unit.

surface clays Clays obtained by open-pit mining.

surface effect The effect caused by the entrainment of secondary air against or parallel to a wall or ceiling when an outlet discharges air against or parallel to the wall or ceiling.

sustainable building A building that provides the specified building performance requirements while minimizing disturbance to and improving the functioning of local, regional, and global ecosystems, both during and after its construction and specified service life.

sustainable development Development that meets the needs of the present without compromising the ability of future generations to meet their own needs.

swing stage A scaffold that is suspended by ropes or cables from a block-and-tackle system attached by roof anchors that can be raised or lowered along a building façade.

switchboard A large single panel or an assembly of panels with switches, overcurrent protection devices, and buses that are mounted on the face or both sides of a panel.

switches Devices to open and close electric circuits or to change connections within circuits.

synthetic fibers Fibers formed by chemical reactions.

synthetic materials Materials formed by the artificial building up of simple compounds.

tackboard A surface used to display announcements and other materials, usually attached with tacks.

tapestry A fabric upon which colored threads are woven by hand to produce a design.

temper designation A specification of the temper or metallurgical condition of an aluminum alloy.

tempered glass A process used to strengthen glass by raising its temperature to near the softening point and then suddenly blowing jets of cold air on both sides to chill it and create surface tension.

tensile bond strength The ability of a mortar to resist forces tending to pull the masonry apart.

tensile stress The stress per unit area of the cross section of a material that resists elongation.

terrazzo A finish-floor material made up of concrete and an aggregate of marble chips that after curing is ground smooth and polished.

test cylinders Cylindrical samples of wet concrete that are cured on site and used to test compressive strength of the hardened concrete.

texture plaster A finish plaster used to produce rough, textured finished surfaces.

thermal break An insulating material or gap, placed between thermal conducting materials in contact to retard the passage of heat or cold.

thermal bridge Formed when a material or assembly is uninsulated from exterior to interior, allowing for conductive heat transfer from the warm side to the cold.

thermal conductance (C) Thermal conductance is the same as thermal conductivity except it is based on a specified thickness of a material rather than on the one inch used for conductivity.

thermal conduction The process of heat transfer through a solid by transmitting kinetic energy from one molecule to the next.

thermal conductivity (k) The rate of heat flow through one square foot of material one inch thick expressed in BTU per hour when a temperature difference of one degree Fahrenheit is maintained between the two surfaces.

thermal convection Heat transmission by the circulation of a liquid or a heated air or gas.

thermal envelope The exterior enclosure that separates conditioned space from the exterior.

thermal mass Process in which materials with a high specific heat capacity store heat from the sun during the day and then discharge that heat by natural heat transfer during the night.

thermal radiation The transmission of heat from a hot surface to a cool one by means of electromagnetic waves.

thermal resistance An index of the rate of heat flow through a material or assembly of materials. It is the reciprocal of thermal conductance (C).

thermal transmittance A measure of heat transfer through an assembly of materials, such as an exterior wall—the reciprocal of the total resistance (R) of the assembly.

thermoplastic Plastics that soften by heating and re-harden when cooled without changing chemical composition.

thermosets Cured plastics that are chemically cross-linked and when heated will not soften but will degrade.

thermostat A temperature-sensitive instrument that controls the flow of electricity to units used to heat and cool spaces in a building.

throw The horizontal or vertical distance an airstream travels after leaving an air outlet before it loses velocity.

tieback anchors Steel anchors grouted into holes drilled in an excavation wall to hold the sheeting, thus reducing the number of braces required.

timber Wood structural members having a minimum thickness of 6 in. (140 mm).

timber joinery The joining of structural wood members using wood joints, such as the mortise and tenon.

topcoat The final coat of paint.

toughness A measure of the ability of a material to absorb energy from a blow or shock without fracturing.

trade association An organization whose membership comprises manufacturers or businesses involved in the production or supply of materials and services in a particular area.

transformer An electrical device used to convert an incoming electric current from one voltage to another voltage.

trap Device used to maintain a water seal against sewer gases that back up the waste pipe. Usually each fixture has a trap.

trowel A hand tool with a flat, broad steel blade used to apply, shape, and spread mortar.

troweling Producing a final smooth finish on freshly poured concrete with a steel-bladed tool after the concrete has been floated.

truss A structural component made of a combination of members, usually in a triangular arrangement, to form a rigid framework often used to support a roof.

tufted carpet Carpet formed by stitching the pile yarn through the backing material.

underpinning Placing a new foundation below an existing foundation

unfibered gypsum A neat gypsum with no additives.

uniform load Any load spread evenly over a large area.

unit-and-mullion system A curtain wall system that is a combination of the stick and unitized methods; the system uses mullions secured to a building structure with preassembled wall panels installed between them.

unitized system A curtain wall system that uses large preassembled wall panels.

universal design A movement that advocates the design of products and environments to be usable by all people, to the greatest extent possible, without the need for adaptation or specialized design.

urban sprawl The rapid and expansive growth of a city in an unplanned and often environmentally problematic fashion.

U-value The overall coefficient of thermal expansion in BTU / (hr. ft2 °F) or W / m² K.

valence The points on an atom to which valences of other elements can bond.

valve A device used to regulate and stop the flow of water.

varnish A transparent coating that dries on exposure to air, providing a protective coating.

vehicle The liquid portion of a paint composed mainly of solvents, resins, or oils.

velvet weave carpet Carpet formed by joining the pile, stuffer, and weft yarns with double warp yarns.

veneer gypsum base A gypsum board product designed to serve as the base for the application of gypsum veneer plaster.

veneer plaster A thin layer of plaster applied over a special veneer gypsum base sheet.

vent pipe Pipe that keeps the water in the trap of the fixture under atmospheric pressure, thus eliminating the chance of its being siphoned out when another fixture is used.

vent stack That part of the soil stack above the highest vent branch.

vents, plumbing Pipes permitting a waste system to operate under atmospheric pressure while preventing sewer gases from entering a building. Plumbing vents allow air to enter and leave the system, preventing water in the traps from being siphoned off.

vertical shear The tendency of one part of a member to move vertically in relationship to its adjacent part.

vinyl composition tiles (VCT) A type of resilient flooring that is used in a variety of commercial applications; made by compressing and heating vinyl resins, plasticizers, stabilizers, fibers, and pigments into a uniform thickness.

viscosity The flow resistance of a liquid under an applied load or pressure.

volatile organic compound Manmade chemicals used and produced in the manufacture of paints, composites, and refrigerants that can potentially harm humans.

volatile organic compounds (VOCs) Compounds released to the atmosphere as a coating dries.

waferboard A panel made of wood wafers, randomly arranged and bonded with a waterproof binder.

warp A variation in a board from a flat, plane condition.

waste pipe Horizontal plumbing pipes that connect a fixture to the soil pipe.

water retention The property of a mortar that prevents the rapid loss of water by absorption into the masonry units.

water-based coatings Coatings formulated with water as the solvent.

water–cement ratio In a concrete or mortar mixture, the ratio of the amount of water (minus that held by the aggregates) to the amount of cement used.

water-vapor transmission rate (WVTR) The steady-state vapor flow in a given time through a given area of a body, normal to specified parallel surfaces, under specific conditions of temperature and humidity at each surface.

watt The unit of measurement of electrical power or rate of work. It is a pressure of one volt flowing at the rate of one ampere.

weatherability The ability of a plastic to resist deterioration due to moisture, ultraviolet light, heat, and chemicals found in the air.

weathering index A value that reflects the ability of clay masonry units to resist the effects of weathering.

weathering steel A steel alloy that forms a natural self-protecting rust.

weep hole Small openings at the bottom of exterior cavity walls to allow moisture in the cavities to drain out.

wet glazing Uses preformed tape or a gunable liquid sealant to set glass in a frame.

whiteboard A type of visual display board that may have a vinyl sheet surface or one of porcelain enamel on steel and the substrate can be particleboard, fiberboard, sheetrock, hardboard, plywood, or plastic honeycomb.

Wilton carpet Carpet formed on a loom capable of feeding yarns of various colors.

winning A term used to describe the mining of clay.

workability (1) Describes the ease or difficulty with which concrete can be placed and worked into its final location. (2) In relation to mortar, the property of freshly mixed mortar that determines the ease and homogeneity with which it can be spread and finished.

wrought alloys Products formed by any of the standard manufacturing processes, such as drawing, rolling, forging, or extruding.

wrought products Steel items produced by means other than casting.

wythe A continuous vertical wall of masonry one unit in thickness.

zoning ordinance Local land use regulations written into municipal law that regulate the degree of building development according to land use.

Index